Introduction to Condensed Matter Theory

Pedro D. Sacramento

Introduction to Condensed Matter Theory

Volume 1

Springer

Pedro D. Sacramento
Instituto Superior Técnico
University of Lisbon
Lisbon, Portugal

ISBN 978-3-031-86250-2 ISBN 978-3-031-86251-9 (eBook)
https://doi.org/10.1007/978-3-031-86251-9

This Springer imprint is published by the registered company Springer Nature Switzerland AG
The registered company address is: Gewerbestrasse 11, 6330 Cham, Switzerland

Preface

This book grew out of lecture notes of several courses taught over the years in the area of Condensed Matter Physics. Specifically, it covers aspects of Thermodynamics, and the subjects lectured in the courses of Statistical Physics, Solid State Physics, Condensed Matter Physics, Phase Transitions and Critical Phenomena, and Topics in Condensed Matter Physics. The first three courses are taught in the Undergraduate Program (first three years of the Physics course at Instituto Superior Técnico, previously part of the Technical University of Lisbon and now part of the University of Lisbon) and the last three courses are taught in the Master's program (last two years of the Physics course).

Due to its comprehensive nature and associated size, the printed book is divided in two volumes but it is intended as a unit. The first volume comprises the first fourteen Chapters and the second volume comprises Chapters 15 to 25.

Several books have been used to support the studies of the various materials covered and have a significant influence on this book. The main references used are

[Callen] H.B. Callen, *Thermodynamics*, Wiley
[Fermi] E. Fermi, *Thermodynamics*, Dover
[Reif] F. Reif, *Fundamentals of Statistical and Thermal Physics*, McGraw-Hill
[Huang] K. Huang, *Statistical Mechanics*, Wiley
[Pathria] R.K. Pathria, *Statistical Mechanics*, Pergamon
[Stanley] E. Stanley, *Introduction to Phase Transitions and Critical Phenomena*, Oxford University Press
[Ma] S.K. Ma, *Modern Theory of Critical Phenomena*, Bemjamin/Cummings
[Kittel] C. Kittel, *Introduction to Solid State Physics*, Wiley
[AshcroftMermin] N.W. Ashcroft and N. D. Mermin, *Solid State Physics*, Harcourt College Publishers
[Ziman] J.M. Ziman, *Principles of the Theory of Solids*, Cambridge University Press
[KittelQTS] C. Kittel, *Quantum Theory of Solids*, Wiley
[Madelung] O. Madelung, *Introduction to Solid-State Theory*, Springer-Verlag
[Gasiorowicz] S. Gasiorowiz, *Quantum Physics*, Wiley
[Merzbacher] E. Merzbacher, *Quantum Mechanics*, Wiley

[BicudoSacramento] P. Bicudo, P. D. Sacramento, M. Cardoso, M. M. Correia, S. F. Martins, R. Monteiro, M. F. Paulos, J. G. Rosa and J. E. Santos, *Mecânica Quântica*, IST Press (in Portuguese)
[Amit] D.J. Amit, *Field Theory, the Renormalization Group and Critical Phenomena*, World Scientific
[Ravndal] Finn Ravndal, *Scaling and Renormalization Groups*, Nordita
[BurkhardtLeeuwen] T.W. Burkhardt and J.M.J. van Leeuwen (editors), *Real-Space Renormalization*, Springer-Verlag
[Reichl] L.E. Reichl, *A Modern Course in Statistical Physics*, University of Texas Press
[Goldenfeld] N. Goldenfeld, *Lectures on Phase Transitions and the Renormalization Group*, Perseus Books Publishing
[ChaikinLubensky] P.M. Chaikin and T.C. Lubensky, *Principles of Condensed Matter Physics*, Cambridge University Press
[Schrieffer] J.R. Schrieffer, *Theory of Superconductivity*, Benjamin/Cummings
[deGennes] P. G. de Gennes, *Superconductivity of Metals and Alloys*, Addison-Wesley
[GuntonDroz] J.D. Gunton and M. Droz, *Introduction to the Theory of Metastable and Unstable States*, Springer-Verlag
[Continentino] M.A. Continentino, *Quantum Scaling in Many-Body Systems*, World Scientific
[Sachdev] S. Sachdev, *Quantum Phase Transitions*, Cambridge University Press
[AraujoSacramento] M.A.N. Araújo and P.D. Sacramento, *Topologia em Matéria Condensada: Uma Introdução*, IST Press (in Portuguese) and its English translation *Topology in Condensed Matter: An Introduction*, World Scientific

Several books with problems have been useful over the years such as [Mihaly] L. Mihály and M.C. Martin *Solid State Physics Problems and Solutions*, Wiley, the collection [PhDQualifying] on *Problems and Solutions of Major American Universities PhD Qualifying Questions and Solutions*, World Scientific, specifically on *Thermodynamics and Statistical Mechanics*, on *Solid State Physics, Relativity and Miscellaneous Topics*, and on *Electromagnetism*, World Scientific, and [Dalvit] D.A.R. Dalvit, J. Frastai, I. Lawrie *Problems on Statistical Mechanics*, Institute of Physics Publishing.

Also, lecture notes from the classes of various teachers and, in general, physics discussions have been used in parts of the book such as by Vitor R. Vieira, Pedro Schlottmann, James D. Gunton, and Theodore W. Burkhardt as well as some personal notes, thesis, or exercises by Iveta Pimentel, Carlos Matos Ferreira, Ana Noronha, Kong-Ju Bock Lee, Nuno Peres, Eduardo Castro, Miguel Gonçalves, Ana Martins, and José Luís Martins.

Discussions over the years with Eduardo Ducla Soares, Dieter Forster, Miguel Araújo, Nuno Peres, Zlatko Tesanovic, Vitalii Dugaev, Eduardo Castro, Pedro Ribeiro, José Carmelo, Dionys Baeriswyl, Pedro Bicudo, Alexander Nersesyan, and more recently with Tilen Cadez, Rubem Mondaini, Hai-Qing Lin, Wing Chi Yu, Nikola Paunkovic, Ricardo Dias, Henrik Johannesson, Natan Andrei, Mucio Continentino, Bruno Mera, and Bruno Amorim are also acknowledged.

The subjects selected are intended so that the book may be used as a textbook on materials in the area of Condensed Matter Physics, up to the Master degree level. Several exercises are proposed at the end of each chapter. In order to keep the mathematical formalism as simple as possible, more advanced topics, that are most easily dealt with using many-body theory, have been avoided or some alternative, less technical, approach has been adopted. More advanced topics and formalism that are not in general covered in this book may be pursued in various references such as

[AGD] A.A. Abrikosov, L.P. Gorkov and I.E. Dzyaloshinski, *Methods of Quantum Field Theory in Statistical Physics*, Dover

[Mahan] G.D. Mahan, *Many-Particle Physics*, Plenum Press

[FetterWalecka] A.L. Fetter and J.D. Walecka, *Quantum Theory of Many-Particle Systems*, Dover

[DoniachSondheimer] S. Doniach and Sondheimer, *Green's Functions for Solid State Physics*, World Scientific

[Fradkin] E. Fradkin, *Field Theories of Condensed Matter Systems*, Addison-Wesley

[PinesNozieres] D. Pines and P. Nozières, *The Theory of Quantum Liquids*, Addison-Wesley

[BernevigHughes] B.A. Bernevig with T. L. Hughes, *Topological Insulators and Topological Superconductors*, Princeton University Press

[Coleman] P. Coleman, *Introduction to Many-Body Physics*, Cambridge University Press

[Hewson] A.C. Hewson, *The Kondo Problem to Heavy Fermions*, Cambridge University Press

Most results presented in the book are derived with some detail but there are some topics that are more advanced or technically more challenging, whose results are presented and discussed without a full derivation and may be considered as better suited as an introduction to a course on Topics in Condensed Matter. Even though the book covers many topics, there are significant omissions, such as a description of liquid systems, including liquid crystals and superfluids, time-dependent phenomena, and a thorough description of systems far from equilibrium (even though some topics are addressed), interface phenomena such as junctions of different materials, spintronics, theory of electronic energy bands, etc. Also, the book is, for the most part, omissive along the direction of the properties and applications of materials and devices, mainly for reasons of space. A brief survey of some materials and their applications is presented in the closing chapter of the book. The reader is encouraged to explore different directions in the wide range of topics in the rich field of Condensed Matter materials.

Lisbon, Portugal
November 2024

Pedro D. Sacramento

Declarations

Competing Interests The author has no competing interests to declare that are relevant to the content of this manuscript.

Contents

Volume 1

Volume 2

Chapter 1
Introduction to Macroscopic Systems

1.1 Thermodynamic Equilibrium States: Irreversibility

In classical mechanics, a description of a system is achieved specifying, for a given instant, the position and velocity of each of its particles. In a system of N particles, there are therefore $6N$ variables. Macroscopic systems contain very large numbers of particles of the order of the Avogadro number, and the problem becomes very complex. It is virtually impossible to describe the individual evolution of the particles that constitute the system.

As it turns out, experience shows that, in order to describe the system, a finite number of parameters is enough. The macroscopic state of the system is characterized by a reduced set of macroscopic variables, related to some average properties of the system. This constitutes the basis for the success of thermodynamics: the ability to describe a system by a reduced set of parameters such as the energy, U; temperature, T; number of particles, N; volume, V; pressure, p; chemical potential, μ; magnetization, M; polarization, P; etc. It is not obvious a priori that this reduced set will be enough to characterize the system, nor which variables are to be chosen. Experiments indicate a way to proceed, but a fully microscopic explanation and description is desirable and needed. This is the goal of a statistical description of the properties of a macroscopic system: through the introduction of some prescriptions, one calculates average values of the properties of the system constituents and establishes a connection to the entities described by thermodynamics and, ultimately, to quantities that are experimentally accessible.

From the point of view of thermodynamics, one introduces postulates or laws that are empirically established. From these laws and some postulates, one draws conclusions about generic properties valid for different systems and materials. Experiment shows that only a rather small number of parameters is enough to describe the (equilibrium) state of a macroscopic system. For instance, it is enough to consider two independent parameters to specify the state of an homogeneous fluid, chosen from p, V, μ, T, U, S, where S is the entropy to be defined later. A

P. D. Sacramento, *Introduction to Condensed Matter Theory*,
https://doi.org/10.1007/978-3-031-86251-9_1

usual choice are the pressure and volume and the equilibrium states of the fluid are represented by points on the $p - V$ plane. As a consequence, the thermodynamic variables in a state of equilibrium have a determined value.

Besides characterizing states of equilibrium, a fundamental goal of a thermodynamic description is also to describe transformations between equilibrium states. This includes processes that are strongly displaced from equilibrium states, and difficult questions such as the characterization of intermediate states arise. If the intermediate states are far from equilibrium, a simple description in terms of a few variables is questionable and a more detailed description may be warranted. If the succession of intermediate states do not deviate much from equilibrium states the description is simplified.

Newton's equations of motion are invariant under time reversal $t \rightarrow -t$ and are reversible. Also, in simple systems and neglecting friction, if a system is isolated, energy is conserved. Writing the energy as a sum of kinetic and potential energy, the possible processes can be described by interchanges of kinetic and potential energy.

On the other hand, macroscopic systems are characterized by states of equilibrium among all the possible states of the system. The notion of an equilibrium state implies therefore a state that, from the macroscopic point of view, looks unchanged as time goes by. Also, if a system is not initially in an equilibrium state, the tendency to equilibrium implies some irreversibility as time goes by. The system tends to find a state of equilibrium and, having found it, remains for an indetermined time in that state. One needs to explain from a microscopic point of view how a system that is basically described by microscopic reversible equations of motion attains irreversibility.

In addition to the fundamental interest of the description of macroscopic systems and their properties, in opposition to simple systems of a few particles, the development of thermodynamics had to do with a practical interest to use the energy contained in macroscopic systems. A delicate issue is the identification of different energy exchange processes as work or heat. Broadly speaking, work is associated with an ordered, controlled, exchange of energy between two systems. Typically it involves changes of energy due to a change of a small number of parameters (such as the height of a weight) that involves some sort of reversibility by changing the direction the variable changes. Other processes, typically distributed among many degrees of freedom, therefore in a somewhat uncontrolled way, are called heat. A simple example is exchanges of kinetic energy among a large number of particles or degrees of freedom. The inherent lack of hability to control how the energy is distributed among the various particles leads to a loss of energy availability, which is clearly associated to a loss to eventually convert the energy into some ordered form (work) that may be used for some practical purpose. This inherent irreversibility, associated with the impossibility to recover the energy dissipated (distributed among a large number of degrees of freedom), is a central point that will be quantified in some way by the formalism. The thermodynamic laws will regulate energy transfer processes and establish criteria for the equilibrium states.

A fundamental problem of thermodynamics is to determine the resulting equilibrium state when two systems are made to interact. If the overall system is isolated, there are no exchanges of any quantity with the outside world, specifically no energy, volume, or number of particles exchanges with the exterior. Any exchanges of these quantities will only take place internally.

Consider an isolated system divided in two subsystems with energies U_1, U_2; volumes V_1, V_2; and numbers of particles N_1, N_2, respectively. Initially the two subsystems are isolated from each other and are separated by a wall that is fixed, not allowing exchange of energy or particles. If we relax some of these restrictions, the systems can exchange volume, energy, and/or particles. The way these conditions are relaxed will determine the final state.

Consider now that the system is actually divided into three subsystems 1, 2, 3. Assume now that we allow the exchange of energy between subsystems 1 and 3 but no exchange of volume or number of particles. There will be a flow of thermal energy between the two subsystems until some equilibrium state is reached. Then we allow that thermal energy can be exchanged between subsystems 2 and 3, and once again we wait long enough so that thermal equilibrium is reached between subsystems 2 and 3. Then it is clear that if we now allow exchange of thermal energy between subsystems 1 and 2 nothing happens: that is, if subsystems 1 and 3 are in thermal equilibrium and subsystems 3 and 2 are in thermal equilibrium, then subsystems 1 and 2 are also in thermal equilibrium. This empirical result is sometimes known as the zeroth law of thermodynamics. Let us loosely introduce the concept of temperature as a quantity that characterizes thermal equilibrium. Specifically, if two or more systems have the same temperature, they are in thermal equilibrium. Reversely, if two systems do not have the same temperature and are put into thermal contact, a heat flow will occur between the two. In order to reach equilibrium, we expect that the flow will be such as to equilibrate the temperatures of the two systems. When the temperature is equal everywhere, the net flow of thermal energy stops.

Two questions need to be answered: (i) Why does the system reach an equilibrium state? And (ii) why does it remain indefinitely in that state? One should note first that equilibrium exists on a macroscopic scale. The key to understand what equilibrium means lies with the microscopic description. In other words, how can we understand an equilibrium macroscopic state, given the constant motion of the microscopic constituents of a large system with the associated very large sequence of microscopic states due to the various individual (or collective) states of the particles?

An example is provided by a simple gas. Consider that the gas is contained in some fraction of the volume of a container. This can be achieved, for instance, by introducing a wall that limits the motion of the particles of the gas to a fraction of the volume (for instance half). Consider now that we remove the wall and that the gas is allowed to expand and fill the entire volume of the container. (To simplify matters, assume the wall is removed perpendicularly to the direction of the expansion of the particles of the gas). Also, for simplicity, consider that the gas is ideal in the sense that we can neglect the interactions between the particles. Therefore the total energy is just kinetic (we also neglect the effect of gravity). Considering elastic collisions between the particles and the walls of the container, the total energy is conserved, and, since the system is isolated, the total number of particles is conserved. As stated above, the equations of motion of each particle are reversible in time. However, the macroscopic system shows irreversibility in a very direct way: the initial state is not observed because at the very least this would require that at some time in the future all the particles would be contained in the initial spatial region (for instance, in the left half of the container). Even though this is not forbidden and is one possible state of the system (actually a very large number of microscopic states of the system), it is not observed. A simple explanation is that it is very unlikely to find all the particles in a fraction of the total volume of the container. A simple estimation goes as follows. Call V the total volume of the container and consider a fraction V' of the total volume. If the particles are free to move and their energy is fully kinetic, all sites in the container are equivalent. Therefore, it must be equally probable to find any given particle in any site (this is a postulate of equal probability). Therefore, the probability to find one particle in volume V' must be proportional to the volume of this region. That is the probability to find one particle in that volume is

$$P(V') = \frac{V'}{V} = p. \tag{1.1}$$

Normalizing the probability, we find that the probability to find the particle in the rest of the container is

$$P(V - V') = \frac{V - V'}{V} = q = 1 - p. \tag{1.2}$$

Assume there are N particles in the container. Then we can find the probability to find N_1 particles in volume V' and $N_2 = N - N_1$ particles in the rest of the container. It is given by

$$P(N_1, N_2) = \frac{N!}{N_1!(N - N_1)!} p^{N_1} q^{N-N_1} = \Omega(N_1). \tag{1.3}$$

The first factor in this expression is the number of different ways to distribute N objects of which N_1 are equal and the remaining N_2 are also equal among themselves, in two separate groups (the two parts of the container). Fixing the value of N_1 (and N_2 since N is fixed), the other factors are just the result that the probability of independent events is just the product of the probabilities of each event separately. The distribution $\Omega(N_1)$ is the binomial or Bernoulli's distribution. It is clearly normalized recalling Newton's binomial

$$(p + q)^N = \sum_{n=0}^{N} \frac{N!}{n!(N - n)!} p^n q^{N-n}. \tag{1.4}$$

The value of the number of particles in one part of the container is therefore a statistical variable with a given probability distribution. We know in general that if a variable can take a set of discrete values, for instance, a variable u that can take a set of values $u_1, \cdots, u_M$, and to each value we may associate a given probability $P(u_1), \cdots, P(u_M)$, then the average value that variable takes is given by

$$\langle u \rangle = \frac{\sum_{i=1}^{M} u_i P(u_i)}{\sum_{i=1}^{M} P(u_i)}. \tag{1.5}$$

If the probability distribution is normalized $\sum_i P(u_i) = 1$, the expression simplifies, as usual. Also, we can calculate the average value of any function of the statistical variable in a similar way

$$\langle f(u) \rangle = \sum_{i=1}^{M} f(u_i) P(u_i). \tag{1.6}$$

We can then calculate the average value of the number of particles in volume V' of the container. This is given by

$$\langle N_1 \rangle = \sum_{N_1=0}^{N} N_1 \frac{N!}{N_1!(N-N_1)!} p^{N_1} q^{N-N_1}. \tag{1.7}$$

Using that

$$N_1 p^{N_1} = p \frac{\partial}{\partial p} \left(p^{N_1} \right),$$

we get

$$\begin{aligned} \langle N_1 \rangle &= p \frac{\partial}{\partial p} \left(\sum_{N_1=0}^{N} \frac{N!}{N_1!(N-N_1)!} p^{N_1} q^{N-N_1} \right) \\ &= p \frac{\partial}{\partial p} (p+q)^N \\ &= Np = N \frac{V'}{V}, \end{aligned} \tag{1.8}$$

as expected. If $V' = V/2$, we expect that on average there are half the number of particles in each half of the container. The interesting point comes out if we calculate the fluctuations around this average value. Recall that we want to see if it is possible to find all the particles in one of the subsystems. This would imply a large fluctuation around the average value since it would require a zero number of particles in one of the subsystems. We can calculate the standard deviation in the usual way. We begin by calculating the variance of the distribution

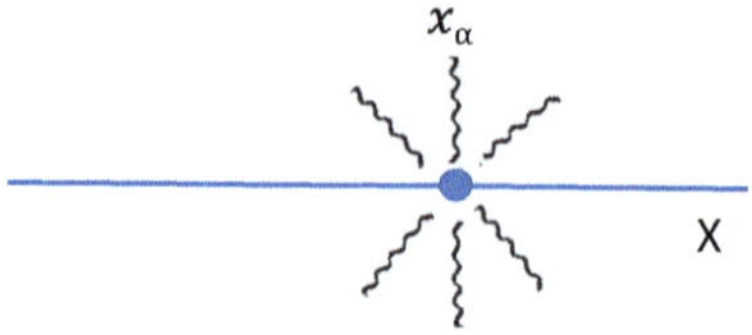

Fig. 1.1 Coupling of a particle moving in a one-dimensional system to α harmonic degrees of freedom of a reservoir

$$\sigma^2 = \sum_{N_1=0}^{N} (N_1 - \langle N_1 \rangle)^2 \, \Omega(N_1). \tag{1.9}$$

A calculation leads to the standard deviation

$$\sigma = \sqrt{Npq}. \tag{1.10}$$

The relative fluctuation around the average value can be calculated and gives

$$\frac{\sigma}{\langle N_1 \rangle} = \frac{1}{\sqrt{N}} \sqrt{\frac{V}{V'} - 1}. \tag{1.11}$$

If the system number of particles is fixed, N fixed, we see that the fluctuation is vanishingly small if $V' >> V/N$. Therefore, we get a very small dispersion if the volume is macroscopically large. In other words the probability decreases very fast as we move away from the average value. Also, the average value is the most probable. Therefore, the average value is the observed one since it is the most probable (a larger number of microscopic configurations that lead to the macroscopic state characterized by a single variable, the number of particles in the subsystem), and since it is much more probable than all the others, the system remains in this most probable state. So it is natural that the system tends to equilibrium, since this is the most probable state, and it remains there since all the other states are much less probable. In other words as the system size N grows, the distribution is extremely sharp around the most probable state.

1.2 Langevin Equation

The origin of irreversibility can be elucidated further considering a system interacting with some external reservoir. In general the Hamiltonian of the system plus reservoir is written as

$$H = H_S + H_{S-R} + H_R. \tag{1.12}$$

To simplify consider one particle in a one-dimensional system. We write

$$
\begin{aligned}
H_S &= \frac{p^2}{2m} + V(x) \\
H_{S-R} &= x \sum_\alpha c_\alpha x_\alpha, \qquad \alpha = 1, \cdots, N \\
H_R &= \sum_\alpha \frac{p_\alpha^2}{2m_\alpha} + \frac{1}{2} m_\alpha \omega_\alpha^2 x_\alpha^2 .
\end{aligned}
\tag{1.13}
$$

We consider that the reservoir is composed of N degrees of freedom associated with particles of different masses, m_α, subject to harmonic potentials of frequencies ω_α associated with oscillations along x_α coordinates. c_α are the couplings to the reservoir variables (Fig. 1.1). Hamilton's equations of motion

$$
\begin{aligned}
\frac{dx}{dt} &= \frac{\partial H}{\partial p}, \qquad & \frac{dx_\alpha}{dt} &= \frac{\partial H}{\partial p_\alpha} \\
\frac{dp}{dt} &= -\frac{\partial H}{\partial x}, \qquad & \frac{dp_\alpha}{dt} &= -\frac{\partial H}{\partial x_\alpha}
\end{aligned}
\tag{1.14}
$$

lead to

$$
\begin{aligned}
\frac{dx}{dt} &= \frac{p}{m}, \qquad & \frac{dx_\alpha}{dt} &= \frac{p_\alpha}{m_\alpha} \\
\frac{dp}{dt} &= -\frac{\partial V}{\partial x} - \sum_\alpha c_\alpha x_\alpha , \qquad & \frac{dp_\alpha}{dt} &= -m_\alpha \omega_\alpha^2 x_\alpha - c_\alpha x .
\end{aligned}
\tag{1.15}
$$

The equations of motion of the degrees of freedom of the reservoir can be solved. Newton's equation is

$$
\frac{d^2 x_\alpha}{dt^2} + \omega_\alpha^2 x_\alpha = -\frac{c_\alpha}{m_\alpha} x . \tag{1.16}
$$

The solution of the homogeneous equation (without the right hand term)

$$
\frac{d^2 x_\alpha^0}{dt^2} + \omega_\alpha^2 x_\alpha^0 = 0 \tag{1.17}
$$

has the form

$$
x_\alpha^0(t) = x_\alpha^0(t_i) \cos \omega_\alpha (t - t_i) + \frac{p_\alpha(t_i)}{m_\alpha \omega_\alpha} \sin \omega_\alpha (t - t_i), \tag{1.18}
$$

with the initial conditions $x_\alpha^0(t_i)$, $p_\alpha(t_i)$ at time t_i. The solution of the full equation is then

$$x_\alpha(t) = x_\alpha^0(t) - \int_{t_i}^{t} dt' \frac{1}{\omega_\alpha} \sin \omega_\alpha (t - t') \frac{c_\alpha}{m_\alpha} x(t'). \tag{1.19}$$

Similarly, Newton's equation for the system may be written as

$$m \frac{d^2 x}{dt^2} = -\frac{\partial V}{\partial x} - \sum_\alpha c_\alpha x_\alpha(t). \tag{1.20}$$

Substituting the solution for the variables of the reservoir, we obtain that

$$\begin{aligned} m \frac{d^2 x}{dt^2} &= -\frac{\partial V}{\partial x} - \sum_\alpha c_\alpha x_\alpha^0(t) + \sum_\alpha \frac{c_\alpha^2}{m_\alpha \omega_\alpha} \int_{t_i}^{t} dt' \sin \omega_\alpha (t - t_i) x(t') \\ &= -\frac{\partial V}{\partial x} - \sum_\alpha c_\alpha x_\alpha^0(t) - \sum_\alpha \frac{c_\alpha^2}{m_\alpha \omega_\alpha^2} \int_{t_i}^{t} dt' \frac{d}{dt} \cos \omega_\alpha (t - t_i) x(t') \\ &= -\frac{\partial V}{\partial x} - \sum_\alpha c_\alpha x_\alpha^0(t) - \frac{d}{dt} \int_{t_i}^{t} dt' \sum_\alpha \frac{c_\alpha^2}{m_\alpha \omega_\alpha^2} \cos \omega_\alpha (t - t_i) x(t') \\ &\quad + \sum_\alpha \frac{c_\alpha^2}{m_\alpha \omega_\alpha^2} x(t). \end{aligned} \tag{1.21}$$

This equation can be written as

$$m \frac{d^2 x}{dt^2} = -\frac{\partial V}{\partial x} + \xi(t) + \Delta \omega_0^2 x(t) - \Gamma \frac{dx(t)}{dt}, \tag{1.22}$$

known as Langevin's equation. The second term on the right hand side is given by

$$\xi(t) = -\sum_\alpha c_\alpha x_\alpha^0(t) \tag{1.23}$$

and

$$\Delta \omega_0^2 = \sum_\alpha \frac{c_\alpha^2}{m_\alpha \omega_\alpha^2}. \tag{1.24}$$

The origin of the fourth term can be understood in the following way. The integrand in the term containing the integral over time,

$$\int_{t_i}^{t} dt' \cos \omega_\alpha (t - t_i) x(t'),$$

is a rapidly oscillating function. The dominant term is therefore when $t' = t$ and the integral is proportional to $x(t)$. This leads to the term proportional to the time derivative of the coordinate $x(t)$. Note that this is a qualitatively new term in the equation of motion, since it is not reversible in time. The irreversibility shows up through this term, which is a term proportional to the velocity and may be recognized as a friction term. This is well-known to lead to irreversibility effects and loss of energy through dissipation.

The term $\xi(t)$ is also new. Note that it depends on the initial conditions of the reservoir. These are not controllable since in general the reservoir has many degrees of freedom. Our ignorance on the specific initial conditions requires that we take some average over the possible values they may take. Taking the average over the possible initial conditions of the reservoir and considering that the dependence is periodic, we can take the average

$$\langle \xi(t) \rangle = \frac{1}{T} \int_{t_i}^{t_i+T} dt' \xi(t'), \tag{1.25}$$

where T is the period. This leads to

$$\langle \xi(t) \rangle = \frac{1}{T} \int_{t_i}^{t_i+T} dt' \left[-\sum_\alpha c_\alpha x_\alpha^0(t_i) \cos \omega_\alpha (t'-t_i) \right.$$
$$\left. - \sum_\alpha c_\alpha \frac{p_\alpha(t_i)}{m_\alpha \omega_\alpha} \sin \omega_\alpha (t'-t_i) \right]. \tag{1.26}$$

Defining $z = t' - t_i$ we can show that

$$\langle \xi(t) \rangle = 0. \tag{1.27}$$

Therefore, this term is a term that fluctuates and has zero average. It is like a random term that results from the coupling of the system to the degrees of freedom of the reservoir. However, this term does not vanish for a given instant. This can be shown considering its correlations. We can show that

$$\langle \xi(t)\xi(t') \rangle \neq 0, \tag{1.28}$$

taking the average over the initial conditions of the reservoir. As will be shown later this correlation is proportional to Γ if $t = t'$ and zero otherwise.

The coupling of the system to the degrees of freedom of the reservoir leads to two terms in the equation of motion of the variable of the system that reflect the irreversibility of the coupling to many degrees of freedom. The term proportional to the velocity is the result of the fast oscillations induced by this coupling. This is true if the frequencies ω_α are large. Effectively, substituting the time dependence of the reservoir variables in the equation of motion for the system and taking the approximation that only the equal time contribution is important is equivalent to eliminating the fast degrees of freedom of the reservoir. This is true if the time evolution of the system is slow compared to the fluctuations of the reservoir.

Chapter 2
Revision of Thermodynamics

2.1 Extensive and Intensive Variables

The energy is an extensive quantity that depends on the size of the system. The larger the system, the larger the energy. Other quantities that are extensive are the number of particles, the volume, and the entropy. Since the energy is extensive, it is assumed that it is only a function of the extensive variables. This leads to

$$U = U(S, X_1, X_2, \cdots, X_t), \tag{2.1}$$

where X_i indicates a set of extensive variables and where we have singled out the entropy, S. We postulate that the energy (or the entropy) is a homogeneous function of degree one:

$$U(\lambda S, \lambda X_1, \cdots, \lambda X_t) = \lambda^1 U(S, X_1, \cdots, X_t). \tag{2.2}$$

Since the natural variables of the energy are the extensive variables, we can write that

$$dU = \frac{\partial U}{\partial S}|dS + \frac{\partial U}{\partial X_1}|dX_1 + \cdots + \frac{\partial U}{\partial X_1}|dX_t. \tag{2.3}$$

Defining the partial derivatives as

$$\begin{aligned} T &= \frac{\partial U}{\partial S}| \\ p_j &= \frac{\partial U}{\partial X_j}|, \end{aligned} \tag{2.4}$$

P. D. Sacramento, *Introduction to Condensed Matter Theory*,
https://doi.org/10.1007/978-3-031-86251-9_2

these are, naturally, intensive variables, since they are ratios of extensive variables. They are therefore homogeneous functions of degree zero, such as

$$T(\lambda S, \lambda X_1, \cdots, \lambda X_t) = \lambda^0 T(S, X_1, \cdots, X_t), \tag{2.5}$$

and, therefore, are independent of a rescaling of the system size. Later it will become clear that T is the temperature. Differentiating equation (2.2) with respect to λ, we get that

$$\begin{aligned}
\frac{\partial U(\lambda S, \cdots)}{\partial(\lambda S)} \frac{\partial(\lambda S)}{\partial \lambda} + \sum_j \frac{\partial U(\cdots)}{\partial(\lambda X_j)} \frac{\partial(\lambda X_j)}{\partial \lambda} &= U(S, \cdots) \\
\frac{\partial U}{\partial S} S + \sum_j \frac{\partial U}{\partial X_j} X_j &= U \\
TS + \sum_j p_j X_j &= U.
\end{aligned} \tag{2.6}$$

Choosing the other variables as the volume and the number of particles and designating the partial derivatives

$$\begin{aligned}
p &= -\frac{\partial U}{\partial V} \\
\mu &= \frac{\partial U}{\partial N},
\end{aligned} \tag{2.7}$$

called pressure and chemical potential, respectively, we get Euler's equation

$$U = TS - pV + \mu N. \tag{2.8}$$

The differential of the energy leads then to

$$dU = TdS + SdT - pdV - Vdp + \mu dN + Nd\mu. \tag{2.9}$$

On the other hand

$$dU = TdS - pdV + \mu dN, \tag{2.10}$$

which implies the Gibbs-Duhem relation

$$SdT - Vdp + Nd\mu = 0. \tag{2.11}$$

As a consequence, this gives that the change of the chemical potential is given by

$$d\mu = -sdT + vdp, \tag{2.12}$$

where $s = S/N$ and $v = V/N$ are intensive quantities. In general we have that

$$SdT + \sum_j X_j dp_j = 0. \tag{2.13}$$

If we have a multicomponent system with different types of particles, we can associate a different chemical potential to each particle species. We define

$$\mu_j = \frac{\partial U}{\partial N_j} \tag{2.14}$$

and the change of energy generalizes to

$$dU = TdS - pdV + \sum_{i=1}^{r} \mu_i dN_i, \tag{2.15}$$

for a r-component system.

The intensive parameters are responsible for the equilibrium conditions between two systems of arbitrary extensions. As said above, these parameters do not depend on the system size. The temperature is associated with thermal equilibrium and heat flux in its absence, pressure with changes of volume, and the chemical potential with particle flux.

2.1.1 Thermometry

We need now to quantify the concept of temperature and thermal equilibrium. If the temperature is the same in two subsystems, we have thermal equilibrium, but if the temperatures are different, the rate at which thermal energy is exchanged depends on the temperatures of the two subsystems, in a way that has to be quantified.

One possible way to quantify temperature is through a gas thermometer. It is known experimentally that the pressure and volume of a dilute gas obey the law $pV \sim T$, as shown in Fig. 2.1. If the temperature of the system is maintained constant, the product pV is approximately constant. One way to maintain the temperature constant is to keep the system in thermal contact with a very large system that has its temperature constant (heat reservoir). One can calibrate the temperature scale by changing its temperature (by thermal contact with another gas at different temperatures) by drawing the various lines at constant temperature in a $p - V$ diagram. For instance, keeping the volume constant only the pressure varies from each line at constant temperature to the next one (isotherms). Therefore we get a dependence law of $p = p(T)$. We can therefore associate a given pressure to a given temperature (having fixed the volume). There are various types of thermometers that operate with different systems and that are suitable in different

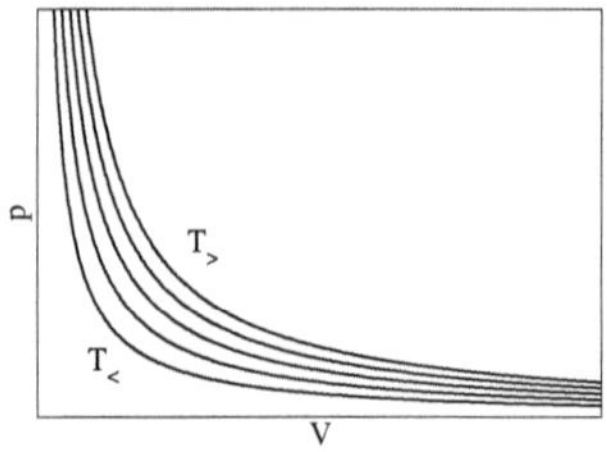

Fig. 2.1 Curves of constant temperature of an ideal gas in a pV diagram

temperature regimes. As it turns out, the gas thermometer is particularly important since it sets a temperature scale designated absolute temperature, as we shall see later. Other examples are:

- Gas thermometer using hydrogen or helium that uses $p(T)$ for a constant V.
- Liquid thermometer using mercury, alcohol, etc. uses volume at constant pressure.
- Electrical resistance thermometer of platinum uses resistivity as a function of temperature
- Thermocouples of various materials such as bismuth-bismuth antimony, iron-constantan, copper-constantan, and chromel-alumel use electric potential difference between the two sides of the metallic-junction, one placed in the cold system (reference) and the other in the hot system.
- Low-temperature thermometers that use as a reference points the ebullition points of N_2, H_2, and He and use electrical resistance of platinum or coal wires. Also, at low temperatures one may use magnetic susceptibility measurements since at low temperatures the susceptibility is inversely proportional to temperature.

When two systems at different temperatures are put into contact, there is thermal energy exchanged between the two systems, until equilibrium is reached. We expect that the net flow of energy goes in the direction of the system with the highest temperature to the system with the lowest temperature. In general the temperature of the system with the highest initial temperature will decrease and vice versa. Therefore if one of the systems is the thermometer, we must guarantee that the measurement process does not affect the temperature of the system to be measured. The thermometer must therefore be small enough that its temperature becomes equal in equilibrium to the temperature of the system to be measured. This is possible if the thermometer is very small as compared to the size of the system to be measured.

2.2 Transformations Between Thermodynamic States

In general, if there is a change of state of a system due to some transformation, there is an exchange of energy with the outside world. The exchange of energy involves in general that some work and/or heat is exchanged in the process. We will use the convention that if in the transformation there is a net influx of heat, this is positive

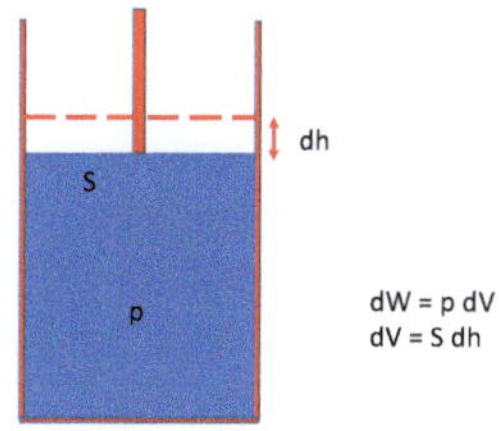

Fig. 2.2 Work that results from the gas pressure on a piston of area S

from the point of view of the system (it receives energy), and if the flux is outward, this is negative (the system loses energy to the outside world). If the system does work on the outside, we take it to be positive, and if the system receives work, this is negative. Note that some books use a different convention and consider that any sort of received energy is positive and any sort of given energy is negative.

Take as an example a gas at pressure p contained in a volume V and a piston of area S, as shown in Fig. 2.2. The force that the gas exerts on the piston is pS. Assume now that, due to this force, the piston is displaced a distance dh from its initial position such that the volume of the gas increases by $dV = Sdh$. Then the work done by the gas on the piston is given by

$$dW = pSdh = pdV. \tag{2.16}$$

Since the change in volume is positive, the work is positive, as is usual in mechanics (note again that the convention of some books is the opposite of the convention in mechanics). This exchange of energy with the outside world is, apparently, easy to quantify since it only involves a change of a reduced number of variables, in this example just the variable h, the perpendicular coordinate to the surface of the piston.

Exchanges of heat are more difficult to quantify since they involve a large number of degrees of freedom. They are associated with the change of the kinetic degrees of freedom of the particles of the system. The exchanges of thermal energy between two systems occur through the complex process of exchange of kinetic energy of the particles of one system with the walls of the container that then are transmitted, also by collisions of the particles of the other system with the wall. This is a complex problem that eventually leads to some sort of thermal equilibrium due to the change of the average kinetic energy due to collisions. Thermal equilibrium is attained since, on average, fast particles in a region of high temperature, when colliding with slower particles in a region of lower temperature, lose part of their kinetic energy (and lower their temperature), while the particles with smaller average velocities increase their energy (and increase their temperature). The complexity of these exchanges due to collisions can only be addressed in terms of average values, as will be seen later on.

The transformations between different states may be of different kinds. It is useful to define some particular processes:

- Quasi-static transformations. If the transformation is very slow, we may consider that it is a succession of equilibrium states. The transformation evolves slow enough that, at each instant, the system has time enough to equilibrate, before being submitted to a new change of parameters. The processes by which a thermodynamic system tends to equilibrium are actually quite fast at the macroscopic scale. The origin lies in the microscopic processes that are ultimately responsible for the equilibration, and these occur at the particle level with time scales that are typically very small, as compared to our scale of one second. The simplification to this type of quasi-static processes is important, since only in equilibrium states are the thermodynamic parameters well-defined. As we will see later in detail, the thermodynamic parameters are defined in terms of some average values of the microscopic properties. If a system is in a state that is far from equilibrium, as it happens if the transformation is fast enough, in general the fluctuations are large and the deviation from an equilibrium state can be large. For instance, the system may be in a strongly inhomogeneous state, and therefore it does not make sense to define the state of the system by a few macroscopic parameters. One may consider some macroscopic parameters that are defined locally, for instance, temperature and pressure that are defined locally, but this leads to an extension to local thermodynamics that is possible, but that we will not pursue here. So we will restrict here, and for the most part, to a thermodynamic description where we only consider slow enough transformations that are close to a sequence of equilibrium states. In the microscopic description that we will also consider later, we will consider more general transformations.

Within this restriction of quasi-static transformations, it seems then reasonable that, in a process where we have a sequence of equilibrium states, therefore the pressure of the system is well-defined throughout the transformation, an expression of the type $dW = pdV$ for the work done by the system would be valid. We will see later that this is not quite so.

- Reversible and irreversible transformations. We define a reversible transformation when, besides being quasi-static, if we reverse the direction of the transformation the system *and* the outside world go through the same sequence of states, but in reverse order. The transformation is irreversible in all other cases. In general, all real transformations are irreversible but one may get as close as possible to a reversible transformation. It is in general possible to revert the sequence of states of a given system but the difficulty is to guarantee the same for the exterior. If there is friction, the transformation is irreversible since, by definition, friction involves an uncontrolled loss of energy that cannot be recovered in an ordered way. As in general in real processes, there is always some sort of friction present; in general the transformations are irreversible. However, as will be shown later, the concept of a reversible transformation is quite important and useful.

  An idealized example of a reversible transformation is a reversible expansion of a gas contained in a cylinder with a piston with negligible friction and that all the walls of the cylinder are such that no exchanges of heat are allowed. These are called adiabatic walls. If we now guarantee that the pressure of the gas and the outside pressure differ by an infinitesimal amount $p_{ext} = p_{int} \pm \delta$, then it is possible to invert the direction of the transformation and the gas and the outside may ideally go through the same sequence of states in reverse order.

A second idealized transformation is a system at a phase transition. For instance, in a mixture of ice and water while the temperature is at the freezing temperature $T = 0$ Celsius, we may change the relative concentrations of ice and water in a reversible way as long as the temperature remains at the critical temperature. The melting of the ice involves changes of the potential energy of the bonds between the water molecules and this supposedly involves few degrees of freedom and may be reversed. The change of the relative concentrations of ice and water is obtained putting, for instance, the system in contact with another one at a slightly higher or lower temperature. There is then a change of the energy potential if we assume the exchange of energy is sufficiently slow and homogeneous that all the energy is used to break or restore the chemical bonds and not to change the kinetic energy of the molecules. On the other hand, the thermal exchange between two systems at different temperatures not at a critical point is clearly irreversible.

- Cyclic transformations are such that the state of the system returns to the initial value.
- Isothermal transformations are such that temperature is constant.
- Adiabatic transformations are such that there is no heat exchange.
- Isobaric transformations are such that pressure is constant.
- Isochoric transformations are such that the volume is constant.

2.2.1 *First Law of Thermodynamics*

The basis of this law is the conservation of energy. In a mechanical system the energy is the sum of the kinetic and the potential energies. If there are no external forces (isolated system), energy is conserved. If the system is not isolated, the energy of the system changes. The change of energy is given by

$$U_B - U_A = -W, \tag{2.17}$$

where A is the initial state, B the final state, and W is the work done by the system. The work done by the system is only a function of A and B and not on the path. This happens because the internal forces are conservative, since we assumed we can write the internal forces as derived from some sort of potential energy.

Consider now a thermodynamic system. For instance, a given amount of water. The equilibrium states A and B are characterized by their pressure that we assume is the atmospheric pressure $p_A = p_B = 1$ atm and different temperatures t_A and t_B such that $t_A < t_B$. We may now consider two separate ways to go from A to B.

(i) We may supply heat to the system by burning a flame. In this way the temperature of the system will rise, and we may perform the transformation

$t_A \rightarrow t_B$. The work done by the outside world on the system is close to zero. Indeed, since the temperature of the system increases, the system will probably expand. Associated with a small change of volume, there is some work involved, but this is small and can be neglected in comparison to the heat exchanged.

(ii) We may now heat up the water by friction. Inserting puddles in the water, we may supply energy to the system such that it will rise its temperature such that the same final temperature is obtained $t_A \rightarrow t_B$. Using the resistance of the water, we need to do work on the system (needed to turn the puddles). The water does the same amount of work $|W|$ as its molecules act on the moving puddles. So the system goes from the same initial state to the same final state performing different amounts of work, which implies that work depends now on the path chosen to go from one state to another.

The energy exchanged in either transformation is of the form

$$\Delta U = Q - W, \tag{2.18}$$

and an infinitesimal transformation obeys the conservation of energy

$$dU + \delta W = \delta Q. \tag{2.19}$$

Both the exchange of thermal energy and the exchange of work depend on the path chosen to go from one state to another. On the other hand, energy is only a function of the change in state. The energy is called a state function. In order to distinguish a state function from changes that depend on the chosen path, we represent their changes differently.

If we consider a cyclic transformation, then $B = A$ and $\Delta U = 0$. In this case $W = Q$. If the system does work ($W > 0$) in order to preserve its energy it must receive the same amount of heat from the exterior ($Q > 0$). The work done by the system in a cyclic transformation is equal to the heat absorbed.

2.2.2 *Heat Capacity and Specific Heat*

If the transformation is quasi-static it seems natural to assume that $\delta W = pdV$. Then the first law may be written as

$$dU + pdV = \delta Q. \tag{2.20}$$

The natural variables of the energy, being an extensive quantity, are the extensive variables. On the other hand, as experiment shows, the thermodynamic state of a homogeneous single component system may be characterized by any two parameters. This means that not all parameters are independent and any two can be chosen. Using this freedom, we may take the energy to be a function of any two parameters, since it is a function of state and the thermodynamic state may be characterized by any two parameters. It is convenient to choose, for instance, that the energy of a state is a function of the extensive quantity volume but also of the intensive quantity temperature. Specifically we may write that $U = U(T, V)$. Actually, any thermodynamic quantity may be written as a function of any two parameters in this case. Therefore we may write that

$$dU = \left(\frac{\partial U}{\partial T}\right)_V dT + \left(\frac{\partial U}{\partial V}\right)_T dV. \tag{2.21}$$

The first partial derivative is calculated at constant volume and the second at constant temperature. Introducing this expression in the first law, we can write that

$$\left(\frac{\partial U}{\partial T}\right)_V dT + \left[\left(\frac{\partial U}{\partial V}\right)_T + p\right] dV = \delta Q. \tag{2.22}$$

We can now use extensively the property that any two functions are a function of any other two functions. Specifically we can consider that $U = U(T, p)$ and $V = V(T, p)$. We may then write that

$$\left[\left(\frac{\partial U}{\partial T}\right)_p + p\left(\frac{\partial V}{\partial T}\right)_p\right] dT + \left[\left(\frac{\partial U}{\partial p}\right)_T + p\left(\frac{\partial V}{\partial p}\right)_T\right] dp = \delta Q. \tag{2.23}$$

The heat capacity at constant X is defined as

$$C_X = \left(\frac{\delta Q}{dT}\right)_X. \tag{2.24}$$

Examples are the heat capacity at constant volume and the heat capacity at constant pressure

$$C_V = \left(\frac{\delta Q}{dT}\right)_V, \qquad C_p = \left(\frac{\delta Q}{dT}\right)_p. \tag{2.25}$$

Considering first constant volume, $dV = 0$, in Eq. (2.22) we get that

$$C_V = \left(\frac{\delta Q}{dT}\right)_V = \left(\frac{\partial U}{\partial T}\right)_V \tag{2.26}$$

and taking constant pressure $dp = 0$ in Eq. (2.23) we get that

$$C_p = \left(\frac{\delta Q}{dT}\right)_p = \left(\frac{\partial U}{\partial p}\right)_T + p\left(\frac{\partial V}{\partial p}\right)_T. \tag{2.27}$$

In the expression for the heat capacity at constant pressure, we can understand that the second term reflects the effect of the work done during the expansion. In the heat capacity at constant volume, it does not show up since the volume is constant.

The heat capacities are extensive quantities (since they are the ratios of an extensive quantity and an intensive quantity) and therefore depend on the size of the system. It is usual to define quantities that are intensive, and therefore characteristic of a given system and transformation, and do not depend on the system size. These are called specific heats and are defined as the ratio of the heat capacity either by the mass of the system, the number of particles, or the number of moles.

2.2.3 The Ideal Gas

As an example of the use of the ideas introduced, let us consider an ideal gas, a diluted system of weakly interacting particles that, to good approximation, may be taken as free particles.

Experimentally it has been determined with very good approximation that

$$pV = \nu RT, \tag{2.28}$$

where ν is the number of moles and R is the specific constant of gases with the value $R = 8.314$ Joules/Kelvin. The temperature scale in Kelvin will be introduced later and is such that it is a centigrade scale where O Celsius $= 273.15$ Kelvin. This equation is an example of an equation of state that relates the thermodynamic variables in an equilibrium state. This experimental law is known as the Boyle-Mariotte or Gay-Lussac law.

Consider now an isothermal quasi-static transformation where the work done by the system as the gas changes its volume from V_1 to V_2 is given by

$$\begin{aligned} W &= \int_{V_1}^{V_2} p dV = \nu RT \int_{V_1}^{V_2} \frac{dV}{V} \\ &= \nu RT \ln \frac{V_2}{V_1} = \nu RT \ln \frac{p_1}{p_2}, \end{aligned} \tag{2.29}$$

since the temperature is constant and therefore pV is constant. We need now to calculate the heat exchanged in such a transformation and the change of internal energy, ΔU. Even though we know the equation of state, we need to know how to calculate the energy of an ideal gas.

The following experiment by Joule, shown in Fig. 2.3, allows to find the energy of an ideal gas. Microscopically we know that, under the assumption that the system is composed of free particles, the energy is simply the sum of their kinetic energies. In the experiment the ideal gas is contained in part of the volume of an adiabatic container. The container is divided in two parts A and B, and initially the gas is

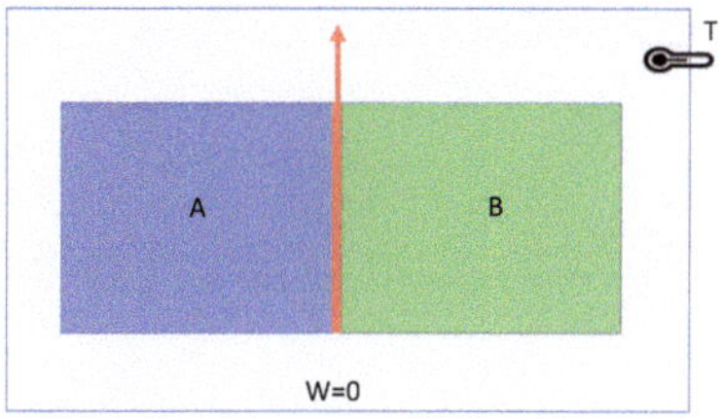

Fig. 2.3 The Joule experiment

contained in part A. There is a wall that separates regions A and B which is then removed perpendicularly. In this way the removal of the wall does no work on the gas. As the wall is removed, the gas expands and fills the entire volume, composed of regions A and B. Measuring the temperature of the gas before and after one finds, within experimental accuracy, that the temperature has not changed. Considering that the energy is a function of $U = U(T, V)$, the experiment shows that the energy is only a function of the temperature. Indeed, using the first law and since the walls are adiabatic, $Q = 0$, and there is no work performed, $W = 0$, we get that $\Delta U = 0$. The volume of the system has changed, and the temperature has remained constant, $\Delta T = 0$. Therefore, the energy is only a function of temperature.

Assume anyway that the energy is a function $U = U(T, V)$. Then we may write that

$$dU = \left(\frac{\partial U}{\partial T}\right)_V dT + \left(\frac{\partial U}{\partial V}\right)_T dV = \left(\frac{\partial U}{\partial T}\right)_V dT, \tag{2.30}$$

which leads to

$$U = C_V T + constant. \tag{2.31}$$

We can therefore write that for an ideal gas

$$\delta Q = pdV + C_V dT. \tag{2.32}$$

Using the equation of state, we get that

$$pdV + Vdp = \nu RdT \tag{2.33}$$

and

$$\delta Q = (C_V + \nu R)\, dT - Vdp. \tag{2.34}$$

We conclude that

$$C_p = C_V + \nu R, \tag{2.35}$$

or the corresponding relation for the molar specific heats

$$c_p = c_V + R. \tag{2.36}$$

This is a particular example of the general relation $c_p > c_V$.

In an adiabatic transformation $\delta Q = 0$, we get that

$$0 = dU + \delta W. \tag{2.37}$$

If the system does work ($W > 0$), the internal energy decreases, and therefore the temperature decreases. Reversely, if the exterior does work on the system ($W < 0$), the internal energy increases and so does the temperature. Therefore an adiabatic expansion leads to a decrease of the temperature and a compression to a temperature increase. Consider one mole of a gas. Then an adiabatic transformation leads to

$$\begin{aligned} c_V dT + p dV &= 0 \\ c_V dT + \frac{RT}{V} dV &= 0 \\ \frac{dT}{T} + \frac{R}{c_V}\frac{dV}{V} &= 0 \\ \ln T + \frac{R}{c_V}\ln V &= constant, \end{aligned} \tag{2.38}$$

which implies that

$$T V^{\frac{R}{c_V}} = constant. \tag{2.39}$$

Defining $\gamma = c_p/c_V = 1 + R/c_V$, we can write that in an adiabatic transformation

$$\begin{aligned} T V^{\gamma-1} &= C \\ p V^{\gamma} &= C' \\ T p^{\frac{1-\gamma}{\gamma}} &= C'', \end{aligned} \tag{2.40}$$

where C, C', C'' are constants. As compared to an isothermal, for which pV is constant, we can easily see that the adiabatics have larger slopes than the isothermal transformations (since $\gamma > 1$), as shown in Fig. 2.4.

A simple application of these results is the decrease of temperature with altitude in the atmosphere. Consider the air is an ideal gas. The temperature decreases with

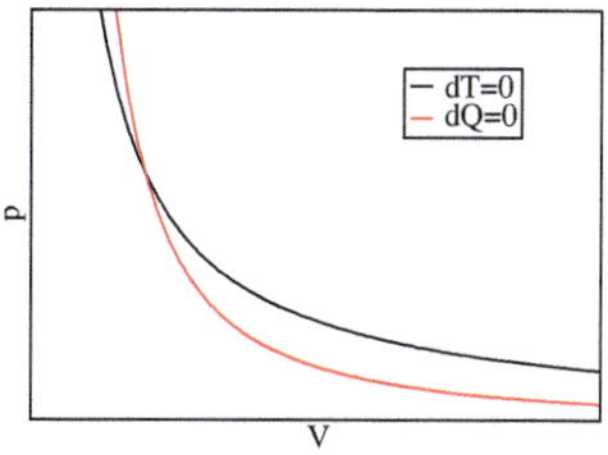

Fig. 2.4 Comparison of isothermal and adiabatic curves of an ideal gas in a pV diagram

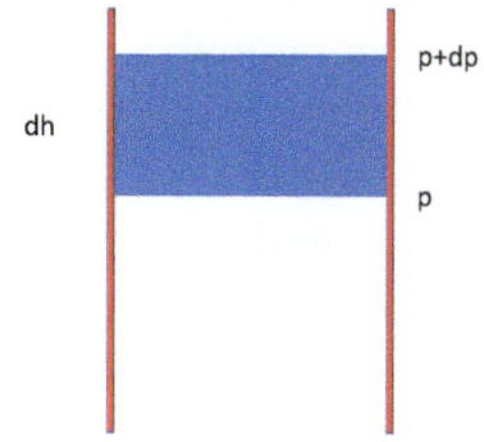

Fig. 2.5 Change of pressure with altitude in the atmosphere

the altitude. The air at lower altitudes is at higher temperature and is transported by convection currents to higher heights. As it moves upward it expands and, since air is a bad heat conductor, we may take the approximation that it does so adiabatically (neglecting the heat exchanges with its neighborhood). Since the expansion is approximately adiabatic, its temperature decreases. Also, as the altitude increases, the pressure decreases.

At a certain height, the pressure is due to the weight of the air molecules at higher altitudes. If we consider a certain height h and a slightly higher altitude $h + dh$, the difference in pressure, dp, is due to the weight of the molecules contained in an infinitesimal volume $dV = Sdh$, where S is an arbitrary area, as illustrated in Fig. 2.5. The weight of these molecules is $\rho g dV$, where ρ is the mass density and g the acceleration of gravity. The pressure difference is $dp = -\rho g S dh/S$. Since $\rho = m/V$ and the mass $m = M\nu$, where M is the molar mass, we can write that

$$\frac{dp}{p} = -\frac{gM}{R}\frac{dh}{T}. \tag{2.41}$$

Using the condition that the transformation is adiabatic, we can write that

$$\frac{dT}{T} = \frac{\gamma - 1}{\gamma}\frac{dp}{p}. \tag{2.42}$$

These results can be combined to give the temperature dependence with the altitude

$$\frac{dT}{dh} = -\frac{\gamma - 1}{\gamma}\frac{gM}{R}. \tag{2.43}$$

Using that for air $\gamma = 7/5$ (as we shall see later) and $M = 28.88g$, we estimate that $dT/dh \sim -9.8$ degrees per km, in good agreement with experiment.

2.3 Second Law of Thermodynamics

The first law of thermodynamics (conservation of energy) expresses the impossibility to build a machine that creates energy. However, it says nothing about the possibility to convert one type of energy into another, specifically heat into work. It is verified experimentally that we can always convert work into heat. We have seen previously one example where through friction we can heat up a system. Another standard example is to convert electrical energy into heat by the flow of an electrical current through a resistance that by Joule's effect dissipates heat.

On the other hand there are limitations on the possibility to convert heat into work. Otherwise it would be, for instance, possible to build a machine that would create work by cooling the environment. Such a machine, if it would exist, would constitute a perpetual motion of the second kind since the amount of heat stored in the surface of the earth is virtually unlimited.

The second law of thermodynamics expresses the fact that such machine cannot be built.

Let us define as a heat source a system that may exchange heat with the universe but that does not exchange work. An example is a mass of water that at temperature T may be considered a heat source if we consider that its volume is basically unchanged as it gives heat. If the system is very large, such that its temperature remains constant, we call it a heat reservoir. As it exchanges heat with another system, the change in temperature is negligible.

Kelvin Postulate A transformation whose only final result is to transform into work heat extracted from a heat reservoir that is at a given temperature is impossible.[1]

This is illustrated in Fig. 2.6. Note that it is not impossible to convert heat into work as long as some other change has occurred in the state of the system. Example: consider an isothermal expansion of an ideal gas that is in thermal contact with a heat reservoir at temperature T. As T is constant, the change in energy vanishes, $\Delta U = 0$. This implies that $W = Q$. The full transformation is such that some amount of heat, Q, is removed from the heat reservoir and injected in the gas that by expanding does work on the outside world. However, the gas now occupies a larger volume and therefore is not in the same state as initially.

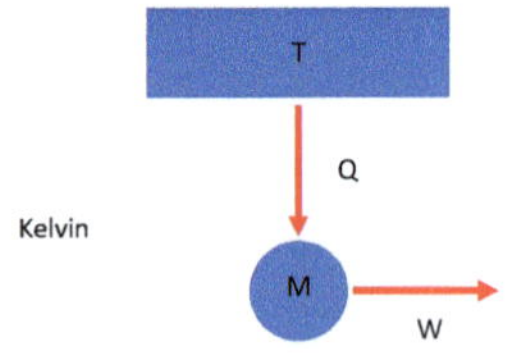

Fig. 2.6 Kelvin postulate: it is impossible that the only outcome of a transformation is to convert heat into work

[1] See also [Fermi].

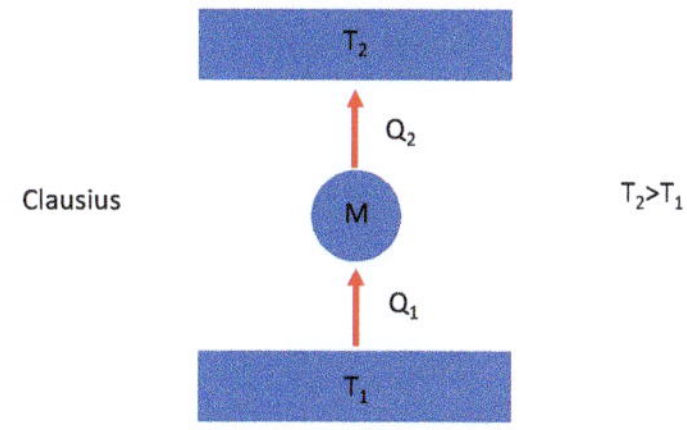

Fig. 2.7 Clausius postulate: it is impossible that the only outcome of a transformation is to transfer heat from a lower temperature reservoir and inject it in a higher temperature reservoir

Clausius Postulate A transformation whose only final result is to transfer heat from a body at a given temperature to another at a higher temperature is impossible. In other words, the spontaneous flux of heat from a system at a lower temperature to another at a higher temperature is not allowed. By definition heat flows spontaneously from higher temperatures to lower temperatures.

This is illustrated in Fig. 2.7. We will now show that the postulates of Kelvin and Clausius are equivalent. We begin by showing that if the postulate of Kelvin were not true, then the postulate of Clausius would not hold.

Suppose that Kelvin's postulate is not valid. Then it is possible to carry out a transformation whose only final result is to completely convert heat into work. This heat can, for instance, be extracted from a heat reservoir at temperature T_1. By friction we can next convert this work into heat and use it to increase the temperature of another system that had an initial temperature T_2 which could be chosen to be such that $T_2 > T_1$. Then the only final result would be to transfer heat from a body at temperature T_1 to another at a temperature $T_2 > T_1$. This would violate Clausius's postulate.

Before showing the equivalence between the two postulates in reverse order, let us define a particular type of cyclic transformation that will be very useful to infer general principles. We infer from the above proof that Kelvin's postulate holds true. As a consequence, in order to convert heat into work and to leave the system invariant (cyclic property of the system), we need to change the state of the outside universe in some way. If we want to exchange heat with the outside world, it is convenient to consider two heat reservoirs at temperatures T_1 and T_2. Specifically we define the so-called Carnot cycle in the following way: consider an ideal gas that is initially in some state, call it A. Consider first the gas in contact with some heat reservoir at temperature T_2 and let the gas expand isothermally until some volume V_B. Then expand the gas adiabatically until a volume V_C. Then compress the gas isothermally in contact with another heat reservoir at temperature $T_1 < T_2$ until a volume V_D is reached. Finally compress the gas adiabatically until the initial state is attained. The sequence of transformations is shown in a $p - V$ diagram in Fig. 2.8. In order to represent the transformations, we have to assume that they are a sequence of equilibrium states (quasi-static transformations).

Along the isothermal transformation at the higher temperature T_2, the ideal gas absorbs heat, Q_2, from the reservoir and does work in equal amounts since the temperature does not change. When it is compressed at temperature T_1 the opposite occurs. The gas gives heat, $-Q_1$, to the reservoir at temperature T_1 and receives

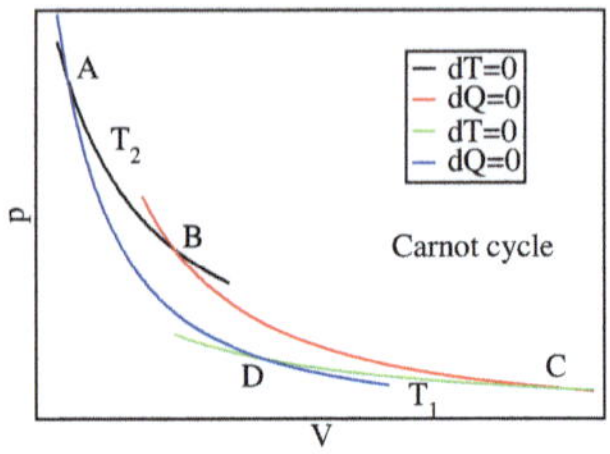

Fig. 2.8 Carnot cycle

work from the exterior again in equal amounts. Going from state B to state C, it does work, and going from state D to state A, it receives work. Since the overall transformation is cyclic, there is no change in internal energy of the gas, and the total work done by the gas equals the total amount of heat received. Therefore

$$Q_2 - Q_1 = W, \tag{2.44}$$

where W is the total amount of work done by the system. Clearly this equals

$$W = W_{AB} + W_{BC} + W_{CD} + W_{DA}. \tag{2.45}$$

Note that W_{AB}, W_{BC} are positive since the gas expands and W_{CD}, W_{DA} are negative since the gas is compressed. If $dW = pdV$ holds for each transformation, we can calculate the work done by the system in each branch of the cycle calculating, for instance,

$$W_{AB} = \int_A^B pdV. \tag{2.46}$$

We can as well calculate the total work, but this is just the area contained by the sequence of transformations. If the cycle is carried clockwise the work is positive. Reversing the sequence of the transformations and going through the cycle in a counterclockwise way, the work is negative and the gas gives heat to the reservoirs (receives heat from the reservoir at the lower temperature and gives heat in a larger amount to the reservoir at the higher temperature).

This cycle is usually represented as in Fig. 2.9 (working clockwise and therefore the gas does work). An amount of heat Q_2 is removed from T_2, an amount of heat Q_1 is injected in T_1, and work W is given to the outside world in each cycle. If the procedure is repeated indefinitely, as long as we keep exchanging heat with the two reservoirs, the system will keeping producing work that can be used to some specified purpose.

An important issue is to determine the efficiency of such a machine. Conservation of energy implies that

$$Q_2 = W + Q_1, \qquad W = Q_2 - Q_1. \tag{2.47}$$

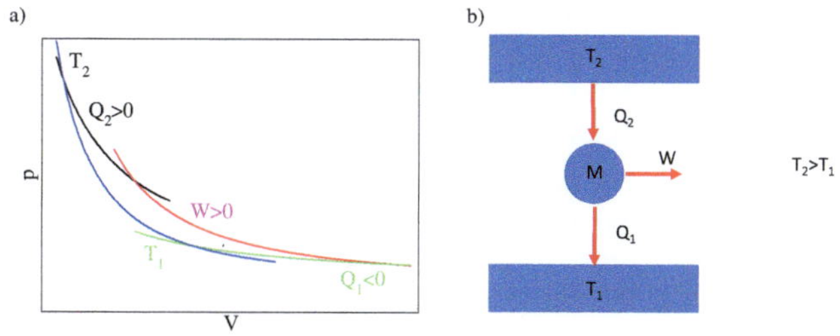

Fig. 2.9 (**a**) Carnot cycle: exchanges of heat and work. (**b**) Schematic representation of energy exchanges on a cycle

The definition of the efficiency of a given machine depends on its purpose. Consider the machine represented above. The outcome of the machine is the work it produces in each cycle. The expenditure is the amount of heat that one needs to provide to operate the machine, meaning in this case the heat Q_2 that is extracted from the reservoir at the higher temperature. The amount of heat Q_1 may be considered as heat that has to be lost. Otherwise heat would be completely converted into work. Then the efficiency of such type of machine, being defined as the ratio of what one gets over what is spent, is obtained as

$$\eta = \frac{W}{Q_2} = \frac{Q_2 - Q_1}{Q_2} = 1 - \frac{Q_1}{Q_2}. \tag{2.48}$$

Note that we can define the same set of transformations for any fluid and not necessarily an ideal gas. This special example will however allow us to obtain simple explicit results later on.

Consider now that the sequence of transformations can be taken as reversible. Therefore, reversing the direction of the cycle we go through the same states for the system and universe in reverse order. As said above, the system receives work instead of producing it.

Let us now show that if the postulate of Clausius is false, then the postulate of Kelvin is also false. Assume that it is possible to transfer an amount of heat Q_2 from a reservoir at a temperature T_1 and inject it into a reservoir at a higher temperature, $T_2 > T_1$, with no other change. This implies that now we can remove the same amount of heat Q_2 from the reservoir at temperature T_2 and use it to produce work W and give some heat Q_1 to the reservoir T_1. Therefore, the only outcome would be to produce work by extracting heat $Q_2 - Q_1$ from a heat reservoir, in this case at temperature T_1. This would violate the postulate of Kelvin. This proves that the two postulates are equivalent.

Consider now a machine (system) working in contact with two reservoirs at temperatures T_1 and T_2 and consider that $T_1 < T_2$. The machine does not have to be necessarily a Carnot machine or reversible. We just need to consider that when in contact with reservoir T_2 (T_1), the machine exchanges an amount of heat Q_2 (Q_1) and the machine does or receives an amount of work W from the exterior. Consider that the exchanges are such that, from the point of view of the reservoir T_2, $Q_2 < 0$ and from the point of view of the system $Q_2 > 0$ (receives heat) and from the point of view of the reservoir T_1, $Q_1 > 0$ and from the point of view of the system $Q_1 < 0$ (loses heat). Consider also a second machine working between the same temperatures T_1 and T_2 but which exchanges energies in the form W', Q'_1, Q'_2. Let us now prove the following theorem:

(i) If the first machine is reversible

$$\frac{Q_2}{Q_1} \geq \frac{Q'_2}{Q'_1}. \tag{2.49}$$

(ii) If the second machine is reversible as well

$$\frac{Q_2}{Q_1} = \frac{Q'_2}{Q'_1}. \tag{2.50}$$

Proof We know that $W = Q_2 - Q_1$ and $W' = Q'_2 - Q'_1$. Let us write with as good precision as desired that

$$\frac{Q_2}{Q'_2} = \frac{N'}{N},$$

where N, N' are two positive integers. Let the second machine perform N' full cycles and the first machine N full cycles in the opposite direction and let it be a reversible machine. The total work, W_t, done by the two machines is

$$W_t = N'W' - NW = Q_{2,t} - Q_{1,t}. \tag{2.51}$$

Similarly

$$\begin{aligned} Q_{1,t} &= N'Q'_1 - NQ_1 \\ Q_{2,t} &= N'Q'_2 - NQ_2 = Q_2\left(N'\frac{Q'_2}{Q_2} - N\right) = 0, \end{aligned} \tag{2.52}$$

which implies that

$$W_t = -Q_{1,t}. \tag{2.53}$$

The reservoir at temperature T_2 suffers no change. The heat absorbed from the reservoir T_1 is $-Q_{1,t}$ and is converted into work W_t. Kelvin's postulate implies that this work has to be negative. Therefore $Q_{1,t} \geq 0$. This implies that $N'Q'_1 \geq NQ_1$ and that

$$\frac{Q_2}{Q_1} \geq \frac{Q'_2}{Q'_1}, \tag{2.54}$$

as wanted.

If the second machine is reversible, we get the opposite relation that

$$\frac{Q'_2}{Q'_1} \geq \frac{Q_2}{Q_1}. \tag{2.55}$$

Therefore, if both machines are reversible, we get an equality.

Recall now that for this type of machine, we have defined the efficiency as $\eta = 1 - Q_1/Q_2$. Similarly, the efficiency of the second machine is given by $\eta' = 1 - Q'_1/Q'_2$. Using that

$$\frac{Q_1}{Q_2} \leq \frac{Q'_1}{Q'_2}, \tag{2.56}$$

we get that the efficiency of the reversible machine is equal or larger than the efficiency of the other machine

$$\eta \geq \eta'. \tag{2.57}$$

We conclude that all reversible machines have the same efficiency, which is the largest possible. Any real machine, being irreversible, is necessarily less efficient. Therefore, in order to have a machine as efficient as possible, it is useful to build a machine that works as close as possible to an ideal reversible one.

The theorem shows that Q_2/Q_1 is the same for all reversible machines that operate between two reservoirs at temperatures T_1 and T_2. Therefore we can write that

$$\frac{Q_2}{Q_1} = f(T_1, T_2), \tag{2.58}$$

where f is some universal function of T_1 and T_2 in the sense that it does not depend on the machine itself. We can now show that this universal function obeys the relation

$$f(T_1, T_2) = \frac{f(T_0, T_2)}{f(T_0, T_1)}, \tag{2.59}$$

where T_0 is some arbitrary temperature. The proof goes as follows: let us name A_1 and A_2 two reversible machines. The first machine operates between two reservoirs at temperatures T_0 and T_1 with which it exchanges amounts of heat Q_0, Q_1 and the second machine operates with two reservoirs at temperatures T_0 and T_2 with which it exchanges heats Q_0 and Q_2. Then we know that

$$\frac{Q_1}{Q_0} = f(T_0, T_1)$$
$$\frac{Q_2}{Q_0} = f(T_0, T_2). \tag{2.60}$$

Therefore it follows that

$$\frac{Q_2}{Q_1} = \frac{f(T_0, T_2)}{f(T_0, T_1)}. \tag{2.61}$$

Consider now that both machines are reversible and one cycle of A_1 and another one in the opposite order of machine A_2. The reservoir at temperature T_0 remains invariant. The reservoir at temperature T_2 gives Q_2 to A_2 and the reservoir at T_1 receives Q_1 from A_1. This can be seen as a cycle that operates between two reservoirs at temperatures T_1 and T_2. Then we know that

$$\frac{Q_2}{Q_1} = f(T_1, T_2),$$

which concludes the proof that

$$f(T_1, T_2) = \frac{f(T_0, T_2)}{f(T_0, T_1)}. \tag{2.62}$$

Since the temperature T_0 is arbitrary, we can take it to be some constant and write that $Kf(T_0, T) = \theta(T)$. Then we can write that

$$\frac{Q_2}{Q_1} = f(T_1, T_2) = \frac{\theta(T_2)}{\theta(T_1)}. \tag{2.63}$$

Let us now define a new temperature scale, called absolute temperature, such that $\theta(T) = T$. (To avoid confusion we should not have used all along T as the temperature, if we do not mean the absolute temperature. From now on T is to be understood as the absolute temperature). Since it is a free scale, we can choose it to be centigrade as before, in the sense that the interval between the boiling and the freezing points of water is divided into 100 values.

We can now show that this absolute scale coincides with the Kelvin scale used in a gas thermometer. Consider the Carnot machine whose system is one mole of an ideal gas. Let it operate between two reservoirs at temperatures T_1 and T_2, as before. Along the isothermal transformations, there is no change of internal energy of the gas, and the heat exchanged is the work of each transformation. Therefore

$$Q_2 = W_{AB} = RT_2 \ln \frac{V_B}{V_A}$$

$$Q_1 = W_{CD} = -RT_1 \ln \frac{V_D}{V_C}. \tag{2.64}$$

States A and D are in the same adiabatic and the same happens with states B and C. Therefore

$$\begin{aligned} T_1 V_D^{\gamma-1} &= T_2 V_A^{\gamma-1} \\ T_1 V_C^{\gamma-1} &= T_2 V_B^{\gamma-1}. \end{aligned} \tag{2.65}$$

This implies that

$$\frac{V_B}{V_A} = \frac{V_C}{V_D}. \tag{2.66}$$

Therefore for a reversible machine, we get that

$$\frac{Q_2}{Q_1} = \frac{T_2}{T_1} \tag{2.67}$$

and the efficiency is

$$\eta = 1 - \frac{T_1}{T_2}. \tag{2.68}$$

The efficiency is an exclusive function of the temperatures.

2.3.1 Thermodynamic Entropy

Let us take a system S on which a cyclic operation is performed. In general we may consider that the system receives or gives heat to a series of reservoirs at temperatures $T_1, T_2, \cdots, T_n$ in the form of $Q_1, Q_2, \cdots, Q_n$. As before if $Q_i > 0$ the system receives heat. We can now prove that [2]

$$\sum_{i=1}^{n} \frac{Q_i}{T_i} \leq 0 \tag{2.69}$$

and the equality holds if the transformation is reversible (Fig. 2.10).

Let us introduce an extra reservoir at a temperature T_0 and let a series of machines operate in n reversible cycles $(C_1, C_2, \cdots, C_n)$ between the n reservoirs at temperatures $T_1, T_2, \cdots, T_n$ and the extra reservoir. Also choose $T_0 > T_i$. The

[2] See also [Fermi].

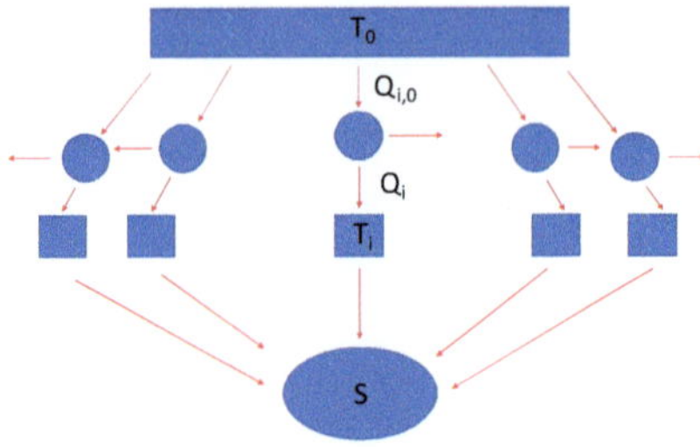

Fig. 2.10 System in contact with a sequence of reservoirs

reservoir at temperature T_0 supplies an amount of heat $Q_{i,0}$ to machine i and the machine supplies an amount of heat Q_i to each reservoir at temperature T_i during the cycle C_i. Since each cycle is operated reversibly we know that

$$Q_{i,0} = \frac{T_0}{T_i} Q_i. \tag{2.70}$$

Consider now a complex cycle such that S goes through one cycle and one cycle of all the C_i. Note that each reservoir supplies S an amount of heat Q_i to S, the same amount it has received from each cycle C_i. The net exchange of heat with the n reservoirs is zero. The reservoir at temperature T_0 gives out an amount of heat

$$Q_0 = \sum_{i=1}^{n} Q_{i,0} = T_0 \sum_{i=1}^{n} \frac{Q_i}{T_i}. \tag{2.71}$$

The final outcome of this complex cyclic operation is that $S + C_1 + C_2 + \cdots + C_n$ receive an amount of heat Q_0 from T_0 and this is converted into work, W, by the n machines. This implies that $W < 0$ which implies that $Q_0 < 0$. Therefore, due to Kelvin's postulate, we conclude that

$$\sum_{i=1}^{n} \frac{Q_i}{T_i} \leq 0, \tag{2.72}$$

otherwise we would transform heat into work with no other change. Once again, if we reverse the direction of operation of the cycles, we would get the opposite inequality. Therefore, if the cycle is reversible, we get that

$$\sum_{i=1}^{n} \frac{Q_i}{T_i} = 0. \tag{2.73}$$

We can now consider an extremal case where the number of reservoirs $n \to \infty$ and that their temperatures are infinitesimally close. We can then write that in general

$$\oint \frac{\delta Q}{T} \leq 0$$
$$\oint \frac{\delta Q}{T} = 0, \qquad (Reversible) \tag{2.74}$$

considering that the system exchanges energy with the reservoirs sequentially. Note that in these expressions T is the temperature of the reservoirs. In general the temperature of the system while it is going through some transformation is not necessarily the same. Of course in the reversible case they are the same since the transformation is slow enough such that the system equilibrates with the reservoir temperatures.

If a system undergoes some reversible transformation from a state A to a state B and back to A, we have just shown that, even though the work or the heat exchanged during this process are in general finite, the cyclic integral of $\delta Q/T$ vanishes. This implies that separating the cyclic integral into the two transformations I from A to B and II from B to A, we get that

$$\oint \frac{\delta Q}{T} = 0 = \left(\int_A^B \frac{\delta Q}{T}\right)_I + \left(\int_B^A \frac{\delta Q}{T}\right)_{II} \tag{2.75}$$

and that the integral is independent of the path chosen since

$$\left(\int_A^B \frac{\delta Q}{T}\right)_I = \left(\int_A^B \frac{\delta Q}{T}\right)_{II}. \tag{2.76}$$

Since this integral is independent of the path chosen, it can be expressed in terms of a function that only depends on the initial and final states: therefore it is a state function. Call is entropy and let us represent it by S

$$\int_A^B \frac{\delta Q}{T} = S(B) - S(A). \tag{2.77}$$

Therefore we can write that for a reversible transformation

$$\frac{\delta Q}{T} = dS. \tag{2.78}$$

One usually says that $1/T$ is an integrating factor in the sense that it converts the variation of an entity that is path dependent, such as heat, into a state function, or an exact differential. The entropy is therefore defined in terms of a differential and therefore an additive constant is to be determined. This constant will be fixed by the third law of thermodynamics. For the moment, the entropy can be defined with respect to some reference state, call it O, as

$$S(A) = \int_O^A dS = \int_O^A \frac{\delta Q}{T}. \tag{2.79}$$

This entropy is called thermodynamic entropy.

Recall that

$$dU = TdS - pdV \tag{2.80}$$

$$dU = \delta Q - \delta W. \tag{2.81}$$

We concluded that for a quasi-static transformation, we should have $\delta W = pdV$. Also we have shown that for a reversible transformation $\delta Q = TdS$. Therefore we will limit ourselves to consider that $\delta W = pdV$ only holds for reversible transformations. If there are no heat exchanges, such as for an adiabatic transformation, then if the transformation is quasi-static, we may still write that $\delta W = pdV$.

Following discussions in the beginning of this chapter, the entropy is an extensive quantity. If a system is divided in two subsystems, the total entropy should be the sum of each subsystem contribution $S = S_1 + S_2$. The same holds for the energy, heat, and work exchanged by the two subsystems with the exterior.

We can now show that in general

$$S(B) - S(A) \geq \int_A^B \frac{\delta Q}{T}, \tag{2.82}$$

where A and B are two states of a given system. We perform a general transformation from A to B and call it I. Next we can return from B to the initial state A through a reversible transformation R. Using a previous result

$$0 \geq \oint \frac{\delta Q}{T} = \left(\int_A^B \frac{\delta Q}{T}\right)_I + \left(\int_B^A \frac{\delta Q}{T}\right)_R = \left(\int_A^B \frac{\delta Q}{T}\right)_I + S(A) - S(B). \tag{2.83}$$

This implies that

$$\int_A^B \frac{\delta Q}{T} \leq S(B) - S(A), \tag{2.84}$$

as wanted. This means that the change of entropy in a given transformation is the smallest one for a reversible transformation.

Consider now an isolated system. In this case there is no exchange of energy with the outside world and $\delta Q = 0$. The previous inequality guarantees that

$$S(B) \geq S(A). \tag{2.85}$$

This is the second law of thermodynamics. It expresses that for an isolated system, the entropy either remains the same or increases, and it cannot decrease in time. This is also an expression of the irreversibility of thermodynamic systems. Any transformation the system goes through must lead to a final state of equal or larger entropy and so the entropy can be used as an arrow of time. The equality holds for a reversible transformation. Therefore, we have a quantitative definition of a reversible transformation. A reversible transformation is such that the entropy of an isolated system remains constant. If the system is composed of several subsystems, the same law implies that a reversible transformation is such that the total entropy remains constant. Any real transformation leads to an increase of the entropy. Note, however, that in a reversible or irreversible transformation, it is possible to decrease the entropy of a given subsystem. The restriction holds for the whole system that may be considered as isolated.

As a corollary, we may conclude that when an isolated system has an entropy that is maximal consistent with a given energy (which is constant since the system is isolated), then it cannot suffer any transformation because it would result in a smaller entropy. Therefore, for a given energy, the state of maximal entropy is thermodynamically the most stable state and therefore the equilibrium state.

An equivalent statement is that the equilibrium state of an isolated system with a given entropy is the one that minimizes the energy. Assume that the energy is not minimal. Then for a given entropy, it is possible to remove energy in the form of work without changing the entropy. (Note that the entropy is associated with heat exchanges. We consider that an exchange of work does not change the entropy.) We are then allowed to give this energy back to the system in the form of heat. This would imply an increase in entropy. Therefore the energy would return to the initial state but the entropy would have increased, and therefore the entropy was not maximal.

As an application of the concept of entropy, let us divide a system in two parts A_1 and A_2. Assume that A_1 is at temperature T_1 and A_2 is at temperature T_2, with $T_1 < T_2$. If the two systems are put in thermal contact, heat will flow from A_2 to A_1. Consider that the two systems can be chosen as heat reservoirs. Then the changes of entropies are

$$\Delta S_1 = \frac{Q}{T_1}, \qquad \Delta S_2 = \frac{-Q}{T_2}, \tag{2.86}$$

since we can take T_1 and T_2 as constants. The total change of entropy, since it is an extensive quantity, is

$$\Delta S = \Delta S_1 + \Delta S_2 = \frac{Q}{T_1} - \frac{Q}{T_2} = Q\left(\frac{1}{T_1} - \frac{1}{T_2}\right) > 0. \tag{2.87}$$

Therefore the flux of heat between two heat reservoirs is irreversible. It will be shown later that the same holds for any two heat sources.

We may now reformulate results previously obtained in terms of the second law in a much easier and straightforward way.

Consider again the diagram of Fig. 2.9b, understood for a general system and not necessarily a Carnot cycle. There is a reservoir at temperature T_2 and another one at temperature T_1, with $T_2 > T_1$. A machine called thermal machine removes heat Q_2 from T_2, inserts heat Q_1 into T_1, and does work W. Kelvin's postulate states that $Q_1 \neq 0$. Applying the second law of thermodynamics, we get that if we consider the possibility that $Q_1 = 0$ the change of entropy of the machine working cyclically (the system returns to its initial state and its change of energy and entropy are zero)

$$\Delta S = -\frac{Q_2}{(T_2} < 0, \tag{2.88}$$

which would be in disagreement with the second law. Therefore Q_1 cannot vanish.

Using that $Q_2 = W + Q_1$, we get that

$$\begin{aligned} \Delta S &= -\frac{Q_2}{T_2} + \frac{Q_1}{T_1} \geq 0 \\ \frac{Q_1}{T_1} &\geq \frac{Q_2}{T_2} \\ \frac{Q_1}{Q_2} &\geq \frac{T_1}{T_2}, \end{aligned} \tag{2.89}$$

as before. Using that $Q_1 = Q_2 - W$, we get that

$$\begin{aligned} -\frac{Q_2}{T_2} + \frac{Q_2 - W}{T_1} &\geq 0 \\ \frac{W}{T_1} &\leq Q_2 \left(\frac{1}{T_1} - \frac{1}{T_2} \right) \\ \eta = \frac{W}{Q_2} &\leq 1 - \frac{T_1}{T_2}, \end{aligned} \tag{2.90}$$

as before.

Reversing the direction of the machine, we get two different types of machines, depending on the purpose. This is illustrated in Fig. 2.11. We can now extract heat

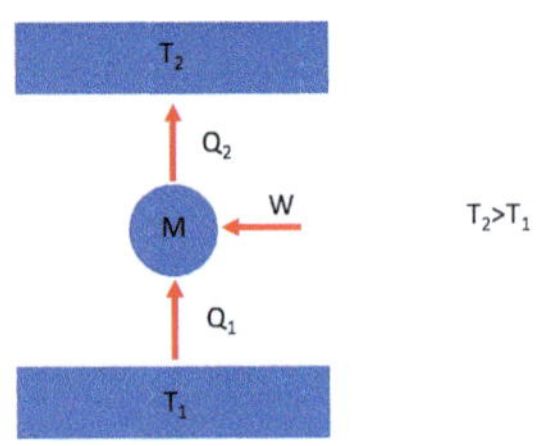

Fig. 2.11 Scheme representing either a refrigerator or a heat pump

from the reservoir at lower temperature and inject heat in the reservoir at the higher temperature, and we have to do work on the system. If we want to cool the reservoir at the lower temperature, this is called a refrigerator. If we want to inject heat into the reservoir at the higher temperature, this is called a heat pump (or heater). The efficiency also depends on the type of machine. In the case of the refrigerator, the efficiency is defined as

$$\eta_r = \frac{Q_1}{W} \tag{2.91}$$

and in the case of the heat pump

$$\eta_h = \frac{Q_2}{W} = 1 + \frac{Q_1}{W} = 1 + \eta_r > 1. \tag{2.92}$$

These two types of machines allow us to rederive Clausius's postulate. Assume we can remove heat from the lower temperature reservoir and insert it in the higher temperature one without doing any work. This would imply that

$$\Delta S = \frac{Q_2}{T_2} - \frac{Q_1}{T_1} = Q\left(\frac{1}{T_2} - \frac{1}{T_1}\right) < 0 \tag{2.93}$$

but it should be greater than zero. So it is impossible. A perfect refrigerator cannot be built.

Using the second law, we can now show that

$$\eta_r \leq \frac{T_1}{T_2 - T_1}. \tag{2.94}$$

The maximal value is obtained for a reversible transformation.

2.3.2 Equation of State and Energy of an Ideal Gas

Two types of experiments on ideal gases allowed to establish the equation of state $pV = \nu RT$ and that the energy is only a function of temperature. We may now show that these two results are related and that the equation of state imposes the result of Joule's experiment.

We know that

$$\begin{aligned} TdS &= dU + pdV \\ dS &= \frac{1}{T}dU + \frac{\nu R}{V}dV \end{aligned}$$

$$= \frac{1}{T}\left(\frac{\partial U}{\partial T}\right)_V dT + \left[\frac{1}{T}\left(\frac{\partial U}{\partial V}\right)_T + \frac{\nu R}{V}\right] dV. \tag{2.95}$$

Considering that the entropy is a function of temperature and volume, we may write that

$$dS = \left(\frac{\partial S}{\partial T}\right)_V dT + \left(\frac{\partial S}{\partial V}\right)_T dV. \tag{2.96}$$

We can now conclude that

$$\begin{aligned} \left(\frac{\partial S}{\partial T}\right)_V &= \frac{1}{T}\left(\frac{\partial U}{\partial T}\right)_V \\ \left(\frac{\partial S}{\partial V}\right)_T &= \frac{1}{T}\left(\frac{\partial U}{\partial V}\right)_T + \frac{\nu R}{V}. \end{aligned} \tag{2.97}$$

Since the entropy is a function of state, second derivatives should be symmetric. Therefore,

$$\left(\frac{\partial}{\partial V}\right)_T \left(\frac{\partial S}{\partial T}\right)_V = \left(\frac{\partial}{\partial T}\right)_V \left(\frac{\partial S}{\partial V}\right)_T. \tag{2.98}$$

Therefore

$$\frac{1}{T}\frac{\partial^2 U}{\partial V \partial T} = \left[-\frac{1}{T^2}\left(\frac{\partial U}{\partial V}\right)_T + \frac{1}{T}\frac{\partial^2 U}{\partial T \partial V}\right], \tag{2.99}$$

which implies that

$$\left(\frac{\partial U}{\partial V}\right)_T = 0, \tag{2.100}$$

since the energy is also a function of state.

2.4 Thermodynamic Potentials and Maxwell Relations

Take an isolated system that is divided in two parts 1 and 2 that may interact with each other. We want to determine the equilibrium conditions. We can use the second law by imposing that the entropy must be stationary (a maximum) when equilibrium is reached. The results are intuitive but the entropy provides a simple, quantitative way to prove our everyday experience.

Recall that[3]

$$dS = \frac{dU}{T} + \frac{p}{T}dV - \frac{\mu}{T}dN \tag{2.101}$$

for a one-component system. Consider first that there are no changes in volume or the number of particles. Then $dV = dN = 0$. In this case we can write that the total change of the entropy of the whole system is

$$dS = dS_1 + dS_2 = \frac{dU_1}{T_1} + \frac{dU_2}{T_2} \geq 0. \tag{2.102}$$

We know that $dU_1 = -dU_2$. Therefore

$$dS = dU_1 \left(\frac{1}{T_1} - \frac{1}{T_2} \right) \geq 0. \tag{2.103}$$

Equilibrium is attained if the temperatures are equal $T_1 = T_2$. This leads to $dS = 0$ and equilibrium.

In general when the two systems are in contact, we have that

$$dS = dU_1 \left(\frac{1}{T_1} - \frac{1}{T_2} \right) + dV_1 \left(\frac{p_1}{T_1} - \frac{p_2}{T_2} \right) - dN_1 \left(\frac{\mu_1}{T_1} - \frac{\mu_2}{T_2} \right). \tag{2.104}$$

Equilibrium is clearly attained if the intensive parameters are equal in the two subsystems. Besides $T_1 = T_2$, if $p_1 = p_2$ and $\mu_1 = \mu_2$ we get $dS = 0$.

The equilibrium conditions depend on the type of processes that are allowed. In a purely mechanical system, we know that the work done by the exterior in a given system satisfies

$$W_{ext} = -\Delta U. \tag{2.105}$$

In thermodynamics we have seen that this expression has to be extended to include heat

$$W = -\Delta U + Q. \tag{2.106}$$

Consider that the system is in contact with a heat reservoir at a constant temperature T and that it suffers a transformation from state A to state B. We know that

$$\int_A^B \frac{\delta Q}{T} \leq S(B) - S(A). \tag{2.107}$$

[3] See also [Reif].

Since the temperature is constant, because the system is in thermal contact with the reservoir and assuming the transformation is slow enough so that this prevails over the transformation, then

$$Q = \int_A^B \delta Q = \int_A^B dQ \leq T\,[S(B) - S(A)]\,. \tag{2.108}$$

This means that there is an upper limit for the heat that the system may receive from the outside world that is a function of T, B, and A. The work obeys

$$\begin{aligned} W &= -U(B) + U(A) + Q \\ W &\leq U(A) - U(B) + T\,[S(B) - S(A)]\,, \end{aligned} \tag{2.109}$$

implying that there is an upper limit for the work that may be delivered by the system, as its state changes from A to B. Let us define a new state function, called Helmholtz free energy, as

$$F = U - TS. \tag{2.110}$$

Then we get that

$$W \leq F(A) - F(B) = -\Delta F. \tag{2.111}$$

Therefore, $-\Delta F$ is the maximal work that the system may deliver. It is called free energy, in the sense that it is the maximal amount of energy available to deliver work. It has a similar role to energy in mechanical systems.

If we suppose that the contact between the system and the outside world is exclusively thermal, $W = 0$, we get that

$$F(B) \leq F(A). \tag{2.112}$$

We conclude that if the system is in thermal contact with a heat reservoir at temperature T and $W = 0$, its free energy decreases. The equilibrium state is such that its free energy is minimized.

In mechanical systems the equilibrium state is attained by minimizing the potential energy. By similarity, the free energy is called a thermodynamic potential, because it also obeys a minimum principle. This free energy is the appropriate thermodynamic potential at constant volume, constant number of particles, and constant temperature. There are energy fluctuations that are allowed because the system is in thermal contact with the heat reservoir, but these are such that temperature is maintained constant.

We may consider other situations where a system is in contact with other types of external systems. One can extend the notion of reservoirs to other types such as a volume reservoir (characterized by a specific pressure) or a particle reservoir (characterized by a constant chemical potential). As another example assume now that the system is in contact with a heat reservoir (constant temperature) and a volume reservoir (constant pressure). This means that fluctuations in energy and volume are allowed. Since pressure is constant, the work done by the system as it changes its volume from an initial state A to a final state B is given by

$$W = p\,[V(B) - V(A)]\,. \tag{2.113}$$

Since any transformation is isothermal, we know that

$$W \leq F(A) - F(B). \tag{2.114}$$

Therefore,

$$p\,[V(B) - V(A)] \leq F(A) - F(B). \tag{2.115}$$

Let us define a new state function[4] $\Phi = F + pV = U - TS + pV$, called Gibbs free energy. We get that

$$\Phi(B) \leq \Phi(A). \tag{2.116}$$

Therefore, the equilibrium is attained when the thermodynamic potential Φ is minimal.

These results can be easily understood in a more formal way. Recall that we postulated that energy is a function of extensive variables. These are called its natural variables. Specifically $U = U(S, V)$ and $dU = TdS - pdV$. Specifically

$$dU = \left(\frac{\partial U}{\partial S}\right)_V dS + \left(\frac{\partial U}{\partial V}\right)_S dV, \tag{2.117}$$

which leads to the results previously considered

$$\begin{aligned} T &= \left(\frac{\partial U}{\partial S}\right)_V \\ -p &= \left(\frac{\partial U}{\partial V}\right)_S. \end{aligned} \tag{2.118}$$

If the extensive variables are constant $dS = dV = 0$ as in an isolated system, this implies that $dU = 0$. The condition of equilibrium for an isolated system is that the energy is a minimum. So far we have only imposed that the state is stationary. We will determine later the conditions that imply a minimum, by studying the

[4] The Gibbs free energy is also often represented as $G = U - TS + pV$.

fluctuations around this stationary point and imposing that these fluctuations are positive.

Since the energy is a state function its second derivatives are symmetric. This leads to

$$\left(\frac{\partial}{\partial V}\right)_S\left(\frac{\partial U}{\partial S}\right)_V = \left(\frac{\partial}{\partial S}\right)_V\left(\frac{\partial U}{\partial V}\right)_S$$
$$\left(\frac{\partial T}{\partial V}\right)_S = -\left(\frac{\partial p}{\partial S}\right)_V . \tag{2.119}$$

This result is one of Maxwell's relations between the thermodynamic variables.

Consider now a system at constant pressure. It is then natural to choose as natural variables the entropy and pressure instead of entropy and volume. The following equality is useful:

$$pdV = d(pV) - Vdp.$$

We may write that

$$dU = TdS - pdV = TdS - d(pV) + Vdp$$
$$d(U + pV) = TdS + Vdp. \tag{2.120}$$

Defining a new function of state as $H = U + pV$, called enthalpy, we obtain that

$$dH = TdS + Vdp. \tag{2.121}$$

This function has S, p as natural variables. This type of transformation where we change the natural variables is called a Légendre transformation. An example is the relation between the Hamiltonian and the Lagrangian of a system. So if a system is such that entropy is constant and pressure is constant, $dS = dp = 0$, this implies that the enthalpy satisfies an extremum principle $dH = 0$, in a way equivalent to the extremum principle of the energy for an isolated system. It turns out it is also a minimum principle. As before we can see that

$$T = \left(\frac{\partial H}{\partial S}\right)_p$$
$$V = \left(\frac{\partial H}{\partial p}\right)_S \tag{2.122}$$

and the corresponding Maxwell relation is

$$\left(\frac{\partial T}{\partial p}\right)_S = \left(\frac{\partial V}{\partial S}\right)_p . \tag{2.123}$$

We can proceed and choose as natural variables T, V, for a system that exchanges thermal energy. Taking steps similar to those for the enthalpy, we get that

$$\begin{aligned} dU &= TdS - pdV = d(TS) - SdT - pdV \\ dF &= -SdT - pdV, \end{aligned} \tag{2.124}$$

where $F = U - TS$ as before. Its natural variables are T and V as wanted. For a system at constant temperature and volume $dT = dV = 0$, we obtain an extremum principle $dF = 0$ (it can be shown that it is a minimum as we saw above). The following relations hold:

$$\begin{aligned} -S &= \left(\frac{\partial F}{\partial T}\right)_V \\ -p &= \left(\frac{\partial F}{\partial V}\right)_T \end{aligned} \tag{2.125}$$

and the corresponding Maxwell relation is

$$\left(\frac{\partial S}{\partial V}\right)_T = \left(\frac{\partial p}{\partial T}\right)_V. \tag{2.126}$$

Finally we consider as natural variables T and p. We obtain that

$$\begin{aligned} dU &= TdS - pdV = d(TS) - SdT - d(PV) + Vdp \\ d\Phi &= -SdT + Vdp, \end{aligned} \tag{2.127}$$

where $\Phi = U - TS + pV$, as before. Its natural variables are T, p as wanted. For a system at constant temperature and pressure $dT = dp = 0$, this new potential becomes a minimum at equilibrium $d\Phi = 0$. The following relations hold

$$\begin{aligned} -S &= \left(\frac{\partial \Phi}{\partial T}\right)_p \\ V &= \left(\frac{\partial \Phi}{\partial p}\right)_T \end{aligned} \tag{2.128}$$

and the corresponding Maxwell relation is

$$-\left(\frac{\partial S}{\partial p}\right)_T = \left(\frac{\partial V}{\partial T}\right)_p. \tag{2.129}$$

Note that $\Phi = \mu N$.

2.5 Response Functions

In order to study the properties of a system, we have to interact with it and to perturb it. The way it responds to the external perturbation contains information on the system. For instance, by exchanging energy with the system, we get information on how energy is distributed through the normal modes (or eigenmodes) of the system.

We have considered the example of heat capacities and specific heats and applied it to the simple case of an ideal gas. We want now to establish some general relations valid for any system and to see other response functions from which we may get complementary information.

As before let us consider the measurement of a heat capacity or specific heat and let us consider that the exchange of heat is carried reversibly. Consider the heat capacities at constant volume and at constant pressure

$$C_V = \left(\frac{\delta Q}{dT}\right)_V = T\left(\frac{\partial S}{\partial T}\right)_V$$
$$C_p = \left(\frac{\delta Q}{dT}\right)_p = T\left(\frac{\partial S}{\partial T}\right)_p . \tag{2.130}$$

We know that taking as variables (T, p)

$$\delta Q = TdS = T\left[\left(\frac{\partial S}{\partial T}\right)_p dT + \left(\frac{\partial S}{\partial p}\right)_T dp\right]$$
$$= C_p dT + T\left(\frac{\partial S}{\partial p}\right)_T dp \tag{2.131}$$

and as a function of (T, V) we can write that

$$\delta Q = TdS = C_p dT + T\left(\frac{\partial S}{\partial p}\right)_T \left[\left(\frac{\partial p}{\partial T}\right)_V dT + \left(\frac{\partial p}{\partial V}\right)_T dV\right]. \tag{2.132}$$

Therefore, we get that

$$C_V = T\left(\frac{\partial S}{\partial T}\right)_V = C_p + T\left(\frac{\partial S}{\partial p}\right)_T \left(\frac{\partial p}{\partial T}\right)_V . \tag{2.133}$$

The derivative of the entropy with respect to pressure is not directly measurable; however using a Maxwell relation, we know that

$$\left(\frac{\partial S}{\partial p}\right)_T = -\left(\frac{\partial V}{\partial T}\right)_p . \tag{2.134}$$

This is directly measurable since it gives the change of volume as temperature is changed keeping pressure fixed. One defines the coefficient of expansion at constant pressure as

$$\alpha = \frac{1}{V}\left(\frac{\partial V}{\partial T}\right)_p . \tag{2.135}$$

We may also rewrite the other derivative in terms of a quantity that is more accessible experimentally. Consider the volume as a function of temperature and pressure. Then we can write an arbitrary change of volume as

$$dV = \left(\frac{\partial V}{\partial T}\right)_p dT + \left(\frac{\partial V}{\partial p}\right)_T dp.$$

Considering now that the volume is constant, $dV = 0$, we can obtain that

$$\left(\frac{\partial p}{\partial T}\right)_V = -\frac{\left(\frac{\partial V}{\partial T}\right)_p}{\left(\frac{\partial V}{\partial p}\right)_T}. \tag{2.136}$$

We define a new coefficient called isothermal compressibility

$$\kappa = -\frac{1}{V}\left(\frac{\partial V}{\partial p}\right)_T . \tag{2.137}$$

We conclude therefore that

$$\left(\frac{\partial p}{\partial T}\right)_V = \frac{\alpha}{\kappa},$$

which finally leads to

$$C_p - C_V = VT\frac{\alpha^2}{\kappa} \tag{2.138}$$

and to the general result that $C_p > C_V$. For liquids and solids $C_p \sim C_V$. For instance, for copper at $T = 25$ Celsius, $p = 1$ atm one gets that $\gamma = C_p/C_V = 10^{-5}$. However, for an ideal gas, as we saw previously $C_p - C_V = \nu R$. As can be checked, we obtain that $\alpha = 1/T$ and $\kappa = 1/p$.

Some useful relations may be obtained. If we choose the entropy to be a function of temperature and volume, we can write that

$$dS = \left(\frac{\partial S}{\partial T}\right)_V dT + \left(\frac{\partial S}{\partial V}\right)_T dV$$

$$= \frac{1}{T} C_V dT + \left(\frac{\partial p}{\partial T} \right)_V dV. \tag{2.139}$$

Integrating, we can obtain an expression for the entropy as

$$\begin{aligned} S(T, V) - S(T_0, V_0) &= \int_{T_0}^{T} dT' \frac{C_V(T', V)}{T'} + \int_{V_0}^{V} dV' \left(\frac{\partial p(T_0, V')}{\partial T} \right)_V \\ &= S(T, V) - S(T_0, V) + S(T_0, V) - S(T_0, V_0). \end{aligned} \tag{2.140}$$

The derivative of the heat capacity with respect to volume may be obtained in terms of the equation of state

$$\begin{aligned} \left(\frac{\partial C_V}{\partial V} \right)_T &= \left(\frac{\partial}{\partial V} \right)_T \left[T \left(\frac{\partial S}{\partial T} \right)_V \right] = T \frac{\partial^2 S}{\partial V \partial T} \\ &= T \frac{\partial^2 S}{\partial T \partial V} = T \left(\frac{\partial}{\partial T} \right)_V \left(\frac{\partial S}{\partial V} \right)_T = T \left(\frac{\partial}{\partial T} \right)_V \left(\frac{\partial p}{\partial T} \right)_V \\ &= T \left(\frac{\partial^2 p}{\partial T^2} \right)_V . \end{aligned} \tag{2.141}$$

Applying to the ideal gas we obtain that the heat capacity at constant volume does not depend on the volume (we will see later that it depends on the number of particles or moles, since it is an extensive quantity).

Another general relation is that starting from $dU = TdS - pdV$ and expressing the entropy in terms of temperature and volume, we can write that

$$dU = C_v dT + \left[T \left(\frac{\partial p}{\partial T} \right)_V - p \right] dV. \tag{2.142}$$

Therefore,

$$\left(\frac{\partial U}{\partial V} \right)_T = T \left(\frac{\partial p}{\partial T} \right)_V - p. \tag{2.143}$$

Applying to the ideal gas and using the equation of state, we obtain again that the energy does not depend on the volume, as shown by Joule's experiment.

2.6 Stability Criteria

We have established various extremum principles that set the conditions to attain equilibrium of a thermodynamic system, either isolated or in contact with different types of reservoirs. We established that for an isolated system at fixed energy, the entropy reaches a maximum as equilibrium is achieved. Also, fixing the entropy, the energy should be a minimum. Imposing that the energy is minimal leads to conditions that characterize the equilibrium of a system. Otherwise the system, even though reaching an extremum state, if slightly perturbed will move away from this state. This is actually of interest since it will allow us to characterize later on conditions for transitions between states of local equilibrium to states of global equilibrium, as happens in phase transitions. For the moment we want to clarify and establish conditions for stable thermodynamic equilibrium.[5] Consider a large isolated system S' divided in two subsystems S and $\tilde{S}$ such that there are ν moles in S and $\tilde{\nu}$ in $\tilde{S}$. We take the situation where the subsystem S is small, that is, $\nu << \tilde{\nu}$. The total number of moles is $\nu' = \nu + \tilde{\nu}$ and

$$U' = \nu u(s, v) + \tilde{\nu}\tilde{u}(\tilde{s}, \tilde{v}), \tag{2.144}$$

where s is the molar entropy and v the molar volume. Similarly $V' = \nu v + \tilde{\nu}\tilde{v}$, $S' = \nu s + \tilde{\nu}\tilde{s}$. Since the total system is isolated, we know that $0 = \nu dv + \tilde{\nu} d\tilde{v}$. Consider now that the total entropy is fixed, $dS' = 0$. This implies that the equilibrium state is obtained by minimizing the energy. Since the system is small $\nu << \tilde{\nu}$, this implies that $|d\tilde{v}| << |dv|$ and $|d\tilde{s}| << |ds|$. If the global system was not isolated, its energy might change according to

$$\Delta U' = \nu \left[du + d^2u + d^3u + \cdots \right] + \tilde{\nu} d\tilde{u}, \tag{2.145}$$

where we neglect the higher-order terms for the energy change of the complementary system. We can write that

$$\begin{aligned} du &= Tds - pdv \\ d^2u &= \frac{1}{2}\left[u_{ss}\,(ds)^2 + 2u_{sv} ds dv + u_{vv}\,(dv)^2 \right] \\ d\tilde{u} &= \tilde{T} d\tilde{s} - \tilde{p} d\tilde{v}, \end{aligned} \tag{2.146}$$

where

$$\begin{aligned} u_{ss} &= \frac{\partial^2 u}{\partial s^2} = \frac{\partial T}{\partial s} \\ u_{sv} &= u_{vs} = \frac{\partial^2 u}{\partial v \partial s} = -\frac{\partial p}{\partial s} = \frac{\partial T}{\partial v} \\ u_{vv} &= \frac{\partial^2 u}{\partial v^2} = -\frac{\partial p}{\partial v}. \end{aligned} \tag{2.147}$$

[5] See also [Callen].

The minimum of energy is obtained imposing $\Delta U' = 0$. The first-order terms imply the well-known conditions that

$$T = \tilde{T}, \qquad p = \tilde{p}. \tag{2.148}$$

In order to have a minimum, the second-order terms must be positive. The first and third terms of d^2u are positive if $u_{ss} \geq 0$ and $u_{vv} \geq 0$. The second term depends on the process chosen since both ds and dv may be either positive or negative. We have therefore to diagonalize this quadratic form. This can be achieved if we choose as natural variables the temperature and the molar volume. We can write the entropy as a function of T, v. This leads to

$$ds = \frac{1}{u_{ss}} \left[dT - u_{sv} dv \right]. \tag{2.149}$$

The second-order term for the change in energy may be written as

$$d^2u = \frac{1}{2} \left[\frac{1}{u_{ss}} (dT)^2 + \left(u_{vv} - \frac{u_{sv}^2}{u_{ss}} \right) (dv)^2 \right]. \tag{2.150}$$

The condition for a minimum may now be written as

$$\begin{aligned} u_{ss} &\geq 0 \\ u_{vv} - \frac{u_{sv}^2}{u_{ss}} &\geq 0. \end{aligned} \tag{2.151}$$

Note that $u_{vv} - \frac{u_{sv}^2}{u_{ss}} = \left(\partial^2 u/\partial v^2\right)_T$. Since the variables that are chosen as independent are T, v and T is taken constant in this derivative, we may write that

$$\left(\frac{\partial^2 u}{\partial v^2}\right)_T = \left(\frac{\partial^2 (u - Ts)}{\partial v^2}\right)_T = \left(\frac{\partial^2 f}{\partial v^2}\right)_T = -\left(\frac{\partial p}{\partial v}\right)_T = \frac{1}{v\kappa_T},$$

where $f = f(T, v) = u - Ts$. Therefore we may rewrite the conditions for equilibrium as

$$\begin{aligned} \left(\frac{\partial T}{\partial s}\right)_v &= \frac{T}{C_V} \geq 0 \\ -\left(\frac{\partial p}{\partial v}\right)_T &= \frac{1}{v\kappa_T} \geq 0. \end{aligned} \tag{2.152}$$

Therefore, if $T \geq 0$ we get that

$$C_V \geq 0 \qquad C_p \geq C_V \geq 0$$

$$\left(\frac{\partial p}{\partial v}\right)_T \leq 0, \qquad \kappa_T \geq 0. \tag{2.153}$$

These results are quite intuitive. A positive specific heat means that if we add heat to a system, its temperature should increase. Also, if we increase the pressure on a system at constant temperature, its volume should decrease.

These results allow us to also establish equilibrium conditions for the thermodynamic potentials. For instance,

$$\begin{aligned}
\left(\frac{\partial^2 F}{\partial T^2}\right)_V &= -\left(\frac{\partial S}{\partial T}\right)_V = -\frac{C_V}{T} \leq 0 \\
\left(\frac{\partial^2 \Phi}{\partial T^2}\right)_p &= -\left(\frac{\partial S}{\partial T}\right)_p = -\frac{C_p}{T} \leq 0 \\
\left(\frac{\partial^2 F}{\partial V^2}\right)_T &= -\left(\frac{\partial p}{\partial V}\right)_T = \frac{1}{V\kappa_T} \geq 0 \\
\left(\frac{\partial^2 \Phi}{\partial p^2}\right)_T &= \left(\frac{\partial V}{\partial p}\right)_T = -V\kappa_T \leq 0.
\end{aligned} \tag{2.154}$$

2.7 Exercises

1. We want to double the pressure of a gas contained in a given volume. Consider the three processes: isothermal, adiabatic, and isochoric. Determine in which of these processes (a) the gas has a larger temperature variation, b) the work performed on the system is maximal, and (c) the heat exchanged with the outside is maximal.
2. Sound propagates in a material alternating compressions and expansions that give rise to longitudinal waves. In the compressed regions, the temperature increases, being lower in the regions that suffer an expansion. However, since air is a bad heat conductor, we may neglect the amount of heat exchanged between those regions as the wave propagates, and the process may be considered adiabatic.

   (a) From the wave equation, the propagation velocity is

   $$v = \sqrt{\frac{dp}{d\rho}},$$

   where p is the pressure and ρ the density, calculate the propagation velocity of the sound waves in air at $T = 0C$. (This result will be obtained later in Eq. (6.186)).

(b) How does the velocity change with temperature? What is its value at $T = 20C$?

3. A mole of an ideal gas goes through a cyclic process: 1. Isothermal expansion from V_A to V_B. 2. Expansion at constant pressure from V_B to V_C. 3. Isothermal compression from V_C to V_D. 4. Compression at constant pressure from V_D to V_A. Take $V_B = 2V_A$ and $V_C = 3V_A$.

   (a) Represent the transformations in a $p - V$ diagram.
   (b) Determine T_C and V_D in terms of the initial parameters.
   (c) In which phases of the cycle does the system absorb heat?
   (d) What type of machine is it?

4. A vertical cylinder, containing ν moles of an ideal gas, is closed with a piston of mass M and area A. The acceleration of gravity is g. The molar specific heat of the gas, c_V, is a constant (independent of temperature). The heat capacities of the piston and cylinder are very small and the friction forces between the piston and the cylinder walls may be neglected. The system is thermically isolated.

   Initially the piston is fixed in a position such that the gas has a volume V_0 and a temperature T_0. The piston is released and after a few oscillations it reaches a final equilibrium position that corresponds to a larger volume of the gas.

   (a) The temperature of the gas increased, decreased, or remained constant?
   (b) The entropy of the gas increased, decreased, or remained constant?
   (c) Calculate the final temperature of the gas in terms of T_0 and V_0 and other parameters mentioned.
   (d) Compare the final volume and the final temperature with the result that would be obtained if the transformation was isentropic (constant entropy).

5. Show that for an ideal gas the enthalpy is only a function of temperature.
6. Show that

$$C_p - C_V = VT\frac{\alpha^2}{K_T},$$

   where α is the thermal expansion coefficient at constant pressure and K_T is the isothermal compressibility.
7. Show that

   (a)

$$\frac{C_p}{C_V} = \frac{K_T}{K_S},$$

   (b)

$$C_V = \frac{TV\alpha^2 K_S}{(K_T - K_s)\, K_T}$$

and

$$C_p = \frac{TV\alpha^2}{(K_T - K_s)},$$

where

$$K_T = -\frac{1}{V}\left(\frac{\partial V}{\partial p}\right)_T$$

and

$$K_S = -\frac{1}{V}\left(\frac{\partial V}{\partial p}\right)_S.$$

8. In a magnetic system we may consider as extensive variables the entropy, S, and the magnetization, M, with intensive variables the temperature, T, and the magnetic field, h. Therefore we may write

$$dU = TdS + hdM. \tag{2.155}$$

   (a) Find the expressions for the enthalpy, $H(S, h)$; Gibbs potential, $\Phi(T, H)$; and Helmholtz potential, $F(T, M)$.
   (b) Show the Maxwell relations

$$\begin{aligned}
\left(\frac{\partial T}{\partial M}\right)_S &= \left(\frac{\partial h}{\partial S}\right)_M \\
\left(\frac{\partial T}{\partial h}\right)_S &= -\left(\frac{\partial M}{\partial S}\right)_h \\
\left(\frac{\partial S}{\partial h}\right)_T &= \left(\frac{\partial M}{\partial T}\right)_h \\
\left(\frac{\partial S}{\partial M}\right)_T &= -\left(\frac{\partial h}{\partial T}\right)_M.
\end{aligned} \tag{2.156}$$

9. Consider again a magnetic system as in the previous problem. Define the susceptibilities

$$\chi_T = \left(\frac{\partial M}{\partial h}\right)_T$$

and

$$\chi_S = \left(\frac{\partial M}{\partial h}\right)_S .$$

Define also,

$$\alpha_h = \left(\frac{\partial M}{\partial T}\right)_h$$

and

$$\alpha_M = \left(\frac{\partial h}{\partial T}\right)_M .$$

Show that the specific heats satisfy

(a)

$$C_h - C_M = T\alpha_h^2 \chi_T ,$$

(b)

$$\frac{C_h}{C_M} = \frac{\chi_T}{\chi_S} .$$

10. In some transformations there is a discontinuity of the entropy and of the volume density of the system in the vicinity of a given temperature, such as in some phase transitions from one phase to another. Associated with the change in entropy, there is heat involved. This is called latent heat, L, of a transition. At the transition the intensive parameters p and T are continuous. Show the Clausius-Clapeyron relation for the slope of a transition line in the $p - T$ diagram:

$$\frac{dp}{dT} = \frac{L}{T(v' - v)}, \tag{2.157}$$

where v', v are the molar volumes on the two sides of the transition.

Chapter 3
Microscopic Description

3.1 Macrostate and Microstate: Thermodynamic Probability

Let us consider a real space description of the state of a system. We divide the volume V of the system in G microcells (small enough that on average are occupied by few particles).[1] The microscopic state of the system, designated microstate, is specified by the set of occupation numbers of each microcell. Experimentally we can only measure properties associated with volumes that contain a large number of these microcells (at least when we consider macroscopic properties of the system). Let us call macrocell a volume that contains g microcells. That is, we divide the volume of the system into G/g macrocells. A macroscopic state, designated macrostate, is then obtained specifying the number of particles in each macrocell. A macrostate contains therefore many microstates.

The probability of a given macrostate is proportional to the sum of the probabilities of all the corresponding microstates. As before, we introduce a postulate that all the microstates are equally probable. If there is translational invariance, we expect that all sites are equally probable, and, therefore, the distribution of the particles in real space should be homogeneous, if there are no external fields that lift the degeneracy. This implies that the probability of a given macrostate is proportional to the number of corresponding microstates. As a consequence, the most probable macrostate, which we identify with the equilibrium state, is the one to which the largest number of microstates corresponds.

The number of microstates that correspond to a given macrostate can be calculated in the following way. Consider n_j particles distributed among g microcells, as shown in Fig. 3.1. The number of ways to distribute them is given by

[1] Here we follow lecture notes by C. Matos Ferreira.

P. D. Sacramento, *Introduction to Condensed Matter Theory*,
https://doi.org/10.1007/978-3-031-86251-9_3

Fig. 3.1 Distribution of n_j molecules through g microcells

$$\Omega_j = \frac{(g-1+n_j)!}{n_j!(g-1)!}. \tag{3.1}$$

This result can be easily understood considering that the microcells can be thought of g boxes (with $g-1$ walls). Then we just have to calculate the number of ways to rearrange $n_j + g - 1$ objects. The numerator is the number of permutations of $n_j + g - 1$ objects, but we have to discount the permutations of n_j objects and the permutations of $g-1$ objects, since they do not correspond to a physically different state. (Recall that the number of permutations of n objects is $n!$).

The number of microstates that correspond to a given macrostate described by a set of occupation numbers of the macrocells (n_j) is given by

$$\Omega\{n_j\} = \prod_{j=1}^{G/g} \Omega_j. \tag{3.2}$$

The equilibrium state is obtained maximizing Ω. Equivalently we may maximize the logarithm. This is obtained as

$$\ln \Omega_j = \ln\left[(g+n_j-1)!\right] - \ln\left[(g-1)!\right] - \ln\left[n_j!\right]. \tag{3.3}$$

Using Stirling's approximation

$$\ln x! \sim x \ln x - x, \qquad (x >> 1),$$

we obtain that

$$\ln \Omega_j = (g+n_j)\ln(g+n_j) - g\ln g - n_j \ln n_j. \tag{3.4}$$

Defining $\langle N_{ij}\rangle = n_j/g$, the average occupation of microcell i of macrocell j, we obtain that

$$\ln \Omega_j = g\left[(1+\langle N_{ij}\rangle)\ln(1+\langle N_{ij}\rangle) - \langle N_{ij}\rangle \ln\langle N_{ij}\rangle\right]$$

$$\ln \Omega = \sum_{j=1}^{G/g} g\left[(1+\langle N_{ij}\rangle)\ln(1+\langle N_{ij}\rangle) - \langle N_{ij}\rangle \ln\langle N_{ij}\rangle\right]$$

$$\ln \Omega = \sum_{i=1}^{G} \left[(1 + \langle N_i \rangle) \ln(1 + \langle N_i \rangle) - \langle N_i \rangle \ln \langle N_i \rangle \right]. \tag{3.5}$$

In order to find an extremum of the number of states, we must add the restriction that the total number of particles is fixed:

$$\sum_{i=1}^{G} \langle N_i \rangle = N. \tag{3.6}$$

3.2 Dilute Limit

Consider first a case where the occupation number of each microcell is very small, $\langle N_i \rangle << 1$. This corresponds to low densities or a very dilute system. In this regime, we get that

$$\ln \Omega = \sum_{i=1}^{G} \langle N_i \rangle \left(1 - \ln \langle N_i \rangle \right). \tag{3.7}$$

Varying the logarithm of the number of microstates, $\delta \ln \Omega = 0$, subject to

$$\sum_{i=1}^{G} \delta \langle N_i \rangle = 0,$$

we get that

$$\sum_{i=1}^{G} \left(\ln \langle N_i \rangle + \alpha \right) \delta \langle N_i \rangle = 0. \tag{3.8}$$

The parameter α is a Lagrange multiplier used to implement the constraint of the total number of particles. Since the variations $\delta \langle N_i \rangle$ are arbitrary, the coefficient of each term of the sum must vanish. This leads to

$$\langle N_i \rangle = e^{-\alpha} \tag{3.9}$$

a constant which can be determined using that

$$\sum_{i=1}^{G} \langle N_i \rangle = N$$

$$\Rightarrow \langle N_i \rangle = \frac{N}{G} = \frac{Na}{V}, \tag{3.10}$$

where a is the volume of each microcell. This also determines α. As expected, the spatial distribution of the particles in the equilibrium state is homogeneous for an isolated system. We can calculate the number of states in equilibrium as

$$\ln \Omega = N\left(1 - \ln \frac{Na}{V}\right)$$
$$\Omega = \left(\frac{v}{a}\right)^N e^N, \tag{3.11}$$

where $v = V/N$ is the volume per particle. Keeping this number constant, we see that Ω grows as the number of particles grows or as the volume of each microcell, a, decreases. In the limit that the volume of each microcell vanishes, the number of states diverges, which indicates that there should be a minimum, finite but small, microcell volume. Otherwise, a finite system with a finite volume and number of particles would have an infinite number of states.

This lower limit for the volume of the microcell is provided by quantum mechanical considerations. Considering the phase space, where we enlarge the space to include the velocities or momenta of the particles, we need six coordinates to characterize each point in this space. The volume of one microcell in phase space can now be written as $\Delta x \Delta p_x \Delta y \Delta p_y \Delta z \Delta p_z$. Heisenberg's uncertainty principle establishes a natural lower limit in this space. Since in quantum mechanics we cannot determine simultaneously a position component and its associated momentum component with infinite precision, the minimum volume of one microcell is of the order of h^3, where h is Planck's constant. This is the consequence of the uncertainty relation

$$\Delta x \Delta p_x \sim h \tag{3.12}$$

and the same for the other spatial components.

3.3 Bose-Einstein, Planck, and Fermi-Dirac Quantum Distributions and Maxwell-Boltzmann Classical Distribution

Consider the expression in the general case when

$$\ln \Omega = \sum_i \left[(1 + \langle N_i \rangle) \ln (1 + \langle N_i \rangle) - \langle N_i \rangle \ln \langle N_i \rangle\right]. \tag{3.13}$$

Let us find again the maximum of the number of microstates by finding the extremum $\delta \ln \Omega = 0$

$$\sum_i \left[\ln(1 + \langle N_i \rangle) + (1 + \langle N_i \rangle) \frac{1}{(1 + \langle N_i \rangle)} - \ln\langle N_i \rangle - \langle N_i \rangle \frac{1}{\langle N_i \rangle} \right] \delta\langle N_i \rangle = 0$$
$$\sum_i \left[\ln \frac{1 + \langle N_i \rangle}{\langle N_i \rangle} \right] \delta\langle N_i \rangle = 0. \tag{3.14}$$

Consider now that the system may exchange both energy and particles with the outside world, being in contact with a heat reservoir (constant temperature) and a particle reservoir (constant chemical potential). Then we have an extremum principle with the restrictions that

$$\sum_i \langle N_i \rangle = N$$
$$\sum_i u_i \langle N_i \rangle = U, \tag{3.15}$$

where N is the average number of particles in the system and U the average energy. We have to introduce two Lagrange multipliers to implement these restrictions. Call them α and β, respectively. We obtain then that

$$\sum_i \left[\ln \frac{1 + \langle N_i \rangle}{\langle N_i \rangle} - \alpha - \beta u_i \right] \delta\langle N_i \rangle = 0. \tag{3.16}$$

Since the fluctuations in the number of particles in each microcell are independent, due to the introduction of the Lagrange multipliers, every term in the sum must vanish. This leads to

$$\langle N_i \rangle = \frac{1}{e^{\alpha + \beta u_i} - 1}. \tag{3.17}$$

This gives the average occupation of the various microcells given α, β (to be determined). This is called the Bose-Einstein distribution. If the total number of particles is not restricted in any way, not even its average value, the restriction on the occupation of each microcell does not exist. Therefore the Lagrange multiplier α is not needed and can be taken to be zero. The Bose-Einstein distribution reduces to the so-called Planck distribution given by

$$\langle N_i \rangle = \frac{1}{e^{\beta u_i} - 1}. \tag{3.18}$$

Recall that these two distributions were obtained considering that the occupation of the microcells by the particles of the system was not limited except that, in the case of the Bose-Einstein distribution, the total number of particles is fixed (or fixed to its average value) and the occupation is such that, in both cases, the total energy (or its average value) is also fixed.

There are particles such that it is not possible that in each microcell, there is more than one particle. Therefore, the numbers of occupation take only two values $N_i = 0, 1$. We have to redo the derivation of the number of microstates corresponding to a given macrostate. There are n_j particles in macrocell j that are to be distributed among g microcells. There are g available microcells for the first particle. There are then $(g-1)$ microcells for the second particle and so on, until for particle j there are $(g-n_j+1)$ available microcells, since in each microcell at most we can place one particle. The number of ways to distribute the n_j particles in the g microcells is therefore

$$(g)(g-1)(g-2)\cdots(g-n_j+1) = \frac{g!}{(g-n_j)!}. \tag{3.19}$$

Considering once again that the n_j particles are identical and permuting them does not lead to a different state, we obtain that

$$\Omega_j = \frac{g!}{(g-n_j)!n_j!}. \tag{3.20}$$

Therefore the total number of states and its logarithm are given by

$$\Omega = \prod_{j=1}^{G/g} \frac{g!}{(g-n_j)!n_j!}$$

$$\ln\Omega = \sum_{i=1}^{G} \left[(\langle N_i\rangle - 1)\ln(1-\langle N_i\rangle) - \langle N_i\rangle \ln\langle N_i\rangle\right]. \tag{3.21}$$

Note that this expression is quite similar to the one for the particles that obey the Bose-Einstein distribution with the two plus signs of the first term replaced by minus signs. Carrying out the same extremum principle and with the additional restrictions that the total number of particles is fixed to some (average) value and the same for the total energy, we get the distribution

$$\langle N_i\rangle = \frac{1}{e^{\alpha+\beta u_i}+1}. \tag{3.22}$$

This is called Fermi-Dirac distribution. Note once again that the only difference to the Bose-Einstein distribution is that the minus sign in the denominator has been replaced by a plus sign. However, as we will show later, this small difference leads to profound differences between systems whose state occupations satisfy the Bose-Einstein or Fermi-Dirac distributions.

In the classical limit, when the occupation of each microcell is on average very small $\langle N_i \rangle << 1$, both the Bose-Einstein and the Fermi-Dirac distributions tend to the same classical distribution

$$\langle N_i \rangle = e^{-(\alpha+\beta u_i)}, \tag{3.23}$$

called Maxwell-Boltzmann distribution. This is the extension of the dilute limit considered before, with the added possibility that there are energy exchanges with a heat reservoir.

3.4 Statistical Entropy and Connection Between Statistical Physics and Thermodynamics

We established that equilibrium is characterized by a macrostate that is obtained by the largest number of microstates. Also, we established that, if a system is isolated and the energy is fixed, the thermodynamic equilibrium state is obtained maximizing the entropy. It is then natural to find a connection between these two maximal principles. Let us for the moment define the so-called statistical entropy as

$$S_s = k_B \ln \Omega, \tag{3.24}$$

where k_B is a constant, called Boltzmann's constant, whose physical significance will be clarified later. Since the number of states satisfies a maximal principle, so does the statistical entropy. Its dependence on Ω follows from the assumption that it is an extensive quantity, in a way similar to the thermodynamic entropy.

Consider that $S_s = f(\Omega)$, a function to be determined. If we divide the system in two parts, we can write that $S_s^1 = f(\Omega_1)$, $S_s^2 = f(\Omega_2)$, and $S_s^{total} = S_s^1 + S_s^2$. If the two subsystems are independent, the total number of states is $\Omega = \Omega_1 \times \Omega_2$. Then

$$f(\Omega) = f(\Omega_1\Omega_2) = f(\Omega_1) + f(\Omega_2). \tag{3.25}$$

If we take a derivative with respect to Ω_1 or with respect to Ω_2, we get the equations

$$\begin{aligned} f'(\Omega)\Omega_2 &= f'(\Omega_1) \\ f'(\Omega)\Omega_1 &= f'(\Omega_2), \end{aligned} \tag{3.26}$$

where f' is the derivative of f. Eliminating $f'(\Omega)$ we get

$$\frac{\Omega_1}{\Omega_2} = \frac{f'(\Omega_2)}{f'(\Omega_1)}, \tag{3.27}$$

which leads to

$$\Omega_1 f'(\Omega_1) = \Omega_2 f'(\Omega_2) = k, \tag{3.28}$$

where k is a constant. Therefore,

$$f'(\Omega) = \frac{k}{\Omega} \tag{3.29}$$

and integrating we get that

$$f(\Omega) = k \ln \Omega. \tag{3.30}$$

The logarithm dependence then justifies the choice Eq. (3.24) as a consequence of the additive property.

The thermodynamic entropy previously studied obeys

$$dS = dS_t = \frac{1}{T}dU + \frac{p}{T}dV - \frac{\mu}{T}dN. \tag{3.31}$$

Considering that an extensive quantity, such as the number of states or the statistical entropy, is a function of the extensive variables, we may write that $\Omega = \Omega(U, V, N)$. Therefore,

$$d\left(\ln \Omega\right) = \left(\frac{\partial \ln \Omega}{\partial U}\right)_{N,V} dU + \left(\frac{\partial \ln \Omega}{\partial V}\right)_{N,U} dV + \left(\frac{\partial \ln \Omega}{\partial N}\right)_{U,V} dN. \tag{3.32}$$

If we want to establish some equivalence between the thermodynamic entropy and the statistical entropy, we need to construct the following correspondences:

$$\begin{aligned}
k_B \left(\frac{\partial \ln \Omega}{\partial U}\right)_{N,V} &= \frac{1}{T} = \left(\frac{\partial S}{\partial U}\right)_{N,V} \\
k_B \left(\frac{\partial \ln \Omega}{\partial V}\right)_{N,U} &= \frac{p}{T} = \left(\frac{\partial S}{\partial V}\right)_{N,U} \\
k_B \left(\frac{\partial \ln \Omega}{\partial N}\right)_{U,V} &= -\frac{\mu}{T} = \left(\frac{\partial S}{\partial N}\right)_{U,V}.
\end{aligned} \tag{3.33}$$

These relations will be used later on to obtain information about thermodynamic systems having performed a statistical calculation of the properties of the system.

This correspondence is restricted so far to an isolated system. We will need to extend this correspondence to systems in contact with different reservoirs. We start

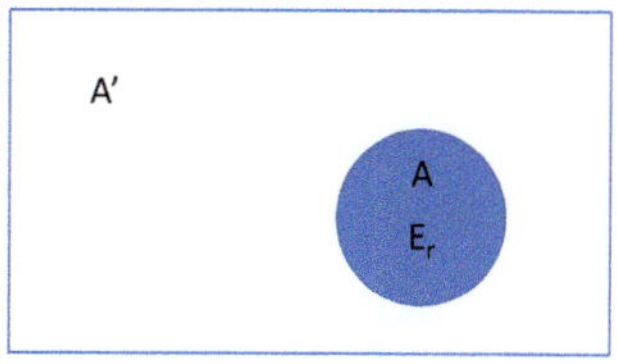

Fig. 3.2 Isolated system with subsystem A and complementary subsystem A' with internal energy exchange

by considering an isolated enlarged system constituted by a system and a reservoir. Let us call the system A and the reservoir A' and consider that the sum of these two systems is isolated from the outside world, as shown in Fig. 3.2. The energies of the states of system A are designated E_r and the energies of the states of the reservoir E'_r.

Consider that A' is a heat reservoir characterized by a given (fixed) temperature T. Since the overall system is isolated, even though we allow for energy exchanges between A and A', we know that

$$E_r + E'_r = E^{(0)}, \tag{3.34}$$

is a total constant energy. By definition the reservoir is very large, consequently

$$\frac{E_r}{E^{(0)}} = 1 - \frac{E'_r}{E^{(0)}} << 1.$$

Let us now fix the state of system A. Compatible with this state, the reservoir A' may take many different microscopic states, compatible with the energy E'_r. Call the number of these states $\Omega'(E'_r)$. Then the probability of a given state r' is

$$P_{r'} \sim \Omega'(E'_r). \tag{3.35}$$

As a consequence the probability of a given state of system A is

$$P_r \sim \Omega'(E'_r) = \Omega'(E^{(0)} - E_r). \tag{3.36}$$

Taking the logarithm, we get

$$\begin{aligned} \ln \Omega'(E'_r) &= \ln \Omega'(E^{(0)}) + \left(\frac{\partial \ln \Omega'}{\partial E'}\right)_{E'=E^{(0)}} (-E_r) + \cdots \\ &= \ln \Omega'(E^{(0)}) - \beta E_r, \end{aligned} \tag{3.37}$$

where $\beta = 1/(k_B T)$. This implies that the probability to find system A in state r, corresponding to a given energy E_r, is a function of its energy and given by

$$P_r \sim e^{-\beta E_r}. \tag{3.38}$$

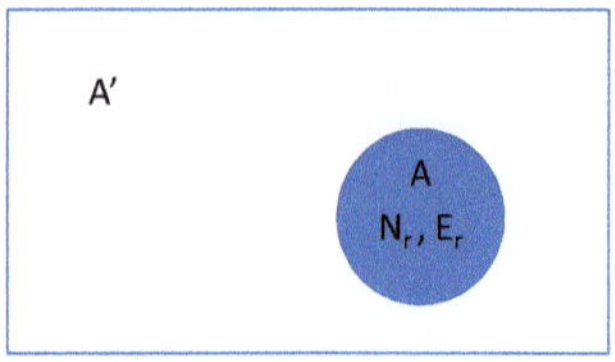

Fig. 3.3 Subsystem A and complementary subsystem A' that may exchange energy and particles

Normalizing the probability, we get that

$$P_r = \frac{e^{-\beta E_r}}{\sum_r e^{-\beta E_r}}. \tag{3.39}$$

The denominator is called the partition function of the system and is written as a sum over states r

$$Z = \sum_r e^{-\beta E_r}. \tag{3.40}$$

Recall that we assumed before that for an isolated system all microstates are equally probable. We see now that if a system is in contact with a heat reservoir, the probability of a given microstate of energy E_r is a function of its energy, and, therefore, the postulate of equally probable microstates does not apply anymore.

We may consider now that system A exchanges both energy and particles with some reservoir that allows the exchange of heat and particles in such a way that the sum of the system with the reservoir is now isolated from the outside world. This case is shown in Fig. 3.3. The states of the system are now characterized by an energy E_s and a particle number N_r. Conservation of energy and particle number imply that

$$\begin{aligned} N_r + N_r' &= N^{(0)} \\ E_s + E_s' &= E^{(0)}. \end{aligned} \tag{3.41}$$

Once again since the reservoir is very large, we can take that $N_r/N^{(0)} << 1$ and $E_s/E^{(0)} << 1$. The probability that the system is in a state with an energy E_s and a number of particles N_r is

$$P_{r,s} \sim \Omega'(N^{(0)} - N_r, E^{(0)} - E_s). \tag{3.42}$$

As before, we can write that

$$\ln \Omega'(N^{(0)} - N_r, E^{(0)} - E_s) = \ln \Omega'(N^{(0)}, E^{(0)}) + \left(\frac{\partial \ln \Omega'}{\partial N'}\right)_{N'=N^{(0)}} (-N_r)$$

$$
\begin{aligned}
&+ \left(\frac{\partial \ln \Omega'}{\partial E'} \right)_{E'=E^{(0)}} (-E_s) + \cdots \\
&= \ln \Omega'(N^{(0)}, E^{(0)}) - \alpha N_r - \beta E_s. \qquad (3.43)
\end{aligned}
$$

Normalizing the probability, we get that

$$
P_{r,s} = \frac{e^{-\alpha N_r - \beta E_s}}{\sum_{r,s} e^{-\alpha N_r - \beta E_s}}, \qquad (3.44)
$$

where the denominator is called the grand-partition function and is written as a sum over energy states and numbers of particles

$$
\mathcal{Z} = \sum_{r,s} e^{-\alpha N_r - \beta E_s}. \qquad (3.45)
$$

We will show later that $\alpha = -\mu/(k_B T)$.

3.5 Kinetic Theory of Gases

A case that illustrates well how a microscopic description of a system of many particles explains its thermodynamic properties is an ideal gas.

An ideal gas (non-interacting particles) in a container of volume V is an apparently simple system to solve. However, as stressed before, the very large number of particles (in general molecules) makes it impossible to follow them individually. A description in terms of averages values is possible and appropriate.[2] We may begin by assuming that space is homogeneous (neglecting the effect of gravity in the molecules). Then a good starting point is to realize that this implies that the particles are distributed homogeneously throughout the volume, since all locations are equivalent and equally probable. Also, if we assume that there are no preferred directions, the motion of the molecules should be locally isotropic, since we expect that the three directions of space and each sense (left, right or up, down) are equivalent. Since the particle's energies are kinetic between (nearly instantaneous) collisions, this means that their velocities are distributed randomly and equally along each direction. We also expect that the molecules move in straight lines at constant velocities until they are perturbed, as illustrated in Fig. 3.4.

There are two sources of perturbations. One is associated with the walls of the container and its geometry conditions to some extent the directions of motion. The other is the infrequent, but possible, collisions between the particles. Even though, due to the low density of the gas, interactions between the molecules are very

[2] Here we follow notes by Ana Noronha.

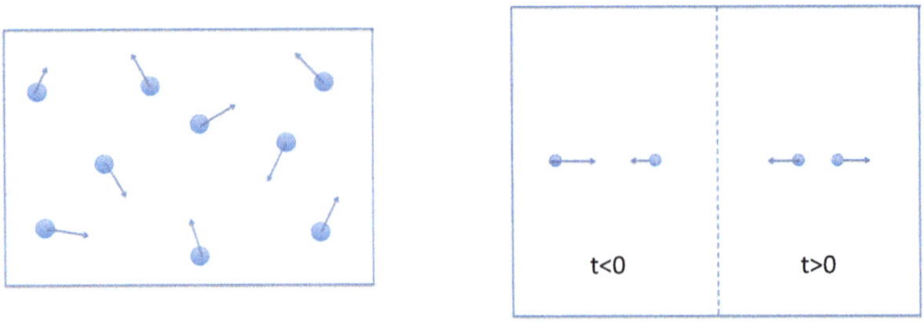

Fig. 3.4 (**a**) Random motion of particles in gas. (**b**) Frontal collision between two particles with different velocities

Fig. 3.5 Change of momentum due to reflection on container wall

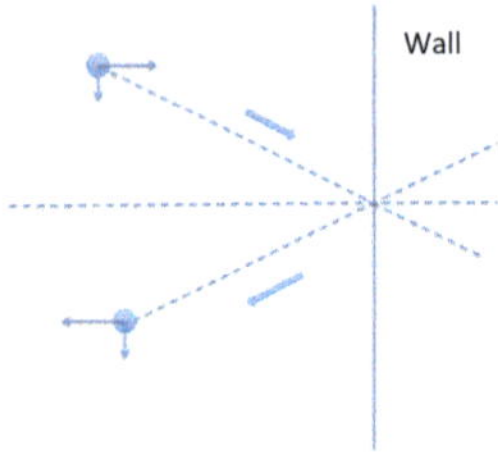

small and negligible, we cannot rule out elastic collisions between the particles. These are the result of collisions due to the repulsive force when two molecules try to occupy the same region of space. The collisions play an important role in some circumstances. For instance, if for some reason the system is in a state away from equilibrium, where one of the parts of the system is at a different temperature or pressure, equilibration requires the molecules to collide, so that exchanges of energy are possible, and on average the faster molecules loose part of their kinetic energy and the slower particles gain energy, so that the system may equilibrate. We will assume that the collisions between the molecules are elastic and instantaneous. When two molecules collide they change their energies and the direction in which they are moving. After the collision they move once again in a straight line, with constant velocity, until the next collision with either another particle or a wall of the container. If the collisions with other molecules are rare, the time between collisions is much larger than the time a collision lasts, and so this collision may be considered instantaneous. In a gas at normal pressure ($p = 1$ atm) and temperature of the order of $T = 20$ C, the average distance between particles if of the order of ten times their radius.

When the molecules collide with the walls of the container, they are reflected back due to the very large mass of the container, as shown in Fig. 3.5. In each collision the change of linear momentum they suffer is transferred to the wall. A transfer of momentum implies a force exerted on the walls and, therefore, pressure. Consider a container wall that is a plane and call the perpendicular direction to the wall x. Assume a molecule moving in the xy plane. When the molecule collides with the wall, its velocity changes from $\mathbf{v} = v_x\mathbf{e}_x + v_y\mathbf{e}_y$ to $\mathbf{v}' = -v_x\mathbf{e}_x + v_y\mathbf{e}_y$.

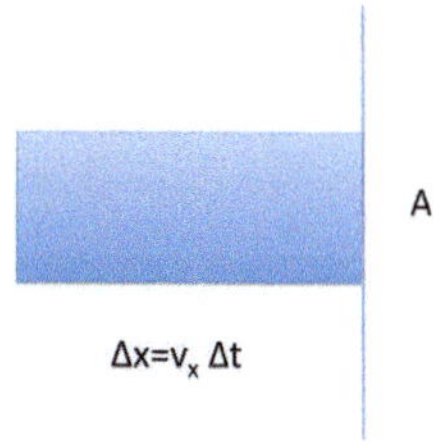

Fig. 3.6 Volume contributing to pressure on area A

Therefore in this collision the molecule suffers a change of momentum given by

$$\Delta \mathbf{p} = -2mv_x\mathbf{e}_x. \tag{3.46}$$

As a consequence, the momentum transferred to the wall is

$$\Delta \mathbf{p}_{col} = 2mv_x\mathbf{e}_x. \tag{3.47}$$

Consider now that the gas has a density $n = N/V$, where N is the total number of particles. In order to calculate the force exerted per unit area on the wall of the container, we need to calculate the total momentum transferred to the wall per unit time on a given area. The total momentum transferred is given by

$$\Delta \mathbf{p} = \langle \Delta \mathbf{p}_{col} n_{col} \rangle, \tag{3.48}$$

where n_{col} is the number of collisions per unit time on that area of the wall. The particles of the gas have different velocities. Consider a subset of these molecules that have a given velocity component v_x. If we take the direction perpendicular to the wall as positive all particles with positive velocity component, which are inside a volume $\Delta V = A\Delta x$, where $\Delta x = v_x \Delta t$, will hit the area A in the wall in the time interval Δt. This is illustrated in Fig. 3.6. The molecules with negative velocity will move away and will not exerted pressure on the area A. Since it is equally probable for the molecules to move to the right or to the left, the number of collisions is given by

$$n_{col} = \frac{1}{2} n A v_x \Delta t, \tag{3.49}$$

assuming that the density does not depend on the velocity. Therefore the average value of the momentum transmitted to the wall is given by

$$\Delta p = \left\langle 2mv_x \frac{1}{2} n A v_x \Delta t \right\rangle. \tag{3.50}$$

Therefore the pressure exerted on the wall (force per unit area) is given as an average value

$$p = nm\langle v_x^2 \rangle. \tag{3.51}$$

Since the three directions of space are equivalent, we expect that

$$\left\langle v_x^2 \right\rangle = \frac{1}{3}\left\langle \left(v_x^2 + v_y^2 + v_z^2 \right) \right\rangle = \frac{1}{3}\langle v^2 \rangle$$

and we conclude that

$$p = \frac{2}{3} n \left\langle \frac{1}{2} m v^2 \right\rangle. \tag{3.52}$$

So pressure is related to the average kinetic energy of each molecule. Since we are using the model of an ideal gas, the total energy is given by

$$U = N \left\langle \frac{1}{2} m v^2 \right\rangle, \tag{3.53}$$

where we consider the molecules as free point particles of mass m. Therefore we get that

$$\begin{aligned} pV &= \frac{2}{3} U \\ p &= \frac{2}{3} u, \end{aligned} \tag{3.54}$$

where $u = U/V$ is the energy density. Comparing with the equation of state $pV = \nu RT = 2/3U$ and recalling that $U = C_V T$, we conclude that

$$C_V = \frac{3}{2} \nu R \tag{3.55}$$

and the molar specific heat $c_V = 3/2R$. Note that from the point of view of thermodynamics, the response functions, such as the heat capacity or specific heat, are coefficients that appear in the theory that are to be measured in an experiment. As we have seen for this simple example, a statistical approach enables to calculate the specific heat from basic principles.

Another consequence of the relations obtained so far is a clear connection between temperature and the average kinetic energy of the molecules of the gas. Since the energy is $U = N\langle \frac{1}{2} m v^2 \rangle$ and is proportional to temperature, this implies the result that temperature is a measure of the average kinetic energy. The higher the temperature, the faster on average will the molecules move in the gas. So as heat is given to a system, with the consequent temperature increase, the supplied energy is absorbed by the molecules and used to increase their average velocities, and therefore the thermal agitation increases, as intuitively it should. These expressions however establish a quantitative relationship between temperature and kinetic energy. Note that so far we are considering the molecules as point particles with no internal structure. If they have internal structure (such as rotations and vibrations), not all energy that is supplied is necessarily converted in average kinetic energy (assumed here as center of mass kinetic energy), but part of it may be used to change the internal structure of the molecules. We will return to this point later.

What happens if the particles have no mass? We know from special relativity that the energy of a relativistic particle may be written as $E = mc^2$ or as $E^2 = p^2c^2 + m_0^2c^4$. If the rest mass term is negligible, we may take that $E = pc$. For instance, in the case of photons $m_0 = 0$ and so the following results apply to the calculation of the pressure exerted by a beam of photons (light) on a wall. Assuming $p = E/c$ and the velocity of the particle is the speed of light, we get that

$$\begin{aligned} \Delta p_{col} &= 2p_x \\ n_{vol} &= \frac{1}{2} n A v_x \Delta t \end{aligned} \tag{3.56}$$

and the pressure is

$$p = n p_x v_x. \tag{3.57}$$

Once again considering that all space directions are equivalent, we can conclude that the equation of state can be written as

$$p = \frac{1}{3} u. \tag{3.58}$$

Note the factor of $1/3$ instead of the factor of $2/3$ we obtained for free particles with mass. Also, this result was obtained using classical mechanics. It turns out that the relation between pressure and the energy density also holds for free quantum particles both for particles that satisfy the Bose-Einstein distribution (bosons) and particles that satisfy the Fermi-Dirac distribution (fermions). These results will be obtained later.

As the molecules move inside the container, they travel an average distance before they collide with other molecules. In general, this average distance is larger than the average distance between particles. We have shown that the average kinetic energy and the temperature are related by

$$\left\langle \frac{1}{2} m v^2 \right\rangle = \frac{U}{N} = \frac{C_V T}{N} = \frac{\nu R T}{N}. \tag{3.59}$$

Using that $N = \nu N_A$ and defining Boltzmann's constant as $k_B = R/N_A$, where N_A is Avogadro's number, the number of particles in a mole, we obtain that

$$\langle v^2 \rangle = \frac{3 k_B T}{m}. \tag{3.60}$$

It is then possible to define a measure of the average velocity of the particles. One defines the quadratic average velocity as

$$\bar{v} = \sqrt{\langle v^2 \rangle} = \sqrt{\frac{3k_B T}{m}}. \tag{3.61}$$

We may now estimate the average distance that a molecule travels between collisions with other molecules. This is called mean free path. Consider that the molecules have a diameter D instead of being point particles. For simplicity, assume one molecule that travels in the container with velocity $\bar{v}$ and that all other molecules are at rest. As it moves along, the molecule will collide with other molecules whose centers are a distance smaller or equal than D. The molecule will collide in a time interval Δt with all other molecules that are contained in a cylinder of base πD^2 and height $\Delta x = \bar{v}\Delta t$. In that volume there are $n\pi D^2 \Delta x$ molecules. If in a length Δx there is a given number of collisions, the average distance between two collisions is given by

$$l = \frac{\Delta x}{n\pi D^2 \Delta x} = \frac{1}{n\pi D^2}. \tag{3.62}$$

This result is just an estimate since neither are the other molecules at rest neither follow the molecules a straight line after each collision.

As expected, the mean free path, l, decreases as the density increases and as the size of the molecules increases. Replacing the numbers for typical gases we get that, for instance, for oxygen at a pressure of 1 atm and room temperature, the mean free path is of the order of 30 times the average distance between molecules. At very low pressures, it can be quite large, of the order of cms. For instance, for hydrogen at a pressure of 10^{-6} atm we get that $l \sim 18$ cm, which may be of the order of the container size. The average time between collisions is $\delta t = l/\bar{v}$, and the number of collisions per unit time is

$$n_{col} = \frac{1}{\Delta t} = \frac{\bar{v}}{l}. \tag{3.63}$$

For instance, for oxygen at a temperature of 27 C, the number of collisions is of the order of $n_{col} \sim 5 \times 10^9$ per second.

Considering the example of an ideal gas, we know that

$$p = nk_B T, \tag{3.64}$$

which implies that

$$l = \frac{k_B}{\pi D^2}\frac{T}{p}. \tag{3.65}$$

The mean free path is therefore proportional to temperature and inversely proportional to pressure.

In the calculation of the mean free path, we assumed a particle moving with the quadratic mean velocity. However, the molecules in the gas have different velocities, and we would like to estimate the number of molecules with a given velocity. For a given total energy U, the sum of the kinetic energies of all molecules is fixed. Therefore no particle can have a kinetic energy larger than the total energy U. Unlike the distribution of the particles in real space, where all locations are equally probable if there are no external fields, the distribution in energy space is clearly not uniform because of the total energy cutoff. Also, it seems intuitive that molecules with very high velocities will be unlikely. On the other hand, we have already established that all directions in space should be equivalent.

Let us call

$$f(v_x, v_y, v_z)dv_x dv_y dv_z,$$

the probability that a molecule has a velocity such that its components are between v_x and $v_x + dv_x$, v_y and $v_y + dv_y$, and v_z and $v_z + dv_z$, and $f(v_x, v_y, v_z)$ is the probability density. Since there is isotropy in the direction of the velocities, the probability density is only a function of the absolute value of the velocity or $f(v^2)$. Since the molecules are free, the velocity components are independent, and we can write that

$$f(v^2) = g_x(v_x)g_y(v_y)g_z(v_z),$$

where, due to isotropy, $g_x = g_y = g_z$. Therefore,

$$f(v^2) = g(v_x)g(v_y)g(v_z). \tag{3.66}$$

Let us define two new functions $G(v_i) = \ln g(v_i)$ and $F(v^2) = \ln f(v^2)$. Then we find that

$$F\left(v_x^2 + v_y^2 + v_z^2\right) = G(v_x) + G(v_y) + G(v_z). \tag{3.67}$$

The solution is

$$G(v_i) = -\alpha v_i^2 + \beta.$$

As a consequence

$$\begin{aligned} g(v_i) &= Ce^{-\alpha v_i^2} \\ f(v^2) &= C^3 e^{-\alpha v^2}. \end{aligned} \tag{3.68}$$

The probability may be normalized using that

$$\int_{-\infty}^{\infty} dv_x \int_{-\infty}^{\infty} dv_y \int_{-\infty}^{\infty} dv_z f(v_x^2 + v_y^2 + v_z^2) = 1$$
$$= C \int_{-\infty}^{\infty} dv_x e^{-\alpha v_x^2} C \int_{-\infty}^{\infty} dv_y e^{-\alpha v_y^2} C \int_{-\infty}^{\infty} dv_z e^{-\alpha v_z^2}. \tag{3.69}$$

Each of these integrals is a Gaussian integral. Define

$$I_n = \int_0^{\infty} dx x^n e^{-\alpha x^2}. \tag{3.70}$$

It can be shown that

$$I_0 = \frac{1}{2}\sqrt{\frac{\pi}{\alpha}}$$
$$I_1 = \frac{1}{2\alpha}$$
$$I_n = \frac{n-1}{2\alpha} I_{n-2}. \tag{3.71}$$

Since each integral above has an integrand that is an even function of the velocity components, we get that $C = \sqrt{\alpha/\pi}$.

We still have to determine the parameter α. Let us calculate the average kinetic energy. In the case of molecules with no internal structure, such as a monoatomic gas, we saw that

$$\left\langle \frac{1}{2} m v^2 \right\rangle = \left\langle \frac{1}{2} m v_x^2 \right\rangle + \left\langle \frac{1}{2} m v_y^2 \right\rangle + \left\langle \frac{1}{2} m v_z^2 \right\rangle = \frac{3}{2} k_B T. \tag{3.72}$$

Since all directions are equivalent, we can write that

$$\left\langle \frac{1}{2} m v_i^2 \right\rangle = \frac{1}{2} k_B T. \tag{3.73}$$

We may now calculate the average value of v_i^2 using the probability density $f(v^2)$ as

$$\langle v_i^2 \rangle = \left(\frac{\alpha}{\pi}\right)^{1/2} \int_{-\infty}^{\infty} dv_i v_i^2 e^{-\alpha v_i^2} = \left(\frac{\alpha}{\pi}\right)^{1/2} 2 I_2 = \frac{1}{2\alpha}. \tag{3.74}$$

We obtain therefore that $\alpha = m/(2k_B T)$. Therefore, the probability distribution function is given by

$$f(v^2) = \left(\frac{m}{2\pi k_B T}\right)^{3/2} e^{-\frac{m}{2k_B T}(v_x^2 + v_y^2 + v_z^2)}. \tag{3.75}$$

We recognize this distribution function as the Maxwell-Boltzmann distribution since it is of the form of the energy of each state (in this case the states of free motion of a particle) divided by $k_B T$. Note that a similar expression was found for the probability for a macrostate of energy E_r. Here the Maxwell-Boltzmann distribution is for the occupations of states of individual particles.

Since the Maxwell-Boltzmann distribution has a Gaussian form, we may now determine the standard deviation for each velocity component. The standard deviation is defined as

$$\sigma = \sqrt{\langle (v_i - \langle v_i \rangle)^2 \rangle}. \tag{3.76}$$

We can obtain that the standard deviation for each velocity component, where the average values are defined with respect to the probability distribution $g(v_i)$, is given by

$$\sigma = \frac{1}{2\sqrt{\alpha}} = \left(\frac{k_B T}{m}\right)^{1/2}. \tag{3.77}$$

This reflects the expected result that the larger the temperature the larger the dispersion in the velocities around the average value.

Since the distribution function only depends on the absolute value of the velocity of each particle, it makes sense to define a new probability density that is the result of integrating over all directions, maintaining the absolute value of the velocity fixed. We want to calculate the probability to find a particle with velocity between v and $v + dv$. It is convenient to use spherical coordinates. These are defined in the usual way (now in velocity space) as

$$\begin{aligned} v_x &= v \sin\theta \cos\varphi \\ v_y &= v \sin\theta \sin\varphi \\ v_z &= v \cos\theta. \end{aligned} \tag{3.78}$$

Also we know that $dv_x dv_y dv_z = v^2 \sin\theta dv d\theta d\varphi$. The probability to find a molecule with velocity between v and $v + dv$ is then

$$dP(v) = F(v)dv = \left(\frac{\alpha}{\pi}\right)^{3/2} e^{-\alpha v^2} v^2 dv \int_0^{\pi} d\theta \sin\theta \int_0^{2\pi} d\varphi. \tag{3.79}$$

This leads to

$$F(v) = 4\pi v^2 \left(\frac{\alpha}{\pi}\right)^{3/2} e^{-\alpha v^2}. \tag{3.80}$$

This is also called the Maxwell-Boltzmann distribution for the absolute value of the velocity and is shown in Fig. 3.7.

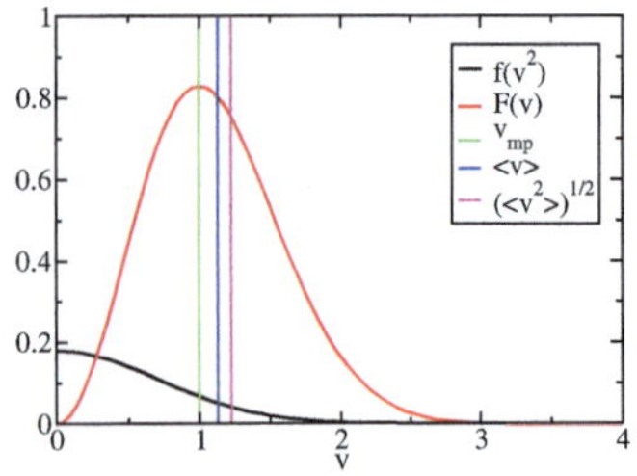

Fig. 3.7 Maxwell-Boltzmann distributions of the velocities

We may now calculate the average value of the absolute of the velocity given by

$$\langle v\rangle = 4\pi \int_0^\infty dv v^3 f(v^2)$$
$$= \frac{2}{\sqrt{\alpha\pi}} = \left(\frac{8k_B T}{\pi m}\right)^{1/2}. \tag{3.81}$$

Also we may calculate the most probable velocity, defined as the maximum of the Maxwell-Boltzmann distribution $F(v) = 4\pi v^2 f(v^2)$. It is easy to see that this is given by

$$v_{mp} = \left(\frac{2k_B T}{m}\right)^{1/2}. \tag{3.82}$$

We find that

$$v_{mp} < \langle v\rangle < \bar{v}. \tag{3.83}$$

We may now recalculate the number of collisions of the particles of the gas with the walls of the container. A naive calculation leads to the estimate that taking z as the direction perpendicular to the wall and considering molecules with velocity v_z along that direction, we need to consider a volume of area A and height $v_z\Delta t$ that contains the molecules with that velocity component that will collide the area A on the wall surface. Out of the particles in that volume $1/6$ will collide with the wall: there are three spatial directions x, y, z, and each velocity component can be positive or negative; hence a probability of $1/6$ that a molecule will travel along the positive z direction. This gives an estimate for the number of collisions per unit area and unit time as

$$\Phi = \frac{1}{6} n\langle v\rangle, \tag{3.84}$$

where n is the particle density.

A full calculation goes as follows. The desired number of particles that collides with area A in a given time interval Δt is

$$\Phi_{v_z} = n v_z A \Delta t. \tag{3.85}$$

The total number of collisions in a unit area and unit time is then

$$\Phi = \int_0^\infty dv_z n v_z g(v_z). \tag{3.86}$$

The integration can be carried out using the result for I_3 and we get that

$$\Phi = n \left(\frac{k_B T}{2\pi m} \right)^{1/2}, \tag{3.87}$$

neglecting the dependence of the density on the velocity, which can be written as

$$\Phi = \frac{1}{4} n \langle v \rangle, \tag{3.88}$$

which is very similar to the naive result obtained above.

3.5.1 Theorem of Equipartition of Energy

The result we obtained earlier, that the average kinetic energy along each direction is given by

$$\left\langle \frac{1}{2} m v_i^2 \right\rangle = \frac{1}{2} k_B T \tag{3.89}$$

is actually valid for other situations. Specifically, in classical statistical mechanics there are various systems for which the average energy of a given degree of freedom is $1/2 k_B T$. This result is called the theorem of equipartition of energy that states the following: consider a system whose energy depends on n variables $q_1, q_2, \cdots, q_n$ and that we may write that the energy is separable in the form

$$E(q_1, q_2, \cdots, q_n) = \epsilon q_i^2 + E'(q_1, \cdots, q_{i-1}, q_{i+1}, \cdots, q_n). \tag{3.90}$$

Then the average value of the term that involves the variable q_i is given by

$$\langle \epsilon q_i^2 \rangle = \frac{1}{2} k_B T. \tag{3.91}$$

The variable q_i can have different physical dimensions such as a generalized coordinate or momentum.

The proof is direct. We saw that the distribution function is of the form of an exponential of the energy in units of $k_B T$. Specifically we just considered the case of an ideal gas where the energy is the kinetic energy of each molecule. However, we showed previously that the same expression holds irrespective of the energy of the individual states of the particles of the system. Therefore

$$\left\langle \epsilon q_i^2 \right\rangle = \frac{\int dq_1 \cdots \int dq_{i-1} \int dq_i \int dq_{i+1} \cdots \int dq_n \epsilon q_i^2 e^{-\frac{1}{k_B T}\left(\epsilon q_i^2 + E'(q_1,\cdots,q_{i-1},q_{i+1},\cdots,q_n)\right)}}{\int dq_1 \cdots \int dq_{i-1} \int dq_i \int dq_{i+1} \cdots \int dq_n e^{-\frac{1}{k_B T}\left(\epsilon q_i^2 + E'(q_1,\cdots,q_{i-1},q_{i+1},\cdots,q_n)\right)}}. \tag{3.92}$$

The term in the variable q_i separates and this gives the simple result

$$\langle \epsilon q_i^2 \rangle = \frac{\int dq_i \epsilon q_i^2 e^{-\frac{1}{k_B T}(\epsilon q_i^2)}}{\int dq_i e^{-\frac{1}{k_B T}(\epsilon q_i^2)}}. \tag{3.93}$$

Using the properties of Gaussian integrals, this gives directly the result that

$$\langle \epsilon q_i^2 \rangle = \frac{1}{2} k_B T. \tag{3.94}$$

We conclude, therefore, that any separable quadratic term in the energy leads to an average energy with the same form. This proof holds if the coordinates are Cartesian-like. Considering the average values in phase space, the proof generalizes to any coordinates. It is left to the exercises to show that

$$\left\langle x_i \frac{\partial H}{\partial x_j} \right\rangle = \delta_{ij} k_B T, \tag{3.95}$$

taking x_i as any of the generalized coordinates, q_i, or conjugated momenta, p_i, and defining the integrations as

$$\langle A \rangle = \frac{\int d\omega A e^{-\beta H}}{\int d\omega e^{-\beta H}}, \tag{3.96}$$

where A is a dynamical variable and the integration is over phase space: $d\omega = dq_1 dq_2 \cdots dp_1 dp_2 \cdots$.

For instance, consider an harmonic oscillator in one dimension. We know that the energy of a particle of mass m under the influence of a spring of spring constant k is written as

$$E(x, v) = \frac{1}{2} m v^2 + \frac{1}{2} k x^2. \tag{3.97}$$

Both terms are quadratic in their variables. The kinetic energy term is quadratic in the velocity, v, and the potential energy term is quadratic in the displacement variable, x. Therefore the theorem of equipartition of energy implies that

$$\langle E \rangle = \left\langle \frac{1}{2} m v^2 \right\rangle + \left\langle \frac{1}{2} k x^2 \right\rangle = \frac{1}{2} k_B T + \frac{1}{2} k_B T = k_B T. \tag{3.98}$$

Note that since the energy is a function of the coordinates and momenta, in general the energy contains terms in phase space, and there is a possible duplication of degrees of freedom. The harmonic oscillator is an example. Even though there is only one coordinate in real space, since both kinetic and potential energies are involved, the number of degrees of freedom in this context actually takes into account real space and momentum space.

3.5.2 *Specific Heat of Gases*

We concluded before that the molar specific heat of an ideal gas is $c_V = 3/2R$. This was shown considering a gas of particles with no internal structure such as a monoatomic gas, where each molecule is composed of a single atom. In general, gases are polyatomic. Each molecule is composed of several atoms that remain confined due to their mutual interactions. In some cases, the molecule is linear, but most of the time the distribution of the atoms inside the molecule is a three-dimensional arrangement.

A molecule of $\mathcal{N}$ atoms has $3\mathcal{N}$ degrees of freedom. We can separate the degrees of freedom in the center of mass coordinates and the motion around the center of mass. Since we are considering a three-dimensional space, there are three degrees of freedom associated with the center of mass. There are left $3\mathcal{N} - 3$ degrees of freedom. If the molecule is linear, there are two degrees of freedom of rotation of the molecule, since rotations around the axis of the molecule leave the molecule invariant. Otherwise we need three degrees of freedom to describe a rotation (such as Euler's angles). We have therefore either $3\mathcal{N} - 5$ or $3\mathcal{N} - 6$ degrees of freedom left for the vibrations of the molecule that originate in the relative displacements of the atoms inside the molecule.

The contribution of the motion of the center of mass to the energy is just kinetic and therefore quadratic in the center of mass momenta components. The energy of rotation is also quadratic and of the form $1/2I\omega^2$ where I is the moment of inertia and ω the frequency of rotation. The vibrations may be approximated by a set of harmonic oscillators that have two contributions from each degree of freedom, since there are terms that result from the kinetic energy and others from the potential energy.

Using the equipartition theorem, we can predict that the average energy

$$U = N\left[\left\langle\frac{1}{2}m\mathbf{v}_{CM}^2\right\rangle + \sum_{i=1}^{n_{rot}}\left\langle\frac{1}{2}I_i\omega_i^2\right\rangle + \sum_{i=1}^{n_{vib}}\left(\left\langle\frac{1}{2}mv_i^2\right\rangle + \left\langle\frac{1}{2}k_ir_i^2\right\rangle\right)\right], \tag{3.99}$$

where $n_{rot} = 2, 3$ and $n_{vib} = 3N - 3 - n_{rot}$ is of the form

$$U = N\left[\frac{3}{2}k_BT + n_{rot}\frac{1}{2}k_BT + 2n_{vib}\frac{1}{2}k_BT\right]. \tag{3.100}$$

In the case of a monoatomic gas, there are no internal degrees of freedom, and only the first term needs to be considered. Therefore, we obtain once again that for a monoatomic gas

$$c_V = \frac{3}{2}\nu R \tag{3.101}$$

taking the derivative of the energy with respect to temperature. A diatomic molecule is necessarily linear, and we get that for a diatomic gas, there are two degrees of freedom of rotation and one of vibration (corresponding to the oscillations of the two atoms in terms of their average distance). We get therefore that the molar-specific heat of a diatomic gas is predicted to be given by

$$c_V = \frac{3}{2}\nu R + 2\frac{1}{2}\nu R + 2\frac{1}{2}\nu R = \frac{7}{2}\nu R. \tag{3.102}$$

An aspect of the prediction is therefore that the specific heat is a constant independent of temperature. It turns out that this is result is not in agreement with experiments.

Consider first a monoatomic gas. The specific heat is in this case the result shown by experiment, but at very low temperatures, the specific heat tends to zero. This is actually a consequence of the third law of thermodynamics that we will consider ahead. But except for this very low temperature disagreement, the equipartition theorem gives a correct result. However, for polyatomic gases, the experimental results do not agree with the theoretical prediction, even for room temperatures. Typically experiments show that at high enough temperatures the result from the equipartition theorem is correct. However, as temperature lowers, the specific heat behaves in a plateau-like manner, as shown in Fig. 3.8 for a diatomic gas. For instance, for a diatomic gas at high temperatures, the specific heat of one mole of gas has the predicted value of $7/2$. At a certain temperature, the specific heat decreases to a value of $5/2$ and maintains this value for temperatures close to room temperature. At lower temperatures, it decreases again to a value of $3/2$, and finally at very low temperatures, it tends to zero as for the monoatomic gas.

The explanation for these results lies beyond the validity of classical mechanics. A quantum mechanical description is needed. Classically the energies of translations, rotations, and vibrations are continuous functions of the variables. A smaller energy is associated with smaller amplitudes. At a certain temperature that

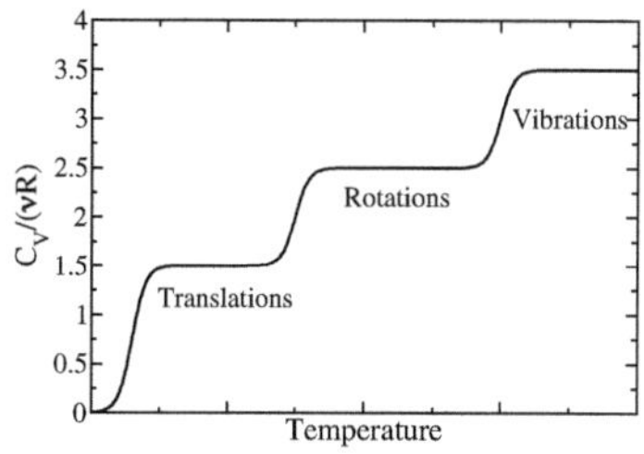

Fig. 3.8 Temperature dependence of the specific heat of a diatomic gas

corresponds to a given average kinetic energy of the molecules, there are collisions between the particles that lead to energy exchanges. These can be absorbed by a given molecule as kinetic energy of the center of mass, or as rotations or vibrations. A larger amount of energy leads to higher momenta or higher frequencies or higher vibration amplitudes. So at high temperatures on average more energy is available to be distributed among these degrees of freedom. As the temperature lowers, there is less and less energy available to be distributed among the various degrees of freedom, but these can always be excited no matter how small the available energy. Therefore all degrees of freedom contribute to the specific heat. However, the transition between $c_V = 7/2$ and $c_V = 5/2$ seems to indicate that some degrees of freedom are no longer available to absorb the energy that results from the molecule collisions. It seems that some degrees of freedom get frozen as the temperature lowers. It turns out that if we freeze the vibrations of the molecules, it is clear how the transition to $5/2$ happens. The reason is that quantum mechanically the energy levels are not continuous but often they are discrete. The energy needed to excite a vibration of the two atoms in the molecule is high, and this energy is only available through collisions if the temperature is high enough. Since the energy levels are discrete, if on a collision the energy transferred is not large enough, the molecule cannot absorb this energy, and therefore it does not contribute to the specific heat. The energies associated with rotations are also discrete (quantized), but are smaller and so at room temperature the energy that results from collisions between the molecules is larger than the difference between the rotation energy levels. Therefore they contribute. If we further lower the temperature, there is no longer enough energy to excite the rotation levels and so the specific heat changes from $5/2$ to $3/2$, since only the kinetic energy of the center of mass can absorb the transferred energy. At very low temperatures, even the translational degrees of freedom get quantized and the same occurs.

This example, among others we will encounter, shows that a classical description is not enough in many systems, even very simple ones such as an ideal gas. In general, a quantum description will be needed for a full description of physical systems. However, a classical description is simpler and still adequate in many situations.

3.6 Gibbs Paradox

The entropy obtained via thermodynamics is well-defined. We can write that for an arbitrary process, since entropy is a function of state, $TdS = dU + pdV$. For an ideal gas we can specify that

$$dS = \frac{\nu c_V(T)}{T} dT + \frac{\nu R}{V} dV. \tag{3.103}$$

Choose a reference state with respect to which we calculate an entropy difference. This reference state has ν_0 moles, volume V_0, and temperature T_0. Also we use the usual notation that $S_0 = \nu_0 s_0$. We may then calculate the entropy $S(T, V, \nu)$. If we consider ν moles of the reference state, its entropy will be νs_0 and its volume $V_0 \nu / \nu_0$. We can then write that

$$S(T, V, \nu) - \nu s_0 = \nu \int_{T_0}^{T} \frac{c_V(T')}{T'} dT' + \nu R \int_{V_0 \frac{\nu}{\nu_0}}^{V} \frac{dV'}{V'}. \tag{3.104}$$

The integral over the volume may be done explicitly, and we get that

$$\begin{aligned} S(T, V, \nu) &= \nu \left[\int_{T_0}^{T} \frac{c_V(T')}{T'} dT' + R \ln \frac{V}{\nu} - R \ln \frac{V_0}{\nu_0} + s_0 \right] \\ &= \nu \left[\int_{T_0}^{T} \frac{c_V(T')}{T'} dT' + R \ln V - R \ln \nu + C' \right], \end{aligned} \tag{3.105}$$

where C' is a constant. If the specific heat is temperature independent, the integration over temperature can be carried out explicitly, and we get that

$$S(T, V, \nu) = \nu \left[c_V \ln \frac{T}{T_0} + R \ln \frac{V}{V_0} - R \ln \frac{\nu}{\nu_0} \right]. \tag{3.106}$$

Consider now a statistical description. We have implicitly used a semiclassical description that when a system is composed of identical particles no new states are obtained by exchanging two or more particles. However, classically all particles are distinguishable, even if identical. Classically we are allowed to label each particle and in principle to follow its motion individually over time (in practice this is not feasible but in theory it should be possible). Let us see one implication of using a classical statistical description that will give further evidence that a quantum description is needed to understand nature.

Consider the apparently simple system of an ideal gas. The entropy can be written as $S = k_B \ln \Omega(N, V, U)$. Let us now calculate the number of states $\Omega(N, V, U)$. Even though we will consider the particles to be distinguishable, we will consider that the individual energy states are actually determined by quantum arguments.

Quantum mechanics reveals itself both at the level of the solutions of the energy levels of each particle individually (the levels are determined by the solution of the single particle Schrödinger's equation) and also at the level of how the particles of a many-body system of free particles are distributed among the available individual particle levels. (This is the origin of the separation of particles into different classes that obey the various distribution functions respecting to bosons or fermions and the classical limit represented by the Maxwell-Boltzmann distribution, in the case of a system at fixed temperature).

Each particle in the gas may be in a given set of one-particle energy levels, ϵ_r. We know that

$$\sum_{r=1}^{3N} \epsilon_r = U.$$

Quantum mechanics tells us that the energy levels of one particle in a cubic volume V of linear dimension L ($V = L^3$) are given by

$$\epsilon_r = \epsilon(n_x, n_y, n_z) = \frac{h^2}{8mL^2}\left(n_x^2 + n_y^2 + n_z^2\right), \tag{3.107}$$

where $n_x, n_y, n_z = 1, 2, 3, \cdots$, h is Planck's constant and m the mass of the particles. Therefore, $\Omega(N, V, U)$ is the number of ways to satisfy the equation

$$\sum_{r=1}^{3N} n_r^2 = \frac{8mV^{2/3}U}{h^2} = U^*. \tag{3.108}$$

Let us then assume that the particles are distinguishable. This implies that if a particle in state i is exchanged with another in state j, the configuration is counted as distinct. Consider a spherical surface in a $3N$-dimensional space of radius $\sqrt{U^*}$. In order to obtain $\Omega(N, V, U)$, we need to determine the function $\Omega(U^*)$ which gives the number of points on the surface with $n_x, n_y, n_z > 0$ for each particle. If the system size is large, the number of points on the surface is close to the number of points in the same sphere, since the number grows fast. Using that the fraction of the volume of a sphere of radius r in a $3N$ dimensional space (N particles in a $3d$ dimensional space) corresponding to the positive values of the quantum numbers is given by

$$V = \frac{1}{2^{3N}} \frac{\pi^{3N/2}}{\left(\frac{3N}{2}\right)!} r^{3N}, \tag{3.109}$$

taking $r = \sqrt{U^*}$ and using Stirling's approximation, the calculation gives that the entropy for a monoatomic gas reads

$$
\begin{aligned}
S &= Nk_B \ln\left[\frac{V}{h^3}\left(\frac{4\pi mU}{3N}\right)^{3/2}\right] + \frac{3}{2}Nk_B \\
&= Nk_B \ln\left[V\left(\frac{2\pi mk_BT}{h^2}\right)^{3/2}\right] + \frac{3}{2}Nk_B.
\end{aligned} \tag{3.110}
$$

We can obtain the energy in the form

$$
U(S, V, N) = \frac{3h^2N}{4\pi mV^{2/3}}e^{\frac{2S}{3Nk_B}-1}. \tag{3.111}
$$

Since temperature obeys

$$
T = \left(\frac{\partial U}{\partial S}\right)_{N,V} = \frac{2}{3}\frac{U}{Nk_B}, \tag{3.112}
$$

we reobtain the result derived above. Note now that the entropy is not extensive: in this expression, we see that the entropy $S \sim \ln N$. This is a first indication that this expression is not correct.

To further see that the expression obtained is not correct, let us consider two different gases that occupy two parts of a container. The first gas has N_1 particles, occupies a volume V_1, and is at temperature T. The second gas is at the same temperature, has N_2 particles, and occupies a volume V_2. The entropy of each gas is initially

$$
S_i = N_ik_B \ln V_i + \frac{3}{2}N_ik_B\left[1 + \ln\left(\frac{2\pi m_ik_BT}{h^2}\right)\right], \tag{3.113}
$$

where $i = 1, 2$. Let us now mix the two gases by removing the wall that separates them. The temperature of the mixture is the same as the initial one. The entropy of the mixture is

$$
S_m = \sum_{i=1}^{2}\left[N_ik_B \ln V + \frac{3}{2}N_ik_B\left[1 + \ln\left(\frac{2\pi m_ik_BT}{h^2}\right)\right]\right], \tag{3.114}
$$

where $V = V_1 + V_2$ is the total volume of the container.

The entropy of the mixture is given by

$$
\begin{aligned}
\Delta S &= S_m - (S_1 + S_2) \\
\Delta S &= k_B\left[N_1 \ln\frac{V_1 + V_2}{V_1} + N_2 \ln\frac{V_1 + V_2}{V_2}\right],
\end{aligned} \tag{3.115}
$$

This expression seems correct. $\Delta S > 0$ as is consistent with an irreversible process, as we discussed before for the expansion of a gas. Consider however that the same gas is initially in the two parts of the container of volume V. In the case the masses of the particles are the same $m_1 = m_2 = m$, the entropy of the mixture takes the form

$$S_m = Nk_B \ln V + \frac{3}{2}Nk_B \left[1 + \ln\left(\frac{2\pi mk_BT}{h^2}\right),\right] \tag{3.116}$$

where $N = N_1 + N_2$. The mixing entropy is equal to the previous case, but the mixing of two identical gases should be reversible and $\Delta S = 0$. This is called Gibbs paradox.

A way to solve the problem was proposed by Gibbs and can be summarized as

$$\begin{aligned} S &\rightarrow S - k_B \ln(N!) \\ S_i &\rightarrow S_i - k_B \ln(N_i!) \\ S_m &\rightarrow S_m - k_B \ln\left((N_1 + N_2)!\right). \end{aligned} \tag{3.117}$$

As can be checked, this leads, using Stirling's approximation, to no change of entropy $\Delta S = 0$, as expected. Tracing back the origin of these changes we realize that it amounts to reduce Ω by $N!$. We have to discount the permutations between the identical N particles since particles are actually indistinguishable. With this correction we obtain that the entropy is given by the Sackur-Tetrode equation

$$S(N, V, U) = Nk_B \ln \frac{V}{N} + \frac{3}{2}Nk_B \left[\frac{5}{3} + \ln \frac{2\pi mk_BT}{h^2}\right]. \tag{3.118}$$

Note that the entropy is now extensive as it should. The resolution of this paradox reinforces the idea that quantum mechanics is necessary to explain a simple system such as an ideal gas.

3.7 Exercises

1. Consider a volume V' in a volume V with N particles, isolated from the exterior.

    (a) What is the probability to find one particle in V'?
    (b) What is the probability, $\Omega(n_1)$, that n_1 particles are in V' and $n_2 = N - n_1$ in $V - V'$?
    (c) Calculate the average value of n_1, $\langle n_1 \rangle$.
    (d) Calculate the relative fluctuation of n_1, $\sigma(n_1)/\langle n_1 \rangle$, where

$$\sigma^2 = \sum_{n_1=0}^{N} (n_1 - \langle n_1 \rangle)^2 \, \Omega(n_1).$$

Study the limits $V' >> V/N$ and $V' << V/N$.

2. Consider a particle that is able to move along one line by hopping between sites distant by a. The probability that the particle moves to the right is p and the probability that it moves to the left is $q = 1 - p$. Evaluate $\langle (\Delta m)^2 \rangle$ where m is the net displacement to the right after N jumps (note that the hops are by a or $-a$) and that $\Delta m = m - \langle m \rangle$.
3. Consider a random walk in a two-dimensional square lattice. Each of the four directions, east, south, west, and north are equally probable. Consider that the lattice constant is a and that when there is a jump from one lattice point to another, the jump is limited to nearest neighbors. What is the probability that after n steps the distance from an initial point in the lattice is smaller than ma?
4. A box is divided in two parts by a wall. The ratio of the volumes of the two compartments is 3:1. The largest volume contains 1000 molecules of the gas Ne and the smaller contains 100 molecules of He. A small hole is punctured on the wall and we wait until equilibrium is reached.

    (a) Determine the average number of molecules of each type in each of the compartments and show that it coincides with the value of N that maximizes the probability associated with the distribution of molecules in the compartments.
    (b) What is the probability to find 1000 molecules of Ne in the larger subsystem and 100 of He in the smaller, when equilibrium is reached?

5. Draw the Fermi-Dirac and Bose-Einstein distributions for three different temperatures 10, 100, 500 K and for energy values between $\mu - 0.5\,\text{eV}$ and $\mu + 0.5\,\text{eV}$.
6. Consider the number of states $\Omega(E^{(0)}, E)$ of a subsystem with variable energy E in contact with a reservoir, such that they constitute an isolated system with total energy $E^{(0)}$. Around equilibrium the number of states may be expanded around its value for the equilibrium case with the average energy, $\langle E \rangle$. Obtain an expression for the energy variance $\langle (E - \langle E \rangle)^2 \rangle$ around equilibrium and compare with Eq. (4.68). Apply the result to the case of an ideal gas.
7. (a) Show that the Maxwell velocity distribution of an ideal gas is given by

$$f(v^2) = \left(\frac{m}{2\pi k_B T} \right)^{3/2} e^{-\frac{1}{2k_B T} m v^2}.$$

    (b) Show that the average value, average square value, and more probable velocity are given by

$$\langle v\rangle = \left(\frac{8k_BT}{\pi m}\right)^{1/2}, \langle v^2\rangle = \frac{3k_BT}{m}, v_{mp} = \left(\frac{2k_BT}{m}\right)^{1/2},$$

respectively.

8. (a) Show the generalized equipartition theorem

$$\left\langle x_i \frac{\partial H}{\partial x_j}\right\rangle = \delta_{ij} k_B T, \tag{3.119}$$

taking x_i as any of the generalized coordinates, q_i, or conjugated momenta, p_i, and considering the integrations over phase space: $d\omega = dq_1 dq_2 \cdots dp_1 dp_2 \cdots$.

   (b) Consider a two-dimensional harmonic oscillator. Show that the theorem holds using Cartesian coordinates, such that $H(x, y, p_x, p_y)$.

   (c) Repeat the same exercise considering polar coordinates where $H(r, \theta, p_r, p_\theta)$. Recall that p_r and p_θ are the momenta conjugate to r and θ, respectively. Also, $d\omega = dr d\theta dp_r dp_\theta$.

9. Show the virial theorem that if we have a set of independent non-interacting particles, in a one-dimensional space, say along the x direction, in the presence of a potential of the type $V \sim x^n$, then the average kinetic energy, $\langle T\rangle$, and the average potential energy, $\langle V\rangle$, satisfy $2\langle T\rangle = n\langle V\rangle$.

10. Show the virial theorem

$$\left\langle \sum_i q_i \frac{dp_i}{dt}\right\rangle = -3Nk_BT,$$

for a system of N particles at temperature T.

   (a) Consider an ideal gas and show, using this result, that $pV = Nk_BT$.

   (b) Consider a system of particles that interact via a potential $u(|\mathbf{r}_1 - \mathbf{r}_2|)$. Show that

$$pV = Nk_BT\left(1 - \frac{2\pi N}{3Vk_BT}\int_0^\infty dr \frac{\partial u(r)}{\partial r} g(r) r^3\right),$$

where $g(r)$ is the probability that the particles are separated by a distance r.

Chapter 4
Classical Statistical Physics

We have seen that in a mechanical classical system a microstate is characterized specifying the positions and velocities of all the particles. Actually, this condition was relaxed to some extent since it is known that, if the particles are identical, it is enough to specify how many particles are in each state of the individual particles. This is actually a semi-classical result: classically, even if identical, all particles are distinguishable. However, from the point of view of quantum mechanics, identical particles are indistinguishable and, as we will see later, exchanging identical particles does not lead to a different microstate. A microstate is only characterized by the distribution of the particles through the microcells. As a consequence, one has to sum over permutations of particles in symmetric (bosons) or anti-symmetric ways (fermions). We have integrated this result before when discussing the equilibrium conditions of a thermodynamic system.

As time goes by, due to the continuous motion of the particles in a thermodynamic system, for instance, a gas, the microstate of the system changes continuously. The properties that are measured in the laboratory correspond to time averages over the very high number of microstates the system passes by.

We want to simulate the various microstates the system goes through calculating averages over a very large number of copies of the system, each one in a different microstate, in the *same* time instant. We introduce the notion of an ensemble of $\mathcal{N}$ copies of the system over which we will take averages to simulate the behavior of the system.

4.1 Phase Space: Liouville's Theorem

The microstates are specified in terms of points in the phase space of the positions and momenta of the particles. Each microstate is represented by a point in this large space of $6N$ coordinates. As said above, in order to simulate the microstates the

P. D. Sacramento, *Introduction to Condensed Matter Theory*,
https://doi.org/10.1007/978-3-031-86251-9_4

system passes by over time, we need to sum over copies of the system representing different states. However, the motion of the point in phase space is restricted to some regions that are consistent with the physical restrictions over the parameters such as constant volume, constant energy, etc.

Let us define a distribution function in phase space such that[1]

$$\rho(\mathbf{x}, \mathbf{p}, t) d^3x d^3p$$

is the number of points in the volume d^3xd^3p. Note that this is a shorthand notation for

$$\begin{aligned} d^3x &= dx_1dy_1dz_1 \cdots dx_Ndy_Ndz_N \\ d^3p &= dp_{x_1}dp_{y_1}dp_{z_1} \cdots dp_{x_N}dp_{y_N}dp_{z_N}, \end{aligned}$$

where N is the number of particles of the system. The average over the copies of the system of a given function $f(\mathbf{x}, \mathbf{p})$ is obtained in the usual way

$$\langle f \rangle = \frac{\int d^3xd^3p f(\mathbf{x}, \mathbf{p}) \rho(\mathbf{x}, \mathbf{p}, t)}{\int d^3xd^3p \rho(\mathbf{x}, \mathbf{p}, t)}. \tag{4.1}$$

Consider a volume ω in phase space. The time rate of change of the number of points inside this volume is

$$\frac{\partial}{\partial t} \int_\omega \rho d\omega,$$

where $d\omega = d^3xd^3p$. The net rate of phase space points that leave the volume ω is also given by

$$\int_\sigma \rho\, (\mathbf{v} \cdot \mathbf{n})\, d\sigma = \int_\omega \nabla \cdot (\rho \mathbf{v})\, d\omega,$$

where σ is the surface of volume ω, $\mathbf{v}$ is the velocity of the points of phase space, and $\mathbf{n}$ is the normal to the surface σ. In this context the divergence operates in a six-dimensional space

$$\nabla \cdot (\rho \mathbf{v}) = \sum_{i=1}^{3N} \left[\frac{\partial}{\partial x_i} \left(\rho \frac{dx_i}{dt} \right) + \frac{\partial}{\partial p_i} \left(\rho \frac{dp_i}{dt} \right) \right]. \tag{4.2}$$

The conservation of the number of points leads to

[1] See also [Pathria].

$$\int_\omega d\omega \nabla \cdot (\rho \mathbf{v}) = -\frac{\partial}{\partial t}\int_\omega d\omega\rho$$

$$\int_\omega d\omega\left[\frac{\partial\rho}{\partial t} + \nabla \cdot (\rho\mathbf{v})\right] = 0$$

$$\frac{\partial\rho}{\partial t} + \nabla \cdot (\rho\mathbf{v}) = 0, \tag{4.3}$$

since ω is arbitrary. Explicitly, the equation of conservation of the density of points can be written as

$$\frac{\partial\rho}{\partial t} + \sum_{i=1}^{3N}\left(\frac{\partial\rho}{\partial x_i}\frac{dx_i}{dt} + \frac{\partial\rho}{\partial p_i}\frac{dp_i}{dt}\right) + \rho\sum_{i=1}^{3N}\left(\frac{\partial\frac{dx_i}{dt}}{\partial x_i} + \frac{\partial\frac{dp_i}{dt}}{\partial p_i}\right) = 0. \tag{4.4}$$

Using Hamilton's equations of motion

$$\frac{dx_i}{dt} = \frac{\partial H(x_i, p_i)}{\partial p_i}$$

$$\frac{dp_i}{dt} = -\frac{\partial H(x_i, p_i)}{\partial x_i} \tag{4.5}$$

and their derivatives

$$\frac{\partial\frac{dx_i}{dt}}{\partial x_i} = \frac{\partial^2 H}{\partial x_i \partial p_i}$$

$$\frac{\partial\frac{dp_i}{dt}}{\partial p_i} = -\frac{\partial^2 H}{\partial p_i \partial x_i} = -\frac{\partial^2 H}{\partial x_i \partial p_i}, \tag{4.6}$$

we can rewrite the equation of probability conservation as

$$\frac{\partial\rho}{\partial t} + \sum_{i=1}^{3N}\left(\frac{\partial\rho}{\partial x_i}\frac{dx_i}{dt} + \frac{\partial\rho}{\partial p_i}\frac{dp_i}{dt}\right) = 0. \tag{4.7}$$

Since the distribution function is of the form $\rho(x_i, p_i, t)$, this equation may be understood as

$$\frac{d\rho}{dt} = 0, \tag{4.8}$$

which means that the local density of points, in the referential of the moving point, is constant. This also means that the fluid of representative points is incompressible. This is called Liouville's theorem.

We may also write this conservation equation as

$$\frac{\partial \rho}{\partial t} = -\left[\rho, H\right]_{PB}, \tag{4.9}$$

where the Poisson bracket is defined as

$$\left[\rho, H\right]_{PB} = \sum_{i=1}^{3N} \left(\frac{\partial \rho}{\partial x_i} \frac{\partial H}{\partial p_i} - \frac{\partial \rho}{\partial p_i} \frac{\partial H}{\partial x_i} \right). \tag{4.10}$$

Let us consider a small volume in phase space

$$\delta \Gamma = \int d\omega = \int d^3x d^3p.$$

The number of points in this volume is given by $\delta N = \rho \delta \Gamma$. We may calculate

$$\frac{d}{dt} \delta N = \left(\frac{d\rho}{dt} \delta \Gamma + \rho \frac{d}{dt} \left(\delta \Gamma\right) \right) = 0, \tag{4.11}$$

being equal to zero because the number of points is conserved. Assuming a small volume, ρ is approximately constant which implies that $d\rho/dt = 0$. Therefore, we have Liouville's theorem that

$$\frac{d}{dt} \left(\delta \Gamma\right) = 0. \tag{4.12}$$

The volume remains constant even though its shape may change.

If we consider a stationary flux of the density of points $\partial \rho / \partial t = 0$, this implies that

$$\left[\rho, H\right]_{PB} = 0. \tag{4.13}$$

Possible solutions are:

- $\rho(\mathbf{x}, \mathbf{p})$ is constant. In this case the average of some function f is given by

$$\langle f \rangle = \frac{1}{\omega} \int d\omega f(\mathbf{x}, \mathbf{p}). \tag{4.14}$$

  This is valid if all the microstates are equivalent. This choice of the distribution function is called the microcanonical ensemble.
- Another possible choice is that $\rho(\mathbf{x}, \mathbf{p}) = \rho\left[H(\mathbf{x}, \mathbf{p})\right]$. This automatically implies that the Poisson bracket vanishes. Among the possible choices, we may take that

$$\rho(\mathbf{x}, \mathbf{p}) \sim e^{-H(\mathbf{x},\mathbf{p})/(k_B T)}. \tag{4.15}$$

This choice leads to the so-called canonical ensemble.

If the system has reached a stationary flux, this means the system is in equilibrium, and, therefore, the average of any physical entity is time independent. If we call this average $\langle f \rangle$ (the average of entity f), this is equal to the time average of f, but is also equal to the average over the ensemble of copies of the system. As said above, in one experiment, an average over time is performed

$$\langle f \rangle_T = \frac{1}{T} \int_{t_0}^{t_0+T} f(\mathbf{x}, \mathbf{p}, t) dt. \tag{4.16}$$

The average over the ensemble is given by

$$\langle f \rangle_{ensemble} = \int \rho(\mathbf{x}, \mathbf{p}) f(\mathbf{x}, \mathbf{p}) d\omega, \tag{4.17}$$

for a given time instant. If the system is ergodic, then the two averages are equal.

Ergodic Theorem: A system is ergodic if the time fraction the state of the system is in a given part of phase space is equal to the fraction of that region, with respect to the whole accessible volume of phase space.

If the system is ergodic, then the averages are equal, due to the postulate of the equiprobabilities of microstates. If time is long enough, the time average of any physical quantity should be the same for all mental copies of the system. These hypothesis imply that the average of the quantity f is given by the average over the ensemble of the time average of f. This time average is independent of the members of the ensemble and therefore $\langle f \rangle$ is the average for long times of f, which is the experimental average over real time.

4.2 Canonical Ensemble

4.2.1 Statistics in the Canonical Ensemble

The canonical ensemble is constructed to describe the interactions between a system and a heat reservoir. Therefore, energy exchanges are allowed between the system and the reservoir such that temperature is constant, T. Consider a system of variable energy. Let us again ask the question: What is the probability that the system will be found in a state characterized by the energy E_r? Call this probability P_r.

Consider the system as a member of a canonical ensemble characterized by $(N, V, T, \cdots)$ such that the energy is distributed by $\mathcal{N}$ identical systems (copies). Let us then study the statistics of the distribution of the energy by the various copies.

Consider that[2]

$$\sum_r n_r = \mathcal{N}$$

$$\sum_r n_r E_r = \epsilon = \mathcal{N}U, \tag{4.18}$$

where n_r is the number of copies with energy E_r, ϵ is the total energy and U is the average energy of the system in contact with the reservoir. The set $\{n_r\}$ is a possible solution of the distribution of the systems by the available states. The number of ways to distribute the systems by the r states is

$$W\{n_r\} = \frac{\mathcal{N}!}{n_1!n_2!n_3!\cdots}. \tag{4.19}$$

The average number of copies in state r is given by

$$\langle n_r \rangle = \frac{\sum'_{\{n_r\}} n_r W\{n_r\}}{\sum'_{\{n_r\}} W\{n_r\}}, \tag{4.20}$$

where the prime over the sums over states means that there are restrictions on the available phase space. In a way similar to before, we may now take the logarithm and obtain that

$$\ln W = \ln \mathcal{N}! - \sum_r \ln (n_r!) \tag{4.21}$$

and using Stirling's approximation, we get that

$$\ln W = \mathcal{N} \ln \mathcal{N} - \sum_r n_r \ln n_r. \tag{4.22}$$

The equilibrium distribution over the copies of the ensemble is now obtained finding the extremum

$$\delta \ln W = 0 = -\sum_r (\ln n_r + 1)\, \delta n_r. \tag{4.23}$$

We must include the restrictions

$$\sum_r \delta n_r = 0$$

[2] See also [Pathria].

$$\sum_r E_r \delta n_r = 0. \tag{4.24}$$

Introducing Lagrange multipliers α, β, we may write that

$$\sum_r [-(\ln n_r + 1) - (\alpha + \beta E_r)]\, \delta n_r = 0.$$

Therefore the equilibrium distribution n_r is given by

$$\frac{n_r}{\mathcal{N}} = \frac{e^{-\beta E_r}}{\sum_r e^{-\beta E_r}} = P_r. \tag{4.25}$$

As before, we rewrite this probability distribution as

$$P_r = \frac{e^{-\beta E_r}}{Z}, \tag{4.26}$$

where Z is the partition function defined as before

$$Z = \sum_r e^{-\beta E_r}. \tag{4.27}$$

Note that we may write that

$$P_r = -\frac{1}{\beta}\frac{\partial}{\partial E_r} \ln \left[\sum_r e^{-\beta E_r}\right] = -\frac{1}{\beta}\frac{\partial}{\partial E_r} \ln Z. \tag{4.28}$$

Also, it is straightforward to write that

$$U = \frac{\sum_r E_r e^{-\beta E_r}}{\sum_r e^{-\beta E_r}} = -\frac{\partial}{\partial \beta} \ln Z. \tag{4.29}$$

The partition function is a generating function from which we may obtain both the probability function and the average energy.

Define now the statistical free energy

$$F_s = -\frac{1}{\beta} \ln Z. \tag{4.30}$$

Its differential is given by

$$d\,(\beta F_s) = -d\,(\ln Z) = U d\beta + \beta \sum_r \frac{e^{-\beta E_r}}{Z} dE_r. \tag{4.31}$$

Therefore, we can write

$$d\left[\beta\left(U-F_s\right)\right]=\beta\left(dU-\sum_r\frac{e^{-\beta E_r}}{Z}dE_r\right). \tag{4.32}$$

Note that the left hand side is a state function. The first term on the right hand side is clearly the change of energy of the system. The second term is the work done by the exterior while changing the energy levels of the system (it is the average of the changes of the energy levels). Therefore, on the right hand side, we have the heat exchanged with the system. Therefore,

$$\beta\left(U-F_s\right)=\beta\delta Q=\frac{\delta Q}{k_BT}=\frac{S}{k_B}. \tag{4.33}$$

Note that, since the left hand side is a state function, we can choose for the right hand side a transformation that is reversible. We conclude therefore that $F_s=F$, the thermodynamic free energy that we have introduced before and whose differential is given by

$$dF=-SdT-pdV+\mu dN. \tag{4.34}$$

Recall that

$$S=-\left(\frac{\partial F}{\partial T}\right)_{N,V},\qquad p=-\left(\frac{\partial F}{\partial V}\right)_{N,T},\qquad \mu=\left(\frac{\partial F}{\partial N}\right)_{V,T}. \tag{4.35}$$

Therefore the following relations should hold

$$\begin{aligned}
S&=k_B\frac{\partial}{\partial T}\left[T\ln Z\right]_{N,V}\\
p&=k_BT\frac{\partial}{\partial V}\left[\ln Z\right]_{N,T}\\
\mu&=-k_BT\frac{\partial}{\partial N}\left[\ln Z\right]_{V,T}.
\end{aligned} \tag{4.36}$$

Since the identification between the statistical and the thermodynamic free energies holds, these relations provide a correspondence between statistical physics using the canonical ensemble and thermodynamics. We obtained before an equivalent connection between statistical mechanics using the microcanonical ensemble and thermodynamics, using, in that instance, the number of microstates as the generating function.

Recall that we obtained before that $U=-\partial/\partial\beta\ln Z$. Therefore, we can write that

$$C_V = \left(\frac{\partial U}{\partial T}\right)_{N,V} = T\left(\frac{\partial S}{\partial T}\right)_{N,V} = -T\left(\frac{\partial^2 F}{\partial T^2}\right)_{N,V} = k_B\beta^2 \frac{\partial^2}{\partial \beta^2}(\ln Z)_{N,V}\,. \tag{4.37}$$

Similarly pressure can be obtained in this framework as

$$p = -\frac{\sum_r \frac{\partial E_r}{\partial V} e^{-\beta E_r}}{\sum_r e^{-\beta E_r}}, \tag{4.38}$$

which leads to the previous result

$$pdV = -\sum_r \frac{e^{-\beta E_r}}{Z} dE_r. \tag{4.39}$$

Finally we may write the entropy as

$$\frac{S}{k_B} = \frac{U-F}{k_B T} = \sum_r P_r\left(\beta E_r + \ln Z\right) = -\sum_r P_r \ln P_r. \tag{4.40}$$

This last expression shows that the entropy can be written exclusively in terms of the probabilities of the copies of the ensemble or in terms of the probabilities of the various states. This is actually a general result for the statistical entropy, which has been obtained here in terms of the canonical ensemble.

4.2.2 Third Law of Thermodynamics

We obtained a statistical description of the entropy of a system

$$S = -k_B \sum_r P_r \ln P_r.$$

Consider now that the reservoir is at zero temperature. Then we have that

$$\begin{aligned} P_r &= \frac{1}{g_0}, \qquad E_r = E_0 \\ P_r &= 0, \qquad E_r \neq E_0. \end{aligned} \tag{4.41}$$

Here E_0 is the smaller energy state, called groundstate energy, g_0 is the degeneracy of the groundstate, and we are considering discrete energy levels to simplify (their origin is quantum mechanical). Then the entropy may be written as

$$S = -k_B \sum_{r=1}^{g_0} \frac{1}{g_0} \ln \frac{1}{g_0} = k_B \ln g_0. \tag{4.42}$$

This is the third law of thermodynamics that states that the groundstate entropy is determined by the degeneracy of the groundstate, that is, the number of states with the minimal energy

$$S(T \to 0) = k_B \ln g_0. \tag{4.43}$$

This result fixes the additive constant that remained to be determined. Most physical systems have groundstates that are nondegenerate. In this case $g_0 = 1$ and

$$S(T \to 0) = 0. \tag{4.44}$$

We may then write the entropy of a state A in relation to the groundstate

$$S_A = S_{T=0} + \int_{T=0}^{T_A} \left(\frac{\partial S}{\partial T}\right)_{V,N} dT$$

$$S_A = S_{T=0} + \int_{T=0}^{T_A} \frac{C_V}{T} dT. \tag{4.45}$$

Since the entropy difference is finite, the integral on the right hand-side is also finite which implies that

$$C_V(T \to 0) \to 0. \tag{4.46}$$

Considering a process at constant pressure we conclude, following similar steps, that

$$C_p(T \to 0) \to 0 \tag{4.47}$$

and similarly for any specific heat at constant X.

We may also consider the change of entropy with respect to other variables, at constant low temperature, in the limit when the temperature tends to zero. For instance, we may show

$$\left(\frac{\partial S}{\partial p}\right)_T \to 0, \quad T \to 0, \tag{4.48}$$

since the entropy is finite and constant as temperature vanishes. Using a Maxwell relation, this leads to

$$-\left(\frac{\partial V}{\partial T}\right)_p \to 0, \quad T \to 0 \tag{4.49}$$

and therefore the coefficient of thermal expansion, α, tends to zero as temperature tends to zero. Similarly using that

$$\left(\frac{\partial S}{\partial V}\right)_T \to 0, \qquad T \to 0, \tag{4.50}$$

we get that

$$\left(\frac{\partial p}{\partial T}\right)_V \to 0, \qquad T \to 0. \tag{4.51}$$

A question that arises is to determine if absolute zero may be reached. Consider, for instance, an ideal gas. In a $p - V$ diagram, we may draw isotherms and adiabatic curves. These adiabatic curves may be taken as reversible and therefore isentropic, which means at constant entropy. We know that as temperature increases, the isotherms are further away from the origin. Also, as the entropy increases, the isentropic curves also move away from the origin. The third law implies that the isotherm at $T = 0$ coincides with the isentropic of minimal entropy. Also, we have seen before that the slopes of the isentropics are higher than those of the isotherms. Therefore, the two types of curves cross each other in such a way that a given isentropic crosses lower and lower temperatures as volume increases.

A simple way to cool a gas is to expand it ($\delta W > 0$) and keep the transformation adiabatic because this implies that $dU < 0$ and the temperature decreases. However, this procedure will not lead us to absolute zero. Consider a given state with an entropy that is larger than the minimal one and at a temperature that is finite. We know that by an adiabatic process, we cannot change from an adiabatic curve to another and therefore we cannot reach absolute zero this way.

An apparent possible way would be to compress the system at constant pressure. In this case $\delta W < 0$ and the system evolves through isotherms of decreasing temperature and therefore $dU < 0$. This implies that $\delta Q < 0$ which is achieved removing heat from the system. This can be achieved by placing the gas in contact with a reservoir at a lower temperature. As the temperature approaches zero, any other reservoir will have higher temperature (such as the neighborhood of the system), and so heat will be transferred to the system instead of removed from it.

We will return to the problem of how to cool a system and reach very low temperatures. In any case, either due to theoretical constraints or technical difficulties, it has not been possible to reach absolute zero even though very small temperatures have been reached.

4.2.3 Partition Function

In most cases the accessible energy levels are degenerate. There are g_r states with the same energy E_r. Taking this into account, we may rewrite the partition function as

$$Z = \sum_r e^{-\beta E_r} = \sum_{E_r} g_r e^{-\beta E_r} \tag{4.52}$$

and we may also write that the probability of a given energy value is

$$P_{E_r} = \frac{g_r e^{-\beta E_r}}{\sum_{E_r} g_r e^{-\beta E_r}}. \tag{4.53}$$

In the continuum limit, we may write that

$$P(E)dE = \frac{g(E)e^{-\beta E} dE}{\int_0^\infty dE g(E) e^{-\beta E}}. \tag{4.54}$$

Consider now that there are N_i particles and g_i states associated with the energy ϵ_i. The number of ways to distribute the N_i identical particles by g_i states is

$$\frac{g_i^{N_i}}{N_i!}.$$

Therefore the number of states of the macrostate is

$$\Omega = \frac{g_1^{N_1}}{N_1!}\frac{g_2^{N_2}}{N_2!}\cdots \tag{4.55}$$

We may now calculate the entropy of such a state. The entropy may be obtained as

$$S = k_B \ln \Omega = -k_B \sum_i N_i \ln \frac{N_i}{g_i} + k_B \ln N. \tag{4.56}$$

We can use that $N_i = NP_i$ to obtain that the entropy is

$$S = -k_B \sum_i N_i \left(\ln \frac{N}{Z} - \frac{\epsilon_i}{k_B T} \right) + k_B N$$

$$S = Nk_B \ln \frac{Z}{N} + \frac{U}{T} + k_B N. \tag{4.57}$$

Now we may obtain the free energy $F = U - TS$. This is given by

$$F = U - T \left(Nk_B \ln \frac{Z}{N} + \frac{U}{T} + k_B N \right)$$

$$F = -k_B T \left(N \ln Z - \ln N! \right)$$

$$F = -k_B T \ln \frac{Z^N}{N!}. \tag{4.58}$$

We conclude that the partition function of a system of N identical particles is given by

$$Z_N = \frac{Z^N}{N!}, \tag{4.59}$$

where Z is here the partition function of one particle, defined as a sum over the one-particle states (and one-particle energies ϵ_i). Note that this result only holds for free systems. If the system is interacting, one cannot write the partition function of the N-particle system as a product over the partition functions of each particle. So (semi)classical systems of identical particles may be described by a partition function of the form

$$Z_N = \frac{1}{N!h^{3N}} \int e^{-\beta H} d\omega. \tag{4.60}$$

The extra factor, $1/N!$, is introduced due to the fact that permutations of the N particles do not lead to different states if they are identical, in the light of a semi-classical description. Also, the volume element in phase-space is adjusted so to satisfy the Heisenberg uncertainty principle.

As an example consider an ideal gas. In this case the Hamiltonian is just a sum of kinetic energies

$$H = \sum_{i=1}^{N} \frac{p_i^2}{2m}. \tag{4.61}$$

The partition function can then be written as

$$\begin{aligned}
Z_N &= \frac{1}{N!h^{3N}} \int \prod_{i=1}^{N} dx_i dp_i e^{-\frac{\beta}{2m}\sum_{i=1}^{N} p_i^2} \\
&= \frac{V^N}{N!h^{3N}} \left[\int_0^\infty 4\pi p^2 dp e^{-\frac{p^2}{2mk_BT}} \right]^N \\
&= \frac{1}{N!} \left[\frac{V}{h^3} \int_0^\infty 4\pi p^2 dp e^{-\frac{p^2}{2mk_BT}} \right]^N \\
&= \frac{1}{N!} \left[\frac{V}{h^3} (2\pi m k_B T)^{3/2} \right]^N \\
Z_N &= \frac{1}{N!} Z^N.
\end{aligned} \tag{4.62}$$

4.3 Equivalence Between Ensembles

So far we have studied in some detail the microcanonical and the canonical ensembles. They are suitable to different physical situations. The first one is a system where its extensive parameters are fixed (such as in an isolated system). In the second one, energy exchanges are allowed, coupling the system to a heat reservoir, such that temperature is fixed, but volume and particle number changes are not allowed. The microscopic statistical description involves calculation of averages of entities that ultimately lead to predictions of macroscopic quantities that may be compared to experimental results for thermodynamic quantities. We have established some correspondences between statistical quantities and thermodynamic quantities, in both the microcanonical and the canonical ensembles. Specifically, we have established that in the microcanonical ensemble, the generating function is the number of microstates Ω (or actually its logarithm)

$$\begin{aligned} S &= k_B \ln \Omega \\ \frac{1}{k_B T} &= \left(\frac{\partial \ln \Omega}{\partial U}\right)_{N,V} \\ \frac{p}{k_B T} &= \left(\frac{\partial \ln \Omega}{\partial V}\right)_{N,U} \\ -\frac{\mu}{k_B T} &= \left(\frac{\partial \ln \Omega}{\partial N}\right)_{U,V}, \end{aligned} \tag{4.63}$$

while in the canonical ensemble the generating function is the partition function (also its logarithm)

$$\begin{aligned} F &= -k_B T \ln Z \\ S &= k_B \frac{\partial}{\partial T} (T \ln Z)_{N,V} \\ \frac{p}{k_B T} &= \left(\frac{\partial \ln Z}{\partial V}\right)_{N,T} \\ -\frac{\mu}{k_B T} &= \left(\frac{\partial \ln Z}{\partial N}\right)_{T,V}. \end{aligned} \tag{4.64}$$

On the other hand, we have seen that equilibrium states (and irreversibility) is attained because these states correspond to sharp maxima of the probability distributions, over the available microstates of a system. Therefore, a state with a given fixed energy or a state with an energy which is some average energy corresponding to a given temperature is almost the same. We will study this in detail in the following. If true, this implies that there is some freedom to select one or the other ensemble to obtain the thermodynamics of the system. To some extent we will be able to derive the same thermodynamic properties, even though the conditions imposed on the different ensembles are different. Specifically, in the canonical ensemble, the energy fluctuates around its average value. So to show the equivalence between the two ensembles, we need to show that the energy fluctuations in the canonical ensemble are very small in the thermodynamic limit, when the system is very large and assuming the system remains in the equilibrium state.

The fluctuations of the energy can be studied looking at the variance of the energy, given by

$$\langle (\Delta E)^2 \rangle = \langle E^2 \rangle - \langle E \rangle^2. \tag{4.65}$$

Note that the average energy is given by

$$U = \langle E \rangle = \frac{\sum_{E_r} E_r g_r e^{-\beta E_r}}{\sum_{E_r} g_r e^{-\beta E-r}} \tag{4.66}$$

and

$$\begin{aligned} \frac{\partial U}{\partial \beta} &= -\frac{\sum_{E_r} E_r^2 g_r e^{-\beta E_r}}{\sum_{E_r} g_r e^{-\beta E-r}} + \frac{[\sum_{E_r} E_r g_r e^{-\beta E_r}]^2}{[\sum_{E_r} g_r e^{-\beta E-r}]^2} \\ &= -\left[\langle E^2 \rangle - \langle E \rangle^2 \right]. \end{aligned} \tag{4.67}$$

Therefore, we obtain

$$\langle (\Delta E)^2 \rangle = -\frac{\partial U}{\partial \beta} = k_B T^2 \frac{\partial U}{\partial T} = k_B T^2 C_V, \tag{4.68}$$

which shows that the fluctuations of the energy are determined by the heat capacity. (Recall that in the canonical ensemble the number of particles and the volume are fixed and, therefore, we get the heat capacity at constant volume.) We may now calculate the relative energy fluctuation taking the ratio of the standard deviation with respect to the average value

$$\frac{\sqrt{\langle (\Delta E)^2 \rangle}}{\langle E \rangle} = \frac{\sqrt{k_B T^2 C_V}}{U} \sim \frac{1}{\sqrt{N}}, \tag{4.69}$$

which tends to zero in the thermodynamic limit (when the size of the system tends to infinity) and, therefore, the number of particles tends to infinity. Therefore, we have shown the expected result that the relative energy fluctuations vanish in the canonical ensemble.

Further insight may be gained if we determine the maximum of the probability distribution which, in this case, is directly related to the energy distribution. The energy probability distribution has a maximum determined by

$$\frac{\partial}{\partial E}\left[g(E)e^{-\beta E}\right]_{E=E^*} = 0. \tag{4.70}$$

This maximum corresponds to the equilibrium value and $E^* = U$ is the average value. We can now expand $\ln ZP(E)$ around the equilibrium state as

$$\begin{aligned}\ln\left[e^{-\beta E}g(E)\right] &= \left(-\beta U + \frac{S}{k_B}\right) + \frac{1}{2}\left(\frac{\partial^2}{\partial E^2}\ln\left(e^{-\beta E}g(E)\right)\right)_{E=U}(E-U)^2 + \cdots \\ &= -\beta(U - TS) - \frac{1}{2k_B T^2 C_V}(E-U)^2 .\end{aligned} \tag{4.71}$$

We have used that $S = \ln g(E)$, the logarithm of the degeneracy of a given energy E, a result that is actually defined in the context of the microcanonical ensemble. In that context, the entropy is the logarithm of the total number of states Ω, for a fixed energy. In this context, the total number of states, for a given energy E, is the degeneracy of that energy, $g(E)$. Therefore the energy probability distribution satisfies

$$P(E) \sim e^{-\beta E}g(E) \sim e^{-\beta F}e^{-\frac{(E-U)^2}{2k_B T^2 C_V}} \cdots \tag{4.72}$$

and is a Gaussian distribution around the average value, the thermodynamic energy U, with a dispersion given by $\sigma = \sqrt{k_B T^2 C_V}$. Note that since both the energy and the heat capacity are extensive, the numerator is proportional to N^2, while the denominator is proportional to N. Therefore, besides an overall factor of N (common to the first exponential of the free energy), the fluctuation involves the intensive quantity-specific heat.

We also need to consider the partition function. This is given by

$$\begin{aligned} Z &= \int_0^\infty dE e^{-\beta E}g(E) \\ &\sim e^{-\beta F}\int_0^\infty dE e^{-\frac{(E-U)^2}{2k_B T^2 C_V}} \\ &\sim e^{-\beta F}\sqrt{2\pi k_B T^2 C_V}. \end{aligned} \tag{4.73}$$

Therefore

$$-k_B T \ln Z = F \sim F - \frac{1}{2}k_B T \ln\left(2\pi k_B T^2 C_V\right). \tag{4.74}$$

While the free energy term is of the order of N, the second term is of the order of $\ln N$ and therefore is negligible.

4.4 Grand Canonical Ensemble

In the grand canonical ensemble, energy and particle number fluctuations are allowed. The ensemble has $\mathcal{N}$ identical copies of the system we want to study. These systems are distributed among the various possible states that the real physical system may be found in. There are restrictions on the distribution of copies of the ensemble[3]

$$\sum_{r,s} n_{r,s} = \mathcal{N}$$
$$\sum_{r,s} n_{r,s} N_r = \mathcal{N}\bar{N}$$
$$\sum_{r,s} n_{r,s} E_s = \mathcal{N}\bar{E}, \tag{4.75}$$

where $n_{r,s}$ is the number of systems in state r, s. The number of states for a given set of these numbers is given as before by

$$W\{n_{r,s}\} = \frac{\mathcal{N}!}{\prod_{r,s}\left(n_{r,s}!\right)}. \tag{4.76}$$

The most probable distribution leads to

$$P_{r,s} = \frac{n_{r,s}}{\mathcal{N}} = \frac{e^{-\alpha N_r - \beta E_s}}{\sum_{r,s} e^{-\alpha N_r - \beta E_r}}. \tag{4.77}$$

The averages may be obtained calculating

$$\bar{N} = -\frac{\partial}{\partial\alpha}\left(\ln\sum_{r,s} e^{-\alpha N_r - \beta E_s}\right)$$
$$\bar{E} = -\frac{\partial}{\partial\beta}\left(\ln\sum_{r,s} e^{-\alpha N_r - \beta E_s}\right). \tag{4.78}$$

These two results suggest to define a generating function as

$$q = \ln\sum_{r,s} e^{-\alpha N_r - \beta E_s} = q\left(\alpha, \beta, \{E_s\}\right) \tag{4.79}$$

[3] See also [Pathria].

the logarithm of the grand partition function. Similarly to the procedure for the canonical ensemble, we obtain that

$$dq = -\bar{N}d\alpha - \bar{E}d\beta - \beta \sum_{r,s} P_{r,s} dE_s$$

$$d\left(q + \alpha\bar{N} + \beta\bar{E}\right) = \beta\left(\frac{\alpha}{\beta}d\bar{N} + d\bar{E} - \sum_{r,s} P_{r,s} dE_s\right). \tag{4.80}$$

As before, the term in parenthesis on the right hand side is the heat exchanged written as

$$\delta Q = d\bar{E} + \delta W - \mu d\bar{N}$$

$$\delta W = -\sum_{r,s} P_{r,s} dE_s$$

$$\mu = -\frac{\alpha}{\beta}$$

$$d\left(q + \alpha\bar{N} + \beta\bar{E}\right) = \beta\delta Q = \frac{dS}{k_B}. \tag{4.81}$$

The potential q is then given by

$$q = \ln \mathcal{Z} = \frac{S}{k_B} - \alpha\bar{N} - \beta\bar{E} = \frac{TS + \mu\bar{N} - \bar{E}}{k_B T} = \frac{pV}{k_B T}. \tag{4.82}$$

We define now a quantity called fugacity as

$$z = e^{-\alpha} = e^{\frac{\mu}{k_B T}}. \tag{4.83}$$

The grand partition function can be written in terms of the fugacity as

$$\begin{aligned}\mathcal{Z} &= \sum_{r,s} z^{N_r} e^{-\beta E_s} \\ &= \sum_{N_r=0}^{\infty} z^{N_r} \sum_{s} e^{-\beta E_s} \\ &= \sum_{N_r=0}^{\infty} z^{N_r} Z_{N_r},\end{aligned} \tag{4.84}$$

where Z_{N_r} is the partition function of a system with fixed number of particles N_r.

The connection between statistical physics and thermodynamics is obtained in a way similar to the other ensembles. We have shown that pressure can be obtained from the grand partition function as

$$p = \frac{k_B T}{V} \ln \mathcal{Z}. \tag{4.85}$$

Other quantities can be obtained in the following ways:

$$\begin{aligned}
N &= z\left(\frac{\partial}{\partial z}q(z,V,T)\right)_{V,T} = k_B T\left(\frac{\partial}{\partial \mu}q(\mu,V,T)\right)_{V,T} \\
U(z,V,T) &= -\left(\frac{\partial}{\partial \beta}q(z,V,T)\right)_{z,V} = k_B T^2\left(\frac{\partial}{\partial T}q(z,V,T)\right)_{z,V} \\
F &= N\mu - pV = Nk_B T\ln z - k_B T\ln q(z,V,T) = -k_B T\ln\frac{\mathcal{Z}(z,V,T)}{z^N} \\
S &= \frac{U-F}{T} = k_B T\left(\frac{\partial q}{\partial T}\right)_{z,V} - Nk_B\ln z + k_B q \\
&= k_B\left(\frac{\partial}{\partial T}Tq(\mu,V,T)\right)_{\mu,V} \\
S &= \left(\frac{\partial(pV)}{\partial T}\right)_{\mu,V} = k_B\left(\frac{\partial(T\ln\mathcal{Z})}{\partial T}\right)_{\mu,V}.
\end{aligned} \tag{4.86}$$

We may now study the fluctuations in the grand canonical ensemble. Both the number of particles, N, and the energy, E, may vary arbitrarily. We want to show now that the relative fluctuations are negligible.

As before, we start by recalling the probability distribution given by

$$P_{r,s} = \frac{z^{N_r}e^{-\beta E_s}}{\mathcal{Z}(z,V,T)}. \tag{4.87}$$

The averages of the number of particles and of its square are given by

$$\begin{aligned}
\langle N\rangle &= \sum_{r,s} N_r P_{r,s} = \frac{1}{\mathcal{Z}}\left(z\frac{\partial}{\partial z}\right)_{V,T}\mathcal{Z} = z\left(\frac{\partial\ln\mathcal{Z}}{\partial z}\right)_{V,T} \\
\langle N^2\rangle &= \sum_{r,s} N_r^2 P_{r,s} = \frac{1}{\mathcal{Z}}\left(z\frac{\partial}{\partial z}\right)^2_{V,T}\mathcal{Z} = \frac{1}{\mathcal{Z}}\left(z\frac{\partial}{\partial z} + z^2\frac{\partial^2}{\partial z^2}\right)_{V,T}\mathcal{Z}.
\end{aligned} \tag{4.88}$$

The variance of the number of particles is given by

$$\langle(\Delta N)^2\rangle = \langle N^2\rangle - \langle N\rangle^2 = \left(z\frac{\partial}{\partial z}\right)^2_{V,T}\ln\mathcal{Z} = \left(z\frac{\partial}{\partial z}\langle N\rangle\right)_{V,T}$$

$$= k_B T \left(\frac{\partial \langle N \rangle}{\partial \mu} \right)_{V,T} . \tag{4.89}$$

The relative fluctuation is therefore given by

$$\frac{\langle (\Delta N)^2 \rangle}{\langle N \rangle^2} = \frac{k_B T}{N^2} \left(\frac{\partial N}{\partial \mu} \right)_{V,T} \tag{4.90}$$

and is identically the same for the relative fluctuation of the particle density $n = N/V$. Recall that $d\mu = vdp - sdT$ where $v = V/N$ and $s = S/N$. Then we get that if the temperature is constant, $dT = 0$, then $d\mu = vdp$. Taking a derivative with respect to v we get

$$\left(\frac{\partial \mu}{\partial v} \right)_T = v \left(\frac{\partial p}{\partial v} \right)_T = -\frac{N^2}{V} \left(\frac{\partial \mu}{\partial N} \right)_{V,T} = V \left(\frac{\partial p}{\partial V} \right)_{N,T} . \tag{4.91}$$

Substituting, we get that the relative fluctuation reads

$$\frac{\langle (\Delta N)^2 \rangle}{\langle N \rangle^2} = -\frac{k_B T}{V^2} \left(\frac{\partial V}{\partial p} \right)_{N,T} = \frac{k_B T}{V} \chi_T , \tag{4.92}$$

where χ_T is the isothermal compressibility. As before, the relative fluctuation is of the order of $1/\sqrt{N}$ and tends to zero in the thermodynamic limit, if the thermal compressibility is finite. In a phase transition, the compressibility may be very high and there are strong fluctuations. Actually, these fluctuations are the precursors of the phase transition that is associated with a change of density between two different phases. This leads to the phenomenon called critical opalescence, associated with the vicinity of a phase transition that involves fluctuations of density.

Finally we can also estimate the energy fluctuations. The averages of the energy and its square are given by

$$\begin{aligned} \langle E \rangle &= \sum_{r,s} E_s P_{r,s} = -\frac{1}{\mathcal{Z}} \left(\frac{\partial}{\partial \beta} \right)_{z,V} \mathcal{Z} = -\left(\frac{\partial \ln \mathcal{Z}}{\partial \beta} \right)_{z,V} \\ \langle E^2 \rangle &= \sum_{r,s} E_s^2 P_{r,s} = \frac{1}{\mathcal{Z}} \left(\frac{\partial^2}{\partial \beta^2} \right)_{z,V} \mathcal{Z} . \end{aligned} \tag{4.93}$$

The fluctuations of the energy are therefore

$$\langle (\Delta E)^2 \rangle = \langle E^2 \rangle - \langle E \rangle^2 = \left(\frac{\partial^2}{\partial \beta^2} \right)_{z,V} \ln \mathcal{Z} = -\left(\frac{\partial \langle E \rangle}{\partial \beta} \right)_{z,V} . \tag{4.94}$$

The relative fluctuation is given by

$$\frac{\langle(\Delta E)^2\rangle}{\langle E\rangle^2} = \frac{k_B T^2}{U^2}\left(\frac{\partial U}{\partial T}\right)_{z,V} \sim \frac{1}{N}, \tag{4.95}$$

which again vanishes in the thermodynamic limit.

The energy fluctuations in the grand canonical ensemble are related with the energy fluctuations in the canonical ensemble. It can be shown that

$$\langle(\Delta E)^2\rangle = k_B T^2 C_V + \langle(\Delta N)^2\rangle \left(\frac{\partial U}{\partial N}\right)_{T,V}^2 . \tag{4.96}$$

4.5 Jaynes Formalism and Classical Information Theory

When we have a situation described in terms of probabilities, there is no complete knowledge of all the variables of the problem. In a statistical problem, there is ignorance, uncertainty, or lack of information that otherwise would allow a deterministic description. When we assign given values of the probabilities to different events, we should use all the information available, since the values of the probabilities depend on the information (Bayes theorem). The importance of a given piece of information is the change of the initial state of uncertainty. Shannon attempted to quantify the notion that we should use all the information we possess and nothing else, to assign values to the probabilities. This is the basis of information theory. It provides an alternative perspective on the reasoning behind a statistical formulation of a classical system.

Four postulates or properties are introduced. Consider an event that may take n possible values. The postulates are:

1. The uncertainty, H, associated with the possible results should be an exclusive function of the probabilities of those results

$$H = H(p_1, \cdots, p_n). \tag{4.97}$$

2. If the events are equally probable, the uncertainty should be a monotonically increasing function of the number of results. In this case we have that $p_i = 1/n$ for all events, i. Therefore

$$H = H\left(\frac{1}{n}, \cdots, \frac{1}{n}\right) = H\left(\frac{1}{n}\right) = \bar{H}(n). \tag{4.98}$$

   This function $\bar{H}$ is monotonically increasing, which reflects the expected result that the larger the possible number of events the larger the uncertainty.
3. If $C = AB$ where A and B are independent events, the uncertainty associated with C is the sum of the uncertainties of events A and B

$$H(C) = H(A) + H(B) = H(AB). \tag{4.99}$$

The uncertainty should be additive.

4. The value of the uncertainty should not depend on how the problem is set. In other words, and choosing a language we used in thermodynamic systems, the uncertainty is a function of state.

With these four properties we are now able to determine the functional form of the uncertainty. A few comments beforehand are useful.

First, the probability distribution itself is some measure of the uncertainty. Consider two probability distributions in terms of some variable such that one of them is broadly peaked around some value and the other one is sharply peaked around some particular value of the variable. It is intuitive that the uncertainty is larger in the case of the first distribution. Simple examples can be constructed with $n = 3$. In a first case take $p_1 = 1$, $p_2 = 0$, $p_3 = 0$. In this case clearly there is no uncertainty. Consider now $p_1 = 0.8$, $p_2 = 0.2$, $p_3 = 0$. Clearly the uncertainty is now finite. Taking $p_1 = p_2 = p_3 = 1/3$ is clearly the case with maximal uncertainty since all outcomes of a measurement are equally probable.

Although the probability distribution is a measure of the uncertainty, it is more convenient to characterize the uncertainty by a single number, instead of specifying the whole set of probabilities.

Consider an event that may take n values with equal probabilities. Consider that this event is the intersection of two independent events one with n_1 possible results and the other with n_2 possible results. Therefore

$$H(n_1 n_2) = H(n_1) + H(n_2). \tag{4.100}$$

Take a derivative with respect to n_1: we get

$$n_2 H'(n_1 n_2) = H'(n_1). \tag{4.101}$$

Take now a derivative with respect to n_2. We get

$$n_1 n_2 H''(n_1 n_2) + H'(n_1 n_2) = 0. \tag{4.102}$$

But this can be written as

$$n H''(n) + H'(n) = 0, \tag{4.103}$$

which leads to

$$\frac{H''}{H'} = -\frac{1}{n}. \tag{4.104}$$

Integrating once we get

$$\ln H'(n) = \ln \frac{K}{n} \tag{4.105}$$

and integrating again we get that

$$H(n) = k \ln n + C. \tag{4.106}$$

Here K and C are constants. If $n = 1$ there is no uncertainty which implies that $C = 0$. Therefore, we get that the uncertainty is given in this case, of equally probable outcomes,

$$H(n) = K \ln n. \tag{4.107}$$

The value of K determines the units of the uncertainty. If we take $K = 1$ and the logarithm in base 2 the units of information are called "bits." If $K = k_B$ Boltzmann's constant and the logarithm is in basis "e," the uncertainty is in Boltzmann units (like the entropy). Note that $H = K \ln n = -K \ln p$ with $p = 1/n$. This is the uncertainty per result which may be interpreted as the average uncertainty per event.

Let us now suppose that we receive some information that allows us to change the probabilities of the various outcomes. That is the probability distribution $\{p\}$ changes to $\{p'\}$. Initially the uncertainty was $H = -K \ln p$, and the uncertainty after the information received is $H' = -K \ln p'$. For the moment, we are restricting to cases where all the events are equally probable: so the number of outcomes has changed. For instance, we may consider a particle in a homogenous system whose volume has changed. If the volume increases (the number of states has increased), so does the uncertainty of the particle location.

We may define $I = H - H' = -K \ln(p/p')$. This is called the information contained in the message that allowed us to change the probabilities. If $I > 0$ there is a reduction of uncertainty. If the information is negative, there is an increase of the uncertainty, as is intuitive. If we calculate the average value of logarithm of the probability

$$\sum_{i=1}^{n} \frac{1}{n} \ln \frac{1}{n} = \ln \frac{1}{n} = \ln p. \tag{4.108}$$

Consider now that we have possible outcomes that are not equally probable. To each event i we may associate a probability p_i. It is convenient to consider the following example: assume we have n balls that we want to distribute in N groups. Therefore a given distribution of the balls leads to n_1 balls in group 1, n_2 balls in group 2 and so on. Clearly $\sum_{i=1}^{N} = n$. Call C the event of distributing the balls and factorize it into two events, an event A which is the choice of one of the groups and an event B which is the distribution of the n_i results in each group. The event A has N possible outcomes. The probability of each outcome is $p_i = n_i/n$. We want to know the average uncertainty per outcome. This we call $H(p_1, \cdots, p_N)$.

The postulates impose that

$$H(C) = H(A) + H(B), \tag{4.109}$$

since A and B are independent. If the event A leads to the result i, the uncertainty per result is $K \ln n_i$. The average uncertainty per outcome relative to the event B is $\sum_{i=1}^{N} p_i K \ln n_i$. Therefore,

$$K \ln n = H(p_1, \cdots, p_N) + \sum_{i=1}^{N} p_i K \ln ni. \tag{4.110}$$

This leads to

$$\begin{aligned} H(p_1, \cdots, p_N) &= K \ln n - \sum_{i=1}^{N} K \ln n_i \\ &= -K \sum_{i=1}^{N} p_i \ln p_i. \end{aligned} \tag{4.111}$$

This is Shannon's formula for the uncertainty.

We may now take into account these ideas to look for an optimal criterion to infer the probabilities of a given set of accessible states of a system. The idea is to maximize the uncertainty (this is similar to the maximization of the entropy we saw before). Usually, in a system there are equations that the variables satisfy and typically the number of unknowns is larger than the number of equations that mirror our knowledge on the system. Let us assume that the information on the system are the average values of some random variables $g_1, g_2, \cdots, g_R$. Therefore, we know $\langle g_r \rangle$, where $r = 1, \cdots, R$. We choose the distribution of probabilities to be normalized. We have therefore $R + 1$ equations

$$\begin{aligned} \sum_i p_i &= 1 \\ \sum_i p_i g_r^{(i)} &= \langle g_r \rangle. \end{aligned} \tag{4.112}$$

The Jaynes formalism makes the proposal that the best way to estimate the probabilities, for a given set of known average values, is to maximize the uncertainty associated with that information. Therefore we need to maximize the uncertainty (called from now on statistical entropy $S = H$) subject to the restrictions of the average values we know. This extremum problem requires the use of Lagrange multipliers to implement the constraints. Let us associate a Lagrange multiplier $\lambda_0 - 1$ with the equation that implements the normalization of the probabilities and Lagrange multipliers λ_r to the other equations. We get

$$dS - (\lambda_0 - 1)\sum_i dp_i - \lambda_r \sum_i g_r^{(i)} dp_i = 0. \tag{4.113}$$

Using that

$$dS = -K \sum_i (\ln p_i + 1)\, dp_i, \tag{4.114}$$

we obtain the equation

$$\sum_i \left(\ln p_i + 1 + \lambda_0 - 1 + \sum_r \lambda_r g_r^{(i)} \right) dp_i = 0. \tag{4.115}$$

We now choose the Lagrange multipliers such that each term in the sum vanishes

$$p_i = e^{-\lambda_0 - \sum_r \lambda_r g_r^{(i)}}. \tag{4.116}$$

Their values are now determined using this form of the probabilities in the equations for the average values that are known. For instance, we may write that

$$e^{\lambda_0} = \sum_i e^{-\sum_r \lambda_r g_r^{(i)}} = Z. \tag{4.117}$$

Z is a sum over states and therefore is like the generalized partition function. (If the various equations that we have to satisfy are conservation laws the variables of the system must satisfy, this construction is called a generalized Gibbs ensemble.) Note that $\lambda_0 = \ln Z$. It is easy to see that

$$\frac{\partial \lambda_0}{\partial \lambda_r} = -\langle g_r \rangle. \tag{4.118}$$

Correlation functions are obtained taking one more derivative

$$\frac{\partial^2 \lambda_0}{\partial \lambda_r \lambda_s} = \langle g_r g_s \rangle - \langle g_r \rangle \langle g_s \rangle. \tag{4.119}$$

4.6 Exercises

1. Show that the solution of Liouville's equation, ρ, is a true probability density, i.e., ρ satisfies the following properties:

$$\rho \geq 0, \int \rho(\Gamma) d\Gamma = 1.$$

2. Consider the one-dimensional motion of a particle of mass m in the presence of a gravitational field. Draw the trajectories of the particle in phase space and verify the validity of Liouville's theorem.
3. Show the validity of Liouville's theorem in the case of an one dimensional elastic collision between two particles. Assume that the particles have equal masses.
4. Obtain the solution of Liouville's equation for a system of harmonic oscillators.
5. Consider a one-dimensional harmonic oscillator.

   (a) The displacement x of the oscillator with respect to its equilibrium position is $x = A\cos(\omega t + \phi)$. Suppose that the exact value of the phase ϕ is unknown and that we only know that it has equal probability to take any value in the interval $0 < \phi < 2\pi$. Determine the probability $P(x)dx$ that x is found in the interval between x and $x + dx$, at a given time t. Express $P(x)$ as a function of A and x.
   (b) Knowing that the energy of the oscillator lies between E and $E + dE$, calculate $P(x)dx$ and express the result as a function of E and x. Relate E with the amplitude and verify that $P(x)dx$ coincides with the one calculated in a).

6. Show that when $T \to 0$, C_V, C_p, $(\partial V/\partial T)_p$, $(\partial S/\partial p)_T$ and $(\partial S/\partial V)_T \to 0$.
7. Consider a system of non-interacting N spins 1/2 in thermal equilibrium in the presence of an external magnetic field **B** and call μ_i the component of the magnetic moment associated with the spin i, in the direction of **B**. Let p and q be the probabilities that any magnetic moment component takes the values μ and $-\mu$, respectively.

   (a) What are the values of p and q when $B = 0$?
   (b) If $B \neq 0$ should we have $p > q$ or $p < q$? Justify.
   (c) Show that the average total magnetic moment of the system, **M**, along the magnetic field direction, is given by $\langle M \rangle = N(p - q)\mu$ and that $\langle (M - \langle M \rangle)^2 \rangle = 4Npq\mu^2$.
   (d) Calculate the relative fluctuation of M.

8. Consider the same system of magnetic moments. The magnetic energy is then $E = -(n_1 - n_2)\mu B$, where n_1 is the number of spins aligned along the field **B** and n_2 the number of spins aligned in the opposite direction.

   (a) Consider the energy interval between E and $E+\delta E$ where $\mu B << \delta E << E$. What is the total number of microstates $\Omega(E)$ in that energy interval?
   (b) Express $\ln \Omega(E)$ as a function of E. Use the Stirling formula.

9. Knowing that the partition function Z of the canonical ensemble and the free energy F are related by $-\beta F = \ln Z$ shows that:

   (a) $\langle p \rangle = \beta^{-1} (\partial \ln Z/\partial V)_\beta$.
   (b) $S = -k_B \beta^2 (\partial \ln Z/\partial \beta)_V$.
   (c) $H = -(\partial \ln Z/\partial \beta)_V + V/\beta\, (\partial \ln Z/\partial V)_\beta$, where $H = U + pV$.
   (d) $G = (\beta^2 - 1)(\partial \ln Z/\partial \beta)_V + \beta^{-1} (\partial \ln Z/\partial V)_\beta$, where $G = H - TS$.

10. In statistical mechanics, we have

$$\left(\frac{\partial \langle E \rangle}{\partial V}\right)_\beta + \beta \left(\frac{\partial p}{\partial \beta}\right)_V = -p$$

and from thermodynamics

$$\left(\frac{\partial U}{\partial V}\right)_T - T \left(\frac{\partial p}{\partial T}\right)_V = -p.$$

Why is it that β cannot be directly proportional to temperature?

11. Consider the following partition function $Z(N, V, T) = f(T)V^N$. Show that the system it describes obeys the equation of state of an ideal gas. ($f(T)$ is an arbitrary function of temperature.)
12. Consider an homogeneous mixture of monoatomic inert gases at temperature T in a container of volume V. Suppose that the mixture is composed by ν_1 moles of gas 1, ν_2 moles of gas 2, $\cdots$, ν_k moles of gas k.

    (a) Derive the equation of states of the mixture starting from the partition function.
    (b) What is the relation between the total pressure p and the pressure p_i that the gas i would exert if it were the only to occupy the volume V at temperature T?

13. Consider an ideal gas of N diatomic molecules, homonuclear, in a container of volume V at temperature T. The Hamiltonian of each of the molecules is

$$H(\mathbf{r}_1, \mathbf{r}_2; \mathbf{p}_1, \mathbf{p}_2) = \frac{1}{2m}\left(p_1^2 + p_2^2\right) + \frac{\alpha}{2}|\mathbf{r}_1 - \mathbf{r}_2|^2,$$

where $\mathbf{r}_1$ and $\mathbf{r}_2$ are the positions of each of the constituent atoms of a molecule and $\mathbf{p}_i$ are the respective momenta, α is a constant.

    (a) Calculate the partition function of the gas.
    (b) Calculate $\langle |\mathbf{r}_1 - \mathbf{r}_2|^2 \rangle$.
    (c) Determine the heat capacity at constant volume.

14. Show the Bohr-van Leeuwen theorem: the phenomenon of diamagnetism (in the sense of a thermodynamic average) does not exist in classical physics.
    Suggestions:

    (a) If $H(\mathbf{q}_i, \mathbf{p}_i)$ is the Hamiltonian of a system of charged particles with charge q_i with no interactions in the absence of an external field, then $H(\mathbf{q}_i; \mathbf{p}_i - q_i\mathbf{A}_i)$ will be the Hamiltonian in the presence of a magnetic field $\mathbf{B} = \nabla \times \mathbf{A}$, where $\mathbf{A}_i$ is the vector potential at position i and c the speed of light.

(b) The magnetization along the magnetic field, $\mathbf{M}$, induced in the system is related with the partition function by

$$\mathbf{M} = k_B T \frac{\partial \ln Z}{\partial \mathbf{B}}.$$

15. Consider a gas of N particles in the extreme relativistic limit, that is, $\epsilon = cp$, that is in equilibrium at temperature T (ϵ and p are respectively the energy and moment of each particle). Determine the internal energy, the entropy, and the specific heat of the gas.
16. Show that the grand canonical partition function of an ideal monoatomic gas may be expressed in the form $\mathcal{Z} = e^{zZ}$ where z is the fugacity Z the partition function per atom.
17. Knowing that the partition function in a isothermal and isobaric ensemble is given by

$$\Delta(p, T, N) = \sum_V Z(N, V, T) e^{-\beta p V},$$

where $Z(N, V, T)$ is the partition function of the canonical ensemble, show that the average volume $\langle V \rangle$ is given by

$$\langle V \rangle = -k_B T \left(\frac{\partial \ln \Delta}{\partial p} \right)_{T,N}.$$

Verify that the thermodynamic potential that corresponds to this ensemble is the Gibbs free energy, $G = k_B T \ln \Delta$.
18. Deduce an equation for the volume fluctuation in an isothermal and isobaric ensemble, i.e., determine

$$\langle (V - \langle V \rangle)^2 \rangle,$$

as a function of the isothermal compressibility.

Chapter 5
Stochastic Processes

5.1 Master Equation and Fokker-Planck Equation

The intrinsic uncertainty of the microstate of a system leads to a statistical description of a system in terms of stochastic variables that fluctuate according to some probability distributions. To simplify, for the moment consider a system that is described by a variable Y. Several probability distributions may be defined that reflect the time evolution of a system together with the values the variable Y may take. The sequence of values may be defined in terms of continuous or discrete variables. One such probability function is defined, $P_1(y_1, t_1)$, as the probability density that the variable Y takes the value y_1 at time t_1.[1] We can as well define $P_2(y_1, t_1; y_2, t_2)$ the probability density that the variable Y takes the value y_1 at time t_1 and the value y_2 at time t_2 and generalize to n events $P_n(y_1, t_1; \cdots ; y_n, t_n)$. It is also convenient to define conditional probabilities, relating the values of the variable at given times with the knowledge that at earlier times the variable took a given sequence of values. To better describe the correlations between the values the variable Y takes as time goes by, we also define $P_{1|1}(y_1, t_1|y_2, t_2)$ as the conditional probability density that variable Y takes the value y_2 at time t_2 knowing that Y took the value y_1 at time t_1 (note the causality). We may also define a generalized conditional probability distribution as

$$P_{k|l}(y_1, t_1; \cdots ; y_k, t_k|y_{k+1}, t_{k+1}; \cdots ; y_{k+l}, t_{k+l}),$$

with the meaning of the conditional probability density that the stochastic variable Y takes the values $(y_{k+1}, t_{k+1}; \cdots ; y_{k+l}, t_{k+l})$ knowing that it took the values $(y_1, t_1; \cdots ; y_k, t_k)$. Also, reduced probability densities are defined integrating over the values of some previous values such as

[1] See also [Reichl].

P. D. Sacramento, *Introduction to Condensed Matter Theory*,
https://doi.org/10.1007/978-3-031-86251-9_5

$$\int P_n(y_1,t_1;\cdots;y_n,t_n)dy_n = P_{n-1}(y_1,t_1;\cdots;y_{n-1},t_{n-1}). \tag{5.1}$$

Since these quantities are probability densities, they are defined positive $P_n \geq 0$. It is convenient to normalize the probability distribution in the usual way:

$$\int P_1(y_1,t_1)dy_1 = 1. \tag{5.2}$$

We are assuming for now that the variable Y is continuous. If the variable takes discrete values, we just need to replace the integrals by sums over the values the variable takes. The sequence of the events expresses the time evolution, and it turns out that the probability distributions obey some integral equations that can be reformulated in terms of differential equations in time. We may define moments of the distribution such as

$$\begin{aligned}&\langle y_1(t_1)y_2(t_2)\cdots y_n(t_n)\rangle\\&\quad=\int dy_1\int dy_2\cdots\int dy_n y_1y_2\cdots y_nP_n(y_1,t_1;y_2,t_2;\cdots;y_n,t_n)\end{aligned} \tag{5.3}$$

that give information about the correlations among the values the variable takes as time evolves.

A process is stationary if, for all values of n, the distributions $P_n(y_i,t_i,i=1,\cdots,n)$ are invariant under a translation in time by τ, for any time τ. As a consequence we can write that $P_1(y_1,t_1) = P_1(y_1)$ is independent of time, and $\langle y_1(t_1)y_2(t_2)\rangle$ only depends on the time difference $|t_1-t_2|$.

It is easy to see that

$$P_{1|1}(y_1,t_1|y_2,t_2) = \frac{P_2(y_1,t_1;y_2,t_2)}{P_1(y_1,t_1)}. \tag{5.4}$$

which leads to

$$P_1(y_2,t_2) = \int P_1(y_1,t_1)P_{1|1}(y_1,t_1|y_2,t_2)dy_1. \tag{5.5}$$

Equation (5.4) can be generalized to give

$$P_{k|l}(y_1,t_1;\cdots;y_k,t_k|y_{k+1},t_{k+1};\cdots;y_{k+l},t_{k+l}) = \frac{P_{k+l}(y_1,t_1;\cdots;y_k+l,t_{k+l})}{P_k(y_1,t_1;\cdots;y_k,t_k)}. \tag{5.6}$$

The conditional probabilities are important in the study of the dynamics of systems if these probabilities have memory effects, that is, if the probabilities are influenced by the sequence of events the system went through as it evolves in time. In the case that the system has no memory of its past, then the treatment of the behavior of the system is simplified. It may happen that this memory is short ranged in the sense that the stochastic variable only keeps memory of its immediate past. Then we get that

$$P_{n-1|1}(y_1, t_1; \cdots; y_{n-1}, t_{n-1}|y_n, t_n) = P_{1|1}(y_{n-1}, t_{n-1}|y_n, t_n). \tag{5.7}$$

This is called a Markov process. A Markov process is completely determined by $P_1(y_1, t_1)$ and by the conditional probability density $P_{1|1}(y_1, t_1|y_2, t_2)$. We know that

$$P_3(y_1, t_1; y_2, t_2; y_3, t_3) = P_1(y_1, t_1)P_{1|1}(y_1, t_1|y_2, t_2)P_{2|1}(y_1, t_1; y_2, t_2|y_3, t_3). \tag{5.8}$$

Considering $n = 3$ as an example and a Markov process, we may write that

$$P_{2|1}(y_1, t_1; y_2, t_2|y_3, t_3) = P_{1|1}(y_2, t_2|y_3, t_3). \tag{5.9}$$

Integrating over y_2 we get that the so-called Chapman-Kolmogorov equation

$$P_{1|1}(y_1, t_1|y_3, t_3) = \int dy_2 P_{1|1}(y_1, t_1|y_2, t_2)P_{1|1}(y_2, t_2|y_3, t_3). \tag{5.10}$$

Since the conditional probability density may be interpreted as a transition probability, the meaning of this equation is that the transition from a value of the variable $Y = y_1$ at time t_1 to the value y_3 at time t_3 may be achieved through an intermediate step through a value y_2 at time $t_1 < t_2 < t_3$, as long as we integrate over all possible values the variable y_2 may take. Clearly, this can be generalized considering as many intermediate times as desired as long as we integrate over all intermediate values of the stochastic variable.

The Chapman-Kolmogorov is an integral equation. We may also obtain a differential equation that governs the probability density distribution. Consider again Eq. (5.5) and choose $t_2 = t_1 + \tau$. We obtain therefore

$$P_1(y_2, t_1 + \tau) = \int dy_1 P_1(y_1, t_1)P_{1|1}(y_1, t_1|y_2, t_1 + \tau). \tag{5.11}$$

Taking $\tau = 0$ implies that

$$P_{1|1}(y_1, t_1|y_2, t_1) = \delta(y_1 - y_2). \tag{5.12}$$

Consider that τ is small. We may write that

$$P_{1|1}(y_1, t_1|y_2, t_1 + \tau) = \delta(y_1 - y_2) - \tau \int dy W_{t_1}(y_1, y)\delta(y_1 - y_2) + \tau W_{t_1}(y_1, y_2). \tag{5.13}$$

We have defined $W_{t_1}(y_1, y_2)$ as the transition probability density per unit time that the system changes from state y_1 to state y_2 in the time interval that goes from $t_1 \to t_1 + \tau$. The meaning of the last term is that $\tau W_{t_1}(y_1, y_2)$ is the probability density that the transition $y_1 \to y_2$ takes place in the time interval τ. The meaning of the first two terms is the probability that there is no transition and $y_2 = y_1$.

Define the time derivative of the probability density in the usual way

$$\frac{dP_1(y_1, t)}{dt} = \lim_{\tau \to 0} \frac{P_1(y_1, t + \tau) - P_1(y_1, t)}{\tau}. \tag{5.14}$$

We need to calculate

$$\lim_{\tau \to 0} \frac{P_{1|1}(y_1, t|y_2, t + \tau) - P_{1|1}(y_1, t|y_2, t)}{\tau}.$$

We obtain that

$$\frac{\partial P_1(y_2, t)}{\partial t} = \int dy_1 P_1(y_1, t) \left[- \int dy W_{t_1}(y_1, y) \delta(y_1 - y_2) + W_{t_1}(y_1, y_2) \right], \tag{5.15}$$

which finally leads to the so-called master equation

$$\frac{\partial P_1(y_2, t)}{\partial t} = \int dy_1 \left[W(y_1, y_2) P_1(y_1, t) - W(y_2, y_1) P_1(y_2, t) \right]. \tag{5.16}$$

This result reflects the expected result that $P_1(y_2, t)$ increases if there are transitions from $y_1 \to y_2$ and decreases if there are transitions in the opposite direction $y_2 \to y_1$. Since we have to sum over the possible states, there is an integration over y_1.

Let us assume that the changes in the variable Y can only take small values. This implies that $W(y', y)$ should decrease fast with $|y - y'|$. We may then write that

$$W(y', y) = W(y'; y - y') = W(y', \xi), \tag{5.17}$$

where $\xi = y - y'$. This implies that the master equation may be written as

$$\frac{\partial P_1(y, t)}{\partial t} = \int d\xi W(y - \xi; \xi) P_1(y - \xi, t) - P_1(y, t) \int d\xi W(y; -\xi). \tag{5.18}$$

Expanding the integrand of the first term in powers of ξ, we obtain

$$\begin{aligned} \frac{\partial P_1(y, t)}{\partial t} &= \int d\xi W(y; \xi) P_1(y, t) - \int d\xi \xi \frac{\partial}{\partial y} W(y; \xi) P_1(y, t) \\ &+ \frac{1}{2} \int d\xi \xi^2 \frac{\partial^2}{\partial y^2} W(y; \xi) P_1(y, t) + \cdots - P_1(y, t) \int d\xi W(y; -\xi). \end{aligned} \tag{5.19}$$

The first and last terms on the right hand side cancel. Neglecting higher order terms it leads to the Fokker-Planck equation for the probability distribution

$$\frac{\partial P_1(y,t)}{\partial t} = -\frac{\partial}{\partial y}\left[\alpha_1(y)P_1(y,t)\right] + \frac{1}{2}\frac{\partial^2}{\partial y^2}\left[\alpha_2(y)P_1(y,t)\right], \tag{5.20}$$

where

$$\alpha_n(y) = \int d\xi \xi^n W(y,\xi) \tag{5.21}$$

are the moments of order n.

5.2 Markov Chains and One-Dimensional Random Walk

A Markov chain of events corresponds to a situation where the variable Y takes a discrete set of values $Y = (y_1, y_2, \cdots, y_l)$ where time takes on a series of discrete values multiples of some time scale τ, where $t = s\tau$ with s an integer and τ the minimal time interval between events.[2] In this case we write that

$$P_1(y_i, 1) = \sum_{j=1}^{l} P_1(y_j, 0) P_{1|1}(y_j, 0|y_i, 1). \tag{5.22}$$

As time evolves from $t = 0 \to t = \tau$ we have that $s = 0 \to s = 1$. We may now define a quantity

$$Q_{ji} = P_{1|1}(y_j, 0|y_i, 1). \tag{5.23}$$

We know that

$$\sum_i Q_{ji} = 1.$$

Using the definition, we can write that $P(1) = P(0)Q$. We can construct a l-component vector $P(s)$. In this case Q is a matrix of dimension $l \times l$. Iterating we can obtain that $P(s) = P(0)Q^s$.

Consider now a particle that moves randomly along one line. The points along the line differ by a distance l and as a time τ passes by the particle jumps to the right with probability p or to the left with probability $q = 1 - p$, as we encountered before. We can write that

---

[2] See also [Reichl].

$$P_1(n_2 l, s\tau) = \sum_{n_1} P_1\left((n_1 l, (s-1)\tau\right) P_{1|1}\left(n_1 l, (s-1)\tau | n_2 l, s\tau\right). \tag{5.24}$$

The transition can be written accordingly as

$$P_{1|1}\left(n_1 l, (s-1)\tau | n_2 l, s\tau\right) = p\delta_{n_2, n_1+1} + q\delta_{n_2, n_1-1}. \tag{5.25}$$

If moving to the right is equally likely as moving to the left, $p = q = 1/2$, we get

$$P_1(nl, s\tau) = \frac{1}{2} P_1\left((n-1)l, (s-1)\tau\right) + \frac{1}{2} P_1\left((n+1)l, (s-1)\tau\right). \tag{5.26}$$

The time evolution of the probability density can be obtained calculating

$$\begin{aligned} &\frac{P_1\left(nl, s\tau\right) - P_1\left(nl, (s-1)\tau\right)}{\tau} \\ &= \frac{l^2}{2\tau}\left[\frac{P_1\left((n+1)l, (s-1)\tau\right) + P_1\left((n-1)l, (s-1)\tau\right) - 2P_1\left(nl, (s-1)\tau\right)}{l^2}\right]. \end{aligned} \tag{5.27}$$

Using that $x = nl$ and $t = s\tau$ and taking the limits $\lim_{l\to 0}$ and $\lim_{\tau\to 0}$ such that $D = l^2/(2\tau)$ is finite, one gets the Fokker-Planck equation in the form of a diffusion equation

$$\frac{\partial P_1(x,t)}{\partial t} = D\frac{\partial^2 P_1(x,t)}{\partial x^2}, \tag{5.28}$$

where D is the diffusion constant.

Let us look for a solution of the Fokker-Planck equation for an initial condition $P_1(x, 0) = \delta(x)$, which describes a particle located at the origin. Define the Fourier transform

$$P_1(k,t) = \int_{-\infty}^{\infty} dx\, P_1(x,t) e^{ikx} \tag{5.29}$$

and its inverse

$$P_1(x,t) = \int_{-\infty}^{\infty} \frac{dk}{2\pi} P_1(k,t) e^{-ikx}. \tag{5.30}$$

We obtain that

$$\frac{\partial P_1(k,t)}{\partial t} = -Dk^2 P_1(k,t). \tag{5.31}$$

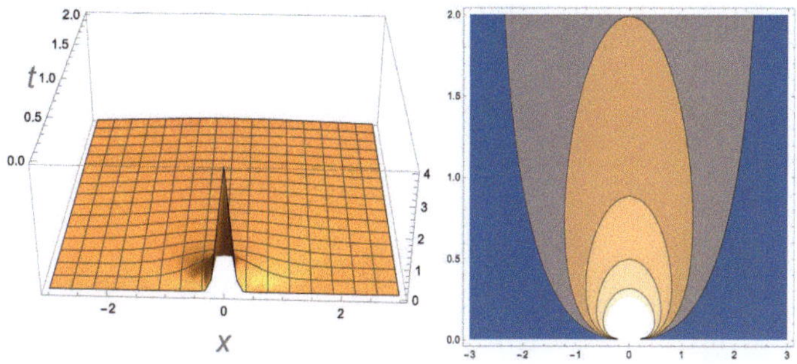

Fig. 5.1 Time evolution of spatial probability distribution with $D = 1$. On the right side a contour plot is shown

The solution is simple and is given by

$$P_1(k,t) = Ae^{-Dk^2 t}. \tag{5.32}$$

Note that

$$P_1(k,0) = \int_{-\infty}^{\infty} dx\, P_1(x,0)e^{ikx} = 1, \tag{5.33}$$

which implies that $A = 1$. We may now calculate the inverse Fourier transform

$$P_1(x,t) = \int_{-\infty}^{\infty} \frac{dk}{2\pi} e^{-Dk^2 t} e^{-ikx} = \frac{1}{\sqrt{4\pi Dt}} e^{-\frac{x^2}{4Dt}} \tag{5.34}$$

using the properties of Gaussian integrals (completing the square). The time dependence of the probability distribution is shown in Fig. 5.1.

We want now to determine how the particle moves in space through time. First we calculate the average value of its location

$$\langle x(t)\rangle = \int_{-\infty}^{\infty} dx x P_1(x,t) = \frac{1}{\sqrt{4\pi Dt}} \int_{-\infty}^{\infty} dx x e^{-\frac{x^2}{4Dt}} = 0, \tag{5.35}$$

since the integrand is an odd function. We may as well calculate the quadratic displacement

$$\begin{aligned}\langle x^2(t)\rangle &= \int_{-\infty}^{\infty} dx x^2 P_1(x,t) \\ &= \frac{1}{\sqrt{4\pi Dt}} \int_{-\infty}^{\infty} dx x^2 e^{-\frac{x^2}{4Dt}}\end{aligned}$$

$$
\begin{aligned}
&= \frac{1}{\sqrt{4\pi Dt}} \left(\frac{-\partial}{\partial \beta} \int_{-\infty}^{\infty} dx e^{-\beta x^2} \right)_{\beta=\frac{1}{4Dt}} \\
&= \frac{1}{\sqrt{4\pi Dt}} \left(\frac{-\partial}{\partial \beta} \sqrt{\frac{\pi}{\beta}} \right)_{\beta=\frac{1}{4Dt}} \\
&= \frac{1}{\sqrt{4Dt}} \frac{1}{2} \frac{1}{\beta^{3/2}} |_{\beta=\frac{1}{4Dt}} \\
&= 2Dt.
\end{aligned}
\tag{5.36}
$$

Therefore, we get

$$
\begin{aligned}
\frac{d}{dt} \langle x(t) \rangle &= 0 \\
\frac{d}{dt} \langle x^2(t) \rangle &= 2D.
\end{aligned}
\tag{5.37}
$$

Note that the standard deviation of the coordinate of the particle in this random walk grows with time like

$$
\sqrt{\langle x^2(t) \rangle} \sim \sqrt{t}. \tag{5.38}
$$

5.3 Liouvillian and BBGKY Hierarchy

The state of a classical system of N particles with $3N$ degrees of freedom may be represented by a vector in phase space $\mathbf{X}^N = (\mathbf{q}^N, \mathbf{p}^N)$ where q_i are generalized coordinates, such as spatial positions $q_i = x_i$, and p_i are the conjugate momenta. We define as before, $\rho(\mathbf{X}^N, t) d\mathbf{X}^N$, the probability that $\mathbf{X}^N$ is between $\mathbf{X}^N$ and $\mathbf{X}^N + d\mathbf{X}^N$, at time t. In contrast to the probabilities considered in the previous sections, which described a series of events in time, as characterized by the evolution of a given variable, the probability distribution gives information about the state at a given time. This probability distribution satisfies Liouville's theorem as expressed in Eqs. (4.8), (4.9), and (4.10). Equation (4.10) may be rewritten as[3]

$$
\frac{\partial \rho^N}{\partial t} = -\hat{H}^N \rho^N \tag{5.39}
$$

defining an operator $\hat{H}^N$. Using the definition of the Poisson bracket, we may write it formally as

[3] See also [Reichl].

$$i\frac{\partial \rho^N}{\partial t} = \hat{L}^N \rho^N, \tag{5.40}$$

where $\hat{L}^N$ is an operator defined by

$$\hat{L}^N = -i\sum_{j=1}^{N}\left(\frac{\partial H^N}{\partial \mathbf{p}_j}\cdot\frac{\partial}{\partial \mathbf{q}_j} - \frac{\partial H^N}{\partial \mathbf{q}_j}\cdot\frac{\partial}{\partial \mathbf{p}_j}\right) \tag{5.41}$$

and is called Liouvillian or Liouville's operator. We may solve Liouville's equation formally to obtain the time evolution of the probability distribution

$$\rho^N(t) = e^{-i\hat{L}^N t}\rho^N(t=0). \tag{5.42}$$

The solution gives full information about the many-particle state. In equilibrium

$$\frac{\partial \rho^N}{\partial t} = 0,$$

which in other words leads to

$$\hat{L}^N \rho^N_{eq} = 0. \tag{5.43}$$

In general we are interested in few particle properties. Their probability distributions may be seen as reduced probability distributions from the full many-particle distribution. In a way similar to the previous sections, integrating the information of the remaining variables simplifies the description of the state of a system. The reduced probability density of s-particles may be obtained as

$$\rho_s(\mathbf{X}_1, \cdots, \mathbf{X}_s, t) = \int d\mathbf{X}_{s+1}\cdots\int d\mathbf{X}_N \rho^N(\mathbf{X}_1, \cdots, \mathbf{X}_N, t). \tag{5.44}$$

These results for the reduced probability densities are related to the results obtained before for the probability distributions of sequences of events of stochastic variables.

Solving for a given problem, defined by a given Hamiltonian, we may calculate average values of physical quantities, either in equilibrium or their time evolution. One and two particle physical quantities are usually the relevant ones, since usually interaction terms involves two-particle scattering. Considering a function of one coordinate of the type

$$O^N_{(1)}(\mathbf{X}^N) = \sum_{i=1}^{N} O(X_i), \tag{5.45}$$

its average is given by

$$\langle O^N_{(1)}(t)\rangle = N\int d\mathbf{X}_1 O(\mathbf{X}_1)\rho_1(\mathbf{X}_1,t). \tag{5.46}$$

If we have functions of two coordinates of the type

$$O^N_{(2)}(\mathbf{X}^N) = \sum_{i<j}^{\frac{1}{2}N(N-1)} O(\mathbf{X}_i,\mathbf{X}_j), \tag{5.47}$$

then their average, if the Hamiltonian is symmetric under the exchange of particles, is given by

$$\langle O^N_{(2)}(t)\rangle = \frac{1}{2}N(N-1)\int d\mathbf{X}_1\int d\mathbf{X}_2 O(\mathbf{X}_1,\mathbf{X}_2)\rho_2(\mathbf{X}_1,\mathbf{X}_2,t). \tag{5.48}$$

These expressions give the time evolution of the averages of one- and two-particle variables, using the time evolution of the probability distributions and similar expressions can be obtained for any quantities involving any number of particles. The time evolution of the distribution function follows Eq. (5.42). Introducing this expression on the time evolution of averages, we get an expression where the initial time distribution appears. Integrating by parts we may write the averages as the initial time average of some time evolved quantity.[4] The time evolution of the averages satisfies a differential equation of the type

$$\frac{\partial O(\mathbf{X}^N,t)}{\partial t} = \mathcal{H}^N O(\mathbf{X}^N,t) = \sum_{i=1}^{N}\left(\frac{d\mathbf{q}_i}{dt}\frac{\partial}{\partial\mathbf{q}_i}+\frac{d\mathbf{p}_i}{dt}\frac{\partial}{\partial\mathbf{p}_i}\right)O(\mathbf{X}^N,t), \tag{5.49}$$

writing $\hat{L}^N = -i\mathcal{H}^N$.

Of course we are assuming that the full distribution is known, which leads to the knowledge of the reduced probability densities and the time dependence of the physical variables. It turns out that the solution for the many-particle system is complex and involves a hierarchy of terms that relates reduced probability distributions of higher and higher numbers of particles. An usual approach involves truncating the hierarchy at some point. Usually one needs some approximate expression for the reduced probability distributions of one and two particles.

Let us define $F_s(\mathbf{X}_1,\cdots,\mathbf{X}_s,t) = V^s\rho_s(\mathbf{X}_1,\cdots,\mathbf{X}_s,t)$ and $F_N(\mathbf{X}_1,\cdots,\mathbf{X}_N,t) = V^N\rho_N(\mathbf{X}_1,\cdots,\mathbf{X}_N,t)$, where V is the volume. Let us consider a rather generic symmetric Hamiltonian that has a kinetic energy part and a term that corresponds to interactions between the particles of a system, written as a sum of pairwise interactions

[4] This is the equivalent of using the Schrödinger or Heisenberg pictures in quantum mechanics.

$$H^N = \sum_{i=1}^{N} \frac{|\mathbf{p}_i|^2}{2m} + \sum_{i<j}^{\frac{1}{2}N(N-1)} \Phi(|\mathbf{q}_i - \mathbf{q}_j|). \tag{5.50}$$

Liouville's theorem for this Hamiltonian gives

$$\frac{\partial \rho^N}{\partial t} = -\left[\sum_i \frac{\mathbf{p}_i}{m} \cdot \frac{\partial}{\partial \mathbf{q}_i} - \sum_{i<j} \frac{\partial \Phi_{ij}}{\partial \mathbf{q}_i} \cdot \frac{\partial}{\partial \mathbf{p}_i} - \sum_{i<j} \frac{\partial \Phi_{ij}}{\partial \mathbf{q}_j} \cdot \frac{\partial}{\partial \mathbf{p}_j} \right] \rho^N. \tag{5.51}$$

Define an operator

$$\mathcal{H}^N(\mathbf{X}_1, \cdots, \mathbf{X}_N) = \sum_{i=1}^{N} \frac{\mathbf{p}_i}{m} \cdot \frac{\partial}{\partial \mathbf{q}_i} - \sum_{i<j} \Theta_{ij}, \tag{5.52}$$

where the operator

$$\Theta_{ij} = \frac{\partial \Phi_{ij}}{\partial \mathbf{q}_i} \cdot \frac{\partial}{\partial \mathbf{p}_i} + \frac{\partial \Phi_{ij}}{\partial \mathbf{q}_j} \cdot \frac{\partial}{\partial \mathbf{p}_j}. \tag{5.53}$$

Then integrating Liouville's equation over the coordinates from $\mathbf{X}_{s+1}$ until $\mathbf{X}_N$ and multiplying by V^s, we get

$$\frac{\partial F_s}{\partial t} + \mathcal{H}^s F_s = V^s \int d\mathbf{X}_{s+1} \cdots \int d\mathbf{X}_N$$
$$\left[- \sum_{i=s+1}^{N} \frac{\mathbf{p}_i}{m} \cdot \frac{\partial}{\partial \mathbf{q}_i} + \sum_{i \le s, s+1 \le j \le N} \Theta_{ij} + \sum_{s+1 \le k \le l} \Theta_{kl} \right] \rho^N(\mathbf{X}_1, \cdots, \mathbf{X}_N, t). \tag{5.54}$$

Note that we can write

$$\int_{vol} \mathbf{p}_i \cdot \frac{\partial}{\partial \mathbf{q}_i} \rho^N d\mathbf{X}_i = \int_{surface} \mathbf{p}_i \rho^N d\mathbf{S}_i. \tag{5.55}$$

Considering that the probability density vanishes on the surface, the surviving term can be written as

$$V^s \int d\mathbf{X}_{s+1} \cdots \int d\mathbf{X}_N \sum_{i \le s, s+1 \le j \le N} \Theta_{ij} \rho^N(\mathbf{X}_1, \cdots, \mathbf{X}_N, t)$$

$$= V^s(N-s) \sum_{i=1}^{s} \int d\mathbf{x}_{s+1} \Theta_{i,s+1} \int d\mathbf{X}_{s+2} \cdots \int d\mathbf{X}_N \rho^N(\mathbf{X}_1, \cdots, \mathbf{X}_N, t)$$

$$= \frac{N-s}{V} \sum_{i=1}^{s} \int d\mathbf{X}_{s+1} \Theta_{i,s+1} F_{s+1}(\mathbf{X}_1, \cdots, \mathbf{X}_{s+1}, t). \tag{5.56}$$

Considering now the thermodynamic limit of a very large system, such that $N \to \infty$ and $V \to \infty$ with $v = V/N$ constant, we get an infinite hierarchy of equations of motion for the probability densities of s particles, known as BBGKY hierarchy[5] that can be written

$$\frac{\partial F_s}{\partial t} + \mathcal{H}^s F_s = \frac{1}{v} \sum_{i=1}^{s} \int d\mathbf{X}_{s+1} \Theta_{i,s+1} F_{s+1}(\mathbf{X}_1, \cdots, \mathbf{X}_{s+1}, t). \tag{5.57}$$

A couple of examples are

$$\begin{aligned} \frac{\partial F_1}{\partial t} + \frac{\mathbf{p}_1}{m} \cdot \frac{\partial F_1}{\partial \mathbf{q}_1} &= \frac{1}{v} \int d\mathbf{X}_2 \Theta_{1,2} F_2(\mathbf{X}_1, \mathbf{X}_2, t) \\ \frac{\partial F_2}{\partial t} &+ \left(\frac{\mathbf{p}_1}{m} \cdot \frac{\partial}{\partial \mathbf{q}_1} + \frac{\mathbf{p}_2}{m} \cdot \frac{\partial}{\partial \mathbf{q}_1} - \Theta_{1,2} \right) F_2 \\ &= \frac{1}{v} \int d\mathbf{X}_3 \left(\Theta_{1,3} + \Theta_{2,3} \right) F_3(\mathbf{X}_1, \mathbf{X}_2, \mathbf{X}_3, t). \end{aligned} \tag{5.58}$$

In order to truncate the hierarchy, one needs to rewrite in some approximate way the F_3 distribution in terms of F_1 and F_2. Similar issues will be considered in later chapters, and explicit truncations will be carried out.

5.4 Dynamics of Lattice Model

Let us consider some microscopic lattice model composed by (micro)cells such that each contains a small number of particles (Fig. 5.2). As we saw before, a macrostate is characterized by larger (macro)cells that are closer in size to some experimental measurement, in number M, over which a distribution of particles is considered. The cells are just some division of the whole system that approaches some continuum distribution of the particles. Let us label the cells by α, β and define the particle concentration (or number of particles in a given cell divided by its volume) on cell α by c_α. The average particle concentration is then given by $c = \sum_{\alpha=1}^{M} c_\alpha / M$.

The grand canonical partition function is[6]

$$Z = Tr_{\{c_\alpha\}} g\{c_\alpha\} e^{-\beta E\{c_\alpha\}}, \tag{5.59}$$

[5] After Bogoliubov, Born, Green, Kirkwood, and Yvon.

[6] See also [GuntonDroz].

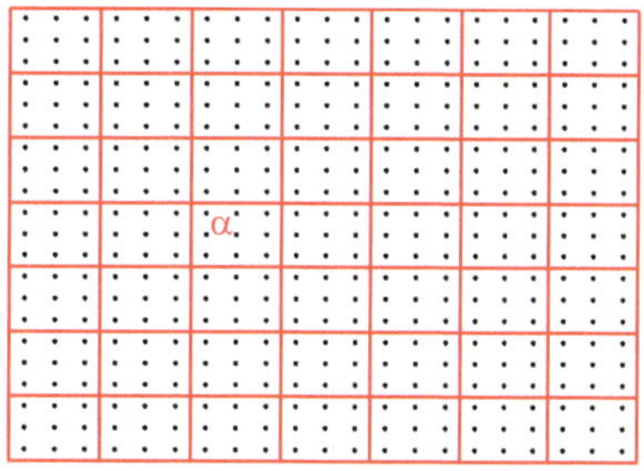

Fig. 5.2 Schematic representation of a lattice of macrocells, α, over a lattice of microcells respecting to the lattice points of the microscopic system (in a real system the number of lattice points in a given macrocell is very large)

where $g\{c_\alpha\}$ is the number of configurations over the microstates compatible with a given macrostate, given a concentration distribution $\{c_\alpha\}$. It is therefore the degeneracy of the configurations of the concentrations on the microcells for a given macrostate. At the level of the macrocells, we may consider that the partition function is given by

$$Z = Tr_{\{c_\alpha\}} e^{-\beta F\{c_\alpha\}}, \tag{5.60}$$

where the effective free energy, that results from the so called coarse graining of the microscopic details up to the macrostate, is given by

$$F\{c_\alpha\} = E\{c_\alpha\} - k_B T \ln g\{c_\alpha\}. \tag{5.61}$$

In the coarse graining procedure, the Hamiltonian (or energy) in the Boltzmann exponential in the partition function is replaced by a free energy-like term as we increase the size of the basic cell unit from the microcell to the macrocell. Increasing further the size of the unit cell, the microscopic details get increasingly washed away, as we tend to a continuum description of the system. In the continuum limit, we may write, considering that the particle concentration is instead the deviation from a homogeneous distribution and expanding, considering that the deviations are small,

$$\begin{aligned} F\{c(\mathbf{x})\} &= \int d\mathbf{x} f\,(c(\mathbf{x})) \\ &= \int d\mathbf{x} \left[\frac{1}{2} a |\nabla c(\mathbf{x})|^2 + V(c(\mathbf{x})) \right], \end{aligned} \tag{5.62}$$

with a some coefficient, and where

$$V(c(\mathbf{x})) = g_2 c^2(\mathbf{x}) + g_4 c^4(\mathbf{x}) + \cdots \tag{5.63}$$

with g_2, g_4 also coefficients.

In the microscopic model, the dynamics of a given lattice model involves the solution of the time evolution of the concentrations at each lattice site, as a result of the Hamiltonian of the system. In general the problem is complicated due to interactions with an external reservoir, the presence of impurities, etc. The problem may be solved generically considering the master equation approach, as before.

5.4.1 Fokker-Planck Equation

We consider the probability of a given concentration distribution, at a given time, $P(\{c\},t)$. Its time evolution can be written formally as

$$\partial_t P(\{c\},t) = \sum_{\{c_\alpha\}}{}' \left[W(\{c\},\{c'\})P(\{c'\},t) - W(\{c'\},\{c\})P(\{c\},t) \right]. \tag{5.64}$$

Here, W is the transition rate between the two concentration distributions, and the sum is constrained such that the probability is conserved

$$\sum_{\{c_\alpha\}} P(\{c\},t) = 1. \tag{5.65}$$

In this description, the details of the dynamics are contained in the transition rate, W. It is expected that the total concentration may be conserved, since it is related to the total number of particles, for instance, atoms or molecules in a given material. The global conservation constraint may be extended to consider that the concentration is also conserved locally in the sense that, if for a given cell $c_\alpha \to c_\alpha +\epsilon$, then the neighboring cell will experience a change $c_{\alpha'} \to c_{\alpha'} - \epsilon$, assuming that other cells remain unchanged. This choice of the transitions between configurations leads to

$$\begin{aligned} W(\{c\},\{c'\}) &= \frac{1}{2} \sum_{\alpha,\alpha'} \prod_{\beta \neq \alpha,\alpha'} \delta(c'_\beta - c_\beta) D_{\alpha,\alpha'} \\ &\times \int_{-\infty}^{\infty} d\epsilon R\left(\{c'\},\{c\}\right) \delta\left(c'_\alpha - c_\alpha - \epsilon\right) \delta\left(c'_{\alpha'} - c_{\alpha'} + \epsilon\right), \end{aligned} \tag{5.66}$$

where $D_{\alpha,\alpha'} = 1$ if α, α' are neighbors, and is zero otherwise, and $R\left(\{c'\},\{c\}\right)$ is some functional of the concentration distribution.

At equilibrium we have that

$$\partial_t P_{eq}(\{c\}) = 0. \tag{5.67}$$

This implies the condition of detailed balance

$$W(\{c\},\{c'\})P_{eq}(\{c'\}) - W(\{c'\},\{c\})P_{eq}(\{c\}) = 0, \tag{5.68}$$

which may be fulfilled by

$$P_{eq}(\{c\}) \sim e^{-\beta F(\{c\})}. \tag{5.69}$$

Therefore,

$$\frac{W(\{c\},\{c'\})}{W(\{c'\},\{c\})} = \frac{P_{eq}(\{c\})}{P_{eq}(\{c'\})} = e^{-\beta\left(F(\{c\})-F(\{c'\})\right)}. \tag{5.70}$$

The transition rate may be obtained in a symmetrized form as

$$W(\{c'\},\{c\}) = e^{\frac{1}{2}\beta[F(\{c\})-F(\{c'\})]}\Omega(\{c'\},\{c\}), \tag{5.71}$$

where Ω is a symmetric function in the concentration configurations and is a function of the changes ϵ.

The master equation, considering the local conservation, may now be written as

$$\partial_t P(\{c\},t) = \frac{1}{2}\sum_{\alpha,\alpha'}{}' D_{\alpha,\alpha'}\int_{-\infty}^{\infty} d\epsilon\Omega(\epsilon)$$
$$\times\left[e^{\frac{\beta}{2}\Delta F}P(\cdots,c_\alpha+\epsilon,c_{\alpha'}-\epsilon,\cdots)-e^{-\frac{\beta}{2}\Delta F}P(\cdots,c_\alpha,c_{\alpha'},\cdots)\right]. \tag{5.72}$$

Here

$$\Delta F = F\left(c_\alpha+\epsilon, c_{\alpha'}-\epsilon\right) - F\left(c_\alpha, c_{\alpha'}\right). \tag{5.73}$$

Expanding now in powers of ϵ leads to a Fokker-Planck equation.

Since $\Omega(\epsilon)$ is symmetric, odd powers of ϵ do not contribute to the expansion. The leading term involves quadratic terms in ϵ. We define the jump rate per site, Γ, as

$$\int_{-\infty}^{\infty} d\epsilon\epsilon^2\Omega(\epsilon) = N_c^{-\left(1+\frac{2}{d}\right)}\Gamma, \tag{5.74}$$

where N_c is the number of sites in a cell. We obtain to second order in the expansion

$$\partial_t P\left(\{c\},t\right) = \frac{\Gamma}{4N_c^{1+\frac{2}{d}}}\sum_{\alpha,\alpha'}{}' D_{\alpha\alpha'}\left(\partial_{c_\alpha}-\partial_{c'_\alpha}\right)$$
$$\times\left[\beta\left(\partial_{c_\alpha}F-\partial_{c'_\alpha}F\right)P\left(\{c\},t\right)+\left(\partial_{c_\alpha}P-\partial_{c'_\alpha}P\right)\right]. \tag{5.75}$$

Using that only neighboring cells are coupled, we get

$$\partial_t P(\{c\},t) = \frac{\Gamma}{2N_c^{1+\frac{2}{d}}} \sum_{\alpha} \partial_{c_\alpha} \left[\beta \left(2d\partial_{c_\alpha} F - \sum_{\langle \alpha\alpha' \rangle} \partial_{c'_\alpha} F \right) P(\{c\},t) \right.$$
$$\left. + \left(2d\partial_{c_\alpha} P(\{c\},t) - \sum_{\langle \alpha\alpha' \rangle} \partial_{c'_\alpha} P(\{c\},t) \right) \right]. \tag{5.76}$$

This equation may be written as a Fokker-Planck equation in the form

$$\partial_t P(\{c\},t) = -\sum_{\alpha} \partial_{c_\alpha} J_\alpha(\{c\},t), \tag{5.77}$$

where the current is given by

$$J_\alpha(\{c\},t) = -\sum_{\beta} \Gamma_{\alpha\beta} \left[\beta P(\{c\},t)\, \partial_{c_\beta} F + \partial_{c_\beta} P(\{c\},t) \right] \tag{5.78}$$

with

$$\Gamma_{\alpha\beta} = \frac{\Gamma}{2N_c^{1+\frac{2}{d}}} \left[2d\delta_{\alpha\beta} - \sum_{\langle \alpha\alpha' \rangle} \delta_{\alpha'\beta} \right]. \tag{5.79}$$

In the continuum limit, we may define the functional derivative through

$$\partial_{c_\alpha} = \left(N_c a_0^d \right) \frac{\delta}{\delta c(\mathbf{x})}. \tag{5.80}$$

The sums over cells are replaced by integrals over positions as

$$\sum_{\alpha} \to \left(N_c a_0^d \right)^{-1} \int d^d x \tag{5.81}$$

and we can write that

$$\Gamma_{\alpha\beta} = -\frac{\Gamma}{2} a_0^{d+2} \nabla^2 \left(\mathbf{x} - \mathbf{x}' \right), \tag{5.82}$$

where we have used a standard expression for a numerical second order derivative. This leads to a Fokker-Planck equation obtained by Langer

$$\partial_t P(c,t) = -\int d^d x \frac{\delta J(c(\mathbf{x},t)}{\delta c(\mathbf{x})}$$
$$J(c(\mathbf{x}),t) = \frac{\Gamma}{2} a_0^{d+2} \nabla^2(\mathbf{x}) \left[\beta P(c,t) \frac{\delta F}{\delta c(\mathbf{x})} + \frac{\delta P(c,t)}{\delta c(\mathbf{x})} \right]. \tag{5.83}$$

Note that the Laplacian appeared as a consequence that we considered that there is a local conservation of the concentration of particles.

5.4.2 Langevin and Fokker-Planck Equations

Equivalently to a description of the time evolution in terms of the evolution of the probability distribution, we may consider dynamical equations for observables directly, such as the Langevin equation we considered before.

Consider we have variables that depend on time that we denote as $A_i(t)$. Due to the coupling of the system to its environment, in general the equations of motion of some dynamical variable is described by a Langevin type equation

$$\partial_t A_i(t) = \varphi_i\left[\{A_j\}, t\right] + \xi_i(t). \tag{5.84}$$

There is a term that is the result of the coupling between the observable and the other observables and a term that is random, the noise term. This type of equation may be obtained considering the coupling of quantities of a given system with some external environment or considering that some variables (degrees of freedom) of a given system act as reservoirs for the remaining internal variables. Typically in some system, there are fast and slow variables, and it is often convenient to integrate out the fast variables leading to Langevin-type equations for the slow variables. In general, as a result of this elimination of variables (either external or fast internal variables), in addition to the coupling between degrees of freedom, as dictated by the Hamiltonian of the system, friction and noise terms appear. The term φ_i is essentially the derivative of the Hamiltonian (or free energy) of the system with respect to the variables of the system.

It is possible to derive the Fokker-Planck equation from the Langevin equations [1]. Here we just give the result that can be written as

$$\partial_t P(\{A_i\}, t) = -\sum_j \frac{\partial}{\partial A_j}\left[\varphi[\{A_i\}]P\right] + \sum_j D_j \frac{\partial^2}{\partial A_j^2} P, \tag{5.85}$$

considering that

$$\begin{aligned} \langle \xi_i(t) \rangle &= 0 \\ \langle \xi_i(t)\xi_j(t') \rangle &= 2D_i\delta_{ij}\delta(t-t') \end{aligned} \tag{5.86}$$

and the set of terms $\{D_i\}$ are independent of the variables. Two possibilities for the dynamics are usually considered, one in which a given variable is not conserved or another one where the variable is conserved (as in the case of the particle concentration we considered above).

The first case where the variable is not conserved is called model A. In this case the Langevin equation is of the type

$$\partial_t c(\mathbf{x}, t) = -\Gamma_0 \frac{\partial F}{\partial c(\mathbf{x}, t)} + \xi(\mathbf{x}, t) \tag{5.87}$$

with

$$\langle \xi(\mathbf{x}, t)\xi(\mathbf{x}', t')\rangle = 2\Gamma_0 \delta(\mathbf{x} - \mathbf{x}')\delta(t - t'). \tag{5.88}$$

The associated Fokker-Planck equation is

$$\partial_t P(c(\mathbf{x}), t) = \Gamma_0 \int d^d x \frac{\partial}{\partial c(\mathbf{x}, t)} \left[\frac{\partial P}{\partial c(\mathbf{x}, t)} + P \frac{\partial F}{\partial c(\mathbf{x}, t)} \right]. \tag{5.89}$$

In the case of a conserved variable dynamics, the so-called model B, the only change is to add the Laplacian we obtained before, $\Gamma_0 \to M\nabla^2$. Explicitly we get for the model B dynamics

$$\partial_t c(\mathbf{x}, t) = -M\nabla^2 \frac{\partial F}{\partial c(\mathbf{x}, t)} + \xi(\mathbf{x}, t), \tag{5.90}$$

with

$$\langle \xi(\mathbf{x}, t)\xi(\mathbf{x}', t')\rangle = 2M\nabla^2(\mathbf{x})\delta(\mathbf{x} - \mathbf{x}')\delta(t - t'). \tag{5.91}$$

The associated Fokker-Planck equation is

$$\partial_t P(c(\mathbf{x}), t) = M \int d^d x \nabla^2(\mathbf{x}) \frac{\partial}{\partial c(\mathbf{x}, t)} \left[\frac{\partial P}{\partial c(\mathbf{x}, t)} + P \frac{\partial F}{\partial c(\mathbf{x}, t)} \right]. \tag{5.92}$$

5.5 Exercises

1. Obtain the Fokker-Planck equation for a system of particles in motion. That is, obtain the Fokker-Planck equation for the probability distribution $P(\mathbf{x}, \mathbf{v}, t)$.
2. Consider the Fokker-Planck equation Eq. (5.20) and take the moments as independent of space. Assume that $P(\pm\infty, t) = 0$ and that a particle is localized initially at a given position $P(y, 0) = \delta(y - y_0)$.

   (a) Use the method of separation of variables, $P(y, t) = T(t)Y(y)$ to find the solution for the distribution.
   (b) Determine the long time stationary state and the expression for the first moment as a function of time.

3. The motion of a Brownian particle moving in one dimension in a fluid at temperature T is described by the Langevin equation Eq. (1.22). Consider that $V = 0$ and neglect the term with $\Delta\omega_0^2$. Assume that $\langle\xi(t)\rangle = 0$ and that $\langle\xi(t)\xi(t')\rangle = C\delta(t - t')$, where C is an unknown constant. Assume that the initial velocity of the particle is $v(0) = v_0$. Rewrite the Langevin equation in terms of the particle velocity at time t.
   (a) Calculate $\langle v(t)\rangle$ and $\langle v^2(t)\rangle$. (Assume that the velocity and the random term are uncorrelated, $\langle v(t)\xi(t)\rangle = 0$).
   (b) After a long time, the particle relaxes to equilibrium and its average kinetic energy satisfies the equipartition theorem. Determine the constant C.

Reference

1. H. Haken, Reviews of Modern Physics **47**, 67 (1975).

Chapter 6
Introduction to Transport Processes

6.1 Elementary Kinetic Theory

The kinetic theory of dilute gases was introduced earlier, in which the random motion of the particles was described statistically. In a dilute system, one assumes free motion of the particles, which is essentially isotropic. However, some restrictions were introduced on the motion of the particles, such as the elastic collisions with the walls of the container and the hard-core collisions with other particles. These restrictions lead to the pressure on the container walls (due to the collisions of the particles with the walls) and that, due to the collisions with other particles, their motion is not a simple straight line, as it would be in empty space. The concept of a mean free path emerged, defined as the mean distance between collisions. Considering molecules of diameter D, as shown in Fig. 6.1, and taking the density of the gas to be n, the mean free path was estimated to be

$$l = \frac{1}{\sqrt{2} n \pi D^2}. \tag{6.1}$$

(The factor $1/\sqrt{2}$ corrects for the three dimensionality of the elastic collisions.) This expression may be seen as the result of considering that the molecules are hard spheres colliding with a scattering cross section $\sigma = \sqrt{2}\pi D^2$. In their erratic motion, the molecules define fluxes through a given area, inside or at the border of the system, which was shown to be

$$\Phi = \frac{1}{4} n \langle v \rangle, \tag{6.2}$$

where $\langle v \rangle$ is the average velocity of the particles, which can be calculated using the Maxwell-Boltzmann distribution.

P. D. Sacramento, *Introduction to Condensed Matter Theory*,
https://doi.org/10.1007/978-3-031-86251-9_6

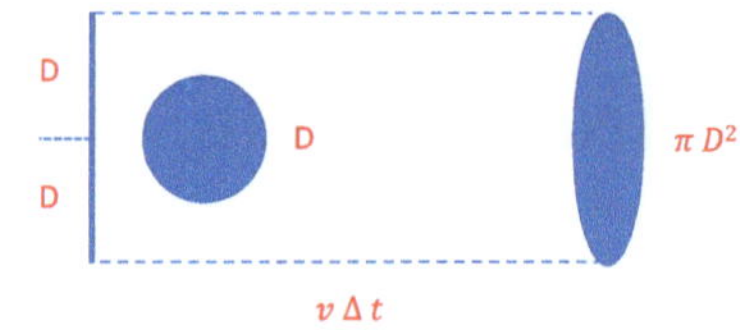

Fig. 6.1 Effective scattering area for the collision of a molecule of diameter D with other molecules

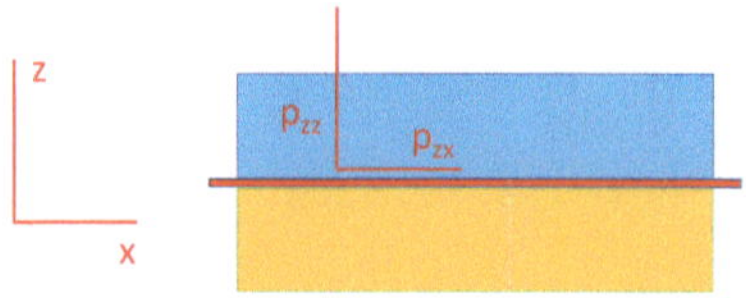

Fig. 6.2 Pressure tensor at the interface between two slabs due to the presence of viscosity

Since pressure is a thermodynamic quantity it makes sense to be defined in macroscopic regions (macrocells), large enough to contain a large number of molecules. In a system in equilibrium, pressure is the same throughout the system. We calculated the pressure the gas exerts on the wall of a container by calculating the linear momentum transferred to a given area of the wall. Due to equilibrium homogeneity, this is also the pressure that is exerted by the molecules on its neighborhood inside the container.

6.1.1 Viscosity

In general, fluids are not ideal, and the particles have interactions among themselves, beyond the hard-core repulsions. Consider a fluid (liquid or gas) and define two axis x and z to define a plane in the fluid.[1] Consider a given value of $z = z_1$. The part of the fluid with $z < z_1$ exerts a force per unit area, $\mathbf{p}_z$, on the other part. The normal component to the plane $xy(z = z_1)$ of the average force per unit area, the z component of $\mathbf{p}_z$, is the pressure of the fluid. Call it $p_{zz} = \bar{p}$. If the fluid is in equilibrium, $p_{zx} = 0$. This is the force exerted by the plane $xy(z = z_1)$ in the direction x, as illustrated in Fig. 6.2).

Consider now a non-equilibrium situation. Consider that $0 \leq z \leq L$. Assume that there is a non-uniform velocity profile, that is, that the fluid is moving in such a way that the velocity is z dependent. For instance, consider that the fluid is at rest for $z = 0$ (for instance, the wall of a tube where the fluid is flowing). The velocity then increases as z grows until it reaches its maximum at $z = L$, call it $v_x = v_0$. This may be achieved, for instance, moving the plane $z = L$ which is in contact with the fluid with a velocity v_0. This establishes a velocity gradient along z (see Fig. 6.3). As a result, tangential forces arise to try to recover the situation of stationary equilibrium. The various planes exert tangential forces, p_{zx}, in the fluid above. If $\partial v_x/\partial z = 0$

[1] See also [Reif].

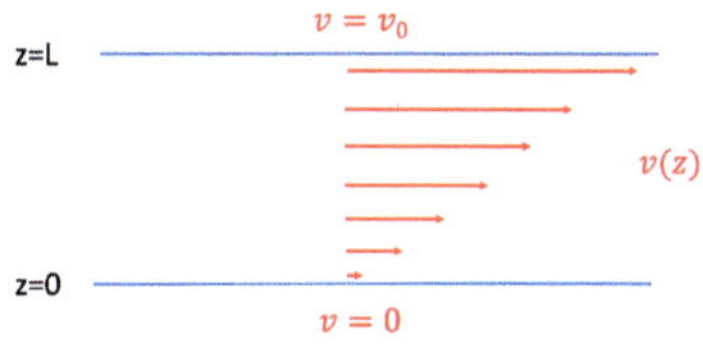

Fig. 6.3 Velocity profile between a plane with zero velocity and a plane close to which the fluid has a finite velocity

then $p_{zx} = 0$. Since in this non-equilibrium case this does not hold and $\partial v_x/\partial z \neq 0$, we expect that

$$p_{zx} = -\eta \frac{\partial v_x}{\partial z}, \tag{6.3}$$

where we have introduced a coefficient η, called viscosity. This relation holds if we assume that the velocity gradient is not very large, such that the response of the system is small and therefore a linear dependence may be assumed. This is a first example of a linear response theory. The coefficient may be obtained from a microscopic description.

Consider again a plane with a given constant value of the coordinate z. Consider two extra planes, one at $z + l$ and another at plane $z - l$, where l is the mean free path. Due to the velocity gradient, the velocity in the various planes is given by $v_x(z)$. Using that pressure is force per unit area and the force is the change of momentum per unit time, we can write that

$$p_{zx} = \frac{1}{4} n \langle v \rangle \left[m v_x(z - l) - m v_x(z + l) \right]. \tag{6.4}$$

We can now expand

$$\begin{aligned} v_x(z + l) &= v_x(z) + \frac{\partial v_x}{\partial z} l + \cdots \\ v_x(z - l) &= v_x(z) - \frac{\partial v_x}{\partial z} l + \cdots \end{aligned} \tag{6.5}$$

Substituting, we get that

$$p_{zx} = \frac{1}{4} n \langle v \rangle m \left(-2 \frac{\partial v_x}{\partial z} l \right) = -\eta \frac{\partial v_x}{\partial z}. \tag{6.6}$$

We conclude that

$$\eta = \frac{1}{2} n \langle v \rangle m l \tag{6.7}$$

is the viscosity. Using the result for the mean free path, we get that

$$\eta = \frac{m}{2\sqrt{2}\pi D^2}\langle v\rangle. \tag{6.8}$$

We conclude that it is independent of the density of the fluid. Since the average velocity depends on temperature, the viscosity also depends on the temperature. For an ideal gas, the equation of state $p = nk_BT$ implies that the viscosity also does not depend on pressure. Since the dilute limit is necessary for the ideal gas approximation to hold, these results apply when $l >> D$.

6.1.2 Thermal Conductivity and Particle Diffusion

In general, non-equilibrium profiles lead to fluxes. For instance, a temperature profile is well-known to lead to a heat flux. As in the previous example, assume now that temperature depends on z. We fix a plane with a given value of z, and we take into account fluxes of particles from two neighboring planes that are displaced up and down with a distance scale set by the mean free path. The energy that crosses the plane at z coming from a lower value of the coordinate z is given by

$$\langle E\rangle_{up} = \frac{1}{4}n\langle v\rangle\bar{\epsilon}(z-l). \tag{6.9}$$

The average energy that crosses the same plane coming from above is similarly given by

$$\langle E\rangle_{down} = \frac{1}{4}n\langle v\rangle\bar{\epsilon}(z+l). \tag{6.10}$$

In these two expressions, $\bar{\epsilon}(z)$ is the average energy per particle at plane z. If the temperature profile is such that temperature increases with z, then so does the average energy per particle. The heat flux along z is given by

$$\begin{aligned}
Q_z &= \frac{1}{4}n\langle v\rangle\left[\bar{\epsilon}(z-l) - \bar{\epsilon}(z+l)\right] \\
&= \frac{1}{4}n\langle v\rangle\left[\left\{\bar{\epsilon}(z) - l\frac{\partial\bar{\epsilon}}{\partial z}\right\} - \left\{\bar{\epsilon}(z) + l\frac{\partial\bar{\epsilon}}{\partial z}\right\}\right] \\
Q_z &= -\frac{1}{2}n\langle v\rangle l\frac{\partial\bar{\epsilon}}{\partial T}\frac{\partial T}{\partial z} \\
Q_z &= -K\frac{\partial T}{\partial z},
\end{aligned} \tag{6.11}$$

which implies that the thermal conductivity, K, is given by

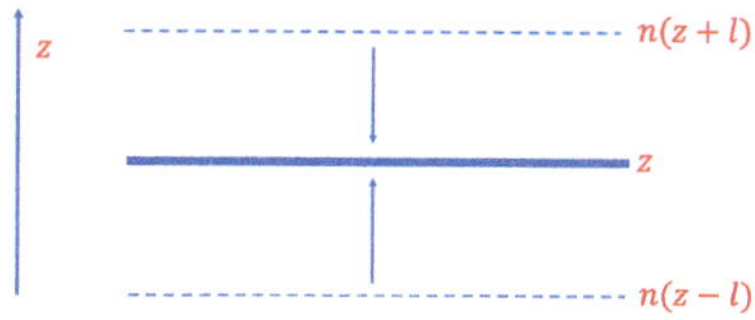

Fig. 6.4 A density profile along a given direction leads to a particle flux

$$K = \frac{1}{2} n \langle v \rangle c l, \tag{6.12}$$

where $c = \partial \bar{\epsilon} / \partial T$ is the specific heat per particle.

Another example is that a density profile leads to a particle flux. Considering that density depends on the coordinate z in a similar way as shown in Fig. 6.4 leads to a particle flux along z given by

$$\begin{aligned} J_z &= \frac{1}{4} \langle v \rangle \left[n(z - l) - n(z + l) \right] \\ &= -\frac{1}{2} \langle v \rangle \frac{\partial n}{\partial z} l \\ J_z &= -D \frac{\partial n}{\partial z}, \end{aligned} \tag{6.13}$$

where the diffusion constant, D, is given by

$$D = \frac{1}{2} \langle v \rangle l. \tag{6.14}$$

The fluxes that are generated by non-equilibrium profiles of some intensive quantity have the physical meaning of some associated physical quantity that flows through a unit area per unit time. It has the same dimensions as a current density and is of the form of the product of the density of the physical quantity being transported times the drift velocity across the surface. Often, these current densities obey conservation laws that will be considered in the next section.

6.1.3 Conservation Laws

The equation of motion of some variable, O, defined in terms of the coordinates of a N-particle system, may be written[2] as in Eq. (5.49)

[2] See also [Reichl].

$$\frac{\partial O(\mathbf{X}^N,t)}{\partial t} = \mathcal{H}^N O(\mathbf{X}^N,t) = \sum_{i=1}^{N}\left(\frac{d\mathbf{q}_i}{dt}\frac{\partial}{\partial \mathbf{q}_i} + \frac{d\mathbf{p}_i}{dt}\frac{\partial}{\partial \mathbf{p}_i}\right) O(\mathbf{X}^N,t), \tag{6.15}$$

where $\mathcal{H}^N$ is a differential operator acting on the N particle space. A general Hamiltonian for the N-particle system may be written as

$$H^N = \sum_{i=1}^{N}\frac{|\mathbf{p}_i|^2}{2m} + \sum_{i<j}^{\frac{1}{2}N(N-1)} \Phi\left(|\mathbf{q}_i - \mathbf{q}_j|\right). \tag{6.16}$$

Using Hamilton's equations, we can write that

$$\begin{aligned} \frac{d\mathbf{q}_i}{dt} &= \frac{\partial H^N}{\partial \mathbf{p}_i} = \frac{\mathbf{p}_i}{m} \\ \frac{d\mathbf{p}_i}{dt} &= -\frac{\partial H^N}{\partial \mathbf{q}_i} = -\sum_{l\neq i}\frac{\partial \Phi\left(|\mathbf{q}_i - \mathbf{q}_j|\right)}{\partial q_i} = \sum_{l\neq i} F_{il}, \end{aligned} \tag{6.17}$$

where the forces satisfy

$$F_{il} = -\frac{\partial \Phi\left(|\mathbf{q}_i - \mathbf{q}_l|\right)}{\partial q_i} = \frac{\partial \Phi\left(|\mathbf{q}_i - \mathbf{q}_l|\right)}{\partial q_l} = -F_{li}. \tag{6.18}$$

Let us apply these expressions to various physical quantities.

Conservation of Particle Density

We begin by considering the particle density. The particle density at point $\mathbf{R}$ is given by

$$n(\mathbf{q}^N, \mathbf{R}) = \sum_{i=1}^{N}\delta(\mathbf{q}_i - \mathbf{R}). \tag{6.19}$$

The equation of motion for the density is given by

$$\frac{\partial n(\mathbf{q}^N,\mathbf{R})}{\partial t} = -\nabla_{\mathbf{R}}\cdot\mathbf{J}^n(\mathbf{p}^N,\mathbf{q}^N;\mathbf{R}), \tag{6.20}$$

where

$$\mathbf{J}^n(\mathbf{p}^N,\mathbf{q}^N;\mathbf{R}) = \sum_{i=1}^{N}\frac{\mathbf{p}_i}{m}\delta(\mathbf{q}_i - \mathbf{R}) \tag{6.21}$$

is the particle current. This is a statement of the conservation of the number of particles. Consider a closed volume, V, and the surface that encloses it, S, and integrate both sides of the equation over the volume. We get that

$$-\int_V d^3R\frac{\partial n(\mathbf{q}^N,\mathbf{R})}{\partial t}=\int_V d^3R\nabla_{\mathbf{R}}\cdot\mathbf{J}^n(\mathbf{p}^N,\mathbf{q}^N;\mathbf{R}). \tag{6.22}$$

Using Gauss's theorem, we can convert the volume integral of the divergence of the particle current by the limiting surface integral as

$$\int_V d^3R\nabla_{\mathbf{R}}\cdot\mathbf{J}^n(\mathbf{p}^N,\mathbf{q}^N;\mathbf{R})=\int_S \mathbf{J}^n(\mathbf{p}^N,\mathbf{q}^N;\mathbf{R})\cdot\mathbf{n}dS, \tag{6.23}$$

where $\mathbf{n}$ is a unit vector, perpendicular at each point to the surface S. This integral is just the flux of particles through the surface. This equation just states that if the number of particles inside the volume changes with time, this is exclusively due to a flux through the surface and that there are no particles appearing or disappearing and, therefore, that the number of particles is conserved, if we sum over the inside and outside of volume V.

Conservation of Momentum

We may now consider the total momentum of the particles of the system. The momentum density is defined by

$$m\mathbf{J}^n(\mathbf{p}^N,\mathbf{q}^N;\mathbf{R})=\sum_{i=1}^{N}\mathbf{p}_i\delta(\mathbf{q}_i-\mathbf{R}). \tag{6.24}$$

As above, we may now obtain the time evolution of this quantity. We get that

$$\frac{\partial}{\partial t}\left(m\mathbf{J}^n(\mathbf{p}^N,\mathbf{q}^N;\mathbf{R})\right)=\sum_{i=1}^{N}\left[\frac{d\mathbf{p}_i}{dt}\delta(\mathbf{q}_i-\mathbf{R})+\mathbf{p}_i\frac{d\mathbf{q}_i}{dt}\cdot\frac{\partial}{\partial\mathbf{q}_i}\delta(\mathbf{q}_i-\mathbf{R})\right]. \tag{6.25}$$

Note that

$$\delta(\mathbf{q}_i-\mathbf{R})=\frac{1}{V}\sum_{\mathbf{k}}e^{i\mathbf{k}\cdot(\mathbf{q}_i-\mathbf{R})}$$

Considering excitations of large wavelengths, which implies small momenta, we can write that the equation of motion may be written once again as a conservation law

$$m\frac{\partial}{\partial t}\mathbf{J}^n(\mathbf{q}^N, \mathbf{R}) = -\nabla_{\mathbf{R}} \cdot \mathbf{J}^p(\mathbf{p}^N, \mathbf{q}^N, \mathbf{R}), \tag{6.26}$$

where the momentum current is defined by

$$\mathbf{J}^p(\mathbf{p}^N, \mathbf{q}^N, \mathbf{R}) = \sum_{i=1}^{N} \frac{\mathbf{p}_i\mathbf{p}_i}{m}\delta(\mathbf{q}_i - \mathbf{R}) + \frac{1}{2}\sum_{i\neq l}(\mathbf{q}_i - \mathbf{q}_l)\, F_{il}\delta(\mathbf{q}_i - \mathbf{R}). \tag{6.27}$$

Conservation of Energy

In an isolated system, energy is conserved, but a subsystem in general exchanges energy with its neighborhood. The energy density at a point $\mathbf{R}$ may be written as

$$\begin{aligned} h(\mathbf{q}^N, \mathbf{p}^N, \mathbf{R}) &= \sum_{i=1}^{N} h_i\delta(\mathbf{q}_i - \mathbf{R}) \\ h_i &= \frac{|\mathbf{p}_i|^2}{2m} + \frac{1}{2}\sum_{j\neq i}\Phi(|\mathbf{q}_i - \mathbf{q}_j|). \end{aligned} \tag{6.28}$$

The equation for the conservation of energy is written as

$$\frac{\partial}{\partial t}h(\mathbf{q}^N, \mathbf{p}^N, \mathbf{R}) = -\nabla_{\mathbf{R}} \cdot \mathbf{J}^h(\mathbf{p}^N, \mathbf{q}^N, \mathbf{R}), \tag{6.29}$$

where the energy current is given by

$$\mathbf{J}^h(\mathbf{p}^N, \mathbf{q}^N, \mathbf{R}) = \sum_{i=1}^{N}\left[h_i\frac{\mathbf{p}_i}{m}\delta(\mathbf{q}_i - \mathbf{R})\right] + \frac{1}{2}\sum_{i\neq j}\left[\frac{\mathbf{p}_i + \mathbf{p}_j}{m} \cdot \mathbf{F}_{ij}(\mathbf{q}_i - \mathbf{q}_j)\delta(\mathbf{q}_i - \mathbf{R})\right]. \tag{6.30}$$

6.1.4 Heat Flow and Diffusion Equation

Let us study in some detail the case of a temperature gradient and the heat flow that is created. This is given by Fick's law

$$\mathbf{Q} = -K\nabla T(\mathbf{r}, t), \tag{6.31}$$

where K is the thermal conductivity and where we consider a general case in which the temperature gradient may depend on time, t. As we saw, there is a conservation law associated with the conservation of energy. Assuming that any energy exchange is in the form of a heat exchange, we can write that the conservation of energy implies a conservation of the heat flux as

$$\nabla \cdot \mathbf{Q} + \frac{\partial}{\partial t}(c\rho T(\mathbf{r}, t)) = 0. \tag{6.32}$$

Here c is the specific heat and ρ the particle density. The corresponding integral equation is obtained easily

$$\int_S \mathbf{Q} \cdot \mathbf{n} dS = -\frac{\partial}{\partial t} \int_V d^3 r c\rho T(\mathbf{r}, t). \tag{6.33}$$

On the right hand side, the integral over the volume gives the energy contained in volume V.

Using Fick's law, we obtain that

$$\nabla \cdot \mathbf{Q} = -K \nabla^2 T = -\frac{\partial}{\partial t} c\rho T(\mathbf{r}, t) \tag{6.34}$$

assuming a homogeneous material such that the thermal conductivity is constant throughout the system. We get the heat diffusion equation

$$\frac{\partial T}{\partial t} = \lambda \nabla^2 T, \tag{6.35}$$

with

$$\lambda = \frac{K}{c\rho}. \tag{6.36}$$

In a stationary regime, the temperature gradient does not depend on time. Then

$$\frac{\partial T}{\partial t} = 0 = \nabla^2 T. \tag{6.37}$$

Let us consider a one-dimensional case to simplify. We write the heat diffusion equation as

$$\frac{\partial T(x,t)}{\partial t} = \lambda \frac{\partial^2 T(x,t)}{\partial x^2}. \tag{6.38}$$

If the regime is stationary, we obtain that

$$\frac{\partial T(x,t)}{\partial x} = C, \tag{6.39}$$

where C is a constant. We may look for a solution using the method of separation of variables. We try a solution in the form $T(x,t) = X(x)A(t)$. Inserting this form of solution in the diffusion equation, we obtain

$$\frac{\partial A}{\partial t}X = \lambda A\frac{\partial^2 X}{\partial x^2}$$

$$\frac{1}{\lambda A}\frac{\partial A}{\partial t} = \frac{1}{X}\frac{\partial^2 X}{\partial x^2} = -k^2, \tag{6.40}$$

where k is another constant since the left hand side is only a function of time and the right hand side is only a function of the space coordinate. For this to hold for all time and coordinate values the two sides must be a constant, independent of both t and x. We can solve the two equations separately. We get that the time equation is

$$\frac{dA}{A} = -k^2\lambda dt$$

which has the solution

$$A(t) = A(0)e^{-k^2\lambda t}. \tag{6.41}$$

The constant $A(0)$ is determined in terms of the initial conditions.

The space part leads to the so-called Helmholtz equation

$$\left(\frac{d^2}{dx^2} + k^2\right)X(x) = 0. \tag{6.42}$$

The solution depends on the boundary conditions. The full solution is written as

$$T(x,t) = A(0)X(x)e^{-k^2\lambda t}, \tag{6.43}$$

where $A(0)$ is determined by the time boundary conditions and the form of $X(x)$ by the space boundary conditions.

As we mentioned above, a stationary solution in a one-dimensional system implies a constant temperature gradient (or for any other quantity that satisfies the diffusion equation). Actually, the system does not have to be one-dimensional as long as the change of temperature only occurs along one direction. A standard example is parallel planes where temperature is constant on each plane, but changes from plane to plane, such as the heat flow across a large wall separating two, or more, regions at different temperatures, such as the inside and the outside of some wall with finite width. The equation

$$\frac{dT(x)}{dx} = C \tag{6.44}$$

has an easy solution in the region between two different temperatures, established at two different points along one line,

$$T(x,t) = T(x_1) + \frac{x - x_1}{x_2 - x_1}\left(T(x_2) - T(x_1)\right). \tag{6.45}$$

Using Fick's law, we can determine the heat flow in the simple way

$$Q = \frac{T_1 - T_2}{\frac{l}{K}}, \tag{6.46}$$

where l is the distance between the points along the line where the temperature takes the values T_1 and T_2. Since heat flows from the higher temperature to the lower temperature, we take here $T_1 > T_2$ and $Q > 0$. We have assumed that the thermal conductivity is a constant, which implies a homogeneous system.

These results may be easily extended to systems that are not homogeneous. As a first example we may consider that the system is composed of different materials as x varies. In general, the different materials may have different widths and different conductivities. If there is heat flow, and considering that the energy transferred is conserved along x, we can conclude that, since the area that heat crosses is constant along x, so is the heat flux, which is the amount of heat that crosses a given area per unit time. In general, the power transmitted (energy per unit time) is conserved as the heat flow takes place, but the flux in general is not conserved, if the area it crosses changes. Standard examples are, for instance, a heat source at the center of a sphere such that it flows radially to the outside, or a wire inside a cylinder, also at a different temperature from the outside environment, such that the heat flow is also radial. Since the area traversed as one moves away from the heat source is changing, the flux changes. These examples will be considered next.

Flow Perpendicular to Parallel Infinite Planes

For the moment, consider that the area crossed is constant. Then if we consider a sequence of n regions with widths l_i and conductivities K_i, the conservation of heat flux leads to

$$Q = \frac{\Delta T_1}{\frac{l_1}{K_1}} = \frac{\Delta T_2}{\frac{l_2}{K_2}} = \cdots = \frac{\Delta T_n}{\frac{l_n}{K_n}}, \tag{6.47}$$

where ΔT_i is the temperature drop in region i. We leave as an exercise to show that we can calculate the heat flux if we measure any two temperatures. For instance, if we know the temperature difference between the first and last points along the line, ΔT, we can obtain that

$$Q = \frac{\Delta T}{R_{eq}}, \tag{6.48}$$

where

$$R_{eq} = \sum_{i=1}^{n} \frac{l_i}{K_i}. \tag{6.49}$$

Radial Flow in a Cylinder

Consider now that in the axis of a cylinder of length l is a thin cylinder of radius r_i, at a temperature T_i. The inner cylinder is surrounded by a cylindrical layer of width $r_o - r_i$. Also, the temperature at the outer surface of the cylindrical layer is T_o. Assuming that $T_i > T_o$, if the cylindrical layer conducts heat (finite thermal conductivity), there will be a heat flow from the inner cylinder to the outer surface. If temperature is uniform along the length of the cylinder and outer surface, the heat flow will take place radially. Due to power conservation, the heat flow times the surface it crosses is constant, at any given distance from the axis. The surface that the heat flow crosses is given by $S = 2\pi rl$ for a surface at distance r from the axis. Therefore the power given by $q = QS$ is given by

$$q = -2\pi rlK\frac{dT(r)}{dr}. \tag{6.50}$$

This equation may be solved. We get that

$$\begin{aligned} \frac{q}{2\pi Kl}\int_{r_i}^{r_o} \frac{1}{r}dr &= -\int_{T_i}^{T_o} dT \\ T_i - T_0 &= \frac{q}{2\pi Kl} \ln \frac{r_i}{r_o} \\ q &= \frac{T_i - T_0}{\frac{\ln \frac{r_i}{r_o}}{2\pi Kl}}. \end{aligned} \tag{6.51}$$

So, if we know the two temperatures T_i and T_o, we may calculate the power transmitted across the cylinder.

Also, we may calculate the temperature profile. We can write that

$$\frac{q}{2\pi Kl}\int_{r_i}^{r} \frac{1}{r'}dr' = -\int_{T_i}^{T} dT'. \tag{6.52}$$

Integrating, we get the temperature profile

$$T(r) = T_i - \frac{T_i - T_0}{\ln \frac{r_o}{r_i}} \ln \frac{r}{r_i}. \tag{6.53}$$

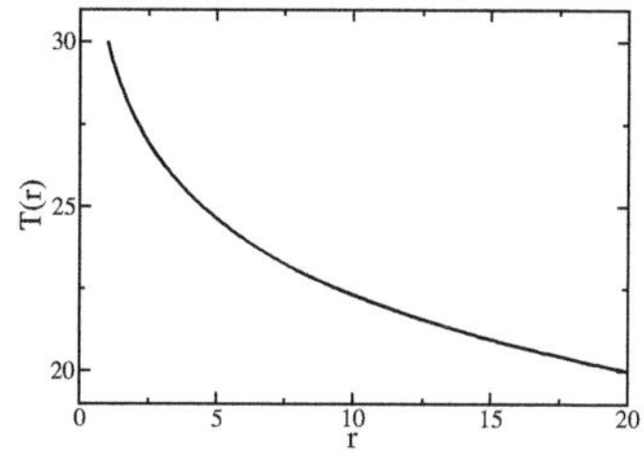

Fig. 6.5 Temperature profile for a cylinder geometry with the parameters $r_i = 1, r_O = 20$ and $t_i = 30, t_o = 20$

This type of profile is shown in Fig. 6.5.

Example of Stationary Three-Dimensional Profile

In a hot day the water temperature at the surface of a swimming pool is at $T_s = 40$ Celsius. All the other walls of the swimming pool are at $T_w = 20$ Celsius. What is the temperature distribution inside the pool?

The swimming pool has sides with lengths a, b, c. The boundary condition is such that $T(x, y, z = c) = T_s$. For convenience, define a new temperature $\bar{T} = T - T_w$. Then in terms of this new temperature all other walls are at $\bar{T} = 0$. We also have that $\bar{T}(x, y, c) = T_s - T_w$.

Recall that the diffusion equation leads to a solution of the form

$$\bar{T}(\mathbf{r}, t) = u(\mathbf{r})A(t), \tag{6.54}$$

where

$$A(t) = A(0)e^{-\lambda k^2 t}, \qquad \left(\nabla^2 + k^2\right) u(\mathbf{r}) = 0. \tag{6.55}$$

The stationary regime is a limit where the temperature does not depend on time. This is obtained considering the limit $t \to \infty$. In this limit the only solution that does not vanish is obtained when $k = 0$. This leads to Laplace's equation

$$\nabla^2 u(\mathbf{r}) = 0. \tag{6.56}$$

Let us look for a solution of the form

$$u(x, y, z) = X(x)Y(y)Z(z). \tag{6.57}$$

Substituting in Laplace equation

$$\frac{1}{X(x)}\frac{d^2X(x)}{dx^2} + \frac{1}{Y(y)}\frac{d^2Y(y)}{dy^2} + \frac{1}{Z(z)}\frac{d^2Z(z)}{dz^2} = 0, \tag{6.58}$$

we may now write that

$$\begin{aligned}\frac{1}{X(x)}\frac{d^2X(x)}{dx^2} &= -\alpha^2\\ \frac{1}{Y(y)}\frac{d^2Y(y)}{dy^2} &= -\beta^2\\ \frac{1}{Z(z)}\frac{d^2Z(z)}{dz^2} &= \gamma^2,\end{aligned} \tag{6.59}$$

with the restriction that $\alpha^2 + \beta^2 = \gamma^2$. Taking into account that temperature $\bar{T}$ vanishes in all walls except on the surface of the pool, we can write that

$$\begin{aligned}X(x) &\sim \sin\alpha x\\ Y(y) &\sim \sin\beta y\\ Z(z) &\sim \sinh\left(\sqrt{\alpha^2+\beta^2 z}\right),\end{aligned} \tag{6.60}$$

with $\alpha a = n\pi$ and $\beta b = m\pi$, where n, m are integers. The general solution of the temperature inside the pool is then

$$u(x, y, z) = T_w + \sum_{n,m=1}^{\infty} A_{nm}\sin(\alpha_n x)\sin(\beta_m y)\sinh(\gamma_{nm}z), \tag{6.61}$$

where

$$\alpha_n = \frac{n\pi}{a}, \qquad \beta_m = \frac{m\pi}{b}, \qquad \gamma_{nm} = \pi\sqrt{\frac{n^2}{a^2}+\frac{m^2}{b^2}}. \tag{6.62}$$

The coefficients A_{nm} are determined using the boundary condition that $\bar{T}(x, y, c) = T_s - T_w$. This translates into

$$T_s - T_w = \sum_{n,m=1}^{\infty} A_{nm}\sin(\alpha_n x)\sin(\beta_m y)\sinh(\gamma_{nm}c). \tag{6.63}$$

The coefficients are obtained using the orthogonality of the sine functions. It leads to

$$A_{nm} = \frac{4(T_s - T_w)}{ab\sinh(\gamma_{nm}c)}\int_0^a dx\int_0^b dy\sin\frac{n\pi x}{a}\sin\frac{m\pi y}{b}. \tag{6.64}$$

Finally, we obtain (Fig. 6.6)

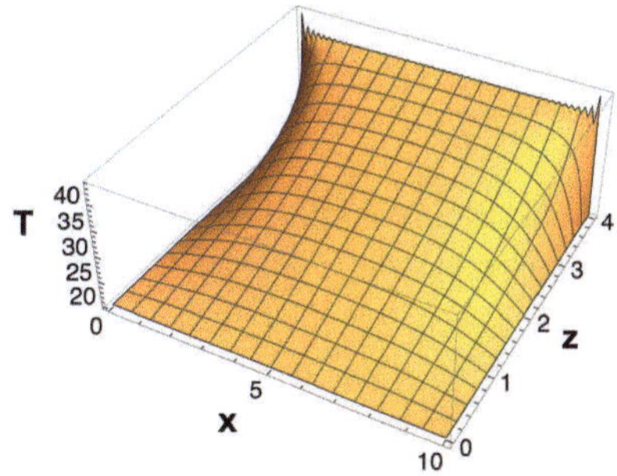

Fig. 6.6 Temperature profile for the swimming pool here described with parameters $a = b = 10$, $c = 4$. The y coordinate is fixed at its half value $a/2$

$$u(x, y, z) = T_w + \sum_{n,m=1,odd}^{\infty} \frac{16(T_s - T_w)}{nm\pi^2 \sinh(\gamma_{nm} c)} \sin\frac{n\pi x}{a} \sin\frac{m\pi y}{b} \sinh(\gamma_{nm} z). \tag{6.65}$$

Example of Time-Dependent Solution

In general a non-stationary problem is more complex. For instance, consider a one-dimensional rod at constant temperature $T(x, t < 0) = T_0$ that at time $t = 0$ is put in thermal contact with two reservoirs at different temperatures. Specifically, the left side of the rod ($x = 0$) is placed in contact with a reservoir at temperature T_l and the right hand side with another reservoir at temperature T_r. Cleary heat will flow from the higher temperature to the lower temperature and, in the infinite time limit, a profile will be established along the rod that will interpolate between the two extrema temperatures.

In general, a solution of this problem can be written as

$$u(\mathbf{r}, t) = u_t(\mathbf{r}, t) + u_s(\mathbf{r}), \tag{6.66}$$

where we have used a general notation u for a solution of a diffusion equation, not necessarily temperature. The first term has the information about the time dependence and it must reflect the initial conditions of the problem. It is called the transient solution. The space boundary conditions are used in the second term of this solution. This is the stationary solution. Therefore the function u_t has homogeneous (zero) space boundary conditions since these are already included in the function u_s. To simplify, consider a one dimensional problem. The diffusion equation is

$$\lambda \frac{\partial^2}{\partial x^2} u_t(x, t) + \lambda \frac{\partial^2}{\partial x^2} u_s(x) = \frac{\partial u_t}{\partial t}. \tag{6.67}$$

When time tends to infinity, the transient solution vanishes, and we are left with the stationary solution, which must obey

$$\frac{d^2}{dx^2} u_s(x) = 0 \tag{6.68}$$

and satisfy the boundary conditions. Clearly,

$$u(x,0) = u_t(x,0) + u_s(x), \qquad u_t(x,0) = u(x,0) - u_s(x), \tag{6.69}$$

which shows that the boundary conditions are absent in u_t.

To be specific, consider that $u_s(x=0)=0$, $u_s(x=L)=20$. We can write that

$$u_t(x,t) = \sum_l A_l \sin\frac{l\pi x}{L} e^{-\lambda\left(\frac{l\pi}{L}\right)^2 t} \tag{6.70}$$

and

$$u_t(x,0) = 20 - u_s(x) = \sum_l A_l \sin\frac{l\pi x}{L}. \tag{6.71}$$

Using the orthogonality of the sine functions we obtain

$$A_l = \frac{2}{L}\int_0^L dx \sin\frac{l\pi x}{L}\left(20 - \frac{20x}{L}\right) = \frac{40}{l\pi}\left(1 - 2(-1)^l\right), \tag{6.72}$$

since the function $u_s(x) = 20x/L$.

6.2 Brownian Motion

Consider a fluid and some particles that may move around inside the fluid. The particles have typically diameters of the order of $D_B \sim 0.2 - 1\,\mu$m. As the particles collide with the molecules of the fluid, they undergo some type of three-dimensional random walk, as schematically shown in Fig. 6.7. The local density of the particles, $n(\mathbf{r}, t)$, is in general not homogeneous. One expects that a particle current will take place from regions of higher density to regions of lower density. There is the

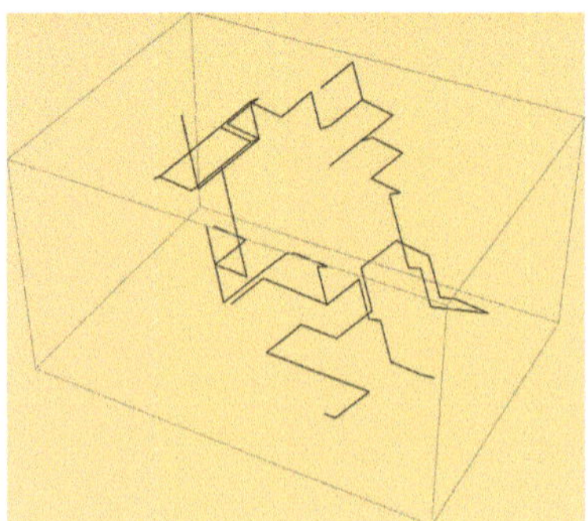

Fig. 6.7 Schematic representation of the motion of a Brownian particle in three-dimensional space resulting from collisions with other particles. Between collisions the particle moves as a free particle, with linear motion. Representation done using Mathematica

conservation law

$$\frac{\partial n}{\partial t} + \nabla \cdot \mathbf{j} = 0, \tag{6.73}$$

where the current obeys a Fick's law as

$$\mathbf{j} = -D\nabla n, \tag{6.74}$$

where D is a diffusion constant. This leads to a diffusion equation for the motion of the particles inside the fluid that is of the type

$$\frac{\partial n}{\partial t} = D\nabla^2 n \tag{6.75}$$

Defining a probability density as

$$n(\mathbf{r}, t) = NP(\mathbf{r}, t) \tag{6.76}$$

with N the total number of those particles, we have that $P(\mathbf{r}, t)d^3r$ has the meaning of the probability of finding a particle at time t in the region between $\mathbf{r}$ and $(\mathbf{r}+d\mathbf{r})$. The probability density satisfies an equivalent diffusion equation

$$\frac{\partial P(\mathbf{r}, t)}{\partial t} = D\nabla^2 P(\mathbf{r}, t) \tag{6.77}$$

This is the three-dimensional version of Eq. (5.28). In a way similar to the one-dimensional case previously considered, we may find a solution when the initial time condition is one particle at some spatial point as

$$P(\mathbf{r}, 0) = \delta\left(\mathbf{r} - \mathbf{r}_0\right) \tag{6.78}$$

which leads to a solution as in Eq. (5.34)

$$P(\mathbf{r}, t) = \frac{1}{(4\pi Dt)^{3/2}} e^{-\frac{(\mathbf{r}-\mathbf{r}_0)^2}{4Dt}}. \tag{6.79}$$

Defining, similarly to the one-dimensional case considered before, the average square displacement as

$$\langle(\mathbf{r} - \mathbf{r}_0)^2\rangle = \int d^3r(\mathbf{r} - \mathbf{r}_0)^2 P(\mathbf{r}, t), \tag{6.80}$$

similar calculations, generalized to three-dimensions, lead to

$$\langle(\mathbf{r} - \mathbf{r}_0)^2\rangle = 6Dt, \tag{6.81}$$

which generalizes the result of Eq. (5.36).

Einstein Relation

It is now our goal to obtain an expression for the diffusion constant, D.

To simplify, assume that the particles that move through the fluid have a spherical shape. As the particles move, they experience a drag force

$$\mathbf{F}_D = -m\xi\mathbf{v}, \tag{6.82}$$

where ξ is the friction coefficient and $\mathbf{v}$ is the particle velocity. Considering the spheres of mass m and radius a and that the fluid has a viscosity η, Stokes law leads to the relation

$$\xi = \frac{6\pi\eta a}{m}. \tag{6.83}$$

This law may be seen as an empirical law or an approximate solution, obtained from the Navier-Stokes equations in the limit of small Reynolds number.

Consider now that the motion of the particles is the result of applying an external force, $\mathbf{F}$. As the particle moves, it is under the action of the external force and the drag force. As is well-known from elementary mechanics, a terminal velocity will in general be achieved, when the two forces balance each other, $\mathbf{F} + \mathbf{F}_D = 0$, and the particle reaches the terminal velocity $\mathbf{v}_T = \mathbf{F}/(m\xi)$. We may associate with the terminal velocity a current density given by $n\mathbf{v}_T$, where n is the particle density. The total particle flux (or current density) is then written as

$$\mathbf{j} = -D\nabla n + \frac{n}{m\xi}\mathbf{F}. \tag{6.84}$$

This leads to the Smoluchowski equation for the particle density

$$\frac{\partial n(\mathbf{r}, t)}{\partial t} = D\nabla^2 n(\mathbf{r}, t) - \frac{1}{m\xi}\nabla \cdot (n(\mathbf{r}, t)\mathbf{F}) \tag{6.85}$$

describing the diffusion in the presence of the external force. This also leads to a Fokker-Planck equation for the probability function as

$$\frac{\partial P(\mathbf{r}, t)}{\partial t} = D\nabla^2 P(\mathbf{r}, t) - \frac{1}{m\xi}\nabla \cdot (P(\mathbf{r}, t)\mathbf{F}). \tag{6.86}$$

In equilibrium the current vanishes: $\mathbf{j}_{eq} = 0$. Consider as a simple example that the external force is the force of gravity: $\mathbf{F} = -mg\mathbf{e}_z$. In thermal equilibrium, the density is a function of altitude as

$$n_{eq}(z) = n(0)e^{-\frac{mgz}{k_B T}}. \tag{6.87}$$

Therefore, we get

$$\nabla n = \frac{\partial n}{\partial z}\mathbf{e}_z = -\frac{mg}{k_B T} n(z)\mathbf{e}_z. \tag{6.88}$$

The vanishing of the current density (or the particle flux) leads to

$$0 = D\frac{mg}{k_B T} n(z)\mathbf{e}_z + \frac{n}{m\xi}(-mg)\mathbf{e}_z. \tag{6.89}$$

This implies the Einstein relation

$$D = \frac{k_B T}{m\xi} \tag{6.90}$$

and determines the diffusion constant in thermal equilibrium. The Einstein expression relates the diffusion with the friction constant (associated with dissipation) and the temperature (associated with thermal fluctuations). It is an example of a relation between quantities that are associated with dissipation and other quantities associated with fluctuations. We will encounter other examples of so-called fluctuation-dissipation theorems in the sense that the two properties are related. Since we have established Eq. (6.81), we can also write that

$$\langle(\Delta\mathbf{r})^2\rangle = \frac{k_B T}{\pi\eta a} t, \tag{6.91}$$

using Stokes law to rewrite the result in terms of the viscosity and temperature. This result has been verified experimentally (Perrin). Since it is an equilibrium result, it is preceded by some transient regime before equilibrium sets in.

Langevin Equation Approach

The Langevin equation may be written as

$$\frac{d\mathbf{v}}{dt} = -\xi\mathbf{v} + \mathbf{A}(t). \tag{6.92}$$

This is an equation for the slow degrees of freedom, obtained by integrating over the fast degrees of freedom. Here ξ is the friction constant and the random, time-dependent, acceleration, results from integrating the fast variables. The one-dimensional case was proposed in Eq. (1.22).

Since the acceleration is a random variable, its average value vanishes. We may calculate the correlation $\langle\mathbf{A}(t_1)\mathbf{A}(t_2)\rangle$ which is finite. We assume that the acceleration is independent of the position and velocity of the particle moving in the fluid.

Call $\mathbf{v}_0 = \mathbf{v}(t = 0)$ and let us average over the distribution of its values. Define $\mathbf{q}(t) = \mathbf{r}(t) - \mathbf{r}(0)$. The velocity equals

$$\mathbf{v} = \frac{d\mathbf{q}}{dt}. \tag{6.93}$$

Rewriting the Langevin equation in terms of **q** and taking the scalar product with **q**, we get

$$\mathbf{q} \cdot \frac{d^2\mathbf{q}}{dt^2} = -\xi \mathbf{q} \cdot \frac{d\mathbf{q}}{dt} + \mathbf{q} \cdot \mathbf{A}. \tag{6.94}$$

Using the identity

$$\mathbf{q} \cdot \frac{d^2\mathbf{q}}{dt^2} = \frac{1}{2}\frac{d^2}{dt^2}\left(q^2\right) - \left(\frac{d\mathbf{q}}{dt}\right)^2 \tag{6.95}$$

we get

$$\frac{1}{2}\frac{d^2q^2}{dt^2} - \left(\frac{d\mathbf{q}}{dt}\right)^2 = -\xi \mathbf{q} \cdot \frac{d\mathbf{q}}{dt} + \mathbf{q} \cdot \mathbf{A}. \tag{6.96}$$

Let us average over initial velocity distribution and acceleration (**A**). We obtain

$$\langle \mathbf{q} \cdot \mathbf{A} \rangle = \langle \mathbf{q} \rangle \cdot \langle \mathbf{A} \rangle = 0, \tag{6.97}$$

since $\langle \mathbf{A} \rangle = 0$. Also,

$$\frac{d^2\langle q^2 \rangle}{dt^2} + \xi \frac{d}{dt}\langle q^2 \rangle = 2\langle v^2 \rangle = 2\frac{3k_BT}{m} \tag{6.98}$$

using the equipartition theorem.

The solution of the Langevin equation is obtained as

$$\langle q^2 \rangle = 6D\left[t - \frac{1}{\xi}\left(1 - e^{-\xi t}\right)\right]. \tag{6.99}$$

For large times $t >> 1/\xi$, we get Einstein's result.

The velocity satisfies

$$\mathbf{v}(t) = e^{-\xi t}\left[\mathbf{v}_0 + \int_0^t ds e^{\xi s}\mathbf{A}(s)\right] \tag{6.100}$$

and its average value

$$\frac{d\langle \mathbf{v} \rangle}{dt} = -\xi \langle \mathbf{v} \rangle. \tag{6.101}$$

Fixing the initial velocity and averaging over random distribution of the acceleration, we get

$$\langle \mathbf{v}(t) \rangle_0 = \mathbf{v}_0 e^{-\xi t}. \tag{6.102}$$

We can obtain that

$$\langle v_i(t_1) v_j(t_2) \rangle_0 = e^{-\xi(t_1+t_2)} \left[v_{0i} v_{0j} + \int_0^{t_1} ds_1 \int_0^{t_2} ds_2 e^{\xi(s_1+s_2)} \langle A_i(s_1) A_j(s_2) \rangle_0 \right]. \tag{6.103}$$

Assume

$$\langle A_i(t_1) A_j(t_2) \rangle = \delta_{ij} \delta(t_1 - t_2) \alpha \tag{6.104}$$

(so-called white noise, since it is independent of frequency). We get

$$\langle v_i(t_1) v_j(t_2) \rangle_0 = \frac{1}{3} \delta_{ij} \left[e^{-\xi(t_1+t_2)} \left(v_0^2 - \frac{\alpha}{2\xi} \right) + \frac{\alpha}{2\xi} e^{-\xi|t_1-t_2|} \right]. \tag{6.105}$$

Take $t_1 = t_2 = t$. Then

$$\langle v^2(t) \rangle_0 = e^{-2\xi t} v_0^2 + \frac{\alpha}{2\xi} \left[1 - e^{-2\xi t} \right]. \tag{6.106}$$

For large t,

$$\langle v^2(t) \rangle \to \frac{3k_B T}{m} \tag{6.107}$$

which implies again that

$$\alpha = \frac{6k_B T \xi}{m}. \tag{6.108}$$

This is also like a fluctuation-dissipation theorem that relates the dissipation, associated with ξ, with the fluctuations of the random force, associated with α.

6.3 Irreversible Processes and Entropy Production

We would like now to revisit the discussion of systems out of equilibrium and to provide further insight.[3]

[3] For further reading see [Callen].

When a system evolves out of equilibrium, there are two types of parameters involved. Those that describe the "forces" that originate the process (using a mechanical language) and those that describe the reply to such forces. Once again let us revisit a system that is divided in two subsystems and let us ignore the outside world (or equivalently, consider that the system is isolated from the outside world). Let us label the total system by S^0 and the two subsystems by S and S'. A given extensive parameter, X, obeys

$$X_k + X_{k'} = X^0, \tag{6.109}$$

where the first term respects to subsystem S and the second term to S'. The equilibrium values are determined by an extremum principle. Considering the energy to be constant, the equilibrium state is achieved maximizing the entropy.

Let us take the quantity (intensive in the entropy representation, since is the ratio of the extensive entropy over an extensive quantity)

$$\mathcal{F}_k = \left(\frac{\partial S^0}{\partial X_k}\right)_{X_k=X_k^0} = \left(\frac{\partial (S+S')}{\partial X_k}\right)_{X_k=X_k^0} = \frac{\partial S}{\partial X_k} - \frac{\partial S'}{\partial X_{k'}} = F_k - F_{k'}. \tag{6.110}$$

If $\mathcal{F}_k = 0$ the system is in equilibrium. If $\mathcal{F}_k \neq 0$ some irreversible process will take place that will tend to take the system to an equilibrium process. The process is by its own nature irreversible since attaining equilibrium is an intrinsically irreversible process.

Some examples previously discussed were

$$X_k = U, \quad , F_k = \frac{\partial S}{\partial U} = \frac{1}{T}, \tag{6.111}$$

$$X_k = V, \quad , F_k = \frac{\partial S}{\partial V} = \frac{p}{T}, \tag{6.112}$$

$$X_k = N, \quad , F_k = \frac{\partial S}{\partial N} = -\frac{\mu}{T}. \tag{6.113}$$

In a way related to before, we define the flux J_k as

$$J_k = \frac{dX_k}{dt}. \tag{6.114}$$

(The concept of flux was associated with a flow of some variable through a surface per unit area and unit time; the definition used in this section is slightly different and is related to a current).

In such processes as the system irreversibly tends to equilibrium, entropy should increase, and, therefore, there is entropy production. This production is given by

$$\frac{dS}{dt} = \sum_k \frac{\partial S}{\partial X_k}\frac{dX_k}{dt} = \sum_k F_k J_k. \tag{6.115}$$

The partition of a system in two subsystems may now be generalized to an arbitrary number of subsystems and in the limit to a continuum of subsystems. This allows the introduction of the definition of a local entropy, defined in the same way as in equilibrium for a macroscopic system. One can extend the definition by considering the local entropy to be a function of the local extensive variables with the same functional dependence as for the equilibrium global entropy. From the expression

$$dS = \sum_k F_k dX_k \tag{6.116}$$

we may define some local entropy per unit volume

$$ds = \sum_k F_k dx_k. \tag{6.117}$$

We may now define the entropy current as

$$\mathbf{J}^s = \sum_k F_k \mathbf{J}_k. \tag{6.118}$$

Using this expression, we can write that

$$\frac{ds}{dt} = \frac{\partial s}{\partial t} + \nabla \cdot \mathbf{J}^s. \tag{6.119}$$

This is not an equation of conservation since there is entropy production (expressed by the term $\frac{ds}{dt}$ that, in general, is non-vanishing). The other parameters satisfy equations of conservation

$$0 = \frac{\partial x_k}{\partial t} + \nabla \cdot \mathbf{J}_k, \tag{6.120}$$

since the other quantities are not produced or destroyed.

The local entropy satisfies

$$\frac{\partial s}{\partial t} = \sum_k F_k \frac{\partial x_k}{\partial t}. \tag{6.121}$$

On the other hand we have that

$$\nabla \cdot \mathbf{J}^s = \nabla \cdot \left(\sum_k F_k \mathbf{J}_k \right) = \sum_k \nabla F_k \cdot \mathbf{J}_k + \sum_k F_k \nabla \cdot \mathbf{J}_k .$$

Therefore, using Eq. (6.120) in the equation for the local entropy Eq. (6.121), we get that Eq. (6.119) becomes

$$\frac{ds}{dt} = \sum_k \nabla F_k \cdot \mathbf{J}_k . \tag{6.122}$$

It is usual to call affinity the gradient of the intensive parameters in the entropy representation. Recall that we are implicitly considering here that there is now a continuum of subsystems, along space directions, and so the definition $\mathcal{F}_k = F_k - F_{k'}$ generalizes to the continuum expression with the gradient operator. In the case of energy transport the affinity is $\mathcal{F}_0 = \nabla \left(\frac{1}{T} \right)$ and in the case of transport of particles of type k the affinity is $\mathcal{F}_k = -\nabla \left(\frac{\mu_k}{T} \right)$.

We may simplify the irreversible process of flow to equilibrium by considering that the process is Markoffian. In other words, the fluxes on a given instant only depend on the values of the affinities in the same instant. As stressed before, this means that the system has no memory. As a consequence we can write that

$$\mathbf{J}_k = \mathbf{J}_k \left(\mathcal{F}_0, \mathcal{F}_1, \cdots ; F_0, F_1, \cdots \right) \tag{6.123}$$

is a function of the affinities $\mathcal{F}$ and the intensive parameters F in a given instant. Each flux depends in general on all affinities, even though we expect that it depends more strongly on the affinity related to the same physical quantity. We may expand the dependence on the fluxes near equilibrium in the form

$$J_k = \sum_j L_{jk} \mathcal{F}_j + \frac{1}{2} \sum_i \sum_j L_{ijk} \mathcal{F}_i \mathcal{F}_j + \cdots \tag{6.124}$$

where we have introduced so called kinetic coefficients

$$L_{jk} = \frac{\partial J_k}{\partial \mathcal{F}_j}$$

$$L_{ijk} = \frac{\partial^2 J_k}{\partial \mathcal{F}_i \partial \mathcal{F}_j} . \tag{6.125}$$

The derivatives are calculated at equilibrium. Therefore, the kinetic coefficients are only function of the intensive parameters and not of the affinities $L_{jk} = L_{jk}(F_0, F_1, \cdots)$. The same happens with the second-order kinetic coefficients L_{ijk}. These kinetic coefficients are called the Onsager coefficients and are symmetric.

As an example consider again Fick's law for the heat flow. We know that

$$\mathbf{Q} = \mathbf{J}_0 = -K\nabla T = KT^2\nabla\left(\frac{1}{T}\right), \tag{6.126}$$

where K is the thermal conductivity and KT^2 is the kinetic coefficient associated with the heat flow, in the description of the entropy production.

Thermoelectric Effects

An important example of non-diagonal kinetic coefficients are thermoelectric and thermomagnetic effects. Specifically, the presence of electric or magnetic fluxes due to thermal inhomogeneities or vice versa. An obvious case and consequence is the simultaneous presence of electrical and thermal currents.

Recalling that

$$ds = \frac{1}{T}du - \frac{\mu}{T}dn \tag{6.127}$$

we get that

$$J_s = \frac{1}{T}J_u - \frac{\mu}{T}J_n, \tag{6.128}$$

where J_s is an entropy current, J_u a thermal current, and J_n a particle current. If the particles are charged, such as electrons, the particle current implies an electrical current. As we saw in the previous section

$$\frac{ds}{dt} = \nabla\left(\frac{1}{T}\right)\cdot\mathbf{J_u} - \nabla\left(\frac{\mu}{T}\right)\cdot\mathbf{J}_n. \tag{6.129}$$

To simplify, consider one-dimensional fluxes, for instance, along x. We can write as in the previous section that $J_k = \sum_l L_{jk}\mathcal{F}_j$ where $\mathcal{F}_j = \nabla F_j$. Explicitly we write

$$\begin{aligned} -J_n &= L_{nn}\nabla\left(\frac{\mu}{T}\right) + L_{un}\nabla\left(\frac{1}{T}\right) \\ J_u &= L_{nu}\nabla\left(\frac{\mu}{T}\right) + L_{uu}\nabla\left(\frac{1}{T}\right). \end{aligned} \tag{6.130}$$

Since the coefficients are symmetric, $L_{un} = L_{nu}$. If we have a reversible process or if there is no work exchanged (or negligible), we can write that $\delta Q = TdS$. Therefore $J_Q = TJ_S$. This implies

$$J_Q = J_u - \mu J_n. \tag{6.131}$$

Using this result we get that

$$\begin{aligned}\frac{ds}{dt} &= \nabla\frac{1}{T}\cdot\mathbf{J}_Q + \nabla\frac{1}{T}\cdot\mu\mathbf{J}_n - \nabla\frac{\mu}{T}\cdot\mathbf{J}_n \\ &= \nabla\frac{1}{T}\cdot\mathbf{J}_Q - \frac{1}{T}\nabla\mu\cdot\mathbf{J}_n.\end{aligned} \tag{6.132}$$

Therefore,

$$\begin{aligned}-J_n &= L_{nn}\frac{1}{T}\nabla\mu + L_{Qn}\nabla\frac{1}{T} \\ J_Q &= L_{nQ}\frac{1}{T}\nabla\mu + L_{QQ}\nabla\frac{1}{T}.\end{aligned} \tag{6.133}$$

Once again $L_{Qn} = L_{nQ}$. If a quantity X is conserved and we have a stationary regime, then the associated current is divergence-free $\nabla\cdot\mathbf{J}_X = 0$. We conclude that since

$$\nabla\cdot\mathbf{J}_Q = \nabla\cdot\mathbf{J}_u - \nabla\cdot(\mu\mathbf{J}_n) \tag{6.134}$$

and since, in a stationary regime $\nabla\cdot\mathbf{J}_u = \nabla\cdot\mathbf{J}_n = 0$, we get

$$\nabla\cdot\mathbf{J}_Q = -\nabla\mu\cdot\mathbf{J}_n. \tag{6.135}$$

We can now combine these results to write that

$$\frac{ds}{dT} = \nabla\left(\frac{1}{T}\right)\cdot\mathbf{J}_Q + \frac{1}{T}\nabla\cdot\mathbf{J}_Q. \tag{6.136}$$

The first term on the right-hand side is the heat flux due to the temperature gradient, and the second term reflects that the appearance of the heat flux contributes to the increase of entropy.

Thermal Conductivity

From Fick's law, we can write that the thermal conductivity can be written as

$$K = -\frac{\partial Q}{\partial\nabla T} \tag{6.137}$$

if the particle current vanishes, $J_n = 0$. We get that

$$0 = L_{nn}\frac{1}{T}\nabla\mu + L_{Qn}\nabla\left(\frac{1}{T}\right). \tag{6.138}$$

This allows us to obtain an expression for the kinetic coefficient

$$L_{Qn} = -\frac{L_{nn}\frac{1}{T}\nabla\mu}{\nabla\left(\frac{1}{T}\right)} = -L_{nn}\frac{1}{T}\frac{\nabla\mu}{-\frac{1}{T^2}\nabla T} = TL_{nn}\frac{\nabla\mu}{\nabla T}, \tag{6.139}$$

where we used that $\nabla(1/T) = -1/T^2\nabla T$. We get then that

$$\begin{aligned} J_Q &= TL_{nn}\frac{\nabla\mu}{\nabla T}\frac{1}{T}\nabla\mu + L_{QQ}\nabla\left(\frac{1}{T}\right) \\ &= \frac{L_{nn}\left(\nabla\mu\right)^2}{\nabla T} - \frac{1}{T^2}L_{QQ}\nabla T. \end{aligned} \tag{6.140}$$

The thermal conductivity is then given by

$$K = -\frac{J_Q}{\nabla T} = \frac{1}{T^2}L_{QQ} - \frac{L_{nn}\left(\nabla\mu\right)^2}{(\nabla T)^2}. \tag{6.141}$$

This can be written as

$$K = \frac{L_{QQ}}{T^2} - \frac{L_{Qn}^2}{T^2 L_{nn}} = \frac{L_{nn}L_{QQ} - L_{Qn}^2}{T^2 L_{nn}}. \tag{6.142}$$

Note that $L_{nn}L_{QQ} - L_{Qn}^2$ is the determinant of the matrix of the Onsager coefficients. If there is no off-diagonal Onsager coefficient, $L_{Qn} = 0$, then we get the usual expression since $L_{QQ}/T^2 = K$.

Another interesting effect is the Thomson effect, which reflects the heat flow when an electric current flows through a temperature gradient.

6.4 Boltzmann Equation: Boltzmann H Theorem

The microscopic description of the dynamics of a group of interacting particles is complex, particularly if the system is not in equilibrium. In this section we will consider an approximate statistical description in terms of a function defined by[4]

$$f(\mathbf{q}, \mathbf{p}, t)d^3q d^3p,$$

which gives the number of particles that have position between $\mathbf{q}$ and $\mathbf{q} + d\mathbf{q}$ and momenta between $\mathbf{p}$ and $\mathbf{p} + d\mathbf{p}$. We have in mind a diluted gas of particles of mass m that interact via a potential with spherical symmetry of the type $\Phi(|\mathbf{q}_i - \mathbf{q}_j|)$. In a way similar to the kinetic theory of an ideal gas, we assume that the average

[4] See also [Reif] and [Reichl].

time between collisions is much larger than the time of each collision. Note that the interacting potential may have a finite range instead of the hard-core collisions previously considered. We will write that

$$f(\mathbf{q}, \mathbf{p}, t) = N\rho_1(\mathbf{q}, \mathbf{p}, t) = \frac{N}{V}F_1(\mathbf{q}, \mathbf{p}, t), \tag{6.143}$$

where N is the number of particles in volume V and where $\rho_1(\mathbf{q}, \mathbf{p}, t)$ has the meaning of a probability.

A simple case is when the interaction potential vanishes, $\phi = 0$, and each particle is subject to an external force $\mathbf{F}(\mathbf{q}, t)$. Consider the particles that are in the interval $\mathbf{q}, \mathbf{q} + d\mathbf{q}$ and in the interval $\mathbf{p}, \mathbf{p} + d\mathbf{p}$, at time t. In a later time $t' = t + dt$, the particles will be located in $\mathbf{q}', \mathbf{q}' + d\mathbf{q}'$ and in $\mathbf{p}', \mathbf{p}' + d\mathbf{p}'$. The new locations are given by

$$\begin{aligned} \mathbf{q}' &= \mathbf{q} + \frac{d\mathbf{q}}{dt}dt = \mathbf{q} + \frac{\mathbf{p}}{m}dt \\ \mathbf{p}' &= \mathbf{p} + \frac{d\mathbf{p}}{dt}dt = \mathbf{q} + \mathbf{F}dt. \end{aligned} \tag{6.144}$$

Since there are no collisions (not even hard-core collisions)

$$f(\mathbf{q}, \mathbf{p}, t)d^3pd^3q = f(\mathbf{q}', \mathbf{p}', t')d^3q'd^3p', \tag{6.145}$$

due to the conservation of the number of particles. We can write that $d^3q'd^3p' = |J|d^3qd^3p$ where $|J|$ is the Jacobian of the change of variables. This can be written as

$$|J| = det \begin{pmatrix} \frac{\partial x'}{\partial x} & \frac{\partial y'}{\partial x} & \frac{\partial z'}{\partial x} & \frac{\partial p_x'}{\partial x} & \frac{\partial p_y'}{\partial x} & \frac{\partial p_z'}{\partial x} \\ \frac{\partial x'}{\partial y} & \frac{\partial y'}{\partial y} & \frac{\partial z'}{\partial y} & \frac{\partial p_x'}{\partial y} & \frac{\partial p_y'}{\partial y} & \frac{\partial p_z'}{\partial y} \\ \frac{\partial x'}{\partial z} & \frac{\partial y'}{\partial z} & \frac{\partial z'}{\partial z} & \frac{\partial p_x'}{\partial z} & \frac{\partial p_y'}{\partial z} & \frac{\partial p_z'}{\partial z} \\ \frac{\partial x'}{\partial p_x} & \frac{\partial y'}{\partial p_x} & \frac{\partial z'}{\partial p_x} & \frac{\partial p_x'}{\partial p_x} & \frac{\partial p_y'}{\partial p_x} & \frac{\partial p_z'}{\partial p_x} \\ \frac{\partial x'}{\partial p_y} & \frac{\partial y'}{\partial p_y} & \frac{\partial z'}{\partial p_y} & \frac{\partial p_x'}{\partial p_y} & \frac{\partial p_y'}{\partial p_y} & \frac{\partial p_z'}{\partial p_y} \\ \frac{\partial x'}{\partial p_z} & \frac{\partial y'}{\partial p_z} & \frac{\partial z'}{\partial p_z} & \frac{\partial p_x'}{\partial p_z} & \frac{\partial p_y'}{\partial p_z} & \frac{\partial p_z'}{\partial p_z} \end{pmatrix}. \tag{6.146}$$

Using that

$$\frac{\partial q_\alpha'}{\partial q_\gamma} = \delta_{\alpha,\gamma}, \qquad \frac{\partial q_\alpha'}{\partial p_\gamma} = \frac{dt}{m}\delta_{\alpha,\gamma}, \qquad \frac{\partial p_\alpha'}{\partial q_\gamma} = \frac{\partial F_\alpha}{\partial q_\gamma}dt, \qquad \frac{\partial p_\alpha'}{\partial p_\gamma} = \delta_{\alpha,\gamma} \tag{6.147}$$

we get

$$|J| = det \begin{pmatrix} 1 & 0 & 0 & \frac{\partial F_x}{\partial x}dt & \frac{\partial F_y}{\partial x}dt & \frac{\partial F_z}{\partial x}dt \\ 0 & 1 & 0 & \frac{\partial F_x}{\partial y}dt & \frac{\partial F_y}{\partial y}dt & \frac{\partial F_z}{\partial y}dt \\ 0 & 0 & 1 & \frac{\partial F_x}{\partial z}dt & \frac{\partial F_y}{\partial z}dt & \frac{\partial F_z}{\partial z}dt \\ \frac{dt}{m} & 0 & 0 & 1 & 0 & 0 \\ 0 & \frac{dt}{m} & 0 & 0 & 1 & 0 \\ 0 & 0 & \frac{dt}{m} & 0 & 0 & 1 \end{pmatrix} = 1 + O(dt^2). \tag{6.148}$$

which implies that, to leading order, $d^3q'd^3p' = d^3qd^3p$ and

$$f(\mathbf{q}, \mathbf{p}, t) = f(\mathbf{q}', \mathbf{p}', t'). \tag{6.149}$$

This equality leads to

$$\begin{aligned} &f\left(\mathbf{q} + \frac{d\mathbf{q}}{dt}dt, \mathbf{p} + \frac{d\mathbf{p}}{dt}, t + dt\right) - f(\mathbf{q}, \mathbf{p}, t) = 0 \\ &\left[\left(\frac{\partial f}{\partial x}\frac{dx}{dt} + \frac{\partial f}{\partial y}\frac{dy}{dt} + \frac{\partial f}{\partial z}\frac{dz}{dt}\right) + \left(\frac{\partial f}{\partial p_x}\frac{dp_x}{dt} + \frac{\partial f}{\partial p_y}\frac{dp_y}{dt} + \frac{\partial f}{\partial p_z}\frac{dp_z}{dt}\right) + \frac{\partial f}{\partial t}\right] \\ &\quad dt = 0. \end{aligned} \tag{6.150}$$

This equation can be written in a shorthand notation as

$$Df = 0, \tag{6.151}$$

where the operator D is defined by

$$Df = \frac{\partial f}{\partial t} + \frac{d\mathbf{q}}{dt} \cdot \frac{\partial}{\partial \mathbf{q}} f + \frac{d\mathbf{p}}{dt} \cdot \frac{\partial}{\partial \mathbf{p}} f. \tag{6.152}$$

This is called the Boltzmann equation in the absence of collisions or the Vlasov equation. In this case we get that

$$\frac{\partial f}{\partial t} = -\frac{d\mathbf{q}}{dt} \cdot \frac{\partial}{\partial \mathbf{q}} f - \frac{d\mathbf{p}}{dt} \cdot \frac{\partial}{\partial \mathbf{p}} f. \tag{6.153}$$

If we reintroduce the potential between the particles, we can write that

$$\frac{\partial f}{\partial t} = -\frac{d\mathbf{q}}{dt} \cdot \frac{\partial f}{\partial \mathbf{q}} - \frac{d\mathbf{p}}{dt} \cdot \frac{\partial f}{\partial \mathbf{p}} + \left(\frac{\partial f}{\partial t}\right)_{collisions}. \tag{6.154}$$

The first term on the right-hand side is due to the particle flux that originates in the free particle equation of motion, and the second term is due to the presence of an external force. We will study next the collisions term.

Let us consider two-body collisions. Consider an Hamiltonian of the type

$$H_2(\mathbf{q}_1, \mathbf{q}_2, \mathbf{p}_1, \mathbf{p}_2) = \frac{\mathbf{p}_1^2}{2m} + \frac{\mathbf{p}_2^2}{2m} + \Phi(|\mathbf{q}_1 - \mathbf{q}_2|). \tag{6.155}$$

It is convenient to use the center of mass coordinates and the relative coordinates

$$\mathbf{q} = \mathbf{q}_2 - \mathbf{q}_1, \quad \mathbf{p} = \frac{1}{2}(\mathbf{p}_2 - \mathbf{p}_1), \tag{6.156}$$

$$\mathbf{P} = \mathbf{p}_1 + \mathbf{p}_2, \quad \mathbf{Q} = \frac{1}{2}(\mathbf{q}_2 + \mathbf{q}_1). \tag{6.157}$$

In terms of these variables the Hamiltonian can be written as

$$H_2(\mathbf{Q}, \mathbf{q}, \mathbf{P}, \mathbf{p}) = \frac{\mathbf{P}^2}{4m} + \frac{\mathbf{p}^2}{m} + \Phi(|\mathbf{q}|). \tag{6.158}$$

Consider elastic collisions where the kinetic energy is conserved

$$\frac{\mathbf{p}_1^2}{2m} + \frac{\mathbf{p}_2^2}{2m} = \frac{(\mathbf{p}_1')^2}{2m} + \frac{(\mathbf{p}_2')^2}{2m}. \tag{6.159}$$

Also, the center of mass momentum is conserved $\mathbf{P} = \mathbf{P}'$. These two conditions imply that, in the center of mass frame, the relative momentum is conserved $|\mathbf{p}| = |\mathbf{p}'| = mg/2$, where g is the absolute value of the relative velocity.

When two particles scatter, we may consider the relative motion of the effective particle with a reduced mass with respect to a scattering center. The collision may be described in the laboratory frame or the center of mass frame. We consider a beam of particles of momentum $\mathbf{p}$ that enters a central field with impact parameter b. In the lab frame we may consider that the incident particle is moving and the second particle is at rest. In the center of mass frame both particles move. After scattering, the incident particle moves along a direction that makes an angle θ_0 with the incident direction and in the center of mass frame the angle is θ. The angular momentum, $L = m/2gb$, is conserved since we consider a central potential.

Let us call I the intensity of the beam and dN/dt the number of scattered particles per second along the solid angle $d\Omega = \sin\theta_0 d\theta_0 d\varphi$. This is given by

$$\frac{dN}{dt} = I\sigma_L(g, \theta_0)\sin\theta_0 d\theta_0 d\varphi = Ibdbd\varphi, \tag{6.160}$$

where $\sigma_L(g, \theta_0)$ is the differential cross section. Therefore,

$$\sigma_L(g,\theta_0) = \frac{b(g,\theta_0)}{\sin\theta_0}\frac{db}{d\theta_0}. \tag{6.161}$$

The scattering may also be described in the center of mass frame. Since the number of particles that are contained in the solid angle, $d\Omega_0 = \sin\theta_0 d\theta_0 d\varphi$, in the lab frame, is equal to the number of particles in the corresponding solid angle in the center of mass frame, $d\Omega_C = \sin\theta d\theta d\varphi$, we get that

$$\sigma_L(g,\theta_0)\sin\theta_0 d\theta_0 d\varphi = \sigma_C(g,\theta)\sin\theta d\theta d\varphi. \tag{6.162}$$

(The angle φ is the same in the two frames.) In the center of mass frame the process $(\mathbf{p}_1, \mathbf{p}_2) \to (\mathbf{p}'_1, \mathbf{p}'_2)$ is identical to the inverse process $(\mathbf{p}'_1, \mathbf{p}'_2) \to (\mathbf{p}_1, \mathbf{p}_2)$ which leads to

$$\sigma_C\left(\mathbf{p}_1, \mathbf{p}_2 \to \mathbf{p}'_1, \mathbf{p}'_2\right) = \sigma_C\left(\mathbf{p}'_1, \mathbf{p}'_2 \to \mathbf{p}_1, \mathbf{p}_2\right). \tag{6.163}$$

Derivation of Boltzmann Equation

Let us call[5]

$$\frac{\partial f}{\partial t}\Delta\mathbf{r}d\mathbf{p}_1$$

the rate of particles scattered off $\Delta V_1 = \Delta\mathbf{r}d\mathbf{p}_1$. The number of particles in volume ΔV_1 at time t with coordinates $\mathbf{p}_1 \to \mathbf{p}_1 + d\mathbf{p}_1$ and $\mathbf{r} \to \mathbf{r} + d\mathbf{r}$ is $f(\mathbf{r}, \mathbf{p}_1, t)\Delta\mathbf{r}d\mathbf{p}_1$.

All the particles with momentum $\mathbf{p}_2$ that are in a cylinder of volume $d\mathbf{q}_2 = 2\pi b db g dt$ will collide with the particles of momentum $\mathbf{p}_1$, during the time interval dt. The number of these particles is

$$f(\mathbf{r}, \mathbf{p}_2, t)2\pi b db g dt d\mathbf{p}_2.$$

The total number of collisions $N\left(\mathbf{p}_1, \mathbf{p}_2 \to \mathbf{p}'_1, \mathbf{p}'_2\right)$, per unit volume in the time interval dt, between particles of momentum $\mathbf{p}_1$ and others of momentum $\mathbf{p}_2$, with the outcome $\mathbf{p}'_1$ and $\mathbf{p}'_2$, is

$$N\left(\mathbf{p}_1, \mathbf{p}_2 \to \mathbf{p}'_1, \mathbf{p}'_2\right) = 2\pi g b db dt f(\mathbf{r}, \mathbf{p}_2, t) f(\mathbf{r}, \mathbf{p}_1, t) d\mathbf{p}_1 d\mathbf{p}_2, \tag{6.164}$$

where we assume that $\mathbf{p}_1$ and $\mathbf{p}_2$ are not correlated, as is valid for a dilute gas and with the assumption of molecular chaos. Similarly,

$$N\left(\mathbf{p}'_1, \mathbf{p}'_2 \to \mathbf{p}_1, \mathbf{p}_2\right) = 2\pi g b db dt f(\mathbf{r}, \mathbf{p}'_1, t) f(\mathbf{r}, \mathbf{p}'_2, t) d\mathbf{p}'_1 d\mathbf{p}'_2. \tag{6.165}$$

In the case of elastic collisions $d\mathbf{p}_1 d\mathbf{p}_2 = d\mathbf{p}'_1 d\mathbf{p}'_2$.

[5] For further reading see [Reichl].

We get that the net increase of particles of momentum $\mathbf{p}_1$ between $\mathbf{p}_1$ and $\mathbf{p}_1 + d\mathbf{p}_1$, per unit volume and unit time, is

$$\left(\frac{\partial f_1}{\partial t}\right)_{collision} d\mathbf{p}_1 = d\mathbf{p}_1 \int d\mathbf{p}_2 \int d\Omega g \sigma_C(\theta, g) \left[f(\mathbf{r}, \mathbf{p}'_1, t) f(\mathbf{r}, \mathbf{p}'_2, t) - f(\mathbf{r}, \mathbf{p}_2, t) f(\mathbf{r}, \mathbf{p}_1, t) \right] . \tag{6.166}$$

We may now insert this result in Boltzmann's equation and obtain

$$\frac{\partial f_1}{\partial t} + \frac{d\mathbf{q}_1}{dt} \cdot \frac{\partial f_1}{\partial \mathbf{q}_1} + \frac{d\mathbf{p}_1}{dt} \cdot \frac{\partial f_1}{\partial \mathbf{p}_1} = \int d\mathbf{p}_2 \int d\Omega g \sigma_C(\theta, g) \left[f(\mathbf{r}, \mathbf{p}'_1, t) f(\mathbf{r}, \mathbf{p}'_2, t) - f(\mathbf{r}, \mathbf{p}_2, t) f(\mathbf{r}, \mathbf{p}_1, t) \right] . \tag{6.167}$$

Comparing the Boltzmann equation with the kinetic equation in the BBGKY hierarchy, Eq. (5.58), we see that in the Boltzmann equation, the integral only depends on f_1, instead of depending on a higher-order correlation function.

The solution of the Boltzmann equation is in general difficult due to the presence of the collision integral. A simplification is often made replacing the collision integral by $-(f - f_0)/\tau$. This is called the relaxation time approximation, and τ is the relaxation time. One assumes implicitly that the true distribution is close to the equilibrium one, f_0, and the term introduced leads to a relaxation to equilibrium, in a time scale fixed by τ.

Considering a system with no external forces the Boltzmann equation is written as

$$\frac{\partial f(\mathbf{q}, \mathbf{p}, t)}{\partial t} + \frac{1}{m} \mathbf{p} \cdot \nabla f(\mathbf{q}, \mathbf{p}, t) = -\frac{1}{\tau} \left(f(\mathbf{q}, \mathbf{p}, t) - f_0(\mathbf{q}, \mathbf{p}, t) \right) . \tag{6.168}$$

Particularizing for a monoatomic classical gas with f_0 given by

$$f_0(\mathbf{q}, \mathbf{p}, t) = \frac{n(\mathbf{q}, \mathbf{t})}{[2\pi m k_B T(\mathbf{q}, \mathbf{t})]^{3/2}} e^{-\frac{|\mathbf{p} - m\mathbf{v}(\mathbf{q}, t)|^2}{2mk_B T(\mathbf{q}, t)}} .$$

and using the conservation laws for the number of particles, momentum and energy, an approximate solution for $f(\mathbf{q}, \mathbf{p}, t)$ of the form $f(\mathbf{q}, \mathbf{p}, t) = f_0(\mathbf{q}, \mathbf{p}, t) + \tau f_1(\mathbf{q}, \mathbf{p}, t) + O(\tau^2)$ can be obtained.[6] In this approximation the density satisfies the equation

$$\frac{\partial (n v_i)}{\partial t} + \frac{1}{m} \frac{\partial (n k_B T)}{\partial x_i} + \sum_j \frac{\partial (n v_i v_j)}{\partial x_j} = 0.$$

[6] See, for instance, [Dalvit] for a detailed derivation.

Boltzmann H Theorem
Boltzmann introduced the function

$$H(t) = \int d\mathbf{q}_1 \int d\mathbf{p}_1 f(\mathbf{q}_1, \mathbf{p}_1, t) \ln f(\mathbf{q}_1, \mathbf{p}_1, t). \tag{6.169}$$

The time dependence is given by

$$\frac{\partial H(t)}{\partial t} = \int d\mathbf{q}_1 \int d\mathbf{p}_1 \frac{\partial f_1}{\partial t} \left[\ln f_1 + 1\right]. \tag{6.170}$$

Consider Boltzmann's equation without external forces and insert it in the previous expression. This leads to

$$\frac{\partial H}{\partial t} = \frac{1}{4} \int d\mathbf{q}_1 \int d\mathbf{p}_1 \int d\mathbf{p}_2 \int d\Omega g\sigma(\theta, g) \left(f_1' f_2' - f_1 f_2\right) \ln \frac{f_1 f_2}{f_1' f_2'}. \tag{6.171}$$

We get Boltzmann H theorem that states

$$\frac{\partial H}{\partial t} \leq 0. \tag{6.172}$$

In equilibrium we get detailed balance in the sense that

$$\ln f_1 + \ln f_2 = \ln f_1' + \ln f_2', \tag{6.173}$$

which implies that $\partial H / \partial t = 0$. In equilibrium and if there are no external forces, the one-particle distribution function has to be independent of the position of the particle and the conserved quantities are momentum, kinetic energy, and the number of particles. Therefore, in equilibrium we can write that

$$\begin{aligned} \ln f_1 &= A + \mathbf{B} \cdot \mathbf{p}_1 + C \frac{\mathbf{p}_1^2}{2m} \\ f_1(\mathbf{p}_1) &= e^{A + \mathbf{B} \cdot \mathbf{p}_1 + C \frac{\mathbf{p}_1^2}{2m}}. \end{aligned} \tag{6.174}$$

The term linear in the momentum leads to a non-vanishing average momentum. This is therefore the stationary solution of Boltzmann equation that we recognize as the Maxwell-Boltzmann distribution.

We can now define a new function called Boltzmann entropy as

$$\begin{aligned} S(t) &= -k_B H(t) \\ &= -k_B \int d\mathbf{q}_1 \int d\mathbf{p}_1 f(\mathbf{q}_1, \mathbf{p}_1, t) \ln f(\mathbf{q}_1, \mathbf{p}_1, t). \end{aligned} \tag{6.175}$$

which either is a constant or increases with time. It is in some sense a confirmation of the second law of thermodynamics. Even though is quite similar to the statistical entropy previously defined (Gibbs), it is only a function of the one-particle distribution, while the previous definition involved the full distribution function of the many-particle ensemble. It is like an entropy defined in terms of a reduced distribution. The statistical proof of the second law for the full entropy is more involved and will not be considered here.

6.4.1 Equations of Motion of Average Values

We may now use the solution of Boltzmann equation to obtain and solve the equations of motion of the average value of some quantity $\chi(\mathbf{r}, \mathbf{p}, t)$. The average value of this quantity is defined in the usual way as[7]

$$\langle \chi(\mathbf{r}, t) \rangle = \frac{\int d^3 p \chi(\mathbf{r}, \mathbf{p}, t) f(\mathbf{r}, \mathbf{p}, t)}{\int d^3 p f(\mathbf{r}, \mathbf{p}, t)} = \frac{1}{n(\mathbf{r}, t)} \int d^3 p \chi(\mathbf{r}, \mathbf{p}, t) f(\mathbf{r}, \mathbf{p}, t), \tag{6.176}$$

where $n(\mathbf{r}, t)$ is the number of particles per unit volume.

Boltzmann equation can be written in a shorthand notation as

$$Df_1 = D_c f_1, \tag{6.177}$$

where D_c is the term that has its origin in the collisions between the particles.

Let us multiply by χ and integrate over the momentum on both sides of the equation. The left hand side gives

$$\int d^3 p_1 Df_1 \chi = \int d^3 p_1 \frac{\partial f_1}{\partial t} \chi + \int d^3 p_1 \frac{\mathbf{p}_1}{m} \cdot \frac{\partial f_1}{\partial \mathbf{r}_1} \chi + \int d^3 p_1 \frac{d\mathbf{p}_1}{dt} \cdot \frac{\partial f_1}{\partial \mathbf{p}_1} \chi \tag{6.178}$$

and the right-hand side gives

$$C(\chi) = \int d^3 p_1 D_c f_1 \chi = \int d^3 p_1 \int d^3 p_2 \int d\Omega \left(f_1' f_2' - f_1 f_2 \right) g \sigma_C \chi. \tag{6.179}$$

The left-hand side can be written as

$$\int d^3 p Df \chi = \frac{\partial}{\partial t} (n \langle \chi \rangle) + \frac{\partial}{\partial x_\alpha} (n \langle p_\alpha \chi \rangle) - n \langle D \chi \rangle. \tag{6.180}$$

[7] See also [Reif].

The collision term on the other hand may be symmetrized yielding

$$\mathcal{C}(\chi) = \frac{1}{2} \int d^3p \int d^3p_1 \int d\Omega f f_1 g \sigma \Delta\chi, \tag{6.181}$$

where $\Delta\chi = \chi' + \chi_1' - \chi - \chi_1$. Considering only cases where χ is conserved in collisions, $\Delta\chi = 0$, implies $\mathcal{C}(\chi) = 0$. In these cases

$$\frac{\partial}{\partial t}\langle n\chi\rangle + \frac{\partial}{\partial x_\alpha}\langle np_\alpha\chi\rangle = n\langle D\chi\rangle. \tag{6.182}$$

If $\chi = m$ this is the equation for the conservation of mass. If $\chi = p_\alpha$ the conservation of momentum and if $\chi = p^2/(2m)$ is the conservation of energy. As an example, we specify $\chi = m$. In this case we get

$$\begin{aligned} \frac{\partial}{\partial t}\langle nm\rangle + \frac{\partial}{\partial x_\alpha}\langle nmp_\alpha\rangle &= 0 \\ \frac{\partial}{\partial t}\langle \rho\rangle + \frac{\partial}{\partial x_\alpha}\langle \rho p_\alpha\rangle &= 0 \\ \frac{\partial \rho}{\partial t} + \nabla \cdot (\rho \mathbf{p}) &= 0, \end{aligned} \tag{6.183}$$

the well-known conservation of mass.

6.5 Exercises

1. (a) Estimate the number of collisions between the molecules in air (taken approximately as ideal) at room temperature and atmospheric pressure, using the expression for the pressure and Eq. (6.7). The viscosity coefficient for air in these conditions is 1.8×10^{-4} in cgs units.
   (b) What is the temperature dependence of the viscosity coefficient?
   (c) If the molecular scattering cross section is $\sigma \sim E_{cm}^2$, where E_{cm} is the center of mass energy of two colliding particles, what is the temperature dependence of η?
2. A closed cylinder of length L and negligible transverse dimension has a distribution of molecules that at some initial time is given by

$$n(x, 0) = n_0 + n_1 \cos\frac{\pi x}{L} + n_3 \cos\frac{3\pi x}{L} + n_5 \cos\frac{5\pi x}{L}.$$

   Calculate $n(x, t)$ as time evolves considering a diffusive process given by Eq. (6.75).

3. Consider a fluid at temperature T in the presence of an electric field $\mathbf{E}(x)$ and that some of its molecules are ionized.

   (a) Using Eqs. (6.84), and (6.85) and using Einstein relation Eq. (6.90), the motion of the ionized particles satisfies the equations[8]

$$\frac{\partial n}{\partial t} + \nabla \cdot \mathbf{j} = 0$$

$$\mathbf{j} = -D\nabla n + \frac{q}{k_B T} Dn\mathbf{E}.$$

   (b) Find the steady-state solution within the canonical ensemble.

4. Boltzmann's equation without external forces can be written as

$$\frac{\partial f_1}{\partial t} + \frac{d\mathbf{q}_1}{dt} \cdot \frac{\partial f_1}{\partial \mathbf{q}_1} = \int d\mathbf{p}_2 \int d\Omega g\sigma(\theta, g)\left(f_1' f_2' - f_1 f_2\right), \tag{6.184}$$

   Use Boltzmann's equation in Eq. (6.170). Assume that the function f and its derivatives vanish at the surface. We may now symmetrize this expression. We may exchange $\mathbf{p}_1$ with $\mathbf{p}_2$ and we may also exchange $\mathbf{p}_1$ with $\mathbf{p}_1'$ and $\mathbf{p}_2$ with $\mathbf{p}_2'$. Show that Eq. (6.171) holds.

5. (a) Considering the mass density ρ and momentum in a fluid, show that, using the equation of conservation Eq. (6.182), we can write that

$$\frac{\partial \rho}{\partial t} + \nabla \cdot (\rho \mathbf{v}) = 0,$$

$$\frac{\partial}{\partial t}(\rho \mathbf{v}) + (\mathbf{v} \cdot \nabla)(\rho \mathbf{v}) + \nabla p = 0, \tag{6.185}$$

   where $\mathbf{v}$ is the velocity and p the pressure.

   (b) A small disturbance in a fluid at rest and initially in equilibrium, with density ρ_0 and pressure p_0, leads to a change in velocity ($\mathbf{v} = 0 \to \mathbf{v}$), in pressure ($p_0 \to p$) and density ($\rho_0 \to \rho$). Using the conservation equations obtain a wave equation for the deviation of the density from equilibrium and obtain an expression for the velocity of the sound waves. Consider that the deviation in pressure takes place adiabatically (constant entropy) and show that the sound velocity is given by

$$v = \sqrt{\left(\frac{dp}{d\rho}\right)_S} \tag{6.186}$$

[8] You may also rederive these equations starting from a Langevin equation [Dalvit].

In the case of air show that

$$v^2 = \gamma \frac{k_B T}{m},$$

where m is the mass of a molecule, k_B is Boltzmann constant, and $\gamma = c_P/c_V$.

Chapter 7
Quantum Statistical Physics

7.1 Statistics of Quantum Systems and Quantum Liouville Theorem

A system is in a given (pure) quantum state, $|\psi\rangle$. We define an operator given by the projector to that state:

$$\rho = |\psi\rangle\langle\psi|. \tag{7.1}$$

Taking the state as normalized, the average of an observable G on that state may be written as

$$\langle G\rangle = \langle\psi|G|\psi\rangle. \tag{7.2}$$

This result is also obtained as

$$\langle G\rangle = Tr\{\rho G\}, \tag{7.3}$$

if we use the cyclic property of the trace. In general, the (mixed) state of the system can be written as a linear combination of states of some (normalized) basis as

$$|\psi\rangle = \sum_n c_n|\phi_n\rangle. \tag{7.4}$$

The operator may be obtained as

$$\rho = \sum_{n,m} c_n c_m^*|\phi_n\rangle\langle\phi_m|. \tag{7.5}$$

P. D. Sacramento, *Introduction to Condensed Matter Theory*,
https://doi.org/10.1007/978-3-031-86251-9_7

The matrix elements are then given by

$$\rho_{nm} = c_n c_m^*. \tag{7.6}$$

The operator, ρ, is called the density matrix operator, and ρ_{nm} are its matrix elements in the basis $|\phi_n\rangle$.

Consider now an ensemble composed of $\mathcal{N}$ identical quantum systems ($\mathcal{N} >> 1$) described by a Hamiltonian H. Including dependence on time, the states of the systems are characterized by states $|\psi^k(t)\rangle$, or by wave functions $\psi^k(\mathbf{R}_i, t)$, where $k = 1, 2, \ldots, \mathcal{N}$ and i labels each particle in the systems, which obey Schrödinger's equation[1]

$$i\hbar\frac{\partial\psi^k(t)}{\partial t} = H\psi^k(t). \tag{7.7}$$

If the states are mixed states, from Schrödinger equation's, we obtain the equation of motion for the coefficients:

$$i\hbar\frac{\partial c_n^k}{\partial t} = \sum_m H_{nm}c_m^k(t), \tag{7.8}$$

where the matrix element $H_{nm} = \int d\omega\phi_n^* H\phi_m$, or in a simplified notation $h_{nm} = \langle\phi_n|H|\phi_m\rangle$. Here $d\omega$ is the elementary volume. The coefficient $c_n^k(t)$ has the usual meaning: it is the probability amplitude that the k system is found in the state ϕ_n, at time t. The normalization of the states implies that $\sum_n |c_n^k(t)|^2 = 1$, for all k values.

Extending the definition of the density operator to include an index for the ensemble copy, k, we get:

$$\rho = \frac{1}{\mathcal{N}}\sum_{k=1}^{\mathcal{N}}|\psi^k\rangle\langle\psi^k|, \tag{7.9}$$

and the matrix elements of this operator are

$$\rho_{nm} = \frac{1}{\mathcal{N}}\sum_{k=1}^{\mathcal{N}}c_m^k(t)c_n^{k*}(t) \tag{7.10}$$

and, therefore, the diagonal matrix elements are given by

[1] See also [Pathria].

$$\rho_{nn} = \frac{1}{\mathcal{N}} \sum_{k=1}^{\mathcal{N}} |c_n^k(t)|^2. \tag{7.11}$$

This diagonal term involves a quantum average and an average over the ensemble of states. It is the probability that a system (a member of the ensemble of copies), chosen randomly over the ensemble, at time t, is in the state ϕ_n, when a measurement is carried out. Obviously, $\sum_n \rho_{nn} = 1$.

The equation of motion of the matrix elements of the density operator is given by

$$\begin{aligned} i\hbar \frac{\partial \rho_{mn}}{\partial t} &= \frac{1}{\mathcal{N}} \sum_{k=1}^{\mathcal{N}} \left[i\hbar \left[\frac{\partial c_m^k}{\partial t} c_n^{k*} + c_m^k \frac{\partial c_n^{k*}}{\partial t} \right] \right] \\ &= (H\rho - \rho H)_{mn}. \end{aligned} \tag{7.12}$$

Therefore, we obtain the quantum Liouville theorem:

$$i\hbar \frac{\partial \rho_{mn}}{\partial t} = [H, \rho]_{mn}. \tag{7.13}$$

In equilibrium, the ensemble is stationary, $\frac{\partial \rho_{mn}}{\partial t} = 0$, and

$$[H, \rho]_{eq} = 0. \tag{7.14}$$

This condition is satisfied if the density operator is a function of the Hamiltonian. If we choose as a basis the wave functions of the Hamiltonian, then the density matrix is diagonal. In any other representation, it is in general non-diagonal, but in equilibrium the evolution of the system, from one state to any other, must be countered by the evolution from the second state to the first one. This is the so-called detailed balance hypothesis.

Let us consider again an operator G. The average value of this operator in the ensemble of states is given by

$$\begin{aligned} \langle G \rangle &= \frac{1}{\mathcal{N}} \sum_{k=1}^{\mathcal{N}} \langle \psi_k | G | \psi_k \rangle \\ &= \sum_{mn} \rho_{mn} G_{nm} = \sum_m (\rho G)_{mm} = Tr\{\rho G\}. \end{aligned} \tag{7.15}$$

In general, if the states are not normalized, we get:

$$\langle G \rangle = \frac{Tr\{\rho G\}}{Tr\{\rho\}}. \tag{7.16}$$

7.1.1 Canonical Ensemble

A possible, and very convenient, choice for the density matrix is the canonical ensemble, where $\rho_{mn} = \rho_n \delta_{mn}$, with $\rho_n = 1/Z_N(\beta) e^{-\beta E_n}$. Here E_n are the energy levels of the system, β is the inverse temperature and $Z_N(\beta)$ is the partition function. The density operator is then given by

$$\begin{aligned}\rho &= \sum_n |\phi_n\rangle \frac{e^{-\beta E_n}}{Z_N(\beta} \langle\phi_n| \\ &= \frac{e^{-\beta E_n}}{Z_N(\beta)} \sum_n |\phi_n\rangle\langle\phi_n| \\ &= \frac{e^{-\beta H}}{Z_N(\beta)}. \end{aligned} \tag{7.17}$$

The average value of an operator is then given by

$$\langle G \rangle = Tr\{\rho G\}, \tag{7.18}$$

with this particular choice of the density matrix operator.

7.1.2 Grand Canonical Ensemble

Another convenient choice is used when the number of particles is not fixed, except on average. This choice is the grand canonical ensemble where

$$\rho = \frac{e^{-\beta(H-\mu N)}}{\mathcal{Z}}. \tag{7.19}$$

where the grand partition function is expressed as

$$\mathcal{Z} = Tr\{e^{-\beta(H-\mu N)}\} = \sum_{n,s} e^{-\beta(E_n-\mu N_s)} \tag{7.20}$$

and μ is the chemical potential. The average value of an operator is in this ensemble obtained as

$$\begin{aligned}\langle G \rangle &= \frac{Tr\{Ge^{-\beta(H-\mu N)}\}}{Tr\{e^{-\beta(H-\mu N)}\}} \\ &= \frac{\sum_{N=0}^{\infty} z^N \langle G \rangle_N Z_N(\beta)}{\sum_{N=0}^{\infty} z^N Z_N(\beta)}, \end{aligned} \tag{7.21}$$

where $z = e^{\beta\mu}$ is the fugacity.

7.2 Semiclassical Treatment and Single-Particle Description

As a first step toward taking into account the effects of a quantum description, let us consider non-interacting particles assuming a single-particle description. The particles satisfy Schrödinger's equation, which determines the eigenvalues and eigenfunctions available to the particles in the system. For the moment let us consider that the quantum nature only affects the energies available to each particle.

7.2.1 *Ideal Gas*

Consider that we have a system of N particles and that its energy is given by $E = \sum_{i=1}^{N} \epsilon_i$. Here ϵ_i are among the energies available to each particle. Let us see that some general conclusions may be reached, which have implications on the thermodynamics of the system.

One particle in a cubic box of side L has a wave function that satisfies

$$\begin{aligned}
\psi(x = 0, L, y, z) &= 0, \\
\psi(x, y = 0, L, z) &= 0, \\
\psi(x, y, z = 0, L) &= 0.
\end{aligned} \tag{7.22}$$

A straightforward solution is

$$\psi = \psi_0 \sin(k_x x) \sin(k_y y) \sin(k_z z), \tag{7.23}$$

where $k_l = n_l \pi / L$, with $n_l = 1, 2, 3, \cdots$ and $l = x, y, z$. The single particle energy is trivially

$$\epsilon(n_x, n_y, n_z) = \frac{h^2}{8mL^2} \left(n_x^2 + n_y^2 + n_z^2 \right). \tag{7.24}$$

Let us call $\Omega(N, V, E)$ the number of solutions of the equation

$$\sum_{R=1}^{3N} n_R^2 = \frac{8mV^{2/3}E}{h^2}, \tag{7.25}$$

where E is the total energy of the N-particle system in the volume $V = L^3$. We see that $\Omega(N, V^{2/3}E)$. This implies that the entropy

$$S(N, V, E) = k_B \ln \Omega(N, V, E) = S(N, V^{2/3}E). \tag{7.26}$$

If the number of particles is conserved and we consider an isentropic process (adiabatic and reversible), the entropy is constant, which implies that $V^{2/3}E$ is constant, which imposes a constraint between the energy and the volume of the system.

Let us calculate, for instance, the pressure using

$$p = -\left(\frac{\partial E}{\partial V}\right)_{N,S} = \frac{2}{3}\frac{E}{V}. \tag{7.27}$$

Since $p \sim E/V$ and since $E \sim V^{-2/3}$, we get that $p \sim V^{-5/3}$. Therefore, $pV^{5/3}$ is constant in an isentropic process. This holds for a monoatomic gas where the only degrees of freedom are the kinetic energy of the translation of the center of mass. This recovers the classical result that $\gamma = 5/3$, for a monoatomic gas.

7.2.2 *Relativistic Gas*

Consider now the single-particle relativistic energy expression:

$$\epsilon^2 = p^2c^2 + m^2c^4. \tag{7.28}$$

Neglecting the mass term, which may be appropriate for light particles, we get that

$$\epsilon = pc = \hbar kc. \tag{7.29}$$

Here $k = |\mathbf{k}|$. The energy takes the form:

$$\epsilon = \frac{hc}{2L}\left(n_x^2 + n_y^2 + n_z^2\right)^{1/2}. \tag{7.30}$$

The number of distinct states, solution of the equation

$$\left(n_x^2 + n_y^2 + n_z^2\right)^{1/2} = \frac{2V^{1/3}}{hc}\epsilon \tag{7.31}$$

we may call $\Omega(1, \epsilon, V)$. For the N particle, we use $\Omega(N, E, V)$ as the number of solutions of the equation:

$$\sum_i \left(n_x^2 + n_y^2 + n_z^2\right)_i^{1/2} = \frac{2V^{1/3}E}{hc}. \tag{7.32}$$

A similar reasoning leads now to a dependence of the entropy as $S(N, V^{1/3}E)$. For an isentropic process, we get therefore that $V^{1/3}E$ is constant. The pressure now follows the law:

$$p = -\left(\frac{\partial E}{\partial V}\right)_{N,S} = \frac{1}{3}\frac{E}{V} \tag{7.33}$$

and in a constant entropy transformation, we get $pV^{4/3}$ constant and $\gamma = 4/3$.

In these arguments we have not considered the effect of the indistinguishability of the particles, that we have shown before is an important point, for instance in the context of the Gibbs paradox.

7.2.3 Single-Particle Partition Function

Using the definition of the partition function in terms of the states j, or its expression in terms of the energy levels as

$$Z = \sum_j e^{-\beta\epsilon_j} = \sum_{\epsilon_i} g_i e^{-\beta\epsilon_i}, \tag{7.34}$$

with

$$\epsilon_j = \frac{h^2}{8m}\left(\frac{n_x^2}{a^2} + \frac{n_y^2}{b^2} + \frac{n_z^2}{c^2}\right), \tag{7.35}$$

(a, b, c are the system dimensions along x, y, z), we can write that

$$Z = \sum_{n_x=1}^{\infty}\sum_{n_y=1}^{\infty}\sum_{n_z=1}^{\infty} e^{-\frac{h^2}{8mk_BT}\left(\frac{n_x^2}{a^2}+\frac{n_y^2}{b^2}+\frac{n_z^2}{c^2}\right)}. \tag{7.36}$$

The three sums factorize. Approximating the sums by integrals, we get that

$$Z = \frac{V}{\lambda^3}, \tag{7.37}$$

where λ is the thermal wavelength given by

$$\lambda = \frac{h}{(2\pi mk_BT)^{1/2}}. \tag{7.38}$$

We may calculate:

$$\ln Z = \ln V + \frac{3}{2}\ln T \frac{3}{2}\ln \frac{2\pi m k_B}{h^2}. \tag{7.39}$$

From the single-particle partition function, we may now calculate the N-particle partition function, using the Gibbs recipe. This leads to the equation of state:

$$\frac{p}{k_B T} = \frac{\partial}{\partial V}(\ln Z_N)_{N,T} = \frac{\partial}{\partial V}\left(\ln \frac{Z^N}{N!}\right)_{N,T} = N\frac{\partial}{\partial V}(\ln Z)_{N,T}\,. \tag{7.40}$$

This gives:

$$\frac{p}{k_B T} = \frac{N}{V}, \tag{7.41}$$

the usual equation of state of the ideal gas. Note that the factor $N!$ does not affect the derivation of the expression for the pressure. Also, we may get that

$$U = N k_B T^2 \left(\frac{\partial \ln Z}{\partial T}\right)_V = \frac{3}{2} N k_B T. \tag{7.42}$$

The Gibbs factor does affect the expression for the entropy, as we saw before.

7.2.4 Harmonic Oscillator in 1d. Quantum Oscillator

First consider a classical harmonic oscillator, with coordinate q and momentum p:

$$E(q, p) = \frac{1}{2} m\omega^2 q^2 + \frac{p^2}{2m}. \tag{7.43}$$

The partition function reads:

$$\begin{aligned} Z &= \frac{1}{h}\int_{-\infty}^{\infty} dq \int_{-\infty}^{\infty} dp e^{-\beta\left(\frac{1}{2}m\omega^2 q^2 + \frac{p^2}{2m}\right)} \\ &= \frac{1}{h}\int_{-\infty}^{\infty} dq e^{-\beta\frac{1}{2}m\omega^2 q^2}\int_{-\infty}^{\infty} dp e^{-\beta\frac{p^2}{2m}} \\ &= \frac{1}{h}\left(\frac{2\pi}{\beta m\omega^2}\right)^{1/2}\left(\frac{2\pi m}{\beta}\right)^{1/2} \\ &= \frac{1}{\beta\hbar\omega}. \end{aligned} \tag{7.44}$$

Considering N oscillators we get $Z_N(\beta) = Z^N(\beta)$. Therefore, using Eqs. (4.35) and (4.36), we get the thermodynamics:

$$\begin{aligned} F &= -k_B T \ln Z_N = N k_B T \ln\left(\frac{\hbar\omega}{k_B T}\right) \\ p &= 0 \\ S &= N k_B \left(\ln\frac{k_B T}{\hbar\omega} + 1\right) \\ U &= N k_B T \\ C_p &= C_V = N k_B. \end{aligned} \tag{7.45}$$

Let us now compare with a quantum oscillator. The energy levels are

$$\epsilon_n = \hbar\omega\left(n + \frac{1}{2}\right), \tag{7.46}$$

with $n = 0, 1, 2, \cdots$. The partition function is

$$Z(\beta) = \sum_{n=0}^{\infty} e^{-\beta\left(n+\frac{1}{2}\right)\hbar\omega} = \frac{e^{-\frac{1}{2}\beta\hbar\omega}}{1 - e^{-\beta\hbar\omega}} = \frac{1}{2\sinh\frac{\beta\hbar\omega}{2}}. \tag{7.47}$$

Considering again a system of N particles, note that the oscillators are localized and therefore are distinguishable. We get that

$$\begin{aligned} Z_N(\beta) &= Z(\beta)^N \\ F &= N k_B T \ln\left(2\sinh\frac{\beta\hbar\omega}{2}\right) = N\left(\frac{1}{2}\hbar\omega + k_B T \ln\left(1 - e^{-\beta\hbar\omega}\right)\right) \\ p &= 0 \\ S &= N k_B \left(\frac{\beta\hbar\omega}{e^{\beta\hbar\omega} - 1} - \ln\left(1 - e^{-\beta\hbar\omega}\right)\right) \\ U &= N\left(\frac{1}{2}\hbar\omega + \frac{\hbar\omega}{e^{\beta\hbar\omega} - 1}\right) \\ C_p &= C_V = N k_B (\beta\hbar\omega)^2 \frac{e^{\beta\hbar\omega}}{\left(e^{\beta\hbar\omega} - 1\right)^2}. \end{aligned} \tag{7.48}$$

Note for instance that $U \neq N k_B T$: the quantum oscillators do not obey the equipartition theorem, except in the high temperature limit. In the low temperature limit, the zero point motion of the quantum oscillator is a clear distinguishing feature with respect to the classical oscillator. In Fig. 7.1 we show results for the entropy,

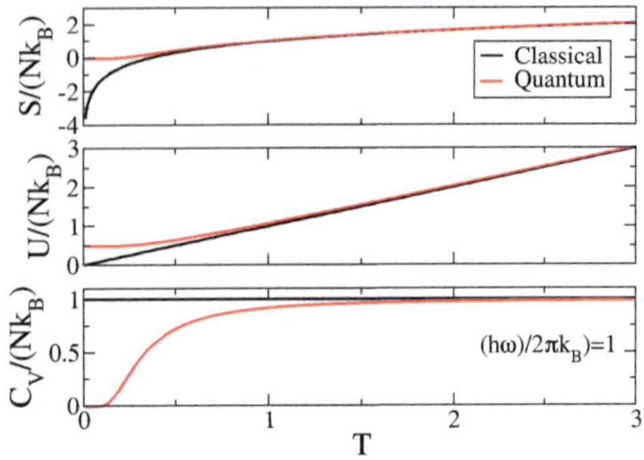

Fig. 7.1 Comparison of the results for the entropy, average energy, and specific heat for a classical and a quantum oscillator, as a function of temperature

average energy, and specific heat comparing the classical and quantum harmonic oscillators, as a function of temperature. In the large temperature limit, the quantum results converge to the classical results. As expected, both in the classical and in the quantum cases, the oscillators do not exert pressure.

7.2.5 Two-Level System: Negative Temperatures

An extreme quantum case is a system with two energy levels. The energies may be written as $E_m = -m\mu H$, with $m = \pm 1$. There is therefore a gap in the spectrum of magnitude $\Delta = 2\,\mu\text{H}$. An example of such a system is a spin-1/2 magnetic moment in a magnetic field.

The partition function is given by

$$Z(\beta) = \sum_m e^{\beta m\mu H} = 2\cosh(\beta\mu H)\,. \tag{7.49}$$

The average magnetic moment is obtained as

$$\langle M\rangle = \langle m\mu\rangle = \mu\frac{\sum_m m e^{-\beta E_m}}{\sum_m e^{-\beta E_m}} = \mu\tanh(\beta\mu H)\,. \tag{7.50}$$

We may also obtain it using

$$\begin{aligned}\langle M\rangle &= \frac{\partial}{\partial(\beta H)}\ln\left(\sum_m e^{\beta m\mu H}\right)\\ &= \frac{\partial}{\partial(\beta H)}\ln(2\cosh(\beta\mu H))\\ &= \mu\tanh(\beta\mu H)\,.\end{aligned} \tag{7.51}$$

In the limit when the applied field is large, $\mu H >> k_B T$, the magnetization saturates, $\langle M \rangle/\mu \to 1$. We may also understand this limit as the limit of a finite field and a vanishing temperature. We expect therefore that only the lowest energy state contributes. In the opposite regime, $\mu H << k_B T$, we get that $\langle M \rangle/\mu \to 0$. This is the high temperature regime.

Let us order the energies of a general system, E_i in increasing order: $E_0 < E_1 < E_2 < \cdots$ with E_0 the lowest energy state available. The Boltzmann factor is written as

$$P_i = \frac{e^{-\beta E_i}}{e^{-\beta E_0} + e^{-\beta E_1} + e^{-\beta E_2} + \dots}$$
$$= \frac{e^{-\beta(E_i - E_0)}}{1 + e^{-\beta(E_1 - E_0)} + e^{-\beta(E_2 - E_0)} + \dots}. \tag{7.52}$$

In the low temperature limit, $T \to 0$, $\beta \to \infty$, since $E_i - E_0 > 0$ (unless $i = 0$), $P_i \to 0$ (unless i is the lowest energy state and then $P_0 = 1$). At zero temperature only the lowest energy state is occupied. On the other hand, in the opposite limit of high temperatures, $\beta \to 0$, all states have the same occupation, $P_i = 1/\Omega$, where Ω is the number of states.

We may now calculate the entropy of the two-level system. Using that $S = -\partial F/\partial T$, we obtain:

$$S = k_B \left(\ln (2 \cosh \beta \mu H) - \frac{\mu H}{k_B T} \tanh \frac{\mu H}{k_B T} \right). \tag{7.53}$$

Using that $F = U - TS$, we may calculate U, the average energy. Taking limits of the entropy, we see that at zero temperature the entropy vanishes (as one would expect from the third law of thermodynamics, since the lowest energy state is not degenerate) and at high temperatures the entropy tends to $S = k_B \ln 2$, since the total number of states for the two-level system is two. At low temperatures the average energy coincides with the lowest energy state and at high temperatures it vanishes, since it tends to the average value of the energy states.

One may also calculate the heat capacity at constant field using

$$C_H = \left(\frac{\partial U}{\partial T} \right)_H = k_B \left(\frac{\mu H}{k_B T} \right)^2 \frac{1}{\cosh^2 \left(\frac{\mu H}{k_B T} \right)}. \tag{7.54}$$

The specific heat vanishes at low and high energies and has a maximum at an energy scale of the order of the gap in the spectrum. The temperature dependence of the thermodynamic quantities is shown in Fig. 7.2.

Since the spectrum is bounded from below and above, there is a maximum energy available to the system. This leads to an interesting consequence. First note that the energy ranges from $-\mu H$ to $+\mu H$. Also, the entropy ranges from zero to $\ln 2$. This maximal value occurs when $U/(\mu H) = 0$. Recalling that the absolute temperature is defined as $T = \partial U/\partial S$ or

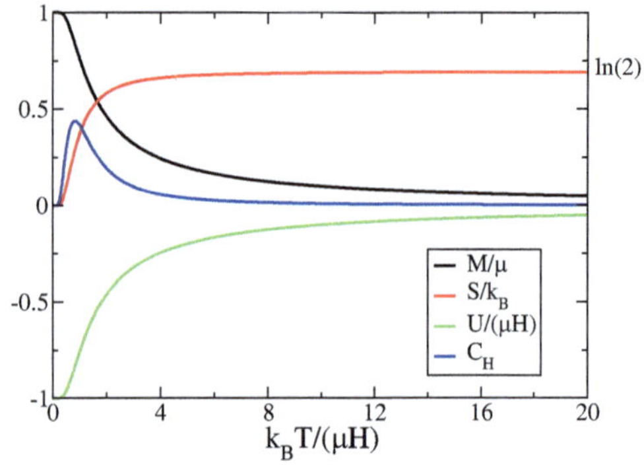

Fig. 7.2 Magnetization, entropy, average energy, and specific heat of a two-level system, as a function of temperature (thermal energy) over the magnetic energy

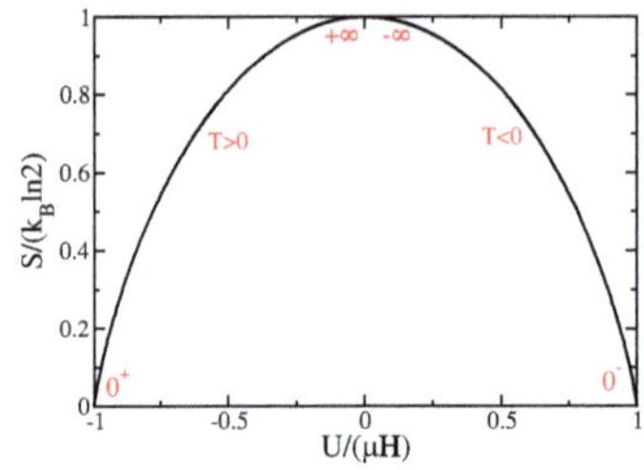

Fig. 7.3 Normalized entropy as a function of the average energy. The inverse of the derivative of the entropy with respect to the energy is proportional to the temperature. In the regime that the energy is negative, the temperature is positive and ranges from zero to infinity. In the regime when the energy is positive (population inversion), the temperature is negative and ranges from zero to negative infinity

$$T = \frac{1}{\frac{\partial S}{\partial U}}, \tag{7.55}$$

we see that while for negative values of U the temperature is positive, since the entropy is increasing as the energy increases, the temperature becomes negative for positive values of U, since as U increases the entropy decreases. At zero positive temperature, the energy is at its lowest value. Increasing the temperature we converge to the regime of equal occupation of the energy levels, and at infinite temperature the average energy vanishes. In this way the energy does not become positive: at most it vanishes. Positive energies are therefore associated with more excited states than those that result from an infinite positive temperature. This suggests therefore that negative temperatures are "hotter" than positive temperatures. Indeed, to excite the positive energy region, one needs what is called population inversion, usually achieved by supplying energy to the system by other means than just thermal excitation. If the system is however in this negative region, it will tend to dissipate energy to lower its energy, and leave the negative temperature regime. The dependence of the entropy on the average energy is shown in Fig. 7.3, together with a qualitative description of the sign of the temperature.

7.3 Identical Particles

Let us now consider in some detail what happens when we have identical particles.

Consider again a system of N particles with no interactions. The Hamiltonian is given by

$$H = \sum_{i=1}^{N} H_i(x_i, p_i). \tag{7.56}$$

The energy, E, of a macroscopic state is just the sum of the energies of the individual particles and the total wave function, ψ_E, factorizes as the product of the individual particle states. These states satisfy Schrödinger's equation of a single particle:

$$H_i u_{\epsilon_i} = \epsilon_i u_{\epsilon_i}. \tag{7.57}$$

where u_{ϵ_i} and ϵ_i are the particle i wave function and energy, respectively. However, we must also to take into account the property that the particles are identical and therefore indistinguishable, as expected from microscopic particles.

Consider that we perform a permutation of particles i and j twice in succession:

$$P_{ij} P_{ij} \psi = \psi. \tag{7.58}$$

This result holds since we expect that $P_{ij}^2 = I$. Therefore, the eigenvalues of the permutation operator of two particles are ± 1:

$$P_{ij} \psi = \pm \psi. \tag{7.59}$$

The eigenvalue $+1$ implies that the wave function is symmetric to the exchange of particles i and j, and in the case of the eigenvalue -1, the wave function is antisymmetric. Assume now that the same holds for any two particles in the system. We may now consider that the wave function is either fully symmetric or fully antisymmetric to the exchange of any two particles:

$$\begin{aligned} \psi_S &= \sum_P P \psi_E \\ \psi_A &= \sum_P \delta_P P \psi_E, \end{aligned} \tag{7.60}$$

where the factor $\delta_P = 1$ is the number of permutations is even and $\delta_P = -1$ if it is odd. An alternative way to write the fully antisymmetric wave function is particularly enlightening. It can be seen that it may be written as a determinant:

$$\psi_A = det \begin{pmatrix} u_1(1) & u_1(2) & \cdots & u_1(N) \\ u_2(1) & u_2(2) & \cdots & u_2(N) \\ \cdots & \cdots & \cdots & \cdots \\ u_N(1) & u_N(2) & \cdots & u_N(N) \end{pmatrix}. \tag{7.61}$$

This is called the Slater determinant. Since if two lines or two columns of the determinant are proportional it vanishes, it follows that one cannot have two particles in the same state. This is a way to express Pauli's exclusion principle for fermions, which states that no two particles may occupy the same single-particle state. Therefore, the antisymmetric wave function is a natural representation for fermionic particles (like electrons). The symmetric wave function is the representation for the other class of particles commonly found in nature, called bosons.

If the particles are identical, the state of the system is actually specified by the number of particles on each single-particle state, that is, by the set $\{n_i\}$, where n_i is the number of particles in state i with energy ϵ_i. If the states are nondegenerate, we may identify the states with the single-particle energies with the restrictions:

$$\sum_i n_i = N$$

$$\sum_i n_i \epsilon_i = E. \tag{7.62}$$

We have to sum over all the permutations of the particles. In classical mechanics each permutation is considered as physically distinct (distinguishable particles). There are $N!/(n_1!n_2!\cdots)$ different ways to distribute the N particles by $n_1, n_2, \cdots$ groups of particles. The Gibbs recipe $1/N!$ in the permutations of the particles gives the statistical weight of each state as

$$W_c\{n_i\} = \frac{1}{n_1!n_2!\cdots}. \tag{7.63}$$

But if the particles are identical (indistinguishable), this weight should be one, since all the permutations should correspond to the same state:

$$W_q\{n_i\} = 1. \tag{7.64}$$

In the case of bosons, the occupation numbers of each state can take any values, $n_i = 0, 1, 2, \cdots$, while for fermions $n_i = 0, 1$.

7.3.1 Real Space Density Matrix

The density matrix operator in the canonical ensemble is given by

$$\rho = \frac{e^{-\beta H}}{Z_N(\beta)}. \tag{7.65}$$

Consider now matrix elements of the density matrix in real space:

$$\langle \mathbf{R}_1, \cdots, \mathbf{R}_N|\rho|\mathbf{R}'_1, \cdots, \mathbf{R}'_N\rangle = \frac{1}{Z_N(\beta)}\langle \mathbf{R}_i|e^{-\beta H}|\mathbf{R}'_i\rangle. \tag{7.66}$$

The partition function may be written as

$$Z_N(\beta) = Tr\left(e^{-\beta H}\right) = \int d^{3N}R\langle \mathbf{R}_1, \cdots, \mathbf{R}_N|e^{-\beta H}|\mathbf{R}_1, \cdots, \mathbf{R}_N\rangle. \tag{7.67}$$

Let us use a simplified notation where $(1, \cdots, N) = (\mathbf{R}_1, \cdots, \mathbf{R}_N)$.

Considering that the particles are free, they only carry kinetic energy, and using periodic boundary conditions, we can write that the single-particle eigenfunctions may be labeled by the momentum, $\mathbf{k}$, as

$$u_{\mathbf{k}}(\mathbf{R}) = \frac{1}{\sqrt{V}}e^{i\mathbf{k}\cdot\mathbf{R}} \tag{7.68}$$

with

$$\mathbf{k} = \frac{2\pi}{V^{1/3}}\mathbf{n}$$

with V the volume of the system and $n_i = 0, \pm 1, \pm 2, \cdots$. The many-body function of the non-interacting system of N particles, for a given set of momenta of the individual particles, $\mathbf{K}$, can be written as

$$\psi_{\mathbf{K}}(1, \cdots, N) = \frac{1}{\sqrt{N!}}\sum_P \delta_P P\{u_{k_1}(1)\cdots u_{k_N}(N)\}. \tag{7.69}$$

Here we define $\delta_P = 1$ for bosons and $\delta_P = (\pm 1)^{[P]}$, where $[P]$ is the order of the permutation, P. The permutation operator acts in the same way either in the position vectors or the momentum vectors of each particle. So we may fix the momenta and just permutate the position vectors of the particles. This leads to

$$\psi_{\mathbf{K}}(1, \cdots, N) = \frac{1}{\sqrt{N!}}\sum_P \delta_P\{u_{k_1}(P1)\cdots u_{k_N}(PN)\}. \tag{7.70}$$

When calculating the matrix elements of the density matrix, we need to sum over the configurations of the individual momenta. It simplifies the calculations to replace the sums by integrals taking the continuum limit as $\sum_{\mathbf{k}} \to \frac{V}{(2\pi)^3}\int d^3k$. The integrals can be carried out exactly since they are Gaussian integrals.

We obtain:

$$\langle 1, \cdots, N|e^{-\beta H}|1', \cdots, N'\rangle = \frac{1}{N!}\left(\frac{m}{2\pi\beta\hbar^2}\right)^{\frac{3N}{2}} \sum_P \delta_P f(P1-1')\cdots f(PN-N'), \tag{7.71}$$

with

$$f(\xi) = e^{-\frac{m}{2\beta\hbar^2}\xi^2}. \tag{7.72}$$

We may now define the thermal wavelength as

$$\lambda = \frac{h}{(2\pi m k_B T)^{1/2}} = \hbar\left(\frac{2\pi\beta}{m}\right)^{1/2}. \tag{7.73}$$

In terms of this quantity, we can rewrite that

$$f(\xi) = e^{-\pi\left(\frac{\xi}{\lambda}\right)^2}. \tag{7.74}$$

The diagonal matrix element of the density matrix is given by

$$\langle \mathbf{R}_1, \cdots, \mathbf{R}_N|e^{-\beta H}|\mathbf{R}_1, \cdots, \mathbf{R}_N\rangle = \frac{1}{N!\lambda^{3N}} \sum_P \delta_P f(P\mathbf{R}_1-\mathbf{R}_1)\cdots f(P\mathbf{R}_N-\mathbf{R}_N), \tag{7.75}$$

where

$$f(\mathbf{R}) = e^{-\pi\frac{R^2}{\lambda^2}}. \tag{7.76}$$

If $P\mathbf{R}_i = \mathbf{R}_i$ (meaning no permutation), we get that $f(0) = 1$. Writing $f_{ij} = f(\mathbf{R}_i - \mathbf{R}_j)$, we see that

$$\sum_P \delta_P \cdots \to 1 \pm \sum_{i<j} f_{ij} f_{ji} + \sum_{i<j<k} f_{ij} f_{jk} f_{ki} \pm \cdots \tag{7.77}$$

In the second term, there is a permutation, in the third two permutations and so on.

If the average distance between the particles is large compared to the thermal wave length $>> \lambda$, which implies that $(V/N)^{1/3} >> \lambda$, we get that $1 >> \lambda n^{1/3}$ and we may approximate $\sum_P \delta_P \sim 1$. In this case we get the classical result

$$Z_N = Tr\left(e^{-\beta H}\right) \sim \frac{1}{N!\lambda^{3N}} \int d^{3N} R = \frac{1}{N!}\left(\frac{V}{\lambda^3}\right)^N. \tag{7.78}$$

In this classical limit

$$\langle \mathbf{R}_1, \cdots, \mathbf{R}_N | e^{-\beta H} | \mathbf{R}_1, \cdots, \mathbf{R}_N \rangle \sim \left(\frac{1}{V} \right)^N . \tag{7.79}$$

The result is just the Nth power of the single-particle result, which indicates that there are no correlations between the particles. In general, there are quantum correlations.

7.3.2 Quantum Correlations

Consider as a simple example, $N = 2$. The previous general expressions simplify to

$$\sum_P \delta_P \cdots = 1 \pm (f(\mathbf{R}_{12}))^2 . \tag{7.80}$$

The diagonal matrix element is given by

$$\langle \mathbf{R}_1 \mathbf{R}_2 | e^{-\beta H} | \mathbf{R}_1 \mathbf{R}_2 \rangle = \frac{1}{2\lambda^6} \left(1 \pm e^{-2\pi \frac{R_{12}^2}{\lambda^2}} \right) . \tag{7.81}$$

The two-particle partition function is given by

$$\begin{aligned} Z_2 &= \frac{1}{2\lambda^6} \int d^3 R_1 \int d^3 R_2 \left(1 \pm e^{-2\pi \frac{R_{12}^2}{\lambda^2}} \right) \\ &= \frac{1}{2} \left(\frac{V}{\lambda^3} \right)^2 \left(1 \pm \frac{1}{V} \int_0^\infty e^{-2\pi \frac{R^2}{\lambda^2}} 4\pi R^2 dR \right) \\ &= \frac{1}{2} \left(\frac{V}{\lambda^3} \right)^2 \left(1 \pm \frac{1}{2^{3/2}} \frac{\lambda^3}{V} \right) , \end{aligned} \tag{7.82}$$

where we used spherical coordinates. Since $\lambda^3 / V << 1$, we get that

$$Z_2 \sim \frac{1}{2} \left(\frac{V}{\lambda^3} \right)^2 . \tag{7.83}$$

Therefore,

$$\langle \mathbf{R}_1 \mathbf{R}_2 | \rho | \mathbf{R}_1 \mathbf{R}_2 \rangle = \frac{1}{V^2} \left(1 \pm e^{-2\pi \frac{R_{12}^2}{\lambda^2}} \right) . \tag{7.84}$$

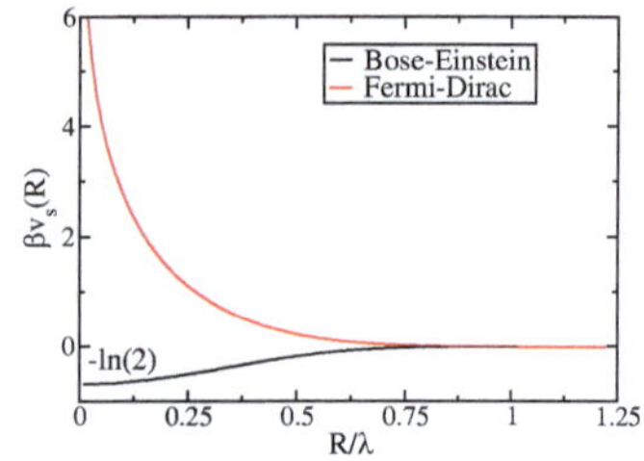

Fig. 7.4 Statistical potential for bosons and fermions as a function of the relative distance between identical particles

If $R_{12} \sim \lambda$, this quantum result is quite different from the classical result $1/V^2$. The plus sign is the proper case for two bosons and the negative sign for fermions. Therefore, the correlation is enhanced for bosons and depleted for fermions. We may define a so-called statistical potential v_S as

$$e^{-\beta v_S} = 1 \pm e^{-2\pi \frac{R^2}{\lambda^2}}. \tag{7.85}$$

As shown in Fig. 7.4, the statistical potential is repulsive in the case of fermions and attractive in the case of bosons.

7.3.3 *Ideal Quantum Gases*

Let us now revisit the discussion of a quantum gas of identical particles, incorporating the various quantum effects. Recall that the partition function is given by

$$Z_N(V,T) = \sum_n e^{-\beta E_n}.$$

The energy of the system is given by $E = \sum_\epsilon n_\epsilon \epsilon$, where n_ϵ is the number of particles with the single-particle energy ϵ. As before $\sum_\epsilon n_\epsilon = N$, the total number of particles. In terms of the occupation numbers, the partition function can be written as

$$Z_N(V,T) = \sum_{\{n_\epsilon\}}{}' g\{n_\epsilon\} e^{-\beta \sum_\epsilon n_\epsilon \epsilon}. \tag{7.86}$$

Here $g\{n_\epsilon\}$ is the statistical weight, and the prime in the summation is a reminder that the sum is restricted, since the total number of particles is N. For bosons the statistical weight is one and for fermions is one if $n_\epsilon = 0, 1$ and is zero otherwise. Therefore, both for bosons and for fermions, the partition function can be written as

$$Z_N(V,T) = \sum_{\{n_\epsilon\}}{}' e^{-\beta \sum_\epsilon n_\epsilon \epsilon}.$$

This is complicated to calculate due to the restriction on the sum.

It would be naturally easier to calculate if the restriction were absent. This is what happens naturally if we calculate instead the grand partition function. We can therefore consider instead the grand canonical ensemble (recall that the thermodynamic averages are independent of the ensemble in the large system limit). We consider therefore the grand partition function:

$$\begin{aligned}\mathcal{Z}(z, V, T) &= \sum_{N=0}^{\infty}\left[z^N \sum_{\{n_\epsilon\}}{}' e^{-\beta \sum_\epsilon n_\epsilon \epsilon}\right] \\ &= \sum_{N=0}^{\infty}\left[\sum_{\{n_\epsilon\}}{}' \prod_\epsilon (ze^{-\beta\epsilon})^{n_\epsilon}\right].\end{aligned} \tag{7.87}$$

We have to sum over all the occupation numbers obeying the restriction $\sum_\epsilon n_\epsilon = N$, and then to sum over N is the same as summing over all values of $n_0, n_1, n_2, \ldots$. Therefore, we can rewrite the grand partition function in a more convenient way:

$$\begin{aligned}\mathcal{Z}(z, V, T) &= \sum_{n_0, n_1, n_2, \ldots}\left[(ze^{-\beta\epsilon_0})^{n_0}(ze^{-\beta\epsilon_1})^{n_1}\ldots\right] \\ \mathcal{Z}(z, V, T) &= \left[\sum_{n_0}(ze^{-\beta\epsilon_0})^{n_0}\right]\left[\sum_{n_1}(ze^{-\beta\epsilon_1})^{n_1}\right]\ldots\end{aligned} \tag{7.88}$$

If the particles are bosons, there are no restrictions on the occupation numbers of the available single-particle states, and taking $n_\epsilon = 0, 1, 2, \ldots$ we obtain:

$$\mathcal{Z}(z, V, T) = \prod_\epsilon \frac{1}{1 - ze^{-\beta\epsilon}}. \tag{7.89}$$

In the case of fermions, the set of occupation numbers is restricted to $n_\epsilon = 0, 1$. Therefore, the grand partition function is given by

$$\mathcal{Z}(z, V, T) = \prod_\epsilon (1 + ze^{-\beta\epsilon}). \tag{7.90}$$

The thermodynamics can be obtained from the knowledge of the grand partition function. For instance, recall that the grand potential $q(z, V, T)$ is given by

$$\begin{aligned}q(z, V, T) &= \frac{pV}{k_B T} \\ &= \ln \mathcal{Z}(z, V, T) = \mp \sum_\epsilon \ln(1 \mp ze^{-\beta\epsilon}).\end{aligned} \tag{7.91}$$

From the expression for the grand potential, we may rederive the various single-particle distributions, namely, the Bose-Einstein and the Fermi-Dirac distribution. Defining $a = \mp 1$, where the upper sign is for bosons and the lower one for fermions, we obtain that

$$
\begin{aligned}
N &= z\left(\frac{\partial q}{\partial z}\right)_{V,T} = \sum_{\epsilon} \frac{1}{z^{-1}e^{\beta\epsilon} + a} \\
U &= -\left(\frac{\partial q}{\partial \beta}\right)_{z,V} = \sum_{\epsilon} \frac{\epsilon}{z^{-1}e^{\beta\epsilon} + a}.
\end{aligned} \tag{7.92}
$$

This implies that

$$
\langle n_\epsilon \rangle = \frac{1}{z^{-1}e^{\beta\epsilon} + a} = \frac{1}{e^{\beta(\epsilon-\mu)} + a} \tag{7.93}
$$

where we recognize the Bose-Einstein and the Fermi-Dirac distribution functions for the single- particle thermal occupations.

We will show that in the case of bosons we can write that

$$
\begin{aligned}
\frac{p}{k_B T} &= \frac{1}{\lambda^3} g_{5/2}(z) \\
\frac{N - N_0}{V} &= \frac{1}{\lambda^3} g_{3/2}(z).
\end{aligned} \tag{7.94}
$$

where N_0 is the number of bosons in the lowest energy single-particle state and

$$
g_n(z) = \frac{1}{\Gamma(n)} \int_0^\infty dx \frac{x^{n-1}}{z^{-1}e^x - 1},
$$

where $\Gamma(n)$ is the gamma function. Due to the boson statistics, the number of bosons in a given state is unrestricted, which leads to a large occupation of the lowest energy single-particle state. At zero temperature one expects that all particles occupy this state, typically a zero-momentum state since the energy is fully kinetic. In the case of fermions, one finds:

$$
\begin{aligned}
\frac{p}{k_B T} &= \frac{g}{\lambda^3} f_{5/2}(z) \\
\frac{N}{V} &= \frac{g}{\lambda^3} f_{3/2}(z).
\end{aligned} \tag{7.95}
$$

Here g is the spin degeneracy and the functions are defined as

$$
f_n(z) = \frac{1}{\Gamma(n)} \int_0^\infty dx \frac{x^{n-1}}{z^{-1}e^x + 1}
$$

It will prove useful to consider the regimes where the fugacity is small. In the regime where z is small, we may find that

$$g_n(z) = \sum_{l=1}^{\infty} \frac{z^l}{l^n}, \tag{7.96}$$

for the case of bosons and

$$f_n(z) = z - \frac{z^2}{2^n} + \frac{z^3}{3^n} - \cdots, \tag{7.97}$$

for the case of fermions.

As concluded before, the single-particle distributions for bosons and fermions lead to rather different behaviors, particularly at low temperatures or large densities. Also, the equations of state, $p/(k_B T)$, and the number density are rather different between the two types of quantum gases. The case of bosons at low temperatures is particularly dominated by the large (macroscopic) occupation of the lowest energy state that will lead to the phenomenon of Bose-Einstein condensation, a phase transition that takes place solely via the effect of the statistical attractive interaction. On the other hand, in the case of fermions, the Pauli exclusion principle takes center stage leading to the existence of the Fermi surface and impenetrable low energy single-particle states. In spite of these significant differences, we will see that, while the expressions for pressure and average energy are different between bosons and fermions, they are related in the same way and in the same way as for a classical system. Specifically we will show that

$$p = \frac{2}{3}\frac{U}{V}, \tag{7.98}$$

as expected of a free system of massive particles. Incidently, we will also show that for photons

$$p = \frac{1}{3}\frac{U}{V}, \tag{7.99}$$

as expected of a free system of massless particles.

In the following, we will specify these results for the boson and fermion gases and also consider the photon gas.

7.4 Boson Gas

The thermodynamics of the Bose gas is contained in the equations:

$$
\begin{aligned}
\frac{pV}{k_B T} &= \ln \mathcal{Z} = -\sum_\epsilon \ln\left(1 - z e^{-\beta\epsilon}\right) \\
N &= \sum_\epsilon \langle n_\epsilon \rangle = \sum_\epsilon \frac{1}{z^{-1} e^{\beta\epsilon} - 1} \\
z &= e^{\mu/(k_B T)}.
\end{aligned} \tag{7.100}
$$

Let us consider a particle confined in a volume V. Let us call $\Sigma(P)$ the number of states accessible to that particle with momentum $p \le P$. Then

$$
\Sigma(P) \sim \int_{p \le P} \int \frac{d^3q\, d^3p}{h^3} = \frac{V}{h^3} \frac{4}{3} \pi P^3.
$$

The number of states between p and $p + dp$ is then

$$
g(p) dp = \frac{d\Sigma(p)}{dp} dp \sim \frac{V}{h^3} 4\pi p^2 dp.
$$

Using that $E = p^2/2m$, we can write that

$$
\Sigma(E) \sim \frac{V}{h^3} \frac{4}{3} \pi (2mE)^{3/2}.
$$

Therefore, the number of states between ϵ and $\epsilon + d\epsilon$ is

$$
a(\epsilon) d\epsilon = \frac{d\Sigma(\epsilon)}{d\epsilon} d\epsilon \sim \frac{V}{h^3} 2\pi (2m)^{3/2} \epsilon^{1/2} d\epsilon.
$$

Then, we can replace:

$$
\sum_\epsilon \to \int d\epsilon\, a(\epsilon) = \int d\epsilon \frac{2\pi V}{h^3} (2m)^{3/2} \epsilon^{1/2}.
$$

In this discussion there is however a subtle point. As we go to a continuous description of the accessible levels, each state now has a zero weight (while it should have a weight one). We will see that the state of $\epsilon = 0$ has important physical consequences. We have therefore to separate this term before taking the continuum limit; otherwise, its special role will be lost. We should actually write that

$$
\begin{aligned}
\frac{p}{k_B T} &= -\frac{1}{V} \ln(1 - z) - \frac{2\pi}{h^3} (2m)^{3/2} \int_0^\infty d\epsilon\, \epsilon^{1/2} \ln(1 - z e^{-\beta\epsilon}) \\
\frac{N}{V} &= \frac{1}{V} \frac{z}{1 - z} + \frac{2\pi}{h^3} (2m)^{3/2} \int_0^\infty d\epsilon\, \epsilon^{1/2} \frac{1}{z^{-1} e^{\beta\epsilon} - 1}.
\end{aligned} \tag{7.101}
$$

If $z << 1$, the first terms are negligible. But if $z \to 1$, the same terms are diverging and therefore cannot be neglected: they are actually terms with an important physical significance.

Let us call N_0 the number of particles in the single-particle ground state with $\epsilon = 0$. Then

$$N_0 = \frac{z}{1-z}.$$

When $z \to 1$ the number of particles $N_0 \to N$ until all the particles of the system are used up. There is a significative fraction of the total number of particles of the system occupying a single quantum state. Clearly, this is only possible since the particles are bosons and there is no restriction on the number of particles that may occupy any state. This phenomenon of a macroscopic occupation of the particles on a single one-particle state is called Bose-Einstein condensation.

Since $N_0 = z/(1-z)$, we conclude that $z = N_0/(N_0+1)$. Therefore the term

$$-\frac{1}{V}\ln(1-z) = -\frac{1}{V}\ln\frac{1}{N_0+1} \sim \frac{1}{V}\ln N_0$$

is negligible when the number of particles (and the volume) is very large. We are left then with the set of equations:

$$\begin{aligned}\frac{p}{k_B T} &= -\frac{2\pi(2mk_B T)^{3/2}}{h^3}\int_0^\infty dx x^{1/2}\ln(1-ze^{-x}) = \frac{1}{\lambda^3}g_{5/2}(z)\\ \frac{N-N_0}{V} &= \frac{2\pi(2mk_B T)^{3/2}}{h^3}\int_0^\infty dx\frac{x^{1/2}}{z^{-1}e^x-1} = \frac{1}{\lambda^3}g_{3/2}(z).\end{aligned} \tag{7.102}$$

Here

$$\lambda = \frac{h}{\sqrt{2\pi mk_B T}},$$

$$g_n(z) = \frac{1}{\Gamma(n)}\int_0^\infty dx\frac{x^{n-1}}{z^{-1}e^x-1},$$

where $\Gamma(n)$ is the gamma function. Note that in the first equation the function $g_{5/2}(z)$ is obtained integrating by parts.[2]

The average energy of the boson gas can be obtained using Eq. (7.92):

$$U = -\left(\frac{\partial q}{\partial\beta}\right)_{z,V} = \frac{3}{2}k_B T\frac{V}{\lambda^3}g_{5/2}(z). \tag{7.103}$$

[2] Note that $g_n(z) = Li_n(z)$, the polylog function.

This means that we obtain a relation which is familiar from the classical ideal gas and the semiclassical treatment that, for free particles with mass,

$$p = \frac{2}{3}\frac{U}{V}. \tag{7.104}$$

Consider now the case when $z << 1$, which implies that $N_0 << N$. Using expansion Eq. (7.96), we get that[3]

$$\frac{pv}{k_B T} = \sum_{l=1}^{\infty} a_l \left(\frac{\lambda^3}{v}\right)^{l-1},$$

with $v = V/N$. Also, we can get that the specific heat is given by

$$\begin{aligned}\frac{C_V}{Nk_B} &= \frac{3}{2}\sum_{l=1}^{\infty} \frac{5-3l}{2} a_l \left(\frac{\lambda^3}{v}\right)^{l-1} \\ &= \frac{3}{2}\left[1 + 0.0884\left(\frac{\lambda^3}{v}\right) + 0.0066\left(\frac{\lambda^3}{v}\right)^2 + \cdots\right].\end{aligned}$$

In the infinite temperature limit ($\lambda \to 0$), we get the classical result that $p \to nk_B T$, and $C_V \to \frac{3}{2}Nk_B$.

7.4.1 Bose-Einstein Condensation

The low temperature limit is more interesting. In this regime λ^3/v is large and, therefore, the expansion used above is not useful. Let us define $N_e = N - N_0$, the number of particles in the excited states. Then

$$N_e = V\frac{(2\pi mk_B T)^{3/2}}{h^3} g_{3/2}(z).$$

Since the average number of particles in a given one-particle state of energy $\epsilon \geq 0$, given by the Bose-Einstein distribution, cannot be negative, we conclude that $z \leq 1$. The maximum value is therefore $z = 1$. The function g_n takes, in this limit, the value

$$g_n(z=1) = \sum_{l=1}^{\infty} \frac{1}{l^n} = \zeta(n),$$

[3] The first coefficients of the expansion are $a_1 = 1$, $a_2 = -0.17678$, $a_3 = -0.0030$, and $a_4 = -0.00011$.

where ζ is the Riemann zeta function. In particular $\zeta(3/2) \sim 2.612$ and $\zeta(5/2) \sim 1.341$. Since the function $g_n(z)$ is monotonically increasing with z, then

$$g_{3/2}(z) \leq g_{3/2}(z=1) \leq \zeta\left(\frac{3}{2}\right).$$

This imposes a restriction on the number of particles that may occupy the excited states, at a certain temperature:

$$N_e \leq V\frac{(2\pi m k_B T)^{3/2}}{h^3}\zeta\left(\frac{3}{2}\right). \tag{7.105}$$

If the total number of particles in the system is $N > N_e^{max}$, the extra particles $N - N_e^{max} = N_0$ have to occupy the ground state, which must have a macroscopic occupation. The condition for Bose-Einstein condensation can then be expressed, keeping N and V constants, as a condition on the temperature. That is, defining a critical temperature, T_c, as the temperature when the macroscopic occupation takes place, we see that, for $T < T_c$, the system is in a different phase. Clearly, the critical temperature is given by

$$T_c = \frac{h^2}{2\pi m k_B}\left(\frac{N}{V\zeta(\frac{3}{2})}\right)^{2/3}. \tag{7.106}$$

For $T < T_c$ the system is divided into two phases: a gas phase with N_e particles and a condensed phase with N_0 particles. For $T < T_c$, the fugacity is constant and given by $z = 1$. Fixing the temperature at the critical point, the Bose-Einstein condensation occurs for a given volume per particle:

$$v_c = \frac{\lambda^3}{\zeta(3/2)} \tag{7.107}$$

(which means that this volume is proportional to $T^{-3/2}$).

7.4.2 Pressure: Specific Heat

Recall that the equation of state is

$$\frac{p}{k_B T} = \frac{1}{\lambda^3}g_{5/2}(z). \tag{7.108}$$

Therefore, since $g_{5/2}(1) = \zeta(5/2)$, we obtain that the pressure, in the condensed phase, $T < T_c$, is given by

$$p(T) = \frac{k_B T}{\lambda^3}\zeta(5/2) = p(T_c)\left(\frac{T}{T_c}\right)^{5/2} = 0.5134 N_e \frac{k_B T}{V}.$$

The pressure at T_c is approximately half the pressure of a classical gas at the same temperature. Also, we conclude that the condensed phase does not exert pressure, and the particles in the excited states are approximately half effective, in comparison with an ideal classical gas. Note that in the regime $T < T_c$, the pressure is independent of volume.

At temperatures higher than the critical point, $T > T_c$, we can write that

$$p(T) = \frac{N}{V} k_B T \frac{g_{5/2}(z)}{g_{3/2}(z)}. \tag{7.109}$$

Using that for $z < 1$

$$g_{3/2}(z) = \frac{\lambda^3}{v} = \frac{N}{V}\lambda^3, \tag{7.110}$$

this expression determines z through

$$g_{3/2}(z) = \frac{N}{V}\frac{h^3}{(2\pi m k_B T)^{3/2}}. \tag{7.111}$$

Returning to the temperature regime in the condensed phase, $T \leq T_c$, we can now obtain an expression for the specific heat:

$$\begin{aligned}\frac{C_V}{N k_B} &= \frac{3}{2}\frac{N}{V}\zeta(5/2)\frac{d}{dT}\left(\frac{T}{\lambda^3}\right)\\ &= \frac{15}{4}\zeta(5/2)\frac{v}{\lambda^3},\end{aligned} \tag{7.112}$$

which means that the specific heat has the low temperature dependence

$$C_V \sim T^{3/2}. \tag{7.113}$$

At the critical point, we get that

$$\frac{C_V(T_c)}{N k_B} = \frac{15}{4}\frac{\zeta(5/2)}{\zeta(3/2)} = 1.925 > \frac{3}{2}, \tag{7.114}$$

where $3/2$ is the classical result. For temperatures above the critical point, the result can be expressed using Eq. (7.103) leading to

$$\frac{C_V(T_c)}{N k_B} = \frac{15}{4}\frac{g_{5/2}(z)}{g_{3/2}(z)} - \frac{9}{4}\frac{g_{3/2}(z)}{g_{1/2}(z)}. \tag{7.115}$$

The specific heat vanishes at zero temperature, which grows and reaches its maximal (finite) value at the critical temperature and then decreases to its asymptotic, finite value, at large temperatures. Note that experimentally a simple bosonic system, such as liquid helium (He^4), displays a specific heat that does not fully agree with the theoretical expectation. There is a transition point at around 2.19 Kelvin indicative of a Bose condensation. However, the specific heat at the transition point actually diverges, unlike the boson gas where it has a finite value. As will be seen later, this indicates that the real system is not actually a gas, and interactions between the particles, however small, must be taken into account.

The expressions for the various quantities can be rewritten in terms of the reduced temperature, $t = T/T_c$ or plotted as function of t using the relation between the fugacity and temperature for $t > 1$. It is natural to separate the regimes above and below the critical temperature.

A convenient expression that relates the fugacity and temperature is

$$\left(\frac{T}{T_c}\right)^{3/2} = \frac{\zeta(3/2)}{g_{3/2}(z)}. \tag{7.116}$$

This expression is shown in Fig. 7.5, together with the behavior for $t < 1$, for which $z = 1$.

The occupation of the ground state and the excited states may be written as

$$\begin{aligned}\frac{N_0}{N} &= 1 - \left(\frac{T}{T_c}\right)^{3/2}, \\ \frac{N_e}{N} &= \left(\frac{T}{T_c}\right)^{3/2}.\end{aligned} \tag{7.117}$$

Their temperature dependence is shown in Fig. 7.5 as well. In the regime when $t > 1$, the number of particles in the ground state is negligible, as discussed above.

The specific heat when $T > T_c$ is written as in Eq. (7.115). When $T < T_c$, it may be written as

$$\frac{C_v}{Nk_B} = \frac{15}{4}\frac{\zeta(5/2)}{\zeta(3/2)}\left(\frac{T}{T_c}\right)^{3/2} \tag{7.118}$$

This is also shown in the same figure.

We may also consider the pressure and rewrite its temperature dependence as

$$\frac{pv}{k_B T} = \frac{\zeta(5/2)}{\zeta(3/2)}\left(\frac{T}{T_c}\right)^{3/2} \tag{7.119}$$

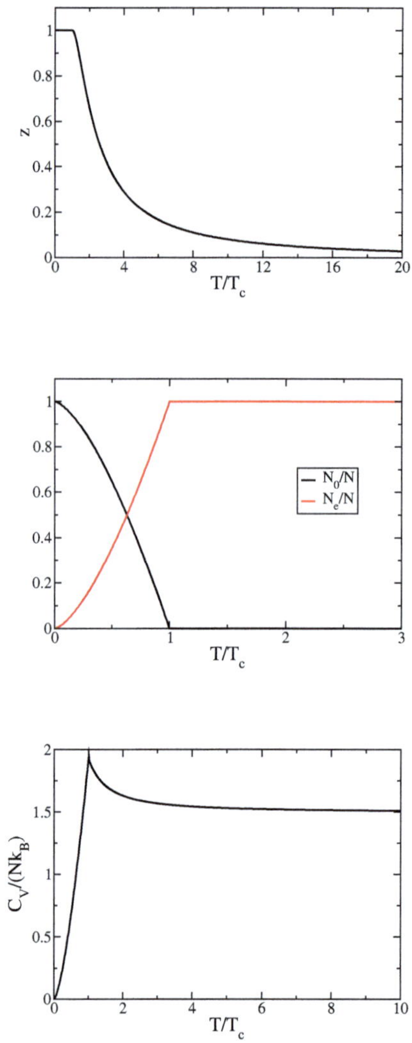

Fig. 7.5 (**a**) Fugacity as a function of the reduced temperature T/T_c. (**b**) Occupation numbers of the ground state and excited states as a function of t. (**c**) Specific heat as a function of the reduced temperature

for $T < T_c$ and

$$\frac{pv}{k_B T} = \frac{g_{5/2}(z)}{g_{3/2}(z)} \tag{7.120}$$

for $T > T_c$. The results are shown in Fig. 7.6.

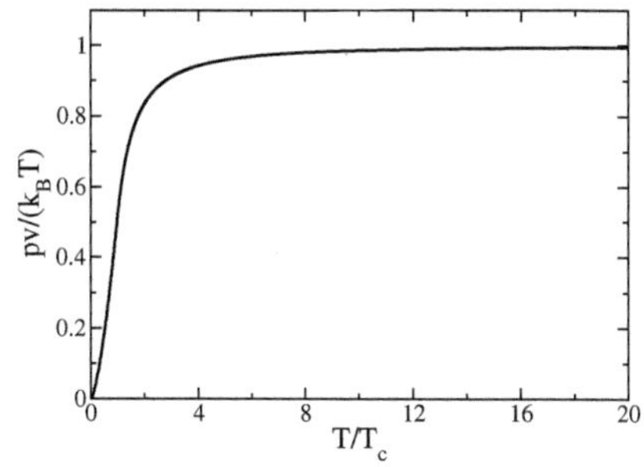

Fig. 7.6 Pressure $(pv/(k_BT))$ as a function of the reduced temperature T/T_c

7.5 Photon Gas

A photon of frequency ν has an energy $\epsilon = h\nu$. The energy of a mode of the electromagnetic field of frequency ν composed of n_ν photons is therefore $E_\nu = n_\nu h\nu$. The total energy of the field is

$$E = \sum_\nu n_\nu h\nu. \tag{7.121}$$

If the system is isolated, the total energy is conserved, but the number of phonons is not conserved, in general. Therefore, the total number of photons, N,

$$N = \sum_\nu n_\nu \tag{7.122}$$

is not fixed. The distribution that gives the number of photons per energy state is Planck's distribution (Bose-Einstein distribution with zero chemical potential, since the number of photons is not fixed):

$$\langle n_\nu \rangle = \frac{1}{e^{\beta h\nu} - 1}. \tag{7.123}$$

The average number of photons in a small volume in phase space d^3xd^3p is

$$d^6N = \langle n_\nu \rangle \frac{d^3xd^3p}{h^3} = \frac{1}{e^{\frac{h\nu}{k_BT}} - 1} \frac{d^3xd^3p}{h^3}. \tag{7.124}$$

Integrating over the volume of real space, we get the number of photons with momentum between $\mathbf{p}$ and $\mathbf{p} + d\mathbf{p}$:

$$\int_V d^6N = \frac{1}{e^{\frac{h\nu}{k_BT}} - 1} \frac{V}{h^3} d^3p. \tag{7.125}$$

Let us define a distribution $f(\mathbf{p})d^3p$ as the number of photons per unit volume, a given momentum, $\mathbf{p}$, and a given polarization. Therefore,

$$f(\mathbf{p}) = \frac{1}{e^{\frac{h\nu}{k_BT}} - 1}\frac{1}{h^3}. \tag{7.126}$$

Using that $\mathbf{p} = \hbar\mathbf{k}$, we get that

$$f(\mathbf{k}) = \frac{1}{e^{\frac{h\nu}{k_BT}} - 1}\frac{1}{(2\pi)^3}. \tag{7.127}$$

Defining $F(k) = 4\pi k^2 f(\mathbf{k})$, and using that $k = 2\pi\nu/c$ which leads to

$$k^2dk = \left(\frac{2\pi}{c}\right)^3 \nu^2 d\nu,$$

we may define a new distribution $n_\nu d\nu$, the average number of photons per unit volume, with frequencies between ν and $\nu + d\nu$, which is given by

$$n_\nu = \frac{1}{e^{\frac{h\nu}{k_BT}} - 1}\frac{8\pi\nu^2}{c^3}. \tag{7.128}$$

The extra factor of 2 appears since there are two polarizations. The energy density for a given frequency is the density of photons times the energy of each photon. It is given by

$$u_\nu = n_\nu h\nu = \frac{8\pi h\nu^3}{c^3}\frac{1}{e^{\frac{h\nu}{k_BT}} - 1}. \tag{7.129}$$

This is called Planck's law for the energy density of the photon field. Also, the results lead to Wien's laws, which state that the frequency for which the energy density is maximal obeys $\nu_m \sim T$ and the value of the distribution at that point scales as $u_{\nu_m} \sim T^3$. The results for the average number of photons and the energy density distribution are shown in Fig. 7.7.

Since the photons have no mass, we have obtained before that the pressure of the photon gas satisfies $p = u/3$ where u is the total energy density integrated over all frequencies:

$$u = \int_0^\infty d\nu u_\nu. \tag{7.130}$$

Performing a change of variable $x = h\nu/(k_BT)$, we obtain that

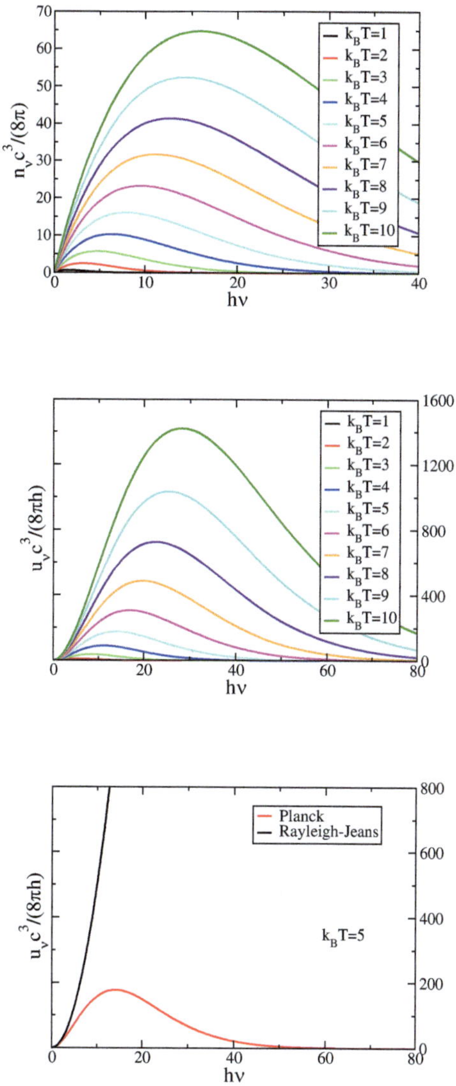

Fig. 7.7 (**a**) Average number of photons, (**b**) average energy density as a function of frequency, ν, for different temperatures. In (**c**) we compare the Planck (quantum case) and the Rayleigh-Jeans (classical case) distributions for a given temperature

$$\begin{aligned} u(T) &= \frac{8\pi h}{c^3}\left(\frac{k_B T}{h}\right)^4 \int_0^\infty dx \frac{x^3 dx}{e^x - 1} \\ &= \frac{4\sigma}{c} T^4 \\ \sigma &= \frac{2\pi h}{c^2}\left(\frac{k_B}{h}\right)^4 \int_0^\infty \frac{x^3 dx}{e^x - 1}. \end{aligned} \tag{7.131}$$

Here, σ is the Stefan-Boltzmann constant. Therefore, the pressure obeys a similar law:

$$p = \frac{4}{3}\frac{\sigma}{c}T^4. \tag{7.132}$$

Experimentally, Planck's distribution was determined considering the energy distribution that is detected from a black-body system. By definition a black-body is a system that absorbs all incident energy and, therefore, the reflection coefficient vanishes. It is, however, allowed to radiate energy. One way to implement such a system is to have a system with a cavity such that there is a small opening that allows any incident energy on the opening to enter the cavity. It is then assumed that this energy remains inside the cavity, since multiple reflections it its internal walls make it unlikely that the (electromagnetic) energy, which goes through the opening, is allowed to leave the cavity. A radiation field is then established in the cavity, and it is possible then that radiation leaks out incoherently with the incident energy.

We expect that some equilibrium situation is established inside the cavity and, therefore, that the distribution of the radiated energy, detected outside the black-body, is characteristic of that equilibrium situation; specifically it is characteristic of the temperature of the cavity region, which should be the same as that of the body itself. The distribution detected is that described by Planck's distribution. In fact, Planck fitted the experimental results to a trial solution. At small frequencies, it is given by Rayleigh-Jeans distribution, which may be understood using classical arguments. At high frequencies it deviates considerably and Planck's reasoning to explain the behavior leads to the conclusion of quantized energy levels. In the classical reasoning, since the radiation field in the cavity is in equilibrium with the walls of the container, the average energy per mode should equal the average energy per mode in the material of the container. Assuming that these modes are oscillations described by harmonic oscillators (even though we know that not all energy modes in a solid are described in this way), using the equipartition theorem we know the average energy per mode as given by $1/2\,\mathrm{k_B T}$. Multiplying by the number of modes per energy interval, derived on Eq. (7.128), we get the Rayleigh-Jeans distribution.

In order to fit the experimental results, Planck realized that the average energy per mode has to be replaced by a distribution that depends both on temperature and the frequency, and this is exactly Planck's distribution for the number of modes in a given frequency (energy) interval. This implied that the energy of each mode is given by multiples, n, of $h\nu$. In Planck's interpretation these are the discrete energy levels of a (quantum) harmonic oscillator, with energies $E_n = nh\nu$, where n is the quantum number. However, returning to the comment that the energy levels of a generic material are not those of harmonic oscillators, Einstein proposed that instead are the energy modes of the radiation field that are given as multiples of $h\nu$. This introduced the concept of the photon and that each mode of the electromagnetic field is actually filled by a given number, n, of energy units, called photons, each with energy $h\nu$ and total energy $E = nh\nu$. Therefore, a given mode with a given frequency ν is actually composed of a very large number of photons. The quantization of a macroscopically occupied mode is therefore basically a continuum of energies, since Planck's constant is very small.

7.6 Fermi-Dirac Gas

The thermodynamics of the Fermi gas is obtained from

$$\frac{pV}{k_BT} = \ln \mathcal{Z} = \sum_{\epsilon} \ln(1 + ze^{-\beta\epsilon})$$

$$N = \sum_{\epsilon} \langle n(\epsilon) \rangle = \sum_{\epsilon} \frac{1}{z^{-1}e^{\beta\epsilon} + 1}. \tag{7.133}$$

As before, z is the fugacity. Note that in the case of fermions, the average number of particles in a given state is never negative and, therefore, there is no restriction on the range of values of the fugacity: $0 \leq z < \infty$. Following similar lines to those taken in the free boson gas, we obtain Eq. (7.95).[4] The average energy is obtained from

$$U = -\frac{\partial}{\partial \beta} (\ln \mathcal{Z})_{z,V} = \frac{3}{2} N k_B T \frac{f_{5/2}(z)}{f_{3/2}(z)}. \tag{7.134}$$

This implies, as for the boson gas, that

$$p = \frac{2}{3}\frac{U}{V}. \tag{7.135}$$

Using that $f_{n-1}(z) = z\frac{\partial}{\partial z} f_n(z)$, we get from the equation for the particle density that

$$\left(\frac{\partial z}{\partial T}\right)_v = -\frac{3}{2}\frac{z}{T}\frac{f_{3/2}(z)}{f_{1/2}(z)}.$$

Then, we obtain that the specific heat is given by

$$\frac{C_V}{Nk_B} = \frac{15}{4}\frac{f_{5/2}(z)}{f_{3/2}(z)} - \frac{9}{4}\frac{f_{3/2}(z)}{f_{1/2}(z)}. \tag{7.136}$$

The free energy and the entropy are given by

[4] Note that $f_n(z) = -Li_n(-z)$, the polylog function.

$$F = N\mu - PV = Nk_BT\left\{\ln z - \frac{f_{5/2}(z)}{f_{3/2}(z)}\right\}$$

$$S = \frac{U-F}{T} = Nk_B\left\{\frac{5}{2}\frac{f_{5/2}(z)}{f_{3/2}(z)} - \ln z\right\}. \tag{7.137}$$

Let us consider the case of small densities or high temperatures. In these cases we have that

$$\frac{n\lambda^3}{g} = \frac{nh^3}{g(2\pi mk_BT)^{3/2}} << 1.$$

Then in the limit of small densities, using Eq. (7.97) in Eq. (7.95), we can obtain the equation of state in powers of $n\lambda^3/g$. The leading term leads, as expected, to the results for the classical ideal gas.

The other limit of high densities or small temperatures is more interesting. Clearly, as $T \to 0$, the fraction $n\lambda^3/g \to \infty$. Also, the occupation number becomes:

$$\begin{aligned}\langle n(\epsilon)\rangle = \frac{1}{e^{\frac{\epsilon-\mu}{k_BT}}+1} &= 1, \epsilon < \mu(T=0)\\ &= 0, \epsilon > \mu(T=0).\end{aligned} \tag{7.138}$$

As we will recall later, $\mu(T=0) = \epsilon_F$ is the Fermi energy, which is the highest one-particle energy state accessible at zero temperature, since the Pauli exclusion principle prevents two fermions to be in the same state.

The density of states is the same as for bosons times a factor of g, the spin degeneracy. At this point is worth recalling that bosons have integer spin values while fermions have half-integer spins. So the minimum spin degeneracy of a fermion is $g = 2$. The total number of particles is then given by

$$\int_0^{p_F} dpa(p) = g\frac{V}{h^3}\int_0^{p_F} dpp^2 = N. \tag{7.139}$$

Then the Fermi momentum is given by

$$p_F = \left(\frac{3N}{4\pi gV}\right)^{1/3} h \tag{7.140}$$

and the Fermi energy is given by

$$\epsilon_F = \frac{p_F^2}{2m} = \left(\frac{6\pi^2 n}{g}\right)^{2/3}\frac{\hbar^2}{2m}. \tag{7.141}$$

This means that the Fermi temperature $T_F = \epsilon_F/k_B$ is a very high temperature, of the order of $10^4 - 10^5$ K, and therefore most experiments are carried out at temperatures that are small compared to the Fermi temperature. The energy of the ground state ($T = 0$) is easily calculated:

$$E(T=0) = \frac{4\pi g V}{h^3} \int_0^{p_F} \frac{p^2}{2m} p^2 dp = \frac{2\pi g V}{5mh^3} p_F^5 = \frac{3}{5} N\epsilon_F. \tag{7.142}$$

Also, the pressure of the ground state is

$$p(T=0) = \frac{2}{3}\frac{E(T=0)}{V} = \frac{2}{5} n\epsilon_F \sim n^{5/3}, \tag{7.143}$$

which is nonzero due to the exclusion principle.

7.6.1 Specific Heat

It is easy to estimate the dependence of the specific heat, before embarking in the solution of the complete expression. The argument is that only the electrons in an interval of energies of the order of $k_B T$ around the Fermi energy may be excited and contribute to the specific heat. These electrons are in a number of the order of T/T_F of the total number. They each absorb an energy of the order of $k_B T$, which means that the energy absorbed by the whole system is of the order of $(NT/T_F)k_B T$. This naturally leads to a specific heat which is of order $C_V \sim Nk_B(T/T_F)$. The first point to note is that the specific heat is linear in the temperature. Also, since in general $T << T_F$, only a small fraction of the electrons contribute to the specific heat (even though all contribute to the conductivity which remained a puzzle before the quantum theory was used to understand metals).

Of course, we can obtain a full understanding of the behavior at small temperatures of the various thermodynamic quantities, solving the above exact and complete equations. As mentioned above, in this regime the fugacity is large and the expansion in powers of z is useless. We can however expand the functions $f_n(z)$ asymptotically in powers of $1/\ln z$. They are expressed as[5]

$$f_{5/2}(z) = \frac{8}{15\sqrt{\pi}} (\ln z)^{5/2} \left(1 + \frac{5\pi^2}{8}\frac{1}{(\ln z)^2} + \cdots\right)$$

$$f_{3/2}(z) = \frac{4}{3\sqrt{\pi}} (\ln z)^{3/2} \left(1 + \frac{\pi^2}{8}\frac{1}{(\ln z)^2} + \cdots\right)$$

[5] See also [Pathria].

$$f_{1/2}(z) = \frac{2}{\sqrt{\pi}} (\ln z)^{1/2} \left(1 - \frac{\pi^2}{24} \frac{1}{(\ln z)^2} + \cdots \right).$$

Inserting the asymptotic expansion of $f_{3/2}(z)$ in the expression for the density, we get that

$$\frac{N}{V} = \frac{g}{\lambda^3} f_{3/2}(z) = \frac{4\pi g}{3} \left(\frac{2m}{h^2}\right)^{3/2} (\mu)^{3/2} \left[1 + \frac{\pi^2}{8} \left(\frac{k_B T}{\mu}\right)^2\right]. \tag{7.144}$$

The chemical potential is obtained from the fugacity as

$$k_B T \ln z = \mu = \epsilon_F \left[1 - \frac{\pi^2}{12} \left(\frac{k_B T}{\epsilon_F}\right)^2\right]. \tag{7.145}$$

We can also obtain the expansion for the energy, the pressure, and the specific heat:

$$\begin{aligned}
\frac{U}{N} &= \frac{3}{5} \epsilon_F \left[1 + \frac{5\pi^2}{12} \left(\frac{k_B T}{\epsilon_F}\right)^2 + \cdots \right] \\
p &= \frac{2}{3} \frac{U}{V} = \frac{2}{5} n \epsilon_F \left[1 + \frac{5\pi^2}{12} \left(\frac{k_B T}{\epsilon_F}\right)^2 + \cdots \right] \\
\frac{C_V}{N k_B} &= \frac{\pi^2}{2} \frac{k_B T}{\epsilon_F} + \cdots .
\end{aligned} \tag{7.146}$$

Also, we can obtain that the free energy and the entropy are given by

$$\begin{aligned}
\frac{F}{N} &= \mu - \frac{pV}{N} = \frac{3}{5} \epsilon_F \left[1 - \frac{5\pi^2}{12} \left(\frac{k_B T}{\epsilon_F}\right)^2 + \cdots \right] \\
\frac{S}{N k_B} &= \frac{\pi^2}{2} \frac{k_B T}{\epsilon_F} + \cdots
\end{aligned} \tag{7.147}$$

which leads, in the zero temperature limit, to a vanishing entropy indicative of a nondegenerate ground state, in agreement with the third law of thermodynamics.

It is easy to show that the relation between the fugacity and temperature may be written as

$$\frac{T}{T_F} = \left(\frac{\frac{4}{3\sqrt{\pi}}}{f_{3/2}(z)}\right)^{2/3}. \tag{7.148}$$

Using this relation we show in Fig. 7.8 the pressure, entropy, average energy, and specific heat as a function of temperature. In the high temperature regime, we reach

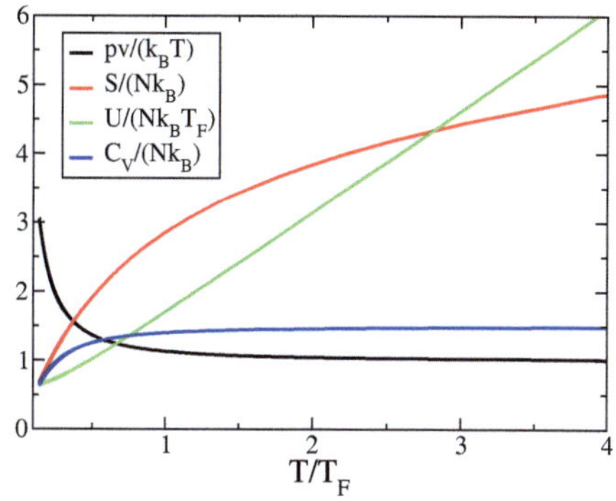

Fig. 7.8 Pressure $(pv/(k_BT))$, entropy $(S/(Nk_B))$, average energy $(U/(Nk_BT_F)$, and specific heat $(C_V/(Nk_B))$ as a function of T/T_F for a Fermi gas

the classical limit, with $pv/(k_BT) \to 1$ and the specific heat per particle reaching $3/2k_B$. At low temperatures. for instance, the entropy and specific heat reach the same behavior and $U/(Nk_BT_F) \to 3/5$, as shown in Eq. (7.142).

7.6.2 *Sommerfeld Expansion*

The dependence of the thermodynamic quantities in the low temperature limit can be obtained, in general, by another method which is useful.

Typically we have to calculate integrals of the form[6]

$$\int_{-\infty}^{\infty} d\epsilon H(\epsilon) f(\epsilon).$$

Here $f(\epsilon)$ is the Fermi function. Let us consider a class of functions such that $H(\epsilon \to -\infty) \to 0$. It is convenient to define

$$K(\epsilon) = \int_{-\infty}^{\epsilon} d\epsilon' H(\epsilon').$$

Then

$$H(\epsilon) = \frac{dK(\epsilon)}{d\epsilon}.$$

We can then write that

$$\int_{-\infty}^{\infty} d\epsilon H(\epsilon) f(\epsilon) = \int_{-\infty}^{\infty} d\epsilon \frac{dK(\epsilon)}{d\epsilon} f(\epsilon)$$

[6] See also [Kittel].

$$= \int_{-\infty}^{\infty} [\frac{d}{d\epsilon}(K(\epsilon) f(\epsilon)] - K(\epsilon)\frac{df(\epsilon)}{d\epsilon}]$$
$$= K(\epsilon) f(\epsilon)|_{-\infty}^{\infty} - \int_{-\infty}^{\infty} d\epsilon K(\epsilon)\frac{df(\epsilon)}{d\epsilon}]$$
$$= \int_{-\infty}^{\infty} d\epsilon K(\epsilon)\frac{-\partial f(\epsilon)}{\partial \epsilon}. \tag{7.149}$$

Expanding around the Fermi energy, we get that

$$K(\epsilon) = K(\mu) + \sum_{n=1}^{\infty} \left(\frac{(\epsilon-\mu)^n}{n!}\right)\left(\frac{d^n K(\epsilon)}{d\epsilon^n}\right)_{\epsilon=\mu}.$$

Then we obtain that

$$\int_{-\infty}^{\infty} d\epsilon H(\epsilon) f(\epsilon)$$
$$= \int_{-\infty}^{\infty} d\epsilon K(\epsilon)\frac{-\partial f(\epsilon)}{\partial \epsilon}$$
$$= \int_{-\infty}^{\infty} d\epsilon \frac{-\partial f(\epsilon)}{\partial \epsilon}[K(\mu) + \sum_{n=1}^{\infty} \left(\frac{(\epsilon-\mu)^n}{n!}\right)]\left(\frac{d^n K(\epsilon)}{d\epsilon^n}\right)_{\epsilon=\mu}$$
$$= K(\mu)\int_{-\infty}^{\infty} d\epsilon \frac{-\partial f(\epsilon)}{\partial \epsilon}\frac{-\partial f(\epsilon)}{\partial \epsilon} + \sum_{n=1}^{\infty}\int_{-\infty}^{\infty} d\epsilon \frac{(\epsilon-\mu)^n}{n!}\left(\frac{d^n K(\epsilon)}{d\epsilon^n}\right)_{\epsilon=\mu}$$
$$= \int_{-\infty}^{\mu} d\epsilon H(\epsilon) + \sum_{n=1}^{\infty}\int_{-\infty}^{\infty} d\epsilon \frac{(\epsilon-\mu)^n}{n!}\frac{-\partial f(\epsilon)}{\partial \epsilon}\left(\frac{d^{n-1} H(\epsilon)}{d\epsilon^{n-1}}\right)_{\epsilon=\mu}. \tag{7.150}$$

Using that

$$\frac{-\partial f(\epsilon)}{\partial \epsilon} = \frac{1}{k_B T}\frac{1}{4\cosh^2 \frac{\epsilon-\mu}{k_B T}},$$

we get:

$$\int_{-\infty}^{\infty} d\epsilon H(\epsilon) f(\epsilon)$$
$$= \int_{-\infty}^{\mu} d\epsilon H(\epsilon) + \sum_{n=1}^{\infty}\int_{-\infty}^{\infty} d\epsilon \frac{(\epsilon-\mu)^{2n}}{(2n)!}\frac{-\partial f(\epsilon)}{\partial \epsilon}\left(\frac{d^{2n-1} H(\epsilon)}{d\epsilon^{2n-1}}\right)_{\epsilon=\mu}.$$

Therefore we arrive at

$$\int_{-\infty}^{\infty} d\epsilon H(\epsilon) f(\epsilon) = \int_{-\infty}^{\mu} d\epsilon H(\epsilon) + \sum_{n=1}^{\infty} a_n (k_B T)^{2n} \left(\frac{d^{2n-1} H(\epsilon)}{d\epsilon^{2n-1}}\right)_{\epsilon=\mu}.$$

Here

$$a_n = \int_{-\infty}^{\infty} dx \frac{x^{2n}}{(2n)!} \left(-\frac{d}{dx}\frac{1}{e^x + 1}\right).$$

Using that $a_1 = (\pi^2/6)$, we get that to leading order in the deviation from the zero temperature result:

$$\int_{-\infty}^{\infty} d\epsilon H(\epsilon) f(\epsilon) = \int_{-\infty}^{\mu} d\epsilon H(\epsilon) + \frac{\pi^2}{6}(k_B T)^2 \left(\frac{d}{d\epsilon} H(\epsilon)\right)_{\epsilon=\mu} + \cdots \tag{7.151}$$

This is the Sommerfeld expansion which shows that, in the Fermi gas, the leading term, in the deviation from the zero temperature result, is quadratic in the temperature.

7.6.3 Fermi Hole

Let us now consider the correlations between two electrons, immersed in the Fermi sea, where one is located at site x and the other at site y, a distance r from each other. If they have opposite spins, they are distinguishable and the wave function of the pair is simply

$$\phi(x, y) = e^{ik_1 \cdot x + ik_2 \cdot y}.$$

Therefore, the probability density is given by

$$g(x, y) = |\phi(x, y)|^2 = 1.$$

However, if they have the same spin, then we have to antisymmetrize the spatial part of the wave function of the pair. We get that

$$\phi(x, y) = \frac{1}{\sqrt{2}} \left(e^{ik_1 \cdot x + ik_2 \cdot y} - e^{ik_2 \cdot x + ik_1 \cdot y}\right)$$

and

$$g(x, y) = 1 - \cos[(k_1 - k_2) \cdot (x - y)].$$

Then

$$\begin{aligned}\langle g(r)\rangle &= \frac{1}{N^2}\sum_{k_1,k_2}(1-\cos[(k_1-k_2)\cdot r]) \\ &= \frac{1}{N^2}\Re\sum_{k_1,k_2}\left(1-e^{i(k_1-k_2)\cdot r}\right) \\ &= 1-\Re\frac{1}{N^2}\frac{V^2}{(2\pi)^6}\int d^3k_1\int d^3k_2 e^{i(k_1-k_2)\cdot r}. \end{aligned} \tag{7.152}$$

Defining

$$F(k_F r) = \frac{V}{N_p(2\pi)^3}\int_{k<k_F} d^3k e^{ik\cdot r}, \tag{7.153}$$

(where N_p is the number of parallel spins and $k_F = p_F/\hbar$, is the Fermi wave vector), we see that

$$\langle g(r)\rangle = 1 - F^2(k_F r).$$

Calculating the integral and using that $V/N_p = 6\pi^2/k_F^3$, we get that

$$F(k_F r) = \frac{3}{(k_F r)^3}\left(\sin(k_F r) - k_F r\cos(k_F r)\right). \tag{7.154}$$

This function goes to zero at infinity and it tends to one at the origin. In the Fig. 7.9 we show the function $\langle g(r)\rangle$. Integrating over space

$$\frac{1}{V}\int d^3r F^2(k_F r) = \frac{1}{N}.$$

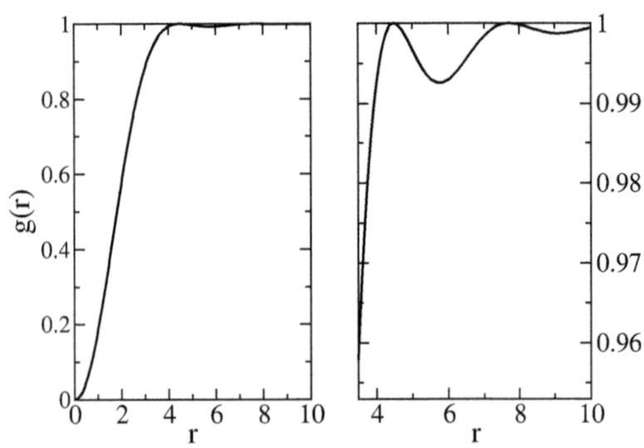

Fig. 7.9 Fermi hole in the correlation function as a function of distance, r, from the location of one electron. In the right a zoom of the convergence to unity is shown

Therefore, the area missing is just one state: this is called the Fermi hole. Due to the exclusion principle, there is a depletion around one electron of exactly one state.

7.7 Exercises

1. (a) Calculate the density of states of free particles with dispersion relations that are linear or quadratic for $d = 1, 2, 3$. That is, show that

$$\begin{pmatrix} & d=1 & d=2 & d=3 \\ E \sim k & const & \sim E & \sim E^2 \\ E \sim k^2 & \sim 1/\sqrt{E} & const & \sim \sqrt{E} \end{pmatrix}. \tag{7.155}$$

   (b) Calculate the temperature dependence of the specific heat at low temperatures for a system of free bosons, for each case.
2. A cylinder is separated in two compartments by a movable piston. Each of the compartments contains an ideal Fermi gas. In one of these, the gas is composed by N particles with spin $1/2$ and in the other N particles with spin $3/2$. Denote by n_1 and p_1 the density and pressure in the first gas and by n_2 and p_2 in the gas 2. Supposing that the molecules of both gases have equal masses, determine which should be the relative density of the two gases in equilibrium in the two limiting cases $T = 0$ and $T \to \infty$.
3. Consider a two-dimensional system. Compare the specific heat of a free system of fermions with that of a free system of bosons and show that they are identical. In one dimension the correspondence holds if the fermion gas is relativistic.
4. Derive an expression for the isothermal compressibility at low temperatures of an ideal Fermi gas.
5. Calculate the diamagnetic susceptibility of a fermion gas and show that in the limit $\hbar \to 0$ this effect disappears, in accordance with the Bohr-Van Leeuwen theorem.
6. Consider a relativistic gas of fermions, as in the case of the electrons in a white dwarf star. Calculate the pressure exerted by the electron gas. Determine the minimal mass so that it will not collapse due to the gravitational attraction (see [Pathria]).
7. At the Bose-Einstein condensation transition line, we get that

$$p v_c^{5/3} = \frac{h^2}{2\pi m} \frac{\zeta(5/2)}{(\zeta(3/2))^{5/3}} \tag{7.156}$$

   is a constant.

   (a) Show that along this transition line, we get that

$$\frac{dp}{dT} = \frac{5}{2}\frac{k_B}{v_c}\frac{\zeta(5/2)}{\zeta(3/2)}. \tag{7.157}$$

(b) Interpret this expression as a Clausius-Clapeyron-like equation as in Eq. (2.157) and obtain an expression for the latent heat of the transition. Consider that $|\Delta v| = v_c$. The reason is that particles that condense formally occupy a zero volume.

8. Consider a reversible adiabatic transformation on the boson gas. This behavior can be obtained fixing the entropy and number of particles.

(a) Using Euler's equation, Eq. (2.8), show that

$$\begin{aligned}\frac{S}{Nk_B} &= \frac{5}{2}\frac{g_{5/2}(z)}{g_{3/2}(z)} - \ln z,\ T \geq T_c,\\ \frac{S}{Nk_B} &= \frac{5}{2}\frac{v}{\lambda^3}\zeta(5/2),\ T \leq T_c.\end{aligned} \tag{7.158}$$

(b) Show that the particles in the condensate do not contribute to the entropy and only the particles in excited states do.

(c) Show that for all temperatures, we get that

$$\begin{aligned} vT^{3/2} &= C\\ pv^{5/3} &= C',\end{aligned} \tag{7.159}$$

both constants. This result is the same as for a classical gas.

(d) Defining as for a classical gas $\gamma = C_p/C_v$, show that

$$\gamma = \frac{5}{3}\frac{g_{5/2}(z)g_{1/2}(z)}{\left(g_{3/2}(z)\right)^2}. \tag{7.160}$$

When $T >> T_c$, we get that $\gamma = 5/3$. For finite temperatures $\gamma > 5/3$. As one approaches the critical point $\gamma \to \infty$. Plot γ as a function of T/T_c.

Chapter 8
Crystalline Structures

Solids are systems of many particles that require, in general, a quantum treatment. Often some form of order is observed. The experimental properties of solid systems are, in general, easily accessible in a laboratory, and any predictions on their properties and behavior can be tested. Usually a very detailed description of their properties is known. In addition to their theoretical interest from a fundamental point of view, many properties are useful in applications.

Solids are composed of atoms that interact in some ways that lead to binding. The atoms themselves are composed of a nucleus and electrons that orbit around the nucleus. The properties of the atoms are determined by the fermionic nature of the electrons and by their mutual interactions. When several atoms are put in proximity, they give rise to larger structures, like molecules, and, in large numbers, may lead to large agglomerates, such as solids or other phases. The agglomerates may have some spatial order, called crystals, or lead to different forms of disordered systems of various degrees, from some imperfections on the regular crystalline structure to fully random spatial distributions of the atoms. These may be identical or mixtures of different types of atoms, or molecules, leading to different types of compounds or alloys. Experimental probes include thermodynamic, transport, optical properties, etc.

In general, the solids have properties that result from the interactions between its constituents, the nuclei and electrons present in the material. The problem is complex and requires some approximation methods to understand its properties, such as perturbation theories, non-perturbative methods, and mean-field theories, in some special cases some form of exact solutions are possible, and often numerical methods are needed for an accurate description. Surprisingly, in some cases a description in terms of independent particles is a good approximation. This holds in particular in the description of the electronic properties of some materials. On the other hand, different types of ordered phases are observed such as magnetic (ferromagnetism, antiferromagnetism, etc.), superconducting, ferroelectric. These

P. D. Sacramento, *Introduction to Condensed Matter Theory*,
https://doi.org/10.1007/978-3-031-86251-9_8

phases require non-perturbative methods to describe the effect of interactions between the particles in the solid. Interactions are also important in nonsolid systems such as in superfluids, liquid crystals, etc.

8.1 Crystal Binding

In general, the binding between the atoms that leads to a regular structure is the result of the Coulomb interactions between the various charges present. The atoms tend to gather in some regular structure, if the energy is lowered with respect to a situation where they are apart. It is usual to define a quantity, called cohesion energy, as the energy that we must supply to a solid to separate its components and place them at rest at infinity. It is, therefore, the difference between the energy of the free atoms and the energy of the solid.

There are different types of bonds[1] such as a van der Waals bond (such as in Ar), ionic bonds (such as in $NaCl$), metallic bond (such as in Na), and covalent bond (such as in C). In the case of the van der Waals bond, the atoms remain neutral, in the ionic bond, one finds that the Cl atoms become ionized with one extra electron Cl^- and the Na atoms loose one electron Na^+, in the metallic bond, the Na atoms loose one electron each that is shared by the whole system and the bond to each ion is broken, and in the covalent bond, electrons are shared between different neighboring atoms. For instance, in Na one finds that the distance between neighboring atoms is of the order of 3.7 Angstrom ($\AA$). The electrons in the $1s$ orbitals are essentially peaked at the nuclei, the $2s$ and $2p$ orbitals have some spatial extension in the intermediate region between two neighboring nuclei, but the $3s$ orbitals are peaked in the intermediate region, leading to a delocalization of the electrons in these atomic orbitals.

In the elements of group 18 of the periodic table, the rare gases with full electronic shells, the binding energies are very weak and of the van der Waals type. The metallic binding, found for instance in the simple metals of groups 1 and 2, is usually weak or of medium size. In the case of the covalent bonds, found for instance in groups 13 and 14, the cohesion energy is strong. Elements of groups 8–12 also have strong bonds (called tight-binding bonds) that, as we shall see later, result from the mixture of some inner orbits in the atoms, in addition to more delocalized, shared, orbitals.

We may note at this point that, under normal pressure conditions, H and He do not lead to a solid structure. The reason is that the quantum zero point motion of the atoms does not allow the existence of the crystal, since the fluctuations of the atom position make a regular structure unstable. One may also note that fluctuations seem to prevent the stability of a strictly two- or one-dimensional structure (stability may

[1] For further reading see [Kittel].

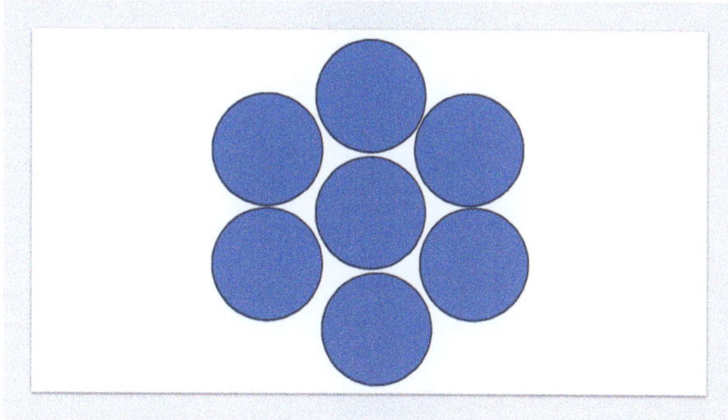

Fig. 8.1 Atomic distribution in a noble gas

be achieved, however, if the structure is immersed in a three-dimensional structure). We will return to this point later.

8.1.1 *Noble Gases*

Since isolated noble gases have full orbitals, the electronic distributions are spherical (Fig. 8.1). Consider two atoms at positions $\mathbf{r}_1$ and $\mathbf{r}_2$. Due to their proximity, one expects induced dipoles on each one, as a result of the displacement of the center of negative charge (the electrons) with respect to the center of positive charges (the nuclei, assumed static). The charge displacements may be considered along the direction, x, which connects the nuclei. Call R the distance between the nuclei, and x_1 and x_2 the displacements of the negative charges with respect to each nucleus, respectively, and call m the reduced mass between the electron and the nucleus. Then, we may write the following Hamiltonian for the kinetic energy and the potential energy, associated with the dipoles seen as a string of force constant C:

$$H_0 = \frac{p_1^2}{2m} + \frac{p_2^2}{2m} + \frac{1}{2}Cx_1^2 + \frac{1}{2}Cx_2^2. \tag{8.1}$$

We consider $R >> x_1, x_2$. There are also the Coulomb interactions between the various charges. This leads to[2]

$$H_1 = \frac{e^2}{R} - \frac{e^2}{R + x_1} - \frac{e^2}{R - x_2} + \frac{e^2}{R + x_1 - x_2}. \tag{8.2}$$

To simplify we consider that the charge of each nucleus is e and each electron distribution has charge $-e$. The Hamiltonian may be approximated by

[2] To simplify notation, in this section and often in the book in the expression for the Coulomb potential, we write e^2 instead of $e^2/(4\pi\epsilon_0)$.

$$
\begin{aligned}
H_1 &= \frac{e^2}{R}\left(1 - \frac{1}{1+\frac{x_1}{R}} - \frac{1}{1-\frac{x_2}{R}} + \frac{1}{1+\frac{x_1-x_2}{R}}\right) \\
&= \frac{e^2}{R}\left(1 - \left(1 - \frac{x_1}{R} + \left(\frac{x_1}{R}\right)^2\right) - \left(1 + \frac{x_2}{R} + \left(\frac{x_2}{R}\right)^2\right)\right. \\
&\quad \left. + \left(1 - \frac{x_1 - x_2}{R} + \left(\frac{x_1 - x_2}{R}\right)^2\right)\right) \\
&\sim -\frac{e^2}{R^3} 2x_1 x_2.
\end{aligned} \tag{8.3}
$$

Therefore, we get that

$$
H = H_0 + H_1 = \frac{p_1^2}{2m} + \frac{1}{2}Cx_1^2 + \frac{p_2^2}{2m} + \frac{1}{2}Cx_2^2 - \frac{2e^2}{R^3}x_1x_2. \tag{8.4}
$$

This Hamiltonian has a quadratic form that may be diagonalized using symmetric and antisymmetric variables defined as

$$
\begin{aligned}
x_s &= \frac{1}{\sqrt{2}}\left(x_1 + x_2\right); \; p_s = \frac{1}{\sqrt{2}}\left(p_1 + p_2\right) \\
x_a &= \frac{1}{\sqrt{2}}\left(x_1 - x_2\right); \; p_a = \frac{1}{\sqrt{2}}\left(p_1 - p_2\right).
\end{aligned} \tag{8.5}
$$

We may write the Hamiltonian in terms of these variables:

$$
H = \frac{1}{2m}\left(p_s^2 + p_a^2\right) + \frac{C}{2}\left(x_s^2 + x_a^2\right) - \frac{e^2}{R^3}\left(x_s^2 - x_a^2\right). \tag{8.6}
$$

Defining the frequencies of two uncoupled harmonic oscillators

$$
\begin{aligned}
\omega_s^2 &= \frac{1}{m}\left(C - \frac{2e^2}{R^3}\right) \\
\omega_a^2 &= \frac{1}{m}\left(C + \frac{2e^2}{R^3}\right),
\end{aligned} \tag{8.7}
$$

the minimum energy of the system is obtained considering the zero point motion of the oscillators leading to a groundstate energy:

$$
E_0 = \frac{\hbar}{2}\left(\omega_s + \omega_a\right). \tag{8.8}
$$

Substituting the expression for the frequencies, we get that

$$E_0 = \frac{\hbar}{2}\sqrt{\frac{C}{m}}\left(\sqrt{1-\frac{2e^2}{CR^3}}+\sqrt{1+\frac{2e^2}{CR^3}}\right). \tag{8.9}$$

Considering that $2e^2$ is small compared to CR^3, we may expand the square roots yielding

$$E_0 = \hbar\omega_0\left(1-\frac{1}{2}\frac{e^4}{C^2R^6}\right). \tag{8.10}$$

The groundstate energy of the two free atoms is $\hbar\omega_0$. Therefore the cohesion energy is given by

$$\Delta E = \frac{1}{2}\hbar\omega_0\frac{e^4}{C^2R^6}, \tag{8.11}$$

which is an attractive energy. This dipolar energy is called the van der Waals energy. Note that if R is small, we may need to include higher-order terms. Also, note that the effect is quantum since we used the zero energy motion of the harmonic oscillators or, alternatively, we may note that the energy is proportional to Planck's constant. In a simplified way, we may say that taking the classical limit ($\hbar \to 0$) the effect vanishes.

This attractive energy decays fast with the distance between the nuclei, since it decays as $1/R^6$. On the other hand, we should note that for small distances, small R, the atomic orbitals overlap significantly which implies two sources of repulsion. One is associated with the repulsive Coulomb interaction between the electronic orbitals, but also there is an effective repulsion that results from Pauli's exclusion principle that tends to prevent the occupation of identical fermionic particles on the same state. As a consequence there is a strong repulsion at short distances.

The short distance repulsive nature can be modeled in various ways such as the Lennard-Jones or Born-Mayer potentials. The Lennard-Jones potential takes the form:

$$U(R) = 4\epsilon\left(\left(\frac{\sigma}{R}\right)^{12}-\left(\frac{\sigma}{R}\right)^{6}\right), \tag{8.12}$$

where the parameters ϵ and σ can be determined from the gas phase. This type of potential is shown in Fig. 8.2. The Born-Mayer potential uses a repulsive potential of the form:

$$\lambda e^{-\frac{R}{\rho}}. \tag{8.13}$$

Considering for instance the Lennard-Jones potential, we see that there is a minimum of the potential at a distance given by $(R_0/\sigma)^6 = 2$, which leads to $R_0/\sigma \sim 1.122$. At the minimum, the potential energy is $U(R_0) = -\epsilon$.

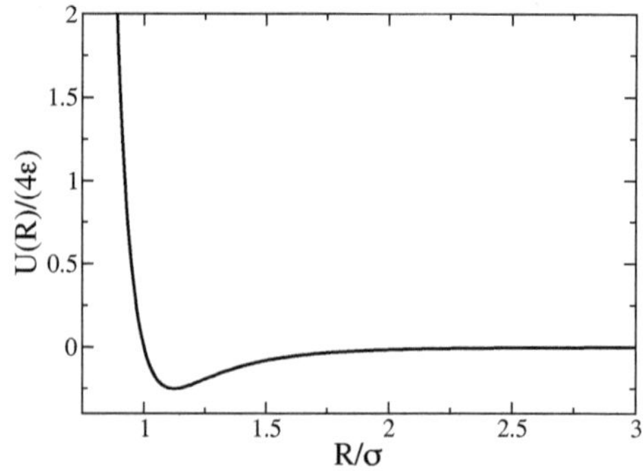

Fig. 8.2 Lennard-Jones potential between two atoms of a noble gas

We may now consider N atoms, instead of just two. Considering the nuclei at fixed positions, as if the lattice were rigid, the total potential energy is written as

$$U = \frac{N}{2} 4\epsilon \left(\sum_j{}' \left(\frac{\sigma}{p_{ij} R} \right)^{12} - \sum_j{}' \left(\frac{\sigma}{p_{ij} R} \right)^{6} \right). \tag{8.14}$$

We have written the distance between nuclei i and j as $R_{ij} = p_{ij} R$, where R is the distance between nearest-neighbors, and the sums exclude the case $i = j$. Both terms in the potential decay fast with distance. Therefore, it is favorable to maximize the number of nearest-neighbors and to minimize the distance R. The atoms tend to be as close to each other as possible, and therefore the lattice will be as closed packed as possible, as illustrated in Fig. 8.1. Possibilities are an fcc lattice or an hcp lattice, as we will show later. The first lattice has a cubic structure with an atom at the center of each square face, and the other has a hexagonal structure in the basis and top.

Considering an fcc lattice, selecting $p_{ij} = 1$ between nearest-neighbors and noticing that each atom has 12 neighbors, we get that

$$\sum_j{}' \frac{1}{p_{ij}^{12}} = 12.13$$

$$\sum_j{}' \frac{1}{p_{ij}^{6}} = 14.45. \tag{8.15}$$

Defining $\bar{R}$ as the distance that minimizes the energy $(dU/dR)_{R=\bar{R}} = 0$, we get that

$$\left(\frac{\bar{R}}{\sigma} \right)^6 = 2 \frac{12.13}{14.45}$$

$$\frac{\bar{R}}{\sigma} = 1.09 < \frac{R_0}{\sigma} = 1.122. \tag{8.16}$$

Comparing with experimental results, one finds a good agreement since the values are obtained as $\bar{R}/\sigma = 1.14(Ne), 1.11(Ar), 1.10(Kr), 1.09(Xe)$.

The result is better for the heavier element, Xe, and becomes less good as the atoms become lighter, since the zero point motion becomes more significative and, therefore, $\bar{R}$ increases due to the oscillations. Replacing the calculated equilibrium value for $\bar{R}$, we find that the minimum energy is given by $U(\bar{R}) = -8.61N\epsilon$, in the approximation that the atoms are at rest (rigid lattice). The atom motion reduces the energy gain and its change is influenced by the mass of the atoms. In addition to changes as we move along the various elements, one also expects an isotope effect. For instance, for Ne^{20} we find $\bar{R} = 4.4644\,Å$ and for Ne^{22} we find $\bar{R} = 4.4559\,Å$.

8.1.2 Ionic Crystals

Consider the example of the ionic crystal $NaCl$. A free sodium atom has filled $n = 1, 2$ shells and one electron in the orbital $(3s)^1$. On the other hand, a chlorine-free atom has filled $n = 1, 2$ shells and 7 electrons in the $n = 3$ shell in the configuration $(3s)^2(3p)^5$. Therefore, the Na has one electron outside a filled shell, and Cl is short of one electron to fill its $3p$ shell. When the two atoms are placed in proximity, the extra electron of sodium is transferred to the chlorine leading to two ions Na^+ and Cl^-, as illustrated in Fig. 8.3. Since they are now charged, there is a Coulomb attraction between the two ions that leads to strong binding. Specifically, the energies involved lead to a two-step process: the first one that results from the ionization of the atoms and then the binding into a solid structure. The energy balances are detailed as

$$Na + 5.14eV \rightarrow Na^+ + e^-$$

$$Cl + e^- \rightarrow Cl^- + 3.61eV$$

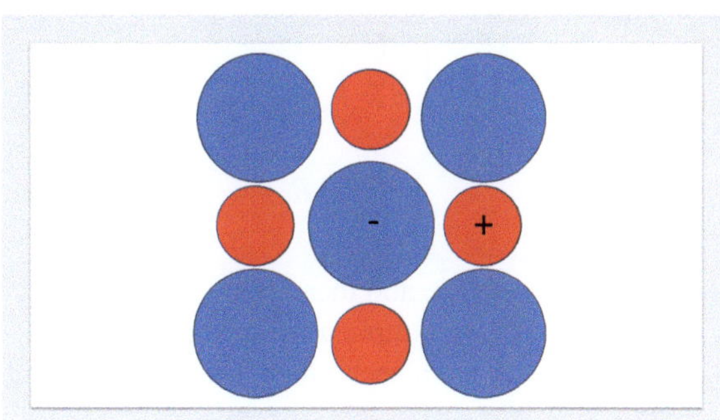

Fig. 8.3 Ions in an ionic crystal

$$Na^+ + Cl^- \rightarrow Na^+Cl^- + 7.9eV. \tag{8.17}$$

First the atoms in a gas phase are ionized, and then the two gases bind to form a solid structure. Therefore, the energy gain with respect to neutral atoms separated in a gas phase is given by $(7.9 - 5.14 + 3.61)\,\text{eV} = 6.4\,\text{eV}$.

The potential energy between two ions is now different from the case of the noble gases, since there is a Coulomb term:

$$U_{ij} = \lambda e^{-\frac{r_{ij}}{\rho}} \pm \frac{q^2}{r_{ij}}, \tag{8.18}$$

where $r_{ij} = p_{ij}R$, the first term is the hardcore repulsion when the atoms are very close, and the dipolar interaction is neglected in comparison with the Coulomb interaction. The total energy is therefore

$$\begin{aligned} U &= N_p \sum_j{}' U_{ij} \\ &= N_p \left(z\lambda e^{-\frac{R}{\rho}} - \frac{q^2}{R}\alpha \right). \end{aligned} \tag{8.19}$$

Here, N_p is the number of pairs of ions, z is the number of nearest-neighbors (since the repulsive term is very short-range, we can safely neglect contributions from further neighbors), and α is called Madelung constant and defined as

$$\alpha = \sum_j{}' \frac{\pm 1}{p_{ij}}. \tag{8.20}$$

The value of this constant depends on the crystalline structure. Since it is a sum of oscillating terms, and due to the long range of the Coulomb interaction, it is in general difficult to solve, even numerically. One may also define some equilibrium distance between the ions, considering the minimum of the total energy. This leads to

$$\begin{aligned} \left(\frac{dU}{dR}\right)_{R=R_0} &= 0 \\ U(R_0) &= -\frac{N_p \alpha q^2}{R_0}\left(1 - \frac{\rho}{R_0}\right). \end{aligned} \tag{8.21}$$

The parameter $\rho < R_0$, and therefore stability of the crystal is ensured ($\alpha > 0$).

Let us take as an example a one-dimensional lattice with charges of alternating signs, distant by R from each other. Therefore, we may write that

$$-\frac{\alpha}{R} = 2\left(-\frac{1}{R} + \frac{1}{2R} - \frac{1}{3R} + \frac{1}{4R} + \dots\right)$$
$$= -\frac{2}{R}\left(1 - \frac{1}{2} + \frac{1}{3} - \frac{1}{4} + \cdots\right). \tag{8.22}$$

Recalling that

$$\ln(1+x) = x - \frac{x^2}{2} + \frac{x^3}{3} - \frac{x^4}{4} + \cdots \tag{8.23}$$

we get that

$$\alpha = 2\ln 2. \tag{8.24}$$

In general, the problem is difficult, since there are convergence problems with the definition of the Madelung constant.

8.1.3 Covalent and Metallic Bonds

The covalent bonds are directional, since they result from the spatial overlap of neighboring orbitals in different atoms. The anisotropy of the orbitals leads to a significant directionality of the arrangement of the orbitals, and therefore the atoms. The overlap is schematically shown in Fig. 8.4. As we will see later, the fermionic nature of the electrons will take an important role through the so-called exchange energy.

In the metallic case, the delocalization of some atomic orbitals leads to a continuum of states occupied by the electrons that have left the vicinity of the atoms to spread all over the material, as shown in Fig. 8.5. This leads, in some first approximation, to a nearly constant electronic density throughout the material. The potential between the neighboring nuclei is in this way lowered (due to the negative charge of the electron cloud), and so the energies of the various electronic states, including the bound states, are lowered: there is therefore an energy gain as a

Fig. 8.4 Charge distributions in covalent bonds

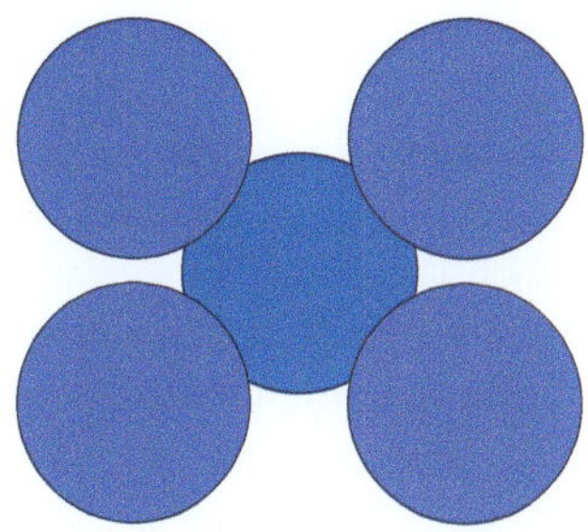

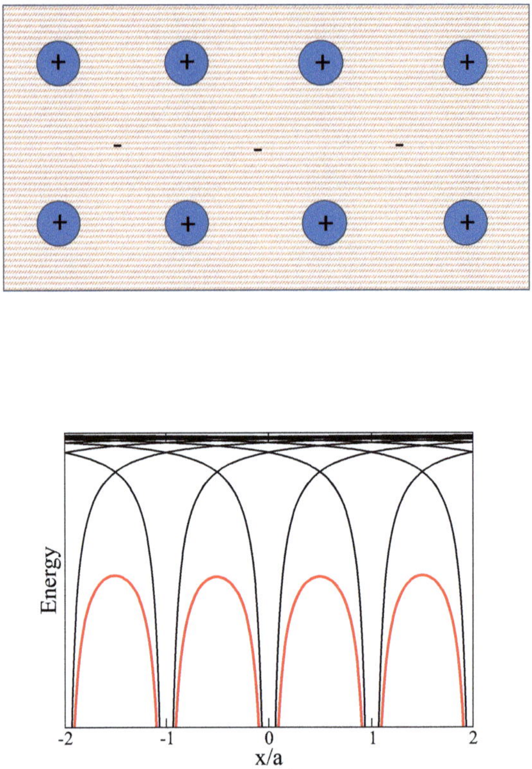

Fig. 8.5 Top: Charge distributions in metallic binding. In the lower panel, the potential felt by each electron is shown. As a result of the sum over the potentials from each ion, the total potential leads to localized (core) states and extended states

result of the overlap between the atomic orbitals and the consequent delocalization of some electrons. This leads to binding and a stable crystalline structure.

8.2 Crystal Structure

In many interesting solid-state systems, the atoms are distributed in a regular way constituting a crystal. A crystal may be defined has some regular distribution of points in a so-called lattice and a basis of atoms at each lattice point, as shown in Fig. 8.6. The lattice has in this case a regular two-dimensional structure to which we may attach a given basis. The basis may be composed of a single entity (like the circles or triangles) or may have some internal structure of more than one entity (such as squares and circles). In a gas there is a continuous translation symmetry which is partially broken when the system crystallizes. The new symmetry is now discrete and defined by the periodic nature of the lattice.

In a three-dimensional system ($3d$), the lattice may be constructed defining some basis vectors, $\mathbf{a}_1, \mathbf{a}_2, \mathbf{a}_3$, taken as linearly independent. Using this basis any lattice point may be written as a linear combination:

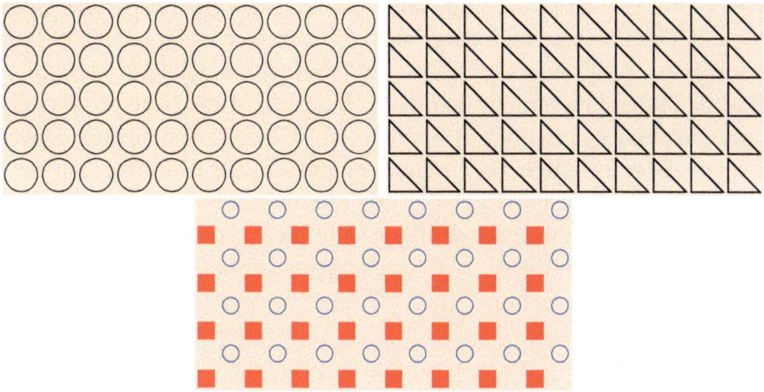

Fig. 8.6 Examples of a lattice as defined by some periodic structure to which a basis is associated, with a single figure or with more than one element

$$\mathbf{l} = l_1\mathbf{a}_1 + l_2\mathbf{a}_2 + l_3\mathbf{a}_3, \tag{8.25}$$

where l_1, l_2, l_3 are integers. The set of points $\mathbf{l}$ defines the lattice. Therefore, the set of basis vectors is chosen as the smaller vectors that describe the lattice.

The discrete translational invariance of the lattice can be expressed by the fact that two points in space differing by a lattice vector, $\mathbf{l}$, are equivalent. That is, two points $\mathbf{r}$ and $\mathbf{r}'$ with

$$\mathbf{r}' = \mathbf{r} + \mathbf{l} \tag{8.26}$$

are equivalent physically and so any physical property should be invariant, minus a translation of $\mathbf{l}$. This clearly assumes an infinite system.

There are limitations to the discrete translational invariance:

- The systems have finite sizes, and therefore there are surface effects associated with the finiteness of the system. However, if the system has a finite macroscopic volume, the number of atoms at the surface is small and the error (deviation from a perfect translational invariance) is small.
- Crystals have defects of different types and so are not perfectly regular.
- Atoms oscillate around their equilibrium positions and so the lattice is not rigid.

8.2.1 Primitive Cell and Bravais Lattices

One may define a primitive cell (or unit cell) as the smallest cell that can be constructed and that obeys the translation properties of the lattice. Different lattices have different sets of atoms in a given unit cell. Examples of unit cells with one atom are for instance crystals of the elements $Al, Cu, Au, Ag, and Fe$. Other lattices have unit cells with a two-atom per cell, such as $NaCl$ and diamond. The number of atoms in a unit cell may be quite large such as in organic or inorganic crystals with unit cells with over 100 atoms, proteins, and DNA with around 10^4 atoms.

A Bravais lattice is a lattice such that it looks the same from all lattice points along any direction from which it is observed. If there are nonequivalent points, then it is not a Bravais lattice. For instance, in $3d$ there are 14 Bravais lattices. If the unit cell has a basis of several atoms, in addition to the set of lattice points defined by the vectors **l**, which define the distribution of unit cells, we also need additional vectors to describe the positions of the atoms within the unit cell. For instance, we need to specify a set of s vectors:

$$\mathbf{r}_j = x_j\mathbf{a}_1 + y_j\mathbf{a}_2 + z_j\mathbf{a}_3, \tag{8.27}$$

for $j = 1, \cdots, s$, where s is the number of atoms in the unit cell, and $|x_j|, |y_j|, |z_j| < 1$. Usually these real numbers are chosen as positive, but is not a requirement.

Even though the unit cell is defined as the minimum size cell that, infinitely repeated defines a given lattice, there are infinite such cells. Taking for instance as a simple example a square lattice, one may define an infinite set of unit cells that satisfy the two requirements: all have the same (minimal) area and the whole space is filled by these cells. A specific choice (possible among infinite) is called the Wigner-Seitz cell. It is chosen such that it has all the same symmetries as any lattice point and therefore is a cell that is invariant under the action of the group of local symmetry operations (such as rotations, inversions, reflection across a plane, mirror symmetries). A simple way to determine the Wigner-Seitz cell is to place its center at a Bravais lattice point, take the half-distance between that point and its neighbors, and define a cell by the region contained in the intersection between the perpendicular planes to the lines that connect the point to the neighbors and that cross the half-distance point. This is illustrated in Fig. 8.7.

As already mentioned, the crystal has therefore a set of symmetries that are associated with the discrete translational symmetry and the local (called point group) symmetries. For instance, in the case of a square lattice, and using that the center of the Wigner-Seitz cell coincides with a lattice point, we find that the cell is invariant under several operations. Specifically, the operations are the identity, a

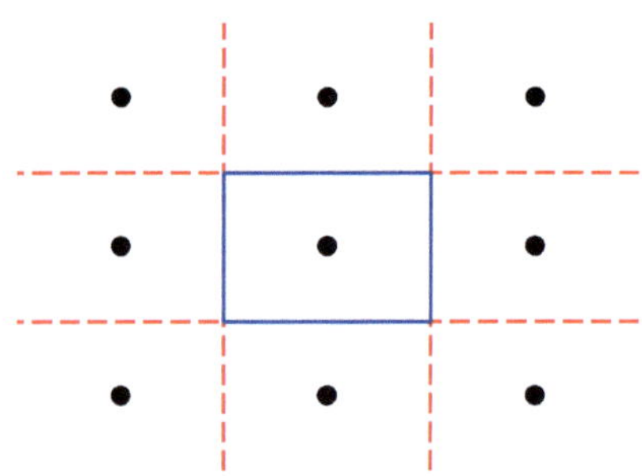

Fig. 8.7 Wigner-Seitz cell of a two-dimensional rectangular lattice

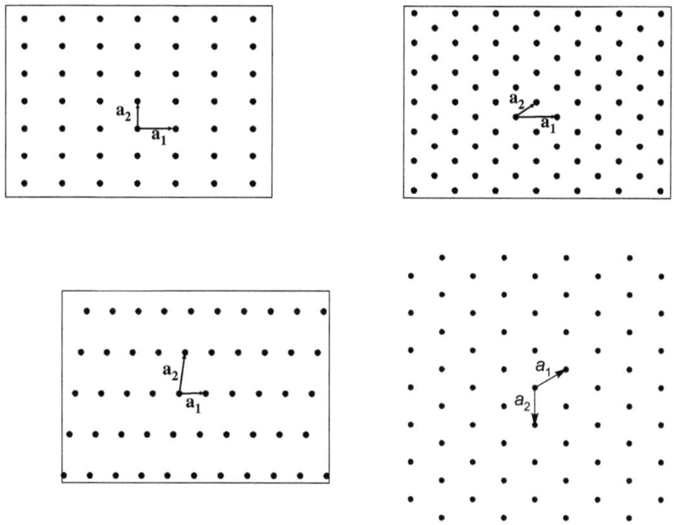

Fig. 8.8 Examples of Bravais lattices in two dimensions: rectangular lattice, centered rectangular lattice, oblique and hexagonal lattices

rotation by $\pi/2$ around the axis perpendicular to the cell and that goes through the center point, four reflection planes, inversion around the center point. If the cell is rectangular, it is invariant with respect to rotations by an angle π; there are two reflection planes, in addition to the identity and inversion symmetry.

The rotations can be labeled as C_n, associated with an angle of rotation $\alpha = 2\pi/n$. Different lattices are invariant under different rotation operations. For instance, $n = 1$ is a trivial rotation as for instance a general oblique lattice is only invariant by a 2π rotation. The case of $n = 2$ applies to a rectangular lattice, $n = 3$ to a triangular lattice and $n = 4$ to a square lattice.

The case $n = 5$ would correspond to a rotation symmetry of a pentagon. However, it is not possible to fill the entire space with pentagons. So the lattice translational symmetry is not satisfied. A hexagonal lattice satisfies rotational invariance with $n = 6$ and also translational invariance. It turns out that, as for $n = 5$, cases with $n \geq 7$ are also impossible.

In Fig. 8.8 we show some two-dimensional Bravais lattices, such as a rectangular lattice (that includes a square lattice if the two sides have equal lengths), a centered rectangular lattice, an oblique lattice, and a hexagonal lattice. An example of a two-dimensional non-Bravais lattice is the honeycomb lattice shown in Fig. 8.9. It can be seen as two interpenetrated hexagonal lattices, such that the unit cell contains one atom of each sublattice. It may also be seen as a hexagonal lattice with a basis. In the figure the basis vectors $\mathbf{a}_1$ and $\mathbf{a}_2$, connecting nearest-neighbors within the same sublattice, and the three nearest-neighbor vectors, $\boldsymbol{\delta}_i$, which connects a given lattice site on a sublattice with the three neighbors located in the other sublattice. This structure is the structure of graphene, which has interesting properties, as will be discussed later.

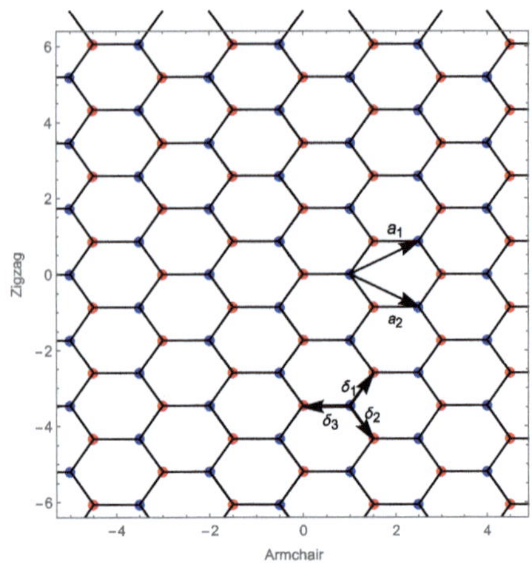

Fig. 8.9 Example of a two-dimensional non-Bravais lattice: honeycomb lattice

These results hold for a lattice with a single type of unit cell. The concept may be generalized defining lattices with two unit cells. These fall into a different class, called quasicrystals, and in some cases it is possible to fill space if one defines two unit cells, or more.

The set of Bravais lattices in $3d$ systems can be characterized as primitive (P), base centered (C), body centered (I), and face centered (F). Let us call a, b, c the lengths of the three basis vectors and α, β, γ the angles between them. The set of 14 lattices is:

- One triclinic lattice, $P, a \neq b \neq c, \alpha \neq \beta \neq \gamma$.
- Two monoclinic lattices, $P, C, a \neq b \neq c, \alpha = \gamma = \pi/2, \alpha, \gamma \neq \beta$.
- Four orthorhombic lattices, $P, C, I, F, a \neq b \neq c, \alpha = \beta = \gamma = \pi/2$.
- Two tetragonal lattices, $P, I, a = b \neq c, \alpha = \beta = \gamma = \pi/2$.
- Three cubic lattices, $P, I, F, a = b = c, \alpha = \beta = \gamma = \pi/2$.
- One trigonal lattice, $P, a = b = c, \alpha = \beta = \gamma < 2\pi/3, \neq \pi/2$.
- One hexagonal lattice, $P, a = b \neq c, \alpha = \beta = \pi/2, \gamma = 2\pi/3$.

Some examples of crystalline structures are shown in Fig. 8.10.

In order to maximize binding, some systems crystallize in such a way that the atoms are as close as possible. There is, therefore, the concept of close packing that is the ideal way to distribute identical spheres in order to maximize the occupied space. This is for instance relevant in the case of an atom with closed shells (and therefore spherical symmetry) such as in the noble gases. There are actually infinite ways to distribute identical spheres in a regular way that maximizes the occupied space. Two examples, found in nature, are the lattices fcc (face-centered cubic) and hcp (hexagonal closed-packed). In both cases we can show that the filling fraction is 0.74.

Considering for instance the hcp structure, shown in Fig. 8.11, in addition to the six corners of the hexagon of the basis, occupied each by a sphere, one places another sphere in the axis above the center of the hexagon. This sphere has *six*

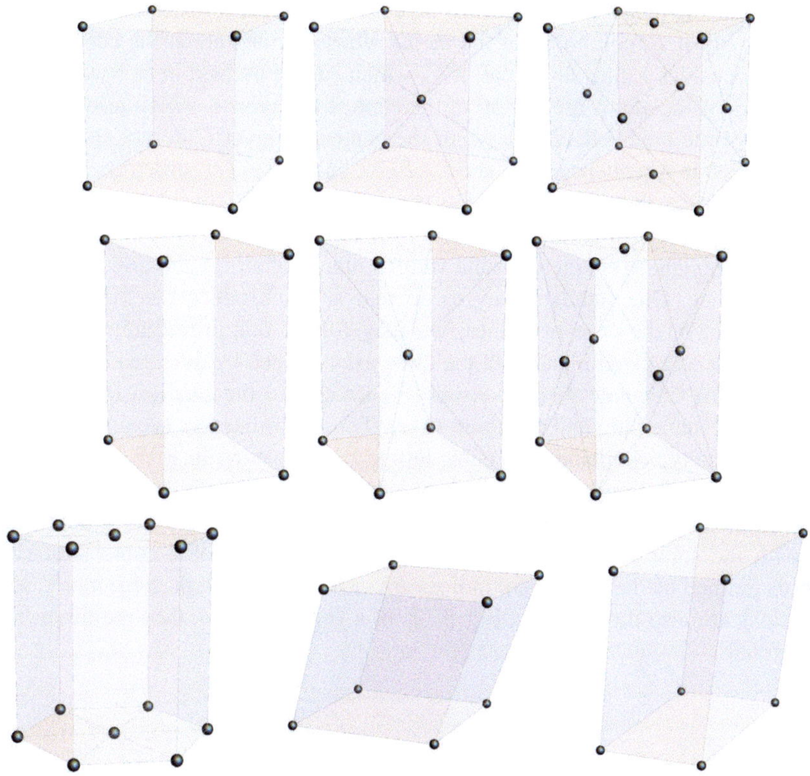

Fig. 8.10 Examples of Bravais lattices in three dimensions: simple cubic, body-centered cubic, face-centered cubic, orthorhombic, body-centered orthorhombic, face-centered orthorhombic, hexagonal, trigonal, and triclinic. Structures done using Mathematica

Fig. 8.11 Hexagonal close-packed three-dimensional lattice. Structure done using Mathematica

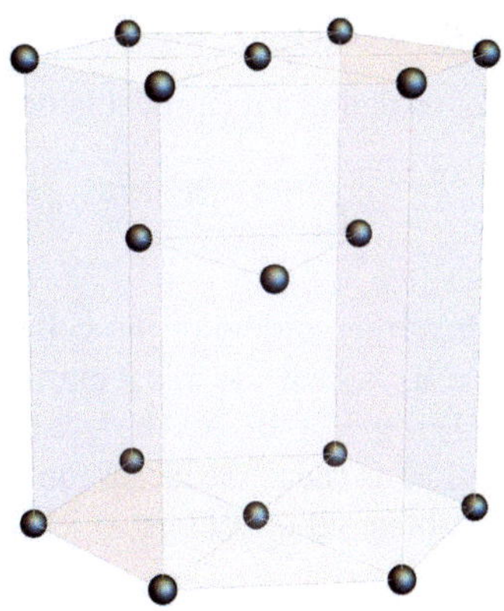

neighbors in the plane. Let us call this layer A. We can next consider another layer of atoms, call it B, composed of the atoms whose centers are at the center points. Next we consider a third layer, call it C, which can be defined in two ways. Either the centers of the atoms are on top of the centers of layer A, which leads to a hcp structure of the type $ABABAB\ldots$, or the centers of layer C do not coincide and are displaced in a sequence $ABCABCABC\ldots$ as in the fcc structure. In the case of the hcp lattice, the primitive cell has two atoms per cell (considering the unit cell of the usual hexagonal lattice). The ratio c/a can be shown to be $\sqrt{8/3} = 1.6333$.

It is convenient to characterize the various planes of atoms and directions across a given lattice. One standard way to do that is to introduce the Miller indices. For instance, in the case of planes, we may follow this procedure: (i) Find the interceptions of a given plane with the three axes defined by the three basis vectors $\mathbf{a}_1, \mathbf{a}_2, \mathbf{a}_3$. (ii) Calculate their inverses. (iii) Reduce to the smallest three integers that have the same ratio and call them (hkl). This set of integers is called the Miller indices. Some typical planes in a cubic lattice are shown in Fig. 8.12.

We may follow some corresponding procedure to define a direction in the crystal. The direction is defined by a set of integers $[u, v, w]$ that are defined as the smaller set with the same ratio $(u\mathbf{a}_1 + v\mathbf{a}_2 + w\mathbf{a}_3)$. In the case of a cubic lattice, the vectors defined by the two sets of indices are orthogonal $(h, k, l) \perp [u, v, w]$. Some directions are also illustrated in Fig. 8.12. In a general lattice, they are not usually orthogonal.

8.2.2 Reciprocal Lattice

Due to the periodicity of the lattice (and to its infinite extension), any property of the system must obey:

$$f(\mathbf{r}) = f(\mathbf{r} + \mathbf{l}). \tag{8.28}$$

Let us define a reciprocal lattice through the basis vectors $(\mathbf{b}_1, \mathbf{b}_2, \mathbf{b}_3)$ where

$$\mathbf{b}_1 = 2\pi \frac{\mathbf{a}_2 \times \mathbf{a}_3}{v}, \mathbf{b}_2 = 2\pi \frac{\mathbf{a}_3 \times \mathbf{a}_1}{v}, \mathbf{b}_3 = 2\pi \frac{\mathbf{a}_1 \times \mathbf{a}_2}{v}, \tag{8.29}$$

where the volume of the cell defined by the three vectors $\mathbf{a}_i$ is given as usual by $v = \mathbf{a}_1 \cdot \mathbf{a}_2 \times \mathbf{a}_3$. In this reciprocal lattice, a generic vector (similarly to the definition of $\mathbf{l}$) is $\mathbf{g} = n_1\mathbf{b}_1 + n_2\mathbf{b}_2 + n_3\mathbf{b}_3$. Note that $\mathbf{b}_i \cdot \mathbf{a}_j = 2\pi\delta_{ij}$.

Let us then consider a function of $\mathbf{r}$, $f(\mathbf{r})$. This function may be expanded in a complete set of vectors which we may choose as being $\mathbf{g}$. The expansion is

$$f(\mathbf{r}) = \sum_{\mathbf{g}} f_{\mathbf{g}} e^{i\mathbf{g}\cdot\mathbf{r}}$$

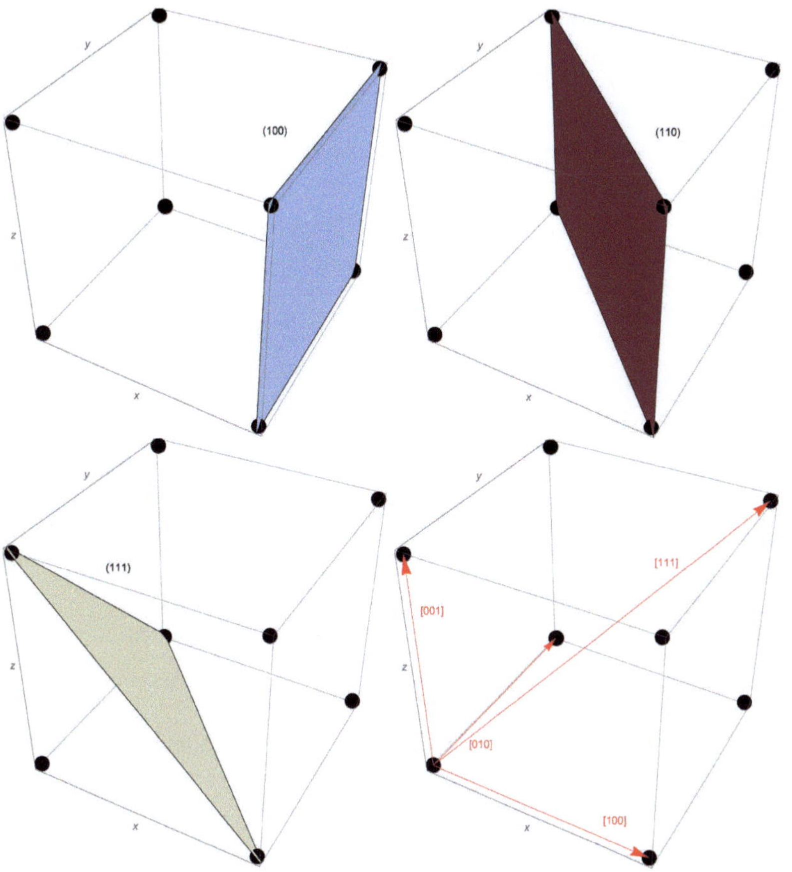

Fig. 8.12 Illustration of some infinite planes in a cubic lattice. The Miller indices are (100), (110), (111). Illustration of some directions in a cubic lattice. The Miller indices are [100], [010], [001], [111]

$$f_{\mathbf{g}} = \frac{1}{V} \int dV f(\mathbf{r}) e^{-i\mathbf{g}\cdot\mathbf{r}}. \tag{8.30}$$

Using that $f(\mathbf{r}+\mathbf{l}) = f(\mathbf{r})$, we obtain that

$$e^{i\mathbf{g}\cdot\mathbf{l}} = 1. \tag{8.31}$$

which implies $\mathbf{g}\cdot\mathbf{l} = 2\pi N$. In an equivalent way, we may develop the function $f(\mathbf{r})$ in a Fourier series:

$$f(\mathbf{r}) = \sum_{\mathbf{k}} f_{\mathbf{k}} e^{i\mathbf{k}\cdot\mathbf{r}}. \tag{8.32}$$

This means that $\mathbf{k} = \mathbf{g}$ and that $f_{\mathbf{k} \neq \mathbf{g}} = 0$. The vectors $\mathbf{g}$ are the terms that contribute to the Fourier series of a function that has the periodicity of the lattice defined by $\mathbf{l} = l_1 \mathbf{a}_1 + l_2 \mathbf{a}_2 + l_3 \mathbf{a}_3$. Clearly the vectors $\mathbf{a}_i$ are characteristic of the particular lattice.

The set of vectors $\mathbf{g}$ defines the reciprocal lattice. One obtains easily that

$$\begin{aligned} \mathbf{g} \cdot \mathbf{a}_1 &= 2\pi n_1 \\ \mathbf{g} \cdot \mathbf{a}_2 &= 2\pi n_2 \\ \mathbf{g} \cdot \mathbf{a}_3 &= 2\pi n_3 \end{aligned} \tag{8.33}$$

and confirms

$$\mathbf{g} \cdot \mathbf{l} = 2\pi \left(l_1 n_1 + l_2 n_2 + l_3 n_3 \right) = 2\pi N. \tag{8.34}$$

The Wigner-Seitz cell of the reciprocal lattice is called the first Brillouin zone. It can be seen easily that the reciprocal lattice of a simple cubic lattice is also a simple cubic lattice. On the other hand, the reciprocal lattice of a bcc lattice is a fcc lattice, and the reciprocal lattice of a fcc lattice is a bcc lattice.

The reciprocal lattice has some interesting and useful properties:

- Each vector of the reciprocal lattice, $\mathbf{g}$, is perpendicular to a set of planes of the real space lattice.
  For example, if we have two sets of points defined by the vectors $\mathbf{l}$ and $\mathbf{l}'$ that satisfy $\mathbf{g} \cdot \mathbf{l} = \mathbf{g} \cdot \mathbf{l}'$, then we get that the reciprocal vector is perpendicular to the difference between the two vectors $\mathbf{l}, \mathbf{l}'$, since $\mathbf{g} \cdot \left(\mathbf{l} - \mathbf{l}' \right) = 0$. The difference defines a plane. An example is $l_1' = l_1 - mn_3$; $l_2' = l_2 - mn_3$; $l_3' = l_3 + m(n_1 + n_2)$.
- If the components of $\mathbf{g}$ do not contain a common factor, then $|\mathbf{g}|$ is inversely proportional to the spacing of the real space planes that are perpendicular to $\mathbf{g}$. The distance between the planes is $d = 2\pi / |\mathbf{g}|$.
- The volume of the unit cell of the reciprocal lattice is inversely proportional to the volume of the unit cell in real space:

$$\mathbf{b}_1 \cdot (\mathbf{b}_2 \times \mathbf{b}_3) = \frac{(2\pi)^3}{\mathbf{a}_1 \cdot (\mathbf{a}_2 \times \mathbf{a}_3)}. \tag{8.35}$$

- The real space lattice is the reciprocal lattice of its reciprocal lattice:

$$2\pi \frac{\mathbf{b}_2 \times \mathbf{b}_3}{\mathbf{b}_1 \cdot (\mathbf{b}_2 \times \mathbf{b}_3)} = \mathbf{a}_1. \tag{8.36}$$

  and cyclic permutations.

8.3 Diffraction

A standard way to determine experimentally the lattice structure of a given material is to use wave diffraction off the material. As is well known, and explicitly recalled later on, a significant diffraction signal is obtained when the wavelength of the incident waves is of the same order as the characteristic dimensions of the structure, in this case the crystal structure. Different types of waves interact differently with the material, and so, depending on which physical quantities we want to measure, different types of radiations are used.

If we are interested in the crystalline structure of the material, a typical length scale is the distance between atoms, call it a, the lattice constant or some average distance. This is a scale of the order of Angstrom.

If we use photons as the incident waves, wavelengths of this order are found if the photon energies are of the order of 10–50 keV, which means using X-rays. The photons will interact mostly with the electrons in the material, and, since these are distributed periodically according to the periodicity of the lattice, one can get information about the lattice structure. If we use neutrons, since these are neutral but have spin, we may expect that these couple to the spins in the material, such as the spins of the nuclei and the spins of the electrons. Since the neutrons have a mass that is much larger than that of the electrons, they will couple preferentially to the spins of the nuclei. The typical suitable energies of the neutrons are of the order of 40–50 meV.

The methods used are not without some difficulties. In addition to the imperfections of the lattices, there are temperature effects that lead to lattice fluctuations and a less sharp diffraction signal. Also, in the case of the spin diffraction, different isotopes have in general different spins that are distributed randomly, with the consequent breaking of the perfect crystal translational symmetry. Also, the spins may be oriented along different directions that may be random, which also leads to a less sharp signal. On the other hand, if there are ordered magnetic structures in the material, this dispersion of the signal is absent, and actually a clear set of peaks in the diffraction pattern allow the determination of the lattice magnetic structure, which may or not coincide with the lattice structure of the material.

In addition to electromagnetic radiation and neutron diffraction, one may also consider electron diffraction with energies of the order of 10–50 eV. The electrons scatter strongly from the material through Coulomb scattering and spin scattering (even though typically smaller with respect to the charge scattering), and, if the material is thick, there are multiple scatterings that complicate the analysis. It is therefore a method that is mainly used to study surfaces.

8.3.1 Bragg's Law

Consider we have a wave with wave vector $\mathbf{k}$ incident on the surface of a material and with amplitude $\psi_{\mathbf{k}}(\mathbf{r})$. As the wave penetrates the material, it will be under the action of a potential of the form $V(\mathbf{r})$. In general, the wave will be partly reflected and partly transmitted, and its wave vector may be changed to a different value $\mathbf{k}'$. As a first approximation, we may use the Born approximation to calculate the scattering amplitude[3]

$$M_{\mathbf{k},\mathbf{k}'} = \frac{1}{V}\int d^3r\,\psi^*_{\mathbf{k}'}(\mathbf{r})V(\mathbf{r})\psi_{\mathbf{k}}(\mathbf{r}). \tag{8.37}$$

In the case of plane waves, we write:

$$\begin{aligned} \psi_{\mathbf{k}}(\mathbf{r}) &= e^{i\mathbf{k}\cdot\mathbf{r}} \\ \psi_{\mathbf{k}'}(\mathbf{r}) &= e^{i\mathbf{k}'\cdot\mathbf{r}}. \end{aligned} \tag{8.38}$$

We may now recall that the potential is periodic and may be expanded as

$$V(\mathbf{r}) = \sum_{\mathbf{g}} V_{\mathbf{g}} e^{i\mathbf{g}\cdot\mathbf{r}}. \tag{8.39}$$

Then

$$\begin{aligned} M_{\mathbf{k},\mathbf{k}'} &= \frac{1}{V}\int d^3r e^{i(\mathbf{k}-\mathbf{k}')\cdot\mathbf{r}}V(\mathbf{r}) \\ &= \sum_{\mathbf{g}} V_{\mathbf{g}}\frac{1}{V}\int d^3r e^{i(\mathbf{k}-\mathbf{k}'+\mathbf{g})\cdot\mathbf{r}} \\ &= \sum_{\mathbf{g}} \delta_{\mathbf{k}',\mathbf{k}+\mathbf{g}} V_{\mathbf{g}} = V_{\mathbf{k}'-\mathbf{k}}. \end{aligned} \tag{8.40}$$

We conclude, therefore, that a transition (diffraction) is only possible if $\mathbf{k}' - \mathbf{k}$ is a reciprocal lattice vector.

Let us consider that the scattering is elastic, which means that there is no energy absorption by the lattice. Then, we have that $|\mathbf{k}| = |\mathbf{k}'|$. Recalling the restriction that the wave vector difference is a reciprocal lattice vector, we can write that

$$\begin{aligned} \mathbf{k}' &= \mathbf{k} + \mathbf{g} \\ (k')^2 &= k^2 + g^2 + 2\mathbf{k}\cdot\mathbf{g} \\ g^2 + 2\mathbf{k}\cdot\mathbf{g} &= 0 \\ \mathbf{k}\cdot\mathbf{g} &= |\mathbf{k}||\mathbf{g}|\cos\alpha \\ |\mathbf{g}| + 2|\mathbf{k}|\cos\alpha &= 0, \end{aligned} \tag{8.41}$$

[3] See also [Kittel].

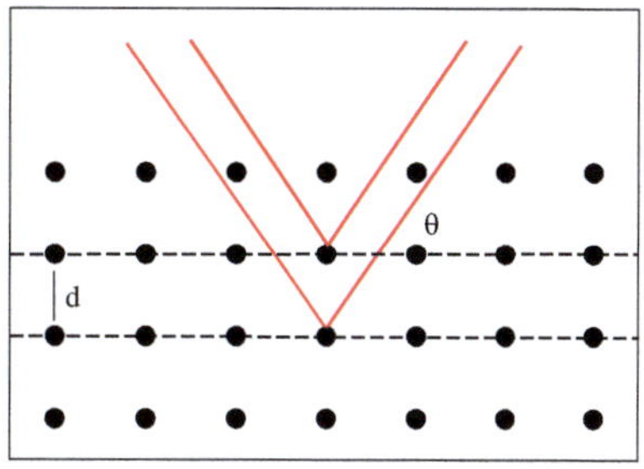

Fig. 8.13 Diffraction of an incident wave on a crystal. d is the separation between the planes along the direction perpendicular to the surface of the material. The angle θ is the angle between the diffracted wave and the crystal planes

where α is the angle between the wave vector and the reciprocal lattice vector. If we define another angle, θ, such that 2θ is the angle between $\mathbf{k}$ and $\mathbf{k}'$, we can see that $\alpha = \pi/2 + \theta$ and so $\cos\alpha = -\sin\theta$. Therefore, we get that

$$|\mathbf{g}| = 2|\mathbf{k}|\sin\theta. \tag{8.42}$$

Recalling that the distance between parallel planes is related to the absolute value of the reciprocal lattice vector, we may write:

$$d = \frac{2\pi N}{|\mathbf{g}|} = \frac{2\pi N}{2|\mathbf{k}|\sin\theta} = \frac{\lambda}{2}\frac{N}{\sin\theta}$$

$$2d\sin\theta = N\lambda. \tag{8.43}$$

Here, λ is the wavelength of the radiation and N is an integer. This relation is called Bragg's law.

We conclude that for a given incident radiation, characterized by the value of the wavelength, and for a given set of planes distant by a value d, there is a set of angles, θ_N, for which the diffraction occurs. Determining the angles for a given orientation of the crystal allows to determine the distances between a set of planes characteristic of a given periodic potential and, therefore, a characteristic atom distribution in the material. Rotating the material we may find other sets of angles corresponding to different values of d. In general, this a complicated procedure if the structure is unknown, but it is possible to carry out this diffraction experiment to determine the crystalline structure. A commonly followed procedure consists in assuming many different structures, predicting the values that follow Bragg's law, and comparing to the experimental results.

As shown in Fig. 8.13, the condition for Bragg law is the constructive interference of the reflected waves from consecutive parallel crystal planes. Since the diffraction amplitude is essentially $|V_{\mathbf{k}'-\mathbf{k}}|^2$, and since the Fourier amplitude of the potential decreases with the difference between the wave vectors, we expect a high scattering amplitude if the reciprocal vector is small and a small intensity if the reciprocal vector is large.

There are other ways to fix the conditions for the existence of diffraction. A way is to express the equation $\Delta\mathbf{k} = \mathbf{g}$ in terms of its projections on the real space basis vectors. Specifically this leads to the so-called Laue equations

$$\begin{aligned}\Delta\mathbf{k}\cdot\mathbf{a}_1 &= 2\pi v_1\\ \Delta\mathbf{k}\cdot\mathbf{a}_2 &= 2\pi v_2\\ \Delta\mathbf{k}\cdot\mathbf{a}_3 &= 2\pi v_3,\end{aligned} \tag{8.44}$$

using that $\mathbf{b}_i\cdot\mathbf{a}_j = 2\pi\delta_{ij}$ with $\mathbf{g} = v_1\mathbf{b}_1 + v_2\mathbf{b}_2 + v_3\mathbf{b}_3$.

Another way is to note that the relation $|\mathbf{g}|^2 + 2\mathbf{k}\cdot\mathbf{g} = 0$ may also be written, changing $\mathbf{g}\to-\mathbf{g}$, as

$$2\mathbf{k}\cdot\mathbf{g} = |\mathbf{g}|^2, \tag{8.45}$$

which is another way of implementing the diffraction condition.

This previous relation may also be written as

$$\mathbf{k}\cdot\left(\frac{1}{2}\mathbf{g}\right) = \left(\frac{\mathbf{g}}{2}\right)^2. \tag{8.46}$$

Once again, $\mathbf{k}$ will be diffracted if it obeys this equation. For instance, $\mathbf{k} = \mathbf{g}/2$ will be diffracted. But this is just the condition for the limits of the Brillouin zone (recall that it is the Wigner-Seitz cell of the reciprocal lattice). Therefore, we reach the conclusion that the limits of the Brillouin zone correspond to the values of the wave vector $\mathbf{k}$ that are in the condition of the Bragg reflection. As an example, take the $1d$ case and call a the lattice constant of a simple one atom cell. So for an incident wave along the direction of the lattice, $a = d$. The condition for diffraction is $2a = N\lambda$. Note that $k = 2\pi/\lambda$. Then, it is easy to see that the limit of the first Brillouin zone is $k = \pi/a$ which implies $\lambda = 2a$.

8.3.2 *Diffraction Intensity: Form Factor and Structure Factor*

Let us now consider in some more detail the transition amplitude that regulates the intensity of the scattered wave. A remark is that the results are only strictly true at zero temperature, as already stressed. The intensity is maximal at zero temperature, and it turns out that at finite temperature the Bragg condition is still satisfied, but the intensity decreases. It will be shown in the next chapter that the intensity is reduced exponentially from the zero temperature result as

$$I \sim |V_{\mathbf{g}}|^2 e^{-W}, \tag{8.47}$$

where W is called the Debye-Waller factor. It increases as the temperature increases. Actually, even at zero temperature it is nonvanishing due to the zero point motion.

We should recall that the real space potential $V(\mathbf{r})$ is actually the result of the interaction of the wave (or particle) with all the lattice points. That is, taking the rigid lattice approximation, the potential at the spatial point $\mathbf{r}$ is given by the sum of the potentials of the atoms located at positions $\mathbf{R_l}$, with V_a the atomic potential:

$$V(\mathbf{r}) = \sum_{\mathbf{l}} V_a(\mathbf{r} - \mathbf{R_l}). \tag{8.48}$$

Therefore, we may write explicitly:

$$M_{\mathbf{k},\mathbf{k}'} = \sum_{\mathbf{l}} \frac{1}{V} \int d^3r e^{i(\mathbf{k}-\mathbf{k}')\cdot\mathbf{r}} V_a(\mathbf{r} - \mathbf{R_l}). \tag{8.49}$$

Defining

$$V_a(\mathbf{k}) = \frac{1}{v_c} \int d^3r e^{-i\mathbf{k}\cdot\mathbf{r}} V_a(\mathbf{r}), \tag{8.50}$$

with v_c the volume of one cell out of N cells ($V = Nv_c$), then, defining $\mathbf{q} = \mathbf{k}' - \mathbf{k}$, we get:

$$M_{\mathbf{q}} = V_a(\mathbf{q}) \frac{1}{N} \sum_{\mathbf{l}} e^{-i\mathbf{q}\cdot\mathbf{R_l}}, \tag{8.51}$$

which can be seen as a product of two factors: a first one called the atomic form factor and the second the structure form factor, characteristic of a given lattice structure. The structure factor reduces to $\delta_{\mathbf{q},\mathbf{g}}$ and so

$$M_{\mathbf{q}} = V_a(\mathbf{g}). \tag{8.52}$$

The lattice dependence is encoded in the set of reciprocal lattice values, $\mathbf{g}$.

Since the periodic potential is associated with the electron density of the material, $n(\mathbf{r})$, the scattering amplitude may equivalently be written as

$$F = \int d^3r n(\mathbf{r}) e^{i(\mathbf{k}-\mathbf{k}')\cdot\mathbf{r}}, \tag{8.53}$$

or, since it is characterized by a given reciprocal lattice vector, as

$$F_{\mathbf{g}} = N \int_{cell} d^3r n(\mathbf{r}) e^{-i\mathbf{g}\cdot\mathbf{r}} = N S_{\mathbf{g}}, \tag{8.54}$$

where $S_{\mathbf{g}}$ is called the structure factor. Consider now a lattice with a basis of s atoms. Using that

$$n(\mathbf{r}) = \sum_{j=1}^{s} n_j(\mathbf{r} - \mathbf{r}_j) \tag{8.55}$$

we can write:

$$\begin{aligned} S_{\mathbf{g}} &= \sum_j \int d^3 r n_j(\mathbf{r} - \mathbf{r}_j) e^{-i\mathbf{g}\cdot\mathbf{r}} \\ &= \sum_j e^{-i\mathbf{g}\cdot\mathbf{r}_j} \int d^3 r n_j(\boldsymbol{\rho}) e^{-i\mathbf{g}\cdot\boldsymbol{\rho}}, \end{aligned} \tag{8.56}$$

where we defined $\boldsymbol{\rho} = \mathbf{r} - \mathbf{r}_j$. Let us now define the atomic form factor as

$$f_j = \int d^3 r n_j(\boldsymbol{\rho}) e^{-i\mathbf{g}\cdot\boldsymbol{\rho}}. \tag{8.57}$$

Then, we finally obtain:

$$S_{\mathbf{g}} = \sum_j f_j e^{-i\mathbf{g}\cdot\mathbf{r}_j}. \tag{8.58}$$

Recalling that we may define $\mathbf{r}_j = x_j\mathbf{a}_1 + y_j\mathbf{a}_2 + z_j\mathbf{a}_3$ and $\mathbf{g} = v_1\mathbf{b}_1 + v_2\mathbf{b}_2 + v_3\mathbf{b}_3$ we can write that

$$S_g(v_1, v_2, v_3) = \sum_j f_j e^{-2\pi i(v_1 x_j + v_2 y_j + v_3 z_j)}. \tag{8.59}$$

We may now consider some examples for different lattices with a basis.

Consider the *bcc* lattice, which applies for instance to metallic Na. In this lattice one finds identical atoms at the locations $x_1 = y_1 = z_1 = 0$ and $x_2 = y_2 = z_2 = 1/2$. Since the atoms are identical, the atomic form factor is independent of j, call it f. Therefore

$$S_g(v_1, v_2, v_3) = f\left(1 + e^{-\pi i(v_1+v_2+v_3)}\right). \tag{8.60}$$

We can conclude that $S_g = 0$ if $v_1 + v_2 + v_3$ is an odd integer and $S_g = 2f$ if the sum is an even integer. There is no diffraction if the sum of the integers is odd, while there is diffraction if the sum is even.

Considering for instance a fcc lattice, we find identical atoms at the locations $x_1 = 0, y_1 = 0, z_1 = 0$; $x_2 = 0, y_2 = 1/2, z_2 = 1/2$; $x_3 = 1/2, y_3 = 0, z_3 = 1/2$; and $x_4 = 1/2, y_4 = 1/2, z_4 = 0$. The structure factor yields:

$$S_g(v_1, v_2, v_3) = f\left(1 + e^{-\pi i(v_2+v_3)} + e^{-\pi i(v_1+v_2)} + e^{-\pi i(v_1+v_3)}\right). \tag{8.61}$$

Therefore, if v_1, v_2, v_3 are all even or all odd, we get that $S_g = 4f$. If one of the integers is odd and two are even or one is even and the other are odd, $S_g = 0$.

8.4 Exercises

1. Suppose that the energy between a potassium atom and a chlorine atom, as a function of the distance R, is given by a Born-Mayer model where, in addition to the Coulomb attraction between the ions K^+ and Cl^-, there is a repulsive energy:

$$E_{rep} = A\exp(-R/\rho),$$

with the parameters $A = 0.6 \times 10^4$ eV and $\rho = 0.33\,\AA$:

(a) What is the cohesive energy of a KCl molecule according to this model?
(b) What is the length of the bond?
(c) In the harmonic approximation, what is the vibration frequency of the molecule?

2. Using the Lennard-Jones potential, calculate the cohesive energies of neon in the structures bcc and fcc. The sums on the lattice are

$$\sum_j p_{ij}^{-12} = 9.11418(bcc)$$

$$= 12.13188(fcc)$$

$$\sum_j p_{ij}^{-6} = 12.2533(bcc)$$

$$= 14.45392(fcc)$$

and $\epsilon = 50 \times 10^{-16} erg$ and $\sigma = 2.74\,\AA$. Determine the lattice constant for both structures.
3. Calculate the cohesive energy of KCl in the cubic lattice of ZnS and in the lattice of $NaCl$. The corresponding Madelung constants are $\alpha_{ZnS} = 1.6381$, $\alpha_{NaCl} = 1.7476$, and assume that $z\lambda = 2.05 \times 10^{-8} erg$, $\rho = 0.326\,\AA$ for both structures. Which structure has the smaller energy? Determine the equilibrium distance between the first neighbors, R_0. The experimental results are $R_0 = 3.147\,\AA$ and the cohesive energy is $= 169.5$ kcal/mol.
4. Which are the symmetry operations of a cube? How many operations are there? Classify them. Do they form a group?
5. Show that a tetragonal Bravais lattice bcc does not exist.

6. Assuming that the atoms are rigid spheres, determine the maximal fraction of the occupied volume for the lattices sc, bcc, fcc, and hcp. Determine the ideal ratio c/a for the crystal hcp.
7. Consider the structure of $CsCl$. The lattice constant is $4.12\,\AA$, all the bond lengths indicated in the figure are equal, and all the angles between the links are identical:

   (a) Which lattice is it and what is the basis of this crystal?
   (b) How many first neighbors (and of what type) have Cs and Cl?
   (c) What is the distance to the first neighbors of Cs?
   (d) What is the angle between the chemical bonds?

8. Consider an fcc lattice. Draw the plane $(02\bar{3})$. Draw the direction $[\bar{1}13]$. What is the distance between planes (100) that go through lattice points? What is the distance between planes (111) that go through lattice points?
9. Consider the conventional cubic cell of a cuprite crystal, Cu_2O. The oxygen atoms are found in the vertices and at the center of the cube. The copper atoms are along a diagonal of the cube, equidistant of the oxygen atoms of the center and those at the vertices. The lattice constant is $a = 4.27\,\AA$:

   (a) Which is the lattice, and what is the basis of this crystal?
   (b) Which are the nearest-neighbors of the atoms of O and Cu, and at what distance are they found?
   (c) The chemistry tables list radius of $1.32\,\AA$ to O^{--} and $0.96\,\AA$ to Cu^{+}. With these radii and a model of rigid spheres, what would be the lattice constant?

10. Determine the reciprocal lattices of the lattices sc, bcc, and fcc and of the hexagonal lattice.
11. Determine the volume of the unit cell of the reciprocal lattice as a function of $\mathbf{a}_1, \mathbf{a}_2, \mathbf{a}_3$ of the direct lattice (real space lattice).
12. X-rays are scattered by the electronic density. In the case of the hydrogen atom in its ground state, the electronic density is given by

$$n(r) = \frac{1}{\pi a_0^3} \exp(-2r/a_0),$$

   where a_0 is Bohr radius. Show that the form factor of hydrogen is

$$f_g = \frac{16}{(4 + (g a_0)^2)^2}.$$

13. An X-ray beam with energy 17.444 keV shines on a sample of polycrystalline aluminum. The aluminum crystals have an fcc structure with lattice constant $a = 4.05\,\AA$:

   (a) Calculate the four smaller diffraction angles (2θ) observed under these conditions.

(b) If a film sensitive to the X-rays is exposed at a distance of 20 cm from the sample, which image of the diffraction figure will be formed? Do not forget to calculate the distances.

14. A neutron beam with energies 0.5 eV shines on a sample of polycrystalline aluminum. The aluminum crystals have an fcc structure with lattice constant $a = 4.05\,\AA$.

(a) Calculate the four smaller diffraction angles (2θ) observed under these conditions using Laue equation.
(b) If the sample is a single crystal, the x-axis parallel to the axis of the beam, the z-axis vertical, and the crystal is initially placed with the cubic axes parallel to xyz, show how to rotate the crystal around the vertical axis to observe diffraction.
(c) Calculate now the six smaller angles using the Laue condition for the conventional cell.
(d) Show that for the angles that did not show up for the primitive cell, the structure factor vanishes.

15. Consider a wave vector $\mathbf{k}$ in one face of the first Brillouin zone:

(a) Show that there is a vector $\mathbf{g}$ of the reciprocal lattice such that $\mathbf{k}+\mathbf{g}$ is the opposing face of the Brillouin zone.
(b) Show that $|\mathbf{k}+\mathbf{g}| = |\mathbf{k}|$.
(c) Show that in first-order perturbation theory in the potential $V(\mathbf{r})$, we obtain that

$$E_{\mathbf{k}} = \frac{\hbar^2}{2m}(\mathbf{k})^2 + V_0 \pm V_{\mathbf{g}}.$$

(d) Consider now another wave vector $\mathbf{k}' = \mathbf{k} + \delta\mathbf{k}$ where $\delta\mathbf{k}$ is small (as compared to $2\pi/a$). Show that at first-order perturbation theory in the potential $V(\mathbf{r})$, we get:

$$E_{\mathbf{k}+\delta\mathbf{k}} = \frac{\hbar^2}{2m}(\mathbf{k}+\delta\mathbf{k})^2 + \frac{\hbar^2}{2m}\mathbf{g}\dot{\delta}\mathbf{k} + V_0 \pm \sqrt{\left(\frac{\hbar^2}{2m}\mathbf{g}\dot{\delta}\mathbf{k}\right)^2 + V_{\mathbf{g}}^2}.$$

(e) Show that $\mathbf{g}\dot{\nabla}_{\mathbf{k}}E_{\mathbf{k}} = 0$ in the face of the Brillouin zone.

16. Consider the wave vector $\mathbf{k} = \frac{2\pi}{a}(1, 1/2, 0)$ for an fcc crystal with lattice constant a:

(a) Show that that wave vector is in the Bragg plane associated with the vector $\mathbf{g}_1 = \frac{2\pi}{a}(2, 0, 0)$.
(b) Show that that vector is in two other Bragg planes and identify their respective vectors.

(c) Suppose that the potential is weak and write the matrix that you have to diagonalize according to the perturbation theory of degenerate states.
(d) Obtain the energy of these states as a function of $V(|\mathbf{g}|)$.

17. Show that the field created by a set of electric dipoles oriented along the axis of a crystal vanishes at a point of a cubic lattice.

Chapter 9
Lattice Vibrations: Phonons

9.1 Dispersion Relation of Lattice Vibrations

The atoms in a solid are not at rest and oscillate around average positions. For instance, a finite temperature leads to these oscillations, and even at zero temperature there is the zero point motion. Also, external perturbations may induce oscillations of the atoms around their equilibrium positions. Therefore, the crystal lattice is not static.[1]

The oscillations of the atoms may be longitudinal or transversal in the sense that if the oscillation can be seen as a propagating wave along a given direction, if the atoms oscillate along that direction the oscillation mode is called longitudinal and if the atoms oscillate in the plane perpendicular to the wave propagation direction, the modes are called transversal, as shown in Fig. 9.1.

In the case of waves that propagate along directions of high symmetry in a crystal, such as considering a cubic lattice along the directions [100], [110], or [111], entire planes of atoms are moving together, and so, even though the system is three-dimensional, we may view the motion as unidimensional.

To simplify, let us then start our analysis considering a one-dimensional crystal. Consider that the centers of the atoms are located a distance a from each other and call a general position along the chain as s (and so the atom is placed along the axis, call it x, at a position $x_s = sa$). Let us call the displacement of atom s with respect to its average position as u_s. At the equilibrium positions, the energy of the system of atoms is minimized, which means that the force on each atom vanishes. Therefore, as we displace the atom from its equilibrium position, we expect a force acting on the atom that tends to bring the atom to its original position. The forces between the atoms are the result of complicated electromagnetic interactions between all the charges in the chain. Later we will try to formulate this situation in

[1] See also [Kittel] and [AshcroftMermin].

P. D. Sacramento, *Introduction to Condensed Matter Theory*,
https://doi.org/10.1007/978-3-031-86251-9_9

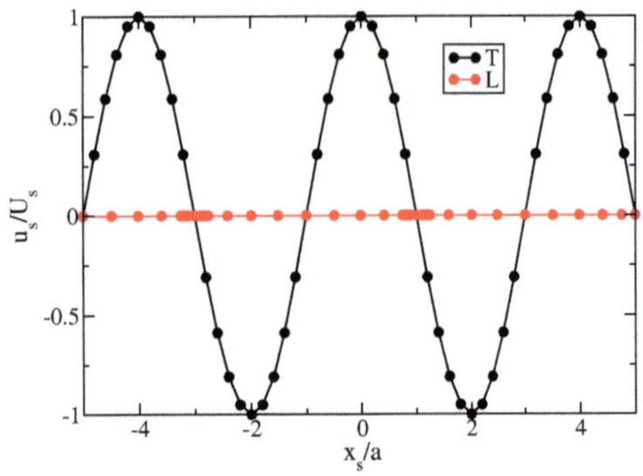

Fig. 9.1 Oscillations of atom positions around equilibrium positions due to the passage of a transverse or longitudinal wave

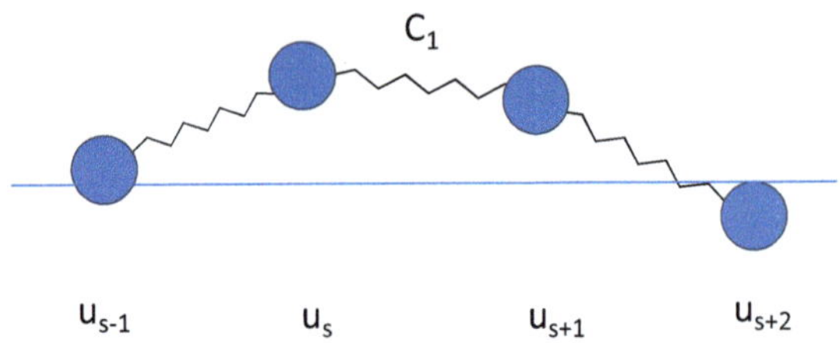

Fig. 9.2 The interaction between neighboring atoms in a one-dimensional lattice is modeled by elastic springs that connect the atoms

terms of the electromagnetic properties of the system. However, let us use a much simplified method that encompasses the physics of the oscillations, approximating the electromagnetic forces by some effective springs, which tend to restore the equilibrium positions of the atoms. This is shown schematically in Fig. 9.2.

Let us write that the force on atom s can be written to leading order as

$$F_s = \sum_p C_p \left(u_{s+p} - u_s\right). \tag{9.1}$$

Considering a small displacement from equilibrium positions, we approximate the force by a linear expression, and therefore the potential energy as a quadratic expression. At this point we neglect the nonlinear terms. The forces are parametrized by the force constants C_p, and we assume that all atoms contribute to the force on a given atom. Clearly, we expect that as the distance, p, between two atoms increases, the force constant C_p should decrease fast. We note that in general the force constants are different if the motion is longitudinal or transversal. From the expression of the force, we may deduce the equations of motion of the atoms. This leads to

$$M\frac{d^2u_s}{dt^2} = \sum_p C_p \left(u_{s+p} - u_s\right), \tag{9.2}$$

where M is the mass of the atoms, considered for the moment to be all of the same type. Let us look for solutions with a given time dependence of the type:

$$u_s = U_s e^{-i\omega t}. \tag{9.3}$$

Substituting in the equation of motion, we get:

$$-\omega^2 M U_s = \sum_p C_p \left(U_{s+p} - U_s\right). \tag{9.4}$$

Let us now look for a spatial dependence of the type:

$$U_s = U e^{iska}. \tag{9.5}$$

Again, substituting we get

$$-\omega^2 M U e^{iska} = \sum_p C_p U \left(e^{iska} e^{ipka} - e^{iska}\right)$$

$$-\omega^2 M = \sum_p C_p \left(e^{ipka} - 1\right). \tag{9.6}$$

For a symmetric lattice, we may take $C_p = C_{-p}$ and we get:

$$\omega^2 M = 2 \sum_{p>0} C_p \left(1 - \cos pka\right). \tag{9.7}$$

Note that in order to have lattice stability, $C_p > 0$. We get therefore that the frequency, ω, of the oscillation depends on k, the wave number. Therefore, the system is dispersive since $\omega = \omega(k)$.

Let us now calculate the group velocity of these oscillations. We may write that

$$M \frac{d\omega^2}{dk} = 2M\omega \frac{d\omega}{dk} = 2a \sum_{p>0} C_p p \sin pka. \tag{9.8}$$

At very small wave vector, $k \sim 0$, we get that the dispersion relation is linear. Therefore, $\omega = v_g k$. This leads to

$$2M v_g^2 k = 2a^2 \sum_{p>0} C_p p^2 k$$

$$v_g(k \sim 0) = \sqrt{\frac{a^2}{M} \sum_{p>0} C_p p^2}. \tag{9.9}$$

On the other hand, focusing on the limits of the first Brillouin zone, $k = \pm\pi/a$, we get that

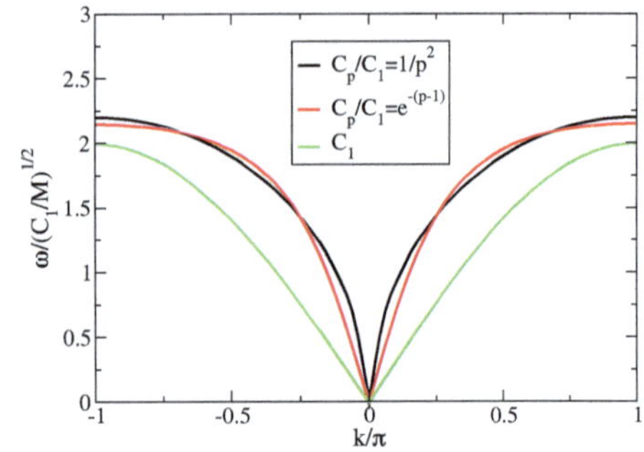

Fig. 9.3 Dispersion relation $w(k)$ for different types of spring couplings between neighbors distant by p sites: $1/p^2$, $exp(-(p-1)$ and limiting case of considering only the coupling between nearest-neighbors: $p = 1$

$$v_g\left(k = \pm\frac{\pi}{a}\right) = \left(\frac{d\omega}{dk}\right)_{k=\pm\pi/a} = 0. \tag{9.10}$$

Note that the frequency is nonvanishing, as can be seen in Fig. 9.3. Therefore, the group velocity vanishes and the wave is stationary. At small wave numbers, the velocity is finite and is associated with the motion of the atoms that we call sound waves.

Since we expect that C_p decreases fast as p increases, it is usual to take the approximation that only nearest-neighbor atoms couple, and so we only keep the term with $p = 1$. Call $C_{p=1} = C$. The dispersion relation then simplifies to

$$\omega = 2\sqrt{\frac{C}{M}}\left|\sin\frac{ka}{2}\right|. \tag{9.11}$$

As above, we only consider the positive square root since the oscillation frequencies should be taken as positive; otherwise, the lattice would not be stable. In Fig. 9.3 we compare the dispersion relations for different dependencies of C_p. Accordingly, for the sound velocity around zero wave number, we get the simplified result:

$$v_s = \frac{d\omega}{dk} = a\sqrt{\frac{C}{M}}. \tag{9.12}$$

We conclude that the heavier the atoms (larger mass, M) the lower the sound velocity and the stiffer, or more tightly bound, the lattice is (larger C) the higher the sound velocity.

The solution we looked for satisfies:

$$\frac{u_{s+1}}{u_s} = e^{ika}. \tag{9.13}$$

Recall that $-\pi \leq ka < \pi$. The meaning is that the wave number, k, is restricted to the first Brillouin zone since other, larger, values of k are redundant. Indeed, recall that $k = 2\pi/\lambda$ which means that $\lambda = 2\pi/k$. Considering the one-dimensional lattice with lattice constant, a, the minimum wavelength allowed is $\lambda = 2a$. Smaller

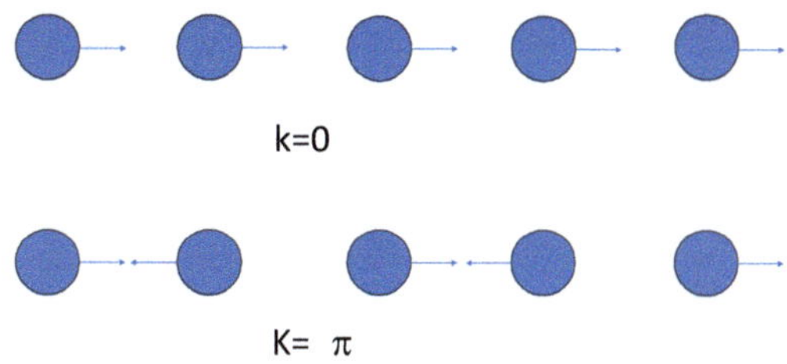

Fig. 9.4 Oscillation modes for zero momentum and Bragg reflection condition

wavelengths have no physical support on the real lattice system. Therefore, $\lambda \geq 2a$. This means that $|k| \leq \pi/a$. Another way to understand is to note that if we take two points in wave number space multiplied by the lattice constant (ka), one inside the first Brillouin zone and one outside, they differ by 2π and therefore it is the same wave motion.

Let us now consider the spatial dependence of the wave and consider $k = \pm\pi/a$. Then

$$\begin{aligned} u_s &= ue^{iska} = ue^{i\pi s} = (-1)^2 u \\ \frac{u_{s+1}}{u_s} &= e^{ika} = e^{i\pi} = (-1). \end{aligned} \tag{9.14}$$

This result implies that neighboring atoms are out of phase (opposite phase). This, of course, leads to a stationary wave, in agreement with the vanishing group velocity. Considering for instance a longitudinal wave, since neighboring atoms are oscillating out of phase, the center of gravity of the atoms is at rest, and therefore, the group velocity vanishes. This is shown in Fig. 9.4. Recall that this result can also be understood as the condition for Bragg reflection for a wave that has a propagation direction perpendicular to the atoms' positions $\theta = \pi/2$ and, since the distance between the atoms is $d = a$, we get $\lambda = 2a$.

From the dispersion relation of the waves propagating through the material, we may obtain information about the force constants in the system. For instance, considering the low wave number limit, measuring the sound velocity gives us information about the force constant, C. Using the more general expression taking into account the several neighbors, Eq. (9.7), we may obtain the force constants, C_p, using

$$M \int_{-\pi/a}^{\pi/a} dk\omega_k^2 \cos rka = 2\sum_{p>0} C_p \int_{-\pi/a}^{\pi/a} dk\,(1 - \cos pka)\cos rka = -2\pi\frac{C_r}{a}, \tag{9.15}$$

which leads to

$$C_p = -\frac{Ma}{2\pi}\int_{-\pi/a}^{\pi/a} dk\omega_k^2 \cos pka \tag{9.16}$$

and allows to determine the force constants, if we measure experimentally the dispersion relation of the oscillations.

9.1.1 Diatomic Chain: Acoustic and Optical Modes

Consider now a system with two types of atoms, distributed regularly in alternate order. Therefore, the unit cell has two atoms, of different types. In the context of lattice vibrations, one relevant parameter that distinguishes the atoms is the mass. Let us call the two masses, M_1 and M_2. Considering a cubic lattice, the lattice constant, a, has now twice the size: it is the distance between two identical atoms. For instance, if we consider a $NaCl$ crystal, the Na and Cl atoms have different masses. The difference between atoms may also arise from different values of the force constants between neighbors. This occurs for instance in diamond.

In three dimensions there are three degrees of freedom per atom and therefore there are six degrees of freedom per unit cell. Therefore, we expect six branches in the space of the wave vector (since the wave vector of a wave is related to its linear momentum, often we will call from now on the wave vector as the momentum of a wave; there are some subtleties associated with the lattice periodicity that will be discussed later). If we have a monoatomic lattice, there are only three degrees of freedom per unit cell. As mentioned above, there is one longitudinal mode and two transverse modes. In general, the dispersion relations of the three modes are different. A generic notation for the three modes is $\omega_L(k), \omega_{T_1}(k), \omega_{T_2}(k)$. In the case of a diatomic lattice, we expect six modes with dispersion relations $\omega_{L_1}(k), \omega_{L_2}(k)$ for the longitudinal modes, and $\omega_{T_1}(k), \omega_{T_2}(k), \omega_{T_3}(k), \omega_{T_4}(k)$ for the four transverse modes.

At small "momenta" ($k \to 0$), the frequency of some modes also tends to zero ($\omega(k \to 0) \to 0$). In a $3d$ space, there are three space directions, and we expect three translational motions, of very long wavelength ($k \sim 1/\lambda \to 0$) that cost very little energy. These modes are called acoustic modes. The remaining oscillation modes have a finite gap at small momenta and are called optical modes. If there are s atoms per cell, in $3d$ we have $3s$ modes. These modes are organized as s longitudinal modes and $2s$ transverse modes. Also, they are organized as 3 acoustic modes and $3(s-1)$ optical modes. These conditions lead to

$$
\begin{gathered}
1LA \; ; \; 2TA \\
(s-1)LO \; ; \; 2(s-1)TO,
\end{gathered}
\tag{9.17}
$$

where A stands for acoustic, O for optical, L for longitudinal, and T for transverse. For instance, if $s = 2$ (the diatomic case), we get $1LA$, $1LO$, $2TA$, and $2TO$.

Let us then consider explicitly the case of a one-dimensional chain of two types of atoms with different masses and a unique force constant, C. The equations of motion for the displacements of the two atom species, u_s and v_s, are

$$
M_1 \frac{d^2 u_s}{dt^2} = C\left(v_s + v_{s-1} - 2u_s\right)
$$

$$M_2 \frac{d^2 v_s}{dt^2} = C\left(u_{s+1} + u_s - 2v_s\right), \tag{9.18}$$

where we only consider nearest-neighbors. Here, the displacements are only allowed along one direction, say the longitudinal direction.

Let us look for solutions of Eq. (9.18) in the form:

$$\begin{aligned} u_s &= ue^{iska - i\omega t} \\ v_s &= ve^{iska - i\omega t}. \end{aligned} \tag{9.19}$$

Substituting we get that

$$\begin{aligned} \omega^2 M_1 u &= C\left(2u - v - ve^{-ika}\right) \\ \omega^2 M_2 v &= C\left(2v - u - ue^{ika}\right). \end{aligned} \tag{9.20}$$

In order to have a nontrivial solution, the determinant of the matrix of the coefficients of the amplitudes u, v must vanish. That is

$$det \begin{pmatrix} \omega^2 M_1 - 2C & C\left(1 + e^{-ika}\right) \\ C\left(1 + e^{ika}\right) & \omega^2 M_2 - 2C \end{pmatrix} = 0. \tag{9.21}$$

Using that $1 - \cos ka = 2\sin^2 ka/2$, we obtain two solutions:

$$\omega_\pm^2 = C\left(\frac{1}{M_1} + \frac{1}{M_2}\right) \pm C\sqrt{\left(\frac{1}{M_1} + \frac{1}{M_2}\right)^2 - \frac{4}{M_1 M_2}\sin^2\frac{ka}{2}}. \tag{9.22}$$

In the limit of small momenta $k \to 0$, the two solutions behave quite differently:

$$\begin{aligned} \omega_+ &= \sqrt{2C\left(\frac{1}{M_1} + \frac{1}{M_2}\right)} \\ \omega_- &= 0. \end{aligned} \tag{9.23}$$

The first solution has a finite energy (finite gap) and is the optical mode, and the second solution vanishes (acoustic mode). In the figure we plot the dispersion relations of the two modes. Near the Brillouin zone limit, the dispersion relations approach quadratically their limiting values. The slope of the acoustic mode gives the sound velocity with a value:

$$v_s = a\sqrt{\frac{C}{2}\frac{1}{M_1 + M_2}}. \tag{9.24}$$

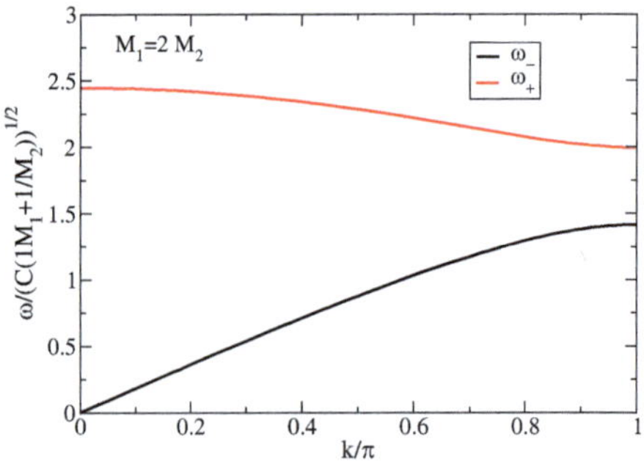

Fig. 9.5 Dispersion relations for the acoustical and optical modes of a diatomic chain

In the limit when $k \to \pi/a$, we get that

$$\omega_{\pm}^2 = C\left(\frac{1}{M_1}+\frac{1}{M_2}\right) \pm C\left(\frac{1}{M_1}-\frac{1}{M_2}\right), \tag{9.25}$$

which takes the values $2C/M_1$ and $2CM_2$, respectively, for $\pm$. The dispersion relations for the two modes are shown in Fig. 9.5.

This analysis was carried out for the longitudinal modes, but a similar analysis could be carried out for the transversal modes: the only difference would be in the values of the force constants. Here we are considering that even though the atoms are distributed along a line ($1d$ spatial distribution), the atoms may oscillate along any direction in a three-dimensional space.

We may now study the vibration modes. The equations of motion are rewritten as

$$\begin{pmatrix} \omega^2 M_1 - 2C & C\left(1+e^{-ika}\right) \\ C\left(1+e^{ika}\right) & \omega^2 M_2 - 2C \end{pmatrix} \begin{pmatrix} u \\ v \end{pmatrix} = 0. \tag{9.26}$$

Using for instance the first row and introducing the dispersion relation, we get for the optical mode, for small momenta:

$$\left(2C\left(1+\frac{M_1}{M_2}\right) - 2C\right)u + 2Cv = 0. \tag{9.27}$$

Therefore, we get:

$$\frac{u}{v} = -\frac{M_2}{M_1} \tag{9.28}$$

which implies that $M_1 u + M_2 v = 0 = M_t x_{CM}$, where $M_t = M_1 + M_2$ and x_{CM} is the center of mass coordinate. The optical mode corresponds to an oscillating dipole of the two atoms in the unit cell, oscillating such that their center of mass is at rest. Since in a ionic crystal, like $NaCl$, the two atoms have opposite electric charges, the two atoms are actually an electric dipole. Its oscillations may be generated by coupling to an external radiation field, and so this mode may be optically excited, hence its name optical mode.

The equation between the wave amplitudes at the two atoms is different in the acoustic case. Considering again the small momentum limit ($k \to 0$), we can write that

$$-2Cu + 2Cv = 0, \tag{9.29}$$

which implies that $u = v$. This corresponds to a translation of the two atoms (and the crystal itself) as a whole. The two atoms move in phase as they oscillate. Since we are in the regime of small momenta, and therefore large wavelengths or large distances, we can take $\lambda >> a$, which means that the lattice structure becomes irrelevant and this gives us the continuum limit, where the lattice structure may be ignored.

We may also analyze what happens in the vicinity of the Brillouin zone limit, $k = \pi/a$. Taking for instance $M_1 > M_2$, we get that the optical mode frequency is given be $\sqrt{2C/M_2}$, and the acoustic mode frequency is $\sqrt{2C/M_1}$. Introducing in the equation for the amplitudes of the optical mode

$$\left(\frac{2C}{M_2}M_1 - 2C\right)u + C\left(1 + e^{-i\frac{\pi}{a}a}\right)v = 0, \tag{9.30}$$

which simplifies to

$$2C\left(\frac{M_1}{M_2} - 1\right)u = 0, \tag{9.31}$$

which means that $u = 0$ and v is undetermined. Therefore, the mass M_1 does not move (it is at rest) and the mass M_2 is free to move. In the case of the acoustic mode, the opposite occurs. The amplitude $v = 0$ and u is undetermined. So the mass M_2 is at rest and the mass M_1 is free to move.

When we have a diatomic lattice (or in general a lattice with a unit cell with several atoms), there are two clear consequences: since the unit cell contains several atoms, the lattice constant related to the size of the unit cell is typically larger than the distance between atoms. Also, there are several oscillation modes as a result of the number of atoms in the unit cell. For instance, in the diatomic case we just considered, there is one acoustic and one optical mode for each type of mode polarization. There is a finite gap between the dispersion relations, both at zero momentum and in the Brillouin zone limit. This last gap appears as a consequence of the different masses of the two atoms. If we start from the diatomic chain description and take $M_1 = M_2$, this gap vanishes. But if the two masses become equal, all the atoms in the lattice are equivalent, and we should recover the monoatomic case. Considering a one-dimensional case for a better understanding, it is clear that the size of the unit cell is now halved: the new lattice constant is half the one of the

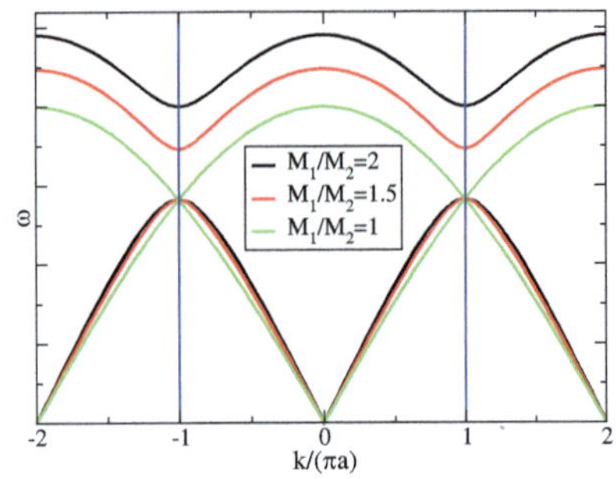

Fig. 9.6 Dispersion relations for different ratios of M_1/M_2. In the case of $M_1 = M_2$, the lattice constant doubles and the Brillouin zone halves

diatomic chain. Therefore, the limit of the Brillouin zone doubles. Also, since there is only one atom in the unit cell, we only have one mode of a given polarization. So the optical mode is now absent and we are left with the acoustic mode.

Let us take the solution for the diatomic lattice and take $M_1 = M_2 = M$. The two solutions of the dispersion relation become now:

$$\omega_+ = 2\sqrt{\frac{C}{M}}\cos\frac{ka}{4}$$
$$\omega_- = 2\sqrt{\frac{C}{M}}\sin\frac{ka}{4}. \tag{9.32}$$

A plot of these solutions is shown in Fig. 9.6. We see the gap closes at $k = \pi/a$. But now this is no longer the limit of the Brillouin zone, since the masses are equal. Recalling the solution of the monoatomic chain, we note that there was only one solution for the dispersion relation and it was of the form of a sine. Actually, if we superimpose that solution, we find that it coincides with the acoustic mode of the diatomic chain with $M_1 = M_2 = M$ and it extends up to $k = 2\pi/a$, recalling that a is the size of the unit cell of the diatomic lattice. Also, we may note that the optical mode of the form of a cosine is just the reflection of the branch for $k > \pi/a$. In conclusion, we see that if we use a Brillouin zone with the edge at $k = \pi/a$, we need to include two modes (the acoustic and the optical one). To describe the limit of same mass, we just need one mode if we consider an "extended" Brillouin zone with limit $k = 2\pi/a$. We will return to this later on when we discuss the electronic energy bands in the crystal.

9.1.2 Wave Momentum and Crystalline Momentum

Associated with the displacement of a given atom from its equilibrium position, u_r, we may define a Fourier amplitude, u_k, of a given one-dimensional lattice with N sites, in the usual way:

$$u_r = \frac{1}{\sqrt{N}} \sum_k u_k e^{ikra}. \tag{9.33}$$

The values of k are determined by the boundary conditions. Let us take periodic boundary conditions, such that the lattice is actually a closed ring, satisfying $u_1 = u_{N+1}$. This implies that $e^{ikNa} = 1$ and $k_n = 2\pi n/(Na)$, with $n = 0, \pm 1, \pm 2, \cdots$. Associated with the displacements and their velocities we obtain the total momentum of the lattice of atoms of mass M, as

$$P = \sum_r M \frac{du_r}{dt}. \tag{9.34}$$

Let us take the wave solution:

$$u_r = u_0 e^{i(kra+\omega t)}, \tag{9.35}$$

which leads to

$$\frac{du_r}{dt} = u_0 i\omega e^{i(rka+\omega t)} \tag{9.36}$$

and associate a momentum to each Fourier component. This is given by

$$\begin{aligned} P_k &= M \sum_r u_0 i\omega e^{i\omega t} e^{irka} \\ &= i\omega M u_0 e^{i\omega t} \sum_{r=1}^{N} e^{irka} \\ &= i\omega M u_0 e^{i\omega t} \frac{1 - e^{iNka}}{1 - e^{ika}} \\ &= i\omega M u_0 e^{i\omega t} \frac{1 - e^{i2\pi n}}{1 - e^{i\frac{2\pi n}{N}}}. \end{aligned} \tag{9.37}$$

This leads to $P_k = 0$, if $k \neq 0$, and $P_k = i\omega M u_0 e^{i\omega t} N$, if $k = 0$. Therefore, the momentum of the vibration mode with a given k (given momentum) vanishes unless $k = 0$. This is the case of a translation in block of the crystal. The oscillations or vibrations we are considering are actually the result of the relative motion of the atoms with respect to each other, and so there is no actual physical momentum being transported. These oscillations are harmonic displacements of the atoms with respect to their equilibrium positions.

Let us now assume that we interact with the solid sending some sort of radiation, such as X-rays or neutrons, with a given wave number (or momentum). The scattering of the incident wave off the lattice was considered in the previous chapter, and it is determined by Bragg's law. This can be written as

$$\mathbf{k}' = \mathbf{k} + \mathbf{g}, \tag{9.38}$$

where $\mathbf{g}$ is a reciprocal lattice vector, $\mathbf{k}$ is the wave vector of the incident wave (for instance a photon) and $\mathbf{k}'$ is the wave vector of the scattered wave. This expression applies when we have elastic scattering. It may happen that the scattering is inelastic and some energy is given to or absorbed from the crystal. The classical harmonic motions of the atoms considered in our description may be quantized in the usual way. The quantized oscillations of the atoms are called phonons. In the case of inelastic scattering, there may be involved a phonon that may be created or absorbed:

$$\begin{aligned} \mathbf{k}' + \mathbf{q} &= \mathbf{k} + \mathbf{g} \\ \mathbf{k}' &= \mathbf{q} + \mathbf{k} + \mathbf{g}, \end{aligned} \tag{9.39}$$

respectively. Using inelastic scattering it is possible to experimentally determine the dispersion relation of the phonons and, therefore, of the oscillation modes of the atoms in the crystal.

9.1.3 Quantization of the Vibrations: Phonons

To simplify notation, several examples we have considered were in the context of a one-dimensional system. In $3d$ the results are just extensions that take into account three independent terms, if we choose appropriately the coordinate system (directions of the oscillation eigenmodes). Let us simplify and consider a three-dimensional Bravais lattice with one atom per unit cell. The Hamiltonian can be written as

$$\begin{aligned} H &= \sum_{l=1}^{N} \frac{M}{2} \left(\frac{d\mathbf{u}_l}{dt} \right)^2 + U\left(\mathbf{R}_1, \cdots, \mathbf{R}_N\right) \\ U\left(\mathbf{R}_1, \cdots, \mathbf{R}_N\right) &= U^0 + \frac{1}{2} \sum_{l,j} \sum_{l',j'} u_l^j G_{ll'}^{jj'} u_{l'}^{j'}. \end{aligned} \tag{9.40}$$

The positions of the atoms are specified by $\mathbf{R}_l = \mathbf{R}_l^0 + \mathbf{u}_l$ given as the equilibrium position, plus the displacement from equilibrium, and U^0 is the energy of the rigid lattice. The quantity $\mathbf{G}$ is the matrix of the second derivatives of the energy with respect to the displacements of atoms in positions l and l', along the directions with components j and j'. For a Bravais lattice, $\mathbf{G}$ is symmetric under the exchange of the two sites, $G_{ll'} = G_{l'l}$. The quantity is also invariant under lattice translations, and therefore we can write that $\mathbf{G}(\mathbf{h})$ where $\mathbf{h} = \mathbf{R}_l^0 - \mathbf{R}_{l'}^0$.

The equations of motion are just the generalization to three dimensions of the cases previously considered. They can be written, for each Fourier component in $3d$ space, as

$$M\omega^2 \mathbf{u}_0(\mathbf{k}) = \mathbf{G}(\mathbf{k})\mathbf{u}_0(\mathbf{k}), \tag{9.41}$$

where

$$\mathbf{G}(\mathbf{k}) = \sum_{\mathbf{h}} e^{i\mathbf{k}\cdot\mathbf{h}} \mathbf{G}(\mathbf{h}). \tag{9.42}$$

Considering a unit cell with one atom, the modes are all acoustic and for each momentum **k** there are three frequencies.

We may also write the Hamiltonian in terms of the coordinates and momenta, its natural variables. The momenta are the canonical conjugates of the coordinates, in the usual way. The Hamiltonian is then written as

$$H = \frac{1}{2}\sum_{\mathbf{k},\lambda} \left(p_{\mathbf{k},\lambda}^2 + \omega_{\mathbf{k},\lambda}^2 q_{\mathbf{k},\lambda}^2 \right), \tag{9.43}$$

using a more general notation, where the coordinates are written as q, and $\lambda = 1, 2, 3$ are the possible polarizations of each eigenmode of the system.

This type of harmonic Hamiltonian, written in terms of independent canonical conjugate variables, can be quantized in a standard way imposing commutation relations as

$$[q_{\mathbf{k},\lambda}, p_{\mathbf{k}',\lambda'}] = i\hbar\delta_{\mathbf{k},\mathbf{k}'}\delta_{\lambda,\lambda'}. \tag{9.44}$$

These quantized modes are the phonons of the 3d system. The energy of a given oscillation mode is then given by

$$\epsilon = \left(n + \frac{1}{2} \right) \hbar\omega_{\mathbf{k}}, \tag{9.45}$$

where n is the number of phonons with a given frequency $\omega_{\mathbf{k}}$.

9.2 Density of States and van Hove Singularities

Let us define the density of states of oscillation modes as the number of frequencies in a given interval, between ω and $\omega + d\omega$. We write this number as $ND(\omega)d\omega$ and this defines the function $D(\omega)$. To simplify, let us start with the case of a chain of atoms. Using periodic boundary conditions, we already showed that the wave number satisfies $k = 2\pi n/L$, where $L = Na$ is the system size and n is an integer. Summing over oscillation modes amounts to summing over the k values. That is, we can write that

$$\sum_k = \sum_n = \frac{L}{2\pi}\int dk = N\frac{1}{\frac{2\pi}{a}}\int dk. \tag{9.46}$$

In $3d$ we get the result:

$$\sum_{\mathbf{k}} = \left(\frac{L}{2\pi}\right)^3 \int d^3k = \frac{N}{v_{BZ}}\int d^3k, \tag{9.47}$$

where we considered a cube of side L and v_{BZ} is the volume of the Brillouin zone.

Going back to $1d$, note that we may write

$$\frac{L}{2\pi}dk = \frac{L}{2\pi}d\omega\frac{dk}{d\omega} = \frac{L}{2\pi}\frac{d\omega}{|v_g|} = ND(\omega)d\omega, \tag{9.48}$$

where we obtain explicitly that the density of states, $D(\omega)$ is

$$D(\omega) = \frac{L}{N}\frac{1}{2\pi}\frac{1}{\left|\frac{d\omega}{dk}\right|}. \tag{9.49}$$

Using the dispersion relation for the $1d$ problem of a monoatomic lattice

$$\omega = 2\sqrt{\frac{C}{M}}\sin\frac{ka}{2}, \tag{9.50}$$

we get that the density of states is obtained as

$$D(\omega) = \frac{a}{2\pi}\frac{1}{a\sqrt{\frac{C}{M}}\cos\frac{ka}{2}} = \frac{1}{\pi}\frac{1}{\sqrt{\omega_0^2-\omega^2}}, \tag{9.51}$$

where $\omega_0 = 2\sqrt{C/M}$. Such a density of states has the behavior shown in Fig. 9.7. At small frequency it is finite, with value $1/(\pi\omega_0)$ and it diverges close to $\omega = \omega_0$.

This behavior of the density of states is to be expected. Recall that at low frequencies, the dispersion relation is linear. Therefore, the group velocity is finite, and as we increase the wave number, the corresponding interval in the frequency range is finite: this leads to a finite number of modes for a given finite range of wave numbers. On the other hand, close to $k = \pi/a$, the group velocity vanishes, since the dispersion relation becomes flat. Therefore, as we change the wave number by a given amount, the change in frequency is decreasing and, in the limit when we approach the Brillouin zone limit, the dispersion relation becomes flat, which implies a finite and increasing number of states, for a decreasing range of wave numbers. Therefore, the density of states diverges at the Brillouin zone limit.

Consider again the $3d$ case. We may write that the sum over states is now of the form:

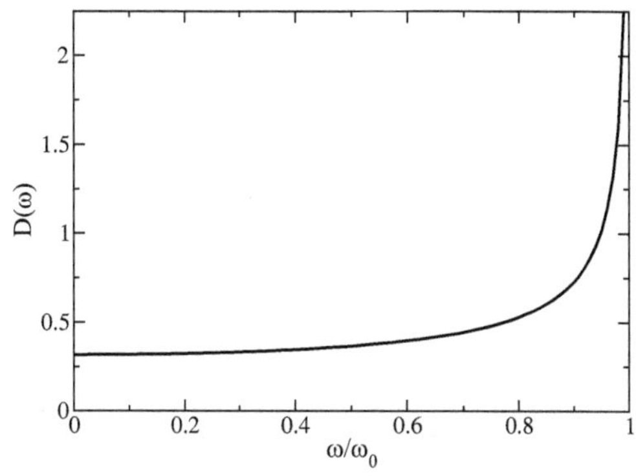

Fig. 9.7 Density of states for a monoatomic lattice

$$\begin{aligned}\sum_{\omega\leq\omega(\mathbf{k})\leq\omega+d\omega} &= \frac{V}{(2\pi)^3}\sum_{pol}\int_{shell} d^3k \\ &= \frac{V}{(2\pi)^3}\sum_{pol}\int dSdk_{\perp}. \end{aligned} \tag{9.52}$$

Here we have summed over the polarizations, S is an equifrequency surface and $\perp$ is the direction perpendicular to this surface. Therefore, by definition of the gradient, we obtain that $d\omega = |\nabla_k \omega| dk_{\perp}$. Using this result we finally arrive at the expression for the density of states:

$$D(\omega) = \frac{V}{N\,(2\pi)^3}\sum_{pol}\int_S dS\frac{1}{|\mathbf{v}_g|}. \tag{9.53}$$

The integral may have singularities at points where the group velocity vanishes. These singularities are called van Hove singularities.

9.3 Thermal Properties

The phonons may be excited if we supply energy to the system. Different oscillation modes may be involved, each with a different dispersion relation and with different occupations, depending on the modes excited. Therefore, the phonons will make a contribution to the specific heat of a crystal.

The total energy absorbed by the phonon system will, therefore, be the sum over all excited modes. In thermal equilibrium, the occupation of a mode with a given frequency, ω, is given by Planck's distribution and is of the form:

$$\langle n(\omega)\rangle = \frac{1}{e^{\beta\hbar\omega}-1}, \tag{9.54}$$

where $\beta = 1/(k_B T)$. This is the distribution to a system of bosons (oscillations in this case) with no conserved number and therefore zero chemical potential. Therefore, the average energy in that mode is given by

$$U(\omega) = \hbar\omega \left(\frac{1}{e^{\frac{\hbar\omega}{k_B T}} - 1} + \frac{1}{2} \right)$$
$$= \frac{\hbar\omega}{2} coth \frac{\hbar\omega}{2k_B T}. \tag{9.55}$$

At high temperatures, we recover the classical result that the average energy is $k_B T$. This result we recognize as the equipartition theorem, which states that the average energy per degree of freedom is $k_B T/2$, and we have two contributions to the Hamiltonian, one from the kinetic energy and one from the potential energy. At low temperatures, the quantum effects are most notable and the result deviates from the classical result, as a reflection of the zero point motion term. In the limit of zero temperature, the average energy is simply $\hbar\omega/2$, as discussed before.

These results apply to a single oscillation mode. Taking into account the many modes that may be involved may be done in different ways.

9.3.1 Einstein Model

Let us simplify the issue and consider that all oscillators are equal and therefore have the same frequency. This is the Einstein model. We may write that the total average energy is

$$U = 3N \frac{\hbar\omega}{2} coth \frac{\hbar\omega}{2k_B T}. \tag{9.56}$$

Therefore, the heat capacity is given by

$$C_V = \left(\frac{\partial U}{\partial T} \right)_V = 3N k_B \left(\frac{\frac{\hbar\omega}{2k_B T}}{sinh \frac{\hbar\omega}{2k_B T}} \right)^2. \tag{9.57}$$

At high temperatures $k_B T >> \hbar\omega$, the heat capacity reduces to the Dulong-Petit law and

$$C_V \sim 3N k_B, \tag{9.58}$$

which is independent of temperature. At low temperatures we get that

$$C_V \sim 3Nk_B \left(\frac{\hbar\omega}{2k_BT}\right)^2 e^{-\frac{\hbar\omega}{k_BT}}. \tag{9.59}$$

This result deviates strongly from the experimental result which shows that the heat capacity (or specific heat) at low temperatures is $C_V \sim T^3$. This deviation is due to the regime of low frequencies or energies (and therefore the low temperature limit is not properly obtained). Indeed, for any temperature there are always modes with energies smaller than the temperature $\hbar\omega < k_BT$, and these modes give an important contribution. Typically a contribution to the specific heat that is exponential at low temperatures holds in a regime where the energies are larger than the temperature. This actually occurs for the optical modes where there is a finite energy gap. The contribution of those modes to the specific heat is well accounted for by the Einstein model, but the acoustic modes require a more careful treatment.

9.3.2 Debye Model

The Einstein model provides a reasonably good description of the optical modes since both the energy gap is finite and the band has a small dispersion. On the other hand, the acoustic mode has a large range that is essentially linear. Close to the Brillouin zone limit, the dispersion relation flattens, but this is a narrow region. In order to have the optical modes excited, we need to couple the system with photons with energies that are typically higher than those associated with thermal energy. So in order to study thermal processes, both the optical modes are not expected to contribute significantly, neither the regime of higher frequencies of the acoustic mode is expected to contribute significantly, unless the temperature is somewhat high. Therefore, at least as far as we are interested in the low temperature regime, it is a good approximation to consider the linear regime as extending over the entire Brillouin zone, and to neglect the contribution from the deviation from this linear regime. However, we must conserve the total number of modes, since this is linked to the number of atoms. A cutoff must be introduced in the wave number, k_D, under the condition that the number of modes is conserved. This is the idea of the Debye model.

In the low-frequency regime, where the dispersion relation is linear, we can write $\omega = v_s k$, with v_s the sound velocity and assuming there is isotropy, which means that the velocity is the same for the various polarizations. In this case the surfaces of constant frequency (or energy and therefore may be called equipotentials in some loose sense) are spheres. The total number of degrees of freedom equals the total number of atoms leading to

$$N = \sum_{\mathbf{k}} = \frac{V}{(2\pi)^3} \int_{k_D} = \frac{V}{(2\pi)^3} \frac{4}{3}\pi k_D^3. \tag{9.60}$$

Note that

$$\frac{4}{3}\pi k_D^3 = v_{BZ} = \frac{N}{V}(2\pi)^3 . \tag{9.61}$$

From the expression for the density of states

$$D(\omega) = \frac{V}{N(2\pi)^3}\sum_{pol}\int_s dS\frac{1}{|v_g|}$$
$$= \frac{V}{N(2\pi)^3}\frac{3}{v_s}4\pi k^2$$
$$D(\omega) = \frac{3}{4\pi k_D^3}3\frac{4\pi k^2}{v_s} = 3\frac{3\omega^2}{\omega_D^3}. \tag{9.62}$$

We have used that $v_g = v_s$, we took into consideration that there are three polarizations and defined $\omega_D = v_s k_D$, the (and Debye wave number or momentum). Later it will also be useful to consider the Debye temperature, Θ_D, defined by $\hbar\omega_D = k_B\Theta_D$.

The total average energy may be written in general as

$$U = \sum_{\mathbf{k},\lambda}\hbar\omega_{\mathbf{k},\lambda}\left(\langle n_{\mathbf{k},\lambda}\rangle + \frac{1}{2}\right)$$
$$= N\int d\omega D(\omega)\left(\langle n(\omega)\rangle + \frac{1}{2}\right)\hbar\omega. \tag{9.63}$$

We get:

$$U = \frac{9N}{2\omega_D^3}\int_0^{\omega_D} d\omega coth\frac{\hbar\omega}{2k_BT}\hbar\omega^3 \tag{9.64}$$

and the heat capacity is

$$C_V = \left(\frac{\partial U}{\partial T}\right)_V = 3Nk_B\int_0^{\omega_D} d\omega\frac{3\omega^2}{\omega_D^3}\frac{\left(\frac{\hbar\omega}{k_BT}\right)^2 e^{\frac{\hbar\omega}{k_BT}}}{\left(e^{\frac{\hbar\omega}{k_BT}}-1\right)^2}. \tag{9.65}$$

Defining a variable $z = \hbar\omega/(k_BT)$, we can rewrite:

$$C_V = 9Nk_B\left(\frac{T}{\Theta_D}\right)^3\int_0^{\frac{\Theta_D}{T}} dz\frac{z^4e^z}{(e^z-1)^2}. \tag{9.66}$$

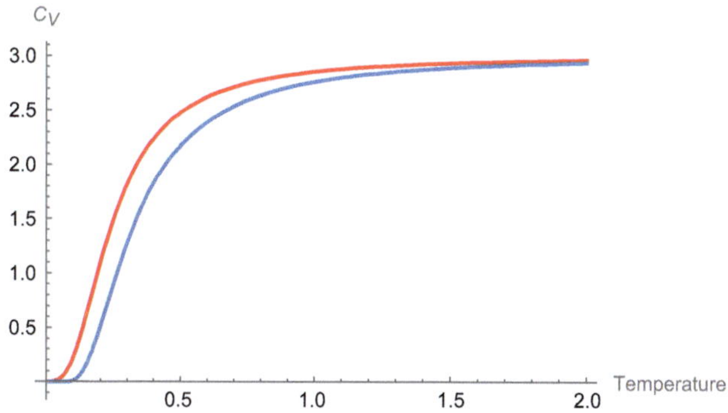

Fig. 9.8 Comparison of the specific heat obtained from the Einstein model (in blue) and the Debye model (in red). In the case of the Einstein model, the temperature ($k_B T$) is scaled by the energy of the oscillation mode ($\hbar\omega$). In the case of the Debye model, the temperature is scaled by the Debye temperature Θ_D. The specific heat is scaled as $C_V/(Nk_B)$

In the limit of low temperatures, $T << \Theta_D$, replacing the upper limit of the integral by ∞, we get that

$$\begin{aligned} C_V &= 9Nk_B \left(\frac{T}{\Theta_D}\right)^3 \frac{4\pi^4}{15} \\ C_V &\sim \left(\frac{T}{\Theta_D}\right)^3, \end{aligned} \tag{9.67}$$

in agreement with the experimental result. In the regime of very high temperatures (even though not strictly correct since we neglect the high-frequency regime where the band becomes flat), we recover the classical result $C_V = 3Nk_B$ (Fig. 9.8).

We can now consider the low-temperature regime of a metal where both the electronic and phonon contributions to the specific heat are taken into account. We find therefore that

$$C_V = \gamma T + AT^3, \tag{9.68}$$

where the linear term is the electronic contribution, as discussed previously.

We have seen that the Fermi temperature is quite high and so room temperature is a small temperature regime as compared to the Fermi temperature. We note that the Debye temperature is lower than the Fermi temperature, but it is still often of the order of several hundred Kelvin. So often the approximation of low temperatures holds for both the electronic contribution (with very good precision) and the phonon contribution. For low temperatures both approximations are quite accurate.

9.4 Debye-Waller Factor

The condition for the diffraction peaks Eq. (8.58) was obtained assuming a static distribution of the lattice atoms. However, the atoms oscillate around their equilibrium positions such that their positions $\mathbf{R} = \mathbf{R}_j + \mathbf{u}$, where $\mathbf{u}$ are the displacements around the lattice regular positions $\mathbf{R}_j$. The atoms oscillate around their equilibrium positions due to temperature effects but also due to their zero point motion at zero temperature. Therefore, the atoms move and the discrete translational invariance is broken. The scattering amplitude is therefore expected to decrease such that

$$I \sim |V_{\mathbf{g}}|^2 e^{-W}, \tag{9.69}$$

where W is called the Debye-Waller factor. If temperature increases W should increase leading to a reduced scattering amplitude.

Each contribution to Eq. (8.58) is written as

$$f_j e^{-i\mathbf{g}\cdot\mathbf{R}_j} \langle e^{-i\mathbf{g}\cdot\mathbf{u}} \rangle, \tag{9.70}$$

where $\langle \cdots \rangle$ is the average over the positions of the atoms, which may be the result of a thermal average or of a quantum average at zero temperature (as a result of the zero point motion). Assuming the oscillations are small, we may expand:

$$\langle e^{-i\mathbf{g}\cdot\mathbf{u}} \rangle = 1 - i\langle \mathbf{g}\cdot\mathbf{u} \rangle - \frac{1}{2}\langle (\mathbf{g}\cdot\mathbf{u})^2 \rangle + \cdots \tag{9.71}$$

Assuming the oscillations are random around the equilibrium positions $\langle \mathbf{g}\cdot\mathbf{u} \rangle = 0$, we are left with the quadratic term, neglecting higher-order terms, which leads to

$$\langle (\mathbf{g}\cdot\mathbf{u})^2 \rangle = g^2 \langle u^2 \rangle \langle \cos^2\theta \rangle = \frac{1}{3} g^2 \langle u^2 \rangle. \tag{9.72}$$

The potential energy associated with the displacement of the atoms from their equilibrium positions is $1/2Cu^2$, where C is the force constant. Using the equipartition theorem, we get that

$$\frac{1}{2} C \langle u^2 \rangle = \frac{3}{2} k_B T. \tag{9.73}$$

Exponentiating back the expansion, and taking the square to obtain the correction to the scattering intensity, we get that

$$I = I_0 e^{-\frac{1}{3} g^2 \langle u^2 \rangle}, \tag{9.74}$$

where I_0 is the scattering intensity disregarding the fluctuations of the atomic positions. The equipartition theorem leads then to

$$I = I_0 e^{-k_B T \frac{g^2}{M\omega^2}}, \tag{9.75}$$

where M is the mass of the atom and ω its oscillating frequency, $\omega^2 = C/M$. Substituting the numbers for a given material, there is a reduction predicted by the Debye-Waller factor, but is not large and experimentally the diffraction peaks are still observed.

At zero temperature the evaluation proceeds in a different way. For instance, the neutron dispersion from a crystal is given by the matrix element $S = \langle 0|e^{ipx}|0\rangle$, where p is a constant. We can show that

$$\langle 0|e^{ipx}|0\rangle = e^{-\frac{p^2}{2}\langle 0|x^2|0\rangle},$$

where $|0\rangle$ is the ground state of the harmonic oscillator with coordinate x, and we are simplifying the calculation to a one-dimensional problem. We may express x in second quantization and expand the exponential in powers of the creation and destruction operators. We can write that

$$\begin{aligned} S &= \langle 0|e^{ip\sqrt{\frac{\hbar}{2\omega}}(a^\dagger + a)}m|0\rangle \\ &= \sum_{n=0}^{\infty} \frac{1}{n!}\left(ip\sqrt{\frac{\hbar}{2\omega}}\right)^n \left(a^\dagger + a\right)^n |0\rangle. \end{aligned} \tag{9.76}$$

Using

$$[a, (a + a^\dagger)^n] = n(a + a^\dagger)^{n-1}$$

and that

$$\langle 0| \left(a^\dagger + a\right)^{2k} |0\rangle = (2k-1)!!\langle 0| \left(a^\dagger + a\right)^2 |o\rangle^k, \tag{9.77}$$

we get:

$$S = \sum_{k=0}^{\infty} \frac{1}{(2k)!!}\left(-p^2\frac{\hbar}{2\omega}\right)^k \langle 0| \left(a^\dagger + a\right)^2 |0\rangle^k. \tag{9.78}$$

Using that $(2k)!! = 2^k k!$, we obtain:

$$S = e^{-\frac{p^2}{2}\langle 0|x^2|0\rangle}, \tag{9.79}$$

which reflects the effect on the scattering amplitude of the quadratic fluctuations of the atom position around its equilibrium value.

Clearly, if the oscillations of the atoms around their equilibrium positions get too large, for instance due to the increase of temperature, there will be a point such that these oscillations, say $\sqrt{\langle u^2 \rangle}$, are of the order of the lattice constant: the lattice structure loses its stability and a phase transition to another structure or to a liquid phase may occur. This is the Lindemann criterion for the melting transition.

9.5 Exercises

1. Consider a chain of identical masses, M, separated by a distance a, and connected to each other by springs with force constant, C. Suppose that a longitudinal wave propagates $u_n = A\cos(\omega t - qna)$:

   (a) Calculate the kinetic and the potential energies of the chain.
   (b) Calculate the average value of the energy per atom (temporal average).

2. Consider a two-dimensional crystal with a square lattice and an atom per unit cell. Suppose that the interactions between atoms may be reduced to springs with force constant C and length a between nearest-neighbors and springs with force constants C' and length $\sqrt{2}a$ between second neighbors:

   (a) Calculate the dispersion relation for longitudinal and transversal phonons in the plane for $\mathbf{k} = (q, 0)$.
   (b) Find the sound velocity.

3. Consider a chain of equal atoms with mass M with couplings between first neighbors. The force constants have values C_1 and C_2, such that each atom is coupled to a neighbor with constant C_1 and to the other with constant C_2. Determine the equations of motion and the dispersion relations. What is the difference to a diatomic chain with atoms of different masses M_1 and M_2 but with equal force constants?
4. Consider transversal vibrations of a square lattice. All the atoms and force constants are equal. Find the equations of motion in the case of coupling between nearest-neighbors and obtain the dispersion relations. Find the first Brillouin zone. Draw the equipotentials (values of the momenta $\mathbf{k}$ with equal energy).
5. The coupling between the atoms and their motion originating in the electrons may have a long-range nature, even though decreasing with distance. An example is a power law (linear) decrease with distance between the atoms, such that the force constants take the form:

$$C_p = A\frac{\sin pk_0 a}{pa},$$

   where we may consider that pa is the distance between planes of atoms, labeled by s and $s + p$. Determine the dispersion relation for the oscillations of the planes. Make a graph and identify any singular behavior. Note that near k_0, the dispersion relation is singular. This property is called Kohn anomaly.

6. Consider a monoatomic chain with N atoms with mass M and consider periodic boundary conditions. Show that the total chain moment for an excitation of momentum k is given by

$$P_k = i\omega M u_0 e^{i\omega t} \sum_{r=0}^{N-1} e^{irka},$$

where u_0 is the amplitude and a is the lattice constant. Show that the physical momentum vanishes unless $k = 0$. Explain the difference between physical and crystalline momentum (phonon momentum).

7. Determine the dispersion relations of the phonons for an fcc lattice with coupling between nearest-neighbors along the propagation directions $(1, 0, 0)$, $(1, 1, 0)$ e $(1, 1, 1)$.
8. At small momenta an optical branch may be approximated by $\omega(\mathbf{q}) = \omega_0 - A|\mathbf{q}|^2$. Calculate the density of states of this mode. Show that it may be written as

$$D(\omega) = \left(\frac{L}{2\pi}\right)^3 \frac{2\pi}{A^{3/2}} (\omega_0 - \omega)^{1/2}$$

for $\omega < \omega_0$ and $D(\omega) = 0$ for $\omega > \omega_0$.

9. Show that in the Debye approximation, the average of the square of the displacement of an atom at absolute zero is $3\hbar\omega_D^2/8\pi^2\rho v_{sound}^3$.
10. Assuming the dispersion relation $\omega = Ak^s$ for waves in a solid (where ω is the frequency and k the wave vector), show that its contribution to the low-temperature specific heat is proportional to $T^{3/s}$.
11. Obtain the density of states $D(\omega)$ for a system in one and two dimensions in the continuum limit. Obtain the specific heat in low dimensions in both cases.
12. Consider the dispersion relation

$$\omega^2 M = 2C(2 - \cos k_x a - \cos k_y a).$$

How many van Hove singularities are there? Classify then according to the type of singularity. Does the density of states diverge at any point?

13. Aluminum crystallizes as a fcc lattice with a lattice constant $4.04\,\mathring{A}$ and a Debye temperature of $428K$. Estimate the melting temperature using the Lindemann criterion ($x_m = 0.2$) and compare with the experimental value $933K$. The atomic weight of Al is 27.
14. Consider the vacuum matrix element

$$S = \langle 0|e^{ipx}|0\rangle.$$

The coordinate x may be written in terms of the creation and destruction operators of the harmonic oscillator as

$$x = \sqrt{\frac{\hbar}{2\omega}}\left(a^{\dagger} + a\right)$$

$$S = e^{-\frac{p^2}{2}\langle 0|x^2|0\rangle}.$$

Show, using that $[a, (a + a^{\dagger})^n] = n(a + a^{\dagger})^{n-1}$, that

$$S = e^{-\frac{p^2}{2}\langle 0|x^2|0\rangle}.$$

Chapter 10
Electronic Energy Bands

To describe the behavior of crystalline solids, we have to take into account the presence of a periodic potential, which is the sum of the (positively charged) nuclei and the (negatively charged) electronic contributions. In addition to the core electrons that remain in the vicinity of nuclei, the remaining electrons may preferentially stay close to the nuclei (such as in ionic crystals), be shared (such as in covalent crystals), or have a high probability to be found in regions between nuclei (such as in metals). For instance, the electrons that are free to move in the crystal "feel" the presence of the periodic attractive Coulomb potential of the ions that constitute the solid (we will see later that the potential is not in general purely Coulombic).

Let us consider simple metals with one free electron per atom, which implies that we assume that the remaining electrons are bound to the atoms in localized states. One example is sodium. There is a distribution of the ions across the material that occupies approximately 15% of the volume. There is therefore a majority of free space through which the delocalized electrons may move. These electrons move through the periodic attractive potential of the ions, but this attractive potential is not strong enough to bind the electrons.

Under these circumstances, taking into consideration the available space to the electrons and the low electronic concentration, we may consider, as a first approximation, that the delocalized electrons do not interact with other delocalized electrons. We may therefore consider the electrons as a non-interacting gas, but in the presence of an external periodic potential, provided by the background of positive charges of the ions. The electrons are then a nearly free system of charges since we ignore the presence of other electrons, except for their effect through the periodic potential. This may be viewed as an external potential greatly simplifying the analysis.

P. D. Sacramento, *Introduction to Condensed Matter Theory*,
https://doi.org/10.1007/978-3-031-86251-9_10

10.1 Bloch Theorem

Let us then, quite generally and without specifying the nature of the external potential, consider one electron, at point $\mathbf{r}$, in a periodic potential. As we discussed before, due to the periodicity of the lattice (and to its supposedly infinite extension, in comparison with the atomic scale), any property of the system, $f(\mathbf{r})$, must obey:

$$f(\mathbf{r}) = f(\mathbf{r} + \mathbf{l}), \tag{10.1}$$

where $\mathbf{l}$ is a lattice vector. This function may be expanded in a complete set of vectors which can be chosen as the reciprocal lattice vectors $\mathbf{g}$. The Fourier decomposition may be written as

$$f(\mathbf{r}) = \sum_{\mathbf{g}} f_{\mathbf{g}} e^{i\mathbf{g}\cdot\mathbf{r}}. \tag{10.2}$$

Let us then consider an electron in a periodic potential. To simplify let us take a one-dimensional system given by the Hamiltonian:

$$H = \frac{\hbar^2}{2m} k^2 + U(x). \tag{10.3}$$

The potential is expanded in the reciprocal lattice basis:

$$U(x) = \sum_{g} U_g e^{igx}$$

$$U(x) = \sum_{g>0} U_g \left(e^{igx} + e^{-igx}\right)$$

$$U(x) = 2 \sum_{g>0} U_g \cos gx, \tag{10.4}$$

where we have chosen $U_0 = 0$ for convenience. Then, Schrödinger's equation is written as

$$H\psi = \epsilon\psi$$

$$\left(\frac{1}{2m} p^2 + U(x)\right) = \left(\frac{1}{2m} p^2 + \sum_{g} U_g e^{igx}\right) \psi(x) = \epsilon\psi(x). \tag{10.5}$$

Expanding the wave function in a Fourier series

$$\psi(x) = \sum_{k} c(k) e^{ikx},$$

with $k = (2\pi n)/L$, where L is the system size and using periodic boundary conditions, and, substituting in Schrödinger's equation, we get that

$$\sum_k \frac{\hbar^2}{2m} k^2 c(k) e^{ikx} + \sum_{gk} U_g c(k) e^{i(k+g)x} = \epsilon \sum_k c(k) e^{ikx}. \tag{10.6}$$

Since the plane waves are independent functions, we can equate the coefficients of a plane wave, and we obtain an equation for the coefficients of the Fourier series, called central equation:

$$(\lambda_k - \epsilon) c(k) + \sum_g U_g c(k - g) = 0, \tag{10.7}$$

where we have defined the kinetic energy $\lambda_k = (\hbar^2 k^2)/(2m)$. Let us now suppose that we are able to solve this equation and that we determine the coefficients $c(k)$. The coefficient $c(k)$ is a function of all the coefficients $c(k - g)$. If we fix k, the solution involves all the values of $k - g$. If a given value of k shows up in the solution of the wave function ψ ($= \sum_k c(k) e^{ikx}$), then all other wave vectors of the expansion of ψ will be of the type $k - g$. Let us then attach the index k to the solution of ψ that contains the wave vector k. Then

$$\psi_k(x) = \sum_g c(k - g) e^{i(k-g)x}$$

$$\psi_k(x) = \left(\sum_g c(k - g) e^{-igx} \right) e^{ikx}$$

$$\psi_k(x) = u_k(x) e^{ikx}, \tag{10.8}$$

where

$$u_k(x) = \sum_g c(k - g) e^{-igx}. \tag{10.9}$$

The function $u_k(x)$ obeys the property that

$$u_k(x + l) = u_k(x). \tag{10.10}$$

It is therefore a periodic function with the period of the underlying lattice. This result is called Bloch theorem. It states that the wave function of the electrons in a periodic lattice factorizes into a plane wave (for a specified wave vector k), which retains the information of a freely propagating electron, as if in empty space, but with a modulation provided by the periodicity of the lattice. Therefore, in the periodic lattice, the electrons are nearly free, since their wave function is very close to the solution of free space (of course keep in mind that so far we have neglected any interactions between the electrons). The electrons are almost free, but in momentum space, the electrons with wave vectors k and $k' = k - g$ are coupled.

Returning to a three-dimensional space, the Bloch theorem can then be written as

$$\psi_{\mathbf{k}}(\mathbf{x}) = u_{\mathbf{k}}(\mathbf{x})e^{i\mathbf{k}\cdot\mathbf{x}}$$
$$u_{\mathbf{k}}(\mathbf{x}) = u_{\mathbf{k}}(\mathbf{x}+\mathbf{l}). \tag{10.11}$$

The phase factor $e^{i\mathbf{k}\cdot\mathbf{x}}$ describes the free propagation of the electrons, and the function $u_{\mathbf{k}}(\mathbf{x})$ describes the modulation of the wave function that results from the periodic potential. Clearly, we may find that

$$\psi_{\mathbf{k}}(\mathbf{x}+\mathbf{l}) = u_{\mathbf{k}}(\mathbf{x}+\mathbf{l})e^{i\mathbf{k}\cdot(\mathbf{x}+\mathbf{l})} = e^{i\mathbf{k}\cdot\mathbf{l}}\psi_{\mathbf{k}}(\mathbf{x}). \tag{10.12}$$

The central equation to be solved is then, back to a one-dimensional system to simplify notation:

$$(\lambda_k - \epsilon)c(k) + \sum_g U_g c(k-g) = 0.$$

We will carry out an approximate solution to exemplify the structure of the solution and capture its main features.

In general, U_g decreases fast as g grows. Let us call $\bar{g}$ the smaller value of g. Let us suppose that $U_{g>\bar{g}} \sim 0$. Also note that $U_{-g} = U_g = U$. To solve the equations (in equal number of the coefficients c), the determinant of the coefficients must be zero:

$$det \begin{pmatrix} \lambda_{k-2\bar{g}} - \epsilon & U & 0 & 0 & 0 \\ U & \lambda_{k-\bar{g}} - \epsilon & U & 0 & 0 \\ 0 & U & \lambda_k - \epsilon & U & 0 \\ 0 & 0 & U & \lambda_{k+\bar{g}} - \epsilon & U \\ 0 & 0 & 0 & U & \lambda_{k+2\bar{g}} - \epsilon \end{pmatrix} = 0. \tag{10.13}$$

Note that the matrix has infinite size, but we are truncating it as an approximation. The solutions are of the form ϵ_{nk}.

Let us consider a value of $k = g/2 = \pi/a$. Then $k^2 = (g/2)^2$ and $(k-g)^2 = (g/2)^2$. This implies that the kinetic energies of the two momenta are the same: $\lambda_k = \lambda_{k-g} = \lambda$. If $c(g/2)$ is important, then $c(-g/2)$ is as well. Taking only these two components, we get the following system of equations to solve:

$$(\lambda - \epsilon)c(g/2) + Uc(-g/2) = 0$$
$$(\lambda - \epsilon)c(-g/2) + Uc(g/2) = 0. \tag{10.14}$$

For all wave vectors k, we neglect all other terms in the sum $\sum_g U_g c(k-g)$ because $U_g << \lambda_k$. But if $k = g/2$, then the symmetric wave vector $k = -g/2$ is equally important. It corresponds to a wave with the same energy moving in the opposite direction. We have therefore to calculate the determinant:

$$det \begin{pmatrix} \lambda - \epsilon & U \\ U & \lambda - \epsilon \end{pmatrix} = 0. \tag{10.15}$$

The solutions are

$$\epsilon_{\pm} = \lambda \pm U = \frac{\hbar^2}{2m}\left(\frac{1}{2}g\right)^2 \pm U. \tag{10.16}$$

There is therefore a gap: $\epsilon_+ - \epsilon_- = 2U$. Note that

$$\frac{c(-g/2)}{c(g/2)} = \frac{\epsilon - \lambda}{U} = \pm 1.$$

Therefore, the wave functions, considering only these two components, read:

$$\psi_{\pm}(x) = C\left(e^{\frac{i}{2}gx} \pm e^{-\frac{i}{2}gx}\right). \tag{10.17}$$

These two wave functions are stationary waves with a group velocity which is zero. There is a gap as shown in Fig. 10.1 that opens up at $k = \pm\pi/a$. There is a first band, a forbidden energy zone, and a second band, due to the presence of the periodic potential. Far from the forbidden energy zone, the bands approach a regime where

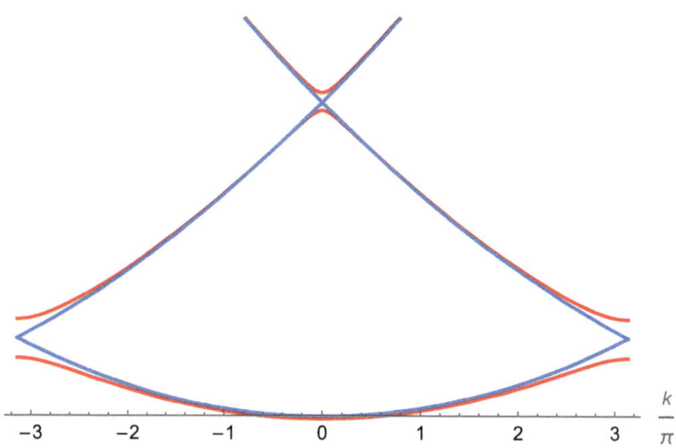

Fig. 10.1 Gap opening at the Brillouin zone edge of a one-dimensional system, as a result of the coupling to the periodic potential. In the case of no potential, the gap vanishes. The potential also opens a gap when the second and third bands cross

the electron is basically free. Neglecting the periodic potential (taking $U = 0$), the bands are parabolic since there is only kinetic energy. The case where U may be neglected and the periodic potential is absent has been considered before in Sect. 7.6. Let us next provide some further results and detail some aspects in a complementary way.

10.2 Free Electron Model

If the electrons were strictly free, their energy would just be the kinetic energy of the type $\epsilon = \hbar^2 k^2/(2m)$; $p = \hbar k$ is the momentum and m is the mass of an electron. Due to the periodic potential, we saw in the previous section that the energy bands deviate from the free electron case, but the difference is particularly significative near the momenta for which gaps open, as a result of the periodic structure. Far from the momenta where the velocity vanishes, the dispersion relation of the electrons is essentially parabolic, as for the free electron case.

Since the electrons are fermions, Pauli's principle applies, and each state can only be occupied by one electron (per spin component). Therefore, if the electron concentration is low, only low energy levels are occupied, and the system behaves essentially as free. For chemical potentials close to energies where the gap is found, there are electrons whose dispersion relation differs significantly from the free case. If the chemical potential lies in the gap, the group velocity vanishes, the system is no longer a metal and a band insulator is found (since one needs a finite excitation energy to delocalize an electron across the energy gap).

Let us then focus on a regime where the electrons may be considered as free. In other words, we are in the regime where the Bloch waves are essentially plane waves and the periodic spatial modulation does not affect much the wave function of the electrons. One may wonder why such an approximation holds in a condensed matter system. One may argue that the solution of Schrödinger equation shows that these states are eigenstates of the independent electron problem (neglecting their mutual interactions). Therefore, the electrons are not scattered because Bloch waves are stationary. Also, the electrons avoid themselves as a result of Pauli's principle, and if the densities are not high, collisions are not frequent. Of course, in a real material, there are impurities, imperfections, and temperature effects that lead to collisions, as evidenced by the experimental observations of a finite mean-free path between collisions and the observation, in general, of Ohm's law.

10.2.1 Energy Levels in a 1d Problem

The problem of a single particle in $1d$ is a simple quantum mechanics problem. Consider a finite system of length L and open boundary conditions, $\psi(0) = \psi(L) = 0$. Schrödinger's equation for a free particle is

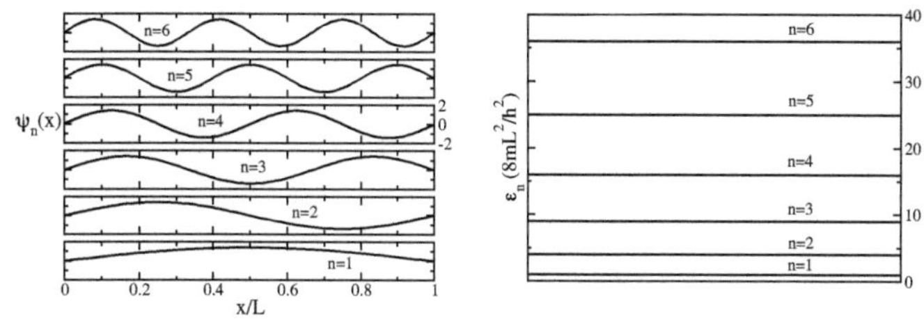

Fig. 10.2 Wave functions and energy levels of a free particle in a finite one-dimensional system

$$H\psi_n = -\frac{\hbar^2}{2m}\frac{d^2\psi_n(x)}{dx^2} = \epsilon_n \psi_n(x). \tag{10.18}$$

The solutions that satisfy the boundary conditions are

$$\psi_n(x) = A_n \sin k_n x, \tag{10.19}$$

with $k_n = n\pi/L$ and the quantized eigenenergies are

$$\epsilon_n = \frac{\hbar^2}{2m}\left(\frac{n\pi}{L}\right)^2. \tag{10.20}$$

The wave functions and eigenenergies are shown in Fig. 10.2. The values of k_n are associated with wavelengths $\lambda_n = 2L/n$ with an increasing number of nodes of the wave functions as the quantum number, n, increases. Since the electrons are fermions, at zero temperature the energy levels are occupied by two electrons, of $\uparrow$ and $\downarrow$ spin components, up to a maximum single-particle energy, called Fermi energy, ϵ_F, associated with the quantum number, n_F. If the number of electrons, N, is even, then $n_F = N/2$. Therefore the Fermi energy is

$$\epsilon_F = \frac{\hbar^2}{2m}\left(\frac{\pi n_F}{L}\right)^2 = \frac{\hbar^2}{2m}\left(\frac{N\pi}{2L}\right)^2. \tag{10.21}$$

At finite temperature, the Fermi-Dirac distribution

$$f(\epsilon) = \frac{1}{e^{\frac{\epsilon-\mu}{k_B T}} + 1} \tag{10.22}$$

deviates from the Heaviside shape around the chemical potential, the discontinuity around the chemical potential ($\mu(T = 0) = \epsilon_F$) is rounded off, and the electrons have a finite occupation of higher energy levels.

10.2.2 Energy Levels in a 3d System

In this case Schrödinger's equation can be written in Cartesian coordinates as

$$-\frac{\hbar^2}{2m}\left(\frac{\partial^2}{\partial x^2}+\frac{\partial^2}{\partial y^2}+\frac{\partial^2}{\partial z^2}\right)\psi_{\mathbf{k}}(\mathbf{r})=\epsilon_{\mathbf{k}}\psi_{\mathbf{k}}(\mathbf{r}). \tag{10.23}$$

Using open boundary conditions on a system of sizes L_x, L_y, L_z, the solutions are

$$\psi_{\mathbf{k}}(\mathbf{r})=A\sin\frac{\pi n_x x}{L_x}\sin\frac{\pi n_y y}{L_y}\sin\frac{\pi n_z z}{L_z}. \tag{10.24}$$

Here, n_x, n_y, n_z are positive integers.

It is more convenient to introduce periodic boundary conditions. This describes an electron injected in a system, described by plane waves instead of stationary waves. It is, for instance, more adequate to describe and understand transport properties. One simulates an infinite system by a very large system that is repeated indefinitely (periodic boundary). Therefore, we use:

$$\psi(x+L_x,y,z)=\psi(x,y,z) \tag{10.25}$$

and similarly along the other directions. Under these boundary conditions, the wave functions are not the stationary sine functions, but plane waves of the form

$$\psi_{\mathbf{k}}(\mathbf{r})=Ae^{i\mathbf{k}\cdot\mathbf{r}}. \tag{10.26}$$

These plane waves are also the eigenfunctions of momentum (generator of translations). In this case we have that

$$k_x=0,\pm\frac{2\pi}{L_x},\pm\frac{4\pi}{L_x},\cdots \tag{10.27}$$

and similarly for k_y, k_z. The energy is then of the type

$$\epsilon_{\mathbf{k}}=\frac{\hbar^2}{2m}\left(k_x^2+k_y^2+k_z^2\right), \tag{10.28}$$

a discrete set of values due to the discretization of the wave vector.

The spacing between the discretized values is then

$$\Delta k_\alpha=\frac{2\pi}{L_\alpha}, \tag{10.29}$$

with $\alpha = x, y, z$. Let us now, to simplify, consider a cubic structure with $L_x = L_y = L_z = L$. The Fermi sphere is the volume that includes all occupied states, at zero temperature. Let us now count the number of states inside the Fermi sphere, of radius k_F. This is given by

$$2 \times \frac{\frac{4}{3}\pi k_F^3}{\left(\frac{2\pi}{L}\right)^3} = \frac{V}{3\pi^2} k_F^3 = N. \tag{10.30}$$

$V = L^3$ is the system volume. The factor of two comes from the spin degeneracy of each electron (recall that N is the number of electrons). This result fixes the Fermi momentum, k_F. This is given by

$$k_F = \left(\frac{3\pi^2 N}{V}\right)^{1/3} \tag{10.31}$$

and is a function of the particle density N/V. The Fermi energy is therefore also a function of the particle density (see Eq. (7.141)):

$$\epsilon_F = \frac{\hbar^2}{2m}\left(3\pi^2 \frac{N}{V}\right)^{2/3}. \tag{10.32}$$

We may also define a Fermi momentum, $p_F = \hbar k_F$ and a Fermi velocity of an electron with an energy equal to the Fermi energy as

$$v_F = \frac{\hbar k_F}{m} = \frac{\hbar}{m}\left(3\pi^2 \frac{N}{V}\right)^{1/3}. \tag{10.33}$$

Also, we may define the Fermi temperature:

$$T_F = \frac{\epsilon_F}{k_B} \sim 10^4 - 10^5\,\text{K}. \tag{10.34}$$

Usual temperatures satisfy $T \ll T_F$, which means that typically systems are found at low temperatures compared to the Fermi temperature, which is the temperature scale associated with the energy scale of the single-particle eigenstates of the problem.

10.2.3 *Density of States*

We may now define the density of states as the number of states in a given energy interval. The number of states is equal to the number of particles (note that we include in this count the spin). Therefore, we may calculate the number of states with energy smaller than ϵ. This is given by

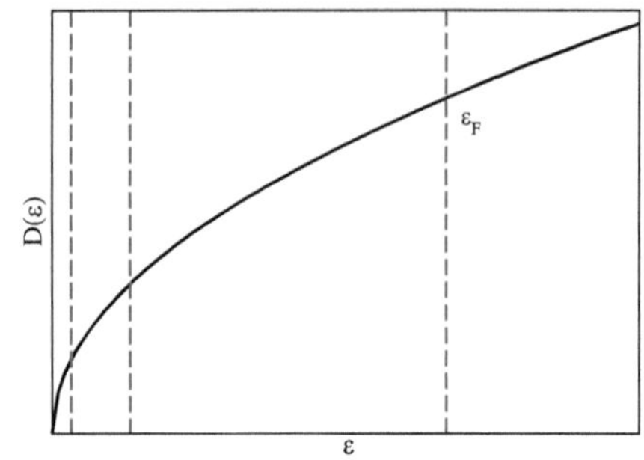

Fig. 10.3 Density of states of $3d$ free electrons. If the Fermi energy, ϵ_F, is large enough, the density of states does not change significantly with energy, as compared to lower energies

$$N(\epsilon) = \frac{V}{3\pi^2}\left(\frac{2m\epsilon}{\hbar^2}\right)^{3/2}. \tag{10.35}$$

The density of states is then simply obtained in $3d$ as

$$D(\epsilon) = \frac{dN(\epsilon)}{d\epsilon} = \frac{V}{2\pi^2}\left(\frac{2m}{\hbar^2}\right)^{3/2}\epsilon^{1/2}. \tag{10.36}$$

The density of states is shown in Fig. 10.3.

10.2.4 Heat Capacity of the Electron Gas

The heat capacity (or specific heat) of the electron gas was discussed before in Sect. 7.6.1. As we supply energy (heat) to a system, we want to calculate the corresponding change in temperature of the system. Classically, we know that if we have a set of harmonic oscillators (or any system with a quadratic energy dispersion, such as a parabolic dispersion relation in terms of momentum), the equipartition theorem shows that the average energy per degree of freedom is a function of temperature and given by $1/2k_BT$. In the case of a three-dimensional free system of N particles, the heat capacity is therefore $C_V = 3N/2k_B$. Experimentally it is observed that the heat capacity is much smaller than the classical result. Indeed, it is of the order of 1% of the classical result. As explained before, due to Pauli's principle, only a fraction of the order of k_BT of the electrons, that is, of the order of NT/T_F, is able to absorb the energy, and lead to an increase of the temperature. Only the electrons in the energy region close to the Fermi energy are able to be promoted to empty states. Each of those electrons absorbs an energy of the order of k_BT. Therefore, the heat capacity

$$C_V = \frac{dQ}{dT}$$

$$\sim \frac{d}{dT}\left(\left(N\frac{T}{T_F}\right)(k_BT)\right)$$

$$\sim Nk_B\left(\frac{T}{T_F}\right). \tag{10.37}$$

We may obtain the specific heat explicitly, to leading order, in a way complementary to the derivation carried out before. The energy at a finite temperature, T, is given by

$$U = \int_{-\infty}^{\infty} d\epsilon D(\epsilon)\epsilon f(\epsilon). \tag{10.38}$$

Using the Sommerfeld expansion, the energy may be approximated as

$$U = \int_{-\infty}^{\mu} d\epsilon D(\epsilon)\epsilon + \frac{\pi^2}{6}(k_BT)^2\left(\mu\frac{dD(\epsilon)}{d\epsilon}|_{\epsilon=\mu} + D(\mu)\right) + \cdots \tag{10.39}$$

We may now use that

$$\frac{dD(\epsilon)}{d\epsilon}|_{\epsilon=\mu} = \frac{1}{2}\frac{1}{\mu}D(\mu). \tag{10.40}$$

To obtain the specific heat, we calculate the derivative of the energy with respect to the temperature. Note that the chemical potential depends on the temperature, as shown in Eq. (7.145). At low temperature it is approximately the Fermi energy. Since the chemical potential depends on temperature, we have to calculate the derivative of the first term of the energy in the upper limit of the integral. Using Eq. (7.145), this leads to

$$C_V = -\frac{\pi^2}{6}k_B^2 T D(\epsilon_F) + \frac{\pi^2}{2}k_B^2 T D(\epsilon_F) = \frac{\pi^2}{3}k_B^2 T D(\epsilon_F). \tag{10.41}$$

Using that

$$D(\epsilon_F) = \frac{3N}{2\epsilon_F} = \frac{3N}{2k_BT_F} \tag{10.42}$$

leads to the final result

$$C_V = \frac{\pi^2}{2}Nk_B\frac{T}{T_F}, \tag{10.43}$$

as in Eq. (7.146).

Therefore, the specific heat is linear in temperature at low temperatures and can be written as

$$C_V = \gamma T \tag{10.44}$$

where

$$\gamma = \frac{\pi^2}{2} N \frac{k_B}{T_F} = \frac{\pi^2}{3} k_B^2 D(\epsilon_F). \tag{10.45}$$

Let us now recall that the Fermi energy is inversely proportional to the electron mass, $\epsilon_F \sim 1/m$. We see then that

$$\gamma \sim \frac{1}{\epsilon_F} \sim m. \tag{10.46}$$

The experimental measurement of the temperature slope of the specific heat allows us to define an effective electronic mass in the solid. That is

$$\frac{\gamma^{exp}}{\gamma^{free}} = \frac{m^{eff}}{m}. \tag{10.47}$$

In general, $m^{eff} \neq m$, where m is the mass of a free electron. A first reason is that electrons are not truly free and interact with the periodic potential of the ions. We will analyze this renormalization of the mass ahead. (Recalling the simple example of a one-dimensional periodic lattice, we found a gap at $k = \pi/a$. In its vicinity the group velocity vanishes which seems to indicate a very large effective mass of the electrons). Also, in general, the ion lattice is not static, and the electrons interact with the vibrations of the lattice, changing their effective mass. We may also understand, to some extent, the effect of the interactions with other electrons as some renormalization of their mass, in addition to other effects to be discussed later.

10.3 Energy Bands: Band Gap and Effective Mass

The nearly free electron model, which ignores the interactions between the electrons but takes into account the interaction with the periodic potential of the lattice, leads to a description of the electrons energies that resembles that of free electrons, if the momenta are small, but with significant differences as we approach the Brillouin zone limit. At least in $1d$, this is the result of our previous analysis. In $3d$ the analysis may be more complicated. Remaining for the moment in the $1d$ case, we saw that in the vicinity of the Brillouin zone limit, $k \sim \pi/a$, a gap opens and the group velocity goes to zero, as obtained through the explicit solution of the central equation. It may also be understood noting that these stationary waves are the result of the Bragg reflection conditions (elastic scattering):

$$(\mathbf{k} + \mathbf{g})^2 = k^2. \tag{10.48}$$

In $1d$ this leads to a solution:

$$k = \pm\frac{1}{2}g = \pm n\frac{\pi}{a}. \tag{10.49}$$

The wave functions that correspond to these solutions are

$$\begin{aligned}\psi_+ &= e^{i\frac{\pi}{a}x} + e^{-i\frac{\pi}{a}x} = 2\cos\frac{\pi}{a}x \\ \psi_- &= e^{i\frac{\pi}{a}x} - e^{-i\frac{\pi}{a}x} = 2i\sin\frac{\pi}{a}x,\end{aligned} \tag{10.50}$$

which are stationary waves with vanishing group velocity. The probability density of the two states is of the type

$$\begin{aligned}\rho_+ &= |\psi_+|^2 \sim \cos^2\frac{\pi}{a}x \\ \rho_- &= |\psi_-|^2 \sim \sin^2\frac{\pi}{a}x.\end{aligned} \tag{10.51}$$

Selecting the origin at one of the lattice positions, we get that state + is peaked at the atom locations while state − is peaked in the intermediate regions, as shown in Fig. 10.4. The state + accumulates electrons with negative charge at the location of the atoms (actually ions of positive opposite charge), and therefore there is an attractive energy that lowers the energy, with respect to a state that accumulates the electrons at intermediate points in the lattice. Hence, the energy gap that opens at the Brillouin zone limit: $E_- - E_+ = E_{gap}$.

An alternative way to calculate the gap is to consider the periodic potential as $U(x) = U\cos 2\pi x/a$ and defining normalized wave functions of the two states as $\psi_+ = \sqrt{2}\cos\pi x/a$ and $\psi_- = \sqrt{2}\sin\pi x/a$, we obtain that the gap is approximately

$$E_{gap} = \int_0^a dx U(x)\left(|\psi_+|^2 - |\psi_-|^2\right) = U. \tag{10.52}$$

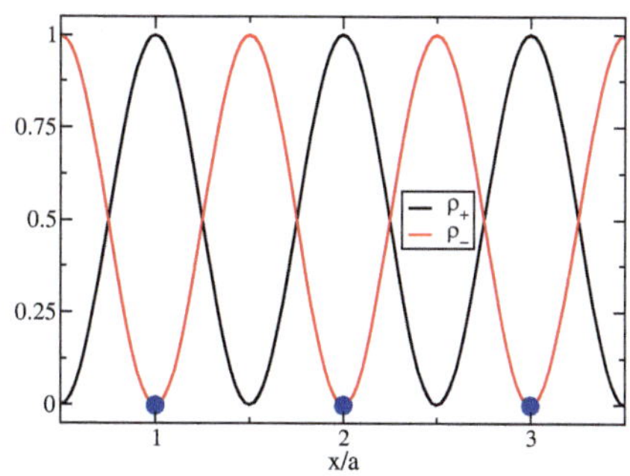

Fig. 10.4 Probability density for the states + and − along a chain. The dots indicate the atoms' positions

Recall that if we have Bloch functions, these obey:

$$\psi_{\mathbf{k}}(\mathbf{r}) = u_{\mathbf{k}}(\mathbf{r})e^{i\mathbf{k}\cdot\mathbf{r}}, \tag{10.53}$$

where $\mathbf{k}$ is considered to be restricted to the first Brillouin zone. We saw that by expanding the wave function in a Fourier series $\psi = \sum_{\mathbf{k}} c(\mathbf{k})e^{i\mathbf{k}\cdot\mathbf{r}}$, the expansion coefficients $c(\mathbf{k})$ couple to the coefficients $c(\mathbf{k} + \mathbf{g})$, and, as a consequence, we may parametrize a Bloch function (solution of the central equation) in terms of $\mathbf{k}$ or in terms of any other $\mathbf{k} + \mathbf{g}$. It is therefore usual to refer the states to a given momentum in the first Brillouin zone. Note that this is different from the case of lattice vibrations (phonons) since in these the momentum is limited to the first Brillouin zone.

The quantity $\mathbf{k}$ is a relevant part in the conservation laws of collisions in a crystal. Let us define $\hbar\mathbf{k}$ as the crystalline momentum of an electron. It is not conserved as the usual momentum in free space since, for instance in a collision with a phonon of wave vector $\mathbf{q}$, with the absorption of a phonon, we have the conservation law:

$$\mathbf{k} + \mathbf{q} = \mathbf{k}' + \mathbf{g}, \tag{10.54}$$

where $\mathbf{k}'$ is the wave vector of the electron after the collision and $\mathbf{g}$ is a reciprocal lattice vector. So the crystalline momentum of an electron is conserved minus a reciprocal lattice vector.

This arbitrariness is related to the freedom to choose different ways to represent the electronic bands that are the solutions of the central equation. We may use different schemes such as the extended zone scheme or the reduced zone scheme. This is the result of the possibility to represent the energy levels in terms of values of the wave number that can be as large as we want, or the possibility to perform translations to the first Brillouin zone, adding or subtracting an appropriate reciprocal lattice vector. This is illustrated in the figure for the case of free electrons (no periodic potential). In this case the energy is simply a parabola in terms of the wave vector. However, we may fold the zones beyond the first Brillouin zone and concentrate the energies inside the first Brillouin zone. This leads to infinite many energy bands in the first Brillouin zone, with increasing energies. These bands are gapless in the sense that there is no gap between the various branches of the energy dispersion relation. In the presence of a periodic potential, gaps open at the edges of the Brillouin zone. The case represented here is a one-dimensional lattice. In higher dimensions the dispersion relations are more complex, as we will see ahead. In general we can write:

$$\epsilon(k_x, k_y, k_z) = \frac{\hbar^2}{2m}\left(\mathbf{k}^* + \mathbf{g}\right)^2, \tag{10.55}$$

where $\mathbf{k}^*$ is a wave vector in the first Brillouin zone and $\mathbf{g}$ is the appropriate reciprocal lattice vector that translates the wave vector $\mathbf{k}$ into the first Brillouin zone.

Near an edge of the Brillouin zone, $k \sim g/2$, we may define the deviation $q = k - g/2$. The energy then satisfies the approximate solution of Eq. (10.13):

$$\epsilon(k) = \frac{1}{2}\left(\lambda_{k-g} + \lambda_k\right) \pm \sqrt{\frac{1}{4}\left(\lambda_{k-g} - \lambda_k\right)^2 + U^2}$$
$$\rightarrow \epsilon_q = \frac{\hbar^2}{2m}\left(\frac{1}{4}g^2 + q^2\right) \pm \sqrt{4\lambda\frac{\hbar^2 q^2}{2m} + U^2}, \tag{10.56}$$

where

$$\lambda = \frac{\hbar^2}{2m}\left(\frac{g}{2}\right)^2.$$

Considering that the deviation from the Brillouin zone edge is small, $\hbar^2 gq/(2m) \ll |U|$, we get:

$$\epsilon_q \sim \frac{\hbar^2}{2m}\left(\frac{1}{4}g^2 + q^2\right) \pm U\left(1 + 2\frac{\lambda}{U^2}\frac{\hbar^2 q^2}{2m}\right). \tag{10.57}$$

Using Eq. (10.16) we can write:

$$\epsilon_q = \epsilon_\pm + \frac{\hbar^2 q^2}{2m}\left(1 \pm \frac{2\lambda}{U}\right). \tag{10.58}$$

Therefore, we have quadratic bands around the edge of the Brillouin zone. The band curvature is positive for the upper band across the gap, and the curvature is negative for the band below the gap. We may define the effective masses, $m^*_\pm$, as

$$\frac{\hbar^2 q^2}{2m}\left(1 \pm \frac{2\lambda}{U}\right) = \frac{\hbar^2 q^2}{2m^*_\pm}$$
$$m^*_\pm = \frac{m}{1 \pm \frac{2\lambda}{U}}. \tag{10.59}$$

Recalling that the potential is attractive, $U = -|U|$, we get that $\epsilon_\pm = \lambda \pm U = \lambda \mp |U|$. Therefore, $\epsilon_+ < \epsilon_-$ and

$$m^*_+ = \frac{m}{1 - \frac{2\lambda}{|U|}} < 0$$
$$m^*_- = \frac{m}{1 + \frac{2\lambda}{|U|}} > 0. \tag{10.60}$$

10.3.1 Metals and Insulators

Let us consider a one-dimensional lattice with a lattice constant, a, and N unit cells (taken to be even), leading to a chain of length $L = Na$. The set of wave numbers is $k = 0, \pm 2\pi/L, \pm 4\pi/L, \ldots, N\pi/L = \pi/a$. The point $k = -\pi/a$ is equivalent to the point $k = \pi/a$, since it can be obtained from the other by subtracting a reciprocal lattice vector, specifically $g = 2\pi/a$. Therefore, the number of k points is N, and is equal to the number of unit cells in real space. Each unit cell contributes with an independent value of k for each energy band. Including the electron spin, we conclude that there are $2N$ orbitals for each energy band.

Consider a simple case, such that each atom contributes with one electron to the extended states in the solid. Since we have one electron per unit cell, the energy band may be half-full (because of the spin degeneracy). If each atom contributes with two electrons to the extended states, then the energy band may be full.

This allows us to establish some naive distinction between different materials. The reason it is a simplified description is that we have in mind a one-dimensional system and in reality we have in general $3d$ structures. The naive reasoning leads us to distinguish materials that may have either partially filled bands or fully occupied bands, depending if the number of electrons per unit cell is odd or even.

Consider very low temperatures or strictly zero temperature. According to Pauli's principle, the ground state is obtained filling up the energy states of the bands, until the number of available electrons is exhausted. If a band is partially filled, this means that a small increase of momentum, or energy, will excite one or more electrons to higher energy states. Considering an arbitrarily small energy transfer, an electron will be excited to an empty state with an energy arbitrarily close to its initial state. Therefore, at first, only electrons that occupy the higher energy levels will be excited. If the band is partially filled, there are available states to excite the electron. This process leads to an imbalance of the total momentum of the full set of electrons, and therefore a current may be generated (we are assuming here a situation where the bands are symmetric with respect to k). If a band is full, recalling that there is a gap to the next energy band, and therefore, to the available electronic states, a finite amount of energy is necessary to excite the electron. If the energy supplied is not enough, the electron is not excited and there is no imbalance of the total momentum and no current is expected. Therefore, if we have a partially filled band, this is supposed to describe a system which may conduct a current, and if the band is filled, it is expected to describe a system that does not conduct. This leads naturally to two classes of systems which we may call metals (conductors) and insulators (called band insulators). There are other types of systems as shown in the figure such as semimetals and semiconductors. In the case of semimetals, the system may be gapless with a full band and an empty band touching at a given set of wave vectors, or there is an energy overlap, such that the top of one band is empty and the bottom of the other band is occupied, in such a way that the spectrum is also gapless. In the case of semiconductors, there is a gap, that is typically small as compared to that found in insulators, such that electrons may be easily excited across the gap, for instance by temperature effects. Also, the addition of impurities may deplete electrons from the top band or add some electrons to it, leading to a doped semiconductor.

Examples of the various types of materials are alkaline metals and noble metals which with one electron per unit cell (or an odd number) are metals. In the case of rare earths, one would expect them to be insulators, since they have two (or an even) number of electrons. However, there is overlap of bands and they are actually metals, even though with a low conductivity. Diamond, silicon, and germanium have two atoms of valence four per cell (meaning four electrons extended per atom) which implies eight electrons per cell, and, since it is an even number, we expect an insulator, which is confirmed by experiments. For instance, carbon has an electronic structure $2s^2 2p^2$, silicon $3s^2 3p^2$, and germanium $4s^2 4p^2$.

Optical absorption may determine the gap values by exciting electrons from one band to another, using photon absorption.

10.4 Equations of Motion in a Crystal

The electrons are in general described by some wave packet that we may take as centered around some given momentum, $\hbar\mathbf{k}$, with a group velocity defined by

$$v_g = \frac{d\omega}{dk} = \frac{1}{\hbar}\frac{d\epsilon}{dk}, \tag{10.61}$$

where $\epsilon = \hbar\omega$ and a one-dimensional case was considered, for simplicity. In general the group velocity is obtained as

$$\mathbf{v}_g = \frac{1}{\hbar}\nabla_{\mathbf{k}}\epsilon(\mathbf{k}). \tag{10.62}$$

In the presence of an electric field, work is performed on the electron. In the time interval δt, the change in energy of the electron is

$$\delta\epsilon = -eEv_g\delta t = \frac{d\epsilon}{dk}\delta k = \hbar v_g \delta k, \tag{10.63}$$

where the velocity is calculated along the direction of the electric field. This means that the wave number changes as

$$\delta k = -\frac{eE}{\hbar}\delta t. \tag{10.64}$$

So, the equation is formally of the type:

$$\hbar \frac{d\mathbf{k}}{dt} = -e\mathbf{E} \tag{10.65}$$

and the external force follows then the law

$$\mathbf{F} = -e\mathbf{E} = \hbar \frac{d\mathbf{k}}{dt}. \tag{10.66}$$

The last term is then the time derivative of the (crystalline) momentum of the electron in the crystal. In free space the momentum is simply taken as $m\mathbf{v}$, but in the crystal the equation of motion involves the crystalline momentum, which includes, in addition to the action of the external field, the effect of the forces due to the ions, through the presence of the periodic potential. If we also apply an external magnetic field, we get the semiclassical expression:

$$\hbar \frac{d\mathbf{k}}{dt} = -e\mathbf{v} \times \mathbf{B} + (-e\mathbf{E}), \tag{10.67}$$

or, more explicitly,

$$\frac{d\mathbf{k}}{dt} = -\frac{e}{\hbar}\mathbf{E} - \frac{e}{\hbar^2}\nabla_{\mathbf{k}}\epsilon \times \mathbf{B}. \tag{10.68}$$

We may derive the equation of motion in a different way.[1] Consider a Bloch wave function:

$$\psi_{\mathbf{k}}(\mathbf{r}) = \sum_{\mathbf{g}} c\,(\mathbf{k}+\mathbf{g})\, e^{i(\mathbf{k}+\mathbf{g})\cdot\mathbf{r}}. \tag{10.69}$$

The expectation value of the electron momentum is given by

$$\begin{aligned} \mathbf{p} &= \langle \mathbf{k}|\,(-i\hbar\nabla)\,|\mathbf{k}\rangle \\ &= \sum_{\mathbf{g}} \hbar\,(\mathbf{k}+\mathbf{g})\,|c\,(\mathbf{k}+\mathbf{g})\,|^2 \\ &= \hbar\left(\mathbf{k} + \sum_{\mathbf{g}} \mathbf{g}|c\,(\mathbf{k}+\mathbf{g})\,|^2\right), \end{aligned} \tag{10.70}$$

where we used that the wave function is normalized $\sum_{\mathbf{g}} |c\,(\mathbf{k}+\mathbf{g})\,|^2 = 1$.

[1] See also [Kittel].

When we apply an external force on the electron, this produces a momentum change in the electron. As a consequence of this change, there is also some momentum transfer from the electron to the lattice. Consider that the electron momentum changed from $\mathbf{k} \rightarrow \mathbf{k} + \Delta\mathbf{k}$ as a result of the external force. To simplify, let us consider that the external force only acts on a single electron, which may be the case if we only have one electron in an otherwise empty band. Assume that the action of the external force is through an impulse like $\mathbf{J} = \int \mathbf{F}dt$. This impulse is the momentum change and may be written

$$\mathbf{J} = \Delta\mathbf{p} = \Delta\mathbf{p}_{el} = \hbar\Delta\mathbf{k} \tag{10.71}$$

if the electron is free and

$$\mathbf{J} = \Delta\mathbf{p} = \Delta\mathbf{p}_{el} + \Delta\mathbf{p}_{l}, \tag{10.72}$$

which includes the contribution to the lattice. But we know that

$$\Delta\mathbf{p}_{el} = \hbar\Delta\mathbf{k} + \sum_{\mathbf{g}} \hbar\mathbf{g}\left(\nabla_{\mathbf{k}}|c\left(\mathbf{k}+\mathbf{g}\right)|^{2} \cdot \Delta\mathbf{k}\right). \tag{10.73}$$

We are left with the calculation of the contribution to the lattice.

When an electron is reflected by the periodic lattice, it transfers momentum to the lattice. An electron with momentum $\hbar\mathbf{k}$ is reflected with momentum $\hbar\left(\mathbf{k}+\mathbf{g}\right)$, and the lattice absorbs a momentum change of $-\hbar\mathbf{g}$. The momentum transferred to the lattice when the state $\psi_{\mathbf{k}}$ is changed in the state $\psi_{\mathbf{k}+\Delta\mathbf{k}}$ is

$$\Delta\mathbf{p}_{l} = -\hbar\sum_{\mathbf{g}} \mathbf{g}\left(\nabla_{\mathbf{k}}|c\left(\mathbf{k}+\mathbf{g}\right)|^{2} \cdot \Delta\mathbf{k}\right), \tag{10.74}$$

which means that the total change of momentum is

$$\Delta\mathbf{p} = \Delta\mathbf{p}_{el} + \Delta\mathbf{p}_{l} = \hbar\Delta\mathbf{k} = \mathbf{J}. \tag{10.75}$$

Therefore, we reach the same conclusion that

$$\mathbf{F} = \hbar\frac{d\mathbf{k}}{dt}, \tag{10.76}$$

which leads to the conclusion that the crystalline momentum encompasses the action of the external force.

10.5 Holes in a Band

The many-body state of the non-interacting electrons in a given state is described by the Slater determinant of the wave functions (eigenstates) of the occupied states. Ignoring the fully occupied bands with lower energies, if there are few electrons occupying a given band, this Slater determinant is a reasonable description, but as the number of electrons in the band increases, this becomes a complex problem. A way around it is to use an occupation number representation and use second quantization methods (see Appendix A). On the other hand, if a band is almost full, with a few non-occupied states, called holes, once again there are a few states (the unoccupied states) that might provide an alternative simple description of the state. The aim is to substitute the representation of the state by the many occupied states by another one that focuses on the empty ones (the holes). This would become a few body problem. This simplified problem contains the same information but in terms of new "variables."

Let us focus first our attention on the simplest problem of a single hole in a band. One way is to consider a fully occupied valence band (such as in an insulator or semiconductor) and a fully empty conduction band, separated by a finite energy gap. As shown above, the curvature of the valence band is taken to be negative (negative effective mass), and the curvature of the conduction band is taken to be positive (positive effective mass). Let us assume that a photon incident on the material excites one electron in the valence band to a state in the conduction band. The energy of the photon is enough to overcome the energy difference between the two states, and, since the velocity of the photon is much higher than that of the electrons, we may assume that there is no momentum change (the transition is vertical in an energy-momentum graph). As a result of the interaction with the photon, there is now a hole in the valence band and one electron in the conduction band. Call $\mathbf{k}_e$ the wave vector of the electron that was excited. Assume that before the photon excited the electron, the system was in equilibrium, with an even number of electrons, such that the total momentum of the valence band was zero. After the electron was removed, the total momentum of the valence band is now

$$\mathbf{k}_h = -\mathbf{k}_e. \tag{10.77}$$

The momentum of the whole valence band may be described by the momentum of a "particle," called hole, with momentum $\mathbf{k}_h$.

If we choose the zero of the energies at the top of the valence band, we can conclude that the energy of the excitation we call hole is just

$$\epsilon_h(\mathbf{k}_h) = -\epsilon_e(\mathbf{k}_e). \tag{10.78}$$

Note that this is the energy of the valence band and does not include the energy of the electron promoted to the conduction band. That depends on the energy difference between the two bands at the momentum $\mathbf{k}_e$.

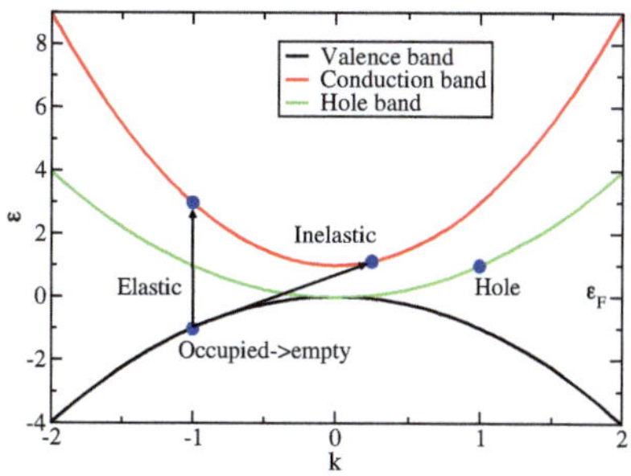

Fig. 10.5 An electron is removed from the valence band and is excited to the conduction band, either with conservation of momentum or with momentum change. The removed electron creates an empty state in the valence band that may be represented by an occupied state in the hole band

Clearly, it costs energy to excite the electron and so the energy of the excitation (the hole) should be positive. The lower the energy of the electron that was excited, the higher the energy we need to supply. The energy of the hole is opposite (symmetrical) to the energy of the missing electron. If the energy band is symmetrical, we get:

$$\epsilon_e(\mathbf{k}_e) = \epsilon_e(-\mathbf{k}_e) = -\epsilon_h(-\mathbf{k}_e) = -\epsilon_h(\mathbf{k}_h). \tag{10.79}$$

We may imagine that the holes constitute now an energy band. The band is then the result of the property that the momentum changes sign and the energy changes sign with respect to the states that a single electron may occupy on the full valence band. Therefore, we get that the slope of the original electron valence band is equal to the corresponding hole band. The velocity satisfies:

$$\mathbf{v}_h = \mathbf{v}_e \tag{10.80}$$

because $\nabla\epsilon_h(\mathbf{k}_h) = \nabla\epsilon_e(\mathbf{k}_e)$. An example of the band structure in the vicinity of a gap that separates the valence band from the conduction band and a representation of the hole band associated with the valence band is shown in Fig. 10.5.

Also, since the curvatures are opposite, we conclude that

$$m_h = -m_e, \tag{10.81}$$

which means that the effective mass of the holes is positive. Note that we can obtain the mass using that $m = \hbar^2/\left((d^2\epsilon)/(dk^2)\right)$ which implies the relation between the sign of the curvature of the energy (second derivative) and the sign of the mass. In general, the system may have bands that are not isotropic, and we get that the effective mass is actually a tensor that can be written as

$$\left(\frac{1}{m^*}\right)_{\mu,\nu} = \frac{1}{\hbar^2}\frac{d^2\epsilon}{dk_\mu dk_\nu}. \tag{10.82}$$

As mentioned before, since the effect of the periodic potential is small far from the Brillouin zone edges, the effective mass is approximately the same as the free electron mass, and only close to the edges a significative change is found. The reason is that only in their neighborhood the amplitude of the waves of the type $e^{i(k-g)x}$ is significative (to simplify consider $1d$). In this regime, there is an enhanced momentum transfer from the electron to the lattice. Near the edge the reflected part is large (this leads to the stationarity condition, $v_g = 0$, at the edge, as discussed at length before), which results from the cancelation of the momentum $\hbar k = \hbar(-1/2g)$ with the momentum $\hbar(1/2g)$. The origin of the negative mass is then interpreted that, as the electron changes its momentum from $\mathbf{k} \to \mathbf{k} + \Delta\mathbf{k}$, the momentum transferred from the electron to the lattice is larger than the momentum transferred to the electron by the external force. This happens in the energy band below the gap. It turns out that the amplitude of the wave $e^{i(k-g)x}$ decreases fast in the band above the gap and the effective mass is positive.

In addition to the sign change that is observed in some bands, if the energy dispersion relation is very flat, the curvature is very small which may lead to very high effective masses (positive or negative). Typically in a good (simple) metal, the effective mass is approximately the mass of the free electron. In some materials, however, there is a large renormalization of the mass, and the effective mass may be very large compared to the free electron mass. Values of the order or 1000 times larger are found in materials called heavy-fermion systems, such as in $UBe_{13}, CeAl_2, CeCu_2Si_2$. We will see later that in these materials, $4f$ and $5f$ electrons are typically responsible for this renormalization. The main reason is associated with very narrow energy bands (and therefore low group velocities or high effective masses). A specific model for this type of bands, called tight-binding model, will be addressed ahead.

Looking at the semiclassical equation of motion under the action of external fields, electric and magnetic, we may obtain that the holes have positive charge instead of the negative charge, $-e$, of the electrons:

$$\begin{aligned}\hbar\frac{d\mathbf{k}_e}{dt} &= -e\left(\mathbf{E} + \mathbf{v}_e \times \mathbf{B}\right)\\ -\hbar\frac{d\mathbf{k}_h}{dt} &= -e\left(\mathbf{E} + \mathbf{v}_h \times \mathbf{B}\right)\\ \hbar\frac{d\mathbf{k}_h}{dt} &= e\left(\mathbf{E} + \mathbf{v}_h \times \mathbf{B}\right). \end{aligned} \tag{10.83}$$

The choice of representation of the state, either in terms of electrons (with an almost full band) or in terms of holes (in this case just one hole), should lead to the same result for physical quantities, such as the charge current density. For instance, if we apply an electric field along a given direction, since an electron has negative charge, it will move in the opposite direction. On the other hand, since the charge of the hole is positive, it will move along the direction of the electric field. Recalling that the charge current density is of the form $J = \rho v$, where ρ here is the charge density which for the electron is negative and for the hole is positive, leads to the expected result that the current density has the same direction of the electric field.

10.6 Brillouin Zones and Fermi Surfaces

There are different ways to represent the electronic energy bands. Usual procedures are the extended zone scheme, where the momentum is allowed to take any values (unrestricted zone), the reduced zone scheme where any momentum value outside the first Brillouin zone is reflected to the first zone, with an appropriate translation by a reciprocal lattice vector, and the periodic zone scheme where the first Brillouin zone is repeated indefinitely outside its range.

The electronic bands are the corresponding eigenvalues of the eigenvectors described by the wave functions:

$$\psi_{n,\mathbf{k}} = e^{i\mathbf{k}\cdot\mathbf{r}} u_{n,\mathbf{k}}(\mathbf{r}) = \sum_{\mathbf{g}} c_n(\mathbf{k}+\mathbf{g}) e^{i(\mathbf{k}+\mathbf{g})\cdot\mathbf{r}}. \tag{10.84}$$

As mentioned before, the first Brillouin zone is the Wigner-Seitz cell of the reciprocal lattice. For instance, if we consider a square lattice in a $2d$ space, the first Brillouin zone is also a square centered around the origin of the momenta, and extends in either direction from $-\pi/a$ to π/a. Recall that the Wigner-Seitz cell is constructed choosing the center at a given point in the reciprocal lattice, take it as the origin, and bisecting the reciprocal lattice vectors to the first neighbors. The second Brillouin zone is defined following a similar procedure bisecting the reciprocal lattice vectors that connect to the second neighbors, and so on for other Brillouin zones as shown in Fig. 10.6. The second Brillouin zone is then the difference between the first Brillouin zone and the larger area that takes into account the second neighbors. This is represented in Fig. 10.7. The first Brillouin zone is connected, but the other Brillouin zones are disconnected, but all have the same area. The first Brillouin zone is a square; the second Brillouin zone is composed of four triangles and the third Brillouin zone of eight triangles.

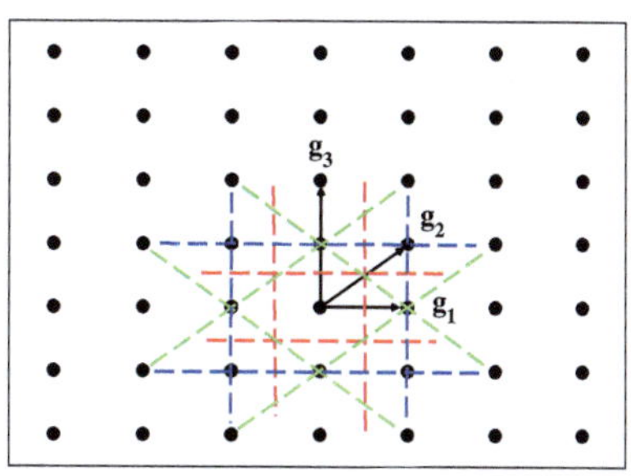

Fig. 10.6 Wigner-Seitz cell and further neighbor cells of a rectangular/square lattice

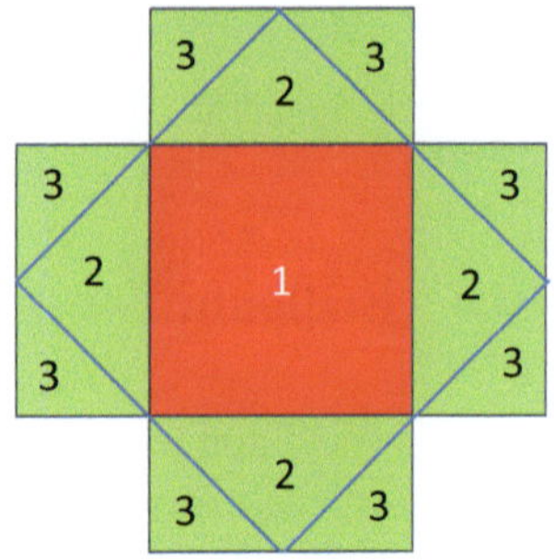

Fig. 10.7 Brillouin zones of a square lattice

Let us now consider a three-dimensional system and the Fermi "sphere." It is a sphere if there is no periodic potential, but the sphere has a distorted shape if we take into account the periodic potential. Most notably if the electronic density is high enough and there are electrons occupying states that approach the edges of the Brillouin zone, the Fermi surface gets distorted.

It is clear that in a square or cubic lattice, if the electronic density is high enough, such that the Fermi momentum is larger than $\sqrt{2}\pi/a$, the Fermi surface will contain the full first Brillouin zone, but will also include parts of the second and third Brillouin zones. At least if, for the moment, we ignore the effect of the periodic potential. Taking this effect into account, we know that the density of states increases near the Brillouin zone edges, more electrons are allowed in states of lower energies, and the Fermi surface may be entirely contained in the first Brillouin zone. But for a higher electron density, the Fermi surface will be divided in several Brillouin zones. In a $3d$ system, and in other Bravais lattices, this often occurs. In general, the first Brillouin zone is not fully occupied, and some higher-order Brillouin zones will have smaller partial occupations, as shown in Fig. 10.8. So the first Brillouin zone will likely be seen as a band with holes, while the other Brillouin zones will be seen as electron bands.

The volume of the Fermi surface was derived previously for the case when the electrons are free. It turns out that the volume does not change when the effect of the periodic potential is added, since the volume is essentially obtained due to a counting of states, and this does not change, since the periodic potential is seen as an external potential. The shape of the Fermi surface changes, but its volume remains constant since a single-electron picture still holds, even though there is the periodic potential. If we consider interactions between electrons, a single-particle description does not hold anymore and a full many-body treatment is needed. It is less obvious, but it can be shown that even with the effect of interactions the volume does not change. This is called Luttinger's theorem.

There is a simple procedure to define the various Brillouin zones which is known as Harrison's construction. Take a set of spheres with radii equal to the Fermi wave vector and center each sphere at a given reciprocal lattice vector. The first zone is the set of points in $\mathbf{k}$ space that are part of a single sphere, the second zone the points that are included in two spheres and so on.

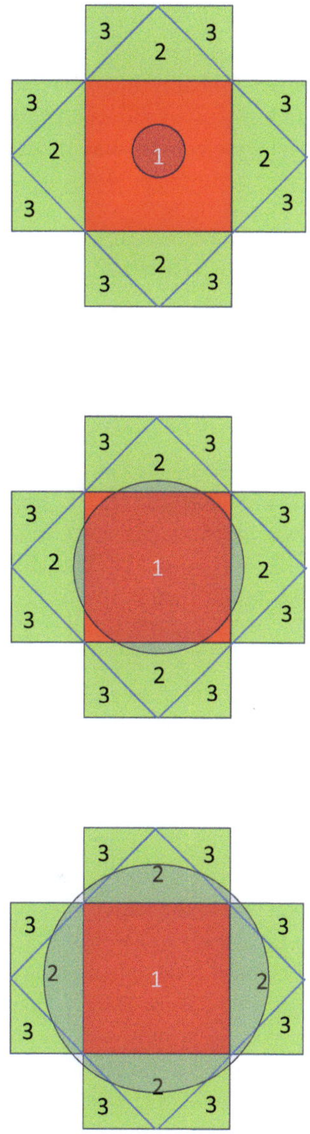

Fig. 10.8 Filling of the Brillouin zones for different size Fermi surfaces of free electrons in a square lattice

Experimentally the Fermi surface may be obtained using the de Haas-van Alphen effect. It is based on the action of a magnetic field on a crystal. Since the Lorentz force does no work, the electrons move along orbits of constant energy. Therefore, electrons on the Fermi surface move along orbits that are located on the Fermi surface. It turns out that these orbits give the larger contribution to the effect due to interferences of orbits at different energies. There is a periodicity in the magnetic response that is associated with the size of the orbits, which allows to determine the shape of the Fermi surface by rotating the orientation of the magnetic field, since the orbits are essentially contained in the plane perpendicular to the external magnetic field. Typically the orbits are of two kinds: they may be closed, if a band is only partially filled and its energies are limited, or they may be open if the orbits are such that the electrons may travel from one Brillouin zone to the next.

10.6.1 Band Structure of Some 3d Materials

In this subsection we will present a simplified description of the band structure and Fermi surfaces of some typical $3d$ systems.[2] In Table 10.1 the periodic table of the elements is shown highlighting the different types of materials. In Tables 10.2, 10.3, 10.4, and 10.5, the electronic configurations of the various elements are shown.

The simplest case is probably the case of monovalent metals which have one free electron per atom. Examples are the alkaline metals (group 1) and noble metals (group 11). The alkaline metals tend to crystalize in a *bcc* structure and the noble metals in an fcc structure. The electronic structures of the alkalines are Li : $[He]2s^1$; Na : $[Ne]3s^1$; K : $[Ar]4s^1$; Rb : $[Kr]5s^1$; Cs : $[Xe]6s^1$, and the electronic structures of the noble metals are Cu : $[Ar]3d^{10}4s^1$; Ag : $[Kr]4d^{10}5s^1$; Au : $[Xe]4f^{14}5d^{10}6s^1$.

The *bcc* lattice has two atoms per cell and the Fermi momentum may be calculated as

$$\frac{k_F^3}{3\pi^2} = n = \frac{2}{a^3},$$

where a is the lattice constant of the underlying simple cubic cell (not the unit cell) which leads to $k_F = 0.620\,(2\pi/a)$. Using that the distance from the center of the Brillouin zone to its edge is $\Gamma N = 2\pi/a\sqrt{(1/2)^2 + (1/2)^2} = 0.707\,(2\pi/a)$, we get that $k_F/\Gamma N = 0.877$. Therefore, the whole Fermi sphere is in the first Brillouin zone and is only slightly distorted from a perfect sphere. Note that the sizes of the gaps may be considerable, and therefore there is a large deviation from a free electron picture, but the Fermi surface is largely insensitive and is almost spherical.

The case of the noble metals is more complex. There are now 11 electrons in the fcc structure which implies a minimum of 6 bands. There are s bands and d bands. The $3d$ bands are less extended than the $4s$ band. So we expect that the higher band occupied is of $4s$ character, and the $3d$ bands have slightly lower energies. Therefore, the Fermi level is expected to lie in a $4s$ band and the $3d$ bands lie below the Fermi level by a finite amount (gap). Since an s band is essentially spherical, we expect that a nearly free electron model is a good description. However, the $3d$

[2] For further reading see [AshcroftMermin].

Table 10.1 The first entry for each element is Z, the number of electrons (and protons) of a neutral free atom. Except for H the group 1 elements are the alkali metals. Group 2 are the alkaline earth metals. Groups 3 through 12 are the transition metals, with the exception of the lanthanides (elements with $Z = 57 - 71$) and the actinides (elements with $Z = 89 - 103$). Group 18 are the noble gases. Group 17 are the halogens. Al, Ga, In, Sn, Tl, Pb, Bi, Po, Nh, Fl, Mc, and Lv are other metals. B, Si, Ge, As, Sb, and Te are called metalloids. Other nonmetals are H, C, N, O, P, S, and Se. Many systems at low enough temperature will order into some magnetic order or enter a state of zero resistance, called superconductivity, as discussed later in this book. nm, non-metallic; mo, magnetic ordering; sc, superconductor; sc*, superconductor under pressure or in thin films; mo*, magnetic order but under pressure or in thin films is a superconductor

1	2	3	4	5	6	7	8	9	10	11	12	13	14	15	16	17	18
1																	2
H																	He
nm																	nm
3	4											5	6	7	8	9	10
Li	Be											B	C	N	O	F	Ne
	sc*											nm	nm	nm	nm	nm	nm
11	12											13	14	15	16	17	18
Na	Mg											Al	Si	P	S	Cl	Ar
												sc	sc*	nm	nm	nm	nm
19	20	21	22	23	24	25	26	27	28	29	30	31	32	33	34	35	36
K	Ca	Sc	Ti	V	Cr	Mn	Fe	Co	Ni	Cu	Zn	Ga	Ge	As	Se	Br	Kr
			sc	sc	mo	mo	mo	mo	mo		sc	sc	sc*	nm	sc*	nm	nm
37	38	39	40	41	42	43	44	45	46	47	48	49	50	51	52	53	54
Rb	Sr	Y	Zr	Nb	Mo	Tc	Ru	Rh	Pd	Ag	Cd	In	Sn	Sb	Te	I	Xe
			sc	sc	sc	sc	sc				sc	sc	sc	sc*	sc*	nm	nm
55	56	71	72	73	74	75	76	77	78	79	80	81	82	83	84	85	86
Cs	Ba	Lu	Hf	Ta	W	Re	Os	Ir	Pt	Au	Hg	Tl	Pb	Bi	Po	At	Ru
sc*	sc*		sc	sc	sc	sc	sc	sc			sc	sc	sc	sc*	nm	nm	nm
87	88	103	104	105	106	107	108	109	110	111	112	113	114	115	116	117	118
Fr	Ra	Lr	Rf	Db	Sg	Bh	Hs	Mt	Ds	Rg	Cn	Nh	Fl	Mc	Lv	Ts	Og
		57	58	59	60	61	62	63	64	65	66	67	68	69	70		
		La	Ce	Pr	Nd	Pm	Sm	Eu	Gd	Tb	Dy	Ho	Er	Tm	Yb		
		sc	mo*	mo	mo		mo	mo	mo	mo	mo	mo	mo	mo	mo		
		89	90	91	92	93	94	95	96	97	98	99	100	101	102		
		Ac	Th	Pa	U	Np	Pu	Am	Cm	Bk	Cf	Es	Fm	Md	No		
			sc	sc	sc												

bands are not far in energy, and their role is mainly felt in optical properties, if the energy supplied is of the order of the small gap to the Fermi level. Since there is one $4s$ electron per atom, the Fermi surface of a half-filled free electron band in an fcc structure is a sphere that is contained in the first Brillouin zone. This Fermi surface approaches the Brillouin zone edge along the direction [111], such that it reaches

Table 10.2 s-wave block of the periodic table. For each element the atomic number, Z, is shown at the top of each entry. The atomic number equals the number of electrons (and protons) in each free atom. In the electronic configuration, the first number is the principal quantum number; the letter indicates the orbital angular momentum ($s : l = 0$; $p : l = 1$; $d : l = 2$; $f : l = 3$) in the usual notation. $1s^2$ means two electrons with $n = 1$ and $l = 0$, $7s^2$ means two electrons in the level $n = 7$ and $l = 0$, and so on. The inner states that correspond to filled energy levels of the noble gas of group 18 of the previous period of Table 10.1 are omitted

1 H $1s$	
3 Li $2s$	4 Be $2s^2$
11 Na $3s$	12 Mg $3s^2$
19 K $4s$	20 Ca $4s^2$
37 Rb $5s$	38 Sr $5s^2$
55 Cs $6s$	56 Ba $6s^2$
87 Fr $7s$	88 Ra $7s^2$

Table 10.3 p-wave block (with the exception of He)

					2 He $1s^2$
5 B $2s^2 2p$	6 C $2s^2 2p^2$	7 N $2s^2 2p^3$	8 O $2s^2 2p^4$	9 F $2s^2 2p^5$	10 Ne $2s^2 2p^6$
13 Al $3s^2 2p$	14 Si $3s^2 2p^2$	15 P $3s^2 2p^3$	16 S $3s^2 2p^4$	17 Cl $3s^2 2p^5$	18 Ar $3s^2 2p^6$
31 Ga $4s^2 2p$	32 Ge $4s^2 2p^2$	33 As $4s^2 2p^3$	34 Se $4s^2 2p^4$	35 Br $4s^2 2p^5$	36 Kr $4s^2 2p^6$
49 In $5s^2 2p$	50 Sn $5s^2 2p^2$	51 Sb $5s^2 2p^3$	52 Ta $5s^2 2p^4$	53 I $5s^2 2p^5$	54 Xe $5s^2 2p^6$
81 Tl $6s^2 2p$	82 Pb $6s^2 2p^2$	83 Bi $6s^2 2p^3$	84 Po $6s^2 2p^4$	85 At $6s^2 2p^5$	86 Rn $6s^2 2p^6$

Table 10.4 *d*-wave block

21	22	23	24	25	26	27	28	29	30
Sc	Ti	V	Cr	Mn	Fe	Co	Ni	Cu	Zn
$3d$	$3d^2$	$3d^3$	$3d^5$	$3d^5$	$3d^6$	$3d^7$	$3d^8$	$3d^{10}$	$3d^{10}$
$4s^2$	$4s^2$	$4s^2$	$4s$	$4s^2$	$4s^2$	$4s^2$	$4s^2$	$4s$	$4s^2$
39	40	41	42	43	44	45	46	47	48
Y	Zr	Nb	Mo	Tc	Ru	Rh	Pd	Ag	Cd
$4d$	$4d^2$	$4d^4$	$4d^5$	$4d^6$	$4d^7$	$4d^8$	$4d^{10}$	$4d^{10}$	$4d^{10}$
$5s^2$	$5s^2$	$5s$	$5s$	$5s$	$5s$	$5s$		$5s$	$5s^2$
57	72	73	74	75	76	77	78	79	80
La	Hf	Ta	W	Re	Os	Ir	Pt	Au	Hg
	$4f^{14}$								
$5d$	$5d^2$	$5d^3$	$5d^4$	$5d^5$	$5d^6$	$5d^9$	$5d^9$	$5d^{10}$	$5d^{10}$
$6s^2$	$6s^2$	$6s^2$	$6s^2$	$6s^2$	$6s^2$		$6s$	$6s$	$6s^2$

Table 10.5 *f*-wave block

58	59	60	61	62	63	64	65	66	67	68	69	70	71
Ce	Pr	Nd	Pm	Sm	Eu	Gd	Tb	Dy	Ho	Er	Tm	Yb	Lu
$4f^2$	$4f^3$	$4f^4$	$4f^5$	$4f^6$	$4f^7$	$4f^7$	$4f^8$	$4f^{10}$	$4f^{11}$	$4f^{12}$	$4f^{13}$	$4f^{14}$	$4f^{14}$
						$5d$	$5d$						$5d$
$6s^2$	$6s^2$	$6s^2$	$6s^2$	$6s^2$	$6s^2$	$6s^2$	$6s^2$	$6s^2$	$6s^2$	$6s^2$	$6s^2$	$6s^2$	$6s^2$
90	91	92	93	94	95	96	97	98	99	100	101	102	103
Th	Pa	U	Np	Pu	Am	Cm	Bk	Cf	Es	Fm	Md	No	Lr
	$5f^2$	$5f^3$	$5f^5$	$5f^6$	$5f^7$	$5f^7$							
$6d^2$	$6d$	$6d$				$6d$							
$7s^2$	$7s^2$	$7s^2$	$7s^2$	$7s^2$	$7s^2$	$7s^2$							

about 0.903 of the distance of the zone center (Γ point) to the hexagonal face. It turns out that the Fermi surface stretches until it touches the *eight* hexagonal faces.

The various orbits can be detected applying a magnetic field, as discussed above. They are also detected measuring the optical conductivity, as discussed later in the book. For instance, for the alkaline metals, in addition to the low-frequency Drude peak associated with the electron translation (to be discussed later), there is a threshold to a peak that is associated with interband transitions, from states in the second Brillouin zone. Something similar occurs in the case of the noble metals with the additional effect of the interband transitions that involve the d electron bands. There is a peak at an energy of the order of $2eV$, associated with the gap between the d bands and the Fermi level, and a peak at the order of $4eV$, associated with a transition to another s band, which is associated with the second Brillouin zone.

Consider now divalent metals in the groups 2 and 12. In group 2 we have $Be(hcp)$: $[He]2s^2$; $Mg(hcp)$: $[Ne]3s^2$; $Ca(fcc)$: $[Ar]4s^2$; $Sr(fcc)$: $[Kr]5s^2$; $Ba(bcc)$: $[Xe]6s^2$. In group 12 we have $Zn(hcp)[Ar]3d^{10}4s^2$; $Cd(hcp)$: $[Kr]4d^{10}5s^2$; $Hg(R)$: $[Xe]4f^{14}5d^{10}6s^2$, where R stands for rhombohedron. It turns out that the d bands of group 12 have energies that are far from the Fermi level (located at the s bands) and essentially do not contribute.

The materials that have a cubic structure are easier to analyze. Since Ca, Sr, Ba have two electrons per cell, we may wonder if they are insulators, as would be expected if the structure were $1d$ and the effect of interactions between the electrons is ignored. We know that the volume contained in the Fermi surface is the same volume as the first Brillouin zone. If we would have truly free electrons, the Fermi surface would intersect the faces of the first Brillouin zone and would cross over to the next Brillouin zone. We would have a complex structure in the first Brillouin zone, and there would be small patches in the second Brillouin zone. Adding the periodic potential, which leads to distortions of the Fermi surface due to the existence of the gaps, it might occur that the increase of the density of states in the first Brillouin zone might accommodate all the electrons, and the second Brillouin zone would become empty, leading indeed to an insulator. It turns out that this does not happen, even though it would be a possibility.

The case of the hcp lattices, which are not of Bravais type, is more involved to carry out a simplified analysis, and a detailed calculation or measurement is required. There are two atoms/unit cell, which leads to $four$ bands, with the additional complication that in the materials Be, Mg, Zn, Cd, we cannot ignore the effect of a small spin-orbit coupling. In the case of trivalent atoms, such as $Al(fcc)$: $[Ne]3s^2 3p^1$, with three electrons per atom, the Fermi surface structure gets more complicated, and it turns out that there are occupied states in the first three Brillouin zones. Tetravalent metals such as $Pb(fcc)$: $[Xe]4f^{14}5d^{10}6s^2 6p^2$ also have a Fermi surface that extends three Brillouin zones. There are four electrons/atom since we may consider the $4f$ and $5d$ electrons as inert, in the sense that the corresponding energy levels are far from the Fermi surface.

The pentavalent elements are nearly insulators but truly are semimetals, with small overlaps between the valence band and the conduction band. Examples are graphite with a structure that is hexagonal with four atoms in each unit cell and As : $[Ar]3d^{10}4s^2 4p^3$; Sb : $[Kr]4d^{10}5s^2 5p^3$; Bi : $[Xe]4f^{14}5d^{10}6s^2 6p^3$ with a structure that is a rhombohedron with two atoms/cell and are therefore close to be insulators. If the structure was simple cubic, they would be good metals, but they have very few free charges due to their actual structures.

Transition metals are somewhat different. They are all metals and are characterized by a varying number of electrons in the d bands, from $3d$ to $4d$ as these bands are filled with electrons, in addition to electrons in an external filled s band, such as $4s^2$ or $5s^2$. The structure changes from fcc to bcc to hcp as the d bands get filled, as we move from Sc to Ni or from Y to Pd. The interesting properties are dominated by the d bands, since these bands are either close to the Fermi level or the Fermi level actually crosses a d band. The properties are different from previous examples where the Fermi level essentially crosses s or p bands, since the d levels are more localized. Therefore, the d bands are in general narrower, as a result of the smaller electron wave functions overlaps between neighboring atoms. A free electron description with parabolic dispersion relation, on top of which we add the effect of the periodic potential, is not a good starting point. Another model that takes into account the fact that the bands are narrow is required. A tight-binding approach is more appropriate, as shown in the next section. As a result, the Fermi surfaces cannot be seen as distorted spheres and the proper description is more complex. As a consequence of narrow bands the density of states is higher at the Fermi level.

Something similar occurs in the rare earths as we move from La to Hf, where the $4f$ level gets filled or as the $5f$ gets filled. In these systems the f levels are more localized, in comparison with the d levels, and are typically further away from the Fermi level. Their more localized nature implies that the effect of interactions between electrons in these levels has a larger role, such as the Coulomb repulsion between electrons that are close by, due to their more localized behavior around a same atom. As we will see, this leads to magnetic moments that lead to interesting magnetic properties.

10.7 Tight-Binding Model

Consider first the case of weakly bound atoms and therefore far away from each other. In this case it is reasonable to look for solutions of the electronic wave functions as a linear combination of atomic orbitals (LCAO). The same situation is achieved if the electrons that are being shared by the atoms are somewhat localized, with spatial extensions that are small compared to the distance, a, between the atoms. As long as these types of orbitals give a significant contribution to the properties of the system, even if there are other more extended orbitals, it is interesting to determine the contribution of the more localized orbitals. Near each atom the wave function is likely similar to that of an electron in an isolated atom.

Let us call $\phi_a(\mathbf{r}-\mathbf{R})$ an atomic orbital centered in $\mathbf{R}$. Let us look for a Bloch orbital of the form[3]

$$\phi_{\mathbf{k}}(\mathbf{r}) = \sum_{\mathbf{R}} e^{i\mathbf{k}\cdot\mathbf{R}} \phi_a(\mathbf{r}-\mathbf{R}). \tag{10.85}$$

In a first approximation let us calculate the average value of the energy:

$$E(\mathbf{k}) = \frac{\int d^3r \phi_{\mathbf{k}}^* \{-\frac{\hbar^2}{2m}\nabla^2 + V(\mathbf{r})\}\phi_{\mathbf{k}}}{\int d^3r \phi_{\mathbf{k}}^* \phi_{\mathbf{k}}}. \tag{10.86}$$

If the atoms are located far from each other

$$\frac{1}{N_a}\int d^3r \phi_{\mathbf{k}}^* \phi_{\mathbf{k}} \sim 1. \tag{10.87}$$

[3] See also [Kittel].

Let us focus on the numerator. We get that, approximately, the energy per particle is

$$\epsilon(\mathbf{k}) \sim \frac{1}{N_a} \sum_{\mathbf{R},\mathbf{R}'} e^{i\mathbf{k}\cdot(\mathbf{R}-\mathbf{R}')} \int d^3r \phi_a^*(\mathbf{r}-\mathbf{R}') \left\{ -\frac{\hbar^2}{2m}\nabla^2 + V(\mathbf{r}) \right\} \phi_a(\mathbf{r}-\mathbf{R}). \tag{10.88}$$

We may now write:

$$\epsilon(\mathbf{k}) = \sum_{\mathbf{h}} e^{i\mathbf{k}\cdot\mathbf{h}} \epsilon_{\mathbf{h}}, \tag{10.89}$$

where we have defined $\mathbf{h} = \mathbf{R} - \mathbf{R}'$. This leads to

$$\epsilon_{\mathbf{h}} = \frac{1}{v_c} \int d^3r \phi_a^*(\mathbf{r}+\mathbf{h}) \left\{ -\frac{\hbar^2}{2m}\nabla^2 + V(\mathbf{r}) \right\} \phi_a(\mathbf{r}). \tag{10.90}$$

The function ϕ_a satisfies:

$$\left\{ -\frac{\hbar^2}{2m}\nabla^2 + V_a(\mathbf{r}) \right\} \phi_a(\mathbf{r}) = \epsilon_a \phi_a(\mathbf{r}), \tag{10.91}$$

where V_a is the potential of an isolated atom at $\mathbf{r} = 0$. We may write:

$$\epsilon_{\mathbf{h}} = \frac{1}{v_c} \int d^3r \phi_a^*(\mathbf{r}+\mathbf{h}) \left\{ -\frac{\hbar^2}{2m}\nabla^2 + V_a(\mathbf{r}) + (V(\mathbf{r}) - V_a(\mathbf{r})) \right\} \phi_a(\mathbf{r}). \tag{10.92}$$

Taking $\mathbf{h} = 0$ we get that

$$\begin{aligned} \epsilon_{\mathbf{h}=0} = & \frac{1}{v_c} \int d^3r \phi_a^*(\mathbf{r}) \left\{ -\frac{\hbar^2}{2m}\nabla^2 + V_a(\mathbf{r}) \right\} \phi_a(\mathbf{r}) \\ & + \frac{1}{v_c} \int d^3r \phi_a^*(\mathbf{r}) \{V(\mathbf{r}) - V_a(\mathbf{r})\} \phi_a(\mathbf{r}). \end{aligned} \tag{10.93}$$

We conclude that

$$\epsilon_0 \sim \epsilon_a \tag{10.94}$$

and

$$\epsilon_{\mathbf{h}} \sim \frac{1}{v_c} \int d^3 \phi_a^*(\mathbf{r}) \{V(\mathbf{r}) - V_a(\mathbf{r})\} \phi_a(\mathbf{r}), \tag{10.95}$$

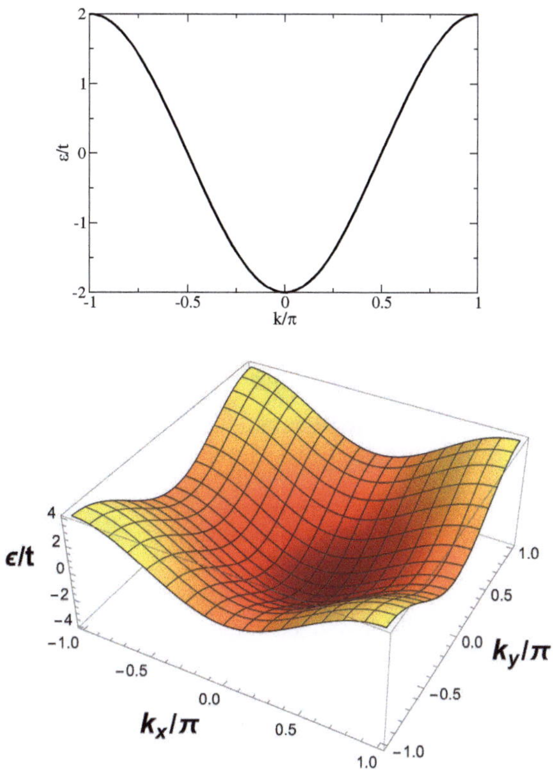

Fig. 10.9 Energy band for a tight-binding model in $1d$ (upper) and $2d$ (lower)

since $\phi_a(\mathbf{r})$ and $V_a(\mathbf{r})$ are very localized in one atom. Note that the definition of $\epsilon_{\mathbf{h}}$ in Eq. (10.90) involves the wave functions of neighboring sites, and is associated with their overlaps mediated by the Hamiltonian of the electronic system. The final expression Eq. (10.95) reveals, therefore, that it may be seen approximately as a potential energy difference contribution at a given site.

We have in a centrosymmetric system:

$$\epsilon(\mathbf{k}) \sim \epsilon_a + e^{i\mathbf{k}\cdot\mathbf{h}}\epsilon_{\mathbf{h}} + e^{-i\mathbf{k}\cdot\mathbf{h}}\epsilon_{-\mathbf{h}} + \cdots \tag{10.96}$$

The vector $\mathbf{h}$ points in a given spatial direction, which in the cubic lattice leads to the nearest-neighbors plus further neighbors. Limiting to first neighbors, we get:

$$\epsilon(\mathbf{k}) = \epsilon_a + 2\tilde{\epsilon}\left(\cos ak_x + \cos ak_y + \cos ak_z\right). \tag{10.97}$$

In three dimensions the band has width $12|\tilde{\epsilon}|$. Depending on the sign of $\tilde{\epsilon}$, the location of the minima and maxima of the band change. Taking $\tilde{\epsilon} < 0$ leads to a minimum of the band at $\mathbf{k} = 0$ and maxima at $k_x, k_y, k_z = \pi/a$. The energy bands for the cases of a $1d$ and a $2d$ system are represented in Fig. 10.9.

If the lattice constant, a, is very large, the atoms are far apart, and the energy levels are the energy levels of a single atom, with degeneracy N, the number of sites in the lattice. As a decreases the levels start to overlap and the degeneracy is lifted. For each atomic level, there is a band of width of the order of $\tilde{\epsilon}$ which reflects the overlap between the atomic orbitals. The model is particularly useful either for a system with very low density or, more realistically, to describe a set of orbitals that are in inner shells of an atom, such as d of f levels, since these are more localized than s states. Since the orbitals associated with d or f levels are more localized, the overlaps are small, and the bands that are created by the overlap of these orbitals are narrow. This implies small group velocities and large effective masses.

10.8 Semiconductors

The materials that we called semiconductors have particular characteristics that we will detail next. Examples are the elements $C(D)$: $2s^2 2p^2$; $Si(D)$: $3s^2 3p^2$; $Ge(D)$: $3d^{10} 4s^2 4p^2$; $Sn(T)$: $4d^{10} 5s^2 5p^2$: $Pb(fcc)$: $4f^{14} 5d^{10} 6s^2 6p^2$ and are therefore tetravalent. D stands for diamond structure and T for tetragonal. They are intrinsic semiconductors that at zero temperature are insulators (with a small gap), which is overcome by temperature effects with some holes in the valence band and a small occupation in the conduction band. The deviation from an insulator may also be achieved adding impurities that may either remove or supply electrons to the bands, leading to what are called extrinsic semiconductors. In Table 10.6 are gathered the energy gaps of some semiconductors.

Consider an intrinsic semiconductor. The chemical potential lies in the gap between the valence band (fully occupied at zero temperature) and the conduction band (fully empty at zero temperature). Call the energy gap $E_g = E_c - E_v$, where E_c

Table 10.6 Energy gaps (direct d or indirect i), in eV, of some semiconductors. HgTe has band overlap and is therefore a semimetal. Adapted with permission from [Kittel]

Material	Type	$T = 0$	$T = 300K$	Material	Type	$T = 0$	$T = 300K$
Diamond	i	5.4		Si	i	1.17	1.11
Ge	i	0.744	0.66	α Sn	d	0	0
InSb	d	0.23	0.17	InAs	d	0.43	0.36
InP	d	1.42	1.27	GaP	i	2.32	2.25
GaAs	d	1.52	1.43	GaSb	d	0.81	0.68
AlSb	i	1.65	1.6	SiC (hex)	i	3	
Te	d	0.33		SnTe	d	0.3	0.18
HgTe	d	−0.30		PbS	d	0.286	0.34-0.37
PbSe	i	0.165	0.27	PbTe	i	0.190	0.29
CdS	d	2.582	2.42	CdSe	d	1.840	1.74
CdTe	d	1.607	1.44	Cu_2O	d	2.172	

is the lower energy of the conduction band and E_v the higher energy of the valence band. The gap may be direct or indirect, depending if the gap is between two points with the same wave vector or with different wave vectors, respectively. The effect of temperature is to weigh the electron density with the Fermi-Dirac distribution, $f(\epsilon)$ where

$$f(\epsilon) = \frac{1}{e^{\frac{\epsilon-\mu}{k_BT}} + 1}. \tag{10.98}$$

The kinetic energy of an electron at the bottom of the conduction band is then written as

$$\epsilon_k = E_c + \frac{\hbar^2 k^2}{2m_e}, \tag{10.99}$$

where m_e is the effective mass of the electrons at the bottom of the conduction band.

Recall that the density of states was obtained as

$$D(\epsilon) = \frac{1}{2\pi^2}\left(\frac{2m_e}{\hbar^2}\right)^{\frac{3}{2}} (\epsilon - E_c)^{\frac{1}{2}}. \tag{10.100}$$

The occupation of the conduction band is due to the temperature effect. We may calculate the density of electrons in the conduction band. It is given by

$$n_{cb} = \int_{E_c}^{\infty} D(\epsilon) f(\epsilon) d\epsilon. \tag{10.101}$$

Let us take small temperatures such that $\epsilon - \mu >> k_BT$. Then, it can be shown that

$$n_{cb} = 2\left(\frac{m_e k_B T}{2\pi\hbar^2}\right)^{\frac{3}{2}} e^{\frac{\mu - E_c}{k_BT}}. \tag{10.102}$$

The density of states for the holes in the valence band can similarly be obtained using that

$$D_h(\epsilon) = \frac{1}{2\pi^2}\left(\frac{2m_h}{\hbar^2}\right)^{\frac{3}{2}} (E_v - \epsilon)^{\frac{1}{2}}. \tag{10.103}$$

The density of states for the valence and conduction bands is shown in Fig. 10.10.

The density of holes is then obtained in a similar way evaluating

$$n_h = \int_{-\infty}^{E_v} D_h(\epsilon) f_h(\epsilon) d\epsilon, \tag{10.104}$$

Fig. 10.10 Density of states of the valence and conduction bands of a pure semiconductor (or insulator)

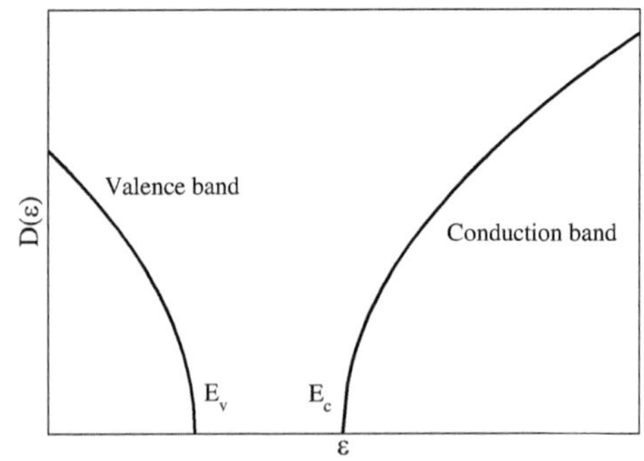

where $f_h(\epsilon) = 1 - f(\epsilon)$. With the same approximation of small temperatures, we get:

$$n_h = 2\left(\frac{m_h k_B T}{2\pi\hbar^2}\right)^{\frac{3}{2}} e^{\frac{E_v-\mu}{k_B T}}. \tag{10.105}$$

Since the charge of the holes is positive, a standard notation for the holes is to represent them by the symbol p (positive) and $n_h = n_p$. The product of the two densities equals

$$n_e n_p = 4\left(\frac{k_B T}{2\pi\hbar^2}\right)^3 (m_e m_h)^{\frac{3}{2}} e^{-\frac{E_g}{k_B T}}. \tag{10.106}$$

Therefore, for a given temperature, the product is fixed. It turns out the same result holds even if the holes and conduction electrons are due to impurities (extrinsic case).

The system is in thermodynamic equilibrium with a photon radiation field with processes of the type photon $\rightarrow$ electron + hole, with a given rate $A(T)$, and processes of the type electron + hole $\rightarrow$ photon, with a rate $B(T)n_e n_p$. The change of the electron density is given by

$$\frac{dn_e}{dt} = A(T) - B(T)n_e n_p = \frac{dn_p}{dt}. \tag{10.107}$$

In equilibrium, the densities do not change in time:

$$\frac{dn_e}{dt} = \frac{dn_p}{dt} = 0, \tag{10.108}$$

which implies that

$$n_e n_p = \frac{A(T)}{B(T)}, \tag{10.109}$$

which is only a function of temperature. In the case of an extrinsic semiconductor, adding impurities may change $n_e + n_p$ but keeping $n_e n_p$ constant.

In the case of an intrinsic semiconductor, naturally $n_e = n_p$, since the electrons in the conduction band are excitations from the valence band that left the corresponding holes. Therefore,

$$n_e = n_p = 2\left(\frac{k_B T}{2\pi\hbar^2}\right)^{\frac{3}{2}} (m_e m_h)^{\frac{3}{4}} e^{-\frac{E_g}{2k_B T}}. \tag{10.110}$$

Also, we may conclude from $n_{cb} = n_p$ that

$$e^{\frac{2\mu}{k_B T}} = \left(\frac{m_h}{m_e}\right)^{\frac{3}{2}} e^{\frac{E_g}{k_B T}}. \tag{10.111}$$

Therefore, the chemical potential has the value:

$$\mu = \frac{1}{2}E_g + \frac{3}{4}k_B T \ln\frac{m_h}{m_e}. \tag{10.112}$$

If the bands are symmetrical, $m_h = m_e$, the chemical potential is at the middle of the gap, $\mu = E_g/2$. The same comes out naturally if $T = 0$.

We will not discuss here the effects that may be achieved when doping semiconductors and, in particular, when two types of doped materials are put into contact, forming heterostructures. Junctions where we mix p-type and n-type doped semicondutors lead to devices with important applications such as transistors.

10.9 Topology of Energy Bands

The topological properties of a system define or condition its properties to some extent. Topology is associated with the idea that different systems can be continuously deformed into each other. For instance, we are interested to determine if Hamiltonians or states may be classified in topological classes. Topological classes group systems that may be continuously transformed into each other, while systems belonging to different classes are not continuously connected. In general, topological properties may be classified by finding numbers, called topological invariants, which do not change under a continuous deformation of the geometry of the system.

In the context of the energy bands of a system, such as in an electronic system, the topological properties may be characterized or quantified by different quantities such as the geometric Berry phase of the bands [1].[4] The Berry phase is a general

---

[4] Here we follow [AraujoSacramento].

concept that will be discussed in the following. Other related quantities are winding numbers and the Chern number. An interesting point is that often these invariants are directly related to physical quantities, and their quantization implies quantization of the physical quantities, under appropriate conditions, such the electric polarization, magnetization, and the Hall conductance.

10.9.1 Berry Phase

Consider a Hamiltonian, $\hat{H}(\boldsymbol{\xi})$, which depends on a set of parameters, $\boldsymbol{\xi}(t)$, that change over time. Schrödinger's equation reads:

$$i\hbar\frac{\partial}{\partial t}|\psi(t)\rangle = \hat{H}(\boldsymbol{\xi})|\psi(t)\rangle\,, \tag{10.113}$$

and gives the time evolution of a state. An equivalent description is obtained considering the action of the time evolution operator, given by the time-ordered exponential of the Hamiltonian. Given some initial state, the action of the time evolution operator provides the state at any later time.

The concept of the Berry phase is associated with an adiabatic evolution such that the parameters are assumed to change very slowly. Assuming this infinitely slow evolution, one may consider the diagonalization of the Hamiltonian at each time, separately, as

$$\hat{H}(\boldsymbol{\xi})|\psi_n(\boldsymbol{\xi})\rangle = E_n(\boldsymbol{\xi})|\psi_n(\boldsymbol{\xi})\rangle\,. \tag{10.114}$$

Under these conditions the evolution of a state may be expressed in terms of the instantaneous eigenstates of the time-evolving Hamiltonian. Considering an eigenstate at $t = 0$ the wave function, $|\psi(t)\rangle$, evolves over time as

$$|\psi(t)\rangle = e^{-\frac{i}{\hbar}\int^t E_n[\boldsymbol{\xi}']dt'}e^{i\gamma_B}|\psi_n[\boldsymbol{\xi}(t)]\rangle\,, \tag{10.115}$$

where, in addition to the dynamical phase in the first exponential, an extra phase, called the Berry phase, γ_B, also appears. Inserting this wave function in Schrödinger equation leads to

$$\frac{\partial\gamma_B}{\partial t} = i\langle\psi_n[\boldsymbol{\xi}(t)]|\,\frac{\partial}{\partial\boldsymbol{\xi}}\,|\psi_n[\boldsymbol{\xi}(t)]\rangle\,\cdot\dot{\boldsymbol{\xi}}\,, \tag{10.116}$$

recalling that the parameters vary with time.

The accumulated Berry phase that results from a path over the space of parameters is the sum of the changes $\delta\gamma_B$ associated with the changes of the parameters $\delta\boldsymbol{\xi}$ over time intervals δt. For instance, if the parameters vary along a given path, C, we get:

$$\gamma_B = i \int_C \langle \psi_n(\boldsymbol{\xi})| \frac{\partial}{\partial \boldsymbol{\xi}} |\psi_n(\boldsymbol{\xi})\rangle \cdot \delta \boldsymbol{\xi} \,. \tag{10.117}$$

There is a formal similarity with the phase accumulated when a charge is coupled to a vector potential. One defines a quantity that may be formally seen as a vector potential, namely, the vector $\mathcal{A}_n(\boldsymbol{\xi})$, known as Berry connection and given by

$$\mathcal{A}_n(\boldsymbol{\xi}) = i \langle \psi_n(\boldsymbol{\xi})| \frac{\partial}{\partial \boldsymbol{\xi}} |\psi_n(\boldsymbol{\xi})\rangle \,. \tag{10.118}$$

Assuming the instantaneous eigenstates as normalized, $\langle \psi_n(\boldsymbol{\xi})|\psi_n(\boldsymbol{\xi})\rangle = 1$, one gets that $Re\langle \psi_n|\frac{\partial}{\partial \boldsymbol{\xi}}\psi_n\rangle = 0$, which implies that, if it is nonvanishing, the Berry connection is real.

The Berry phase is not gauge invariant and a suitable change of phase of the wave function may render it null. This is in agreement with the fact that the vector potential in electromagnetism is not gauge invariant. Changing the Berry connection, therefore, does not lead to an invariant Berry phase. However, it is natural to expect that the curvature of the Berry connection may be gauge invariant, by similarity to the magnetic field. We may define a gauge invariant quantity, called Berry curvature, as

$$\boldsymbol{\Omega} = \nabla_{\boldsymbol{\xi}} \times \mathcal{A}_n(\boldsymbol{\xi}) \,, \tag{10.119}$$

Considering a closed path in the space of parameters, we expect that the associated Berry phase will be an invariant, since it is expected to be related to the flux of the Berry curvature. The Berry curvature can be obtained as

$$\Omega_{\mu\nu} = i \left[\langle \frac{\partial \psi_n}{\partial \xi^\mu} | \frac{\partial \psi_n}{\partial \xi^\nu} \rangle - \langle \frac{\partial \psi_n}{\partial \xi^\nu} | \frac{\partial \psi_n}{\partial \xi^\mu} \rangle \right] = (\boldsymbol{\Omega})_\gamma \epsilon_{\mu\nu\gamma} \,. \tag{10.120}$$

The relevance of the Berry phase in the context of energy bands of a solid may be seen considering that, by applying a field, we may imagine that an electron will move across the Brillouin zone, as time goes by. For instance, in a one-dimensional system, applying an electric field will move the electron across the Brillouin zone, and due to its periodicity, closed trajectories in momentum space may be obtained. The geometrical Berry phase, in this context called Zak phase, has an expression that results from integrating over the Brillouin zone the change of the wave function. The Zak phase is an invariant and its properties will be useful in the classification of the bands.

In general, a Berry phase does not have a topological nature and it may take any real values. However, the presence of some symmetries, like inversion, chiral, or charge conjugation symmetries, ensures a quantization of the Berry phase. It is convenient to use a discretization of the space of parameters to calculate the Berry phase. This is calculated $mod(2\pi)$ and a band that has zero Berry phase is called trivial and a band that has a Berry phase of π is called topological. The Berry phase of a given band may be calculated dividing the Brillouin zone considering N discrete points, $k \in [0, 2\pi]$, with momenta taking the values $k_1, \ldots, k_N$. Assuming that the various bands are separated by finite energy gaps, the band's Berry phase γ_B may be obtained defining the link variable $U(k_l) = \varphi^*(k_l)\varphi(k_{l+1})$, and summing over k_l as

$$\gamma_B = -i \sum_l \log U(k_l), \tag{10.121}$$

where $\varphi(k_l)$ are the eigenstates of the Hamiltonian in momentum space. This procedure can be generalized if there are degenerate points between the bands (therefore the band gap vanishes) taking a link variable as $U(k_l) = det\mathbf{U}(k_l)$, involving the determinant of the matrix:

$$\mathbf{U}_{ij}(k_l) = \varphi_i^*(k_l)\varphi_j(k_{l+1}), \tag{10.122}$$

where $1 \leq i, j \leq N$ run over the eigenstates and $\varphi_i(k_l)$ is the eigenstate at momentum k_l of the ith band. In this case the system has nontrivial topology if the Berry phase of a group of bands is nontrivial. These bands, while having degenerate points, have finite gaps to other bands of higher and lower energies.

The Two-Level System

As an example of the existence of a Berry phase, let us consider a simple model that consists of two levels (or two bands). The Hamiltonian may be written as[5]

$$\hat{H}(\boldsymbol{\xi}) = \mathbf{h}(\boldsymbol{\xi}) \cdot \boldsymbol{\tau} + h_0 \tau_0, \tag{10.123}$$

where $\boldsymbol{\tau}$ denotes the three Pauli matrices in orbital space, and τ_0 is the identity matrix, since the Pauli matrices and the identity constitute a basis in the space of 2×2 matrices. In terms of the spherical coordinates (h, θ, ϕ), the eigenfunctions of (10.123) read:

$$|\psi_-\rangle = \begin{pmatrix} \sin\frac{\theta}{2} \\ -e^{i\phi}\cos\frac{\theta}{2} \end{pmatrix} \quad \text{and} \quad |\psi_+\rangle = \begin{pmatrix} \cos\frac{\theta}{2} \\ e^{i\phi}\sin\frac{\theta}{2} \end{pmatrix}. \tag{10.124}$$

The Berry connection for each level is given by

$$\mathcal{A}_- = i\langle\psi_-|\nabla_{\boldsymbol{\xi}}|\psi_-\rangle = -\cos^2\frac{\theta}{2}\,\nabla_{\boldsymbol{\xi}}\phi\,, \tag{10.125}$$

[5] For further reading see [AraujoSacramento].

$$\mathcal{A}_+ = i\langle\psi_+|\nabla_\xi|\psi_+\rangle = -\sin^2\frac{\theta}{2}\,\nabla_\xi\phi\,, \tag{10.126}$$

If the parameters are the spherical coordinates themselves, then the Berry connection has two components:

$$\mathcal{A}_\theta = 0\,, \mathcal{A}_\phi = -\cos^2\frac{\theta}{2} \qquad \text{for level} -, \tag{10.127}$$

$$\mathcal{A}_\theta = 0\,, \mathcal{A}_\phi = -\sin^2\frac{\theta}{2} \qquad \text{for level} +. \tag{10.128}$$

We can see that, for the lower level,[6]

$$\mathcal{A}_\theta = h\mathcal{A}\cdot\boldsymbol{u}_\theta = 0\,, \tag{10.129}$$

$$\mathcal{A}_\phi = h\sin\theta\mathcal{A}\cdot\boldsymbol{u}_\phi = -\cos^2\frac{\theta}{2}\,. \tag{10.130}$$

which leads to

$$\mathcal{A} = -\frac{1}{2h}\,\text{cotg}\,\frac{\theta}{2}\,\boldsymbol{u}_\phi \qquad (\text{level }-). \tag{10.131}$$

This is similar to the vector potential for a magnetic monopole, at the origin, $\boldsymbol{h} = 0$. Taking the curl in $\boldsymbol{h}$ space,

$$\nabla_h \times \mathcal{A} = \boldsymbol{\Omega}^- = \frac{1}{2h^2}\hat{\boldsymbol{h}}\,. \tag{10.132}$$

The total flux through the sphere is $\mp 2\pi$, for the two energy bands. This is schematically represented in Fig. 10.11. The existence or absence of this quantized flux (in units of 2π) clearly distinguishes different types of bands. As the parameters change, the area covered on the sphere of unit radius depends on the Hamiltonian or its eigenstates.

10.9.2 Winding Number

Consider a one-dimensional tight-binding model where the matrix elements connecting the lattice sites alternate between two values, t_1 and t_2. This implies that the unit cell then has two types of sites (a and b)[2]. The wave function has amplitudes (a_n, b_n) in the n-th cell:

[6] Recall that $\nabla\psi = ((\partial\psi/\partial h), (1/h)(\partial\psi/\partial\theta), 1/(h\sin\theta)(\partial\psi/\partial\phi))$.

Fig. 10.11 Schematic representation of a monopole at the origin with an outgoing flux

$$\psi(n) = \begin{pmatrix} a_n \\ b_n \end{pmatrix} . \tag{10.133}$$

The Hamiltonian's matrix elements read:[7]

$$\langle a, n|\hat{H}|b, n\rangle = t_1 , \qquad \langle b, n|\hat{H}|a, n+1\rangle = t_2 . \tag{10.134}$$

In the case of an infinite system with periodic boundary conditions, the Hamiltonian states may be parametrized by the momentum values. The corresponding Hamiltonian matrix of the simplified SSH model is

$$H(k) = \begin{pmatrix} 0 & t_1 + t_2^* e^{-ik} \\ t_1^* + t_2 e^{ik} & 0 \end{pmatrix} . \tag{10.135}$$

The eigenvalue problem is written as

$$H(k) \begin{pmatrix} a \\ b \end{pmatrix} = E \begin{pmatrix} a \\ b \end{pmatrix} \tag{10.136}$$

and the wave function is a Bloch wave:

$$\psi(n) = e^{ikn} \begin{pmatrix} a \\ b \end{pmatrix} . \tag{10.137}$$

The eigenenergies $E(k) = \pm|t_1 + t_2^* e^{-ik}|$ and the eigenstates of the Hamiltonian matrix may be written as

[7] Here we follow [AraujoSacramento].

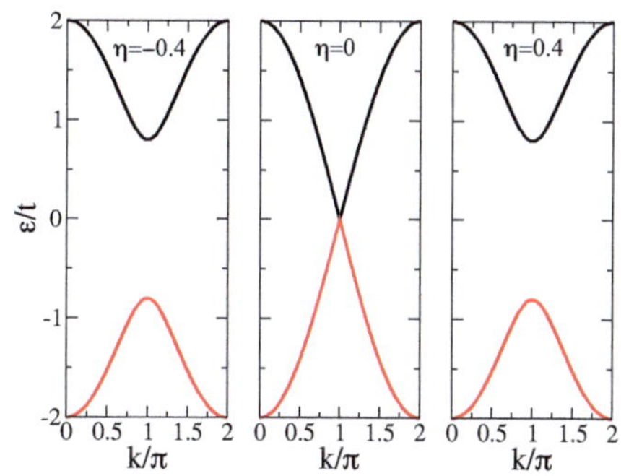

Fig. 10.12 Energy bands of the SSH model for different values of the dimerization η, defined in the text

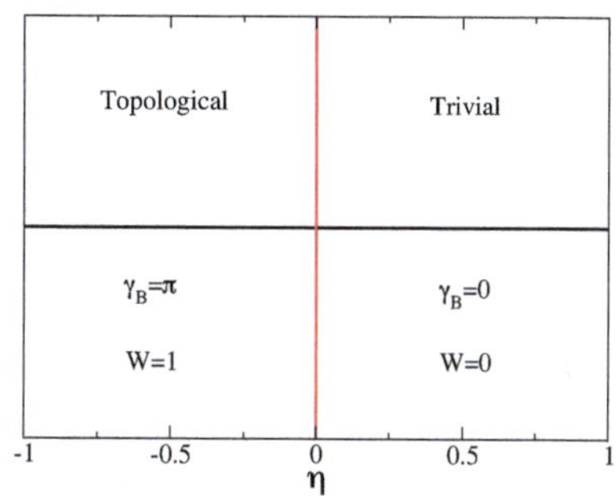

Fig. 10.13 Phase diagram of the SSH model where we take $t_1 = t(1+\eta)$, $t_2 = t(1-\eta)$ and we take $t = 1$. Represented are the winding number, W, and the Berry phase, γ_B, in the two phases

$$\psi_k^+ = \frac{1}{\sqrt{2}} \begin{pmatrix} \frac{z_k}{|z_k|} \\ 1 \end{pmatrix}$$
$$\psi_k^- = \frac{1}{\sqrt{2}} \begin{pmatrix} 1 \\ -\frac{z_k^*}{|z_k|} \end{pmatrix}. \tag{10.138}$$

Here $z_k = t_1 + t_2^* e^{-ik}$. We may parametrize $t_1 = t(1+\eta)$, $t_2 = t(1-\eta)$, and take $t = 1$ as the unit of energy. In Fig. 10.12 we show the two energy bands for different values of η. When $\eta = 0$ ($t_1 = t_2$), the spectrum is gapless. For $\eta \neq 0$ the spectrum has a gap.

Two qualitatively different cases, as represented in Fig. 10.13, emerge related to the number of times that the matrix element $H_{12}(k)$ travels counterclockwise around the origin of the complex plane, as k runs over all its domain. The number of times this occurs is called the winding number. If $|t_2| > |t_1|$, as k takes on values $0 \to 2\pi$, the complex number $t_1 + t_2^* e^{-ik}$ makes one turn around the origin of the complex plane. If $|t_2| < |t_1|$, this winding number is zero. It is the winding of the phase of the complex number z_k that accumulates a variation equal to a multiple of 2π. We can then define the topological invariant, known as winding number, as

$$W = \frac{1}{2\pi i} \int_0^{2\pi} dk \frac{d \log(t_1 + t_2^* e^{-ik})}{dk}. \tag{10.139}$$

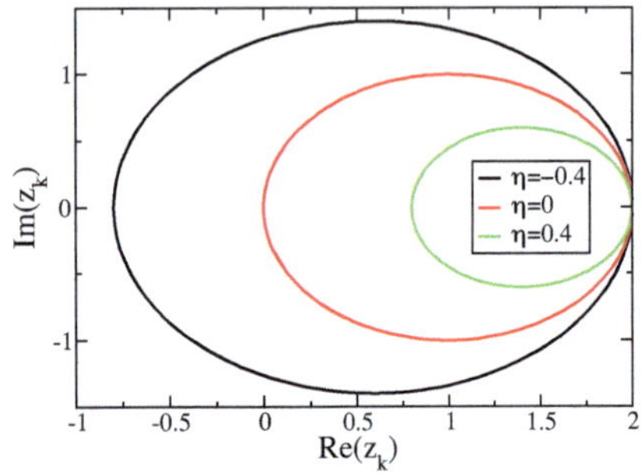

Fig. 10.14 $Im(z_k)$ vs. $Re(z_k)$ for different dimerization, η, values. In the topological phase $\eta < 0$, the origin of the complex plane is included in the parametric plot. In the trivial phase, the origin is not included and in the gapless case the origin is crossed

The separation between the two phases at $|t_1| = |t_2|$ is characterized by the closing of the energy gap between the two bands and that the matrix element $H_{12}(k)$ intercepts the origin. The various cases are shown in Fig. 10.14.

To simplify consider now $t_1 = 0$. The Berry phase associated with, for instance, the state with positive energy may be calculated as

$$\begin{aligned}\gamma_B &= \frac{i}{2}\int_0^{2\pi} dk \begin{pmatrix} \frac{t_2 e^{ik}}{|t_2|} & 1 \end{pmatrix} \frac{d}{dk} \begin{pmatrix} \frac{t_2^* e^{-ik}}{|t_2|} \\ 1 \end{pmatrix} \\ &= \frac{i}{2}\int_0^{2\pi} dk\,(-i) = \pi . \end{aligned} \tag{10.140}$$

It is easy to see that if t_1 is finite and $t_2 = 0$, the Berry phase vanishes. The Berry phase therefore also distinguishes the topological from the trivial regimes, as the winding number does.

Consider now a finite large system, and with open boundary conditions. For an eigenstate with energy E, we can write:

$$\begin{aligned} E a_n &= t_1 b_n + t_2^* b_{n-1} \\ E b_n &= t_2 a_{n+1} + t_1^* a_n \end{aligned} . \tag{10.141}$$

Take first a semi-infinite chain, $n \geq 0$ and let us look for a solution with $E = 0$. This implies that

$$a_n = \left(-\frac{t_1^*}{t_2}\right)^n a_0 , \qquad b_n = 0 , \tag{10.142}$$

which holds if $|t_2| > |t_1|$. This means that the chain starts with a link which has a smaller magnitude (weak link). This state is localized at the edge. When the crystal ends with a weak link, an edge state appears that decreases exponentially toward the bulk of the chain.

We may also consider the end of the crystal at the cell $n = L$. We then obtain, for $n \leq L$:

$$a_n = 0\,, \qquad b_n = \left(-\frac{t_1}{t_2^*}\right)^{L-n} b_L\,, \tag{10.143}$$

so, the state with $E = 0$ lives on sublattice b. In a very long finite chain, $0 \leq n \leq L \to \infty$, such that there is no overlap between the edges, there exist two exponentially decaying edge states supported on different sublattices.

The topological phase is then characterized both by the existence of edge states and a nonvanishing quantized winding number or a nontrivial quantized Berry phase. This is called bulk-edge correspondence. The topological properties are reflected both in the bulk states and in the edges of a finite (or semi-infinite) chain.

10.9.3 Polarization

Classically, the dipole moment (or polarization) of a system of charges, q_i, located at points $\mathbf{r}_i$ with respect to some origin, is given by

$$\mathbf{P} = \sum_i q_i \mathbf{r}_i \tag{10.144}$$

Consider a one-dimensional system, to simplify. Quantum mechanically we may expect that the polarization of an electronic system is given by[8]

$$P = e \int dx\, x|\phi(x)|^2\,. \tag{10.145}$$

The electron coordinate, x, is continuous and we calculate the electric polarization of a given orbital ϕ. Note that if the system is infinite and the wave function is not localized, the polarization is not well-defined. In particular, if we have an infinite system with periodic boundary conditions, the definition requires some form of localized function, so that the polarization is meaningful. Let $\phi(x-n)$ denote a Wannier state localized at site n of a chain with $N \to \infty$ sites. A Bloch wave associated with these localized wave functions can be written as

$$\psi_k(x) = \frac{1}{\sqrt{N}} \sum_n e^{ikn}\phi(x-n) = e^{ikx}u_k(x)\,, \tag{10.146}$$

where

$$u_k(x) = \frac{1}{\sqrt{N}} \sum_n e^{ik(n-x)}\phi(x-n)\,. \tag{10.147}$$

[8] Here we follow [AraujoSacramento].

Note that

$$u_k^*(x) i \partial_k u_k(x) = \frac{1}{N} \sum_{n,m} e^{ik(n-m)} (x-n)\phi(x-n)\phi^*(x-m) \,. \tag{10.148}$$

Summing over all momenta, k, in the Brillouin zone, and using that $\sum_k e^{ik(n-m)} = N\delta_{n,m}$, we get:

$$\frac{1}{N} \sum_k u_k^*(x) i \partial_k u_k(x) = \frac{1}{N} \sum_n (x-n)|\phi(x-n)|^2 \,, \tag{10.149}$$

Integrating expression (10.149) over x, we conclude that

$$\begin{aligned} P &= e \int dx \int_{-\pi}^{\pi} \frac{dk}{2\pi} u_k^*(x) i \partial_k u_k(x) \\ P &= \frac{e}{2\pi} \int_{-\pi}^{\pi} \mathcal{A}_k dk \,, \end{aligned} \tag{10.150}$$

The integral over the Berry connection is the Zak phase.

The Berry connection is therefore related to the polarization of Wannier states of an energy band. Multiplying the latter by $e/(2\pi)$, we obtain the electric polarization per unit cell, P. Returning to the case of the SSH model, the Berry (Zak) phase takes two values depending if we are in the trivial or topological phases. Therefore, the polarization is given by $P = 0, e/2$ (we are using units of lattice constant $a = 1$). Therefore, in the trivial phase, and choosing x with its origin on the left side of the unit cell, say at the atom position A, we get that a vanishing polarization implies that the wave function is centered at the A atom locations. In the topological phase, a polarization of $e/2$ means that the wave function is centered on the B atom locations. The result that there are only two possible (quantized) values for the center of the wave function results from the inversion symmetry of the SSH chain. Indeed, we may choose the center of inversion symmetry to be located at the A atom location (or B atom location as well). As argued by Zak [3], if there is inversion symmetry, the Bloch wave functions may be chosen such that the Wannier functions are even or odd around either of the two centers of symmetry, located at the A or the B atom, and therefore the locations $x = 0$ and $x = 1/2$ (in units of the lattice constant). Therefore, the wave functions, selecting $n = 0$, satisfy $\phi(-x) = \pm\phi(x)$, which implies that the polarization vanishes selecting the center of inversion around the origin. Selecting the center of inversion around the symmetry point $1/2$, we get that $\phi(-x+1) = \pm\phi(x)$, which implies that the polarization equals $P = e/2$.

The result (10.150) can be generalized to three-dimensional space by proceeding in an analogous fashion. The polarization along the Cartesian axes $j (= 1, 2, 3)$ is given by

$$P_j = e \int d^3r \int \frac{d^3k}{(2\pi)^3} u^*_{\mathbf{k}}(\mathbf{r}) i \partial_{k_j} u_{\mathbf{k}}(\mathbf{r}) \,. \tag{10.151}$$

In general, the polarization of a periodic system is ill-defined. Since the charge distribution is periodic in space, the electric dipole moment is not well-defined and depends on the choice of unit cell. One way around the problem is to define the expectation value of the position operator, $\hat{x}$, on a given state $|\psi_0\rangle$ of a charge as [4]

$$\langle \hat{x} \rangle = \frac{L}{2\pi} \Im \ln \langle \psi_0 | e^{i \frac{2\pi}{L} \hat{x}} | \psi_0 \rangle , \tag{10.152}$$

where L is the system size and periodic boundary conditions are assumed. With this definition, the periodicity is implemented and the average value is obtained, modulo L. This definition of the position leads to a definition for the polarization that also implements the periodicity:

$$P = \lim_{L \to \infty} \frac{e}{2\pi} \Im \ln \langle \psi_0 | e^{i \frac{2\pi}{L} \hat{x}} | \psi_0 \rangle , \tag{10.153}$$

where e is the electron charge.

10.9.4 Chern Number

The Chern number invariant is defined as the Berry phase obtained from the line integral of the connection around a two-dimensional Brillouin zone, or from the integral of the curvature over the Brillouin zone:

$$C = \frac{1}{2\pi} \oint_{BZ} \mathcal{A} \cdot d\mathbf{k} = \frac{1}{2\pi} \int dk_x \, dk_y \, \Omega_{k_x, k_y} \,, \tag{10.154}$$

recalling Stokes theorem. In the case of a two-band system, the Chern number may also be written as

$$C = \frac{1}{2\pi} \int_{\text{h space}} \frac{\boldsymbol{h}}{2h^3} \cdot d\boldsymbol{S} = \frac{1}{4\pi} \int_{\text{k space}} \frac{\partial \hat{\boldsymbol{h}}}{\partial k_x} \times \frac{\partial \hat{\boldsymbol{h}}}{\partial k_y} \cdot \hat{\boldsymbol{h}} \, dk_x dk_y \,, \tag{10.155}$$

(recall Eq. (10.132)). It is then understood that C counts the number of times that the integration domain $\int dk_x \, dk_y$ covers the sphere in $\boldsymbol{h}$ space. This expression is obtained using the following results:

$$\frac{dS}{h^2} = \sin\theta d\theta d\phi$$

$$\sin\theta = \frac{\partial \hat{\boldsymbol{h}}}{\partial \theta} \times \frac{\partial \hat{\boldsymbol{h}}}{\partial \phi} \cdot \hat{\boldsymbol{h}}$$

$$dk_x dk_y = \left| \frac{\partial k_x}{\partial \theta}\frac{\partial k_y}{\partial \phi} - \frac{\partial k_x}{\partial \phi}\frac{\partial k_y}{\partial \theta} \right| d\theta d\phi. \tag{10.156}$$

We have seen from the example of the SSH model that the winding number may characterize the topology of one-dimensional systems. Likewise, the Chern number may be used as a topological invariant in two-dimensional systems. There is also a bulk-edge correspondence, and the Chern number is also related to the number of edge states at the border of a finite two-dimensional system. As for the one-dimensional case, in a topological phase considering open boundary conditions, edge states appear as a signature of the nontrivial topological nature.

The definition and calculation of the Chern number is expected to meet some problems since we are in some sense projecting a sphere (over which the vector **h** moves) onto a plane. It seems that to get some nontrivial result some sort of singularity is unavoidable. Using periodic boundary conditions, the momentum space is actually a torus in two dimensions.

To determine the Chern number, it is necessary, for instance, to solve Eq. (10.154) which applies Stokes theorem. However, as the Brillouin zone is a torus, it has no boundary.[9] So if the Berry curvature would be a well-behaved function, one would always obtain a zero Chern number, which is precisely what happens for non-topological phases. Nevertheless, the procedure breaks down if the Berry connection presents singularities. The origin of such problematic points lies on the impossibility of picking a global gauge that guarantees a smooth and single-valued wave function. Instead, it is necessary to define distinct gauges in different regions of the Brillouin zone and separately apply the Stokes theorem to the different patches.

One way to understand the problem is to consider, for instance, the wave functions (10.124). We see that they are not single-valued in all space. Indeed, ψ_- is indeterminate at the north pole ($\theta = 0$) because the angle ϕ is not determined there. For an analogous reason, ψ_+ is indeterminate at the south pole. One could attempt to solve the problem by multiplying any of those wave functions by $e^{-i\phi}$, which means adopting a different gauge. Then again, that would only transfer the indeterminacy to the opposite pole. When the Chern number is nonzero, all the sphere is covered, and from the previous paragraph, we conclude that it is not possible to adopt a gauge where the wave function is continuous and single-valued all over the Brillouin zone.

Consider, then, the insulator with states in the lower band given by ψ_-, and that they are occupied (if the chemical potential is located in the gap between the two bands). Consider that at a point $\mathbf{k}^*$, there is a north pole, $\boldsymbol{h}(\mathbf{k}^*) = (0, 0, h_z > 0)$. In the vicinity of $\mathbf{k}^*$, h_x, and h_y vanish linearly. In going around that point, ψ_- accumulates a phase 2π coming from the factor $e^{i\phi}$ in the wave function, the same

[9] For further reading see [BernevigHughes]. See also Master thesis of Ana Silva, Instituto Superior Técnico, 2014.

one that caused the indeterminacy. We may then think about calculating the Chern number from the line integral (10.154) around the Brillouin zone by decomposing the latter into plaquettes and summing over all of them:

$$C = \frac{1}{2\pi} \oint_{ZB} \mathcal{A} \cdot d\mathbf{k} = \frac{1}{2\pi} \sum_{\square} \oint_{\square} \mathcal{A} \cdot d\mathbf{k}\,. \tag{10.157}$$

Here, $\square$ denotes a plaquette. The line integrals cancel along the sides of neighboring plaquettes. But there is a plaquette that contains the point $\mathbf{k}^*$. For that one, the line integral yields $\pm 2\pi$. If there are $|C|$ such points in the Brillouin zone, the summation (10.157) yields the Chern number, an integer number that may be used to classify different classes. This allows the calculation of C in a numerically very efficient way [5]. Discretizing the Brillouin zone as $k_\mu = 2\pi j/N$, with $j = 1, \ldots, N$, and $\mu = x, y$, defining the link variable, $U_\mu(\mathbf{k})$, for the link $\delta\mathbf{k}_\mu$ (with $|\delta\mathbf{k}_\mu| = 2\pi/N$) oriented along the μ direction from the point $\mathbf{k}$ may be defined as

$$U_\mu(\mathbf{k}) = \frac{\langle n(\mathbf{k})|n(\mathbf{k} + \delta\mathbf{k}_\mu)\rangle}{|\langle n(\mathbf{k})|n(\mathbf{k} + \delta\mathbf{k}_\mu)\rangle|}\,. \tag{10.158}$$

From the lattice field strength, defined as

$$F_{xy}(\mathbf{k}) = \ln\left(U_x(\mathbf{k})U_y(\mathbf{k} + \delta\mathbf{k}_x)U_x(\mathbf{k} + \delta\mathbf{k}_y)^{-1}U_y(\mathbf{k})^{-1}\right), \tag{10.159}$$

with $-\pi < -iF_{xy}(\mathbf{k}) \leq \pi$, a gauge-invariant expression for the Chern number is obtained:

$$C_n = \frac{1}{2\pi i} \sum_{\mathbf{k}} F_{xy}(\mathbf{k}). \tag{10.160}$$

Dirac-Like Hamiltonian

An interesting and relevant case is that of a Hamiltonian such that its dispersion relation is linear in the momentum. Consider the Hamiltonian[10]

$$\hat{H} = (k_x, k_y, h_z) \cdot \boldsymbol{\tau}\,, \tag{10.161}$$

where $h_z = const$ is a so-called Dirac mass term. Then, the Berry curvature, $\boldsymbol{\Omega}$,

$$\Omega^-_{k_x,k_y} = \frac{h_z}{2h^3} = (\boldsymbol{\Omega}^-)_z\,. \tag{10.162}$$

[10] For further reading see [AraujoSacramento].

Integrating the Berry curvature over the Brillouin zone (k_x, k_y), we get:

$$\begin{aligned}\int \Omega^{-}_{k_x,k_y} dk_x\, dk_y &= \frac{h_z}{2}\int \frac{dk_y\, dk_y}{(k_x^2+k_y^2+h_z^2)^{3/2}}\\ &= \frac{h_z}{2}\int_0^{\infty} \frac{2\pi r dr}{(r^2+h_z^2)^{3/2}}\\ &= \frac{\pi h_z}{2}\int_0^{\infty} \frac{dt}{(t+h_z^2)^{3/2}}\\ &= \pi \cdot \mathrm{sgn}\,(h_z)\ ,\end{aligned} \tag{10.163}$$

where the domain of integration $\int dk_x\, dk_y$ has been extended to infinity. The flux of the monopole through a half-sphere is π and gives a half-integer contribution to the Chern number. Since the latter must be integer, there must be an even number of Dirac-like points in the Brillouin zone.

As a final remark, the topological properties of energy bands are often a result of the existence or not of some symmetries of the Hamiltonian. Some of these symmetries are briefly mentioned in Appendix B.

10.10 Semimetals

Band theory explains the difference between electronic systems whose spectrum is gapless, such as metals, in the sense that the chemical potential is placed inside a band, and systems that are gapped, such as insulators or semiconductors, in the sense that the chemical potential is located in an energy region between two bands, with a finite energy interval and with no available extended states. Other possibilities to classify a given system are possible, such as superconductors that will be considered later, both of gapped and gapless type, and systems that are classified as semimetals. One example of semimetals are systems where the gap between bands vanishes and the chemical potential is located at the energy value that coincides with the higher energy value of the lower band and the lower energy point of the upper band. Notable examples of this type of systems are two-dimensional graphene and three-dimensional Weyl semimetals. These two cases will be considered next. It is also possible that the bands overlap in energy space, in the sense that the lower energy value of the "upper" band is smaller than the higher energy values of the "lower" band. In this case the chemical potential may be located in such a way that, at zero temperature there are hole states at the top of the lower energy band and only a small fraction of the states of the upper band are occupied. This type of systems is also called a semimetal.

10.10.1 Energy Bands of Graphene

Graphene is made of carbon atoms sitting at the sites of a honeycomb lattice. The honeycomb lattice is not a Bravais lattice. It can be described as a triangular lattice with a basis of two sites (A and B) as shown in Fig. 8.9. We may also see it as two interpenetrating triangular lattices.[11] The closest neighbors to an atom A are B atoms and vice versa. Because of this property, the lattice is said to be bipartite.

Considering a finite size system, there are different ways to cut the edges of the two-dimensional lattice. Two usual choices are the so-called armchair and zigzag edges, as also shown in Fig. 8.9. It is convenient to define different unit cell basis vectors depending on the orientation of the cut and defining the x-axis as the direction along the edge.

One possible definition for the two basis vectors, useful for the case of an armchair edge, is

$$\mathbf{a}_1 = \frac{a}{2}\left(3, \sqrt{3}\right); \qquad \mathbf{a}_2 = \frac{a}{2}\left(3, -\sqrt{3}\right), \tag{10.164}$$

where we call a the distance between nearest-neighbors, one on sublattice A and the other on sublattice B. These vectors connect two neighboring sites on the same lattice, either A or B. The three nearest neighbor points are reached by the vectors:

$$\boldsymbol{\delta}_1 = \frac{a}{2}\left(1, \sqrt{3}\right), \qquad \boldsymbol{\delta}_2 = \frac{a}{2}\left(1, -\sqrt{3}\right), \qquad \boldsymbol{\delta}_3 = -a\,(1, 0)\,. \tag{10.165}$$

In the case of a zigzag edge, we may define instead a pair of basis vectors as

$$\mathbf{a}_1 = a\,(1, 0)\,, \qquad \mathbf{a}_2 = \frac{a}{2}\left(1, -\sqrt{3}\right), \tag{10.166}$$

where here a is the lattice constant of the unit cell, the distance between two sites of the same sublattice.

Given the basis vectors, any point on a given sublattice may be reached taking combinations of the basis vectors:

$$\mathbf{R} = m\mathbf{a}_1 + n\mathbf{a}_2, \tag{10.167}$$

where m, n are integers.

Consider for now the contribution of nearest-neighbor terms to a hopping-like Hamiltonian. This Hamiltonian may be written in second quantization formalism, as reviewed in Appendix A:

[11] For further reading see Eduardo V Castro, N. M. R. Peres, J. M. B. Lopes dos Santos, F. Guinea, and A. H. Castro Neto, Chap. 4 of *Strongly Correlated Systems, Coherence and Entanglement*, edited by J. M. P. Carmelo, J. M. B. Lopes dos Santos, V. R. Vieira and P. D. Sacramento.

$$H_0 = -t \sum_{\mathbf{R},\boldsymbol{\delta},\sigma} c^{\dagger}_{\mathbf{R},\sigma} c_{\mathbf{R}+\boldsymbol{\delta},\sigma} \tag{10.168}$$

for the "kinetic term," where t is the hopping amplitude. In first quantization, this just means that the Hamiltonian is described by a matrix that has matrix elements that describe the overlaps between local states at locations $\mathbf{R}$ and $\mathbf{R}+\boldsymbol{\delta}$, such that the spin component, σ, is preserved. In this case the matrix elements have the amplitude t. The operator $c^{\dagger}_{\mathbf{R},\sigma}$ adds one electron with spin projection σ, at site $\mathbf{R}$ and the operator $c_{\mathbf{R}+\boldsymbol{\delta},\sigma}$ removes one electron with the same spin projection from site $\mathbf{R}+\boldsymbol{\delta}$, where $\boldsymbol{\delta}$ connects a site with its neighbor. The process of removing one electron from one site and adding it at a different site is a representation of the physical process where an electron moves from one site to its neighbor, with a given probability amplitude, here given by t.

Distinguishing the sites on the two sublattices, we may write that

$$H_0 = -t \sum_{\mathbf{R},\sigma} \left(a^{\dagger}_{\sigma}(\mathbf{R}) b_{\sigma}(\mathbf{R}) + a^{\dagger}_{\sigma}(\mathbf{R}) b_{\sigma}(\mathbf{R}-\mathbf{a}_1) + a^{\dagger}_{\sigma}(\mathbf{R}) b_{\sigma}(\mathbf{R}-\mathbf{a}_2) + h.c. \right). \tag{10.169}$$

Here we use the notation that at sites that are of type A, the operator $c = a$ and at sites B, the operator $c = b$. This notation highlights the two-sublattice nature of the problem by introducing two types of operators.

Associated with the real space lattice, we may introduce the reciprocal space lattice, with basis vectors defined as usual by $\mathbf{b}_i \cdot \mathbf{a}_j = 2\pi\delta_{ij}$, leading in the case of the armchair edge to

$$\mathbf{b}_1 = \frac{2\pi}{3a}\left(1, \sqrt{3}\right) \qquad \mathbf{b}_2 = \frac{2\pi}{3a}\left(1, -\sqrt{3}\right), \tag{10.170}$$

or for the zigzag edge choice of basis vectors to

$$\mathbf{b}_1 = \frac{2\pi}{a}\left(1, \frac{1}{\sqrt{3}}\right) \qquad \mathbf{b}_2 = -\frac{2\pi}{a}\left(0, \frac{2}{\sqrt{3}}\right). \tag{10.171}$$

Carrying out a Fourier transform of the operators, and specifying for the zigzag edge, we can write the Hamiltonian in momentum space as

$$H = \sum_{\mathbf{k},\sigma} \psi^{\dagger}_{\mathbf{k},\sigma} H_{\mathbf{k}} \psi_{\mathbf{k},\sigma}, \tag{10.172}$$

where

$$\psi_{\mathbf{k},\sigma} = \begin{pmatrix} a_{\mathbf{k},\sigma} \\ b_{\mathbf{k},\sigma} \end{pmatrix}, \tag{10.173}$$

in terms of the amplitudes on the two sublattices, and where the Hamiltonian matrix is given by

$$H_{\mathbf{k}} = -t \begin{pmatrix} 0 & s_{\mathbf{k}} \\ s_{\mathbf{k}}^* & 0 \end{pmatrix}, \tag{10.174}$$

with

$$s_{\mathbf{k}} = 1 + e^{i\mathbf{k}\cdot\mathbf{a}_1} + e^{i\mathbf{k}\cdot\mathbf{a}_2}. \tag{10.175}$$

The Hamiltonian is easily diagonalized leading to the energy levels:

$$E_{\mathbf{k}} = \pm t\sqrt{3 + 2\left(\cos(k_x a) + 2\cos(\frac{k_x a}{\sqrt{2}})\cos(\sqrt{3}\frac{k_y a}{2})\right)} = \pm t|s_{\mathbf{k}}|. \tag{10.176}$$

In the case of the armchair edge basis definition, we can write that

$$E_{\mathbf{k}} = \pm t\sqrt{3 + f(\mathbf{k})} - t' f(\mathbf{k}), \tag{10.177}$$

including next-nearest neighbors, linked by the hopping amplitude t', and where

$$f(\mathbf{k}) = 2\cos(\sqrt{3}k_y a) + 4\cos\left(\frac{\sqrt{3}}{2}k_y a\right)\cos\left(\frac{3}{2}k_x a\right). \tag{10.178}$$

Note that if $t' = 0$, the spectrum is symmetric around zero energy, and therefore the coupling to next-nearest neighbors breaks the particle-hole symmetry (see Appendix B for a discussion of this and some other symmetries).

In Fig. 10.15 we show the energy bands as a function of the momenta (k_x, k_y). At a set of points in the Brillouin zone, the energy difference between the two bands vanishes and the spectrum is gapless. These points are called the Dirac cones. There are two inequivalent Dirac points written as

$$\mathbf{K} = \frac{2\pi}{3a}\left(1, \frac{1}{\sqrt{3}}\right) \qquad \mathbf{K}' = \frac{2\pi}{3a}\left(1, -\frac{1}{\sqrt{3}}\right) \tag{10.179}$$

in the definition for the armchair edge, or

$$\mathbf{K} = -\frac{2\pi}{3a}(2, 0) \qquad \mathbf{K}' = \frac{2\pi}{3a}(2, 0) \tag{10.180}$$

in the definition for the zigzag edge.

The Hamiltonian matrix

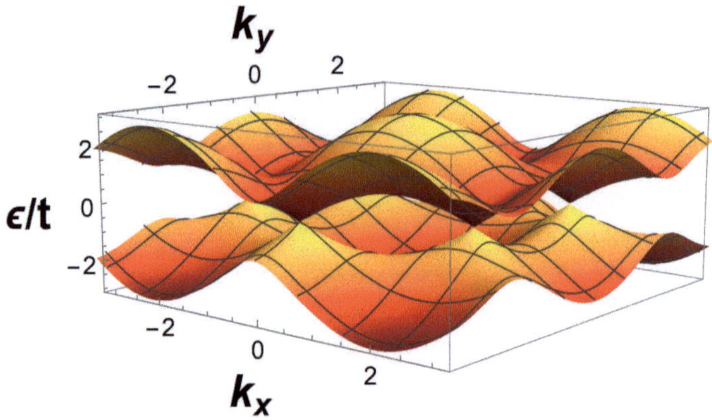

Fig. 10.15 Energy bands of graphene as obtained for the case of the parametrization for armchair edges. Note the appearance of the gapless Dirac nodes

$$H_{\mathbf{k}} = -t \begin{pmatrix} 0 & s_{\mathbf{k}} \\ s_{\mathbf{k}}^* & 0 \end{pmatrix} \tag{10.181}$$

may be linearized around the Dirac points. We may redefine:

$$\begin{aligned} k_x &= \frac{1}{2}q_x + \frac{\sqrt{3}}{2}q_y \\ k_y &= -\frac{\sqrt{3}}{2}q_x + \frac{1}{2}q_y, \end{aligned} \tag{10.182}$$

with $\mathbf{q} = \mathbf{k} - \mathbf{K}$ leading to

$$H_{\mathbf{k}} = v_F \begin{pmatrix} 0 & k_x - ik_y \\ k_x + ik_y & 0 \end{pmatrix}. \tag{10.183}$$

Also, we may linearize around the other Dirac point considering $\mathbf{q} = \mathbf{k} - \mathbf{K}'$ and

$$\begin{aligned} k_x &= -\frac{1}{2}q_x - \frac{\sqrt{3}}{2}q_y \\ k_y &= \frac{\sqrt{3}}{2}q_x - \frac{1}{2}q_y, \end{aligned} \tag{10.184}$$

leading to

$$H_{\mathbf{k}} = v_F \begin{pmatrix} 0 & k_x + ik_y \\ k_x - ik_y & 0 \end{pmatrix}. \tag{10.185}$$

Therefore, we may write:

$$
\begin{aligned}
H_{\mathbf{K}} &= v_F \boldsymbol{\sigma} \cdot \mathbf{p} \\
H_{\mathbf{K}'} &= v_F \boldsymbol{\sigma}^* \cdot \mathbf{p},
\end{aligned} \tag{10.186}
$$

where $\mathbf{p}$ is the momentum and we are taking units $\hbar = 1$. Therefore, near each Dirac point, the spectrum is linear.

Independently of the choice of the basis vectors, we may expand the energy dispersions in the vicinity of the Dirac cones. Using, for instance, the basis for the armchair edge, we write:

$$
E_{\pm}(\mathbf{q}) = 3t' \pm v_F |\mathbf{q}| + O(|\mathbf{q}|^2), \tag{10.187}
$$

where the Fermi velocity $v_F = 3ta/2$. It turns out that its magnitude is of the order of $c/300$, where c is the velocity of light in vacuum. So the electrons have a dispersion relation that is similar to that of light, or of a relativistic massless particle, but with a velocity that is two orders of magnitude smaller than a relativistic speed.

The density of states associated with the dispersion relation in graphene is in general complicated, but in the vicinity of the Dirac cones, where the spectrum is linear, it can be obtained easily. Let us consider only nearest-neighbor hopping. The Dirac cone projects to a circle of radius $q = \sqrt{q_x^2 + q_y^2}$ on the two-dimensional plane. The disk has an area

$$
A = \pi q^2 = \frac{\pi E^2}{v_F^2}. \tag{10.188}
$$

The area of the unit cell in real space is given by

$$
A_c = |\mathbf{a}_1 \times \mathbf{a}_2| = \frac{3\sqrt{3}a^2}{2}. \tag{10.189}
$$

Considering that there are two inequivalent Dirac cones and two spin components, we obtain that the number of states with energy smaller than E leads to

$$
N(E) = \frac{4\frac{\pi E^2}{v_F^2}}{\frac{2\pi^2}{A_c}}. \tag{10.190}
$$

The density of states is then obtained as

$$
\rho(E) = \frac{\partial N(E)}{\partial E} = \frac{2A_c|E|}{\pi v_F^2} \tag{10.191}
$$

and is linear in energy in the vicinity of the Dirac points and, therefore, vanishes at the Fermi surface (set of Dirac cone points). Note that the behavior is different from that of a usual quadratic energy dispersion. This leads to different behavior at the Fermi surface and therefore different thermodynamic and transport properties.

10.10.2 Weyl Semimetals

It was argued in the discussion of the energy bands of graphene that the electrons behave as massless Dirac-like fermions in two dimensions, since near the Dirac points the spectrum is linear in the momentum. Recalling the equation proposed by Dirac, to describe a relativistic particle of mass m satisfying the energy-momentum relation $E^2 = p^2c^2 + m^2c^4$ and formally similar to Schrödinger's equation for a nonrelativistic particle

$$i\hbar\frac{\partial\psi}{\partial t} = H\psi, \tag{10.192}$$

implies that the Hamiltonian has a matrix structure which involves matrices that have a minimal dimension 4×4. The Hamiltonian was obtained as

$$H = \boldsymbol{\alpha}\cdot\mathbf{p}c + \beta mc^2, \tag{10.193}$$

where $\boldsymbol{\alpha}, \boldsymbol{\beta}$ are matrices. These Hermitian matrices obey the relations $\alpha_i^2 = 1$, $\beta^2 = 1$, $\alpha_i\alpha_j + \alpha_j\alpha_i = \delta_{ij}$, and $\alpha_i\beta + \beta\alpha_i = 0$.

Different definitions of these 4×4 matrices satisfy these restrictions. Dirac chose a representation in terms of Pauli matrices:

$$\boldsymbol{\alpha} = \begin{pmatrix} 0 & \boldsymbol{\sigma} \\ \boldsymbol{\sigma} & 0 \end{pmatrix} = \sigma_1 \otimes \boldsymbol{\sigma} \tag{10.194}$$

and

$$\beta = \begin{pmatrix} 1 & 0 \\ 0 & -1 \end{pmatrix} = \sigma_3 \otimes 1. \tag{10.195}$$

It is convenient to define other associated matrices as $\gamma^0 = \beta$, and $\gamma^i = \beta\alpha^i$, with $i = 1, 2, 3$, the so-called Dirac γ-matrices. In terms of these matrices, the Dirac equation is written as

$$i\hbar\gamma^\mu\partial_\mu\psi = mc\psi, \tag{10.196}$$

where $\mu = 0, 1, 2, 3$, and the matrices obey the anti-commutation relations $\{\gamma^\mu, \gamma^\nu\} = 2\eta^{\mu\nu}$. The matrices are represented by

$$\gamma^0 = \begin{pmatrix} 1 & 0 \\ 0 & -1 \end{pmatrix} = \sigma_3 \otimes 1 \tag{10.197}$$

and

$$\gamma^i = \begin{pmatrix} 0 & \sigma_i \\ -\sigma_i & 0 \end{pmatrix} = i\sigma_2 \otimes \sigma_i . \tag{10.198}$$

In this representation the matrix γ^0 has a diagonal form and the other matrices have an off-diagonal form. Also, in the Hamiltonian, the mass term has a diagonal form, and the terms associated with the momentum have an off-diagonal form. Taking units where $c = 1$, we can write then that the energy eigenvalue equation is

$$\begin{pmatrix} m & \boldsymbol{\sigma} \cdot \mathbf{p} \\ \boldsymbol{\sigma} \cdot \mathbf{p} & -m \end{pmatrix} \begin{pmatrix} \Phi_A \\ \Phi_B \end{pmatrix} = E \begin{pmatrix} \Phi_A \\ \Phi_B \end{pmatrix}, \tag{10.199}$$

where Φ_A, Φ_B are two-component spinors.

Another possible choice is to define the matrices as

$$\begin{aligned} \boldsymbol{\alpha} &= (\sigma_3 \otimes \sigma_1, \sigma_3 \otimes \sigma_2, \sigma_3 \otimes \sigma_3) \\ \beta &= -\sigma_1 \otimes 1. \end{aligned} \tag{10.200}$$

With this choice the mass term is off-diagonal and the momentum part is diagonal.

Yet another choice of matrices, proposed by Weyl, is

$$\gamma^0 = \begin{pmatrix} 0 & 1 \\ 1 & 0 \end{pmatrix} \tag{10.201}$$

and

$$\gamma^i = \begin{pmatrix} 0 & \sigma_i \\ -\sigma_i & 0 \end{pmatrix}, \tag{10.202}$$

where all matrices are off-diagonal. This leads to the Weyl equation

$$\begin{pmatrix} -m & i\partial_t - \boldsymbol{\sigma} \cdot \mathbf{p} \\ i\partial_t + \boldsymbol{\sigma} \cdot \mathbf{p} & -m \end{pmatrix} \begin{pmatrix} \psi_L \\ \psi_R \end{pmatrix} = 0, \tag{10.203}$$

where ψ_L, ψ_R are different two-component spinors.

At zero momentum the four-component spinor of the Dirac equation separates in the positive and negative energy solutions. The Dirac equation also has an interesting limit when we consider a nonrelativistic regime, where only the positive energies are significative and the four-component spinor is reduced to a two-component spinor. This satisfies a generalized Schrödinger like equation, which includes the spin of the electron, the so-called Pauli equation.

Another interesting limit of the Dirac equation is obtained if the particle is massless, $m = 0$. Its character is clearer if one uses the Weyl representation. We see that in this limit the four-component spinors decouple in two two-component spinors leading for instance to the decoupled equations of motion:

$$
\begin{aligned}
(i\partial_t - \boldsymbol{\sigma}\cdot\mathbf{p})\,\psi_R &= 0 \\
(i\partial_t + \boldsymbol{\sigma}\cdot\mathbf{p})\,\psi_L &= 0,
\end{aligned}
\tag{10.204}
$$

which leads to two energy eigenvalues

$$
E = \pm\boldsymbol{\sigma}\cdot\mathbf{p}, \tag{10.205}
$$

respectively for the right and left movers.

Some condensed matter electronic systems have energy dispersions that are of this type and, since the system is gapless, are called Weyl semimetals. Typically the gap between bands closes at a few points, called Weyl points, in a way somewhat similar to the Dirac points in graphene. The Weyl semimetals are three-dimensional systems with at least a pair of bands, and focusing on these two bands, the Hamiltonian may be written as[12]

$$
H = \mathbf{h}\cdot\boldsymbol{\sigma} + h_0\sigma_0 = h_\mu\sigma_\mu. \tag{10.206}
$$

Designating the Weyl point as **k**, a point in the Brillouin zone, in order to satisfy the condition of a Weyl point, we can impose that

$$
h_\mu(\mathbf{k}+\mathbf{q}) = \hbar\mathbf{v}_\mu\cdot\mathbf{q}, \tag{10.207}
$$

where **q** is the deviation from the Weyl point. In the three-dimensional space, we have that $\mathbf{q} = (q_x, q_y, q_z)$. Taking $\hbar = 1$ to simplify the notation, we see that the spectrum in the vicinity of the Weyl point is

$$
E(\mathbf{q}) = \mathbf{v}_0\cdot\mathbf{q} \pm \sqrt{\sum_{i=1}^{3}(\mathbf{v}_i\cdot\mathbf{q})^2}. \tag{10.208}
$$

We may define the chirality of a given Weyl point as $C = sgn\,(\mathbf{v}_1\cdot\mathbf{v}_2\times\mathbf{v}_3)$.

Some general considerations on the Weyl semimetals may be used to extract interesting information about the properties of these systems. Consider the action

[12] For further reading see [AraujoSacramento].

of the symmetry operations parity and time reversal. (See the brief description in Appendix B.) The effect of the parity operation is such that $H(\mathbf{p}) \rightarrow H(-\mathbf{p})$. Therefore, applying this operation to $H(\mathbf{k}+\mathbf{q}) = \mathbf{v}_\mu \cdot \mathbf{q}\sigma_\mu$, we get $H(-\mathbf{k}-\mathbf{q})$. Changing the sign of $\mathbf{q}$ this leads to $H(-\mathbf{k}+\mathbf{q}) = -\mathbf{v}_\mu \cdot \mathbf{q}\sigma_\mu$. So if the system is invariant under the parity transformation, we get that

$$H(\mathbf{k}+\mathbf{q}) = \mathbf{v}_\mu \cdot \mathbf{q}\sigma_\mu, \qquad H(-\mathbf{k}+\mathbf{q}) = -\mathbf{v}_\mu \cdot \mathbf{q}\sigma_\mu, \tag{10.209}$$

which means that the points $\pm\mathbf{k}$ have opposite chiralities and, therefore, cannot be equivalent in the Brillouin zone.

Consider now the action of time reversal. Since we do not consider spin here (the spinor degeneracy is associated with the two-band structure), the operation of time reversal on the Hamiltonian reduces to the operation of complex conjugation $H \rightarrow H^*$ and change of the sign of the momenta. This implies that $H(\mathbf{k}+\mathbf{q}) = \mathbf{v}_\mu \cdot \mathbf{q}\sigma_\mu$, under the action of time reversal, becomes $H^*(\mathbf{k}-\mathbf{q})$. If the system is invariant under the time reversal operation, then we get $H(-\mathbf{k}+\mathbf{q}) = -\mathbf{v}_\mu \cdot \mathbf{q}\sigma_\mu^*$. This implies that the points $\pm\mathbf{k}$ have the same chirality (recall that $\sigma_2 = \sigma_y$ is purely imaginary). So, if both symmetries are present, parity and time reversal, this would imply that $\mathbf{v}_2 = 0$ and then we get that $C = 0$ and the Weyl point would not exist. So in order for the Weyl point to exist, one of the symmetries has to be broken. If it exists then it turns out that its existence is robust and the argument is associated with the existence of band degeneracies.

Let us first recall a result known for one-dimensional systems with no internal degrees of freedom. The equation of motion of the electrons is given by the Schrödinger equation associated with the existence of the external periodic potential due to the ion background. The eigenvalue problem involves the Laplacian, a second-order differential operator, and therefore there are two solutions. For a given energy, there are two solutions and so at most two points in momentum space. Changing the sign of the momentum, we get in general the same energy, which implies that there are no band crossings.

Let us now consider we have several bands in a three-dimensional system and let us focus on a pair of bands. Let us call their energy dispersions $E_{n,+}(\mathbf{k})$, $E_{n,-}(\mathbf{k})$. In the subspace of these two bands, we can write an effective Hamiltonian like

$$H_e = \sum_{\mathbf{k}} \psi_{\mathbf{k}}^\dagger H(\mathbf{k}) \psi_{\mathbf{k}}. \tag{10.210}$$

The Hamiltonian matrix may be written as

$$H(\mathbf{k}) = f(\mathbf{k})1 + \sum_{j=1}^{3} g_j(\mathbf{k})\sigma_j, \tag{10.211}$$

which leads to a spectrum

$$E_{\pm} = f(\mathbf{k}) \pm \sqrt{\sum_{j=1}^{3} g_j^2(\mathbf{k})}. \tag{10.212}$$

There is degeneracy between the two bands at points $\mathbf{k}_0$ such that $g_j(\mathbf{k}_0) = 0$ for all $j = 1 - 3$. In three dimensions it may be possible since we may adjust three momentum components. If the zeros exist, then they come in pairs. We may expand the Hamiltonian around each of these degeneracy points as

$$H(\mathbf{k}) = E_{\mathbf{k}_0} + \hbar \mathbf{v}_0 \cdot (\mathbf{k} - \mathbf{k}_0) 1 + \sum_{j=1}^{3} \hbar \mathbf{v}_j \cdot (\mathbf{k} - \mathbf{k}_0) \sigma_j + \cdots \tag{10.213}$$

if $\mathbf{v}_j$ are orthogonal. This is then like an anisotropic Weyl point. It has a 2×2 structure that is linear in momentum, and there is a zero mass (due to its gaplessness). Note that in general in two dimensions, there are only two momentum components of $\mathbf{k}$ that can be freely varied which means it is impossible to find solutions to three equations, in general. Unless there are additional symmetries, the two levels do not cross.

In the three-dimensional case, one expects that it is in general possible to adjust the location of the Weyl point, if it exists, which attests to its robustness to changes in the parameters. The result can be further strengthened noting that the Hamiltonian around the Weyl point is of the type of a monopole at the Weyl location leading to a Berry curvature:

$$\boldsymbol{\Omega} = \frac{\mathbf{h}}{2h^3} = \frac{(\mathbf{v}_1 \cdot \mathbf{q}, \mathbf{v}_2 \cdot \mathbf{q}, \mathbf{v}_3 \cdot \mathbf{q})}{2h^3}. \tag{10.214}$$

We may define the chiral charge of the Weyl point as

$$C = sgn\left(\int \boldsymbol{\Omega} \cdot d\mathbf{S}\right), \tag{10.215}$$

integrated over h-space or over the Brillouin zone. Its topological nature ensures the robustness of the Weyl point, if the conditions for its existence hold. Since each Weyl point is like a monopole of Berry flux, one expects that the number of Weyl points should be even. Also, we may expect that Berry flux will diverge from a positive chirality Weyl point and converge to a negative chirality Weyl point.

In the vicinity of a Weyl point, the density of states has a quadratic dependence on the energy. Taking $\mathbf{v}_0 = 0$, the density of states is given by

$$\rho(E) = \int \frac{d^3q}{(2\pi)^3}\delta\left(\sqrt{\sum_{j=1}^{3}(\mathbf{v}_j \cdot \mathbf{q})^2} - E\right). \tag{10.216}$$

Changing variables to $\xi_i = \mathbf{v}_i \cdot \mathbf{q}$ we get, using that the Jacobian of the transformation is

$$J(\partial q/\partial \xi) = |\mathbf{v}_1 \cdot \mathbf{v}_2 \times \mathbf{v}_3|, \tag{10.217}$$

we can write

$$\rho(E) = \frac{1}{|\mathbf{v}_1 \cdot \mathbf{v}_2 \times \mathbf{v}_3|}\int \frac{d^3\xi}{(2\pi)^3}\delta\left(\sqrt{\sum_{j=1}^{3}\xi_j^2} - E\right). \tag{10.218}$$

Changing to spherical coordinates $r^2 = \sum_{j=1}^{3}\xi_j^2$, $d^3\xi = 4\pi r^2 dr$, we obtain that

$$\rho(E) = \frac{E^2}{2\pi^2\hbar^3|\mathbf{v}_1 \cdot \mathbf{v}_2 \times \mathbf{v}_3|}. \tag{10.219}$$

10.11 Exercises

1. Consider free particles in two dimensions in a rectangle of sides L_x and L_y, and with boundary conditions of an infinite barrier, that is, $\psi(0, y) = \psi(L_x, y) = 0$, $\psi(x, 0) = \psi(x, L_y) = 0$:
   (a) Calculate the Fermi wave vector and the Fermi energy.
   (b) Calculate the system energy as a function of temperature.
2. Suppose that you may apply the free electron model to aluminum. Aluminum crystallizes in an fcc structure with three valence electrons per atom and lattice constants $a = 4.05\,\mathring{A}$:
   (a) What is the Fermi wave vector?
   (b) What is the Fermi energy in eV?
   (c) What is the density of states at the Fermi level in number of states $/eV$?
3. Show that the kinetic energy of a gas of N three-dimensional gas of free electrons at $T = 0\,\mathrm{K}$ is given by

$$U_0 = \frac{3}{5}N\epsilon_F.$$

Calculate the pressure exerted by the electron gas on the walls of the container where they are found. Calculate the gas compressibility.

4. Draw the energy bands for an fcc crystal with zero potential along the directions $(q, 0, 0)$ and (q, q, q).
5. Consider a three-dimensional lattice of free electrons described by the band $\epsilon_{\mathbf{k}} = \hbar^2 k^2/2m$ and reduce the momentum vectors to the first Brillouin zone subtracting an adequate reciprocal lattice vector, **g**. Assume a simple cubic lattice and draw the energies along the axis k_x, that is, the direction [100] for the first 6 bands (choose units such that $\hbar^2 = 2m$). What happens if the periodic potential is turned on? Indicate schematically.
6. Consider a square lattice with a crystalline potential given by either:

(a)

$$U(x, y) = -2U\left(\cos\frac{2\pi x}{a} + \cos\frac{2\pi y}{a}\right),$$

or

(b)

$$U(x, y) = -4U\cos\frac{2\pi x}{a}\cos\frac{2\pi y}{a}.$$

Use the central equation, to leading order, to calculate the gaps of the energy spectrum at the point $\mathbf{k} = (\pi/a, \pi/a)$ and compare the effects of the two potentials.

7. Consider a particle that moves along one dimension. Consider two consecutive potential barriers with values smaller than the total energy of the particle and localized at the points $x = a$ and $x = b$:

   (a) Determine the reflection and transmission amplitude coefficients of the set of two barriers, as a function of the coefficients of each barrier.
   (b) Interpret geometrically the results obtained in terms of the multiple reflections and transmissions between the two barriers.
   (c) Under which conditions does the reflection coefficient vanish?

8. Kronig-Penney model: Consider an electron moving in a one-dimensional space under the action of a potential, like a periodic well of depth U_0, such that each barrier has a width b and the potential barriers are distant by a. The potential extends from $-\infty$ to $+\infty$. Solve Schrödinger equation and obtain the eigenvalue equation. Consider the limit $U_0 \to \infty$, $b \to 0$ with $U_0 bam/\hbar^2$ finite. Discuss the graphical solution. Determine in an approximate way the value of the first gap.
9. A two-dimensional metal in the shape of a square lattice has two conduction electrons per atom. In the approximation of nearly free electrons, make a careful

scheme of the energy surfaces of the electrons and of the holes. For the electrons choose a scheme such that the Fermi surface is represented as closed.

10. In the presence of a lattice potential, the wave functions are of the type[13]

$$\left(\frac{\mathbf{p}^2}{2m} + U(\mathbf{r})\right) e^{i\mathbf{k}\cdot\mathbf{r}} u_{\mathbf{k}}(\mathbf{r}) = \epsilon_{\mathbf{k}} e^{i\mathbf{k}\cdot\mathbf{r}} u_{\mathbf{k}}(\mathbf{r}),$$

and may be rewritten as

$$\left(\frac{(\mathbf{p}+\hbar\mathbf{k})^2}{2m} + U(\mathbf{r})\right) u_{\mathbf{k}}(\mathbf{r}) = \epsilon_{\mathbf{k}} u_{\mathbf{k}}(\mathbf{r}).$$

At $\mathbf{k} = 0$ an approximate solution may be written as $\psi_0(\mathbf{r}) = u_0(\mathbf{r})$. At finite momentum we may take as approximate solution:

$$\psi_{\mathbf{k}}(\mathbf{r}) = e^{i\mathbf{k}\cdot\mathbf{r}} u_0(\mathbf{r})$$

which is a proper solution if we drop the term $\mathbf{k}\cdot\mathbf{p}$ in the equation for the periodic function. One may then treat the term $\mathbf{k}\cdot\mathbf{p}$ as a perturbation.

Use second-order perturbation theory to find the result:

$$\epsilon_n(\mathbf{k}) = \epsilon_n(0) + \frac{\hbar^2 k^2}{2m} + \frac{\hbar^2}{m^2}\sum_j \frac{(| < n0|\mathbf{k}\cdot\mathbf{p}|j0 > |)^2}{\epsilon_n(0) - \epsilon_j(0)},$$

for the n band of a cubic crystal where the sum is extended to all orbitals $\psi_{j,\mathbf{k}}$ for $\mathbf{k} = 0$ and show that the effective mass at this point is

$$\frac{m}{m*} = 1 + \frac{2}{m}\sum_j{}' \frac{| < n0|\mathbf{p}|j0 > |^2}{\epsilon_n(0) - \epsilon_j(0)}.$$

Apply the result to the case of the top of the conduction band of a semiconductor.

11. Calculate the density of states of the one-dimensional tight-binding model.
12. Apply an electric field on a solid:

    (a) Solve the equation of motion of a free electron in the solid in the presence of the electric field.

[13] For further reading see [Kittel].

(b) If the electron is part of a band composed of atomic orbitals with reduced spatial extension, the tight-binding model describes in an approximate way the dispersion relation of the electron in the form:

$$\epsilon(k) = -2t\cos(ka),$$

considering a one-dimensional system with lattice constant a and where t is the hopping amplitude between orbitals in neighboring atoms. Describe the motion of the electron due to the action of a small electric field. Explain the appearance of oscillations (called Bloch oscillations) in the properties of the system (consider the system pure, with no impurities).

(c) Solve the same problem for a semiconductor and calculate the oscillation frequency assuming that $a = 50$ Å and that the electric field has an amplitude $E = 5 \times 10^4$ V cm^{-1}.

(d) In the case of a metal, the average time between collisions of the electrons with impurities is of the order of 10^{-13} s. Discuss the possibility to observe the Bloch oscillations in this case.

13. The Wannier functions of a band are defined in terms of the Bloch functions of the same band as

$$w(\mathbf{r} - \mathbf{r}_n) = \frac{1}{\sqrt{N}} \sum_{\mathbf{k}} e^{-i\mathbf{k}\cdot\mathbf{r}_n} \psi_{\mathbf{k}}(\mathbf{r}),$$

where $\mathbf{r}_n$ is a lattice point considering N atoms on a line with lattice constant a.

(a) Show that the Wannier functions around different lattice points n, m are orthogonal:

$$\int dV w^*(\mathbf{r} - \mathbf{r}_n) w(\mathbf{r} - \mathbf{r}_m) = 0,$$

if $n \neq m$.

(b) Show that for $\psi_{\mathbf{k}} = \frac{1}{\sqrt{N}} e^{ikx} u_0(x)$, the Wannier wave function is given by

$$w(x - x_n) = u_0(x) \frac{\sin \pi (x - x_n)/a}{\pi (x - x_n)/a},$$

14. Consider the Kitaev model defined by the Hamiltonian matrix in momentum space:

$$\begin{pmatrix} \epsilon_k - \mu & -2i\Delta \sin k \\ 2i\Delta \sin k & -\epsilon_k + \mu \end{pmatrix}, \tag{10.220}$$

with $\epsilon_k = -2t \cos k$. The transitions lines occur for $\mu = 2$ and $k = \pi$, for $\mu = -2$ and $k = 0$ and for $\Delta = 0$ and $\cos k = -\mu/(2t)$. Therefore, for $\Delta = 0$ and $\mu = 0$, the transition occurs at $k = \pi/2$. Calculate the winding numbers of the various phases as we change the chemical potential and the parameter Δ.

15. A toy model that involves a transition between two topological regimes with Chern numbers $C = 2$ and $C = -2$ is the Hamiltonian:

$$\begin{aligned} h_x &= \cos k_x + \cos k_y \\ h_y &= \cos k_x - \cos k_y \\ h_z &= t_2 \sin k_x \sin k_y. \end{aligned} \tag{10.221}$$

The two regimes are obtained changing the sign of t_2. The gap closes at the points $\boldsymbol{k} = (\pi/2, \pi/2), (\pi/2, -\pi/2)$, and equivalent points. Calculate the Chern numbers in the two regimes.

References

1. See, for instance, D. Xiao, M.-C. Chang and Q. Niu, Reviews of Modern Physics **82**, 1959 (2010).
2. This model is a simplified version of the SSH model, W. P. Su, J. R. Schrieffer, and A. J. Heeger, Physical Review Letters 42, 1698 (1979).
3. J. Zak, Physical Review Letters **62**, 2747 (1989).
4. R. Resta, Physical Review Letters 80, 1800 (1998).
5. T. Fukui, Y. Hatsugai and H. Suzuki, Journal of the Physical Society of Japan **74**, 1674 (2005).

Chapter 11
Linear Response Theory

The action of a time-dependent perturbation on a system is described by the solution of the time-dependent Schrödinger equation

$$i\hbar \frac{\partial \psi(t)}{\partial t} = H\psi(t). \tag{11.1}$$

The solution can be written formally as

$$|t\rangle = U(t, t_0)|t_0\rangle, \tag{11.2}$$

in terms of the unitary time evolution operator, $U(t, t_0)$, between an initial time, t_0, and a given instant, t. This operator satisfies the relations

$$\begin{aligned} U(t_0, t_0) &= 1 \\ U(t_2, t_0) &= U(t_2, t_1)U(t_1, t_0). \end{aligned} \tag{11.3}$$

Taking $t_1 = t_2 - \delta t$, where δt is small, it can be shown that the operator satisfies an evolution equation

$$i\hbar \frac{\partial U(t)}{\partial t} = H(t)U(t), \tag{11.4}$$

which naturally leads back to Schrödinger equation.

If the Hamiltonian is time independent, $H(t) = H$, the solution is straightforward

$$|t\rangle = e^{-\frac{i}{\hbar}H(t-t_0)}|t_0\rangle. \tag{11.5}$$

P. D. Sacramento, *Introduction to Condensed Matter Theory*,
https://doi.org/10.1007/978-3-031-86251-9_11

If the Hamiltonian is time dependent, the solution can be written as

$$
\begin{aligned}
U(t,t_0) &= 1 - \frac{i}{\hbar}\int_{t_0}^{t} dt' H(t')U(t',t_0) \\
&= 1 + \sum_{n=1}^{\infty}\left(-\frac{i}{\hbar}\right)^n \int_{t_0}^{t} dt_1 \int_{t_0}^{t_1} dt_2 \cdots \int_{t_0}^{t_{n-1}} dt_n H(t_1)H(t_2)\cdots H(t_n)
\end{aligned}
\tag{11.6}
$$

and involves a time-ordered product (recall that the same operator in general does not commute at different times). This solution is not very helpful since it involves infinite terms. Each term involves the total Hamiltonian of the system, which includes the original Hamiltonian plus the time-dependent perturbation. Even if this extra term is small, the series has to be summed since the terms of increasing order are not of smaller amplitude. It is therefore convenient to obtain an expansion that only involves the perturbation. Using the fact that this term is small, we may hope to truncate the series at some desired order.

The standard Schrödinger equation is written in the so-called Schrödinger picture. In the standard approach, the time dependence is contained in the quantum states and the operators are in general time independent. For instance, the time-dependent matrix element of an operator A can be written as $\langle t|A|t\rangle$. Using that $|t\rangle = U|t_0\rangle$ we can write that

$$
\langle t|A_S|t\rangle = \langle t_0|U^\dagger A_S U|t_0\rangle = \langle t_0|A_H|t_0\rangle,
\tag{11.7}
$$

where we have added the designation A_S to emphasize that the (time-independent) operator is in the Schrödinger picture, and we have defined a time-dependent operator A_H, where H stands for Heisenberg picture. This operator is defined by

$$
A_H = U^\dagger A_S U.
\tag{11.8}
$$

If the Hamiltonian is time independent, $H(t) = H$, we can write that

$$
A_H(t) = e^{\frac{i}{\hbar}Ht} A_S e^{-\frac{i}{\hbar}Ht}
\tag{11.9}
$$

(taking $t_0 = 0$, to simplify). We obtain therefore that

$$
\begin{aligned}
\frac{dA_H}{dt} &= \frac{i}{\hbar} H e^{\frac{i}{\hbar}Ht} A_S e^{-\frac{i}{\hbar}Ht} \\
&- \frac{i}{\hbar} e^{\frac{i}{\hbar}Ht} A_S H e^{-\frac{i}{\hbar}Ht} \\
&+ e^{\frac{i}{\hbar}Ht} \frac{\partial A_S}{\partial t} e^{-\frac{i}{\hbar}Ht},
\end{aligned}
\tag{11.10}
$$

including the possibility that the operator in the Schrödinger picture is explicitly time dependent. Interestingly, this equation can be written in a way that is formally equivalent to the Ehrenfest theorem for the time average of an operator

$$i\hbar\frac{d}{dt}\langle t|A_S|t\rangle = \langle t|[A_S, H]|t\rangle + i\hbar\left\langle\frac{\partial A_S}{\partial t}\right\rangle. \tag{11.11}$$

Indeed, we can write that the operator in the Heisenberg picture obeys

$$i\hbar\frac{dA_H(t)}{dt} = [A_H(t), H(t)] + i\hbar\frac{\partial A_H}{\partial t}. \tag{11.12}$$

Consider now that to a given Hamiltonian H_{0S} we add a perturbation V_S, intended to be small. The time evolution of a state in Schrödinger picture is governed by the full Hamiltonian $H_S = H_{0S} + V_S$ as

$$i\hbar\frac{\partial}{\partial t}|t\rangle_S = H_S|t\rangle_S. \tag{11.13}$$

In the spirit of time-dependent perturbation theory, assume that we know the solution of

$$i\hbar\frac{\partial}{\partial t}|t\rangle_S = H_{0S}|t\rangle_S \tag{11.14}$$

and that we want to know the effect of the extra term, V_S, assumed to be small compared to H_{0S}. One defines a unitary operator

$$\bar{U} = e^{\frac{i}{\hbar}H_{0S}t} \tag{11.15}$$

that leads to a state in the so-called interaction picture

$$|t\rangle_I = \bar{U}|t\rangle_S \tag{11.16}$$

and an operator

$$A_I = \bar{U}A_S\bar{U}^{-1}. \tag{11.17}$$

At the initial time, $t = 0$, the interaction picture coincides with the Schrödinger picture and if the extra term $V_S = 0$, the interaction picture coincides with the Heisenberg picture. Therefore, it is also called intermediate picture.

Note that $H_{0S} = H_{0I} = H_0$. We conclude that

$$i\hbar\frac{\partial}{\partial t}|t\rangle_I = e^{\frac{i}{\hbar}H_0t}V_S|t\rangle_S \tag{11.18}$$

and

$$i\hbar\frac{\partial}{\partial t}|t_I = V_I|t\rangle_I. \tag{11.19}$$

Also, we may obtain

$$i\hbar\frac{d}{dt}A_I = [A_I, H_0] + i\hbar\frac{\partial A_I}{\partial t}. \tag{11.20}$$

Using that $|t\rangle_S = U_S|t_0\rangle_S$ and defining the operator U_I as

$$|t\rangle_I = U_I|t_0\rangle_I, \tag{11.21}$$

we get

$$U_I = \bar{U}U_S. \tag{11.22}$$

We may calculate the time evolution of the state in the interaction picture

$$i\hbar\frac{\partial}{\partial t}|t\rangle_I = V_I(t)U_I(t)|t_0\rangle_I. \tag{11.23}$$

We obtain therefore

$$i\hbar\frac{\partial}{\partial t}U_I(t) = V_I(t)U_I(t). \tag{11.24}$$

The solution of this equation follows similar steps to those for the evolution operator U_S. Using that $U_I(t_0, t_0) = 1$ we obtain

$$U_I(t, t_0) = 1 + \sum_{n=1}^{\infty}\left(-\frac{i}{\hbar}\right)^n \int_{t_0}^{t} dt_1 \int_{t_0}^{t_1} dt_2 \cdots \int_{t_0}^{t_{n-1}} dt_n V_I(t_1)V_I(t_2)\cdots V_I(t_n). \tag{11.25}$$

This is also a time-ordered product. A significant improvement is that it involves powers of the interaction, assumed small. Therefore, we may truncate the series to a few terms, since the solution is expected to converge fast.

11.1 Susceptibilities and Linear Response Theory

Let us consider now that we apply an external field to a system and we want to study the response of the system to the external probe. Let us then apply a field which in general will be time dependent. We have then to solve Schrödinger equation or some equivalent alternative way. Recall that we can define an operator in the so-called Heisenberg representation as[1]

$$A_H(t) = U^{-1}(t) A_S U(t). \tag{11.26}$$

Let us now consider that at a given time, t_i, we apply an external field such that the Hamiltonian is given by

$$\mathcal{H}_H(t) = \mathcal{H} + \delta\mathcal{H}, \tag{11.27}$$

where, as an example, $\delta\mathcal{H}$ is the result of applying a magnetic field, $\mathbf{h}(\mathbf{r}, t)$ like

$$\delta\mathcal{H} = -\int d^3r \mathbf{h}(\mathbf{r}, t) \cdot \mathbf{S}(\mathbf{r}); \tag{11.28}$$

here $\mathbf{S}(\mathbf{r})$ is the spin density operator. The magnetization is therefore given by

$$\mathbf{m}(\mathbf{r}, t) =< \mathbf{S}_H(\mathbf{r}, t) >= Tr\left(\rho_{\mathcal{H}} \mathbf{S}_H(\mathbf{r}, t)\right). \tag{11.29}$$

Here $\rho_{\mathcal{H}} = \rho^{eq} = \frac{1}{Z} e^{-\beta\mathcal{H}}$ is the canonical distribution. Then

$$\mathbf{S}_H(\mathbf{r}, t) = U^{-1}_{\mathcal{H}+\delta\mathcal{H}}(t, t_i) \mathbf{S}_S(\mathbf{r}) U_{\mathcal{H}+\delta\mathcal{H}}(t, t_i). \tag{11.30}$$

If $\delta\mathcal{H} = 0$ and $\mathcal{H}$ is independent of time, the evolution operator is simply given by

$$U_{\mathcal{H}}(t, t_i) = e^{-\frac{i}{\hbar}\mathcal{H}(t-t_i)}.$$

It is convenient to introduce the so-called interaction representation where

$$\begin{aligned} \mathbf{S}_I(\mathbf{r}, t) &= e^{\frac{i}{\hbar}\mathcal{H}(t-t_i)} \mathbf{S}_S(\mathbf{r}) e^{-\frac{i}{\hbar}\mathcal{H}(t-t_i)} \\ U^{-1}_{\mathcal{H}+\delta\mathcal{H}}(t, t_i) &= e^{-\frac{i}{\hbar}\mathcal{H}(t-t_i)} U_I(t, t_i). \end{aligned} \tag{11.31}$$

We obtain

$$U_I(t, t_i) = 1 - \frac{i}{\hbar} \int_{t_i}^{t} dt' \delta\mathcal{H}_I(t') U_I(t', t_i). \tag{11.32}$$

Iterating to first order gives

$$U_I(t, t_i) = 1 - \frac{i}{\hbar} \int_{t_i}^{t} dt' \delta\mathcal{H}_I(t') + \ldots \tag{11.33}$$

[1] Here we follow [BicudoSacramento].

The magnetization can then be written as

$$\begin{aligned}\mathbf{m}(\mathbf{r}, t) &= Tr\left\{\frac{1}{Z}e^{-\beta\mathcal{H}}U^{-1}_{\mathcal{H}+\delta\mathcal{H}}\mathbf{S}_S U_{\mathcal{H}+\delta\mathcal{H}}\right\} \\ &= Tr\left\{\frac{1}{Z}e^{-\beta\mathcal{H}}U_I^{-1}\mathbf{S}_I U_I\right\}.\end{aligned} \tag{11.34}$$

Replacing the expression for the evolution operator U_I and its inverse to linear order in the external field, we obtain

$$\mathbf{m}(\mathbf{r}, t) =< \mathbf{S}_I(\mathbf{r}, t) >_{eq} + \frac{i}{\hbar}\int_{t_i}^{t} dt' < [\delta\mathcal{H}_I(t'), \mathbf{S}_I(\mathbf{r}, t)] >_{eq} . \tag{11.35}$$

Taking the fluctuation of the magnetization with respect to the equilibrium value without external field, we get the response of the system to the external stimulus. The p component is given by

$$\delta m^p(\mathbf{r}, t) = \int_{t_i}^{\infty}\int d^3r' \sum_l \chi^{pl}(\mathbf{r}, \mathbf{r}'; t - t')h^l(\mathbf{r}', t'), \tag{11.36}$$

with, for instance in Cartesian coordinates, $p = x, y, z$. We have introduced the susceptibility $\chi^{pl}(\mathbf{r}, \mathbf{r}'; t - t')$, which is the response function which relates the external field to the response of the system which, in this example, is the change in the magnetization. The susceptibility is given by

$$\chi^{pl}(\mathbf{r}, \mathbf{r}'; t - t') = \frac{i}{\hbar}\theta(t - t') < [S_I^p(\mathbf{r}, t), S_I^l(\mathbf{r}', t')] >_{eq} . \tag{11.37}$$

The subscript eq means that the average values are to be evaluated using the Boltzmann weight. The θ (step) function reflects the causality of the problem: there is no response before the external field is switched on. Note that in the time integration we can extend the lower limit to $-\infty$, since before t_i the perturbation is zero anyway. In equilibrium, $\delta\mathcal{H} = 0$ and, therefore, the operators in the interaction and in the Heisenberg representations are equal. Therefore, we may also rewrite

$$\chi^{lm}(\mathbf{r}, \mathbf{r}'; t - t') = \frac{i}{\hbar}\theta(t - t') < [S_H^l(\mathbf{r}, t), S_H^m(\mathbf{r}', t')] >_{eq} . \tag{11.38}$$

It is easy to see that, as expected,

$$\chi^{lm}(\mathbf{r}, \mathbf{r}'; t - t') = \frac{\partial m_l(\mathbf{r}, t)}{\partial h_m(\mathbf{r}', t')}. \tag{11.39}$$

11.1.1 Kramers-Kronig Relations

Let us write

$$\chi^{lm}(\mathbf{r},t;\mathbf{r}',t') = 2i\theta(t-t')\chi''_{lm}(\mathbf{r},t;\mathbf{r}',t')$$
$$\chi''_{lm}(\mathbf{r},t;\mathbf{r}',t') = \frac{1}{2\hbar} < [S^l_H(\mathbf{r},t), S^m_H(\mathbf{r}',t')] >_{eq} . \tag{11.40}$$

Assuming that the system is uniform in space and time, we get that

$$\chi_{lm}(\mathbf{r},t;\mathbf{r}',t') = \chi_{lm}(\mathbf{r}-\mathbf{r}';t-t')$$
$$\chi''_{lm}(\mathbf{r},t;\mathbf{r}',t') = \chi''_{lm}(\mathbf{r}-\mathbf{r}';t-t'). \tag{11.41}$$

Taking $t_i \to -\infty$ and performing a Fourier transform, we can write

$$\chi_{ij}(\mathbf{k},\omega) = \int d(t-t') \int d^3(x-x') e^{-i\mathbf{k}\cdot(\mathbf{r}-\mathbf{r}')+i\omega(t-t')} \chi_{ij}(\mathbf{r}-\mathbf{r}';t-t') \tag{11.42}$$

and

$$\delta m_i(\mathbf{k},\omega) = \sum_j \chi_{ij}(\mathbf{k},\omega) h_j(\mathbf{k},\omega). \tag{11.43}$$

To simplify notation, let us focus now our attention on the time dependence of the susceptibilities and let us drop the matrix indices. We write then $\chi(t-t') = 2i\theta(t-t')\chi''(t-t')$. We can write the following sequence of identities using that the function is causal

$$\begin{aligned}
\chi(\omega) &= 2i \int_0^\infty d(t-t') e^{i\omega(t-t')} \chi''(t-t') \\
&= 2i \int_0^\infty dt e^{i\omega t} \chi''(t) \\
&= 2i \int_0^\infty dt e^{i\omega t} \int_{-\infty}^{\infty} \frac{d\omega'}{2\pi} e^{-i\omega' t} \chi''(\omega') \\
&= 2i \int_{-\infty}^{\infty} \frac{d\omega'}{2\pi} \chi''(\omega') \int_0^\infty dt e^{-i(\omega'-\omega-i\eta)t}, \ \eta \to 0^+ \\
&= 2i \int_{-\infty}^{\infty} \frac{d\omega'}{2\pi} \chi''(\omega') \frac{1}{i(\omega'-\omega-i\eta)}
\end{aligned}$$
$$\chi(\omega) = \int_{-\infty}^{\infty} \frac{d\omega'}{\pi} \frac{\chi''(\omega')}{\omega'-\omega-i\eta}. \tag{11.44}$$

Using that

$$\frac{1}{x \pm i\eta} = P\frac{1}{x} \mp i\pi\delta(x), \tag{11.45}$$

where P stands for principal part and $\delta(x)$ is the Dirac delta function written as

$$P\frac{1}{x} = \lim_{\eta\to 0^+} \frac{x}{x^2+\eta^2}$$

$$\delta(x) = \lim_{\eta\to 0^+} \frac{1}{\pi}\frac{\eta}{x^2+\eta^2},$$

we can write that

$$\chi(\omega) = P\int \frac{d\omega'}{\pi}\frac{\chi''(\omega')}{\omega'-\omega} + i\chi''(\omega). \tag{11.46}$$

If $\chi''(\omega)$ is real (as we will show later), then

$$\begin{aligned} Im\chi(\omega) &= \chi''(\omega) \\ Re\chi(\omega) &= \chi'(\omega) = P\int \frac{d\omega'}{\pi}\frac{\chi''(\omega')}{\omega'-\omega}, \end{aligned} \tag{11.47}$$

which implies a relation between the real and the imaginary parts of the dynamical susceptibility $\chi(\omega)$. This relation is the result of the causality of the response function. This constitutes one of the Kramers-Kronig relations that relate the real and imaginary parts of the frequency part of a causal function.

Since $\chi(t) = 2i\theta(t)\chi''(t)$ we can write that

$$\chi(\omega) = 2i\int_0^\infty dt\chi''(t)e^{i\omega t}. \tag{11.48}$$

Taking the frequency as a complex variable, we can write that $\omega = \omega_r + i\omega_i$. Then we conclude that

$$|e^{i(\omega_r+i\omega_i)t}| = |e^{i\omega_r t}e^{-\omega_i t}| = e^{-\omega_i t}.$$

Therefore, if $\chi(\omega)$ exists for real frequency ($\omega = \omega_r$), then it also exists in the upper complex plane ($\omega_i > 0$). Therefore, $\chi(\omega)$ is analytical in the upper half-plane. Considering a contour, $C = C_1 + C_2$, in the complex plane that includes the real axis, C_1, and a semicircle of infinite radius in the upper half-plane, C_2, we may write

$$\frac{1}{2\pi i}\oint_C \frac{\chi(z')}{z'-z}dz' = \chi(z). \tag{11.49}$$

If $\chi(z)$ decays at least as fast as $1/z$ as $|z| \to \infty$, then the integral C_2 vanishes and we obtain that

$$\chi(z) = \frac{1}{2\pi i} \int_{-\infty}^{\infty} d\omega' \frac{\chi(\omega')}{\omega' - z}, \tag{11.50}$$

where ω' is the real axis. Taking $z = \omega + i\eta$ and $\eta \to 0$ we obtain

$$\frac{1}{2\pi i} P \int_{-\infty}^{\infty} d\omega' \frac{\chi(\omega')}{\omega' - \omega} + \frac{1}{2}\chi(\omega) = \chi(\omega) \tag{11.51}$$

and

$$\chi(\omega) = -iP \int_{-\infty}^{\infty} \frac{d\omega'}{\pi} \frac{\chi(\omega')}{\omega' - \omega}. \tag{11.52}$$

Writing, as before, $\chi(\omega) = \chi'(\omega) + i\chi''(\omega)$, we arrive at the Kramers-Kronig relations between the real and imaginary parts of the susceptibility

$$\chi'(\omega) = P \int_{-\infty}^{\infty} \frac{d\omega'}{\pi} \frac{\chi''(\omega')}{\omega' - \omega}$$

$$\chi''(\omega) = -P \int_{-\infty}^{\infty} \frac{d\omega'}{\pi} \frac{\chi'(\omega')}{\omega' - \omega}. \tag{11.53}$$

This shows that both functions χ' and χ'' are real functions.

11.1.2 Dynamical Structure Factor

When a beam of particles is sent to a target, it interacts with its particles. The scattering may be used to obtain information about the properties of the constituents of the target, such as its charge density or spin density correlations.

Consider a beam of particles of mass M scattering off a target, composed of particles at locations $\mathbf{r}_j$, through an interaction Hamiltonian

$$H_{int}(\mathbf{r}) = \sum_j V(\mathbf{r} - \mathbf{r}_j). \tag{11.54}$$

Using Fermi's golden rule, the cross section in Born approximation is given by

$$\frac{1}{V}\frac{d^2\sigma(i \to f)}{d\Omega d\omega} = \frac{1}{V}\frac{k_f}{k_i}\left(\frac{M}{2\pi}\right)^2 |\langle \mathbf{k}_f f|H_{int}|i\mathbf{k}_i\rangle|^2 \delta(\hbar\omega - E_f + E_i). \tag{11.55}$$

Before the collision of a given incident particle with the target, the state of the particle is given by $|\mathbf{k}_i\rangle$ and its energy is ϵ_i and the state of the target is $|i\rangle$ and its energy is E_i. After the collision, the particle is in a state $|\mathbf{k}_f\rangle$ with energy ϵ_f and the target is in state $|f\rangle$ with energy E_f. Conservation of energy implies

$$\epsilon_i + E_i = \epsilon_f + E_f. \tag{11.56}$$

We also define

$$\hbar\omega = \epsilon_i - \epsilon_f = E_f - E_i. \tag{11.57}$$

The matrix element may be written as

$$\begin{aligned}\langle \mathbf{k}_f f|H_{int}|i\mathbf{k}_i\rangle &= \left\langle \mathbf{k}_f f|\sum_j V(\mathbf{r}-\mathbf{r}_j)|i\mathbf{k}_i\right\rangle \\ &= \left\langle \mathbf{k}_f f|\sum_j \int \frac{d^3k'}{(2\pi)^3} V(\mathbf{k}')e^{i\mathbf{k}'\cdot(\mathbf{r}-\mathbf{k}')}|i\mathbf{k}_i\right\rangle \\ &= \int \frac{d^3k'}{(2\pi)^3} V(\mathbf{k}')\left\langle \mathbf{k}_f|e^{i\mathbf{k}'\cdot\mathbf{r}}|\mathbf{k}_i\right\rangle\left\langle f|\sum_j e^{-i\mathbf{k}'\cdot\mathbf{r}_j}|i\right\rangle. \end{aligned} \tag{11.58}$$

The matrix element due to the incident particle can be calculated easily, for instance, introducing a decomposition of the identity

$$\begin{aligned}\langle \mathbf{k}_f|e^{i\mathbf{k}'\cdot\mathbf{r}}|\mathbf{k}_i\rangle &= \int d^3r'\langle \mathbf{k}_f|e^{i\mathbf{k}'\cdot\mathbf{r}}|\mathbf{r}'\rangle\langle \mathbf{r}'|\mathbf{k}_i\rangle \\ &= \int d^3r' e^{i\mathbf{k}'\cdot\mathbf{r}'}\langle \mathbf{k}_f|\mathbf{r}'\rangle\langle \mathbf{r}'|\mathbf{k}_i\rangle \\ &= (2\pi)^3\delta(\mathbf{k}'-\mathbf{k}_f+\mathbf{k}_i), \end{aligned} \tag{11.59}$$

where we have used that $\langle \mathbf{r}|\mathbf{k}\rangle = e^{i\mathbf{k}\cdot\mathbf{r}}$. Let us now define $\mathbf{k} = \mathbf{k}_i - \mathbf{k}_j$. We conclude then

$$\langle \mathbf{k}_f f|H_{int}|i\mathbf{k}_i\rangle = V(\mathbf{k})\left\langle f|\sum_j e^{i\mathbf{k}\cdot\mathbf{r}_j}|i\right\rangle. \tag{11.60}$$

Therefore, the differential cross section of the transition $i \to f$ is given by

$$\frac{1}{V}\frac{d^2\sigma(i\to f)}{d\Omega d\omega} = \frac{1}{V}\frac{k_f}{k_i}\left(\frac{M}{2\pi}\right)^2\delta(\hbar\omega - E_f + E_i)|V(\mathbf{k})|^2|\langle f|\sum_j e^{i\mathbf{k}\cdot\mathbf{r}_j}|i\rangle|^2. \tag{11.61}$$

In a scattering experiment we may control both $\mathbf{k}_i$ and $\mathbf{k}_f$, but in general we do not know the states $|i\rangle$ and $|f\rangle$ of the target. We must, therefore, perform an average over the initial states of the target. A convenient choice is to perform a thermodynamical average assuming the target is initially in thermal equilibrium. We may consider for instance the canonical ensemble and take that the density matrix of the target is given by

$$\rho_i = \frac{e^{-\beta E_i}}{Z}.$$

We must also carry out a sum over all the possible final states of the target. The desired cross section is given by

$$\begin{aligned}
\frac{1}{V}\frac{d^2\sigma}{d\Omega d\omega} &= \sum_{i,f} \rho_i \frac{d^2\sigma(i \to f)}{d\Omega d\omega} \\
&= \frac{1}{V}\frac{k_f}{k_i}\left(\frac{M}{2\pi}\right)^2 |V(\mathbf{k})|^2 \sum_{i,f} \delta(\hbar\omega - E_f + E_i)\rho_i |\langle f|\sum_j e^{i\mathbf{k}\cdot\mathbf{r}_j}|i\rangle|^2 \\
&= \frac{1}{2\pi\hbar}\frac{k_f}{k_i}\left(\frac{M}{2\pi}\right)^2 |V(\mathbf{k})|^2 \bar{S}_T(\mathbf{k}, \omega).
\end{aligned} \tag{11.62}$$

Using that

$$2\pi\hbar\delta(\hbar\omega - E_f + E_i) = \int dt e^{-i\left(\omega - \frac{E_f - E_i}{\hbar}\right)t},$$

we get that

$$\bar{S}_T(\mathbf{k}, \omega) = \sum_{i,f} \frac{1}{V}\rho_i \int dt e^{-i\left(\omega - \frac{E_f - E_i}{\hbar}\right)t} \left\langle i|\sum_l e^{-i\mathbf{k}\cdot\mathbf{r}_l}|f\right\rangle \left\langle f|\sum_j e^{i\mathbf{k}\cdot\mathbf{r}_j}|i\right\rangle. \tag{11.63}$$

Note that in the Heisenberg picture we may write

$$\mathbf{r}_j(t) = e^{\frac{i}{\hbar}Ht}\mathbf{r}_j(t=0)e^{-\frac{i}{\hbar}Ht}. \tag{11.64}$$

Therefore, we may write

$$\bar{S}_T(\mathbf{k}, \omega) = \frac{1}{V}\int_{-\infty}^{\infty} dt e^{-i\omega t} \left\langle \sum_l e^{-i\mathbf{k}\cdot\mathbf{r}_l} \sum_j e^{i\mathbf{k}\cdot\mathbf{r}_j(t)} \right\rangle. \tag{11.65}$$

Note that the particle density at point $\mathbf{r}$ at time t is given by

$$n(\mathbf{r},t)=\sum_i \delta(\mathbf{r}-\mathbf{r}_i(t)). \tag{11.66}$$

Its Fourier transform is given by

$$n(\mathbf{k},t)=\frac{1}{\sqrt{V}}\int d^3re^{-i\mathbf{k}\cdot\mathbf{r}}n(\mathbf{r},t)=\frac{1}{\sqrt{V}}\sum_i e^{-i\mathbf{k}\cdot\mathbf{r}_i(t)}. \tag{11.67}$$

Introducing $\sum_f |f\rangle\langle f|=1$ leads to

$$\begin{aligned}\bar{S}_T(\mathbf{k},\omega)&=\int_{-\infty}^{\infty}dte^{-i\omega t}\langle n(\mathbf{k},0)n(-\mathbf{k},t)\rangle\\&=\int_{-\infty}^{\infty}dte^{-i\omega t}\int\frac{d^3r}{\sqrt{V}}\int\frac{d^3r'}{\sqrt{V}}e^{-i\mathbf{k}\cdot\mathbf{r}}e^{i\mathbf{k}\cdot\mathbf{r}'}\langle n(\mathbf{r},t=0)n(\mathbf{r}',t)\rangle,\end{aligned} \tag{11.68}$$

which gives information about the density-density correlation functions in space-time, or in momentum-energy. This quantity is called dynamical structure factor.

11.1.3 Dissipation

Let us calculate the energy dissipated in the system due to the presence of the external perturbation. The coupling of the external source to the system degrees of freedom may be written as

$$H_{int}=-\sum_i A_iF_i(t), \tag{11.69}$$

where we consider that the external sources, represented by the terms $F_i(t)$, are taken as classical and the internal variables are represented by the Hermitian operators A_i. The energy dissipated in the system is given by

$$W=\int_{-\infty}^{\infty}dt\left\langle\frac{\partial H}{\partial t}\right\rangle_{n-e}=-\int_{-\infty}^{\infty}dt\sum_i\left\langle A_i\frac{\partial F_i(t)}{\partial t}\right\rangle=\int_{-\infty}^{\infty}dt\sum_i\left\langle\frac{\partial A_i}{\partial t}\right\rangle F_i(t). \tag{11.70}$$

The averages are taken in a non-equilibrium system due to the presence of the time-dependent perturbation. We have integrated by parts assuming that the sources vanish at $t=\pm\infty$. In linear response theory we know that

$$\langle A_i(t)\rangle_{n-e}=\langle A_i(t)\rangle_{eq}+\sum_j\int_{-\infty}^{\infty}dt'\chi_{ij}(t-t')F_j(t'). \tag{11.71}$$

The average in equilibrium is time independent. Therefore,

$$
\begin{aligned}
W &= \int_{-\infty}^{\infty} dt \int_{-\infty}^{\infty} dt' \sum_{ij} F_i(t) \frac{\partial}{\partial t} \chi_{ij}(t-t') F_j(t') \\
&= \int_{-\infty}^{\infty} dt \int_{-\infty}^{\infty} dt' \sum_{ij} F_i(t) \int_{-\infty}^{\infty} \frac{d\omega}{2\pi} e^{-i\omega(t-t')} (-i\omega) \chi_{ij}(\omega) F_j(t') \\
&= \sum_{ij} \int_{-\infty}^{\infty} \frac{d\omega}{2\pi} F_i^*(\omega)(-i\omega)\chi_{ij}(\omega) F_j(\omega).
\end{aligned} \tag{11.72}
$$

Since $F_i(t)$ is real, it implies that $F_i^*(\omega) = F_i(-\omega)$. Making the substitutions $\omega \to -\omega$ and $i \leftrightarrow j$, we may as well write that

$$
W = \sum_{ij} \int_{-\infty}^{\infty} \frac{d\omega}{2\pi} F_i^*(\omega) i\omega \chi_{ji}(-\omega) F_j(\omega), \tag{11.73}
$$

which leads to

$$
W = \sum_{ij} \int_{-\infty}^{\infty} \frac{d\omega}{2\pi} F_i^*(\omega) i\omega \frac{1}{2} \left(-\chi_{ij}(\omega) + \chi_{ji}(-\omega)\right) F_j(\omega). \tag{11.74}
$$

We may now write

$$
\begin{aligned}
\chi_{ji}(-\omega) &= P \int_{-\infty}^{\infty} \frac{d\omega'}{\pi} \frac{\chi''_{ji}(\omega')}{\omega' + \omega} + i\chi''_{ji}(-\omega) \\
&= -P \int_{-\infty}^{\infty} \frac{d\omega'}{\pi} \frac{\chi''_{ji}(-\omega')}{\omega' - \omega} + i\chi''_{ji}(-\omega) \\
&= -P \int_{-\infty}^{\infty} \frac{d\omega'}{\pi} \frac{\chi''_{ij}(\omega')}{\omega' - \omega} + i\chi''_{ji}(-\omega).
\end{aligned} \tag{11.75}
$$

This implies that

$$
\frac{1}{2}\left(-\chi_{ij}(\omega) + \chi_{ji}(-\omega)\right) = \frac{1}{2}\left(-i\chi''_{ij}(\omega) + i\chi''_{ji}(-\omega)\right) = -i\chi''_{ij}(\omega).
$$

Therefore, we arrive at the result

$$
W = \sum_{ij} \int_{-\infty}^{\infty} \frac{d\omega}{2\pi} F_i^*(\omega) \omega \chi''_{ij}(\omega) F_j(\omega), \tag{11.76}
$$

which shows that the dissipation is controlled by the imaginary part of the susceptibility, as expected.

Consider now that $F_i = a_i e^{i\omega t}$. We obtain then

$$\sum_{ij} a_i^* \omega \chi''_{ij}(\omega) a_j \geq 0. \tag{11.77}$$

Therefore,

$$\omega \chi''_{ii} \geq 0. \tag{11.78}$$

The definition of susceptibility may be extended to two operators as

$$\chi''_{AB}(t - t') = \frac{1}{2\hbar} \langle [A(t), B(t')] \rangle. \tag{11.79}$$

It follows that

$$\chi''_{AB}(t - t') = -\chi''_{BA}(t' - t) \tag{11.80}$$

and

$$\chi''_{AB}(\omega) = -\chi''_{BA}(-\omega). \tag{11.81}$$

If the two operators A and B are Hermitian,

$$\begin{aligned} \left(\chi''_{AB}(t - t')\right)^* &= \frac{1}{2\hbar} \langle [A(t), B(t')] \rangle^* \\ &= \frac{1}{2\hbar} Tr \left[\rho [A(t), B(t')]\right]^\dagger \\ &= \frac{1}{2\hbar} Tr \left[[A(t), B(t')]^\dagger \rho^\dagger\right] \\ &= \frac{1}{2\hbar} Tr \left[\rho [B(t'), A(t)]\right] \\ &= \chi''_{BA}(t' - t). \end{aligned} \tag{11.82}$$

Also,

$$\begin{aligned} \left(\chi''_{AB}(\omega)\right)^* &= \int d(t - t') e^{-i\omega(t-t')} \left(\chi''_{AB}(t - t')\right)^* \\ &= \int d(t' - t) \chi''_{BA}(t' - t) e^{i\omega(t'-t)} \\ &= \chi''_{BA}(\omega). \end{aligned} \tag{11.83}$$

Therefore,

$$\left(\chi''_{AB}(\omega)\right)^* = \chi''_{BA}(\omega) = -\chi''_{AB}(-\omega). \tag{11.84}$$

If the two operators are the same, $B = A$, we get that χ''_{AA} is real, as we saw above.

11.1.4 Fluctuation-Dissipation Theorem

The dissipation is associated with the imaginary part of the susceptibility which, as we found before, is given by a commutator of two operators

$$\chi''_{AB}(t-t') = \frac{1}{2\hbar}\langle [A(t), B(t')]\rangle. \tag{11.85}$$

On the other hand, a perturbation leads to a response in the system that may be measured in a dispersion experiment. Typically this involves correlation functions of the type

$$\bar{S}_{AB}(t-t') = \langle A(t)B(t')\rangle. \tag{11.86}$$

The two types of correlation functions are related leading to a relation between dissipation in a system and its response in a scattering experiment.

In thermal equilibrium the density matrix, ρ, is given by the Boltzmann exponential and we obtain

$$\begin{aligned}
\bar{S}_{AB}(t-t') &= \frac{1}{Z} Tr e^{-\beta H} A(t)B(t') \\
&= \frac{1}{Z} Tr e^{-\beta H} A(t) e^{\beta H} e^{-\beta H} B(t') \\
&= \frac{1}{Z} Tr e^{-\beta H} e^{\frac{i}{\hbar}Ht} A(0) e^{-\frac{i}{\hbar}Ht} e^{\beta H} e^{-\beta H} B(t') \\
&= \frac{1}{Z} Tr e^{-\beta H} B(t') A(t+i\beta\hbar) \\
\bar{S}_{AB}(t-t') &= \bar{S}_{BA}(t'-t-i\beta\hbar).
\end{aligned} \tag{11.87}$$

Carrying a Fourier transform we get

$$\bar{S}_{AB}(\omega) = \int_{-\infty}^{\infty} d(t-t') e^{i\omega(t-t')} \bar{S}_{BA}(t'-t-i\beta\hbar). \tag{11.88}$$

Making a change of variables $t' - t \to t$, and then another transformation $t \to t + i\beta\hbar$, we get

$$\begin{aligned}\bar{S}_{AB}(\omega) &= e^{\beta\hbar\omega} \int_{-\infty}^{\infty} dt e^{-i\omega t} \bar{S}_{BA}(t) \\ &= e^{\beta\hbar\omega} \bar{S}_{BA}(-\omega).\end{aligned} \tag{11.89}$$

Note that

$$\chi''_{AB}(t-t') = \frac{1}{2\hbar}\langle [A(t), B(t')] \rangle = \frac{1}{2\hbar}\left[\bar{S}_{AB}(t-t') - \bar{S}_{BA}(t'-t)\right]. \tag{11.90}$$

In terms of their Fourier transforms, this can be expressed

$$\begin{aligned}\chi''_{AB}(\omega) &= \frac{1}{2\hbar}\left[\bar{S}_{AB}(\omega) - \bar{S}_{BA}(-\omega)\right] \\ \chi''_{AB}(\omega) &= \frac{1}{2\hbar}\left(1 - e^{-\beta\hbar\omega}\right)\bar{S}_{AB}(\omega),\end{aligned} \tag{11.91}$$

which relates the susceptibility and the scattering cross section of the system degrees of freedom.

Consider now the fluctuation of an operator around its mean value, $\delta A = A - \langle A \rangle$. We may define a correlation function of the fluctuations of two operators A and B and take their symmetrized product

$$S_{AB}(t-t') = \frac{1}{2}\langle \{\delta A(t), \delta B(t')\} \rangle. \tag{11.92}$$

It is straightforward to see that

$$S_{AB}(t-t') = \frac{1}{2}\left(\bar{S}_{AB}(t-t') + \bar{S}_{BA}(t'-t)\right), \tag{11.93}$$

plus a possible constant term. In Fourier space we obtain

$$\begin{aligned}S_{AB}(\omega) &= \frac{1}{2}\left(\bar{S}_{AB}(\omega) + \bar{S}_{BA}(-\omega)\right) \\ &= \frac{1}{2}\left(1 + e^{-\beta\hbar\omega}\right)\bar{S}_{AB}(\omega).\end{aligned} \tag{11.94}$$

Using the relation to the imaginary part of the susceptibility, we get

$$\begin{aligned}S_{AB}(\omega) &= \frac{1}{2}\left(1 + e^{-\beta\hbar\omega}\right) 2\hbar \frac{1}{1 - e^{-\beta\hbar\omega}} \chi''_{AB}(\omega) \\ S_{AB}(\omega) &= \hbar \coth\frac{\beta\hbar\omega}{2} \chi''_{AB}(\omega).\end{aligned} \tag{11.95}$$

This relation is called the fluctuation-dissipation theorem since it relates the fluctuations expressed in $S_{AB}(\omega)$ with the dissipation of the system expressed in $\chi''_{AB}(\omega)$.

11.2 Conductivity: Kubo Formulas

Linear response theory may also be used to determine the current of a system of particles in the presence of an external field. An example is the charge current due to the presence of an electric field. The electric field may be the result of a potential gradient or the result of a time-varying vector potential.

In the presence of a vector potential, the minimal coupling of a particle of charge e_i to the electromagnetic field leads to the change in the linear momentum $\mathbf{p} \to \mathbf{p} - e_i\mathbf{A}$. The Hamiltonian of a system of free particles in the presence of such an external field is written as

$$H = \sum_i \frac{1}{2m}\mathbf{p}_i^2 \to \sum_i \frac{1}{2m}\left[\mathbf{p}_i - e_i\mathbf{A}\right]^2 . \tag{11.96}$$

Expanding, we get

$$H = \sum_i \frac{1}{2m}\left[\mathbf{p}_i^2 - e_i\left(\mathbf{p}_i \cdot \mathbf{A} + \mathbf{A} \cdot \mathbf{p}_i\right) + (e_i\mathbf{A})^2\right]. \tag{11.97}$$

Neglecting terms that are quadratic in the vector potential, the perturbation due to the external field is

$$\delta H = \sum_i \frac{-e_i}{2m}\left(\mathbf{p}_i \cdot \mathbf{A} + \mathbf{A} \cdot \mathbf{p}_i\right). \tag{11.98}$$

The charge current operator is given in the usual way by

$$\mathbf{J}(\mathbf{x},t) = \frac{1}{2}\sum_i \left[e_i\frac{(\mathbf{p}_i - e_i\mathbf{A}(\mathbf{x},t))}{m}\delta(\mathbf{x}-\mathbf{x}_i) + \delta(\mathbf{x}-\mathbf{x}_i)e_i\frac{(\mathbf{p}_i - e_i\mathbf{A}(\mathbf{x},t))}{m}\right]. \tag{11.99}$$

The current operator may be separated in two terms

$$\mathbf{J}(\mathbf{x},t) = \mathbf{j}_p(\mathbf{x}) + \mathbf{j}_d(\mathbf{x}), \tag{11.100}$$

the paramagnetic current

$$\mathbf{j}_p(\mathbf{x}) = \frac{1}{2}\sum_i \left[e_i\frac{\mathbf{p}_i}{m}\delta(\mathbf{x}-\mathbf{x}_i) + \delta(\mathbf{x}-\mathbf{x}_i)e_i\frac{\mathbf{p}_i}{m}\right] \tag{11.101}$$

and the diamagnetic current

$$\mathbf{j}_d(\mathbf{x}) = -\frac{Ne^2}{m}\mathbf{A}(\mathbf{x}, t), \tag{11.102}$$

where we have considered that all charges are equal and N is the number of particles.

Experimentally, one measures the current due to the total electric field, which is the sum of the external plus the induced internal field. In linear response the current is proportional to the electric field and is given by

$$J_\alpha(\mathbf{r}, t) = \sum_\beta \sigma_{\alpha\beta}(\mathbf{q}, \omega)E_\beta(\mathbf{r}, t), \tag{11.103}$$

taking the electric field as

$$E_\beta(\mathbf{r}, t) = E_\beta(0)e^{i\mathbf{q}\cdot\mathbf{r}-i\omega t} \tag{11.104}$$

and where $\sigma_{\alpha\beta}(\mathbf{q}, \omega)$ is the dynamical conductivity. The conductivity is complex $\sigma_{\alpha\beta}(\mathbf{q}, \omega) = Re\left(\sigma_{\alpha\beta}(\mathbf{q}, \omega)\right) + iIm\left(\sigma_{\alpha\beta}(\mathbf{q}, \omega)\right)$. In general we have that

$$J_\alpha(\mathbf{r}, t) = \sum_\beta \int d^3r' \int_{-\infty}^{\infty} dt' \sigma_{\alpha\beta}(\mathbf{r} - \mathbf{r}'; t - t')E_\beta(\mathbf{r}', t'). \tag{11.105}$$

The average value of the current operator is given by

$$\langle \mathbf{J}(\mathbf{x}, t)\rangle = \langle \mathbf{j}_p(\mathbf{x}, t)\rangle - \frac{Ne^2}{m}\mathbf{A}(\mathbf{x}, t). \tag{11.106}$$

Writing the Hamiltonian as $H + \delta H$, where δH is the term linear in the vector potential, we may write that the average of the paramagnetic current in terms of the equilibrium states (where $\delta H = 0$ and, therefore, we are using the Heisenberg picture) is

$$\begin{aligned}
\langle \mathbf{j}_H(\mathbf{x}, t)\rangle &= \langle U_{H+\delta H}^{-1}\mathbf{j}_S U_{H+\delta H}\rangle \\
&= \langle \bar{U}^{-1}\mathbf{j}_I(\mathbf{x}, t)\bar{U}\rangle \\
&= \left\langle \left(1 + \frac{i}{\hbar}\int_{t_0}^{t} dt'\delta H_I(t')\right)\mathbf{j}_I(\mathbf{x}, t)\left(1 - \frac{i}{\hbar}\int_{t_0}^{t} dt''\delta H_I(t'')\right)\right\rangle \\
&\simeq \langle \mathbf{j}_I(\mathbf{x}, t)\rangle + \left\langle \frac{i}{\hbar}\int_{t_0}^{t} dt'\langle[\delta H_I(t'), \mathbf{j}_I(\mathbf{x}, t)]\right\rangle,
\end{aligned} \tag{11.107}$$

where we have dropped the index p in the paramagnetic current to simplify the notation and have neglected terms of order $(\delta H_I)^2$. Recalling that the averages are taken over the equilibrium states, we write explicitly

$$\mathbf{j}(\mathbf{x},t) = \langle \mathbf{j}_I(\mathbf{x},t)\rangle_{eq} + \frac{i}{\hbar}\int d^3x' \int_{t_0}^{t} dt' \sum_{\beta} A_\beta(\mathbf{x}',t')\langle [j_I^\beta(\mathbf{x}',t'), \mathbf{j}_I(\mathbf{x},t)]\rangle_{eq}. \tag{11.108}$$

Considering that the current vanishes in the equilibrium state and writing

$$J_\alpha(\mathbf{x},t) = \int d^3x' \int_{t_0}^{t} dt' \sum_{\beta} R_{\alpha\beta}(\mathbf{x}-\mathbf{x}'; t-t')A_\beta(\mathbf{x}',t'), \tag{11.109}$$

we obtain the Kubo formula

$$R_{\alpha\beta}(\mathbf{x}-\mathbf{x}'; t-t') = -\frac{i}{\hbar}\left\langle [j_I^\alpha(\mathbf{x},t), j_I^\beta(\mathbf{x}',t')\right\rangle_{eq} - \delta_{\alpha\beta}\frac{Ne^2}{m}\delta(\mathbf{x}-\mathbf{x}')\delta(t-t'). \tag{11.110}$$

A time-varying vector potential generates a time-dependent electric field in the usual way $\mathbf{E} = -\partial\mathbf{A}/\partial t$. Considering a wave of frequency ω, this implies that $\mathbf{E} = i\omega\mathbf{A}$. We can write that the current is given by

$$J_\alpha(\omega) = \sigma_{\alpha\beta}(\omega)E_\beta(\omega) = i\omega\sigma_{\alpha\beta}(\omega)A_\beta(\omega) = R_{\alpha\beta}(\omega)A_\beta(\omega). \tag{11.111}$$

We finally obtain the Kubo formula for the conductivity

$$\sigma_{\alpha\beta}(\mathbf{q},\omega) = \frac{1}{\hbar\omega V}\int_{-\infty}^{t} dt' e^{i\omega(t-t')}\langle [j_\alpha^\dagger(\mathbf{q},t), j_\beta(\mathbf{q},t')]\rangle + i\frac{ne^2}{m\omega}\delta_{\alpha\beta}, \tag{11.112}$$

which can be rewritten as

$$\sigma_{\alpha\beta}(\mathbf{q},\omega) = \frac{1}{\hbar\omega V}\int_{0}^{\infty} dt e^{i\omega t}\langle [j_\alpha^\dagger(\mathbf{q},t), j_\beta(\mathbf{q},0)]\rangle + i\frac{ne^2}{m\omega}\delta_{\alpha\beta}. \tag{11.113}$$

11.3 Exercises

1. Consider the interaction picture. Defining

$$|t\rangle_I = \bar{U}|t\rangle_S = e^{\frac{i}{\hbar}H_{0,S}t}|t\rangle_S,$$

where the Hamiltonian is written in the Schrödinger description as $H_S = H_{0,S} + V_S$, and defining that

$$A_I = \bar{U}A_S\bar{U}^{-1},$$

show that

$$i\hbar\frac{\partial}{\partial t}U_I(t) = V_I(t)U_I(t),$$

with U_I the time evolution operator in the interaction picture defined by $|t\rangle_I = U_I(t)|t_0\rangle_I$.

2. In an X-ray or neutron scattering experiment, a beam of particles with momentum $\mathbf{k}_i$ and energy ϵ_i is impinged on a system and the scattered particles are collected with final momentum $\mathbf{k}_f$ and energy ϵ_f. In this process the dispersed particle of mass M collides with particles or excitations of the system that change their energetic state from E_i to E_f. The cross section of the collision is given by Fermi's golden rule

$$\frac{1}{V}\frac{d^2\sigma(i\to f)}{d\Omega d\omega} = \frac{1}{V}\frac{k_f}{k_i}\left(\frac{M}{2\pi}\right)^2 |\langle k_f f|H_{int}|ik_i\rangle|^2\delta(\hbar\omega - E_f + E_i),$$

where H_{int} is the interaction term between the beam and the system. Consider that

$$H_{int} = \sum_j V(\mathbf{r}-\mathbf{r}_j).$$

(a) Show that

$$\frac{1}{V}\frac{d^2\sigma(i\to f)}{d\Omega d\omega}$$
$$= \frac{1}{V}\frac{k_f}{k_i}\left(\frac{M}{2\pi}\right)^2 |V(\mathbf{k})|^2\delta(\hbar\omega - E_f + E_i)|\langle f|\sum_j e^{-i\mathbf{k}\cdot\mathbf{r}_j}|i\rangle|^2,$$

where $V(\mathbf{k})$ is the Fourier transform of the interaction potential.

(b) Performing the thermodynamic average on the initial states of the solid, i, and carrying out a summation over the final states, f, show that

$$\frac{1}{V}\frac{d^2\sigma(i\to f)}{d\Omega d\omega} = \frac{1}{h}\frac{k_f}{k_i}\left(\frac{M}{2\pi}\right)^2 |V(\mathbf{k})|^2\bar{S}_T(\mathbf{k},\omega),$$

where

$$\bar{S}_T(\mathbf{k},\omega) = \int_{-\infty}^{\infty} dte^{-i\omega t}\langle n(\mathbf{k},0)n(-\mathbf{k},t)\rangle.$$

This factor is called dynamical structure factor. Suggestion: use that

$$h\delta(\hbar\omega - E_f + E_i) = \int dt e^{-i\left(\omega - \frac{E_f - E_i}{\hbar}\right)t},$$

and that

$$n(\mathbf{r}, t) = \sum_{i=1}^{N} \delta(\mathbf{r} - \mathbf{r}_i(t)).$$

3. Show that the Kramers-Kronig relations for a response function $\alpha(\omega) = \alpha'(\omega) + i\alpha''(\omega)$ can be written as

$$\alpha'(\omega) = \frac{2}{\pi} P \int_0^\infty \frac{s\alpha''(s)}{s^2 - \omega^2} ds$$

$$\alpha''(\omega) = -\frac{2\omega}{\pi} P \int_0^\infty \frac{\alpha'(s)}{s^2 - \omega^2} ds.$$

4. The Kubo formula for the dynamic conductivity is given by

$$\sigma_{\alpha\gamma}(\mathbf{q}, \omega) = \frac{1}{\omega V} \int_0^\infty dt e^{i\omega t} \langle [j_\alpha^\dagger(\mathbf{q}, t), j_\gamma(\mathbf{q}, 0)] \rangle + \frac{ne^2}{m\omega} i\delta_{\alpha,\gamma}.$$

Using that

$$j_\alpha^\dagger(\mathbf{q}, t) = e^{iHt} j_\alpha(\mathbf{q}, 0) e^{-iHt}$$

and assuming we know the eigenstates of the Hamiltonian

$$H|n\rangle = E_n|n\rangle,$$

show that the real part of the conductivity can be written as

$$Re\sigma_{\alpha\gamma} = \frac{1}{\omega V Z} \sum_{n,m} \left(e^{-\beta E_n} - e^{-\beta E_m}\right) Re\{\langle n|j_\alpha^\dagger(\mathbf{q}, 0)|m\rangle\langle m|j_\gamma(\mathbf{q}, 0)|n\rangle$$

$$\int_0^\infty dt e^{i(\omega + E_n - E_m)t},$$

where Z is the partition function and β the inverse temperature. Therefore, show that we obtain

$$Re\sigma_{\alpha\gamma}(0, 0) = \frac{\pi\beta}{VZ} \sum_{n,m} e^{-\beta E_n} \langle n|j_\alpha(0, 0)|m\rangle\langle m|j_\gamma(0, 0)|n\rangle \delta(E_n - E_m).$$

Chapter 12
Electromagnetic Response

12.1 Dielectrics and Metals

A simplistic description of the difference between insulators and metals is associated with two possible extreme cases where the electrons are either bound or free to move around the system, respectively. An obvious insulating limit is obtained if we consider a gas of neutral atoms, far from each other, such that all electrons are bound to their nuclei. As the atoms approach each other, one may expect that the wave functions start to overlap and the electrons may be transferred from one atom to another leading to possible extended states, assuming the atoms have some regular distribution in a solid structure (we will focus here on solid systems and consider, for now, periodic structures). We have seen before that the periodicity of the atoms distribution leads to the possibility of bands and gaps between them, originating in the Bragg reflections. A consequence of the gap opening due to the band structure provides a separation between metals and insulators (called band insulators) as the bands are partially filled or completely filled, respectively, as illustrated in Fig. 12.1.

In order to obtain information about any system properties, we must interact with it and study its response. The behavior of the system is affected by changing various external parameters such as temperature, pressure, chemical potential, magnetic field, or electrical field. In this chapter we are interested on the effects of applying an electromagnetic field on a solid. The effects of a static magnetic field will be considered in some detail in the following chapter.

To start with, let us apply a static electric field along a given direction. If the electrons are free, they will accelerate and lead to current transport that increases indefinitely as time goes by (in the ideal case of no collisions) and gives rise to a conductivity, σ, that goes to infinity as $t \to \infty$. If the electrons are bound, the electric field will displace the electrons with respect to the position of the nuclei and will lead to a finite polarization and a dielectric response, characterized by the dielectric constant ϵ. If instead of a static electric field we apply an electric field that is oscillating in time, the electrons will oscillate back and forth, and the difference

P. D. Sacramento, *Introduction to Condensed Matter Theory*,
https://doi.org/10.1007/978-3-031-86251-9_12

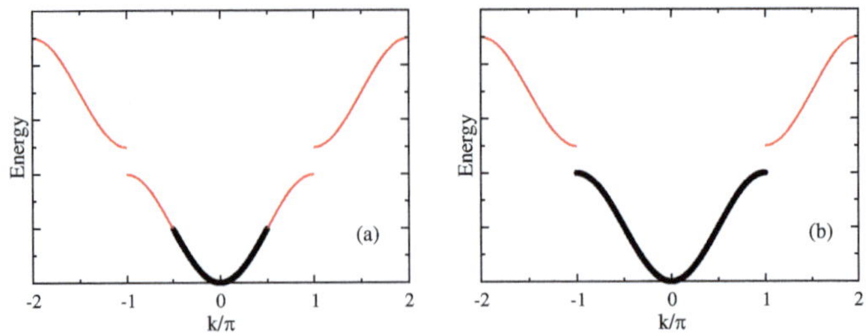

Fig. 12.1 (**a**) The lower band is not full, leading to a metal. (**b**) Lower band is full, and, due to the gap to the next band, the system is a (band) insulator

between the insulator and the conductor becomes less sharp. Both the conductivity and the dielectric constant are now a function of the frequency of oscillation of the electric field and are not fully independent. We may rewrite the conductivity as a function $\sigma(\omega)$ and the dielectric constant as $\epsilon(\omega)$.

Let us recall Maxwell's equations

$$\begin{aligned} \nabla \cdot \mathbf{D} &= \rho \\ \nabla \cdot \mathbf{B} &= 0 \\ \nabla \times \mathbf{H} &= \mathbf{J} + \frac{\partial \mathbf{D}}{\partial t} \\ \nabla \times \mathbf{E} &= -\frac{\partial \mathbf{B}}{\partial t}. \end{aligned} \tag{12.1}$$

These equations are supplemented by the relations

$$\begin{aligned} \mathbf{D} &= \epsilon \mathbf{E} \\ \mathbf{B} &= \mu \mathbf{H} \end{aligned} \tag{12.2}$$

between the electric displacement vector, $\mathbf{D}$, and the electric field, the magnetic induction, $\mathbf{B}$, and the magnetic field, $\mathbf{H}$, where μ is the magnetic permeability. The displacement vector is related to the polarization of the medium as

$$\mathbf{D} = \epsilon_0 \mathbf{E} + \mathbf{P}, \tag{12.3}$$

which implies that the relative dielectric constant is

$$\epsilon_r = \frac{\epsilon}{\epsilon_0} = 1 + \chi_e, \tag{12.4}$$

where we have defined the dielectric susceptibility , χ_e, as $\mathbf{P} = \epsilon_0 \chi_e \mathbf{E}$. In addition, we define the current, $\mathbf{J}$, through the equation that reflects the charge conservation

$$\nabla \cdot \mathbf{J} + \frac{\partial \rho}{\partial t} = 0 \tag{12.5}$$

and in a normal metal we have Ohm's law $\mathbf{J} = \sigma \mathbf{E}$.

A plane wave is a solution of Maxwell's equations. Choosing $\mathbf{E} = \mathbf{E}_0 e^{i\mathbf{k}\cdot\mathbf{r} - i\omega t}$ we can write that

$$\nabla \times \mathbf{H} = (\sigma - i\omega\epsilon_0)\,\mathbf{E} = -i\omega\epsilon_0 \left(1 + i\frac{\sigma}{\epsilon_0\omega}\right)\mathbf{E}. \tag{12.6}$$

In this expression we have neglected the polarization part by assuming that $\chi_e \sim 0$. Therefore, in a simplified way we may consider that the system is a dielectric with a complex frequency-dependent dielectric constant, written as

$$\epsilon_r(\omega) = 1 + i\frac{\sigma}{\epsilon_0\omega}, \tag{12.7}$$

where the imaginary part is proportional to the conductivity (considered here frequency independent), or, equivalently, we may consider the system as a conductor with a complex frequency-dependent conductivity given by

$$\sigma(\omega) = -i\omega\epsilon_0\,(\epsilon_r(\omega) - 1)\,. \tag{12.8}$$

Recall now that the index of refraction is defined as $n = c/v$, where v is the light velocity in a given medium and c the velocity in vacuum. Using that $v = \omega/k$ we obtain that

$$n(\omega) = \frac{ck}{\omega}. \tag{12.9}$$

Recalling that $n = \sqrt{\epsilon_r\mu_r}$, $(\mu_r = \mu/\mu_0)$ we get that the momentum is given by

$$k^2 = \frac{\mu_r\omega^2}{c^2}\epsilon_r(\omega) = \frac{\mu_r\omega^2}{c^2}\left(1 + i\frac{\sigma}{\epsilon_0\omega}\right). \tag{12.10}$$

This implies that the momentum has a real and an imaginary part, $k = \alpha + i\beta$, and that in general any wave will be attenuated in the material.

Using that $k^2 = \omega^2\mu_r\epsilon_r(\omega)/c^2$ we may conclude the following:

- If ϵ is real (which happens in general if $\sigma = 0$) and if $\epsilon > 0$, then k is real and a wave propagates inside the material. If $\epsilon < 0$ then k is imaginary and the wave is attenuated exponentially away from the surface.
- If ϵ is complex, in general waves are attenuated.
- If $\epsilon = 0$, for which $\chi_e = -1$, the effective potential due to a charge diverges. Recall that inside a dielectric, an external potential, for instance due to a test charge, is renormalized as V_{ext}/ϵ. Therefore, if ϵ vanishes the effective potential

diverges. The physical consequence is that, even if the potential is vanishingly small, it is possible to have a finite effective potential. This means an instability of the system and leads to normal oscillation modes.

- If $\epsilon \rightarrow \infty$ the effective potential vanishes. Considering the potential on one charge as due to neighboring charges (such as on one atom due to the neighboring atoms), the forces between the atoms become vanishingly small. This implies the possibility of a structural instability.

Simple Microscopic Model

A simple microscopic model used to calculate the dielectric constant consists in regarding the motion of the electrons, taken as independent particles of mass m and charge $-e$, as that of forced and damped harmonic oscillators. The classical (one-dimensional-like) equation of motion for each electron is written as

$$m\left(\frac{d^2}{dt^2}+\lambda\frac{d}{dt}+\omega_0^2\right)x=F. \tag{12.11}$$

Here λ is the damping coefficient, ω_0 is the frequency of the oscillator, x is the electron displacement from its equilibrium position, and F is the external force applied along a given direction, here the x-axis. Looking for a solution of the type $e^{-i\omega t}$ we find that

$$m\left(-\omega^2-i\lambda\omega+\omega_0^2\right)x_\omega=F_\omega. \tag{12.12}$$

Writing that $x_\omega=\alpha(\omega)F_\omega$, where α is the polarizability, we get that

$$\alpha(\omega)=\frac{1}{m\left(\omega_0^2-\omega^2-i\omega\lambda\right)}. \tag{12.13}$$

Using this result we may obtain an expression for the dielectric constant:

$$\epsilon(\omega)=1+\frac{P_\omega}{\epsilon_0E_\omega}=1-\frac{nex_\omega}{\epsilon_0E_\omega}=1+\frac{ne^2}{\epsilon_0}\alpha(\omega), \tag{12.14}$$

where n is the electron density. Therefore,

$$\epsilon_r(\omega)=1+\frac{\omega_p^2}{\omega_0^2-\omega^2-i\omega\lambda}, \tag{12.15}$$

where

$$\omega_p^2=\frac{ne^2}{\epsilon_0m} \tag{12.16}$$

is the plasma frequency, as we will explain later.

Separating the dielectric constant in real and imaginary parts $\epsilon(\omega) = \epsilon'(\omega) + i\epsilon''(\omega)$ we get that

$$\begin{aligned}
\epsilon_r'(\omega) &= 1 + \omega_p^2 \frac{\left(\omega_0^2 - \omega^2\right)}{\left(\omega_0^2 - \omega^2\right)^2 + \omega^2\lambda^2} \\
\epsilon_r''(\omega) &= \omega_p^2 \frac{\omega\lambda}{\left(\omega_0^2 - \omega^2\right)^2 + \omega^2\lambda^2}.
\end{aligned} \tag{12.17}$$

The same result is obtained if we consider that ω_0 is the gap for transitions between bands.

Using the simplified relation between the conductivity and the dielectric constant we obtain

$$\begin{aligned}
\sigma(\omega) &= -i\omega\epsilon_0 \left(\epsilon_r(\omega) - 1\right) \\
&= -i\omega\epsilon_0 \frac{\omega_p^2}{\omega_0^2 - \omega^2 - i\omega\lambda}.
\end{aligned} \tag{12.18}$$

Separating the conductivity in the real and imaginary parts $\sigma(\omega) = \sigma'(\omega) + i\sigma''(\omega)$ we obtain

$$\begin{aligned}
\sigma'(\omega) &= \epsilon_0\omega^2\lambda \frac{\omega_p^2}{\left(\omega_0^2 - \omega^2\right)^2 + \omega^2\lambda^2} \\
\sigma''(\omega) &= -\epsilon_0\omega\omega_p^2 \frac{\left(\omega_0^2 - \omega^2\right)}{\left(\omega_0^2 - \omega^2\right)^2 + \omega^2\lambda^2}.
\end{aligned} \tag{12.19}$$

Let us now consider that the electrons are free. This can be particularized taking $\omega_0 = 0$ (no harmonic force on the electron). In this case we get that

$$\begin{aligned}
\alpha(\omega) &= \frac{1}{m\omega} \frac{-1}{\omega + i\lambda} \\
\epsilon_r'(\omega) &= 1 - \omega_p^2 \frac{1}{\omega^2 + \lambda^2} \\
\epsilon_r''(\omega) &= \frac{\omega_p^2}{\omega} \frac{\lambda}{\omega^2 + \lambda^2} \\
\sigma'(\omega) &= \epsilon_0\lambda \frac{\omega_p^2}{\omega^2 + \lambda^2} \\
\sigma''(\omega) &= \epsilon_0\omega\omega_p^2 \frac{1}{\omega^2 + \lambda^2}.
\end{aligned} \tag{12.20}$$

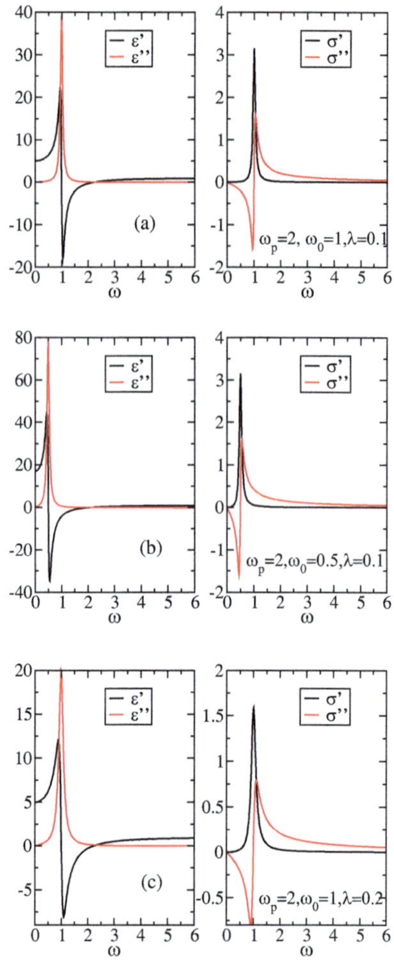

Fig. 12.2 Real and imaginary parts of relative dielectric function and conductivity as a function of frequency, for (**a**) $\omega_p = 2, \omega_0 = 1, \lambda = 0.1$, (**b**) same parameters but $w_0 = 0.5$, and (**c**) same parameters but $\lambda = 0.2$. The conductivity is multiplied by $4\pi/\epsilon_0$

Note that λ has dimensions of frequency. Therefore, it may also be written as the inverse of a time scale, as $\lambda = 1/\tau$. This time scale has the interpretation of an average time between collisions that lead to damping of the electrons. Several examples for the real and imaginary parts of the dielectric function and conductivity are shown in Figs. 12.2 and 12.3.

An alternative way to present the expression for the conductivity, in terms of the scattering time between collisions, is

$$
\begin{aligned}
\sigma'(\omega) &= \sigma_0 \frac{1}{1+(\omega\tau)^2} \\
\sigma''(\omega) &= \sigma_0 \frac{\omega\tau}{1+(\omega\tau)^2},
\end{aligned}
\tag{12.21}
$$

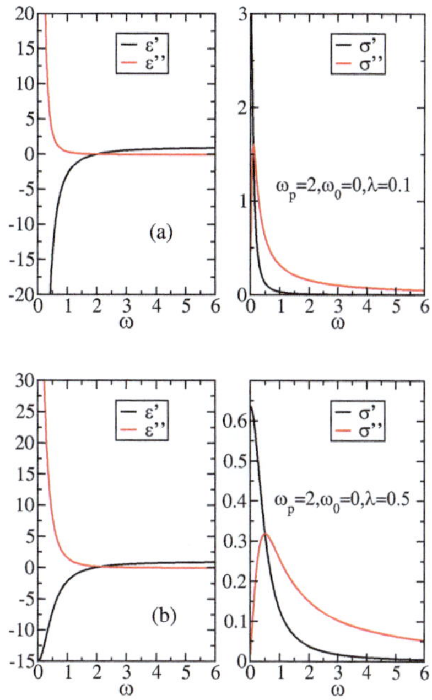

Fig. 12.3 Real and imaginary parts of relative dielectric function and conductivity as a function of frequency, for (**a**) $\omega_p = 2$, $\omega_0 = 0$, $\lambda = 0.1$ and (**b**) same parameters but $\lambda = 0.5$. The conductivity is multiplied by $4\pi/\epsilon_0$

where

$$\sigma_0 = \frac{ne^2\tau}{m} \tag{12.22}$$

is the d.c. (zero frequency) conductivity of a gas of non-interacting free electrons subject to a drag force due to collisions with imperfections, phonons, etc. This term is called Drude conductivity. Using Eqs. (12.21) and (12.22) we can show the sum rule

$$\int_0^\infty d\omega\sigma'(\omega) = \frac{\pi ne^2}{2m}. \tag{12.23}$$

Matthiessen's Rule

Experimentally it is found that the contributions to the (inverse) average time between collisions just add in an appropriate manner. Specifically, Matthiessen's rule states that the contributions from impurities or imperfections, τ_i, just add to the phonon contribution, τ_p,

$$\frac{1}{\tau} = \frac{1}{\tau_i} + \frac{1}{\tau_p}. \tag{12.24}$$

Since the resistivity is the inverse of the conductivity, we may also conclude that

$$\rho = \rho_i + \rho_p, \tag{12.25}$$

for the impurity and phonon contributions. The impurity/imperfections contribution is essentially temperature independent at low temperatures, except if the impurities are magnetic leading to the Kondo effect, as will be discussed later. On the other hand, the phonon contribution depends on temperature.

Let us consider the phonon contribution and consider first that the temperature is small. In this case only phonons of small energies are excited, so ω is small and so is the momentum q. The scattering of one electron with momentum k_F, along a given direction, off a phonon, leads to a deviation of the electron motion to a direction that makes an angle θ with the incident direction, say, the direction of the applied electric field that gives rise to the current in the metal. The deviation from the incident direction implies that the projection of the momentum of the electron along the incident direction decreases by $\delta k_F = k_F(1-\cos\theta)$. Since the momentum of the phonon is small, the deviation can be approximated by $\hbar\delta k_F \sim \hbar k_F\theta^2/2$. In each collision the momentum decreases a fraction $\theta^2/2$. We may estimate that after r collisions the momentum projection vanishes leading to

$$\hbar k_F\left(1-\frac{\theta^2}{2}\right)^r \sim \hbar k_F\left(1-r\frac{\theta^2}{2}\right) \sim 0. \tag{12.26}$$

Since $\theta \sim q/k_F$ we get that

$$r \sim \frac{2k_F^2}{q^2}. \tag{12.27}$$

The energy of the phonons obeys $\hbar\omega = \hbar v_s q \sim k_B T$ which implies that $r \sim T^{-2}$. The number of collisions with the phonons is proportional to their density,

$$n_p \sim \int_0^{\omega_D} d\omega \frac{w^2}{e^{\frac{\hbar\omega}{k_B T}}-1}. \tag{12.28}$$

At small temperatures $t << \theta_D$ we get that

$$n_p \sim T^3. \tag{12.29}$$

We can now conclude that

$$\frac{1}{\tau} \sim \frac{n_p}{r} \sim T^5 \tag{12.30}$$

and, therefore, the phonon contribution to the resistivity at small temperatures is

$$\rho_p \sim \frac{1}{\tau} \sim T^5. \tag{12.31}$$

At higher temperatures, we get that the phonon density is of the order of the temperature. Also, at higher temperatures the phonon momentum is high, which means that, on average, one or few collisions are enough to produce a significant change of the momentum of the incident electron. So we may take $r \sim 1$. This leads to the result that, at high temperatures,

$$\rho_p \sim \frac{1}{\tau} \sim n_p \sim T. \tag{12.32}$$

12.1.1 Ideal Conductor: Drude Peak and Plasma Frequency

A very pure material is a system for which the friction coefficient is very small (small λ) or the time between collisions is very long (large τ). A good conductor is therefore approximated if we consider $\lambda \to 0$ (or $\tau \to \infty$). It is convenient to consider this limit in the expression of the polarizability. Using that

$$\lim_{\lambda\to 0} \frac{1}{\omega + i\lambda} = P\left(\frac{1}{\omega}\right) - i\pi\delta(\omega), \tag{12.33}$$

we obtain that the polarizability is given by

$$\alpha(\omega) = -\frac{1}{m\omega}\left(P\left(\frac{1}{\omega}\right) - i\pi\delta(\omega)\right). \tag{12.34}$$

The conductivity is given by

$$\begin{aligned} \sigma'(\omega) &= \epsilon_0 \pi \omega_p^2 \delta(\omega) \\ \sigma''(\omega) &= \epsilon_0 \frac{\omega_p^2}{\omega} \end{aligned} \tag{12.35}$$

and the relative dielectric constant by

$$\begin{aligned} \epsilon_r'(\omega) &= 1 - \frac{\omega_p^2}{\omega^2} \\ \epsilon_r''(\omega) &= \frac{\pi\omega_p^2}{\omega}\delta(\omega). \end{aligned} \tag{12.36}$$

Several comments should be made at this point:

- Consider first the real part of the conductivity. Equation (12.20) shows that, if the damping term is finite (λ finite), the real part has a Lorentzian shape centered at zero frequency and with a width given by λ. This is called the Drude peak of the conductivity and is a characteristic of a conductor. As the material becomes a better metal and $\lambda \to 0$, the peak sharpens around zero frequency and, in the ideal limit, becomes a Dirac-$\delta(\omega)$ infinite d.c. conductivity, as expected of a pure conductor.
- In the same limit of a very pure conductor, the real part of the dielectric function Eq. (12.36) has a singular behavior. Near zero frequency it diverges, but it is negative. As it turns out, it is negative for frequencies $\omega < \omega_p$. Since $\epsilon < 0$, this implies that any wave of this frequency range, when incident on the system, is totally reflected (meaning that it is attenuated exponentially inside the material). In the frequency range $\omega > \omega_p$ the dielectric function is positive, and since in the limit $\lambda \to 0$ the imaginary part vanishes, waves propagate in the material. The frequency point $\omega = \omega_p$ is special since $\epsilon = 0$. This corresponds to oscillatory normal modes called plasma oscillations. Considering that the nuclei are fixed (due to their large mass and interactions that lead to the regular structure of the solid), only the electrons oscillate around their average positions. This displacement leads to the polarization that ultimately leads to a non-trivial dielectric function. As a result, we have a relative motion of the negative charges with respect to the positive charges and the origin of the plasma oscillations.

12.1.2 Atomic Contribution to Dielectric Function

In general, in a dielectric there are different types of oscillatory motion, with different natural frequencies, ω_j. Each oscillator contributes additively to the polarization of the system, leading straightforwardly to a dielectric function of the form

$$\epsilon_r(\omega) = 1 + \frac{ne^2}{\epsilon_0 m} \sum_j \frac{f_j}{\omega_j^2 - \omega^2 - i\omega\lambda_j}, \tag{12.37}$$

where λ_j is the damping coefficient of oscillator j and f_j is the number of electrons with natural frequency ω_j. Considering that there are z electrons per molecule (and n is the number of molecules per unit volume), there is a sum rule $\sum_j f_j = z$. A schematic representation with all $f_j = 1$ is shown in Fig. 12.4.

We may consider that in general the damping coefficients are small $\lambda_j \sim 0$. Therefore, $\epsilon_r(\omega)$ has a small imaginary part. Neglecting the imaginary part if the frequency is not very close to one of the resonant natural frequencies, the fraction

$$\frac{1}{\omega_j^2 - \omega^2}$$

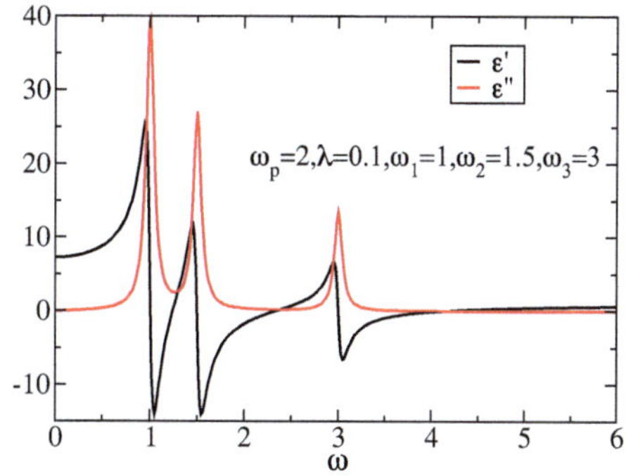

Fig. 12.4 Real and imaginary parts of relative dielectric function as a function of frequency, for $\omega_p = 2$, $\lambda = 0.1$, for a sequence of values of ω_0

is positive if $\omega < \omega_j$ and is negative if $\omega > \omega_j$. Therefore, close to the resonance, $\omega = \omega_j$, the fraction changes sign from positive to negative, as the frequency increases. For very small frequency, smaller than any of the natural frequencies of the oscillators, $\epsilon_r(\omega) > 1$. As the frequency increases, the frequency ω will cross each natural frequency in turn. Each time a crossing occurs negative contributions appear that decrease the dielectric function. When $\omega > \omega_j$ for all j, the sum is necessarily negative and $\epsilon_r(\omega) < 1$. For large enough frequency we get that

$$\epsilon_r(\omega) = 1 - \frac{\omega_p^2}{\omega^2}, \tag{12.38}$$

where

$$\omega_p^2 = \frac{nze^2}{\epsilon_0 m} \tag{12.39}$$

is the plasma frequency. Using that the refraction index $n(\omega) = \sqrt{\epsilon_r(\omega)}$, where we have considered that $\mu_r \sim 1$, and replacing Eq. (12.38) we can write that

$$\omega n(\omega) = \sqrt{\omega^2 - \omega_p^2}. \tag{12.40}$$

Since $n(\omega) = ck/\omega$ we get the dispersion relation

$$\omega^2 = \omega_p^2 + c^2k^2. \tag{12.41}$$

Since ω_p is finite, this dispersion relation has a (large) gap. These excitations require a large energy to be excited of the order of several electron Volt.

If the frequency is very close to one of the natural frequencies of the oscillators, the damping cannot be neglected. At the resonance energy the contribution to the dielectric function is purely imaginary and the imaginary part has a peak. This signals energy dissipation, and, since the imaginary part is positive, it signals resonant absorption. Clearly, if one of the natural frequencies is vanishingly small, we recover the Drude peak found previously.

In general the dielectric function is obtained as the sum of various contributions: the electronic, ionic, and dipolar. The electronic and ionic contributions are of similar type. The ionic contribution arises since the inner electronic orbitals are also polarizable and the ions are not strictly static. The dipolar contribution has its origin in the possible existence of permanent dipolar moments in an atom or molecule. Their motion in the presence of a perturbation is different, if we assume that the dipolar moment has a fixed magnitude and at most rotates rigidly in the presence of an orienting electric field. In this case its contribution to the real part of the dielectric function has a maximal value at zero frequency and then it decreases as frequency increases. There is no plasma oscillation mode in this case, and therefore the characteristic resonance through the natural oscillator frequency is absent.

12.2 Transverse and Longitudinal Modes

An electromagnetic wave is a transverse mode of propagation. From the wave equation we have that

$$\begin{aligned} \frac{\partial^2 \mathbf{D}}{\partial t^2} &= \frac{1}{\mu}\nabla^2 \mathbf{E} \\ \mathbf{E} &\sim e^{-i\omega t} e^{i\mathbf{k}\cdot\mathbf{r}} \\ \mathbf{D} &= \epsilon(\omega, \mathbf{k})\mathbf{E}. \end{aligned} \tag{12.42}$$

This leads to

$$c^2 k^2 = \epsilon_r(\omega, \mathbf{k})\mu_r \omega^2. \tag{12.43}$$

This is equivalent to the previous relation $\omega n(\omega) = ck$ (since $n(\omega) = \sqrt{\epsilon_r(\omega)\mu_r}$).

Let us consider the plasma of the electrons and ions (considered static) in a solid. The ions contribute to the dielectric constant and change the vacuum contribution ($\epsilon = 1$) to a different value, call it $\epsilon(\infty)$. The dielectric function of the electrons in the presence of the ions is therefore corrected to

$$\epsilon(\omega) = \epsilon(\infty) - \frac{ne^2}{m\omega^2}. \tag{12.44}$$

Therefore, we get that

$$\epsilon(\omega) = \epsilon(\infty)\left(1 - \frac{\tilde{\omega}_p^2}{\omega^2}\right), \tag{12.45}$$

where the plasma frequency is changed to

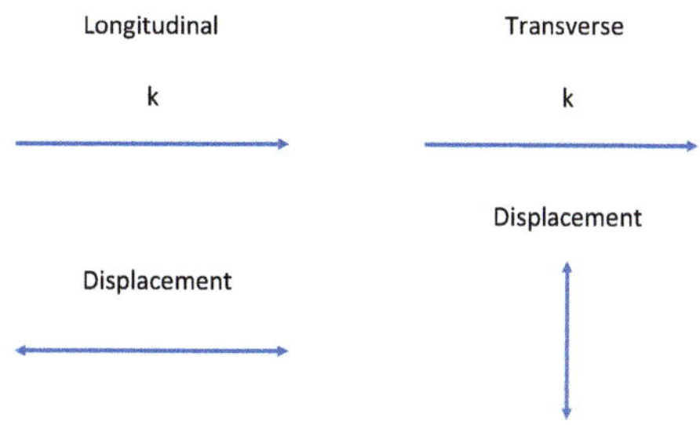

Fig. 12.5 Longitudinal and transverse modes of the charge displacements

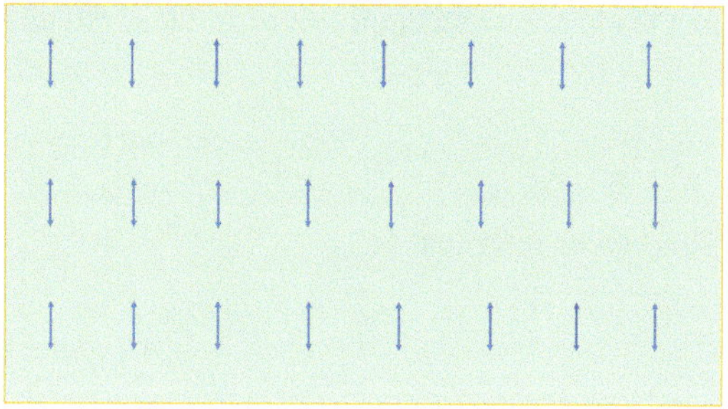

Fig. 12.6 Charge longitudinal collective displacements (plasmons)

$$\tilde{\omega}_p^2 = \frac{ne^2}{\epsilon(\infty)m}. \tag{12.46}$$

As before, if the frequency is smaller than the plasma frequency the wave is fully reflected since the momentum becomes purely imaginary ($k^2 < 0$). The dispersion relation of the transverse wave is given by

$$\omega^2 = \tilde{\omega}_p^2 + \frac{c^2k^2}{\epsilon_r(\infty)}. \tag{12.47}$$

It is also possible to have longitudinal excitations in the system, as illustrated in Fig. 12.5 and detailed in Fig. 12.6. These will be discussed next.

12.2.1 Plasmons

The normal longitudinal modes are given by the zeros of the dielectric function. Since they are spontaneous oscillations, they do not require the stimulus of an external field and are the result of mutual interactions between the plasma constituents. In this context this is achieved at the frequency $\omega_L = \tilde{\omega}_p$. A straightforward picture is obtained considering a solid where the ions are static at given positions (such as the crystal structure of the material) and where, at an initial stage, the electrons' average positions correspond to centers of local neutrality. Assume some internal perturbation that induces a displacement of the electrons from their equilibrium average positions. To simplify, we neglect here spatial modulations and assume a zero momentum, uniform displacement of the electrons. We can view the system as two slabs, one charged positively with the ions and one charged negatively with the electrons that are displaced with respect to one another. It is convenient to think of both slabs as having finite thickness. The relative displacement induces a surface charge, for instance in the ions, since they are assumed static. The surface charges have different signs and are given by $\sigma = \pm neu$, where u is the displacement of the electron slab with respect to the ion slab. As a consequence, an electric field arises inside the slab, $E = 4\pi neu$. This electric field acts on the electrons with a force that tends to restore equilibrium and leads to an equation of motion

$$nm\frac{d^2u}{dt^2} = -neE = -4\pi n^2e^2u. \tag{12.48}$$

This can be rewritten as

$$\frac{d^2u}{dt^2} + \omega_p^2 u = 0, \tag{12.49}$$

where ω_p is the plasma frequency. Clearly the solution is an oscillatory motion around the equilibrium position. Also, it is a longitudinal wave. It is also clear that, since $\epsilon = 0$ this implies $\mathbf{D} = 0$ and a field that opposes the polarization $\mathbf{E} = -\mathbf{P}/\epsilon_0$, as expected.

These modes may be quantized and are called plasmons. Their dispersion relation will be obtained later on.

12.3 Screening

If we insert a test charge in a collection of electric charges, the system responds to this perturbation through a polarization that creates a field that adds up to the external field. The total field acts locally to create the polarization, and, therefore, the solution of the problem is obtained self-consistently: the system of charges is polarized due to the total local potential which is the sum of the external field and the field due to the polarization. The local potential is given by the external potential divided by the dielectric function.

Consider an external potential of the type $\delta V_{ext}(\mathbf{q}) = Ve^{i\mathbf{q}\cdot\mathbf{r}}$. We consider a static perturbation (zero frequency) or a perturbation that acts very slowly on the system of charges. A similar form is assumed for the polarization $\delta\Phi(\mathbf{q}) = \Phi e^{i\mathbf{q}\cdot\mathbf{r}}$ and the total potential $\delta U(\mathbf{q}) = ue^{i\mathbf{q}\cdot\mathbf{r}}$. Since the dielectric function relates the external with the local potential, we can write that

$$\delta U(\mathbf{q}) = \frac{\delta V_{ext}(\mathbf{q})}{\epsilon_r(\mathbf{q}, \omega = 0)}. \tag{12.50}$$

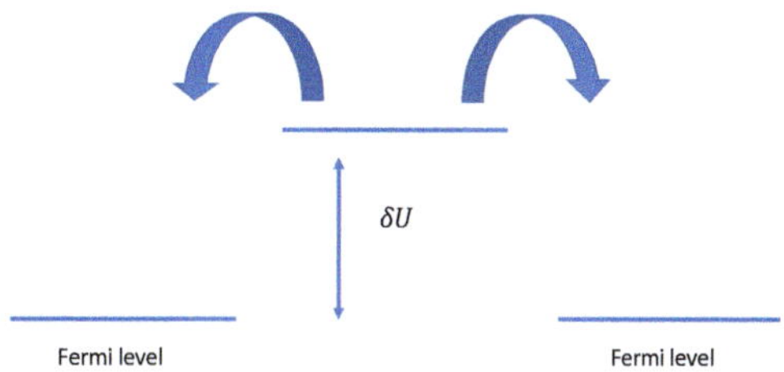

Fig. 12.7 A local perturbation of amplitude δU changes the Fermi level leading to a charge depletion due to the electron motion to equilibrate the chemical potential

12.3.1 Thomas-Fermi Approximation

We further simplify the perturbation assuming that in a small region the potential energy is changed with respect to neighboring regions.[1] Therefore, the chemical potential is changed locally. Assume this change is positive. Since the perturbation is very slow, there is a flow of charge from this region to the neighboring regions to equilibrate the chemical potential, as shown schematically in Fig. 12.7. As a consequence there is an electron charge depletion at the same region. The depletion in the electron density is given approximately by

$$\delta n(\mathbf{r}) = -D(\epsilon_F)\delta U(\mathbf{r}), \tag{12.51}$$

where $D(\epsilon_F)$ is the density of states per volume of the electronic system, at the Fermi level. This is the Thomas-Fermi approximation. A change in the electronic density induces a change in the potential energy via Poisson's equation

$$\nabla^2 (\delta\Phi) = -\frac{1}{\epsilon_0}(-e)\delta\rho, \tag{12.52}$$

where $\delta\rho = -e\delta n(\mathbf{r})$. Therefore,

$$\nabla^2 (\delta\Phi) = -\frac{1}{\epsilon_0}e^2\delta n(\mathbf{r}) = \frac{1}{\epsilon_0}e^2 D(\epsilon_F)\delta U(\mathbf{r}) = q_s^2\delta U(\mathbf{r}). \tag{12.53}$$

Using the plane wave solution we get that

$$-q^2\delta\Phi = q_s^2\delta U(\mathbf{r})$$

$$\delta\Phi = -\frac{q_s^2}{q^2}\delta U$$

$$\delta U = V + \delta\Phi = V - \frac{q_s^2}{q^2}\delta U$$

[1] See also [Kittel].

$$\delta U\left(1+\frac{q_s^2}{q^2}\right) = V$$

$$\delta U = \frac{V}{1+\frac{q_s^2}{q^2}}. \tag{12.54}$$

This final result implies that the dielectric function is given, in this limit, by

$$\epsilon_r(q) = 1+\frac{q_s^2}{q^2}, \tag{12.55}$$

where $q_s^2 = e^2 D(\epsilon_F)/\epsilon_0$.

Consider now a test charge, such as an electron, at a given point in space. The potential exerted on any other electron at a distance r from the test charge is given by

$$\delta V = \frac{e^2}{4\pi\epsilon_0}\frac{1}{r}. \tag{12.56}$$

Taking the Fourier transform gives the potential

$$\delta V(\mathbf{q}) = \frac{e^2}{\epsilon_0 q^2}. \tag{12.57}$$

The total potential is given by

$$\delta U(\mathbf{q}) = \frac{\delta V(\mathbf{q})}{\epsilon_r(\mathbf{q},0)} = \frac{e^2}{\epsilon_0 q^2}\frac{1}{1+\frac{q_s^2}{q^2}}. \tag{12.58}$$

This leads to the results

$$\delta U(\mathbf{q}) = \frac{e^2}{\epsilon_0}\frac{1}{q^2+q_s^2}$$

$$\delta U(\mathbf{r}) = \frac{e^2}{4\pi\epsilon_0}\frac{e^{-q_s r}}{r}. \tag{12.59}$$

The effective potential is therefore not Coulombic. It does not decrease as the inverse of distance but much faster, exponentially with distance. $1/q_s$ (screening length) gives the distance scale over which the potential decays. This is the screening effect one observes in a conductor. For instance in copper the screening length if of the order of $0.55\mathring{A}$ and in Al it is of the same order. This is rather small distances, of the order of the Bohr radius, which means screening is very effective in reducing the spatial extent of the Coulomb interaction. It turns out that in an insulator the behavior is different, as will be shown later. In an insulator the decay is Coulombic.

12.3.2 Mott Transition

Consider a system of protons and an equal number of electrons and single out one of these electrons. The potential energy between each proton and this electron is screened and of the type

$$U(r) = -\frac{e^2}{4\pi\epsilon_0 r}e^{-q_s r}, \tag{12.60}$$

where r is the distance between the electron and a given proton. Recall that $q_s^2 = e^2 D(\epsilon_F)/\epsilon_0 = 3.939n^{1/3}/a_0$, where n is the electronic density and a_0 is the Bohr radius. Note that as the density increases, q_s also increases. If q_s is large, the potential energy is small, unless r is small. Therefore, the potential is short ranged. As a consequence, even though attractive it may not lead to a bound state between the electron and the proton. Therefore, the electron may be delocalized since the proton is not able to bind the electron. In three dimensions a minimum attractive potential is needed to create a bound state. It can be shown that there is a bound state that is $q_s < 1.19/a_0$. If the bound state occurs, then the system is expected to be an insulator: the electrons will be bound to their respective protons. A sufficiently small q_s occurs if the density is small enough, or the atoms are sufficiently far apart. Clearly, if the atoms are very far from each other, the system is insulator. As the atoms approach, their relative distances decrease, the density increases, and, assuming a simple cubic structure ($n = 1/a^3$), the system will have a transition from insulator to a metal when the distance between the atoms (lattice constant) is $a < a_c = 2.78a_0$. A more detailed calculation leads Mott to the prediction that $a_c = 4.5a_0$.

12.3.3 Screening and Phonons

Phonons were introduced before as the vibrational modes of the atoms of a lattice structure, which result from harmonic motions around equilibrium positions. The model used considered that the atoms are connected by springs, with given force constants that are adjusted phenomenologically, for instance measuring the sound velocity, or different types of deformations of the solid. Ultimately, the interactions between the atoms have an electromagnetic origin. We will next try to establish a connection between the dielectric response of a solid and its vibrations.[2]

Consider, to simplify, longitudinal acoustic phonons. As discussed above, the longitudinal normal modes are obtained from the zeros of the real part of the dielectric function $\epsilon(\omega_L) = 0$. We will consider the total dielectric function due to the sum of the electronic and ionic contributions. We will consider the

[2] See also [Kittel].

adiabatic approximation since the ions move slowly with respect to the electrons. The interactions between the charges of the ions are screened by the electrons. We will use the Thomas-Fermi approximation for the dielectric function of the electrons

$$\epsilon_{r,e}(\omega, \mathbf{k}) = 1 + \frac{q_s^2}{k^2}, \tag{12.61}$$

We consider that the ions are far from each other and that they move independently. We use $\epsilon(\omega, 0)$ with the mass of the ions, M, leading to the total dielectric function

$$\epsilon_r(\omega, k) = 1 - \frac{ne^2}{\epsilon_0 M\omega^2} + \frac{q_s^2}{k^2}. \tag{12.62}$$

We are interested in studying the regime where both the frequency and momenta are small, where the dispersion relation of the phonons is linear with a slope given by the sound velocity. In the regime where both k and ω are small, we get that the normal modes are obtained imposing

$$\epsilon_r(\omega, k) \sim -\frac{ne^2}{\epsilon_0 M\omega^2} + \frac{q_s^2}{k^2} = 0. \tag{12.63}$$

More specifically, multiplying Eq. (12.62) by ω^2, we can write that

$$\omega^2 = \frac{ne^2}{\epsilon_0 M}\frac{k^2}{k^2 + q_s^2}. \tag{12.64}$$

Taking small wave vector, $k << q_s$, this gives the dispersion relation

$$\omega^2 = \frac{ne^2}{\epsilon_0 M q_s^2}k^2 = \frac{m}{3M}v_F^2 k^2. \tag{12.65}$$

Therefore, the sound velocity is given by

$$v_s = \left(\frac{m}{3M}\right)^{1/2} v_F. \tag{12.66}$$

There is a reasonably good agreement with the experimental result for the sound velocity of simple metals, like the alkalines. For instance, for potassium the calculated value is about $v_s = 1.8 \times 10^5$ cm/s, while the experimental value is 2.2×10^5 cm/s.

At high frequencies the Thomas-Fermi approximation for the electrons is not good. It is a better approximation to write in the zero momentum limit

$$\epsilon_r(\omega, 0) = 1 - \frac{ne^2}{\epsilon_0 M\omega^2} - \frac{ne^2}{\epsilon_0 m\omega^2}. \tag{12.67}$$

The eigenmodes are then just the plasma frequency where the electronic mass is replaced by the reduced mass of the electron and the ion.

12.4 Polaritons

As we have seen previously, if the unit cell of a lattice has more than one atom, in general there are, in addition to the acoustic phonons, higher energy branches called optical phonons. Considering two atoms per cell, while the acoustic phonons describe in phase motion of the atoms of the unit cell, the optical phonons describe motion where the atoms oscillate with opposite phases. Both the acoustical and the optical phonons in general have longitudinal and transverse modes.

If electromagnetic waves are incident on the material, one expects that they may polarize the ions and excite transverse oscillation modes. Since the speed of light is very high and much higher than the speed of sound, the dispersion relation $\omega = ck$ will not intersect the acoustical branch. We expect, therefore, that the photon will interact mainly with the optical phonon, since it will become degenerate with this mode at very small momentum but finite energy. This will lead to some perturbation of the motion of the ions, which requires degenerate perturbation theory involving two modes, and leads to the opening of a gap (not related to the periodicity of the lattice). The eigenmodes that result from the coupling of the photons and the phonons are called polaritons (Fig. 12.8).

It is actually easier to consider this problem classically, in the sense that electromagnetism contains the root for this coupling.[3] Recall that we can write that

$$c^2k^2E = \omega^2 \left(E + P/\epsilon_0\right). \tag{12.68}$$

Let us call ω_T the zero momentum energy of the transverse optical phonon. In general, the dispersion relation near zero momentum is flat, and, since the

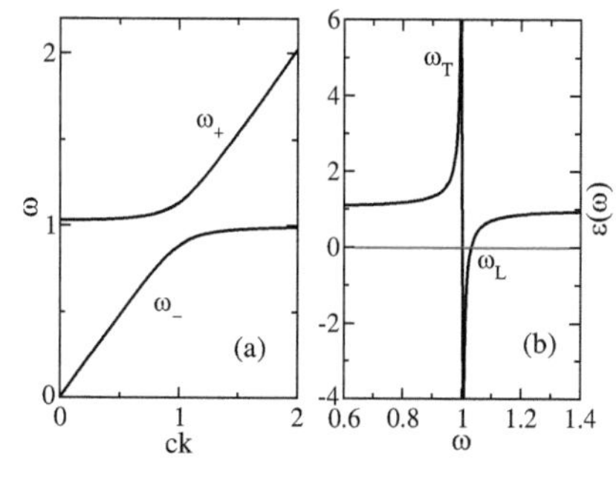

Fig. 12.8 (**a**) Dispersion relations of the polariton modes. (The frequency is in units of ω_T). (**b**) Dielectric function as a function of frequency. In the gap between ω_- and ω_+ the dielectric function is negative, corresponding to the regime of no transmissivity of the photons

[3] See also [Kittel].

momentum region where the coupling occurs is very narrow, we may consider that the frequency is a constant and independent of momentum, k. If some force is applied on the atoms, these will oscillate around their equilibrium positions giving rise to some polarization, $P = nqu$, where u is the displacement. The equation of motion for the polarization is of the type

$$\frac{d^2P}{dt^2} + \omega_T^2 P = F,$$

(where F is the force) which leads to

$$-\omega^2 P + \omega_T^2 P = \frac{nq^2}{M} E. \tag{12.69}$$

Here n is the number of pairs of ions of effective charge q and reduced mass, M, per unit volume. Let us neglect the electronic contribution to the polarization and focus on the ionic contribution. Putting together Maxwell's equation for the polarization and the forced harmonic oscillator equation we get that

$$\begin{aligned}
-\left(c^2k^2 - \omega^2\right) E + \frac{\omega^2}{\epsilon_0} P &= 0 \\
\frac{nq^2}{M} E - \left(\omega_T^2 - \omega^2\right) P &= 0.
\end{aligned} \tag{12.70}$$

A non-trivial solution is obtained if the determinant of the coefficients of the electric field and the polarization vanishes. (We may think of the electric field as the "photon" and the polarization as the "phonon.") The determinant is given by

$$det \begin{pmatrix} \omega^2 - c^2k^2 & \frac{\omega^2}{\epsilon_0} \\ \frac{nq^2}{M} & \omega^2 - \omega_T^2 \end{pmatrix} = 0.$$

The solution leads to two modes

$$\omega_\pm^2 = \frac{1}{2}\left(c^2k^2 + \omega_T^2 + \frac{nq^2}{\epsilon_0 M} \pm \sqrt{\left(c^2k^2 + \omega_T^2 + \frac{nq^2}{\epsilon_0 M}\right)^2 - 4c^2k^2\omega_T^2}\right). \tag{12.71}$$

At zero momentum, $k = 0$, the two branches give

$$\begin{aligned}
\omega_+^2 &= \omega_T^2 + \frac{nq^2}{\epsilon_0 M} \\
\omega_-^2 &= 0.
\end{aligned} \tag{12.72}$$

On the other hand, in the limit when the momentum is large, the lower mode has the limit $\omega_- = \omega_T$. Therefore, a gap appears in the spectrum. At low momenta the lower branch is very similar to a photon, while the upper branch is similar to an optical phonon. As the momentum increases the two branches result from a strong mixing of the photon and phonon and at larger momenta the lower branch is similar to the optical phonon and the upper branch is similar to the photon mode.

The dielectric function is given by

$$\epsilon_r(\omega) = 1 + \frac{P}{\epsilon_0 E} = 1 + \frac{nq^2}{\epsilon_0 M} \frac{1}{\omega_T^2 - \omega^2}. \tag{12.73}$$

Including the core electronic contribution it is renormalized to

$$\epsilon(\omega) = \epsilon(\infty) + \frac{nq^2/M}{\omega_T^2 - \omega^2}. \tag{12.74}$$

Consider now the case when $\omega = 0$. We get that

$$\epsilon(0) = \epsilon(\infty) + \frac{nq^2}{M\omega_T^2}. \tag{12.75}$$

This means we can rewrite

$$\begin{aligned} \epsilon(\omega) &= \epsilon(\infty) + (\epsilon(0) - \epsilon(\infty)) \frac{\omega_T^2}{\omega_T^2 - \omega^2} \\ &= \frac{\omega_T^2 \epsilon(0) - \omega^2 \epsilon(\infty)}{\omega_T^2 - \omega^2} \\ &= \epsilon(\infty) \left(\frac{\omega_L^2 - \omega^2}{\omega_T^2 - \omega^2} \right). \end{aligned} \tag{12.76}$$

The dielectric function diverges when $\omega = \omega_T$. This means a structural instability resulting from a vanishing force between the system charges. It is a signature of some sort of order in the distribution of the electric charges, such as ferroelectric order. The dielectric function vanishes when $\omega = \omega_L$. When the frequency is in the range $\omega_T < \omega < \omega_L$ the dielectric function is negative, the waves are attenuated which means there are no propagating modes, which is consistent with the existence of a gap in the spectrum. In this interval the reflection coefficient approaches one. When the frequency vanishes we get the Lyddane-Sachs-Teller (LST) relation

$$\frac{\omega_L^2}{\omega_T^2} = \frac{\epsilon(0)}{\epsilon(\infty)}. \tag{12.77}$$

12.5 Optical Processes

The relation between the dielectric function and the index of refraction allows experimental access using optical methods. As we have shown before, in general the index of refraction has a real and an imaginary part

$$n(\omega) = \sqrt{\epsilon + i\frac{\sigma}{\epsilon_0 \omega}} = n_r + in_i. \tag{12.78}$$

Recall that the current is given by

$$\mathbf{J} = -i\epsilon_0 \omega n^2(\omega)\mathbf{E}. \tag{12.79}$$

The power dissipated by Joule's effect is given by

$$\frac{1}{2}Re\,(\mathbf{J}\cdot\mathbf{E}) = \frac{1}{2}Re\left(-i\epsilon_0\omega n^2(\omega)\right)|\mathbf{E}|^2. \tag{12.80}$$

Using that

$$n^2(\omega) = (n_r + in_i)^2 = n_r^2 - n_i^2 + 2in_r n_i$$

and defining the absorption coefficient, η, by the ratio between the dissipated power and the power transmitted, we get

$$\eta = \frac{\frac{1}{2}Re\,(\mathbf{J}\cdot\mathbf{E})}{\sqrt{\frac{\epsilon}{\mu}}|\mathbf{E}|^2} = \frac{n_i\omega}{c}. \tag{12.81}$$

A measurement of the absorption coefficient directly reads the imaginary part of the index of refraction.

On the other hand, we may also measure the reflection part. Considering normal incidence on the material the reflected part over the incident part, R, is given by

$$R = \left|\frac{1-n}{1+n}\right|^2 = \frac{(n_r-1)^2 + n_i^2}{(n_r+1)^2 + n_i^2}. \tag{12.82}$$

A combined measurement of the absorption and reflection coefficients gives information about the real and imaginary parts of the index of refraction and the dielectric function. Indeed (taking $\mu_r = 1$)

$$\begin{aligned} \epsilon_r'(\omega) &= n_r^2 - n_i^2 \\ \epsilon_r''(\omega) &= 2n_r n_i. \end{aligned} \tag{12.83}$$

However, the real and imaginary parts of the dielectric function are not independent and do not need to be both measured. The reason is that the dielectric function is a response function and does satisfy Kramers-Kronig relations, which imply that the imaginary and real parts are related, if we know the whole frequency dependency of one of the functions.

Recall that $\epsilon_r - 1 = \chi_e$, where χ_e is a susceptibility that relates the polarization with a given electric field. Therefore, it is a response function that satisfies Kramers-Kronig relations.

$$\chi(\omega) = -iP \int_{-\infty}^{\infty} \frac{d\omega'}{\pi} \frac{\chi(\omega')}{\omega' - \omega}. \tag{12.84}$$

Separating the dielectric function in real and imaginary parts we get that

$$\begin{aligned} \epsilon_r'(\omega) - 1 &= P \int_{-\infty}^{\infty} \frac{d\omega'}{\pi} \frac{\epsilon_r''(\omega')}{\omega' - \omega} \\ \epsilon_r''(\omega) &= -P \int_{-\infty}^{\infty} \frac{d\omega'}{\pi} \frac{\epsilon_r'(\omega') - 1}{\omega' - \omega}. \end{aligned} \tag{12.85}$$

These can be rewritten as

$$\begin{aligned} \epsilon_r'(\omega) - 1 &= \frac{2}{\pi} P \int_0^{\infty} d\omega' \frac{\omega' \epsilon_r''(\omega')}{(\omega')^2 - \omega^2} \\ \epsilon_r''(\omega) &= -\frac{2\omega}{\pi} P \int_0^{\infty} d\omega' \frac{\epsilon_r'(\omega')}{(\omega')^2 - \omega^2}, \end{aligned} \tag{12.86}$$

or

$$\begin{aligned} n_r^2(\omega) - n_i^2(\omega) &= \frac{2}{\pi} P \int_0^{\infty} d\omega' \frac{\omega' 2n_r(\omega') n_i(\omega')}{(\omega')^2 - \omega^2} \\ 2n_r(\omega) n_i(\omega) &= -\frac{2\omega}{\pi} P \int_0^{\infty} d\omega' \frac{n_r^2(\omega') - n_i^2(\omega')}{(\omega')^2 - \omega^2}. \end{aligned} \tag{12.87}$$

In addition to the "classical" electromagnetic response of incident radiation on a material, there are also quantum effects due to the interaction of the photons with the energy bands of the electrons. In general, the electronic levels are organized in bands, separated by finite gaps. Since the velocity of photons is very high, the momentum transfer between a photon and an electron is rather small, for a given finite energy transfer. Therefore, with good approximation we may consider that these processes involve no momentum transfer. They are however responsible for transitions between bands and only occur if the energy of an absorbed photon equals the energy difference between different electronic energy bands. Excitations of electrons within the same band can have arbitrarily small energies and small or finite momenta and do not involve absorption of a single photon.

12.5.1 Excitons

Assume that we supply a finite energy and finite momentum so that an electron gains energy and momentum and is able to occupy an empty state, leaving behind an empty state. This is a particle-hole excitation.[4] This is a picture of the process in momentum (energy) space. Consider however a similar process in real space. Assume that by sending energy into the system an electron is excited from a given state leaving behind a hole. We may assume that the electron absorbs a photon and is promoted to an excited band. Since the hole and the electron have opposite charges, there will be some attraction between the two, and they will eventually tend to recombine to form some sort of bound state, called exciton. Consider the possibility that this process occurs but that the electron has not returned to its initial state. If the binding is weak, the pair will be loosely bound and the electron and the hole will move together in a pair that extends over space in a way similar to a hydrogen-like atom. This is called a Mott-Wannier exciton and is characterized by an extended wave function. Since this is expected to be a bound state, its energy is negative with respect to the lower edge of the band to which the electron was promoted, by the absorption of the photon. We expect therefore energy levels close to the lower edge of the excited band, due to the presence of these excitons. Another possibility is that the system strongly binds the excited electron to the atom it originated from. This happens if the system has a strong ionic character and each electron is strongly bound to atomic orbitals, which implies an insulator to begin with. Bound states between the electron and the hole it left behind are also possible, but are typically strongly bound. These are called Frenkel excitons and have wave functions that are rather localized. These two limiting cases of very weakly bound and strongly bound excitons lead to a wide range of binding energies that can be as small as 1meV or as large as 1 eV.

The Mott-Wannier excitons can be well modeled by a simple attractive potential between the members of the pair as

$$U(r) = -\frac{e^2}{4\pi\epsilon r}. \tag{12.88}$$

The problem has a straightforward solution as the hydrogen atom and leads to energy levels as

$$E_n = E_g - \frac{\mu e^4}{2\hbar^2(4\pi\epsilon)^2}\frac{1}{n^2}, \tag{12.89}$$

with respect to the energy gap to the excited band and where here μ is the reduced mass of the electron and hole. For instance, for Cu_2O we have that $\epsilon = 10$, $\mu =$

[4] See also [Kittel].

0.7 m, which gives a binding energy of about 95 meV, as compared to the gap to the excited band which is about 2.17 eV.

The strongly coupled nature of the Frenkel excitons suggests that the electrons and their pair could move between nearest atoms, as in the tight-binding model, giving rise to some narrow band below the excited band. Consider a crystal with N atoms and to simplify, consider a one-dimensional arrangement of atoms. Denote the ground state of the system as $\psi_g = u_1 u_2 \cdots u_N$, where u_i are one-particle Wannier wave functions localized at atom i. Consider now that at site j the electron is in an excited state v_j. The wave function is changed to $\phi_j = u_1 u_2 \cdots u_{j-1} v_j u_{j+1} \cdots u_N$. Consider now a Hamiltonian that essentially is just kinetic energy. It will be possible for the electron to stay at the same atom or to change its location either to the left or the right of site j. We consider then that the action of the Hamiltonian on the wave function ϕ_j is of the type

$$H\phi_j = \epsilon\phi_j + T\left(\phi_{j-1} + \phi_{j+1}\right), \tag{12.90}$$

where T is the probability amplitude that the electron, under the action of the Hamiltonian, changes its location. Clearly, the wave function ϕ_j is not an eigenfunction of the Hamiltonian since the electron in the excited state v_j may delocalize. In order to diagonalize the Hamiltonian, we may choose instead a wave function of the Bloch form

$$\psi_k = \sum_j e^{ijka}\phi_j. \tag{12.91}$$

Applying the Hamiltonian to this trial wave function we get that

$$\begin{aligned} H\psi_k &= \sum_j e^{ijka} H\phi_j \\ &= \sum_j e^{ijka}\left(\epsilon\phi_j + T\phi_{j-1} + T\phi_{j+1}\right) \\ &= \sum_j e^{ijka}\left(\epsilon + Te^{ika} + Te^{-ika}\right)\phi_j \\ &= \left(\epsilon + 2T\cos(ka)\right)\psi_k, \end{aligned} \tag{12.92}$$

which is therefore an eigenfunction of the Hamiltonian with eigenvalue

$$E_k = \epsilon + 2T\cos(ka). \tag{12.93}$$

The bandwidth is determined by T, which is small if the electron-hole pair is strongly bound. An example of this type of exciton is found in ionic crystals such as KBr.

12.6 Exercises

1. The static electric conductivity of a metal is given by Eq. (12.22). Estimate the average time between collisions for an electron in different metals, such as copper.
2. Consider a system described by a set of oscillators subject to a force F. The equation of motion of the total displacement of the system, x, is

$$\sum_j M_j \left(\frac{d^2}{dt^2} + \rho_j \frac{d}{dt} + \omega_j^2 \right) x = F.$$

We define the response function $\alpha(\omega)$ as

$$x(\omega) = \alpha(\omega) F(\omega),$$

where F is the real part of $F(\omega)e^{-i\omega t}$ and x the real part of $x(\omega)e^{-i\omega t}$.

(a) Show that

$$\alpha(\omega) = \sum_j \frac{f_j}{\omega_j^2 - \omega^2 - i\omega\rho_j}$$

and $f_j = 1/M_j$.

(b) Show that the following sum rule should hold

$$\sum_j f_j = \frac{2}{\pi} \int_0^\infty \omega \alpha''(\omega) d\omega,$$

where $\alpha(\omega) = \alpha'(\omega) + i\alpha''(\omega)$.

3. The real part of the frequency-dependent dielectric function for free electrons is given by Eq. (12.36). Show that metals fully reflect radiations with frequencies smaller than the plasma frequency. Calculate the maximum wave length of radiation that propagates in several metals, such as sodium.
4. A plasma occupies an infinite half-plane ($z > 0$) and the remaining half-plane ($z < 0$) is vacuum. The solutions of Laplace equation $\nabla^2 \phi = 0$ are given by[5]

$$\phi_p(x, z \geq 0) = A \cos kx e^{-kz}, \qquad \phi_0(x, z \leq 0) = A \cos kx e^{kz}.$$

[5] See [Kittel].

These expressions satisfy the condition that the tangential component of the electric field is continuous. Show that imposing that the perpendicular component of the displacement vector, **D**, is continuous implies that, at the surface of the plasma, $\epsilon_r(\omega) = -1$. Therefore, one obtains the Stern-Ferrell result

$$\omega_s^2 = \frac{1}{2}\omega_p^2,$$

for the frequency of a surface oscillation of the plasma.

5. Consider a plane wave of polarized light of frequency ω that propagates in a solid and that the dielectric function is given by Eq. (12.17).
   (a) Calculate the refraction index $n(\omega)$.
   (b) Make a scheme of the real and imaginary parts of $n(\omega)$ in the limit of small densities ($\omega_p/\omega_0 << 1$ and $\omega_p/\lambda << 1$).
   (c) At resonance ($\omega = \omega_0$), calculate the attenuation length in the solid in the limit of small densities.
6. (a) Determine the range of photon wave lengths for which $NaCl$ reflects the incident radiation due to the existence of polariton modes. (static dielectric constant $\epsilon = 5.90$, lattice constant $a = 5.6\mathring{A}$, atomic mass of $Na = 23$, and atomic mass of $Cl = 35.5$).
   (b) Consider a solid with $\epsilon(0) = 5.9$, $\epsilon(\infty) = 2.25$ and zero reflectivity, $R = 0$ in Eq. (12.82), for $\lambda = 3.06\mu m$ and $\epsilon(\omega)$ real. Determine ω_L and ω_T.
7. Taking the limit of weak dissipation in Eq. (12.13) using Eq. (12.33), we obtain the expression Eq. (12.34) for the polarizability. The case of the vicinity of the gap in a semiconductor may be obtained approximately making the replacement

$$\delta(\omega) \rightarrow \frac{1}{2}\delta(\omega - \omega_g)$$

in the polarizability. This leads to the expression

$$\epsilon_r''(\omega) = \frac{ne^2}{2\epsilon_0 m\omega}\pi\delta(\omega - \omega_g).$$

Show that the real part of the dielectric permittivity can be written as

$$\epsilon_r'(\omega) = 1 + \frac{\omega_p^2}{\omega_g^2 - \omega^2},$$

where ω_p is the plasma frequency.

8. In addition to the sum rule Eq. (12.23), there are other sum rules that prove useful. One example is the sum rule

$$\int_0^\infty \frac{a\sigma'(\omega)}{\omega^2} d\omega = \epsilon_r(\omega = 0) - 1,$$

where σ' is the real part of the conductivity. Show this result and determine a. Recall that the Kramers-Kronig relations imply that

$$\epsilon_r'(\omega) = \frac{2}{\pi} P \int_0^\infty \frac{\omega' \epsilon_r''(\omega')}{(\omega')^2 - \omega^2} d\omega' + 1.$$

Chapter 13
Effects of Magnetic Fields

An external magnetic field interacts with a solid in two ways. There is an orbital effect associated with the vector potential. This term changes the Hamiltonian of the system via the minimal coupling and affects the electronic orbits inducing cyclotron motion. This contribution leads to a diamagnetic response of the system.

There is also a coupling with the magnetic moment due to the spins of the constituents such as the electrons and the nuclear magnetic moment. Also, there is a coupling with the magnetic moment due to the orbital motion of orbits. These two contributions lead to a paramagnetic response of the system.

In general, each electron has orbital and spin angular momenta that are coupled to yield a total angular momentum. Each atom has several electrons whose angular momenta add up to give a total angular momentum, which determines the total magnetic moment.

13.1 Magnetic Moments

We will consider first a situation in which the magnetic moments are localized. Later in this section the magnetic moments will be associated with the spin of electrons in a metal. In this case we will focus for now on electrons that are delocalized, and, therefore, the energy has a term that results from the coupling of the spin magnetic moment with an external magnetic field and another term that is the kinetic energy of the electron.

P. D. Sacramento, *Introduction to Condensed Matter Theory*,
https://doi.org/10.1007/978-3-031-86251-9_13

13.1.1 Langevin and Brillouin Paramagnetism

Each atom has a possible magnetic moment, whose origin will be discussed later in the book. Before we determine the magnetic moment of a given atom, let us consider a magnetic dipole, $\boldsymbol{\mu}$, and an external magnetic field, $\mathbf{H}$. The energy of interaction of the dipole with the field is given by $E = -\boldsymbol{\mu} \cdot \mathbf{H}$. If we consider N dipoles localized in space, the total energy of the system is

$$E = -\sum_{i=1}^{N} \boldsymbol{\mu}_i \cdot \mathbf{H}.$$

Let us choose for simplicity the direction of the uniform external field as defining the z axis, $\mathbf{H} = H\mathbf{e}_z$. The energy is written as

$$E = -\sum_{i=1}^{N} \mu H \cos\theta_i = -\mu H \sum_{i=1}^{N} \cos\theta_i . \tag{13.1}$$

Here θ_i are the angles that each dipole makes with the direction of the magnetic field. Then the partition function of the system is given by

$$\begin{aligned} Z_N(\beta) &= \sum_{\{\theta_i\}} e^{\beta\mu H \sum_{i=1}^{N} \cos\theta_i} \\ &= \sum_{\{\theta_i\}} \prod_{i=1}^{N} e^{\beta\mu H \cos\theta_i} \\ &= \prod_{i=1}^{N} \sum_{\theta_i} e^{\beta\mu H \cos\theta_i} \\ &= (Z(\beta))^N , \end{aligned} \tag{13.2}$$

where the single dipole partition function is given by

$$Z(\beta) = \sum_{\theta} e^{\beta\mu H \cos\theta} . \tag{13.3}$$

The partition function of the system of N independent, non-interacting dipoles is the product of the partition functions of each system. Therefore, the magnetization is given by

$$M_z = \left\langle \sum_{i=1}^{N} \mu \cos\theta_i \right\rangle = \frac{1}{\beta} \frac{\partial}{\partial H} \ln Z_N(\beta)$$

$$= \frac{N}{\beta}\frac{\partial}{\partial H}\ln Z(\beta) = N\langle\mu\cos\theta\rangle. \tag{13.4}$$

All that is left is the calculation of the average value of $\cos\theta$. The result depends on whether the dipoles are classical or quantum. In the classical case, we obtain that the partition function

$$Z(\beta) = \int_0^{2\pi} d\varphi \int_0^{\pi} d\theta \sin\theta e^{\beta\mu H\cos\theta} = 4\pi\frac{\sinh\beta\mu H}{\beta\mu H}. \tag{13.5}$$

Therefore, the magnetization per dipole is given by

$$m_z = \mu L(\beta\mu H), \tag{13.6}$$

where the Langevin function $L(x)$ is given by

$$L(x) = \coth x - \frac{1}{x}.$$

For small temperatures $k_B T \ll \mu H$, the Langevin function is approximately unity and the magnetization per dipole is approximately μ, as expected. In the opposite limit of high temperatures, the Langevin function is approximately $L(x) \sim x/3$ and the susceptibility is of the Curie type

$$\chi = \frac{C}{T}, C = N\frac{\mu^2}{3k_B}. \tag{13.7}$$

Here C is called Curie constant.

In the quantum case the partition function is calculated as

$$Z(\beta) = \sum_{m=-J}^{J} e^{\beta g\mu_B Hm}.$$

We are considering a spin-J dipole, $\mu = |\boldsymbol{\mu}| = g\mu_B J$, where g is the Landé factor, and $\mu_B = \frac{e\hbar}{2m}$ is the Bohr magneton. Therefore, the components of the dipole operator are given by the $(2J+1)$ projections of the spin operator, $\mathbf{J}$: $\mu_z = g\mu_B m$ where $m = -J, \ldots, J$. Defining $x = \beta(g\mu_B J)H$ we obtain that the partition function is given by

$$Z(\beta) = \frac{\sinh(2J+1)\frac{x}{2J}}{\sinh\frac{x}{2J}}. \tag{13.8}$$

The magnetization per dipole is therefore given by

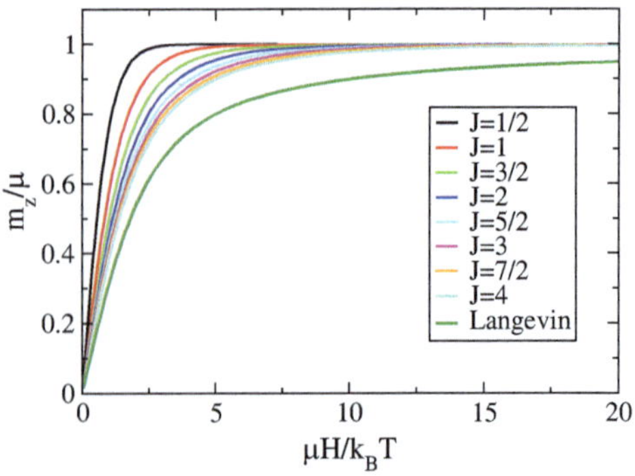

Fig. 13.1 Magnetization normalized by the saturation value at zero temperature as a function of the magnetic energy over the thermal energy, for different spin values from the quantum limit ($S = 1/2$) to the classical Langevin regime ($S \to \infty$)

$$m_z = \mu B_J(x), \tag{13.9}$$

where the Brillouin function is

$$B_J(x) = \left(1 + \frac{1}{2J}\right) \coth\left[\left(1 + \frac{1}{2J}\right)x\right] - \frac{1}{2J} \coth\left[\frac{1}{2J}x\right]. \tag{13.10}$$

In the limit $x \gg 1$ (which corresponds either to low temperatures or high magnetic field), $B_J(x) \to 1$. In the opposite limit

$$B_J(x) \sim \frac{1}{3}\left(1 + \frac{1}{J}\right)x + \ldots$$

Therefore, in this limit the susceptibility is again of the Curie type, but with a different Curie constant. The susceptibility is given by

$$\chi = \frac{C_J}{T}, C_J = \frac{g^2 \mu_B^2 J(J+1)}{3k_B}. \tag{13.11}$$

For instance, $C_{\frac{1}{2}} = \frac{\mu^2}{k_B}$ and $C_\infty = C_{classical} = \frac{\mu^2}{3k_B}$. We conclude therefore that $C_{\frac{1}{2}} = 3C_{classical}$. Also note that as expected $B_\infty(x) = L(x)$.

In Fig. 13.1 we show the magnetization for the classical (Langevin) and quantum cases (Brillouin) as a function of the ration of the magnetic energy over the thermal energy. In Fig. 13.2 we show the Curie susceptibility for different spin values.

13.1.2 Van Vleck Paramagnetism

Let us now consider an atom with no magnetic moment in its ground state, represented by the ket $|0\rangle$. Consider now one excited state $|s\rangle$, such that $\langle s|\mu_z|0\rangle \neq 0$. Let us consider the energy of the magnetic moment in the presence of the magnetic field, as a perturbation. Call $\Delta = E_s - E_0$. The perturbed ground state is

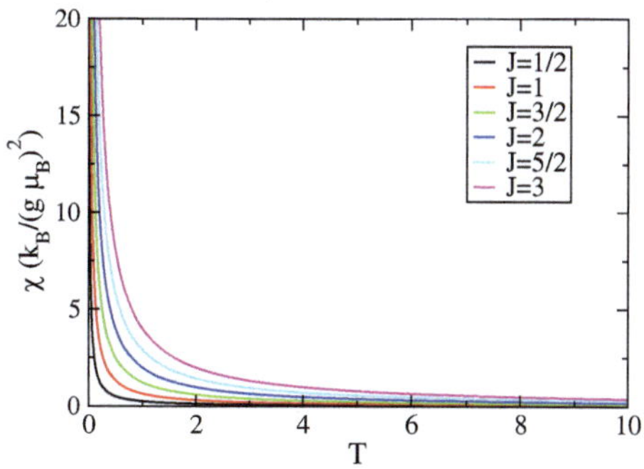

Fig. 13.2 Curie susceptibility for various spin values

$$|0\rangle' = |0\rangle + \frac{B}{\Delta}\langle s|\mu_z|0\rangle|s\rangle$$

and the perturbed excited state is given by

$$|s\rangle' = |s\rangle - \frac{B}{\Delta}\langle 0|\mu_z|s\rangle|0\rangle.$$

The magnetic moment of each perturbed state is

$$\begin{aligned}\langle 0'|\mu_z|0'\rangle &= \frac{2B}{\Delta}|\langle s|\mu_z|0\rangle|^2 \\ \langle s'|\mu_z|s'\rangle &= -\frac{2B}{\Delta}|\langle s|\mu_z|0\rangle|^2.\end{aligned} \tag{13.12}$$

In the limit when the temperature is high, as compared to the gap between the ground state and the first excited state, we find

$$\frac{N_0}{N_s} = e^{\frac{\Delta}{k_B T}} \sim 1 + \frac{\Delta}{k_B T}.$$

Then $N_0 - N_s = N_s\Delta/(k_B T)$. In this high temperature regime, $N_s \sim N/2$, and, therefore,

$$M \sim \frac{2B|\langle s|\mu_z|0\rangle|^2}{\Delta}\frac{N\Delta}{2k_B T}. \tag{13.13}$$

Then the susceptibility is given by

$$\chi = N\frac{|\langle s|\mu_z|0\rangle|^2}{k_B T}, \tag{13.14}$$

which is independent of Δ and inversely proportional to the temperature.

In the opposite low temperature limit, $\Delta \gg k_B T$, all atoms are in the ground state. Therefore, the susceptibility is given by

$$\chi = \frac{2N|\langle s|\mu_z|0\rangle|^2}{\Delta}, \tag{13.15}$$

which is independent of the temperature.

13.1.3 Magnetic Permeability Tensor

Let us consider a magnetic material where there are localized magnetic dipoles. The magnetic moment and the angular momentum are related by $\mathbf{m} = -\gamma \mathbf{s}$, where γ is the gyromagnetic ratio and $\mathbf{s}$ the spin angular momentum. Applying a magnetic field along z, for instance, a torque (force moment) arises on the magnetic moment that is given by[1]

$$\boldsymbol{\tau} = \mathbf{m} \times \mathbf{B}_0 = \mu_0 \mathbf{m} \times \mathbf{H}_0 = -\mu_0 \gamma \mathbf{s} \times \mathbf{H}_0 \,. \tag{13.16}$$

Since the torque is equal to the time variation of the angular momentum, we obtain

$$\frac{d\mathbf{m}}{dt} = -\mu_0 \gamma \mathbf{m} \times \mathbf{H}_0 \,. \tag{13.17}$$

The magnetization also obeys the same equation of motion,

$$\frac{d\mathbf{M}}{dt} = -\mu_0 \gamma \mathbf{M} \times \mathbf{H} \,, \tag{13.18}$$

where $\mathbf{H}$ is the internal magnetic field. Let us consider, now, the effect of adding a small magnetic field, $\mathbf{h}$, that changes with time. The total field and total magnetization are then given by

$$\begin{aligned} \mathbf{H}_t &= H_0 \mathbf{e}_z + \mathbf{h} \,, \\ \mathbf{M}_t &= M_s \mathbf{e}_z + \mathbf{M} \,, \end{aligned} \tag{13.19}$$

where M_s is the saturation magnetization, due to the external field, and the additional magnetization is taken in the plane xy.

The equations of motion for the magnetization components are given, in the limit of a small changing field, by

[1] Here we follow [AraujoSacramento].

$$\frac{dM_x}{dt} = -\omega_0 M_y + \omega_m h_y ,$$
$$\frac{dM_y}{dt} = \omega_0 M_x - \omega_m h_x ,$$
$$\frac{dM_z}{dt} = 0 , \tag{13.20}$$

where $\omega_0 = \mu_0 \gamma H_0$ and $\omega_m = \mu_0 \gamma M_s$. We obtain that the components satisfy

$$\frac{d^2 M_x}{dt^2} + \omega_0^2 M_x = \omega_m \frac{dh_y}{dt} + \omega_0 \omega_m h_x ,$$
$$\frac{d^2 M_y}{dt^2} + \omega_0^2 M_y = -\omega_m \frac{dh_x}{dt} + \omega_0 \omega_m h_y . \tag{13.21}$$

If the field **h** has a time dependence of the form $e^{i\omega t}$, then we obtain

$$\left(\omega_0^2 - \omega^2\right) M_x = \omega_0 \omega_m h_x + i\omega\omega_m h_y ,$$
$$\left(\omega_0^2 - \omega^2\right) M_y = -i\omega\omega_m h_x + \omega_0 \omega_m h_y . \tag{13.22}$$

These relations may be condensed in a matrix form,

$$M = \chi H = \begin{pmatrix} \chi_{xx} & \chi_{xy} & 0 \\ \chi_{yx} & \chi_{yy} & 0 \\ 0 & 0 & 0 \end{pmatrix} h , \tag{13.23}$$

where the matrix elements are given by

$$\chi_{xx} = \chi_{yy} = \frac{\omega_0 \omega_m}{\omega_0^2 - \omega^2} ,$$
$$\chi_{xy} = -\chi_{yx} = \frac{i\omega\omega_m}{\omega_0^2 - \omega^2} . \tag{13.24}$$

The frequency dependence of the magnetic susceptibility is shown in Fig. 13.3. This implies that the magnetic permeability tensor defined by $B = \mu_t h$ is given by

$$\mu_t = \begin{pmatrix} \mu & i\kappa & 0 \\ -i\kappa & \mu & 0 \\ 0 & 0 & \mu_0 \end{pmatrix} . \tag{13.25}$$

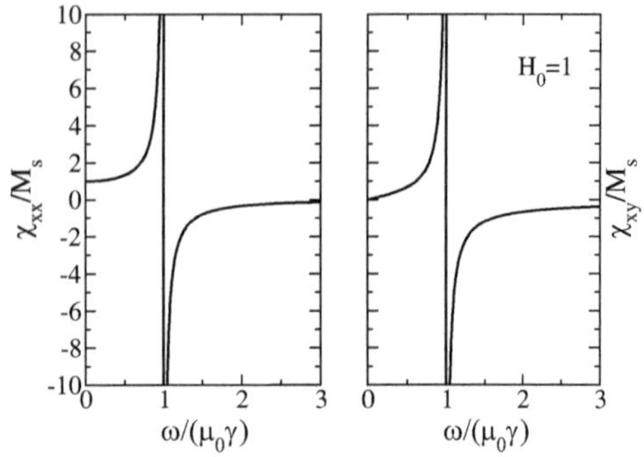

Fig. 13.3 Diagonal and off-diagonal components of the magnetic susceptibility tensor as a function of frequency

The elements of the permeability tensor are given by

$$\mu = \mu_0 (1 + \chi_{xx}) = \mu_0 \left(1 + \chi_{yy}\right) = \mu_0 \left(1 + \frac{\omega_0 \omega_m}{\omega_0^2 \omega^2}\right),$$

$$\kappa = -i \mu_0 \chi_{xy} = i \mu_0 \chi_{yx} = \mu_0 \frac{\omega \omega_m}{\omega_0^2 - \omega^2}. \tag{13.26}$$

A material that has a permeability of this form is called gyrotropic. The matrix is antisymmetric due to the presence of the magnetic field. Changing the direction of the magnetic fields, the signal of μ remains invariant but κ changes sign. These results assume that there are no losses.

13.1.4 Pauli Paramagnetism

Consider a gas of electrons. The energy of each electron in the presence of an external magnetic field has two components: the magnetic field affects the orbital motion and the spin of the electron interacts with the external field as a magnetic dipole. Let us now concentrate only on the interaction of the spin with the external field (Zeeman term). We will consider the other term shortly.

The energy of the free particle corrected with the Zeeman term is[2]

$$\epsilon = \frac{p^2}{2m} - \boldsymbol{\mu}^* \cdot \mathbf{B}. \tag{13.27}$$

Here $\boldsymbol{\mu}^*$ is the magnetic moment. The asterisk is to not confuse with the chemical potential, denoted as usual by μ. Since we are considering the electron gas, the spin is $1/2$. There are two groups of particles corresponding to the two spin projections

[2] See, for instance, [Pathria].

on the quantization direction (which we take as the direction of the magnetic field, once again assumed uniform). The two groups of particles have energies

$$\text{(i)} \qquad \epsilon = \frac{p^2}{2m} - \mu^* B$$

$$\text{(ii)} \qquad \epsilon = \frac{p^2}{2m} + \mu^* B. \tag{13.28}$$

Here $B = |\mathbf{B}|$ and $\mu^* = |\boldsymbol{\mu}^*|$. At $T = 0$ all levels are occupied up to the Fermi level, ϵ_F. The number of particles in each group is then

$$N_+ = \frac{4\pi V}{3h^3}\left(2m(\epsilon_F + \mu^* B)\right)^{\frac{3}{2}}$$

$$N_- = \frac{4\pi V}{3h^3}\left(2m(\epsilon_F - \mu^* B)\right)^{\frac{3}{2}}. \tag{13.29}$$

The magnetic moment of the electron gas in the presence of the magnetic field is

$$M = \mu^*(N_+ - N_-) = \frac{4\pi V}{3h^3}\mu^*(2m)^{\frac{3}{2}}\left[(\epsilon_F + \mu^* B)^{\frac{3}{2}} - (\epsilon_F - \mu^* B)^{\frac{3}{2}}\right]. \tag{13.30}$$

The susceptibility is then given by

$$\chi(B \to 0) = \lim_{B\to 0}\frac{M}{VB} = \frac{4\pi(\mu^*)^2}{h^3}\epsilon_F^{1/2} = \chi_0(T = 0). \tag{13.31}$$

Using that

$$\epsilon_F = \left(\frac{6\pi^2 n}{g}\right)^{\frac{2}{3}}\frac{\hbar^2}{2m},$$

(n is the electron density) and that $g = 2$, we get that, at $T = 0$,

$$\chi_0 = \frac{3}{2}\frac{n(\mu^*)^2}{\epsilon_F}. \tag{13.32}$$

At finite small temperature T, the calculation is a bit more involved and one gets

$$\chi = \chi_0\left[1 - \frac{\pi^2}{12}\left(\frac{k_B T}{\epsilon_F}\right)^2\right]. \tag{13.33}$$

Since the susceptibility is positive, this is a paramagnetic contribution.

13.2 One Electron in External Fields

In the presence of external electric and magnetic fields, an electron will have both longitudinal (along the electric field) and transverse motions. A simple description may be taken considering the classical equations of motion, leading to the explanation of the classical Hall effect where a transverse potential difference is induced due to the magnetic field. A transverse motion may also be induced in the absence of a magnetic field due to quantum effects, such as due to a Berry curvature of the electronic bands in a solid. These effects are considered in this section.

13.2.1 Classical Equations of Motion

Let us begin by considering a simple model for a dielectric, assuming that it is composed by electrons, bound harmonically and isotropically to a given center of force, and let us apply an electric field and a magnetic field, taking into account the presence of friction on the motion of the electrons.[3] The equation of motion of one of these electrons, of charge q and mass m, is written, in a classical approximation, as

$$m\frac{d^2\mathbf{r}}{dt^2} = q\mathbf{E} + q\frac{d\mathbf{r}}{dt} \times \mathbf{B} - m\nu_c\frac{d\mathbf{r}}{dt} - m\omega_0^2\mathbf{r}\,. \tag{13.34}$$

Looking for a solution of the type $\mathbf{r} \propto \mathbf{r}_0 e^{i\omega t}$, we obtain

$$-m\omega^2\mathbf{r} = q\mathbf{E} + i\omega q\mathbf{r} \times \mathbf{B} - m\nu_c i\omega\mathbf{r} - m\omega_0^2\mathbf{r}\,. \tag{13.35}$$

The relation between the coordinates and the electric field may be written in a matrix form,

$$A\begin{pmatrix} x \\ y \\ z \end{pmatrix} = q\begin{pmatrix} E_x \\ E_y \\ E_z \end{pmatrix}, \tag{13.36}$$

where the matrix A is given by

$$\begin{pmatrix} m(\omega_0^2-\omega^2)+m\nu_c i\omega & -qi\omega B_z & qi\omega B_y \\ qi\omega B_z & m(\omega_0^2-\omega^2)+m\nu_c i\omega & -qi\omega B_x \\ -qi\omega B_y & qi\omega B_x & m(\omega_0^2-\omega^2)+m\nu_c i\omega \end{pmatrix}. \tag{13.37}$$

[3] See, for instance, [AraujoSacramento].

The electric dipolar moment is given by $\mathbf{p} = q\mathbf{r}$, the polarization $\mathbf{P} = n_q q\mathbf{r} = \epsilon_0\chi_e\mathbf{E}$, and also $\epsilon = \epsilon_0(1 + \chi_e)$. From here we obtain that the dielectric tensor is given by

$$\epsilon = \epsilon_0 \left(1 + \frac{n_q q^2}{\epsilon_0} A^{-1}\right). \tag{13.38}$$

The inverse of the matrix A is given by

$$A^{-1} = \frac{1}{a(a^2 - \tilde{B}^2)} \begin{pmatrix} a^2 - \tilde{B}_x^2 & -\tilde{B}_x\tilde{B}_y + ia\tilde{B}_z & -\tilde{B}_x\tilde{B}_z - ia\tilde{B}_y \\ -\tilde{B}_x\tilde{B}_y - ia\tilde{B}_z & a^2 - \tilde{B}_y^2 & -\tilde{B}_y\tilde{B}_z + ia\tilde{B}_x \\ -\tilde{B}_x\tilde{B}_z + ia\tilde{B}_y & -\tilde{B}_y\tilde{B}_z - ia\tilde{B}_x & a^2 - \tilde{B}_z^2 \end{pmatrix}, \tag{13.39}$$

where $a = A_{11} = A_{22} = A_{33}$ and $\tilde{B}_j = B_j q\omega$. We can see that in the absence of a magnetic field the matrix A is diagonal. If the magnetic field is non-vanishing, it is antisymmetric, due to the presence of terms that break time-reversal invariance that results from the magnetic field. Considering for instance a magnetic field along the z axis, and a transverse wave ($E_z = 0$) propagating along this direction, the matrix structure of the dielectric tensor ensures that the index of refraction has two distinct eigenvalues, which lead to two different indices of refraction and the Faraday effect.

13.2.2 Classical Hall Effect

The equation of motion for the velocity of the electron in the absence of the harmonic potential ($\omega_0 = 0$) and in the presence of a magnetic field, B, along the z direction, may be written as

$$\begin{aligned} m\left(\frac{d}{dt} + \frac{1}{\tau}\right) v_x &= -e\left(E_x + Bv_y\right) \\ m\left(\frac{d}{dt} + \frac{1}{\tau}\right) v_y &= -e\left(E_y - Bv_x\right) \\ m\left(\frac{d}{dt} + \frac{1}{\tau}\right) v_z &= -eE_z, \end{aligned} \tag{13.40}$$

where the friction term was written in terms of the average time between collisions, τ. In the stationary state, $d/dt = 0$, we get

$$\begin{aligned} v_x &= -\frac{e\tau}{m} E_x - \omega_c \tau v_y \\ v_y &= -\frac{e\tau}{m} E_y + \omega_c \tau v_x \end{aligned}$$

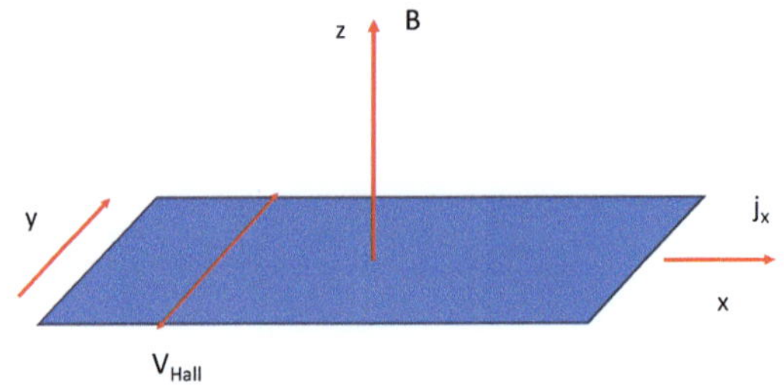

Fig. 13.4 Hall geometry: a current flows along the x direction in a plane under the simultaneous presence of a perpendicular magnetic field along the z direction. Due to the Lorentz force a potential difference is induced in the transverse, y, direction

$$v_z = -\frac{e\tau}{m} E_z, \tag{13.41}$$

with the cyclotron frequency $\omega_c = (eB)/m$.

Consider now a two-dimensional conductor such that the magnetic field is perpendicular to the plane of the conductor. The dimension along the x axis is large and the dimension along the y axis is finite. This is called a Hall geometry, shown in Fig. 13.4. Due to the perpendicular magnetic field and applying an electric field along the x direction, the Lorentz force acts on the electrons and deviates their trajectory, such that they move along the y direction. There is an accumulation of electrons on one edge of the sample that acts as a repulsive potential as more electrons are deviated towards that edge. In the stationary state, the motion along the y direction stops and $v_y = 0$. In this regime we get that

$$\frac{e\tau}{m} E_y = \omega_c \tau v_x. \tag{13.42}$$

Also, in the same regime we get

$$v_x = -\frac{e\tau}{m} E_x. \tag{13.43}$$

Combining these expressions,

$$E_y = -\omega_c \tau E_x = -\frac{eB\tau}{m} E_x. \tag{13.44}$$

As a result, due to the presence of the perpendicular magnetic field, an electric field appears along the y direction. This leads to the appearance of an electric potential difference between the two edges of the conductor. This is called the Hall voltage. It is usual to define the ratio

$$R_H = \frac{E_y}{j_x B}, \tag{13.45}$$

called the Hall coefficient, and where j_x is the x component of the current density. Using that $j_x = ne^2 \tau E_x / m$ we obtain

$$R_H = -\frac{1}{ne}, \tag{13.46}$$

which is only a function of the carrier density and their charge. If the carriers are electrons, the Hall coefficient is negative, and if the carriers are holes, the Hall coefficient is positive. Depending on the type of carriers, the Hall potential difference changes sign.

Let us return to the equation of motion for the velocity of a charge q in the presence of the combined action of an electric and a magnetic field:

$$m\frac{d\mathbf{v}}{dt} = q\mathbf{E} + q\mathbf{v}\times\mathbf{B} - \frac{m}{\tau}\mathbf{v}. \tag{13.47}$$

Looking for a solution of the type $\mathbf{v} \sim \mathbf{v}_0 e^{i\omega t}$, we get that

$$\begin{pmatrix} j_x \\ j_y \\ j_z \end{pmatrix} = n_q q \begin{pmatrix} v_x \\ v_y \\ v_z \end{pmatrix} = n_q q^2 U^{-1} \begin{pmatrix} E_x \\ E_y \\ E_z \end{pmatrix}, \tag{13.48}$$

where the matrix U is given by

$$\begin{pmatrix} mi\omega + \frac{m}{\tau} & -qB & 0 \\ qB & mi\omega + \frac{m}{\tau} & 0 \\ 0 & 0 & mi\omega + \frac{m}{\tau} \end{pmatrix}. \tag{13.49}$$

At zero frequency, static case, the conductivity for the electron system, $q = -e$, may be written as

$$\sigma = \sigma_0 \begin{pmatrix} \frac{1}{1+(\omega_c\tau)^2} & -\frac{\omega_c\tau}{1+(\omega_c\tau)^2} & 0 \\ \frac{\omega_c\tau}{1+(\omega_c\tau)^2} & \frac{1}{1+(\omega_c\tau)^2} & 0 \\ 0 & 0 & 1 \end{pmatrix}, \tag{13.50}$$

where $\sigma_0 = ne^2\tau/m$ is the Drude conductivity, as found before in Eq. 12.22. Also, note that comparing the static transverse conductivity, $\sigma_x = \sigma_y$, in the previous expression with the real part of the longitudinal conductivity of Eq. 12.21, the role of the frequency ω is taken by the cyclotron frequency, ω_c.

13.2.3 Berry Curvature, Anomalous Velocity, and Hall Effect

The Berry curvature gives a contribution to the equation of motion of an electron in the presence of an external electric field, as a correction to the usual term considered before. An electric field, $\boldsymbol{E}$, may be described by means of a vector potential, $\boldsymbol{A}(t)$, such that $\boldsymbol{E} = -\partial_t \boldsymbol{A}$. The Hamiltonian,

$$\hat{H}(\mathbf{k}, t) = \hat{H}\left[\mathbf{k} - \frac{e}{\hbar}\mathbf{A}(t)\right], \tag{13.51}$$

changes slowly in time if the field is weak. The velocity operator is given by[4]

$$\hat{\boldsymbol{v}}(\mathbf{k}, t) = \frac{1}{\hbar}\partial_{\mathbf{k}}\hat{H}\left[\mathbf{k} - \frac{e}{\hbar}\mathbf{A}(t)\right]. \tag{13.52}$$

Let us now calculate the average value of the velocity operator in the state evolved adiabatically from one initial state, defined as an instantaneous eigenstate of the Hamiltonian. Equation (10.115) is an approximation, since as the state evolves, even if very slowly, there is an overlap with the other instantaneous eigenstates of the Hamiltonian. Therefore, there is a superposition of terms with time-dependent coefficients for each instantaneous eigenstate. The time dependence of these coefficients is obtained substituting the wave function in Schrödinger equation, as usually done in standard time-dependent perturbation theory. Assuming that at some initial instant the system is in some particular eigenstate m, integrating the equations for the time-dependent coefficients leads to the time-evolved wave function as

$$\begin{aligned} |\psi_m(\mathbf{k}, t)\rangle &= e^{i\gamma_B(t)} e^{-\frac{i}{\hbar}\int^t E_m(t')dt'} \\ &\times \left[|u_m(\mathbf{k}, t)\rangle - i\hbar \sum_{j(\neq m)} \frac{\langle u_j(\mathbf{k}, t)|\partial_t u_m(\mathbf{k}, t)\rangle}{E_m(t) - E_j(t)} |u_j(\mathbf{k}, t)\rangle \right]. \end{aligned} \tag{13.53}$$

It turns out we need to calculate

$$\begin{aligned} \langle \psi_m(\mathbf{k}, t)|\hat{\boldsymbol{v}}(\mathbf{k}, t)|\psi_m(\mathbf{k}, t)\rangle &= \langle u_m(\mathbf{k}, t)|\hat{\boldsymbol{v}}(\mathbf{k}, t)|u_m(\mathbf{k}, t)\rangle \\ &- i\ [\langle \partial_{\mathbf{k}} u_m(\mathbf{k}, t)|\partial_t u_m(\mathbf{k}, t)\rangle - c.c\ ], \end{aligned} \tag{13.54}$$

since we may use the chain rule as

$$|\partial_t u_m(\mathbf{k}, t)\rangle = \frac{e}{\hbar}\boldsymbol{E} \cdot \partial_{\mathbf{k}}|u_m(\mathbf{k}, t)\rangle, \tag{13.55}$$

to relate the time derivative with a momentum derivative. Using that for $j \neq m$,

$$\hbar\langle u_m|\hat{\boldsymbol{v}}(\mathbf{k}, t)|u_j\rangle = \left(E_j - E_m\right)\langle u_m|\partial_{\mathbf{k}} u_j\rangle = \left(E_m - E_j\right)\langle \partial_{\mathbf{k}} u_m|u_j\rangle, \tag{13.56}$$

we can obtain that

[4] For further reading see D. Xiao, M.-C. Chang and Q. Niu, Reviews of Modern Physics **82**, 1959 (2010) and [AraujoSacramento].

$$\langle v_i(t)\rangle = (\boldsymbol{v}_m)_i - \frac{ie}{\hbar}E_j\left[\langle\partial_{k_i}u_m|\partial_{k_j}u_m\rangle - c.c\right]. \tag{13.57}$$

We may now recall the definition of the curvature vector,

$$(\boldsymbol{\Omega})_\mu\epsilon_{ij\mu} = \Omega^{(m)}_{ij} = i\langle\partial_{k_i}u_m|\partial_{k_j}u_m\rangle - c.c\,, \tag{13.58}$$

and rewrite (13.57) as

$$\langle v_i(t)\rangle = (\boldsymbol{v}_m)_i - \frac{ie}{\hbar}(\boldsymbol{E}\times\boldsymbol{\Omega})_i\,,$$

or, going back to vectorial notation,

$$\langle\boldsymbol{v}(\mathbf{k},t)\rangle = \boldsymbol{v}_m(\mathbf{k},t) - \frac{e}{\hbar}\boldsymbol{E}\times\boldsymbol{\Omega}^{(m)}\,. \tag{13.59}$$

The second term goes under the designation of anomalous velocity and is perpendicular to the electric field. The Berry curvature may therefore produce a current perpendicular to the applied field, after performing the integration over occupied states. This is a Hall-like current induced by the topology and in the absence of a magnetic field. It is one example of an anomalous Hall effect, in the sense that the effect exists without an applied external magnetic field. An anomalous Hall effect also appears due to magnetization effects.

Variation of Electric Polarization

The previous discussion allows us another way around the problem that in a periodic system the electric polarization is ill defined. The insight is to realize that variations of polarization are well-defined. Using arguments from electrostatics, we may recall that the polarization and the charge density are related by

$$\nabla\cdot\mathbf{P}(\mathbf{r}) = -\rho(\mathbf{r}). \tag{13.60}$$

The charge conservation equation relates the charge density and the charge current

$$\frac{\partial\rho(\mathbf{r})}{\partial t} + \nabla\cdot\mathbf{j} = 0. \tag{13.61}$$

Taking a time derivative we get that

$$\nabla\cdot\left(\frac{\partial\mathbf{P}}{\partial t} - \mathbf{j}\right) = 0. \tag{13.62}$$

Considering two states at times $t = 0$ and $t = T$, we may relate the polarization difference between the two states with the current through the system, as the system evolves adiabatically in time. Apart from a term with zero divergence, we may write that

$$\Delta \mathbf{P} = \int_0^T dt \mathbf{j}. \tag{13.63}$$

Using that the adiabatic current that results from some time-dependent perturbation is given by Xiao et al. [1]

$$j = -\sum_n \int_{BZ} \frac{dk}{2\pi} \Omega^n_{kt}, \tag{13.64}$$

where we used Eqs. (13.53) and (13.54), we get that the α component of the change in polarization may be written as

$$\Delta P_\alpha = e \sum_n \int_0^T dt \int_{BZ} \frac{d\mathbf{k}}{(2\pi)^d} \Omega^n_{k_\alpha t}. \tag{13.65}$$

The change in polarization is therefore written in terms of a Berry-like curvature for the level n with wave function u_n

$$\Omega^n_{kt} = i \left(\langle \frac{\partial u_n}{\partial k} | \frac{\partial u_n}{\partial t} \rangle - \langle \frac{\partial u_n}{\partial t} | \frac{\partial u_n}{\partial k} \rangle \right) \tag{13.66}$$

and therefore the polarization difference is well-defined in this way.

13.3 Landau Levels

Let us consider a free electron in a homogeneous magnetic field, aligned along the z axis. In minimal coupling the free Hamiltonian is replaced by

$$H = \frac{1}{2m} (\mathbf{p} + e\mathbf{A})^2 . \tag{13.67}$$

Here $\mathbf{A}$ is the vector potential and $-e$ is the electron charge. We may choose a gauge such that $\nabla \cdot \mathbf{A} = 0$. Choosing a uniform field like $\mathbf{B} = B\mathbf{e}_z$, we can use a Landau gauge as $\mathbf{A} = (0, Bx, 0)$ or $\mathbf{A} = (-By, 0, 0)$ or also $\mathbf{A} = B/2(-y, x, 0) = 1/2\mathbf{B} \times \mathbf{r}$, called symmetric gauge.

Diagonalization of Hamiltonian: Solution of Schrödinger Equation
We can obtain the energy levels of the particle in the magnetic field solving, for instance, Schrödinger equation

$$\frac{1}{2m} (-i\hbar\nabla + e\mathbf{A})^2 \psi = \epsilon \psi, \tag{13.68}$$

which can be expanded to

$$-\frac{\hbar^2}{2m}\nabla^2\psi - \frac{i\hbar e}{m}\mathbf{A}\cdot\nabla\psi + \frac{e^2}{2m}\mathbf{A}^2\psi = \epsilon\psi. \tag{13.69}$$

Selecting, for instance a Landau gauge $\mathbf{A} = (0, Bx, 0)$, we get that Schrödinger equation is written as

$$\frac{\partial^2\psi}{\partial x^2} + \frac{\partial^2\psi}{\partial y^2} + \frac{\partial^2\psi}{\partial z^2} + \frac{2ieB}{\hbar}x\frac{\partial\psi}{\partial y} - \frac{e^2B^2}{\hbar^2}x^2\psi + \frac{2m\epsilon}{\hbar^2}\psi = 0. \tag{13.70}$$

Taking a solution of the form

$$\psi(x, y, z) = e^{i(k_y y + k_z z)}u(x),$$

we get that

$$-\frac{\hbar^2}{2m}\frac{d^2u(x)}{dx^2} + \frac{1}{2}m\left(\frac{eB}{m}x + \frac{\hbar k_y}{m}\right)^2 u(x) = \epsilon' u(x). \tag{13.71}$$

Here $\epsilon' = \epsilon - \frac{\hbar^2}{2m}k_z^2$. This equation is the equation for a harmonic oscillator of frequency $\omega_B = \frac{eB}{m}$, centered at the point $x_0 = -\frac{1}{\omega_B}\frac{\hbar k_y}{m}$. Therefore, the energy, besides the kinetic energy along the z direction, is that of the levels of the harmonic oscillator: $\epsilon_n' = \hbar\omega_B(n + \frac{1}{2})$. Therefore,

$$\epsilon = \frac{\hbar^2}{2m}k_z^2 + \hbar\omega_B\left(n + \frac{1}{2}\right). \tag{13.72}$$

As in the classical case, the motion is a straight line along the field direction plus a circular motion in the perpendicular plane. However, due to the quantum nature of the problem, the states in the perpendicular plane are now quantized. The energy in the plane xy is quantized in units of the cyclotron frequency, ω_B. The wave functions of the lowest Landau levels are shown in Fig. 13.5.

Due to the liberty of placing the center of each harmonic oscillator in any point along the x direction, the harmonic oscillator levels are degenerate. The degeneracy is given by

$$g(B) = L_x L_y \frac{eB}{h}, \tag{13.73}$$

where we have used periodic boundary conditions for the plane waves along the y direction.

If we use instead the symmetric gauge, Schrödinger equation can be written as

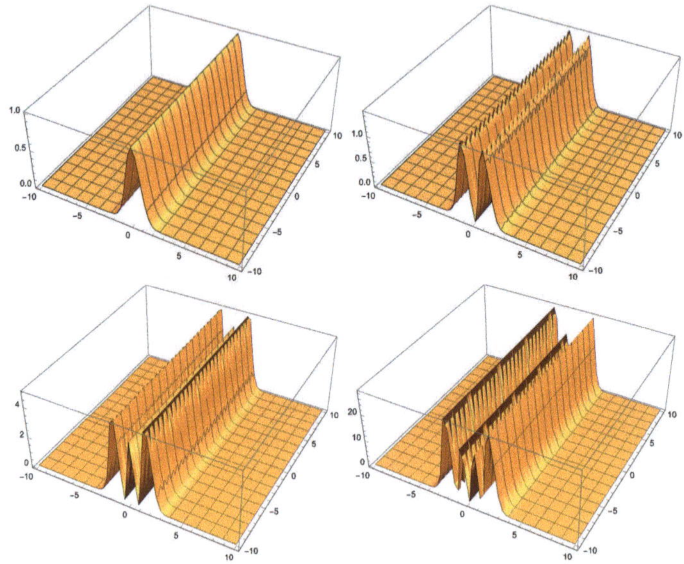

Fig. 13.5 Absolute value squared of the wave functions of the first Landau levels ($n = 0, 1, 2, 3$) using the Landau gauge

$$-\frac{\hbar^2}{2m_e}\nabla^2\psi - \frac{i\hbar eB}{2m_e}\left(-y\frac{\partial}{\partial x} + x\frac{\partial}{\partial y}\right)\psi + \frac{e^2B^2}{8m_e}\left(x^2 + y^2\right)\psi = \epsilon\psi \qquad (13.74)$$

(it is convenient to write the mass of the electron as m_e). Noting that the z component of the angular momentum can be written as

$$L_z = i\hbar\left(x\frac{\partial}{\partial y} - y\frac{\partial}{\partial x}\right), \qquad (13.75)$$

we can rewrite Schrödinger equation as

$$-\frac{\hbar^2}{2m_e}\nabla^2\psi + \frac{eB}{2m}L_z\psi + \frac{e^2B^2}{8m}\left(x^2 + y^2\right)\psi = \epsilon\psi. \qquad (13.76)$$

It is convenient to use cylindrical coordinates $x = \rho\cos\phi$, $y = \rho\sin\phi$, $z = z$ and to choose a solution of the form

$$\psi(z, \rho, \phi) = e^{im\phi}e^{ikz}u(\rho). \qquad (13.77)$$

We can then reformulate Schrödinger equation as

$$\frac{d^2u(\rho)}{d\rho^2} + \frac{1}{\rho}\frac{du}{d\rho} - \frac{m^2}{\rho^2}u - \frac{e^2B^2}{4\hbar^2}\rho^2 u(\rho) + \left(\frac{2m_e\epsilon}{\hbar^2} - \frac{eBm}{\hbar} - k^2\right)u(\rho) = 0. \tag{13.78}$$

Defining a new variable as

$$x = \sqrt{\frac{eB}{2\hbar}}\rho \tag{13.79}$$

and looking for a solution of the form

$$u(x) = x^{|m|}e^{-\frac{x^2}{2}}G(x), \tag{13.80}$$

we get an equation for the function $G(x)$ as

$$\frac{d^2G}{dx^2} + \left(\frac{2|m|+1}{x} - 2x\right)\frac{dG}{dx} + (\lambda - 2 - 2|m|)\,G = 0, \tag{13.81}$$

with

$$\lambda = \frac{4m_e}{eB\hbar}\left(\epsilon - \frac{\hbar^2k^2}{2m_e}\right) - 2m. \tag{13.82}$$

Defining further a new variable $y = x^2$, we get

$$\frac{d^2G(y)}{dy^2} + \left(\frac{|m|+1}{y} - 1\right)\frac{dG}{dy} + \frac{\lambda - 2 - 2|m|}{4y}G = 0. \tag{13.83}$$

This equation may be recognized as the equation for the radial part of the hydrogen atom. Therefore,

$$G(y) = L_{n_r}^{|m|}(y), \tag{13.84}$$

the associated Laguerre polynomial. Here, $n_r = 0, 1, 2, 3, \ldots$. The energy levels are

$$\epsilon = \frac{\hbar^2k^2}{2m_e} + \frac{eB\hbar}{2m_e}\left(2n_r + 1 + |m| + m\right) \tag{13.85}$$

and the wave functions are

$$\psi = e^{im\phi}e^{ikz}\left(\left(\frac{eB}{2\hbar}\right)^{1/2}\rho\right)^{|m|} e^{-\frac{1}{2}\frac{eB}{2\hbar}\rho^2}L_{n_r}^{|m|}\left(\frac{eB}{2\hbar}\rho^2\right). \tag{13.86}$$

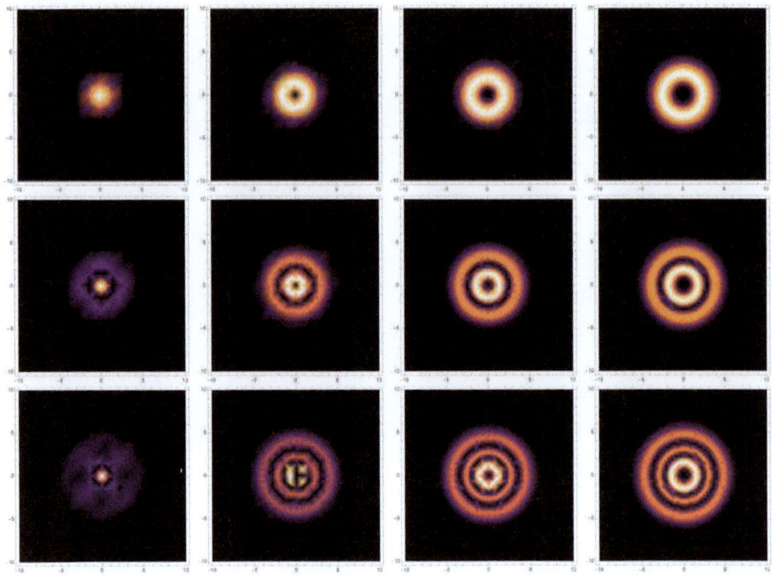

Fig. 13.6 Absolute value squared of the wave functions of the first Landau levels using the symmetric gauge. In the first row $n = 0$ and $m = 0, 1, 2, 3$. In the second row $n = 1$ and the same values of m and in the third row $n = 2$ and the same values of m

Since the wave functions are gauge dependent, the wave functions are different from those in the Landau gauge and are shown in Fig. 13.6. However, the energy levels are not gauge dependent, and it can be verified that the spectrum is the same.

Note that applying a magnetic field introduces a scale in the problem that is called magnetic length $l = \sqrt{\frac{\hbar}{eB}}$, in terms of which all quantities are measured, independently of the gauge used.

The lowest energy state is obtained taking $n_r = 0$ and $m \leq 0$. In this case the unnormalized wave function leads to

$$\psi_{LLL} = e^{-i|m|\phi} e^{ikz} \left(\frac{1}{\sqrt{2}} \frac{\rho}{l} \right)^{|m|} e^{-\frac{1}{4}(\frac{\rho}{l})^2}, \tag{13.87}$$

since

$$L_0^{|m|} = 1. \tag{13.88}$$

Defining $z = x + iy$, where x, y are the Cartesian coordinates of the plane, we can write that the normalized wave function is

$$\psi_{LLL} = e^{ikz} \frac{1}{\sqrt{2\pi l^2 2^{|m|} |m|!}} (z^*)^{|m|} e^{-\frac{1}{4}\left(\frac{|z|}{l}\right)^2}. \tag{13.89}$$

In the literature it is often considered that the magnetic field points in the $-z$ direction. With this choice, the term in the Hamiltonian that is proportional to the z component of the angular momentum has a negative sign and the spectrum has the form

$$\epsilon = \frac{\hbar^2 k^2}{2m_e} + \frac{eB\hbar}{2m_e} \left(2n_r + 1 + |m| - m\right), \tag{13.90}$$

which implies that the lowest energy level has $m = |m|$, which means the values of m are positive. This leads to a wave function of the lowest Landau level in the form

$$\psi_{LLL} = e^{ikz} \frac{1}{\sqrt{2\pi l^2 2^m m!}} z^m e^{-\frac{1}{4}\left(\frac{|z|}{l}\right)^2}. \tag{13.91}$$

Diagonalization of Hamiltonian: Operator Formalism

As we saw using the Landau gauge, the wave functions in the plane perpendicular to the magnetic field are those of the harmonic oscillator along the x direction and plane waves in the y direction. This suggests that we may use an operator formalism to diagonalize the Hamiltonian, instead of solving directly Schrödinger equation.

The velocity operator can be written in terms of the canonical momentum operator as

$$\mathbf{v} = \frac{1}{m} \left(\mathbf{p} + e\mathbf{A}(\mathbf{r})\right) \tag{13.92}$$

and the velocity components do not commute. For instance,

$$[v_x, v_y] = -i\frac{\hbar\omega_B}{m}. \tag{13.93}$$

Note that the commutator is gauge independent. We may now define the operators

$$\begin{aligned} \Pi &= \frac{ml}{\hbar}\frac{v_x - iv_y}{\sqrt{2}} \\ \Pi^\dagger &= \frac{ml}{\hbar}\frac{v_x + iv_y}{\sqrt{2}}, \end{aligned} \tag{13.94}$$

which obey harmonic-oscillator commutation relations like those of lowering and raising operators:

$$[\Pi, \Pi^\dagger] = 1. \tag{13.95}$$

The role of the coordinate and momentum operators in a one-dimensional harmonic oscillator is taken by the x and y velocity components. Note that the velocity operator mixes momentum and spatial coordinates. The Hamiltonian is of the form

$$H = \frac{1}{2}mv^2 = \hbar\omega_B \left(\Pi^\dagger \Pi + \frac{1}{2} \right) + \frac{\hbar^2 k_z^2}{2m}, \tag{13.96}$$

as expected of an harmonic oscillator, with the spectrum

$$\epsilon_n(k_z) = \left(n + \frac{1}{2} \right) \hbar\omega_B + \frac{\hbar^2 k_z^2}{2m}. \tag{13.97}$$

Magnetic Translation Group

The classical trajectory of a particle in a magnetic field, projected to the plane perpendicular to the magnetic field, is a circle of a given radius, R. This radius is a function of the magnetic field, charge, and mass of the particle, such that the particle is orbiting around a given center in space, characterized by coordinates, X and Y, with an angular frequency given by ω_B. Defining complex variables $z = x + iy$ and $v = v_x + iv_y$ the classical equations of motion on the plane can be written as $z(t) = Z + Re^{-i\omega_B t}$ and $v(t) = -i\omega_B Re^{-i\omega_B t}$, where $Z = X + iY$. These equations lead to equations for the coordinates of the center of the orbit as[5]

$$\begin{aligned} X &= x + \frac{v_y}{\omega_B} \\ Y &= y - \frac{v_x}{\omega_B}. \end{aligned} \tag{13.98}$$

In the quantum case, where the quantities are now operators, we can see that

$$[X, v_x] = [X, v_y] = [Y, v_x] = [Y, v_y] = 0 \tag{13.99}$$

and, consequently,

$$[X, H] = [Y, H] = 0, \tag{13.100}$$

which means that the center of rotation is a constant of motion. On the other hand,

$$[X, Y] = -il^2. \tag{13.101}$$

Note that we are considering that the magnetic field is aligned along $-z$ which implies the commutation relation $[v_x, v_y] = i\hbar\omega_B/m$.

It is now convenient to define the operators

$$T_x = -m\omega_B Y = m\left(v_x - \omega_B y\right)$$

[5] See also G. F. Giuliani and G. Vignale, *Quantum Theory of the Electron Liquid*, Cambridge University Press.

$$T_y = m\omega_B X = m\left(v_y + \omega_B x\right). \tag{13.102}$$

These operators satisfy the following commutation relations:

$$[T_x, x] = [T_y, y] = -i\hbar \tag{13.103}$$

and

$$[T_x, y] = [T_y, x] = 0. \tag{13.104}$$

These commutation relations are of the type of those that we find between the momentum and position operators in a system with no applied magnetic field (see Appendix C). Therefore, the operators T_i, with $i = x, y$, take the role of the translation generators in the presence of a magnetic field. However, unlike the components of the translation generators in the absence of a magnetic field that commute with each other, here

$$[T_x, T_y] = -i\frac{\hbar^2}{l^2} \neq 0. \tag{13.105}$$

Even though both operators T_x and T_y commute with the Hamiltonian, since they do not commute with each other, this means that we may label the Hamiltonian eigenstates with the eigenvalues of one of the translational generators, but not both. For instance, denoting the eigenvalues of T_y as $\hbar k_y$ we may conclude that

$$T_y|\epsilon_n, k_y\rangle = \hbar k_y|\epsilon_n, k_y\rangle, \tag{13.106}$$

since the eigenstates of the two operators, H and T_y, may be chosen as common eigenstates. However, the action of T_x is non-trivial.

The translation operator by a vector $\mathbf{a}$ may be written as

$$t(\mathbf{a}) = e^{\frac{i}{\hbar}\mathbf{T}\cdot\mathbf{a}}. \tag{13.107}$$

The different commutation relations between the translational generators lead to an algebra for the translation operators that is different from the case without magnetic field. The algebra leads to a different group called the magnetic translation group.

The translation along one of the spatial directions may be defined in the usual way, say along y, and the non-trivial nature is reflected when a translation along x is considered. For instance, using periodic boundary conditions in the y direction we get as usual, defining the translational operator across the system length L_y as $t(\mathbf{a} = L_y\mathbf{u}_y) = e^{\frac{i}{\hbar}T_y L_y} = 1$, leads to the usual quantization $k_y = n_y(2\pi)/L_y$, where n_y is an integer.

The action of a translation along x is more complex. Consider first a displacement of the type $\mathbf{a} = 2\pi l^2/L_y\mathbf{u}_x$. It can be seen that

$$t(2\pi l^2/L_y\mathbf{u}_x)|\epsilon_n, \frac{2\pi}{L_y}n_y\rangle = |\epsilon_n, \frac{2\pi}{L_y}(n_y+1)\rangle, \tag{13.108}$$

changing k_y and maintaining the energy while displacing the x coordinate by $\Delta x = 2\pi l^2/L_y$. If the system size along the x direction is L_x, the maximal number of displacements is $N_x = L_x/\Delta x$, which is

$$N_x = \frac{L_x L_y}{2\pi l^2}. \tag{13.109}$$

Since the energy is maintained, N_x is also the maximal number of k_y values allowed on each energy level and, therefore, its degeneracy, $g = N_x$. This is given by

$$g = \frac{eBA}{h}, \tag{13.110}$$

as previously shown, where $A = L_x L_y$ is the area perpendicular to the magnetic field.

More generally we have that the translation operators obey the composition law

$$t(\mathbf{a})t(\mathbf{b}) = e^{\frac{i}{2l^2}(\mathbf{a}\times\mathbf{b})\cdot\mathbf{u}_z}t(\mathbf{b}+\mathbf{a}) \tag{13.111}$$

and satisfy the magnetic algebra

$$t(\mathbf{a})t(\mathbf{b}) = e^{-i\frac{(\mathbf{a}\times\mathbf{b})\cdot\mathbf{u}_z}{l^2}}t(\mathbf{b})t(\mathbf{a}). \tag{13.112}$$

Note that if

$$\frac{(\mathbf{a}\times\mathbf{b})\cdot\mathbf{u}_z}{l^2} = 2\pi n, \tag{13.113}$$

with n an integer, the translation operators commute. Defining an area A by the two vectors **a** and **b**, the condition may be seen as a condition for the magnetic flux across the area A as

$$\Phi = BA = \Phi_0 n, \tag{13.114}$$

where the quantity

$$\Phi_0 = \frac{h}{e} \tag{13.115}$$

is called the quantum of flux (to be discussed later). So if the magnetic flux is a multiple of the flux quantum, the magnetic translation operators have an algebra that is similar to that in the absence of magnetic field.

13.4 Landau Diamagnetism

To study the response of the system to the external field due to the presence of the vector potential, we have to calculate the average magnetization. Let us consider the grand partition function. Recall that is given by

$$\ln \mathcal{Z} = \sum_{\epsilon} \ln\left(1 + ze^{-\beta\epsilon}\right)$$

$$\epsilon = \frac{e\hbar B}{m}\left(j + \frac{1}{2}\right) + \frac{p_z^2}{2m}. \tag{13.116}$$

Converting the sum over the z component momenta into an integral, we get that

$$\ln \mathcal{Z} = \int_{-\infty}^{\infty} dp_z \frac{L_z}{h}\left(\sum_{j=0}^{\infty}\left(L_x L_y \frac{eB}{h}\right) \ln\left(1 + ze^{-\frac{\beta e\hbar B}{m}(j+\frac{1}{2})}e^{-\beta\frac{p_z^2}{2m}}\right)\right). \tag{13.117}$$

Here as before $z = e^{\frac{\mu}{k_B T}}$ is the fugacity.

Let us consider first the case when $z \ll 1$. Expanding the logarithm in powers of z, we get that

$$\ln \mathcal{Z} = \frac{zVeB}{h^2}\left(\frac{2\pi m}{\beta}\right)^{\frac{1}{2}} \frac{1}{2\sinh\frac{\beta e\hbar B}{2m}}.$$

Then the magnetization is given by

$$M = \frac{1}{\beta}\frac{\partial}{\partial B}\ln \mathcal{Z} = \frac{zV}{\lambda^3}\mu\left[\frac{1}{\sinh x} - x\frac{\cosh x}{\sinh^2 x}\right]. \tag{13.118}$$

Here $x = \beta\mu\beta$, $\mu = \frac{eh}{4\pi m}$ and the thermal wave length is $\lambda = \frac{h}{(2\pi m k_B T)^{1/2}}$. Note that the magnetization is negative, and therefore we get a magnetization that opposes the external magnetic field: this is diamagnetism. Also note that this is a quantum phenomenon: if we let $h \to 0$, the magnetization vanishes, in agreement with the Bohr-van Leeuwen theorem. In the limit of $\mu B \ll k_B T$ we get again a Curie-like expression where the susceptibility is now

$$\chi = -\frac{z\mu^2}{3\lambda^3 k_B T}.$$

The case when $z \gg 1$ is more involved. In the case when $\mu B \ll k_B T \ll E_F$ one gets that the susceptibility is given by

$$\chi = -\frac{1}{2}\frac{n\mu^2}{\epsilon_F}.$$

This result is such that

$$|\chi_{diamagnetic}| = \frac{1}{3}|\chi_{Pauli}|. \quad (13.119)$$

13.5 Quantum Hall Effect

13.5.1 Integer Quantum Hall Effect

As we discussed before, in the Hall effect one typically considers a two-dimensional system to which we apply a perpendicular magnetic field which, by the Lorentz force that acts on the moving charges (originated by the application of a longitudinal electric field), leads to a deflection of the charges to the transverse direction, such that an electric potential difference emerges in the direction perpendicular to the current and the magnetic field. The static ($\omega = 0$) conductivity and resistivity tensors in the plane have the form

$$\sigma = \frac{\sigma_0}{1+(\omega_c\tau)^2}\begin{pmatrix} 1 & -\omega_c\tau \\ \omega_c\tau & 1 \end{pmatrix} \quad (13.120)$$

and

$$\rho = \begin{pmatrix} \rho_0 & \frac{B}{ne} \\ -\frac{B}{ne} & \rho_0 \end{pmatrix}, \quad (13.121)$$

where B is the magnetic field intensity, ne is the charge density, m is the mass of the particles, $\omega_c = eB/m$, and τ is the average time between collisions. The conductivity $\sigma_0 = (ne^2\tau)/m$ is the conductivity in the absence of the magnetic field and $\rho_0 = 1/\sigma_0$.

In the presence of a very high magnetic field, it was found experimentally that, in certain conditions in electronic two-dimensional systems, the conductivity tensor has the form

$$\sigma = \begin{pmatrix} 0 & -i_{LL}e^2/h \\ i_{LL}e^2/h & 0 \end{pmatrix} \quad (13.122)$$

and the resistivity tensor has the inverse form

$$\rho = \begin{pmatrix} 0 & h/i_{LL}e^2 \\ -h/i_{LL}e^2 & 0 \end{pmatrix}, \quad (13.123)$$

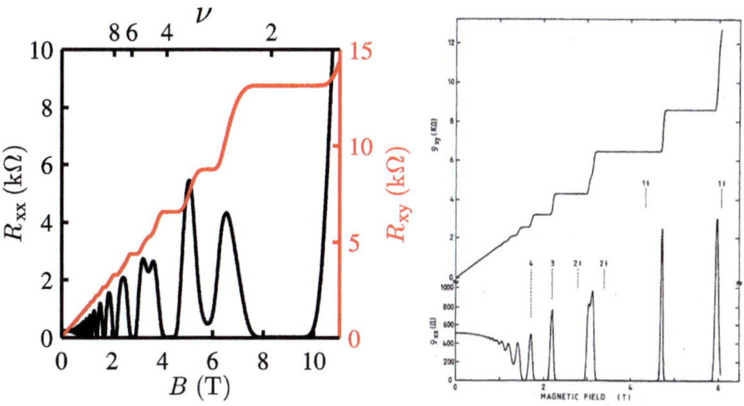

Fig. 13.7 Integer quantum Hall effect. On the left panel scheme of the Hall resistance plateaus and on the right panel experimental results for $GaAs - Al_xGa_{1-x}As$ heterostructures at low temperatures (Reproduced with permission from M E Suddards, A Baumgartner, M Henini, and C J Mellor, New Journal of Physics **14**, 083015 (2012) and from Klaus von Klitzing, Reviews of Modern Physics **58**, 519 (1986))

where i_{LL} is an integer. This is shown in Fig. 13.7, where it can be seen that there are regimes where the transverse resistivity has a plateau, and that in that regime the longitudinal resistivity vanishes [2]. This quantization of the transverse resistivity constitutes the quantum Hall effect.

If the system is strictly two-dimensional and we apply a perpendicular magnetic field, then there is no motion along the direction of the magnetic field and the energy levels are those of a harmonic oscillator (recall that if we use the symmetric gauge, $\mathbf{A} = \frac{1}{2}\mathbf{B}\times\mathbf{r}$, the states are parametrized by the Landau level and by the component of the orbital angular momentum and, also, are localized in the two spatial directions). Since the degeneracy of each Landau level is given by g, we define the filling of the Landau levels by the quantity

$$\nu = n/g, \tag{13.124}$$

where n is the electronic density (number of electrons per unit area). If $n = i_{LL}(g/A)$, where i_{LL} is an integer, then the system has i_{LL} levels completely full. Inserting this electronic density in the expression for the transverse resistivity, we obtain that

$$\frac{B}{ne} = \frac{h}{i_{LL}e^2}. \tag{13.125}$$

In other words, the quantized value of the transverse resistivity (or conductivity) is explained by the quantization of the Landau levels. However, for other values of the

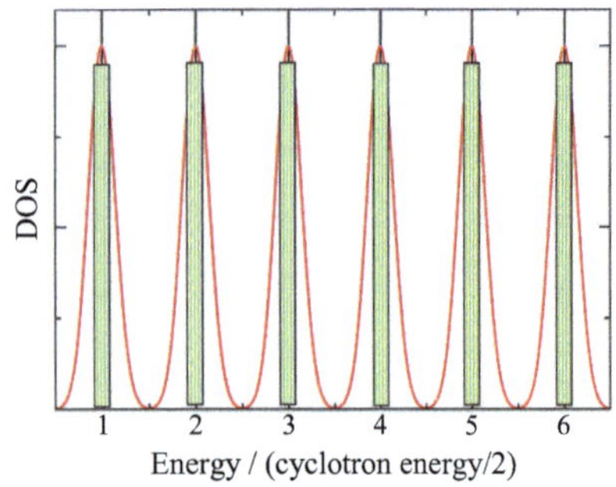

Fig. 13.8 In the absence of disorder the energy levels are the Landau levels. Introducing disorder, states appear in the gaps between the original Landau levels. Some of these states are extended (in the crossed regions near the original Landau levels) and the remaining states are localized

magnetic field, such that the last Landau level is not completely full, the expression does not justify the existence of the observed plateau.

Disorder Effect

In the Hall geometry, consider a relativistic transformation to a frame of reference that is moving with a velocity $\mathbf{v} = \mathbf{E} \times \mathbf{B}/B^2$. In this frame of reference the electric field vanishes and the magnetic field is approximately the same, given by $\mathbf{B}' = \mathbf{B}/\gamma$, with $\gamma \sim 1$ if the velocity is small compared to the speed of light, c. Then the current, which in general is given by (see the stationary regime of Eq. (13.47))

$$\mathbf{j} = \sigma_0 \mathbf{E} - \frac{\sigma_0}{ne} \mathbf{j} \times \mathbf{B}, \tag{13.126}$$

has the direction and adequate value if the Landau level is full. That is, the current is transverse. However, experimentally this situation is only observed at the plateaus. If the system is invariant under translations, then the passage to this frame of reference should always be allowed, and, therefore, to describe the experiment it is necessary that the translational invariance must be broken. This happens if we consider that there are impurities in the system. What happens is shown in Fig. 13.8. Without the presence of impurities, the density of states has localized delta function peaks at the energies that correspond to the Landau levels. The states are extended along the x direction and localized along y, with at least an extended state that carries the Hall current.

As disorder is introduced, the Landau levels become bands of delocalized states with states localized in the regions between bands. The presence of the electric field and impurities lifts the degeneracy of the states. The localized states do not transport current and, if the chemical potential is on the localized states, a variation of the chemical potential (or of the magnetic field that changes the degeneracy of each level) does not lead to a change in the conductivity: this is the origin of the plateau. These are regimes such that the states that are occupied do not contribute to the conductivity. When the chemical potential reaches the delocalized levels, one leaves the plateau regime, until the chemical potential crosses the next Landau level and moves again to a region of localized states (the gap region between the Landau levels in the regime without the disorder). The vanishing longitudinal resistivity (and conductivity) results from the gap in the system, due to the quantization of the Landau levels and the localized states originated by the impurities. When the chemical potential is found on a localized state, the longitudinal conductivity vanishes.

Some comments should be made at this point:

(i) A two-dimensional system with no magnetic field present and in the presence of disorder is a system where localization takes place, and, therefore, all states are localized. This will be shown later in the book. Such is due to interference effects between the incident electronic waves and the waves reflected by the impurities. What the quantum Hall effect shows is that, in order to explain the effect, the introduction of the impurities is necessary and that the conductivity is non-vanishing in general (outside the plateaus). In other words, the presence of the magnetic field counteracts the localization effect caused by the impurities, since by breaking the time-reversal invariance the interference effect does not occur fully. In other words, the effect of the magnetic field is to reduce the kinetic energy, due to the restriction that the electrons are conditioned to follow the Landau orbitals.

(ii) The case of a single impurity that is modeled via a repulsive or attractive local potential, of the Dirac delta form, has been studied, and the conclusion is that the local potential gives origin to a localized state (which does not carry current), but with the interesting property that the remaining delocalized states carry a current excess, which compensates exactly the lost current due to the localized state.

(iii) An important experimental result is that the quantum Hall effect is very robust. That is, it is observed in several materials, shapes, and disorder amounts, which suggests some universality. In other words it raises the possibility that one may choose particular geometry shapes, suitable for calculations, to attempt a better understanding of the phenomenon. This independence on the material and the accuracy of the measured values (of the order of one part per million) has been used to establish new standards of current, in addition to the value of the fine structure constant.

(iv) The universality of the phenomenon suggests that the phenomenon may have some macroscopic coherence.

(v) The interpretation of the phenomenon does not take into account any interactions between the electrons, and it is enough to adopt a single-particle nature and take into account the fermionic nature of the electrons. At densities close to the complete filling of the Landau levels, the excitations have high energies, of the order of ω_C. Both the kinetic energy (that is effectively eliminated by the restriction that the orbits are Landau levels) and the potential energy between the electrons are small, in comparison with the magnetic energy.

Gauge Invariance and Quantization

These ideas led Laughlin [3] to search for a universal principle that would regulate the systems behavior. The ribbon geometry shown in Fig. 13.9 was considered. The system is traversed by a current, and, due to the perpendicular magnetic field, a Hall

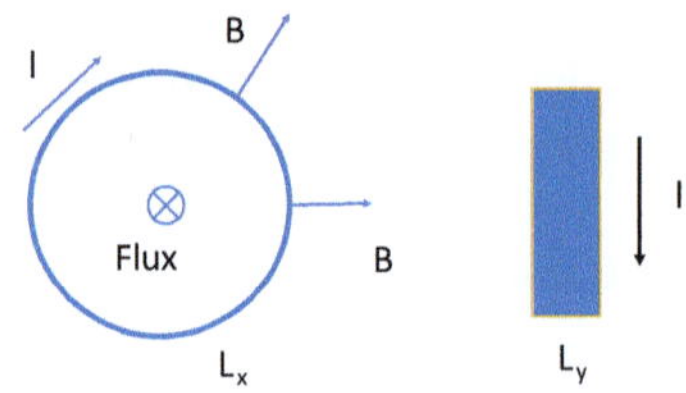

Fig. 13.9 Laughlin geometry. A ribbon of length L_x and width L_y is traversed by a current I along the x direction and a magnetic field, B, is applied perpendicularly to the ribbon at each point. A magnetic flux, Φ, is also applied through the ring

tension appears between the edges of the ribbon with length L_x and width L_y. The current is given by

$$I = \frac{\partial E}{\partial \Phi} = \frac{1}{L_x}\frac{\partial E}{\partial A}, \tag{13.127}$$

where Φ is the magnetic flux and E the energy. If the states are extended along the system and we change the vector potential A, then the change of the vector potential, which leads to a change of the wave function as given by the Dirac-Feynman phase $e^{(i/\hbar)S} = e^{ieAx/\hbar}$, where S is the action that results from the minimal coupling between the electron current and the vector potential, implies that the vector potential (or its change) must be of the type

$$A = i_A \frac{h}{eL_x}, \tag{13.128}$$

where i_A is an integer, since the magnetic flux must be a multiple of the flux quantum, Φ_0, in order to leave the system invariant as we go around the length of the ribbon.

In this geometry it is more convenient to use a Landau gauge as $\mathbf{A} = (-By, 0, 0)$. Applying a uniform electric field along the y direction, one needs to add to the Hamiltonian a term as $V = eE_0 y$. Solving Schrödinger equation following similar steps as before and selecting a wave function of the form $\psi = e^{i(k_x x + k_z z)}u(y)$ leads to a solution along the y direction that can be seen as a harmonic oscillator with frequency ω_B centered around a point given by

$$y_0 = \frac{1}{\omega_B}\left(\frac{\hbar k_x}{m} - \frac{E_0}{B}\right).$$

If we change the vector potential by a value ΔA, the effect of this variation is to change the center of the harmonic oscillators by $y_0 \to y_0 - \frac{\Delta A}{B}$. The energy of the system changes linearly with ΔA since

$$\epsilon_{n,k} = \left(n + \frac{1}{2}\right)\hbar\omega_B + eE_0 y_0 + \frac{1}{2}m\left(\frac{E_0}{B}\right)^2 . \tag{13.129}$$

The displacement of the center of a harmonic oscillator can be seen as a multiple of $L_y\Phi_0/\Phi_B$, where Φ_0 is the flux quantum and Φ_B is the total magnetic flux across the ribbon due to the magnetic field B. Displacing y_0 by L_y leads to the result that Φ_B/Φ_0 is an integer.

Considering a flux change by a flux quantum, the system tends to return to its original state. The increase in energy results from a charge transfer of i_A electrons (one per each occupied Landau level) from one side of the ribbon to the other edge. The corresponding current is given by

$$I = \frac{i_A e\Delta V}{\Delta\phi} = \frac{i_A e^2 \Delta V}{h}. \tag{13.130}$$

13.5.2 Quantum Hall Effect and Chern Number

The quantization of the Hall conductivity is then the result of gauge invariance, a general principle. Therefore, it is a way to explain the universality of the phenomenon. On the other hand, it is possible to calculate the Hall conductivity directly using the Kubo formula. In the regime of localized states, with an exponential decay from given spatial locations, one expects that the conductivity is insensitive to changes in boundary conditions. Under the assumption that the ground state of the system is nondegenerate (for a completely filled Landau level) and that there is a finite gap to the excited states, it has been shown that [4]

$$\sigma = \frac{e^2}{h}\int_0^{2\pi}\int_0^{2\pi} d\theta d\varphi \frac{1}{2\pi i}\left(< \frac{\partial\phi_0}{\partial\varphi} | \frac{\partial\phi_0}{\partial\theta} > - < \frac{\partial\phi_0}{\partial\theta} | \frac{\partial\phi_0}{\partial\varphi} > \right), \tag{13.131}$$

where ϕ_0 is the ground state. Here θ and ϕ are the phase angles used in the twisted boundary conditions along the x and y directions of the finite system. That is, instead of periodic boundary conditions, one takes $\phi_0(0, 0) = \phi_0(L_x, 0)e^{i\theta L_x} = \phi_0(0, L_y)e^{i\varphi L_y}$. This expression for the Hall conductivity may be understood in a different way that shows that the result is associated with a topological invariant, the Chern number. Parametrizing the space of variables in terms of the components of the momentum (k_x, k_y), we can see that Eq. (13.131) may be understood as an example of Eq. (10.120). Therefore, it can be verified that the Hall conductivity for a filled band, n, is

$$\sigma_{yx} = \frac{e^2}{h}\int \Omega^{(n)}_{k_x,k_y}\, dk_x\, dk_y = \frac{e^2}{h}C^{(n)}, \tag{13.132}$$

which confirms that the conductivity can be understood in terms of the topological invariant, the Chern number.

The Berry curvature, $\Omega_{\mu\nu}^{(n)}$, and as a consequence the expression for the Hall conductivity, may be written as a sum over states which does not involve derivatives of the wave function.[6] After taking the derivative of the Schrödinger equation,

$$\hat{H}(\boldsymbol{R})|\psi_n(\boldsymbol{R})\rangle = E_n(\boldsymbol{R})|\psi_n(\boldsymbol{R})\rangle\,, \tag{13.133}$$

with respect to $\boldsymbol{R}$, we can then either left-multiply by $\langle\psi_n(\boldsymbol{R})|$, thereby obtaining Hellmann-Feynman theorem,

$$\langle\psi_n(\boldsymbol{R})|\frac{\partial \hat{H}}{\partial R^\nu}|\psi_n(\boldsymbol{R})\rangle = \frac{\partial E_n(\boldsymbol{R})}{\partial R^\nu}\,, \tag{13.134}$$

or left-multiply by $\langle\psi_{n'}(\boldsymbol{R})|$, obtaining

$$\langle\psi_{n'}(\boldsymbol{R})|\frac{\partial \hat{H}}{\partial R^\nu}|\psi_n(\boldsymbol{R})\rangle = (E_n - E_{n'})\left\langle\psi_{n'}|\frac{\partial \psi_n}{\partial R^\nu}\right\rangle\,, \qquad n \neq n'\,. \tag{13.135}$$

This leads to the equations:

$$|\frac{\partial \psi_n}{\partial R^\mu}\rangle = \sum_{n'(\neq n)} \frac{1}{E_n - E_{n'}}\langle\psi_{n'}|\frac{\partial \hat{H}}{\partial R^\mu}|\psi_n\rangle\,|\psi_{n'}\rangle + \alpha_{n,\mu}|\psi_n\rangle\,, \tag{13.136}$$

$$|\frac{\partial \psi_n}{\partial R^\nu}\rangle = \sum_{n'(\neq n)} \frac{1}{E_n - E_{n'}}\langle\psi_{n'}|\frac{\partial \hat{H}}{\partial R^\nu}|\psi_n\rangle\,|\psi_{n'}\rangle + \alpha_{n,\nu}|\psi_n\rangle\,, \tag{13.137}$$

where α is pure imaginary, so $\alpha^* = -\alpha$. Inserting (13.136)–(13.137) into (10.120) and simplifying the notation, $|n\rangle \equiv |\psi_n\rangle$, we obtain

$$\Omega_{\mu\nu}^{(n)} = i \sum_{n'(\neq n)} \frac{\langle n|\frac{\partial \hat{H}}{\partial R^\mu}|n'\rangle\langle n'|\frac{\partial \hat{H}}{\partial R^\nu}|n\rangle - \langle n|\frac{\partial \hat{H}}{\partial R^\nu}|n'\rangle\langle n'|\frac{\partial \hat{H}}{\partial R^\mu}|n\rangle}{(E_n - E_{n'})^2}\,. \tag{13.138}$$

There are some conditions for the finiteness of the Chern number and, therefore, for the Hall conductivity. In particular, the Chern number is conditioned by symmetry properties. Consider the effect of the time-reversal operation, $\mathcal{T}$, and its effect on the Berry curvature. The transformation $\boldsymbol{k} \longrightarrow -\boldsymbol{k}$ implies $\partial\boldsymbol{h}(-\boldsymbol{k})/\partial k_j = -\partial\boldsymbol{h}(\boldsymbol{k})/\partial k_j$, so, the product of derivatives does not change sign. After the rotation by T, only one component of $\boldsymbol{h}$ changes sign. Hence, we get for the Berry curvature

[6] Here we follow [AraujoSacramento].

$$\mathcal{T}: \frac{1}{h^2}\frac{\partial \boldsymbol{h}}{\partial k_x} \times \frac{\partial \boldsymbol{h}}{\partial k_y} \cdot \hat{\boldsymbol{h}} \longrightarrow -\frac{1}{h^2}\frac{\partial \boldsymbol{h}}{\partial k_x} \times \frac{\partial \boldsymbol{h}}{\partial k_y} \cdot \hat{\boldsymbol{h}}. \tag{13.139}$$

We conclude that time reversal inverts the Chern number, $\mathcal{T} : C \longrightarrow -C$. Therefore, a time-reversal invariant system has $C = -C = 0$. As a consequence, time-reversal symmetry has to be broken to have a finite Chern number and a finite Hall conductivity.

13.5.3 Quantum Hall Effect in Graphene

The orbital effect of the magnetic field in a continuum system has been considered using the minimal coupling approach, $\mathbf{p} \rightarrow \mathbf{p} - e\mathbf{A}$, where $\mathbf{A}$ is the vector potential. In real space this translates to $-i\nabla \rightarrow -i\nabla - e\mathbf{A}$. On a lattice we may in general consider the Peierls substitution, where the hopping amplitude is replaced as $t_{ij} \rightarrow t_{ij}e^{ie\int_{R_i}^{R_j}\mathbf{A}\cdot d\mathbf{r}}$ (where we are taking $\hbar = 1$). Let us focus, however, on the vicinity of the Dirac points in graphene, where the spectrum is linear and a continuum approximation is valid.[7]

We consider a perpendicular homogeneous magnetic field and let us choose a Landau gauge $\mathbf{A} = (-y, 0, 0)\, B$. The magnetic field introduces the magnetic length scale given by $l_B = \sqrt{\hbar/(eB)}$. The energy dispersion in graphene is linear, of the order of $v_F q$, and q may be seen as the inverse of the characteristic length scale of the problem, in this case the magnetic length, l_B. Therefore, the spectrum introduces an energy scale which, recalling the circular motion of the orbits in the presence of a perpendicular magnetic field, can be seen as a scale like the cyclotron energy of the problem that obeys

$$\omega_B \sim \frac{v_F}{l_B} \sim \sqrt{B}. \tag{13.140}$$

Note that the linearity of the energy spectrum leads to a dependence of the cyclotron energy that differs from the usual dependence $\omega_B \sim B$, associated with an energy dispersion that is quadratic in the momentum.

In addition to the different dispersion relation of graphene, the parameters that characterize its dispersion relation lead to significantly different properties. The aspect we focus here has to do with the response to the effect of the magnetic field. Consider, for example, a large magnetic field of the order of $B \sim 10T$. In a usual system, this leads to a cyclotron energy of the order of $k_B \times 10\,\mathrm{K}$. However, in graphene the same field gives rise to a cyclotron energy of the order of $k_B \times 1000\,\mathrm{K}$, which is a large scale that implies that room temperature is a relatively

[7] For further reading see A.H. Castro Neto, F. Guinea, N.M.R. Peres, K.S. Novoselov, and A.K. Geim, Reviews of Modern Physics **81**, 109 (2009) and [AraujoSacramento].

small temperature, and we may expect a quantum Hall effect at finite temperatures, instead of being just a small temperature effect. Also, the Zeeman energy associated with such a magnetic field is small and may be safely discarded.

The problem may be solved explicitly considering the minimal coupling on the Hamiltonian matrix Eq. (10.183), transformed to real space

$$H_{\mathbf{K}}(\mathbf{r}) = v_F \begin{pmatrix} 0 & \frac{\hbar}{i}\partial_x + eyB - \hbar\partial_y \\ \frac{\hbar}{i}\partial_x + eyB + \hbar\partial_y & 0 \end{pmatrix} \tag{13.141}$$

and similarly for the other Dirac point

$$H_{\mathbf{K}'}(\mathbf{r}) = v_F \begin{pmatrix} 0 & \frac{\hbar}{i}\partial_x + eyB + \hbar\partial_y \\ \frac{\hbar}{i}\partial_x + eyB - \hbar\partial_y & 0 \end{pmatrix}. \tag{13.142}$$

Transforming to momentum space along the x axis, defining the basis states as

$$\psi_k(\mathbf{r}) = e^{ikx} \begin{pmatrix} c_1\Phi_1^A(y) \\ c_2\Phi_2^B(y) \end{pmatrix}, \tag{13.143}$$

we find the eigenvalue problem

$$v_F\frac{\hbar}{l_B} \begin{pmatrix} 0 & l_Bk - \frac{y}{l_B} \mp l_B\partial_y \\ l_Bk - \frac{y}{l_B} \pm l_B\partial_y & 0 \end{pmatrix} \begin{pmatrix} c_1\Phi_1^A \\ c_2\Phi_2^B \end{pmatrix} = E \begin{pmatrix} c_1\Phi_1^A \\ c_2\Phi_2^B \end{pmatrix}, \tag{13.144}$$

where the two cases are in the vicinity of the two Dirac points. In this way we rescale the distance coordinate y with the magnetic length.

It is now convenient to define a new variable $\tilde{y} = y/l_B - l_Bk$ which implies $\partial_{\tilde{y}} = l_B\partial_y$. Let us introduce the harmonic oscillator like operators

$$\begin{aligned} a &= \frac{1}{\sqrt{2}}\left(\tilde{y} + \partial_{\tilde{y}}\right) \\ a^\dagger &= \frac{1}{\sqrt{2}}\left(\tilde{y} - \partial_{\tilde{y}}\right). \end{aligned} \tag{13.145}$$

Introducing the states of the harmonic oscillator, we know that

$$\begin{aligned} a^\dagger\Phi_n &= \sqrt{n+1}\Phi_{n+1} \\ a\Phi_n &= \sqrt{n}\Phi_{n-1}. \end{aligned} \tag{13.146}$$

Defining

$$\begin{pmatrix} c_1\Phi_1^A \\ c_2\Phi_2^B \end{pmatrix} = \begin{pmatrix} \Phi_n^A(\tilde{y}) \\ \pm\Phi_{n+1}^B(\tilde{y}) \end{pmatrix}, \tag{13.147}$$

for the **K** Dirac point and

$$\begin{pmatrix} c_1\Phi_1^A \\ c_2\Phi_2^B \end{pmatrix} = \begin{pmatrix} \Phi_{n+1}^A(\tilde{y}) \\ \pm\Phi_n^B(\tilde{y}) \end{pmatrix}, \tag{13.148}$$

for the **K**′ Dirac point, we get the energy eigenvalues

$$E_{\mathbf{K}} = E_{\mathbf{K}'} = \pm v_F\sqrt{2}\frac{\hbar}{l_B}\sqrt{n+1}. \tag{13.149}$$

These energy levels are doubly degenerate, with the additional double degeneracy due to the spin of the electrons. Compare this solution with the one obtained for a quadratic dispersion relation shown in Eq. (13.72).

There is in addition a zero energy solution for each Dirac point. Verifying by inspection, we find that the state

$$\begin{pmatrix} 0 \\ \Phi_0(\tilde{y}) \end{pmatrix} \tag{13.150}$$

for the Dirac point **K**, and the state

$$\begin{pmatrix} \Phi_0(\tilde{y}) \\ 0 \end{pmatrix} \tag{13.151}$$

for the Dirac point **K**′ have zero energy. The Landau levels for a usual parabolic band and for a Dirac band are compared in Fig. 13.10.

The variable $\tilde{y}$ is a function of k, the momentum along the x direction, and, therefore, the Landau level center moves across the sample as k changes, as for the usual Hall effect. We have that each electron level is twice degenerate due to the two Dirac points. We also have two zero energy levels, one in electron levels and one in hole levels. If the chemical potential is located between the Landau levels n and $n+1$, we have, for each spin component, $(2n+1)$ occupied levels. The one

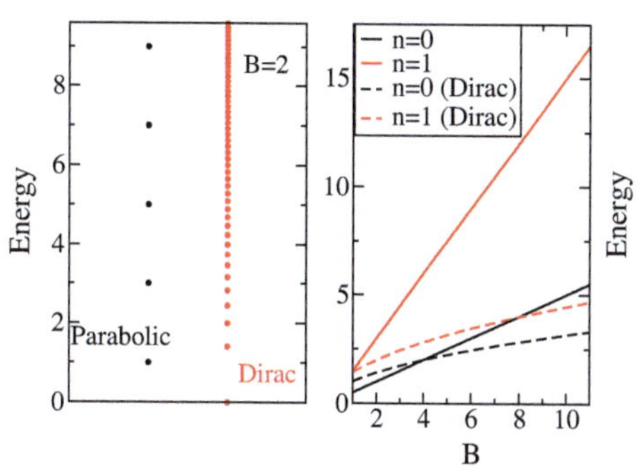

Fig. 13.10 Landau levels for a parabolic band and near a Dirac cone. On the left panel, the magnetic field is fixed and the sequence of levels is compared. On the right panel, the first two levels are considered as a function of B

comes from the zero energy level and the factor of 2 comes from the Dirac cone degeneracy. If we now include the spin degeneracy, we have $2(2n+1)$ occupied levels that can be written as $4(n+1/2)$. Using the Laughlin argument we have that the Hall conductivity should be of the type

$$\sigma_{xy} = \pm 2(2n+1)\frac{e^2}{h}, \tag{13.152}$$

since these are the occupied levels that are transferred from one edge of the sample to the other.

13.5.4 Haldane Model: Anomalous Hall Insulator

An explicit realization of an anomalous Hall insulator was proposed by Haldane [5]. The model is based on the honeycomb lattice and is an extension of the simpler model for graphene.

As an example let us take the choice of basis suited for the zigzag boundary. Consider an A site at position $\mathbf{R}$. Let $t = 1$ be the hopping amplitude to the neighboring B sites at positions $\mathbf{R} - \boldsymbol{\delta}_3$, $\mathbf{R} - \boldsymbol{\delta}_1$, $\mathbf{R} - \boldsymbol{\delta}_2$. We may take, as primitive cell, the one made up of atoms A and B at positions $\mathbf{R}$ and $\mathbf{R} - \boldsymbol{\delta}_3$, respectively. Haldane modified graphene's Hamiltonian in such a way as to obtain finite and symmetrical masses, h_z, at the Dirac points. A way to achieve this is to introduce an intra-sublattice hopping matrix element as

$$\begin{aligned}\langle \mathbf{R}|\hat{H}|\mathbf{R}+\boldsymbol{\delta}_2-\boldsymbol{\delta}_3\rangle &= \langle \mathbf{R}|\hat{H}|\mathbf{R}+\boldsymbol{\delta}_1-\boldsymbol{\delta}_2\rangle = \langle \mathbf{R}|\hat{H}|\mathbf{R}+\boldsymbol{\delta}_2-\boldsymbol{\delta}_1\rangle \equiv -it_2\,,\\ \langle \mathbf{R}|\hat{H}|\mathbf{R}+\boldsymbol{\delta}_2-\boldsymbol{\delta}_1\rangle &= \langle \mathbf{R}|\hat{H}|\mathbf{R}+\boldsymbol{\delta}_1-\boldsymbol{\delta}_3\rangle = \langle \mathbf{R}|\hat{H}|\mathbf{R}+\boldsymbol{\delta}_3-\boldsymbol{\delta}_2\rangle \equiv it_2\,.\end{aligned} \tag{13.153}$$

These next-nearest-neighbor hopping terms are $\pm\, it_2$, depending on whether the electron deviates to the right or left.

The Haldane model tight-binding Hamiltonian may be generalized and can be written as

$$H = -t_1 \sum_{\langle i,j\rangle} c_i^\dagger c_j - t_2 \sum_{\langle i,j\rangle} e^{-i\phi_{ij}} c_i^\dagger c_j + M \sum_i \eta_i c_i^\dagger c_i + \text{H.c.}\,, \tag{13.154}$$

where $c_i = A, B$ are defined in the two triangular sublattices that form the honeycomb lattice. The first term t_1 represents a standard real nearest-neighbor hopping that links the two triangular sublattices. The next term represents a complex next-nearest-neighbor hopping $t_2 e^{-i\phi_{ij}}$ acting within each triangular sublattice with a phase ϕ_{ij} that has opposite signs $\phi_{ij} = \pm\phi$ in the two sublattices. (The previous discussion corresponds to $\phi = \pi/2$.) This term breaks time-reversal symmetry and

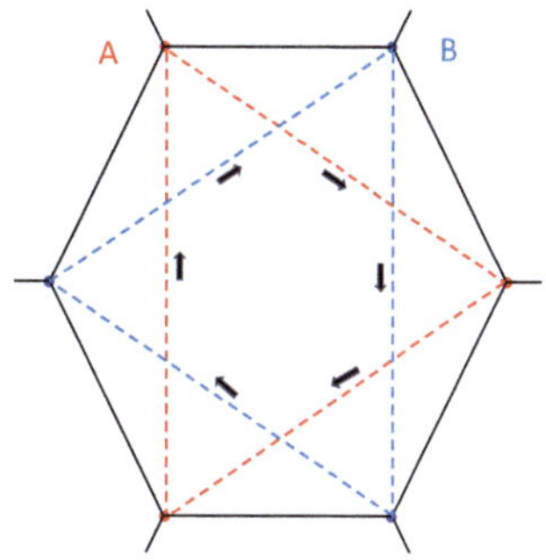

Fig. 13.11 Unit cell of the Haldane model with the addition of second-neighbor hoppings between atoms in the same sublattice. The hopping has a phase which is positive along the arrows indicated in the figure

opens a non-trivial topological gap at the Dirac points. Since it breaks time-reversal symmetry, there is a possibility of a non-vanishing Chern number. Note however that there is zero total magnetic flux in a unit cell and so effectively the magnetic field vanishes. The last term represents a staggered potential ($\eta_i = \pm 1$). It breaks inversion symmetry and opens a trivial gap at the Dirac points. The unit cell of the Haldane model is represented in Fig. 13.11.

We may also consider the Hamiltonian matrix in momentum space as $H(\mathbf{k}) = \boldsymbol{h}(\mathbf{k}) \cdot \boldsymbol{\tau}$, where $\boldsymbol{\tau}$ acts in sublattice space (A and B). For $h_z = 0$, graphene (with $t_2 = 0$) enjoys chiral symmetry, parity (implemented by the matrix $P = \tau_x$), and time-reversal symmetry. The Hamiltonian is given by

$$\begin{aligned} H(\boldsymbol{k}) &= 2t_2 \cos\phi \sum_i \cos(\boldsymbol{k} \cdot \boldsymbol{b}_i) I \\ &+ t_1 \sum_i (\cos(\boldsymbol{k} \cdot \boldsymbol{a}_i)\sigma_1 + \sin(\boldsymbol{k} \cdot \boldsymbol{a}_i)\sigma_2) \\ &+ \left(M - 2t_2 \sin\phi \sum_i \sin(\boldsymbol{k} \cdot \boldsymbol{b}_i) \right) \sigma_3. \end{aligned} \tag{13.155}$$

Here t_1 is the hopping between nearest neighbors between one sublattice and the other, M is the mass term, and the lattice vectors are

$$\boldsymbol{a}_1 = (1, 0), \boldsymbol{a}_2 = (-1/2, \sqrt{3}/2), \boldsymbol{a}_3 = (-1/2, -\sqrt{3}/2) \tag{13.156}$$

and

$$\boldsymbol{b}_1 = \boldsymbol{a}_2 - \boldsymbol{a}_3, \boldsymbol{b}_2 = \boldsymbol{a}_3 - \boldsymbol{a}_1, \boldsymbol{b}_3 = \boldsymbol{a}_1 - \boldsymbol{a}_2. \tag{13.157}$$

There are topological regions characterized by non-vanishing Chern numbers, ± 1, for $|t_2/t_1| < 1/3$, that lead to non-vanishing Hall conductances. As a function of the phase ϕ and the gap magnitude, M, the non-trivial phases occur if $|M/t_2| < 3\sqrt{3}|\sin\phi|$. The phase diagram is shown in Fig. 13.12.

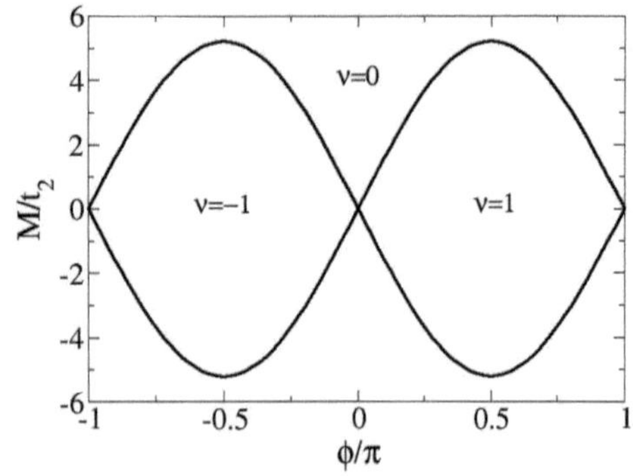

Fig. 13.12 Phase diagram of the Haldane model

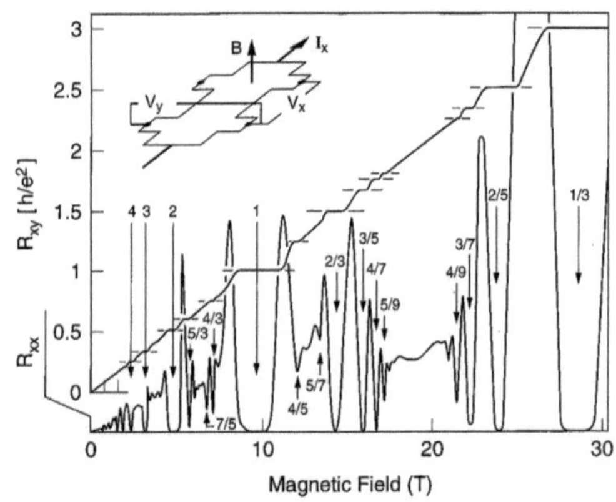

Fig. 13.13 Experimental results for the fractional quantum Hall effect (Reproduced with permission from Horst L. Stormer, Daniel C. Tsui, and Arthur C. Gossard, Reviews of Modern Physics **71**, S298 (1999))

13.5.5 Fractional Quantum Hall Effect

Two years after the discovery of the integer quantum Hall effect, the fractional quantum Hall effect was discovered. Considering even higher magnetic fields, very low temperatures, and very pure systems, it was found that the Hall conductivity is quantized and presents plateaus for the Hall conductivity of the type

$$\sigma = f\frac{e^2}{h}, \tag{13.158}$$

where $f \sim \nu$ is a rational fraction of the type p/q, where p and q are integers and that, as before, ν is the filling factor of the Landau levels. In the first experiments it was found that q is odd. It turns out that the experimental behavior is very similar to that of the integer effect, as shown in Fig. 13.13.

In the Hall resistivity plateau regimes, the longitudinal resistivity is very small. The only apparent difference with respect to the integer effect are the fractionary values of the Hall conductivity, associated with the fractionary values of ν. This property suggests that the Fermi level is also localized in a gap. The proof that in the integer case the Hall conductivity is a topological invariant shows that, if the ground state is nondegenerate, then the quantization should lead to an integer. The fact that the conductivity is quantized but with a fractionary value implies that the state must be degenerate. Also, the observation that the fraction f is related with ν suggests that, in addition to the fact that the magnetic field quantizes the orbits in Landau levels, since we are now in a regime where the Landau level is not completely full (let us consider the case $\nu < 1$), the gap does not occur between the Landau levels but probably results from the interactions between the electrons. Once again, the apparent macroscopic coherence of the result suggests a correlated state between the electrons, which as we shall see, leads to fractional charges. Since the Landau level is not full, the Coulomb energy between the electrons is now an important energy scale in the problem, because the kinetic energy is confined, due to the magnetic field. In the case of the integer effect, the relevant excitations are those across the gap between the Landau levels that involve much higher energies than those associated with the interactions between the electrons, and, therefore, the interactions may be neglected. The existence of the phenomenon then suggests that it is favorable for the electrons to correlate in such a way as to give origin to a gap for the excitations. In this way the system is formally equivalent to the integer effect, with the difference that the origin for the excitation gap are the Coulomb interactions between the electrons, and not directly the magnetic field. Similarly, one needs to consider the presence of disorder to break the spatial translational invariance, but, since the gaps are now much smaller, it is necessary that other terms do not destroy the effect. As a consequence, the systems have to be more pure and the temperature must be smaller for the effect to be observed.

Incompressibility

The explicit numerical solution of a problem of three electrons that are moving in a two-dimensional space and under the influence of an external perpendicular magnetic field, strong enough that the electrons are in the lowest Landau level, allowed Laughlin [6] an understanding of the state that the electronic system is in. It is convenient in this case to use the symmetric gauge. In this gauge the vector potential is written as $\mathbf{A} = \frac{1}{2}\mathbf{B} \times \mathbf{r}$. The single-particle states are given by

$$|m, n> = \frac{1}{\sqrt{2^{m+n+1}\pi m! n!}} e^{\frac{1}{4}(x^2+y^2)} \left(\frac{\partial}{\partial x} + i\frac{\partial}{\partial y} \right)^m \left(\frac{\partial}{\partial x} - i\frac{\partial}{\partial y} \right)^n e^{-\frac{1}{2}(x^2+y^2)}, \tag{13.159}$$

where we are using units such that the magnetic length and the cyclotron frequency are 1. The energy is given as before where n is the Landau level and the state $|m, n>$ is also an eigenstate of the angular momentum with value $m - n$. If the magnetic field is high enough, then the electrons are in the lowest Landau level $n = 0$ and the states are given by the simple expression

$$|m> = \frac{1}{\sqrt{2^{m+1}\pi m!}} z^m e^{-\frac{1}{4}|z|^2}, \tag{13.160}$$

where $z = x + iy$. Introducing a Coulomb interaction between the electrons, Laughlin concluded that the interelectron distance has some preferential values that change discontinuously as the pressure in the system changes. This leads to the conclusion that the system is incompressible, the electrons avoid each other but the magnetic field imposes that these orbit around each other, and there is therefore some balance between the Coulomb repulsion and the effect of the magnetic field. This correlated state may then serve as a basis for the many-electron state and possibly explain the similarities with a state (called superfluid state to be discussed ahead) such that the electrons may move almost with no dissipation, under certain favorable conditions. Also, Laughlin determined numerically that the excitation energies are of the order of $0.03e^2/l$.

Laughlin Wave Function

Laughlin introduced a variational wave function to describe the state of the system in the form

$$\psi = \prod_{j<k} f(z_j - z_k) e^{-\frac{1}{4}\sum_l |z_l|^2}. \tag{13.161}$$

The wave function must be contained in the lowest Landau level and therefore the function f must be an analytical function of z_j. In addition, the wave function must be antisymmetric and therefore the function f must be odd. The conservation of the angular momentum implies that $\prod_{j<k} f(z_j - z_k)$ must be a homogeneous polynomial of degree M, where M is the total angular momentum. These conditions restrict the wave function to

$$\psi_m(z_1, \ldots, z_N) = \prod_{j<k} (z_j - z_k)^m e^{-\frac{1}{4}\sum_l |z_l|^2}, \tag{13.162}$$

where m must be odd. Using the analogy of the quantum problem to a classical problem, we may write that

$$|\psi_m|^2 = e^{-\beta\Phi}, \tag{13.163}$$

where Φ is the thermodynamic potential of the equivalent classical problem. Choosing arbitrarily $\beta = 1/m$, it is concluded that the classical potential is given by

$$\Phi = -\sum_{j<k} 2m^2 \ln |z_j - z_k| + \frac{1}{2} m \sum_l |z_l|^2. \tag{13.164}$$

The equivalent classical problem is that of N particles of charge m that interact with each other via a logarithmic potential (as the Coulomb interaction in two dimensions) being immersed in a medium with uniform charge density of opposite sign, $\sigma = (2\pi l^2)^{-1}$.

This classical (plasma) system has been studied in detail and it is known that for $m < 70$ it is in a liquid phase and that for higher values of m it is in a solid phase. Since the state of lowest energy of the plasma is a state where locally the charge density is compensated, the electronic system should have a density $\sigma_m = (m2\pi l^2)^{-1}$. When the densities are high, the preferential state will have $m = 1$ (the case of the integer effect) and as we lower the density the stable states correspond to $m = 3, 5, \ldots$. (As we lower the density, it may be favorable for the electronic system to crystallize in a Wigner crystal, instead of remaining liquid. This type of crystallization will be considered in the next chapter. Gapless excitations will appear in this case.) In the liquid regime the system is found in a new incompressible state. Laughlin confirmed the success of the variational wave functions comparing with numerical results for systems with small numbers of particles, and he found that the agreement is excellent, particularly for small values of m.

Excitations with Fractional Charges

As the system orders into the new state, the degeneracy is lifted, leading to a nondegenerate ground state, and therefore, similar arguments used by Laughlin in the integer effect apply. Laughlin proposed that the excitations may be obtained inserting a flux tube at a position z_0 and changing adiabatically the flux by a flux quantum. Again, due to the gauge invariance, as we introduce an integer number of flux quanta the Hamiltonian returns to an eigenstate. However, the energy transmitted to the system by the change of the magnetic flux leads to an excited state and, therefore, to elementary excitations. The effect of changing the flux has a consequence on the one-particle functions

$$z^k e^{-\frac{1}{4}|z|^2} \to z^{k\pm 1} e^{-\frac{1}{4}|z|^2},$$

that is, the angular momentum is changed by a unit. This can be understood by transforming the state at the nearest site. The procedure is similar to the motion of the particle along the system (recall that the typical size of the state with the angular momentum m is $\sqrt{2m}$). In this operation a charge is transported across the system. This charge is the average charge per state. In the case of the many-particle system described by the Laughlin variational wave function, the result of this operation is

$$\psi_m^{+z_0} = e^{-\frac{1}{4}\sum_l |z_l|^2} \prod_i (z_i - z_0) \prod_{j<k} (z_j - z_k)^m, \tag{13.165}$$

for the excitation called quasihole and

$$\psi_m^{-z_0} = e^{-\frac{1}{4}\sum_l |z_l|^2} \prod_i \left(2\frac{\partial}{\partial z_i} - z_0^*\right), \tag{13.166}$$

for the excitation called quasielectron. Once again, the agreement with the energies calculated numerically for small systems is good. In a similar way, Laughlin established a connection between the quantum system and a classical equivalent system comparing $|\psi^{+z_0}|^2 = e^{-\beta\Phi'}$ obtaining that

$$\Phi' = \Phi - 2m \sum_l \ln |z_l - z_0|. \tag{13.167}$$

Therefore, the effective potential Φ' describes the plasma interacting with a charge localized at point z_0. To minimize the energy, the plasma compensates the additional charge localized at z_0. Since the plasma is actually composed of unitary charges (instead of charges of value m, defined arbitrarily due to the definition $\beta = 1/m$), the true accumulated charge is $1/m$. Therefore, the charge of the excitation is fractionary and given by $1/m$. Since the charge of the state is $1/m$, this means that as we change the flux by a flux quantum, the charge transported across the system (being the charge of each state) is then $1/m$, which leads, in a way similar to the integer effect, to a contribution to the Hall conductivity that is fractionary. In this interpretation, the fractional effect is very similar to the integer effect, with the important difference that the charges that are transported are fractionary.

If we now consider that we vary the factor ν, it is to be expected that quasiholes or quasielectrons may appear. In these circumstances it is possible that the quasiparticles may themselves condensate in states of quasiparticles that, in a way similar to the integer effect, give origin to gaps that stabilize again the Hall conductivity in plateaus. This process leads to a hierarchy of states that allowed the interpretation of other fractions that are observed experimentally. This construction becomes quite involved and does not explain all the experimental observations. For instance, it is restricted to cases when the denominator of the fractions is odd. However, it has been found some cases where the denominator of the fractions is even, such as the fraction $5/2$.

Composite Fermions

With this problem in mind, Jain [7] introduced a description of the excitations of the system that is in better agreement with the exact energy values that allows to justify other fractions observed experimentally and that allows a more direct parallelism with the integer case. The initial idea was to consider that each electron is fractionalized in m parts, since the excitations have charge e/m. To distinguish the particles it was considered that there are m different types, labeled by λ. At the end of the calculation the positions of each particle are $z_{j,\lambda} = z_j$, for the particles that are the constituents of electron j. The idea is to describe the fractional quantum Hall effect in a way as close as possible to the integer case. That is, to use Fermi statistics as the dominant mechanism, together with the existence of a gap at the Fermi level. Therefore, the constituent particles are fermions. That is why m must be odd. The particle density must be equal to the electron density. Calling p_λ the filling factor of the λ particle levels, we must have that $e_\lambda p_\lambda = ep$, where p is the filling factor of the electrons since the density is given by $Be_\lambda p_\lambda / h$. Then we have that

$$p = \frac{1}{\sum_{\lambda=1}^{m} p_\lambda^{-1}}. \tag{13.168}$$

This expression shows that even when the factors p_λ are integers, in general the factor p is fractionary. The states are then classified by the numbers p_λ of the occupied levels of the fictitious particles.

Using the same arguments of gauge invariance, it is possible to conclude that the Hall conductivity is quantized according to $\sigma = pe^2/h$, as wanted. The excitations are then obtained considering variations (holes or quasiparticles) in the numbers p_λ. Therefore, if $m = 1$ we have the state of the integer Hall effect and the particles have charge e. In the case that the set of values of $p_\lambda = [1, 1, 1, \ldots]$, we have Laughlin state where all the particles have charge e/m. The hole excitation is easily obtained creating a hole in the state that respects for instance to $\lambda = 1$. The fully filled state is given by

$$\chi_1 = \begin{vmatrix} 1 & 1 & 1 & \cdots \\ z_1 & z_2 & z_3 & \cdots \\ z_1^2 & z_2^2 & z_3^2 & \cdots \\ \cdots & \cdots & \cdots & \cdots \end{vmatrix} e^{-\frac{1}{4}\sum_i |z_i|^2}$$

$$\chi_1 = \prod_{j<j} (z_j - z_k) e^{-\frac{1}{4}\sum_i |z_i|^2}.$$

The state respecting to a quasihole is equal to the excitation proposed by Laughlin and is written in this language as

$$\chi_1^+ = \begin{vmatrix} z_1 & z_2 & z_3 & \cdots \\ z_1^2 & z_2^2 & z_3^2 & \cdots \\ z_1^3 & z_2^3 & z_3^3 & \cdots \\ \cdots & \cdots & \cdots & \cdots \end{vmatrix} e^{-\frac{1}{4}\sum_i |z_i|^2},$$

which may be written as $\chi_1^+ = \prod_k z_k \chi_1$. On the other hand, the excitation that corresponds to a quasielectron is written in a way different from Laughlin as

$$\chi_1^+ = \begin{vmatrix} z_1^* & z_2^* & z_3^* & \cdots \\ 1 & 1 & 1 & \cdots \\ z_1 & z_2 & z_3 & \cdots \\ \cdots & \cdots & \cdots & \cdots \end{vmatrix} \chi_1^2,$$

from which one obtains energy values that are in better agreement with the results obtained by exact diagonalization of small systems than the Laughlin excitations. On the other hand, we may also consider states of the type $[p_1, 1, 1, \cdots]$. These states correspond to sequences of the type

$$p = \frac{p_1}{(m-1)p_1 + 1},$$

or states that lead to even denominators.

It is more convenient to interpret the Jain states considering that the Laughlin states

$$\chi_{1/m} = \prod_{j<k} (z_j - z_k)^m e^{-\frac{1}{4}\sum_i |z_i|^2}$$

can be written as

$$\chi_{1/m} = \prod_{j<k} (z_j - z_k)^{m-1} \chi_1. \tag{13.169}$$

This expression may be interpreted as attaching $(m-1)$ flux quanta to each electron. For instance, in the case of $m = 3$ the interpretation is that to each electron one adds two flux quanta that move together with the electron. Then each electron is under the action of an effective magnetic field

$$B_{eff} = B - 2\pi(m-1)n_e, \tag{13.170}$$

in units such that the quantum of flux is 2π and where n_e is the electronic density. Since the filling factor for the electrons is given by $\nu = 2\pi n_e/B$, we may define an effective filling factor for the composite fermions as $p = 2\pi n_e/B_{eff}$. Then we have that $p^{-1} = \nu^{-1} - (m-1)$. In the case of $m = 3$ we obtain that $\nu = p/(2p+1)$. In terms of these composite fermions, the ground state may be represented with the lowest level filled with composite fermions and where one excitation of the quasihole type is represented and obtained simply by promoting one composite fermion to the level immediately above. Recall that the gaps between the accessible states to the composite fermions have origin in the interactions and not on the cyclotron energy. In the same way, we may represent the creation of two excitations promoting two composite fermions to a higher level and leaving two holes, as in the usual interpretation in terms of excitations on a band creating particle-hole pairs. The agreement between this theory and the numerical results is considerably improved as compared with the excitations proposed by Laughlin.

Fractional Statistics

This interpretation of the quantum Hall effect in terms of composite fermions has interesting consequences. Consider a charged particle orbiting around a flux tube, Φ, at a distance r. The particle is then subjected to a vector potential $A_\varphi = \Phi/2\pi r$. We may eliminate the vector potential performing a gauge transformation of the type

$$A'_i = A_i - \partial_i \Lambda = 0,$$

where $\Lambda = \Phi\varphi/2\pi$. The wave function of the particle transforms as

$$\psi' = e^{iq\Phi\varphi/2\pi}\psi.$$

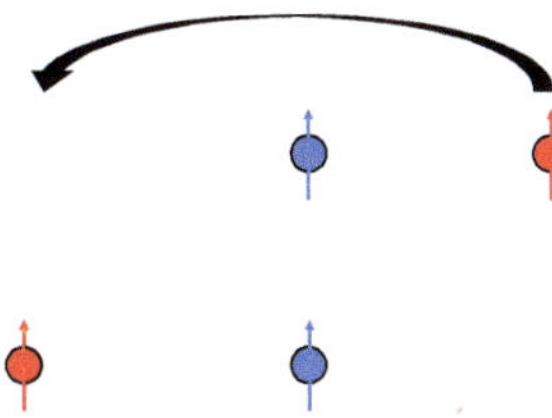

Fig. 13.14 Permutation of two composite fermions. The arrows represent the fluxes attached to each fermion

Since the wave function is periodic with period 2π, we have that

$$\psi'(\varphi + 2\pi) = e^{iq\Phi}\psi'(\varphi).$$

The allowed solutions are of the type $\psi' \sim e^{im\varphi}$ where m =integer $+ \, q\Phi/2\pi$.

Let us then consider two composite fermions orbiting around each other recalling that each has a flux attached. Let us consider that the two composite fermions exchange positions by rotating one around the other by an angle π, as shown in Fig. 13.14. Separating the center of mass motion from the relative motion, it is easy to see that the result of the permutation of the two sets of fermion-flux quanta may be written as

$$\psi(R, \theta; r, \varphi + \pi) = e^{2\pi i\Delta}\psi(R, \theta; r, \varphi), \tag{13.171}$$

where $\Delta = q\Phi/2\pi$.

In the case of bosons we get that $\Delta = 0$; if $\Delta = 1/2$, we have the usual case of fermions, but in general we have a phase that interpolates between the two cases, and these particles are called anyons. In the case of composite particles we also have to consider the effect of the circulation of the particles around the flux tubes. Each of the particles goes through a half-turn around the flux tube attached to the other. Therefore, in addition to the usual statistical phase, we have to consider the phase associated with the circulation around a flux tube. It is clear that if the charged particles are fermions, an additional minus sign appears so that the spin of the anyon is then s_a =integer $+ \, \Delta + s$ where s is the intrinsic spin of the particle.

In order to calculate the effect of the permutation of the excitations, we may explicitly consider the Laughlin wave function[8]. Let us then consider the fractional quantum Hall effect and a filling factor $\nu = n/m$. As we saw in the case of $n = 1$, the charge of the quasiparticles is given by $e^* = e/m$. To determine the charge of the quasiparticles in an alternative way, we may calculate the change of phase that results from the circulation of one excitation along a trajectory of radius R. Consider the excitation that corresponds to a quasihole $\psi_m^{+z_0}$. The change of phase γ is then

$$\gamma = \frac{e^*}{\hbar}\int \mathbf{A} \cdot d\mathbf{l} = 2\pi\frac{e^*}{e}\frac{\phi}{\phi_0}. \tag{13.172}$$

[8] Here we follow D. Arovas, J. R. Schrieffer and F. Wilczek, Physical Review Letters **53**, 722 (1984).

This is a Berry phase and satisfies

$$\frac{d\gamma(t)}{dt} = i < \psi(t)|\frac{d\psi(t)}{dt} > . \tag{13.173}$$

Considering the wave function of the Laughlin excitation, we obtain, as expected, that

$$\gamma = -2\pi\nu\frac{\phi}{\phi_0}, \tag{13.174}$$

which implies, as we know, that $e^* = \nu e$. Consider now two excitations of the quasihole type localized at z_a and z_b. Then the state is given by

$$\psi_m^{z_a,z_b} = \prod_i (z_i - z_a)(z_i - z_b)\psi_m . \tag{13.175}$$

Consider now to transport z_a along a closed path of radius R. If z_b is outside the circle, the analysis done for one excitation stands. However, if z_b is contained in the circle then an additional phase given by

$$\Delta\gamma = 2\pi\nu \tag{13.176}$$

appears. When $\nu = 1$ the additional phase is given by $\Delta\gamma = 2\pi$ and the phase associated with the permutation of the quasiparticles (half-circle) is given by $\Delta\gamma/2 = \pi$ corresponding to fermionic statistics. If the factor ν is fractionary, then the statistics is also fractionary.

13.6 Strain and Synthetic Magnetic Fields

The presence of a magnetic field is manifested on the orbital degrees of freedom via minimal coupling or a Peierls factor. In this case, it is manifested by its effect on the phase of the hopping between different locations, for instance, in a tight-binding model. Strain on a lattice typically changes the hopping amplitudes and, under appropriate conditions, may be seen as an effective vector potential and, consequently, an effective magnetic field applied on the system. This occurs for instance in the case of graphene. It turns out that under appropriate conditions, the effective magnetic field may be very large, even larger than the highest real magnetic fields available.

Let us consider a two-dimensional system on the plane xy. We may consider that there is an in-plane deformation of the type $\mathbf{u} = (u_x, u_y)$. The lattice vectors are then displaced to new values $\mathbf{R} \rightarrow \mathbf{R} + \mathbf{u}$. We may also consider a displacement of the atoms into the perpendicular direction z and write an expression for the displacement of the atoms as $\mathbf{d} = (u_x, u_y, h)$ such that[9]

$$\mathbf{R} \to \mathbf{R}^{(0)} + \mathbf{d}(\mathbf{R}). \tag{13.177}$$

We consider nearest-neighbor vectors, $\boldsymbol{\delta}$, such that the unstrained lattice vectors satisfy

$$|\mathbf{R}^{(0)} - \left(\mathbf{R}^{(0)} + \boldsymbol{\delta}\right)| = |\boldsymbol{\delta}| = a. \tag{13.178}$$

Under the action of strain, we get

$$|\mathbf{R}^{(0)} + \mathbf{d}(\mathbf{R}) - \left(\mathbf{R}^{(0)} + \mathbf{d}(\mathbf{R} + \boldsymbol{\delta}) + \boldsymbol{\delta}\right)| = | - \boldsymbol{\delta} + \mathbf{d}(\mathbf{R}) - \mathbf{d}(\mathbf{R} + \boldsymbol{\delta})|. \tag{13.179}$$

The hopping amplitudes are changed into

$$t_{\boldsymbol{\delta}} = t\{| - \boldsymbol{\delta} + \mathbf{d}_a(\mathbf{R}) - \mathbf{d}_b(\mathbf{R} + \boldsymbol{\delta})|\}, \tag{13.180}$$

between two sites a and b. Considering that the displacements $\mathbf{u}$ are small and neglecting any displacement along the z direction, this leads to

$$t_{\boldsymbol{\delta}} \to t_{\boldsymbol{\delta}} - t_{\boldsymbol{\delta}}\beta \left(\hat{\boldsymbol{\delta}} \cdot \nabla\right) \mathbf{u} \cdot \hat{\boldsymbol{\delta}}, \tag{13.181}$$

where

$$\hat{\boldsymbol{\delta}} = \frac{\boldsymbol{\delta}}{|\boldsymbol{\delta}|} \tag{13.182}$$

and $\beta = -\partial \ln t / \partial \ln a$, considering only nearest neighbors and that the hopping amplitude of the unstrained lattice is the same between the various first neighbors, distant by a.

Let us consider the case of graphene and consider the choice of basis that is convenient for the case of a zigzag edge. The nearest-neighbor vectors are

$$\begin{aligned}
\boldsymbol{\delta}_1 &= (0, 1)\, \tilde{a} \\
\boldsymbol{\delta}_2 &= \left(\frac{\sqrt{3}}{2}, -\frac{1}{2}\right) \tilde{a} \\
\boldsymbol{\delta}_2 &= \left(-\frac{\sqrt{3}}{2}, -\frac{1}{2}\right) \tilde{a},
\end{aligned} \tag{13.183}$$

[9] We follow unpublished personal notes by Eduardo Castro.

where $\tilde{a}$ is the distance between two points on the same sublattice. We obtain

$$\begin{aligned} t_{\delta_1} &= t - t\beta u_{yy} \\ t_{\delta_2} &= t - t\beta\left(\frac{3}{4}u_{xx} - \frac{\sqrt{3}}{4}u_{xy} - \frac{\sqrt{3}}{4}u_{yx} + \frac{1}{4}u_{yy}\right) \\ t_{\delta_3} &= t - t\beta\left(\frac{3}{4}u_{xx} + \frac{\sqrt{3}}{2}u_{xy} + \frac{1}{4}u_{yy}\right), \end{aligned} \tag{13.184}$$

where $u_{a,b} = \partial_a u_b$, with $a, b = x, y$.

Consider a strain such that $\mathbf{u} = \left(0, cy^2 + D\right)$. In this case we have that $u_{yy} = 2cy$ and

$$\begin{aligned} t_{\delta_1} &= t - t\beta 2cy \\ t_{\delta_2} &= t - t\beta\frac{1}{2}cy \\ t_{\delta_3} &= t - t\beta\frac{1}{2}cy. \end{aligned} \tag{13.185}$$

At low energies the Hamiltonian may be written as

$$H_{\mathbf{K}} = v_F \int d\mathbf{r}\left(a^\dagger(\mathbf{r})\; b^\dagger(\mathbf{r})\right)\begin{pmatrix} 0 & \partial_x - i\partial_y \\ \partial_x + i\partial_y & 0 \end{pmatrix}\begin{pmatrix} a(\mathbf{r}) \\ b(\mathbf{r}) \end{pmatrix}, \tag{13.186}$$

in the vicinity of the Dirac point $\mathbf{K}$ and in the vicinity of the other Dirac point we get

$$H_{\mathbf{K}'} = v_F \int d\mathbf{r}\left(a^\dagger(\mathbf{r})\; b^\dagger(\mathbf{r})\right)\begin{pmatrix} 0 & -\partial_x - i\partial_y \\ -\partial_x + i\partial_y & 0 \end{pmatrix}\begin{pmatrix} a(\mathbf{r}) \\ b(\mathbf{r}) \end{pmatrix}. \tag{13.187}$$

The term due to the strain, which shifts the value of the hopping amplitude, may be written in the form

$$H_A = \int d\mathbf{r}\left(A(\mathbf{r})a^\dagger(\mathbf{r})b(\mathbf{r}) + A^*(\mathbf{r})b^\dagger(\mathbf{r})a(\mathbf{r})\right), \tag{13.188}$$

where

$$A(\mathbf{r}) = \sum_{\delta}\delta t(\mathbf{r})e^{-i\delta\cdot\mathbf{k}}. \tag{13.189}$$

At the two Dirac points we get

$$A(\mathbf{r}) = \frac{3}{4} t\beta \left(-u_{xx} + u_{yy} - 2iu_{xy}\right), (\mathbf{K})$$
$$A(\mathbf{r}) = \frac{3}{4} t\beta \left(-u_{xx} + u_{yy} + 2iu_{xy}\right), (\mathbf{K}') \tag{13.190}$$

(note that in this expression $u_{xy} = 1/2(\partial_x u_y + \partial_y u_x)$). The strain term contribution to the Hamiltonian at the Dirac point $\mathbf{K}$ may be written as

$$H_A = v_F e \int d\mathbf{r} \psi^\dagger(\mathbf{r}) \boldsymbol{\sigma} \cdot \mathbf{A}_{el}(\mathbf{r}) \psi(\mathbf{r}), \tag{13.191}$$

where as before $v_F = 3at/(2\hbar)$ and

$$e\mathbf{A}_{el} = \frac{\hbar\beta}{a} \left(\frac{1}{2}(u_{xx} - u_{yy}), -u_{xy} \right). \tag{13.192}$$

The magnetic field is given by

$$\mathbf{B} = \nabla \times \mathbf{A}_{el} = B_z \mathbf{e}_z = \frac{\hbar\beta}{ae} c\mathbf{e}_z, \tag{13.193}$$

a constant magnetic field along z. Since time-reversal symmetry is preserved by the strain, the total magnetic field is zero, and the contributions from the two Dirac points cancel. Estimates of the magnitude of this pseudo-magnetic field yield a field of around $10T$, if we consider a strain of about 20% in a ribbon of around 100 nm.

13.7 Exercises

1. The energy of an electric dipole in the presence of an external electric field is formally equivalent to the energy of a magnetic dipole in the presence of an external magnetic field.
   (a) If the orientation of the dipoles is such that they are not aligned with the external field, the electric dipole and the magnetic dipole will tend to reorient to minimize the energy. Compare the equations of motion of the two types of dipoles.
   (b) If the deviation from the ground state is due to temperature, calculate the polarization of a gas of N permanent independent electric dipoles of dipole moment $\mathbf{p}$ and show that it satisfies a Curie law at small electric fields.
2. The nuclei of the atoms that constitute a crystalline lattice have spin 1. Each nucleus may have the spin quantum numbers $m = 1, 0, -1$. These quantum numbers measure the projection of the nuclear spin along one of the crystalline axes of the solid. Since the charge distribution of the nucleus does not have a

spherical symmetry but instead an ellipsoidal symmetry, the energy of a nucleus depends on the orientation of its spin with respect to the existing electric field at the point where it is located. In this way a nucleus in a spin state corresponding to $m = -1$ or to $m = 1$ has energy $E = \pm e$ and in the spin state $m = 0$ it has energy $E = 0$. Suppose that the solid is in thermal equilibrium at temperature T.

(a) Determine an expression for the nuclear contribution to the internal energy of the solid and express it as a function of temperature.
(b) Determine an expression for the entropy of the solid. Study the limits $T \to 0$ and $T \to \infty$.
(c) Sketch the nuclear specific heat as a function of T and calculate explicitly its dependence on temperature. How does C_V change with T when $T \to \infty$?

3. Explain the method of cooling a system by adiabatic demagnetization. Using this method temperatures of the order of $10^{-3} K$ may be reached. The method relies on the result that as the magnetic field is increased the entropy of a system of spins decreases.
4. Some organic molecules have an excited triplet state ($S = 1$) with an energy $k_B \Delta$ that is larger than the energy of the singlet ground state ($S = 0$).

(a) Find an expression for the average value of the magnetic moment $\langle \mu \rangle$ in the presence of a magnetic field, B.
(b) Show that the susceptibility for $T >> \Delta$ is approximately independent of Δ.
(c) Explain how this system may be cooled down by adiabatic magnetization (not adiabatic demagnetization).

5. A metal with magnetic permeability $\mu = 1$ is placed under the action of a uniform magnetic field $\mathbf{B} = B_0 \hat{z}$. The conduction electrons may be considered as a gas of free electrons with density n and average time between collisions, τ. Derive an expression for the resistivity tensor.
6. If we have a circularly polarized wave propagating in a material and a magnetic field is applied along the propagation direction, since the magnetic field gives origin to electronic motion that rotates in the perpendicular plane, the two rotation motions will interfere (Faraday effect).

(a) Using the equation of motion of the electrons under the combined effects of the wave electric field and the external magnetic field and considering that the electrons are bound to the atoms via an elastic potential,

$$m \frac{d^2 \mathbf{r}}{dt^2} = -e\mathbf{E} - e\mathbf{v} \times \mathbf{B} - m\omega_0^2 \mathbf{r}$$

and that it depends on the direction of the polarization rotation, calculate the refraction index. Therefore, the wave numbers for the two polarizations are different.

(b) Consider a wave that has a superposition of right and left polarizations, and specify to the case of linear polarization. Show that the direction of polarization changes as the wave propagates in the dielectric.

7. Consider a metallic bar with an electric current in the direction x that is the result of a potential difference V_x applied at its edges. In the middle point of the bar, two side contacts, A and B, are used to measure the Hall voltage, V_H, in the direction y. If the two contacts are slightly misaligned, the measurement of V_H will have a systematic error. Show that that error may be canceled if two measurements are carried out, one with the magnetic field in the direction z and another with the field in the direction $-z$.
8. In a given semiconductor there are 10^{13} donors per cm^3 with an ionization energy E_d of 1 meV and an effective mass $0.01m$.

(a) Estimate the conduction electron concentration at the temperature 4 K.
(b) What is the value of the Hall coefficient? Assume that there are no acceptor atoms present and that $E_g \gg k_B T$.

9. A two-dimensional system with one type of charge carriers in the presence of a perpendicular magnetic field has a resistivity tensor given by Eq. (13.121). Rewrite this expression as

$$\rho = \begin{pmatrix} \rho_0 & -R_H B \\ R_H B & \rho_0 \end{pmatrix}, \tag{13.194}$$

where R_H is the Hall coefficient.
Consider now that we have two types of charge carriers with the same density, n, opposite charges (e and $-e$), masses m_1, m_2, and relaxation rates τ_1, τ_2, respectively.

(a) Calculate the magnetoresistance $\Delta\rho = \rho(B) - \rho(B = 0)$, where B is the magnetic field.
(b) Determine the condition for a zero Hall effect.

10. Compare the wave functions of an electron in a plane in a homogeneous perpendicular magnetic field using the Landau gauge and the symmetric gauge. In particular, plot the absolute value squared of a few of the lowest energy eigenfunctions, as function of distance scaled by the magnetic length.
11. The Sommerfeld condition

$$\oint \frac{ds}{\lambda} = n,$$

where λ is the de Broglie wavelength and the integration is carried out along a close circuit, together with the de Broglie relation leads to

$$\oint p ds = nh,$$

which can be generalized to electrons as

$$\oint pds = (n + 1/2)h.$$

(a) Using the equation of motion of an electron in the presence of an external magnetic field (neglect the Zeeman effect), show that the flux through a closed electron orbit is given by

$$\Phi_n = (n + 1/2)\Phi_0. \tag{13.195}$$

(b) The real space closed orbit has a corresponding orbit in momentum space. Show that the orbit in momentum space has an area that satisfies

$$S_n = (n + 1/2)\frac{2\pi e}{\hbar}B, \tag{13.196}$$

where B is the magnetic field.

12. We can write the interaction of an electron with an external magnetic field, using the symmetric gauge as

$$H_{int} = \mu_B (\mathbf{L} + g\mathbf{S}) \cdot \mathbf{B} + \frac{e^2}{8m}B^2 \left(x^2 + y^2\right). \tag{13.197}$$

Using first-order perturbation theory, and considering only the effect of the second term, proportional to B^2, obtain an expression for the magnetic susceptibility and show that it leads to a diamagnetic contribution.
Estimate the magnitude of the molar susceptibility and show that it is small.

13. Consider a disc of a material containing free electrons, of radius r_2 and with a circular hole of radius $r_1 < r_2$. We apply a perpendicular magnetic field of amplitude B_0 through the material and that is zero outside the annular region. We also apply a magnetic flux, Φ, through the hole [8].
The combined effect of the flux Φ and the magnetic field $\mathbf{B} = B_0\mathbf{e}_z$ leads to a vector potential directed along the azimuthal direction $\mathbf{A} = A\mathbf{e}_\theta$ where we are using cylindrical coordinates (r, θ, z) and in the symmetric gauge

$$A = \frac{1}{2}B_0 r + \frac{\Phi}{2\pi r}. \tag{13.198}$$

The eigenstates, $\psi_{m,n_r}(\mathbf{r})$, are centered around some point, r_m, and the quantization of flux can be written as

$$B_0\pi r_m^2 + \Phi = m\Phi_0, \tag{13.199}$$

where Φ_0 is the flux quantum, m is the angular momentum quantum number, and n_r is the Landau level index.

(a) Show that the current associated with a given eigenstate may be written as

$$I_{m,n_r} = \frac{e}{m^*} \int_0^\infty dr |\psi_{m,n_r}(\mathbf{r})|^2 \left(\frac{m\hbar}{r} - eA(r) \right) . \qquad (13.200)$$

(b) Show that the current may be approximated by

$$I_{m,n_r} \sim \frac{e^2 B_0}{m^*} \int_0^\infty dr |\psi_{m,n_r}(\mathbf{r})|^2 (r_m - r) . \qquad (13.201)$$

(c) As a consequence, the currents near $r = r_2$ and near $r = r_1$ are different. Considering the cyclotron motion due to the presence of a perpendicular magnetic field, explain why the currents along the two edges have different signs.

14. Calculate the Chern numbers of the various phases of the Haldane anomalous Hall insulator shown in Fig. 13.12.
15. In two-dimensional systems, the existence of only one component of the angular momentum implies no quantization of the angular momentum and so it can take any fractional value. Going around a circle and changing the angle by 2π for integer angular momentum, or by 4π for half-integer angular momentum, in a three-dimensional space leads back to the same state, but in a two-dimensional space it can lead to an arbitrary phase between the initial and final states [9].
We may define on a plane exchanges of particles by windings (for instance counter-clockwise) associated with a given arbitrary phase $e^{i\theta}$. The exchanges are not characterized by the permutation group but by the braid group, constructed by the operator T_i acting between particles i and $i + 1$.

(a) Show that the exchanges may be defined by the conditions

$$\begin{aligned} T_i T_j &= T_j T_i , & |i - j) > 1, \\ T_i T_j T_i &= T_j T_i T_j , & |i - j| = 1, \end{aligned} \qquad (13.202)$$

and that $T_i^{-1} \neq T_i$.

(b) Consider a winding between two anyons by an angle φ. Then the wave function acquires a phase $e^{i\theta\varphi/\pi}$. Show that the relative angular momentum, $L_z = -i\hbar\varphi$, has a fractional shift with respect to the integer value found in three-dimensional systems.

References

1. D. Xiao, M.-C. Chang and Q. Niu, Reviews of Modern Physics **82**, 1959 (2010).
2. More accurate experimental results have been obtained more recently. See, for instance, Fig. 2.6 of the PhD thesis of Benedikt Friess, https://mediatum.ub.tum.de/doc/1219842/1219842.pdf that has been published by Springer in https://link.springer.com/book/10.1007/978-3-319-33536-0.
3. R.B. Laughlin, Physical Review B **23**, 5632 (1981).
4. D.J. Thouless, M. Kohmoto, M.P. Nightingale and M. den Nijs, Physical Review Letters **49**, 405 (1982); Q. Niu, D.J. Thouless and Y.S. Wu, Physical Review B **31**, 3372 (1985).
5. F. D. M. Haldane, Physical Review Letters **61**, 2015 (1988).
6. R.B. Laughlin, Physical Review B **27**, 3383 (1983).
7. J.K. Jain, Physical Review B **40**, R8079 (1989); Physical Review B **41**, 7653 (1990).
8. See, B. I. Halperin, Physical Review B **25**, 2185 (1982).
9. See, M. Greiter and F. Wilczek, arXiv:2210.02530.

Chapter 14
Microscopic Description of Interactive Quantum Systems

In this chapter we will consider, in some detail, the effect of interactions in a quantum system. We will limit ourselves to the lowest order of approximation, with focus on variational methods, such as the Hartree and Hartree-Fock approximations. One of the interesting consequences of interactions in a many-body system is the appearance of new phases of matter, such as magnetic or superconducting phases, to be considered in later chapters. These are in general the result of non-perturbative effects of the interactions and the simplest method to describe them is to use a mean-field approach (such as self-consistent Hartree and Hartree-Fock approximations). The need for non-perturbative methods, such as variational methods, arises already at the level of a single atom. We will review the approach usually taken to understand a simple atom such as He, stressing the need to use methods beyond perturbation theory. An outcome of this study is the understanding of how a magnetic moment arises in an atom and how it is related to the Coulomb potential and to the symmetry properties of wave functions, under the exchange of identical particles.

14.1 Bose Liquid

Let us consider first a Bose liquid. Unlike a Bose gas, where the particles are non-interacting, we will now consider the effect of interactions, hence a liquid. We are interested in the low-energy excitations which we expect are those corresponding to low momenta, or equivalently long distances. These are therefore the sound waves of the system. The energy dispersion relation is then of the form $\omega = uk$ or $\epsilon = up$, where u is the sound velocity. This only holds at low energies. As the momentum increases, the energy ϵ will no longer be linear in p. At large momenta (small distances) we should recover the free particle spectrum where $\epsilon \sim p^2$.

P. D. Sacramento, *Introduction to Condensed Matter Theory*,
https://doi.org/10.1007/978-3-031-86251-9_14

14.1.1 Interacting Bosons

Let us then consider a system of bosons with weak interactions. The system is described by an Hamiltonian of the form

$$\begin{aligned} H &= H_0 + H_{int} \\ H_0 &= \sum_{\mathbf{k}} \epsilon_{\mathbf{k}} a^{\dagger}_{\mathbf{k}} a_{\mathbf{k}} \\ H_{int} &= \frac{1}{2}\frac{1}{\Omega^2} \int d^3 r_1 \int d^3 r_2 V(\mathbf{r}_1 - \mathbf{r}_2) a^{\dagger}(\mathbf{r}_1) a(\mathbf{r}_1) a^{\dagger}(\mathbf{r}_2) a(\mathbf{r}_2). \end{aligned} \tag{14.1}$$

Here $\epsilon_{\mathbf{k}} = \frac{k^2}{2M}$ and Ω is the volume. This model may describe He^4, a system of weakly coupled bosons where the number of bosons is conserved.

Let us define

$$a(\mathbf{r}) = \sum_{\mathbf{k}} e^{i\mathbf{k}\cdot\mathbf{r}} a_{\mathbf{k}}.$$

Then we can rewrite the interaction term introducing the Fourier transform of the interaction potential as

$$H_{int} = \frac{1}{2} \sum_{k_1, k_1', k_2, k_2'} \delta_{k_1+k_2, k_1'+k_2'} V(k_1' - k_1) a^{\dagger}_{k_1} a^{\dagger}_{k_2} a_{k_2'} a_{k_1'} + \frac{1}{2} N V(r=0). \tag{14.2}$$

Here $N = \sum_{\mathbf{k}} a^{\dagger}_k a_k$.

If $V = 0$, we have seen that the system will show Bose-Einstein condensation at $T < T_c$ and all the particles will be in the state of $k = 0$, at $T = 0$. If $V \neq 0$, then some particles will occupy excited states ($\mathbf{k} \neq 0$). There will be N_e particles in the excited states and N_0 particles in the ground state, with $N_e << N_0$.

Let us now look at the structure of the terms of the Hamiltonian. They represent particles in different states and are involved in different processes like:

(i) Interactions in the ground state: $a^{\dagger}_0 a^{\dagger}_0 a_0 a_0$. This leads to a contribution $\frac{1}{2} V(k = 0) N_0^2$.
(ii) One particle out of the ground state: $a^{\dagger}_0 a^{\dagger}_k a_k a_0$ and $a^{\dagger}_k a^{\dagger}_0 a_0 a_k$. This leads to $N_0 V(k=0) \sum_k' a^{\dagger}_k a_k$. The $'$ excludes the momentum $k = 0$.
(iii) Exchange of one particle in the ground state with another one in a different state: $a^{\dagger}_k a^{\dagger}_0 a_k a_0$ and $a^{\dagger}_0 a^{\dagger}_k a_0 a_k$. This leads to $N_0 \sum_k' V(k) a^{\dagger}_k a_k$.
(iv) Both particles in the ground state or both in an excited state: $a^{\dagger}_0 a^{\dagger}_0 a_k a_{-k}$ or $a^{\dagger}_k a^{\dagger}_{-k} a_0 a_0$. This leads to $\frac{1}{2} N_0 \sum_k' V(k)(a_k a_{-k} + a^{\dagger}_k a^{\dagger}_{-k})$. Since N_0 is very large, we may approximate $\sqrt{N_0+2}\sqrt{N_0+1} \sim N_0$ or $\sqrt{N_0}\sqrt{N_0-1} \sim N_0$, respectively.

Other terms do not include particles in the ground state. Since N_0 is large, these are the dominant terms due to the N_0 prefactor. We get then that the Hamiltonian is written as

$$H = \sum_k \epsilon_k a_k^\dagger a_k + \frac{1}{2} N_0^2 V_0 + N_0 V_0 \sum_k{}' a_k^\dagger a_k$$
$$+ N_0 \sum_k{}' V(k) a_k^\dagger a_k + \frac{1}{2} N_0 \sum_k{}' V(k)(a_k a_{-k} + a_k^\dagger a_{-k}^\dagger) + \cdots \tag{14.3}$$

Keeping only these terms is equivalent to considering only quadratic terms in a_k, $a_k^\dagger$. Using that $N = N_0 + \sum_k' a_k^\dagger a_k$ we can rewrite the Hamiltonian

$$H = \frac{1}{2} N^2 V_0 + \sum_k{}' (\epsilon_k + NV(k))\, a_k^\dagger a_k + \frac{1}{2} N \sum_k{}' V(k)(a_k a_{-k} + a_k^\dagger a_{-k}^\dagger) + \cdots \tag{14.4}$$

14.1.2 Diagonalization of Interacting Hamiltonian

Let us diagonalize the system using two possible methods.

Equations of Motion

The equations of motion of the operators are given by

$$i \frac{\partial}{\partial t} a_k^\dagger = [a_k^\dagger, H] = -\left(\epsilon_k + NV(k)\right) a_k^\dagger - NV(k) a_{-k}$$
$$i \frac{\partial}{\partial t} a_{-k} = [a_{-k}, H] = \left(\epsilon_k + NV(k)\right) a_{-k} + NV(k) a_k^\dagger. \tag{14.5}$$

Taking the Fourier transform we get that

$$[\omega + \epsilon_k + NV(k)] a_k^\dagger + NV(k) a_{-k} = 0$$
$$-NV(k) a_k^\dagger + [\omega - \epsilon_k - NV(k)] a_{-k} = 0. \tag{14.6}$$

Taking the determinant to be zero we get

$$\omega^2 - (\epsilon_k + NV(k))^2 + (NV(k))^2 = 0. \tag{14.7}$$

The solution is therefore

$$\omega = \sqrt{\epsilon_k^2 + 2NV(k)\epsilon_k}. \tag{14.8}$$

In the limit when $k \to 0$ we get

$$\omega_{k\to 0} = \sqrt{\frac{NV_0}{M}}k.$$

In the limit when $k >> 1$ we get

$$\omega_{k>>1} = \epsilon_k\left(1 + \frac{NV(k)}{\epsilon_k}\right) \sim \epsilon_k + NV(k).$$

Therefore, when $V = 0$, the dispersion at small momenta is quadratic in k. However, when there is an interaction between the bosons $V \neq 0$, the dispersion is linear in the momentum (note that $V_0 > 0$, meaning a repulsive potential). Any infinitesimal potential changes qualitatively the nature of the spectrum, as shown in Fig. 14.1.

Bogoliubov-Valatin Transformation

The equations of motion couple $a_k^\dagger$ and a_{-k}. Let us then define new operators $\alpha_k^\dagger$ and α_k such that

$$H = \sum_k \lambda_k \alpha_k^\dagger \alpha_k + const. \tag{14.9}$$

With this choice we find

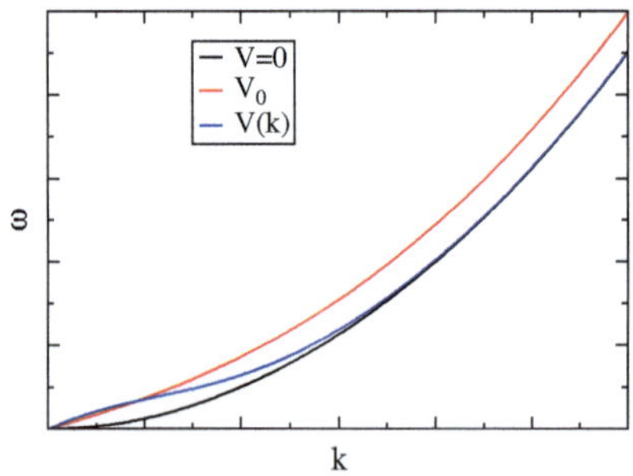

Fig. 14.1 Comparison of dispersion relations for the excitations without interactions ($V = 0$), with a constant interaction independent of momentum (V_0), and with an interaction ($V(k)$) that reflects the repulsive (Born-Mayer-like) and attractive parts (van der Waals) of the two-body interaction

$$[\alpha_k^\dagger, H] = -\lambda_k \alpha_k^\dagger$$

$$[\alpha_k, H] = \lambda_k \alpha_k,$$

and, since the new operators are also bosons, they obey

$$[\alpha_k, \alpha_{k'}^\dagger] = \delta_{k,k'}.$$

The Hamiltonian may be written in a convenient way like

$$\begin{aligned}
H &= \frac{1}{2}N^2 V_0 + \frac{1}{2}\sum_k{}' H_k \\
H_k &= \omega_0 \left(a_k^\dagger a_k + a_{-k}^\dagger a_{-k}\right) + \omega_1 \left(a_k a_{-k} + a_{-k}^\dagger a_k^\dagger\right) \\
\omega_0 &= \epsilon_k + NV(k) \\
\omega_1 &= NV(k).
\end{aligned} \tag{14.10}$$

Let us then consider the linear combinations

$$\begin{aligned}
\alpha_k &= u_k a_k - v_k a_{-k}^\dagger \\
\alpha_k^\dagger &= u_k a_k^\dagger - v_k a_{-k}.
\end{aligned} \tag{14.11}$$

The commutation relations between the new operators lead to

$$[\alpha_k, \alpha_k^\dagger] = u_k^2 - v_k^2 = 1.$$

Calculating the following commutator

$$\begin{aligned}
[\alpha_k^\dagger, H_k] &= u_k(-\omega_0 a_k^\dagger - \omega_1 a_{-k}) - v_k(\omega_0 a_{-k} + \omega_1 a_k^\dagger) \\
&= -\lambda_k (u_k a_k^\dagger - v_k a_{-k}),
\end{aligned} \tag{14.12}$$

implies that

$$\begin{aligned}
\omega_0 u_k + \omega_1 v_k &= \lambda_k u_k \\
\omega_1 u_k + \omega_0 v_k &= -\lambda_k v_k.
\end{aligned} \tag{14.13}$$

These equations have a solution if the determinant of the matrix

$$\begin{pmatrix} \omega_0 - \lambda_k & \omega_1 \\ \omega_1 & \omega_0 + \lambda_k \end{pmatrix}$$

vanishes. The solutions are given by

$$\lambda_k^2 = (\epsilon_k + NV(k))^2 - (NV(k))^2 .$$

Therefore, in the limit $k \to 0$, we get as before

$$\lambda_{k\to 0} = \left(\frac{NV_0}{M}\right)^{1/2} k.$$

14.1.3 Excitations

Let us consider, to be specific, that the interaction potential between the bosons is of the Lennard-Jones or Born-Mayer type with a van der Waals attraction. At low momenta the spectrum of the interacting Bose liquid is linear, like that of phonons in a solid. As the momentum grows, there is a local minimum in the dispersion relation which is related to the minimum in the potential. At high momenta the energy dispersion converges to that of free bosons, as previously discussed.

In the neighborhood of the local minimum, located around $1\mathring{A}^{-1} - 2\mathring{A}^{-1}$, a new type of quasi-particles appears, which were called rotons by Landau. The average distance between the atoms in He^4 in the liquid is about $r \sim (2\pi)/(2\mathring{A}^{-1}) \sim 3.1\mathring{A}$. The structure factor of the liquid shows a high probability of finding two atoms at a distance of about $3\mathring{A}$. There is therefore short-range order in the liquid.

Superfluidity

Let us now consider a particle of mass M and velocity $\mathbf{v}$ that is moving through the Bose liquid.[1] As it moves through the liquid it will interact with the elementary excitations of the liquid. There is conservation of energy and momentum expressed by

$$\begin{aligned} \frac{1}{2}Mv^2 &= \frac{1}{2}M(v')^2 + \omega_k \\ M\mathbf{v} &= M\mathbf{v}' + \mathbf{k}, \end{aligned} \tag{14.14}$$

where $\mathbf{k}$ and ω_k are the momentum and energy of the elementary excitation of the Bose liquid, respectively. We can then get that

$$(M\mathbf{v} - \mathbf{k})^2 = M^2(v')^2 = M^2v^2 - 2M\mathbf{v}\cdot\mathbf{k} + k^2$$

[1] See, for instance, [KittelQTS].

$$0 = \omega_k - \mathbf{v}\cdot\mathbf{k} + \frac{k^2}{2M}. \tag{14.15}$$

The smallest velocity that satisfies this condition is

$$v_c = \min\left(\frac{\omega_k}{k} + \frac{k}{2M}\right) > \min\frac{\omega_k}{k}. \tag{14.16}$$

If the potential $V = 0$ the critical velocity vanishes (since the dispersion is quadratic). However, if $V \neq 0$ then the ratio of ω_k/k is finite. This implies that the critical velocity is nonzero. If the velocity of the incident particle is $v < v_c$, there is no loss of energy to the medium and the particle moves without losing energy. The system becomes a superfluid. Note that if the system of Bosons is a gas (non-interacting), it cannot be superfluid. Note however that there are other excitations which we have not considered, which have a different nature from the elementary excitations we have paid attention up to now, called vortices, that decrease the value of the critical velocity by about an order of magnitude. The vortices require a lower threshold to be excited, but, still below the renormalized critical velocity, there will be superfluidity.

First and Second Sound

The coefficient of the dispersion relation with respect to the momentum is the usual sound (called first sound) such that $\omega = c_1 k$ where c_1 is the velocity of sound. There are however other excitations, which are constituted of waves of energy density, that are of the form $\omega = c_2 k$, which are called second sound. Let us determine the velocity of propagation of these waves.

Let us assume that there are no Umklapp processes, involving a reciprocal lattice vector, and that the mean free path of the phonons is smaller than the wave length of the second sound. Let us define

$$f(k, x) d^3k d^3x,$$

the number of phonons in an elementary volume in phase space. Then the rate of change of the density of phonons is given by the Boltzmann like equation

$$\frac{df}{dt} = \frac{\partial f}{\partial t} + \mathbf{v}\cdot\nabla f = \Delta_c f,$$

which is the rate of change of f due to the collisions. Now

$$\mathbf{v} = c_1 \frac{\mathbf{k}}{|\mathbf{k}|}.$$

Using this we get

$$\frac{\partial f}{\partial t} + c_1 \frac{\mathbf{k}}{|\mathbf{k}|} \cdot \nabla f = \Delta_c f. \tag{14.17}$$

The total momentum and the total energy of the liquid are conserved. They are given by

$$\begin{aligned} \mathbf{P}(\mathbf{r}) &= \int d^3k \mathbf{k} f \\ U(\mathbf{r}) &= \int d^3k c_1 k f. \end{aligned} \tag{14.18}$$

Since these quantities are conserved

$$\begin{aligned} \frac{d\mathbf{P}}{dt} &= 0 = \int d^3k \mathbf{k} \Delta_c f \\ \frac{dU}{dt} &= 0 = \int d^3k c_1 k \Delta_c f. \end{aligned} \tag{14.19}$$

Multiplying Eq. (14.17) by k_μ and integrating over the momentum we get that the equation of motion for f gives

$$\frac{\partial}{\partial t} \int d^3k k_\mu f + \int d^3k \frac{c_1}{k} k_\mu \sum_\nu k_\nu \frac{\partial f}{\partial x_\nu} = \int d^3k k_\mu \Delta_c f = 0$$

and, therefore,

$$\frac{\partial}{\partial t} P_\mu + \sum_\nu \frac{\partial}{\partial x_\nu} c_1 \int d^3k \frac{k_\mu k_\nu}{k} f = 0. \tag{14.20}$$

Due to the conservation of total momentum the collision term does not contribute, see Eq. (14.19).

Multiplying now Eq. (14.17) by $c_1 k$ and integrating over the momentum we get

$$\frac{\partial}{\partial t} \int d^3k c_1 k f + \int d^3k c_1^2 \sum_\nu k_\nu \frac{\partial f}{\partial x_\nu} = \int d^3k c_1 k \Delta_c f = 0$$

and, therefore,

$$\frac{\partial U}{\partial t} + c_1^2 \sum_\nu \frac{\partial}{\partial x_\nu} P_\nu = 0. \tag{14.21}$$

Note again that due to the conservation of energy the collision term does not contribute, see again Eq. (14.19).

If we now consider a situation close to equilibrium $f \sim f^0$, we can simplify

$$c_1 \int d^3k \frac{k_\mu k_\nu}{k} f \sim \delta_{\mu\nu} c_1 \int d^3k \frac{k_\mu^2}{k} f,$$

and identify as the dominant contribution. This is approximately given by

$$\frac{1}{3}\delta_{\mu\nu} c_1 \int d^3k \frac{k^2}{k} f \sim \frac{1}{3}\delta_{\mu\nu} U.$$

Then we get that

$$\frac{\partial}{\partial t} P_\mu + \frac{1}{3}\frac{\partial U}{\partial x_\mu} = 0. \tag{14.22}$$

Taking now the time derivative in the Eq. (14.21), we get

$$\frac{\partial^2 U}{\partial t^2} - \frac{c_1^2}{3}\sum_\nu \frac{\partial^2 U}{\partial x_\nu^2} = 0. \tag{14.23}$$

We obtain, therefore, an equation for the energy that is a wave equation with a velocity of propagation $c_2 = c_1/\sqrt{3}$. These energy density waves are called second sound.

14.2 Solid State Physics and the Fermi Liquid

The electrons of the atoms in a solid have orbitals that overlap. The electrons are therefore shared between the ions, some of them eventually becoming strongly delocalized to become conduction electrons. The microscopic Hamiltonian of the system is quite complex due to the strong electromagnetic interactions between the charges of the system. Clearly, some simplification has to be done to obtain the energy states.

The Hamiltonian[2] of a system of electrons with mass m at positions $\mathbf{r}_k$ and ions with mass M_i at positions $\mathbf{R}_i$ is in general of the form[3]

[2] To simplify notation, in the expression for the Coulomb potential we write e^2 instead of $e^2/(4\pi\epsilon_0)$. In Sects. 14.2, 14.3, 14.4, and 14.5 this replacement should be made.

[3] See also [Madelung].

$$
\begin{aligned}
H &= H_{el} + H_{ion} + H_{el-ion} \\
H_{el} &= H_{el,kinetic} + H_{el-el} = \sum_k \frac{p_k^2}{2m} + \frac{1}{2}\sum_{k,k'}{}' \frac{e^2}{|\mathbf{r}_k - \mathbf{r}_{k'}|} \\
H_{ion} &= H_{ion,kinetic} + H_{ion-ion} = \sum_i \frac{p_i^2}{2M_i} + \frac{1}{2}\sum_{i,i'}{}' V_{ion}(\mathbf{R}_i - \mathbf{R}_{i'}) \\
H_{el-ion} &= \sum_{k,i} V_{el-ion}(\mathbf{r}_k - \mathbf{R}_i).
\end{aligned} \tag{14.24}
$$

V_{ion} is the interaction between two ions and V_{el-ion} is the interaction between one electron and one ion. Taking into account the motion of the ions around their equilibrium positions, we may separate the terms involving the ion coordinates as

$$
\begin{aligned}
H_{ion,ion} &= H^0_{ion-ion} + H_{ph} \\
H_{el,ion} &= H^0_{el-ion} + H_{el-ph}.
\end{aligned} \tag{14.25}
$$

Here H^0 respects to the equilibrium positions of the ions and the label ph is the part due to the phonons. We can now make approximations on the terms of the Hamiltonian.

Consider first the case of the electrons. As a first step we can write that the full electron Hamiltonian is approximated by

$$
H_{el} = -\sum_k \frac{\hbar^2}{2m}\nabla_k^2 + \frac{1}{2}\sum_{k,k'}{}' \frac{e^2}{|\mathbf{r}_k - \mathbf{r}_{k'}|} + H_+. \tag{14.26}
$$

This Hamiltonian respects only to the electrons. The positive charges of the ions are replaced by a positively charged background, H_+ (required to maintain charge neutrality and stabilize the system). This is the so-called jellium model. It can be improved, if instead of considering a uniform background, we implement the symmetries of the lattice of ions. This amounts, as shown before, to replace the free electron spectrum by the band structure due to the periodicity of the lattice. Clearly, the important new step is now the interactions between the electrons, in comparison with the nearly free electron model previously discussed.

We can as well make some approximations for the Hamiltonian of the ions. Take as Hamiltonian

$$
H_{ion} = -\sum_i \frac{\hbar^2}{2M_i}\nabla_i^2 + \frac{1}{2}\sum_{i,i'}{}' V_{ion}(\mathbf{R}_i - \mathbf{R}_{i'}) + H_-. \tag{14.27}
$$

In the same way we replace the interaction of the ions with the electrons by an effective negative medium. The solution of this Hamiltonian gives the dynamics of the lattice. For consistency $H_+ + H_- = 0$.

We have effectively separated the dynamics of the electrons and of the ions in a first approximation. This constitutes the basis of the adiabatic approximation of Born-Oppenheimer. The electrons have masses which are far smaller than those of the ions. Therefore, since the strength of the forces is similar, the electrons move much faster than the ions. The ions respond slowly to any change of configuration of the electrons, but the electrons follow any changes of the configuration of the ions. The ions respond to an average of the configurations of the electrons and the wave functions of the electrons can be parametrized by the distribution of the ions. This implies that we can write that

$$(H_{el} + H_{el-ion})\psi = E_{el}\psi. \tag{14.28}$$

The overall wave function Ψ is such that

$$\Psi = \psi(\mathbf{r}_1, \cdots, \mathbf{r}_N; \mathbf{R}_1, \cdots, \mathbf{R}_N)\varphi(\mathbf{R}_1, \cdots, \mathbf{R}_N). \tag{14.29}$$

In the part of the wave function associated with the electrons, the coordinates of the ions are just parameters. We can write

$$\begin{aligned} H\Psi &= (H_{el} + H_{ion} + H_{el-ion})\psi\varphi \\ &= \psi(H_{ion} + E_{el})\varphi - \sum_i \frac{\hbar^2}{2M_i}\left(\varphi\nabla_i^2\psi + 2\nabla_i\varphi \cdot \nabla_i\psi\right). \end{aligned} \tag{14.30}$$

If we neglect the second term that mixes the electrons and nuclei, since the ratios of the electron mass to the nuclei masses are small, we get an approximate equation for the motion of the ions

$$(H_{ion} + E_{el})\varphi = E\varphi, \tag{14.31}$$

where E_{el} is obtained from the solution for the electrons.

Let us then focus our attention on the electron problem. This is given by the solution of

$$(H_{el} + H_{el-ion}^0)\psi = E_{el}\psi, \tag{14.32}$$

where we have neglected the interactions of the electrons with the phonons, valid in regimes where few phonons are excited. Usually, in solid state physics the interactions between the conduction electrons are neglected and only the free electron problem in the presence of a periodic potential is solved. However, the interactions between the electrons may be important. Also, the restrictions on the phase space due to the Pauli exclusion principle have to be taken into account.

Let us then consider that the interactions between the electrons are weak enough so that they do not destroy the concept of the Fermi surface, even if there may be (and surely are) changes of the energy levels. Also, there may be some changes to the equilibrium Fermi-Dirac distribution, due to the interactions. This is the approximation taken by Landau, that the concept of a Fermi surface remains. The Fermi liquid is then assumed to have overall characteristics that are similar to those of the Fermi gas, with excitations that are fermions, even though with different energy levels.

Let us then assume that the excitations of the Fermi liquid are of the same type as those of the Fermi gas. That is, the concept of the Fermi surface prevails. Actually, at $T = 0$ the discontinuity of the occupation number at the Fermi surface has a different (smaller) amplitude. The discontinuity is smaller because the interactions between the electrons renormalize parameters in the theory. Therefore, even at $T = 0$ there is some depletion of the levels close to the Fermi surface.

Consider the case of the Fermi gas at finite temperature. Excitations arise that transfer particles from the levels inside the Fermi surface to states above the Fermi surface. This implies that the excited states (due to the thermal excitation) differ from the equilibrium distribution, due to the excitation of particles. In the case of the Fermi liquid, there are also changes in the Fermi-Dirac distribution that represents the excitations. However, due to the interactive character of the particles (and of the excitations themselves), the energy of the excitations involves interactive terms in the density of the particles. Due to the interactions, the excitations have now a finite lifetime since they may scatter off other excitations. Therefore, they are actually quasi-particles. Only very close to the Fermi surface does the lifetime of the excitations tend to infinite. In a Fermi liquid, the excitations close to the Fermi surface are well-defined and have a character similar to the excitations of the Fermi gas. The theory is therefore limited to energies close to the Fermi surface.

14.2.1 Landau's Fermi Liquid Theory: Renormalized Fermi Gas

The distribution of free electrons on states in thermodynamic equilibrium is governed by the Fermi-Dirac distribution[4]

$$n^0(\mathbf{p}) = n^0(\epsilon) = \frac{1}{1 + e^{\frac{\epsilon-\mu}{k_B T}}} \tag{14.33}$$

and is only a function of the energy, ϵ, of a given state, the temperature, T, and the chemical potential, μ. At zero temperature, it is a step function at the chemical potential, with all lower-energy state occupied and all higher states empty. At a finite

[4] For further reading see [PinesNozieres].

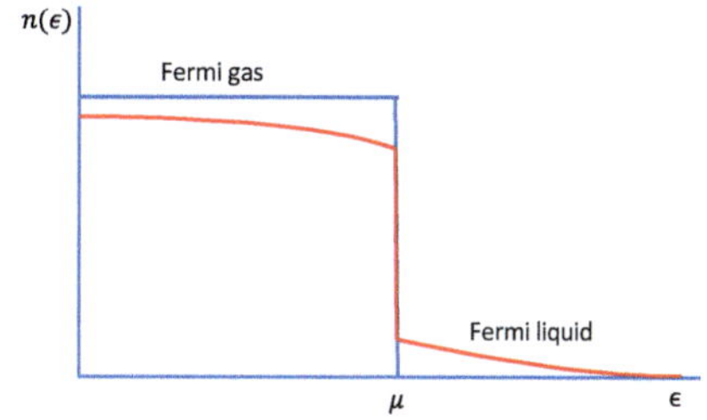

Fig. 14.2 Schematic comparison of Fermi-Dirac distribution for a Fermi gas and a Fermi liquid

temperature, the distribution gets distorted due to the excitation of states above the chemical potential (and consequent depletion at energies lower than the chemical potential). We found before that, at low temperatures, the specific heat is linear in temperature and the magnetic susceptibility is independent of T. We wish now to address the issue when there are interactions present. Interactions may lead to significative changes of the system, and, as discussed in the following chapters, several types of phases may appear such as charge density waves and magnetic and superconducting phases. However, there are also some situations in which the system retains some characteristics of the free system, with some renormalized parameters that are the result of the effect of the interactions. These systems are called normal Fermi liquids, in the sense that they differ in a quantitative way from a free system but are qualitatively similar. One significant feature is that they retain the concept of a Fermi surface, as a discontinuity in the distribution of the electrons over the available states, however with a jump in the occupation number that is not unity, such as schematically shown in Fig. 14.2. We may define a system called non-Fermi liquid as a system where such a discontinuity is absent.

The total energy of a system of N free electrons in a volume Ω can be written as

$$E = \sum_{\mathbf{p}} n_{\mathbf{p}} \frac{p^2}{2m}. \tag{14.34}$$

The total electron density determines the Fermi energy or momentum as

$$\frac{N}{\Omega} = \frac{1}{3\pi^2} \left(\frac{p_F}{\hbar} \right)^3. \tag{14.35}$$

Also, at zero temperature,

$$\mu = E_0(N+1) - E_0(N) = \frac{\partial E_0}{\partial N}, \tag{14.36}$$

which leads to $\mu = p_F^2/(2m)$, the Fermi energy.

At zero temperature we may consider excitations changing the distribution with respect to the equilibrium one as

$$\delta n_{\mathbf{p}} = n_{\mathbf{p}} - n_{\mathbf{p}}^{0}, \tag{14.37}$$

such that the total energy changes as

$$E - E_0 = \sum \frac{p^2}{2m} \delta n_{\mathbf{p}}, \tag{14.38}$$

and the free energy changes as

$$F - F_0 = \sum_{\mathbf{p}} \left(\frac{p^2}{2m} - \mu \right) \delta n_{\mathbf{p}}. \tag{14.39}$$

Since we are considering zero temperature the only term that contributes to the free energy, with respect to the energy, is the one associated with the chemical potential. If the total number of particles is conserved then

$$\sum_{\mathbf{p}} \delta n_{\mathbf{p}} = 0. \tag{14.40}$$

Consider now an interacting system under the condition of being a normal Fermi liquid. Then, if the deviation from the equilibrium distribution is small, we may attempt a Landau expansion as

$$E[n_{\mathbf{p}}] = E_0 + \sum_{\mathbf{p}} \epsilon_{\mathbf{p}} \delta n_{\mathbf{p}} + O\left(\delta n_{\mathbf{p}}^2\right). \tag{14.41}$$

Here the term of second order in the deviations is the result of interactions between the excitations. This leads to a renormalization of the energies, and the modes of the system are no longer strictly free particles, but are called quasiparticles. There is a one-to-one correspondence between the free particles and the quasiparticles, in the case the change is small. Therefore, the energies get renormalized and are called $\epsilon_{\mathbf{p}}$. The velocity of the quasiparticles is the group velocity of these renormalized bands and is given by

$$\mathbf{v}_{\mathbf{p}} = \nabla_{\mathbf{p}} \epsilon_{\mathbf{p}}. \tag{14.42}$$

Particularizing for the Fermi momentum (which we assume may be defined at the Fermi surface), the absolute value of the velocity can be written as

$$v_{p_F} = \frac{p_F}{m^*}, \tag{14.43}$$

which defines the effective mass, m^*, which in general differs from the free electron mass. Note that it is usual in this context to define the effective mass in terms of its relation to the velocity instead of the alternative definition as

$$\frac{1}{m^*} = \frac{\partial^2 \epsilon_{\mathbf{p}}}{\partial p^2}. \tag{14.44}$$

Landau proposed that the free energy deviation with respect to the lowest energy state may be written as

$$\begin{aligned} F - F_0 &= E - E_0 - \mu(N - N_0) \\ &= \sum_{\mathbf{p}} \left(\epsilon_{\mathbf{p}} - \mu\right) \delta n_{\mathbf{p}} + \frac{1}{2} \sum_{\mathbf{p},\mathbf{p}'} f_{\mathbf{p},\mathbf{p}'} \delta n_{\mathbf{p}} \delta n_{\mathbf{p}'} + O\left(\delta n_{\mathbf{p}}^3\right). \end{aligned} \tag{14.45}$$

Note that the first two terms are of the same order. Indeed, $(\epsilon_{\mathbf{p}} - \mu)$ is of order $\delta n_{\mathbf{p}}$, which implies that the first term is of the same order as the interaction term. The origin of the deviation from the free case results from the interactions and not temperature. Clearly, temperature will have a similar effect, but even at zero temperature there is a deviation from the Fermi-Dirac distribution, and the interactions lead to a population of states above the Fermi energy. A similar change of distribution was found in the problem of interacting bosons. In the regimes where $\delta n_{\mathbf{p}} \neq 0$, the deviation $(\epsilon_{\mathbf{p}} - \mu)$ is of the same order since we are assuming that the interactions do not have a profound effect on the distribution. The total number of excitations, $\sum_{\mathbf{p}} |\delta n_{\mathbf{p}}|$, should be small if the theory is a good approximation.

The parameters that encode the effect of the interactions are expected to be symmetric, $f_{\mathbf{p},\mathbf{p}'} = f_{\mathbf{p}',\mathbf{p}}$, since they can be obtained from the second derivative of the energy or of the free energy. Under the assumption of a normal Fermi liquid, we also expect that these f parameters and the energy dispersion relation of the quasiparticles (first derivative of the energy with respect to the deviation from the equilibrium distribution) should be continuous across the Fermi surface. Since the deviation is small (small changes around equilibrium) and if we consider small energies, we effectively remain close to the Fermi surface. This implies that, to leading order, the parameters $f_{\mathbf{p},\mathbf{p}'}$ are calculated at the Fermi surface. Therefore, it only depends on the directions of the momenta $\mathbf{p}$ and $\mathbf{p}'$ and on the spin directions σ, σ'. If there is no magnetic field, the system is invariant under time reversal. Therefore,

$$f_{\mathbf{p},\sigma;\mathbf{p}',\sigma'} = f_{-\mathbf{p},-\sigma;-\mathbf{p}',-\sigma'}. \tag{14.46}$$

If the system is invariant under spatial inversion ($\mathbf{p} \to -\mathbf{p}$), then we get

$$f_{\mathbf{p},\sigma;\mathbf{p}',\sigma'} = f_{\mathbf{p},-\sigma;\mathbf{p}',-\sigma'} \tag{14.47}$$

and only depends on the relative orientations of the spins.

There are two cases: when the spins are parallel and when the spins are antiparallel. We define different sets of parameters as

$$f^{\uparrow\uparrow}_{\mathbf{p},\mathbf{p}'} = f^{s}_{\mathbf{p},\mathbf{p}'} + f^{a}_{\mathbf{p},\mathbf{p}'}$$
$$f^{\uparrow\downarrow}_{\mathbf{p},\mathbf{p}'} = f^{s}_{\mathbf{p},\mathbf{p}'} - f^{a}_{\mathbf{p},\mathbf{p}'}, \tag{14.48}$$

where s stands for symmetric and a stands for antisymmetric. If the system is isotropic, $f^{s}_{\mathbf{p},\mathbf{p}'}$ and $f^{a}_{\mathbf{p},\mathbf{p}'}$ only depend on the angle ξ between the momenta $\mathbf{p}$ and $\mathbf{p}'$. Then we may expand these parameters in terms of Légendre polynomials

$$f^{s}_{\mathbf{p},\mathbf{p}'} = \sum_{l=0}^{\infty} f^{s}_{l} P_l(\cos\xi)$$
$$f^{a}_{\mathbf{p},\mathbf{p}'} = \sum_{l=0}^{\infty} f^{a}_{l} P_l(\cos\xi). \tag{14.49}$$

It is convenient to introduce some reduced coefficients as

$$F^{s(a)}_{l} = \nu(0) f^{s(a)}_{l}, \tag{14.50}$$

where $\nu(0)$ is the density of states at the Fermi surface and given by

$$\nu(0) = \frac{\Omega m^* p_F}{\pi^2 \hbar^3}. \tag{14.51}$$

The effect of the f functions (or parameters) may now be incorporated defining new energies that are called "local" quasiparticle energies. To see their meaning let us consider the state of the system as defined by the set of deviations $\{\delta n_{\mathbf{p}'}\}$ to which we add one quasiparticle with momentum $\mathbf{p}$. The free energy of this quasiparticle is defined as

$$(\bar{\epsilon}_{\mathbf{p}} - \mu) = (\epsilon_{\mathbf{p}} - \mu) + \sum_{\mathbf{p}'} f_{\mathbf{p},\mathbf{p}'} \delta n_{\mathbf{p}'}. \tag{14.52}$$

Define now a distribution

$$\bar{n}^0_{\mathbf{p}} = n^0\left(\bar{\epsilon}_{\mathbf{p}} - \mu\right). \tag{14.53}$$

Recall that n^0 is the equilibrium Fermi-Dirac distribution and

$$n^0_{\mathbf{p}} = n^0\left(\epsilon_{\mathbf{p}} - \mu\right). \tag{14.54}$$

Define the deviations

$$\delta\bar{n}_{\mathbf{p}} = n_{\mathbf{p}} - \bar{n}^0_{\mathbf{p}}$$

$$\delta n_{\mathbf{p}} = n_{\mathbf{p}} - n^0_{\mathbf{p}}. \tag{14.55}$$

This implies that

$$\begin{aligned}\delta n_{\mathbf{p}} - \delta \bar{n}_{\mathbf{p}} &= \left(n_{\mathbf{p}} - n^0_{\mathbf{p}}\right) - \left(n_{\mathbf{p}} - \bar{n}_{\mathbf{p}}\right) \\ &= \bar{n}^0_{\mathbf{p}} - n^0_{\mathbf{p}} = \frac{\partial n^0_{\mathbf{p}}\left(\epsilon_{\mathbf{p}} - \mu\right)}{\partial \epsilon_{\mathbf{p}}}\left(\bar{\epsilon}_{\mathbf{p}} - \epsilon_{\mathbf{p}}\right).\end{aligned} \tag{14.56}$$

Using 14.52 we arrive at

$$\delta \bar{n}_{\mathbf{p}} = \delta n_{\mathbf{p}} - \frac{\partial n^0}{\partial \epsilon_{\mathbf{p}}} \sum_{\mathbf{p}'} f_{\mathbf{p},\mathbf{p}'} \delta n_{\mathbf{p}'}. \tag{14.57}$$

When $T \to 0$

$$\frac{\partial n^0}{\partial \epsilon_{\mathbf{p}}} = -\delta\left(\epsilon_{\mathbf{p}} - \mu\right) \tag{14.58}$$

and, as a consequence, $\delta n_{\mathbf{p}}$ and $\delta \bar{n}_{\mathbf{p}}$ are only nonzero very close or on the Fermi surface.

Consider an isotropic Fermi liquid at zero temperature. Associated with symmetric and antisymmetric spin parts define

$$\begin{aligned}\delta n_{\mathbf{p},\pm} &= \delta n^s_{\mathbf{p}} \pm \delta n^a_{\mathbf{p}} \\ \delta \bar{n}_{\mathbf{p},\pm} &= \delta \bar{n}^s_{\mathbf{p}} \pm \delta \bar{n}^a_{\mathbf{p}}.\end{aligned} \tag{14.59}$$

We can find an expansion for the coefficients, on the Fermi surface, using a spherical harmonic expansion

$$\delta n^{s,a}_{\mathbf{p}} = \sum_{l,m} \delta\left(\epsilon_{\mathbf{p}} - \mu\right) \delta n^{s.a}_{l,m} Y_{lm}(\theta, \phi), \tag{14.60}$$

where θ, ϕ are the angles of the momentum $\mathbf{p}$. Using these expressions in 14.57 and 14.49 and using the spherical harmonics addition theorem we get

$$\begin{aligned}\delta \bar{n}^s_{l,m} &= \left(1 + \frac{F^s_l}{2l+1}\right) \delta n^s_{l,m} \\ \delta \bar{n}^a_{l,m} &= \left(1 + \frac{F^a_l}{2l+1}\right) \delta n^a_{l,m}.\end{aligned} \tag{14.61}$$

We may now obtain information on some equilibrium properties. The specific heat is obtained as

$$C_V = \left(\frac{\partial E}{\partial T}\right)_N = \left(\frac{\partial F}{\partial T}\right)_\mu . \tag{14.62}$$

Recalling the free energy expansion

$$F - F_0 = \sum_{\mathbf{p}} \left(\epsilon_{\mathbf{p}} - \mu\right) \delta n_{\mathbf{p}} + \frac{1}{2} \sum_{\mathbf{p},\mathbf{p}'} f_{\mathbf{p}\mathbf{p}'} \delta n_{\mathbf{p}} \delta n_{\mathbf{p}'}, \tag{14.63}$$

we can show that the first term is of order T^2, while the second term is of order T^4 and is therefore negligible at small temperatures. Indeed, recall that

$$\delta n_{\mathbf{p}} = n^0_{\mathbf{p}}(T, \mu) - n^0_{\mathbf{p}}(0, \mu), \tag{14.64}$$

where

$$n^0_{\mathbf{p}}(T, \mu) = \frac{1}{1 + e^{\frac{\bar{\epsilon}_{\mathbf{p}} - \mu}{k_B T}}} . \tag{14.65}$$

Then we get

$$\int \delta n_{\mathbf{p}} p^2 dp \sim T^2. \tag{14.66}$$

Using the Sommerfeld expansion we can show that

$$F(T) - F(0) = \frac{\pi^2}{6} \nu(0) \left(k_B T\right)^2 , \tag{14.67}$$

which leads to a linear specific heat

$$C_V = \frac{m^* p_F}{3\hbar^2} k_B^2 T, \tag{14.68}$$

and is not affected by the interactions, to leading order.

On the other hand, the spin susceptibility is affected by the interactions. Applying a magnetic field to the electron system there are the orbital and Zeeman effects due to the electron spin. Let us focus on this effect. The energy of a spin-1/2 in an external magnetic field is $-g\mu_B \sigma H$, where g is the Landé factor and the Bohr magneton is $\mu_B = e\hbar/(2mc)$. The energy splitting of the bands of $\uparrow$ and $\downarrow$ spins influences the distributions:

$$n_{\mathbf{p},\sigma}(H) = n^0(\bar{\epsilon}_{\mathbf{p}} - \mu - g\mu_B \sigma H), \tag{14.69}$$

for the two spin projections, $\sigma = \pm 1$. As we saw before, this is the origin of the Pauli paramagnetism.

This leads to

$$\delta \bar{n}_{\mathbf{p},\sigma} = -g\mu_B \sigma H \frac{\partial n^0}{\partial \epsilon_{\mathbf{p}}}, \tag{14.70}$$

since

$$\delta \bar{n}_{\mathbf{p}} = -d\mu \frac{\partial n^0}{\partial \epsilon_{\mathbf{p}}}.$$

The magnetization is given by

$$M = \sum_{\mathbf{p}} g\mu_B \sigma \delta n_{\mathbf{p},\sigma}. \tag{14.71}$$

We have that

$$\delta n_{\mathbf{p},\sigma} = \frac{\delta \bar{n}_{\mathbf{p},\sigma}}{1 + F_0^a} = -\frac{g\mu_B \sigma H}{1 + F_0^a} \frac{\partial n^0}{\partial \epsilon_{\mathbf{p}}}. \tag{14.72}$$

Taking $g = 2$ we get

$$M = \mu_B^2 \frac{\nu(0)}{1 + F_0^a} H \tag{14.73}$$

and a spin susceptibility

$$\chi_s = \frac{M}{\Omega H} = \frac{m^* p_F}{\pi^2 \hbar^3} \frac{\mu_b^2}{1 + F_0^a}, \tag{14.74}$$

which is then changed due to the presence of the interactions, parametrized by F_0^a, in this case. This result shows that $1 + F_0^a > 0$, or else, if it vanishes, an instability arises, which may lead to some magnetic order, such as ferromagnetism.

Wiedemann-Franz Law and Wilson Ratio

The Fermi liquid theory applies to situations where the interacting system has a behavior similar to a free Fermi gas, with appropriately renormalized parameters that take into account the effect of interactions. We may expect that some ratios of physical quantities may lead to results that are to some extent universal, if the renormalized parameters cancel such that the expressions are characteristic of the free fermion problem.

It was shown before that the zero frequency real part of the conductivity of fermions with no interactions is given by the Drude expression

$$\sigma = \frac{ne^2\tau}{m}, \tag{14.75}$$

where n is the particle density, τ is the average time between collisions with impurities or imperfections in the system, and m is the electron mass. Also, we showed before that the thermal conductivity of the Fermi gas can be written as

$$K = \frac{1}{3}Cvl, \tag{14.76}$$

where C is the heat capacity, v the carrier velocity at the Fermi level, and l is the mean free path between collisions. Assuming that the relevant charges are those at the Fermi level, and assuming all move with the Fermi velocity, we have that $l = v_F\tau$. The heat capacity was shown to be

$$C = \frac{\pi^2}{2}Nk_B\frac{T}{T_F}, \tag{14.77}$$

where N is the number of particles and T_F the Fermi temperature. It is then easy to show that

$$\frac{K/T}{\sigma V} = \frac{\pi^2 k_B^2}{3e^2}, \tag{14.78}$$

which is a number that only depends on fundamental constants. This number is the Lorenz number and this independence of the systems properties, such as τ or m, the effective mass of the electrons, is called the Wiedemann-Franz law. The same result is expected to hold for a system described by the Fermi liquid theory. Even though each quantity is different from the result for the Fermi gas, in the ratio the system dependent parameters cancel. One may regard any deviation from this law as a sign of deviation from the applicability of the Fermi liquid hypothesis, at least from the point of view of the system response to charge and heat transport.

Another ratio that we may consider is the so-called Wilson ratio, defined in the limit when temperature goes to zero by

$$R_W = \frac{4\pi^2}{3}\frac{\chi}{\gamma}, \tag{14.79}$$

where χ is the magnetic susceptibility and $C = \gamma T$. In a Fermi liquid the heat capacity is linear in temperature and γ is a constant. In the case of a Fermi liquid (or non-interacting fermionic system) the Wilson ratio defined this way equals one in units such that $k_B = g\mu_B = 1$. Recall that the Pauli susceptibility at zero temperature is given by

$$\chi(T=0) = \frac{3}{2}\frac{n(\mu^*)^2}{\epsilon_F}, \tag{14.80}$$

where μ^* is the electron magnetic moment and that

$$\gamma = \frac{\pi^2}{2} N \frac{k_B}{T_F}. \tag{14.81}$$

We can then show that the Wilson ratio $R_W = 1$.

A clear source of a different behavior between an interacting and a free system is the possibility of a phase transition, which naturally leads to a behavior that is qualitatively different from that of a non-interacting system. Also, a deviation from the linear in temperature behavior of the specific heat leads to a different Wilson ratio. It turns out that some systems have a Fermi liquid-like behavior but with a different Wilson ratio. One example is a Heisenberg chain that has a Fermi-like behavior at low temperatures, but with a Wilson ratio $R_W = 2$. Another example is the Kondo problem that has a Fermi liquid-like behavior at low temperatures but the Wilson ratio is also $R_W = 2$.

14.2.2 Wigner Crystal

The expectation is that if an electronic system does not have a large density, we would expect that the approximation of a non-interacting system would hold. When the density is high we expect that the effect of the interactions between the electrons, specifically the Coulomb interaction, would lead to a significant effect. It turns out however that, if the density is quite small and the electron gas is diluted, the effect of the interactions cannot be ignored. If the density is very small and the temperature is very low we may expect that the electrons have small kinetic energies and the effect of the repulsion between the electrons may tend to create a state where the electrons are almost frozen, in equilibrium positions. This state would be like a crystal, the Wigner crystal.

It is usual to parametrize the electronic density defining the Wigner-Seitz radius, r_s, the radius of a sphere that contains on average one electron. Usually it is defined as a dimensionless parameter, with the distance defined in units of the Bohr radius. If the electrons form a lattice r_s is of the order of the distance between the electrons.

The Coulomb potential energy between the electrons can be written as

$$U = \frac{1}{2}\sum_{i \neq j} \frac{e^2}{|\mathbf{R}_i - \mathbf{R}_j|}, \tag{14.82}$$

where $\mathbf{R}_i$, $\mathbf{R}_j$ are the lattice positions and we are omitting the contribution from the positive charge background that provides the charge neutrality. The potential energy density is of the order of

$$u = U/\Omega \sim \frac{1}{\Omega}\Omega^{2/3} \sim \frac{1}{\Omega^{1/3}} \sim \frac{1}{r_s}. \tag{14.83}$$

If the electrons form such a lattice, we have to take into account the oscillations of the electron positions around the equilibrium positions. The rigidity of the structure leads to oscillations whose frequencies are a function of the force constants that result from the structure. The force constants are proportional to the second derivative of the potential energy and therefore are of the order of $1/r_s^3$. This implies that the oscillation frequencies are of the order of $\sqrt{1/r_s^3}$, which means that the oscillation energies, with equal contributions from potential energy and kinetic energy, are of the order of $1/r_s^{3/2}$.

When the density is small and therefore r_s is large, taking into account that the potential energy is of order $1/r_s$ and the kinetic energy is of the order of $1/r_s^{3/2}$ implies that the potential energy is dominant over the kinetic energy. Therefore, the crystalline structure, if it exists, is dominated by the Coulomb potential energy over the kinetic energy. We stress again that temperature effects have to be small. The full solution of the Wigner crystal limit is therefore complex due to the significant role of the Coulomb interaction, and in general a numerical solution is required.

14.3 Origin of Magnetic Moments

In order to understand the magnetic response of a system, it is important to understand how the magnetic moments of each individual atom are determined.[5] It turns out that interactions between the electrons in an atom have an important role. The simplest example is a set of two electrons, such as in a helium atom (Fig. 14.3).

14.3.1 Helium Atom

The helium atom is composed of a nucleus with two protons and two electrons. Since the nucleus has a much larger mass than the electrons, it is usually a good

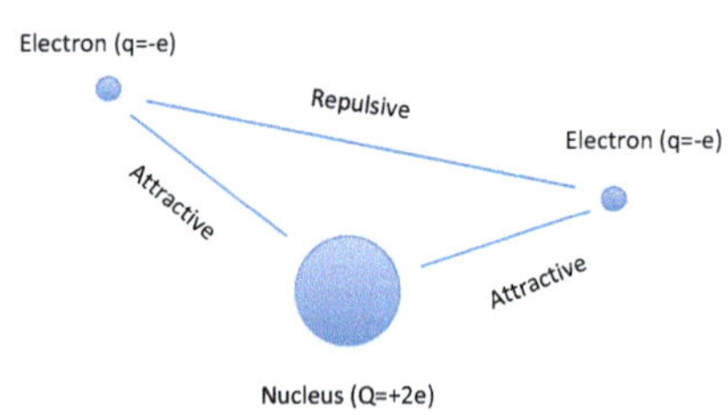

Fig. 14.3 Scheme of an helium atom

[5] For further reading see [Gasiorowicz] and [BicudoSacramento].

approximation to consider that the nucleus is at rest. Neglecting relativistic effects, such as the spin-orbit interaction, and the effects of the current due to the motion of one electron on the other, we may write the Hamiltonian as

$$H = \frac{1}{2m}\mathbf{p_1}^2 + \frac{1}{2m}\mathbf{p_2}^2 - \frac{Ze^2}{r_1} - \frac{Ze^2}{r_2} + \frac{e^2}{|\mathbf{r_1} - \mathbf{r_2}|}. \tag{14.84}$$

This can be written as

$$H = H^{(1)} + H^{(2)} + V, \tag{14.85}$$

where

$$H^{(i)} = \frac{1}{2m}\mathbf{p_i}^2 - \frac{Ze^2}{r_i} \tag{14.86}$$

and the interaction between the electrons,

$$V = \frac{e^2}{|\mathbf{r_1} - \mathbf{r_2}|}, \tag{14.87}$$

may be considered as a perturbation, on the separate motion of each electron around the nucleus. Neglecting the effect of the perturbation, each subsystem is equivalent to a hydrogenoid atom with charge $Z = 2$. In this case, the wave function of the system may be written as a product of the individual wave functions in the form

$$u(\mathbf{r_1}, \mathbf{r_2}) = \phi_{n_1 l_1 m_1}(r_1)\, \phi_{n_2 l_2 m_2}(r_2), \tag{14.88}$$

being solutions of the simplified Schrödinger equation

$$[H^{(1)} + H^{(2)}]\, u(\mathbf{r_1}, \mathbf{r_2}) = E\, u(\mathbf{r_1}, \mathbf{r_2}), \tag{14.89}$$

with eigenvalue

$$E_{n_1 n_2} = E_{n_1} + E_{n_2}, \tag{14.90}$$

where $E_n = -\frac{mc^2}{2}\frac{(Z\alpha)^2}{n^2}$. Here, $\alpha = e^2/(\hbar c)$ is the fine structure constant. The lowest energy state is $E_{11} = 2E_1 = -mc^2(2\alpha)^2$ and the first excited state is $E_{12} = E_{21} = E_1 + E_2$.

Actually, since the electrons are identical, the wave functions have to be antisymmetric. The wave function has a spatial part and a spin part. Two possibilities arise: the spatial part is symmetric to the interchange of the two electrons and the spin part is antisymmetric or the other way around.

The ground state is special since the spatial part of the wave function is necessarily symmetric which implies a spin singlet

$$u_0(\mathbf{r_1}, \mathbf{r_2}) = \phi_{100}(\mathbf{r_1})\phi_{100}(\mathbf{r_2})\, \chi_{singlet}, \tag{14.91}$$

where, as usual,

$$\chi_{singlet} = \frac{1}{\sqrt{2}}(\chi_\uparrow^{(1)}\chi_\downarrow^{(2)} - \chi_\downarrow^{(1)}\chi_\uparrow^{(2)}), \tag{14.92}$$

or, in other (Dirac) notation,

$$\chi_{singlet} = \frac{1}{\sqrt{2}}(|\uparrow\downarrow\rangle - |\downarrow\uparrow\rangle). \tag{14.93}$$

Clearly, the spatial part is symmetric to the change of the coordinates of the two electrons and so the spin part is antisymmetric.

The first excited state has two possible solutions

$$u_1{}^{(s)}(\mathbf{r_1}, \mathbf{r_2}) = \frac{1}{\sqrt{2}}\,[\phi_{100}(\mathbf{r_1})\phi_{2lm}(\mathbf{r_2}) + \phi_{2lm}(\mathbf{r_1}), \phi_{100}(\mathbf{r_2})]\, \chi_{singlet} \tag{14.94}$$

$$u_1{}^{(t)}(\mathbf{r_1}, \mathbf{r_2}) = \frac{1}{\sqrt{2}}[\phi_{100}(\mathbf{r_1})\phi_{2lm}(\mathbf{r_2}) - \phi_{2lm}(\mathbf{r_1})\phi_{100}(\mathbf{r_2})]\, \chi_{triplet}, \tag{14.95}$$

where

$$\chi_{triplet} = \begin{cases} \chi_\uparrow^{(1)}\chi_\uparrow^{(2)} \\ \frac{1}{\sqrt{2}}(\chi_\uparrow^{(1)}\chi_\downarrow^{(2)} + \chi_\downarrow^{(1)}\chi_\uparrow^{(2)}) \\ \chi_\downarrow^{(1)}\chi_\downarrow^{(2)} \end{cases}. \tag{14.96}$$

Let us now add the interaction between the two electrons as a perturbation. Using time independent perturbation theory, the leading correction to the energy of the ground state is given by

$$\Delta E_0 = \int \mathrm{d}^3 r_1\, \mathrm{d}^3 r_2\, u_0^*(\mathbf{r_1}, \mathbf{r_2})\, \frac{e^2}{|\mathbf{r_1} - \mathbf{r_2}|}\, u_0(\mathbf{r_1}, \mathbf{r_2}). \tag{14.97}$$

$$\Delta E_0 = \int \mathrm{d}^3 r_1\, \mathrm{d}^3 r_2\, |\phi_{100}(\mathbf{r_1})|^2\, \frac{e^2}{|\mathbf{r_1} - \mathbf{r_2}|}\, |\phi_{100}(\mathbf{r_2})|^2. \tag{14.98}$$

This expression has a clear (classical-like) physical meaning. Noting that $e|\phi_{100}(\mathbf{r_i})|^2$ is the charge electronic density at point $\mathbf{r_i}$, this is the standard Coulomb interaction between two charge densities at different spatial points, $\mathbf{r_1}$ and $\mathbf{r_2}$. Since the wave function ϕ_{100} is known, we may calculate the energy correction. The result is

$$\Delta E_0 = \frac{5}{8}\frac{Ze^2}{a_0} = \frac{5}{4}Z\Big(\frac{1}{2}mc^2\alpha^2\Big). \tag{14.99}$$

Taking $Z = 2$ we get that the perturbed energy $E'_{11} \simeq -74, 8$ eV which may be compared with the experimental value $E_{exp} = -78, 975$eV. The agreement is not good. One reason is that the potential the nucleus exerts on one electron is partially shielded (screened) by the motion of the other electron. In some sense, the effective nucleus charge is smaller. This effect becomes even more important when the atom has more electrons.

14.3.2 Direct and Exchange Terms: Heisenberg Interaction

Let us now consider the correction to the first excited state. Since the perturbation commutes with L_z, we may select any value of the orbital projection. Specifically, we may choose $m = 0$. The energy correction is given by

$$\begin{aligned}\Delta E_1^{(s,t)} &= \int \mathrm{d}^3r_1 \int \mathrm{d}^3r_2\, u_1^{(s,t)*}(\mathbf{r_1}, \mathbf{r_2})\, V\, u_1^{(s,t)}(\mathbf{r_1}, \mathbf{r_2}) \\ &= \frac{1}{2}e^2 \int \mathrm{d}^3r_1 \int \mathrm{d}^3r_2\, [\phi_{100}(\mathbf{r_1})\phi_{210}(\mathbf{r_2}) \pm \phi_{210}(\mathbf{r_1})\phi_{100}(\mathbf{r_2})]^* \\ &\quad\times \frac{1}{|\mathbf{r_1} - \mathbf{r_2}|}[\phi_{100}(\mathbf{r_1})\phi_{210}(\mathbf{r_2}) \pm \phi_{210}(\mathbf{r_1})\phi_{100}(\mathbf{r_2})],\end{aligned} \tag{14.100}$$

where we have the two possibilities, singlet and triplet,

$$\begin{aligned}\Delta E_1^{(s,t)} &= e^2 \int \mathrm{d}^3r_1 \int \mathrm{d}^3r_2\, |\phi_{100}(\mathbf{r_1})|^2|\phi_{210}(\mathbf{r_2})|^2 \frac{1}{|\mathbf{r_1} - \mathbf{r_2}|} \\ &\pm e^2 \int \mathrm{d}^3r_1 \int \mathrm{d}^3r_2\, \phi^*_{100}(\mathbf{r_1})\phi^*_{210}(\mathbf{r_2}) \frac{1}{|\mathbf{r_1} - \mathbf{r_2}|}\phi_{210}(\mathbf{r_1})\phi_{100}(\mathbf{r_2}).\end{aligned} \tag{14.101}$$

Note that since the Coulomb potential does not depend on spin, the overlap in spin space is trivial and chosen to be unity, due to the normalization of the spin wave functions. The $\pm$ signs are for the singlet and triplet spin states (symmetric and antisymmetric spatial wave functions, respectively). At this point it is worthwhile to stress that, even though the unperturbed Hamiltonian and the perturbation do not depend on the spin, the energy correction depends. Also note that the raising of the degeneracy of the two spin states by the perturbation is originated by a Coulomb-like energy interaction, as shown schematically in Fig. 14.4.

The first term of the energy correction is of the same form as the correction to the ground state energy. It may be understood as an interaction between two charge densities. It is called a direct term. However, the second term has no classical interpretation. Note that it is not possible to rewrite the wave functions as two

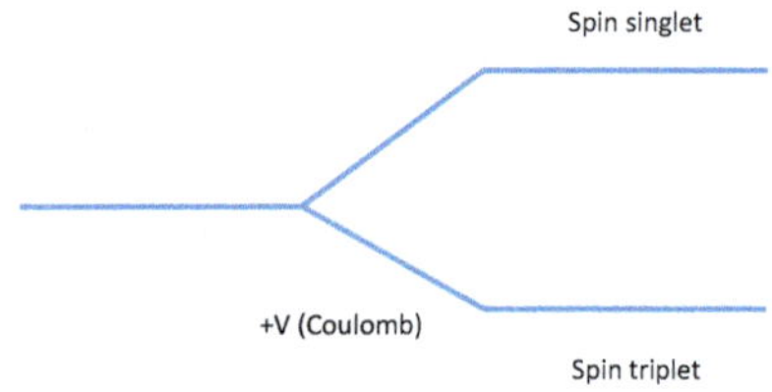

Fig. 14.4 Lifting of the degeneracy between the triplet and singlet spin terms due to the Coulomb repulsion

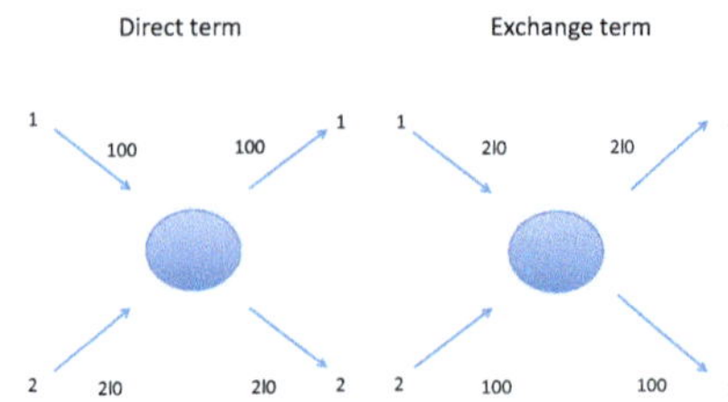

Fig. 14.5 Direct and exchange terms in the scattering of two electrons

density distributions. For the same set of quantum numbers, one wave function is associated with one electron and the complex conjugate with the other electron. It is called an exchange term and has no classical correspondence. The two types of terms are illustrated in Fig. 14.5.

In general we may write that

$$\begin{aligned}\Delta E_{n,l}^{(s)} &= J_{n,l} + K_{n,l},\\ \Delta E_{n,l}^{(t)} &= J_{n,l} - K_{n,l}.\end{aligned} \tag{14.102}$$

Again, even though the perturbation does not depend on the spin, the energies of the two spin states are different. This is a consequence of the symmetry of the wave function. Here

$$J_{n,l} = e^2 \int \mathrm{d}^3 r_1 \int \mathrm{d}^3 r_2 \, |\phi_{100}(\mathbf{r_1})|^2 |\phi_{nl0}(\mathbf{r_2})|^2 \frac{1}{|\mathbf{r_1} - \mathbf{r_2}|} \tag{14.103}$$

and

$$K_{n,l} = e^2 \int \mathrm{d}^3 r_1 \int \mathrm{d}^3 r_2 \, \phi_{100}^*(\mathbf{r_1}) \phi_{nl0}^*(\mathbf{r_2}) \frac{1}{|\mathbf{r_1} - \mathbf{r_2}|} \phi_{nl0}(\mathbf{r_1}) \phi_{100}(\mathbf{r_2}). \tag{14.104}$$

It is easy to see that $J_{n,l} > 0$. If $K_{n,l} > 0$ then

$$J_{n,l} - K_{n,l} < J_{n,l} + K_{n,l} \Leftrightarrow \Delta E_{n,l}^{(t)} < \Delta E_{n,l}^{(s)}. \tag{14.105}$$

It can be shown that is so. Therefore, the system prefers to be in a triplet state. The perturbation lifts the degeneracy between the two spin states.

This may be understood from a simple argument. The spin triplet is associated with a spatial wave function that is antisymmetric. Therefore, there is a small probability that the electrons are both in the neighborhood of the nucleus. The electrons are attracted to the nucleus but repel each other. It seems reasonable, therefore, that it will be favorable energetically if the electrons are far apart. In this way, the repulsive energy is minimized, while the attraction to the nucleus is still in place. A symmetric wave function would imply a higher probability to find the electrons near the nucleus and a larger repulsive contribution to the energy of the system. Therefore, in this simple case, the spin takes the maximal value.

A similar result holds also in atoms with a larger number of electrons. This constitutes the first Hund's rule which states that the electrons organize themselves in such a way as to maximize the spin.

If we couple two spins, $\mathbf{s_1}, \mathbf{s_2}$, the total spin, $\mathbf{S}$, satisfies

$$\mathbf{S}^2 = \mathbf{s_1}^2 + \mathbf{s_2}^2 + 2\mathbf{s_1}.\mathbf{s_2}. \tag{14.106}$$

Acting on the states $|s_1 s_2 S m\rangle$ we may rewrite the equation replacing the diagonal operators by their eigenvalues as

$$s(s+1)\hbar^2 = \frac{3}{4}\hbar^2 + \frac{3}{4}\hbar^2 + 2\mathbf{s_1} \cdot \mathbf{s_2},$$

which implies

$$\frac{2\mathbf{s_1}.\mathbf{s_2}}{\hbar^2} = s(s+1) - \frac{3}{2} = \begin{cases} \frac{1}{2} \text{ triplet } (s=1) \\ -\frac{3}{2} \text{ singlet } (s=0) \end{cases}. \tag{14.107}$$

Therefore,

$$\frac{1}{2}\sigma_1.\sigma_2 = \begin{cases} \frac{1}{2} \text{ triplet} \\ -\frac{3}{2} \text{ singlet} \end{cases}. \tag{14.108}$$

and we may rewrite

$$\Delta E_{n,l} = J_{n,l} - \frac{1}{2}(1 + \sigma_1.\sigma_2) K_{n,l}. \tag{14.109}$$

This way to rewrite the energy correction is chosen to highlight that there is a spin dependence and suggests some effective interaction between the spins of the two electrons. This type of expression has the general designation of Heisenberg exchange interaction. As will be seen later, this is a form that is supposed to be the basis for the magnetic properties of interacting spin systems.

14.3.3 Variational Method: Screening

A better approximation for the energy levels of the atom may be attempted using a variational method. A simple choice of the variational wave function for the ground state is to select

$$\psi(\mathbf{r_1}, \mathbf{r_2}) = \phi_{100}(\mathbf{r_1})\phi_{100}(\mathbf{r_2}), \tag{14.110}$$

where

$$\Big(\frac{\mathbf{p}^2}{2m} - \frac{Z^* e^2}{r}\Big)\phi_{100}(\mathbf{r}) = \epsilon\, \phi_{100}(\mathbf{r}), \tag{14.111}$$

with $\epsilon = -\frac{1}{2}mc^2(Z^*\alpha)^2$, are the eigenfunctions of a hydrogenoid atom with charge Z^*, taken as the variational parameter. Minimizing with respect to the variational parameter, the average value of the Hamiltonian in this state given by

$$\begin{aligned}\langle\psi|H|\psi\rangle = \int \mathrm{d}^3 r_1 \int \mathrm{d}^3 r_2 \phi^*_{100}(\mathbf{r_1})\phi^*_{100}(\mathbf{r_2})\Big(\frac{\mathbf{p_1^2}}{2m} + \frac{\mathbf{p_2^2}}{2m} - \frac{Ze^2}{r_1} - \frac{Ze^2}{r_2} + \\ + \frac{e^2}{|\mathbf{r_1} - \mathbf{r_2}|}\Big)\phi_{100}(\mathbf{r_1})\phi_{100}(\mathbf{r_2})\end{aligned} \tag{14.112}$$

leads to

$$\langle\psi|H|\psi\rangle = -\frac{1}{2}mc^2\alpha^2\Big(4ZZ^* - 2Z^{*2} - \frac{5}{4}Z^*\Big). \tag{14.113}$$

Minimizing determines the value of the effective charge of the nucleus

$$Z^* = Z - \frac{5}{16}, \tag{14.114}$$

which leads to a slightly improved value of the ground state energy

$$E_0 \leq -\frac{1}{2}mc^2\alpha^2\Big[2\Big(Z - \frac{5}{16}\Big)^2\Big] = -77,38\ \text{eV}. \tag{14.115}$$

A way to improve the variational method and approach further the exact ground state energy is to increase the size of the variational space, for instance, selecting trial wave functions with larger freedom by increasing the number of variational parameters. An alternative way is Hartree's method, where the functional form of the trial wave functions itself is to be determined.

14.3.4 Hartree Method

Consider an atom with Z electrons. Neglecting the motion of the nucleus, the Schrödinger equation may be written as

$$\left(\sum_{i=1}^{Z} \frac{\mathbf{p_i}^2}{2m} - \frac{Ze^2}{r_i} + \sum_{i>j} \frac{e^2}{|\mathbf{r_1} - \mathbf{r_2}|} \right) \psi(\mathbf{r_1}, \mathbf{r_2}, \cdots, \mathbf{r_Z}) = E\psi(\mathbf{r_1}, \mathbf{r_2}, \cdots, \mathbf{r_Z}). \tag{14.116}$$

This is a complex problem that in general requires a numerical solution. Due to the arbitrarily large number of electrons, screening is important and a perturbative method will not yield good results.

Let us choose a variational method where the trial wave function has the form

$$\psi(\mathbf{r_1}, \mathbf{r_2}, \cdots, \mathbf{r_Z}) = \phi_1(\mathbf{r_1})\phi_1(\mathbf{r_2}) \cdots \phi_1(\mathbf{r_Z}), \tag{14.117}$$

where the functions $\phi_i(\mathbf{r_i})$ are to be determined. The average value of the Hamiltonian in this state is given by

$$\begin{aligned} \langle\psi|H|\psi\rangle = &\sum_{i=1}^{Z} \int d^3r_i \phi_i^*(\mathbf{r_i}) \left(-\frac{\hbar^2}{2m}\nabla_i^2 - \frac{Ze^2}{r_i} \right) \phi_i(\mathbf{r_i}) \\ &+ e^2 \sum_{i<j} \sum_{j} \int d^3r_i \int d^3r_j \frac{|\phi_i(\mathbf{r_i})|^2 |\phi_j(\mathbf{r_j})|^2}{|\mathbf{r_i} - \mathbf{r_j}|}. \end{aligned} \tag{14.118}$$

If a given set of functions $\phi_i(\mathbf{r_i})$ minimizes the average value of the Hamiltonian in state $|\psi\rangle$, then if we consider a shift of the wave function in the form $\phi_i(\mathbf{r_i}) \to \phi_i(\mathbf{r_i}) + \lambda f_i(\mathbf{r_i})$, where λ is a small parameter and $f_i(\mathbf{r_i})$ is an arbitrary function, the average value of the Hamiltonian differs from the minimal value by terms that are at least of the order of λ^2. In other words, the term linear in λ must vanish since we are expanding around a minimum. Also, the shift must keep the normalization of the wave functions. This implies that

$$\int d^3r_i |\phi_i(\mathbf{r_i}) + \lambda f_i(\mathbf{r_i})|^2 = 1. \tag{14.119}$$

To leading order in λ this leads to

$$\int d^3r_i \left[\phi_i^*(\mathbf{r_i}) f_i(\mathbf{r_i}) + \phi_i(\mathbf{r_i}) f_i^*(\mathbf{r_i}) \right] = 0. \tag{14.120}$$

The shift of the wave functions may now be inserted in $\langle\psi|H|\psi\rangle$. The condition for minimum is that the term linear in λ vanishes, subject to the restriction 14.120.

The kinetic energy term is written as

$$\lambda \sum_i \int d^3r_i \left[f_i^*(\mathbf{r_i}) \left[-\frac{\hbar^2}{2m} \nabla_i^2 \phi_i(\mathbf{r_i}) \right] + c.c. \right],$$

where we have integrated by parts twice, imposing that $f_i(\mathbf{r_i})$ must vanish at infinity and where $c.c.$ stands for complex conjugate.

The Coulomb interaction with the nucleus is written as

$$-\lambda \sum_i \int d^3r_i \left[f_i^*(\mathbf{r_i}) \frac{Ze^2}{r_i} \phi_i(\mathbf{r_i}) + c.c. \right].$$

Finally, the term that describes the interactions between the electrons:

$$\lambda e^2 \sum_{i>j} \sum_j \int d^3r_i \int d^3r_j \frac{1}{|\mathbf{r_i} - \mathbf{r_j}|} \times$$

$$\left[f_i^*(\mathbf{r_i})\phi_i(\mathbf{r_i}) + f_i(\mathbf{r_i})\phi_i^*(\mathbf{r_i}) \right] |\phi_j(\mathbf{r_j})|^2 + \left[f_j^*(\mathbf{r_j})\phi_j(\mathbf{r_j}) + f_j(\mathbf{r_j})\phi_j^*(\mathbf{r_j}) \right] |\phi_i(\mathbf{r_i})|^2.$$

Collecting the various linear terms and introducing the restriction with a Lagrange multiplier, $-\epsilon_i$, we obtain

$$\sum_i \int d^3r_i f_i^*(\mathbf{r_i}) \left[-\frac{\hbar^2}{2m} \nabla_i^2 \phi_i(\mathbf{r_i}) - \frac{Ze^2}{r_i} \phi_i(\mathbf{r_i}) \right]$$

$$+ e^2 \sum_{i \neq j} \sum_j \int d^3r_i \int d^3r_j f_i^*(\mathbf{r_i}) \frac{|\phi_j(\mathbf{r_j})|^2}{|\mathbf{r_i} - \mathbf{r_j}|} \phi_i(\mathbf{r_i})$$

$$- \epsilon_i \int d^3r_i f_i^*(\mathbf{r_i}) \phi_i(\mathbf{r_i}) = 0 \tag{14.121}$$

and an equivalent equation that is the complex conjugate. Since the function $f_i^*(\mathbf{r_i})$ is arbitrary, this implies that the integrand vanishes. This leads to the Hartree equations

$$\left[-\frac{\hbar^2}{2m} \nabla_i^2 - \frac{Ze^2}{r_i} + e^2 \sum_{j \neq i} \int d^3r_j \frac{|\phi_j(\mathbf{r_j})|^2}{|\mathbf{r_i} - \mathbf{r_j}|} \right] \phi_i(\mathbf{r_i}) = \epsilon_i \phi_i(\mathbf{r_i}). \tag{14.122}$$

The first thing to note about these equations is that they are eigenvalue equations, where the Lagrange multiplier is the eigenvalue and the wave functions are the eigenfunctions. The second aspect is that these are self-consistent equations for the wave functions. If the interaction between the electrons is neglected, the Lagrange multipliers are simply the eigenenergies of the electrons, in the presence of a nucleus of charge Ze. We expect therefore a similar meaning for the parameters ϵ_i, when the electron repulsion is taken into account. The term due to this interaction is interesting. Within this approximation, it may be interpreted as the mean field due to the electrons $j \neq i$, on electron i. Note that it is the sum over all other electrons of a term that is the Coulomb potential, due to some charge density at point $\mathbf{r_j}$ and then summed over all $j \neq i$. Therefore, Hartree's variational approach leads to some sort of mean-field approximation, where the equation for the wave functions of a given electron i involves all the remaining electrons interacting with electron i, but where the interactions among the j electrons do not appear explicitly, in contrast with the Hamiltonian of the full system.

To some extent, Hartree's approximation reduces the many-body interacting problem to a single-particle problem in the presence of an effective potential

$$V_{eff}^i(\mathbf{r_i}) = -\frac{Ze^2}{r_i} + e^2 \sum_{j \neq i} \int d^3 r_j \frac{|\phi_j(\mathbf{r_j})|^2}{|\mathbf{r_i} - \mathbf{r_j}|}, \tag{14.123}$$

although self-consistent, since the potential itself depends on the solution of the wave functions of the other electrons. This is typical of a mean-field approximation as will be seen in later sections.

If the effective potential has spherical symmetry, we may integrate over the solid angle and the problem simplifies to a problem with a central potential. We may therefore focus only on the radial part and introduce the usual set of quantum numbers characteristic of the hydrogen atom.

14.3.5 Hartree-Fock Equations

The description we have just considered is, however, incomplete. Since the electrons are fermions their wave function should be completely antisymmetric. The trial wave function should be taken as the Slater determinant of the trial functions $\phi_i(\mathbf{r_j})$. The minimization of the average value of the Hamiltonian in the Slater determinant leads to a set of equations that have an extra term, due to the exchange between the electrons. This term is called the Fock term and the new set of equations are called Hartree-Fock equations that can be written as

$$\begin{aligned}\left[-\frac{\hbar^2}{2m}\nabla_i^2 - \frac{Ze^2}{r_i} + e^2 \sum_{j \neq i} \int d^3 r_j \frac{|\phi_j(\mathbf{r_j})|^2}{|\mathbf{r_i} - \mathbf{r_j}|}\right] \phi_i(\mathbf{r_i}) \\ - e^2 \sum_{j \neq i} \int d^3 r_j \frac{\phi_j^*(\mathbf{r_j})\phi_i(\mathbf{r_j})}{|\mathbf{r_i} - \mathbf{r_j}|} \phi_i(\mathbf{r_i}) \\ = \epsilon_i \phi_i(\mathbf{r_i}).\end{aligned} \tag{14.124}$$

The solution of these equations is in general more complex than the Hartree equations. In the context of atoms, the improvement that results from the addition of the Fock term is not very significative. Often an analysis of the solutions using the Hartree approximation and taking into account Pauli's exclusion principle lead to a good description of the energy levels of atoms. This problem of atomic physics is beyond the scope of this book and will not be elaborated further. In some cases their solution in the context of solid state physics is possible and later on we will consider some examples of their solution.

14.3.6 Hund Rules and Magnetic Moments in a Crystal

To a good approximation, the single-particle states of an atom with Z electrons can be characterized by the same quantum numbers as those of a hydrogen-like atom, with a nuclear charge $|Ze|$. The distribution of the electrons among the available single-particle states requires a calculation of the level degeneracies. The quantum numbers are traditionally taken as n, l, m_l, m_s, where n is the principal quantum number, $l = 0, \cdots, n-1$ is the angular momentum, $m_l = -l, -l+1, \cdots, l-1, l$ is the angular momentum projection along some quantization axis, and $m_s = -1/2, 1/2$ is the spin angular momentum projection of each electron. In the absence of external fields, the energy eigenvalues are only a function of n. The degeneracy of each energy level is $g = 2n^2$ since

$$\sum_{l=0}^{n-1} 2\,(2l+1) = 2n^2.$$

Each orbital with a given pair of quantum numbers n, l has $2(2l+1)$ states. If all m_l states are occupied with one electron, the particle distribution is given by

$$\sum_{m_l=-l}^{l} |R_{nl}(r)|^2 |Y_{lm_l}(\theta,\varphi)|^2 = \frac{2l+1}{4\pi}$$

(a constant) and therefore is spherically symmetric. Given an atom with Z electrons the ground state of the atom is obtained distributing the electrons among the available single-particle states appropriately, as exemplified in Fig. 14.6.

The problem of the determination of the energy states is in general complex but some empirical rules are helpful. These rules are summarized in Hund's rules.[6] Let us consider a free atom where all energy levels are filled, except a higher energy one which is partially filled. Assume this last orbital has a given angular momentum, l. In addition to the potential energy due to the nucleus, there are other

[6] See also [Kittel] and [BicudoSacramento].

Fig. 14.6 Standard example of occupation of energy levels of hydrogen-like atom, with the levels $n = 1, 2$ fully occupied

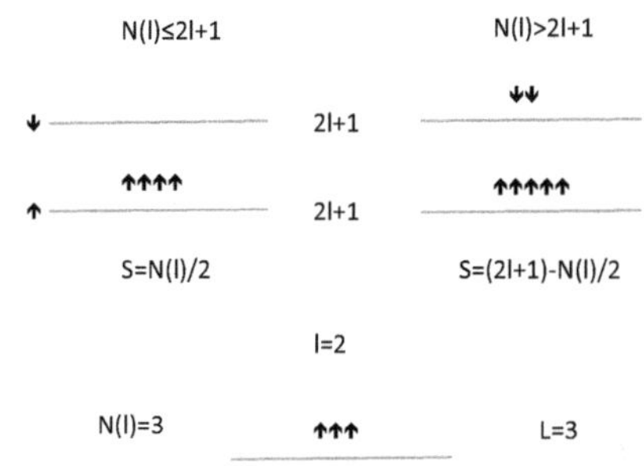

Fig. 14.7 Examples of Hund's rules. $N(l)$ is the number of electrons in orbital l

contributions. As we have seen above, there is the interaction between the electrons. In a first approximation we are assuming that the Hartree and Fock terms are a small enough perturbation that, even though shifts the energy levels, does not lead to level crossings and mixings, such that we are still able to label the states with the quantum numbers of the hydrogen atom. However, there are other corrections to the hydrogen atom itself that are important as the number of electrons increases, such as the spin-orbit interaction.

Hund first rule: The first empirical rule has already been discussed. The electrons organize themselves to be in a state of maximal total spin, S (total spin is a spin that is the sum of all the individual spins of each electron). Note that the total spin of a filled orbital vanishes, due to Pauli's principle.

Hund second rule: The total angular momentum, L, has the maximal value that is compatible with the first rule and with Pauli's principle.

Hund third rule: the total angular momentum, J, where $\mathbf{J} = \mathbf{L} + \mathbf{S}$, is given by $J = |L - S|$ if the orbital is less than half-full, and $J = L + S$ if the orbital is more than half-full. This rule is the result of the spin-orbit interaction written as $\lambda_{SO}(\mathbf{S} \cdot \mathbf{L})$. If $\lambda_{SO} > 0$, the spin and angular momentum minimize the energy if they are parallel, and, if $\lambda_{SO} < 0$, they tend to be anti-parallel. If the orbital is less than half-full the first case occurs and vice versa if the band is more than half-full. As examples we will consider the rare-earth materials of the lanthanide and actinide families and the iron group elements (some are shown in Fig. 14.7).

First, recall that the various orbitals with different l values are denoted as $s, p, d, f, \cdots$ when $l = 0, 1, 2, 3, \cdots$, respectively. The degeneracy of each orbital, taking into account the spin degeneracy, is $2, 6, 10, 14, \cdots$, respectively. For instance, the lanthanides have in common a set of filled levels that share with the rare gas Xe. In addition, they have an electronic configuration such that there are two electrons in the $n = 5, l = 0$ orbital and six electrons in the $n = 5, l = 1$ orbital. That is, they have a configuration $5s^2 5p^6$. The various elements along the lanthanide series include the elements La^{57}; Ce^{58}; Pr^{59}; Nd^{60}; $\cdots$; Yb^{70}; Lu^{71}. Their electronic configurations include as well the following electrons:

$$(5d, 6s^2); (4f^2, 6s^2); (4f^3, 6s^2); (4f^4, 6s^2); \cdots ; (4f^{14}, 6s^2); (4f^{14}, 5d, 6s^2).$$

The main difference between the various lanthanides is the filling of the $4f$ orbital. In comparison with the $n = 5$ and $n = 6$ orbitals, the $4f$ orbital is of shorter range and more internal. The actinides differ by the filling of the $5f$ orbital and a similar property may be invoked. When several atoms are brought together, such as in the solid state, the spatial extent of these internal partially filled orbitals is typically smaller than the distance between the atoms. Therefore, the neighborhood of each atom does not affect in an important way the electrons that occupy these orbitals.

The case of the iron group elements is different. For instance, considering the elements Fe; Co; Ni; Cu and the electronic configurations of the higher energy orbitals $(3d^6, 4s^2)$; $(3d^7, 4s^2)$; $(3d^8, 4s^2)$; $(3d^10, 4s)$, a similar trend is verified. In this case they differ by the filling of the $3d$ orbital. Note that in Cu this orbital is completely filled and, therefore, is expected to be inert. An important difference with respect to the lanthanides or the actinides is that the spatial extent of the $3d$ orbital is similar to the radius of the atom. Therefore, the presence of neighboring atoms will affect the electrons of the $3d$ orbital in a significant way.

As mentioned above, a filled orbital has no total spin and, therefore, no spin magnetic moment. The possible contribution to the spin, and total magnetic momentum of the atom, is originated in the partially filled orbitals. As an exercise let us use Hund rules to calculate the angular momenta of various atoms. As atoms are brought together, their valencies will change, depending on the neighboring atoms.

Consider for instance Mn. Its partially filled level has the configuration $(3d^5, 4s^2)$. The two electrons in the $4s$ orbital have $S = 0, L = 0$. The electrons in the $3d$ orbital have $S = 5/2, L = 0$. The same happens if the valence state is Mn^{2+}, where the configuration is simply $3d^5$. Therefore $J = S = 5/2$. Another example is Ce^{3+} with electronic configuration $4f^1$. It is easy to see that $S = 1/2$, $L = 3$ and $J = 5/2$. For instance, Fe leads to $S = 2$, $L = 2$ and $J = 4$. The same result holds for a divalent ion, where the two $4s$ electrons are lost.

A given total angular momentum leads as usual to a finite magnetic moment given by

$$\boldsymbol{\mu} = -g\mu_B \mathbf{J}, \tag{14.125}$$

where g is the Landé factor

$$g = 1 + \frac{J(J+1) + S(S+1) - L(L+1)}{2J(J+1)} = \frac{3}{2} + \frac{S(S+1) - L(L+1)}{2J(J+1)} \tag{14.126}$$

and μ_B is the Bohr magneton. The energy of a magnetic moment in an external magnetic field **B** is given by

$$U = -\boldsymbol{\mu} \cdot \mathbf{B} = m_J g \mu_B B, \tag{14.127}$$

where m_J is the projection of the total angular momentum along the direction of the applied magnetic field. We may define the effective number of Bohr magnetons as

$$p = g\sqrt{J(J+1)}. \tag{14.128}$$

This number can be measured directly through the magnetic susceptibility at high temperatures. As we have shown previously, the susceptibility at high temperatures follows Curie's law $\chi = C/T$, where T is the temperature and C is Curie's constant given by

$$C = \frac{Np^2\mu_B^2}{3k_B}. \tag{14.129}$$

The experimental results for the effective number of Bohr magnetons when the atom is in the solid state are somewhat unexpected. In the case of rare earths, the agreement between the calculated value and the experimental value is quite good. The calculated value uses Hund rules, whose validity holds for an isolated atom. In the solid state the atomic states give place to different Bloch states together with more atomic-like states, associated with the orbitals deep inside the atom cores. As discussed above, the orbitals responsible for the magnetic moment of the rare earths are quite confined around each atom, which explains the good agreement with the experimental value. In the case of the iron group elements, the agreement is much worse. It turns out that a good fit of the experimental results is obtained if we take that

$$p = 2\sqrt{S(S+1)}. \tag{14.130}$$

This expression is obtained from the general case if $L = 0$, which implies $J = S$ and $g = 2$. We conclude that in the crystal the orbital angular momentum is quenched to zero, due to the effects of the neighborhood of an atom.

In the solid state the outer orbitals overlap significantly and their shape changes considerably. In a metal some electrons become completely delocalized in extended states, leaving behind ions whose charge distributions create a crystalline field that acts on the remaining electrons of the neighboring lattice sites. Due to the lattice structure, the crystalline field is in general non-homogeneous and not spherically symmetric. As a consequence, it tends to distort the orbitals. Also, energetic considerations imply that some orientations are preferred and degeneracies are lifted due to the crystalline field. The energies involved may be competitive with the energies associated with Hund rules. Typically the first and second Hund rules are very robust, but the competition between the spin-orbit interaction and the crystalline field has to be taken into account. For instance, for the elements Fe, Co, Ni, typically the crystalline field is more important than the spin-orbit interaction, but as we move to atoms with a higher number of electrons, such as Ru, Rh, Pd or Os, Ir, Pt, the spin-orbit interaction becomes dominant over the crystalline field. The outcome is still non-trivial since the energies are comparable.

As a simple example of the effect of the crystal on the orbitals, consider a planar structure such that along a given direction (call it z), there is a positive electric field, both above and below the plane. In a free atom the three p-orbitals ($l = 1$) are degenerate. However, in the planar structure it is clearly preferable energetically that the electrons are oriented along the z direction. The orbital p_z has a lower energy with respect to the degenerate p_x, p_y orbitals and the three-degeneracy is partially lifted by the field.

Consider as an example a crystal with orthorhombic symmetry, such that the neighboring ions produce an electrostatic potential on a given site in the form[7]

$$e\varphi = Ax^2 + By^2 - (A + B)z^2. \tag{14.131}$$

Note that this potential satisfies Laplace's equation $\nabla^2\varphi = 0$. The $l = 1$ wave functions in free space can be written as

$$u_x = xf(r), u_y = yf(r), u_z = zf(r). \tag{14.132}$$

Note that $L^2 u_i = l(l+1)u_i = 2u_i$, in units of $\hbar^2$, and where $i = x, y, z$. Now it is easy to show, by symmetry, that

$$\langle u_x|e\varphi|u_y\rangle = \langle u_x|e\varphi|u_z\rangle = \langle u_y|e\varphi|u_z\rangle = 0. \tag{14.133}$$

On the other hand, the diagonal matrix elements do not vanish. For instance,

$$\langle u_x|e\varphi|u_x\rangle = \int dxdydz|f(r)|^2\left[Ax^4 + By^2x^2 - (A + B)x^2z^2\right] = A(I_1 - I_2), \tag{14.134}$$

where $I_1 = \int dxdydz|f(r)|^2x^4$ and $I_2 = \int dxdydz|f(r)|^2x^2y^2$. Similarly,

$$\begin{aligned} \langle u_y|e\varphi|u_y\rangle &= B(I_1 - I_2) \\ \langle u_z|e\varphi|u_z\rangle &= -(A + B)(I_1 - I_2). \end{aligned} \tag{14.135}$$

As a consequence, the electrostatic field lifts the degeneracy between the three orbitals.

We may now calculate the average value of L_z in the three orbitals. Using that $L_z = -i\hbar(x\partial/\partial y - y\partial/\partial x)$ it is easy to see that

$$\langle u_x|L_z|u_x\rangle = \langle u_y|L_z|u_y\rangle = \langle u_z|L_z|u_z\rangle = 0 \tag{14.136}$$

[7] See also [Kittel].

and the off-diagonal matrix elements such as $\langle u_x|L_z|u_y\rangle$ do not vanish. Since the degeneracy is lifted by the crystalline field, the average value of the angular momentum is taken on each orbital separately leading to a vanishing value and therefore quenching of the angular momentum. If the degeneracy is not lifted, for instance in a cubic crystal the electrostatic field $e\varphi = 0$, the average value of the angular momentum has to be evaluated in a state that is a linear combination of the three degenerate states. Even though the diagonal terms vanish, there is a contribution from the off-diagonal matrix elements that, therefore, leads to a non-vanishing average value of the angular momentum.

Note however that even if there is cubic symmetry quenching may occur. As we saw a distortion of the orbitals leads to a lifted degeneracy. This may lead to a lower energy state if the occupation of the degenerate levels is such that the energy is smaller. It may be then energetically favorable for the system to distort spontaneously. This is called a Jahn-Teller effect. As a consequence, there may be again quenching of the angular momentum.

14.4 Hartree-Fock Approximation for a Many-Body Problem

Let us consider again the electron Hamiltonian in the jellium model[8]

$$H = -\sum_k \frac{\hbar^2}{2m}\nabla_k^2 + \frac{1}{2}\sum_{k,k'}{}' \frac{e^2}{|\mathbf{r}_k - \mathbf{r}_{k'}|} + H_+$$

$$H = \sum_k H_k + \sum_{k,k'}{}' H_{kk'}. \tag{14.137}$$

Here we are considering an extended system, such as a crystal composed of a very large number of atoms and electrons. If there were no interactions between the electrons, the problem would be separable in the problem of a collection of one-electron problems (independent electron approximation). Adding the interactions we may solve the problem approximately, for instance using the Hartree-Fock approximation. The procedure is similar to the one used in the case of the helium atom. A significant difference is that the wave functions of the non-interacting basis suitable to the problem are not the atomic localized eigenfunctions, but wave functions that are extended and that satisfy Bloch's theorem. To leading approximation we will take plane waves. For a better description one may replace the plane waves by the eigenfunctions that result from the presence of the periodic potential.

[8] Here we follow [BicudoSacramento].

As in the case of an atom we may ignore the antisymmetry and derive the Hartree equations for the many-body electron system

$$\left(-\frac{\hbar^2}{2m}\nabla^2 + V(r) + e^2 \sum_{k\neq j} \int dV' \frac{|\varphi_k(\mathbf{r}')|^2}{|\mathbf{r}-\mathbf{r}'|}\right) \varphi_j(\mathbf{r}) = E_j \varphi_j(\mathbf{r}). \tag{14.138}$$

Antisymmetrizing the single-particle wave functions leads to the Hartree-Fock equations

$$\left[-\frac{\hbar^2}{2m}\nabla^2 + V(\mathbf{r})\right] \varphi_j(\mathbf{r}) + e^2 \sum_{k\neq j} \int dV' \frac{|\varphi_k(\mathbf{r}')|^2}{|\mathbf{r}-\mathbf{r}'|} \varphi_j(\mathbf{r})$$
$$- e^2 \sum_{k\neq j;spins=} \int dV' \frac{\varphi_k^*(\mathbf{r}')\varphi_j(\mathbf{r}')}{|\mathbf{r}-\mathbf{r}'|} \varphi_k(\mathbf{r}) = E_j \varphi_j(\mathbf{r}). \tag{14.139}$$

In the last term it is explicitly stressed that the spins are equal due to the orthogonality of the states in both real and spin spaces.

Let us now rederive the same results using second quantization (the formalism of second quantization is reviewed in Appendix A). The interacting term of the Hamiltonian is written as

$$H_{int} = \frac{1}{2} \sum_{ij}{}' \frac{e^2}{|\mathbf{r}_i - \mathbf{r}_j|} c_i^\dagger c_i c_j^\dagger c_j. \tag{14.140}$$

Introducing Fourier components

$$c_i = \frac{1}{\sqrt{N}} \sum_k e^{ik\cdot r_i} c_k,$$

we get that the Hamiltonian can be written as

$$H = \sum_{k,\sigma} \epsilon_k c_{k,\sigma}^\dagger c_{k,\sigma} + \sum_q \frac{2\pi}{q^2} e^2 \sum_{kk'} \sum_{\sigma\sigma'} c_{k+q,\sigma}^\dagger c_{k-q,\sigma'}^\dagger c_{k',\sigma'} c_{k,\sigma}. \tag{14.141}$$

We have introduced the Fourier transform of the Coulomb potential

$$\int d^3x e^{-i\mathbf{k}\cdot\mathbf{r}} \frac{e^{-\alpha|\mathbf{r}|}}{|\mathbf{r}|} = \frac{4\pi}{k^2+\alpha^2}.$$

We have introduced, for regularization, a small constant α, which at the end is taken to zero, to calculate the Fourier transform of the Coulomb potential. Note that the Fourier transform diverges as $q \to 0$, due to the long range nature of the potential. This divergence has important consequences, as we will see. In Eq. (14.141) and later expressions the regularization α is not shown explicitly, but it should be kept in mind.

The Hamiltonian can now be diagonalized if we take a mean-field approach (we will see that the mean-field approach leads to the Hartree-Fock approximation). It is actually convenient to use the real space description. In terms of the field operators the Hamiltonian is written as

$$H = \int d^3x \psi^\dagger(x) \left[\frac{p^2}{2m} + v(x) \right] \psi(x) + \frac{1}{2} \int d^3x \int d^3y \psi^\dagger(x) \psi^\dagger(y) \\ \times V(x-y) \psi(y) \psi(x). \tag{14.142}$$

Let us expand the field operators in terms of the wave functions of the free problem

$$\psi(x) = \sum_j c_j \varphi_j(x),$$

where the wave functions are the solutions of

$$\left[\frac{p^2}{2m} + v(x) \right] \varphi_j(x) = \epsilon_j \varphi_j(x).$$

Using that the field operators anticommute

$$\{\psi(x), \psi^\dagger(x')\} = \delta(x - x'),$$

we conclude, using that $\{c_j, c_k^\dagger\} = \delta_{j,k}$,

$$\sum_j \varphi_j(x) \varphi_j^*(x') = \delta(x - x').$$

If the potential is constant, then it is clear that the wave functions are plane waves and we recover the previous case.

Let us consider the equation of motion of the field operator

$$\begin{aligned} i \frac{\partial \psi(x)}{\partial t} &= [\psi(x), H] \\ &= \left(\frac{p^2}{2m} + v(x) \right) \psi(x) + \frac{1}{2} \int d^3y \psi^\dagger(y) V(x-y) \psi(y) \psi(x) \\ &\quad + \frac{1}{2} \int d^3y \psi^\dagger(y) V(y-x) \psi(y) \psi(x) \\ &= \left(\frac{p^2}{2m} + v(x) \right) \psi(x) + \int d^3y \psi^\dagger(y) V(x-y) \psi(y) \psi(x), \end{aligned} \tag{14.143}$$

if the potential is symmetric.

The equation has a non-linear term that results from the interacting term of the Hamiltonian. The solution of the equation becomes feasible if the equation is linearized. This linearization may be achieved in different ways. One way consists of using a mean-field approximation

$$
\begin{aligned}
i\frac{\partial\psi(x)}{\partial t} &\sim \left[\frac{p^2}{2m}+v(x)+\int d^3yV(x-y)\langle\psi^\dagger(y)\psi(y)\rangle\right]\psi(x) \\
&\quad -\int d^3yV(x-y)\langle\psi^\dagger(y)\psi(x)\rangle\psi(y). \qquad (14.144)
\end{aligned}
$$

We have replaced

$$
\psi^\dagger(y)\psi(y)\psi(x) \to \langle\psi^\dagger(y)\psi(y)\rangle\psi(x) - \langle\psi^\dagger(y)\psi(x)\rangle\psi(y),
$$

which amounts to replacing two of the operators by some average (c-number) value. Note that the average values of the type $\langle\psi\rangle$ or $\langle\psi^\dagger\rangle$ vanish, because they do not conserve the number of particles. The minus sign of the second term appears because of the anticommutation relations of the field operators

$$
\psi(y)\psi(x) = -\psi(x)\psi(y).
$$

The average is over some state (such as the ground state) or a thermal average, depending on the problem.

Another way to linearize the equation of motion is to implement a mean-field approximation on the Hamiltonian. The purpose is to replace the interacting (quartic) term by a non-interacting-like (quadratic) term. The non-linear term is then replaced by a linear term.

Let us consider a variable A that fluctuates around an average value $\langle A\rangle$ (the same reasoning also applies considering an operator and its average value). Let us consider the fluctuations $\delta A = A - \langle A\rangle$ and take them to be small. Then we can write that

$$
A = \langle A\rangle + (A - \langle A\rangle) = \langle A\rangle + \delta A. \qquad (14.145)
$$

Let us then consider the product of two fluctuating variables. Using the same decomposition we can write that, within a mean-field approximation,

$$
\begin{aligned}
AB &= (\langle A\rangle + \delta A)(\langle B\rangle + \delta B) \\
&= \langle A\rangle\langle B\rangle + \langle A\rangle\delta B + \delta A\langle B\rangle + \delta A\delta B \\
&\sim \langle A\rangle\langle B\rangle + \langle A\rangle(B - \langle B\rangle) + (A - \langle A\rangle)\langle B\rangle \\
AB &\sim \langle A\rangle B + A\langle B\rangle - \langle A\rangle\langle B\rangle. \qquad (14.146)
\end{aligned}
$$

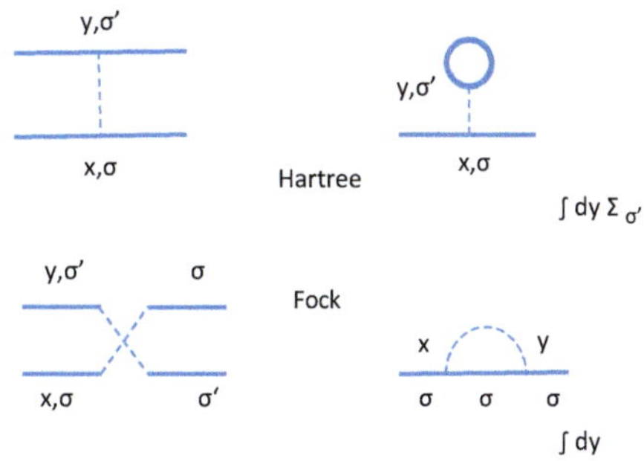

Fig. 14.8 Hartree and Fock terms in the scattering of two electrons and their effect on the mean-field equations of motion

Returning to the Hamiltonian, A and B may be taken as the products of two operators. The decomposition leads to terms that have at most two operators (two operators are replaced by some average), leading naturally to linear equations of motion.

It is easy to see that in the decoupled equation of motion, the first term is the direct Hartree term and the second term is the exchange, or Fock term, represented in Fig. 14.8. Expanding the field operators in terms of the eigenfunctions of the free problem, we get

$$\begin{aligned} i\frac{\partial\psi(x)}{\partial t} &= \sum_j \epsilon_j c_j \varphi_j(x) \\ &= \left[\frac{p^2}{2m} + v(x) + \int d^3y V(x-y) \sum_i \varphi_i^*(y)\varphi_i(y)\langle c_i^\dagger c_i\rangle\right] \sum_j c_j\varphi_j(x) \\ &\quad - \int d^3y V(x-y) \sum_i \varphi_i^*(y)\varphi_i(x)\langle c_i^\dagger c_i\rangle \sum_j c_j\varphi_j(y). \end{aligned} \tag{14.147}$$

We have used that $\langle c_i^\dagger c_j\rangle = \delta_{ij}\langle c_i^\dagger c_i\rangle$. The sum $\sum_i \langle c_i^\dagger c_i\rangle$ is a sum over the occupied sites at zero temperature. Comparing the coefficients of the operator c_j we get

$$\begin{aligned} \epsilon_j\varphi_j(x) &= \left[\frac{p^2}{2m} + v(x) + \sum_{i,occupied} \int d^3y V(x-y)\varphi_i^*(y)\varphi_i(y)\right]\varphi_j(x) \\ &\quad - \sum_{i,occupied,=spin} \int d^3y V(x-y)\varphi_i^*(y)\varphi_i(x)\varphi_j(x), \end{aligned} \tag{14.148}$$

which are again the Hartree-Fock equations. As before, we note that only parallel spins are correlated in this approximation.

14.4.1 Koopman's Theorem

Let us discuss the meaning of these equations. Consider a system of N electrons with ground state energy, E_N, and a state with one less particle with energy E^l_{N-1}. We will now show that, if N is large enough,

$$E_N - E^l_{N-1} = \epsilon_l \tag{14.149}$$

and can be understood as the ionization energy of the system. Indeed,

$$\begin{aligned}
\langle\phi|H|\phi\rangle - \langle\phi_l|H|\phi_l\rangle &= \sum_{m,occup} \langle m| \left(\frac{p^2}{2m} + v(\mathbf{r})\right) |m\rangle \\
&+ \frac{1}{2} \sum_{m,p,occup} (\langle mp|V|mp\rangle - \langle mp|V|pm\rangle) \\
&- \text{(same with l missing)} \\
&= \langle l| \left(\frac{p^2}{2m} + v\right) |l\rangle + \frac{2}{2} \sum_p (\langle lp|V|lp\rangle - \langle lp|V|pl\rangle) \\
&= \epsilon_l .
\end{aligned} \tag{14.150}$$

This result only holds if we assume that the states are not changed when we remove an electron. This occurs in the solid state since the number of electrons is very large and removing one electron has little influence on the available states of the remaining electrons. This result is called Koopman's theorem. This does not occur in an atom where removing one electron significantly changes the states of the system.

14.4.2 Hartree and Fock Contributions to the Energy

Let us now look at the consequences of the Hartree-Fock approximation and let us specify to the jellium model. The potential is then

$$v(x) = -\int d^3y|e|\frac{\rho_0^+(y)}{|x-y|},$$

using a simplified notation such that $\mathbf{r} = x$. Here, the charge density of the positive background is uniform and given by

$$\rho_0^+(y) = \frac{N|e|}{\Omega},$$

where, as before, N is the number of electrons. Since we have a uniform background charge, the wave functions are plane waves

$$\varphi_k(x) = \frac{1}{\sqrt{\Omega}} e^{ik\cdot x}. \tag{14.151}$$

Then the Hartree equation reduces to

$$\left[\frac{\hbar^2 k^2}{2m} - \frac{Ne^2}{\Omega}\int d^3y \frac{1}{|x-y|} + \frac{(N-1)e^2}{\Omega}\int d^3y \frac{1}{|x-y|}\right]\frac{1}{\sqrt{\Omega}}e^{ik\cdot x}$$
$$= \epsilon_k \frac{1}{\sqrt{\Omega}} e^{ik\cdot x}. \tag{14.152}$$

In the thermodynamic limit the free particle energy is a solution, since the interacting term scales as $1/\Omega^{1/3}$ and $\epsilon_k = \hbar^2 k^2/(2m)$.

Consider now the Fock term. Let us consider again plane waves. The Fock term gives

$$-\sum_{l,occupied}\int d^3y \frac{1}{\Omega} e^{i(k_j-k_l)\cdot y} V(x-y) \frac{1}{\sqrt{\Omega}} e^{ik_l\cdot x}.$$

Therefore, the plane waves Eq. (14.151) are a solution of the equation

$$\begin{aligned}
\epsilon_{k_j} &= \frac{\hbar^2 k_j^2}{2m} - \frac{1}{\Omega}\sum_{k_l}\int d^3y e^{i(k_j-k_l)\cdot(y-x)} V(x-y) \\
&= \frac{\hbar^2 k_j^2}{2m} - \frac{1}{\Omega}\sum_{l,occupied}{}' G(k_j-k_l),
\end{aligned}$$

where

$$G(k) = \frac{4\pi e^2}{k^2}.$$

Defining the z axis parallel to $\mathbf{k}_j$ and using spherical coordinates, we can calculate the integral and obtain the exchange energy due to the Fock term as

$$E_{exchange} = -\frac{e^2}{\pi}\left(\frac{k_F^2 - k_j^2}{2k_j}\ln\left|\frac{k_F + k_j}{k_F - k_j}\right| + k_F\right). \tag{14.153}$$

In Fig. 14.9 we show the exchange energy as a function of momentum. We see that at the Fermi momentum the derivative is infinite. This means that the density of states vanishes at the Fermi energy. This must be wrong. There are two reasons

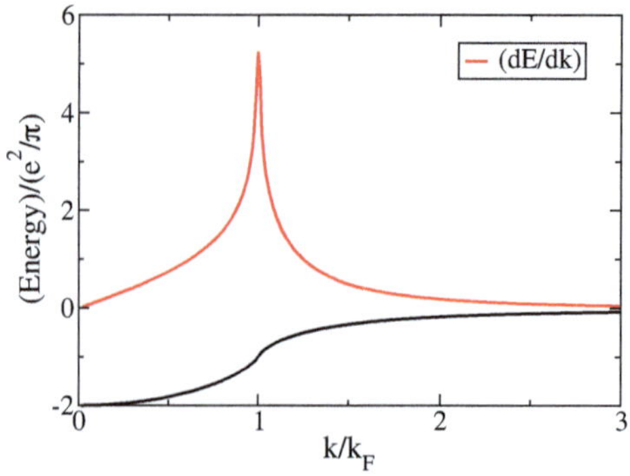

Fig. 14.9 Exchange energy in the Hartree-Fock approximation. Note the divergence of its derivative at the Fermi momentum, with consequent (incorrect) vanishing density of states

for the incorrect result: the first is that plane waves are not a good solution of the interacting problem, and second, in this approximation, the particles interact via the Coulomb potential while, in reality, there is screening of the potential between two charges due to the effect of the other charges.

Even though screening has to be included for a proper description of the system, the approximation does yield the existence of binding of the electronic system in the presence of the ion background. We leave as an exercise to show that the energy of the system per atom (in Rydberg) is given within the Hartree-Fock approximation by

$$\epsilon = \frac{2.21}{r_s^2} - \frac{0.916}{r_s}, \tag{14.154}$$

where $r_s = r_0/r_H$ (r_0 is the average distance between the atoms and r_H is the Bohr radius). We see therefore that, within the Hartree-Fock approximation, there is binding between the electrons and there are bound states.

14.5 Charge and Spin Density Waves

Before we turn to the problem of taking into account the screening effect, let us look for other solutions of the Hartree-Fock equations. In particular, we will consider non-homogeneous distributions of charge and spin.

14.5.1 Charge Density Waves (CDW)

Let us consider the one-dimensional case. The Hamiltonian is once again

$$H = \sum_{k,\sigma} \epsilon_k c^\dagger_{k,\sigma} c_{k,\sigma} + \sum_q{}' \left(\sum_{kk'} \sum_{\sigma\sigma'} \frac{2\pi e^2}{q^2} c^\dagger_{k+q,\sigma} c_{k'-q,\sigma'} c_{k',\sigma'} c_{k,\sigma} \right). \tag{14.155}$$

The wave vector $q = 2k_F$ connects the two Fermi points. Let us consider a reduced (mean-field) Hamiltonian, where in the interaction term we replace a product of two operators by its mean value, in the form

$$\begin{aligned} H_{red} &= \sum_{k,\sigma} \epsilon_k c^\dagger_{k,\sigma} c_{k,\sigma} \\ &+ V(2k_F)\left(\sum_{k,\sigma}\langle c^\dagger_{k+2k_F,\sigma} c_{k-2k_F,\sigma}\rangle \sum_{k',\sigma'} c^\dagger_{k'-2k_F,\sigma'} c_{k',\sigma'} + (h.c.)\right) \\ &- V(2k_F)\left(\sum_{k,\sigma}\langle c^\dagger_{k+2k_F,\sigma} c_{k'-2k_F,\sigma}\rangle \sum_{k',\sigma'} \langle c^\dagger_{k'-2k_F,\sigma'} c_{k',\sigma'}\rangle\right). \end{aligned} \tag{14.156}$$

Here $h.c.$ stands for Hermitian conjugate. Let us define the parameter

$$\Delta = V(2k_F)\sum_{k,\sigma}\langle c^\dagger_{k+2k_F,\sigma} c_{k-2k_F,\sigma}\rangle. \tag{14.157}$$

Then we can rewrite the reduced Hamiltonian like

$$\begin{aligned} H_{red} &= \sum_{k,\sigma}\left(\epsilon_k c^\dagger_{k,\sigma} c_{k,\sigma} + \Delta c^\dagger_{k-k_F,\sigma} c_{k+k_F,\sigma} + \Delta^* c^\dagger_{k+k_F,\sigma} c_{k-k_F,\sigma}\right) \\ &- \frac{1}{V(2k_F)}|\Delta|^2. \end{aligned} \tag{14.158}$$

This problem is diagonalizable since it is quadratic (or the equations of motion are linear). Perform a Bogoliubov-Valatin transformation defining new operators through

$$\begin{aligned} c_{k-k_F,\sigma} &= \cos\theta_k \alpha_{k,\sigma} - \sin\theta_k \beta_{k,\sigma} \\ c_{k+k_F,\sigma} &= \sin\theta_k \alpha_{k,\sigma} + \cos\theta_k \beta_{k,\sigma}. \end{aligned}$$

Then the kinetic energy term is replaced by

$$\sum_{k,\sigma} \epsilon_k c^\dagger_{k,\sigma} c_{k,\sigma} \to \sum_{k>0,\sigma}{}' \left(\epsilon_{k-k_F,\sigma} c^\dagger_{k-k_F,\sigma} c_{k-k_F,\sigma} + \epsilon_{k+k_F,\sigma} c^\dagger_{k+k_F,\sigma} c_{k+k_F,\sigma}\right).$$

Substituting the transformation we obtain terms of the type

$$\alpha^\dagger_{k,\sigma}\alpha_{k,\sigma},\ \beta^\dagger_{k,\sigma}\beta_{k,\sigma},\ \alpha^\dagger_{k,\sigma}\beta_{k,\sigma},\ \beta^\dagger_{k,\sigma}\alpha_{k,\sigma}.$$

We obtain therefore that the Hamiltonian can be written as

$$H = \sum_{k,\sigma} \left(\frac{1}{2} \left[(\epsilon_{k-k_F} + \epsilon_{k+k_F}) - \sqrt{(\epsilon_{k-k_F} - \epsilon_{k+k_F})^2 + 4\Delta^2} \right] \alpha^\dagger_{k,\sigma} \alpha_{k,\sigma} \right)$$
$$+ \sum_{k,\sigma} \left(\frac{1}{2} \left[(\epsilon_{k-k_F} + \epsilon_{k+k_F}) + \sqrt{(\epsilon_{k-k_F} - \epsilon_{k+k_F})^2 + 4\Delta^2} \right] \beta^\dagger_{k,\sigma} \beta_{k,\sigma} \right). \tag{14.159}$$

There are extra crossed terms that vanish if we choose

$$\tan 2\theta_k = -\frac{2\Delta}{\epsilon_{k+k_F} - \epsilon_{k-k_F}}.$$

The α operators create excitations with energies in a band which has lower energy than the excitations created by the β operators. At the momentum k_F there is a gap in the spectrum due to the appearance of a non-zero value for the parameter Δ. The gap at the Fermi momentum is 2Δ. This parameter corresponds to a so-called charge density wave of momentum $2k_F$. It is called an order parameter since it vanishes in the absence of the CDW and becomes finite in its presence. The notion of an order parameter to distinguish two phases will be considered later in the context of phase transitions.

If the system is at half-filling then $2k_F = \pi/a$. In this case the elementary reciprocal lattice vector is such that $g = 2\pi/a = 4k_F$. Let us now assume that the periodicity of the system, instead of being a, doubles to $2a$, due to some distortion of the system. Then the lattice constant goes from $a \to 2a$, the reciprocal lattice vector halves $g \to g/2$, and, since $2k_F = g/2$, then $k_F \to \pi/(2a)$ is now the limit of the Brillouin zone, due to the doubling in the periodicity of the lattice. Therefore, at the border of the new Brillouin zone there is now a stationary wave with a velocity that vanishes. The gap opens and the system lowers its energy to accommodate the vanishing velocity at the border of the filled band (since the dispersion relations flatten to zero). We might expect therefore that, at least at half-filling, it should be favorable for the system to dimerize (double its lattice constant).

However, let us analyze the equation of state for the order parameter Δ. In terms of the average values of the new operators, we get that

$$\Delta = V(2k_F) \sum_{k,\sigma} \sin 2\theta_k \left(\langle \alpha^\dagger_{k,\sigma} \alpha_{k,\sigma} \rangle - \langle \beta^\dagger_{k,\sigma} \beta_{k,\sigma} \rangle \right). \tag{14.160}$$

At $T = 0$ the occupancy of the upper band is zero and the lower band is full. Therefore

$$\Delta = V(2k_F) \sum_{k,\sigma} \sin 2\theta_k = 2V(2k_F) \sum_k \frac{\tan 2\theta_k}{\sqrt{1 + \tan^2 2\theta_k}}$$

$$\Delta = -4\Delta V(2k_F)\sum_k \frac{1}{\sqrt{(\epsilon_{k-k_F} - \epsilon_{k+k_F})^2 + 4\Delta^2}}.$$

Linearizing around the Fermi level like

$$\epsilon_{k\pm k_F} = v_F\hbar k_F \pm v_F\hbar k,$$

(for $|k| < k_c$) we get that

$$\Delta = -\Delta V(2k_F)\int_{-k_c}^{k_c} dk \frac{1}{\sqrt{(v_f\hbar k)^2 + \Delta^2}}$$

$$\Delta \sim -2\Delta V(2k_F)\frac{2\pi}{av_F\hbar}\ln\frac{v_f\hbar k_c}{\Delta}. \tag{14.161}$$

Therefore, if we take $\Delta > 0$ on the right-hand side then $\Delta < 0$ on the left-hand side. Therefore, $\Delta = 0$. There is no charge density wave (CDW). It is not favorable energetically. Only appears as an excited state, but not at $T = 0$.

However, if we add the jellium we may have a CDW due to a deformation of the lattice. The electronic instability may cause a lattice instability. We need therefore to couple the electrons to the phonons. In this case we may have a metal-insulator transition due to the gap originated from the CDW.

Let us consider that there is indeed a lattice deformation that tends to decrease the electronic energy, due to the creation of a gap at the Fermi momentum, k_F. Let us now add the lattice energy due to the lattice deformation.[9] This deformation involves an energy increase due to the elasticity of the lattice. Let us call the deformation δ. The equilibrium deformation is given by the solution of

$$\frac{d}{d\delta}(E_e + E_l) = 0, \tag{14.162}$$

where E_e is the electronic energy and E_l is the elastic energy of the lattice. Let us consider a tension of the form $\delta\cos(2k_F x)$ (we consider here, for illustrative purposes, the simple case of a one-dimensional deformation, as we have been considering). The average elastic energy per unit length is given by

$$E_l = \frac{1}{2}C\delta^2\langle\cos^2(2k_F x)\rangle = \frac{1}{4}C\delta^2, \tag{14.163}$$

where C is the linear force constant of the metal. We have now to estimate the electronic contribution.

[9] See also [Kittel].

The potential felt by the electrons due to the ions is given by

$$U(x) = 2A\delta \cos(2k_F x).$$

Recall that from the central equation we have that

$$\epsilon = \frac{1}{2}\left(\lambda_{k-g} + \lambda_k\right) \pm \sqrt{\frac{1}{4}\left(\lambda_{k-g} - \lambda_k\right)^2 + U^2}. \tag{14.164}$$

Expanding around $\tilde{k} = k - g/2$, we get that near the border of the Brillouin zone,

$$\epsilon_k = \frac{\hbar^2}{2m}\left(\frac{1}{4}g^2 + \tilde{k}^2\right) \pm U\left(1 + 2\frac{\lambda}{U^2}\frac{\hbar^2\tilde{k}^2}{2m}\right). \tag{14.165}$$

Using that around that point

$$\lambda = \frac{\hbar^2}{2m}\left(\frac{1}{2}g\right)^2, \tag{14.166}$$

we get that

$$\epsilon_k = \frac{\hbar^2}{2m}\left(k_F^2 + \tilde{k}^2\right) \pm \sqrt{4\frac{\hbar^2 k_F^2}{2m}\frac{\hbar^2\tilde{k}^2}{2m} + A^2\delta^2}, \tag{14.167}$$

(recall that $g = 2k_F$). Defining

$$x_{\tilde{k}} = \frac{\hbar^2\tilde{k}^2}{m}, x_F = \frac{\hbar^2 k_F^2}{m}, x = \frac{\hbar^2\tilde{k}k_F}{m},$$

we may calculate (taking the smaller solution)

$$\frac{d\epsilon_k}{d\delta} = -\frac{A^2\delta}{\left(x_F x_{\tilde{k}} + A^2\delta^2\right)^{1/2}}. \tag{14.168}$$

Since $d\tilde{k}/\pi$ is the number of orbitals per unit length, we can obtain that

$$\begin{aligned}\frac{dE_e}{d\delta} &= \frac{2}{\pi}\int_0^{k_F} d\tilde{k}\frac{d\epsilon_k}{d\delta} \\ &= -\frac{2A^2\delta}{\pi}\int_0^{k_F}\frac{d\tilde{k}}{\left(x_{\tilde{k}}x_F + A^2\delta^2\right)^{1/2}}\end{aligned}$$

$$= -\frac{2A^2\delta}{\pi}\frac{k_F}{x_F}\int_0^{k_F}\frac{dx}{\left(x^2 + A^2\delta^2\right)^{1/2}}$$

$$= -\frac{2A^2\delta}{\pi}\frac{k_F}{x_F}\sinh^{-1}\frac{x_F}{A\delta}. \tag{14.169}$$

Adding the electronic and the lattice contributions we get that

$$\frac{1}{2}C\delta - \frac{2A^2 m\delta}{\pi\hbar^2 k_F}\sinh^{-1}\frac{\hbar^2 k_F^2}{mA\delta} = 0, \tag{14.170}$$

or, equivalently,

$$\frac{\hbar^2 k_F^2}{mA\delta} = \sinh\left(\frac{\hbar^2 k_F \pi C}{4mA^2}\right). \tag{14.171}$$

Expanding for small deformations we get

$$|A|\delta \sim \frac{2\hbar^2 k_F^2}{m}e^{-\frac{\pi\hbar^2 k_F C}{4mA^2}}. \tag{14.172}$$

Defining $W = \hbar^2 k_F^2/(2m)$ the bandwidth of the conduction band, using that $D(0) = 2m/(\pi\hbar^2 k_F)$ is the density of states at the Fermi level, and identifying $V = 2A^2/C$ the effective electron-electron interaction, we can rewrite

$$|A|\delta \sim 4We^{-\frac{1}{D(0)V}}. \tag{14.173}$$

Note that the equation for the equilibrium deformation of the lattice is non-analytic in the "interaction." As we will see in examples ahead, this is a common characteristic of various phase transitions. An example of this type of distortion occurs in TaS_3.

14.5.2 Spin Density Waves (SDW)

Let us now consider another possible solution of the Hartree-Fock equations. This time let us look for a density wave in the spin degrees of freedom (SDW). Take again the Hamiltonian

$$H = \sum_{k,\sigma}\epsilon_k c^\dagger_{k,\sigma}c_{k,\sigma} + V(2k_F)\sum_q{}' \left(\sum_{kk'}\sum_{\sigma\sigma'} c^\dagger_{k+q,\sigma}c_{k'-q,\sigma'}c_{k',\sigma'}c_{k,\sigma}\right),$$

where we have approximated the interaction by its value at $q = 2k_F$. Let us now consider different spins. We consider now a different factorization of the interaction term of the form

$$H_{red} = \sum_{k,\sigma} \epsilon_k c^\dagger_{k,\sigma} c_{k,\sigma} - V(2k_F) \left(\sum_{kk'} \sum_{\sigma} \langle c^\dagger_{k+2k_F,\sigma} c_{k',-\sigma} \rangle c^\dagger_{k'-2k_F,-\sigma} c_{k,\sigma} + h.c. \right).$$

Note the minus sign in the factorized term. Let us now define the order parameter

$$\Delta = -2V(2k_F) \sum_k \langle c^\dagger_{k+2k_F,\uparrow} c_{k,\downarrow} \rangle.$$

Then the reduced Hamiltonian may be written as

$$\begin{aligned} H_{red} = \sum_k \Big(& \epsilon_{k-k_F} c^\dagger_{k-k_F,\uparrow} c_{k-k_F,\uparrow} + \epsilon_{k+k_F} c^\dagger_{k+k_F,\uparrow} c_{k+k_F,\uparrow} \\ & + \epsilon_{k-k_F} c^\dagger_{k-k_F,\downarrow} c_{k-k_F,\downarrow} + \epsilon_{k+k_F} c^\dagger_{k+k_F,\downarrow} c_{k+k_F,\downarrow} \Big) \\ & + \Delta c^\dagger_{k-k_F,\downarrow} c_{k+k_F,\uparrow} + \Delta c^\dagger_{k+k_F,\uparrow} c_{k-k_F,\downarrow}. \end{aligned} \tag{14.174}$$

Once again this Hamiltonian can be diagonalized performing a Bogoliubov-Valatin rotation. Define

$$\begin{aligned} c_{k-k_F,\uparrow} &= \cos\theta_k \alpha_{k,\uparrow} - \sin\theta_k \beta_{k,\downarrow} \\ c^\dagger_{k+k_F,\uparrow} &= \sin\theta_k \alpha_{k,\downarrow} + \cos\theta_k \beta^\dagger_{k,\uparrow}. \end{aligned}$$

We obtain in the same way as before that

$$\Delta = 2\Delta V(2k_F) \frac{2\pi}{\hbar v_F a} \ln \frac{\hbar k_c v_F}{\Delta}. \tag{14.175}$$

The case $\Delta = 0$ is again a solution. But there is now a second solution given by

$$1 = 2V(2k_F) \frac{2\pi}{\hbar v_F a} \ln \frac{\hbar k_c v_F}{\Delta}. \tag{14.176}$$

This gives a non-trivial solution which is non-analytic in the potential and therefore cannot be obtained by perturbation theory and can be expressed as

$$\Delta = \hbar k_c v_F e^{-\frac{2\hbar k_F v_c}{V(2k_F)}}. \tag{14.177}$$

Clearly as $V \to 0$ then $\Delta \to 0$. This collective mode, called spin density wave (SDW), is now stable. The charge distribution is homogeneous (without the ion background or in the jellium model) but the spin distribution is non-homogeneous. The energy difference is

$$\Delta E = E(\Delta) - E(0) = -\frac{\Delta^2}{4v_F \hbar k_F}(1 + 2\ln\frac{v_F \hbar k_F}{\Delta})$$

(Overhauser, 1961). In one dimension the SDW is the ground state. In three dimensions it is less favorable (however, for instance chromium has a SDW ground state).

14.6 Coulomb Interaction: Lindhard Function

Let us revisit the effect of the interactions between the electrons and consider second-order perturbation theory to see the effect of the Coulomb potential on the electron gas, in any number of dimensions. Consider the correction to the energy of two particles in states 1 and 2 in second order

$$\epsilon^{(2)}_{1,2} = -\sum_{3,4} \frac{2m}{\hbar^2} \frac{\langle 1, 2|V|3, 4\rangle\langle 3, 4|V|1, 2\rangle}{(k_3^2 + k_4^2) - (k_1^2 + k_2^2)}, \tag{14.178}$$

where

$$\langle 1, 2|V|3, 4\rangle = \frac{e^2}{\epsilon_0 \Omega q^2}\delta_{k_1+k_2,k_3+k_4}\delta_{\mathbf{q}=\mathbf{k}_1-\mathbf{k}_3=\mathbf{k}_4-\mathbf{k}_2}$$

and Ω is the volume. We can expand the above expression to obtain that

$$\begin{aligned}
\epsilon^{(2)}_{1,2} &= -i\frac{2m}{\hbar^2}\left(\frac{e^2}{\epsilon_0\Omega q^2}\right)^2 \sum_{\mathbf{q}} \frac{1}{q^4}\frac{1}{(\mathbf{k}_1-\mathbf{q})^2 + (\mathbf{k}_2+\mathbf{q})^2 - \mathbf{k}_1^2 - \mathbf{k}_2^2} \\
&= -\frac{m}{\hbar^2}\left(\frac{e^2}{\epsilon_0\Omega q^2}\right)^2 \sum_{\mathbf{q}} \frac{1}{q^4}\frac{1}{q^2 + \mathbf{q}\cdot(\mathbf{k}_2-\mathbf{k}_1)} \\
&= -\frac{m}{\hbar^2}\left(\frac{e^2}{\epsilon_0\Omega q^2}\right)^2 \frac{\Omega}{(2\pi)^3}2\pi\int_0^\infty dq q^2\int_{-1}^1 d\mu\frac{1}{q^4}\frac{1}{q^2}\frac{1}{1+\mu\frac{|\mathbf{k}_1-\mathbf{k}_2|}{q}} \\
&= -\frac{m}{\hbar^2}\left(\frac{e^2}{\epsilon_0\Omega q^2}\right)^2 \frac{\Omega}{(2\pi)^2}\frac{1}{|\mathbf{k}_1-\mathbf{k}_2|}\int_0^\infty dq\frac{q}{q^4}\ln|\frac{q+|\mathbf{k}_1-\mathbf{k}_2|}{q-|\mathbf{k}_1-\mathbf{k}_2|}|.
\end{aligned} \tag{14.179}$$

The integrating function behaves as q^{-3} (and the integral as q^{-2}) and therefore diverges as $q \to 0$. Each order in the perturbation expansion is infinite. The cause is the long range nature of the Coulomb potential ($q \to 0$ is problematic). Actually what happens is that the electronic system screens the interaction. The interaction between two electrons far apart is screened by the other electrons, as seen before. The Coulomb potential is altered to a Yukawa-like potential of the form

$$\frac{e^2}{4\pi\epsilon_0 r} \to \frac{e^2}{4\pi\epsilon_0 r} e^{-\frac{r}{l_s}}, \tag{14.180}$$

where we have introduced a length l_s which is the screening length (we will see later that $l_s \sim v_F\sqrt{\frac{m}{ne^2}}$). Once again this result is non-analytical and cannot be obtained in perturbation theory. We must use another method. The idea was considered before (for instance in the context of the Thomas-Fermi approximation) but here we want to consider a more general and systematic way to handle the self-consistent nature of the problem.

Let us consider a time-dependent perturbation (recall the linear response theory previously discussed). Consider then[10]

$$\delta U(\mathbf{r}, t) = u e^{-i\mathbf{q}\cdot\mathbf{r}+i\omega t} e^{\alpha t}. \tag{14.181}$$

Here $\alpha = 0^+$. The states are perturbed to

$$\psi_{\mathbf{k}} = |\mathbf{k}\rangle + b_{\mathbf{k}-\mathbf{q}}(t)|\mathbf{k}-\mathbf{q}\rangle, \tag{14.182}$$

where

$$|\mathbf{k}\rangle = e^{i\mathbf{q}\cdot\mathbf{r}-i\epsilon(\mathbf{k})t/\hbar} \tag{14.183}$$

is a plane wave. The coefficients in the expansion on other states are given by

$$b_{\mathbf{k}-\mathbf{q}}(t) = \frac{\langle \mathbf{k}-\mathbf{q}|\delta U|\mathbf{k}\rangle}{\epsilon(\mathbf{k}) - \epsilon(\mathbf{k}-\mathbf{q}) + \hbar\omega - i\hbar\alpha} = \frac{u e^{i\omega t} e^{\alpha t}}{\epsilon(\mathbf{k}) - \epsilon(\mathbf{k}-\mathbf{q}) + \hbar\omega - i\hbar\alpha}. \tag{14.184}$$

The change in the electronic density due to the perturbation is

$$\begin{aligned}\delta\rho(\mathbf{r}, t) &= e\frac{1}{\Omega}\sum_{\mathbf{k}}\left(|\psi_{\mathbf{k}}(\mathbf{r}, t)|^2 - 1\right)\\ &= e\frac{1}{\Omega}\sum_{\mathbf{k}}\left(b^*_{\mathbf{k}-\mathbf{q}}(t)e^{i\mathbf{q}\cdot\mathbf{r}} + e^{-i\mathbf{q}\cdot\mathbf{r}}b_{\mathbf{k}-\mathbf{q}}(t)\right)\end{aligned}$$

[10] For further reading see [Ziman].

$$= e\frac{1}{\Omega}\sum_{\mathbf{k}}\left(\frac{ue^{-i\omega t}e^{\alpha t}e^{i\mathbf{q}\cdot\mathbf{r}}}{\epsilon(\mathbf{k})-\epsilon(\mathbf{k}-\mathbf{q})-\hbar\omega-i\hbar\alpha}\right.$$
$$\left.+\frac{ue^{i\omega t}e^{\alpha t}e^{-i\mathbf{q}\cdot\mathbf{r}}}{\epsilon(\mathbf{k})-\epsilon(\mathbf{k}-\mathbf{q})-\hbar\omega+i\hbar\alpha}\right). \tag{14.185}$$

If it is a real perturbation, instead of a complex one, then we have to sum the terms with $\omega \to -\omega$ and $\mathbf{q} \to -\mathbf{q}$. It gives

$$\delta\rho(\mathbf{r},t) = eu\frac{1}{\Omega}\sum_{\mathbf{k}}\left(\frac{e^{-i\omega t}e^{\alpha t}e^{i\mathbf{q}\cdot\mathbf{r}}}{\epsilon(\mathbf{k})-\epsilon(\mathbf{k}-\mathbf{q})-\hbar\omega-i\hbar\alpha}\right.$$
$$\left.+\frac{e^{i\omega t}e^{\alpha t}e^{-i\mathbf{q}\cdot\mathbf{r}}}{\epsilon(\mathbf{k})-\epsilon(\mathbf{k}+\mathbf{q})+\hbar\omega-i\hbar\alpha}+c.c.\right)$$
$$= ee^{\alpha t}ue^{i\omega t-i\mathbf{q}\cdot\mathbf{r}}$$
$$\frac{1}{\Omega}\sum_{\mathbf{k},occupied}\left(\frac{1}{\epsilon(\mathbf{k})-\epsilon(\mathbf{k}+\mathbf{q})+\hbar\omega-i\hbar\alpha}\right.$$
$$\left.+\frac{1}{\epsilon(\mathbf{k})-\epsilon(\mathbf{k}-\mathbf{q})-\hbar\omega+i\hbar\alpha}\right)$$
$$+\,c.c. \tag{14.186}$$

Inserting the Fermi functions to give a thermodynamic weight to the occupied states, we get

$$\delta\rho(\mathbf{r},t) = ee^{\alpha t}u\frac{1}{\Omega}\sum_{\mathbf{k}}\frac{f(\epsilon(\mathbf{k}))-f(\epsilon(\mathbf{k}+\mathbf{q}))}{\epsilon(\mathbf{k})-\epsilon(\mathbf{k}+\mathbf{q})+\hbar\omega-i\hbar\alpha}e^{i\omega t-i\mathbf{q}\cdot\mathbf{r}}+c.c. \tag{14.187}$$

Recall now that Poisson's equation implies that

$$\nabla^2(\delta\phi) = \frac{e}{\epsilon_0}\delta\rho. \tag{14.188}$$

Take

$$\delta\phi = \phi e^{i\omega t-i\mathbf{q}\cdot\mathbf{r}}e^{\alpha t}. \tag{14.189}$$

Then we get

$$\phi = u\frac{e^2}{\epsilon_0 q^2}\frac{1}{\Omega}\sum_{\mathbf{k}}\frac{f(\epsilon(\mathbf{k}))-f(\epsilon(\mathbf{k}+\mathbf{q}))}{\epsilon(\mathbf{k})-\epsilon(\mathbf{k}+\mathbf{q})+\hbar\omega-i\hbar\alpha}. \tag{14.190}$$

We should now add ϕ to u in δU. In this way the calculation becomes self-consistent. We have that the external potential plus the polarization of the medium gives the total potential. The total charge density is the external charge term plus the induced charge. We really have

$$-i\mathbf{q}\cdot\mathbf{D}(\mathbf{q},\omega)=\rho_{ext} \tag{14.191}$$

and that the induced charge is related to the polarization through

$$i\mathbf{q}\cdot\mathbf{P}=\rho_{induced}. \tag{14.192}$$

Defining the dielectric function in the usual way we get

$$\mathbf{D}(\mathbf{q},\omega)=\epsilon_0\mathbf{E}(\mathbf{q},\omega)+\mathbf{P}(\mathbf{q},\omega)=\epsilon(\mathbf{q},\omega)\mathbf{E}(\mathbf{q},\omega). \tag{14.193}$$

Then writing the external potential as

$$\delta V=Ve^{i\omega t-i\mathbf{q}\cdot\mathbf{r}}e^{\alpha t}+c.c. \tag{14.194}$$

we get

$$u=\frac{V}{\epsilon_r(\mathbf{q},\omega)} \tag{14.195}$$

and the dielectric function is obtained through the self-consistent solution of

$$u=V+u\frac{e^2}{\epsilon_0 q^2}\frac{1}{\Omega}\sum_{\mathbf{k}}\frac{f(\epsilon(\mathbf{k}))-f(\epsilon(\mathbf{k}+\mathbf{q}))}{\epsilon(\mathbf{k})-\epsilon(\mathbf{k}+\mathbf{q})+\hbar\omega-i\hbar\alpha} \tag{14.196}$$

and is given by

$$\epsilon_r(\mathbf{q},\omega)=1-\frac{e^2}{\epsilon_0 q^2}\frac{1}{\Omega}\sum_{\mathbf{k}}\frac{f(\epsilon(\mathbf{k}))-f(\epsilon(\mathbf{k}+\mathbf{q}))}{\epsilon(\mathbf{k})-\epsilon(\mathbf{k}+\mathbf{q})+\hbar\omega-i\hbar\alpha}. \tag{14.197}$$

This is the so-called Lindhard function.

Different regimes can be considered leading, in some cases, to a rederivation of results previously discussed.

14.6.1 Static Properties

Consider the static uniform case. Take $\omega=0$ and $\mathbf{q}\to 0$. In this limit

$$\epsilon_r(\mathbf{q}\to 0,\omega=0)=1-\lim_{\mathbf{q}\to 0}\frac{1}{\Omega}\sum_{\mathbf{k}}\frac{e^2}{\epsilon_0 q^2}\frac{f(\epsilon(\mathbf{k}))-f(\epsilon(\mathbf{k}+\mathbf{q}))}{\epsilon(\mathbf{k})-\epsilon(\mathbf{k}+\mathbf{q})+-i\hbar\alpha}. \tag{14.198}$$

We can expand $\epsilon(\mathbf{k}+\mathbf{q}) = \epsilon(\mathbf{k}) + \mathbf{q}\cdot\nabla_{\mathbf{k}}\epsilon(\mathbf{k})$. Also using that

$$\frac{1}{x-i\eta} = P\frac{1}{x} + i\pi\delta(x), \tag{14.199}$$

($\eta \to 0$) we conclude that the imaginary part is zero and in this case we can take $\alpha = 0$ in Eq. (14.198). Using that

$$f(\epsilon(\mathbf{k}+\mathbf{q})) = f(\epsilon(\mathbf{k})) + \mathbf{q}\frac{\partial f}{\partial \epsilon}\cdot\nabla_{\mathbf{k}}\epsilon(\mathbf{k}), \tag{14.200}$$

we obtain that the dielectric function is given by

$$\epsilon_r(\mathbf{q}\to 0, \omega = 0) = 1 - \frac{e^2}{\epsilon_0 q^2}\lim_{\mathbf{q}\to 0}\frac{1}{\Omega}\sum_{\mathbf{k}}\frac{-\frac{\partial f}{\partial \epsilon}\mathbf{q}\cdot\nabla_{\mathbf{k}}\epsilon(\mathbf{k})}{-\mathbf{q}\cdot\nabla_{\mathbf{k}}\epsilon(\mathbf{k})} = \lim_{\mathbf{q}\to 0}\frac{1}{\Omega}\sum_{\mathbf{k}}\frac{\partial f}{\partial \epsilon_{\mathbf{k}}}. \tag{14.201}$$

The sum over the momentum can be converted into an integral over the energies. Also, at low enough temperatures, $(\partial f)/(\partial\epsilon) \sim -\delta(\epsilon - \epsilon_F)$. We get then that

$$\epsilon_r(\mathbf{q}\to 0, \omega = 0) = 1 + \frac{e^2}{\epsilon_0 q^2}N(\epsilon_F), \tag{14.202}$$

where $N(\epsilon_F)$ is the density of states per volume at the Fermi level. The dielectric function diverges at zero momentum. This result can be recognized as the Thomas-Fermi approximation previously discussed. In the static case ($\omega = 0$) taking the limit $q \to 0$ means long distances and long times. We can imagine the electrons as filling the Fermi sea at each site in the lattice. In equilibrium, the chemical potential (Fermi level) is homogeneous. Imagine now that at a certain site $\mathbf{r}$ we apply a perturbation that increases the energy locally. Then at that the point the chemical potential is higher. There will be a flow of the electrons from that site to the neighboring sites, to decrease the density at that site and try to homogenize the system. Since the electrons flow from that site, there is an electronic deficiency at that point as discussed before, which through Poisson's equation leads to a change in the potential and the Thomas-Fermi approximation for the dielectric function. As we saw before, the form of the dielectric function, as for the Thomas-Fermi result, leads to screening of a test charge and a Yukawa-like potential.

Consider now the regime of very high temperatures $T >> T_F$. A classical treatment is therefore justified. Then

$$\delta n = n_0 e^{-\frac{\delta U(\mathbf{r})}{k_B T}} - n_0 = -n_0\frac{\delta U(\mathbf{r})}{k_B T}. \tag{14.203}$$

Using again Poisson's equation we obtain

$$\nabla^2(\delta\phi) = \frac{e^2 n_0}{\epsilon_0 k_B T}\delta U(\mathbf{r}), \tag{14.204}$$

which implies that in this limit

$$\lambda^2 = \frac{e^2 n_0}{\epsilon_0 k_B T}. \tag{14.205}$$

This is the Debye-Huckel approximation.

Let us return to the Lindhard function.

$$\epsilon_r(\mathbf{q}, \omega) = 1 - \frac{e^2}{\epsilon_0 q^2} \frac{1}{\Omega} \sum_{\mathbf{k}} \frac{f(\epsilon_{\mathbf{k}}) - f(\epsilon_{\mathbf{k}+\mathbf{q}})}{\epsilon(\mathbf{k}) - \epsilon(\mathbf{k}+\mathbf{q}) + \hbar\omega - i\hbar\alpha}$$

$$\epsilon_r(\mathbf{q}, 0) = 1 - \frac{e^2}{\epsilon_0 q^2} \frac{1}{\Omega} \sum_{\mathbf{k}} \frac{f(\epsilon_{\mathbf{k}}) - f(\epsilon_{\mathbf{k}+\mathbf{q}})}{\epsilon(\mathbf{k}) - \epsilon(\mathbf{k}+\mathbf{q})}.$$

Take the case of three dimensions and $\omega = 0$. The difference in the momentum summation can be simplified. Make a translation in the second term $\mathbf{k} \to \mathbf{k} - \mathbf{q}$ and assuming isotropy for free electrons we can write that $\epsilon(\mathbf{k} - \mathbf{q}) = \epsilon(\mathbf{k} + \mathbf{q})$. Then we can write that

$$\epsilon_r(\mathbf{q}, 0) = 1 - \frac{e^2}{\epsilon_0 q^2} 2 \frac{1}{\Omega} \sum_{\mathbf{k}} \frac{f(\epsilon_{\mathbf{k}})}{\epsilon_{\mathbf{k}} - \epsilon_{\mathbf{k}+\mathbf{q}}}. \tag{14.206}$$

The sum can be carried out

$$2\frac{1}{\Omega} \sum_{\mathbf{k}} \frac{f(\epsilon_{\mathbf{k}})}{\epsilon_{\mathbf{k}} - \epsilon_{\mathbf{k}+\mathbf{q}}} = \frac{4m}{\hbar^2 (2\pi)^3} \int d^3k \frac{f(\epsilon_{\mathbf{k}})}{k^2 - (\mathbf{k}+\mathbf{q})^2}$$

$$= -\frac{4m}{\hbar^2 (2\pi)^3} 2\pi \int_0^{k_F} dk k^2 \int_{-1}^{1} d\mu \frac{1}{2kq\mu + q^2}.$$

The integrations can be performed and yield for the dielectric function, using that $N(\epsilon_F) = (m)/\hbar^2 (2\pi^2)$,

$$\epsilon_r^{(3d)}(\mathbf{q}, 0) = 1 + \frac{e^2}{\epsilon_0 q^2} N(\epsilon_F) k_F \left(\frac{1}{2} + \frac{1}{2qk_F} (k_F^2 - \frac{q^2}{4}) \ln |\frac{2k_F + q}{2k_F - q}| \right). \tag{14.207}$$

Consider now the two-dimensional case. Here we mean a three-dimensional system but arranged in layers. The dispersion relation of the free electrons is taken as

$$\epsilon_{\mathbf{k}} = \frac{\hbar^2}{2m} (k_x^2 + k_y^2). \tag{14.208}$$

The vector $\mathbf{q}$ is also contained in the xy plane. This is valid for $|k_z| < \frac{1}{2} K$ where K is some cutoff that arises since the spacing in the z direction is very small and therefore the quantization along the z direction is such that only very high momenta states are allowed. The motion along z is frozen.

In this case we can write that

$$\epsilon_{\mathbf{k}} - \epsilon_{\mathbf{k}+\mathbf{q}} = -\frac{\hbar^2}{2m}(2\mathbf{k} \cdot \mathbf{q} + q^2). \tag{14.209}$$

Then we can write

$$\begin{aligned} 2\frac{1}{\Omega}\sum_{\mathbf{k}} \frac{f(\epsilon_{\mathbf{k}})}{\epsilon_{\mathbf{k}} - \epsilon_{\mathbf{k}+\mathbf{q}}} &= -\frac{4m}{\hbar^2(2\pi)^3} K \int_0^{k_F} dk k \int_0^{2\pi} d\theta \frac{1}{2kq\cos\theta + q^2} \\ &= -\frac{mK}{2\hbar^2\pi^3}\frac{1}{2q}\int_0^{k_F} dk \frac{2\pi}{\sqrt{\left(\frac{q}{2k}\right)^2 - 1}}\theta(q - 2k). \end{aligned}$$

Using that the density of states is $N(\epsilon) = \frac{mK}{\hbar^2(2\pi)^2}$ we can obtain

$$\begin{aligned} \epsilon_r^{(2d)}(\mathbf{q}, 0) &= 1 + \frac{e^2}{\epsilon_0 q^2} N(\epsilon_F), q < 2k_F \\ &= 1 + \frac{e^2}{\epsilon_0 q^2} N(\epsilon_F)(1 - \frac{1}{q}\sqrt{q^2 - 4k_F^2}), q > 2k_F. \end{aligned} \tag{14.210}$$

Finally, let us consider the one-dimensional case. In this case

$$\epsilon_k = \frac{\hbar^2 k_z^2}{2m}. \tag{14.211}$$

Again we consider a system quasi-one-dimensional where $|k_x|, |k_y| < \frac{1}{2}K$. Like before, we analyze the sum over the momenta

$$\begin{aligned} 2\frac{1}{\Omega}\sum_{\mathbf{k}} \frac{f(\epsilon_{\mathbf{k}})}{\epsilon_{\mathbf{k}} - \epsilon_{\mathbf{k}+\mathbf{q}}} &= 2\frac{1}{\hbar^2(2\pi)^3} K^2 \int_{-k_F}^{k_F} dk_z \frac{2m}{k_z^2 - (k_z + q)^2} \\ &= -4m\frac{K^2}{\hbar^2(2\pi)^3 2q} \ln|\frac{2k_F + q}{2k_F - q}.| \end{aligned}$$

Using that the density of states at the Fermi level is

$$N(\epsilon_F) = \frac{2K^2 m}{\hbar^2(2\pi)^3 k_F}, \tag{14.212}$$

we get

$$\epsilon_r^{(1d)}(\mathbf{q}, 0) = 1 + \frac{e^2}{\epsilon_0 q^2} N(\epsilon_F)\frac{k_F}{q} \ln|\frac{2k_F + q}{2k_F - q}|. \tag{14.213}$$

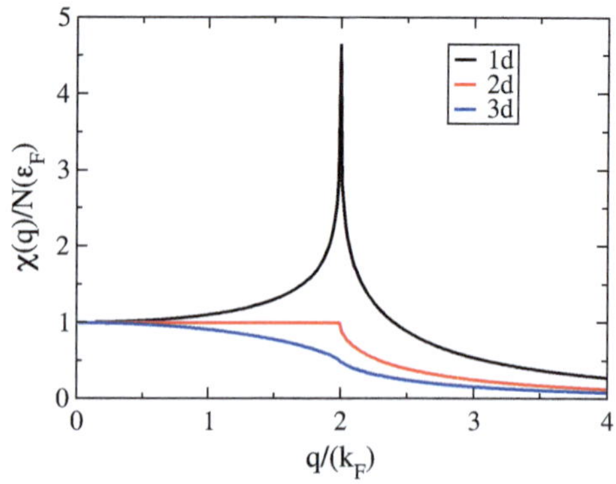

Fig. 14.10 Susceptibility $\chi(q)$, normalized by the density of states at the Fermi level, for the $1d$, $2d$, $and3d$ cases, as a function of q/k_F

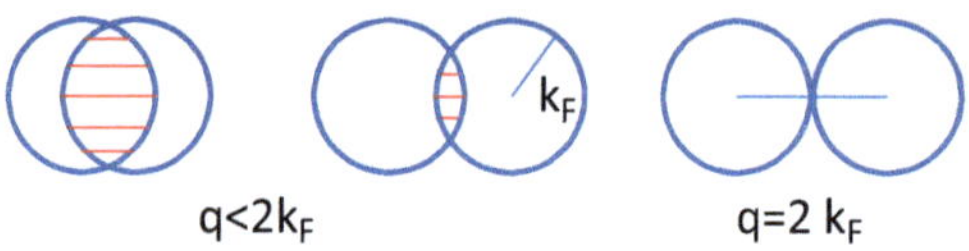

Fig. 14.11 When $q < 2k_F$ there is an overlap between the two Fermi surfaces that results from the states at momentum $\mathbf{k}$ and $\mathbf{k} + \mathbf{q}$. For $q > 2k_F$ the two Fermi surfaces no longer overlap and there is no cancellation in the numerator of the Lindhard function. $q = 2k_F$ is the transition point between the two regimes

In the three cases of $1d$, $2d$, and $3d$ the dielectric function can be written as

$$\begin{aligned}\epsilon_r(\mathbf{q}, 0) &= 1 + \frac{e^2}{\epsilon_0 q^2}\chi(q) \\ \chi(0) &= N(\epsilon_F) \\ \chi(q) &\geq 0.\end{aligned} \tag{14.214}$$

In Fig. 14.10 we show the so-called susceptibility $\chi(q)$ for the three cases, as a function of the momentum. The point $q = 2k_F$ is special. In $1d$ the susceptibility diverges at this point, indicative of the instability previously mentioned. In $2d$ the susceptibility is continuous, but its derivative is discontinuous. Only in three dimensions the behavior of the susceptibility is smooth.

It is easy to understand why $q = 2k_F$ is a special point. As a perturbation is applied there is an exchange of momentum q between the electrons. We may imagine that the perturbation shifts the Fermi sphere of the electrons by an amount which is the momentum transferred by the external perturbation. At small momentum many of the electrons do not contribute to the process because, due to exclusion principle, there is a large overlap of the Fermi surfaces, and most states are occupied. As the momentum increases, the volume that contributes increases and when the special point $q = 2k_F$ is reached, all electrons contribute, as illustrated in Fig. 14.11. After this point no qualitative change occurs and the susceptibility decreases since the momentum exchanged keeps increasing.

Let us then consider a phonon with momentum $\mathbf{q}$. The electrons follow the motion of the ions and therefore the interaction between the ions (positive charges immersed in the negative medium provided by the electrons) is screened by the electrons. As we have seen, the screening is proportional to $1/\epsilon$. Therefore, the forces between the ions are screened by the electrons to F/ϵ. The electronic medium changes the coupling constants between the ions. Therefore, the phonon spectrum is changed in an important way when $q = 2k_F$. This is the so-called Kohn effect. In the special case of $d = 1$ the dielectric function diverges, and, therefore, the effective force between the ions vanishes. The forces are completely screened when $q = 2k_F$. We expect therefore a strong instability in the lattice structure and the system may have a structural phase transition, because the ions are no longer bound. The outcome is that the lattice becomes dimerized: this is the so-called Peierls instability. If we distort the crystal to form dimers (bound states of two ions), the periodicity doubles, as discussed above; the lattice constant is now $2a$; this implies that the Brillouin zone halves, the dispersion relation is changed, and $q \neq 2k_F$. The force between the ions is restored since the screening is not fully effective anymore and this leads in general to the stabilization of the lattice. There is in general the so-called Peierls distortion (not necessarily dimerization).

The case of an insulator is slightly different. The treatment is also different from the one we followed for the metal, because now we cannot simply take the conduction electrons to be free. We want to consider the case when the insulator originated in the band structure, which implies taking into account the periodic potential (even though it is also possible to originate an insulator solely due to the effect of interactions leading to the so-called Mott insulator). Let us then consider that the gap arises due to the Bragg reflections (clearly the result depends now on the particular lattice structure). The dielectric function is written as

$$\epsilon_r(\mathbf{q},\omega) = 1 - \frac{e^2}{\epsilon_0 q^2}\frac{1}{\Omega}\sum_{\mathbf{k},\mathbf{g}} \frac{|<\mathbf{k}|e^{i\mathbf{q}\cdot\mathbf{r}}|\mathbf{k}+\mathbf{q}+\mathbf{g}>|^2\left(f(\epsilon_{\mathbf{k}}) - f(\epsilon_{\mathbf{k}+\mathbf{q}+\mathbf{g}})\right)}{\epsilon_{\mathbf{k}} - \epsilon_{\mathbf{k}+\mathbf{q}+\mathbf{g}} + \hbar\omega - i\hbar\alpha}. \tag{14.215}$$

Consider now the static case $\omega = 0$. The term $\mathbf{g} = 0$ corresponds to two energies in the same band. Since we are considering an insulator, either both bands are empty or both are full. Therefore, it does not contribute. Among the reciprocal vector values $\mathbf{g} \neq 0$, the most relevant is the one such that

$$\epsilon_{\mathbf{k}+\mathbf{q}+\mathbf{g}} - \epsilon_{\mathbf{k}} \sim E_{gap}. \tag{14.216}$$

Let us consider small momentum $\mathbf{q}$ (the case of $q = 0$ does not contribute). The leading term is proportional to q^2. This term cancels the same factor in the denominator and therefore the divergence disappears. We get that

$$\epsilon_r(\mathbf{q} \to 0, 0) = 1 + \frac{ne^2\hbar^2}{\epsilon_0 m(E_{gap})^2}. \tag{14.217}$$

It is finite: an insulator cannot screen completely. Note that the dielectric function can be written as

$$\epsilon_r(\mathbf{q} \to 0, 0) = 1 + \left(\frac{\hbar\omega_p}{E_{gap}}\right)^2. \tag{14.218}$$

Here

$$\omega_p = \sqrt{\frac{ne^2}{\epsilon_0 m}} \tag{14.219}$$

is called the plasma frequency, as discussed before.

14.6.2 Dynamic Properties

Consider now the momentum- and frequency-dependent dielectric function. In general it has a real and an imaginary part

$$\epsilon(q, \omega) = \epsilon_1(q, \omega) + i\epsilon_2(q, \omega). \tag{14.220}$$

Using the expression obtained before for the dielectric function, and using that

$$\frac{1}{\epsilon_k - \epsilon_{k+q} + \hbar\omega - i\hbar\alpha} = P\frac{1}{\epsilon_k - \epsilon_{k+q} + \hbar\omega} + i\pi\delta(\epsilon_k - \epsilon_{k+q} + \hbar\omega), \tag{14.221}$$

we obtain that the real part is given, at $T = 0$, by

$$\begin{aligned}\epsilon_{r,1}(q, \omega) = 1 + \frac{q_{TF}^2}{q^2}\{\frac{1}{2} + \frac{k_F}{4q}&\left[\left(\frac{(\hbar\omega + \frac{\hbar^2 q^2}{2m})^2}{(\hbar q v_F)^2} - 1\right)\ln|\frac{\hbar\omega - v_F\hbar q + \frac{\hbar^2 q^2}{2m}}{\hbar\omega + v_F\hbar q + \frac{\hbar^2 q^2}{2m}}|\right.\\ &\left. - \left(\frac{(\hbar\omega - \frac{\hbar^2 q^2}{2m})^2}{(\hbar q v_F)^2} - 1\right)\ln|\frac{\hbar\omega - v_F\hbar q - \frac{\hbar^2 q^2}{2m}}{\hbar\omega + v_F\hbar q - \frac{\hbar^2 q^2}{2m}}|\right]\},\end{aligned} \tag{14.222}$$

where the Thomas-Fermi momentum is given by

$$q_{TF}^2 = \frac{3\omega_p^2}{v_F^2} = 3\frac{ne^2}{\epsilon_0 m v_F^2}. \tag{14.223}$$

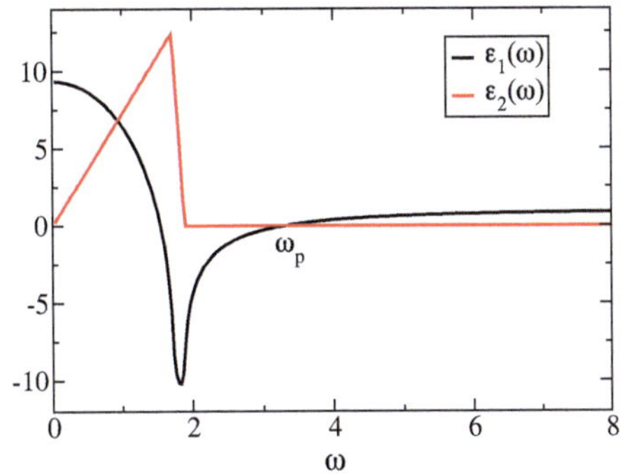

Fig. 14.12 Real part ($\epsilon_{r,1}$) and imaginary part ($\epsilon_{r,2}$) of the relative dielectric function as a function of frequency for low momentum. At high frequencies, the real part has a zero that corresponds to the plasma frequency. In this regime, the imaginary part, $\epsilon_{r,2}$ vanishes as well

The imaginary part is given by

$$\begin{aligned}\epsilon_{r,2}(q,\omega) &= \frac{\pi}{2}\frac{q_{TF}^2}{q^2}\frac{\omega}{qv_F}, 0 \leq \hbar\omega \leq v_F\hbar q - \frac{\hbar^2 q^2}{2m} \\ &= \frac{\pi}{4}\frac{q_{TF}^2}{q^2}\frac{k_F}{q}\left(1 - \frac{(\hbar\omega - \frac{\hbar^2 q^2}{2m})^2}{(\hbar q v_F)^2}\right), \hbar q v_F - \frac{\hbar^2 q^2}{2m} \leq \hbar\omega \\ &\leq \hbar q v_F + \frac{\hbar^2 q^2}{2m} \\ &= 0, \hbar\omega \geq \hbar q v_F + \frac{\hbar^2 q^2}{2m}.\end{aligned} \tag{14.224}$$

Note that this is a continuous function of the frequency.

In the limit $q \to 0$, the real part of the dielectric function diverges. For small values of the momentum q, we show in Fig. 14.12 the real and imaginary parts of the dielectric function, as a function of the frequency ($q << q_{TF}, k_F$), for some representative parameters. The imaginary part vanishes at zero frequency, as we saw before. The real part is finite in the static case, then it decreases, goes through zero, and becomes negative, and at high frequencies it becomes positive again, crossing the real axis at the plasma frequency ω_p. When the real part becomes negative, the imaginary part is finite. Then the imaginary part decreases to zero, and at the plasma frequency, both the real and the imaginary parts vanish. Therefore, the dielectric function vanishes at the plasma frequency. Therefore, the electric field is finite even if the applied external field is zero. There is an excitation of the system which is called plasmon where an internal electric field is spontaneously created by any infinitesimal fluctuation in the system, as we discussed before.

Since we are in a regime where the imaginary part is zero, we can take $\alpha = 0$ in the expression for the dielectric function. We obtain then that

$$\epsilon_r(q,\omega) = 1 - \frac{e^2}{\epsilon_0 q^2}\frac{1}{\Omega}\sum_k f(\epsilon_k)\left(\frac{1}{\epsilon_k - \epsilon_{k+q} + \hbar\omega} + \frac{1}{\epsilon_k - \epsilon_{k-q} - \hbar\omega}\right). \tag{14.225}$$

Let us consider small q and large ω. Then

$$\epsilon_r(q,\omega) \sim 1 - \frac{e^2}{\epsilon_0 q^2}\frac{1}{\Omega}\sum_k f(\epsilon_k)\left(\epsilon_{k+q} - \epsilon_k + \epsilon_{k-q} - \epsilon_k\right) \sim 1$$
$$- \frac{e^2}{\epsilon_0 q^2}\frac{\hbar^2 q^2}{m}\frac{1}{\Omega}\sum_k f(\epsilon_k). \tag{14.226}$$

We finally get that

$$\epsilon_r(q,\omega) \sim 1 - \frac{\omega_p^2}{\omega^2}, \tag{14.227}$$

where ω_p is the plasma frequency. Clearly when $\omega = \omega_p$ we obtain that $\epsilon(q,\omega) = 0$.

Let us take again the dielectric function given by

$$\epsilon_r(q,\omega) = 1 - \frac{e^2}{\epsilon_0 q^2}\frac{1}{\Omega}\sum_k f(\epsilon_k)\frac{2(\epsilon_k - \epsilon_{k+q})}{(\epsilon_k - \epsilon_{k+q})^2 - (\hbar\omega)^2}$$
$$= 1 - \frac{\omega_p^2}{\omega^2} + \frac{e^2}{\epsilon_0 q^2(\hbar\omega)^4}\frac{1}{\Omega}\sum_k f(\epsilon_k)2(\epsilon_k - \epsilon_{k+q})^3 + O\left(\frac{1}{\omega^6}\right).$$

Considering small momenta, converting the sum over the momenta as an integral, and performing the integral, we can obtain that

$$\epsilon_r(q,\omega) = 1 - \frac{\omega_p^2}{\omega^2}\left(1 + \frac{1}{\omega^2}\frac{3}{5}(qv_F)^2\right). \tag{14.228}$$

The zeros of the dielectric function are obtained from the equation $\epsilon(q,\omega_p(q)) = 0$ and are given by

$$\omega_p^4(q) - \omega_p^2(q)\omega_p^2 - \frac{3}{5}\omega_p^2(qv_F)^2 = 0. \tag{14.229}$$

The solutions are

$$\omega^2 = \frac{\omega_p^2}{2}\left(1 + \sqrt{1 + \frac{12}{5}\frac{(qv_F)^2}{\omega_p^2}}\right). \tag{14.230}$$

Expanding, we get that for small momenta

$$\omega = \omega_p + \frac{3}{10}\frac{(qv_F)^2}{\omega_p}. \tag{14.231}$$

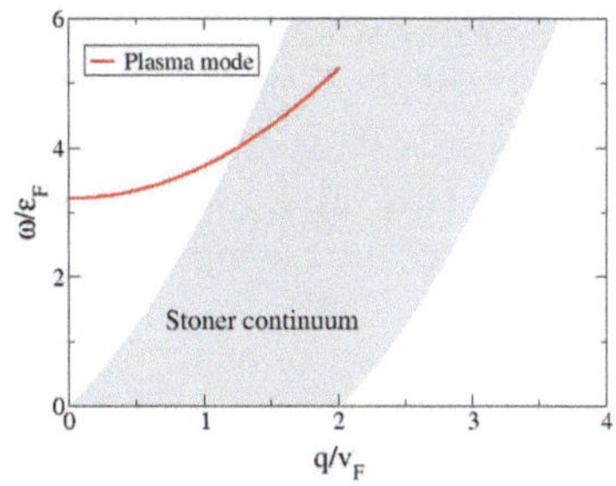

Fig. 14.13 Dispersion relation of the longitudinal plasma modes and Stoner continuum of the particle-hole excitations. The plasma mode is well-defined until it enters the Stoner continuum where it becomes unstable to the creation of particle-hole pairs

At zero momentum the frequency is just the plasma frequency. As the momentum increases, the eigenenergies increase quadratically with the momentum. These excitations are called plasmon modes. They are well-defined very-high-energy excitations of the system. They are stable for small momentum values. However, there are other excitations in the system since there are particle-hole excitations that are allowed. Since these are two-particle excitations they form a continuum. As shown in Fig. 14.13 they have very low energy at small momentum and are therefore favorable energetically, while the plasmons remain frozen. When the dispersion relation of the plasmons meets the Stoner continuum of excitations, the plasmons may decay into particle-hole excitations and are no longer stable excitations.

Random-Phase Approximation

Let us summarize some of the results we reached. First notice that the factor $e^2/(\epsilon_0 q^2)$ in Eq. (14.214) is the Fourier transform of the Coulomb potential, $V(q)$. The susceptibility $\chi(q)$ defined in the same expression is actually the change in the charge density caused by the external perturbation (see Eq. (14.187)). Therefore,

$$\delta\rho = -e\chi(q,\omega)\delta u = e\frac{1}{\Omega}\sum_k \frac{f(\epsilon_k) - f(\epsilon_{k+q})}{\epsilon_k - \epsilon_{k+q} + \hbar\omega - i\hbar\alpha}\delta u. \tag{14.232}$$

If we do not consider the self-consistent nature of the problem we get that

$$\delta\rho_{HF} = -e\chi(q,\omega)\delta V_{ext} \tag{14.233}$$

and we do not get screening. If however we consider the perturbation in a self-consistent way, this leads to the so-called random phase approximation (RPA) like

$$\delta\rho_{RPA} = -e\frac{\chi(q,\omega)}{1 + \frac{e^2}{\epsilon_0 q^2}\chi(q,\omega)}\delta V_{ext}. \tag{14.234}$$

In this way we get the proper screening property of the system of charges.

Linear response theory implies causality of the response function and involves a density-density correlation function, of the quantity in study, such as in Eqs. (11.37,11.38). Applying a time-dependent perturbation $\delta U(\mathbf{r}, t) = ue^{-i\mathbf{q}\cdot\mathbf{r}+i\omega t}e^{\alpha t}$, with $\alpha \to 0^+$, we get a change in the charge density, $\delta\rho(\mathbf{r}, t)$ that satisfies Poisson equation. Taking the external potential with amplitude V, and defining $\mathbf{D}(\mathbf{q}, \omega) = \epsilon(\mathbf{q}, \omega)\mathbf{E}(\mathbf{q}, \omega)$ and using the usual definition $u = V/\epsilon_r(\mathbf{q}, \omega)$, we obtained the Lindhard function written as

$$\epsilon_r = 1 - V(q)\Pi, \tag{14.235}$$

where $V(q)$ is the Fourier transform of the Coulomb potential

$$V(q) = \frac{e^2}{\epsilon_0 q^2} \tag{14.236}$$

and $\Pi = -\chi(q)$, also called polarization, is given by

$$\Pi = \frac{1}{\Omega}\sum_{\mathbf{k}} \frac{f(\epsilon(\mathbf{k})) - f(\epsilon(\mathbf{k}+\mathbf{q}))}{\epsilon(\mathbf{k}) - \epsilon(\mathbf{k}+\mathbf{q}) + \hbar\omega - i\hbar\alpha}, \tag{14.237}$$

while the electric polarization is the response to an applied electric field through $P = \epsilon_0\chi_e E$, as introduced before. Combining Eqs. (14.187), (14.197), and (14.214) the previous relations follow. Note that $\chi_e(q) = V(q)\chi(q)$.

As discussed before, when $\epsilon \to 0$ (or $1/\epsilon \to \infty$), we find the normal modes of the system, since we may get a finite response (finite u), even if V is infinitesimal. For instance, at small q and at the plasma frequency, we get that $\epsilon(\mathbf{q}, \omega)$ is zero (both real and imaginary parts), and, as we showed, we get a finite electric field without applied electric field giving rise to the appearance of plasmons. On the other hand, if $\epsilon \to \infty$ and therefore $u \to 0$, the forces between charges vanish and we get a structural instability and a consequent phase transition that is generally associated with some divergent susceptibility, if it is of second order. As we showed before, the condition of $\epsilon \to 0$ not only provides the condition for the plasma frequency (to leading order in $1/\omega$), but going to the next order also provides the dispersion relation of the plasmons up to the merging into the Stoner continuum.

An alternative way to discuss the results is to adopt a description that is focused on the dynamical structure factor, a density-density correlation. Let us again consider the response of the electron gas to an external charge density $\rho_{ext}(\mathbf{x}, t)$

$$H_{ext} = \int d^3x \int d^3x' \rho(\mathbf{x}) V(\mathbf{x} - \mathbf{x}') \rho_{ext}(\mathbf{x}', t). \tag{14.238}$$

The total charge density is then

$$\begin{aligned} \langle \rho_{tot}(\mathbf{q}, \omega) \rangle &= \rho_{ext}(\mathbf{q}, \omega) + \langle \rho_{ind}(\mathbf{q}, \omega) \rangle \\ &= \frac{\rho_{ext}(\mathbf{q}, \omega)}{\epsilon_r(\mathbf{q}, \omega)}. \end{aligned} \tag{14.239}$$

The total Hamiltonian is therefore $H_{tot} = H + H_{ext}$ and the induced charge density is

$$\begin{aligned}\langle \rho_{ind}(\mathbf{x}, t)\rangle &= \langle \psi(t)|\rho_{\mathbf{x}}|\psi(t)\rangle \\ &= \int_{-\infty}^{\infty} dt' \int d^3x' \int d^3x'' K(\mathbf{x}-\mathbf{x}'; t-t') V(\mathbf{x}'-\mathbf{x}'')\rho_{ext}(\mathbf{x}'', t'),\end{aligned} \tag{14.240}$$

where, by linear response theory, there is causality and

$$K(\mathbf{x}-\mathbf{x}'; t-t') = -i\theta(t-t')\langle[\rho(\mathbf{x}, t), \rho(\mathbf{x}', t')]\rangle. \tag{14.241}$$

In momentum space we get

$$\begin{aligned}\langle \rho_{ind}(\mathbf{q}, t)\rangle &= \int_{-\infty}^{\infty} dt' K(\mathbf{q}, t-t') V(\mathbf{q})\rho_{ext}(\mathbf{q}, t') \\ K(\mathbf{q}, t-t') &= -i\theta(t-t')\langle[\rho(\mathbf{q}, t), \rho(-\mathbf{q}, t')]\rangle,\end{aligned} \tag{14.242}$$

assuming spatial translational invariance. Taking the time Fourier transform we can write

$$\langle \rho_{ind}(\mathbf{q}, \omega)\rangle = V(\mathbf{q})K(\mathbf{q}, \omega)\rho_{ext}(\mathbf{q}, \omega), \tag{14.243}$$

which leads to

$$\frac{1}{\epsilon_r(\mathbf{q}, \omega)} = 1 + V(\mathbf{q})K(\mathbf{q}, \omega). \tag{14.244}$$

In this expression we see now that if $\epsilon \to 0$, then the kernel diverges, $K(\mathbf{q}, \omega) \to \infty$. This kernel is a density-density correlation function (dynamic structure factor). It is like a susceptibility and so we obtain the result that the normal modes, associated to instabilities, are associated to a divergent quantity, a susceptibility like quantity. This will be discussed in detail in the following chapters.

In the RPA the kernel may be written as

$$K(\mathbf{q}, \omega) = \frac{\Pi(\mathbf{q}, \omega)}{1 - V(\mathbf{q})\Pi(\mathbf{q}, \omega)}. \tag{14.245}$$

Indeed we may now conclude that

$$\frac{1}{\epsilon_r} = 1 + VK = 1 + \frac{V\Pi}{1 - V\Pi} = \frac{1 - V\Pi + V\Pi}{1 - V\Pi} = \frac{1}{1 - V\Pi} \tag{14.246}$$

and we recover that $\epsilon_r(\mathbf{q}, \omega) = 0$ is the same as $V(\mathbf{q})\Pi(\mathbf{q}, \omega) = 1$. Here, it can be shown that, at zero temperature,

$$\Pi(\mathbf{q}, \omega) = \sum_{p<p_F;|p+q|>p_F} \left(\frac{1}{\hbar\omega - \hbar\omega_q(p) + i\hbar\alpha} - \frac{1}{\hbar\omega + \hbar\omega_q(p) + i\hbar\alpha} \right), \tag{14.247}$$

where $\hbar\omega_q(p) = \epsilon(p+q) - \epsilon(p)$. The condition of $1/\epsilon_r \to \infty$ then leads to solutions that contain the plasmon dispersion relation but also the Stoner continuum.

14.7 Exercises

1. Diagonalize the Hamiltonian

$$H = w_a a^\dagger a + w_b b^\dagger b + V\left(b^\dagger a + a^\dagger b\right).$$

Consider the operators as being both bosonic and both fermionic.

Use two solution methods:

(a) Obtain the equation of motion of the operators a and b and their Hermitian conjugates. Verify that the time evolution involves a linear combination of the operators. Look for a solution in the form $a, b \sim e^{i\omega t}$ and determine the possible frequency values.

(b) Define new operators as

$$\alpha = ua + vb,$$

$$\alpha^\dagger = ua^\dagger + vb^\dagger.$$

The operators α and $\alpha^\dagger$ are bosonic and fermionic, respectively, and therefore

$$u^2 + v^2 = 1.$$

The condition that the Hamiltonian is diagonalized, that is,

$$H = \omega\alpha^\dagger\alpha,$$

implies that

$$[\alpha, H] = \omega\alpha.$$

Use this equation to determine the spectrum and the eigenstates of the Hamiltonian. The relation between the original operators and the operators that diagonalize the Hamiltonian is called Bogoliubov-Valatin transformation.

2. Consider a helium atom and the interaction term between the two electrons as a perturbation.
   (a) Calculate the correction to the ground state energy using first order perturbation theory.
   (b) Calculate the same correction using the Ritz variational method choosing as trial function the product of two wave functions of the hydrogen atom and taking as variational parameter the nucleus charge. Calculate the effective charge of the nucleus and interpret the result. Compare the result of a) with the experimental result.
   (c) Show that the integrals

$$K_{n,l} = \frac{e^2}{4\pi\epsilon_0} \int \mathrm{d}^3 r_1 \int \mathrm{d}^3 r_2\, \phi^*_{100}(\mathbf{r_1})\phi^*_{nl0}(\mathbf{r_2}) \frac{1}{|\mathbf{r_1} - \mathbf{r_2}|} \phi_{nl0}(\mathbf{r_1})\phi_{100}(\mathbf{r_2}),$$

   are positive if $l = n - 1$.
   (d) Determine the spin state of the electrons in the ground state and in the first excited state.

3.(a) The electronic configuration of carbon is $1s^2 2s^2 2p^2$. Show that the valence state can be written as ${}^3P_0$ where we are using the notation ${}^{2S+1}L_J$ with S the spin, L the orbital angular momentum, and J the total angular momentum and the set $L = 0, 1, 2, 3, \cdots$ is represented by $S, P, D, F, \cdots$.
   (b) Show that in the case of nitrogen ($1s^2 2s^2 2p^3$) the valence state is ${}^4S_{3/2}$.
   (c) Show that in the case of Mn^{2+} we may have $J = 5$ and in which conditions.

4. The orbital magnetic moment is given by

$$\mu_L = \frac{e\hbar}{2m}\sqrt{L(L+1)}$$

   and the spin magnetic moment is given by

$$\mu_S = \frac{e\hbar}{m}\sqrt{S(S+1)}.$$

   The total magnetic moment is given by

$$\boldsymbol{\mu} = \boldsymbol{\mu}_L + \boldsymbol{\mu}_S.$$

   This vector is not parallel to $\mathbf{J} = \mathbf{L} + \mathbf{S}$. The projection of this vector in the direction of $\mathbf{J}$ is called μ_J. Show that

$$\mu_J = \frac{e\hbar}{2m}\sqrt{J(J+1)}g,$$

where the Landé factor g is given by

$$g = 1 + \frac{J(J+1) + S(S+1) - L(L+1)}{2J(J+1)}.$$

5. Consider three operators A, B, and C. Show that

$$[A, BC] = [A, B]C - B[C, A] = \{A, B\}C - B\{C, A\}.$$

6. Consider a ring where free electrons may move. We apply a magnetic field that gives origin to a magnetic flux Φ across the ring. The ring may be considered as a one-dimensional system with periodic boundary conditions.

   (a) Determine how the boundary conditions are changed due to the presence of the magnetic flux. (Recall that in the presence of a magnetic field, a term appears that gives origin to a phase in the wave function of the form $e^{-\frac{ie}{\hbar}\int d\mathbf{l}\cdot\mathbf{A}}$ where $\mathbf{A}$ is the vector potential.)

   (b) Performing a local gauge transformation show that in the context of the tight-binding model this can be written as

$$H = -\sum_{i,j,\sigma} t_{i,j} c_{i,\sigma}^{\dagger} c_{j,\sigma},$$

   where

$$t_{i,j} = t e^{-\frac{ie}{\hbar}a(i-j)A}.$$

7(a) Use the formalism of second quantization and write the Hamiltonian of a gas of free electrons. Write now an expression for the Hamiltonian subtracted by the ground state energy. Defining

$$\epsilon(\mathbf{k}) = E(\mathbf{k}) - \mu,$$

where $E(\mathbf{k})$ is the kinetic energy of the electrons, and separating the expression for the energy of the system subtracted by the energy of the ground state in two terms, one that respects to the absolute values of the momentum $k > k_F$ and another that respects to momenta such that $k < k_F$, where k_F is the Fermi momentum, identify the contributions that respect to the particles and to the holes. That is, show that

$$E - E_0 = \sum_{k>k_F,\sigma} \epsilon(\mathbf{k}) n_{\mathbf{k},\sigma} + \sum_{k<k_F,\sigma} |\epsilon(\mathbf{k})|(1 - n_{\mathbf{k},\sigma}),$$

identify the various terms, and discuss them. Repeat the same type of exercise for the total momentum of the electron gas.

(b) Finally, discuss the elementary excitations of the system. Discuss also the excitation spectrum.

8. A bosonic system interacts in the form of a density-density potential in the form

$$H = \sum_k \epsilon_k c_k^\dagger c_k + \int d^3r \int d^3r' V(\mathbf{r} - \mathbf{r}') n(\mathbf{r}) n(\mathbf{r}').$$

Obtain the Hartree-Fock equations for this system.

9. Consider a one-dimensional electronic system described by the Hamiltonian

$$H = \sum_{k,\sigma} \epsilon_k c_{k,\sigma}^\dagger c_{k,\sigma} + \sum_q{}' \left(\sum_{kk'} \sum_{\sigma\sigma'} \frac{e^2}{2\epsilon_0 q^2} c_{k+q,\sigma}^\dagger c_{k'-q,\sigma'}^\dagger c_{k',\sigma'} c_{k,\sigma} \right). \tag{14.248}$$

Assume that the interaction terms, $V(q) = (e^2/(\epsilon_0 q^2))$, that respect to an exchange of momentum $q = 2k_F$ are the most important.

(a) Rewrite the Hamiltonian isolating these terms.

(b) We want to use a mean-field approximation to diagonalize the Hamiltonian. Defining an order parameter as

$$\Delta = V(2k_F) \sum_{k,\sigma} \langle c_{k+2k_F,\sigma}^\dagger c_{k,\sigma} \rangle, \tag{14.249}$$

obtain the expression for the Hamiltonian in mean field.

(c) Obtain the equations of motion of the fermionic operators.

Recall that $[A, BC] = [A, B]C - B[C, A] = \{A, B\}C - B\{C, A\}$ and

$$i\frac{d}{dt}A(t) = [A, H].$$

(d) If the parameter Δ has a non-vanishing value, what does it represent physically? Hint: carry out the Fourier transformation into real space.

10. Consider the Lindhard function

$$\epsilon_r(\mathbf{q}, \omega) = 1 - \frac{e^2}{\epsilon_0 q^2} \frac{1}{\Omega} \sum_{\mathbf{k}} \frac{f(\epsilon_{\mathbf{k}}) - f(\epsilon_{\mathbf{k}-\mathbf{q}})}{\epsilon_{\mathbf{k}} - \epsilon_{\mathbf{k}-\mathbf{q}} + \hbar\omega - i\hbar\alpha}.$$

(a) Consider $\omega = 0$, $\mathbf{q} \to 0$. Show that

$$\epsilon_r(\mathbf{q} \to 0, 0) = 1 + \frac{e^2}{\epsilon_0 q^2} N(\epsilon_F),$$

where $N(\epsilon_F)$ is the density of states per volume at the Fermi level.

(b) Consider now $\omega = 0$ but arbitrary moment. Show that you may write that

$$\epsilon_r(\mathbf{q}, 0) = 1 - \frac{e^2}{\epsilon_0 q^2} \frac{1}{\Omega} \sum_{\mathbf{k}} \frac{f(\epsilon_{\mathbf{k}})}{\epsilon_{\mathbf{k}} - \epsilon_{\mathbf{k}-\mathbf{q}}},$$

considering free electrons and that the energy is even. Converting the sum into an integral show that

$$\epsilon_r(\mathbf{q}, 0) = 1 + \frac{e^2}{\epsilon_0 q^2} \frac{m}{2\pi^2 \hbar^2} k_F \left(\frac{1}{2} + \frac{1}{2qk_F} (k_F^2 - \frac{q^2}{4}) \log |\frac{2k_F + q}{2k_F - q}| \right).$$

(c) Consider now large ω and small q. Show that

$$\epsilon_r(\mathbf{q}, \omega) = 1 - \frac{\omega_p^2}{\omega^2},$$

where ω_p is the plasma frequency.

(d) Consider an insulator. In this case the Lindhard function is generalized to include contributions from several bands

$$\epsilon_r(\mathbf{q}, \omega) = 1 - \frac{e^2}{\epsilon_0 q^2} \frac{1}{\Omega} \sum_{\mathbf{k},\mathbf{g}} \frac{|\langle \mathbf{k}| e^{i\mathbf{q}\cdot\mathbf{r}} |\mathbf{k} + \mathbf{q} + \mathbf{g}\rangle|^2 \left(f(\epsilon_{\mathbf{k}}) - f(\epsilon_{\mathbf{k}+\mathbf{q}+\mathbf{g}}) \right)}{\epsilon_{\mathbf{k}} - \epsilon_{\mathbf{k}+\mathbf{q}+\mathbf{g}} + \hbar\omega - i\hbar\alpha},$$

where $\mathbf{g}$ are reciprocal lattice vectors. Consider in particular the contribution from the smaller reciprocal lattice vector. Show that

$$\epsilon_r(\mathbf{q} \to 0, 0) = 1 + \left(\frac{\hbar\omega_p}{\epsilon_{gap}} \right)^2,$$

where ϵ_{gap} is the energy difference between the valence band and the conduction band.

Start by showing that

$$\frac{1}{2}[[H, e^{i\mathbf{q}\cdot\mathbf{r}}], e^{-i\mathbf{q}\cdot\mathbf{r}}] = -\frac{\hbar^2 q^2}{2m},$$

considering a free Hamiltonian, H.

Considering the matrix element of this expression in an eigenstate of the Hamiltonian (defined by $H|s\rangle = E_s|s\rangle$), show that

$$\frac{\hbar^2 q^2}{2m} = \sum_n (E_n - E_s) |\langle n| e^{i\mathbf{q}\cdot\mathbf{r}} |s\rangle|^2,$$

where $|n\rangle$ is the complete set of eigenstates of H. Apply this expression considering that the states $|s\rangle$ and $|n\rangle$ may be identified with the states $|\mathbf{k}\rangle$ and $|\mathbf{k}+\mathbf{q}+\mathbf{g}\rangle$, respectively.

11. Lindhard's function can be written as

$$\epsilon_r(\mathbf{q},\omega) = 1 - \frac{e^2}{\epsilon_0 q^2}\frac{1}{\Omega}\sum_{\mathbf{k}} \frac{f(\epsilon_{\mathbf{k}}) - f(\epsilon_{\mathbf{k}-\mathbf{q}})}{\epsilon_{\mathbf{k}} - \epsilon_{\mathbf{k}-\mathbf{q}} + \hbar\omega - i\hbar\alpha}.$$

(a) Consider a static perturbation ($\omega = 0$). Obtain an expression for the dielectric function in the limit of small momenta. Show that the Coulomb potential of a point charge is screened. That is, show that the bare Coulomb potential is replaced in a metal by the Yukawa potential.
(b) Consider now a strictly two-dimensional system. The Coulomb potential in real space is of the type $V(r) = a/r$, where r is the distance in the plane from the point charge. Show that the Fourier transform is of the form $V(q) = 2\pi a/q$ where q is the absolute value of the two-dimensional momentum.

$$\int_0^{2\pi} \frac{p dx}{p^2 + q^2\cos(x)^2} = \frac{2\pi}{\sqrt{p^2+q^2}},$$

$$\int_0^{2\pi} \frac{p\cos(x)dx}{p^2 + q^2\cos(x)^2} = 0.$$

(c) Note that the dielectric function can be written as

$$\epsilon(q,\omega) = 1 - V(q)\Pi(q,\omega),$$

where we can write that

$$\Pi(q,\omega) = \frac{1}{\Omega}\sum_{\mathbf{k}} \frac{f(\epsilon_{\mathbf{k}}) - f(\epsilon_{\mathbf{k}-\mathbf{q}})}{\epsilon_{\mathbf{k}} - \epsilon_{\mathbf{k}-\mathbf{q}} + \hbar\omega - i\hbar\alpha}.$$

Derive the momentum dependence of the dielectric function in the case of $\omega = 0$ and small momenta. Is there screening as in the three-dimensional case?
(d) Obtain the dispersion relation of the plasmons in the limit of small momenta in the two-dimensional problem. You may expand in powers of the frequency. Show that the plasmons are gapless.

MIX
Papier aus verantwortungsvollen Quellen
Paper from responsible sources
FSC® C105338

If you have any concerns about our products,
you can contact us on
ProductSafety@springernature.com

In case Publisher is established outside the EU,
the EU authorized representative is:
Springer Nature Customer Service Center GmbH
Europaplatz 3, 69115 Heidelberg, Germany

Printed by Libri Plureos GmbH
in Hamburg, Germany

Introduction to Condensed Matter Theory

Pedro D. Sacramento

Introduction to Condensed Matter Theory

Volume 2

Springer

Pedro D. Sacramento
Instituto Superior Técnico
University of Lisbon
Lisbon, Portugal

ISBN 978-3-031-86250-2 ISBN 978-3-031-86251-9 (eBook)
https://doi.org/10.1007/978-3-031-86251-9

This Springer imprint is published by the registered company Springer Nature Switzerland AG
The registered company address is: Gewerbestrasse 11, 6330 Cham, Switzerland

Preface

This book grew out of lecture notes of several courses taught over the years in the area of Condensed Matter Physics. Specifically, it covers aspects of Thermodynamics, and the subjects lectured in the courses of Statistical Physics, Solid State Physics, Condensed Matter Physics, Phase Transitions and Critical Phenomena, and Topics in Condensed Matter Physics. The first three courses are taught in the Undergraduate Program (first three years of the Physics course at Instituto Superior Técnico, previously part of the Technical University of Lisbon and now part of the University of Lisbon) and the last three courses are taught in the Master's program (last two years of the Physics course).

Due to its comprehensive nature and associated size, the printed book is divided in two volumes but it is intended as a unit. The first volume comprises the first fourteen Chapters and the second volume comprises Chapters 15 to 25.

Several books have been used to support the studies of the various materials covered and have a significant influence on this book. The main references used are

[Callen] H.B. Callen, *Thermodynamics*, Wiley
[Fermi] E. Fermi, *Thermodynamics*, Dover
[Reif] F. Reif, *Fundamentals of Statistical and Thermal Physics*, McGraw-Hill
[Huang] K. Huang, *Statistical Mechanics*, Wiley
[Pathria] R.K. Pathria, *Statistical Mechanics*, Pergamon
[Stanley] E. Stanley, *Introduction to Phase Transitions and Critical Phenomena*, Oxford University Press
[Ma] S.K. Ma, *Modern Theory of Critical Phenomena*, Bemjamin/Cummings
[Kittel] C. Kittel, *Introduction to Solid State Physics*, Wiley
[AshcroftMermin] N.W. Ashcroft and N. D. Mermin, *Solid State Physics*, Harcourt College Publishers
[Ziman] J.M. Ziman, *Principles of the Theory of Solids*, Cambridge University Press
[KittelQTS] C. Kittel, *Quantum Theory of Solids*, Wiley
[Madelung] O. Madelung, *Introduction to Solid-State Theory*, Springer-Verlag
[Gasiorowicz] S. Gasiorowiz, *Quantum Physics*, Wiley
[Merzbacher] E. Merzbacher, *Quantum Mechanics*, Wiley

[BicudoSacramento] P. Bicudo, P. D. Sacramento, M. Cardoso, M. M. Correia, S. F. Martins, R. Monteiro, M. F. Paulos, J. G. Rosa and J. E. Santos, *Mecânica Quântica*, IST Press (in Portuguese)
[Amit] D.J. Amit, *Field Theory, the Renormalization Group and Critical Phenomena*, World Scientific
[Ravndal] Finn Ravndal, *Scaling and Renormalization Groups*, Nordita
[BurkhardtLeeuwen] T.W. Burkhardt and J.M.J. van Leeuwen (editors), *Real-Space Renormalization*, Springer-Verlag
[Reichl] L.E. Reichl, *A Modern Course in Statistical Physics*, University of Texas Press
[Goldenfeld] N. Goldenfeld, *Lectures on Phase Transitions and the Renormalization Group*, Perseus Books Publishing
[ChaikinLubensky] P.M. Chaikin and T.C. Lubensky, *Principles of Condensed Matter Physics*, Cambridge University Press
[Schrieffer] J.R. Schrieffer, *Theory of Superconductivity*, Benjamin/Cummings
[deGennes] P. G. de Gennes, *Superconductivity of Metals and Alloys*, Addison-Wesley
[GuntonDroz] J.D. Gunton and M. Droz, *Introduction to the Theory of Metastable and Unstable States*, Springer-Verlag
[Continentino] M.A. Continentino, *Quantum Scaling in Many-Body Systems*, World Scientific
[Sachdev] S. Sachdev, *Quantum Phase Transitions*, Cambridge University Press
[AraujoSacramento] M.A.N. Araújo and P.D. Sacramento, *Topologia em Matéria Condensada: Uma Introdução*, IST Press (in Portuguese) and its English translation *Topology in Condensed Matter: An Introduction*, World Scientific

Several books with problems have been useful over the years such as [Mihaly] L. Mihály and M.C. Martin *Solid State Physics Problems and Solutions*, Wiley, the collection [PhDQualifying] on *Problems and Solutions of Major American Universities PhD Qualifying Questions and Solutions*, World Scientific, specifically on *Thermodynamics and Statistical Mechanics*, on *Solid State Physics, Relativity and Miscellaneous Topics*, and on *Electromagnetism*, World Scientific, and [Dalvit] D.A.R. Dalvit, J. Frastai, I. Lawrie *Problems on Statistical Mechanics*, Institute of Physics Publishing.

Also, lecture notes from the classes of various teachers and, in general, physics discussions have been used in parts of the book such as by Vitor R. Vieira, Pedro Schlottmann, James D. Gunton, and Theodore W. Burkhardt as well as some personal notes, thesis, or exercises by Iveta Pimentel, Carlos Matos Ferreira, Ana Noronha, Kong-Ju Bock Lee, Nuno Peres, Eduardo Castro, Miguel Gonçalves, Ana Martins, and José Luís Martins.

Discussions over the years with Eduardo Ducla Soares, Dieter Forster, Miguel Araújo, Nuno Peres, Zlatko Tesanovic, Vitalii Dugaev, Eduardo Castro, Pedro Ribeiro, José Carmelo, Dionys Baeriswyl, Pedro Bicudo, Alexander Nersesyan, and more recently with Tilen Cadez, Rubem Mondaini, Hai-Qing Lin, Wing Chi Yu, Nikola Paunkovic, Ricardo Dias, Henrik Johannesson, Natan Andrei, Mucio Continentino, Bruno Mera, and Bruno Amorim are also acknowledged.

The subjects selected are intended so that the book may be used as a textbook on materials in the area of Condensed Matter Physics, up to the Master degree level. Several exercises are proposed at the end of each chapter. In order to keep the mathematical formalism as simple as possible, more advanced topics, that are most easily dealt with using many-body theory, have been avoided or some alternative, less technical, approach has been adopted. More advanced topics and formalism that are not in general covered in this book may be pursued in various references such as

[AGD] A.A. Abrikosov, L.P. Gorkov and I.E. Dzyaloshinski, *Methods of Quantum Field Theory in Statistical Physics*, Dover

[Mahan] G.D. Mahan, *Many-Particle Physics*, Plenum Press

[FetterWalecka] A.L. Fetter and J.D. Walecka, *Quantum Theory of Many-Particle Systems*, Dover

[DoniachSondheimer] S. Doniach and Sondheimer, *Green's Functions for Solid State Physics*, World Scientific

[Fradkin] E. Fradkin, *Field Theories of Condensed Matter Systems*, Addison-Wesley

[PinesNozieres] D. Pines and P. Nozières, *The Theory of Quantum Liquids*, Addison-Wesley

[BernevigHughes] B.A. Bernevig with T. L. Hughes, *Topological Insulators and Topological Superconductors*, Princeton University Press

[Coleman] P. Coleman, *Introduction to Many-Body Physics*, Cambridge University Press

[Hewson] A.C. Hewson, *The Kondo Problem to Heavy Fermions*, Cambridge University Press

Most results presented in the book are derived with some detail but there are some topics that are more advanced or technically more challenging, whose results are presented and discussed without a full derivation and may be considered as better suited as an introduction to a course on Topics in Condensed Matter. Even though the book covers many topics, there are significant omissions, such as a description of liquid systems, including liquid crystals and superfluids, time-dependent phenomena, and a thorough description of systems far from equilibrium (even though some topics are addressed), interface phenomena such as junctions of different materials, spintronics, theory of electronic energy bands, etc. Also, the book is, for the most part, omissive along the direction of the properties and applications of materials and devices, mainly for reasons of space. A brief survey of some materials and their applications is presented in the closing chapter of the book. The reader is encouraged to explore different directions in the wide range of topics in the rich field of Condensed Matter materials.

Lisbon, Portugal
November 2024

Pedro D. Sacramento

Declarations

Competing Interests The author has no competing interests to declare that are relevant to the content of this manuscript.

Contents

Volume 1

Volume 2

Chapter 15
Phase Transitions

15.1 Collective Phenomena: Symmetry Breaking

Some of the dramatic effects of the interactions between particles are collective phenomena. In certain circumstances, even though the interactions may be weak, long-range correlations are established, leading in many cases to phase transitions. In general, in a phase transition, there is a symmetry change in the system that may be discrete or continuous. In general, the high-temperature phases are symmetrical phases (disordered), in which all the symmetries of the system are preserved. As one lowers the temperature, there occur spontaneous symmetry breakings in relation to the disordered phase. Such changes may occur in different ways, leading to sharp phase transitions (first-order transitions) or continuous (second order and higher orders) being possible however that derivatives of the free energy may be discontinuous (see ahead). Many of these transitions are conditioned by general theorems. We will consider in this chapter some examples of phase transitions and refer some theorems that characterize the various possible transitions.

The following are examples of phase transitions:

- At $T = 0$ and at a pressure of 1 atm, water becomes unstable and separates into two phases, a solid one (ice) and a liquid one (liquid water), that coexist in equilibrium.
- At a temperature of $T = 2.19$ Kelvin and at a pressure of 1 atm, liquid helium separates in two phases I and II.
- Structural transformations in solids such as change from tetragonal to cubic transitions.
- Evaporation in liquids.
- Magnetic transitions.
- Electric transitions.
- Superconductivity and superfluidity.

P. D. Sacramento, *Introduction to Condensed Matter Theory*,
https://doi.org/10.1007/978-3-031-86251-9_15

In general, at a phase transition, there are changes in the values of some variables. A particular type of variable that provides us a useful description of a phase transition is called order parameter. Typically, it is defined such that in one phase (say a high-temperature disordered phase), it vanishes, while in another phase (typically a lower-temperature phase, called ordered), it takes a finite value. The behavior of such a type of variable allows a distinction between phases. A possible classification of the phase transitions focuses on the types of possible changes of different variables, including those of the order parameters.

A transition is called of first order if the thermodynamic potentials are continuous across the transition, but their first derivatives are discontinuous. For instance, the chemical potential is continuous and its derivatives are discontinuous:

$$\left(\frac{\partial \mu}{\partial T}\right)_p = -s$$
$$\left(\frac{\partial \mu}{\partial p}\right)_T = v. \tag{15.1}$$

Therefore, the entropy per particle and the volume per particle (inverse density) are discontinuous. In the particular case of the order parameters, in this type of transition, there is a discontinuity in the order parameter when the transition occurs.

A transition is called second order if the thermodynamic potentials and their derivatives are continuous but their second derivatives are discontinuous (such as specific heats and susceptibilities). In particular, in the case of an order parameter, typically the change from a zero value to a finite value occurs in a continuous way and, at exactly the phase transition point, the value of the order parameter is still zero, and successively for higher-order transitions.

In most cases of interest, the classifications into the first two types of transitions is enough for our purposes. Usually, at a second-order phase transition, the discontinuities of the second derivatives are infinite. In some cases, they turn out to be finite. As will be discussed later, the approach to the critical point, where the second-order transition occurs, involves large fluctuations of the correlations in the system that allow, in the thermodynamic limit, that the system is able to probe the phase space of variables and change its properties, such as the symmetry of the system.

A typical example of a first-order transition is the transition between a liquid and its vapor. While to some approximation a gas (at least a dilute one) is a non-interacting system, interactions in a liquid are not negligible. Therefore, the role of interactions has to be taken into account. A simple way to introduce these interactions leads to the van der Waals gas.

15.2 van der Waals Gas: Virial Expansion

Let us consider a monoatomic gas with N atoms of mass m. Consider the system in thermodynamic equilibrium at a temperature T occupying a volume V and consider the gas to have a small density $n = N/V$. A classical statistics description is adequate for our purposes in this context.

Take as Hamiltonian of the system[1]

$$H = K + U$$
$$K = \frac{1}{2m}\sum_{j=1}^{N}\mathbf{p}_j^2$$
$$U = \frac{1}{2}\sum_{j=1}^{N}\sum_{k=1}^{N} u_{jk}(j \neq k). \tag{15.2}$$

Let us choose as the two-body interaction potential the Lennard-Jones potential given by

$$u(r) = u_0\left(\left(\frac{r_0}{r}\right)^{12} - 2\left(\frac{r_0}{r}\right)^{6}.\right) \tag{15.3}$$

Let us approximate the potential by

$$u(r) = \infty, r < r_0$$
$$u(r) = -u_0\left(\frac{r_0}{r}\right)^{6}, r > r_0. \tag{15.4}$$

We can then calculate the partition function given by

$$Z = \frac{1}{N!}\int \dots \int e^{-\beta(K+U)}\frac{d^3p_1 \dots d^3p_N d^3r_1 \dots d^3r_N}{h^{3N}}$$
$$Z = \frac{1}{h^{3N}N!}\int \dots \int e^{-\beta K(p_1,\dots,p_N)}d^3p_1 \dots d^3p_N \int \dots \int e^{-\beta U(r_1,\dots,r_N)}d^3r_1 \dots d^3r_N. \tag{15.5}$$

The gaussian integrals are straightforward and are given by

$$\int_{-\infty}^{\infty} d^3p e^{-\frac{\beta}{2m}\mathbf{p}^2} = \left(\frac{2\pi m}{\beta}\right)^{3/2}. \tag{15.6}$$

[1] See also [Pathria].

Then, the partition function is reduced to

$$Z = \frac{1}{N!}\left(\frac{2\pi m}{h^2\beta}\right)^{\frac{3}{2}N} Z_U. \tag{15.7}$$

We are left to do the integration over the space coordinates included in

$$Z_U = \int ... \int e^{-\beta U(r_1,...,r_N)} d^3r_1...d^3r_N. \tag{15.8}$$

This is in general difficult to calculate. Clearly, if $U \to 0$ or if $\beta \to 0$ $(T \to \infty)$, the calculation is straightforward and simply gives that $Z_U = V^N$. However, if the density is small, we can approximate the integral by a series of powers in the density.

We can write that the average potential energy is given by

$$\bar{U} = \frac{\int e^{-\beta U} U d^3r_1...d^3r_N}{\int e^{-\beta U} d^3r_1...d^3r_N} = -\frac{\partial}{\partial\beta}\ln Z_U.$$

Integrating, we get that

$$\begin{aligned} \ln Z_U(\beta) &= N \ln V - \int_0^\infty \bar{U}(\beta')d\beta' \\ Z_U(\beta = 0) &= V^N. \end{aligned} \tag{15.9}$$

The average potential energy can be written as

$$\bar{U} = \frac{1}{2}N(N-1)\bar{u} \sim \frac{1}{2}N^2\bar{u},\ N \gg 1,$$

where $\bar{u}$ is the average potential energy per particle given by

$$\bar{u} = \frac{\int e^{-\beta u} u d^3r}{\int e^{-\beta u} d^3r}.$$

We assume that the motion of a pair of atoms is not correlated with the motion of the other atoms. The probability of the relative positional vector between two atoms, **R**, is proportional to

$$P(\mathbf{R}) \sim e^{-\beta u(\mathbf{R})} d^3R.$$

Note that we can write that

$$\bar{u} = -\frac{\partial}{\partial\beta}\ln\int e^{-\beta u} d^3R.$$

If $u \sim 0$, then the exponential is close to 1. Then,

$$\int e^{-\beta u} d^3R = \int \left[1 + (e^{-\beta u} - 1)\right] d^3R = V + I(\beta) = V\left(1 + \frac{I(\beta)}{V}\right), \tag{15.10}$$

where

$$I(\beta) = \int \left(e^{-\beta u} - 1\right) d^3R = \int_0^{\beta} \left(e^{-\beta u} - 1\right) 4\pi R^2 dR. \tag{15.11}$$

This integral is expected to be much smaller than V. Then,

$$\begin{aligned}\bar{u} &= -\frac{\partial}{\partial \beta}\left[\ln V + \ln\left(1 + \frac{I(\beta)}{V}\right)\right] \\ &= -\frac{1}{V}\frac{\partial I(\beta)}{\partial \beta}.\end{aligned} \tag{15.12}$$

Therefore,

$$\bar{U} = -\frac{1}{2}\frac{N^2}{V}\frac{\partial I(\beta)}{\partial \beta}.$$

We obtain then that the partition function is given by

$$\ln Z_U(\beta) = N \ln V + \frac{1}{2}\frac{N^2}{V} I(\beta). \tag{15.13}$$

The pressure is therefore given by

$$\begin{aligned}p &= \frac{1}{\beta}\frac{\partial \ln Z}{\partial V} \\ \frac{p}{k_B T} &= \frac{N}{V} - \frac{1}{2}\left(\frac{N}{V}\right)^2 I(\beta).\end{aligned} \tag{15.14}$$

The expansion for the pressure is of the form

$$\frac{p}{k_B T} = n + B_2(T)n^2 + B_3(T)n^3 + \cdots \tag{15.15}$$

called the virial expansion. The coefficient $B_2(T) = -\frac{1}{2}I(\beta)$. Using the simplified form of the potential Eq. (15.4), we can calculate this coefficient, considering a generalized power law decay with exponent s:

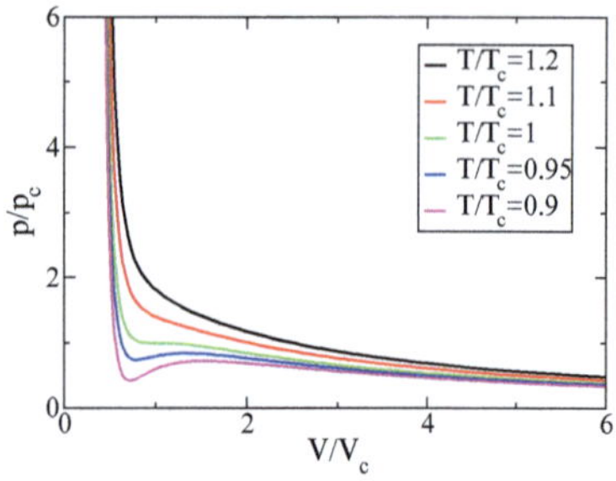

Fig. 15.1 Isotherms of the van der Waals gas: reduced pressure as a function of the reduced volume (ratio to the values at the critical point) for different reduced temperatures

$$
\begin{aligned}
B_2(T) &= -2\pi \int_0^\infty (e^{-\beta u} - 1)R^2 dR \\
u(R) &= \infty, R < R_0 \\
u(R) &= -u_0 \left(\frac{R_0}{R}\right)^s, R > R_0 \\
B_2(T) &= b' - \frac{a'}{k_B T},
\end{aligned}
\tag{15.16}
$$

where $b' = 2\pi/3R_0^3$ and $a' = 3/(s-3)b'u_0$. For $s > 3$, the integral is convergent, which is the case of the van der Waals attraction ($s = 6$). We can rewrite the equation in the form known as the van der Waals equation:

$$
\left(p + \frac{a}{v^2}\right)(v - b) = RT. \tag{15.17}
$$

Here, $a = N_A^2 a'$, $b = N_A b'$, $v = V/\nu$, and $R = N_A k_B$, where N_A is Avogadro's number and ν is the number of moles.

The isotherms of van der Waals equation are well-known[2] and represented in Fig. 15.1. At large temperatures, they are like those of the ideal gas. As the temperature lowers, they remain of the same type for temperatures $T > T_c$, and for temperatures $T < T_c$, they change qualitatively, since the dependence of the pressure as a function of volume is no longer monotonically decreasing. The critical temperature T_c is the value of the temperature that separates the two regimes. As is well-known from thermodynamics, if $(\partial p)/(\partial v)|_T < 0$, the system is in thermodynamic equilibrium, but otherwise the system is in an unstable state. Therefore, along an isotherm, at a temperature $T < T_C$, there is a region where the system is in an unstable state, and therefore one expects some sort of phase transition to a stable state (or at least a metastable state). Indeed, if a system is characterized by the intensive coordinates T, p, the thermodynamic function that characterizes

[2] For further reading see [Callen].

its equilibrium state and displacements from equilibrium is the Gibbs free energy $G = U - TS + pV = \mu N$. As is well-known,

$$d\mu = -sdT + vdp.$$

Along an isotherm $dT = 0$ and, therefore, $d\mu = vdp$. The equilibrium state is determined minimizing the chemical potential given as

$$\mu_X - \mu_O = \int_O^X vdp,$$

where O is a reference point and X is a running point along the isotherm. It is also convenient to plot the isotherm reversing the axis, with v as a function of p, as shown in Fig. 15.2.

It is possible to see (integrating the van der Waals equation along an isotherm) that points C and F (represented in Fig. 15.3) are such that the chemical potential $\mu_C = \mu_F$ and, therefore, the points where the phase transition takes place. Point C (and points A, B) is representative of a phase with low values of the volume per particle (high density) and the point F (and points G, H) is representative of a phase with high values of v (low density). Considering, for instance, point B (and point E),

Fig. 15.2 Isotherm of the van der Waals gas at $T < T_c$ in (**a**) a diagram $p - V$ and (**b**) in a diagram $V - p$

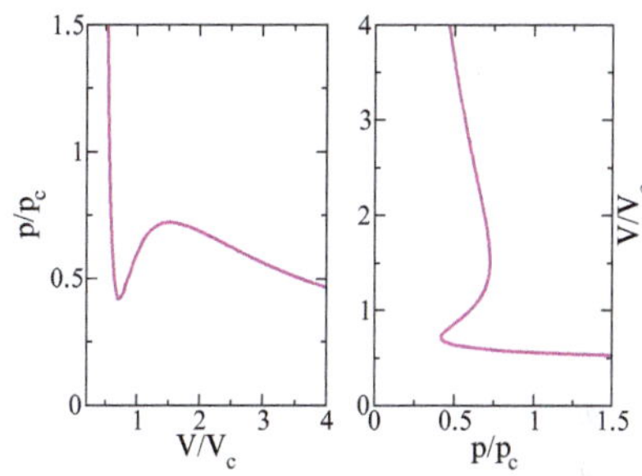

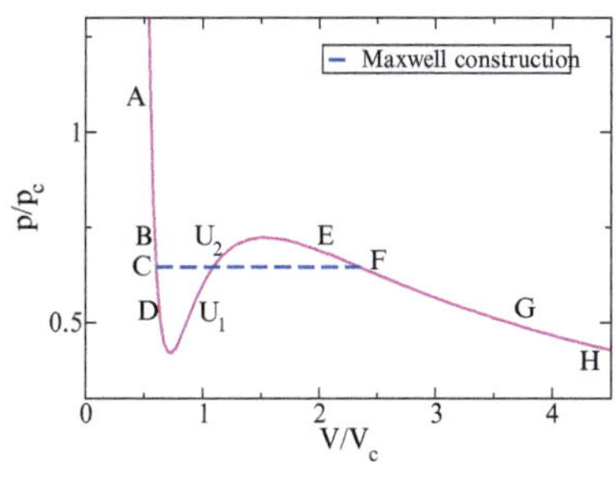

Fig. 15.3 Isotherm of the van der Waals gas at $T < T_c$. Points A, B are in the liquid phase and points G, H are in the gas phase. At point C, the transition takes place to the coexisting phase at point F. D and E are in metastable phases. Points U_1, U_2 represent unstable phases. The dashed line is the Maxwell construction

we see that the isotherm has three solutions (including a point in the unstable regime U_2). Comparing free energies, one can show that the free energy at point B is lower than at point E (and the free energy at point F is lower than at point C) indicating that the stable state is the one represented by point B and that the state represented by point E is metastable (a state that satisfies the conditions for thermodynamic equilibrium but with a free energy that is not the lowest). The state at point U_1 is also unstable. The unstable points are therefore excluded from the isotherm, and the calculation of the chemical potential allows to select the most stable states leading to the so-called Maxwell construction. There is therefore a phase transition between two phases, which occurs when the chemical potential has two solutions (in this case, two phases coexist; in other systems, more phases may coexist).

The van der Waals equation is usually regarded as a simple model for the liquid-gas transition. If the transition occurs at a temperature $T < T_c$, there is a discontinuity in the density of the system (and in other thermodynamic quantities) and the change of phase is abrupt (first order). If the temperature increases and approaches T_c, the discontinuity decreases in magnitude until, in the limit $T = T_c$, the discontinuity vanishes and the transition is continuous (second-order transition). For temperatures above the critical temperature, there is no true phase transition and only a crossover. When v is large, we have that $-(\partial p/\partial v)_T$ is small and the phase is compressible, a characteristic of a gas. When v is small, we have that $-(\partial p/\partial v)_T$ is large and the phase is incompressible, a characteristic of a liquid. Therefore, the simple model that leads to the van der Waals gas describes qualitatively a liquid-gas phase transition.

The van der Waals equation may be written as

$$pV^3 - (pb + RT)\,V^2 + aV - ab = 0, \tag{15.18}$$

a third-order equation. The critical point (inflexion point) is defined by

$$\begin{aligned} \left(\frac{\partial p}{\partial V}\right)_c &= 0 \\ \left(\frac{\partial^2 p}{\partial V^2}\right)_c &= 0. \end{aligned} \tag{15.19}$$

We may write that

$$\begin{aligned} p &= \frac{RT}{V-b} - \frac{a}{V^2} \\ \frac{\partial p}{\partial V} &= -\frac{RT}{(V-b)^2} + 2\frac{a}{V^3} \\ \frac{\partial^2 p}{\partial V^2} &= 2\frac{RT}{(V-b)^3} - 6\frac{a}{V^4}. \end{aligned} \tag{15.20}$$

This leads to

$$\frac{RT_c}{(V_c-b)^2} = \frac{2a}{V_c^3}$$
$$\frac{2RT_c}{(V_c-b)^3} = 6\frac{a}{V_c^4}, \tag{15.21}$$

which implies that the variables at the critical point are

$$V_c = 3b; \qquad p_c = \frac{a}{27b^2}; \qquad T_c = \frac{8}{27}\frac{a}{Rb}. \tag{15.22}$$

As a consequence, the critical point is completely characterized by the parameters of the van der Waals equation. Defining parameters renormalized by the values at the critical point

$$\mathcal{P} = \frac{p}{p_c}; \qquad \mathcal{V} = \frac{V}{V_c}; \qquad \mathcal{T} = \frac{T}{T_c}, \tag{15.23}$$

we can rewrite the van der Waals equation in an universal way, taking the same form for all substances as

$$\left(\mathcal{P} + \frac{3}{\mathcal{V}^2}\right)\left(\mathcal{V} - \frac{1}{3}\right) = \frac{8}{3}\mathcal{T}. \tag{15.24}$$

The isotherms are represented in Fig. 15.1.

Recalling Eq. (2.143) and rewriting

$$p = \frac{RT}{V-b} - \frac{a}{V^2}, \tag{15.25}$$

we obtain that

$$\left(\frac{\partial U}{\partial V}\right)_T = \frac{a}{V^2} \neq 0, \tag{15.26}$$

in contrast to the ideal gas case. Integrating, we get that

$$U = -\frac{a}{V} + f(T) \tag{15.27}$$

and, therefore, $U = U(T, V)$, reflecting the presence of the attractive potential between the particles we considered in the derivation of the van der Waals equation.

For temperatures smaller than T_c, we expect discontinuities due to the first-order transition. The example of the discontinuity in the volume per particle (or density) has been discussed. There is also a discontinuity in the entropy that leads to a latent heat at the transition. We can write that

$$ds = \left(\frac{\partial s}{\partial v}\right)_T dv. \tag{15.28}$$

Comparing two isotherms, one respecting to a temperature T and another respecting to a temperature $T+\Delta T$, selecting two points on the lower temperature isotherm, for instance, say C and F with the same chemical potential and therefore representatives of the transition in the Maxwell construction, we may write

$$\Delta s = s_F - s_C = \int \left(\frac{\partial p}{\partial T}\right)_v dv, \tag{15.29}$$

where we have used a Maxwell relation. Therefore, if ΔT is small, we obtain that the change in entropy is approximately the area between the two curves divided by the temperature difference. Associated with the change in entropy, we obtain a heat flux, called latent heat, which is given by

$$L = T\left(s_F - s_C\right). \tag{15.30}$$

Phase F is associated with a less ordered regime and therefore a higher entropy. Therefore, the latent heat is positive. This is the heat released as we cross from a phase less ordered (larger entropy, s) to another phase that is more ordered (smaller entropy, s). In the present case, as we change from a gas phase to the liquid phase, heat is released. When the opposite occurs, to transform from a liquid to a gas, we need to supply heat, and therefore heat is absorbed by the system.

The idea may be generalized to other situations such as the fusion heat (solid-liquid), vaporization heat (liquid-gas), and sublimation heat (solid-gas).

15.3 Gibbs Phase Rule

Let us consider a system with r components. The chemical potentials in each phase, i, are

$$\mu^i = \mu^i(T, p, x_1^i, x_2^i, \ldots, x_{r-1}^i), \tag{15.31}$$

where $i = 1, \ldots, M$, with M the number of phases, and where x_α^i are the fractions of component α in phase i. Therefore,

$$\sum_{\alpha=1}^{r} x_\alpha^i = 1. \tag{15.32}$$

The total number of independent variables is $2 + M(r - 1)$, since T, p may always be chosen as independent variables, plus $(r - 1)$ independent components for each of the M phases. There are $M - 1$ equations for each component of the type

$$\mu_\alpha^i = \mu_\alpha^j = \cdots = \mu_\alpha^M. \tag{15.33}$$

Therefore, there are $r(M - 1)$ equations between the variables. This leads that the total number of independent variables, f, is

$$f = (2 + M(r - 1)) - (r(M - 1)) = r - M + 2. \tag{15.34}$$

This is called the Gibbs phase rule. Its meaning is that it provides the number of thermodynamic degrees of freedom needed to specify the thermodynamic state of a system with r components in a state of coexistence of M phases.

The simplest example is a system with one component ($r = 1$) in a given phase ($M = 1$). In this case, $f = 2$, which means it is enough to specify any two variables to characterize the equilibrium state of a system. For instance, selecting intensive variables T, p, μ, any two of them are sufficient, since there is a relation between them such as $d\mu = -sdT + vdp$.

Another example is the case of one component and two phases coexisting. This corresponds to $r = 1$, $M = 2$ and $f = 1$. This means that one variable (a line of points) characterizes the coexistence of these two phases. Clearly, $r = 1$, $M = 3$ implies $f = 0$, which means that three phases of a system of one component only coexist at one point (triple point). The case $r = 1$ and $M = 4$ has no solution and therefore four phases cannot coexist. Increasing the number of components $r > 1$, the phase diagrams become increasingly complex. For instance, a binary alloy $r = 2$ allows coexistence regions that are surfaces. Taking $r = 2$, $M = 1$ leads to $f = 3$, and three variables are needed to specify a point in the phase diagram. Taking for $r = 2$ two phases $M = 2$ allows $f = 2$, and therefore a surface of coexistence.

15.4 Lee-Yang Theory

Let us consider the analytic behavior of the thermodynamic quantities.[3] The thermodynamic quantities are obtained taking derivatives of the grand partition function (in the grand canonical ensemble). For instance, the pressure or the specific volume can be obtained as

$$p(z, T) = k_B T \lim_{V\to\infty} \left(\frac{1}{V} \ln \mathcal{Z}\right)$$

[3] For further reading see [Pathria].

$$\frac{1}{v(z,T)} = \lim_{V\to\infty}\left(\frac{1}{V}\left(\frac{\partial \ln \mathcal{Z}}{\partial \ln z}\right)_{V,T}\right)$$
$$= \lim_{V\to\infty}\left(\frac{\partial}{\partial \ln z}\left(\frac{1}{V}\ln \mathcal{Z}\right)\right)_{V,T}. \tag{15.35}$$

If the derivative and the limit can be interchanged, then

$$\frac{1}{v(z,T)} = \frac{1}{k_B T}\left(\frac{\partial p(z,T)}{\partial \ln z}\right)_T. \tag{15.36}$$

Let us write $\mathcal{Z}(z, V, T)$, which is a polynomial of degree N_m, in terms of its zeros in the complex plane:

$$\mathcal{Z}(z, V, T) = \sum_{N=0}^{N_m} Q_N(V,T) z^N = \prod_{k=1}^{N_m}\left(1 - \frac{z}{z_k}\right), \tag{15.37}$$

where

$$\sum_{N=0}^{N_m} Q_N(V,T) z_k^N = 0.$$

The zeros of the grand partition function are $z_k = z_k(V, T)$ with $k = 1, 2, ..., N_m$ and are a function of the interactions between the particles through the partition function Q_N. The solutions are, in general, complex. If some are real, then they have to be negative because the partition function is positive definite. Since the complex roots of an algebraic equation with real coefficients must occur in pairs, this implies that if z_k is a root, then z_k^* is also a root. This implies that the distribution of roots is symmetrical around the real axis. The number of roots N_m is equal to the maximum number of particles allowed in the system. This is clearly proportional to the volume of the system. If the volume increases, then the number of roots will also increase, proportionally. Therefore, if $V \to \infty$, the number of roots $N_m \to \infty$. Therefore, the zeros may be represented by a distribution $g(z)$.

If the limiting distribution is such that some zeros converge to the real axis for positive values, at one or more points z_c, approaching the real axis as much as we want (but strictly never reaching it because there can be no positive zeros), then the analytic behavior of the function

$$F(z,T) = \lim_{V\to\infty}\left(\frac{1}{V}\ln \mathcal{Z}\right) = \frac{p(z,T)}{k_B T},$$

and of any other function derived from this one may become singular at the points $z = z_c$, along the positive real axis. The presence of such a singularity would

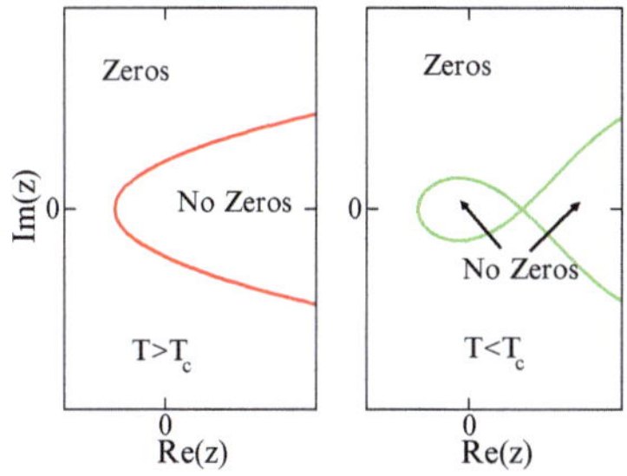

Fig. 15.4 Schematic representation of Lee-Yang zeros in the complex plane. In the left panel is represented a typical situation of a single phase above some critical regime: there are no zeros in the vicinity of the positive real axis. In the right panel, the vicinity around a point in the positive real axis has a non-vanishing density of states in the thermodynamic limit, signaling a singularity and a phase transition

imply the appearance of a phase transition in the system. For a given value of the volume V and for each physical value of z (real and positive), the function $\mathcal{Z}(z)$ increases monotonically with z. This implies that $\ln \mathcal{Z}$ and $\frac{1}{V} \ln \mathcal{Z}$ also increase monotonically with z.

These results lead to the first theorem of Yang and Lee:

(i) The limiting function $F(z, T)$ exists for all the real and positive values of z.
(ii) It is a continuous function, which increases monotonically with z.
(iii) It is independent of the shape of the system (unless the area grows faster than $V^{2/3}$).

For any given value of a finite volume, the function $\mathcal{Z}(z)$ is an analytical function of z, while $\frac{1}{V} \ln \mathcal{Z}$ is analytical everywhere except at the points z_k, where $\mathcal{Z} = 0$. Since none of the z_k is to be found on the real positive axis, the function $\frac{1}{V} \ln \mathcal{Z}$ and its derivatives are analytical in the whole positive and real axis. This implies that for all physical values of z, there can be no singularities in the thermodynamic functions, if the volume is finite.

This then leads to the second theorem of Yang and Lee:
If when $V \to \infty$ there are zeros that approach arbitrarily the real positive axis, then there is a phase transition, since any neighborhood, which can be chosen as small as we want, will always contain zeros. In Fig. 15.4, we illustrate schematically the physical situations of a disordered phase and a point that separates two phases.

15.5 One-Dimensional Ising Model

A model that was early proposed to describe a spontaneous magnetization was the Ising model. This model is a special case of so-called Heisenberg type models, which are models for interacting spin systems. These types of models reflect the exchange interactions between localized systems described by their internal angular momenta (spin). They are of the type

$$H = -\sum_{ij} \sum_{\alpha=x,y,z} J_{ij}^{\alpha} S_i^{\alpha} S_j^{\alpha} - \sum_{i} \sum_{\alpha} h_i^{\alpha} S_i^{\alpha}, \tag{15.38}$$

where $\mathbf{S}_i$ is a spin operator located at site i of a d-dimensional lattice and $\mathbf{h}_i$ is an external magnetic field applied at site i. If $J_{ij}^{\alpha} = J_{ij}$ with i, j nearest neighbors, this leads to the isotropic Heisenberg model; if $J_{ij}^{x} = J_{ij}^{y} = J$ and $J_{ij}^{z} = 0$, this is the XX model; if $J_{ij}^{x} \neq J_{ij}^{y}$ and $J_{ij}^{z} = 0$ is called the XY model; finally if $J_{ij}^{x} = J_{ij}^{y} = 0$ and $J_{ij}^{z} = J$, this is the Ising model. In the Heisenberg case, the order parameter is a vector moving in $3d$ (3 components); in the XX, XY models, it is a vector confined to a plane (2 components); and in the Ising model, it is scalar (1 component).

Ising proposed to solve the $1D$ Ising model to find ferromagnetism, choosing all $J_i > 0$, but it turns out, as we will see, that the model has no symmetry breaking at a finite temperature. However, in higher dimensions, there is a phase transition that has proven quite important and useful in various domains.

Let us concentrate for the time being on the $1D$ case (the two-dimensional case will be considered in a later chapter). The Ising model in one dimension can be written as[4]

$$H = -\sum_{i=1}^{N-1} J_i S_i S_{i+1}. \tag{15.39}$$

We have chosen a chain with N sites and, for convenience, the interaction depends on the site (at the end, we will take $J_i = J$). For the moment, we switch off the external magnetic field. The variables $S_i = \pm 1$. The partition function is simply written as

$$\begin{aligned} Z_N &= Z_N(J_1, J_2, ..., J_{N-1}) \\ &= \sum_{S_1=-1}^{+1} \sum_{S_2=-1}^{+1} ... \sum_{S_N=-1}^{+1} e^{\sum_{i=1}^{N-1} K_i S_i S_{i+1}} \\ &= \sum_{\{S_i\}} e^{\sum_{i=1}^{N-1} K_i S_i S_{i+1}}. \end{aligned} \tag{15.40}$$

Here, $K_i = \beta J_i$. We have 2^N spin configurations. We can solve the problem establishing a recurrence relation:

$$Z_{N+1} = \sum_{S_1=-1}^{+1} \sum_{S_2=-1}^{+1} ... \sum_{S_N=-1}^{+1} e^{\sum_{i=1}^{N-1} K_i S_i S_{i+1}} \sum_{S_{N+1}=-1}^{+1} e^{K_N S_N S_{N+1}}. \tag{15.41}$$

[4] See [Stanley].

The last sum can be performed easily. We get the recurrence relation:

$$Z_{N+1} = 2Z_N \cosh K_N. \tag{15.42}$$

Using that $Z_1 = 2$ (since there are only two states) and taking the case that $J_i = J$, we get

$$Z_N = 2^N \cosh^{N-1} K. \tag{15.43}$$

The correlation function

$$\langle S_k S_{k+r} \rangle = \frac{1}{Z_N} \sum_{\{S\}} S_k S_{k+r} e^{\sum_{i=1}^{N-1} K_i S_i S_{i+1}} = \Gamma_k(r), \tag{15.44}$$

is such that

$$Z_N \Gamma_k(r) = \frac{\partial}{\partial K_k} \frac{\partial}{\partial K_{k+1}} \cdots \frac{\partial}{\partial K_{k+r-1}} Z_N, \tag{15.45}$$

since $S_i^2 = 1$. Clearly, $\Gamma_k(1) = \tanh K_k$ and

$$\Gamma_k(r) = \prod_{i=1}^{r} \tanh K_{k+i-1}. \tag{15.46}$$

Once again taking $J_i = J$, we get that

$$\Gamma_k(r) = \tanh^r K. \tag{15.47}$$

Let us now consider adding the external magnetic field. The Hamiltonian is once again

$$\beta H = -k_B T J \sum_{i=1}^{N-1} S_i S_{i+1} - k_B T h \sum_{i=1}^{N}. \tag{15.48}$$

Let us define the matrix

$$U(S_i, S_{i+1}) = -k_B T J S_i S_{i+1} - \frac{1}{2} k_B T h (S_i + S_{i+1})$$

and the matrix

$$f(S_i, S_{i+1}) = e^{-\beta U(S_i, S_{i+1})}.$$

Then, the partition function may be written as

$$Z_N(T,h) = \sum_{S_1=-1}^{+1} \dots \sum_{S_N=-1}^{+1} f(S_1,S_2)f(S_2,S_3)\dots f(S_N,S_1). \tag{15.49}$$

We also define the transfer matrix between two consecutive sites in the lattice:

$$T = \begin{pmatrix} e^{J+h} & e^{-J} \\ e^{J} & e^{J-h} \end{pmatrix}.$$

Then, the partition function can be conveniently written as

$$Z_N(T,h) = \sum_{S_1=-1}^{+1} \left(T^N\right)_{S_1S_1} = Tr\left(T^N\right). \tag{15.50}$$

We have therefore to find the eigenvalues of the transfer matrix. These are given by

$$\lambda_\pm = e^J \cosh h \pm \left(e^{2J}\sinh^2 h + e^{-2J}\right)^{1/2} \tag{15.51}$$

and, as a consequence,

$$Z_N(T,h) = \lambda_+^N + \lambda_-^N. \tag{15.52}$$

The free energy per site is then given by

$$g(T,h) = -k_BT \lim_{N\to\infty} \frac{1}{N} \ln Z_N = -k_BT \ln \lambda_+. \tag{15.53}$$

This leads to

$$g(T,h) = -k_BT\{J + \ln[\cosh h + \sqrt{\sinh^2 h + e^{-4J}}\}. \tag{15.54}$$

Now, the magnetization is given by

$$\langle S\rangle = -\left(\frac{\partial g}{\partial h}\right)_T = k_BT\frac{\sinh h + \frac{1}{2}\frac{2\sinh h\cosh h}{\sqrt{\sinh^2 h+e^{-4J}}}}{\cosh h + \sqrt{\sinh^2 h + e^{-4J}}}. \tag{15.55}$$

Therefore, in the limit when the magnetic field goes to zero, the magnetization also vanishes, and there is no spontaneous magnetization at a finite temperature. The Ising model in $1D$ does not show spontaneous magnetization. In higher dimensions, it does, as shown by Onsager for the $2D$ case.

The result of the absence of spontaneous magnetization in the one-dimensional Ising model may be understood in the following way. In the ground state, all the spins are parallel. The system may choose one of two degenerate states and the spins may all point up or all point down. To fix ideas, let all of them point up. Let us now consider excited states by flipping some spins. Consider a region of length L with spins down in the background of up spins. Since the model only couples nearest-neighbor spins, there is a cost of energy with respect to the ground state that only involves the spins at the frontiers between the region of L spins and the rest of the system: the energy does not depend on L. So, at any finite temperature, the system may be arbitrarily disordered. It is therefore natural that there is no spontaneous magnetization at finite temperature.

Consider now a two-dimensional Ising model. Again consider that in the ground state, all spins point up, and now consider a region of down spins. These define an area A and its frontier is composed of L spins. Using similar arguments, the excitation energy does not depend on the area, but it depends on L. In $1d$ the frontier was just composed of two sites. However, for the same area A, there are different values of L that are allowed. The partition function may be estimated to be of the form

$$Z \sim g_L e^{-2a\beta L}. \tag{15.56}$$

Here, g_L is the number of blocks with a given perimeter L. This typically grows like a power law $g_L \sim b^L$. Here, a and b are some constants. Let us now increase L. The exponential factor decreases fast and for a finite temperature tends to zero as L grows. At small temperatures, β is large and therefore large regions with turned spins are not favorable. Therefore, these spin configurations do not contribute to the partition function. However, at large temperatures, $\beta \to 0$, the exponential factor is finite, and the degeneracy factor, g_L, grows fast with L. Therefore, these configurations where the spins are disordered give an important contribution to the partition function and the system is probably disordered. There are therefore two regimes: one at low temperature that may be ordered and one at high temperature that is disordered. It turns out, as we will see later, that there is a critical temperature that separates the two regimes. The cases of $1d$ and $2d$ are shown in Fig. 15.5.

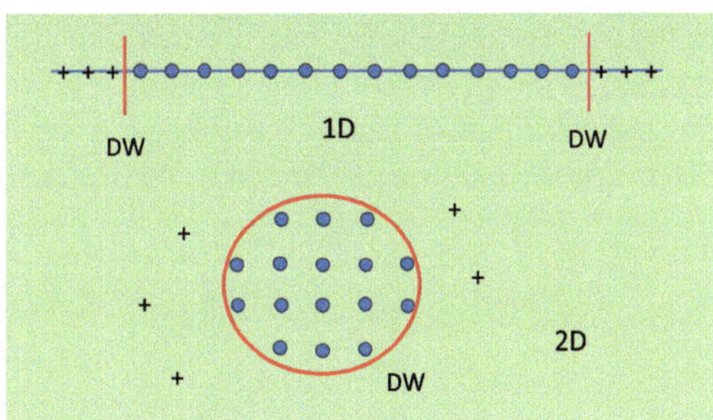

Fig. 15.5 Schematics of the domain walls separating a phase with the spins aligned in one direction (represented by +) and a phase where the spins are aligned in the opposite direction (represented by a small circle). In $1d$ the domain wall is just two points, while in $2d$ it has a length L

15.6 Second-Order Phase Transition in a Binary Alloy

Let us consider an alloy with two types of atoms A and B, in a *bcc* lattice with $2N$ available sites. The *bcc* lattice can be seen as two interpenetrated simple cubic lattices. Let us call them a and b. We have N atoms of type A and N atoms of type B distributed among the $2N$ sites.[5]

Let us define an order parameter P such that

$$N_A^a = \frac{1}{2}(1+P)N$$
$$N_A^b = \frac{1}{2}(1-P)N. \tag{15.57}$$

If $P = 1$, then $N_A^a = N, N_A^b = 0$, and if $P = -1$, then $N_A^a = 0, N_A^b = N$. Of course,

$$N_B^a = \frac{1}{2}(1-P)N$$
$$N_B^b = \frac{1}{2}(1+P)N. \tag{15.58}$$

There are sum rules, since $N_A^a + N_B^a = N_A^b + N_B^b = N$, where N_F^f is the number of F atoms in the f sublattice. For the two extreme cases $P = \pm 1$, the order is perfect. In the case $P = 0$, the lattice is fully disordered since we have a homogeneous distribution with $N_A^a = N_A^b = N/2$.

Let us now consider the general case and distribute the atoms among the lattice sites. Let us assume that all sites are occupied (no double occupancy allowed) and therefore a certain site in a certain sublattice is occupied, either by an A atom or by a B atom. Let us write the energy of the system as

$$E = N_{AA}U_{AA} + N_{BB}U_{BB} + N_{AB}U_{AB}, \tag{15.59}$$

where N_{FG} is the number of links between an F atom and a G atom and U_{FG} is the interaction energy between the same two atoms. Note that there are *eight* neighbors in the *bcc* lattice. The probability that one A atom in the lattice a has a link AA is equal to the probability that an atom A occupies a neighbor site in the lattice b, times the number of nearest neighbors. A similar reasoning may be carried out to determine the number of BB links and of AB links. We can get that

$$N_{AA} = 8\left[\frac{1}{2}(1+P)N\right]\left[\frac{1}{2}(1-P)N\right] = 2(1-P^2)N$$

[5] See also [Kittel].

$$N_{BB} = 8\left[\frac{1}{2}(1+P)N\right]\left[\frac{1}{2}(1-P)N\right] = 2(1-P^2)N$$

$$N_{AB} = 8N\left[\frac{1}{2}(1+P)\right]^2 + 8N\left[\frac{1}{2}(1-P)\right]^2 = 4(1+P^2)N. \quad (15.60)$$

Also, the energy can be written as

$$\begin{aligned} E &= E_0 + 2NP^2U \\ E_0 &= 2N\,(U_{AA} + U_{BB} + 2U_{AB}) \\ U &= 2U_{AB} - U_{AA} - U_{BB}. \end{aligned} \quad (15.61)$$

To determine the equilibrium situation at a certain temperature, we need to minimize the free energy $F = U - TS$. Let us then calculate the entropy. The number of states is given by

$$\Omega = \left(\frac{N!}{[\frac{1}{2}(1+P)N]![\frac{1}{2}(1-P)N]!}\right)^2, \quad (15.62)$$

which is the number of ways of distributing N atoms between the two sublattices such that $N_A^a = \frac{1}{2}(1+P)N$ and so on. The entropy is now obtained as $S = k_B \ln \Omega$ as usual. Therefore, the entropy is obtained as

$$S = 2Nk_B \ln 2 - Nk_B\left((1+P)\ln(1+P) + (1-P)\ln(1-P)\right), \quad (15.63)$$

where we have used Stirling's approximation, as usual ($\ln n! = n \ln n - n$, when $n \gg 1$). Clearly, when $P = \pm 1$, the entropy vanishes, and when $P = 0$, the entropy is maximal and is given by $S = 2Nk_B \ln 2$. Minimizing the free energy $F = U - TS$ with respect to the order parameter P, we get the equation of state (equilibrium condition):

$$4NP_0U + Nk_BT \ln \frac{1+P_0}{1-P_0} = 0. \quad (15.64)$$

There are two solutions: (i) the trivial solution $P_0 = 0$ that corresponds to the disordered (high temperature) phase and (ii) there is a nontrivial solution $P_0 \neq 0$. Considering the regime where $P_0 \sim 0$, we can expand the equation of state and obtain that

$$4NP_0U + 2Nk_BT_cP_0 = 0.$$

The critical temperature is given by $T_c = -\frac{2U}{k_B}$. At this temperature, the order parameter becomes non-vanishing. There is a solution if the potential is attractive

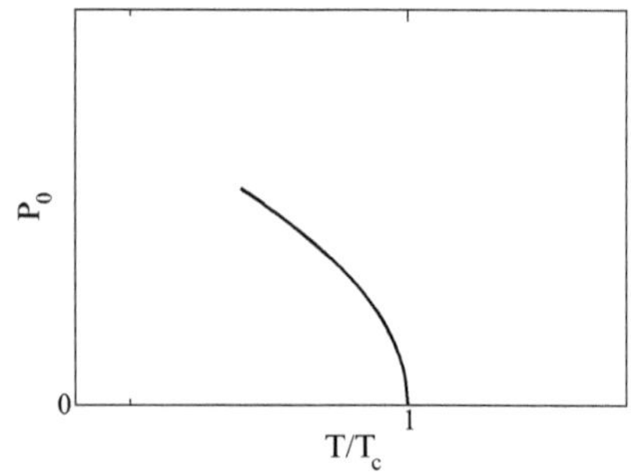

Fig. 15.6 Behavior of the order parameter P_0 for a binary alloy near the critical point

$U < 0$ $(2U_{AB} - U_{AA} - U_{BB} < 0)$. There is therefore a second-order (continuous) transition at the critical temperature.

For values of the temperature smaller than T_c, the value of the order parameter increases. We may expand to the next order:

$$\ln \frac{1 + P_0}{1 - P_0} = 2P_0 + \frac{2}{3} P_0^3. \tag{15.65}$$

Substituting in Eq. (15.64), we get that

$$P_0 \sim (T_c - T)^{1/2}, \tag{15.66}$$

since $U = -k_B T_c/2$. Therefore, the order parameter deviates from the high-temperature zero value as a power law in the temperature difference to the critical temperature. This behavior is illustrated in Fig. 15.6. We will get back to this point later.

15.7 ϕ^4 Theory

15.7.1 Representation of the Ising Model by Functional Integrals

Let us recall the gaussian integral:

$$\int_{-\infty}^{\infty} dx e^{-\frac{1}{4a^2}x^2 + sx} = \sqrt{4\pi a^2} e^{a^2 s^2}.$$

This expression can be generalized

$$\int_{-\infty}^{\infty} \prod_{i=1}^{N} dx_i e^{-\frac{1}{4}x_i V_{ij}^{-1} x_j + S_i x_i} = C e^{S_i V_{ij} S_j}, \tag{15.67}$$

where C is a constant. This transformation is often called the Hubbard-Stratonovich transformation.

Let us consider again the Ising model. The Hamiltonian is of the type

$$H = -\sum_{ij} J_{ij} S_i S_j - \sum_i h_i S_i. \tag{15.68}$$

The partition function is

$$\begin{aligned} Z\{H_i\} &= \sum_{\{S_i\}} e^{-\beta H\{S_i\}} \\ &= \sum_{\{S_i\}} e^{\sum_{ij} K_{ij} S_i S_j + \sum_i H_i S_i} \\ &= \sum_{\{S_i\}} \int_{-\infty}^{\infty} \prod_{i=1}^{N} d\phi_i e^{-\frac{1}{2}\phi_i K_{ij}^{-1} \phi_j + (\phi_i + H_i) S_i} \\ &= \int_{-\infty}^{\infty} \prod_{i=1}^{N} d\phi_i e^{-\frac{1}{4}(\phi_i - H_i) K_{ij}^{-1} (\phi_j - H_j)} \sum_{\{S_i\}} e^{\phi_i S_i}. \end{aligned} \tag{15.69}$$

having defined $K_{ij} = \beta J_{ij}$ and $H_i = \beta h_i$. The final sum is easy to carry out:

$$\sum_{\{S_i\}} e^{\phi_i S_i} = \prod_i (2 \cosh \phi_i) = C e^{\sum_i \ln(\cosh \phi_i)}. \tag{15.70}$$

Therefore, the partition function can be rewritten as

$$Z\{H_i\} = C \int_{-\infty}^{\infty} \prod_{i=1}^{N} d\phi_i e^{-\frac{1}{4}(\phi_i - H_i) K_{ij}^{-1} (\phi_j - H_j)} e^{\sum_i \ln(\cosh \phi_i)}. \tag{15.71}$$

Expanding

$$\ln \cosh x = \frac{1}{2} x^2 - \frac{1}{12} x^4 + \cdots ,$$

we get that the partition function takes the form

$$Z\{H_i\} \sim \int \mathcal{D}\phi e^{-\frac{1}{4}(\phi_i - H_i) K_{ij}^{-1} (\phi_j - H_j)} e^{\sum_i \left(\frac{1}{2}\phi_i^2 - \frac{1}{12}\phi_i^4 + \cdots\right)}. \tag{15.72}$$

where we defined $\mathcal{D}\phi = \prod_{i=1}^{N} d\phi_i$. We have replaced the partition function as a sum over (classical) spin configurations by a functional integral over classical (bosonic-like) fields with an effective action that is given by an expansion in powers of this bosonic field. Since we truncate the expansion to fourth order in powers of the field ϕ, this is called ϕ^4 theory. We will see later that, in most cases, it is enough to truncate the expansion to this order. (The whole treatment is classical, since the Ising model only involves one spin component and we do not need to use the nontrivial commutation relations of the spin operators. Also, we may just consider the spins as the classical variables $S_i = \pm 1$ as we have been considering. If some magnetic field in a transverse direction, such as along the xy plane is added, the problem becomes quantum if we are considering spin operators, as will be discussed in a later chapter. The transverse field couples to a different spin operator component and so the quantum nature naturally emerges. Also, the designation of bosonic field is an abuse of language, but is often used).

We can establish some exact relations between averages of the spin variables, such as the magnetization, and averages of the auxiliar bosonic fields. Consider the exact expression for the partition function in terms of the ϕ fields, Eq. (15.71). Since the partition function is the function of the magnetic fields, H_i, taking the derivative of its logarithm leads to the average value of the spin, the magnetization. We get that

$$\frac{\partial \ln Z}{\partial H_l} = M_l = \frac{1}{2}\left(K^{-1}(\langle\phi\rangle - H)\right)_l, \tag{15.73}$$

where we have used that the matrix K is symmetric ($K_{ij} = K_{ji}$) and $\langle\phi_i\rangle$ is the average value of the bosonic field at site i. Therefore, we get the exact relation

$$\langle\phi_i\rangle = H_i + 2\sum_j K_{ij} M_j, \tag{15.74}$$

where $M_j = \langle S_j\rangle$. So any conclusions for the average value of the field ϕ can be related to conclusions for the magnetization. In particular, nonzero solutions for the average of the ϕ field, in the absence of the external magnetic field, translate to nontrivial solutions for the physical magnetization.

We may also establish exact relations for any higher derivatives of the partition function and, in particular, establish a relation between, for instance, correlation functions of two bosonic field and magnetic susceptibilities. Defining a connected correlation function between two fields as

$$\langle\phi_l\phi_m\rangle_c = \langle\phi_l\phi_m\rangle - \langle\phi_l\rangle\langle\phi_m\rangle, \tag{15.75}$$

we can find the relation

$$\langle\phi\phi\rangle_c = K + K\langle SS\rangle_c K, \tag{15.76}$$

seen as a matrix equation and, therefore, we have to sum over spatial variables, since the matrix K is off-diagonal in space (for instance, by coupling nearest neighbors in the simplest Ising model).

Even though we are focusing here on the Ising model, the procedure is similar if we consider the Heisenberg model. The difference is that the bosonic fields, instead of being scalars (one component since the spins only have one component), become vectors.

15.7.2 Mean-Field Theory

The partition function is then of the type

$$Z \sim \int \mathcal{D}\phi e^{-\frac{H(\phi)}{T}}.$$

Let us consider that in general the action (argument of the Boltzmann factor)

$$\frac{H(\phi)}{T} = \int d^d x[a_0 - B(x)\phi(x) + R(T)\phi^2(x) + u(T)\phi^4(x) + S(T)\left(\nabla\phi(x)\right)^2 + \cdots], \tag{15.77}$$

where we have added a gradient term to allow for non-homogeneous field configurations. This type of expansion is called the Landau theory for a second-order phase transition. It assumes that near the phase transition, the order parameter is small, and an expansion in its powers provides a good description of the transition and of the ordered phase in its vicinity. To obtain the thermodynamics, we have to calculate the partition function. In general, this is very complicated. If $u(T) = 0$ and all higher-order terms vanish as well, we may solve exactly the problem: this is the gaussian model. However, this is not in good agreement with the experimental results. We have therefore to develop some approximation methods. In this case, we are interested in looking for solutions that correspond to values of $\langle\phi\rangle \neq 0$, that is, to phase transitions.

The term $-B(x)\phi(x)$ corresponds to the action of an external field (if $\phi = M$, B is the magnetic field). Since the system has to be invariant if we make the transformation $\phi \rightarrow -\phi$ and, at the same time, $B \rightarrow -B$, there can be no odd terms like $\phi^3, \phi^5, \ldots$. The meaning of $R(T)$ will become clear momentarily. The coefficient $u(T)$ is the amplitude of the interaction part. Finally, $S(T)$ is the coefficient that results of the possible spatial variation of the field. We assume that the variations are smooth and, therefore, we neglect higher-order derivatives. In the case of uniform solutions, this term vanishes. If a uniform solution is preferred physically, then the presence of this term means that it costs energy to introduce a spatial modulation if $S(T) > 0$. We will take $S(T) = 1$.

To calculate the partition function, let us expand the action around its minimum (actually saddle point), since it will give the largest contribution to the partition function:

$$H(\phi) = H(\bar{\phi}) + \frac{1}{2}\left(\frac{\partial^2 H}{\partial \phi^2}\right)_{\bar{\phi}} (\phi - \bar{\phi})^2 + \frac{1}{3!}\left(\frac{\partial^3 H}{\partial \phi^3}\right)_{\bar{\phi}} (\phi - \bar{\phi})^3 + \cdots$$

where

$$\left(\frac{\partial H}{\partial \phi}\right)_{\bar{\phi}} = 0. \tag{15.78}$$

Then, we obtain that

$$\begin{aligned} Z &\sim e^{-\frac{H(\bar{\phi})}{T}} \int \mathcal{D}(\phi - \bar{\phi}) e^{-\frac{1}{T}\frac{1}{2}\left(\frac{\partial^2 H}{\partial \phi^2}\right)_{\bar{\phi}}(\phi-\bar{\phi})^2 - \frac{1}{T}\frac{1}{3!}\left(\frac{\partial^3 H}{\partial \phi^3}\right)_{\bar{\phi}}(\phi-\bar{\phi})^3 - \frac{1}{T}\frac{1}{4!}\left(\frac{\partial^4 H}{\partial \phi^4}\right)_{\bar{\phi}}(\phi-\bar{\phi})^4 + \cdots} \\ &\sim e^{-\frac{H(\bar{\phi})}{T}} \int \mathcal{D}x e^{-\frac{1}{2T}H''x^2} \left[1 - \frac{x^3}{3!T}H''' + \frac{1}{2!}\left(\frac{x^3}{3!T}H'''\right)^2 + \cdots\right] \\ &\quad \left[1 - \frac{1}{4!T}x^4 H^{iv} + \frac{1}{2!}\left(\frac{x^4}{4!T}H^{iv}\right)^2 + \cdots\right][1 + \cdots]. \end{aligned} \tag{15.79}$$

The leading order constitutes the mean-field approximation. At this order, we get that

$$Z^{(0)} \sim e^{-\frac{H(\bar{\phi})}{T}}. \tag{15.80}$$

The next order includes the gaussian fluctuations around the saddle-point solution. It is given by

$$Z^{(1)} \sim e^{-\frac{H(\bar{\phi})}{T}} \int \mathcal{D}x e^{-\frac{1}{2T}H''x^2}. \tag{15.81}$$

For the moment, let us focus our attention on the mean-field approximation. The equation of state is written as

$$\int d^3x[-B(x) + 2R(T)\bar{\phi}(x) + 4u(T)\bar{\phi}^3(x) - 2\nabla^2\bar{\phi}(x)]\delta\phi = 0, \tag{15.82}$$

where we have varied the action with respect to a variation of the field and where we have integrated by parts the gradient term like

$$\delta\left(\nabla\phi \cdot \nabla\phi\right) = 2\nabla\phi \cdot \delta(\nabla\phi) = -2\nabla^2\phi\delta\phi.$$

The equation of state reduces therefore to

$$2R(T)\bar{\phi}(x) + 4u(T)\bar{\phi}^3(x) - 2\nabla^2\bar{\phi}(x) = B(x), \tag{15.83}$$

since the variation $\delta\phi$ is arbitrary.

Let us take, for simplicity, a uniform field $B(x) = B$. Then, we expect that the field is uniform $\bar{\phi}(x) = \bar{\phi}$. The equation of state reduces in this case to

$$2R(T)\bar{\phi} + 4u(T)\bar{\phi}^3 = B.$$

If $B = 0$, there are two solutions. The trivial one $\bar{\phi} = 0$ and a second solution

$$\bar{\phi} = \pm\left(\frac{-R(T)}{2u(T)}\right)^{1/2}.$$

Clearly, this implies that this second solution only exists if $R(T) < 0$. If $T > T_c$, the trivial solution is the chosen one. Therefore, we expect that we can write $R(T) = a(T - T_c)$. If $T < T_c$, the coefficient $R(T) < 0$ and the two solutions are possible. If $T > T_c$, the coefficient $R(T) > 0$ and only the trivial solution is possible.

We have to determine when $T < T_c$, which of the two solutions is the most stable. It is easy to see that the nontrivial solutions are the most stable. Close to T_c, the coefficient is a constant $u(T_c)$. Therefore, close to the critical temperature, we have

$$\bar{\phi} \sim (T_c - T)^{1/2}. \tag{15.84}$$

Defining a critical exponent β as

$$\bar{\phi} \sim (T_c - T)^{\beta},$$

we see that in mean field, this exponent is $\beta = 1/2$. At the critical point $R(T_c) = 0$, we get from the equation of state that $4u(T_c)\bar{\phi}^3 = B$. Defining another critical exponent like

$$\bar{\phi} \sim B^{1/\delta},$$

we get in the mean-field theory that $\delta = 3$.

We can also study the behavior of the susceptibility near the critical point. The susceptibility at constant temperature is defined by $\chi_T = (\partial\bar{\phi})/(\partial B)_T$. Using the equation of state, we get that

$$[2R(T) + 12U(T)\bar{\phi}^2]\left(\frac{\partial\bar{\phi}}{\partial B}\right)_T = 1.$$

Considering $T > T_c$ ($\bar{\phi} = 0$), we get

$$\chi_T = (\partial\bar{\phi})/(\partial B)_T = \frac{1}{2R(T)} \sim \frac{1}{T - T_c} \sim (T - T_c)^{-\gamma}, \tag{15.85}$$

which means that the critical exponent $\gamma = 1$ in mean field. If we consider a temperature below the critical temperature $T < T_c$ and using the nontrivial solution for $\bar{\phi}$, we get the same exponent (but a different prefactor).

It turns out, however, that the exponents observed experimentally are not in agreement with the mean-field predictions. The reason is that the fluctuations, which are neglected in mean-field theory, are important and not negligible near the critical point. Actually, the fluctuations are fundamental near the transition point, since they are needed to explore regions of phase space that correspond to a description of the other phase. This is necessary in order for the system to make the transition to a different phase, which may be more favorable as the parameters are changed. A proper treatment of the fluctuations is, therefore, necessary and will be considered later. But first let us consider the possibility of including higher-order terms in the expansion of the free energy.

15.7.3 Tricritical Point

Since the average on the bosonic field relates to the magnetization, we may change the notation and consider that the free energy is an expansion in powers of the magnetization, call it m.

Let us now consider an expansion of the Landau free energy, F_L, in the magnetization around a critical point up to the next order:

$$\frac{F_L}{k_B T} = \int d^d x \left(c\,(\nabla m)^2 + a(T - T_c)m^2 + um^4 + vm^6 - Hm \right), \tag{15.86}$$

with coefficients $c, a, v > 0$.

(a) Using that $\nabla m \cdot \nabla m = -m\nabla^2 m$, we get the equation of state:

$$\frac{\delta F_L}{\delta m} = 0 = -2c\nabla^2 m + 2a(T - T_c)m + 4um^3 + 6vm^5 - H, \tag{15.87}$$

which can be rewritten as

$$2a(T - T_c)m + 4um^3 - 2c\nabla^2 m + 6vm^5 = H. \tag{15.88}$$

Take now $H = 0$, neglect the term that comes from the m^6 term (since $m^5 \ll m^3$), and choose $\nabla^2 m = 0$. Then, the problem reduces to the case previously solved and leads to

$$2a(T - T_c)m + 4um^3 = 0, \tag{15.89}$$

which can be factorized in terms of its solutions:

$$m\left(2a(T - T_c) + 4um^2\right) = 0, \tag{15.90}$$

as we saw before.

We can write

$$\beta F_L^{MF} = V\left(a(T - T_c)m^2 + um^4\right). \tag{15.91}$$

In the high-temperature phase, $T > T_c$, the magnetization vanishes, $m = 0$. Considering both regimes, we get

$$\begin{aligned} \beta F_L^{MF} &= 0, && T > T_c \\ &= V\left(a(T - T_c)(-2a)\frac{T - T_c}{4u} + u\frac{4a^2(T - T_c)^2}{16u^2}\right), && T < T_c, \end{aligned} \tag{15.92}$$

which simplifies in the ordered phase to

$$\frac{F_L^{MF}}{k_B T} = -\frac{V}{2}\frac{a^2(T - T_c)^2}{u}. \tag{15.93}$$

We may calculate the specific heat recalling that

$$C = -T\frac{\partial^2 F}{\partial T^2}. \tag{15.94}$$

Considering that $F_L \sim -T(T - T_c)^2 \sim -T_c(T - T_c)^2$, in the vicinity of the critical point, we obtain that the specific heat is a constant (independent of temperature), which implies the critical exponent:

$$\alpha = 0. \tag{15.95}$$

Consider now a perturbation in the magnetization profile at the origin:

$$F_L = \int d^d x \left(-cm\nabla^2 m + a(T - T_c)m^2 + um^4 + vm^6 - \lambda m(x)\delta(x)\right). \tag{15.96}$$

The equation of state is written as

$$-2c\nabla^2 m + 2a(T - T_c)m + 4um^3 = \lambda\delta(x). \tag{15.97}$$

Let us choose c such that $2c = 1$. We can write

$$\left(\nabla^2 - r_o^2\right) m(x) = -\lambda\delta(x), \tag{15.98}$$

where $r_0^2 = 2a(T - T_c)$. The solution of this equation may be obtained considering the Fourier transform:

$$m(x) = \int \frac{d^d k}{(2\pi)^d} m(k) e^{ikx}. \tag{15.99}$$

Substituting, we get

$$m(k) = \frac{\lambda}{k^2 + r_0^2}, \tag{15.100}$$

and calculating the inverse transform, we get

$$m(x) = \int \frac{d^d k}{(2\pi)^d} \frac{\lambda}{k^2 + r_0^2} e^{-ikx}$$

$$m(x) \sim \frac{1}{x^{-(2-d)} e^{-\frac{x}{\xi}}}, \tag{15.101}$$

where $\xi = 1/r_0$. The result for the magnetization, $m(k)$, shows that $\eta = 0$ defined as the exponent of $k^{2+\eta}$ and since

$$\xi \sim \frac{1}{(2a)^{1/2}(T - T_c)^{1/2}} \sim \frac{1}{(T - T_c)^{\nu}}, \tag{15.102}$$

this implies that

$$\xi = \frac{1}{2}. \tag{15.103}$$

When $u > 0$, we find therefore a second-order transition as shown in Fig. 15.7. When $T > T_c$, the state with the lowest energy is the one with $m = 0$. Increasing the parameter m leads to a free energy that is higher than that of $m = 0$, which means that the stable state is the disordered phase (or if there is a magnetic field, a paramagnetic phase). When $T < T_c$, a state with finite m has a lower free energy and is preferable with respect to the trivial, $m = 0$, solution. The free energy has a local maximum at $m = 0$ and two symmetric local minima (if $H = 0$) at finite values. At $T = T_c$, there is a transition from the trivial to the nontrivial solution, which happens continuously. This is a second-order phase transition.

Consider now that $u < 0$. Since the Landau theory assumes an expansion around the trivial solution, close to the transition point, the highest term coefficient has

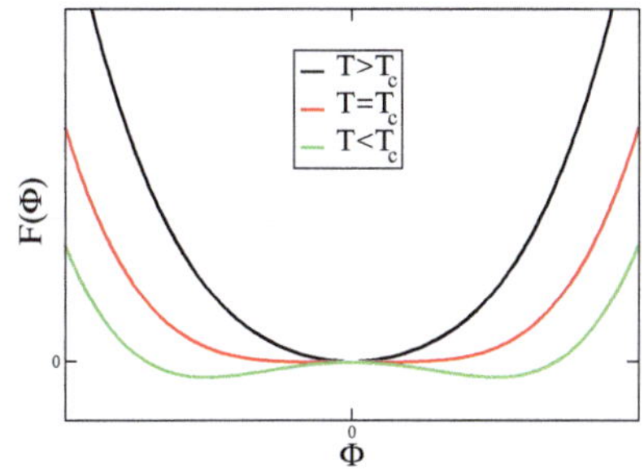

Fig. 15.7 Free energy as a function of the average value of the order parameter in the case of a second-order phase transition: the three regimes are shown for $T > T_c, T = T_c, T < T_c$

to be positive; otherwise, a solution with infinite value of m would be the most stable. We may take $u < 0$ provided we include a higher-order term, say m^6, with a positive coefficient $v > 0$. In this case, there is a temperature, T_1, to be determined, such that if $T > T_1$ (high-temperature regime), the system is disordered and we expect that $m = 0$ is the most stable solution. As we lower the temperature, there is the possibility that a nonzero value of m becomes the preferred solution. Call this temperature T_1. Note that is not T_c: this is the critical temperature of the second-order phase transition that is found if $u > 0$. For $T < T_1$, we find that finite values of m are indeed the preferred solution, with a lower free energy. It turns out that at $T = T_1$, there are three solutions with identical free energies. One is the disordered solution with $m = 0$. The other are symmetric solutions at a finite value of m, and so the transition is not continuous. At $T = T_1$, there are three solutions that differ by a finite amount of the magnetization. At the transition, there is a discontinuity of the magnetization and so is classified as a first-order phase transition. Recall that if $H = 0$, the free energy for $T \geq T_1$ vanishes, since $m = 0$ is one of the solutions.

We have therefore two conditions at the point of the first-order transition. One is the equation of state, obtained from the stationarity of the free energy, $\partial F/\partial m = 0$, and the other the value of the free energy at the transition:

$$\begin{aligned} 2a(T_1 - T_c)m + 4um^3 + 6vm^5 &= 0 \\ a(T_1 - T_c)m^2 + um^4 + vm^6 &= 0. \end{aligned} \tag{15.104}$$

These equations may be solved to yield

$$\begin{aligned} m^2 &= -\frac{u}{2v} \\ T_1 - T_c &= \frac{u^2}{4av}. \end{aligned} \tag{15.105}$$

Note that $T_1 > T_c$. As a consequence, as we lower the temperature from high values, the first-order transition takes place first. There is therefore a high-temperature phase that is separated from an ordered phase by a first-order transition. This is shown in Fig. 15.8. If we consider $u > 0$ as before, the sixth-order term in the free energy is

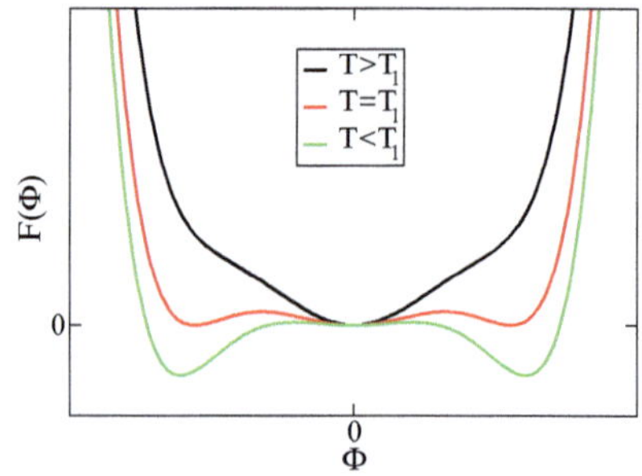

Fig. 15.8 Free energy as a function of the average value of the order parameter in the case of a first-order phase transition: the three regimes are shown for $T > T_1, T = T_1, T < T_1$

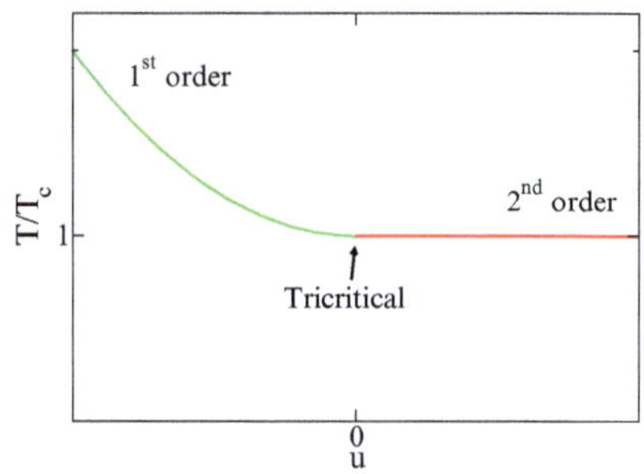

Fig. 15.9 Phase diagram as a function of u representing the second-order phase transition for $u > 0$, the first-order transition for $u < 0$, and the tricritical point for $u = 0$

not relevant, and we find the second-order transition we mentioned. At $u = 0$, the first-order and the second-order transition lines meet at what is called a tricritical point. This point has different exponents in relation to those of the finite $u > 0$, second-order transition. In the plane u, T there is a horizontal line at constant $T = T_c$ that separates the two phases, and at $u < 0$ the transition line changes quadratically with u. This is shown in Fig. 15.9.

Consider $u = 0$ and $T \sim T_c$. We get, reintroducing the magnetic field,

$$2a(T - T_c)m + 6vm^5 = H. \tag{15.106}$$

If $H = 0$, we get

$$m \sim (T - T_c)^{\frac{1}{4}}, \tag{15.107}$$

which implies a different critical exponent:

$$\beta = \frac{1}{4}. \tag{15.108}$$

The free energy behaves as

$$\frac{F_L}{k_B T} \sim V\left(a(T - T_c)m^2 + vm^6\right)$$

$$\sim V\left(a(T-T_c)(T-T_c)^{\frac{1}{2}}+v(T-T_c)^{\frac{3}{2}}\right)$$
$$\sim (T-T_c)^{\frac{3}{2}}. \tag{15.109}$$

This leads to the specific heat exponent, $\alpha = 1/2$

$$C \sim \frac{\partial^2 F_L}{\partial T^2} \sim (T-T_c)^{-\frac{1}{2}}. \tag{15.110}$$

Taking a derivative of the equation of state in the presence of magnetic field with respect to H, we get

$$\left(2a(T-T_c)+30vm^4\right)\frac{\partial m}{\partial H} = 1. \tag{15.111}$$

This leads to the susceptibility

$$\chi \sim \frac{1}{T-T_c} \tag{15.112}$$

and $\gamma = 1$. Also, the magnetization behaves as

$$m \sim H^{\frac{1}{5}} \tag{15.113}$$

and $\delta = 5$. Introducing a local magnetic field leads to a non-homogeneous magnetization that satisfies the equation:

$$\left(-2c\nabla^2+2a(T-T_c)+30vm^4\right)\delta m = \delta H, \tag{15.114}$$

which can be rewritten as before:

$$\left(\nabla^2-\frac{1}{\xi^2}\right)\delta m = -\frac{1}{2c}\delta H, \tag{15.115}$$

with

$$-\frac{1}{\xi^2} = -\frac{1}{2c}\left(2a(T-T_c)+30vm^4\right). \tag{15.116}$$

Therefore,

$$\xi \sim \frac{1}{(T-T_c)^{\frac{1}{2}}}, \tag{15.117}$$

which implies $\eta = 0$, $\nu = 1/2$.

15.7.4 Ginzburg Criterion

Let us now analyze the influence of fluctuations around the mean-field solution. The functional is taken to be of the type

$$\frac{1}{2}|\nabla\phi|^2 + \frac{1}{2}a\phi^2 + \frac{b}{4!}\phi^4 + \cdots$$

where $a \sim (T - T_c)$. Varying with respect to ϕ and taking a uniform solution, we get as before $a\phi_0 + b/6\phi_0^3 = 0$. The solution is $\phi_0^2 = -6a/b$. The fluctuations are given by

$$\langle\phi(x)\phi(x')\rangle = \int \frac{d^dk}{(2\pi)^d}\frac{e^{ik(x-x')}}{k^2+a}.$$

The local field fluctuation is given by

$$\langle\phi^2(x)\rangle = \int \frac{d^dk}{(2\pi)^d}\frac{1}{k^2+a} \sim a^{\frac{d-2}{2}} \sim (T - T_c)^{\frac{d-2}{2}}. \tag{15.118}$$

Then, the relative fluctuation is given by

$$\frac{\langle\phi^2\rangle}{\langle\phi\rangle^2} \sim \frac{(T-T_c)^{\frac{d}{2}-1}}{(T-T_c)} \sim (T-T_c)^{\frac{d}{2}-2} \sim (T-T_c)^{\frac{d-4}{2}}. \tag{15.119}$$

If the space dimension is $d > d_c$, where $d_c = 4$, the fluctuations are not important. If on the other hand $d < d_c$, the fluctuations cannot be neglected. If $T - T_c$ is not too small, then the ratio is finite. Only when $T \to T_c$, there are problems. The Ginzburg criterion establishes therefore a region around the critical point where fluctuations are important, if $d < d_c$.

Local Field Fluctuations

Let us now look with some more care at the result for the local field fluctuation, Eq. (15.118). The partition function to the second order in the fluctuations may be written as

$$Z = \int D\phi e^{-\int d^dx\left((\nabla\phi(x))^2 + a\phi^2(x)\right)}. \tag{15.120}$$

Expanding in a Fourier series, we define

$$\phi(x) = \frac{1}{\sqrt{V}}\sum_k e^{-ikx}\phi(k). \tag{15.121}$$

The gradient term is therefore written as

$$(\nabla\phi)^2 = -\frac{1}{V}\sum_{k,k'} e^{-ikx}e^{-ik'x}\phi(k)\phi(k')kk', \tag{15.122}$$

which, using that

$$\int d^d x e^{-ikx}e^{-ik'x} = V\delta_{k',-k},$$

leads to

$$\int d^d x\,(\nabla\phi)^2 = V\int \frac{d^d k}{(2\pi)^d}k^2\phi(k)\phi(-k). \tag{15.123}$$

We may write that

$$\langle\phi(x)\phi(x')\rangle = \frac{1}{V}\sum_{k,k'} e^{-ikx}e^{-ik'x'}\langle\phi(k)\phi(k')\rangle. \tag{15.124}$$

Using the Fourier decomposition, we may also obtain that

$$\int d^d x\phi^2(x) = V\int \frac{d^d k}{(2\pi)^d}\phi(k)\phi(-k) \tag{15.125}$$

and the partition function, to the second order in the fluctuations, may be written as

$$Z = \int D\phi e^{-V\int \frac{d^d k}{(2\pi)^d}(k^2+a)\phi(k)\phi(-k)}. \tag{15.126}$$

We may now calculate the correlation between the Fourier components of the fluctuations:

$$\langle\phi(k)\phi(k')\rangle = \frac{\int D\phi\phi(k)\phi(k')e^{-V\int \frac{d^d k}{(2\pi)^d}(k^2+a)\phi(k)\phi(-k)}}{\int D\phi e^{-V\int \frac{d^d k}{(2\pi)^d}(k^2+a)\phi(k)\phi(-k)}}. \tag{15.127}$$

Recalling that

$$\int_{-\infty}^{\infty} d\phi e^{-\frac{1}{2}\phi\alpha\phi+\lambda\phi} = \left(\frac{2\pi}{\alpha}\right)^{1/2} e^{\frac{1}{2}\frac{\lambda^2}{\alpha}}, \tag{15.128}$$

we can write that the partition function is

$$Z = \int \left(\prod_{l=1}^{N} d\phi_l \right) e^{-\frac{1}{2}\sum_{i,j}\phi_i \alpha_{ij} \phi_j + \sum_i \lambda_i \phi_i}$$
$$= \prod_{p=1}^{N} \left(\frac{2\pi}{\alpha_p} \right)^{1/2} e^{\frac{1}{2}\sum_p \frac{\lambda_p^2}{\alpha_p}}, \tag{15.129}$$

in the basis that diagonalizes the matrix α, and we considered that α_{ij} are real. In general, we can write that

$$Z = (2\pi)^{N/2} \left(\frac{1}{det\alpha} \right)^{1/2} e^{\frac{1}{2}\sum_{ij} \lambda_i \alpha_{ij}^{-1} \lambda_j}$$
$$= e^{-\frac{1}{2}Tr(\ln(\frac{\alpha}{2\pi})) + \frac{1}{2}\sum_{ij} \lambda_i \alpha_{ij}^{-1} \lambda_j}. \tag{15.130}$$

A correlation function is now defined by

$$\langle \phi_i \phi_j \rangle = \frac{1}{Z} \int \left(\prod_{l=1}^{N} d\phi_l \right) \phi_i \phi_j e^{-\frac{1}{2}\sum_{m,n}\phi_m \alpha_{mn} \phi_n + \sum_m \lambda_m \phi_m}. \tag{15.131}$$

It can be calculated taking derivatives with respect to the source terms, λ_i, λ_j, as usual, leading to

$$\langle \phi_i \phi_j \rangle|_{\lambda_m = 0} = \frac{1}{2} \left(\alpha_{ij}^{-1} + \alpha_{ji}^{-1} \right). \tag{15.132}$$

If the matrix is symmetric, it simplifies to

$$\langle \phi_i \phi_j \rangle|_{\lambda_m = 0} = \alpha_{ij}^{-1}. \tag{15.133}$$

Applying these results to the Fourier components, we get

$$\langle \phi(k) \phi(k') \rangle = \delta_{k',-k} \frac{1}{V} \frac{1}{k^2 + a}. \tag{15.134}$$

If we consider long-range fluctuations ($k \to 0$), we see that if $a \to 0$ (which means approaching the critical point), the fluctuation diverges. Integrating in a d-dimensional space leads to the result of Eq. (15.118).

15.8 Blume-Capel Model

Consider the following system of classical spins defined by the Hamiltonian:

$$H = -J \sum_{\langle i,j \rangle} S_i S_j + \Delta \sum_j S_j^2. \tag{15.135}$$

The spins are spin-1 like and $S_i = \pm 1, 0$. Let us also look at the problem in Fourier space. Define

$$S_k = \frac{1}{\sqrt{N}} \sum_i S_i e^{ikr_i}, \tag{15.136}$$

where r_i are the positions of the spin variables in real space.

The energy of a homogeneous system with $S_i = \pm 1$ is given by

$$E = -J \frac{1}{2} N q + \Delta N, \tag{15.137}$$

where q is the number of nearest neighbors. The energy vanishes if $S_i = 0$. If $\Delta > \frac{1}{2} q J$, the ground state has $S_i = 0$, and if $\Delta < \frac{1}{2} q J$, the ground state has $S_i = \pm 1$.

At finite temperature, the problem is complex. We may approximate a solution using mean field. The partition function is given by

$$\begin{aligned} Z &= \sum_{\{S_i\}} e^{-\beta H} \\ &= \sum_{\{S_i\}} e^{\beta J \sum_{\langle i,j \rangle} S_i S_j - \beta \Delta \sum_i S_i^2} \\ &= \sum_{\{S_i\}} e^{K \sum_{\langle i,j \rangle} S_i S_j - D \sum_i S_i^2}, \end{aligned} \tag{15.138}$$

where $K = \beta J$ and $D = \beta \Delta$. We may now consider the problem as a set of spins in interaction with some average field caused by the other spins as

$$e^{K(S_i + \cdots + S_q) S_0 - D S_0^2 + h S_0} \rightarrow e^{(Kqm + h) S_0 - D S_0^2},$$

where an external field, $h = \beta H$, has been added and $m = \langle S \rangle$ is the average spin value at a given site in the system. Here, we are implicitly assuming a homogeneous solution. Clearly, the model is now reduced to a single spin in the presence of some effective field and

$$m = \frac{\sum_S S e^{(Kqm + h) S - D S^2}}{\sum_S e^{(Kqm + h) S - D S^2}}. \tag{15.139}$$

The average spin may be written self-consistently as

$$m = \frac{e^{Kqm+h} - e^{-(Kqm+h)}}{e^{Kqm+h} + e^{-(Kqm+h)} + e^{D}}. \tag{15.140}$$

In the vicinity of the transition, which occurs between a vanishing average spin and a finite average, we may consider that $|Kqm + h| << 1$. Expanding for small values, we get that

$$m = \frac{2}{2+e^D}(Kqm+h)\left(1 - C(Kqm+h)^2 + \cdots\right), \tag{15.141}$$

where

$$C = \frac{1}{2+e^D} - \frac{1}{6}. \tag{15.142}$$

Let us take $h = 0$. Taking $T < T_c$, the regime for which m does not vanish, we get that

$$1 = \frac{2}{2+e^D}Kq\left(1 - C(Kqm)^2 + \cdots\right). \tag{15.143}$$

We may get from the linearized equation the critical temperature:

$$\frac{Jq}{k_B T_c} = 1 + \frac{1}{2}e^{\frac{\Delta}{k_B T_c}}. \tag{15.144}$$

Solving for m near the critical temperature, we get, as expected from a mean-field approach, that

$$m^2 \sim (T_c - T), \tag{15.145}$$

which implies that the critical exponent $\beta = 1/2$.

There is a tricritical point defined by

$$C = \frac{1}{2+e^D} - \frac{1}{6} = 0, \tag{15.146}$$

since in this case, the cubic term of the equation of state vanishes. This leads to $e^D = 4$. Therefore, the temperature of the tricritical point is given by

$$\begin{aligned} \frac{Jq}{k_B T_t} &= 3 \\ \frac{\Delta}{k_B T_t} &= \ln 4. \end{aligned} \tag{15.147}$$

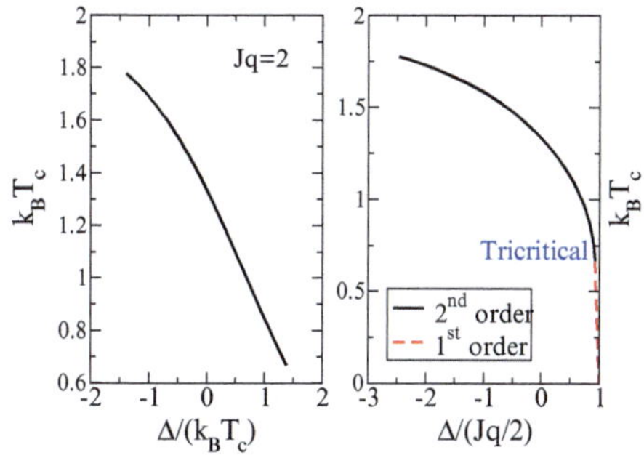

Fig. 15.10 Phase diagram of the Blume-Capel model

In its vicinity, we get that $\beta = 1/4$ and $\delta = 5$, while away from the tricritical point, we get that $\delta = 3$. The phase diagram is shown in Fig. 15.10.

15.9 Classical Monte Carlo Method

Problems in statistical physics involve many-body systems, which imply many degrees of freedom. The thermal average of a given physical quantity, A, that depends on a continuous variable, x, may be written as

$$\langle A(x)\rangle = \frac{\int d^{3N}x A(x) e^{-\frac{H(x)}{k_BT}}}{\int d^{3N}x e^{-\frac{H(x)}{k_BT}}}, \tag{15.148}$$

considering, for example, that we have N particles and in a three-dimensional space, we have $3N$ coordinates. There is a very large number of configurations that must be taken into account. Some approximation has to be taken, such that reduces the amount of work to be carried out by selecting the most important contributions to the sum of the configurations. One method is called the Monte Carlo method.

Out of the many configurations, we may select a subset. One possible way is to just randomly select configurations (simple sampling) and hoping that, as the number of random configurations increases, the average over these configurations approaches the correct value, when all configurations are taken into account. Another possibility is to select some weighted sampling of the configurations in order to select a better, more convergent, set of configurations.

We replace the integral over all values the $3N$ coordinates take, by a summation like

$$\overline{A(x)} = \frac{\sum_{l=1}^{M} e^{-\frac{H(x_l)}{k_BT}} A(x_l)}{\sum_{l=1}^{M} e^{-\frac{H(x_l)}{k_BT}}}, \tag{15.149}$$

where x_l is a given configuration selected randomly. At the end, the true average is obtained taking the limit when $M \to \infty$, but the idea is to select a finite number of configurations.

Let us choose the sequence of configurations not really at random but selecting the configurations in the phase space that may have a larger contribution, at a given temperature. Let us then consider configurations chosen with a probability $P(x_l)$. We can write

$$\overline{A(x)} = \frac{\sum_{l=1}^{M} e^{-\frac{H(x_l)}{k_B T}} A(x_l) \frac{P(x_l)}{P(x_l)}}{\sum_{l=1}^{M} e^{-\frac{H(x_l)}{k_B T}} \frac{P(x_l)}{P(x_l)}}. \tag{15.150}$$

If we choose the sequence of x_l with the probability $P(x_l)$ instead of randomly, then

$$\overline{A(x)} = \frac{\sum_{l'=1}^{M} e^{-\frac{H(x_l')}{k_B T}} A(x_l') \frac{1}{P(x_l')}}{\sum_{l'=1}^{M} e^{-\frac{H(x_l')}{k_B T}} \frac{1}{P(x_l')}}. \tag{15.151}$$

Let us then choose

$$P(x_l') = e^{-\frac{H(x_l')}{k_B T}}. \tag{15.152}$$

This leads to the expression for the average of the quantity A as

$$\overline{A(x)} = \frac{\sum_{l'=1}^{M} A(x_l')}{M}. \tag{15.153}$$

The question is then how to implement that the sampling of the configurations is given by the Boltzmann distribution probability.

The problem may be solved using the so-called Metropolis algorithm. It goes as follows: (i) Let us consider a given configuration x_l. (ii) We obtain a new configuration x_{l+1} through the transition probability $W(x_l \to x_{l+1})$. (iii) W is then chosen such that, in the limit when $M \to \infty$, we get that the distribution tends to the equilibrium one:

$$P(x_l) \to P_{eq}(x_l) = \frac{1}{Z} e^{-\frac{H(x_l)}{k_B T}}. \tag{15.154}$$

(iv) It is a sufficient condition that

$$P_{eq}(x_l) W(x_l \to x_l') = P_{eq}(x_l') W(x_l' \to x_l). \tag{15.155}$$

(v) This implies that

$$\frac{W(x_l \to x_l')}{W(x_l' \to x_l)} = e^{-\frac{\delta H}{k_B T}}, \tag{15.156}$$

where $\delta H = H(x_l') - H(x_l)$. (vi) The condition (v) does not specify W completely. Two usual choices are

$$W(x_l \to x_l') = \frac{1}{2}\left(1 - \tanh\frac{\delta H}{2k_B T}\right) = \frac{e^{-\frac{\delta H}{k_B T}}}{1 + e^{-\frac{\delta H}{k_B T}}}, \tag{15.157}$$

or

$$\begin{aligned} W(x_l \to x_l') &= e^{-\frac{\delta H}{k_B T}}, \delta H > 0 \\ W(x_l \to x_l') &= 1, \delta H \le 0. \end{aligned} \tag{15.158}$$

Let us take as an example the Ising model in two dimensions in a magnetic field:

$$H = -J \sum_{\langle m,n \rangle} S_m S_n - h \sum_m S_m, \tag{15.159}$$

where $S_m = \pm 1$ are classical variables and m labels a given site in the lattice of N sites. The average spin (magnetization) may be calculated as

$$S = \frac{1}{N}\sum_m \langle S_m \rangle. \tag{15.160}$$

Let us try a different algorithm for the choice of the sequence of configurations. In the two-dimensional lattice, a given site m may be labeled by two indices i, j that run along the two directions. A given site i, j has as first neighbors $i+1, j$ and $i-1, j$ along the x-direction and $i, j+1$ and $i, j-1$ along the y-direction. Let us now, at a given site, perform a spin flip: $S_{i,j} \to -S_{i,j}$. This leads to the Hamiltonian change:

$$\begin{aligned} \delta H =& -J\left(-S_{i,j}S_{i+1,j} - S_{i,j}S_{i,j+1} - S_{i,j}S_{i-1,j} - S_{i,j}S_{i,j-1}\right) - h\left(-S_{i,j}\right) \\ &+J\left(S_{i,j}S_{i+1,j} + S_{i,j}S_{i,j+1} + S_{i,j}S_{i-1,j} + S_{i,j}S_{i,j-1}\right) + h\left(S_{i,j}\right) \\ =& \, 2JS_{i,j}\left(S_{i,j+1} + S_{i,j-1} + S_{i+1,j} + S_{i-1,j}\right) + 2hS_{i,j}. \end{aligned} \tag{15.161}$$

Consider now the interval on the real line between zero and unity and divide it into two parts, one with length $e^{-\frac{\delta H}{k_B T}}$ (A) and the remaining one with length $1 - e^{-\frac{\delta H}{k_B T}}$ (B). (i) Generate a random number between zero and unity. (ii) If the number falls in region A, the spin flip is accepted. If the random number falls in region B, the spin flip is rejected. (iii) In any case, include the configuration that is selected in (ii) in the calculation of the average. This procedure may now be repeated going over all the lattice points in a given sequence. We may go through the lattice sequentially, or we may avoid nearest neighbors jumping to second neighbors or further neighbors. Going over the whole lattice points in some way is called a Monte Carlo step. The procedure is then repeated many times. Typically, a few thousand Monte Carlo steps give a good estimate of the average value of any physical quantity. However, the vicinity of a phase transition requires in general many Monte Carlo steps.

15.10 Goldstone Bosons: Anderson-Higgs Mechanism

In this section, we deviate from the specific context of a condensed matter system but adopt a description that is more general, in the context of field theory.[6]

Global Gauge Invariance
Let us consider a Lagrangian (Lagrangian density) of the form

$$\mathcal{L}(\phi(x), \partial^{\mu}\phi(x)) = \partial_{\mu}\phi^{*}\partial^{\mu}\phi - V(\phi^{*}\phi), \tag{15.162}$$

where $x = (ct, \mathbf{r})$ and the four-derivative $\partial_{\mu} = (\partial_t/c, \nabla_{\mathbf{r}})$, or the Hamiltonian:

$$H = \int d^3x[\Pi^{*}\Pi + \nabla\phi^{*} \cdot \nabla\phi + V(\phi^{*}\phi)], \tag{15.163}$$

where $\Pi = (\partial\phi)/(\partial t)$. Let us consider the transformation $\phi(x) \to U\phi(x)$ where $U = e^{-i\alpha}$ is a global gauge transformation (α is a real constant). Both the Lagrangian and the Hamiltonian are invariants under the transformation. This is therefore a symmetry of the problem.

Let us now consider the equation of motion of the theory:

$$\left(\frac{\partial^2}{\partial t^2} - \nabla^2\right)\phi = -\frac{\partial V}{\partial\phi^{*}}.$$

A possible solution corresponds to a constant and uniform solution $\phi(x) = \phi_0$, as long as ϕ_0 is a minimum of the potential V. Then, if $\phi_0 \neq 0$, clearly the solution is not invariant under the gauge transformation. The solution of the equations of motion does not respect the original symmetry of the problem. There is therefore a spontaneous symmetry breaking. (In condensed matter systems, examples of a spontaneous symmetry breaking are, for instance, that the rotation symmetry is broken in a ferromagnet and the translation symmetry is broken in a crystal). The

[6] For further reading see K. Huang, *Quarks, leptons and gauge fields*, World Scientific.

state of lowest energy, represented by ϕ_0, is infinitely degenerate since the phase α is arbitrary.

Let us then consider as an example

$$V(\phi^*\phi) = \mu^2\phi^*\phi + \lambda(\phi^*\phi)^2 + C.$$

The equation of motion is then

$$\left(\frac{\partial^2}{\partial t^2} - \nabla^2 + \mu^2\right)\phi = -2\lambda(\phi^*\phi)\phi.$$

(The coefficient $\lambda > 0$ for the theory to be stable). If $\mu^2 > 0$, there is only one minimum: $\phi = 0$. However, if $\mu^2 < 0$, there are two symmetrical minima located at values $\phi = \pm\phi_0 \neq 0$. Let us then consider the case when $\phi_0 \neq 0$ and, for convenience, rewrite the potential as

$$V(\phi^*\phi) = \lambda\left(\phi^*\phi - \phi_0^2\right)^2.$$

The state of lowest energy corresponds to

$$(\phi^*\phi)(x) = \phi_0^2.$$

This implies that there is an invariance, because we can choose

$$\phi(x) = \phi_0 e^{i\alpha_0}.$$

Let us analyze the low-energy states close to the minimum energy state and parametrize them like

$$\phi(x) = [\phi_0 + \eta(x)]\, e^{i\alpha(x)},$$

where η and α are real. The Lagrangian can be rewritten as

$$\begin{aligned}
\mathcal{L} &= \partial_\mu\phi^*\partial^\mu\phi - \lambda\left(\phi^*\phi - \phi_0^2\right)^2 \\
&= \partial_\mu[(\phi_0+\eta)e^{-i\alpha}]\partial^\mu[(\phi_0+\eta)e^{i\alpha}] - \lambda[(\phi_0+\eta)e^{-i\alpha}[\phi_0+\eta]e^{i\alpha} - \phi_0^2]^2 \\
&= \partial^\mu\eta\partial_\mu\eta - \lambda(2\phi_0-\eta)^2\eta^2 + (\phi_0+\eta)^2\partial^\mu\alpha\partial_\mu\alpha.
\end{aligned} \tag{15.164}$$

Let us now consider η small. We obtain that

$$\mathcal{L} = \left[\partial^\mu\eta\partial_\mu\eta - 4\lambda\phi_0^2\eta^2\right] + \left[\phi_0^2\partial^\mu\alpha\partial_\mu\alpha\right] + O(\eta^3). \tag{15.165}$$

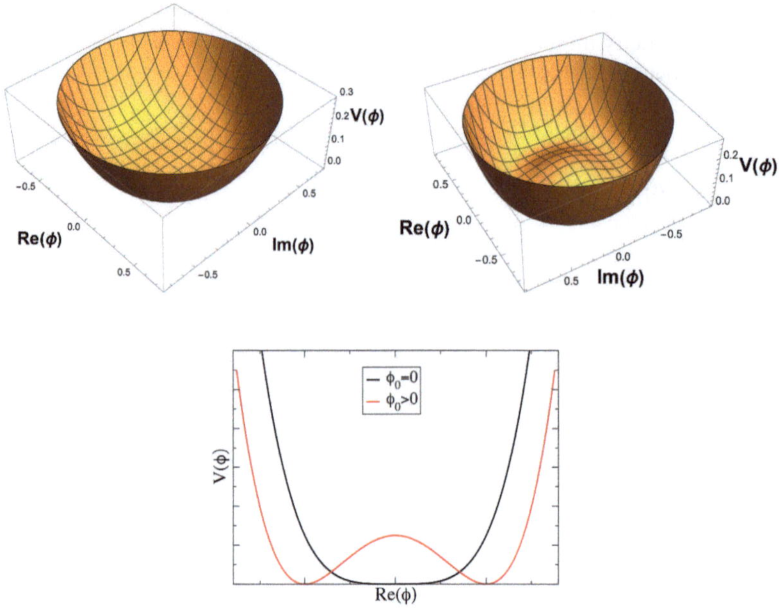

Fig. 15.11 Form of potential in the cases of no symmetry breaking (left panel) and with symmetry breaking (right panel). In the first case, the potential has a minimum at the origin. In the second case, the potential has an infinite number of minima (due to the presence of the gapless mode α). Deviating from the valley of minima, it costs energy due to the presence of the gapped η mode. In the lower panel, a cut along the real axis is shown for the two cases

The first term corresponds to free particles with a mass $2\phi_0\sqrt{\lambda}$, the second term corresponds to massless free particles,[7] and the third term includes interactions that we neglect at this point.

The form of the potential is shown in Fig. 15.11. The η mode corresponds to fluctuations perpendicular to the "valley" that circles the origin. The mode α corresponds to moving around, along the minimum of the valley, which therefore costs no energy.[8] Each value of α_0 gives a possible vacuum state (minimum energy state), which is therefore degenerate. The amplitude of transition between the various values of α_0 vanishes when the volume is infinite (orthogonal catastrophe).

[7] One way to understand the presence or absence of a mass is achieved considering the relativistic expression for the energy as $E^2 = p^2c^2+m^2c^4$.. The term with the derivatives leads to an equation of motion that has as solutions plane waves that satisfy the relativistic relation for the energy and momentum with zero mass. Including in the Lagrangian a term that is quadratic in the field leads to an extra term in the equation of motion, such that looking for a plane wave solution leads to the relativistic equation for a "particle" with finite mass.

[8] It is also usual to identify a massless theory as a theory with gapless modes, and a theory with a mass a theory which has modes with a gap, hence the finite energy necessary to excite such modes.

We obtain the so-called Goldstone theorem: the spontaneous breaking of a global continuous symmetry implies the existence of a particle with no mass. The α mode is called the Goldstone boson. As examples, we have the spin waves in a ferromagnet, which we will study later on, or the phonons in a crystal, therefore energy modes that have zero energy when the momentum goes to zero.

Local Gauge Invariance

Let us now consider a local gauge invariance. In a local gauge transformation, we have that

$$\phi(x) \to U(x)\phi(x),$$

where $U(x) = e^{-i\alpha(x)}$. In this case,

$$\partial^\mu \phi(x) \to U(x)\partial^\mu \phi(x) + \phi(x)\partial^\mu U(x)$$

and the Lagrangian is no longer invariant. To get an invariant Lagrangian, we introduce a gauge field $A^\mu(x)$ such that

$$A^\mu(x) \to A^\mu(x) + \frac{1}{e}\partial^\mu \alpha(x).$$

Here, e is called the "charge" (by similarity to the gauge field in electromagnetism). It is convenient to define the covariant derivative:

$$D^\mu \phi(x) = \left[\partial^\mu + ieA^\mu(x)\right]\phi(x),$$

which transforms in the appropriate way under the local gauge transformation:

$$D^\mu \phi(x) \to U(x) D^\mu \phi(x).$$

Then, defining a new Lagrangian, where we replace

$$\partial^\mu \to D^\mu,$$

this Lagrangian for the fields ϕ remains invariant. Considering now a free term for the gauge field, the Lagrangian that results from the sum of the ϕ field contribution with that of the gauge field can be written as

$$\begin{aligned}
\mathcal{L} &= -\frac{1}{4}F^{\mu\nu}F_{\mu\nu} + (D^\mu\phi)^*(D_\mu\phi) - V(\phi^*\phi) \\
F^{\mu\nu}(x) &= \partial^\mu A^\nu(x) - \partial^\nu A^\mu(x) \\
V(\phi^*\phi) &= \lambda\left(\phi^*\phi - \phi_0^2\right)^2,
\end{aligned} \tag{15.166}$$

for $\phi_0 \neq 0$. This theory is usually called "scalar" electrodynamics, due to its similarities to the usual electrodynamics of charges. The full Lagrangian is invariant under the transformation:

$$\begin{aligned} A^\mu(x) &\to A^\mu(x) + \partial^\mu \omega(x) \\ \phi(x) &\to e^{-ie\omega(x)}\phi(x) \\ \phi^*(x) &\to e^{ie\omega(x)}\phi^*(x). \end{aligned} \tag{15.167}$$

The equations of motion are given by

$$\begin{aligned} \partial_\mu F^{\mu\nu} &= j^\nu \\ D_\mu D^\mu \phi &= -\frac{\partial V}{\partial \phi^*} \\ (D_\mu D^\mu \phi)^* &= -\frac{\partial V}{\partial \phi} \\ j^\nu &= ie\left[\phi^* D^\mu \phi - (D^\mu \phi)^* \phi\right] \\ \partial_\nu j^\nu &= 0. \end{aligned} \tag{15.168}$$

The low-energy solution is given by the class of functions $A^\mu(x) = 0$ and $\phi(x) = \phi_0 e^{i\alpha_0}$.

Let us study the fluctuations around this solution. Let us consider a gauge such that $\phi(x)$ is real: $\phi(x) = \rho(x)$. The equations of motion become

$$\begin{aligned} \partial_\mu F^{\mu\nu} &= -2e^2 A^\nu \phi^* \phi \\ \left(\partial^\mu + ieA^\mu\right)\left(\partial_\mu + ieA_\mu\right)\rho &= 2\lambda\rho\left(\phi_0^2 - \rho^2\right). \end{aligned} \tag{15.169}$$

Since

$$\partial_\nu \partial_\mu F^{\mu\nu} = 0,$$

this implies that

$$\partial_\mu A^\mu(x) = 0,$$

(if $\rho(x) \neq 0$). Let us consider now that we write

$$\rho(x) = \phi_0 + \eta(x)$$

and let us consider η and A small. Then, the equations become

$$\left(\frac{\partial^2}{\partial t^2} - \nabla^2 + 2e^2\phi_0^2\right) A^\mu = 0$$

$$\left(\frac{\partial^2}{\partial t^2} - \nabla^2 + 4\lambda\phi_0^2\right) \eta = 0. \tag{15.170}$$

The first equation corresponds to a spin 1 particle with mass $e\sqrt{2}\phi_0$ and the second equation to a spin 0 particle with mass $2\sqrt{\lambda}\phi_0$. Note that there is no massless particle. The original fields A^1, A^2, ϕ, ϕ^* are replaced by A^1, A^2, A^3, η. This is the so-called Anderson-Higgs mechanism. The spontaneous breaking of a local gauge symmetry "gives" mass to the longitudinal part of the gauge field (the Goldstone boson gives mass to the gauge field).[9]

We may also understand the essence of the existence of Goldstone modes using some arguments that are closer to a condensed matter description.[10] A general enough model may be taken as a field $\mathbf{s}(\mathbf{r})$ coupled linearly to an external source $\mathbf{h}(\mathbf{r})$. One may think of a spin field coupled to a magnetic field. The Helmholtz free energy is a function of the external field and may be written as

$$F(\mathbf{h}) = -\frac{1}{\beta} \ln Tr \left[e^{-\beta\left(H - \int d^d r \mathbf{h}(\mathbf{r})\cdot\mathbf{s}(\mathbf{r})\right)}\right]. \tag{15.171}$$

As before, we may obtain the average of the field using the functional derivative:

$$-\delta_{\mathbf{h}(\mathbf{r})} F(\mathbf{h}) = \langle \mathbf{s}(\mathbf{r}) \rangle = \mathbf{m}(\mathbf{r}). \tag{15.172}$$

Performing a Légendre transform, we may introduce the Gibbs free energy as

$$\Gamma(\mathbf{m}) = F(\mathbf{h}) + \int d^d r \mathbf{h}(\mathbf{r}) \cdot \mathbf{s}(\mathbf{r}). \tag{15.173}$$

We get that

$$\delta_{\mathbf{m}(\mathbf{r})} \Gamma(\mathbf{m}) = \mathbf{h}(\mathbf{r}), \tag{15.174}$$

[9] A standard example is the Meissner effect in superconductivity. The Hamiltonian is invariant under the $U(1)$ gauge transformations since the number of particles is conserved. However, the Cooper pairs break this symmetry. This implies that the gauge field (electromagnetic field) acquires a mass. This is the reason for the expulsion of the magnetic field inside the superconductor since the effect of the mass in the gauge field is a finite penetration length. It is easy to see that if in the equation of motion of a free field we add a mass term and look for a plane wave solution, the momentum acquires an imaginary part with the consequence of a finite penetration length. Actually, in the case of superconductivity, it is called dynamical symmetry breaking because there is no Higgs field to start with. We will study these phenomena in a later chapter.

[10] Here we follow personal notes by Pedro Ribeiro.

since the fields **m** and **h** are conjugate. Also,

$$\delta_{m_\alpha(\mathbf{r})}\delta_{m_\beta(\mathbf{r}')}\Gamma(\mathbf{m}) = (\chi)^{-1}_{\beta\alpha}(\mathbf{r}', \mathbf{r})$$
$$\delta_{h_\beta(\mathbf{r}')}\delta_{h_\alpha(\mathbf{r})}F(\mathbf{h}) = -\chi_{\beta\alpha}(\mathbf{r}', \mathbf{r}). \tag{15.175}$$

We are interested in considering a spontaneous symmetry breaking in the absence of an external field (or external source). For instance, we want to consider a situation such that out of the several components of the vector field **m**, only one component, say the longitudinal component, call it 1, becomes non-vanishing as a phase transition occurs. We may then expect Goldstone modes in the remaining transverse directions. If there is no external field, we know that

$$\delta_{\mathbf{m}(\mathbf{r})}\Gamma(\mathbf{m}) = 0. \tag{15.176}$$

Let us consider that there is some continuous symmetry that acts on the order parameter by transforming it as

$$\mathbf{m} \to R_s\mathbf{m}. \tag{15.177}$$

As a consequence, the Gibbs free energy remains invariant under such a transformation $\Gamma(\mathbf{m}) = \Gamma(R_s\mathbf{m})$. We may consider a transformation that leaves the direction 1 invariant, but that in general changes the other components of the order parameter (for instance, a rotation around the preferred direction selected by the spontaneous symmetry breaking). So we may write that for an infinitesimal transformation of amplitude, ϵ

$$R_s\mathbf{m} = \mathbf{m} + \epsilon\delta\mathbf{m}. \tag{15.178}$$

We may take

$$\delta m_\alpha = t_{\alpha\beta}m_\beta, \tag{15.179}$$

where $\beta = 1, 2, \ldots, n$, with n the number of components. Considering the effect of the symmetry operation on Eq. (15.176) and imposing that the zero field condition holds order by order up to the order ϵ leads to

$$\sum_\beta \int d^d r' \delta_{m_\beta(\mathbf{r}')}\delta_{m_\alpha(\mathbf{r})}\Gamma(\mathbf{m})\delta m_\beta(\mathbf{r}') = 0. \tag{15.180}$$

This leads to

$$\sum_{\beta,\gamma} \int d^d r' \,(\chi)^{-1}_{\beta\alpha}(\mathbf{r}', \mathbf{r})t_{\beta\gamma}m_\gamma = 0. \tag{15.181}$$

The order parameter is of the type

$$m_\gamma(\mathbf{r}') = m\delta_{\gamma,1}, \tag{15.182}$$

which leads to

$$\sum_\beta (\chi(\mathbf{k}=0))^{-1}_{\alpha\beta} t_{\beta 1} m = 0, \tag{15.183}$$

which implies that for the transverse directions, $\beta = 2, \ldots, n$ the momentum space inverse susceptibility has zero eigenvalues. These are the Goldstone modes found in the transverse directions.

15.11 Hohenberg-Mermin-Wagner Theorem

Let us consider once again the classical Heisenberg model:

$$H = -J \sum_{<i,j>} \mathbf{S}_i \cdot \mathbf{S}_j$$

and let us consider that the spin vector has n components like

$$\mathbf{S} = (S_1, S_2, \ldots, S_n),$$

with $|\mathbf{S}| = 1$. As before, for $n = 1$, we have the Ising model; for $n = 2$, the $X - Y$ model; for $n = 3$, the Heisenberg model; and for $n = \infty$, the so-called spherical model. The partition function is obtained integrating

$$Z = \int d^N \Omega e^{-\frac{H}{k_B T}}.$$

Consider now the continuum limit and rewrite the partition function as

$$Z = \int D\mathbf{S} e^{-\int d^d x \left[(\nabla \mathbf{S})^2 + uS^2 + v(S^2)^2\right]}.$$

Note that $(\nabla\mathbf{S})^2 = (\nabla S_1)^2 + \cdots + (\nabla S_n)^2$. For $n = 2, 3, \ldots$, we have excitations in the low-temperature phase that are of the spin-wave type where $|\mathbf{S}|$ =const, but $\nabla\mathbf{S} \neq 0$. The Hamiltonian for the spin waves (this will be treated again later in detail) is

$$\begin{aligned} H_{spin\text{-}waves} &= J' \int d^d x (\nabla S)^2 \\ &= J'' \sum_{\mathbf{q}} q^2 |S_{\mathbf{q}}|^2. \end{aligned} \tag{15.184}$$

Using the equipartition theorem, we can establish that

$$J''q^2\langle|S_{\mathbf{q}}|^2\rangle = \frac{1}{2}.$$

Therefore,

$$\langle(\delta\mathbf{S}(\mathbf{r}))^2\rangle \sim \int d^d q\langle|S_{\mathbf{q}}|^2\rangle \sim \int_0^{\Lambda} dq q^{d-1}\frac{1}{q^2}.$$

This integral diverges when $q \sim 0$, if $d \leq 2$. We showed before that if $d > 4$, the mean-field solution should work well. This is the upper critical dimension for the phase transition to an ordered phase. However, if $d \leq 2$, the fluctuations should destroy the ordered phase, since the fluctuations diverge, by the simple above argument. Therefore, for $d \leq 2$, there should not be a phase transition. If $n > 2$, then $d = 2$ is the lower critical dimension. But, as we will see later, if $n = 2$ and $d = 2$, there is a transition of another type in which there is no spontaneous magnetization in the low-temperature phase, but there is long-range order of topological defects (vortices).

Let us now show that the above argument is correct and that indeed $d = 2$ is the lower critical dimension: that is, for dimensions equal or lower, there is no phase transition at finite temperature. This is the essence of the so-called Hohenberg-Mermin-Wagner, which we will next prove [1].

Let us first consider the so-called Bogoliubov inequality:

$$\frac{1}{2}\langle\{A, A^{\dagger}\}\rangle\langle[[C, H], C^{\dagger}]\rangle \geq k_B T|\langle[C, A]\rangle|^2.$$

Let us prove this inequality. First, define

$$(A, B) = \sum_{i,j}{}' \langle i|A|j\rangle^*\langle|B|j\rangle\frac{W_i - W_j}{E_j - E_i}.$$

Here,

$$W_i = \frac{e^{-\beta E_i}}{Tre^{-\beta H}}.$$

It is easy to show that

$$\frac{W_i - W_j}{E_j - E_i} > 0.$$

Also

$$(A, A) \leq \frac{1}{2}\beta\langle\{A, A^\dagger\}\rangle$$
$$(A, A)(B, B) \geq |(A, B)|^2. \tag{15.185}$$

Choosing $B = [C^\dagger, H]$, then

$$(A, B) = \langle[C^\dagger, A^\dagger]\rangle$$
$$(B, B) = \langle[C^\dagger, [H, C]]\rangle. \tag{15.186}$$

This shows the inequality.

Let us now apply it to the Heisenberg model:

$$H = -\sum_{\mathbf{R},\mathbf{R}'} J(\mathbf{R} - \mathbf{R}')\mathbf{S}(\mathbf{R}) \cdot \mathbf{S}(\mathbf{R}') - h\sum_{\mathbf{R}} S_z(\mathbf{R})e^{-i\mathbf{q}\cdot\mathbf{R}}.$$

Here, we are taking that $J(0) = 0$, $J(\mathbf{R}) = J(-\mathbf{R})$. Also the interaction J has a finite range so that

$$\sum_{\mathbf{R}} R^2|J(\mathbf{R})|$$

is convergent. Let us choose

$$C = S_+(\mathbf{k}), A = S_-(-\mathbf{k} - \mathbf{q}).$$

Then, the Bogoliubov inequality leads to

$$\langle\{S_+(\mathbf{k} + \mathbf{q}), S_-(-\mathbf{k} - \mathbf{q})\}\rangle \leq N^2 k_B T S_z^2$$
$$\frac{1}{\frac{1}{N}\sum_{k'}[J(\mathbf{k}' - J(\mathbf{k}' - \mathbf{q})]\,\langle S_z(-\mathbf{k}')S_z(\mathbf{k}') + \frac{1}{4}\{S_+(\mathbf{k}'), S_-(-\mathbf{k}')\}\rangle + \frac{N}{2}hS_z}, \tag{15.187}$$

where

$$S_z = \frac{1}{N}\sum_{\mathbf{R}}\langle S_z(\mathbf{R})\rangle e^{i\mathbf{q}\cdot\mathbf{R}}.$$

Using that

$$\sum_{\mathbf{k}'}\langle S_i(-\mathbf{k}')S_i(\mathbf{k}')\rangle = N\sum_{\mathbf{R}}\langle S_i(\mathbf{R})S_i(\mathbf{R})\rangle = N^2\langle S_i(\mathbf{R}_0)S_i(\mathbf{R}_0)\rangle$$

and summing Eq. (15.187) over $\mathbf{q}$, we obtain that

$$S(S+1) > 2k_BTS_z^2\frac{1}{N}\sum_{\mathbf{k}}\frac{1}{hS_z + S(S+1)\sum_{\mathbf{R}} R^2J(\mathbf{R})k^2}. \tag{15.188}$$

When the volume of the system goes to infinity, we get that the sum over the momenta is replaced by an integral like

$$S(S+1) > \frac{2k_BTS_z^2}{\rho}\int_{1stzone}\frac{d^dk}{(2\pi)^d}\frac{1}{S(S+1)\sum_{\mathbf{R}} R^2J(\mathbf{R})k^2 + hS_z}. \tag{15.189}$$

Here, $1/\rho$ is the volume per spin. Denoting $\bar{\omega} = S(S+1)\sum_{\mathbf{R}} R^2J(\mathbf{R})$, we can write that

$$S_z^2 < \frac{\rho S(S+1)}{2k_BT}\frac{1}{\int_{1stzone}\frac{d^dk}{(2\pi)^d}\frac{1}{\bar{\omega}k^2+hS_z}}. \tag{15.190}$$

The first Brillouin zone can be taken as a sphere of radius k_0. Using that

$$\frac{1}{\pi}\int_0^{k_0} dk\frac{1}{k^2+a} = \frac{1}{\pi}\frac{1}{\sqrt{a}}\arctan\frac{k_0}{\sqrt{a}}, d=1$$
$$\frac{1}{2\pi}\int_0^{k_0} dk\frac{k}{k^2+a} = \frac{1}{4\pi}\ln\frac{k_0^2+a}{a}, d=2$$
$$\frac{1}{2\pi^2}\int_0^{k_0} dk\frac{k^2}{k^2+a} = \frac{1}{2\pi^2}\{k_0 - \sqrt{a}\arctan\frac{k_0}{\sqrt{a}}\}, d=3. \tag{15.191}$$

We have therefore that in the limit that $h \to 0$

$$S_z < const\frac{h^{1/3}}{T^{2/3}}, d=1$$
$$S_z < \frac{const}{T^{1/2}}\frac{1}{|\ln h|^{1/2}}, d=2$$
$$S_z^2 < \frac{\rho S(S+1)}{2k_BT}\frac{2\pi^2\bar{\omega}}{k_0}, d=3. \tag{15.192}$$

In the first two cases, $S_z \to 0$ as $h \to 0$. In the case of $d = 3$, the magnetization is independent of h and nothing is concluded. The conclusion proves the theorem: in $d = 1, 2$, there is no spontaneous magnetization.

Hohenberg-Mermin-Wagner Theorem and Graphene
One of the unexpected properties of graphene is that it is not supposed to be stable. Indeed, the Hohenberg-Mermin-Wagner theorem, applied to a two-dimensional

system, leads to the conclusion that the fluctuations should destroy the crystalline long-range order. A clue to the solution of the problem is that the system is not strictly two-dimensional, but it is instead a two-dimensional system immersed in a three-dimensional space and, therefore, oscillations in the third spatial direction are possible.

A simple, and classical, consideration similar to that used in a two-dimensional magnetic system is that in the harmonic approximation, fluctuations diverge. Recalling the argument, we expect the appearance of terms in the Hamiltonian that are of the type

$$H_h = \frac{\rho}{2} \sum_{\mathbf{k}} k^2 n_{\mathbf{k}}^2, \tag{15.193}$$

where ρ is the stiffness of the harmonic oscillations and $n_{\mathbf{k}}$ are the variables associated with the density modes of a given momentum. The classical argument uses the equipartition theorem to estimate the average value of the quadratic fluctuations leading to

$$\left\langle \frac{\rho}{2} k^2 n_{\mathbf{k}}^2 \right\rangle \sim k_B T \tag{15.194}$$

and therefore

$$\langle n_{\mathbf{k}}^2 \rangle \sim \frac{T}{\rho k^2}. \tag{15.195}$$

Taking the Fourier transform, the leading term to the real space fluctuations of the density may be calculated using

$$\langle n(\mathbf{r})^2 \rangle \sim \int d^2k \frac{1}{k^2} \sim \int dk \frac{k}{k^2} \sim \int_{k_0}^{\Lambda} \frac{dk}{k} \sim \ln \frac{\Lambda}{k_0}. \tag{15.196}$$

There is therefore an infrared divergence (when $k_0 \to 0$), which leads to long-range fluctuations that diverge and lead to the instability of the lattice. This harmonic approximation is corresponding to the spin-wave theory previously discussed. The result may also be shown with more care using a Bogoliubov-type inequality.

While the harmonic fluctuations destroy the long-range order (in this case, the positional order of the atoms of the lattice), strong nonharmonic terms may stabilize the lattice structure. This occurs through rippling and crumbling effects in the third, perpendicular direction, which stabilize the structure in the case of graphene and other two-dimensional systems, that have attracted considerable attention recently.

15.12 First-Order Phase Transitions

15.12.1 Metastable States

In addition to equilibrium states, characterized by certain stability criteria, there are other thermodynamic states of interest. One example are unstable states, which are stationary states such that any small deviation implies that the system evolves to a different state. Another example are metastable states that are stable if the deviation is small but that are not the most stable states in a system. Larger deviations from the stationary condition of a metastable state may lead to an evolution of the state, away from the metastable state into a state with a lower free energy, such as another metastable state or a stable state, in the sense of the lowest free energy. The deviations may have origin in coupling to some external environment, in the form of thermal fluctuations, or, at low temperatures, quantum fluctuations may lead to the evolution from a metastable state to another, eventually more stable state. In quantum mechanics, a related situation is tunneling between different states (for instance, spatial states).

An example of a system that has the various types of states was considered before, in the form of the van der Waals gas. This system was studied in some detail and stable, metastable, and unstable states were identified. We used the stability criterion that $(\partial p/\partial V)_T < 0$ in equilibrium, which allowed us to realize that there are regions where this condition is violated (unstable states) and also some regimes where this condition is satisfied. In addition, some solutions of the equation of state have different free energies that allow to separate the stable from the metastable states (Maxwell construction).

Using the minimization of the Gibbs free energy (or chemical potential) as a criterion to identify the stable equilibrium state, the isothermal of stable states was found, leading to a regime where a mixture of the two phases is found. This mixture corresponds to the first-order transition between the two phases with different densities (or volume per particle). Starting from high temperatures, larger than a critical temperature where the width of this coexistence region vanishes, and allowing the temperature to decrease past the critical temperature, T_c, the coexistence region of the first-order transition develops continuously (starting at the critical point of a second-order phase transition) for $T < T_c$. In this coexistence region, we can separate the region of metastable states from that of unstable states, and the separation between the two regions is called the spinodal line. Considering the chemical potential as a function of the concentration for a given temperature $T < T_c$, the spinodal points are the points where $(\partial \mu/\partial c)_T = 0$. The regime where $(\partial \mu/\partial c)_T < 0$ is the unstable regime and when $(\partial \mu/\partial c)_T > 0$ is the stable regime. In Fig. 15.12, we show a phase diagram for the pressure or temperature as a function of the volume (such as in the van der Waals equation). In the high-temperature regime, the system is in a disordered phase. Lowering the temperature or pressure at constant volume, we first cross the critical line, and we enter the regime where first-order transitions take place. In this regime, the spinodal line separates the

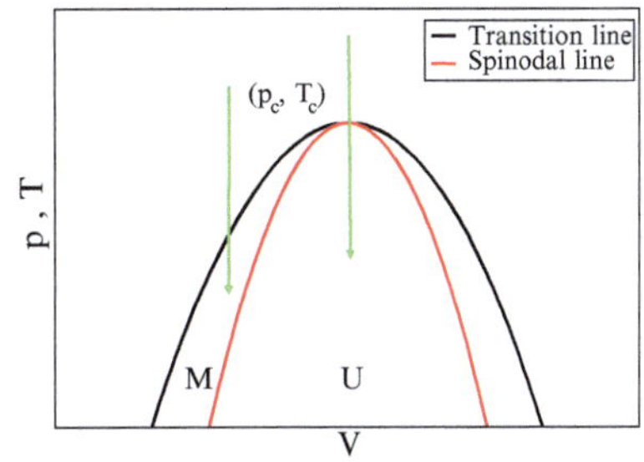

Fig. 15.12 Schematic representation of the spinodal line separating the metastable (M) and unstable (U) regimes

metastable from the unstable regimes. Starting from the high-temperature phase, we may quench the system to either the metastable or the unstable regimes, represented by the arrows. The full dynamics after the quench is complicated and will not be addressed here (see a detailed discussion in [GuntonDroz]).

Here, we will be interested in the dynamics of a possible evolution from a metastable state to the stable state. The system will evolve to minimize its free energy. This is accomplished if fluctuations lead to the creation of so-called droplets of the stable phase. Droplets are regions in real space where the system has the parameters that are associated with the stable phase (for instance, a concentration that is close to that of the concentration of the stable phase). The droplet is therefore a small region of the stable phase immersed in the larger region of the metastable phase. The process of the appearance of these droplets may be called a nucleation of the stable state. As we will show next, if the droplet that is created is large enough, it is favorable that it grows, and eventually the region will expand and the whole system will change its state. If the droplet is not large enough, it will shrink and its size will converge to zero. If it is growing, and is large enough, the process of its growth can be understood as interface motion, where one phase becomes dominant over the other. This will be considered next. A signature of this growth may be identified measuring the structure factor that measures correlations as a function of distance or as a function of momentum. As the droplet grows in size, the momentum associated with the length scale of the droplets decreases, and the maximal value of the momentum structure factor is displaced to smaller momentum values, as time increases.

15.12.2 Tunneling Rate of a Single Particle

Before turning to the study of the droplets and their possible growth in the context of a thermodynamic system, it is useful to consider first the related problem of the tunneling of a particle over (or through) some potential barrier.

Consider a quantum particle in a potential $V(x)$, where, to simplify, we consider a one-dimensional problem (two typical cases are shown in Fig. 15.13). The probability amplitude that starting from a given spatial location, x_i at a given time

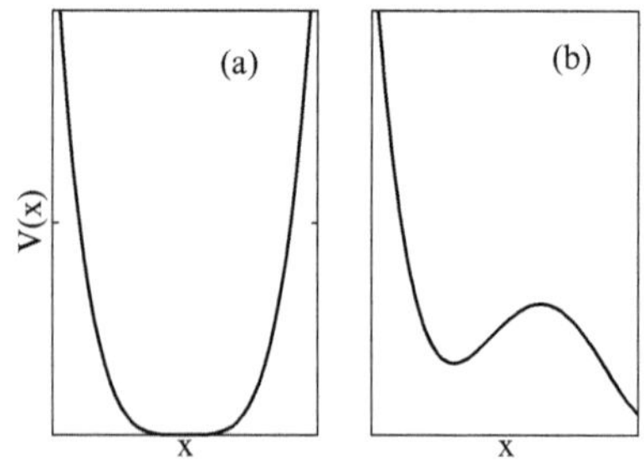

Fig. 15.13 One-dimensional potentials: (**a**) potential with a single minimum. (**b**) Potential with a local minimum, which is not a global minimum

t_i it will be found at x_f at time t_f is given by

$$\langle x_f | e^{-\frac{i}{\hbar}H(t_f - t_i)} | x_i \rangle, \tag{15.197}$$

where the Hamiltonian is

$$H = \frac{p^2}{2m} + V(x), \tag{15.198}$$

with m as the particle's mass. The time interval may be divided in many steps, in number N, that can be taken very large, such that the evolution operator is divided in many steps each corresponding to the time evolution over an infinitesimal time lapse, $\Delta t = (t_f - t_i)/N$. Inserting decompositions of unity in terms of the positions and momenta, we can obtain the path integral expression:[11]

$$\langle x_f | e^{-\frac{i}{\hbar}H(t_f - t_i)} | x_i \rangle = \int DxDpe^{\frac{i}{\hbar}S}, \tag{15.199}$$

where the action (integral over time of the Lagrangian, written as a function of the coordinates and momenta) is given by

$$S = \int_{t_i}^{t_f} \left(p\frac{dx}{dt} - H(x(t), p(t), t) \right) dt \tag{15.200}$$

and where the limit $N \to \infty$ was taken and

$$Dx = \prod_{k=1}^{N-1} dx_k; \qquad Dp = \prod_{k=0}^{N-1} \frac{dp_{k+1}}{2\pi\hbar}. \tag{15.201}$$

[11] Note that this procedure is somewhat similar to the Chapman-Kolmogorov equation.

The integral over the momenta may be carried out exactly, since the Hamiltonian is quadratic in the momentum, leading to

$$\langle x_f | e^{-\frac{i}{\hbar}H(t_f - t_i)} | x_i \rangle = \int Dx e^{\frac{i}{\hbar}\int_{t_i}^{t_f}\left(\frac{1}{2}m(\frac{dx}{dt}(t))^2 - V(x(t))\right)dt}, \tag{15.202}$$

where the Lagrangian is now written in terms of the coordinates and velocities.

It is convenient to define imaginary time as $\tau = it$, where t is real time. Also, let us define $T = i(t_f - t_i)$ and consider that the time evolution occurs between positions $x_i = x(-T/2)$ and $x_f = x(T/2)$. The probability amplitude of the transition between the initial and final position is written as

$$\langle x_f | e^{-\frac{1}{\hbar}HT} | x_i \rangle = \int Dx e^{-\frac{1}{\hbar}S}, \tag{15.203}$$

where the action (now called the Euclidean action, defined over imaginary time) is[12]

$$S = \int_{-T/2}^{T/2} d\tau \left[\frac{1}{2}\left(\frac{dx}{d\tau}\right)^2 + V \right], \tag{15.204}$$

where, to simplify, we considered that the mass is unity. Recall that the integral over Dx is understood to be over all functions $x(\tau)$ with the boundary conditions that are fixed by the initial and final positions. Both in the real time and in the imaginary time descriptions, the trajectories that contribute the most to the transition amplitude are those that render the action stationary, $\delta S = 0$. These are the classical trajectories (small $\hbar$ limit), $\bar{x}$, and are, therefore, from the infinite quantum trajectories those that give the stationary condition of the action (and therefore the Euler-Lagrange equations). Also, note that in the imaginary time Euclidean description, the action is the integral over some Lagrangian where $V \to -V$. Therefore, in Euclidean time, the Hamiltonian or energy is of the type

$$E = \frac{1}{2}\left(\frac{d\bar{x}}{d\tau}\right)^2 - V(\bar{x}). \tag{15.205}$$

This is illustrated in Fig. 15.14 for the case of a potential with a local minimum.

A case of interest here is a potential $V(x)$ that has some local minimum at a given position, say $x = 0$, but which is not the global minimum of the potential. Between the location of the minimum at $x = 0$ and the global minimum, there is a potential barrier of some finite amplitude. If a particle is located near $x = 0$ or in its vicinity, classically it will not be able to cross the potential barrier. If the particle is coupled

[12] For further reading see Curtis G. Callan, Jr. and Sidney Coleman, Physical Review D **16**, 1762 (1977) and S. Coleman, "The uses of instantons", 1977 Lectures at International School of Subnuclear Physics: Ettore Majorana.

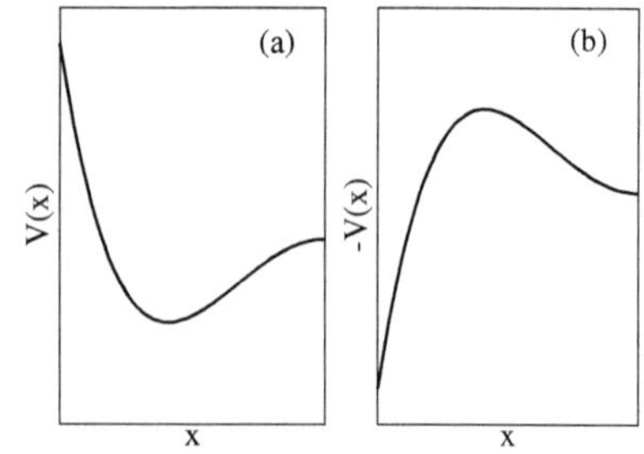

Fig. 15.14 (**a**) One-dimensional potential with a local minimum and (**b**) its symmetric that appears in the Euclidean action

to some external reservoir, it may receive enough energy to overcome the potential barrier (such as a result of thermal fluctuations associated with energy exchanges with the heat reservoir). As is well-known, the particle may also cross the barrier via quantum tunneling, and it may be found at some position x_f in the classically forbidden region, such as the position of the global potential energy minimum.

The stationary points (stationary, classical, trajectory $\bar{x}(t)$) of the action satisfy

$$\frac{\delta S}{\delta \bar{x}} = -\frac{d^2\bar{x}}{d\tau^2} + \frac{dV}{d\bar{x}} = 0. \tag{15.206}$$

We may now expand the action around its stationary point (in general, there may be various stationary points but let us assume there is a single stationary point). The expansion may be carried out considering that any trajectory can be written as

$$x(\tau) = \bar{x}(\tau) + \delta x, \tag{15.207}$$

where δx is the deviation from the classical trajectory. The stationary condition implies that first-order terms, in the deviation from the stationary trajectory, vanish. The leading order term is just the action replaced by the classical trajectory, and the next term is the one that results from the quadratic deviation from the classical solution. The second derivative of the action (at the stationary point) leads to an operator whose eigenfunctions and eigenvalues satisfy

$$-\frac{d^2x_n}{d\tau^2} + \frac{d^2V}{dx^2}x_n = \lambda_n x_n. \tag{15.208}$$

The deviation from the classical trajectory may be expanded in terms of the eigenfunctions of the second derivative operator leading to

$$x(\tau) = \bar{x}(\tau) + \sum_{n=1}^{\infty} c_n x_n(\tau), \tag{15.209}$$

where c_n are the expansion coefficients. In order to satisfy the boundary conditions, the classical trajectory has to satisfy these conditions, and it also imposes that

$$\int_{-T/2}^{T/2} d\tau x_n(\tau) x_m(\tau) = \delta_{nm}$$

$$x_n\left(\pm\frac{T}{2}\right) = 0. \tag{15.210}$$

Summing over trajectories is therefore the same as summing over the coefficients of the expansion. Therefore, we may interpret

$$Dx \sim \prod_n dc_n. \tag{15.211}$$

Up to second order, we get therefore that

$$\langle x_f|e^{-\frac{1}{\hbar}HT}|x_i\rangle \sim e^{-\frac{1}{\hbar}S(\bar{x})} \prod_n \lambda_n^{-1/2} [1 + O(\hbar)]$$

$$\sim e^{-\frac{1}{\hbar}S(\bar{x})} N \left[det\left[-\partial_\tau^2 + \frac{d^2V}{dx^2}(\bar{x})\right]\right]^{-1/2}, \tag{15.212}$$

since all the integrals are gaussian and assuming that all the eigenvalues are positive. N is some normalization factor that it is convenient to introduce here even though its specific value is not needed. Note, however, that not all eigenvalues are positive. Indeed, there is a zero eigenvalue with an eigenfunction that is of the type

$$x_1 \sim \frac{d\bar{x}}{d\tau}. \tag{15.213}$$

We will return to this point shortly.

Single-Well Potential

Consider a single-well potential as shown in Fig. 15.13, with a minimum at $x = 0$, with value $V(x) = 0$. Consider that the particle is initially at that position and that the final point of the trajectory is the same: $x_i = x_f = 0$. The solution of Eq. (15.206) is then $\bar{x}(\tau) = 0$, which leads to $S = 0$. Using the notation $\omega^2 = d^2V/dx^2(0)$, we can write that, for this potential and trajectory,

$$\langle 0|e^{-\frac{1}{\hbar}HT}|0\rangle = N\left[det\left(-\partial_\tau^2 + \omega^2\right)\right]^{-1/2}. \tag{15.214}$$

We need now to evaluate the determinant.

The eigenvalue problem of Eq. (15.208) is of the type

$$\left(-\partial_\tau^2 + W(\tau)\right)\psi = \lambda\psi. \tag{15.215}$$

The operator acting on functions that vanish at $\pm T/2$ has an eigenvalue λ_n if the eigenfunctions satisfy the boundary conditions $\psi_{\lambda_n}(\pm T/2) = 0$, as required by the physical condition of the problem we are solving. Let us now consider functions $\psi_\lambda(\tau)$ such that they are solutions of Eq. (15.215) but with boundary conditions at the initial time:

$$\begin{aligned}\psi_\lambda(-T/2) &= 0\\ \partial_\tau\psi_\lambda(-T/2) &= 1.\end{aligned} \tag{15.216}$$

It can be shown that

$$det\left[\frac{-\partial_\tau^2 + W^{(1)} - \lambda}{-\partial_\tau^2 + W^{(2)} - \lambda}\right] = \frac{\psi_\lambda^{(1)}\left(\frac{T}{2}\right)}{\psi_\lambda^{(2)}\left(\frac{T}{2}\right)}. \tag{15.217}$$

We may now define a normalization as

$$\frac{det\left(-\partial_\tau^2 + W\right)}{\psi_0\left(\frac{T}{2}\right)} = 2\pi\hbar N^2, \tag{15.218}$$

since it is independent of the function W. We conclude that

$$N\left[det\left(-\partial_\tau^2 + W\right)\right]^{-1/2} = (2\pi\hbar)^{-1/2}\left[\psi_0\left(\frac{T}{2}\right)\right]^{-1/2}. \tag{15.219}$$

We may now apply this result to the case when $W = \omega^2$ (which, incidentally, is the case of an harmonic oscillator). A function that satisfies Eq. (15.215) and the boundary conditions Eq. (15.216) is

$$\psi_0(\tau) = \frac{1}{\omega}\sinh\left[\omega\left(\tau + \frac{T}{2}\right)\right]. \tag{15.220}$$

Therefore,

$$\psi_0\left(\frac{T}{2}\right) = \frac{1}{\omega}\sinh(\omega T). \tag{15.221}$$

In the limit that $T \to \infty$, we get that

$$N\left[det\left(-\partial_\tau^2 + \omega^2\right)\right]^{-1/2} = \left(\frac{\omega}{\pi\hbar}\right)^{1/2} e^{-\frac{\omega T}{2}}. \tag{15.222}$$

Considering the eigenstates of the Hamiltonian

$$H|n\rangle = E_n|n\rangle, \tag{15.223}$$

we can write that

$$\langle x_f|e^{-\frac{1}{\hbar}HT}|x_i\rangle = \sum_n e^{-\frac{1}{\hbar}E_nT}\langle x_f|n\rangle\langle n|x_i\rangle. \tag{15.224}$$

Taking the limit $T \to \infty$, we see that the dominant term is that of the ground state corresponding to the eigenvalue E_0. Therefore, comparing with the result of Eq. (15.222), we get that $E_0 = \hbar\omega/2$, as expected of an harmonic oscillator.

Double-Well Potential: Instanton Solutions

Consider now a symmetric double-well potential with minima at $x = \pm a$, with zero potential, shown in Fig. 15.15. As before, call $d^2V(x)/dx^2(x = \pm a) = \omega^2$. Consider trajectories such that $x_i = -a, x_f = a$, or $x_i = -a, x_f = -a$. The trajectories have zero energy, which implies that

$$\frac{dx}{d\tau} = \sqrt{2V}. \tag{15.225}$$

In the case that the trajectory starts at $x_i = -a$ and ends at $x_f = a$, we may consider infinite ways to go from that initial point to that final point. The simplest is such that x interpolates in a monotonic way from x_i to x_f. We can write that

$$S_0 = \int d\tau\left[\frac{1}{2}\left(\frac{dx}{d\tau}\right)^2 + V\right] = \int d\tau\left(\frac{dx}{d\tau}\right)^2 = \int_{-a}^{a} dx\sqrt{2V}. \tag{15.226}$$

This solution is called an instanton and is such that $dx/d\tau > 0$ and does not vanish at any point. At large times, the coordinate approaches a and can be expanded as

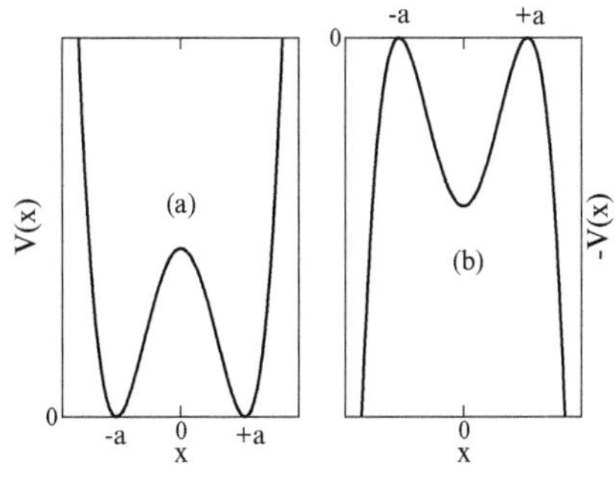

Fig. 15.15 Double-well symmetric potential and its symmetric

$dx/d\tau = \omega(a - x)$ which implies

$$(a - x) = e^{-\omega\tau}. \tag{15.227}$$

This is a solution that is well localized in time with a scale $1/\omega$, which means that the value of the coordinate changes fast in (Euclidean) time from x_i to x_f, with the time scale $1/\omega$. Note that we may also consider the opposite change, considering that the coordinate changes from x_f to x_i (an anti-instanton) and therefore we may consider a sequence of instantons and anti-instantons, in a given number. The only restriction is that we start at a given position and end at a given position, which may be the same potential minimum or the symmetric minimum. In general, we need to sum over all possible numbers of instantons and anti-instantons but, for the moment, let us consider the contribution of one instanton, which applies to the case that we start from $x_i = -a$ and end in $x_f = a$.

Recall that an eigenfunction with zero eigenvalue can be found as

$$x_1 = S_0^{-1/2}\frac{d\bar{x}}{d\tau}, \tag{15.228}$$

where we have added the appropriate normalization, considering that the eigenfunction is normalized in the interval $[-a, a]$. Therefore,

$$\frac{d^2x_1}{d\tau^2} = \frac{d^2V(\bar{x})}{dx^2}x_1. \tag{15.229}$$

Since the instanton is well localized, it may be characterized by the location in time where the change between the two position points is maximal. Call it $\tau = \tau_1$. Recall that in the gaussian integration over the fluctuations around the classical trajectory, we may replace the integration over the coordinates by the integration over the coefficients: c_n. We may now write that for the instanton,

$$dx = \frac{dx}{d\tau}d\tau = x_1 dc_1 = \frac{d\bar{x}}{d\tau_1}d\tau_1 = x_1 S_0^{1/2} d\tau_1. \tag{15.230}$$

As a consequence, we can replace

$$(2\pi\hbar)^{-1/2}\, dc_1 = \left(\frac{S_0}{2\pi\hbar}\right)^{1/2} d\tau_1. \tag{15.231}$$

Therefore,

$$Dx = \prod_n (2\pi\hbar)^{-1/2}\, dc_n = \left[\prod_{n\neq 1} (2\pi\hbar)^{-1/2}\, dc_n\right]\left[\left(\frac{S_0}{2\pi\hbar}\right)^{1/2} d\tau_1\right]. \tag{15.232}$$

The integration over τ_1 is just the whole Euclidean time interval, T. Therefore, we arrive at

$$\langle a|e^{-\frac{1}{\hbar}HT}|-a\rangle = NT\left(\frac{S_0}{2\pi\hbar}\right)^{1/2} e^{-S_0/\hbar}\left[det'\left(-\partial_\tau^2+\frac{d^2V}{dx^2}\right)\right]^{-1/2}, \tag{15.233}$$

where in the determinant we omit its zero eigenvalue, since it has already been taken care of, integrating over τ_1. This is the result for one instanton.

We now have to sum over all possible combinations of instantons and anti-instantons. This can be done considering that all the instantons and anti-instantons are very localized in time, which means that they do not interfere. We may consider a sequence of times $\tau_1, \tau_2, \ldots$, which are the locations of the instantons and anti-instantons, such that T is taken as much larger than the size of each instanton, $1/\omega$. If we have n separated instantons, in the first approximation, they do not interfere and we can take that $S = nS_0$. The single-well solution was

$$\left(\frac{\omega}{\pi\hbar}\right)^{1/2} e^{-\frac{\omega T}{2}}. \tag{15.234}$$

It is usual to write the solution of the double-well with n instantons as

$$\left(\frac{\omega}{\pi\hbar}\right)^{1/2} e^{-\frac{\omega T}{2}} K^n, \tag{15.235}$$

where K is to be determined. It can be determined considering the case of one instanton. It is given by

$$K = \left(\frac{S_0}{2\pi\hbar}\right)^{1/2}\left[\frac{det\left(-\partial_\tau^2+\omega^2\right)}{det'\left(-\partial_\tau^2+\frac{d^2V}{dx^2}\right)}\right]^{1/2}. \tag{15.236}$$

If we have n instantons, we have to integrate over the times, which leads to

$$\int_{-T/2}^{T/2} d\tau_1 \int_{-T/2}^{\tau_1} d\tau_2 \cdots \int_{-T/2}^{\tau_{n-1}} d\tau_n = \frac{T^n}{n!}. \tag{15.237}$$

We have n even if $-a \to -a$ and n is odd if $-a \to a$, which leads to

$$\langle -a|e^{-\frac{1}{\hbar}HT}|-a\rangle = \left(\frac{\omega}{\pi\hbar}\right)^{1/2} e^{-\frac{\omega T}{2}} \sum_{n,even} \frac{\left(Ke^{-S_0/\hbar}T\right)^n}{n!} \tag{15.238}$$

and similarly if the end point is $x_f = a$, taking only n odd. The final expression is

$$\langle \pm a| e^{-\frac{HT}{\hbar}} | -a\rangle = \left(\frac{\omega}{\pi\hbar}\right)^{1/2} e^{-\frac{\omega T}{2}} \frac{1}{2}\left[e^{Ke^{-S_0/\hbar}T} \mp e^{-Ke^{-S_0/\hbar}T}\right]. \qquad (15.239)$$

This suggests that we have two energy states:

$$E_\pm = \frac{1}{2}\hbar\omega \pm \hbar K e^{-\frac{S_0}{\hbar}}. \qquad (15.240)$$

Note that the second term is exponentially small with respect to the first one.

Evolution from a Metastable State

The case of a metastable state, as illustrated in Fig. 15.16, is qualitatively different. An example of a metastable state is a trajectory that tunnels through a barrier or the particle has a trajectory that somewhere satisfies

$$\frac{d\bar{x}}{d\tau} = 0, \qquad (15.241)$$

as found in a trajectory where the particle bounces from one point to another and returns to the original position. The bounce was considered above but as a result of a sum of trajectories (for instance an instanton and one anti-instanton) such that for each $\bar{x}(\tau)$ is monotonic. If $d\bar{x}/d\tau$ has a zero, then x_1 has a node and therefore is not the lowest eigenvalue eigenfunction (which should have no nodes). Since the eigenvalue associated with x_1 is zero, this means that there must be an eigenvalue that is negative. This means that this solution is not a minimum but a saddle point such that the integration over the coefficient c_0 diverges. Note that such an eigenvalue is not part of the Hamiltonian spectrum, and therefore this is an "unstable" state. From the point of view of the physical potential V, if the particle crosses over to a region where $x > \sigma$, it will flow to higher values of x. From the point of view of the effective $-V$ potential, the point at $x = 0$ is unstable and the bounce from the point at $x = \sigma$ returns the particle to the origin, keeping bouncing until it escapes the metastable region of the potential V. This negative eigenvalue implies that K is imaginary or that the energy has an imaginary part as evidenced

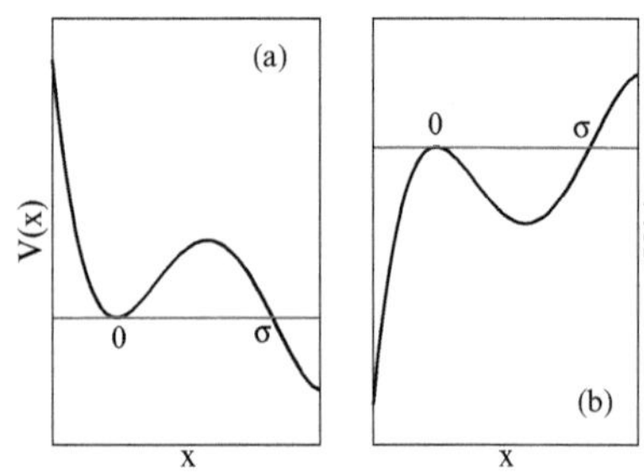

Fig. 15.16 Potential for a metastable (or unstable) state

by Eq. (15.240). Note that in the case K is imaginary, the exponentially small term is now relevant. In this case, we obtain that the imaginary part of the energy

$$ImE = \hbar|K|e^{-\frac{S_0}{\hbar}}, \tag{15.242}$$

which is related to the width of the unstable state. This situation has to be dealt with some care, however, since this energy is not an eigenvalue of the original Hamiltonian. Indeed, the proper solution has to be obtained by a suitable analytic continuation, but we will refrain here from the full derivation.

We will also consider qualitatively this type of solution later, when we consider the growth of a stable phase inside a metastable or an unstable phase. In that problem, we will see that the free energy acquires an imaginary part that will be related to the decay rate of the unstable state (or growth of the stable state).

The problem of non-Hermitian Hamiltonians has attracted great attention. It is briefly discussed in Appendix D.

15.12.3 Nucleation

Consider the Ising model with N spins under the action of a magnetic field h. For small temperatures $T < T_c$ with $T \sim 0$, nearly all spins take the value $+1$, if $h > 0$. Therefore, the magnetization $M > 0$. Also, a collection of atoms or molecules on a lattice with a distribution of concentrations, such that each site is occupied or empty, is formally equivalent to an Ising model with spins $S_i = \pm 1$. The role of the magnetic field is taken by a chemical potential. The following discussion holds for both types of problems, or other equivalent ones that may be characterized by variables that take two values.

Using the spin language, consider now that we change the magnetic field such that the temperature is maintained constant, decreasing the magnetic field through zero and to negative values. If $h < 0$ and the temperature is small, the magnetization is negative with most spins taking the value -1. If we change the value of the magnetic field from positive to negative, infinitely slowly, the magnetization may retain a positive value even if $h < 0$ but small, and the system will be in a metastable state until eventually fluctuations will lead to a transition to the stable, negative magnetization solution.

As mentioned before, the process takes place through the formation of droplets of the stable phase inside the metastable phase, which eventually grow in size and eventually fill the entire space with the stable phase.[13] In the beginning of this process of nucleation of the droplets, we expect that these are small and very dilute,

[13] For further reading see [GuntonDroz].

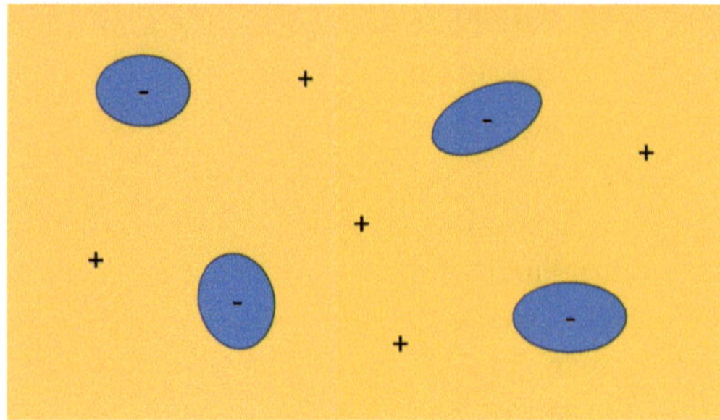

Fig. 15.17 Dilute small droplets of the stable phase in a background of the unstable phase

like a non-interacting gas of droplets as shown schematically in Fig. 15.17. The droplets will have a typical size, l, and the number of droplets with a given size is of the order of

$$n_l = Ne^{-\beta\epsilon_l}. \tag{15.243}$$

Here, N is some normalization, and the energy, ϵ_l, in the Boltzmann factor is the formation free energy of a droplet of size l (with l spins). We expect that is composed of a bulk (volume) term and a surface term:

$$\epsilon_l = 2hl + \sigma l^{\frac{d-1}{d}}. \tag{15.244}$$

The first term is associated with the energy that is needed to flip the l spins in the droplet (from the positive $S = 1$ to the negative $S = -1$). The second, surface, term is proportional to the surface tension, $\sigma > 0$, and we are assuming that the droplet has an approximate d-dimensional spherical shape.

If $h > 0$, the number of droplets, n_l, is a monotonically decreasing function of the droplet size, and it is maximal at $l = 0$, which means of course that any droplet will shrink to zero size. If $h < 0$, there is a minimum such that $(dn_l/dl)_{l_c} = 0$, and beyond this point, for $l > l_c$, the number of droplets increases. This means that if a droplet has a size such that $l > l_c$, it will grow and expand. The point where n_l is minimum is also the point where

$$\frac{\partial \epsilon_l}{\partial l}_{l=l_c} = 0 = 2h + \sigma\frac{d-1}{d}l^{-1/d}. \tag{15.245}$$

This means that

$$l_c = \left(\frac{\sigma(d-1)}{2d|h|}\right)^d, \tag{15.246}$$

at which point the free energy, ϵ_l, has a maximum.

15.12.4 Growth of Droplets and Interface Dynamics

We may view the change of size of the droplet over time as the motion of the interface between the inside region (with the stable phase) and the outside region (with the metastable phase). If the droplet is large enough, we only need to consider its neighborhood and we may consider the interface as planar, separating the two regimes that may be approximated by filling the whole space on the two sides of the interface. This type of description applies to intermediate to long time regimes, when the size of the droplet is already macroscopic.[14] We will consider later oscillations of the planar interface and a droplet that is approximately spherical.

Planar Interface

We may consider some variable (like a concentration of particles or spin density) defined in a continuum, represented by $c(\mathbf{r}, z)$, where z is the direction perpendicular to the interface and $\mathbf{r}$ are the transversal coordinates (parallel to the interface). Also, we may consider that $c(\mathbf{r}, z)$ interpolates between two asymptotic values, when $z \to -\infty$ and $z \to \infty$, which are characteristic of the two phases.

Consider a generic non-homogeneous system described by a free energy:

$$F(\{c(\mathbf{x})\}) = \int d^d x \left[\frac{1}{2}|\nabla c|^2 - \frac{1}{2}\tau c^2 + \frac{1}{4!}g c^4 - hc\right], \tag{15.247}$$

with $\tau > 0$ (which corresponds to $T < T_c$). The equation of state for a "concentration" that only changes along z leads to

$$\frac{\delta F}{\delta c(\mathbf{x})} = -\frac{\partial^2 c}{\partial z^2} + \frac{\partial}{\partial c}\left[-\frac{1}{2}\tau c^2 + \frac{1}{4!}g c^4 - hc\right] = -\frac{\partial^2 c}{\partial z^2} - \tau c + \frac{g}{3!}c^3 - h = 0. \tag{15.248}$$

If $h = 0$, a solution is given by

$$c_{cl}(\mathbf{r}, z) = m_{z_0}(z) = m_0 \tanh\left[\left(\frac{\tau}{2}\right)^{1/2}(z - z_0)\right], \tag{15.249}$$

where

$$m_0 = \left(\frac{6\tau}{g}\right)^{1/2}. \tag{15.250}$$

This type of solution in shown in Fig. 15.18. The solution depends on a parameter, z_0, which can be chosen freely. Its physical significance is the location of the change of the classical solution from one asymptotic behavior to the other. This type of solution is called a kink or a domain wall. The width of the transition region is

[14] For further reading see [GuntonDroz].

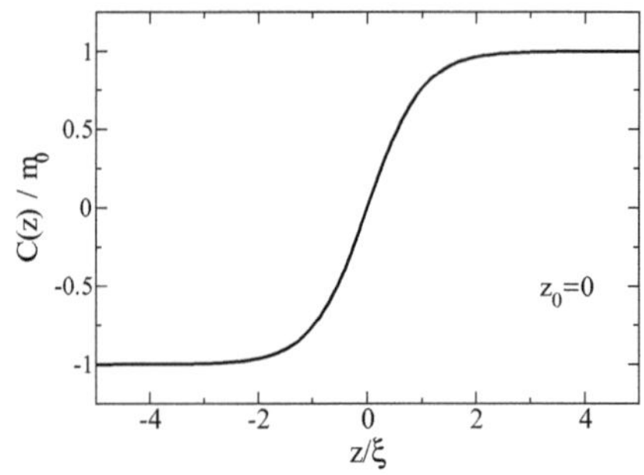

Fig. 15.18 Planar interface between the regimes of Eq. (15.249)

$\xi \sim 1/\sqrt{\tau}$, centered around $z = z_0$. Therefore, the interface is very narrow if τ is large, which means that $T \ll T_c$. Since z_0 is arbitrary, this means that the location of the interface is arbitrary. This is associated with the translational invariance of the Hamiltonian. Clearly, selecting a particular value of z_0, and therefore a particular spatial location for the interface, breaks the continuous translational invariance. This suggests that we may expect a Goldstone-like mode, of zero energy, associated with the breaking of the continuous symmetry. A similar result (but in the time domain) was encountered before in the case of the instanton solution, where a zero mode was also found.

The solution Eq. (15.249) is the lowest energy possibility. In general, the interface may have fluctuations along the transverse direction.

Spherical Droplet

In the case of a spherical droplet, shown in Fig. 15.19, the interface is a sphere, and we have one phase in the interior of the sphere and another phase in the exterior. We consider a sphere with radius R_0 and a phase with a magnetization (or any other physical variable) m_- in the interior and a magnetization m_+ outside. Since we are considering interfaces that are very sharp, we may neglect the spatial dependence of c far from the interface leading to the equation of state:

$$\frac{\partial}{\partial c}\left[-\frac{1}{2}\tau c^2 + \frac{1}{4!}gc^4 - hc\right] = 0$$

$$-\tau c + \frac{g}{6}c^3 = h. \tag{15.251}$$

It is easy to see that

$$c = m_0 + \frac{h}{2\tau}, \tag{15.252}$$

is a solution. We may write then

$$m_\pm = \pm m_0 + \frac{h}{2\tau}, \tag{15.253}$$

Fig. 15.19 Spherical droplet of radius R of a phase m_- inside a background of phase m_+

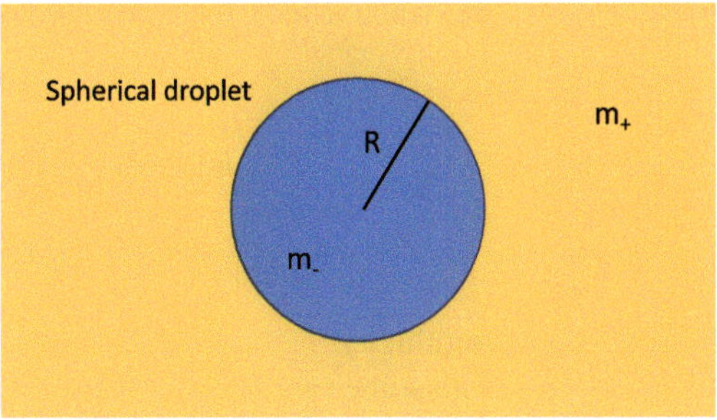

where $m(R = 0) = m_-$ and $m(R \to \infty) = m_+$, where R is the distance to the center of the spherical droplet. We may then write

$$m_{R_0}(R) \sim \frac{1}{2}(m_+ + m_-) + \frac{1}{2}(m_+ - m_-) \tanh\left[\left(\frac{\tau}{2}\right)^{1/2}(R - R_0)\right]$$
$$m_{R_0}(R) = \frac{h}{2\tau} + m_0 \tanh\left[\left(\frac{\tau}{2}\right)^{1/2}(R - R_0)\right]. \tag{15.254}$$

We may now estimate the free energy of the droplet. This is given by

$$F_{dr}(R_0) = F(m_{R_0}(R)) - F(m_+) = F^0_{dr}(R_0) + hF^1_{dr}(R_0) + O(h^2). \tag{15.255}$$

At zero magnetic field, we saw that the energy of the droplet is given by a surface contribution $F^0_{dr}(R_0) = \sigma\Sigma_d(R_0)$, where the surface of the spherical droplet is given by

$$\Sigma_d(R_0) = \frac{2\pi^{d/2}}{\Gamma(d/2)} R_0^{d-1}, \tag{15.256}$$

(for $d = 3$, it is just $4\pi R_0^2$). The correction term is proportional to the volume of the droplet:

$$F^1_{dr}(R_0) = 2m_0 V_d, \tag{15.257}$$

where

$$V_d = \frac{2\pi^{3/2}}{d\Gamma(d/2)} R_0^d, \tag{15.258}$$

(in $d = 3$ is just $4\pi R_0^3/3$). Therefore, the free energy of the droplet has a surface term and a volume term, as we saw before and is given by

$$F_{dr}(R_0) = \sigma\Sigma_d(R_0) + 2m_0 V_d h. \tag{15.259}$$

As we saw before, when $h < 0$, the free energy has an extremum, which leads to a critical radius:

$$R_{0,c} = \frac{\sigma(d-1)}{2m_0|h|}, \tag{15.260}$$

such that for $R > R_{0,c}$, the droplet grows.

Free Energy Singularity

Let us analyze with some care the behavior of the free energy, considering a spherical droplet of the stable phase in the metastable or unstable phase background.[15] In this section, there will be some overlap with previously discussed topics.

The partition function is written as

$$Z = \int D[c]\, e^{-F\{c\}}. \tag{15.261}$$

If the magnetic field $h > 0$, we denote the stable phase by m_+, and if $h < 0$, we denote the stable phase by m_-. The free energy has two extrema for $c_\pm = m_\pm$ and if $h > 0$, for instance, the ratio of the free energies is such that

$$\frac{e^{-F(m_-)}}{e^{-F(m_+)}} \to 0, \tag{15.262}$$

in the thermodynamic limit. Therefore, the partition function is dominated by the solution $c_+ = m_+$. Taking into account fluctuations around the classical solution $c(\mathbf{x}) = m_+ + \hat{c}(\mathbf{x})$, we get

$$Z = Z_0 = e^{-F(m_+)} \int D\left[\hat{c}\right] e^{-\frac{1}{2}\int d^3x \hat{c}(\mathbf{x}) M_0 \hat{c}(\mathbf{x}) + O(\hat{c}^3)}, \tag{15.263}$$

where

$$\begin{aligned} M_0 &= \frac{\delta^2 F}{\delta c(\mathbf{x})^2}_{c=m_+} \\ &= -\nabla^2 - \tau + \frac{1}{2} g m_+^2, \end{aligned} \tag{15.264}$$

in the low-temperature regime, as we saw earlier.

The eigenfunctions of the M_0 operator, defined by

$$M_0 \varphi_j^0(\mathbf{x}) = \omega_j^0 \varphi_j^0(\mathbf{x}), \tag{15.265}$$

[15] For further reading see [GuntonDroz].

may be used to expand the fluctuations:

$$\hat{c}(\mathbf{x}) = \sum_j \eta_j^0 \varphi_j^0(\mathbf{x}). \tag{15.266}$$

Keeping only quadratic terms in the fluctuations, the integral is gaussian and may be performed leading to

$$Z_0 = e^{-F(m_+)} \prod_j \left(\frac{2\pi}{\omega_j^0} \right)^{1/2}, \tag{15.267}$$

similar to the problem previously considered. Again, we assume that all the eigenvalues, ω_j^0, are positive, which is the case if $h > 0$.

Let us now change the magnetic field, such that it becomes $h < 0$, but this is done in such a way that the system has $\langle c \rangle > 0$. This occurs, for instance, as a result of a sudden quench of the magnetic field to a different, negative, value. Therefore, the system is now in a metastable state, while the new equilibrium state is characterized by a uniform solution with $c = m_-$. The solution $c = m_+$ is now a local maximum in the sense that is has a higher free energy than the stable $c = m_-$ state. (There is some interaction between the spins that tends to align them; when $h > 0$, a preferred direction is imposed that selects the m_+ solution. Reversing the sign of h sets the opposite direction as preferred, and it will be favorable for the spins to rotate to the m_- state). As discussed at length, fluctuations may lead to the appearance of a droplet of the stable phase, and so we must take into account droplet contributions to the partition function. It turns out that they give a finite contribution. Let us call again the droplet solution as $m(\mathbf{R})$, which means that we must take into account solutions of the type

$$c(\mathbf{x}) = m(\mathbf{R}) + \hat{c}(\mathbf{x}). \tag{15.268}$$

This means that the partition function has the contribution from the metastable solution, Z_0, to which we must add at least the contribution from one droplet, Z_1, leading to

$$Z = Z_0 + Z_1 + \cdots \tag{15.269}$$

and other contributions that we will ignore, for simplicity. We write that

$$Z_1 = e^{-F(m)} \int D\left[\hat{c}\right] e^{-\frac{1}{2}\int d^3x \hat{c}(\mathbf{x}) M \hat{c}(\mathbf{x}) + O(\hat{c}^3)}, \tag{15.270}$$

with

$$M = \frac{\delta^2 F}{\delta \hat{c}(\mathbf{x})^2}_c = m$$
$$= -\nabla^2 - \tau + \frac{1}{2} g m^2. \tag{15.271}$$

Taking into account previous results, we expect that the matrix M will have a lowest eigenvalue that is negative, a set of d-degenerate zero modes and the remaining are positive eigenvalues. Therefore, the action has a saddle point. The negative eigenvalue implies that Z_1 is imaginary. The zero modes are treated separately and lead to a contribution that is proportional to the volume of the system. Therefore, $Z_1 = i V Z_2$. This implies

$$Z = Z_0 \left[1 + i V \frac{Z_2}{Z_0} \right] \tag{15.272}$$

and the free energy density is given by

$$f = \frac{F}{V} = \frac{1}{V} \ln Z = \frac{1}{V} \ln Z_0 + \frac{1}{V} \ln \left[1 + i V \frac{Z_2}{Z_0} \right] \tag{15.273}$$

and finally

$$f = \frac{1}{V} \ln Z_0 + i \frac{Z_2}{Z_0}. \tag{15.274}$$

We conclude that the droplet configuration contributes even for infinite volume.

Recall that the droplet configuration may be written as

$$m(R - R_0) = \frac{h}{2\tau} + m_0 \tanh \left[\left(\frac{\tau}{2} \right)^{1/2} (R - R_0) \right], \tag{15.275}$$

such that $m(R - R_0) \rightarrow m_+$, for large values of R. The free energy difference between the droplet configuration and the equilibrium solution $\Delta F(m) = F(m) - F(m_+)$ is

$$\Delta F(m) = 4\pi \int_0^{\infty} dR R^2 \left[\frac{1}{2} (\nabla m)^2 - \frac{\tau}{2} \left(m^2 - m_+^2 \right) \right.$$
$$\left. + \frac{g}{4!} \left(m^4 - m_+^4 \right) - h (m - m_+) \right]. \tag{15.276}$$

At low temperatures (large τ), the zero field part is sharp around $R = R_0$, which implies

$$\Delta F(m) = 4\pi \left[4\sqrt{2} R_0^2 \frac{\tau^{3/2}}{g} - \frac{2\sqrt{2}}{\sqrt{3}} |h| \frac{\tau^{1/2}}{g^{1/2}} R_0^3 \right], \tag{15.277}$$

where the first term is a surface term and the second term is a volume term. This leads to a surface tension:

$$\sigma = 4\sqrt{2}\frac{\tau^{3/2}}{g}, \tag{15.278}$$

the critical radius is given by

$$R_{0,c} = \frac{4\tau}{\sqrt{3}|h|g^{1/2}}, \tag{15.279}$$

and the free energy difference is finite at the critical radius, meaning it must be taken into account:

$$\Delta F(R_0 = R_{0,c}) = \frac{256}{9}\pi\left(\frac{\tau^{7/2}}{|h|^2 g^2}\right). \tag{15.280}$$

Since the droplet configuration has spherical symmetry, we may use spherical coordinates and write the Laplacian operator in these coordinates. Also, we may use Eq. (15.275) to obtain that

$$\begin{aligned} M &= -\frac{d^2}{dR^2} - \frac{2}{R}\frac{d}{dR} + \frac{l(l+1)}{R^2} - \tau + \tau V(R) - hW(R) \\ V(R) &= 3\tanh^2\left[\left(\frac{\tau}{2}\right)^{1/2}(R - R_{0,c})\right] \\ W(R) &= -\frac{\sqrt{6}}{4}\left(\frac{g}{\tau}\right)^{1/2}\tanh\left[\left(\frac{\tau}{2}\right)^{1/2}(R - R_{0,c})\right], \end{aligned} \tag{15.281}$$

where l is the angular momentum and takes the values $l = 0, 1, 2, \cdots$. If τ is large or $h = 0$, the term $W(R)$ is negligible. We diagonalize

$$M\varphi_{n,l}(R) = \omega_{n,l}\varphi_{n,l}(R), \tag{15.282}$$

where in the limit considered

$$M = -\frac{d^2}{dR^2} - \frac{2}{R}\frac{d}{dR} + \frac{l(l+1)}{R^2} - \tau + \tau V(R). \tag{15.283}$$

The effective potential is $V_{eff} = \tau(V(R) - 1)$, which has a sharp minimum for $R \sim R_{0,c}$. The lowest eigenvalues correspond to states that are located in the neighborhood of the critical radius. Recall that $R_{0,c} \sim 1/|h|$; therefore, in the neighborhood of the critical radius, $1/R \sim |h|$, $1/R^2 \sim |h|^2$. If l is small, we may approximate the eigenvalue equation by

$$\left[-\frac{d^2}{dR^2} - \tau + 3\tau\tanh^2\left[\left(\frac{\tau}{2}\right)^{1/2}(R - R_0)\right]\right]\varphi_{n,l}(R) = \omega_{n,l}\varphi_{n,l}(R). \tag{15.284}$$

This simplified equation has two bound states that we may call

$$\varphi_{0,l}(R) \sim 1/\cosh^2\left[\left(\frac{\tau}{2}\right)^{1/2}(R-R_0)\right] \sim \frac{dm}{dR}, \tag{15.285}$$

with an eigenvalue close zero $\omega_{0,l} \sim 0$, and a solution

$$\varphi_{1,l}(R) \sim \frac{\sinh\left[\left(\frac{\tau}{2}\right)^{1/2}(R-R_0)\right]}{\cosh^2\left[\left(\frac{\tau}{2}\right)^{1/2}(R-R_0)\right]}, \tag{15.286}$$

with an energy $\omega_{1,l} \sim 3\tau/4$. It turns out that

$$\omega_{0,l} \sim \frac{l(l+1)-2}{2-R_{0,c}^2}, \tag{15.287}$$

which leads to $\omega_{0,1} = 0$, as expected, but also leads to

$$\omega_{0,0} = -\frac{1}{R_{0,c}^2} = -\frac{3|h|^2 g}{16\tau^2} < 0, \tag{15.288}$$

confirming the existence of a negative eigenvalue and an imaginary contribution to the droplet free energy. The zero mode is replaced by a factor that is proportional to the volume of the system, and it can be shown that

$$Z_1 = e^{-F(m_+)}e^{-\Delta F(m)}\Gamma\left(\frac{2\pi}{\omega_{0,0}}\right)^{1/2}\prod_{n,l}{}'\left(\frac{2\pi}{\omega_{n,l}}\right)^{\frac{2l+1}{2}}. \tag{15.289}$$

In the product, we exclude the negative eigenvalue and the zero modes, and the power of $2l+1$ is the result of the degeneracy of the angular momentum modes. The volume factor is given by

$$\Gamma = V\left(\frac{32\pi^{1/2}2^{1/4}}{3}\right)^3\left(\frac{\tau^{7/4}}{g|h|}\right)^3. \tag{15.290}$$

15.13 Exercises

1. Consider the Ising model in a two-dimensional square lattice of lattice constant a. We divide the system in cells of size $L \times L$ where $L = la$. Associate a variable σ_α to cell α defined by the average value of the Ising variables at the sites contained in the cell α. A microstate of the system is specified by the set

of variables σ_i at the sites of the original lattice. The state of the system may also be characterized by the set of variables σ_α.

Rewrite the partition function of the system in terms of the variables σ_α and identify the argument of each exponential in the expression for the partition function.

2. Equation (6.11) gives the heat flux (flow of heat per unit area and unit time) when there is a temperature gradient. Consider a three-dimensional system with a surface where some sudden change of temperature takes place. Assume the three-dimensional system is a body of water that is in the liquid state with a temperature slightly above the freezing point. The change in temperature at the surface of the water is negative and given by ΔT. Therefore, the water freezes from the surface with an ice thickness that increases with time. Assume that the ice does not change its temperature and the entire heat flux from the water to the surface is converted into freezing latent heat. Show that the thickness grows as a square root of time and find an expression for the coefficient relating the thickness with the time dependence.
3. Derive an expression in terms of the order parameter $P(T)$ for the specific heat of a binary alloy, with N atoms of type A and N atoms of type B in two sublattices a and b, each with N available sites. We may write that

$$N_A(a) = \frac{1}{2}(1+P)\,N;\; N_A(b) = \frac{1}{2}(1-P)\,N,$$

and complementary for N_B.
4. Consider the gaussian model. It is obtained neglecting in the Ginzburg-Landau functional the powers in the order parameter that are higher than the second order. Obtain the critical exponents of the gaussian model. Consider the disordered and ordered phases.
5. Consider an integral of the form

$$I(s) = \int dz g(z) e^{sf(z)}.$$

(a) Show that

$$I(s) \sim \sqrt{2\pi}\, g(z_0) e^{sf(z_0)} \frac{e^{i\alpha}}{\sqrt{|sf''(z_0)|}},$$

where $z = z_0 + \delta e^{i\alpha}$, with small δ, where z_0 is the value that maximizes f and f'' is the second derivative of f. (See, for instance, the book *Mathematical Methods for Physicists*, Arfken).

(b) Show that $s! \sim \sqrt{2\pi s} s^s e^{-s}$.

6. Consider the Ginzburg-Landau functional and that $\phi(r)$ is not necessarily uniform and introduce the term in the field gradient.

(a) Rewrite the functional in Fourier space.
(b) Using a self-consistent mean-field approximation, replace the functional by an expression that is quadratic in the fields. That is, show that the correlation function $G(q)$, Fourier transform of $G(x, x') = \langle \phi(x)\phi(x') \rangle$, may be approximated by

$$TG^{-1}(q) = r + q^2 + 12u\langle \phi^2 \rangle.$$

(c) Show that the susceptibility may be written as

$$\chi^{-1} = TG^{-1}(q = 0).$$

(d) Calculate the critical exponent of the susceptibility for dimension d. Particularize for $d = 4$.
(e) Calculate the critical exponent of the correlation length for dimension d. Particularize for $d = 4$.
(f) If ϕ is a vector with n-components, show that the self-consistent approximation is exact in the limit when $n \to \infty$ (spherical model).

7. Consider a ferromagnetic interaction between Ising spins such that all sites are equally coupled, independently of the their distance: $J_{ij} = g/N$, where g is a constant and N is the number of lattice sites. Determine the critical temperature in the thermodynamic limit, and show that a transition takes place for any dimension. Use the mean-field theory and comment on the applicability of this approximation.
8. Consider the Ising model with nearest-neighbor coupling given by J. The system is half-infinite, that is, it has a surface defined by the plane $r = 1$. The next layer is labeled by $r = 2$ and so on for the following planes. Using the mean-field approximation:

(a) Derive a set of equations that determine m_r, the magnetization of the layer r, self-consistently.
(b) Show that for a temperature slightly higher than the critical temperature, T_c, and in the presence of a small magnetic field applied on the layer $r = 1$, the equations have the solution

$$m_r = m_1 \exp\{-(r-1)/\xi\}.$$

In this expression, ξ is the correlation length that diverges as

$$\xi \sim |T - T_c|^{-\nu}.$$

Calculate m_1, ξ, ν.

9. Solve numerically, using the Monte Carlo method, the Ising model in one and two dimensions (in a square lattice). Calculate

$$M = \sum_i \langle S_i \rangle$$

and

$$\chi = \sum_i \sum_j \langle S_i S_j \rangle,$$

as a function of temperature, where i and j denote lattice sites. Consider a lattice of 100 spins in the one-dimensional case and a lattice 6×6 in the two-dimensional case. Use periodic boundary conditions: $S(i) = S(i + N)$ where N is the number of spins in a given direction.

10. Consider a Ginzburg-Landau functional of the type

$$H = \frac{1}{2} r \phi^2 + u \phi^4 + C \phi^3 - h \phi,$$

where, as usual, $r = at$, $t = (T - T_c)/T_c$ and a and u are positive.

Show that the system has a first-order transition and show graphically H as a function of ϕ for different characteristic temperatures.

11. Consider a Lagrangian density

$$L = \frac{1}{2} \partial_\mu \phi \partial^\mu \phi - U(\phi),$$

where $\mu = 0, 1$ (one time dimension and one spatial dimension). $U(\phi)$ is an arbitrary function of ϕ but does not contain derivatives of ϕ.

The energy of a field configuration of ϕ is given by

$$E = \int dx \left(\frac{1}{2} (\partial_0 \phi)^2 + \frac{1}{2} (\partial_1 \phi)^2 + U(\phi) \right).$$

Consider that $U(\phi) = \frac{\lambda}{2} \left(\phi^2 - a^2 \right)^2$.

(a) Consider time-independent solutions. Show that the condition of conservation of energy, $\delta E = 0$, is equivalent to the problem of the motion of one particle of mass $m = 1$ in the presence of a potential $V = -U$. Make a graphic of $U(\phi)$ and of $V = -U(\phi)$.

(b) Consider zero energy trajectories. Which possibilities are there? Show that there are solutions of the type

$$x = \pm \int_{\phi_0}^{\phi} d\phi' \frac{1}{\sqrt{2U(\phi')}},$$

with ϕ_0 arbitrary.

(c) In terms of the field ϕ, the same trajectory satisfies the equation:

$$\frac{1}{2}\,(\partial_1\phi)^2 = U(\phi).$$

Show that the solution is of the type $\phi = a\tanh(\mu x)$. Calculate the energy of the field.

(d) Consider now fluctuations around the found solution. Show that the equation for the eigenmodes of the oscillations is given by

$$-\frac{d^2\psi_n}{dx^2} + U''(\phi)\psi_n = \omega_n^2\psi_n.$$

Show that $d\phi/dx$ is a solution with eigenvalue zero and comment its physical significance. (Recall a translational invariance of the problem.). Interpret then the meaning of the stationary solution that you found.

Reference

1. N. D. Mermin and H. Wagner, Physical Review Letters **17**, 1133 (1966); P. C. Hohenberg, Physical Review **158**, 383 (1967).

Chapter 16
Quantum Magnetism

Spin systems are described in terms of spin operators that have commutation relations that are neither those of bosons nor those of fermions. However, as we will see, in some circumstances, their excitations are boson-like. Also, later on, we will see that in one dimension, their spectrum may have a fermion-like dispersion (or like hard core bosons). One of the reasons for this duality is that one can represent spin operators by bilinear representations in terms of either bosons or fermions. We will not elaborate at this point on the various possible representations of the spin operators, but will throughout this book refer to different representations, as needed.

The magnetic systems we will consider may either be of a localized (insulator) type or itinerant magnetic systems, associated with extended electronic states. At low temperatures, spin systems may order magnetically in different manners, as shown in Fig. 16.1. We will focus mainly on the ferromagnetic and the antiferromagnetic orders, since these are the most commonly treated ones.

In the case of localized spins, we consider that they interact by some exchange overlap, that is eventually originated in the localized electron orbitals, that may overlap between the atoms where the spins are defined. In the case of itinerant magnetism, it originates from the magnetic moments associated with extended conduction electron orbitals and where the interactions between the electrons, of a Coulombic nature, determine the magnetic properties.

We consider first the case of localized magnetic moments. We want to see the effect on the magnetic moments of several atoms brought together. The goal is to understand a solid of many atoms where each one has partially filled orbitals and, therefore, a possible local magnetic moment.

P. D. Sacramento, *Introduction to Condensed Matter Theory*,
https://doi.org/10.1007/978-3-031-86251-9_16

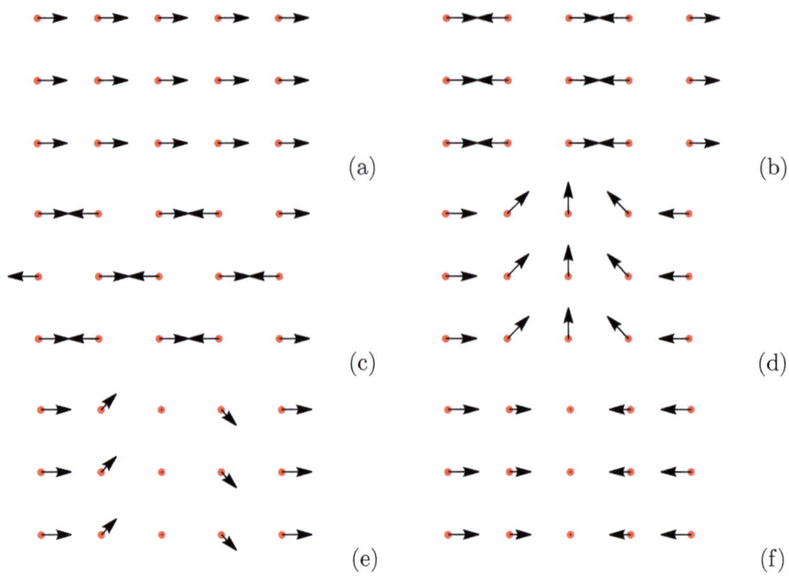

Fig. 16.1 Examples of magnetic structures in a two-dimensional system. The generalization to three dimensions is straightforward. The cases of (**a**) a ferromagnet, (**b**) an antiferromagnet with momentum $(\pi, 0)$, (**c**) an antiferromagnet with momentum (π, π), (**d**) a helix, (**e**) a transverse spin density wave, and (**f**) a longitudinal spin density wave are shown

16.1 Hydrogen Molecule

As a first example, we consider two hydrogen atoms brought together to form a hydrogen molecule.[1]

To simplify things further, we consider the ionized hydrogen molecule where one electron has been removed, H_2^+. Let us call $\mathbf{R}$ the coordinate of one proton with respect to the other and $\mathbf{r}$ the coordinate of the electron with respect to the middle point between the two protons. The Schrödinger equation may be written as.[2]

$$\left(-\frac{\hbar^2}{2M}\nabla_R^2 - \frac{\hbar^2}{2m}\nabla_r^2 - \frac{e^2}{|\mathbf{r}-\mathbf{R}/2|} - \frac{e^2}{|\mathbf{r}+\mathbf{R}/2|} + \frac{e^2}{R}\right)\psi(\mathbf{r},\mathbf{R}) = E\psi(\mathbf{r},\mathbf{R}). \tag{16.1}$$

Here, $M = M_p/2$, where M_p is the mass of each proton. Since the mass of the protons is much larger than the mass of the electron, the nuclei move much slower than the electrons, and using the Born-Oppenheimer approximation, we may neglect the motion of the nuclei. The electrons move in an effective field of the nuclei, for a

[1] For further reading see [Gasiorowicz] and [BicudoSacramento].

[2] In this section, to simplify notation, in the Coulomb potential, we write e^2 instead of $e^2/(4\pi\epsilon_0)$.

given set of positions of the nuclei (in the case of the hydrogen molecule for a given value of R, the distance between the protons). As the nuclei change their relative positions, the electrons are assumed to move fast enough to adapt instantaneously to the new nuclei configuration, as discussed before. Specifically, in the case of the hydrogen molecule, the distance R may be considered as a parameter, and we may neglect the kinetic energy of the nuclei.

Neglecting the kinetic energy, we get a simplified electronic Hamiltonian written as

$$\begin{aligned} H_0 u_0(\mathbf{r}, \mathbf{R}) &= \Big(\frac{p_e^2}{2m} - \frac{e^2}{|\mathbf{r} - \mathbf{R}/2|} - \frac{e^2}{|\mathbf{r} + \mathbf{R}/2|} + \frac{e^2}{R}\Big) u_0(\mathbf{r}, \mathbf{R}) \\ &= \epsilon_0(R) u_0(\mathbf{r}, \mathbf{R}). \end{aligned} \tag{16.2}$$

The first term is the kinetic energy of the motion of the electron, the next two terms are the attractive interactions between the electron and the protons, and the last term is the Coulomb repulsion between the protons, assumed static. The problem of the electron and one proton is just the hydrogen atom. We may attempt a variational approach where we use two ground state eigenfunctions of the hydrogen atom, specifically the functions

$$\psi_1(\mathbf{r}, \mathbf{R}) = \Big(\frac{1}{\pi a_0^3}\Big)^{1/2} e^{|\mathbf{r} - \mathbf{R}/2|/a_0} \tag{16.3}$$

and

$$\psi_2(\mathbf{r}, \mathbf{R}) = \Big(\frac{1}{\pi a_0^3}\Big)^{1/2} e^{|\mathbf{r} + \mathbf{R}/2|/a_0}, \tag{16.4}$$

which may be interpreted as the two possibilities that either the electron is attached to one proton or the other.

As a trial solution for the ionized hydrogen molecule, we may consider a symmetric and an antisymmetric linear combination of these two wave functions. Since the Hamiltonian obeys the symmetry ($\mathbf{p}_e \to -\mathbf{p}_e,\ \mathbf{r} \to -\mathbf{r},\ \mathbf{R} \to -\mathbf{R}$), we may use $\psi_1(\mathbf{r}, \mathbf{R})$ and $\psi_2(\mathbf{r}, \mathbf{R})$:

$$\begin{aligned} \psi_s(\mathbf{r}, \mathbf{R}) &= C_+(R)[\psi_1(\mathbf{r}, \mathbf{R}) + \psi_2(\mathbf{r}, \mathbf{R})], \\ \psi_a(\mathbf{r}, \mathbf{R}) &= C_-(R)[\psi_1(\mathbf{r}, \mathbf{R}) - \psi_2(\mathbf{r}, \mathbf{R})], \end{aligned} \tag{16.5}$$

The normalizations are defined by

$$\frac{1}{C_\pm} = \langle \psi_1 \pm \psi_2 | \psi_1 \pm \psi_2 \rangle. \tag{16.6}$$

Defining

$$S(R) = \int d^3r \; \psi_1(\mathbf{r}, \mathbf{R}) \psi_2(\mathbf{r}, \mathbf{R}), \tag{16.7}$$

we can calculate

$$\begin{aligned} \langle H_0 \rangle_{s,a} &= \frac{1}{2[1 \pm S(R)]} \langle \psi_1 \pm \psi_2 | H_0 | \psi_1 \pm \psi_2 \rangle \\ &= \frac{1}{2[1 \pm S(R)]} [\langle \psi_1 | H_0 | \psi_1 \rangle + \langle \psi_2 | H_0 | \psi_2 \rangle \pm \langle \psi_1 | H_0 | \psi_2 \rangle \pm \langle \psi_2 | H_0 | \psi_1 \rangle] \\ &= \frac{\langle \psi_1 | H_0 | \psi_1 \rangle \pm \langle \psi_1 | H_0 | \psi_2 \rangle}{1 \pm S(R)}. \end{aligned} \tag{16.8}$$

The calculation leads to

$$\begin{aligned} \langle H_0 \rangle_{s,a} = \frac{1}{1 \pm S(R)} \Bigg[& E_1 + \frac{e^2}{R} \left(1 + \frac{R}{a_0} \right) e^{-\frac{2R}{a_0}} \\ & \pm \left(E_1 + \frac{e^2}{R} \right) S(R) \mp \frac{e^2}{a_0} \left(1 + \frac{R}{a_0} \right) e^{-\frac{R}{a_0}} \Bigg]. \end{aligned} \tag{16.9}$$

The energy E_1 is the energy of a hydrogen atom. The normalization $S(R) \sim e^{-R/a_0}$. The plot of the two energy solutions in Fig. 16.2 shows that, as a function of R, the even solution corresponds to a bound state and the odd solution leads to an unbound state. Therefore, the symmetric solution is favored energetically. However, the agreement with the exact solution is not very good, particularly when the distance between the protons is small, small R. As $R \to 0$, the wave function should tend to the ionized helium atom, He^+.

Let us now consider the neutral hydrogen molecule, as shown in Fig. 16.3. We have now two electrons and two protons. We denote the two electrons as 1 and 2 and the two protons as a and b. The distance between the protons is denoted R_{ab}. The distance between the electrons is r_{12}. The distance from electron i to proton α

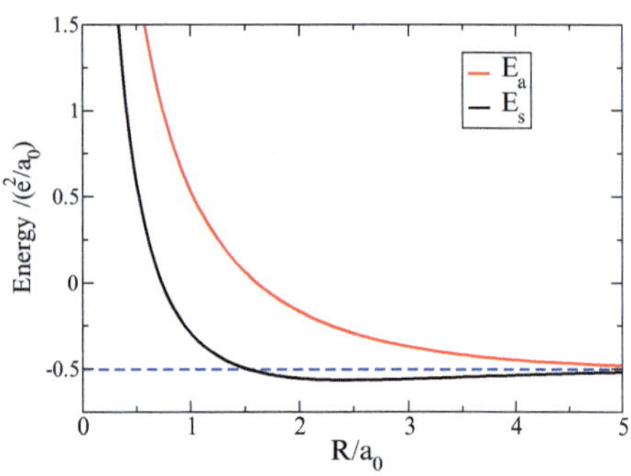

Fig. 16.2 Energies of the symmetric and antisymmetric wave functions of Eq. (16.9)

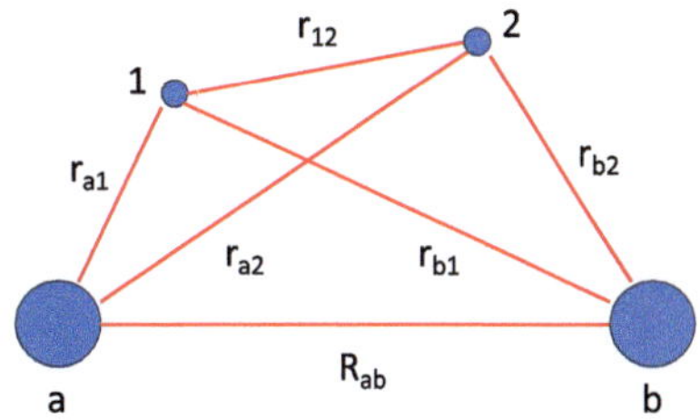

Fig. 16.3 Scheme of the neutral hydrogen molecule, with two protons, a and b, and two electrons 1 and 2

is $r_{\alpha i}$, where $i = 1, 2$ and $\alpha = a, b$. The Hamiltonian may be written as

$$H = H_1 + H_2 + \frac{e^2}{r_{12}} + \frac{e^2}{R_{ab}}, \tag{16.10}$$

with

$$H_i = \frac{\mathbf{p}_i^2}{2m_e} - \frac{e^2}{r_{ai}} - \frac{e^2}{r_{bi}}. \tag{16.11}$$

The previous example of the ionized hydrogen molecule is described by the Hamiltonian

$$\bar{H}_i = H_i + \frac{e^2}{R_{ab}}, \tag{16.12}$$

and the total Hamiltonian may be written as

$$H = \bar{H}_1 + \bar{H}_2 + \frac{e^2}{r_{12}} - \frac{e^2}{R_{ab}}. \tag{16.13}$$

We may now use as a variational wave function the one for the ionized hydrogen molecule:

$$\psi_s(\mathbf{r_1}, \mathbf{r_2}) = \frac{1}{2(1 + S(R_{ab}))}\big[\psi_a(\mathbf{r_1}) + \psi_b(\mathbf{r_1})\big]\big[\psi_a(\mathbf{r_2}) + \psi_b(\mathbf{r_2})\big]\chi_{singlet}. \tag{16.14}$$

Each electron is therefore associated with both protons. Note that the spin part is a singlet since the orbital part is symmetric. The calculation of the energy leads to a binding energy of 2.68 eV and a separation of 0.85 Å, while the experimental values are 4.75 eV and 0.74 Å, respectively. The agreement may be improved with a different choice of variational function. The choice of a molecular orbital is not the most adequate, particularly for small internuclear distances. A better choice is called the valence bond method (or Heitler-London method), which is constructed by a superposition of atomic orbitals. The details are beyond this book and in general a numerical variational method is required. The important point to recall is that the symmetric nature of the spatial part of the wave function implies a singlet spin part.

16.2 Wigner Theorem and Lieb and Mattis Theorem

The results of the previous section seem to indicate that in the hydrogen molecule, the lowest energy state has spin zero. The same result is obtained if we consider a system with two electrons not necessarily bound to two protons, but that are free to move in a d-dimensional space. These electrons have kinetic energy and may interact with each other via an arbitrary potential, which we may assume is symmetric under the exchange of the two electrons coordinates. Then, it can be shown, with some generality, that the ground state of this system has a spatial wave function that is symmetric and, therefore, the spin part is antisymmetric. This is called the Wigner theorem.

The theorem may be shown in the following way.[3] Consider a Hamiltonian of the general type:

$$H = \frac{\mathbf{p}_1^2}{2m} + \frac{\mathbf{p}_2^2}{2m} + V(\mathbf{r}_1, \mathbf{r}_2). \tag{16.15}$$

Choose the potential to be real and symmetric. The two electrons have spin $s = 1/2$. Therefore, the pair has total spin $S = 0, 1$. We want now to show that the ground state has $S = 0$. Since the potential is real, we may take the wave functions of the Hamiltonian to be real. The ground state satisfies $H\psi_0 = E_0\psi_0$. We want to show that a $S = 0$ wave function written as

$$\psi_0 = \phi_s(\mathbf{r}_1, \mathbf{r}_2)\left[| \uparrow\downarrow\rangle - | \downarrow\uparrow\rangle\right], \tag{16.16}$$

where ϕ_s is symmetric, is the ground state. A $S = 1$ wave function may be written as

$$\psi_{S=1} = \phi_{as}(\mathbf{r}_1, \mathbf{r}_2)\chi_{S=1}, \tag{16.17}$$

where ϕ_{as} is antisymmetric. Let us now suppose that the lowest energy state has $S = 1$. Consider then the trial function:

$$\tilde{\psi}_0 = |\phi_{as}|\left[| \uparrow\downarrow\rangle - | \downarrow\uparrow\rangle\right]. \tag{16.18}$$

[3] For further reading see E. Lieb and D. Mattis, Physical Review **125**, 164 (1962).

This trial wave function has $S = 0$ and is antisymmetric (since the absolute value of the antisymmetric spatial wave function is symmetric). Pauli's principle is therefore obeyed. We may now calculate

$$\begin{aligned} E_{S=1} &= \langle \psi_{S=1} | H | \psi_{S=1} \rangle \\ \tilde{E}_{S=0} &= \langle \tilde{\psi}_0 | H | \tilde{\psi}_0 \rangle . \end{aligned} \tag{16.19}$$

Explicitly,

$$\begin{aligned} E_{S=1} &= \int d^3 r_1 \int d^3 r_2 \phi^*_{as}(\mathbf{r_1}, \mathbf{r_2}) H \phi_{as}(\mathbf{r_1}, \mathbf{r_2}) \\ \tilde{E}_{S=0} &= \int d^3 r_1 \int d^3 r_2 |\phi^*_{as}(\mathbf{r_1}, \mathbf{r_2})| H |\phi_{as}(\mathbf{r_1}, \mathbf{r_2})| . \end{aligned} \tag{16.20}$$

Clearly, $E_{S=1} = \tilde{E}_{S=0}$. This implies that there is always an eigenfunction with $S = 0$ with an energy that is, at least, as low as the lowest energy with $S = 1$. But $\tilde{\psi}_0$ is not an eigenfunction, unless the potential is quite special, such as an infinitely repulsive core in one dimension. Therefore, we may use it as a trial function in a variational procedure, from which we expect to find some lower energy average value $E(S = 0) < E(S = 1)$, because the variational method will lead to a smaller energy. Therefore, the lowest energy has $S = 0$.

This result may be extended to an arbitrary number of electrons on a one-dimensional system. It may also be shown that the ground state has zero total spin. This is called the Lieb and Mattis theorem.

These results seem to indicate a tendency for antiparallel alignment of spins between neighboring points. This contrasts to some extent with the result expressed by Hund's rule where the spin of an atom is maximized. (Recall, however, that the spin also vanished for the ground state of the helium atom. In that case, the wave function of the two electrons had a symmetrical spatial part and, consequently, spin zero). Also, we stress again that the coupling between the spins has a Coulombic origin in conjunction with symmetry properties of the electron wave functions, as in the case of Hund's rules. Classically, there is also an interaction between magnetic moments, which is due to a magnetic dipolar interaction: a magnetic dipole gives origin to a magnetic field that is felt by another magnetic moment. However, this interaction is rather weak and does not explain in general the magnitude of the magnetic interactions found experimentally.

In reality, magnetic systems are found with different types of magnetic ordering, including ferromagnetic systems, in which the spins in neighboring sites are parallel. One possible way is to consider some super-exchange mechanism in which the spins interact via some intermediate electronic orbitals, located at a nonmagnetic site that is interstitial between the two spins. One of the spins will tend to be antiparallel with one orbital in the interstitial site. If Hund's rules apply to the interstitial atom, another electron will tend to be parallel, which in turn will cause

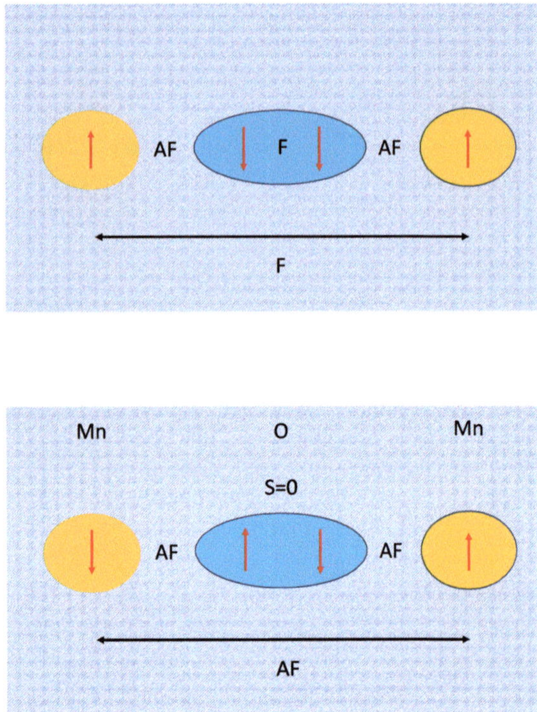

Fig. 16.4 Super-exchange mechanisms leading to ferromagnetic or antiferromagnetic couplings between magnetic moments in atoms that are mediated by some interstitial site. In (**a**) the interstitial site that may have a magnetic moment and in (**b**) the intermediate site that has no magnetic moment. An example is the antiferromagnetic coupling between Mn atoms via a nonmagnetic O atom

the other spin to be antiparallel to the electron spin and, therefore, parallel to the first spin. This is schematically shown in Fig. 16.4a. In Fig. 16.4b, an example of an antiferromagnetic coupling is considered via a nonmagnetic atom, such as oxygen. The mediation between local magnetic moments may also be achieved through the spin density of conduction electrons. This leads to an oscillating coupling that may be ferromagnetic or antiferromagnetic between different atoms. The oscillation is a function of the Fermi momentum of the conduction electrons. Another mechanism that leads to ferromagnetism is the Stoner mechanism that we will consider later in this chapter. The nature of the coupling between systems depends sensitively on the overlap between wave functions of nearby atoms, and many different phases are found experimentally, including non-collinear phases, spiral structures, conical forms, etc. Also, systems may have phases that change as a function of temperature between ferromagnetic and antiferromagnetic, as an example. For instance, some of the lanthanides have a ferromagnetic type ordering at low temperatures, as temperature increases there is a transition to some form of antiferromagnetic ordering until a transition to a paramagnetic regime is attained at higher temperatures. Due to the somewhat localized nature of the magnetic orbitals, the exchange couplings arise due to the coupling to their environment. Their phases are referred in Table 16.1.

Table 16.1 Temperature ranges of the magnetic phases, ferromagnetic (F), antiferromagnetic (AF), and paramagnetic (P) of some lanthanides

$Atom$	F	AF	P
Gd	$0-293K$	$-$	$>293K$
Tb	$0-220K$	$220-230K$	$>230K$
Dy	$0-88K$	$88-176K$	$>176K$
Ho	$0-19K$	$19-130K$	$>130K$
Er	$0-19K$	$19-85K$	$>85K$
Tm	$0-32K$	$32-57K$	$>57K$

16.3 Ferromagnetism: Mean-Field Theory

Let us first focus on a ferromagnetic system, in which the spins tend to align with each other. A simple model that describes exchange interactions between localized magnetic momenta is the Heisenberg model that can be written as

$$H = \frac{1}{2} J \sum_{i,j} \mathbf{S}_i \cdot \mathbf{S}_j. \tag{16.21}$$

The sum extends over nearest-neighbor spins located on a discrete lattice. The coupling between the spins is parametrized by the exchange interaction J. This model describes magnetism in an insulating system. If $J > 0$, it describes an antiferromagnetic system, and if $J < 0$, it describes a ferromagnetic system. The description of magnetism in a conducting system is studied later.

To simplify, we begin by considering a simplified model such that the interaction is limited to one spin component (Ising model), such as the z-th component:

$$H = \frac{1}{2} J \sum_{i,j} S_i^z S_j^z - g\mu_B B \sum_i S_i^z, \tag{16.22}$$

where we have added the effect of an external magnetic field, B, along the z-th direction. Here, g is the Landé factor and μ_B is the Bohr magneton. This model can be treated in a mean-field approximation where fluctuations are neglected. Specifically, like before, we rewrite the exchange part of the Hamiltonian as

$$\frac{1}{2} J \sum_{i,j} \left[\left(S_i^z - \langle S_i^z \rangle\right) \left(S_j^z - \langle S_j^z \rangle\right) + \langle S_j^z \rangle S_i^z + \langle S_i^z \rangle S_j^z - \langle S_i^z \rangle \langle S_j^z \rangle \right]. \tag{16.23}$$

In the mean-field approximation, the first term is neglected, since it is proportional to the square of the fluctuations around the mean value of the spin, at a given point in space. Within this approximation, the effect of the interactions between the spins can be understood as an effective magnetic field on each spin. In addition to the external magnetic field, there is an exchange term that is the result of the neighboring spins.

The Hamiltonian may then be written as

$$H = -\left(g\mu_B B - J\sum_{j=1}^{z}\langle S_j^z\rangle\right)\sum_i S_i^z - \frac{J}{2}z\sum_i \langle S_i^z\rangle^2, \tag{16.24}$$

where z is the coordination number (number of first neighbors). Equivalently, we may write

$$H = -g\mu_B B_{eff}\sum_i S_i^z + C, \tag{16.25}$$

where C is a constant. Note that it is straightforward to generalize these expressions to a rotationally symmetric expression.

Consider now, to simplify, that the system is composed of spins $S = 1/2$. Recalling the results obtained for a paramagnetic system in the presence of an external field, the magnetization can be written as

$$N = N\mu \tanh\left(\frac{\mu B_{eff}}{k_B T}\right). \tag{16.26}$$

N is the number of spins, $\mu = g\mu_B S$. Consider now that the external field vanishes, $B = 0$. The magnetization satisfies the equation of state:

$$M = N\mu \tanh\left(\frac{\mu\lambda M}{k_B T}\right). \tag{16.27}$$

Here, $\lambda = B_{eff}/M$. Considering a ferromagnetic system, for which $J = -|J|$ and using the translational invariance of the system, $M_j = M/N$, where M_j is the magnetization at site j, we see that

$$B_{eff} = \frac{|J|z}{g\mu_B}\frac{M}{N}, \lambda = \frac{|J|z}{g\mu_B N}. \tag{16.28}$$

The equation of state has a trivial solution, $M = 0$, that corresponds to the paramagnetic phase. There is also a nontrivial solution with finite M. At high temperatures, the only solution is the trivial one, since the thermal energy completely misaligns the spins. At zero temperature, the spins are all parallel and the magnetization is finite (and maximal). Between the two extreme regimes, there is a transition from zero magnetization to finite magnetization. This may occur as a crossover or as a transition. It turns out that there is a transition at a finite temperature, which is called critical temperature, T_c. Let us assume that the transition is continuous, in the sense that the magnetization raises from zero continuously. Therefore, near the critical temperature, it is small.

We may linearize the equation of state:

$$M = N\mu \frac{\mu\lambda M}{k_B T_c} = \frac{N\mu^2\lambda}{k_B T_c} M. \tag{16.29}$$

Within mean field, the critical temperature is given by

$$k_B T_c = N\mu^2\lambda = g\mu_B z S^2 |J|. \tag{16.30}$$

Define $m = M/(N\mu)$. This leads to

$$m = \tanh\left(\frac{T_c}{T} m\right). \tag{16.31}$$

This equation may be solved by the graphical method. Since the slope at the origin of the tanh dependence is T_c/T, if $T > T_c$, there is only one solution, $m = 0$. However, if $T_c < T$, there are two solutions: one where $m = 0$ and one for which m is finite, showing that there is a nontrivial phase with a spontaneous magnetization, which results from the exchange effective field, due to the neighboring spins (Figs. 16.5 and 16.6).

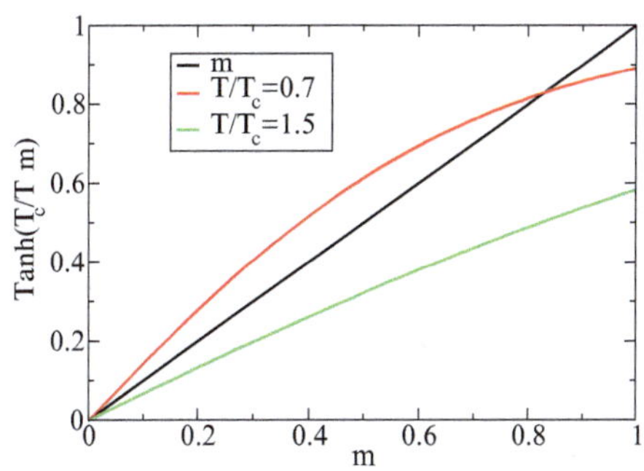

Fig. 16.5 Intersections of m and $\tanh\left(\frac{T_c}{T} m\right)$ for temperatures larger and smaller than the critical temperature, T_c. When $T > T_c$, the curves only intersect at $m = 0$. When $T < T_c$, the curves intersect twice showing the existence of a finite solution for m

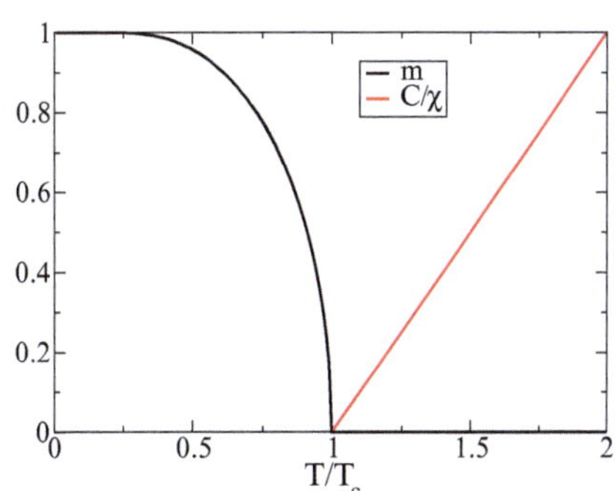

Fig. 16.6 Normalized magnetization m and inverse susceptibility multiplied by the Curie constant as a function of temperature normalized to the critical point

In the paramagnetic regime, $T > T_c$, and restoring the external field, B, we can write that

$$M = \chi_p (B + \lambda M), \tag{16.32}$$

where $\chi_p = C/T$ is the standard Curie paramagnetic susceptibility, where C is Curie's constant. The Curie-Weiss susceptibility in the limit, when the external field is small, can be written as

$$\chi = \frac{M}{B} = \frac{\chi_p}{1 - \lambda\chi_p} = \frac{C}{T - T_c}, \tag{16.33}$$

where $T_c = \lambda C$.

At this point, several remarks are in order: (i) The susceptibility diverges at the finite temperature, T_c, instead of at zero temperature, as in the paramagnetic phase. (ii) The susceptibility may be expanded as

$$\begin{aligned} \chi &= \chi_p \left(1 + \lambda\chi_p + (\lambda\chi_p)^2 + \cdots\right) \\ &= \frac{C}{T} + \frac{C}{T}\lambda\frac{C}{T} + \frac{C}{T}\lambda\frac{C}{T}\lambda\frac{C}{T} + \cdots \end{aligned} \tag{16.34}$$

Truncating the sum at any finite order, the divergence of the susceptibility occurs at zero temperature. Only summing the infinite series will the divergence temperature shift to the finite critical temperature, T_c. (iii) Recalling that $\lambda \sim |J|$, this shows that the result is non-perturbative in the coupling between the spins. (iv) In the case when the spin coupling is of a vectorial nature, it can be easily shown that the critical temperature is given by

$$k_B T_c = \frac{1}{3} g\mu_B z S(S+1)|J|. \tag{16.35}$$

(v) The effective magnetic field that results from the exchange interaction, Eq. (16.28), can be quite large, of the order of 10^3 Tesla in strong ferromagnets, leading to high critical temperatures. (vi) The mean-field approximation leads to the conclusion that the most stable solution is a uniform, homogeneous phase, where the magnetic moments are ordered and aligned along a given direction. Considering a finite system, even though large compared to a microscopic scale, it is not favorable energetically to have all magnetic moments aligned. If the whole system has all the magnetic moments aligned, this magnetization will give rise to magnetic field lines that escape to the external region of the magnetic material: the magnet gives rise to magnetic field lines that have to close and that diverge from the north pole and converge to the south pole of the magnet. The magnetic energy associated with these field lines is finite. It is energetically favorable if no magnetic field lines leave the material. This is achieved if the magnetic moments near the surface are parallel to the surface of the material. Therefore, not all magnetic moments point

Table 16.2 Critical exponents γ and β for some magnetic materials

$Materials$	γ	β	$T_c(Kelvin)$
$Co, CrBR_3, Gd, Fe, Ni, CrO_2$	1.21 − 1.63	0.34 − 0.42	32.56 − 1388

in the same direction. It turns out that the magnetic moments order in domains, such that in each domain, the magnetic moments all point in the same direction but, moving from one magnetic domain to a neighboring one, the magnetization direction changes. This costs some energy due to the various frontiers between the domains, but it is energetically favorable with respect to a single magnetic domain, due to the leakage of magnetic field lines in this case. These frontiers give origin to what is called domain walls. These can have different widths as one crosses from one magnetization direction to another. (vii) The dependence of the susceptibility near the critical point gives a critical exponent $\gamma = 1$, as expected from mean-field theory and discussed at length in the previous chapter. As also discussed in that chapter, the experimental behavior deviates from the mean-field prediction. Also, the mean-field prediction for the critical exponent $\beta = 1/2$ is not observed experimentally. In Table 16.2, some experimental results are summarized showing the deviations from the mean-field theory.

16.4 Ferrimagnetism and Antiferromagnetism

Some magnetic systems show a zero temperature magnetization that does not reach the saturation value expected, using Hund's rules to determine the spin value of the magnetic moment. An example is magnetite,[4] Fe_3O_4. The iron atoms are involved in two types of bonds FeO and Fe_2O_3. The electronic configuration of oxygen is $2s^22p^4$ and of iron is $3d^64s^2$. A s-orbital is full with two electrons, a p-orbital with six electrons, and a d-orbital with ten electrons. In a bond FeO, the two $4s$ electrons are transferred to the p-orbital of the oxygen atom. The iron atom is therefore in the valence state Fe^{2+} with a configuration $3d^6$. Hund's rules lead to $S = 2$ and $L = 2$ but are expected to be quenched by the crystal field.

In the bonds Fe_2O_3, the four s-electrons of the iron atoms are transferred to the p-orbitals of the oxygens plus one d-electron of each iron atom. The valence state is therefore Fe^{3+}, and the electronic configuration of each iron atom is $3d^5$. This leads to $S = 5/2$ and $L = 0$. Therefore, we expect a spin $(1 \times 2 + 2 \times 5/2) = 7$. However, the zero temperature magnetization corresponds to a spin $S = 2.05$. In the unit cell of the lattice structure of magnetite, there are two types of lattice sites. There are A sites with tetrahedral symmetry, filled with Fe^{3+} ions, and B sites with octahedral symmetry, filled with Fe^{3+} and Fe^{2+} ions in equal amounts. The interactions between the various spins are such that their exchange

[4] See also [Kittel].

integrals favor antiparallel orientations. Specifically, J_{AA}, J_{AB}, J_{BB}, are all of an antiferromagnetic origin. The $S = 5/2$ of the A sites are compensated by the $S = 5/2$ spins of the B sites. We are therefore left with the $S = 2$ spins of the remaining B sites, explaining the observed experimental value.

It turns out that this explanation lies on the fact that the largest exchange constant is the interaction between the A and B spins: $|J_{AB}| > |J_{AA}|, |J_{BB}|$. It is energetically favorable that, at low temperatures, the A spins are parallel, the same with the B spins, and that the A and B spins are antiparallel with respect to each other. Even though some of the orientations individually do not satisfy the alignment suggested by the exchange couplings (some bonds are frustrated), the overall configuration minimizes the energy.

The exchange fields satisfy

$$\begin{aligned} \mathbf{B}_A &= -\lambda \mathbf{M}_A - \mu \mathbf{M}_B \\ \mathbf{B}_B &= -\mu \mathbf{M}_A - \nu \mathbf{M}_B, \end{aligned} \tag{16.36}$$

where λ, μ, ν are positive constants and the minus signs indicate that each coupling has an antiferromagnetic nature. The energy of the interaction of the magnetic moments with the exchange magnetic fields is given by

$$\begin{aligned} U &= -\frac{1}{2}\left(\mathbf{B}_A \cdot \mathbf{M}_A + \mathbf{B}_B \cdot \mathbf{M}_B\right) \\ &= \frac{1}{2}\lambda M_A^2 + \mu \mathbf{M}_A \cdot \mathbf{M}_B + \frac{1}{2}\nu M_B^2. \end{aligned} \tag{16.37}$$

The first and third terms are clearly positive. However, the second term favors an antiparallel alignment between the A and B spins. There is a trivial solution, $M_A = M_B = 0$, that leads to a vanishing energy. A more favorable phase is obtained if the energy is negative. This implies

$$\frac{1}{2}\left(\lambda M_A^2 + \nu M_B^2\right) < \mu M_A M_B, \tag{16.38}$$

if we take $\mathbf{M}_a \cdot \mathbf{M}_B = -M_A M_B$.

To simplify, let us take $\lambda = \nu = 0$. In this case, we get

$$\begin{aligned} \mathbf{B}_A &= -\mu \mathbf{M}_B \\ \mathbf{B}_B &= -\mu \mathbf{M}_A. \end{aligned} \tag{16.39}$$

Applying an external magnetic field, B_a, and following similar steps to the ferromagnetic case, we can write, in the paramagnetic phase, that

$$\begin{aligned} M_A T &= C_A\left(B_a - \mu M_B\right) \\ M_B T &= C_B\left(B_a - \mu M_A\right), \end{aligned} \tag{16.40}$$

where C_A, C_B are the Curie constants of the spins of sites A and B, respectively, and T is the temperature.

Consider now a vanishing external field $B_a = 0$. A nontrivial solution is obtained near the critical point if

$$T_c = \mu\sqrt{C_A C_B} \tag{16.41}$$

and we get that

$$M_A = -\frac{\sqrt{C_A}}{\sqrt{C_B}} M_B. \tag{16.42}$$

The susceptibility is given by

$$\chi = \frac{M_A + M_B}{B_a} = \frac{(C_A + C_B)\,T - 2\mu C_A C_B}{T^2 - T_c^2} \tag{16.43}$$

and diverges at the critical temperature T_C, similar to a ferromagnet.

An interesting case occurs if $C_A = C_B = C$. In this case, $M_A = -M_B$ and is called the antiferromagnetic case. Defining $T_c = \mu C = T_N$, called the Néel temperature, the susceptibility is reduced to

$$\chi = \frac{2C}{T + T_N}. \tag{16.44}$$

It is no longer divergent. As one enters the antiferromagnetic phase, the susceptibility has a cusp, but no divergence. Also, the total magnetization at zero temperature vanishes. The magnetization of each sublattice is finite however, but since the spins are oriented in opposite directions, the total magnetization vanishes.

An example of an antiferromagnet is MnO. In Fig. 16.7, the neutron scattering data at a temperature smaller than the Néel temperature ($T_N = 118K$) and at room temperature are shown. The extra well-defined scattering peaks in the antiferromagnetic phase show the presence of antiferromagnetic order together with the evidence of a double size magnetic cell, such as the (111) peak. The spins are aligned in the (111) plane, and adjacent planes are aligned antiparallel, as shown by the scheme of the magnetic structure.

Considering a temperature smaller than the Néel temperature, it turns out that the susceptibility depends on the orientation of the applied magnetic field with respect to the direction of the anti-aligned spins. If the magnetic field is applied perpendicular to the direction of the spins, both the A and B spins will tend to align along the direction of the applied (small) magnetic field. This costs energy, since the exchange energy is no longer minimized. If each magnetization A and B approaches the direction of the applied field by an angle ϕ, the angle between each magnetic moment is 2ϕ. Denoting $M = |M_a| = |M_B|$, we get that the energy has the form

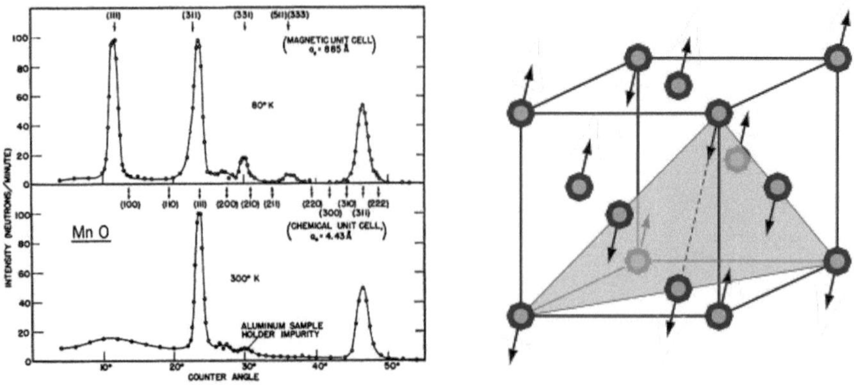

Fig. 16.7 Neutron scattering data on antiferromagnet MnO and magnetic structure. Reproduced with permission from C. G. Shull and J. Samuel Smart, Physical Review **76**, 1256 (1949) and from Andrew L. Goodwin, Matthew G. Tucker, Martin T. Dove, and David A. Keen, Physical Review Letters **96**, 047209 (2006)

$$\begin{aligned} U &= \mu \mathbf{M}_A \cdot \mathbf{M}_B - \mathbf{B}_a \cdot (\mathbf{M}_A + \mathbf{M}_B) \\ &= -\mu M^2 \left(1 - \frac{1}{2}(2\phi)^2\right) - 2B_a M\phi. \end{aligned} \tag{16.45}$$

Searching for an equilibrium solution, we may find the angle that minimizes the energy. This leads to

$$\frac{dU}{d\phi} = 0 = 4\mu M^2\phi - 2B_a M \tag{16.46}$$

and therefore an equilibrium value

$$\phi = \frac{B_a}{2\mu M}. \tag{16.47}$$

The perpendicular susceptibility is therefore a constant at low temperatures:

$$\chi_{\perp} = \frac{2M\phi}{B_a} = \frac{1}{\mu}. \tag{16.48}$$

On the other hand, the parallel susceptibility obtained applying an external field along the direction of the magnetic moments,costs no exchange energy and the susceptibility vanishes. This happens if the magnetic moments of the A and B sites change orientation by the same angle. The susceptibility of an antiferromagnet as a function of temperature is shown in Fig. 16.8.

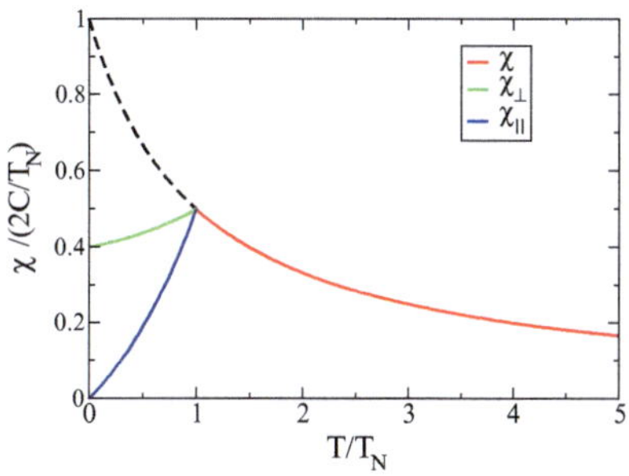

Fig. 16.8 Susceptibility of an antiferromagnet as a function of temperature. In the regime $T/T_N > 1$, the susceptibility is given by Eq. (16.44). The dashed curve is just the extension of the same curve: it diverges when $T = -T_N$. In the regime $T/T_N < 1$, we show some schematic plots of the perpendicular and parallel susceptibilities

16.5 Excitations in Magnetic Systems

16.5.1 Ferromagnetic Systems

Let us consider once again the Heisenberg model and consider for now the one-dimensional case:

$$H = -2J \sum_{i=1}^{N} \mathbf{S}_i \cdot \mathbf{S}_{i+1}.$$

Let us consider here the classical model where we take the spins as classical vectors and let us take $J > 0$. The ground state is obtained if all the spins are parallel. In this case, $\mathbf{S}_i \cdot \mathbf{S}_{i+1} = S^2$, and the energy of the ground state is given by $E_0 = -2NJS^2$. We want now to study the excitations around this minimum energy state.[5]

(i) Consider, for instance, an excitation in which one spin is turned "down." This state has an energy $E = E_0 + 8JS^2$. This excitation is a single-particle (single spin) excitation.
(ii) It is also possible to construct an excitation that requires much less energy if we divide the fluctuation over all spins in the system. The two types of excitations are considered in Fig. 16.9.

Let us consider the collective excitation that is constituted by a wave that describes the change of the orientation of the spins, in the plane perpendicular to their preferential direction (which was assumed as pointing "up"). The terms in the energy that involve the spin i are

$$-2J\mathbf{S}_i \cdot (\mathbf{S}_{i+1} + \mathbf{S}_{i-1}).$$

[5] See also [Kittel].

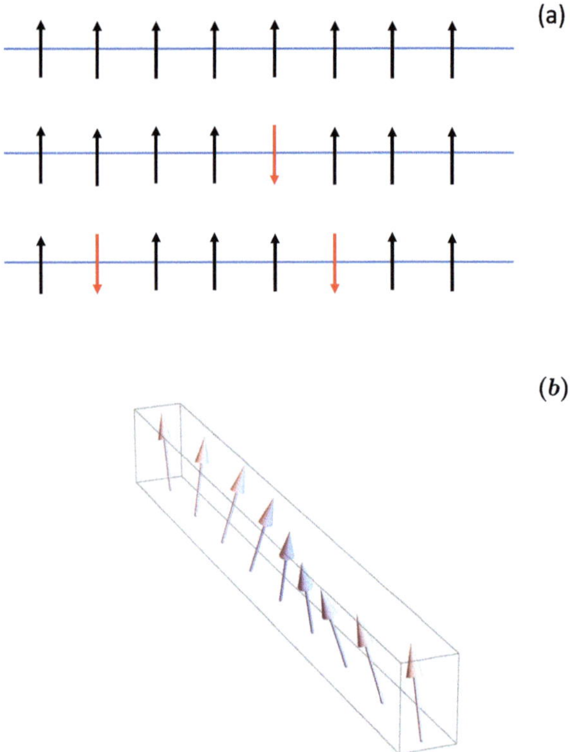

Fig. 16.9 Excitations of a one-dimensional ferromagnetic spin problem. (**a**) Single-particle excitations. In the top line is an example of the fully polarized spin configuration found in the ground state. In the following line, one of the spins is flipped (changes direction), and in the following line, two spins have reversed their orientations. (**b**) Spin-wave collective excitation seen as a small deviation of each spin from its ground state orientation. The spin component deviation is divided among all the spins

The magnetic moment at site i is $\boldsymbol{\mu}_i = -g\mu_B \mathbf{S}_i$. Therefore, we can rewrite that the contribution to the energy due to the site i is

$$- \boldsymbol{\mu}_i \cdot \left[\left(\frac{-2J}{g\mu_B} \right) (\mathbf{S}_{i+1} + \mathbf{S}_{i-1}) \right] \sim -\boldsymbol{\mu}_i \cdot \mathbf{B},$$

where $\mathbf{B}$ is the effective magnetic field acting on the spin i. From electromagnetism, we know that

$$\hbar \frac{d\mathbf{S}_i}{dt} = \boldsymbol{\mu}_i \times \mathbf{B}_i .$$

We get, therefore, the equation of motion for the spins in the form

$$\frac{d\mathbf{S}_i}{dt} = \frac{2J}{\hbar} (\mathbf{S}_i \times \mathbf{S}_{i+1} + \mathbf{S}_i \times \mathbf{S}_{i-1}) . \tag{16.49}$$

If the amplitude of the excitation is small, then $S_i^x, S_i^y << S$, and we can linearize the equation of motion, taking $S_i^z = S$ and neglecting terms of the order of $S^x S^y$:

$$\frac{dS_i^z}{dt} = 0$$
$$\frac{dS_i^x}{dt} = \frac{2JS}{\hbar}\left(2S_i^y - S_{i+1}^y - S_{i+1}^y\right)$$
$$\frac{dS_i^y}{dt} = -\frac{2JS}{\hbar}\left(2S_i^x - S_{i+1}^x - S_{i+1}^x\right). \tag{16.50}$$

Let us look for solutions of the type

$$S_p^x = ue^{i(pka-\omega t)},$$

$$S_p^y = ve^{i(pka-\omega t)}.$$

Substituting these solutions in the equations of motion, we get that

$$-i\omega u = \frac{2JS}{\hbar}\left(2 - e^{-ika} - e^{ika}\right)v = \frac{4JS}{\hbar}(1-\cos ka)v$$
$$-i\omega v = -\frac{2JS}{\hbar}\left(2 - e^{-ika} - e^{ika}\right)u = -\frac{4JS}{\hbar}(1-\cos ka)u. \tag{16.51}$$

To get a nontrivial solution, the determinant of the matrix

$$\begin{pmatrix} i\omega & \frac{4JS}{\hbar}(1-\cos ka) \\ -\frac{4JS}{\hbar}(1-\cos ka) & i\omega \end{pmatrix},$$

must vanish. This leads to the dispersion relation:

$$\hbar\omega = 4JS(1-\cos ka). \tag{16.52}$$

The solution corresponds to $v = -iu$, which is a circular precession around the z axis. At low momenta, the dispersion relation is quadratic $\omega \sim k^2$. It saturates at π/a with a zero velocity. To be more precise,

$$\hbar\omega \sim (2JSa^2)k^2,$$

at small momenta.

Let us now quantize these waves called spin waves. They are bosonic-like (classical) and let us quantize them by using the spectrum of the harmonic oscillator:

$$\epsilon_k = \left(n_k + \frac{1}{2}\right)\hbar\omega_k.$$

(In this approximation, the spin waves do not interact and, therefore, the linearization leads to independent mode that we have quantized. Later on, we will solve the problem with more care). Since these excitations are bosons, their thermodynamic occupation number obeys the Bose-Einstein distribution (with a zero chemical potential since their number is not conserved) like

$$\langle n_k \rangle = \frac{1}{e^{\frac{\hbar\omega_k}{k_B T}} - 1}.$$

The total number of these quantized spin waves (called magnons) is given by

$$\sum_k n_k = \int d\omega D(\omega) \langle n(\omega) \rangle.$$

Here, the density of states per unit volume is given by

$$D(\omega) = \left(\frac{1}{2\pi} \right)^3 4\pi k^2 \frac{dk}{d\omega}.$$

Using that $\hbar\omega \sim (2JSa^2)k^2$, we obtain

$$D(\omega) = \frac{1}{4\pi^2} \left(\frac{\hbar}{2JSa^2} \right)^{3/2} \omega^{1/2}.$$

The total number of magnons is therefore

$$\begin{aligned} \sum_k n_k &= \frac{1}{4\pi^2} \left(\frac{\hbar}{2JSa^2} \right)^{3/2} \int_0^\infty d\omega \frac{\omega^{1/2}}{e^{\beta\hbar\omega} - 1} \\ &= \frac{1}{4\pi^2} \left(\frac{k_B T}{2JSa^2} \right)^{3/2} \int_0^\infty dx \frac{x^{1/2}}{e^x - 1}. \end{aligned}$$

The integral is equal to $4\pi^2 0.0587$. Therefore, we obtain Bloch's law that gives that

$$\frac{\sum_k n_k}{NS} = \frac{\Delta M}{M(0)} \sim \left(\frac{k_B T}{JS} \right)^{3/2}, \tag{16.53}$$

for the deviation of the magnetization from the ground state value.

Note that the result obtained from the mean-field approximation is different and not correct. Starting from Eq. (16.31), we may expand it near zero temperature. At zero temperature, the magnetization is saturated and $m \to 1$. Defining the deviation from perfect alignment as $\delta m = 1 - m$, we can see that we obtain

$$\delta m = 2e^{-2\frac{T_c}{T}} \tag{16.54}$$

and, therefore, the deviation from saturation is exponential and not a power law, as given by Bloch's law. In the same way that near the critical temperature thermal fluctuations are important, invalidating the mean-field theory approach, near zero temperature quantum fluctuations are important. This suggests a full quantum treatment.

16.5.2 Holstein-Primakoff Transformation

Again consider the Heisenberg Hamiltonian:

$$H = -J \sum_{\mathbf{j},\boldsymbol{\delta}} \mathbf{S_j} \cdot \mathbf{S_{j+\delta}} - 2\mu_0 B_0 \sum_{\mathbf{j}} S_{\mathbf{j},z}, \tag{16.55}$$

with $\mu_0 = 2\mu_B/2$, B_0 is the external magnetic field, and $\boldsymbol{\delta}$ connects each site, $\mathbf{j}$, to its neighbors. The total spin of the system is such that $\mathbf{S}^2 = (\sum_j \mathbf{S}_j)^2$ (we have replaced $\mathbf{j}$ by j to simplify the notation) and $S_z = \sum_j S_{jz}$. The ground state of the ferromagnetic system is such that

$$\mathbf{S}^2|0\rangle = NS(S+1)|0\rangle$$

$$S_z|0\rangle = NS|0\rangle.$$

It is a state completely ordered where all spins are aligned parallel. This is the ground state even if the magnetic field vanishes. The direction of alignment is then arbitrary and fixing a given direction breaks the initial rotational symmetry.

Let us now introduce a bosonic description of the spin operators. First, let us define the raising and lowering spin operators like

$$S_j^+ = S_{jx} + iS_{jy} = \sqrt{2S}\sqrt{1 - \frac{a_j^\dagger a_j}{2S}}\, a_j$$

$$S_j^- = S_{jx} - iS_{jy} = \sqrt{2S}\, a_j^\dagger \sqrt{1 - \frac{a_j^\dagger a_j}{2S}}. \tag{16.56}$$

The operators $a_j^\dagger$ and a_j are bosonic operators satisfying

$$\left[a_j, a_l^\dagger\right] = \delta_{jl}.$$

Using that $S_z^2 = S(S+1) - S_x^2 - S_y^2$, introducing the above expressions for the ladder operators, and using that $[a^\dagger a, a] = -a$, we can show that

$$S_{jz} = S - a_j^\dagger a_j. \tag{16.57}$$

To verify that this bosonic representation is appropriate, we have to show that the operators S_j^+, S_j^-, S_{jz} defined in Eqs. (16.56), and (16.57) obey the commutation relations of the spin operators. It is easy to verify that, for instance, $[S^+, S^-] = 2S^z$ using the above definitions in terms of the bosonic operators.

Let us now consider the Fourier transform of the bosonic operators:

$$b_k = \frac{1}{\sqrt{N}} \sum_j e^{ik\cdot x_j} a_j$$

$$b_k^\dagger = \frac{1}{\sqrt{N}} \sum_j e^{-ik\cdot x_j} a_j^\dagger.$$

These definitions lead to a commutation relation $[b_k, b_{k'}^\dagger] = \delta_{k,k'}$ which shows that they are bosons as well. We can then write the spin operators in terms of these momentum space bosonic operators like

$$\begin{aligned} S_j^+ &= \sqrt{2S}\left[a_j - \frac{1}{4S} a_j^\dagger a_j a_j + \cdots\right] \\ &= \sqrt{\frac{2S}{N}}\left[\sum_k e^{-ik\cdot x_j} b_k - \frac{1}{4SN}\sum_{kk'k''} e^{i(k-k'-k'')\cdot x_j} b_k^\dagger b_{k'} b_{k''} + \cdots\right] \\ S_j^- &= \sqrt{\frac{2S}{N}}\left[\sum_k e^{ik\cdot x_j} b_k^\dagger - \frac{1}{4SN}\sum_{kk'k''} e^{i(k+k'-k'')\cdot x_j} b_k^\dagger b_{k'}^\dagger b_{k''} + \cdots\right] \\ S_j^z &= S - a_j^\dagger a_j = S - \frac{1}{N}\sum_{kk'} e^{i(k-k')\cdot x_j} b_k^\dagger b_{k'}. \end{aligned} \tag{16.58}$$

Also, the deviation from the maximal magnetization is given by

$$\delta S_z = NS - \sum_j S_{jz} = \sum_j a_j^\dagger a_j = \sum_k b_k^\dagger b_k. \tag{16.59}$$

The full quantum expression obtained using the Holstein-Primakoff transformation shows the previous hypothesis that the displacement of the magnetization from its saturation is due to the bosonic terms, here those that result from the transformation. Therefore, Bloch's law is also obtained from the full quantum treatment.

Let us now rewrite the Hamiltonian in terms of these bosonic operators. It yields, considering that there are z nearest neighbors,

$$H = -JNzS^2 - 2\mu_0 B_0 NS + H_0 + H_1. \tag{16.60}$$

Here,

$$H_0 = -JzS\sum_k \left(\gamma_k b_k b_k^\dagger + \gamma_{-k} b_k^\dagger b_k - 2b_k^\dagger b_k\right) + 2\mu_0 B_0 \sum_k b_k^\dagger b_k, \tag{16.61}$$

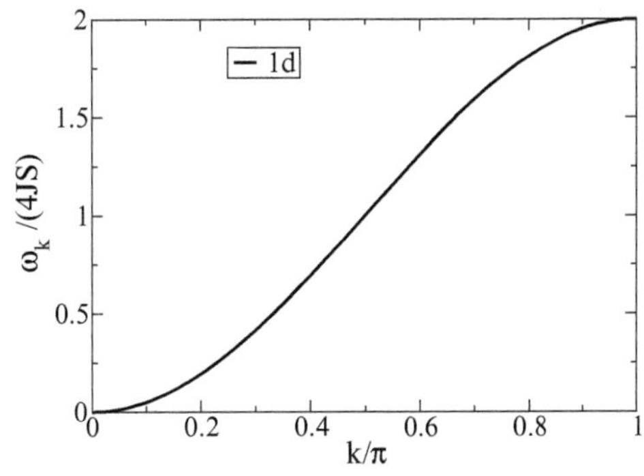

Fig. 16.10 Dispersion relation of the spin-wave excitations of a one-dimensional ferromagnetic system

where

$$\gamma_k = \frac{1}{z}\sum_\delta e^{ik\cdot\delta}.$$

Note that $\sum_k \gamma_k = 0$. If there is a center of symmetry, then $\gamma_k = \gamma_{-k}$. In this case, we get that

$$H_0 = \sum_k n_k \omega_k, \tag{16.62}$$

where the dispersion relation of the excitations (spin waves) is

$$\omega_k = 2JSz(1-\gamma_k) + 2\mu_0 B_0. \tag{16.63}$$

It is shown for a one-dimensional system in Fig. 16.10 in zero magnetic field. The term H_1 contains all higher-order terms in the boson operators, cubic and higher. These terms describe the interactions between the spin waves and can be systematically included. We will not pursue this here.

Writing the dispersion relation as $\omega_k = Dk^2$, where D is called the spin stiffness, we get that, if the magnetic field is zero, $D = 2SJa^2$. The total energy is given by

$$\begin{aligned} U &= \sum_k \langle n_k\rangle \omega_k = \sum_k \omega_k \frac{1}{e^{\beta\omega_k}-1} = \frac{1}{(2\pi)^3}\int d^3k Dk^2 \frac{1}{e^{\beta Dk^2}-1} \\ &= \frac{(k_BT)^{5/2}}{4\pi^2 D^{3/2}}\int_0^{x_m} dx x^{3/2}\frac{1}{e^x-1}. \end{aligned} \tag{16.64}$$

This means that $U \sim T^{5/2}$ and that the specific heat is $C_V \sim T^{3/2}$.

16.5.3 Antiferromagnetic Systems

Consider now the case of antiferromagnetism described by the Heisenberg Hamiltonian, but now with a coupling constant with opposite sign: it is now favorable classically for the spins to align in opposite directions. There are now two interpenetrating sublattices, such that the spins on each lattice are preferably aligned, while spins in different lattices are antiparallel. Therefore, the nearest neighbors of each spin belong to the other sublattice.

The Hamiltonian can now be written conveniently as

$$H = J\sum_{j\delta} \mathbf{S}_j \cdot \mathbf{S}_{j+\delta} - 2\mu_0 H_A \sum_j S^a_{jz} + 2\mu_0 H_A \sum_j S^b_{jz}. \tag{16.65}$$

We have divided the system in sublattices A and B and assigned the labels a and b to the spins belonging to the two sublattices, respectively. Here, $J > 0$ to have antiferromagnetism. H_A is a magnetic field on the A sublattice, and a field $-H_A$ is applied on the B sublattice.

Let us now define two sets of bosonic operators, one associated with the spins a and the other set to the spins b. Then, we define

$$\begin{aligned}
S^+_{aj} &= \sqrt{2S}\left(1 - \frac{1}{2S}a^\dagger_j a_j\right)^{1/2} a_j \\
S^-_{aj} &= \sqrt{2S}a^\dagger_j\left(1 - \frac{1}{2S}a^\dagger_j a_j\right)^{1/2} \\
S^+_{bl} &= \sqrt{2S}b^\dagger_l\left(1 - \frac{1}{2S}b^\dagger_l b_l\right)^{1/2} \\
S^-_{bl} &= \sqrt{2S}\left(1 - \frac{1}{2S}b^\dagger_l b_l\right)^{1/2} b_l \\
S^z_{aj} &= S - a^\dagger_j a_j \\
S^z_{bl} &= -S + b^\dagger_l b_l.
\end{aligned}$$

Note that the spins of the B lattice point in the opposite direction, and, therefore, the bosons describe deviations from the opposite direction. The presence of bosons associated with the A sites describes the lowering of the z-component of the magnetization. Therefore, lowering the spin component is associated with creating a boson. On the other hand, for the B sites, the presence of bosons is associated with the increase of the spin component: therefore, the spin raising operator is associated with the creation of a boson. Like before, we define the Fourier transforms of the bosonic fields:

$$
\begin{aligned}
c_k &= \frac{1}{\sqrt{N}} \sum_j^a e^{ik\cdot x_j} a_j \\
c_k^\dagger &= \frac{1}{\sqrt{N}} \sum_j^a e^{-ik\cdot x_j} a_j^\dagger \\
d_k &= \frac{1}{\sqrt{N}} \sum_l^b e^{-ik\cdot x_l} b_l \\
d_k^\dagger &= \frac{1}{\sqrt{N}} \sum_l^b e^{ik\cdot x_l} b_l^\dagger .
\end{aligned}
\tag{16.66}
$$

Substituting these definitions in terms of the bosons in the expressions for the spin operators and keeping only the quadratic terms in the bosonic operators, we get that, similar to the ferromagnetic case, the Hamiltonian can be written as

$$
H = -2NzJS^2 - 4N\mu_0 H_A S + H_0 + H_1, \tag{16.67}
$$

where, as before, H_1 is the term that contains the interactions between the magnons and

$$
H_0 = 2JzS \sum_k \left[\gamma_k \left(c_k^\dagger d_k^\dagger + c_k d_k\right) + \left(c_k^\dagger c_k + d_k^\dagger d_k\right)\right] + 2\mu_0 H_A \sum_k \left(c_k^\dagger c_k + d_k^\dagger d_k\right). \tag{16.68}
$$

Here,

$$
\gamma_k = \frac{1}{z} \sum_\delta e^{ik\cdot\delta} = \gamma_{-k}, \tag{16.69}
$$

if there is a center of symmetry, as before.

To diagonalize the Hamiltonian, let us make a Bogoliubov-Valatin transformation. Define

$$
\begin{aligned}
\alpha_k &= u_k c_k - v_k d_k^\dagger \\
\alpha_k^\dagger &= u_k c_k^\dagger - v_k d_k \\
\beta_k &= u_k d_k - v_k c_k^\dagger \\
\beta_k^\dagger &= u_k d_k^\dagger - v_k c_k .
\end{aligned}
\tag{16.70}
$$

As before, using that the new operators are also bosons, we get that $u_k^2 - v_k^2 = 1$. Comparing with our solution of the phonons in the Bose liquid, we get that, making the replacements,

$$
a_k \to c_k; \qquad a_k^\dagger \to c_k^\dagger; a_{-k} \to d_k; a_{-k}^\dagger \to d_k^\dagger
$$

and making the replacements

$$\omega_0 \to 2JzS + 2\mu_0 H_A; \qquad \omega_1 \to 2JzS\gamma_k.$$

This leads then to

$$\omega_k^2 = (\omega_e + \omega_A)^2 - \omega_e^2 \gamma_k^2, \tag{16.71}$$

with $\omega_e = 2JzS$ and $\omega_A = 2\mu_0 H_A$. In terms of the new operators, we can write the Hamiltonian like

$$H_0 = -N(\omega_e + \omega_A) + \sum_k \omega_k \left(\alpha_k^\dagger \alpha_k + \beta_k^\dagger \beta_k + 1\right)$$

and the full Hamiltonian is then written as

$$H = -2NzJS(S+1) - 4\pi\mu_0 H_A \left(S + \frac{1}{2}\right) + \sum_k \omega_k \left(n_k + \frac{1}{2}\right) + H_1. \tag{16.72}$$

Here, the sum over the momentum has a weight 2 for each momentum value. If the magnetic field vanishes and we consider small momenta, we get

$$\omega_k \sim (4\sqrt{3} J S a)k, \tag{16.73}$$

and, therefore the dispersion relation is linear in k. It is shown for a one-dimensional system in Fig. 16.11 in the zero external field.

It is interesting to calculate the zero temperature magnetization of each sublattice. For instance,

$$\begin{aligned} S_a^z &= \sum_j S_{aj}^z = NS - \sum_k c_k^\dagger c_k \\ &= NS - \sum_k \left(u_k^2 \alpha_k^\dagger \alpha_k + v_k^2 \beta_k \beta_k^\dagger + \text{terms not diagonal}\right). \end{aligned}$$

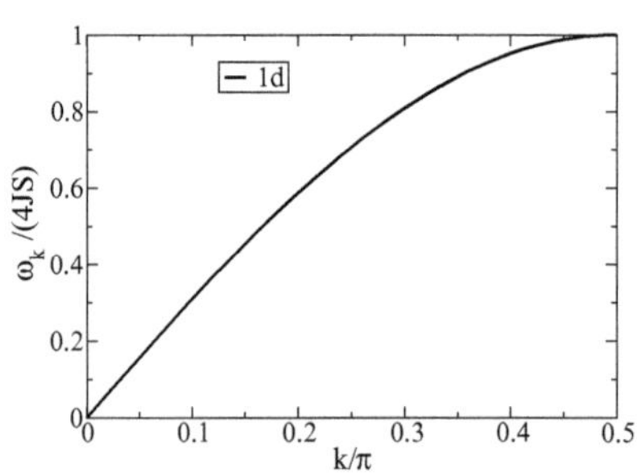

Fig. 16.11 Dispersion relation of the spin-wave excitations of a one-dimensional antiferromagnetic system

Consider then $T = 0$ and $n_k = 0$. Then, the magnetization deviation is given by

$$\begin{aligned}\Delta S^z &= NS - \langle S^z \rangle = \sum_k v_k^2 \\ &= \frac{1}{2} \sum_k (\cosh 2\chi_k - 1) \\ &= \frac{1}{2} \sum_k \left(\frac{1}{\sqrt{1-\gamma_k^2}} - 1 \right).\end{aligned}$$

The change of magnetization per unit volume is

$$\Delta S^z = -\frac{N}{2} + \frac{1}{2(2\pi)^3} \int d^3k \frac{1}{\sqrt{1-\gamma_k^2}} = 0.078N \neq 0. \tag{16.74}$$

This numerical result is for the cubic lattice. The quantum result is different from the classical case, since the sublattice magnetization is not fully saturated. This is due to quantum fluctuations that are nontrivial due to the singlet spin nature of the ground state.

Low-Temperature Specific Heat

The dispersion relations of the low-energy excitations of a ferromagnet ($\omega \sim k^2$) and of an antiferromagnet ($\omega \sim k$) are different. As a consequence, their low-temperature behavior is different, such as the specific heat. Let us consider that

$$\omega = \gamma k^\alpha. \tag{16.75}$$

The energy can be written as

$$U = \int d^d k \gamma k^\alpha \frac{1}{e^{\beta \gamma k^\alpha} - 1} = \int d\omega D(\omega) \omega \frac{1}{e^{\beta\omega} - 1}, \tag{16.76}$$

where $D(\omega)$ is the density of states and the distribution is the Planck distribution with $\beta = 1/(k_B T)$. We can calculate

$$\frac{d\omega}{dk} = \alpha \gamma k^{\alpha-1} \sim \omega^{\frac{\alpha-1}{\alpha}}. \tag{16.77}$$

Rewriting the expression for the energy, we get

$$U \sim \int dk k^{d-1} k^\alpha \frac{1}{e^{\beta\gamma k^\alpha} - 1}. \tag{16.78}$$

Table 16.3 Temperature dependence of the low-temperature specific heat of a bosonic system, with a dispersion relation $\omega \sim k^{\alpha}$ in a space of dimension d. The top line is the case of a ferromagnet and the lower line is the case of an antiferromagnet

ω	$d = 1$	$d = 2$	$d = 3$
k^2	$T^{1/2}$	T	$T^{3/2}$
k	T	T^2	T^3

We conclude that the density of states has the form

$$D(\omega) \sim \omega^{\frac{d-1}{\alpha}} \omega^{\frac{1-\alpha}{\alpha}} \tag{16.79}$$

and

$$U \sim \int d\omega \omega^{\frac{d}{\alpha}} \frac{1}{e^{\beta\omega} - 1}. \tag{16.80}$$

Rescaling the frequency in terms of the temperature, we get for the energy

$$\begin{aligned} U &\sim T \int dx \frac{1}{e^x - 1} x^{d/\alpha} T^{d/\alpha} \\ &\sim T^{\frac{d}{\alpha}}. \end{aligned} \tag{16.81}$$

We can summarize the results for the temperature dependence of the low-temperature specific heat in Table 16.3.

16.6 Stoner Theory of Itinerant Magnetism

16.6.1 Hubbard Model and Motivation

We saw that in simple metals, the free electron model, plus periodic potential and plus Coulomb potential, describes in a qualitatively correct way the conduction bands. Actually, besides the possibility that ordered phases may appear at low temperatures, such as SDW, CDW, or superconductivity (to be discussed in the next chapter), the description of the Fermi liquid is very similar to the description of the Fermi gas.

However, transition metals or rare earths behave in a different way because, besides the conduction bands, they have d or f bands partially filled. The states from which these bands originate lead to overlap integrals between neighboring atoms that are smaller and, therefore, lead to narrow energy bands. In particular, in the case of f bands, the overlap integral is rather small and suggests a localized description of the f electrons. In the case of d electrons, the extension of the orbitals is no longer negligible, and, therefore, we have to take into consideration the possibility of delocalization of the electrons. However, since the probability that one electron is moving from one atom to another neighboring atom is small, we may consider that, on average, one electron spends more time on the atom locations jumping occasionally with a probability amplitude, t, called hopping integral. We might expect, therefore, that the correlations of the electrons with the other electrons on the same atom are more important than the correlations with the electrons sitting on other atoms.

According to Hund's rule (Pauli's exclusion principle), the Coulomb interactions between the electrons of the same atom are such that it is favorable to maximize the spin state (if all electrons have the same spin component then, by Pauli's principle, they cannot be at the same point in space and, as a consequence, the Coulomb repulsion between them is minimized).

(i) A first consequence is that it seems favorable for a given atom to preserve a fixed value of the spin and, therefore, each atom may be characterized by a certain spin, which seems to indicate that this may lead (at least from the magnetic point of view) to a description in terms of a Heisenberg-like Hamiltonian.
(ii) A second consequence is that to satisfy Hund's rule, it seems to be favorable that, if an electron hops to an atom, it will be better if that electron has a spin component that is parallel to the preexisting spin of the atom. That is, there appears an effective attractive interaction for electrons of the same spin and repulsive for electrons of opposite spins: this tends to stabilize the spin of the atom.

Let us then consider a model (called the Hubbard model) in which the attention is focused on these narrow bands of d or f electrons, whose properties are dominant in the behavior of these materials. To simplify, let us consider that they have $l = 0$. The true orbital nature of these materials is actually relevant, but we will not consider this more realistic case.

The Hubbard model is defined by

$$H = \sum_{\mathbf{k},\sigma} \epsilon_{\mathbf{k}} c^{\dagger}_{\mathbf{k},\sigma} c_{\mathbf{k},\sigma} + \frac{1}{2} \sum_{\mathbf{k}_1,\mathbf{k}_2,\mathbf{k}'_1,\mathbf{k}'_2} \sum_{\sigma_1,\sigma_2} \left\langle k_1, k_2 \middle| \frac{1}{r} \middle| k'_1, k'_2 \right\rangle c^{\dagger}_{\mathbf{k_1},\sigma_1} c^{\dagger}_{\mathbf{k_2},\sigma_2} c_{\mathbf{k_2}',\sigma_2} c_{\mathbf{k_1}',\sigma_1} . \tag{16.82}$$

Let us define the Wannier functions:

$$\phi(\mathbf{r}) = \frac{1}{\sqrt{N}} \sum_{\mathbf{k}} \psi_{\mathbf{k}}$$

$$\psi_{\mathbf{k}} = \frac{1}{\sqrt{N}} \sum_{i} e^{i\mathbf{k}\cdot\mathbf{R}_i} \phi(\mathbf{r} - \mathbf{R}_i).$$

Here, $\mathbf{R}_i$ is the position of atom i. Then, defining

$$c_{\mathbf{k},\sigma} = \frac{1}{\sqrt{N}} \sum_{i} e^{i\mathbf{k}\cdot\mathbf{R}_i} c_{i,\sigma}$$

and its Hermitian conjugate, we can rewrite the Hamiltonian as

$$H = \sum_{i,j} \sum_{\sigma} T_{ij} c^{\dagger}_{i,\sigma} c_{j,\sigma} + \frac{1}{2} \sum_{i,j} \sum_{k,l} \sum_{\sigma,\sigma'} \left\langle i, j | \frac{1}{r} | k, l \right\rangle c^{\dagger}_{i,\sigma} c^{\dagger}_{j,\sigma'} c_{l,\sigma'} c_{k,\sigma},$$

where

$$T_{ij} = \frac{1}{N} \sum_{k} \epsilon_k e^{i\mathbf{k}\cdot(\mathbf{R}_i - \mathbf{R}_j)}$$

$$\langle i, j | \frac{1}{r} | k, l \rangle = \frac{e^2}{4\pi\epsilon_0} \int d^3x \int d^3x' \frac{\phi^*(\mathbf{r} - \mathbf{R}_i)\phi(\mathbf{r} - \mathbf{R}_k)\phi^*(\mathbf{r}' - \mathbf{R}_j)\phi(\mathbf{r}' - \mathbf{R}_l)}{|\mathbf{r} - \mathbf{r}'|}.$$

Since the Wannier functions are localized in space, let us only consider the Coulomb interaction matrix element on the same lattice site. In the same way, let us only consider the hopping between nearest neighbors. These approximations lead to the Hubbard model:

$$H = -t \sum_{\mathbf{r},\boldsymbol{\delta}} \sum_{\sigma} c^{\dagger}_{\mathbf{r},\sigma} c_{\mathbf{r}+\boldsymbol{\delta},\sigma} + U \sum_{\mathbf{r}} n_{\mathbf{r},\uparrow} n_{\mathbf{r},\downarrow}, \tag{16.83}$$

a tight-binding Hamiltonian with local, Coulombic, interactions added. Only the electrons of opposite spins may be at the same site ($l = 0$). The sum over $\boldsymbol{\delta}$ is the sum over the nearest neighbors.

We may also write the Hamiltonian in momentum space. It is given by

$$H = \sum_{\mathbf{p},\sigma} \epsilon_{\mathbf{p}} c^{\dagger}_{\mathbf{p},\sigma} c_{\mathbf{p},\sigma} + \frac{U}{N} \sum_{\mathbf{p},\mathbf{p}',\mathbf{q}} c^{\dagger}_{\mathbf{p}+\mathbf{q},\uparrow} c_{\mathbf{p},\uparrow} c^{\dagger}_{\mathbf{p}'-\mathbf{q},\downarrow} c_{\mathbf{p}',\downarrow}. \tag{16.84}$$

The Hilbert space of the model is the tensor product of *four* states per site of the lattice: these are the empty singlet state, two sites that correspond to the single occupancy of the site with spin up or with spin down and the singlet doubly occupied state.

The spin operator is defined as

$$\mathbf{S}(\mathbf{r}) = \frac{\hbar}{2} c^{\dagger}_{\sigma}(\mathbf{r}) \boldsymbol{\sigma}_{\sigma,\sigma'} c_{\sigma'}(\mathbf{r}) = \boldsymbol{\sigma}_{el}(\mathbf{r}),$$

where $\boldsymbol{\sigma}$ is a Pauli matrix. The number operator is given by

$$n(\mathbf{r}) = \sum_{\sigma} n_{\sigma}(\mathbf{r}) = \sum_{\sigma} c^{\dagger}_{\sigma}(\mathbf{r}) c_{\sigma}(\mathbf{r}).$$

Let us consider the operator

$$\sum_{\mathbf{r}} (\mathbf{S}(\mathbf{r}))^2 = \sum_{\mathbf{r},a=x,y,z} S^a(\mathbf{r})S^a(\mathbf{r}) = \sum_{\mathbf{r},a} \frac{1}{4}\left(c^\dagger_\alpha(\mathbf{r})\sigma^a_{\alpha,\beta}c_\beta(\mathbf{r})\right)\left(c^\dagger_\gamma(\mathbf{r})\sigma^a_{\gamma,\delta}c_\delta(\mathbf{r})\right).$$

Using that

$$\sum_a \sigma^a_{\alpha,\beta}\sigma^a_{\gamma,\delta} = 2\delta_{\alpha,\gamma}\delta_{\beta,\delta} - \delta_{\alpha,\beta}\delta_{\gamma,\delta},$$

we obtain

$$\sum_{\mathbf{r}} (\mathbf{S}(\mathbf{r}))^2 = \sum_{\mathbf{r}} \left(\frac{1}{4}n(\mathbf{r}) - \frac{3}{2}n_\uparrow(\mathbf{r})n_\downarrow(\mathbf{r})\right).$$

Therefore, we may rewrite the interacting term in the Hamiltonian:

$$U\sum_{\mathbf{r}} n_\uparrow(\mathbf{r})n_\downarrow(\mathbf{r}) = -\frac{2U}{3}\sum_{\mathbf{r}} (\mathbf{S}(\mathbf{r}))^2 + \frac{N_e U}{6}.$$

Here, N_e is the number of electrons. Then, we may rewrite the Hubbard model as

$$H = -t\sum_{\mathbf{r},\boldsymbol{\delta},\sigma}\left(c^\dagger_\gamma(\mathbf{r})c_\delta(\mathbf{r}+\boldsymbol{\delta}) + c^\dagger_\gamma(\mathbf{r}+\boldsymbol{\delta})c_\delta(\mathbf{r})\right) - \frac{2U}{3}\sum_{\mathbf{r}} (\mathbf{S}(\mathbf{r}))^2, \tag{16.85}$$

plus an irrelevant constant. This Hamiltonian is obviously invariant under $SU(2)$ rotations of the spin. For $U > 0$, the energy is minimized if the total spin state at each site is maximized (Hund's rule). It is therefore natural to expect that it may be favorable for the system to be in a state or magnetic order, as more favorable as U is larger. Note, however, that at this point, it is not clear what type of order will emerge, since at this point, it is not clear what the interaction is (if any) between neighboring spins.

16.6.2 Susceptibility and Stoner Criterion

Let us consider an external field:

$$H_{ext} = -\int d^3x \mathbf{H}(x,t)\cdot\boldsymbol{\sigma}(x).$$

Here, $\boldsymbol{\sigma}(x) = \sum_{\text{electrons}} \delta(x - x_{el})\boldsymbol{\sigma}_{el}$, is the spin density of the electrons at a site, $x = \mathbf{r}$, in three-dimensional space. We want to calculate.[6]

$$\langle \boldsymbol{\sigma}(x,t) \rangle = \langle \psi(t) | \boldsymbol{\sigma}(x) | \psi(t) \rangle .$$

We can use the linear response theory previously discussed:

$$\langle \sigma_i(x,t) \rangle = \langle \sigma_i(x,t) \rangle_0 + \sum_j \int dt' \int d^3x' \chi_{ij}(x-x', t-t') H_j(x', t'), \tag{16.86}$$

where the susceptibility is given by

$$\chi_{ij}(x-x', t-t') = i\theta(t-t') \langle [\sigma_i^I(x,t), \sigma_j^I(x',t')] \rangle . \tag{16.87}$$

The spin operator can be written as

$$\boldsymbol{\sigma}(x) = \sum_q e^{iq\cdot x} \sum_{p,\alpha,\beta} c^\dagger_{p+q,\alpha}(t) \boldsymbol{\sigma}_{\alpha,\beta} c_{p,\beta}(t),$$

where the creation and destruction operators are written in the interaction representation. Define as usual the ladder operators $\sigma^{\pm} = \frac{1}{2}\left(\sigma_x \pm i\sigma_y\right)$ and therefore

$$\sigma^+(x) = \sum_q e^{iq\cdot x} \sum_p c^\dagger_{p+q,\uparrow} c_{p,\downarrow}$$

$$\sigma^-(x) = \sum_q e^{iq\cdot x} \sum_p c^\dagger_{p+q,\downarrow} c_{p,\uparrow} .$$

The susceptibility is diagonal and isotropic. Therefore, $\chi_{ij} = 2\chi^{-+}\delta_{ij}$. It is enough to calculate the transverse susceptibility that is given by

$$\chi^{-+}(x-x', t-t') = i\theta(t-t') \langle [\sigma_I^-(x,t), \sigma_I^+(x',t')] \rangle$$

$$\chi^{-+}(x,t) = \sum_{p,q} e^{iq\cdot x} \chi^{-+}(p,q,t),$$

where

$$\chi^{-+}(p,q,t) = i\theta(t) \left\langle \left[c^\dagger_{p+q,\downarrow}(t) c_{p,\uparrow}(t), \sigma^+(0,0) \right] \right\rangle . \tag{16.88}$$

[6] For further reading see [DoniachSondheimer].

In order to calculate the susceptibility, let us obtain its equation of motion:

$$\begin{aligned} i\frac{\partial}{\partial t}\chi^{-+}(p,q,t) &= -\delta(t)\left\langle\left[c^\dagger_{p+q,\downarrow}c_{p,\uparrow},\sigma^+(0,0)\right]\right\rangle \\ &\quad +i\theta(t)\left\langle\left[i\frac{\partial}{\partial t}\left(c_{p+q,\downarrow}(t)c_{p,\uparrow}(t)\right),\sigma^+(0,0)\right]\right\rangle \\ &= -\delta(t)\langle[c_{p+q,\downarrow}c_{p,\uparrow},\sigma^+(0,0)]\rangle \\ &\quad + i\theta(t)\left\langle\left[\left[c^\dagger_{p+q,\downarrow}c_{p,\uparrow},H\right],\sigma^+(0,0)\right]\right\rangle. \end{aligned} \tag{16.89}$$

In this case, the Hamiltonian is the Hubbard model:

$$H=\sum_{p,\sigma}\epsilon_p c^\dagger_{p,\sigma}c_{p,\sigma}+\frac{U}{N}\sum_{p,p',q}c^\dagger_{p+q,\uparrow}c_{p,\uparrow}c^\dagger_{p'-q,\downarrow}c_{p',\downarrow}.$$

Calculating the commutator, we obtain the equation of motion like

$$\begin{aligned} i\frac{\partial}{\partial t}\chi^{-+}(p,q,t) &= -\delta(t)\left\langle\left[c^\dagger_{p+q,\downarrow}c_{p,\uparrow},\sigma^+(0,0)\right]\right\rangle \\ &\quad +i\theta(t)\left\langle\left[-(\epsilon_{p+q}-\epsilon_p)c^\dagger_{p+q,\downarrow}c_{p,\uparrow},\sigma^+(0,0)\right]\right\rangle \\ &\quad +i\theta(t)\left\langle\left[\frac{U}{N}\sum_{p',q'}c^\dagger_{p+q,\downarrow}c_{p-q',\uparrow}c^\dagger_{p'-q',\downarrow}c_{p',\downarrow},\sigma^+(0,0)\right]\right\rangle \\ &\quad -i\theta(t)\left\langle\left[\frac{U}{N}\sum_{p',q'}c^\dagger_{p'+q',\uparrow}c_{p',\uparrow}c^\dagger_{p+q-q',\downarrow}c_{p,\uparrow},\sigma^+(0,0)\right]\right\rangle. \end{aligned} \tag{16.90}$$

Consider now the Hartree-Fock approximation. The products of *four* operators are replaced by products of two operators times the average value of the other two operators. Recall also that

$$\left\langle c^\dagger_{p,\alpha}c_{p',\beta}\right\rangle=\delta_{p,p'}\delta_{\alpha,\beta}f_{p,\alpha},$$

where f is the Fermi function. Then, the term in U is simplified to

$$\frac{U}{N}\sum_{p'}\left((f_{p,\uparrow}-f_{p+q,\downarrow}\right)c^\dagger_{p+p'+q,\downarrow}c_{p+p',\uparrow}+\left(f_{p',\downarrow}-f_{p',\uparrow}\right)c^\dagger_{p+q,\downarrow}c_{p,\uparrow}.$$

Considering that

$$\sigma^+(0,0) = \sum_{p,q} c^\dagger_{p+q,\uparrow} c_{p,\downarrow},$$

we have that

$$\left\langle \left[c^\dagger_{p+q,\downarrow} c_{p,\uparrow}, \sigma^+(0,0) \right] \right\rangle = f_{p+q,\downarrow} - f_{p,\uparrow}.$$

Putting everything together, we get that the equation of motion is

$$\begin{aligned} &\left[i\frac{\partial}{\partial t} + (\tilde{\epsilon}_{p+q,\uparrow} - \tilde{\epsilon}_{p,\downarrow}) \right] \chi^{-+}(p,q,t) \\ &= -\delta(t)\left(f_{p+q,\downarrow} - f_{p,\uparrow}\right) - \left(f_{p+q,\downarrow} - f_{p,\uparrow}\right)\frac{U}{N}\sum_{p'} \chi^{-+}(p',q,t), \end{aligned} \tag{16.91}$$

where

$$\tilde{\epsilon}_{p,\sigma} = \epsilon_p - \frac{U}{N}\sum_{p'} f_{p',-\sigma}.$$

Let us take Fourier transforms:

$$\chi(\omega) = \int_{-\infty}^{\infty} dt e^{i\omega t} \chi(t).$$

Then,

$$\chi^{-+}(p,q,\omega) = \frac{\left(f_{p+q,\downarrow} - f_{p,\uparrow}\right)\left(1 + \frac{U}{N}\chi^{-+}(q,\omega)\right)}{\omega + \tilde{\epsilon}_{p+q,\uparrow} - \tilde{\epsilon}_{p,\downarrow}}, \tag{16.92}$$

where we have defined $\chi(q) = \sum_p \chi(p,q)$. Summing over **p**, we obtain that

$$\chi^{-+}(q,\omega) = \frac{\Gamma^{-+}(q,\omega)}{1 - U\Gamma^{-+}(q,\omega)}, \tag{16.93}$$

where

$$\Gamma^{-+}(q,\omega) = \frac{1}{N}\sum_p \frac{f_{p+q,\downarrow} - f_{p,\uparrow}}{\omega - (\tilde{\epsilon}_{p,\downarrow} - \tilde{\epsilon}_{p+q,\uparrow}) + i\eta}, \tag{16.94}$$

where $\eta \sim 0$ (compare with Eq. (14.197).

In the paramagnetic state $f_{p,\uparrow} = f_{p,\downarrow} = f_p$, $\tilde{\epsilon}_{p,\uparrow} = \tilde{\epsilon}_{p,\downarrow} = \tilde{\epsilon}_p$, and $\Gamma^{-+}(q,\omega) = \Gamma(q,\omega)$. Let us consider $\omega = 0$. In this case,we get

$$\chi(q,0) = \frac{\Gamma(q,0)}{1 - U\Gamma(q,0)}. \tag{16.95}$$

If $1 = U_c\Gamma(q,0)$, the susceptibility diverges. There is therefore a magnetic instability and a spontaneous magnetization appears in the system. Let us look closely at the expression:

$$\Gamma(q,\omega=0) = \frac{1}{N}\sum_p \frac{f_p - f_{p+q}}{\epsilon_{p+q} - \epsilon_p}.$$

In the limit, when the momentum goes to zero, we obtain

$$\begin{aligned}\lim_{q\to 0}\Gamma(q,\omega=0) &= \frac{1}{N}\sum_p \lim_{q\to 0}\frac{-q\cdot\frac{\partial\epsilon_p}{\partial p}\frac{\partial f_p}{\partial\epsilon_p} + \cdots}{q\cdot\frac{\partial\epsilon_p}{\partial p} + \cdots} \\ &= \frac{1}{N}\sum_p\left(-\frac{\partial f_p}{\partial\epsilon_p}\right).\end{aligned}$$

At zero temperature, we get

$$\Gamma(q=0,\omega=0) = N(\epsilon_F),$$

where as usual the density of states is defined as $N(\epsilon) = \frac{1}{N}\sum_p \delta(\epsilon - \epsilon_p)$. We get therefore Stoner's criterion for the appearance of magnetic order:

$$U_c N(\epsilon_F) = 1. \tag{16.96}$$

Since we have been looking at an instability at zero momentum, this is a ferromagnetic instability.

For a parabolic band, $\Gamma(q)$ decreases monotonically with $|q|$. If we increase U, the instability occurs first at $q = 0$. For other bands, the instability may occur first at other values of the momentum. This then leads to a SDW (Overhauser). For instance, consider a dispersion relation like

$$\epsilon_p = -\epsilon_0\left(\cos p_1 a + \cos p_2 a + \cos p_3 a\right).$$

For a moment $\mathbf{q} = \mathbf{Q} = \frac{1}{2}\mathbf{G}$ where $\mathbf{G}$ is a reciprocal lattice vector $\mathbf{G} = \frac{2\pi}{a}(1,1,1)$ and therefore for the vector $\mathbf{Q} = \frac{\pi}{a}(1,1,1)$, we get that

$$\epsilon_{\mathbf{p}+\mathbf{Q}} = -\epsilon_{\mathbf{p}}.$$

In this case, we can see that

$$\Gamma(q=Q,\omega=0) = -\int_{-3\epsilon_0}^{\epsilon_F} d\epsilon \frac{N(\epsilon)}{2\epsilon} + \int_{-\epsilon_F}^{3\epsilon_0} d\epsilon \frac{N(-\epsilon)}{2\epsilon}$$

$$= -\int_{-3\epsilon_0}^{\epsilon_F} d\epsilon \frac{N(\epsilon)}{\epsilon}.$$

Take the lower limit as very large. When $\epsilon_F = 0$, there is a divergence. This happens when the band is at half-filling. At half-filling, the instability leads to antiferromagnetism. This result is independent of U. There is no critical U required, in contrast to the ferromagnetic instability.

16.6.3 Ferromagnetic Regime

In this regime, the occupancy of the spin-up states is different from the occupation of the down-spin states. The energies in mean field become

$$\tilde{\epsilon}_{p,\sigma} = \epsilon_p - U\langle n_{-\sigma}\rangle.$$

The difference between the two spin projection energies is

$$\tilde{\epsilon}_{p,\uparrow} - \tilde{\epsilon}_{p,\downarrow} = U\langle n_\uparrow - n_\downarrow\rangle = \Delta.$$

Then,

$$\langle n_\sigma\rangle = \frac{1}{N}\sum_p \frac{1}{e^{\beta(\tilde{\epsilon}_{p,\sigma}-\mu)}+1}.$$

It yields, considering Δ small,

$$\langle n_\uparrow\rangle = \frac{1}{2}n + \frac{\Delta}{2N}\sum_p \frac{\partial f}{\partial \epsilon_p}$$

$$\langle n_\downarrow\rangle = \frac{1}{2}n - \frac{\Delta}{2N}\sum_p \frac{\partial f}{\partial \epsilon_p}.$$

Therefore, we get, as before,

$$\Delta = U\Delta N(\epsilon_F) \to UN(\epsilon_F) = 1.$$

Let us consider now

$$\lim_{q\to 0}\Gamma^{-+}(q,\omega) = \lim_{q\to 0}\frac{1}{N}\sum_p \frac{f_{p,\uparrow}-f_{p+q,\downarrow}}{\omega+(\epsilon_{p+q}-\epsilon_p+\Delta)}$$
$$= \frac{\Delta/U}{\omega+\Delta}.$$

This implies that

$$\lim_{q\to 0}\chi^{-+}(q,\omega) = \frac{\Delta/U}{\omega}. \qquad (16.97)$$

Therefore, when $\omega \to 0$, there is a divergence. This is the Goldstone mode that signals the instability at zero frequency, due to the spontaneous symmetry breaking of the rotational invariance.

In general, at finite momenta, we can determine the instabilities of the system looking for divergencies of the susceptibility. These instabilities are the spin waves of the problem. At small momenta, the dispersion relation is of the type $\omega = Dq^2$. Also, there is a continuum of excitations like in the charge part. This is the Stoner continuum. At small momenta, the spin waves are well-defined and are stable. At zero momentum, there is a gap of Δ to the longitudinal excitations, in the ferromagnetic regime. When the dispersion relation of the spin waves meets the continuum, the spin waves are no longer stable and may decay into the other particle-hole excitations. The analysis is obtained from the complete expressions Eqs. (16.93) and (16.94). The schematic behavior is shown in Fig. 16.12.

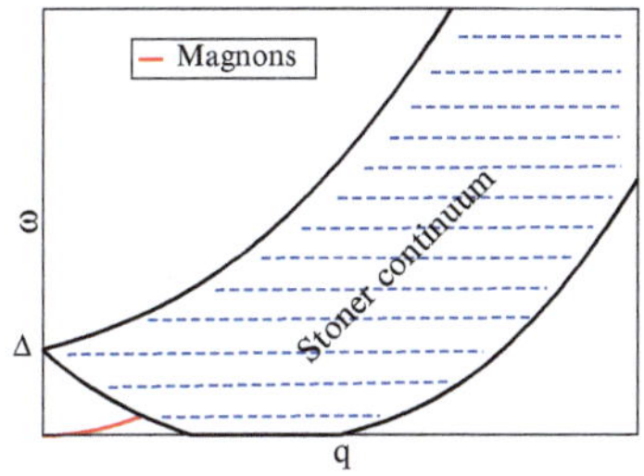

Fig. 16.12 Excitations of a ferromagnetic system. There is a gap to the longitudinal excitations and a continuum of excitations, which result from particle-hole processes (Stoner continuum), like in the charge part. There is also a gapless branch of spin-wave transverse excitations that eventually merges into the continuum, as the momentum increases

16.6.4 The t-J Model

It is then to expect that for $U >> t$, the model may be expressed in terms of an interacting magnetic model. Consider once again the Hubbard Hamiltonian:

$$H = -t \sum_{<i,j>,\sigma} \left(c^{\dagger}_{i,\sigma} c_{j,\sigma} + c^{\dagger}_{j,\sigma} c_{i,\sigma} \right) + U \sum_{i} n_{i,\uparrow} n_{i,\downarrow}$$

and let us consider $U/t >> 1$ and $1 - n_c << 1$ (close to half-filling). Let us then rewrite the Hamiltonian like

$$\begin{aligned}
H &= T_h + T_d + T_{mix} + V \\
T_h &= -t \sum_{<i,j>,\sigma} (1 - n_{i,-\sigma}) c^{\dagger}_{i,\sigma} c_{j,\sigma} (1 - n_{j,-\sigma}))) + h.c. \\
T_d &= -t \sum_{<i,j>,\sigma} n_{i,-\sigma} c^{\dagger}_{i,\sigma} c_{j,\sigma} n_{j,-\sigma} + h.c. \\
T_{mix} &= -t \sum_{<i,j>,\sigma} n_{i,-\sigma} c^{\dagger}_{i,\sigma} c_{j,\sigma} (1 - n_{j,-\sigma}) + h.c. \\
&\quad -t \sum_{<i,j>,\sigma} (1 - n_{i,-\sigma}) c^{\dagger}_{i,\sigma} c_{j,\sigma} n_{j,-\sigma} + h.c. \\
V &= U \sum_{i} n_{i,\uparrow} n_{i,\downarrow}.
\end{aligned}$$

The term T_h describes the transport of holes, and T_d describes the transport of doubly occupied sites. These two terms do not mix different bands. Note that the hoppings of the holes or doubly occupied sites are restricted. The term T_h describes the processes such that one electron of spin projection σ from a singly occupied j site can only hop to a i site that is empty. On the other hand, T_d describes processes where an electron from a doubly occupied site j hops to a site i that is already occupied by an electron of opposite spin projection. In contrast, T_{mix} mixes different bands (keep in mind that U is very large and therefore the bands are separated by a large gap).

Let us now consider a unitary transformation:

$$H_{eff} = e^{iS} H e^{-iS} = H + i[S, H] + \cdots$$

and let us choose S such that T_{mix} is eliminated to the first order. We want to remove processes that mix different bands. We choose therefore

$$i[S, T_h + T_d + V] = -T_{mix},$$

or

$$S = \sum_{n,m} |n\rangle \frac{\langle n|T_{mix}|m\rangle}{i(\epsilon_n - \epsilon_m)} \langle m|,$$

where $|n\rangle$ and $|m\rangle$ are eigenstates of $T_h + T_d + V$ with eigenvalues ϵ_n and ϵ_m, respectively.

We do not know ϵ_n exactly (except in one dimension; actually in one dimension, we also know the eigenenergies of the full Hubbard model, using a method called Bethe ansatz, but we will not consider it here. In a later chapter, the method is illustrated in the case of the Heisenberg chain of spin-1/2, where the solution is easier). But for large U, we have that $(\epsilon_n - \epsilon_m) = \pm U + O(t)$. Therefore,

$$\begin{aligned} S &= -i\frac{t}{U} \sum_{<i,j>,\sigma} n_{i,-\sigma} c^{\dagger}_{i,\sigma} c_{j,\sigma} (1 - n_{j,-\sigma}) \\ &\quad + i\frac{t}{U} \sum_{<i,j>,\sigma} (1 - n_{i,-\sigma}) c^{\dagger}_{i,\sigma} c_{j,\sigma} n_{j,-\sigma} \\ H_{eff} &= T_h + T_d + T_{mix} + V + i[S, T_h + T_d + T_{mix} + V] \\ &= T_h + T_d + V + i[S, T_{mix}]. \end{aligned}$$

If we now consider a subspace with no double occupancy, the Hamiltonian can be written as

$$H_{eff} = T_h + i[S, T_{mix}] = T_h + J \sum_{<i,j>} \left(\mathbf{S}_i \cdot \mathbf{S}_j - \frac{1}{4} \right), \tag{16.98}$$

for the case of spin 1/2 ($n_c = 2$) and where we have defined the exchange constant $J = (4t^2)/U$. This is called $t - J$ model. The terms of order t^3/U^2, $(1 - n_c)t^2/U$ and $(1 - n_c)^2 t$ have been neglected.

This model has two terms. The first one is a hopping term of the holes and the other is a Heisenberg term. Note that the motion of the electrons is strongly restricted by the projectors $(1 - n_{i,-\sigma})$. If we are at half-filling, the model reduces to the antiferromagnetic Heisenberg model.

16.7 Exercises

1. Holstein-Primakoff transformation: Consider the following representation for the spin operators:

$$S^+ = S_x + iS_y = \sqrt{2S}\sqrt{1 - \frac{a^\dagger a}{2S}}a$$

$$S^- = S_x - iS_y = \sqrt{2S}a^\dagger\sqrt{1 - \frac{a^\dagger a}{2S}},$$

where the operators a and $a^\dagger$ are bosonic operators. These operators obey the commutation relations:

$$[a, a^\dagger] = 1$$

$$[a, a] = [a^\dagger, a^\dagger] = 0.$$

(a) Calculate S_z^2 and S_z and show that

$$S_z = S - a^\dagger a.$$

(b) Verify that the bosonic representation satisfies the commutation relation:

$$[S^+, S^-] = 2S^z.$$

2. Consider the isotropic antiferromagnetic Heisenberg model given by the Hamiltonian:

$$H = J \sum_{<i,j>} S_i^z S_j^z + \frac{J}{2} \sum_{<i,j>} \left(S_i^+ S_j^- + S_i^- S_j^+ \right),$$

with $J > 0$ and focus on a single link between nearest-neighbor spins.

(a) Show that the exchange energy is minimized if these are in a spin singlet (total spin zero).
(b) However, show that the minimum is not obtained simply by considering that the spins point in opposite directions. That is, show that the state $|S\rangle_i|-S\rangle_j$ is not an eigenstate.

3. Some magnetic systems may be described by a Heisenberg ferromagnetic Hamiltonian of the form

$$H = J_z \sum_{<i,j>} S_i^z S_j^z + \frac{J_\pm}{2} \sum_{<i,j>} \left(S_i^+ S_j^- + S_i^- S_j^+ \right),$$

where we consider nearest-neighbor exchange terms in a d-dimensional space and $\mathbf{S}$ are spin operators. Consider a material where the Heisenberg interaction is anisotropic, $J_\pm \neq J_z$.

(a) Show that if $J_z > J_\pm$, the energy of the spin waves is finite for $k = 0$.
(b) Show that for $J_z < J_\pm$, the state in which all spins are parallel is not the ground state. Calculate the energy of the state for which one spin has the opposite direction with respect to all other spins.

4. (a) We want to determine the magnetic properties of a given material. We may measure its magnetic susceptibility as a function of temperature and carry out neutron dispersion to determine its magnetic structure. How can you distinguish, using each of the methods, if the system is paramagnetic, ferromagnetic, or antiferromagnetic?
   (b) Assume now that we have instead experimental equipment to measure specific heat at low temperatures. Considering the system to be three dimensional, how can you distinguish the type of magnetic order? The density of states is $D(\omega) \sim k^2 dk/d\omega$. The dispersion relation of the magnons is of the form $\omega(k) \sim k^2$ (ferromagnetic) or $\omega(k) \sim k$ (antiferromagnetic).
   (c) The measurement of the specific heat of EuO in the regime of low temperatures shows that it is proportional to $T^{3/2}$. Determine the dispersion relation of the excitations, $\omega(k) \sim k^\alpha$, consistent with this observation.
5. In a gas of conduction electrons at $T = 0$ in a weak magnetic field, B, the electronic concentration of $\uparrow$ spins, N_+, and of $\downarrow$ spins, N_-, is given in Eq. (13.29), and the magnetization is given by Eq. (13.30). Adding the effect of interactions between the electrons, we may approximate the exchange energy between the electrons assuming that only electrons with parallel spins interact among themselves, with energy $-V$, where $V > 0$. How is the magnetization modified?
6. The Hubbard model describes a set of electrons with a kinetic energy of the tight-binding type and with local interactions between electrons with different spins of the form

$$H = -t \sum_{\mathbf{r},\boldsymbol{\delta}} \sum_{\sigma} c^\dagger_{\mathbf{r},\sigma} c_{\mathbf{r}+\boldsymbol{\delta},\sigma} + U \sum_{\mathbf{r}} n_{\mathbf{r},\uparrow} n_{\mathbf{r},\downarrow}.$$

   (a) Consider the Hartree-Fock approximation written as a mean-field theory in the fermionic operators. Show that the Fock term (exchange term) vanishes.
   (b) The spin operator is defined as

$$\mathbf{S}(\mathbf{r}) = \frac{\hbar}{2} c^\dagger_\sigma(\mathbf{r}) \boldsymbol{\sigma}_{\sigma,\sigma'} c_{\sigma'}(\mathbf{r}),$$

where $\boldsymbol{\sigma}$ is a Pauli matrix. The number operator is given by

$$n(\mathbf{r}) = \sum_{\sigma} n_{\sigma}(\mathbf{r}) = \sum_{\sigma} c_{\sigma}^{\dagger}(\mathbf{r}) c_{\sigma}(\mathbf{r}).$$

Consider the operator

$$\begin{aligned}\sum_{\mathbf{r}} (\mathbf{S}(\mathbf{r}))^2 &= \sum_{\mathbf{r}, a=x,y,z} S^a(\mathbf{r}) S^a(\mathbf{r}) \\ &= \sum_{\mathbf{r}, a} \frac{1}{4} \left(c_{\alpha}^{\dagger}(\mathbf{r}) \sigma_{\alpha,\beta}^a c_{\beta}(\mathbf{r}) \right) \left(c_{\gamma}^{\dagger}(\mathbf{r}) \sigma_{\gamma,\delta}^a c_{\delta}(\mathbf{r}) \right).\end{aligned}$$

Using that

$$\sum_{a} \sigma_{\alpha,\beta}^a \sigma_{\gamma,\delta}^a = 2\delta_{\alpha,\delta}\delta_{\beta,\gamma} - \delta_{\alpha,\beta}\delta_{\gamma,\delta},$$

show that

$$\sum_{\mathbf{r}} (\mathbf{S}(\mathbf{r}))^2 = \sum_{\mathbf{r}} \left(\frac{3}{4} n(\mathbf{r}) - \frac{3}{2} n_{\uparrow}(\mathbf{r}) n_{\downarrow}(\mathbf{r}) \right)$$

and that

$$U \sum_{\mathbf{r}} n_{\uparrow}(\mathbf{r}) n_{\downarrow}(\mathbf{r}) = -\frac{2U}{3} \sum_{\mathbf{r}} (\mathbf{S}(\mathbf{r}))^2 + \frac{N_e U}{2},$$

where N_e is the number of electrons. Show then that one may expect some form of magnetic order in the system.

7. A magnetic impurity of spin S, localized at the point $\mathbf{R}_1$ in a matrix of conduction electrons, interacts with the electrons in the form of an interaction between the impurity spin and the electronic spin density at the site of the impurity. The Hamiltonian is

$$H = \sum_{k,\sigma} \epsilon_k c_{k,\sigma}^{\dagger} c_{k,\sigma} - J\mathbf{S}(\mathbf{R}_1) \cdot \sum_{\sigma,\sigma'} c_{\mathbf{R}_1,\sigma}^{\dagger} \mathbf{s}_{\sigma,\sigma'} c_{\mathbf{R}_1,\sigma'}.$$

The interaction is antiferromagnetic. Consider now that there are two magnetic impurities at the points $\mathbf{R}_1$ and $\mathbf{R}_2$. Each of the impurities interacts with the electrons in the same way. Suppose that we may neglect the direct interaction between the impurities. However, the impurities interact in an indirect way via the conduction electrons.

(a) Consider the effect of the interaction term on the energy of the system in the second-order perturbation theory. Use as unperturbed states the states of the free electrons. Consider the spins of the impurities as classical spins. Write the interaction term in momentum space. Convert the sums over the momenta as integrals. As an intermediate step, show that the correction to the energy may be written as

$$mJ^2\mathbf{S}_1\cdot\mathbf{S}_2\frac{V}{(2\pi)^6}P\int_0^{k_F}d^3k\int_{k_F}^{\infty}d^3k'\frac{e^{-i(\mathbf{k}-\mathbf{k}')\cdot\mathbf{r}}}{k^2-(k')^2}+c.c.$$

(b) Calculating the integrals (for instance, in the complex plane), show that the correction term to the energy is written as

$$4mJ^2\mathbf{S}_1\cdot\mathbf{S}_2\frac{V^2}{(2\pi)^3}k_F^4F(2k_Fr),$$

where

$$F(x)=\frac{x\cos x-\sin x}{x^4}.$$

The interaction oscillates with a period given by $2k_F$, and, as a consequence, the effective interaction may be ferromagnetic or antiferromagnetic. This interaction is called the RKKY interaction (Ruderman-Kittel-Kasuya-Yosida).

8. Consider a spin-S problem on a honeycomb lattice defined by the unit cell vectors Eq. (10.164). The unit vectors connecting nearest neighbors on the two sublattices may be defined as in Eq. (10.165). The second-nearest neighbors, connecting sites on the same sublattice, may be written as

$$\begin{aligned}\delta_1^2&=a\left(1,\sqrt{3}\right)\\ \delta_2^2&=\frac{a}{2}\left(-3,-\sqrt{3}\right)\\ \delta_3^2&=\frac{a}{2}\left(3,-\sqrt{3}\right).\end{aligned}\qquad(16.99)$$

The Hamiltonian is given by

$$H=-J\sum_{\langle i,j\rangle_1}\mathbf{S}_i\cdot\mathbf{S}_j+\sum_{\langle i,j\rangle_2}\mathbf{D}_{ij}\cdot\left[\mathbf{S}_i\times\mathbf{S}_j\right]+B\sum_i S_i^z.\qquad(16.100)$$

The first term describes the (ferromagnetic, $J>0$) interaction between nearest-neighbor spins. The second term describes a Dzyaloshinskii-Moriya interaction between second neighbors, and the last term is an external magnetic field along the z-direction. Consider two second neighbors i, j. These two points may also be reached through the common nearest-neighbor point of sites i and j. Site i is connected to that nearest-neighbor site by a vector $\boldsymbol{\delta}_i$ and site j is connected to that nearest-neighbor site by a vector $\boldsymbol{\delta}_j$. The Dzyaloshinskii-Moriya term is then defined by $\mathbf{D}_{ij} = Dd_{ij}\hat{\mathbf{z}}$, where $d_{ij} = 2/\sqrt{3}\boldsymbol{\delta}_i \times \boldsymbol{\delta}_j \cdot \hat{\mathbf{z}} = \pm 1$, where $\hat{\mathbf{z}}$ is the unit vector along the perpendicular direction to the lattice.

(a) Using spin-wave theory to the lowest order using the Holstein-Primakoff transformation, write the Hamiltonian in terms of the bosonic operators, a_i. Show that the Hamiltonian may be written as

$$H = 2JS \sum_{\langle i,j\rangle_1} \left(a_i^\dagger a_i - a_i^\dagger a_j\right) - i2SD \sum_{\langle i,j\rangle_2} d_{ij} a_i^\dagger a_j - B \sum_i a_i^\dagger a_i, \tag{16.101}$$

plus some constant.

(b) Since we have two sublattices in the honeycomb lattice, there are two types of bosonic operators. The same happens in momentum space leading to two bands labeled by operators $a(\mathbf{k})$ and $b(\mathbf{k})$. Show that in momentum space, the Hamiltonian may be written as

$$H = \sum_{\mathbf{k}} \left(a^\dagger(\mathbf{k})\ b^\dagger(\mathbf{k})\right) H(\mathbf{k}) \begin{pmatrix} a(\mathbf{k}) \\ b(\mathbf{k}) \end{pmatrix}, \tag{16.102}$$

where the Hamiltonian matrix may be written as

$$H(\mathbf{k}) = h_I I + \mathbf{h}(\mathbf{k}) \cdot \boldsymbol{\sigma}, \tag{16.103}$$

where $\boldsymbol{\sigma}$ are the Pauli matrices. Determine the vector $\mathbf{h}(\mathbf{k})$.

(c) Using Eq. (10.170), we may define the Brillouin zone. Calculate the Chern number of the two bands, summing over plaquettes covering the Brillouin zone, using Eq. (10.160). This shows that the magnon bands have a topological nature [1], if $D \neq 0$. Consider both $D > 0$ and $D < 0$.

Reference

1. See, for instance, P. A. McClarty, Annual Review of Condensed Matter Physics **13**, 171 (2022).

Chapter 17
Superconductivity

17.1 Zero Resistance and Meissner Effect

In 1911, Kamerlingh Onnes found a sudden drop of the resistance of Hg at very low temperatures. At about 4.2 Kelvin, the resistance is apparently zero as shown in Fig. 17.1. In 1933, the Meissner effect was discovered: a similar material penetrated by a magnetic field completely expels the magnetic field lines from its interior (the magnetic induction vanishes inside the material: $B = 0$), as the temperature is lowered beyond the point where the resistance vanishes.

A set of properties later discovered shed some light into the mechanism that gives rise to such spectacular phenomenon, called superconductivity, due to its infinite conductivity:

(i) Above the critical temperature, the materials are conductors, but the best metals do not become superconductors.
(ii) The materials that are good magnets also do not become superconductors.
(iii) At low temperatures, the specific heat changes from a behavior typical of a conductor, with the electronic and phonon contributions, to a regime where it becomes exponentially small at very low temperatures. This is indicative of the presence of a gap in the energy spectrum, as seen, for instance, in a two-level system.
(iv) The temperatures at which the system becomes a superconductor are typically very small, usually below 20 Kelvin.
(v) An important clue was provided by the discovery of the isotope effect. Substituting some atoms by different isotopes, their mass is changed. It turns out that the critical temperature is changed, indicating that, somehow, the ions are involved in the process.
(vi) A large magnetic field destroys the superconducting state.

P. D. Sacramento, *Introduction to Condensed Matter Theory*,
https://doi.org/10.1007/978-3-031-86251-9_17

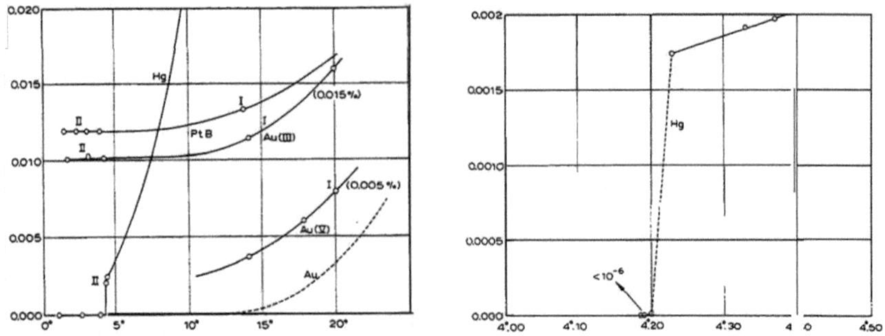

Fig. 17.1 Resistance (normalized to the value at zero Celsius) of several metals as a function of temperature. At very low temperatures, the resistance of Hg drops suddenly to zero ($<10^{-6}$) as also shown in the right panel (Reproduced with permission from the Kamerlingh Onnes Nobel lecture (1913))

The properties of different superconductors are summarized in Table 17.1. Note that the highest critical temperature is that of Nb reaching the value of 9.5 K. A list of some materials with higher critical temperatures is shown in Table 17.2.

It should be noted that the above set of properties are common to the so-called conventional superconductors. Other superconductors do not follow those properties, but we will focus here in explaining the conventional ones, for which a theory has been constructed with great success. A discussion of other systems will be presented later in the book.

The list of the above properties seems to suggest the following picture: interactions must have an important role since very good conductors do not become superconductors. It is expected that in very good conductors interactions are not important since resistance is low. The existence of a gap suggests some form of bound state between the electrons that must be weak, since the critical temperatures are usually rather low. Magnetism and magnetic fields compete with superconductivity. The influence of the lattice suggests that a possible mechanism for the creation of bound states involves the lattice vibrations, that is, the phonons. Indeed, in order to create a bound state between the electrons, one expects an attractive interaction between the electrons. The direct Coulomb interaction is repulsive, even though it should be screened. Some other interaction must yield the attractive force. It has been proposed that this mechanism involves the exchange of phonons between the electrons. If bound states between the electrons exist, one expects that these extend in space, since their energy should be small. On the other hand, since the electrons are fermions, a pair of electrons is like a boson. Therefore, we may expect at low temperatures some sort of boson condensation and some sort of superfluid state, with no resistance. Note, however, that this picture is somewhat simplified because a pair of electrons only seems like a boson far away from it. This picture is oversimplified since the radius of a pair is relatively large (of the order of a thousand Angstrom).

Table 17.1 Parameters of elemental superconductors: transition temperature $T_c(K)$, critical magnetic field at $T = 0K$ in gauss, energy gap at $T = 0K$ in meV (Adapted with permission from [Kittel])

3	4	5	6	7	8	9	12	13	14
								Al 1.140 105 3.4	
	Ti 0.39 100	V 5.38 1420 16					Zn 0.875 53 2.4	Ga 1.091 51 3.3	
	Zr 0.546 47	Nb 9.50 1980 30.5	Mo 0.92 95 2.7	Tc 7.77 1410	Ru 0.51 70	Rh 0.0003 .049	Cd 0.56 30 1.5	In 3.4035 293 10.5	Sn 3.722 309 11.5
	Hf 0.12	Ta 4.483 830 14	W 0.012 1.07	Re 1.4 198	Os 0.655 65	Ir 0.14 19	Hg 4.153 412 16.5	Tl 2.39 171 7.35	Pb 7.193 803 27.3
La 6.00 1100 19									
	Th 1.368 1.62	Pa 1.4							

Table 17.2 Critical temperatures (in Kelvin) of some higher-temperature superconductors (Adapted with permission from [Kittel])

Material	T_c	Material	T_c
Nb_3Sn	18.05	Nb_3Ge	23.2
Nb_3Al	17.5	NbN	16.0
V_3Si	17.1	V_3Ga	16.5
La_3In	10.4	$PbMo_{5.1}S_6$	14.4

17.1.1 London Theory

A simple, early, theory assumes that the superconductor is composed of two fluids, a normal fluid and a superfluid, with densities n_n and n_s and velocities v_n and v_s, respectively. The electronic density of the material is $n = n_n + n_s$. The current of the normal part, $\mathbf{J}_n = -en_n\mathbf{v}_n$, follows the usual Ohm's law

$$\mathbf{J}_n = \sigma_n \mathbf{E}, \tag{17.1}$$

under the presence of an electric field **E**. This law indicates a dissipative current due to the presence of friction with impurities, imperfections, phonons, etc. The superfluid part does not follow Ohm's law since it has infinite conductivity. The motion of the particles of the superfluid follows Newton's law instead. The particles are accelerated due to the presence of an external force. That is, the force is proportional to the acceleration, while Ohm's law describes a set of particles whose force is proportional to the velocity. For the superfluid part, we postulate that,[1] defining the current as $\mathbf{J}_s = -en_s\mathbf{v}_s$,

$$\frac{d\mathbf{J}_s}{dt} = -en_s\mathbf{a}_s = \frac{e^2 n_s}{m}\mathbf{E}. \tag{17.2}$$

Using that $\mathbf{E} = -\partial\mathbf{A}/\partial t$, this shows that the current is proportional to the vector potential, **A**, instead of proportional to the electric field.

Maxwell's equation

$$\nabla \times \mathbf{E} = -\frac{\partial \mathbf{B}}{\partial t}, \tag{17.3}$$

can be written as

$$\nabla \times \mathbf{J}_s = -\frac{n_s e^2}{m}\mathbf{B}. \tag{17.4}$$

Another Maxwell equation is written as

$$\nabla \times \mathbf{B} = \mathbf{J} = (\mathbf{J}_n + \mathbf{J}_s)\,, \tag{17.5}$$

where we have neglected the displacement current. Since we expect the superfluid current to be much larger than the normal current, we also neglect its contribution. Taking the curl of this equation, we get that

$$\nabla \times (\nabla \times \mathbf{B}) = \nabla \times \mathbf{J}_s. \tag{17.6}$$

Using that $\nabla \cdot \mathbf{B} = 0$, we get

$$\nabla^2\mathbf{B} = \frac{n_s e^2}{m}\mathbf{B} = \frac{1}{\lambda_L^2}\mathbf{B}, \tag{17.7}$$

[1] For further reading see [deGennes].

where, λ_L, the London penetration depth, is given by

$$\lambda_L = \left(\frac{m}{n_s e^2}\right)^{1/2}. \tag{17.8}$$

To understand why it is called penetration depth, let us consider a superconductor in contact with vacuum. Let us call z the direction perpendicular to the surface of the superconductor and place the surface at $z = 0$. To simplify, we assume that the superconductor surface is flat and is contained in the infinite plane xy. Consider that the field in the vacuum is parallel to the surface, $\mathbf{B} = B\mathbf{e}_x$, for instance. We have then to solve the set of equations

$$\begin{aligned} \nabla \times \mathbf{B} &= \mathbf{J}_s \\ \nabla^2 \mathbf{B} &= \frac{1}{\lambda_L^2}\mathbf{B} \\ \nabla \cdot \mathbf{B} &= 0, \end{aligned} \tag{17.9}$$

near the surface. If the field is tangencial and along the x direction, then the current is along the y direction $\mathbf{J}_s = J_s \mathbf{e}_y$. The first equation then simplifies to

$$\frac{dB}{dz} = J_s. \tag{17.10}$$

Also,

$$\begin{aligned} \frac{dJ_s}{dz} &= \frac{n_s e^2}{m} B \\ \frac{d^2 B}{dz^2} &= \frac{B}{\lambda_L^2}. \end{aligned} \tag{17.11}$$

This leads to an exponentially decreasing profile for the amplitude of the magnetic field inside the superconductor with a scale given by the London penetration depth

$$B(z) = B(z = 0)e^{-\frac{z}{\lambda_L}}. \tag{17.12}$$

Deep inside the superconductor, the magnetic field, B, vanishes: this is the result expected from the Meissner effect. This is shown in Fig. 17.2.

The penetration depth is a function of the superfluid density. At very low temperature, the density is basically the superfluid density. For Al and Sn, the electron mass is very close to the free electron mass and $\lambda_L \sim 500\mathring{A}$. For transition metals, the effective mass is larger, and the penetration length increases, $\lambda_L \sim 2000\mathring{A}$. Also, for systems with low electronic density, the penetration length increases.

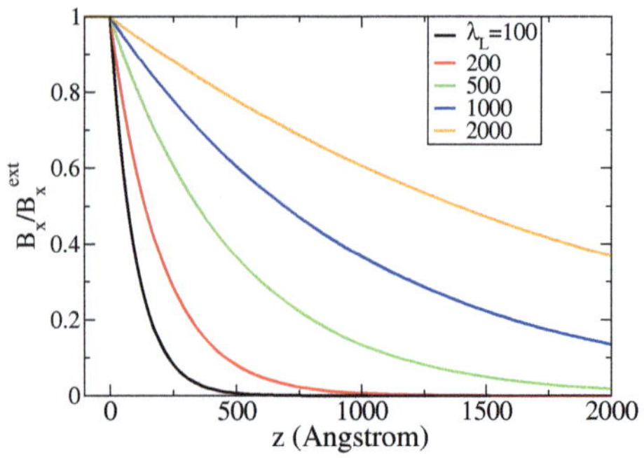

Fig. 17.2 Exponential decrease of a parallel magnetic field component inside a superconductor due to the Meissner effect. The decay length scale is the penetration depth, λ_L, in $\mathring{A}$

17.1.2 Thermodynamics in the Presence of External Field

Consider a superconductor and a magnet that is brought from infinity into the neighborhood of the superconducting material.[2] Let us call the field created by the magnet on the superconductor, B_a. Assume that this applied field is approximately constant in the region of the superconductor. If the field is sufficiently small, such that the system remains superconducting, Meissner effect implies that the magnetic induction inside the superconductor vanishes. Since $B = \mu_0(H + M)$, $B = 0$ implies that $M = -B_a$ (we will use H or B_a interchangeably). This means the superconductor is a perfect diamagnet.

The work done on the superconductor as we bring the magnet from infinity ($B_a = 0$) until a given point, where $B_a \neq 0$, is given by

$$W = -\int_0^{B_a} \mathbf{M} \cdot d\mathbf{B}_a. \tag{17.13}$$

The change in free energy is given by $dF = -\mathbf{M} \cdot d\mathbf{B}_a$ which leads to

$$dF_s = B_a dB_a. \tag{17.14}$$

This can be integrated easily to yield

$$F_s(B_a) - F_s(0) = \frac{B_a^2}{2}. \tag{17.15}$$

If the applied field is large enough, the system will suffer a transition from the superconducting phase to the normal phase (conductor). At the transition, the free energies of the two phases are equal: $F_s(B_{ac}) = F_N(B_{ac})$, where B_{ac} is the critical field that causes the transition to occur. On the other hand, in a normal metal (nonmagnetic), the magnetic susceptibility is small. We may then approximate

[2] See also [deGennes].

$F_N(B_{ac}) = F_N(0)$. This holds if the critical field is not large. Therefore,

$$F_N(B_{ac}) = F_s(B_{ac}) = F_s(0) + \frac{B_{ac}^2}{2} = F_N(0). \tag{17.16}$$

This implies that the change in the free energy, *in zero field*, is given by

$$\Delta F = F_N(0) - F_s(0) = \frac{B_{ac}^2}{2}. \tag{17.17}$$

As expected, the free energy of the normal phase is larger than in the superconducting phase. This difference is called the condensation energy and shows that the system becomes more stable by changing phase into the superconducting regime. Also, it is related to the critical magnetic field, showing an interesting connection to the magnetic response of the system.

We may also study the entropy of the two phases. We know that

$$S_N = -\frac{dF_N}{dT}, \; S_s = -\frac{dF_s}{dT}. \tag{17.18}$$

The difference in entropy between the two phases is given by

$$S_N - S_s = -\frac{d}{dT}(F_N - F_s) = -H_c\frac{dH_c}{dT}. \tag{17.19}$$

The normal phase is a higher-temperature phase and it is expected to have higher entropy. Therefore, the entropy difference should be positive. This implies that the critical magnetic field should decrease with temperature. This is to be expected: at zero field, there is a finite critical temperature at which superconductivity emerges. A sufficiently large applied magnetic field destroys superconductivity. Therefore, as temperature increases, the superconducting state becomes less robust, and a smaller magnetic field is enough to destroy superconductivity.

If there is an entropy difference between the normal and the superconducting phase in the presence of a magnetic field, there is a latent heat at the transition. The latent heat is given by

$$L = T(S_N - S_s) = -TH_c\frac{dH_c}{dT} > 0. \tag{17.20}$$

Phenomenologically, it is known that, approximately

$$H_c(T) = H_c(0)\left(1 - \frac{T^2}{T_c^2}\right). \tag{17.21}$$

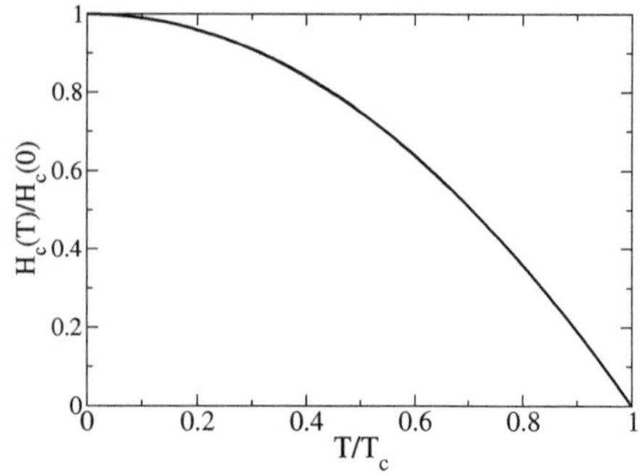

Fig. 17.3 Critical magnetic field as a function of temperature

This is shown in Fig. 17.3. We may also calculate the change in the specific heat at the transition

$$\begin{aligned} C_N - C_s &= T \frac{d}{dT} (S_N - S_s)_{T=T_c} \\ &= -\frac{T}{2} \left(\frac{dH_c}{dT} \right)^2_{T=T_c} . \end{aligned} \tag{17.22}$$

There is a discontinuity at the transition. Since there is a latent heat, the transition from the normal phase to the superconducting phase is of first order if the magnetic field is finite. At zero field, the discontinuity vanishes, which suggest a second-order phase transition near the critical temperature T_c. (It turns out that the transition is actually weakly first order, the discontinuity is very small, and we will assume the transition to be second order).

It is easy to see that

$$\left(\frac{dH_c(T)}{dT} \right)_{T=T_c} = -2 \frac{H_c(0)}{T_c}, \tag{17.23}$$

which allows an estimate of the critical field at zero temperature from the slope of dependence of the critical field near the critical point $T = T_c$.

17.2 Cooper Pairs

Consider again the Fermi liquid. The Coulomb interactions between the electrons are repulsive in the vacuum. However, consider now the electrons in a solid. As we saw, the interactions get renormalized in the medium. Let us assume that the interactions between the electrons and the lattice may cause an effective attractive interaction between the electrons. Such is possible through, for instance, the electron-phonon interaction, as we shall see later on. For the moment, let us assume that there is an attractive interaction between the electrons. Let us consider two electrons plus the electrons that occupy the Fermi sea.

The electrons have coordinates $\mathbf{r}_1$ and $\mathbf{r}_2$ and interact via an attractive potential. To simplify, let us consider states where the center of mass of the two electrons is at rest. Then the wave function of the pair is

$$\psi(\mathbf{r}_1, \mathbf{r}_2) = \psi(\mathbf{r}_1 - \mathbf{r}_2) = \sum_{\mathbf{k}} g(\mathbf{k}) e^{i\mathbf{k}\cdot(\mathbf{r}_1 - \mathbf{r}_2)}.$$

Here, $g(\mathbf{k})$ is the probability amplitude that one electron has momentum $\mathbf{k}$ (and that the other has momentum $-\mathbf{k}$). Pauli's principle implies that $g(\mathbf{k}) = 0$ if $k < k_F$. The Fermi sea is filled and we are considering two electrons above the Fermi sea. The Schrödinger equation is therefore

$$-\frac{\hbar^2}{2m}\left(\nabla_1^2 + \nabla_2^2\right)\psi(\mathbf{r}_1, \mathbf{r}_2) + V(\mathbf{r}_1, \mathbf{r}_2)\psi(\mathbf{r}_1, \mathbf{r}_2) = \left(E + \frac{\hbar^2 k_F^2}{m}\right)\psi(\mathbf{r}_1, \mathbf{r}_2).$$

In terms of the momentum amplitude, we get that

$$\frac{\hbar^2}{2m}k^2 g(\mathbf{k}) + \sum_{\mathbf{k}'} g(\mathbf{k}') V_{\mathbf{k},\mathbf{k}'} = (E + 2E_F) g(\mathbf{k}),$$

where the matrix element is given by

$$V_{\mathbf{k},\mathbf{k}'} = \frac{1}{L^3}\int d^3r\, V(\mathbf{r}) e^{-i(\mathbf{k}-\mathbf{k}')\cdot\mathbf{r}},$$

($\mathbf{r} = \mathbf{r}_1 - \mathbf{r}_2$). Considering that the two electrons have opposite momenta ($\mathbf{k}$ and $-\mathbf{k}$), the effect of the interaction is to scatter the pair to another state such that they remain with opposite momenta $\mathbf{k}'$ and $-\mathbf{k}'$. This is shown schematically in Fig. 17.4. In the top panel, the scattering is shown in momentum space. In the bottom panel, the scattering is shown in real space: two electrons moving with opposite momenta interact and shift their momenta such that they keep orbiting each other giving rise to a bound state, as will be shown next.

Consider now that $V < 0$ and let us look for bound states (solutions with negative energy). To be specific, let us consider an interaction of the form

$$\begin{aligned} V_{\mathbf{k},\mathbf{k}'} &= -\frac{V}{L^3}, \frac{\hbar^2 k^2}{2m} < E_F + \hbar\omega_D, \frac{\hbar^2 (k')^2}{2m} < E_F + \hbar\omega_D \\ &= 0, \text{ otherwise.} \end{aligned} \tag{17.24}$$

Here, ω_D is the Debye frequency of the phonons or any other cutoff frequency representing the maximum exchange energy available, and the volume of the system is L^3, assuming that the system is cubic. Implicitly, $\hbar\omega_D$ is the energy

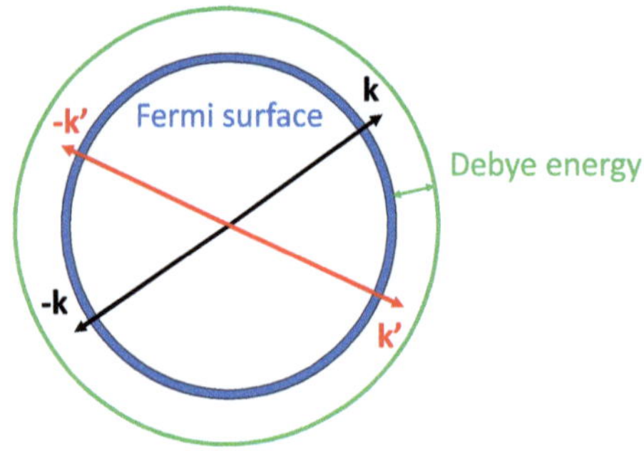

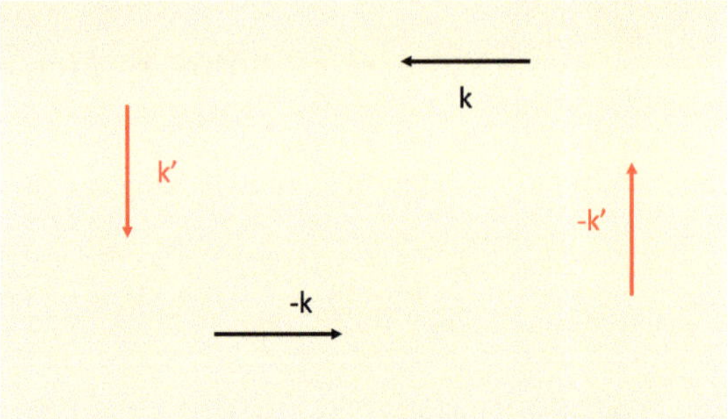

Fig. 17.4 Scattering between two electrons via an attractive interaction that leads to scattering between a pair with opposite momenta **k** and $-\mathbf{k}$ to a new set of momenta $\mathbf{k}'$ and $-\mathbf{k}'$. In the top panel, the scattering is schematically shown in momentum space and on the bottom panel in real space

scale exchanged in the scattering of the electrons. Since we are assuming that the mechanism for the attractive interaction is the exchange of phonons between the electrons, $\hbar\omega_D$ is the Debye energy. If the mechanism were another, then it would represent the energy exchanged through the excitations responsible for the attractive interaction.

Using the simplified potential, we obtain that

$$\left(-\frac{\hbar^2k^2}{2m}+E+2E_F\right)g(\mathbf{k})=-\frac{V}{L^3}\sum_{\mathbf{k}'}g(\mathbf{k}')=C$$

and

$$E_F<\frac{\hbar^2(k')^2}{2m}<E_F+\hbar\omega_D.$$

Summing over **k**,

$$\sum_{\mathbf{k}}g(\mathbf{k})=\sum_{\mathbf{k}}\frac{1}{-E+\frac{\hbar^2k^2}{2m}-2E_F}\frac{V}{L^3}\sum_{\mathbf{k}'}g(\mathbf{k}').$$

We finally arrive at

$$1 = \frac{V}{L^3} \sum_{\mathbf{k}} \frac{1}{-E + \frac{\hbar^2 k^2}{2m} - 2E_F}. \tag{17.25}$$

Defining $\zeta' = \frac{\hbar^2 (k')^2}{2m} - E_F$ and recalling that the density of states is given by

$$N(\zeta') = \frac{1}{(2\pi)^3} 4\pi (k')^2 \frac{dk'}{d\zeta'},$$

we can rewrite that

$$1 = V \int_0^{\hbar\omega_D} d\zeta' N(\zeta') \frac{1}{2\zeta' - E}.$$

Since in general $\hbar\omega_D << E_F$, we can approximate the density of states by its value at the Fermi surface, $N(0)$. Then the integral gives

$$\begin{aligned} 1 &= VN(0) \int_0^{\hbar\omega_D} d\zeta \frac{1}{2\zeta - E} \\ &= \frac{VN(0)}{2} \ln \frac{E - 2\hbar\omega_D}{E}. \end{aligned}$$

This yields in the limit when $VN(0) << 1$ (in the weak coupling limit)

$$E \sim -2\hbar\omega_D e^{-\frac{2}{VN(0)}}. \tag{17.26}$$

Several points are of relevance: First, as anticipated, the solution is of negative energy and there is a bound state. Second, the energy of the bound state is non-perturbative in the potential. Third, whatever the value of the potential, there is a bound state. This does not happen in free space as you may recall from elementary quantum mechanics (except in one dimension, which is not our concern here). In free space ($3d$), there is a minimum value for the potential required for the occurrence of a bound state. Fourth, the wave function $\psi(\mathbf{r}_1, \mathbf{r}_2)$ has to be antisymmetric since the electrons are fermions. With our simplified choice $g(\mathbf{k})$ only depends on $k = |\mathbf{k}|$. This implies that the spatial part of the wave function is symmetric. This implies, on the other hand, that the spin part of the wave function must be antisymmetric. Therefore, the two electrons are in a spin singlet state.

Let us now calculate the mean square radius of a pair of electrons (called Cooper pair). It is given by

$$\rho^2 = \frac{\int d^3 r r^2 |\psi(\mathbf{r}_1 - \mathbf{r}_2)|^2}{\int d^3 r |\psi(\mathbf{r}_1 - \mathbf{r}_2)|^2}.$$

This expression can be written as

$$\rho^2 = \frac{\sum_{\mathbf{k}} |\nabla_{\mathbf{k}} g(\mathbf{k})|^2}{\sum_{\mathbf{k}} |g(\mathbf{k})|^2}$$

$$= \frac{N(0) \left(\frac{\partial \zeta}{\partial k}\right)^2_{\zeta=0} \int_0^\infty d\zeta \left(\frac{\partial g}{\partial \zeta}\right)^2}{N(0) \int_0^\infty d\zeta g^2}.$$

Recall that

$$g = \frac{C}{E - 2\zeta}$$

and that

$$\frac{dk}{d\zeta}_{\zeta=0} = \frac{1}{\hbar v_F}.$$

Therefore, we get for the mean square radius of the Cooper pair the result

$$\rho = \frac{2}{\sqrt{3}} \frac{\hbar v_F}{E}. \tag{17.27}$$

17.3 Electron-Phonon Interaction and Attractive Interaction

From the point of view of the electrons, the lattice acts as an external potential. The interaction may be written as

$$H_{int} = \int d^3 r V(\mathbf{r}) n(\mathbf{r}), \tag{17.28}$$

where $V(\mathbf{r})$ is the potential due to the lattice and $n(\mathbf{r})$ is the electron density at point $\mathbf{r}$. In second quantization and considering that the potential due to the lattice is an infinite sum of contributions from ions located at points $\mathbf{R}_l$, we can write

$$H_{int} = \sum_l \int d^3 r V(\mathbf{r} - \mathbf{R}_l) c^\dagger(\mathbf{r}) c(\mathbf{r}), \tag{17.29}$$

(we have omitted the spin dependence, for simplicity). Introducing a Fourier decomposition of the electron operators

$$c(\mathbf{r}) = \sum_{\mathbf{k}} e^{i\mathbf{k}\cdot\mathbf{r}} c_{\mathbf{k}}$$

and writing that the positions of the ions are given by a vector that refers to the regular lattice position, $\mathbf{l}$, and a small displacement around the equilibrium position, $\mathbf{y}_l$, as $\mathbf{R}_l = \mathbf{l} + \mathbf{y}_l$, we can write that

$$H_{int} = \sum_{\mathbf{k},\mathbf{k}',l} c^\dagger_{\mathbf{k}} \langle \mathbf{k}| V(\mathbf{r} - \mathbf{l} - \mathbf{y}_l)|\mathbf{k}'\rangle c_{\mathbf{k}'}. \tag{17.30}$$

The matrix element is the Fourier transform of the potential.

We may then write the full Hamiltonian as $H = H_0 + H_{int}$ where

$$H_0 = \sum_{\mathbf{k}} \epsilon_{\mathbf{k}} c^\dagger_{\mathbf{k}} c_{\mathbf{k}} + \sum_{\mathbf{q},\lambda} \hbar w_{\mathbf{q}\lambda} a^\dagger_{\mathbf{q}\lambda} a_{\mathbf{q}\lambda} \tag{17.31}$$

is the sum of a free electron part and a set of harmonic oscillators, where λ are the polarizations. Making a shift of variables $\mathbf{r} \to \mathbf{r} + \mathbf{l} + \mathbf{y}_l$, we get that

$$\begin{aligned} H_{int} &= \sum_{\mathbf{k},\mathbf{k}',l} \int d^3 r e^{-i\mathbf{k}\cdot\mathbf{r}} V(\mathbf{r} - \mathbf{l} - \mathbf{y}_l) e^{i\mathbf{k}'\cdot\mathbf{r}} c^\dagger_{\mathbf{k}} c_{\mathbf{k}'} \\ &= \sum_{\mathbf{k},\mathbf{k}',l} e^{i(\mathbf{k}'-\mathbf{k})\cdot(\mathbf{l}+\mathbf{y}_l)} \int d^3 r e^{-i(\mathbf{k}-\mathbf{k}')\cdot\mathbf{r}} V(\mathbf{r}) c^\dagger_{\mathbf{k}} c_{\mathbf{k}'} \\ &= \sum_{\mathbf{k},\mathbf{k}',l} e^{i(\mathbf{k}'-\mathbf{k})\cdot(\mathbf{l}+\mathbf{y}_l)} V_{\mathbf{k}-\mathbf{k}'} c^\dagger_{\mathbf{k}} c_{\mathbf{k}'}. \end{aligned} \tag{17.32}$$

Since the displacement of the ions from their equilibrium positions is small, we may expand

$$\begin{aligned} e^{i(\mathbf{k}'-\mathbf{k})\cdot\mathbf{y}_l} &= 1 + i(\mathbf{k}' - \mathbf{k}) \cdot \mathbf{y}_l \\ &= 1 + \frac{i}{\sqrt{N}} (\mathbf{k}' - \mathbf{k}) \cdot \sum_{\mathbf{q}} e^{i\mathbf{q}\cdot\mathbf{l}} \mathbf{y}_{\mathbf{q}}. \end{aligned} \tag{17.33}$$

The interaction term may then be split into two terms: one due to the coupling to the regular lattice and one due to the coupling with the phonons. The term due to the regular lattice is written as

$$\begin{aligned} H_{Bloch} &= \sum_{\mathbf{k},\mathbf{k}',l} e^{i(\mathbf{k}'-\mathbf{k})\cdot\mathbf{l}} V_{\mathbf{k}-\mathbf{k}'} c^\dagger_{\mathbf{k}} c_{\mathbf{k}'} \\ &= N \sum_{\mathbf{k},\mathbf{g}} V_{\mathbf{g}} c^\dagger_{\mathbf{k}+\mathbf{g}} c_{\mathbf{k}}, \end{aligned} \tag{17.34}$$

where $\mathbf{g}$ is a reciprocal lattice vector. This is the usual term that reflects the influence of the lattice on the free electron bands, which leads to the band structure of the nearly free electron model. Far from the borders of the Brillouin zone, its influence may be neglected. In order words, if we are far from Bragg reflections, we may neglect its effect. The second term is the electron-phonon interaction term that most concerns us here. It may be written as

$$\begin{aligned} H_{e-p} &= \frac{i}{\sqrt{N}} \sum_{\mathbf{k},\mathbf{k}',l,\mathbf{q}} e^{i(\mathbf{k}'-\mathbf{k}+\mathbf{q})\cdot\mathbf{l}} (\mathbf{k}'-\mathbf{k})\cdot\mathbf{y_q} V_{\mathbf{k}-\mathbf{k}'} c^{\dagger}_{\mathbf{k}} c_{\mathbf{k}'} \\ &= i\sqrt{N}\sum_{\mathbf{k},\mathbf{k}'} (\mathbf{k}'-\mathbf{k})\cdot\mathbf{y}_{\mathbf{k}-\mathbf{k}'} V_{\mathbf{k}-\mathbf{k}'} c^{\dagger}_{\mathbf{k}} c_{\mathbf{k}'} \\ &= i\sum_{\mathbf{k},\mathbf{k}',\lambda} \left(\frac{N\hbar}{2M\omega_{\mathbf{k}-\mathbf{k}',\lambda}}\right)^{1/2} (\mathbf{k}'-\mathbf{k})\cdot\boldsymbol{\lambda} V_{\mathbf{k}-\mathbf{k}'} \left(a^{\dagger}_{\mathbf{k}'-\mathbf{k},\lambda} + a_{\mathbf{k}-\mathbf{k}'.\lambda}\right) c^{\dagger}_{\mathbf{k}} c_{\mathbf{k}'} . \end{aligned} \tag{17.35}$$

The dot product selects modes that are longitudinal (polarization and momentum difference parallel). Neglecting the Bloch term yields Fröhlich Hamiltonian

$$H = \sum_{\mathbf{k},\sigma} \epsilon_{\mathbf{k}} c^{\dagger}_{\mathbf{k}\sigma} c_{\mathbf{k}\sigma} + \sum_{\mathbf{q}} \hbar\omega_{\mathbf{q}} a^{\dagger}_{\mathbf{q}} a_{\mathbf{q}} + \sum_{\mathbf{k},\mathbf{k}',\sigma} M_{\mathbf{k}\mathbf{k}'} \left(a^{\dagger}_{-\mathbf{q}} + a_{\mathbf{q}}\right) c^{\dagger}_{\mathbf{k}\sigma} c_{\mathbf{k}',\sigma}, \tag{17.36}$$

with

$$M_{\mathbf{k}\mathbf{k}'} = i\sqrt{\frac{N\hbar}{2M\omega_{\mathbf{q}}}} |\mathbf{k}'-\mathbf{k}| V_{\mathbf{k}-\mathbf{k}'} . \tag{17.37}$$

Here, $\mathbf{q} = \mathbf{k} - \mathbf{k}'$.

In the momentum representation, we defined the matrix element $V_{\mathbf{k},\mathbf{k}'}$. This is the matrix element of the electron-electron interaction where, in the initial time (I), the electrons are in states ($\mathbf{k}$, $\uparrow$) and ($-\mathbf{k}$, $\downarrow$) and in the final time (II) the electrons are in states ($\mathbf{k}'$, $\uparrow$) and ($-\mathbf{k}'$, $\downarrow$). In addition to the Coulomb repulsion term, the matrix element contains two terms resulting from the interaction with the phonons. The interaction with the phonons implies that an electron may create a phonon or absorb a phonon. An effective interaction between two electrons is obtained considering second-order terms by considering that one phonon emitted by one electron is absorbed by another electron.[3] This leads to the two terms represented in Fig. 17.5. The process then involves the exchange of a phonon between two electrons. The matrix element has the structure

$$\langle I|V|II\rangle = \int d^3r_1 \int d^3r_2 e^{-i\mathbf{k}\cdot(\mathbf{r}_1-\mathbf{r}_2)} V(\mathbf{r}_1-\mathbf{r}_2) e^{i\mathbf{k}'\cdot(\mathbf{r}_1-\mathbf{r}_2)}$$

[3] See also [deGennes].

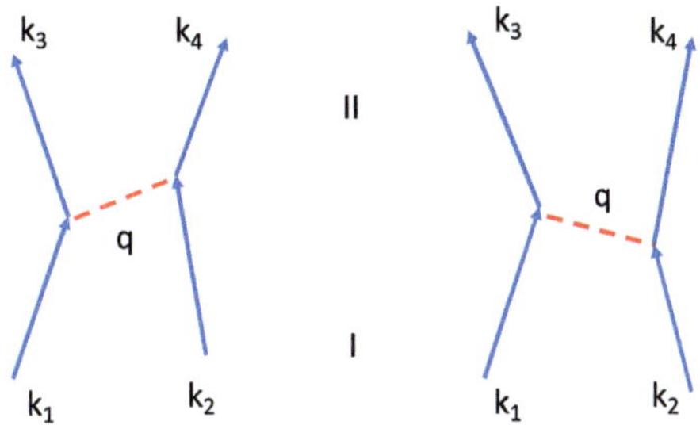

Fig. 17.5 Scattering between electrons with different momenta due to the interchange of an excitation, like a phonon. Here, we are considering $\mathbf{k}_1 = \mathbf{k}$, $\mathbf{k}_2 = -\mathbf{k}$, $\mathbf{k}_3 = \mathbf{k}'$, $\mathbf{k}_4 = -\mathbf{k}'$, and a phonon of momentum $\mathbf{q}$

$$= \int d^3r V(\mathbf{r}) e^{i\mathbf{q}\cdot\mathbf{r}} = V(\mathbf{q}),$$

where $\mathbf{q} = \mathbf{k}' - \mathbf{k}$ is the relative momentum. As a function of time, two processes may occur: In process, (i) the electron with momentum $\mathbf{k}$ emits a phonon with momentum $-\mathbf{q}$ and stays in a state with momentum $\mathbf{k} + \mathbf{q}$. At a later time, the phonon is absorbed by the electron with momentum $-\mathbf{k}$, which then stays in a state with momentum $-\mathbf{k}-\mathbf{q}$. In the other process, (ii) first, the electron with momentum $-\mathbf{k}$ emits a phonon with momentum $\mathbf{q}$ and stays in a state with momentum $-\mathbf{k}-\mathbf{q}$. Later, the other electron with momentum $\mathbf{k}$ absorbs the phonon and stays in a state with momentum $\mathbf{k} + \mathbf{q}$. The initial and final states of both processes are the same, but the intermediate steps are different. The initial state has energy $E_I = 2\zeta_k$ where $\zeta_k = \frac{\hbar^2 k^2}{2m} - E_F$. The final state has energy $E_{II} = 2\zeta_{k'}$. The two intermediate states have also equal energies $E_i = E_{ii} = \zeta_{k'} + \zeta_k + \hbar\omega_q$. Here, ω_q is the energy of the phonon of momentum q.

To second order, the matrix element of the transition between the two states is given by

$$\langle I|H_{el-ph}|II\rangle = \sum_j \langle I|V|j\rangle \frac{1}{2}\left(\frac{1}{E_{II} - E_j} + \frac{1}{E_I - E_j}\right)\langle j|V|II\rangle$$
$$= \frac{|W_q|^2}{\hbar}\left(\frac{1}{\omega - \omega_q} - \frac{1}{\omega + \omega_q}\right). \tag{17.38}$$

The sum is over the intermediate states; we defined $W_q = \langle j|V|II\rangle$, assumed that $W_q^* = \langle I|V|j\rangle$, and defined $\hbar\omega = \zeta_{k'} - \zeta_k$. We obtain finally that

$$\langle I|H|II\rangle = V_q + \frac{2|W_q|^2}{\hbar}\frac{\omega_q}{\omega^2 - \omega_q^2}. \tag{17.39}$$

The first term is the direct repulsive Coulomb repulsion between the electrons, and the second term is the contribution to the electron-electron interaction mediated by the electron-phonon interaction. If $\omega < \omega_q$, it is possible that the total interaction is attractive. Since ω_q is the energy exchanged, which is originated in the phonons, it depends on the mass of the ions. This leads to the known isotope effect in superconductivity.

Let us reformulate the problem of the attractive interaction due to the coupling to the phonons in a simple way. Recall Eq. (12.62) which describes the dielectric function of the combined system of the ions and the electrons. For the ions, we consider the plasma modes, assuming that the atoms do not overlap and only a local contribution is required. For the electrons, we use the Thomas-Fermi approximation, independent of frequency, because the electrons move much faster than the ions, almost instantaneously. The expression can be written as

$$\epsilon_r(k, \omega) = 1 - \frac{\omega_i^2}{\omega^2} + \frac{q_s^2}{k^2}. \tag{17.40}$$

This can be written as

$$\epsilon_r(k, \omega) = \frac{\omega^2(k^2 + q_s^2) - k^2\omega_i^2}{\omega^2 k^2}. \tag{17.41}$$

The normal modes are obtained taking $\epsilon = 0$. This implies that

$$\omega_k^2(q_s^2 + k^2) = \omega_i^2 k^2 \tag{17.42}$$

and, therefore,

$$\omega_k^2 = \omega_i^2 \frac{k^2}{k^2 + q_s^2}. \tag{17.43}$$

We can now rewrite the dielectric function in terms of the normal modes leading to

$$\epsilon_r(k, \omega) = \frac{k^2 + q_s^2}{k^2} \frac{\omega^2 - \omega_k^2}{\omega^2}, \tag{17.44}$$

where the plasma frequency of the ions was eliminated. The effective interaction is therefore given by

$$V_{eff}(k, \omega) = \frac{V(k, \omega)}{\epsilon_r(k, \omega)} = \frac{e^2}{\epsilon_0(k^2 + q_s^2)} \frac{\omega^2}{\omega^2 - \omega_k^2}. \tag{17.45}$$

This result once again shows that there are frequency regimes ($\omega < \omega_k$) for which an effective attractive interaction is possible. Note that without screening, $\lambda = 0$, the normal mode frequency is just the plasma frequency of the ions and the dielectric constant is always negative. In Fig. 17.6a, we plot the dielectric function as a function of momentum and frequency, showing the two regimes when it is positive ($\omega > \omega_k$) and when it is negative ($\omega < \omega_k$). In Fig. 17.6b, we plot cuts

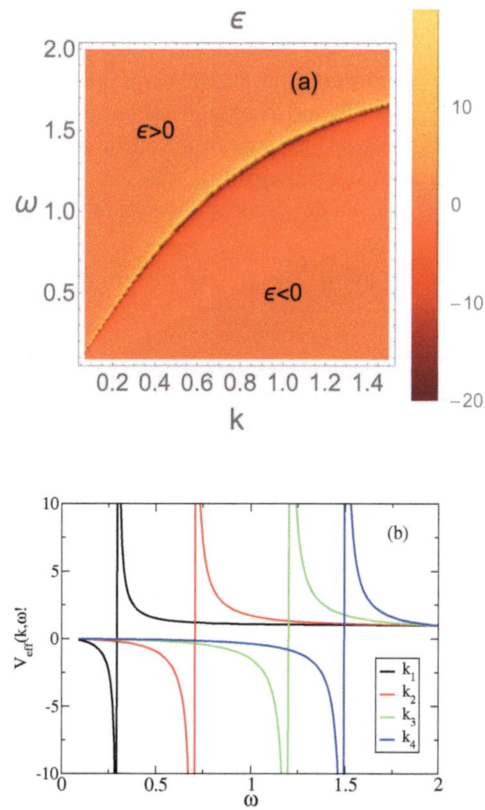

Fig. 17.6 (**a**) Relative dielectric function as a function of k, ω. (**b**) Effective interaction for constant momenta, $k_1 = 0.15$, $k_2 = 0.375$, $k_3 = 0.75$, $k_4 = 1.125$, as a function of frequency, in units of (e^2/ϵ_0). Both plots display the regimes when the effective interaction is attractive. We have taken $\lambda = 1$, $\omega_i = 2$

of the effective interaction, V_{eff} at constant momentum and as a function of ω, showing the regimes where the interaction is attractive.

17.4 BCS Theory

We have seen that the system is unstable to the creation of pairs of electrons of opposite momenta and opposite spins. Actually, we have not considered the case in which the momentum of the center of mass is finite. It can however be shown that the zero-momentum case is particularly favorable, requiring a smaller attractive potential.

We can therefore expect that the ground state is predominantly composed by pairs of electrons $(\mathbf{k}, \uparrow)$, $(-\mathbf{k}, \downarrow)$. We may therefore attempt a wave function for the system of the form

$$|G\rangle = \prod_k \left(u_k + v_k^* c_{k,\uparrow}^\dagger c_{-k,\downarrow}^\dagger \right) |0\rangle. \tag{17.46}$$

Here, $|0\rangle$ is the vacuum state and u_k (v_k) are the probability amplitudes of not having (having) one pair of electrons in the state of momentum k. We may interpret this wave function as a variational wave function. Normalizing the wave function implies $u_k^2 + v_k^2 = 1$ (since there is no other possibility). The probability amplitudes may be determined, in the spirit of the variational method, fixing their values to minimize the ground state energy.

Let us consider the Hamiltonian of the interacting electronic system as being[4]

$$H = \sum_{k,\sigma} \epsilon_k c_{k,\sigma}^\dagger c_{k,\sigma} + \sum_{k,k'} V_{kk'} c_{k,\uparrow}^\dagger c_{-k,\downarrow}^\dagger c_{-k',\downarrow} c_{k',\uparrow}, \tag{17.47}$$

and let us not consider other terms for not being relevant to the pairing mechanism. If the electrons pair up, we expect that $\langle c_{k,\uparrow}^\dagger c_{-k,\downarrow}^\dagger \rangle$ and its conjugate will have nonvanishing values in the superconducting phase. Note that these average values do not conserve the number of particles. It is natural then to use the grand canonical ensemble. This amounts to making the replacement $\epsilon_k \to \epsilon_k - \mu$.

Since the problem is interacting, the calculation of the ground state energy and higher energy levels is not straightforward. It is therefore convenient to approximate the solution of the problem. Let us then use mean-field theory and write

$$c_{-k,\downarrow} c_{k,\uparrow} = \langle c_{-k,\downarrow} c_{k,\uparrow} \rangle + (c_{-k,\downarrow} c_{k,\uparrow} - \langle c_{-k,\downarrow} c_{k,\uparrow} \rangle)$$

and the corresponding for the Hermitian conjugate. These equations are of the type of Eq. (14.145) having selected the relevant operator as a pair of electrons. The mean-field Hamiltonian is given by

$$\begin{aligned} H_{MF} &= \sum_{k,\sigma} (\epsilon_k - \mu) c_{k,\sigma}^\dagger c_{k,\sigma} \\ &\quad - \sum_k \left(\Delta_k c_{k,\uparrow}^\dagger c_{-k,\downarrow}^\dagger + \Delta_k^* c_{-k,\downarrow} c_{k,\uparrow} - \Delta_k^* \langle c_{-k,\downarrow} c_{k,\uparrow} \rangle \right). \end{aligned} \tag{17.48}$$

Introducing the parameter

$$\Delta_k = -\sum_{k'} V_{k'k} \langle c_{-k',\downarrow} c_{k',\uparrow} \rangle \tag{17.49}$$

allows to decouple the Hamiltonian along the line of Eq. (14.146). Other decouplings are also possible, but here we limit the Hamiltonian to its most relevant form in regard to the superconducting pairing. The parameter Δ_k is to be evaluated self-consistently in the ground state if we consider zero temperature, or if considering finite temperatures, the average is to be understood as an ensemble average. Recall

[4] Here we follow [BicudoSacramento].

that in general a quartic Hamiltonian may be decoupled in different ways, as we saw when we discussed charge- and spin-density waves.

The variational wave function depends on the parameters $u_k and v_k$, and the Hamiltonian is parametrized by Δ_k, which in turn is a function of the amplitudes $u_k and v_k$ self-consistently. The various parameters may be determined in different, equivalent, ways. In general, the energy is given by $E(\lambda) = \langle H(\lambda)\rangle$ if it depends on a set of parameters (λ). (The states also depend in general on the parameters). Using the Hellman-Feynman theorem,

$$\frac{\partial}{\partial \lambda}\langle H(\lambda)\rangle = \left\langle \frac{\partial H}{\partial \lambda}\right\rangle, \tag{17.50}$$

we may minimize the average to determine the ground state energy and the state itself. This approach is suggested for the ground state energy in Exercise 5. It leads to a set of equations that relate the probability amplitudes and the parameter Δ_k. The same approach may be followed at any temperature and chemical potential value, taking the ensemble averages, leading to equations that are formally equivalent.

Since in the mean-field approach the Hamiltonian is reduced to a noninteracting problem in the presence of a mean field, determined self-consistently, we may diagonalize the problem in a way similar to the treatments for the weakly interacting Bose gas or an antiferromagnet. Considering the equation of motion of $c_{k,\uparrow}$ leads to[5]

$$i\hbar\frac{\partial c_{k,\uparrow}}{\partial t} = [c_{k,\uparrow}, H] = (\epsilon_{k,\uparrow} - \mu)c_{k,\uparrow} - \Delta_k c^{\dagger}_{-k,\downarrow}. \tag{17.51}$$

Therefore, the operators $c_{k,\uparrow}$ and $c^{\dagger}_{-k,\downarrow}$ are coupled. It makes sense physically: a pair of momenta $(k, \uparrow), (-k, \downarrow)$ is broken either annihilating one electron $(k, \uparrow)$ or creating an electron $(-k, \downarrow)$. The mean-field Hamiltonian may be diagonalized performing a Bogoliubov-Valatin transformation. Appropriate linear combinations[6] may be written as

$$\begin{aligned} \alpha_k &= u_k c_{k,\uparrow} - v_k^* c^{\dagger}_{-k,\downarrow} \\ \beta_{-k} &= u_k c_{-k,\downarrow} + v_k^* c^{\dagger}_{k,\uparrow} \end{aligned} \tag{17.52}$$

and their Hermitian conjugates. Imposing the new operators as fermionic implies

$$\left\{\alpha_k, \alpha^{\dagger}_{k'}\right\} = \left\{\beta_k, \beta^{\dagger}_{k'}\right\} = \delta_{k,k'} \tag{17.53}$$

[5] Using that $[A, BC] = \{A, B\}C - B\{C, A\}$.

[6] For further reading see [FetterWalecka] and [BicudoSacramento].

and that all other anti-commutators vanish. This condition leads to the restriction that

$$|u_k|^2 + |v_k|^2 = 1. \tag{17.54}$$

In the variational method, this condition resulted from the normalization of the wave function, and here, it results from the imposition of fermionic anti-commutation relations of the Bogoliubov operators. The equations for the new fermionic operators may be inverted to yield

$$\begin{aligned} c_{k,\uparrow} &= u_k \alpha_k + v_k^* \beta_{-k}^\dagger \\ c_{-k,\downarrow} &= u_k \beta_{-k} - v_k^* \alpha_k^\dagger . \end{aligned} \tag{17.55}$$

This allows to rewrite the effective mean-field Hamiltonian in terms of the Bogoliubov operators. Keeping only terms that can be written as number operators, of the type $\alpha_k^\dagger \alpha_k$, $\beta_{-k}^\dagger \beta_{-k}$, and imposing that other terms are absent imply taking

$$2u_k v_k \tilde{\epsilon}_k - \Delta_k u_k^2 + \Delta_k^* v_k^2 = 0$$

to annihilate the unwanted crossed terms. The Bogoliubov operators describe the eigenmodes of the system. Their meaning may be confirmed showing that, for instance,

$$\alpha_k |G\rangle = 0 \tag{17.56}$$

and that the same result holds for β_{-k}. Since these operators destroy the excitations of the system, they annihilate the ground state which, by definition, has no excitations.

We can rewrite the mean-field Hamiltonian as

$$\begin{aligned} H_{MF} = &\sum_k \alpha_k^\dagger \alpha_k \left(\tilde{\epsilon}_k (|u_k|^2 - |v_k|^2) + \Delta_k u_k^* v_k + \Delta_k^* u_k v_k^* \right) \\ &+ \sum_k \beta_{-k}^\dagger \beta_{-k} \left(\tilde{\epsilon}_k (|u_k|^2 - |v_k|^2) + \Delta_k u_k^* v_k + \Delta_k^* u_k v_k^* \right) \\ &+ H_0, \end{aligned} \tag{17.57}$$

where the constant term is given by

$$H_0 = \sum_k \Delta_k^* u_k v_k^* (\langle \alpha_k^\dagger \alpha_k \rangle + \langle \beta_{-k}^\dagger \beta_{-k} \rangle) + \sum_k (2|v_k|^2 \tilde{\epsilon}_k - \Delta_k v_k u_k^* - 2\Delta_k^* u_k v_k^*)$$

and ($\tilde{\epsilon}_k = \epsilon_{k,\sigma} - \mu$).

The Bogoliubov-Valatin transformation in Eq. (17.55) allows to obtain an explicit expression for the parameter Δ_k. Using its definition, we may find that

$$\begin{aligned}\Delta_k &= -\sum_{k'} V_{kk'} \langle c_{-k',\downarrow} c_{k',\uparrow} \rangle \\ &= -\sum_{k'} V_{kk'} \left\langle \left(u_{k'}\beta_{-k'} - v_{k'}^{*}\alpha_{k'}^{\dagger} \right) \left(u_{k'}\alpha_{k'} + v_{k'}^{*}\beta_{-k'}^{\dagger} \right) \right\rangle \\ &= -\sum_{k'} V_{kk'} \left\langle \left(u_{k'}v_{k'}^{*}\beta_{-k'}\beta_{-k'}^{\dagger} - v_{k'}^{*}u_{k'}\alpha_{k'}^{\dagger}\alpha_{k'} \right) \right\rangle . \end{aligned} \tag{17.58}$$

We get therefore that

$$\Delta_k = -\sum_{k'} V_{kk'} u_{k'} v_{k'}^{*} [1 - 2f(E_{k'})], \tag{17.59}$$

establishing its relation with the parameters u_k and v_k. (Note that the parameters of the Bogoliubov-Valatin transformations are yet to be determined. They will be shown to be equivalent to the probability amplitudes considered in the variational approach.) The energy dispersion, E_k, is obtained from the diagonalized Hamiltonian as

$$E_k = \tilde{\epsilon}_k (|u_k|^2 - |v_k|^2) + \Delta_k u_k^{*} v_k + \Delta_k^{*} u_k v_k^{*}, \tag{17.60}$$

and f is the Fermi function.[7]

Often, u_k and v_k may be chosen as real (this is not possible if an external magnetic field is applied, as we will see later). Since $u_k^2 + v_k^2 = 1$, this implies that we may choose $u_k = \cos \chi_k$ and $v_k = \sin \chi_k$. Then the condition for the amplitudes u and v can be rewritten as

$$2u_k v_k \tilde{\epsilon}_k - \Delta_k u_k^2 + \Delta_k v_k^2 = 0$$

$$\tan 2\chi_k = \frac{\Delta_k}{\tilde{\epsilon}_k}. \tag{17.62}$$

This implies that

$$\sin 2\chi_k = \pm \frac{\Delta_k}{E_k} = 2u_k v_k$$

[7] Note that here

$$f(E) = \frac{1}{1 + e^{E/(k_B T)}}. \tag{17.61}$$

$$\cos 2\chi_k = \pm \frac{\tilde{\epsilon}_k}{E_k} = u_k^2 - v_k^2,$$

From these equations, we are able to determine the energy eigenvalues and the parameters $u_k and v_k$ as

$$E_k = \left(\tilde{\epsilon}_k^2 + \Delta_k^2\right)^{1/2}$$
$$u_k^2 = \frac{1}{2}\left(1 + \frac{\tilde{\epsilon}_k}{E_k}\right)$$
$$v_k^2 = \frac{1}{2}\left(1 - \frac{\tilde{\epsilon}_k}{E_k}\right).$$

The momentum dependence of the parameters is shown in Fig. 17.7. The diagonalized Hamiltonian may be written as

$$H_{MF} = \sum_k E_k \left(\alpha_k^\dagger \alpha_k + \beta_{-k}^\dagger \beta_{-k}\right) + H_0. \tag{17.63}$$

If $\tilde{\epsilon}_k = 0$, then $E_k = \Delta_k$, and there is a gap in the excitation spectrum. Substituting the previous expressions in the equation for the gap function, we get that

$$\Delta_k = -\sum_{k'} \frac{V_{kk'}\Delta_{k'}}{2E_{k'}} \left(1 - 2f(E_{k'})\right). \tag{17.64}$$

This is called the gap equation. One possible solution of the gap equation is that the gap vanishes. Since it is related to the average value of a Cooper pair, this is a trivial solution for which there is no spontaneous formation of Cooper pairs and the system is not in the superconducting phase. This is the typical high-temperature solution. A nontrivial solution, which describes some superconducting phase, is obtained if the gap is nonvanishing. We may interpret then this quantity as an order parameter

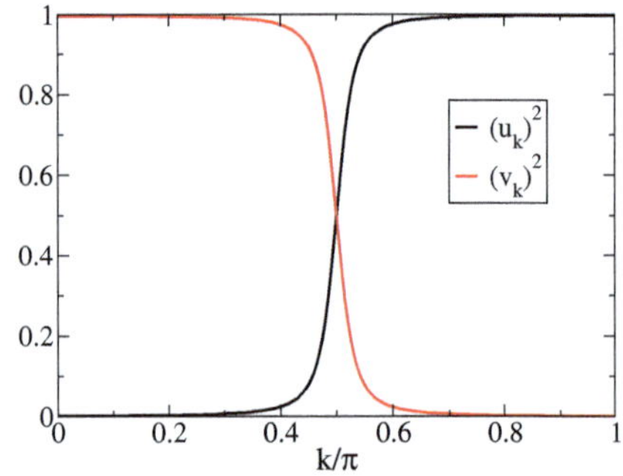

Fig. 17.7 Momentum dependence of u_k^2 (probability to find a particle excitation) and v_k^2 (probability to find a hole excitation) for a tight-binding model with vanishing chemical potential, as a function of the absolute value of the momentum

for the superconducting phase. A possible solution of the gap equation corresponds to the simple uniform case and, therefore, to phase coherence over large distances (recall that, in general, the gap function is a complex number) and a field that is well-defined in space—therefore, a state with macroscopic coherence.

This mean-field approach to a superconductor has been proposed by Bardeen, Cooper, and Schrieffer in 1957 and is called BCS theory.

17.4.1 Critical Temperature

Let us take the gap near zero, $\Delta_k \to 0$, and therefore $E_k = \tilde{\epsilon}_k$. Then the gap equation becomes

$$\Delta_k = -\sum_{k'} \frac{V_{kk'}\Delta_{k'}}{2|\tilde{\epsilon}_{k'}|}(1 - 2f(|\tilde{\epsilon}_{k'}|)) .$$

Let us choose a simple form for the interaction, like $V_{kk'} = -V$ if $|\tilde{\epsilon}_k|, |\tilde{\epsilon}_{k'}| < \hbar\omega_D$ and $V_{kk'} = 0$, otherwise. Then the gap equation is rewritten as

$$\Delta_k = V \sum_{k',|\tilde{\epsilon}_{k'}|<\hbar\omega_D} \Delta_{k'} \frac{1 - 2f(|\tilde{\epsilon}_{k'}|)}{2|\tilde{\epsilon}_{k'}|}. \tag{17.65}$$

If $\Delta_k \to 0$, then it is independent of k and gives

$$1 = VN(0) \int_0^{\hbar\omega_D} d\epsilon \frac{1 - 2f(\epsilon)}{\epsilon}.$$

The Fermi function is in this case calculated at the critical temperature $T = T_c$. Defining $x = \epsilon/(k_B T_c)$ and integrating by parts, we obtain that

$$\frac{1}{VN(0)} = \ln \frac{\hbar\omega_D}{k_B T_c}(1 - 2f(\hbar\omega_D)) + 2\int_0^{\frac{\hbar\omega_D}{k_B T_c}} dx \ln x \frac{d}{dx}\left(\frac{1}{e^x + 1}\right).$$

Neglecting $f(\hbar\omega_D) << 1$, we get

$$\frac{1}{VN(0)} = \ln\left(A\frac{\hbar\omega_D}{k_B T_c}\right),$$

where

$$\ln A = 2\int_0^{\infty} dx \ln x \frac{d}{dx}\left(\frac{1}{e^x + 1}\right)$$

($A = \frac{2\gamma}{\pi} \sim 1.13$ where γ is Euler's constant). We obtain finally that

$$k_B T_c = 1.13\hbar\omega_D e^{-\frac{1}{VN(0)}}. \quad (17.66)$$

Once again, let us emphasize that the result is non-perturbative in the potential.

17.4.2 Gap

Consider once again the gap equation

$$\Delta_k = -\sum_{k'} V_{kk'} \frac{\Delta_{k'}}{2E_{k'}}[1 - 2f(E_{k'})]$$

and the same form for the potential, as above. Then the gap is independent of the momentum, and we get

$$1 = N(0)V \int_0^{\hbar\omega_D} \frac{d\epsilon}{E}[1 - 2f(E)]$$
$$E = (\epsilon^2 + \Delta^2)^{1/2}.$$

Consider now that $\Delta, k_B T_c << \hbar\omega_D$. Then we obtain that[8]

$$\frac{1}{N(0)V} = \sinh^{-1}\frac{\hbar\omega_D}{\Delta} + 2\int_0^{\infty} d\epsilon \frac{\partial f(E)}{\partial \epsilon} \sinh^{-1}\frac{\epsilon}{\Delta}.$$

Integrating by parts, we get as before

$$\frac{1}{N(0)V} \sim \ln\frac{2\hbar\omega_D}{\Delta} + F\left(\frac{\Delta}{k_B T}\right).$$

Using the expression for the critical temperature, we may write that

$$\Delta(T) = \frac{2}{1.13} k_B T_c e^{F\left(\frac{\Delta}{k_B T}\right)}. \quad (17.67)$$

If Δ is finite, then when $T \to 0$, the Fermi-Dirac function tends to zero and $F \to 0$. Then

$$\Delta(0) = 2\hbar\omega_D e^{-\frac{1}{N(0)V}} = 1.77 k_B T_c$$

[8] See also [deGennes].

and

$$\frac{\Delta(T)}{\Delta(0)} = e^{F\left(\frac{\Delta}{k_B T}\right)}.$$

The ratio starts at one at zero temperature and decreases to zero at the critical temperature. Near T_c, we get

$$\frac{\Delta(T)}{\Delta(0)} = 1.74\left(1 - \frac{T}{T_c}\right)^{1/2}.$$

In our notation of the critical exponents, we get $\beta = 1/2$ for the dependence of the order parameter, here taken as the gap function, as a function of the temperature difference to the critical temperature. This is obviously the mean-field result, as expected, since the whole treatment was based upon a mean-field approximation of the Hamiltonian. At the same time, this result suggests a description of the superconducting phase transition by a functional, in which the free energy is expanded in powers of the order parameter. This is actually possible, as we shall see later, and it is called the Ginzburg-Landau functional. Anticipating the later treatment, it can be obtained expanding the gap function in powers of Δ and equalizing it to the stationary condition for the free energy.

17.4.3 Thermodynamics

The Hamiltonian was obtained as

$$H = \sum_{k,\alpha} E_k \gamma_{k,\alpha}^\dagger \gamma_{k,\alpha} + H_0,$$

changing the notation for the fermionic operators as $\gamma_{k,+} = \alpha_k$ and $\gamma_{k,-} = \beta_{-k}$. The average energy is given by

$$U = \langle H \rangle = \sum_{k,\alpha} E_k f(E_k) + H_0 = 2\sum_k E_k f(E_k) + H_0.$$

The specific heat per unit volume is then obtained from

$$c_V = \frac{2}{V}\sum_k E_k \left(\frac{\partial f(E_k)}{\partial T}\right)_V.$$

Using that

$$\frac{\partial f(E)}{\partial T} = \frac{\partial f(E)}{\partial E}\frac{\partial E}{\partial T} + \frac{\partial f}{\partial T},$$

we obtain

$$c_V = -\frac{2}{VT} N(0) \int_{-\infty}^{\infty} d\epsilon \left(E^2 + \frac{\beta}{2} \frac{\partial \Delta^2}{\partial \beta} \right) \frac{\partial f(E)}{\partial E}. \tag{17.68}$$

Using that

$$\frac{\partial f(E)}{\partial E} = -\frac{1}{k_B T} f(E)(1 - f(E)),$$

the specific heat is given by

$$c_V = 2k_B \beta^2 N(0) \int_0^{\infty} d\epsilon f(E)(1 - f(E)) \left[E^2 - T\Delta \frac{d\Delta}{dT} \right].$$

At low temperatures, $\Delta \frac{d\Delta}{dT}$ is negligible, and the specific heat is obtained from

$$\begin{aligned} c_V &\sim 2\beta^2 k_B N(0) \Delta^2 \int_0^{\infty} d\epsilon e^{-\beta\sqrt{\epsilon^2+\Delta^2}} \\ &\sim 2\beta^2 k_B N(0) \Delta^2 e^{-\beta\Delta} \int_0^{\infty} d\epsilon e^{-\beta \frac{\epsilon^2}{2\Delta}}. \end{aligned}$$

We see that at low temperatures the specific heat is exponentially small due to the presence of the gap in the spectrum. The difference between the superconducting and the normal phase specific heats is

$$c_s - c_n = k_B N(0) \beta_c^2 \left(\frac{d\Delta^2}{d\beta} \right)_{T_c} = 10.2 k_B^2 T_c N(0).$$

It is therefore discontinuous at the critical temperature.

17.5 Bogoliubov-de Gennes Equations

Let us consider now an alternative way to solve the problem that is easily generalizable to more complex situations (such as the inclusion of magnetic fields, the addition of disorder effects, etc.). Let us consider a description in real space, and let us consider creation and destruction operators in the real space.

The Hamiltonian of the electrons with an interaction may now be written as

$$
\begin{aligned}
H = &\int d^3r \sum_{\alpha} \psi^{\dagger}(\mathbf{r},\alpha)\left(\frac{(\mathbf{p}-e\mathbf{A})^2}{2m} + U_0(\mathbf{r})\right)\psi(\mathbf{r},\alpha) \\
&- \frac{1}{2}V\int d^3r \sum_{\alpha,\beta} \psi^{\dagger}(\mathbf{r},\alpha)\psi^{\dagger}(\mathbf{r},\beta)\psi(\mathbf{r},\beta)\psi^{\dagger}(\mathbf{r},\alpha),
\end{aligned}
\tag{17.69}
$$

considering a local interaction between the electrons and considering a local potential $U_0(\mathbf{r})$ and the eventual action of an external magnetic field ($\mathbf{A}$ is the vector potential). Let us then consider a local pairing of electrons with opposite spins (we are considering a local attractive interaction that corresponds to a Fourier transform independent of the momentum, as before). Then the mean-field Hamiltonian is given by

$$
\begin{aligned}
H_{MF} = &\int d^3r \Bigg[\sum_{\alpha} \psi^{\dagger}(\mathbf{r},\alpha)H_e(\mathbf{r})\psi(\mathbf{r},\alpha) + U(\mathbf{r})\psi^{\dagger}(\mathbf{r},\alpha)\psi(\mathbf{r},\alpha) \\
&+ \Delta(\mathbf{r})\psi^{\dagger}(\mathbf{r},\uparrow)\psi^{\dagger}(\mathbf{r},\downarrow) + \Delta^*(\mathbf{r})\psi(\mathbf{r},\downarrow)\psi(\mathbf{r},\uparrow)\Bigg],
\end{aligned}
\tag{17.70}
$$

plus a constant, where

$$
\begin{aligned}
\Delta(\mathbf{r}) &= -V\langle\psi(\mathbf{r},\downarrow)\psi(\mathbf{r},\uparrow)\rangle = V\langle\psi(\mathbf{r},\uparrow)\psi(\mathbf{r},\downarrow)\rangle \\
U(\mathbf{r}) &= -V\langle\psi^{\dagger}(\mathbf{r},\uparrow)\psi(\mathbf{r},\uparrow)\rangle = -V\langle\psi^{\dagger}(\mathbf{r},\downarrow)\psi(\mathbf{r},\downarrow)\rangle \\
H_e(\mathbf{r}) &= \frac{1}{2m}(-i\hbar\nabla - e\mathbf{A})^2 + U_0(\mathbf{r}) - E_F.
\end{aligned}
\tag{17.71}
$$

This Hamiltonian can be diagonalized. Let us then write

$$
\begin{aligned}
\psi(\mathbf{r},\uparrow) &= \sum_n \left(\gamma_{n,\uparrow}u_n(\mathbf{r}) - \gamma^{\dagger}_{n,\downarrow}v_n^*(\mathbf{r})\right) \\
\psi(\mathbf{r},\downarrow) &= \sum_n \left(\gamma_{n,\downarrow}u_n(\mathbf{r}) + \gamma^{\dagger}_{n,\uparrow}v_n^*(\mathbf{r})\right).
\end{aligned}
\tag{17.72}
$$

The new operators are fermionic as well. Since this transformation diagonalizes the Hamiltonian, we can write it as

$$
H_{MF} = E_g + \sum_{n,\alpha} \epsilon_n \gamma^{\dagger}_{n,\alpha}\gamma_{n,\alpha},
\tag{17.73}
$$

where E_g is the energy of the ground state and (n,α) are the excitations of the system with energies ϵ_n (we are here neglecting the Zeeman term). It is then obvious that the following relations hold

$$[H_{MF}, \gamma_{n,\alpha}] = -\epsilon_n \gamma_{n,\alpha}, \tag{17.74}$$

$$[H_{MF}, \gamma_{n,\alpha}^\dagger] = \epsilon_n \gamma_{n,\alpha}^\dagger. \tag{17.75}$$

Let us calculate the commutators

$$[\psi(\mathbf{r}, \uparrow), H_{MF}] = [H_e + U]\psi(\mathbf{r}, \uparrow) + \Delta(\mathbf{r})\psi^\dagger(\mathbf{r}, \downarrow), \tag{17.76}$$

$$[\psi(\mathbf{r}, \downarrow), H_{MF}] = [H_e + U]\psi(\mathbf{r}, \downarrow) - \Delta^*(\mathbf{r})\psi^\dagger(\mathbf{r}, \uparrow). \tag{17.77}$$

Introducing the transformation of the fields $\psi(\mathbf{r}, \alpha)$ in terms of the operators of the excitations of the system $\gamma_{n,\alpha}$ and its Hermitian conjugate, we obtain, comparing the coefficients of the same operators, the Bogoliubov-de Gennes equations:

$$\begin{aligned} [H_e + U]u(\mathbf{r}) + \Delta(\mathbf{r})v(\mathbf{r}) &= \epsilon u(\mathbf{r}) \\ -[H_e^* + U]v(\mathbf{r}) + \Delta^*(\mathbf{r})u(\mathbf{r}) &= \epsilon v(\mathbf{r}). \end{aligned} \tag{17.78}$$

The functions $u(\mathbf{r})$ and $v(\mathbf{r})$ are the eigenfunctions of these equations and ϵ are the eigenvalues. We have reduced the problem to the solution of an eigensystem. This way to solve the problem is quite useful in many situations. The solution of these equations allows the calculation of all physical quantities, even in complex situations. In general, these equations will have to be solved numerically.

Note that if there is no superconductivity, $\Delta(\mathbf{r})$ vanishes everywhere, and the equations decouple in two equations, one for the wave function $u(\mathbf{r})$ and another for the wave function $v(\mathbf{r})$. These are just Schrödinger-like equations for each wave function. In this approximation, they are just equations for a free particle in an external potential, such that the wave functions have energies that are symmetric. We may see that the wave functions $u(\mathbf{r})$ are particle-like solutions and the wave functions $v(\mathbf{r})$ are hole-like solutions. The Bogoliubov-de Gennes equations may be seen as a generalization of Schrödinger equations to a superconductor, which involves a spinor-like structure with two components, the u and v parts, such that it reduces to a simple particle-hole description in the nonsuperconducting (high-temperature) phase. There is therefore some doubling of the degrees of freedom in this representation, since the normal system may just be described in terms of particles or holes.

17.6 Flux Quantization

Applying an external electric field to a superconductor particles moves without resistance. Currents established in a closed circuit in the normal, nonsuperconducting, state remain finite without apparent decay, when the system makes the transition to the superconducting phase.[9] Another characteristic of (a class of) superconductors is the Meissner effect which implies that, in the presence of an external magnetic field, the field lines are expelled from the material and, therefore, the superconductors are perfect diamagnets (called type-*I* superconductors). Some superconductors (called of type-*II*) allow the penetration of magnetic field lines in their bulk, giving origin to currents that circulate around the lines of magnetic field (vortices), to create magnetic fields inside the material that shield the external applied magnetic field. The phase coherence implies that any magnetic field lines, which penetrate the material, must have the magnetic flux quantized.

Indeed, if we recall the Aharonov-Bohm effect and consider a geometry, where we have a superconducting ring and inside the hole of the ring magnetic field lines are allowed, since the wave function of the pairs of electrons (Cooper pairs) is a well-defined wave function (the wave function is related to the gap function), if we take the circulation of the pair around the ring, the wave function must be single valued. But the circulation of a pair of charge $q = 2e$ along a given path accumulates a phase of the type

$$\frac{q}{\hbar}\Phi, \tag{17.79}$$

where Φ is the flux of the magnetic field that threads the area defined by the trajectory. The wave function is invariant if

$$\frac{q}{\hbar}\Phi = 2\pi n. \tag{17.80}$$

Therefore, the magnetic flux should be given by

$$\Phi = \frac{2\pi\hbar}{q}n, \tag{17.81}$$

where n is an integer. The flux is quantized in units of the quantum of flux

$$\Phi = \Phi_0 n, \tag{17.82}$$

where $\Phi_0 = h/2e$.

17.7 Tunneling in Macroscopic Systems: Josephson Effect

Tunneling is a well-known property of microscopic systems. However, the possibility to characterize a superconducting state by a coherent macroscopic wave function, as expressed by Eq. (17.46), leads to the possibility that a macroscopic system may

[9] Here we follow [BicudoSacramento].

have a behavior that, in some sense, can be interpreted as tunneling through a barrier. Something similar occurs in Bose condensates. Indeed, it is also possible to observe interference phenomena as in microscopic systems.

The wave function that represents the condensate, ψ, is the probability amplitude of the electron pairs in a given space position. The pair concentration $n = \psi^*\psi = n_s^*$ may be assumed as constant, in the simplest case when there is translational invariance (no magnetic field or impurities). We may then write that

$$\psi = n^{\frac{1}{2}} e^{i\theta}, \tag{17.83}$$

since the density $n = n_s^*$ is the pair density. Under the assumption that the superconducting state is some condensate (or coherent) state, we expect that the phase of the probability amplitude is the same everywhere.

Let us now consider a superconductor divided in two (semi-infinite) regions separated by an insulator of small width. One expects that there may be some tunneling of the unpaired electrons through the insulator. In general, we may expect that in the two regions the wave functions, ψ_1 and ψ_2, may describe superconductors with different densities, n_1 and n_2, and that the phases of the wave functions, θ_1 and θ_2, may be different. The probability amplitude to find a given pair has a phase that is characteristic of the full many-body state, on each side. We may also expect that there may be tunneling of the electron pairs across the junction. This effect is called Josephson effect.

Let us consider that[10]

$$\begin{aligned} \psi_1 &= n_1^{\frac{1}{2}} e^{i\theta_1} \\ \psi_2 &= n_2^{\frac{1}{2}} e^{i\theta_2}, \end{aligned} \tag{17.84}$$

where in general the phase may be different across the insulating junction. Considering that tunneling may occur across the junction, we may write two Schrödinger equations for the amplitudes as

$$\begin{aligned} i\hbar \frac{\partial \psi_1}{\partial t} &= \hbar T \psi_2 \\ i\hbar \frac{\partial \psi_2}{\partial t} &= \hbar T \psi_1, \end{aligned} \tag{17.85}$$

where $\hbar T$ represents the transfer of electron pairs across the junction, considering the zero-temperature case where all electrons are paired. Defining $\delta = \theta_2 - \theta_1$ the phase difference between the two superconductors (a macroscopic parameter since

[10] See also [Kittel] and [BicudoSacramento].

there is phase coherence), we obtain that

$$
\begin{aligned}
\frac{\partial n_1}{\partial t} &= 2T\,(n_1 n_2)^{\frac{1}{2}} \sin\delta \\
\frac{\partial n_2}{\partial t} &= -2T\,(n_1 n_2)^{\frac{1}{2}} \sin\delta \\
\frac{\partial \theta_1}{\partial t} &= -T\left(\frac{n_2}{n_1}\right)^{\frac{1}{2}} \cos\delta \\
\frac{\partial \theta_2}{\partial t} &= -T\left(\frac{n_1}{n_2}\right)^{\frac{1}{2}} \cos\delta .
\end{aligned}
\tag{17.86}
$$

Since the current is proportional to the density variation, we have that

$$J = J_0 \sin\delta \tag{17.87}$$

and, therefore, is function of the phase difference between the probability amplitudes in the two superconductors. It depends therefore on a parameter associated with a macroscopic quantity, as mentioned above, since it describes a state of many particles.

Consider now two Josephson junctions, as shown in Fig. 17.8, and that we apply a perpendicular magnetic field through the circuit. Even in the absence of the junctions, a Aharonov-Bohm effect arises. The phase circulation is affected by the flux of the magnetic field. As seen before, we should have that

$$\hbar \nabla\theta = q\mathbf{A}, \tag{17.88}$$

and, therefore, the presence of the magnetic field implies a phase that depends on space. In this case, the charge of the particles is given by $q = -2e$, since the particles being transported along the circuit are Cooper pairs. The phase difference must be a multiple of 2π, since the amplitude is well-defined. Therefore,

$$\int_C \nabla\theta \cdot d\mathbf{l} = 2\pi m. \tag{17.89}$$

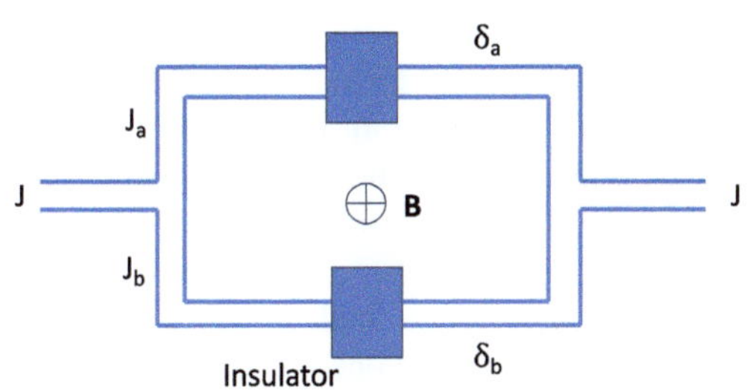

Fig. 17.8 Two Josephson junctions in parallel in the presence of an external perpendicular magnetic field. The current is a function of the phase difference δ_0 between the two superconductors and the magnetic flux

As a consequence, the flux is quantized as

$$\Phi = \int_C \mathbf{A} \cdot d\mathbf{l} = \frac{h}{2e} m, \tag{17.90}$$

where C is the closed contour and m is integer.

Returning to the Josephson junctions, we obtain that the phase circulation along the closed circuit must be given by

$$\int_C \nabla\theta \cdot d\mathbf{l} = \frac{2e}{\hbar}\Phi. \tag{17.91}$$

Let us call δ_a the phase difference between the two superconductors across a junction and δ_b across the other junction, taken both along the same direction. We have therefore

$$\delta_b - \delta_a = \frac{2e}{\hbar}\Phi, \tag{17.92}$$

and we may write

$$\delta_a = \delta_0 - \frac{e}{\hbar}\Phi$$

$$\delta_b = \delta_0 + \frac{e}{\hbar}\Phi. \tag{17.93}$$

We want now to connect two leads to these two junctions to allow passage of current. So the two junctions act in parallel. The total current is given by the sum of the currents through each junction. Therefore,

$$J = J_0 \sin\left(\delta_0 - \frac{e}{\hbar}\Phi\right) + J_0 \sin\left(\delta_0 + \frac{e}{\hbar}\Phi\right). \tag{17.94}$$

The current is then a function of the external magnetic flux and oscillates as a function of the magnetic field

$$J = 2J_0 \sin\delta_0 \cos\frac{e\Phi}{\hbar}. \tag{17.95}$$

These oscillations are the result of the interference between the two junctions and therefore result from the interference between two macroscopic systems. The absolute value of the current has maxima when $e\Phi/(\hbar) = n\pi$. This occurs when $\Phi = n\Phi_0^p$, where $\Phi_0^p = h/(2e)$ is the quantum of flux in a superconductor. The charge is that of a Cooper pair. The system we are considering allows a very accurate measurement of very small fields, of the order of changes of the magnetic flux through the junction of one flux quantum, by detecting the maxima in the Josephson current. This device is called a SQUID (superconducting quantum interference device).

17.8 Ginzburg-Landau Phenomenological Theory

The absence of resistivity and similarity to Bose condensation, with the condensation to a zero-energy (minimal energy) zero-momentum state, suggested a coherent nature of the superconducting state even before the introduction of the microscopic BCS theory. Ginzburg-Landau introduced a description of the state in terms of some order parameter, nonvanishing in the coherent, superconducting, phase that was proposed to be the wave function of the coherent condensate. This was the approach taken in the previous section.

Assuming a second-order phase transition to the superconducting state, a functional that describes the system close to the transition to the coherent phase, with respect to the normal phase described by F_n, was introduced in the form[11]

$$F_s = F_n + \alpha(T)|\psi|^2 + \frac{1}{2}\beta|\psi|^4 + \gamma\left[|\frac{\partial\psi}{\partial x}|^2 + |\frac{\partial\psi}{\partial y}|^2 + |\frac{\partial\psi}{\partial z}|^2\right] + \cdots, \tag{17.96}$$

where ψ is the wave function of the condensate, which the microscopic theory allows to understand as the probability amplitude of Cooper pairs. Therefore, $|\psi|^2 = n_s^*$ is the density of the particles that contribute to the condensate (recall the London two-fluid model and the discussion of the Josephson effect). The BCS theory allows to understand that at zero temperature $n_s^* = n_s/2$, since the Cooper pairs have two electrons in the condensate. This functional is defined at each spatial location, $\mathbf{r}$. Note that in light of the microscopic theory the order parameter may also be chosen as the amplitude of the gap parameter, $\Delta(\mathbf{r})$, since it is finite in the ordered phase and vanishes in the normal, nonsuperconducting, phase. We may replace the wave function by the gap function in the functional integral.

In the presence of a magnetic field, the Ginzburg-Landau functional can be written as

$$F_{sh} = F_{n0} + \alpha|\psi|^2 + \frac{1}{2}\beta|\psi|^4 + \frac{1}{2m^*}|\left(-i\hbar\nabla - e^*\mathbf{A}\right)\psi|^2 + \frac{1}{2\mu}h^2, \tag{17.97}$$

neglecting higher-order terms in the derivatives of the wave function. $\mathbf{A}$ is the vector potential of the magnetic field, m^* is the mass of the particles in the condensate (which BCS theory implies $m^* = 2m$), $e^* = 2e$ is the charge of the particles, and

$$\mathbf{h} = \nabla \times \mathbf{A} \tag{17.98}$$

is the magnetic field.

[11] See also, for further reading, [FetterWalecka].

The equilibrium equations of state are obtained varying the total free energy Ginzburg-Landau functional with respect to the condensate wave function and the vector potential. Defining

$$\mathcal{F}_{sh} = \int d^3r\, F_{sh}(\mathbf{r}) \tag{17.99}$$

and considering the equilibrium condition $\delta\mathcal{F}_{sh} = 0$,

$$\begin{aligned}\delta\mathcal{F}_{sh} &= \int d^3r \left\{ (\delta\psi^*) \left[\alpha\psi + \beta|\psi|^2\psi + \frac{1}{2m^*}\left(-i\hbar\nabla - e^*\mathbf{A}\right)^2 \psi \right] + c.c. \right\} \\ &+ \int d^3r\, (\delta\mathbf{A}) \cdot \left\{ \frac{1}{\mu}\nabla\times\mathbf{h} + \frac{1}{2m^*} e^*\psi^*\left(-i\hbar\nabla - e^*\mathbf{A}\right)\psi + c.c. \right\} \\ &= 0. \end{aligned} \tag{17.100}$$

Since the variations are arbitrary, they imply the equations of state

$$\begin{aligned} &\frac{1}{2m^*}\left(\frac{\hbar}{i}\nabla - e^*\mathbf{A}\right)^2 \psi + \alpha\psi + \beta|\psi|^2\psi = 0 \\ &\frac{e^*\hbar}{2m^* i}\left(\psi^*\nabla\psi - \psi\nabla\psi^*\right) - \frac{(e^*)^2}{m^*}|\psi|^2\mathbf{A} = \nabla\times\mathbf{h} = \mathbf{j}_s, \end{aligned} \tag{17.101}$$

where $\mathbf{j}_s$ is the superconducting current.

The wave function (or the gap function) is in general a complex quantity $\psi = \psi_0 e^{i\theta}$ with an amplitude and a phase, as discussed in the context of the Josephson effect, and a natural conclusion from the definition of the gap function in the microscopic theory. The appearance of the superconducting phase is in general associated with a nonvanishing amplitude. In the BCS theory in its simplified version, the phase coherence appears simultaneously. In some systems, it has been argued that, as one approaches the superconducting phase, there is already a preformed amplitude, but the phase is incoherent. Therefore, the average of the wave function vanishes, and the transition to the superconducting phase in those systems is ultimately associated with the coherence of the phase.

Consider first the appearance of a nonvanishing amplitude, and let us ignore for the moment the importance of the phase. Considering no magnetic field, a homogeneous solution of the equation for the wave function is then

$$|\psi_\infty|^2 = -\frac{\alpha}{\beta} = \frac{|\alpha|}{\beta}. \tag{17.102}$$

Recall that, as usual in the Landau theory, the parameter α is linear in the temperature difference to the critical temperature, with $\alpha < 0, T < T_c$, and the subscript ∞ is used to recall that the solution is homogeneous and the same throughout the bulk of the system, all the way to ∞. Recalling the relation between the free energy of the superconducting phase and that of the normal phase, we can write

$$F_{n0} - F_{s0} = \frac{1}{2\mu} H_c^2 = \frac{\alpha^2}{2\beta}. \tag{17.103}$$

We may now find a solution of the equations when there is an interface between the outside and the superconductor, located at $z = 0$, where z is the direction perpendicular to a planar interface. In zero magnetic field (zero vector potential), we may solve the equation

$$-\frac{\hbar^2}{2m^*}\frac{d^2\psi}{dz^2} + \alpha\psi + \beta|\psi|^2\psi = 0, \tag{17.104}$$

for positive values of z, corresponding to the inside of the superconductor region and taking $\psi(z = 0) = 0$. We find a solution of the type

$$\psi(z) = |\psi_\infty| \tanh\left(\frac{z}{\sqrt{2}\xi}\right), \tag{17.105}$$

where ξ is the coherence length of the wave function, which measures the scale over which the condensate forms inside the superconductor. Using Eq. (17.102), it is given by

$$\xi^2 = \frac{\hbar^2}{2m^*|\alpha|}. \tag{17.106}$$

Since α is proportional to the temperature deviation from the critical temperature, we obtain that the critical exponent $\nu = 1/2$, as expected.

Recall the expression for the magnetic penetration length obtained by London, given by

$$\lambda_L^2 = \frac{m^*}{n_s^*(e^*)^2}. \tag{17.107}$$

As we saw above, we can write that

$$n_s^* = |\psi_\infty|^2 = \frac{|\alpha|}{\beta}. \tag{17.108}$$

This allows us to define a parameter that is the ratio of the two lengths, the penetration length and the coherence length, given by

$$\kappa = \frac{\lambda}{\xi} = \frac{m^*}{\hbar e^*}\sqrt{2\beta} = \frac{\sqrt{2/\mu}e^*}{\hbar} H_c \lambda_L^2. \tag{17.109}$$

Let us now define new variables with convenient rescalings:

$$\mathbf{r}' = \frac{\mathbf{r}}{\lambda}$$

$$\mathbf{h}' = \frac{\mathbf{h}}{\sqrt{2\mu} H_c}$$

$$\mathbf{A}' = \frac{\mathbf{A}}{\sqrt{2} H_c \lambda}$$

$$\psi' = \frac{\psi}{|\psi_\infty|}. \tag{17.110}$$

In terms of these variables, the free energy is rewritten as

$$F_{sh} = F_{n0} + \frac{1}{4\pi} H_c^2 \left[\frac{1}{2} - |\psi'|^2 + \frac{1}{2}|\psi'|^4 + |\left(\frac{1}{i\kappa}\nabla' - \mathbf{A}' \right) \psi'|^2 + \left(h'\right)^2 \right]. \tag{17.111}$$

Redefining as well the free energy as $F' = F/(H_c^2)$, we get

$$F'_{sh} = F'_{n0} + \frac{1}{2} - |\psi'|^2 + \frac{1}{2}|\psi'|^4 + |\left(\frac{1}{i\kappa}\nabla' - \mathbf{A}' \right) \psi'|^2 + \left(h'\right)^2. \tag{17.112}$$

This leads to the equations of state

$$\left(\frac{1}{i\kappa}\nabla' - \mathbf{A}' \right)^2 \psi' = \psi' \left(1 - |\psi'|^2 \right) \tag{17.113}$$

and

$$\frac{1}{2}\left[(\psi')^* \left(\frac{1}{i\kappa}\nabla' - \mathbf{A}' \right) \psi' + \psi' \left(-\frac{1}{i\kappa}\nabla' - \mathbf{A}' \right) (\psi')^* \right] = \nabla' \times \mathbf{h}'. \tag{17.114}$$

Note that the equations only depend on the parameter κ.

The equations may now be solved near the planar interface between a normal system, or vacuum, and a superconductor. The solutions depend on the value of κ. If $\kappa << 1$, which implies that the penetration length is much smaller than the coherence length, we find that any external magnetic field decays fast toward the interior of the superconductor and the wave function amplitude slowly converges to its bulk value, $|\psi_\infty|$. This is the behavior characteristic of the so-called type-I superconductors.

On the other hand, if $\kappa >> 1$ and therefore the penetration length is much larger than the coherence length, the magnetic field decays slowly toward the interior of the superconductor, and the bulk value, $|\psi_\infty|$, is reached in a short distance. These

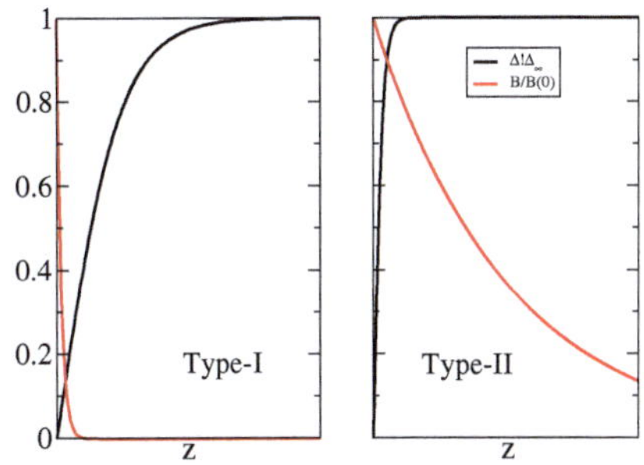

Fig. 17.9 Type-*I* ($\kappa < 1$) and type-*II* ($\kappa > 1$) superconductors

are called type-II superconductors. The two types of superconductors are shown in Fig. 17.9.

In type-I superconductors, λ is small, and the Meissner effect expels the magnetic field. Since ξ is large, the gain in condensation energy is small. This means a positive energy associated with the interface. In other words, it is favorable to minimize the contact surface between the superconductor and the exterior.

In type-II superconductors, λ is large and ξ is small. So there is a gain to condensate the system, and there is no significant energy cost, since the magnetic field changes little near the surface. So there is a negative energy associated with the interface. In other words, it is favorable to maximize the contact surface with the exterior of the superconductor.

Consider then a superconductor in the presence of an external magnetic field. If the superconductor is of type I, we expect that the Meissner effect leads to the expulsion of the magnetic field from the interior of the superconductor. This indeed occurs if the superconductor has a cigar-like shape aligned along the direction of the field lines. There is a small bending of the magnetic field lines in order to have $\mathbf{B} = 0$ inside the superconductor. However, if the system is perpendicular to the field lines, in order to bend the field lines, a large magnetic energy is required. There is therefore a balance between the condensation energy gain and the loss due to the magnetic energy. It is reasonable then to expect that at some point it is energetically favorable that the magnetic field penetrates the superconductor, at least in some regions.

In type-I superconductors, one expects that the magnetic field penetrates the material in regions of finite transversal extent and the system has alternating regions that are normal, where the magnetic field lines penetrate the material, and superconducting regions that fully expel the field lines. Since the surface energy of the interfaces between the two types of regions is positive, it is favorable to minimize these surfaces. On the other hand, if the material is a type-II superconductor, the surface energy is negative, and so it is favorable to maximize the contact surface. Therefore, for a given magnetic field applied to the material and therefore for a given surface area and a given magnetic flux, it is favorable to have as many regions as possible through which the field lines penetrate. However, as we saw above, the magnetic flux is quantized, which means that we may expect as many lines as the

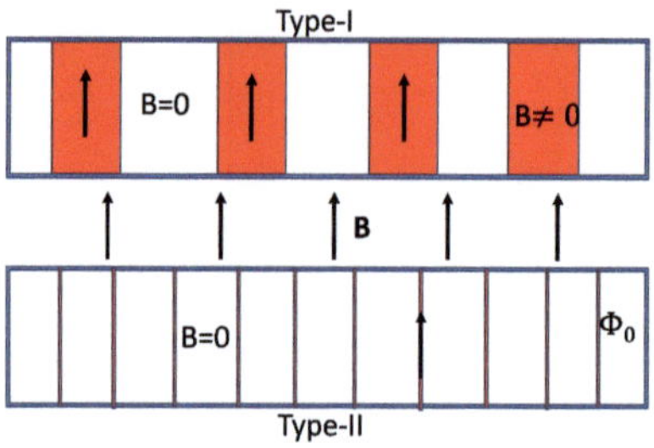

Fig. 17.10 In type-*I* (positive surface energy) and type-*II* (negative surface energy) superconductors, the magnetic field penetrates differently inside a superconducting material. In type-*I* superconductors, the magnetic field penetrates through regions of finite spatial extent. In type-*II* superconductors, the magnetic field penetrates through vortices, each containing a quantum of magnetic flux

total magnetic flux divided by the quantum flux. The penetration of the magnetic field lines is shown schematically in Fig. 17.10.

These lines through which a quantum of magnetic flux penetrates the superconductor are called vortex lines. Through these vortices, the magnetic field penetrates, and there is some spatial extent, transverse to the direction of the vortex, of size λ, the penetration length. If the magnetic field is very small, we expect the density of vortices to be small. A vortex line appears in the material when the field is large enough so that a quantum of flux penetrates the superconductor. A simplified view is that inside the vortex there is a magnetic field and outside the vortex the magnetic field vanishes (due to the Meissner effect). So, we may think that the interior of the vortex is like a normal region and outside we find the superconducting region. Thinking in terms of the wave function of the condensate and considering now that the magnetic field increases considerably, such that the vortex density is very large, there is a point beyond which the system is no longer a superconductor. The reason is that, since the interior of a vortex is essentially a normal region, when the vortices are close together, and the coherence lengths start to overlap between neighboring vortices, superconductivity is lost.

So in type-II superconductors, we expect two different characteristic magnetic fields. In type I, there was a single magnetic field which was the field necessary to destroy superconductivity, H_c. In type II, we can define a lower critical field, H_{c1}, which is the field that gives rise to one flux quantum inside the system, beyond which there is magnetic field inside and the Meissner phase no longer takes place. But the system is still superconducting until a second, in general much larger, field destroys superconductivity, as a result of the overlaps of the vortices. This upper critical field is called H_{c2}. A typical phase diagram is shown in Fig. 17.11.

The two critical fields may be estimated considering that

$$\Phi_0 = H_{c1}\pi\lambda^2, \tag{17.115}$$

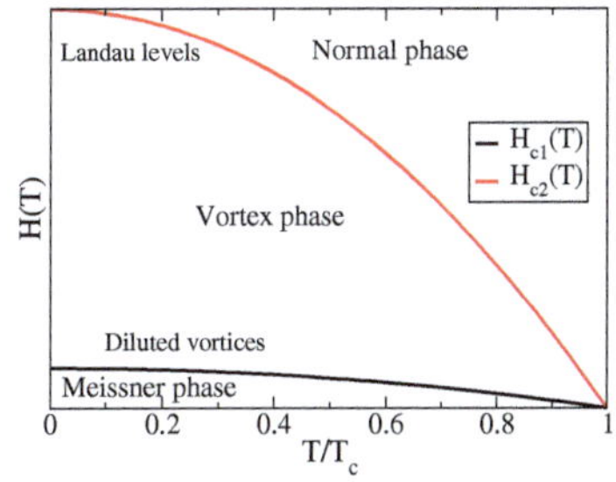

Fig. 17.11 Typical phase diagram of a type-*II* superconductor. At low external magnetic field, H, the B field lines are expelled from the interior leading to the Meissner phase limited above by $H_{c1}(T)$. For larger fields, vortices appear giving rise to the vortex phase limited above by $H_{c2}(T)$. Close to $H_{c2}(0)$, some superconductors may show a Landau-level quantization if κ is very large

which determines the first critical field and

$$\Phi_0 = H_{c2}\pi\xi^2. \tag{17.116}$$

This means that their ratio can be written as

$$\frac{H_{c2}}{H_{c1}} \sim \left(\frac{\lambda}{\xi}\right)^2 \sim \kappa^2. \tag{17.117}$$

It turns out that in type-II superconductors κ may be very large. Typically. H_{c1} is of the order of a few hundreds of gauss, while H_{c2} can range from $1T$ to hundreds of Tesla. For instance, in some nonconventional superconductors with high critical temperatures, $\kappa > 100$.

In type-II superconductors, we expect a large penetration length and a small coherence length. In addition to an intrinsic effect of a superconductor with physical parameters that lead to that result, we see from Eqs. (17.106) and (17.107) that increasing the effective mass of the charges or decreasing their density leads to an enhancement of λ and a decrease of ξ. One way to increase the effective mass of the carriers is to add disorder to a system, leading to an increased effective mass. So we may expect that introducing disorder, or considering dirty superconductors, may be a way to turn the system into a type-II superconductor. Another way, which may be intrinsic, is to consider systems with few charge carriers, which also increases λ. Type-II superconductors are indeed useful because they may support much higher magnetic fields than type-I superconductors, due to the gradual penetration of the magnetic field, while maintaining superconducting order. This is relevant, for instance, in allowing the system to carry larger currents. These currents create a magnetic field in the material itself and, if large enough, will destroy the superconductivity, and the state of no dissipation of energy due to the zero resistance is lost. Also, note that due to the magnetic field, there are vortices in the system. While there is no resistance to the charge transport, the motion of the vortices is resistive, and therefore, some way of pinning the vortices is needed. In that respect, disorder also helps, since it tends to pin the vortices at the locations of the imperfections in the system.

Consider now that we are inside a superconducting phase and that the temperature is small enough that the wave function of the condensate is finite. Let us see now the effect of the phase of the condensate writing again that

$$\psi = \psi_0 e^{i\theta} \tag{17.118}$$

and taking ψ_0 as fixed in space. So any variation of the condensate in space is only due to any contribution from the phase. The free energy reduces to

$$F_s = F_n + \alpha\psi_o^2 + \frac{1}{2}\beta\psi_0^4 + \frac{\hbar^2}{2m^*}\psi_0^2\left(\nabla\theta\cdot\nabla\theta\right)^2 . \tag{17.119}$$

Considering that the amplitude is frozen, the free energy is then only a constant plus a term of the type

$$F_s \sim J\left(\nabla\theta\right)^2 . \tag{17.120}$$

There is a striking similarity between this expression and the Hamiltonian of the classical XX spin model. Indeed, this Hamiltonian may be written as

$$\begin{aligned} H &= -J\sum_{i,j}\cos\left(\theta_i - \theta_j\right) = -J\sum_{i,j}\cos\left(\Delta\theta\right)_{i,j} \\ &= -J\sum_{i,j}\left(1 - \frac{1}{2}\left(\Delta\theta\right)_{i,j}^2 + \cdots\right). \end{aligned} \tag{17.121}$$

17.9 Vortices in a Superconductor

The magnetic field lines in the vortex, and associated magnetic flux, induce a charge current that circulates around the center of the vortex. These supercurrents shield the penetrating magnetic field outside the vortex and give origin to the Meissner effect. Associated with this current, the wave function acquires a phase, the circulation of which gives rise to the current. So in the presence of a vortex (and in general in the presence of a magnetic field), the condensate wave function cannot be taken as real, and a phase has to be included, which is not position-independent. The current is associated with the vector potential, and this is in general nonhomogeneous, and certainly is not homogeneous in the presence of a vortex, since its circulation around the vortex determines the magnetic flux contained in a given closed trajectory around the vortex center.

Let us then consider one vortex in a superconductor.[12] Consider large λ and small ξ, as expected of a strongly type-II superconductor. Since λ is large, the magnetic field decays slowly toward the superconducting region and, since ξ is small, the normal region has a small core. Take $\xi \sim 0$.

The energy can be written as

$$E = \int_{r>\xi} d^3r \frac{1}{2} \left[\mathbf{h}^2 + \lambda_L^2 \left(\nabla \times \mathbf{h} \right)^2 \right]. \tag{17.122}$$

Here, λ_L is the London penetration length

$$\lambda_L = \left(\frac{m^*}{n_s (e^*)^2} \right)^{1/2}. \tag{17.123}$$

The expression for the energy can be understood as follows: The free energy is written as

$$\mathcal{F} = \int F_s d^3r + \int d^3r \frac{1}{2} m v^2 n_s + \int d^3r \frac{h^2}{2\mu}, \tag{17.124}$$

where the first term is the energy of the superconducting state in the absence of vortices and, therefore, no kinetic energy associated with the currents that shield the magnetic field, the second term is the kinetic energy of the free noninteracting particles that constitute the condensate with zero resistance, and the last term is the magnetic energy. We can write that

$$\begin{aligned} \mathbf{j}_s(\mathbf{r}) &= n_s e^* \mathbf{v}(\mathbf{r}) \\ \nabla \times \mathbf{h} &= \mathbf{j}_s. \end{aligned} \tag{17.125}$$

Taking $T = 0$, we get the energy expression considered above.

The equilibrium solution is obtained varying the energy. This leads to London equation

$$\mathbf{h} + \lambda_L^2 \nabla \times (\nabla \times \mathbf{h}(\mathbf{r})) = 0. \tag{17.126}$$

Adding now Maxwell's equation

$$\nabla \times \mathbf{h} = \mathbf{j}_s \tag{17.127}$$

leads to

$$\nabla \times \mathbf{j}_s = -\frac{n_s e^2}{m} \mathbf{h} \tag{17.128}$$

---

[12] For further reading see [deGennes].

and to

$$\mathbf{j} = -\frac{ne^2}{m}\mathbf{A}, \tag{17.129}$$

where we have approximated $\mathbf{j}_n = 0$; the contribution of the normal fluid is negligible. In Landau's gauge, $\nabla \cdot \mathbf{A} = 0$ and $\nabla \cdot \mathbf{j} = 0$, as well.

Consider now that we have a vortex. London equation 17.126 holds for points far from the vortex location $|\mathbf{r}| > \xi$. However, near the vortex, we must include the effect of the flux due to the presence of the vortex. The equation has to be modified to

$$\mathbf{h} + \lambda_L^2 \nabla \times (\nabla \times \mathbf{h}) = \Phi_0 \mathbf{e}_z \delta(\mathbf{r}). \tag{17.130}$$

Here, the vortex is located at the origin and z is the direction of the magnetic field lines. The term that was added guarantees that integrating the equation on a surface, S, perpendicular to the field lines gives

$$\int_S \mathbf{h} \cdot d\mathbf{S} + \lambda_L^2 \oint \nabla \times \mathbf{h} \cdot d\mathbf{l} = \Phi_0, \tag{17.131}$$

where $\mathbf{l}$ is tangent to the border of the surface S and we have used Stokes theorem.

Recalling that

$$\mathbf{j}(\mathbf{r}) = \nabla \times \mathbf{h}, \tag{17.132}$$

we know that for $r >> \lambda_L$ the current vanishes, and so the integral over the area of the field is just the quantum of flux, as required. On the other hand, if $\xi << r << \lambda_L$, the field h is small, and the integral over the area is negligible (also, the area is small). The flux contained in that small area is just a fraction $\Phi_0 r^2/\lambda^2$. We get in this region that

$$\lambda_L^2 2\pi r |\nabla \times \mathbf{h}| = \Phi_0, \tag{17.133}$$

which implies that

$$|\nabla \times \mathbf{h}| = \frac{\Phi_0}{2\pi\lambda_L^2}\frac{1}{r}. \tag{17.134}$$

Using that $|\nabla \times \mathbf{h}| = -dh/dr$, we get

$$h = \frac{\Phi_0}{2\pi\lambda_L^2}\left(\ln\frac{\lambda_L}{r} + C\right) \tag{17.135}$$

if $\xi << r << \lambda_L$. The full solution is

$$h(r) = \frac{\Phi_0}{2\pi\lambda_L^2} K_0\left(\frac{r}{\lambda_L}\right). \tag{17.136}$$

At large distances $r >> \lambda_L$, it reduces to

$$h(r) = \frac{\Phi_0}{2\pi\lambda_L^2}\sqrt{\frac{\pi\lambda_L}{2r}} e^{-\frac{r}{\lambda_L}}. \tag{17.137}$$

We may now calculate the energy of the vortex approximating the whole space by the region inside the core of the vortex, of size ξ, as

$$\begin{aligned} E &= \frac{\lambda_L^2}{2}\int_{\sim\xi} d\mathbf{S}\cdot\mathbf{h}\times(\nabla\times\mathbf{h}) \\ &= \lambda_L^2\pi\xi h(\xi)|\nabla\times\mathbf{h}(\xi)| \\ &= \frac{\Phi_0^2}{4\pi\lambda_L^2}\ln\left(\frac{\lambda_L}{\xi}\right). \end{aligned} \tag{17.138}$$

Here, we have integrated by parts the second term of Eq. (17.122) to obtain a surface term.

Consider now that we have two vortices in the system. This case will allow us to see if there are interactions between the two vortices. We consider one vortex located at $\mathbf{r}_1 = (x_1, y_1)$ and the other vortex at location $\mathbf{r}_2 = (x_2, y_2)$ and the magnetic field is aligned along z, as before. The coherence length is the same at both locations, and call the field at site 1 as $\mathbf{h}_1$ and at site 2 as $\mathbf{h}_2$. The equation for the field is

$$\mathbf{h} + \lambda_L^2\nabla\times(\nabla\times\mathbf{h}) = \mathbf{\Phi}_0\left[\delta(\mathbf{r}-\mathbf{r}_1) + \delta(\mathbf{r}-\mathbf{r}_2)\right], \tag{17.139}$$

where $\mathbf{\Phi}_0 = \Phi_0\mathbf{e}_z$. The total field is given by

$$\mathbf{h}(\mathbf{r}) = \mathbf{h}_1(\mathbf{r}) + \mathbf{h}_2(\mathbf{r}). \tag{17.140}$$

The solutions are of the type

$$\mathbf{h}_i(\mathbf{r}) = \frac{\Phi_0}{2\pi\lambda_L^2} K_0\left(\frac{r - r_i}{\lambda_L}\right). \tag{17.141}$$

The energy of the two vortex system may be approximated by the contributions over the two surfaces of the vortices

$$E = \frac{\lambda_L^2}{2}\int_{|\mathbf{r}-\mathbf{r}_i|=\xi} \mathbf{h}\times(\nabla\times\mathbf{h})\cdot d\mathbf{S}. \tag{17.142}$$

This leads to

$$E = \frac{\lambda_L^2}{2} \int (d\mathbf{S}_1 + d\mathbf{S}_2) \cdot (\mathbf{h}_1 + \mathbf{h}_2) \times (\nabla \times \mathbf{h}_1 + \nabla \times \mathbf{h}_2) \, . \tag{17.143}$$

There are several terms to consider. Two are of the type of the single-vortex case leading to a contribution to the total energy that is basically the sum of the energies of the two separate vortices, E_1:

$$\frac{\lambda_L^2}{2} \left[\int d\mathbf{S}_1 \cdot \mathbf{h}_1 \times (\nabla \times \mathbf{h}_1) + \int d\mathbf{S}_2 \cdot \mathbf{h}_2 \times (\nabla \times \mathbf{h}_2) \right] = 2E_1 . \tag{17.144}$$

Four other terms are basically zero since $\xi << \lambda_L$:

$$\int (\mathbf{h}_1 + \mathbf{h}_2) \cdot [(\nabla \times \mathbf{h}_1) \times d\mathbf{S}_2 + (\nabla \times \mathbf{h}_2) \times d\mathbf{S}_1] \sim 0. \tag{17.145}$$

Finally, there is one term that reflects the interaction between the two vortices. Its contribution to the energy is

$$U_{12} = \frac{\lambda_L^2}{2} \int [\mathbf{h}_1 \times (\nabla \times \mathbf{h}_2) \cdot d\mathbf{S}_2 + \mathbf{h}_2 \times (\nabla \times \mathbf{h}_1) \cdot d\mathbf{S}_1] \, . \tag{17.146}$$

Since $|\nabla \times \mathbf{h}_i| \sim 1/(|r - r_i|)$ (for $|r - r_i| << \lambda_L$), we can write that

$$h_{12} = h_1(r_2) = h_2(r_1) = \frac{\Phi_0}{2\pi \lambda_L^2} K_0 \left(\frac{r_1 - r_2}{\lambda_L} \right) . \tag{17.147}$$

Therefore,

$$U_{12} = \Phi_0 h_{12} > 0 \tag{17.148}$$

is a repulsive force.

This result is consistent with previous considerations, where we concluded that the surface energy is negative and it is convenient energetically to maximize the contact surface between the normal and the superconducting regions. We may therefore anticipate that if we look for vortex-like solutions in a type-I superconductor, the vortices will attract each other to give rise to a collapse of the vortices, which may lead to extended, finite regions, through which the magnetic field penetrates the material. The repulsive energy has the following limits:

$$\begin{aligned} U_{12} &\sim \frac{1}{\sqrt{r_{12}}} e^{-\frac{r_{12}}{\lambda_L}} , \qquad && r_{12} >> \lambda_L \\ U_{12} &\sim \ln \left(|\frac{\lambda_L}{r_{12}}| \right) , \qquad && r_{12} << \lambda_L . \end{aligned} \tag{17.149}$$

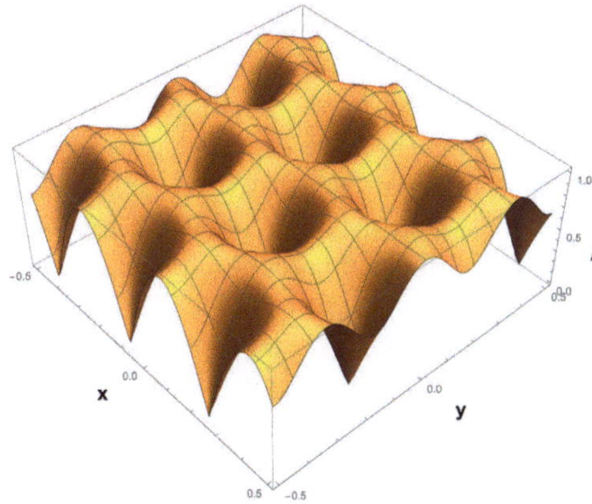

Fig. 17.12 Abrikosov lattice in real space. The order parameter has zeros at the centers of the vortices, organized in a triangular lattice structure due to their mutual repulsive interactions

Due to the repulsive nature of the vortices, they tend to form a regular periodic structure called the Abrikosov lattice. It turns out to be a triangular lattice. It is shown in Fig. 17.12.

In the case of superfluids of atoms with no charge, there is no coupling to the electromagnetic field. Therefore, we get that

$$\psi^* \mathbf{v}_s \psi = \frac{n_s}{m} \left(\hbar \nabla \theta(\mathbf{r}) \right). \tag{17.150}$$

Therefore, the velocity is

$$\mathbf{v}_s = \frac{\hbar}{m} \nabla \theta \tag{17.151}$$

and

$$\nabla \times \mathbf{v}_s = \frac{\hbar}{m} \nabla \times (\nabla \theta) = 0, \tag{17.152}$$

and so the flow is irrotational. In a superfluid, there is in general a superfluid part and a normal part ($T > 0$). We obtained that $\nabla \times \mathbf{v}_s = 0$. It remains to determine the rotational of the velocity of the normal part of the superfluid.

For instance, if we have a container with the superfluid and we impose some rotation (let us say a cylinder rotating around its axis), since the normal part is resistive, due to the friction with the container walls, it starts to rotate, while the superfluid part does not rotate. In the frame that rotates with the cylinder, we get that the Hamiltonian can be written as

$$H'(r', p') = H(r', p') - \boldsymbol{\omega} \cdot \mathbf{L}(r', p'). \tag{17.153}$$

If the frequency of rotation is large enough, there will be a point where it is energetically favorable that the superfluid rotates. This is similar to the case of the superconductor, where the role of the rotation angular momentum is taken by the magnetic field. The free energy in that case is of the type $F' = F - \mathbf{m} \cdot \mathbf{H}$. Considering the angular momentum of the rotation as a function of the rotation frequency, we expect that at low frequencies the angular momentum of the superfluid part vanishes (like the Meissner phase in a superconductor for small enough external magnetic field, H). Increasing the frequency, we may expect the appearance of some vortices (circulating currents), which correspond to the mixed phase of a type-II superconductor. Vortices start to appear ($H > H_{c1}$) and finally a normal fluid only, with no superfluid component, that corresponds to the normal phase of a superconductor, when the external magnetic field is strong enough ($H > H_{c2}$).

If vortices appear due to the rotational motion, then the superfluid contribution to the velocity is now of the type

$$\nabla \times (\nabla\theta) \neq 0, \tag{17.154}$$

which implies a singularity in the phase θ. This is the vortex. Therefore, the circulation of the velocity along a closed loop is nonvanishing:

$$\oint \mathbf{v}_s \cdot d\mathbf{l} = 2\pi\chi. \tag{17.155}$$

Taking a radius r, this leads to

$$2\pi r v_s = 2\pi\chi, \tag{17.156}$$

which means that

$$v_s = \frac{\chi}{r}. \tag{17.157}$$

The quantity is of the form

$$\chi = n\frac{\hbar}{m}, \tag{17.158}$$

where n is an integer. A vortex of lowest vorticity is therefore the case with $n = 1$.

Let us now determine the energy cost of adding one vortex to the system. This is given by

$$\begin{aligned}\Delta E &= \int dV \frac{1}{2}\rho_s v_s^2 \\ &= \frac{1}{2}\rho_2 L_z \int dr 2\pi r v_s^2\end{aligned}$$

$$= L_z \rho_s \pi \chi^2 \int \frac{dr}{r}$$

$$\Delta E = L_z \pi \rho_s \frac{\hbar^2}{m^2} \ln \frac{R}{a}. \tag{17.159}$$

Here, L_z is the size of the vortex along the z direction. So the energy per length is just the result divided by L_z (as considered before for the vortex energy in the superconductor). R is the radius of the container, and a is a cutoff that may be seen as of the order of the atomic distance of the superfluid. We see that the energy ΔE diverges with the system size, unlike in the superconductor where there was an upper cutoff, the penetration length, λ.

If we now calculate the angular momentum, L, we get that

$$L = \int dV \rho_s v_s r = \rho_s \chi \int dV = Lz\pi R^2 \frac{\hbar}{m} \rho_s. \tag{17.160}$$

The condition for the appearance of a vortex is then that the change in energy due to the rotation becomes zero, which means that $L\omega = \Delta E$. We get a critical frequency, ω_c, such that for larger rotation frequencies vortices appear:

$$\omega > \omega_c = \frac{\hbar}{mR^2} \ln \frac{R}{a}. \tag{17.161}$$

Note that, as in the superconductor, the energy depends on the vorticity as $(n\Phi_0)^2$; the same occurs in the superfluid where the energy depends on χ^2, which also leads to a dependence on n^2. So it is energetically favorable to have two separate vortices (energy is twice that of a single vortex, plus the interaction term which is small), while a vortex with vorticity 2 has an energy which is four times larger than that of a vortex. This is another argument that leads to the conclusion that it is not energetically favorable to condense (group) vortices into a larger vortex with larger vorticity (circulation).

Caroli, de Gennes, and Matricon Energy States

The nature of the quantum low-energy states in the presence of a single vortex was solved long ago. There are localized bound states in the vortex core, in addition to states in the continuum. The lowest-energy states for a vortex were obtained analytically by Caroli, de Gennes, and Matricon (CGM). The energy states may be obtained analytically in an approximate way.[13] The energy of the lowest eigenvalues is of the form

$$E_n \sim \frac{\mu \Delta^2}{E_F}. \tag{17.162}$$

[13] See, for instance, [de Gennes].

These are localized states at the vortex core. Moreover, it was obtained that the density of states is $N(\epsilon) \sim N(0)\xi^2$, with the interpretation that the vortex line is equivalent to a normal region of radius ξ, confirming our previous discussion. Since the low-lying states are bound to the vortex, the specific heat should be linear in T, as in a normal conductor.

To obtain the solution, one can start from the Bogoliubov-de Gennes equations in real space

$$\begin{aligned}(H_e + U(\mathbf{r})\, u(\mathbf{r}) + \Delta(\mathbf{r})v(\mathbf{r}) &= \epsilon u(\mathbf{r}) \\ -\left(H_e^* + U(\mathbf{r}\right) v(\mathbf{r}) + \Delta^*(\mathbf{r})u(\mathbf{r}) &= \epsilon v(\mathbf{r}),\end{aligned} \tag{17.163}$$

where

$$H_e = \frac{1}{2m}\left(-i\hbar\nabla - e\mathbf{A}\right)^2 - \mu. \tag{17.164}$$

The magnetic field penetrates along a line. Considering a two-dimensional superconductor, the magnetic field is aligned along the perpendicular direction, say, z-direction. There is therefore cylindrical symmetry with respect to this axis, and so it is convenient to use cylindrical coordinates $\rho, \varphi, andz$. We can use a gauge that fixes the pair potential as $\Delta(\mathbf{r}) = \Delta(\rho)e^{-i\varphi}$. At the origin of the coordinates, the pair potential vanishes $\Delta(\rho = 0) = 0$, and far from the vortex line, the pair potential reaches a given finite value, Δ_∞, the equilibrium value in the bulk of the superconductor. The distance scale over which the order parameter reaches its asymptotic value is the coherence length, ξ. Considering a strongly type-II superconductor, the magnetic field penetration length, λ, is taken as very large, much larger than the coherence length.

The general solution is obtained solving numerically the BdG equations solved using the decompositions [1]

$$\begin{aligned}u^i(\rho,\varphi) &= \sum_{\mu,j} c^i_{\mu,j} e^{i\varphi(\mu-\frac{1}{2})}\phi_{j,\mu-\frac{1}{2}}(\rho) \\ v^i(\rho,\varphi) &= \sum_{\mu,j} d^i_{\mu,j} e^{i\varphi(\mu+\frac{1}{2})}\phi_{j,\mu+\frac{1}{2}}(\rho)\end{aligned} \tag{17.165}$$

and choosing a gauge such that $\Delta(\rho,\varphi) = \Delta(\rho)e^{-i\varphi}$. Here, $\mu = \pm\frac{1}{2}, \pm\frac{3}{2}, \pm\frac{5}{2}, \cdots$ is the planar angular momentum, j is the number of the zero of the Bessel functions in a disc of radius R, and the normalized functions $\phi_{j,m}$ constitute a complete set over the zeros and are defined by

$$\phi_{j,m} = \frac{\sqrt{2}}{R J_{m+1}(\alpha_{j,m})} J_m\left(\alpha_{j,m}\frac{\rho}{R}\right). \tag{17.166}$$

Here, J_m is the Bessel function of order m and $\alpha_{j,m}$ is the jth zero of the Bessel function J_m. The functions $\phi_{j,m}$ by construction are zero at the border of the disc $\rho = R$. In the basis of the functions $\phi_{j,m}$, we have to solve a set of equations for the c and d coefficients of the form

$$\sum_{\mu',j'} \begin{pmatrix} T^{-}_{\mu,j;\mu',j'} & \Delta_{\mu,j;\mu',j'} \\ \Delta^{T}_{\mu,j;\mu',j'} & T^{+}_{\mu,j;\mu',j'} \end{pmatrix} \begin{pmatrix} c_{\mu',j'} \\ d_{\mu',j'} \end{pmatrix} = E \begin{pmatrix} c_{\mu,j} \\ d_{\mu,j} \end{pmatrix}.$$

Here, the various components are diagonal in the angular momentum $T^{\pm}_{\mu,j;\mu',j'} = \delta_{\mu,\mu'} T^{\pm}_{\mu,;j,j'}$ and the same for the off-diagonal terms. The various terms are given by

$$T^{\pm}_{\mu;j,j'} = \mp \left(\frac{\hbar^2}{2m} \left(\frac{\alpha_{j,\mu\pm 1/2}}{R} \right)^2 - E_F \right) \delta_{j,j'}$$

$$\mp \frac{1}{2m} \int_0^R d\rho \rho \phi_{j,\mu\pm 1/2}(\rho) e^2 A^2_{\varphi}(\rho) \phi_{j',\mu\pm 1/2}(\rho)$$

$$- \frac{1}{2m} \int_0^R d\rho \rho \phi_{j,\mu\pm 1/2}(\rho) 2e\hbar(\mu \pm 1/2) \frac{A_{\varphi}(\rho)}{\rho} \phi_{j',\mu\pm 1/2}(\rho)$$

and

$$\Delta_{\mu;j,j'} = \int_0^R d\rho \rho \phi_{j,\mu-1/2}(\rho) \Delta(\rho) \phi_{j',\mu+1/2}(\rho). \tag{17.167}$$

The physical quantities are obtained in the usual way.

The vector potential is given by Maxwell's equations

$$\nabla \times \mathbf{B} = \nabla \times \nabla \times \mathbf{A} = \mathbf{J}_{total}, \tag{17.168}$$

which, in the Coulomb gauge ($\nabla.\mathbf{A} = 0$), is given by

$$\nabla^2 \mathbf{A} = -\mathbf{J}_{total}. \tag{17.169}$$

The current density originated in the supercurrents is obtained self-consistently by

$$\mathbf{J}(\mathbf{r}) = \frac{e\hbar}{im} \sum_i f(E_i) u_i^*(\mathbf{r}) \left[\nabla - \frac{ie}{\hbar} \mathbf{A}(\mathbf{r}) \right] u_i(\mathbf{r})$$

$$+ [(1 - f(E_i))] v_i(\mathbf{r}) \left[\nabla - \frac{ie}{\hbar} \mathbf{A}(\mathbf{r}) \right] v_i^*(\mathbf{r}) - c.c. \tag{17.170}$$

In the cylindrical basis, the order parameter is obtained as

$$\Delta(\rho) = V \sum_{\mu,i(|E_i|\leq\omega_D)} \bar{u}_{\mu,i}(\rho)\bar{v}_{\mu,i}(\rho)\left(1 - 2f(E_{\mu,i})\right), \tag{17.171}$$

with $u_{\mu,i}(\rho,\varphi) = e^{i\varphi(\mu-1/2)}\bar{u}_{\mu,i}(\rho)$ and $v_{\mu,i}(\rho,\varphi) = e^{i\varphi(\mu+1/2)}\bar{v}_{\mu,i}(\rho)$. Also, the current is given by

$$j_\varphi(\rho) = -\frac{2|e|\hbar}{m} Re \sum_{\mu,i} \left(f(E_{\mu,i})\bar{u}_{\mu,i}(\rho)^2 \left((\mu - 1/2)\frac{1}{\rho} + \frac{|e|}{\hbar} A_\varphi(\rho) \right) \right.$$
$$\left. + \ (1 - f(E_{\mu,i}))\bar{v}_{\mu,i}(\rho)^2 \left(-(\mu + 1/2)\frac{1}{\rho} + \frac{|e|}{\hbar} A_\varphi(\rho) \right) \right).$$

The BdG equations are then solved self-consistently. The energy levels show bound states as predicted by Caroli, de Gennes, and Matricon (CGM) together with states in the continuum. Both the coherence length and the penetration length are obtained self-consistently. At low temperatures, $T < T_c/k_F\xi$, where k_F is the Fermi momentum and ξ the coherence length, the quantum limit is reached, and the various physical quantities show oscillations with wave length $\sim 1/k_F$. The eigenfunctions also display the same oscillatory behavior. As the temperature lowers, the size of the vortex shrinks (Kramer and Pesch effect) and decreases until a minimum value is reached. The lowest eigenvalues corresponding to the bound states of CGM are not particle-hole symmetric, as evidenced by the local density of states (LDOS).

A more general solution may be found taking the order parameter in the form

$$\Delta(\mathbf{r}) = \Delta(\rho)e^{-in\varphi}. \tag{17.172}$$

This form describes a magnetic flux equal to n flux quanta ($\Phi = n\Phi_0 = n\frac{h}{2e}$). The wave functions u_i and v_i are expanded in a similar way:

$$u_i(\mathbf{r}) = \sum_{\mu,j} c^i_{\mu,j}\phi_{j,\mu-1/2}e^{i(\mu-1/2)\varphi}, \tag{17.173}$$

$$v_i(\mathbf{r}) = \sum_{\mu,j} d^i_{\mu,j}\phi_{j,\mu-1/2+n}e^{i(\mu-1/2+n)\varphi}. \tag{17.174}$$

The set of values of the angular momentum is given by $\mu = \pm(2l+1)/2$ where $l = 0, 1, 2, \cdots$. It is important to note that the symmetry of the BdG equations

$$u_i(\mathbf{r}) \to v_i^*(\mathbf{r}), \tag{17.175}$$

$$v_i(\mathbf{r}) \to -u_i^*(\mathbf{r}), and \tag{17.176}$$

$$E_i \to -E_i \tag{17.177}$$

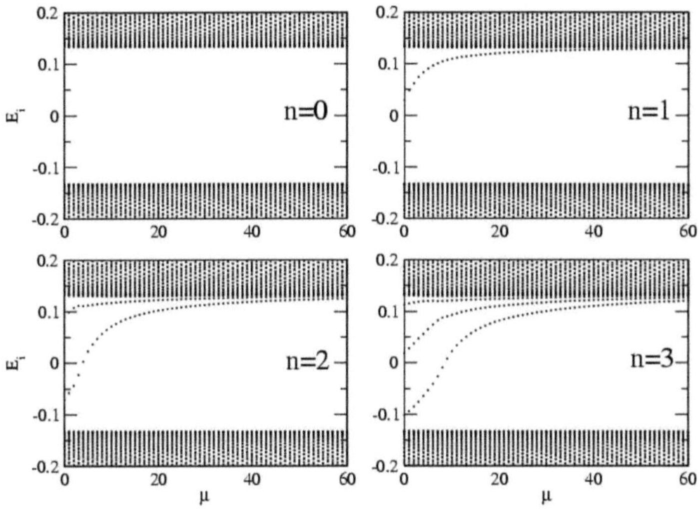

Fig. 17.13 Set of energy eigenvalues as a function of angular momentum, μ, for different vorticities $n = 0, 1, 2, 3$ (Reproduced with permission from Marco Cardoso, Pedro Bicudo, and Pedro D Sacramento, Journal of Physics Condensed Matter **18**, 8623 (2006))

allows to reduce the solution to the positive values of μ. We may obtain the eigenvectors and eigenvalues for negative values of μ using the above symmetry. The terms $(\mu - 1/2)$ and $(\mu - 1/2 + n)$ may be rewritten in a more symmetrical way like $(\mu' - n/2)$ and $(\mu' + n/2)$ if μ' is half-odd integer if n is odd and μ' is integer if n is even.

In Fig. 17.13, we show the numerical solution of the energy eigenvalues as a function of angular momentum, $\mu > 0$, for different vorticities. In the case of $n = 0$, there is no vortex, and the spectrum is just a continuum of states above the superconducting gap (s-wave superconductor). For $n = 1, 2, 3, \cdots, n$, bands of bound states appear inside the gap.

High-Field Regime

Since the critical magnetic field, $H_{c2}(T = 0)$, can be very high in unconventional type-II superconductors, it may be necessary to consider the Landau-level structure of the normal phase eigenstates. This implies that the superconducting state is obtained pairing electron states that take into account the effect of the large magnetic field. Indeed, observation in some superconductors, like high-T_c cuprates, $A15$ superconductors, boro-carbides, many organic superconductors, etc., of oscillations in the magnetic susceptibility, as the magnetic field is varied (the de Haas-van Alphen effect), have confirmed this idea. In conventional type-II systems (like Nb), this effect is expected to be negligible. This is so because, within the BCS theory, the scale of the cyclotron splitting between LLs near $H_{c2}(0)$, $\omega_c \sim \omega_{c2}(0) \equiv eH_{c2}(0)/m^*$, is set by the condensation energy, $\sim T_{c0}^2/E_F$, and should be much smaller than either the thermal smearing, $\sim T$, or the gap $\Delta(T, H)$, the scale for both being T_{c0}.

Including the Landau-level (LL) structure of the normal state leads to a set of Bogoliubov-de Gennes (BdG) equations for the LL quantized quasiparticles in the presence of the gap function $\Delta(\mathbf{r})$, describing the Abrikosov vortex lattice.

The field operators are written as

$$\begin{aligned}\psi_\uparrow(\mathbf{r}) &= \sum_{\nu,\mathbf{q}} \left(u^\nu_\mathbf{q}(\mathbf{r})\gamma_{\nu,\mathbf{q},\uparrow} - v^{*\nu}_\mathbf{q}(\mathbf{r})\gamma^\dagger_{\nu,\mathbf{q},\downarrow} \right)\\ \psi_\downarrow(\mathbf{r}) &= \sum_{\nu,\mathbf{q}} \left(u^\nu_\mathbf{q}(\mathbf{r})\gamma_{\nu,\mathbf{q},\downarrow} + v^{*\nu}_\mathbf{q}(\mathbf{r})\gamma^\dagger_{\nu,\mathbf{q},\uparrow} \right)\end{aligned} \tag{17.178}$$

and

$$\epsilon_\beta = \sum_{\sigma,\nu,\mathbf{q}} \epsilon^\nu_\mathbf{q} \gamma^\dagger_{\nu,\mathbf{q},\sigma}\gamma_{\nu,\mathbf{q},\sigma}. \tag{17.179}$$

Here, $\gamma^\dagger_{\nu,\mathbf{q},\sigma}$ creates a quasiparticle in the level ν, with momentum $\mathbf{q}$ and spin σ. The problem is diagonal in $\mathbf{q}$ (vectors of the magnetic Brillouin zone) due to the periodicity of the Abrikosov vortex lattice. The amplitudes u and v are expanded in the Landau basis (n is the Landau level) as

$$\begin{aligned}u^\nu_\mathbf{q}(\mathbf{r}) &= \sum_n u^\nu_{\mathbf{q}n}\phi_{\mathbf{q}n}(\mathbf{r})\\ v^{*\nu}_\mathbf{q}(\mathbf{r}) &= \sum_n v^{*\nu}_{\mathbf{q}n}\phi_{-\mathbf{q}n}(\mathbf{r}).\end{aligned} \tag{17.180}$$

Here, $\phi_{\mathbf{q}n}(\mathbf{r})$ are the eigenfunctions of the magnetic translation group in the Landau gauge ($A_x = -Hy, A_y = 0, A_z = 0$) belonging to the nth Landau level. The amplitudes $u^\nu_{\mathbf{q}n}$ and $v^\nu_{\mathbf{q}n}$ are the solutions of the BdG equations, and $\epsilon^\nu_\mathbf{q}$ are the energy eigenvalues. The BdG equations to be solved are

$$\begin{aligned}\epsilon_n u^\nu_{\mathbf{q}n} + \sum_m \Delta_{nm}(\mathbf{q}) v^\nu_{\mathbf{q}m} &= \epsilon^\nu_\mathbf{q} u^\nu_{\mathbf{q}n}\\ -\epsilon_n v^\nu_{\mathbf{q}n} + \sum_m \Delta^*_{nm}(\mathbf{q}) u^\nu_{\mathbf{q}m} &= \epsilon^\nu_\mathbf{q} v^\nu_{\mathbf{q}n},\end{aligned} \tag{17.181}$$

where $\epsilon_n = (n + 1/2) - \mu/\hbar\omega_c$ (μ is the chemical potential) and $\Delta_{nm}(\mathbf{q})$ is the matrix element of the order parameter $\Delta(\mathbf{r})$ between electronic states $(\mathbf{q}, n)$ and $(-\mathbf{q}, m)$.

If the field is high, the splitting of the Landau levels is large. Close to the transition to the normal phase, the amplitude of the order parameter is small, and it is natural that the dominant contribution is to only take the diagonal terms of the matrix Δ_{nm}. In the diagonal approximation (DA), these equations are easily solved

$$\epsilon_{\mathbf{q}}^{\nu} = \pm\sqrt{\epsilon_n^2 + |\Delta_{nn}(\mathbf{q})|^2} \equiv \epsilon_{\mathbf{q}}^{n}$$
$$u_{\mathbf{q}n} = \frac{1}{\sqrt{2}}\left(1 + \frac{\epsilon_n}{\epsilon_{\mathbf{q}}^{\nu}}\right)^{1/2}$$
$$v_{\mathbf{q}n} = \frac{1}{\sqrt{2}}\left(1 - \frac{\epsilon_n}{\epsilon_{\mathbf{q}}^{\nu}}\right)^{1/2}, \tag{17.182}$$

(and therefore, $\nu \equiv n$). In this case,

$$u_{\mathbf{q}n}^{\nu} v_{-\mathbf{q}n}^{*\nu} = \frac{\Delta_{nn}(\mathbf{q})}{2\epsilon_{\mathbf{q}}^{n}}. \tag{17.183}$$

As Δ grows, the off-diagonal terms become increasingly important, and the problem has to be solved numerically.

The solution of the energy of the lowest band above the Fermi level, in the DA, as a function of $\mathbf{q}$ shows that the energy of the lowest band above the Fermi level is given by the gap function $|\Delta_{nn}(\mathbf{q})|$. This is given by

$$\Delta_{nn}(\mathbf{q}) = \frac{\Delta}{\sqrt{2}}\frac{(-1)^n}{2^{2n}n!}\sum_k e^{2ikq_ya-(q_x+\pi k/a)^2l^2} H_{2n}\left[\sqrt{2}(q_x + \pi\frac{k}{a})l\right], \tag{17.184}$$

where H are Hermite polynomials. The gap $|\Delta_{nn}(\mathbf{q})|$ has zeros in the magnetic Brillouin zone (associated with the magnetic translation group, previously discussed) in momentum points (q_j), which are in direct correspondence with the positions of the vortices in real space (z_i), such that $q_jl = z_i/l$. The gapless points grow considerably in number as n_c grows. Including off-diagonal terms, small gaps open in the spectrum. However, throughout the Brillouin zone, the numerical prefactors are very small, and the spectrum is similar almost everywhere to the one obtained in the DA. Therefore, due to the high magnetic field present, the spectrum has gapless points, even though the superconductor is of the usual type considered in the simpler BCS theory here described. The spectra in the diagonal approximation, taking into account off-diagonal terms, are shown in Fig. 17.14, for a typical case. As we will see next, there are other possibilities for the pairing of the Cooper pairs that allow gapless spectra without the need to apply a magnetic field.

As mentioned before, in an s-wave superconductor, isolated vortices contribute a linear term to the low-temperature specific heat. This is due to the core states which have a density of states which can be interpreted as resulting from a normal cylinder, of radius the coherence length. In the high-field regime, the low-T specific heat has been calculated also yielding a linear specific heat $C_V = \gamma T$ but where $\gamma(H)$ is a nonlinear function of the magnetic field.

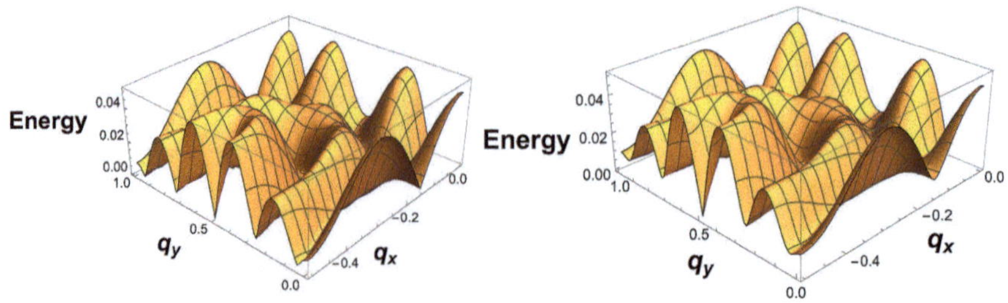

Fig. 17.14 Energy bands in the high-field regime, considering that the chemical potential is located near $n_c = 10$ and integrating over $|m - n_c| = 4$ Landau levels (taken as the Debye energy range) in (left) diagonal approximation and (right) considering the effect of the off-diagonal terms due to the coupling to the other LL

17.10 Electron Pairing Matrix

In the discussion of the BCS theory, arguments were presented that justified the choice that, in the Cooper pairs, the two electrons are in a spin singlet state and that have opposite momenta. So in the pair, one electron has spin $\uparrow$, and the other electron has spin $\downarrow$, and their momenta are $\mathbf{k}$ and $-\mathbf{k}$, which implies that the center of mass momentum vanishes. Also, to simplify, it was chosen that the pairing is just a constant, independent of momentum.

The symmetry of the pairing allows other choices. Since the electrons are fermions, their wave function is antisymmetric. The choice of a spin singlet implies a symmetric spatial part. On the other hand, symmetry allows to choose a spin triplet wave function which opens the door to an antisymmetric spatial part. Actually, the pairing resulting from the effective attractive interaction mediated in the basic theory by phonons (actually there may be other possibilities such as magnetic excitations and others to mediate the pairing between the electrons) implies that there may be some temporal dependence in the interaction, assumed instantaneous in the theory we have presented here. This means that the wave function actually has three parts: the spatial and spin parts already considered but also a temporal part. In the same way as the spatial and spin parts, being maybe symmetrical or antisymmetrical, the temporal part may also be taken as symmetrical or antisymmetrical. This increases the number of choices with just the overall restriction that the product of the three parts is antisymmetric. In particular, choosing the temporal part as antisymmetric allows the possibility that both the spin and spatial parts are symmetric or antisymmetric. In the following, we will take that the temporal part is symmetric and further simplify the treatment by assuming there is no time dependence and the interactions are instantaneous.

We can write the interaction responsible for superconducting pairing as follows:[14]

[14] For further reading see, for instance, [AraujoSacramento].

$$\hat{V} = \sum_{\mathbf{k},\mathbf{k}'} V_{\sigma_1\sigma_2\sigma_3\sigma_4}(\mathbf{k}, \mathbf{k}')\hat{a}^{\dagger}_{\mathbf{k},\sigma_1}\hat{a}^{\dagger}_{-\mathbf{k},\sigma_2}\hat{a}_{-\mathbf{k}',\sigma_3}\hat{a}_{\mathbf{k}',\sigma_4} \tag{17.185}$$

where $\hat{a}^{\dagger}_{\mathbf{k},\sigma_1}$ creates an electron with momentum $\mathbf{k}$ and spin projection σ_1. We are restricting the pairing to the case when the total momentum of a Cooper pair vanishes, as this is the most energetically favorable in a homogeneous system. Therefore, the interaction scatters a pair from momentum $\mathbf{k}$ to a momentum $\mathbf{k}'$. It has the following properties:

$$V_{\sigma_1\sigma_2\sigma_3\sigma_4}(\mathbf{k}, \mathbf{k}') = -V_{\sigma_2\sigma_1\sigma_3\sigma_4}(-\mathbf{k}, \mathbf{k}'), \tag{17.186}$$

$$V_{\sigma_1\sigma_2\sigma_3\sigma_4}(\mathbf{k}, \mathbf{k}') = -V_{\sigma_1\sigma_2\sigma_4\sigma_3}(\mathbf{k}, -\mathbf{k}'), \tag{17.187}$$

$$V_{\sigma_1\sigma_2\sigma_3\sigma_4}(\mathbf{k}, \mathbf{k}') = V^*_{\sigma_4\sigma_3\sigma_2\sigma_1}(\mathbf{k}', \mathbf{k}). \tag{17.188}$$

In the spirit of the mean-field approach, we can now decouple the interaction and introduce the anomalous coupling terms in (17.185) as

$$\hat{V}_{eff} = \sum_{\mathbf{k},\mathbf{k}'} \Big[V^*_{\sigma_1\sigma_2\sigma_3\sigma_4}(\mathbf{k}, \mathbf{k}')\langle \hat{a}^{\dagger}_{\mathbf{k}',\sigma_4}\hat{a}^{\dagger}_{-\mathbf{k}',\sigma_3}\rangle \hat{a}_{-\mathbf{k},\sigma_2}\hat{a}_{\mathbf{k},\sigma_1}$$
$$+ V_{\sigma_1\sigma_2\sigma_3\sigma_4}(\mathbf{k}, \mathbf{k}')\langle \hat{a}_{-\mathbf{k}',\sigma_3}\hat{a}_{\mathbf{k}',\sigma_4}\rangle \hat{a}^{\dagger}_{\mathbf{k},\sigma_1}\hat{a}^{\dagger}_{-\mathbf{k},\sigma_2}\Big], \tag{17.189}$$

whereby we define

$$\Delta_{\sigma_1\sigma_2}(\mathbf{k}) \equiv \sum_{\mathbf{k}'} V_{\sigma_1\sigma_2\sigma_3\sigma_4}(\mathbf{k}, \mathbf{k}')\langle \hat{a}_{-\mathbf{k}',\sigma_3}\hat{a}_{\mathbf{k}',\sigma_4}\rangle . \tag{17.190}$$

It is possible to see, then, that $\Delta_{\sigma_2\sigma_1}(\mathbf{k}) = -\Delta_{\sigma_1\sigma_2}(-\mathbf{k})$ or

$$\hat{\Delta}(\mathbf{k}) = -\hat{\Delta}^T(-\mathbf{k}), \tag{17.191}$$

where the matrix $\hat{\Delta}$ has spin indices. The equality (17.191) follows immediately from the fact that we are pairing up fermionic operators:

$$\hat{\psi}_\alpha \Delta_{\alpha\beta}\hat{\psi}_\beta = -\hat{\psi}_\beta \Delta_{\alpha\beta}\hat{\psi}_\alpha = -\hat{\psi}\Delta^T\hat{\psi}, \tag{17.192}$$

(where the indices denote space or spin variables), and this implies the antisymmetry property $\Delta = -\Delta^T$.

If the Cooper pair is a singlet, then $\hat{\Delta}$ is an even function of $\mathbf{k}$: $\Delta_{\sigma_2\sigma_1}(\mathbf{k}) = -\Delta_{\sigma_1\sigma_2}(\mathbf{k})$; therefore, it is antisymmetric in the spin indices. We write

$$\Delta_{\sigma_2\sigma_1}(\mathbf{k}) = \begin{pmatrix} 0 & \psi(\mathbf{k}) \\ -\psi(\mathbf{k}) & 0 \end{pmatrix}_{\uparrow,\downarrow} = \psi(\mathbf{k}) i\sigma_y \quad \text{, with } \psi(\mathbf{k}) \text{ even.} \tag{17.193}$$

If the Cooper pair is a triplet, then $\hat{\Delta}$ in an odd function of $\mathbf{k}$ and the matrix $\hat{\Delta}$ is symmetric in the spin indices: $\Delta_{\sigma_1\sigma_2}(\mathbf{k}) = \Delta_{\sigma_2\sigma_1}(\mathbf{k})$. We write

$$\hat{\Delta}(\mathbf{k}) = \hat{\Delta}^T(\mathbf{k}) = i\boldsymbol{d}(\mathbf{k}) \cdot \boldsymbol{\sigma}\sigma_y \tag{17.194}$$

$$= \begin{pmatrix} -d_x(\mathbf{k}) + id_y(\mathbf{k}) & d_z(\mathbf{k}) \\ d_z(\mathbf{k}) & d_x(\mathbf{k}) + id_y(\mathbf{k}) \end{pmatrix}_{\uparrow,\downarrow}, \quad \text{with } \boldsymbol{d}(\mathbf{k}) \text{ odd}. \tag{17.195}$$

If time-reversal symmetry is present, then

$$i\sigma_y \hat{\Delta}^*(-\mathbf{k})(-i\sigma_y) = \hat{\Delta}(\mathbf{k}).$$

This implies that the functions $\psi(\mathbf{k})$ and $\boldsymbol{d}(\mathbf{k})$ are real. Conversely, adding imaginary parts implies a system that breaks time-reversal invariance.

In the continuum limit, ignoring for now that the system is actually a lattice with some symmetry spatial group behind it, the functions $\psi(\mathbf{k})$ and $\boldsymbol{d}(\mathbf{k})$ may be expanded in a basis of spherical harmonics and therefore in the basis of angular momentum. The symmetric function $\psi(\mathbf{k})$ has components in the even angular momenta $l = 0, 2, 4, \cdots$, and the antisymmetric function $\boldsymbol{d}(\mathbf{k})$ has components in the odd angular momenta $l = 1, 3, 5, \cdots$. A similar angular momentum expansion may be carried out in the presence of a lattice symmetry. Typically, an expansion in powers of the momentum is replaced by an expansion with trigonometric functions of the momentum (and its multiples).

The simplest case, considered before when the BCS theory was presented, is the one when the spin part is a singlet and the function $\psi(\mathbf{k})$ is a constant, which means $l = 0$. These are called s-wave superconductors. Other possibilities are obtained choosing $l = 2, 4, \cdots$. While many superconductors are of the traditional class, other superconductors, called nonconventional, fall in the class of $l = 2$, d-wave superconductors, such as the high-temperature cuprate superconductors discovered in 1986. In the class of the triplet superconductors, the simplest case corresponds to $l = 1$, p-wave superconductors. These superconductors are harder to find in nature, even though some examples seem to be in this class, such as Sr_2RuO_4, UPt_3, and $Cu_xBi_2Se_3$. Other materials, called pnictides, are multiband systems that seem to be well described by s-wave pairing but such that the pairing function has opposite signs on the two bands.

In some cases, the pairing is such that at some points or lines it may be possible that at the Fermi surface there are gapless points, because there may be zeros in the pairing function. An example is a two-dimensional d-wave superconductor with a pairing of the type

$$\psi(\mathbf{k}) = \Delta\left(\cos(k_x a) - \cos(k_y a)\right), \tag{17.196}$$

called $d_{x^2-y^2}$ pairing. If the normal state is described by some tight-binding band on a square lattice, we find that the Fermi surface is defined by $\cos(k_x a) +$

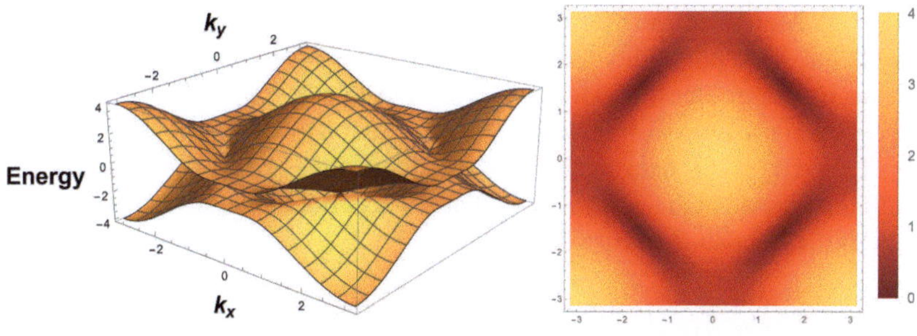

Fig. 17.15 Energy band for a $d_{x^2-y^2}$ superconductor showing that the spectrum has gapless points. These are found when the gap function has zeros on the Fermi surface. In the left panel, the particle and hole bands are shown. In the right panel, the positive energy band is shown highlighting the gapless points

$\cos(k_ya) - E_F = 0$. Intersecting the Fermi surface with the zeros of $\psi(\mathbf{k}) = \Delta\left(\cos(k_xa) - \cos(k_ya)\right)$, we find four gapless points at $k_x = k_y = \pi/(2a)$ and symmetric points. The energy band is shown in Fig. 17.15.

17.11 Goldstone and Higgs Modes

In the superconducting phase, the complex order parameter (in the microscopic description) or the condensate wave function (in the macroscopic description) acquires a finite value. Since they are complex, they have both an amplitude and a phase. The Landau-Ginzburg free energy is a function of the real and imaginary parts of the order parameter, and the general picture is that it has a minimum in the region of parameter space where the equilibrium order parameter takes place. In the real part, or in the imaginary part, of the order parameter, one may consider deviations from the point corresponding to the equilibrium value. The ordered state breaks $U(1)$ invariance, and so there is a global phase that takes a particular value. Changing the value of the phase does not affect the free energy. As a consequence, we may change the phase, keeping the amplitude of the equilibrium order parameter, with no change in the free energy. This is the so-called Goldstone mode associated with the breaking of the continuous symmetry. We may also consider deviations of the amplitude of the order parameter away from the equilibrium value. These deviations are perpendicular to the direction of the Goldstone mode and in general cost a finite energy. This is called the Higgs mode.

As a function of the momentum of the excitations around the equilibrium solution, we expect therefore gapless phase modes (Goldstone modes) and gapped amplitude modes (Higgs modes). It turns out that the Higgs mode has a minimum finite energy, that is, the energy of breaking a pair as given by the BCS theory, therefore is immersed in the continuum of quasiparticle excitations, and is difficult to observe. Moreover, they have no charge, electric dipole moment or magnetic moment, but some possibility to observe these moments has been achieved through coupling to charge density waves. On the other hand, the Goldstone modes in a superconductor acquire a finite energy due to its coupling to the electromagnetic field via the Higgs mechanism, and this energy is of the order of the plasma energy. Therefore, become high energy modes that are also not easily accessible.

Recall the Landau-Ginzburg functional

$$f(\psi) = f_0 + \alpha|\psi(\mathbf{r})|^2 + \frac{\beta}{2}|\psi(\mathbf{r})|^4 + \frac{1}{2m^*}|\left(-i\nabla - e^*\mathbf{A}\right)\psi(\mathbf{r})|^2, \tag{17.197}$$

where $\alpha = \alpha_0(T - T_c)$ and $\mathbf{A}$ is the vector potential. In the ordered phase, $\alpha < 0$, where the global $U(1)$ symmetry is spontaneously broken. Let us write the order parameter in terms of its amplitude and phase, taking into account that the amplitude may deviate from its equilibrium value by $H(\mathbf{r})$, a small amplitude deviation:

$$\psi(\mathbf{r}) = (\psi_0 + H(\mathbf{r}))\, e^{i\theta(\mathbf{r})}. \tag{17.198}$$

The value ψ_0 is the ground state amplitude value and $\theta(\mathbf{r})$ are the phase fluctuations. Inserting this decomposition in the expression for the free energy and recalling that $|\psi_0|^2 = -\alpha/\beta$, we get

$$\begin{aligned} f &= f_0 + \frac{1}{2m^*}(\nabla H)\cdot(\nabla H) - 2\alpha H^2 \\ &+ \frac{(e^*)^2}{2m^*}\left(\mathbf{A} - \frac{1}{e^*}\nabla\theta\right)\cdot\left(\mathbf{A} - \frac{1}{e^*}\nabla\theta\right)(\psi_0 + H)^2 . \end{aligned} \tag{17.199}$$

We see that the H (amplitude) mode is massive with a mass of the order of $\sqrt{-\alpha}$ and the θ phase mode is gapless (as expected of a Goldstone mode). However, the phase variable θ appears associated with the vector potential. The phase may be absorbed defining an effective vector potential as

$$\mathbf{A}' = \mathbf{A} - \frac{\nabla}{e^*}. \tag{17.200}$$

In this way, the phase mode is absorbed in the longitudinal components of the electromagnetic field (along the vector potential and recalling the Landau/London transverse gauge $\mathbf{q}\cdot\mathbf{A}$). The free energy is written

$$f = f_0 - 2\alpha H^2 + \frac{1}{2m^*}(\nabla H)^2 + \frac{(e^*)^2\psi_0^2}{2m^*}\left(\mathbf{A}'\right)^2 + \frac{(e^*)^2\psi_0}{m^*}\left(\mathbf{A}'\right)^2 H + \cdots \tag{17.201}$$

The photons acquire a mass of the order of

$$\sqrt{\frac{(e^*)^2\psi_0^2}{m^*}}, \tag{17.202}$$

via the Anderson-Higgs mechanism. The result also reveals that there is no linear coupling between $\mathbf{A}$ and H. But there is a nonlinear coupling of the type A^2H, which involves a coupling between two photons and a Higgs mode.

The amplitude mode may also be seen on the equation for the current since

$$\mathbf{j} = -\frac{\partial f}{\partial \mathbf{A}} = -\frac{ie^*}{2m^*}\left(\psi^\dagger \nabla \psi - \left(\nabla \psi^\dagger\right)\psi\right) - \frac{(e^*)^2}{m^*}\mathbf{A}\psi^\dagger\psi. \tag{17.203}$$

Using the decomposition $\psi = (\psi_0 + H)\,e^{i\theta}$ leads to a contribution to the current, in addition to that obtained from the ground state wave function, ψ_0, which is given by

$$\mathbf{j} = -\frac{2e^*\psi_0}{m^*}\mathbf{A}H. \tag{17.204}$$

It is also of the London form, proportional to the vector potential.

17.12 Metals, Insulators, and Superconductors

At zero temperature, the distinction between metals and insulators is usually clear, such as the absence or existence of a gap where the chemical potential is located, leading to finite or vanishing conductivity, respectively. This classification is particularly clear in the context of band theory as applied to noninteracting electronic systems.

When we consider interacting problems, the energy spectrum becomes more involved, and the appearance of gaps may be the result of the lattice or the effect of interactions. Even if experimentally it may be possible to determine the nature of the system as metal or insulator, it is useful to introduce a criterion to determine, from a theoretical calculation, the nature of the system. Particularly, often, a numerical calculation is needed to solve the problem.

In addition to the distinction between metals and conductors (or semiconductors, semimetals, etc.), there is also the issue of identifying a superconductor. One way is the measurement of the resistivity at low temperatures and its vanishing (or infinite conductivity). However, from the theoretical point of view, this result is very similar to an ideal conductor. A better characterization of a superconductor is through the Meissner effect and in general the electromagnetic response of the system. In general, it is important to study the momentum-frequency response of the system to an external perturbation and to take appropriate limits and their sequence, as we approach zero momentum and/or zero frequency. The different limits in general do not commute, and the differences may be used to distinguish metals from conductors and from superconductors. The general procedure relies on the linear response theory, previously described.

Let us begin with a simple criterion to distinguish a metal from an insulator considering the ground state of a many-body system. Diagonalizing a system with N_a sites (taken as large) and with N electrons, we may calculate the energy, E, and two quantities:

$$\begin{aligned}\mu_+ &= E(N+1) - E(N)\\ \mu_- &= E(N) - E(N-1).\end{aligned} \tag{17.205}$$

The criterion is that if $\mu_+ = \mu_-$ the system is a conductor and if $\mu_+ > \mu_-$ the system is an insulator. In other words, in the first case, the spectrum is gapless, since there is a continuum of states as we add or subtract one particle from the system, and in the second case, there is a gap, and so a finite energy is required to add one particle to the system. Using this criterion, it was determined that in the case of the one-dimensional Hubbard model, in general, the system is in a metallic state. However, at half-filling (one electron per site) for the Hubbard interaction $U = 0$, we get that $\mu_+ = \mu_-$, indicative of a metallic state, while for any nonzero interaction, one finds $\mu_+ > \mu_-$, indicative of an insulating state. This means that there is no critical interaction strength, U_c, beyond which there is a metal-insulator transition. In the case of a two-dimensional Hubbard model, such a phase transition (Mott-Hubbard transition) occurs at a finite U_c.

Let us now turn to the case of a superconductor.[15] The main ideas are already in London's theory. The superconducting, non-dissipative current is proportional to the vector potential and satisfies[16]

$$\mathbf{j} = -\frac{n_s e^2}{mc}\mathbf{A}. \tag{17.206}$$

If the current is conserved, this leads to $\nabla \cdot \mathbf{A} = 0$, a transverse field. There is also a contribution to the current, called paramagnetic current, that satisfies

$$\mathbf{j}^p(\mathbf{r}) = -\frac{e\hbar}{2mi}\sum_{j=1}^{N_s}\int d^2r_1\cdots d^3r_{N_s}\left(\psi_s^*\nabla_j\psi_s - \psi_s\nabla_j\psi_s^*\right)\delta(\mathbf{r}_j-\mathbf{r}). \tag{17.207}$$

[15] For further reading see [Schrieffer], Theory of Superconductivity; D. J. Scalapino, S. R. White, and S. C. Zhang, Physical Review Letters **68**, 2830 (1992); Douglas J. Scalapino, Steven R. White, and Shoucheng Zhang, Physical Review B **47**, 7995 (1993). Also, we follow private notes by Nuno M.R. Peres.

[16] In this section, we use gaussian CGS units.

This current vanishes if there is no applied vector potential. A main idea of London is that if a weak vector potential field is applied, the superconducting wave function, ψ_s, is unaffected to first order, and therefore, it is rigid with respect to the external perturbation. Therefore, the paramagnetic current to leading order still vanishes. The non-vanishing contribution to the current, called diamagnetic current, may be written as

$$\mathbf{j}_s(\mathbf{r}) = -\sum_{j=1}^{N_s} \frac{e^2}{mc}\mathbf{A}(\mathbf{r}) \int \psi_s^* \psi_s \delta(\mathbf{r}_j - \mathbf{r}) d^3 r_1 \cdots d^3 r_{N_s} = -\frac{n_s e^2}{mc}\mathbf{A}(\mathbf{r}), \tag{17.208}$$

which is not zero in the superconductor. In a normal metal, the paramagnetic current does not vanish, and the paramagnetic and diamagnetic currents cancel each other. In an insulator, the paramagnetic current does not vanish since there are interband matrix elements that may be large.

As mentioned before, we may recall the minimal coupling that results from the action of a vector potential which leads to the expression for the average current

$$\langle \mathbf{j} \rangle = -\langle ne\mathbf{v} \rangle = -\frac{ne}{m}\left\langle \mathbf{p} + \frac{e}{c}\mathbf{A} \right\rangle = -\frac{ne^2}{mc}\mathbf{A}, \tag{17.209}$$

since $\langle \mathbf{p} \rangle = 0$, due to the rigidity of ψ_s, and therefore, we have perfect diamagnetism, reconfirming Eq. (17.206). Also, recalling that $\nabla \cdot \mathbf{B} = 0$, we see that the magnetic field is transverse, and therefore, applying a magnetic field only affects the transverse excitations of the system.

The paramagnetic current may be understood from a term added to the Hamiltonian (or Lagrangian) that is linear in the vector potential, since the current is obtained from the Hamiltonian as a functional derivative with respect to the vector potential. Specifically, we recover the usual coupling of the type

$$H^p = -\frac{1}{c}\int \sum_{\mu} j^p_\mu(x) A_\mu(x) d^3 r = -\frac{1}{2}\int \left(\mathbf{j}^p(x) \cdot \mathbf{A}(x) - \rho_e(x) c\varphi(x)\right) d^3 r, \tag{17.210}$$

where $\mu = 1, 2, 3, 0$, $A_\mu(x) = (\mathbf{A}(x), c\varphi(x))$ with $x = (\mathbf{r}, t)$ and t is time. The linear dependence on the vector potential of the diamagnetic term suggests a quadratic dependence of the perturbing Hamiltonian, leading to

$$H^d = -\frac{e}{2mc^2}\int \rho_e(x) \sum_{i=1}^{3} A_j^2(x) d^3 r, \tag{17.211}$$

where $\rho_e(x)$ is the electronic density at point $\mathbf{r}$ and time t.

Linear Response and Gauge Invariance
The space- and time-dependent perturbation leads to a response of the system that may be written as

$$J_\mu(x) = -\frac{e}{4\pi}\sum_\nu \int K_{\mu\nu}(\mathbf{r}, t; \mathbf{r}', t')A_\nu(\mathbf{r}', t')d^3r'dt'. \tag{17.212}$$

The explicit behavior of the response will be considered next.

It is convenient at this point to make some general remarks. A first remark is that London's equation

$$j_i(x) = -\frac{ne^2}{mc}A_i(x), \tag{17.213}$$

with $i = x, y, z = 1, 2, 3$, is not gauge-invariant. But the choice $\nabla \cdot \mathbf{A} = 0$ imposes the rule that we should only consider the transverse part, fixing the gauge. (Actually, any longitudinal part of the vector potential couples to longitudinal excitations, which implies that the paramagnetic terms in general will not vanish. The rule is that they are not considered.)

If there is translational invariance (in space and time), it is convenient to transform to momentum and frequency space. If $A_\mu(q) = (\mathbf{A}(q), c\varphi(q))$, we can define a reciprocal space current kernel and

$$J_\mu(q) = -\frac{c}{4\pi}\sum_\nu K_{\mu\nu}(q)A_\nu(q), \tag{17.214}$$

where $q = (\mathbf{q}, q_0 = \omega)$. We will see later that we can write

$$K_{\mu\nu}(q) = \frac{4\pi}{c^2}R_{\mu\nu}(q) + \frac{1}{\lambda_L^2}\delta_{\mu,\nu}\left(1 - \delta_{\nu,0}\right), \tag{17.215}$$

where λ_l is the London penetration length. The Meissner effect requires that the transverse part of $K_{\mu\nu}$ remains finite taking a static situation ($q_0 = \omega = 0$) and taking the long distance limit obtained taking $\mathbf{q} \to 0$. This guarantees the magnetic field is expelled from the interior of the superconductor.

The gauge transformation $A_\mu(q) \to A_\mu(q)'iq_\mu\Lambda(q)$ when introduced in Eq. (17.214), and imposing gauge invariance for the current leads to

$$\sum_{\nu=0}^{3} K_{\mu\nu}(q)q_\nu = 0. \tag{17.216}$$

Considering the static case $q_0 = 0$, it simplifies to

$$\sum_j K_{ij}q_j = 0. \tag{17.217}$$

On the other hand, charge conservation

$$\nabla \cdot \mathbf{j} + \frac{\partial \rho}{\partial t} = 0 \tag{17.218}$$

leads to

$$\sum_{\mu=0}^{3} q_\mu j_\mu(q) = 0. \tag{17.219}$$

Inserting in the same Eq. (17.214), we get, taking again the static case,

$$\sum_i q_i K_{ij} = 0. \tag{17.220}$$

These restrictions impose that the kernel should take the form

$$K_{ij}(\mathbf{q}, 0) = \left(\delta_{ij} - \frac{q_i q_j}{q^2}\right) K(q^2). \tag{17.221}$$

The last factor should have only a dependence on the modulus of $\mathbf{q}$, due to rotational invariance of the wave function, ψ_s. The form of the kernel results from the condition of being purely transverse, in agreement with the transverse gauge used, $\mathbf{q}\cdot\mathbf{A}(q) = 0$. Also, the Meissner effect imposes a finite response in the long-distance, long-wavelength limit, and so we get that

$$K(q^2) > 0, \qquad q^2 \to 0. \tag{17.222}$$

Lattice Description

It is useful to consider now a lattice version of the theory, instead of the continuum description considered up to here. The minimal coupling is now introduced through a Peierls factor, and an expansion for weak fields leads naturally to the separation into paramagnetic and diamagnetic terms.

The kinetic energy of a tight-binding model may be written as

$$H_k = -t \sum_{\langle i,j\rangle,\sigma} \left(c^\dagger_{i,\sigma} c_{j,\sigma} + c^\dagger_{j,\sigma} c_{i,\sigma}\right), \tag{17.223}$$

using a simplified notation since the space is in general d-dimensional. For a vector potential of the form

$$A_x(l, t) = A_x(q) e^{iq\cdot l - i\omega t}, \tag{17.224}$$

the Peierls factor is

$$t \rightarrow t e^{-\frac{ie}{\hbar c} A_x \cdot (R_i - R_j)_x}. \tag{17.225}$$

This leads to a change of the kinetic term as follows:

$$H_{k,A} = -t \sum_{l,\sigma} \left(e^{\frac{ie}{\hbar c} A_x(l)} c^\dagger_{l+x,\sigma} c_{l,\sigma} + e^{-\frac{ie}{\hbar c} A_x(l)} c^\dagger_{l,\sigma} c_{l+x,\sigma} \right). \tag{17.226}$$

Expanding the Peierls factors up to second order in the weak vector potential, we get that

$$\begin{aligned}
H_{k,A} &= H_k - \sum_l \left(\frac{e}{\hbar c} j^p_x(l) A_x(l) + \frac{1}{2} \left(\frac{e}{\hbar c}\right)^2 K_x(l) A^2_x(l) \right) \\
j^p_x(l) &= it \sum_\sigma \left(c^\dagger_{l+x,\sigma} c_{l,\sigma} - c^\dagger_{l,\sigma} c_{l+x,\sigma} \right) \\
K_x(l) &= -t \sum_\sigma \left(c^\dagger_{l+x,\sigma} c_{l,\sigma} + c^\dagger_{l,\sigma} c_{l+x,\sigma} \right).
\end{aligned} \tag{17.227}$$

The total current density has therefore two terms, one paramagnetic and the other diamagnetic that result from the functional derivative of the perturbed Hamiltonian with respect to the vector potential

$$J_x(l) = -\frac{\delta H_{k,A}}{\delta A_x(l)} = \frac{e}{\hbar c} j^p_x(l) + \left(\frac{e}{\hbar c}\right)^2 K_x(l) A_x(l), \tag{17.228}$$

where the second term can be recognized as the diamagnetic term, proportional to the vector potential. Note that this derivation is general for the tight-binding model and is not related to a superconductor at this point.

As mentioned above, the perturbation is space- and time-dependent, and we need to use linear response theory to evaluate the response of the system to such a perturbation. Linear response theory leads to the full current expression

$$\langle J_x(l,t) \rangle = \langle j_x(l,t) \rangle - \frac{i}{\hbar} \int_{-\infty}^{t} dt' \langle [j_x(l,t), H_i(t']) \rangle, \tag{17.229}$$

where H_i is the extra term that results from the coupling to the external vector potential.

Since $\langle j^p_x(l) \rangle = 0$, only the diamagnetic current contributes

$$\langle j_x(l,t) \rangle = \left(\frac{e}{\hbar c}\right)^2 \langle K_x(l) \rangle A_x(l,t). \tag{17.230}$$

The commutator can be expanded as

$$[j_x(l,t), H_i(t')] = \left(\frac{e}{\hbar c}\right)^2 \left[j_x^p(l,t), (-)\sum_j j_x^p(j,t') \right] A_x(j,t'). \tag{17.231}$$

Considering the usual case of adiabatic switching of the perturbation, we may write

$$A_x(l,t) = \frac{cE_x(l,t)}{i(\omega + i\delta)}, \tag{17.232}$$

where $\delta \rightarrow 0 and \delta > 0$. The imaginary part added to the frequency, ω, guarantees a time dependence of the perturbation of the form $e^{i\omega t}e^{\delta t}$, which vanishes when $t \rightarrow -\infty$.

Assuming space translational invariance, it is convenient to define

$$j_x^p(-q,t) = \sum_l j_x^p(l,t) e^{-iq\cdot l}. \tag{17.233}$$

Separating in Fourier components both in space and in time, we can write that

$$\langle J_x(l,t)\rangle = J_x e^{iq\cdot l - i\omega t}. \tag{17.234}$$

We may now average over the spatial position, l, and obtain that

$$\begin{aligned} J_x &= \left(\frac{e}{\hbar c}\right)^2 \frac{\langle K\rangle}{dN} \frac{ce_x}{i(\omega + i\delta)} \\ &+ \left(\frac{e}{\hbar c}\right)^2 \frac{ce_x}{i(\omega+i\delta)} \frac{1}{N} \frac{i}{\hbar} \int_0^\infty dt' e^{i\omega t'} \langle [j_x^p) q, 0), j_x^p(-q,-t')]\rangle; \end{aligned} \tag{17.235}$$

here, e_x is the electric field density, d is the spatial dimension, and N is the number of sites. Using Ohm's law that defines the conductivity as the relation between the induced current density and the electric field density, we may write that

$$\sigma(q,\omega) = \left(\frac{e}{\hbar c}\right)^2 c \frac{\frac{\langle K\rangle}{dN} + \frac{\Lambda_{xx}(q,\omega)}{\hbar N}}{i(\omega + \imath\delta)}. \tag{17.236}$$

The evaluation of Λ_{xx} is complicated and is most conveniently carried out using many-body methods that are beyond the scope of this book. But it is informative to continue formally the analysis.

Recall first that

$$\frac{1}{i(\omega + i\delta)} = \frac{-i\omega}{\omega^2 + \delta^2} - \pi\delta(\omega). \tag{17.237}$$

Therefore, the conductivity has real and imaginary parts given by

$$\begin{aligned}\sigma(q,\omega) &= \sigma'(q,\omega) + i\sigma''(q,\omega)\\ \sigma'(q,\omega) &= \pi c\left(\frac{e}{\hbar c}\right)^2 D\delta(\omega) + \sigma_r(q,\omega)\\ D &= -\left(\frac{\langle K\rangle}{dN} + \frac{\Lambda'_{xx}(q,0)}{\hbar N}\right)\\ \sigma_r &= \left(\frac{e}{\hbar c}\right)^2 \frac{\Lambda''_{xx}(q,\omega)}{\hbar N}\frac{\omega}{\omega^2+\delta^2}.\end{aligned} \tag{17.238}$$

Here, D is the Drude peak and σ_r is called the regular part of the conductivity. The Drude peak associated with an ideal conductor is defined through the limit $\mathbf{q} = 0, \omega \to 0$.

Consider now units such that $\hbar = c = 1$, define intensive quantities per particle, and rescale K_x by the spatial dimension. The existence of a Drude peak at zero frequency implies a zero resistance state. Let us rewrite it as

$$\frac{D_s}{\pi e^2} = \langle -K_x\rangle - \Lambda_{xx}(q \to 0, \omega = 0), \tag{17.239}$$

a finite value. In the Meissner effect, we consider a static situation, $\omega = 0$, and a transverse gauge potential, $\mathbf{q}\cdot\mathbf{A}(\mathbf{q}, \omega = 0) = 0$. Consider small momenta and consider a two-dimensional system, as example. We have seen that we may write that

$$\langle j_\alpha(\mathbf{q})\rangle = f(q)\left(\delta_{\alpha\beta} - \frac{q_\alpha q_\beta}{q^2}\right)A_\beta(q). \tag{17.240}$$

The finite stiffness of the superfluid or superconductor implies that (due to the Meissner effect)

$$f(q \to 0) = -\frac{1}{4\pi\lambda^2} = -\frac{D_s}{\pi} \tag{17.241}$$

is finite. Therefore,

$$\frac{D_s}{\pi e^2} = \langle -K_x\rangle - \Lambda_{xx}(q_x = 0, q_y \to 0, \omega = 0) = \left(\frac{n_s}{m}\right), \tag{17.242}$$

where n_s is the superfluid density; recall Eqs. (12.16, 12.35) for a normal conductor. Also, as required by gauge invariance,

$$\langle -K_x\rangle - \Lambda_{xx}(q_x \to 0, q_y = 0, \omega = 0) = 0. \tag{17.243}$$

We get then the following criteria: if a system is a superconductor both D, the Drude peak, and D_s, the superconducting stiffness, are finite. In a metal, only the Drude peak is finite, $D \neq 0$, and $D_s = 0$. In an insulator, both the Drude peak and the stiffness vanish, $D = D_s = 0$. These results hold at zero temperature and without disorder effects, in a clean system.

Using Eq. (17.238) and considering $q \to 0$, we can write

$$\sigma'(o, \omega) = \pi c \left(\frac{e}{\hbar c}\right)^2 D\delta(\omega) + \sigma_r(0, \omega). \tag{17.244}$$

Considering, for instance, a single-band tight-binding model, it may be shown that $\Lambda_{xx}(0, \omega) = 0$. Therefore, we get that

$$D = -\frac{\langle K \rangle}{dN}, \tag{17.245}$$

where K is the kinetic energy and the sum rule

$$\int_0^\infty d\omega\sigma'(\omega) = -\frac{\pi c^2}{2\hbar^2}\frac{\langle K \rangle}{dN}. \tag{17.246}$$

In general, $K = \delta^2 H/\delta A^2$, where H is the Hamiltonian and A a vector potential.

As we have seen before, adding disorder (or finite temperature) the δ function in the conductivity becomes a Lorentzian with a width that is related to the collisions or relaxation rate in the system. Therefore, we get that the Drude peak is no longer of the type of a δ function singularity and $D = 0$, even though the conductivity is still finite at zero frequency.

17.13 Exercises

1. In a superconductor, the electrons are paired due to an attractive interaction between the electrons, which results from interactions with the crystal lattice. The energy spectrum of a superconductor is in the simplest cases given by

$$E(\mathbf{k}) = \pm\sqrt{(\epsilon(\mathbf{k}) - E_F)^2 + \Delta^2},$$

where $\epsilon(\mathbf{k})$ is the dispersion relation of the free electrons, E_F is the Fermi energy, and Δ is a parameter that is related to the energy needed to break a pair of electrons.

(a) Show that the system has a gap.
(b) Assuming that the Fermi level is not near a van Hove singularity, calculate the density of states assuming that $\Delta << E_F$. Compare with the density of states of the free electrons.

2. A current is established in a ring in the superconducting state. In a given amount of time (say, of the order of 1 year), no change in the current is observed, within a given, small, precision. Estimate the experimental limit for the resistivity of the material in the superconducting state.
3. In the context of BCS, consider the effective Hamiltonian

$$H_{eff} = \int d^3r \sum_{\alpha=\uparrow,\downarrow} \psi^\dagger(\mathbf{r}\alpha)\frac{(\mathbf{p}-e\mathbf{A})^2}{2m}\psi(\mathbf{r}\alpha)$$
$$+ \int d^3r[\Delta^*(\mathbf{r})\psi(\mathbf{r}\downarrow)\psi(\mathbf{r}\uparrow) + \Delta(\mathbf{r})\psi^\dagger(\mathbf{r}\uparrow)\psi^\dagger(\mathbf{r}\downarrow)]$$
$$+\ constant.$$

Identify the constant and determine the form of $\Delta(\mathbf{r})$.

Defining the quasiparticle operators as

$$\psi(\mathbf{r}\uparrow) = \sum_n \left(\gamma_{n\uparrow}u_n(\mathbf{r}) - \gamma^\dagger_{n\downarrow}v^*_n(\mathbf{r})\right)$$
$$\psi(\mathbf{r}\downarrow) = \sum_n \left(\gamma_{n\downarrow}u_n(\mathbf{r}) + \gamma^\dagger_{n\uparrow}v^*_n(\mathbf{r})\right),$$

obtain the Bogoliubov-de Gennes equations

$$H_e u(\mathbf{r}) + \Delta(\mathbf{r})v(\mathbf{r}) = \epsilon u(\mathbf{r})$$
$$-H^*_e v(\mathbf{r}) + \Delta^*(\mathbf{r})u(\mathbf{r}) = \epsilon v(\mathbf{r}),$$

where

$$H_e = \frac{1}{2m}(-i\hbar\nabla - e\mathbf{A})^2 - E_F$$
$$H^*_e = \frac{1}{2m}(i\hbar\nabla - e\mathbf{A})^2 - E_F.$$

Hint: The diagonalized Hamiltonian is written as

$$H = E_g + \sum_{n,\sigma} \epsilon_n \gamma^\dagger_{n\sigma}\gamma_{n\sigma}.$$

(a) Show that

$$[H, \gamma_{n\sigma}] = -\epsilon_n\gamma_{n\sigma}$$
$$[H, \gamma^\dagger_{n\sigma}] = \epsilon_n\gamma^\dagger_{n\sigma}.$$

(b) Calculate the commutator

$$[H, \psi].$$

(c) Show that

$$[\psi(\mathbf{r}, \uparrow), H] = H_e\psi(\mathbf{r}, \uparrow) + \Delta(\mathbf{r})\psi^\dagger(\mathbf{r}, \downarrow)$$

$$[\psi(\mathbf{r}, \downarrow), H] = H_e\psi(\mathbf{r}, \downarrow) - \Delta^*(\mathbf{r})\psi^\dagger(\mathbf{r}, \uparrow).$$

(d) Compare the coefficients of the operators γ and $\gamma^\dagger$, and obtain the Bogoliubov-de Gennes equations.

These equations are defined in real space and allow the solution of problems that are not translational invariant, such as external magnetic fields or effects of disorder.

4. Calculate for a superconductor

$$\sum_\sigma \langle c^\dagger_{\mathbf{k},\sigma} c_{\mathbf{k},\sigma} \rangle.$$

(a) Considering $T = 0$
(b) Considering T is finite

5. Consider the effective Hamiltonian

$$H = \sum_{\mathbf{k},\sigma}(\epsilon_k - \mu)c^\dagger_{\mathbf{k},\sigma}c_{\mathbf{k},\sigma} - \sum_{\mathbf{k}}\left(\Delta_{\mathbf{k}}c^\dagger_{\mathbf{k},\uparrow}c^\dagger_{-\mathbf{k},\downarrow} + \Delta_{\mathbf{k}}c_{-\mathbf{k},\downarrow}c_{\mathbf{k},\uparrow}\right)$$

and a variational state as

$$|GS\rangle = \prod_{\mathbf{k}}\left(u_{\mathbf{k}} + v_{\mathbf{k}}c^\dagger_{\mathbf{k},\uparrow}c^\dagger_{-\mathbf{k},\downarrow}\right)|0\rangle.$$

(a) Show that the normalization of the state implies

$$u_k^2 + v_k^2 = 1.$$

(b) Using the variational method and minimizing in order to u_k and v_k, show that we obtain the same condition of the Bogoliubov-Valatin transformation for the functions u and v

$$2u_k v_k(\epsilon_k - \mu) = \Delta_k\left(u_k^2 - v_k^2\right)$$

and therefore the same solutions.

6. Recall that

$$\nabla \times \mathbf{h} = \mathbf{J}$$

and Stokes theorem to obtain Ampère's law.

Consider a long wire of radius a and a current, I, flowing through. The magnetic field outside the wire is given by $H = I/(2\pi r)$, where r is the distance to the axis. Consider that the wire is a type-I superconductor and that the critical magnetic field is H_c.

(a) Calculate the critical current for which the system becomes a normal metal at the surface.
(b) In the conditions of (a), there is now a sheet where the system is normal. Show that the situation is unstable. (Hint: Consider that the normal sheet has width b. Show that the magnetic field at the surface between the superconducting part of the wire and the normal part of the wire is larger than the critical magnetic field.)
(c) However, we know that for very small values of r the magnetic field inside the wire becomes infinitesimally small. So, in that region (certainly for $r = 0$), the magnetic field must be smaller than the critical field. A model by London assumes that for some $r_1 < a$ the system is in a mixed state with regions superconducting and regions a normal metal. Also, the model assumes that $H(r < r_1) = H_c$. For $r > r_1$, the system is in the normal phase. Therefore, there is current flowing both for $r < r_1$ and $r > r_1$. Using that

$$I(r) = 2\pi r H_c;\ J(r) = \frac{2H_c}{r},$$

that the electric field is independent of r and assuming that only a part r/r_1 of the mixed region is resistive, implying that

$$J(r < r_1) = \frac{r_1}{r\rho} E,$$

calculate the resistance of the wire as a function of the current, I.

See M. Tinkham's "*Introduction to Superconductivity*" (Dover).

7. Discuss the sentence: Good metals are bad superconductors.
8. Calculate the energy of a vortex in a superconductor and in a superfluid, and discuss the differences and similarities.

Reference

1. F. Gygi, M. Schlüter, Phys. Rev. B **43**, 7609 (1991).

Chapter 18
Renormalization Group

18.1 Scaling: Renormalization Group

Near the critical point of a second-order phase transition, there is a divergency of various functions. The susceptibilities diverge at the critical point, typically with a power law divergency. For instance, in magnetism $\chi = (\delta M)/(\delta H) \to \infty$, so that in the limit $\delta H \to 0$, the magnetization deviation, δM, is finite. In general, in the neighborhood of the critical point, the various thermodynamic quantities have power law behavior, even if they are not divergent. Near the phase transition, the system displays large fluctuations in the order parameter. These fluctuations nucleate a symmetry change and enable that the order parameter spontaneously acquires a nonvanishing average value.

The extension of the spatial correlations of the order parameter defines a scale in the problem, called correlation length (ξ). As we get close to the critical point, the correlation length increases and tends to infinity, reflecting that the long-range fluctuations are significative. In this regime, all lengths are important, which implies a scale invariance of the system or scaling. As we get close to the critical point, the spatial correlations extend over blocks of ever increasing size that are such that, if we perform a scale transformation, the system displays a typical length of a point further away to the critical point. At the critical point ($\xi \to \infty$), there is scale invariance.

As we approach the critical point, the short-range details lose importance, since are the long-range correlations ($q \to 0$) that lead to the transition and to the symmetry breaking. The loss of importance of the short-distance details leads to the concept of universality: different systems that may differ in the microscopic details, but that have equivalent behaviors on large scales, will have the same critical behavior. As we get close to the critical point, the system is described by blocks of larger and larger size, which characterize the universality class to which the system belongs. Characterizing the system by interacting blocks requires that we specify the couplings between the blocks in relation to the initial coupling constants of the

P. D. Sacramento, *Introduction to Condensed Matter Theory*,
https://doi.org/10.1007/978-3-031-86251-9_18

problem. We will have therefore to renormalize the constants as we approach the critical point. Eventually, as we construct blocks, new interactions, with coupling constants not present in the initial problem, appear in the renormalization procedure. In general, the appearance of these new couplings complicates the problem.

The process of approach to the critical point is carried out eliminating, in succession, short-distance (or high momenta) degrees of freedom and determining recursion relations between the interaction parameters. As a remark, this is the usual procedure in condensed matter as a model is chosen to describe a particular phenomenon: the complete description of a many-body system is extremely complex, and, in general, an equivalent model is used, which describes the low-energy regime corresponding to a given phenomenon. The process of choice of an adequate low-energy effective Hamiltonian may then be obtained in the same way, eliminating high-energy degrees of freedom.

As said above, the thermodynamic quantities in general have power laws close to the critical point. We determined the exponents for the ϕ^4 theory (continuous version of the Ising model) using mean-field theory, but the critical exponents are not accurate. This is known because some models have exact analytical solutions or because of numerical solutions or because of experimental measurements. Even though mean-field theory gives a good description of the phase diagram, close to the critical point, it fails. We have therefore to look more carefully into the problem and take into account the fluctuations. As a guide, we will take the scaling invariance of the problem.

Widom introduced the hypothesis that the thermodynamic functions separate into a regular part and, in a singular part, that is relevant close to the critical point. Also, he assumed that the singular part is a generalized homogeneous function. As we will see, homogeneity explains the relations between the exponents.

18.1.1 Homogeneous Functions

A function is homogeneous if $F(\lambda x) = \lambda^p F(x)$, where λ is a scale factor and p is an exponent. Taking $\lambda = x^{-1}$, we get $F(1) = x^{-p} F(x)$. This implies

$$F(x) = Cx^p$$

is a power. Consider now a function of two variables. Then the function is homogeneous if

$$F(\lambda x, \lambda y) = \lambda^p F(x, y).$$

Choosing, for instance, $\lambda = y^{-1}$, we can write that

$$F\left(\frac{x}{y}, 1\right) = y^{-p} F(x, y).$$

This is now a function of only one variable. Defining $f(z) = F(z, 1)$, we can write that

$$F(x, y) = y^p f\left(\frac{x}{y}\right) = x^p \tilde{f}\left(\frac{y}{x}\right),$$

where we have defined similarly $\tilde{f}(z) = F(1, z)$.

Let us now consider a generalized homogeneous function like[1]

$$F(\mu^u x, \mu^v y) = \mu^p F(x, y).$$

Choosing $\lambda = \mu^p$, we get that

$$F(\lambda^a x, \lambda^b y) = \lambda F(x, y), \tag{18.1}$$

where $a = u/p$ and $b = v/p$. Choosing $\lambda = y^{-1/b}$, we get

$$F(x, y) = y^{\frac{1}{b}} f\left(\frac{x}{y^{\frac{a}{b}}}\right).$$

Consider then the singular part of the Gibbs function. Let us assume it is a generalized homogeneous function

$$G(\lambda^a t, \lambda^b B) = \lambda G(t, B), \tag{18.2}$$

having defined $t = (T - T_c)/T_c$ and B the magnetic field. Therefore, the magnetization satisfies

$$\lambda^b M(\lambda^a t, \lambda^b B) = \lambda M(t, B).$$

Selecting $B = 0$ and choosing $\lambda = (-t)^{-1/a}$, we get that

$$M(t, 0) \sim (-t)^\beta, \tag{18.3}$$

where $\beta = \frac{1-b}{a}$.

Let us now choose $t = 0$. Choosing now $\lambda = B^{-1/b}$, we get that

$$M(0, B) \sim B^{\frac{1}{\delta}}, \tag{18.4}$$

where $\delta = \frac{b}{1-b}$. Using that $\chi = (\partial M)/(\partial B)$, we get

$$\lambda^{2b} \chi(\lambda^a t, \lambda^b B) = \lambda \chi(t, B).$$

[1] For further reading see [Ravndal].

Choosing $\lambda = t^{-1/a}$ and taking $B = 0$, we get

$$\chi(t, 0) \sim t^{-\gamma}, \tag{18.5}$$

where $\gamma = \frac{2b-1}{a} = \gamma'$. Solving in terms of a and b, we get

$$\begin{aligned} a &= \frac{1}{\beta}\frac{1}{\delta+1} \\ b &= \delta\frac{1}{\delta+1}. \end{aligned} \tag{18.6}$$

This implies the so-called Griffiths relation:

$$\gamma = \gamma' = \beta(\delta - 1). \tag{18.7}$$

We can as well obtain the exponents of the specific heat taking a derivative of the free energy with respect to t and taking $B = 0$:

$$\lambda^{2a} C_B(\lambda^a t, 0) = \lambda C_B(t, 0).$$

Choosing $\lambda = t^{-1/a}$, we obtain

$$C_B(t, 0) \sim t^{-\alpha}, \tag{18.8}$$

where $\alpha = 2 - \frac{1}{a} = \alpha'$. We obtain therefore another relation between the critical exponents

$$\alpha + \beta(\delta + 1) = 2. \tag{18.9}$$

The Widom hypothesis leads to relations between the exponents but not to a direct way of calculating them.

18.1.2 Thermodynamic Inequalities

Let us recall some thermodynamic relations suitable for a magnetic system,[2]

$$\begin{aligned} dU &= TdS + HdM \\ dF &= -SdT + HdM \\ dG &= -SdT - MdH, \end{aligned} \tag{18.10}$$

[2] For further reading see [Stanley].

where $F = U - TS$ and $G = U - TS - MH$ (note that we often use B or H for the magnetic field). We may define specific heats at constant M or constant H:

$$C_M = \left(\frac{\partial U}{\partial T}\right)_M = T\left(\frac{\partial S}{\partial T}\right)_M$$
$$C_H = \left(\frac{\partial U}{\partial T}\right)_H = T\left(\frac{\partial S}{\partial T}\right)_H . \tag{18.11}$$

The isothermal susceptibility is

$$\chi_T = \left(\frac{\partial M}{\partial H}\right)_T = -\left(\frac{\partial^2 G}{\partial H^2}\right)_T . \tag{18.12}$$

It satisfies

$$\chi_T \left(C_H - C_M\right) = T\alpha_H^2, \tag{18.13}$$

where

$$\alpha_H = \left(\frac{\partial M}{\partial T}\right)_H . \tag{18.14}$$

Since the specific heats are positive, we get the inequality

$$C_H \geq T\frac{\left(\frac{\partial M}{\partial T}\right)_H^2}{\chi_T}. \tag{18.15}$$

Recall the definitions of the exponents in terms of $t = (T - T_c)/T_c$ as

$$C_H \sim (-t)^{-\alpha'},\ \chi_T \sim (-t)^{-\gamma'},\ \left(\frac{\partial M}{\partial T}\right)_H \sim (-t)^{\beta-1}. \tag{18.16}$$

Note now that if we have two functions given by some power laws, as $f(x) \sim x^{\lambda}$, $g(x) \sim x^{\varphi}$, and we consider $x << 1$ if we have $f(x) \leq g(x)$, this implies that $\lambda \geq \varphi$. Therefore, we get the inequality

$$-\alpha' \leq 2(\beta - 1) + \gamma', \tag{18.17}$$

which leads to the Rushbrooke inequality

$$\alpha' + 2\beta + \gamma' \geq 2. \tag{18.18}$$

This holds at $H = 0$ and $T \to T_c$. It is interesting that the exact solution of the Ising model obeys this expression as an equality. Also, mean-field theory satisfies this relation as an equality.

Considering now $T = T_c$ and a small magnetic field $H \to 0$, we define $M \sim H^{1/\delta}$, $\chi_T \sim H^{1/\delta - 1}$. Defining $c_H \sim H^{-\varphi} \sim M^{-\varphi\delta}$ at $T = T_c$ and $S(H) \sim -H^{\psi} \sim M^{\psi\delta}$, this leads to

$$- H^{\psi-1} \sim \left(\frac{\partial S}{\partial H}\right)_T = \left(\frac{\partial M}{\partial T}\right)_H, \tag{18.19}$$

(using a Maxwell relation). We get therefore that

$$H^{-\varphi} \geq H^{2(\psi-1)-(1/\delta-1)}, \tag{18.20}$$

which leads to the Coopersmith inequality

$$\varphi + 2\psi - \frac{1}{\delta} \geq 1. \tag{18.21}$$

Using thermodynamic relations, we may obtain other inequalities such as the Griffiths and others

$$\begin{aligned} \beta(\delta+1) &\geq 2-\alpha' \\ \gamma' &\geq \beta(\delta-1) \\ \gamma(\delta') &\geq (2-\alpha)(\delta-1), \end{aligned} \tag{18.22}$$

the Buckingham, Gunton, and Fisher inequalities

$$\begin{aligned} d\frac{\delta-1}{\delta+1} &\geq 2-\eta \\ \frac{d\gamma'}{2\beta+\gamma'} &\geq 2-\eta \\ (2-\eta)\nu &\geq \gamma \\ d\alpha' &\geq 2-\eta_E \\ \frac{d\delta\varphi}{1+\xi+\varphi\delta} = \frac{d\varphi}{\varphi+\psi} &\geq 2-\eta_E, \end{aligned} \tag{18.23}$$

and the Josephson inequalities

$$\begin{aligned} d\nu' &\geq 2-\alpha' \\ d\nu &\geq 2-\alpha. \end{aligned} \tag{18.24}$$

18.1.3 Dimensional Analysis and Hyperscaling

Consider the functional integral

$$Z = \int d\Phi e^{-\frac{1}{2}\Phi K \phi + \eta \Phi} = \frac{1}{\sqrt{det K}} e^{\frac{1}{2}\eta K^{-1}\eta}. \tag{18.25}$$

The two-field correlator is obtained calculating

$$\begin{aligned} \langle \Phi\Phi \rangle &= \frac{1}{Z}\frac{\partial^2}{\partial \eta^2}\int d\Phi e^{-\frac{1}{2}\Phi K \Phi + \eta\Phi} \\ &= K^{-1} + \left(\eta K^{-1}\right)^2, \end{aligned} \tag{18.26}$$

as we saw earlier. Taking the source term $\eta \to 0$, it simplifies to

$$\langle \Phi\Phi \rangle = K^{-1}. \tag{18.27}$$

Consider now

$$Z = \int d\Phi e^{-H(\Phi,\eta)}, \tag{18.28}$$

with

$$\begin{aligned} H(\Phi,\eta) &= \int dx \left(-\frac{1}{2}\Phi\nabla^2\Phi + \frac{1}{2}r_0^2\Phi^2 - \eta\Phi\right) \\ &= \int dk \left(\frac{1}{2}(k^2 + r_0^2)\Phi(k)\Phi(-k) - \eta(-k)\Phi(k)\right). \end{aligned} \tag{18.29}$$

In this case, we see that

$$K^{-1} = \frac{1}{k^2 + r_0^2}. \tag{18.30}$$

Therefore,

$$\langle \Phi(k)\Phi(-k) \rangle = \frac{1}{k^2 + r_0^2}. \tag{18.31}$$

Taking the inverse Fourier transform, we get that

$$\langle \Phi^2(r) \rangle = \int d^3k e^{ikr} \frac{1}{k^2 + r_0^2}$$

$$= \frac{1}{r} e^{-\frac{r}{r_0}}$$

$$= \frac{1}{r^p} e^{-\frac{r}{r_0}}, \tag{18.32}$$

with $p = 1$. In general, one defines $p = d - 2 + \eta$. In $d = 3$, it is $p = 1 + \eta$. This implies that in the example here considered, defined by Eq. (18.29), we get that $\eta = 0$.

Let us now consider the susceptibility

$$\chi = \frac{1}{V} \frac{\partial M}{\partial H}, \tag{18.33}$$

where $M = -\partial G/\partial H$. Defining, as before, a connected correlation function

$$\Gamma(r) = \langle m(r)m(0)\rangle - \langle m(r)\rangle\langle m(0)\rangle, \tag{18.34}$$

the susceptibility may be obtained as

$$\chi = \frac{1}{k_B T} \int d^3 r \Gamma(r). \tag{18.35}$$

Note that as $t \to 0$

$$\Gamma(r) \to \frac{1}{r^p} e^{-\frac{r}{\xi}} \tag{18.36}$$

and that $\xi \sim |t|^{-\nu}$.

The coherence length ξ has dimensions of length, L. Since $G/(k_B T)$ has no dimensions, we get that defining the free energy per volume $g = G/V$, its dimensions are $[g] = L^{-d}$ in d space dimensions. The dimensions of the correlation function are $[\Gamma(r)] = L^{2-d-\eta} = [\langle m(0)\rangle^2]$. Therefore, $[M/V] = L^{(2-d-\eta)/2}$ and $[k_B T \chi] = L^{2-\eta}$. Since $M = -\partial G/\partial H$, we get that $[H/(k_B T)] = [g]/[M/V] = L^{(2+d-\eta)/2}$. Taking $L \to \xi$ and $\xi \sim t^{-\nu}$, we get the following set of scalings:

$$\begin{pmatrix} & \text{Scaling} & \text{Definition} \\ \frac{G}{k_B T V} & \nu d & 2-\alpha \\ \frac{M}{V} & -\nu\frac{2-d-\eta}{2} & \beta \\ k_B T \chi & -\nu(2-\eta) & -\gamma \\ \frac{H}{k_B T} & \nu\frac{2+d-\eta}{2} & \beta\delta \end{pmatrix}. \tag{18.37}$$

The relations that involve the spatial dimension, d, are called hyperscaling relations. We see that we recover the Josephson and Fisher inequalities as equalities. We get the relations

$$
\begin{aligned}
2-\alpha &= \nu d \\
\beta &= -\nu \frac{2-d-\eta}{2} \\
\gamma &= \nu(2-\eta) \\
\beta\delta &= \nu \frac{2+d-\eta}{2}.
\end{aligned}
\tag{18.38}
$$

18.1.4 Wilson Spin Blocks: Renormalization Group

Consider as an example a set of N spins in a d-dimensional lattice given by the partition function[3]

$$
Z(\mathbf{K}, N) = \sum_{\{S_i\}} e^{-H(\mathbf{K},\{S_i\},N)}. \tag{18.39}
$$

The vector $\mathbf{K} = (K_1, K_2, \cdots, K_N)$ has as components the Hamiltonian coupling constants defined as

$$
H(\mathbf{K}, \{S_i\}, N) = K_0 + K_1 \sum_{<i,j>}^{N} S_i S_j + K_2 \sum_{<<i,j>>}^{N} S_i S_j + K_3 \sum_{i,j,k}^{N} S_i S_j S_k + \cdots \tag{18.40}
$$

where K_1 couples nearest neighbor spins and K_2 couples second neighbor spins. As before, the Boltzmann factor β was absorbed in the definition of the coupling constants.

Let us consider the N spins, and let us divide them in blocks, for instance, of 3×3 spins in $d = 2$, as illustrated in Fig. 18.1. The total number of spins in one block of size L in a d-dimensional space is in general L^d. The number of blocks is then NL^{-d}. We may define the total spin of the block I as

$$
S_I' = \frac{1}{L^d} \sum_{i \varepsilon I} S_i .
$$

Then, $-1 \leq S_I' \leq 1$. Define a spin normalized to unity $S_I' \to S_I$, imposing $S_I = \pm 1$. Let us call σ_I the individual spins in the block I. Then the partition

[3] For further reading see [Ravndal] and [ChaikinLubensky].

Fig. 18.1 Blocks of size $L(=3)$ in a two-dimensional square lattice

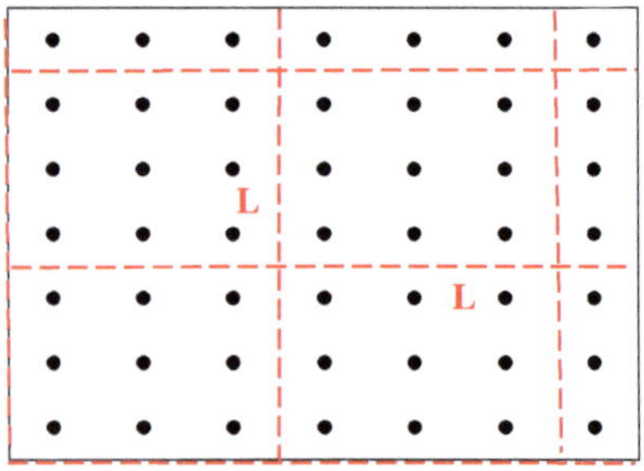

function can be written formally as

$$
\begin{aligned}
Z(\mathbf{K}, N) &= \sum_{\{S_I\},\{\sigma_I\}} e^{-H(\mathbf{K},\{S_I,\sigma_I\},N)} \\
&= \sum_{\{S_I\}} e^{-H(\mathbf{K}_L,\{S_I\},NL^{-d})} \\
&= Z(\mathbf{K}_L, NL^{-d}),
\end{aligned} \tag{18.41}
$$

where we have summed over the individual spins inside each block. $\mathbf{K}_L$ is an effective interaction between the spins of the blocks. The new coupling constants are a function of the previous ones like

$$\mathbf{K}_L = \mathbf{T}(\mathbf{K}). \tag{18.42}$$

The partition function remains the same since the integration was exact.

Let us now assume that the functional form of the partition function remains unaltered (even though with other parameter values). Then,

$$
\begin{aligned}
g(\mathbf{K}) &= \lim_{N\to\infty} \frac{1}{N} \ln Z(\mathbf{K}, N) \\
&= \lim_{N\to\infty} \frac{1}{N} \ln Z(\mathbf{K}_L, NL^{-d}) = L^{-d} g(\mathbf{K}_L).
\end{aligned} \tag{18.43}
$$

We may repeat the same procedure n times. There will be a recursion relation like

$$\mathbf{K}_{nL} = \mathbf{T}(\mathbf{K}_{(n-1)L}),$$

where L is the scale factor of each transformation. If $L > 1$, the distances, measured in terms of the new scale, are smaller. Then the correlation length decreases, and it corresponds to stepping away from the critical point. However, if we are exactly at the critical point, where the correlation length diverges, the scale factor leaves the correlation length invariant, and, therefore, we remain at the critical point. That is, at the critical point, the scale transformation leaves the system invariant. This implies that the coupling constants are invariant. Therefore, the transformation between the coupling constants reaches a fixed point—that is,

$$\mathbf{K}^* = \mathbf{T}(\mathbf{K}^*). \tag{18.44}$$

We can therefore identify a fixed point of the coupling constant transformation law as a critical point.

We have, therefore, due to the scale transformation, a series of transformations between the coupling constants, whose trajectories, in the space of parameters, may be followed. The fixed points of those trajectories define possible critical points. This sequence of transformations constitutes the so-called renormalization group (Wilson).

As an example, let us consider the case of two coupling constants $\mathbf{K} = (K_1, K_2)$. The fixed point is defined by (K_1^*, K_2^*). In general, the transformation $\mathbf{T}$ that relates the coupling constants is nonlinear. Let us consider a neighborhood of a fixed point, and let us linearize the equations of the transformation.

Let us call $\delta\mathbf{K}_L = (\mathbf{K}_L - \mathbf{K}^*)$ and $\delta\mathbf{K} = (\mathbf{K} - \mathbf{K}^*)$. Then, taking both deviations as small, we can write

$$\delta\mathbf{K}_L = A\delta\mathbf{K},$$

where the matrix is defined as

$$A = \begin{pmatrix} \frac{\partial K_{1L}}{\partial K_1} & \frac{\partial K_{1L}}{\partial K_2} \\ \frac{\partial K_{2L}}{\partial K_1} & \frac{\partial K_{2L}}{\partial K_2} \end{pmatrix},$$

where the derivatives are calculated at a fixed point. The eigenvectors of A can be written as

$$\delta\mathbf{u}_L = \Lambda\delta\mathbf{u}.$$

The eigenvalues are expressed in the matrix

$$\Lambda = \begin{pmatrix} \lambda_1 & 0 \\ 0 & \lambda_2 \end{pmatrix}$$

and

$$\delta\mathbf{u} = \begin{pmatrix} \delta u_1 \\ \delta u_2 \end{pmatrix}.$$

Then,

$$\begin{aligned} \delta u_{nL,1} &= (\lambda_1)^n \, \delta u_1 \\ \delta u_{nL,2} &= (\lambda_2)^n \, \delta u_2. \end{aligned} \tag{18.45}$$

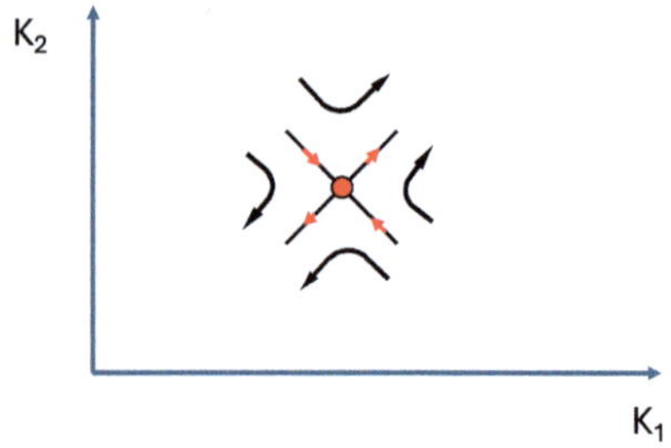

Fig. 18.2 Flow under the scaling transformation in the vicinity of a fixed point, indicated in red. Along one direction, the trajectory flows outward (relevant), and along the other direction, it flows inward (irrelevant)

If the eigenvalue is larger than 1, the point moves away from the fixed point, and if the eigenvalue is smaller than 1, the point moves closer. Therefore, directions in parameter space for which the eigenvalue is larger than 1 are called relevant. The others are called irrelevant. In Fig. 18.2, we illustrate the behavior of points in the space of the parameters K_1 and K_2 near a fixed point, with the motion along one direction as moving away from the fixed point and the opposite around the other direction.

For instance, we expect that the temperature is a relevant parameter (clearly, if the temperature is not at the critical point, the system is not critical). So, we will call u_1 as temperature and u_2 as reserve for other parameters. The irrelevant parameters converge to the fixed point with the iteration procedure. It is the relevant parameters that characterize the class of universality.

Let us now write the free energy in terms of the parameters u_i (eigenvectors). The singular part is then written as

$$g_s(\delta u_1, \delta u_2, \delta u_3, \cdots) = L^{-d} g_s(\lambda_1 \delta u_1, \lambda_2 \delta u_2, \lambda_3 \delta u_3, \cdots).$$

This result is similar to the scaling hypothesis of Widom. Identifying

$$\begin{aligned} \delta u_1 &= t = \frac{T_c - T}{T_c} \\ \delta u_2 &= B, \end{aligned} \tag{18.46}$$

we can write that

$$g(t, B) = \frac{1}{\lambda} g(\lambda^a t, \lambda^b B). \tag{18.47}$$

Note that the magnetic field is also a relevant perturbation and therefore in this case both eigenvalues are larger than 1. Taking now $\lambda = L^d$, we get that

$$\begin{aligned} \lambda_1 &= \left(L^d\right)^a \to a = \frac{\ln \lambda_1}{d \ln L} = \frac{1}{\beta} \frac{1}{\delta + 1} \\ \lambda_2 &= \left(L^d\right)^b \to b = \frac{\ln \lambda_2}{d \ln L} = \delta \frac{1}{\delta + 1}. \end{aligned} \tag{18.48}$$

We can also obtain that

$$\frac{1}{a} = 2 - \alpha = d\frac{\ln L}{\ln \lambda_1}$$
$$\frac{1}{b} = 1 + \frac{1}{\delta} = d\frac{\ln L}{\ln \lambda_2}. \quad (18.49)$$

Determining the eigenvalues of the transformation allows to calculate the critical exponents.

Let us now consider that we may write that

$$\frac{d\mathbf{K}_L}{d\tilde{t}} = \mathbf{B}(\mathbf{K}_L),$$

where $\tilde{t} = \ln L$, and let us linearize the equation around the fixed point given by $\mathbf{B}(\mathbf{K}^*) = 0$. We obtain

$$\frac{d\mathbf{K}_L}{d\tilde{t}} = M(\mathbf{K}_L - \mathbf{K}^*), \quad (18.50)$$

where $M = (\partial B_i)/(\partial K_j)$ calculated at the fixed point. Let us now consider the eigenvectors and eigenvalues of the matrix M given by

$$M \cdot \mathbf{e}_K^R = z_k \mathbf{e}_K^R$$
$$\mathbf{e}_K^L \cdot M = z_k \mathbf{e}_K^L. \quad (18.51)$$

Then the projections of $(\mathbf{K}_L - \mathbf{K}^*)$ on the eigenvectors

$$v_{kL} = \mathbf{e}_K^L \cdot (\mathbf{K}_L - \mathbf{K}^*)$$

satisfy

$$\frac{dv_{kL}}{d\tilde{t}} = \mathbf{e}_K^L \cdot \frac{d\mathbf{K}_L}{dt}$$
$$= \mathbf{e}_K^L \cdot M \cdot (\mathbf{K}_L - \mathbf{K}^*)$$
$$= z_k \mathbf{e}_K^L \cdot (\mathbf{K}_L - \mathbf{K}^*)$$
$$= z_k v_{kL}. \quad (18.52)$$

Integrating, we obtain that

$$v_{kL} = v_k e^{z_k \ln L} = v_k L^{z_k}. \quad (18.53)$$

Therefore, if $z_k > 0$, the parameter is relevant, and if $z_k < 0$, the parameter is irrelevant. The exponents z_k are called anomalous dimensions of the eigenvector directions. Then,

$$z_1 = z_t = ad = \frac{d}{\beta(\delta + 1)}$$

$$z_2 = z_B = \frac{d\delta}{\delta + 1}. \tag{18.54}$$

The ideas of scaling and the renormalization group method will be applied in the next section to calculate the critical exponents in a second-order phase transition. Also, later, it will be applied to study the Berezinskii-Kosterlitz-Thouless transition and the scaling theory of localization.

18.1.5 The Critical Exponents of the ϕ^4 Model

As an example of the description of the critical region and, in particular, the calculation of critical exponents using the renormalization procedure, let us consider the model for which we carried out explicitly the mean-field analysis. Consider again the ϕ^4 model

$$Z = \int \mathcal{D}\phi e^{-H(\phi)},$$

where the action (argument of the Boltzmann exponential) can be written as

$$\begin{aligned} H(\phi) &= \frac{1}{2}\int_0^\Lambda d^d q \left(q^2 + r\right)\phi(q)\phi(-q) \\ &+ \frac{u}{4!}\int_0^\Lambda d^d q_1 \int_0^\Lambda d^d q_2 \int_0^\Lambda d^d q_3 \int_0^\Lambda d^d q_4 \\ &\quad \phi(q_1)\phi(q_2)\phi(q_3)\phi(q_4)\delta(q_1 + q_2 + q_3 + q_4). \end{aligned} \tag{18.55}$$

To simplify notation in the integrals over d-dimensional space, we only indicate that the absolute value of the momenta has a cutoff called Λ. Also, the q_i argument of the fields is a shorthand notation for a d-dimensional vector.

Let us eliminate the degrees of freedom of the high momenta and try to reconstruct the initial Hamiltonian, renormalizing the parameters in a suitable way. Let us start by taking $u = 0$ and considering only the gaussian term. We separate the integration domain in two parts:

$$\int_0^\Lambda d^d q \to \int_0^{\Lambda/l} d^d q + \int_{\Lambda/l}^\Lambda d^d q.$$

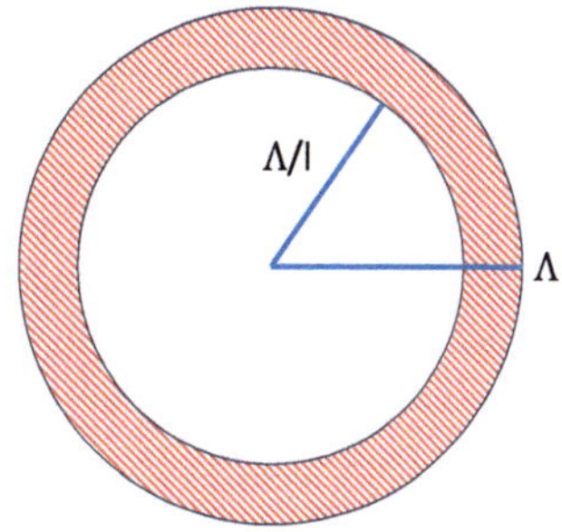

Fig. 18.3 Illustration of the shell of momenta that are integrated over with a modulus of the momenta in the range between Λ/l and Λ. Λ is the energy cutoff of the order of the inverse of the lattice constant

Here, $l \geq 1$ is a scaling factor. This is illustrated in Fig. 18.3 for the case of a three-dimensional space. Let us perform only the second integration, which corresponds to the high momenta contribution. The quadratic term of the action leads to

$$\int_0^{\Lambda} d^d q \frac{1}{2}\left(q^2+r\right)\phi(q)\phi(-q) = \int_0^{\Lambda/l} d^d q \frac{1}{2}\left(q^2+r\right)\phi(q)\phi(-q) + \int_{\Lambda/l}^{\Lambda} d^d q \frac{1}{2}\left(q^2+r\right)\phi(q)\phi(-q). \tag{18.56}$$

The second integral just gives a constant which has no physical meaning. So, we are left with only the first term. To remain in the initial space, we have to rescale the momenta. Define $q' = lq$. At the same time, rescale the fields like

$$\phi'(q') = z^{-1}\phi(q).$$

We obtain in terms of the rescaled momenta and fields that

$$\int_0^{\Lambda/l} d^d q \frac{1}{2}\left(q^2+r\right)\phi(q)\phi(-q) = \int_0^{\Lambda} d^d q' \frac{1}{2}\left(l^{-2}(q')^2+r\right) z^2 l^{-d}\phi'(q')\phi'(-q'). \tag{18.57}$$

To have a problem equal to the initial one, we redefine

$$\begin{aligned} r' &= l^2 r \\ z &= l^{\frac{d+2}{2}}. \end{aligned} \tag{18.58}$$

With these transformations, we recover the initial problem. We had to redefine the temperature difference from the critical point, and we renormalized the fields.

The renormalization of the fields may be interpreted as a renormalization of the magnetic field. If we had a term of the type

$$\int dx B\phi(x) = B\phi(q=0) = Bz\phi'(q'=0),$$

it would be natural to redefine a new magnetic field $B' = zB = l^{\frac{d+2}{2}} B$.

To summarize, in the gaussian model ($u = 0$), the temperature difference and the magnetic field are renormalized

$$\begin{aligned} r' &= l^2 r \\ B' &= l^{\frac{d+2}{2}} B. \end{aligned} \tag{18.59}$$

The fixed point of the transformation is, therefore, $r^* = 0$ and $B^* = 0$. The eigenvalues of the transformation are $\lambda_r = l^2$ and $\lambda_B = l^{1+\frac{d}{2}}$. Both λ_r and λ_B are larger than 1, and therefore, both are relevant, as discussed above. Using that,

$$\begin{aligned} a &= \frac{\ln \lambda_r}{d \ln l} = \frac{\ln l^2}{d \ln l} = \frac{2}{d} = \frac{1}{\beta} \frac{1}{\delta + 1} \\ b &= \frac{\ln \lambda_B}{d \ln l} = \frac{\ln l^{1+\frac{d}{2}}}{d \ln l} = \frac{2+d}{2d} = \delta \frac{1}{\delta + 1}. \end{aligned} \tag{18.60}$$

We conclude

$$\begin{aligned} \delta &= \frac{d+2}{d-2} \\ \beta &= \frac{d-2}{4} \\ \alpha &= 2 - \frac{d}{2} \\ \gamma &= 1 \\ \nu &= \frac{1}{2}, \end{aligned} \tag{18.61}$$

where we have used that $\alpha = 2 - 1/a$, $\gamma = \beta(\delta - 1)$, and $\nu = (2 - \alpha)/d$.

Let us now consider the quartic term. Let us again eliminate the high momenta. We have an integral of the type

$$\int \mathcal{D}\phi e^{-H_0(\phi) - V}.$$

Next, we split

$$H_0 = \tilde{H}_0 + H_0 = \int_0^{\Lambda/l} \cdots + \int_{\Lambda/l}^{\Lambda} \cdots.$$

We have to calculate an integral of the type

$$Tre^{-H_0-V}=Tre^{-H_0}\frac{Tre^{-H_0-V}}{Tre^{-H_0}}=Z_0\langle e^{-V}\rangle_0=Z_0e^W.$$

We want to replace the initial problem by a rescaled one of the form

$$\int\mathcal{D}\phi' e^{-H_{eff}(\phi')},$$

where $H_{eff}=\tilde{H}_0+$ (term to be discovered). We can progress further by noting that writing

$$e^W=\langle e^{-\lambda V}\rangle_0$$

and taking $\lambda=0$, we get $W=0$. Taking a derivative with respect to λ, we get

$$\begin{aligned}W'e^W&=-\langle e^{-\lambda V}V\rangle_0\\(W''+(W')^2)e^W&=\langle e^{-\lambda V}V^2\rangle_0\\W''&=\langle V^2\rangle_0-\langle V\rangle_0^2\\Tre^{-H_0-V}&=Z_0e^{-\lambda\langle V\rangle_0+\frac{\lambda^2}{2}(\langle V^2\rangle_0-\langle V\rangle_0^2)}.\end{aligned}\tag{18.62}$$

As an example of the description of the critical region and, in particular, the calculation of critical exponents using the renormalization procedure, let us consider the model 18.55. Let us then consider the integration of the quartic term. The term $\tilde{H}_0$ of the effective Hamiltonian is given by

$$\frac{1}{2}l^{-d}z^2\left(l^{-2}(q')^2+r\right),$$

after rescaling the momenta and renormalizing the fields. The remaining term in the effective Hamiltonian is more complex. Let us call A and B the integration in the lower momenta and in the shell close to the cutoff, respectively. The terms that include a B part have to be integrated. It is trivial that

$$(A+B)^4=A^4+4A^3B+6A^2B^2+4AB^3+B^4.$$

The average value of an odd number of fields vanishes. We are left with the terms

$$A^4, 6A^2B^2, B^4.$$

In the first term, no field is integrated; in the second term, two fields are integrated; and in the final one, all fields are integrated. The vertices u can be represented as

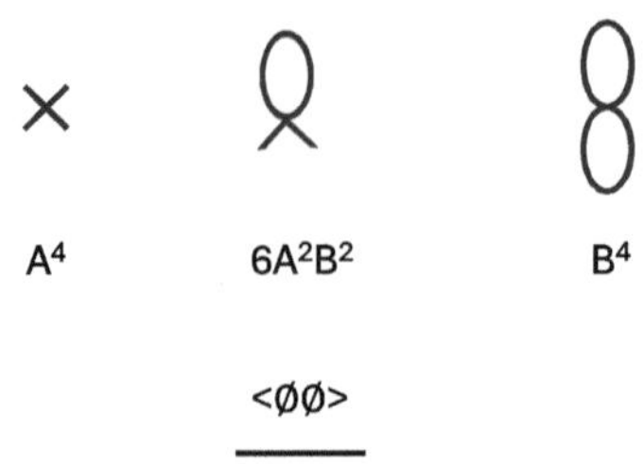

Fig. 18.4 Graphical representation of the terms contributing to the leading order in the interaction. An interaction vertex is represented by a cross. If momenta are integrated over, the lines emerging from the vertex are connected by a line. The solid line represents the average over two fields $\langle\phi(q_1)\phi(q_2)\rangle$

in Fig. 18.4. A vertex is represented as a cross, with each line representing a given momentum variable, q_i. If two momenta are integrated over, those external lines are connected. If all momenta are integrated over, the lines are connected, two by two.[4]

The term B^4 just gives a constant and can be dropped out. The A^4 term is given by

$$\frac{u}{4!}z^4 l^{-3d}\int_0^{\Lambda}\phi'(q_1')\phi'(q_2')\phi'(q_3')\phi'(q_4')\delta(q_1'+q_2'+q_3'+q_4')d^dq_1'd^dq_2'd^dq_3'd^dq_4'.$$

The term that remains is of the form

$$6\frac{u}{4!}\int_0^{\Lambda/l}d^dq_1\int_0^{\Lambda/l}d^dq_2\int_{\Lambda/l}^{\Lambda}d^dq_3\int_{\Lambda/l}^{\Lambda}d^dq_4\delta(q_1+q_2+q_3+q_4)$$
$$\langle\phi(q_1)\phi(q_2)\phi(q_3)\phi(q_4)\rangle_0.$$

The fields that depend on q_1 and q_2 are not integrated over the high momenta. Using that

$$\langle\phi(q_3)\phi(q_4)\rangle_0=\frac{1}{q_3^2+r}\delta(q_3+q_4)$$

(since the integration is gaussian), we obtain that the term gives

$$6\frac{u}{4!}\int_0^{\Lambda/l}d^dq_1\phi(q_1)\phi(-q_1)\int_{\Lambda/l}^{\Lambda}d^dq_3\frac{1}{q_3^2+r}.$$

[4] The representation in Fig. 18.4 and ahead in Fig. 18.5 is merely a graphical representation that is convenient. Some readers may recognize the similarities with some diagrammatic representation and associated diagrammatic rules. Here, we do not follow such route since it is beyond the level intended for this book.

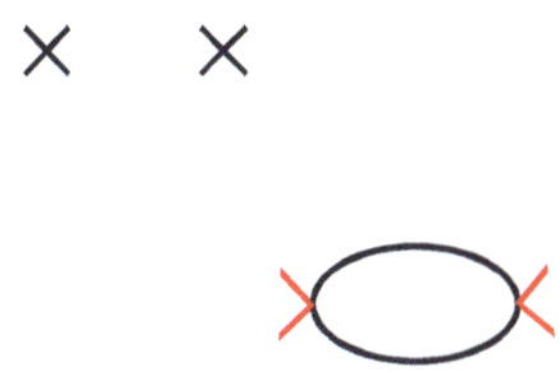

Fig. 18.5 Graphical representation of the terms contributing to the second-order perturbation theory, u^2, that lead to a renormalization of the coupling constant, u. The solid line represents the average over two fields $\langle\phi(q_1)\phi(q_2)\rangle$. We show when no momenta are integrated over (lines in red) or four momenta are integrated

Rescaling the momenta and the fields, this reduces to

$$z^2 l^{-d} 6\frac{u}{4!}\int_0^\Lambda d^d q_1 \phi(q_1)\phi(-q_1)\int_{\Lambda/l}^\Lambda d^d q_3 \frac{1}{q_3^2+r}.$$

Putting the terms together, we get

$$z^2 l^{-d}\left[l^{-2}(q')^2 + r + \frac{1}{2}u\int_{\Lambda/l}^\Lambda d^d q_3 \frac{1}{q_3^2+r}\right] = (q')^2 + r'. \tag{18.63}$$

The renormalization factor z is the same (at this order) as in the gaussian model. Therefore, we obtain

$$r' = l^2\left[r + \frac{u}{2}\int_{\Lambda/l}^\Lambda d^d q_3 \frac{1}{q_3^2+r}\right]. \tag{18.64}$$

We are left now with the renormalization of the coupling constant, u.

The renormalization of the coupling constant is obtained calculating the second-order perturbation theory term

$$\begin{aligned}
&= \left(\frac{u}{4!}\right)^2 \int_0^{\Lambda/l} dq_1 dq_2 \int_{\Lambda/l}^\Lambda dq_3 dq_4 \int_0^{\Lambda/l} dp_1 dp_2 \int_{\Lambda/l}^\Lambda dp_3 dp_4 \\
&\quad \langle\phi(q_1)\phi(q_2)\phi(q_3)\phi(q_4)\phi(p_1)\phi(p_2)\phi(p_3)\phi(p_4)\rangle \\
&\quad \delta(q_1+q_2+q_3+q_4)\delta(p_1+p_2+p_3+p_4) \\
&= \left(\frac{u}{4!}\right)^2 \int_0^{\Lambda/l} dq_1 dq_2 \int_{\Lambda/l}^\Lambda dq_3 dq_4 \int_0^{\Lambda/l} dp_1 dp_2 \int_{\Lambda/l}^\Lambda dp_3 dp_4 \\
&\quad \phi(q_1)\phi(q_2)\phi(p_1)\phi(p_2)\langle\phi(q_3)\phi(q_4)\phi(p_3)\phi(p_4)\rangle\delta(q_1+q_2+p_1+p_2).
\end{aligned} \tag{18.65}$$

Using that the average value is of the type

$$\langle \phi(q_3)\phi(q_4)\phi(p_3)\phi(p_4)\rangle = \frac{1}{q_3^2+r}\frac{1}{q_4^2+r}\delta(q_3+p_3)\delta(q_4+p_4), \tag{18.66}$$

plus all other pairings, two by two, we are left with integrations of the type

$$\int_{\Lambda/l}^{\Lambda} dq_3 \frac{1}{q_3^2+r}\frac{1}{(p_1+p_2-q_3)^2+r}.$$

Renormalizing the momenta and rescaling the fields in the same way as before, we get

$$u' = z^4 l^{-3d}\left[u - \frac{3}{2}u^2\int_{\Lambda/l}^{\Lambda} dq \frac{1}{(q^2+r)^2}\right]. \tag{18.67}$$

The procedure is illustrated in Fig. 18.5. There are 36 ways to obtain an integral of the type considered above, and we have taken $p_1 + p_2 = 0$ to simplify. For each vertex, there are *six* ways to select two lines out of the four lines. There are two ways to connect the lines from the two vertices, and there is a factor of 1/2 due to the second-order perturbation theory. Summarizing our results, we obtain the recursion relations

$$r' = l^2\left[r + \frac{u}{2}\int_{\Lambda/l}^{\Lambda}\frac{d^dq}{q^2+r}\right]$$

$$u' = l^{4-d}\left[u - \frac{3}{2}u^2\int_{\Lambda/l}^{\Lambda}\frac{d^dq}{(q^2+r)^2}\right]. \tag{18.68}$$

When $d = 4$, the coupling is dimensionless.

Let us now consider an infinitesimal scale transformation of the type $l = 1 + \delta l$. We have integrals of the type

$$\int d^dq f(q) = \int dq S_d q^{d-1} f(q).$$

With the infinitesimal scale transformation, we can write that

$$\Lambda - \frac{\Lambda}{l} = (l-1)\frac{\Lambda}{l} = \frac{\delta l}{l}\Lambda.$$

Then, using that δl is small, we obtain the transformations

$$r' = l^2\left[r + \frac{u}{2}\frac{S_d}{(2\pi)^d}\Lambda^d\frac{1}{\Lambda^2+r}\frac{\delta l}{l}\right]$$

$$r' = (1+2\delta l)\left[r + \frac{u}{2}\frac{S_d}{(2\pi)^d}\Lambda^{d-2}\frac{1}{1+\frac{r}{\Lambda^2}}\frac{\delta l}{l}\right]$$

$$\frac{\frac{r'-r}{\Lambda^2}}{\delta l} \sim \frac{2r}{\Lambda^2} + \frac{1}{2}\frac{S_d}{(2\pi)^d}u\Lambda^{d-4}\frac{1}{1+\frac{r}{\Lambda^2}}. \tag{18.69}$$

Note that both r/Λ^2 and $u\Lambda^{d-4}$ are dimensionless. Let us do the following rescalings:

$$\frac{r}{\Lambda^2} \to r$$
$$\frac{S_d}{(2\pi)^d}u\Lambda^{d-4} \to u. \tag{18.70}$$

Then we obtain that

$$l\frac{dr}{dl} = 2r + \frac{1}{2}\frac{u}{1+r}. \tag{18.71}$$

The other equation is (defining $\epsilon = 4-d$)

$$u' = (1+\epsilon\delta l)\left[u - \frac{3}{2}u^2\frac{S_d}{(2\pi)^d}\Lambda^d\frac{\delta l}{l}\frac{1}{(\Lambda^2+r)^2}\right].$$

This leads to

$$\frac{\delta u}{\delta l} \sim \epsilon u - \frac{3}{2}u^2\frac{S_d}{(2\pi)^d}\Lambda^d\frac{1}{l}\frac{1}{(\Lambda^2+r)^2}.$$

Taking $l = 1$ in the right-hand side, we get that

$$l\frac{du}{dl} = \epsilon u - \frac{3}{2}\frac{u^2}{(1+r)^2}. \tag{18.72}$$

Let us now look for the fixed points of the trajectories of r and u, as l changes. Clearly, the trivial (gaussian) fixed point $r^* = 0, u^* = 0$ is also a fixed point of these equations. Another fixed point is given by

$$r^* = -\frac{1}{6}\epsilon$$
$$u^* = \frac{2}{3}\epsilon \tag{18.73}$$

to leading order in ϵ. Defining $\tilde{t} = \ln l$, we can rewrite the evolution equations as

$$\frac{dr}{d\tilde{t}} = 2r + \frac{1}{2}\frac{u}{1+r}$$
$$\frac{du}{d\tilde{t}} = \epsilon u - \frac{3}{2}\frac{u^2}{(1+r)^2}. \tag{18.74}$$

Recalling our definitions

$$\frac{d\mathbf{K}_L}{d\tilde{t}} = \mathbf{B}(\mathbf{K}_L)$$
$$= M \cdot (\mathbf{K}_L - \mathbf{K}_L^*)$$
$$M_{ij} = \left(\frac{\partial B_i}{\partial K_{Lj}}\right)_{\mathbf{K}_L = \mathbf{K}_L^*} \tag{18.75}$$

leads to

$$B_r = 2r + \frac{1}{2}\frac{u}{1+r}$$
$$B_u = \epsilon u - \frac{3}{2}\frac{u^2}{(1+r)^2}. \tag{18.76}$$

In our case, we get

$$M_{rr} = \frac{\partial B_r}{\partial r} = 2 - \frac{1}{2}\frac{u^*}{(1+r^*)^2}$$
$$M_{ru} = \frac{\partial B_r}{\partial u} = \frac{1}{2}\frac{1}{1+r^*}$$
$$M_{ur} = \frac{\partial B_u}{\partial r} = 3\frac{(u^*)^2}{(1+r^*)^3}$$
$$M_{uu} = \frac{\partial B_u}{\partial u} = \epsilon - 3\frac{u^*}{(1+r^*)^2}. \tag{18.77}$$

The matrix to be diagonalized is therefore

$$M = \begin{pmatrix} 2 - \frac{1}{2}\frac{u^*}{(1+r^*)^2} & \frac{1}{2}\frac{1}{1+r^*} \\ 3\frac{(u^*)^2}{(1+r^*)^3} & \epsilon - 3\frac{u^*}{(1+r^*)^2} \end{pmatrix}.$$

To leading order, this reduces to

$$M \sim \begin{pmatrix} 2 - \frac{1}{2}u^* & \frac{1}{2}(1-r^*) \\ 0 & \epsilon - 3u^* \end{pmatrix}.$$

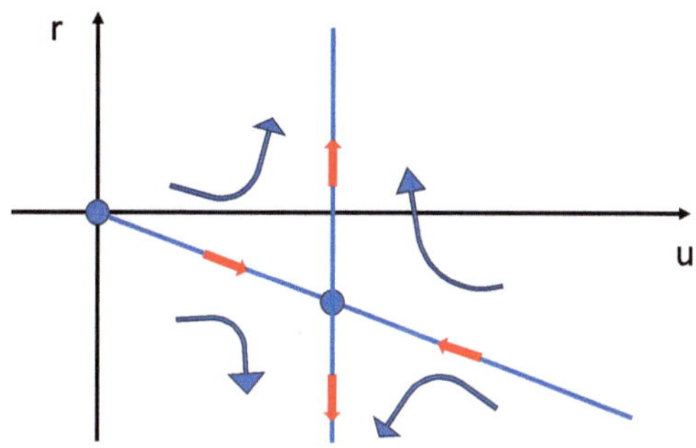

Fig. 18.6 Schematic flow of the trajectories of the parameters r and u in the vicinity of the fixed points: the Gaussian fixed point at $r^* = u^* = 0$ and the nontrivial fixed point at $r^* = -\epsilon/6, u^* = 2/3\epsilon$

The eigenvalues are

$$z_1 = 2 - \frac{1}{2}u^* = 2 - \frac{1}{3}\epsilon = z_T$$
$$z_2 = \epsilon - 3u^* = -\epsilon. \tag{18.78}$$

This means that the first parameter is relevant and the second one is irrelevant (for $\epsilon > 0, d < 4$).

In the gaussian fixed point, $z_1 = z_T = 2$ and $z_2 = \epsilon$. When $d > 4$ ($\epsilon < 0$), this implies that $u^* < 0$, which is nonphysical because u must be positive for the theory to be stable (for negative u, the theory diverges at infinity). The matrix of the gaussian fixed point is given by

$$M = \begin{pmatrix} 2 & \frac{1}{2} \\ 0 & \epsilon \end{pmatrix}.$$

The temperature eigenvector is always relevant, but for $d > 4$, the second eigenvalue is irrelevant, and, therefore, any trajectory along the direction of this eigenvector will converge to the gaussian fixed point. For $d < 4$, the gaussian point is always unstable since both eigenvectors become relevant. We are then left with the other fixed point. For $d > 4$ ($\epsilon < 0$), the temperature eigenvector is relevant, but the other eigenvector is irrelevant, which means that any trajectory along the direction of this second eigenvector will converge to the fixed point. The general structure of the flow is shown in Fig. 18.6.

We can now calculate the critical exponents for the other fixed point. Using the equations previously established, we get that

$$z_T = 2 - \frac{1}{3}\epsilon = \frac{d}{\beta(\delta+1)} = \frac{1}{\nu} = \frac{d}{2-\alpha}. \tag{18.79}$$

This means

$$\alpha = \frac{\epsilon}{6}$$

$$\nu = \frac{1}{2} + \frac{1}{12}\epsilon. \tag{18.80}$$

Also, we can show that $\delta = 3+\epsilon$. In the table below, we compare the results of the ϵ expansion to leading order with the experimental results. We see that the agreement is quite good and much more accurate than the mean-field results. Considering terms to order ϵ^2 improves somewhat the agreement, but going to higher orders the agreement is worse.

	$d = 4$	$d < 4$	$\epsilon = 1$	$Exp.(d = 3)$
α	0	$\frac{\epsilon}{6}$	0.17	0–0.2
β	$\frac{1}{2}$	$\frac{1}{2} - \frac{1}{6}\epsilon$	0.33	0.3–0.4
δ	3	$3 + \epsilon$	4.0	$4 - 5$
γ	1	$1 + \frac{1}{6}\epsilon$	1.17	$\sim$1.3
η	0	0	0.0	$\sim$0.1
ν	$\frac{1}{2}$	$\frac{1}{2} + \frac{\epsilon}{12}$	0.58	0.6–0.7

18.2 Real Space Renormalization

We want now to carry out an elimination of degrees of freedom in real space. As we approach a critical point, the characteristic length scales become larger, and microscopic details become less important. So, as we approach the critical point, we may define larger and larger unit cells, which contain increasing numbers of degrees of freedom. We may then consider a sequence of transformations where we move from the original set of degrees of freedom, say, spins on a lattice, to blocks that contain several spins, in such a way that we replace the original degrees of freedom by a smaller set, those corresponding to the behavior of a given block. As we repeat this procedure, the real space scale increases. Also, as we replace the original spin degrees of freedom by the new ones, the Hamiltonian of the new spin variables may also be changed with respect to the original one. Both because we get terms of similar type with different coupling constants and because new terms, which were not in the original Hamiltonian, may appear. In general, an infinite set of new terms may be generated as we repeat the renormalization procedure. One expects, however, that only a few of those terms will be relevant in the approach to the critical point.[5]

Let us consider the free energy per particle of a system of N spins defined as

$$f = \lim_{N\to\infty} \frac{1}{N} \ln Z, \tag{18.81}$$

[5] For further reading see [BurkhardtLeeuwen] and [ChaikinLubensky].

where we take the partition function as

$$Z = \sum_{\sigma} e^{H(\sigma)}, \tag{18.82}$$

where the factor of $-1/\beta$ has been absorbed in the Hamiltonian. The spins are represented by the σ variables.

At each step of the renormalization procedure, we introduce new spin variables denoted σ', associated with each block of the original spins. The renormalization procedure leads to a transformation of the type

$$e^{H'(\sigma')} = \sum_{\sigma} P(\sigma', \sigma) e^{H(\sigma)}. \tag{18.83}$$

We have that

$$\sum_{\sigma'} P(\sigma', \sigma) = 1, \tag{18.84}$$

so that Z is the same. The Hamiltonian may be expressed as a sum over different terms, $\Omega_\alpha(\sigma)$, indexed by α, that in general runs to infinity

$$H(\sigma) = \sum_{\alpha} k_\alpha \Omega_\alpha(\sigma). \tag{18.85}$$

Even if the original Hamiltonian has a finite number of terms describing the couplings, k_α, between the spin variables, as mentioned above, the renormalization generates arbitrary sets of terms in the new spin variables.

Consider now blocks of spins and that the effective interactions between the block spins are a function of the original couplings—that is, consider $\mathbf{k}' = \mathbf{k}'(\mathbf{k})$. We start with a state of N spins, and after a first step, we have N' spins related to the block volume, of side $L \equiv b$, as $b^d = N/N'$. Also, the coupling constants k_α are changed to new coupling constants k'_α.

The exact transformation in Eq. (18.83) is in general difficult to implement exactly, particularly as the sequence of reduction of spin variables (or enlargement of the blocks) carries on. Kadanoff proposed to replace

$$H(\sigma) \to H(\sigma) + V(\sigma), \tag{18.86}$$

such that the procedure can be evaluated, step by step. This is an approximate method but can be carried out with appropriate choices of $V(\sigma)$. Then,

$$e^{H'(\sigma')} = \sum_{\sigma} P(\sigma', \sigma) e^{H(\sigma)+V(\sigma)}. \tag{18.87}$$

Now, Z is not invariant. Indeed,

$$\begin{aligned}\sum_{\sigma'} e^{H'(\sigma')} &= \sum_{\sigma}\sum_{\sigma'} P(\sigma',\sigma)e^{H(\sigma)+V(\sigma)} \\ &= \sum_{\sigma} e^{H(\sigma)+V(\sigma)} \\ &= \sum_{\sigma} e^{H(\sigma)}\,(1+V(\sigma)+\cdots)\,. \end{aligned} \tag{18.88}$$

Therefore, the partition function becomes

$$Z(H') = Z(H+V) = Z(H)\,(1+\langle V\rangle_H)\,. \tag{18.89}$$

If $\langle V\rangle_H = 0$, this implies that $Z(H') \geq Z(H)$ and $F(H') \leq F(H)$. Consider now that we perform infinite transformations. If at each step $H^{(n)} \to H^{(n+1)}$ we choose the potential such that $\langle V^{(n)}\rangle_{H^{(n)}} = 0$, this gives us a lower bound for the free energy. Also, we may choose $V(\sigma)$ such that we may evaluate the sum

$$\sum_{\sigma} P(\sigma',\sigma)e^{H(\sigma)+V(\sigma)}. \tag{18.90}$$

In some cases, the procedure is easy to implement, and an exact solution may be found. One example is the $1d$ Ising model.

One-Dimensional Ising Model

Consider a set of spins (classical variables here with values $\sigma = \pm 1$) along a chain with a given lattice constant, a. Take $b = 2$, which means that we take a block with two spins, that is, we want to eliminate one every other spin, as shown in Fig. 18.7. For instance, consider a spin at a given site n, and eliminate all spins at sites $\cdots, n-3, n-1, n+1, n+3, \cdots$. The partition function may be written as

$$Z = \sum_{\cdots,\sigma_{n-1},\sigma_n,\sigma_{n+1},\sigma_{n+2},\cdots} \cdots e^{\bar{k}(\sigma_{n-1},\sigma_n)} e^{\bar{k}(\sigma_n,\sigma_{n+1})} e^{\bar{k}(\sigma_{n+1},\sigma_{n+2})} \cdots\,. \tag{18.91}$$

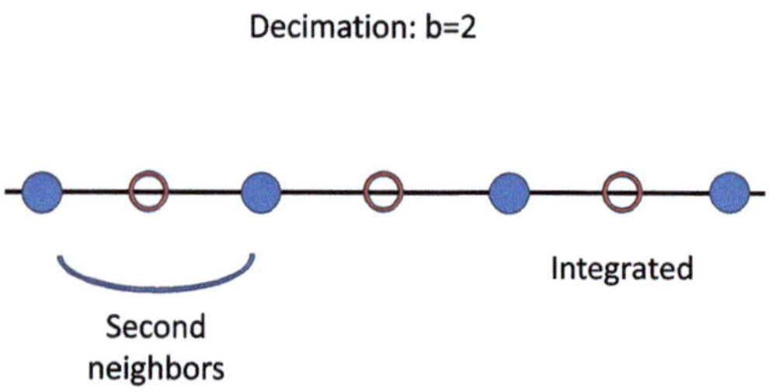

Fig. 18.7 Decimation of the one-dimensional Ising model in which every other spin is eliminated ($b = 2$)

Any group of three consecutive sites, say, $n, n+1, n+2$, contributes to the partition function with the product of the matrices

$$\begin{pmatrix} e^k & e^{-k} \\ e^{-k} & e^k \end{pmatrix}\begin{pmatrix} e^k & e^{-k} \\ e^{-k} & e^k \end{pmatrix} = \begin{pmatrix} e^{2k}+e^{-2k} & 2 \\ 2 & e^{2k}+e^{-2k} \end{pmatrix}. \tag{18.92}$$

This multiplication allows the elimination of site $n+1$. The matrix elements, σ and σ', between any two neighbors may be written as

$$\begin{pmatrix} e^k & e^{-k} \\ e^{-k} & e^k \end{pmatrix}_{\sigma,\sigma'} = \cosh k + \sigma\sigma' \sinh k. \tag{18.93}$$

We get that

$$\begin{aligned} \cosh k' + \sinh k' &= 2\cosh(2k) \\ \cosh k' - \sinh k' &= 2. \end{aligned} \tag{18.94}$$

Therefore, we obtain that

$$\tanh k' = \tanh^2 k, \tag{18.95}$$

as the exact relation between the initial coupling between the spins and the renormalized one after eliminating every other spin.

This expression was obtained for $b = 2$, but it can be generalized to any block size leading to

$$\tanh k' = \tanh^b k. \tag{18.96}$$

Therefore, we have the transformation

$$k' = \tanh^{-1}\left(\tanh^b k\right) = R^{(b)}(k). \tag{18.97}$$

The transformation has two fixed points with $k' = k = k^*$ that satisfy the equations

$$\begin{aligned} \tanh k &= 0, \qquad (T \to \infty) \\ \tanh k &= 1, \qquad (T = 0). \end{aligned} \tag{18.98}$$

Since the $|\tanh|$ is less than one, applying the transformation leads to a smaller value of k. Therefore, the $T = 0$ fixed point is unstable, and the $T \to \infty$ fixed point is stable. Therefore, we recover the result that the system becomes disordered. This is shown in Fig. 18.8.

Fig. 18.8 Flow of the parameter k of the one-dimensional Ising model. The low-temperature fixed point is unstable, and the flow is to the infinite temperature fixed point, leading to a disordered system

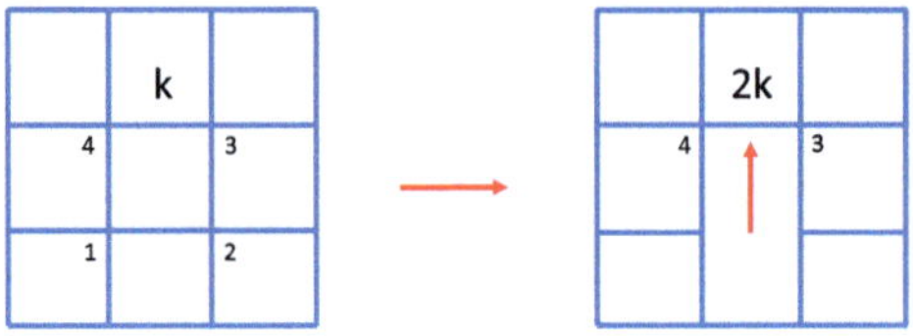

Fig. 18.9 Illustration of the method of bond-moving. One of the bonds is moved in a way that coalesces with a neighboring bond that has now a coupling that has doubled

Higher-Dimensional Systems: Bond-Moving

Higher-dimensional systems are in general more complicated, and several choices for the construction of blocks have been considered. Also, in general, new types of interactions may arise as a result of the choice of blocks. For instance, in a two-dimensional square lattice, we may take as a cell a collection of four sites (four spins) that form a square with sides the lattice constant of the square lattice. Consider now $b = 2$, and imagine a transformation that eliminates every other cell, in both directions. As these alternating cells are eliminated, new terms of the type of interactions between second neighbors arise, which were not present in the original Hamiltonian and which complicate the iteration procedure.

One convenient way to decimate the system of spins that has been used is called bond-moving. This is represented in Fig. 18.9. We consider a unit cell and label the sites from 1 to 4, for instance. The idea is to eliminate, for instance, the interaction between spins 1 and 2 (in the horizontal axis) and move the bond such that the interaction between spins 3 and 4 is now $2k$. The potential term that may be added to the Hamiltonian to reproduce this change is

$$V(\sigma) = k\,(\sigma_3\sigma_4 - \sigma_1\sigma_2)\,. \tag{18.99}$$

Using translational invariance, we can see that this transformation satisfies the condition

$$\langle V\rangle_H = 0. \tag{18.100}$$

It can be seen easily that the energy of one state with all spins parallel is not changed by the bond-moving operation. Therefore, at zero temperature, if the system is ferromagnetic, the transformation is exact.

Another type of transformation is designated Migdal-Kadanoff transformation. As an example, consider the Ising model in d-dimensions (hypercubic lattice). The Migdal transformation, for an isotropic spin system, is written as

$$k' = R^{(b)}(b^{d-1}k). \tag{18.101}$$

As we have seen, it is exact for $d = 1$. Consider, for instance, $d = 2, b = 3$. The procedure starts using a bond-moving method. Consider a square cell of size $3a$ with 16 sites. The horizontal bonds are moved up and down, and the vertical bonds are moved right and left. Recalling that the cell has neighboring cells where the same bond-moving is performed, we get a square of lateral size $3a$ (as before), with 12 spins and with interactions between the spins (that are only located at the border of the cell, since the spins in the interior have been eliminated) that has a magnitude of $3k$. Now, the edges of this enlarged cell may be viewed as $1d$ Ising chains. We may now consider the transformation used in the $1d$ case, taking $b = 3$ and reducing the set of four spins along each edge to just two, by eliminating two of the spins. This leads to a $1d$ decimation of the type

$$k' = \tanh^{-1}\left(\tanh^3(3k)\right). \tag{18.102}$$

This is illustrated in Fig. 18.10.

Consider now anisotropic systems with different coupling constants, depending on the space direction, as $k_1, k_2, \cdots, k_d$ in a d-dimensional space. The procedure is now similar to the isotropic case. The Migdal-Kadanoff transformation can be taken in the following steps: (i) k_1 is not moved and all the others move. For instance, taking $d = 2$, we have for a similar lattice of size $3a$ along x the interactions with value k_1 between the sites, and along y, we get that the couplings are now $3k_2$. (ii) We may now decimate the coupling k_1 leading to a coupling $\tilde{k}_1$ given by $\tilde{k}_1 = R^{(3)}(k_1)$. In general, we would have $\tilde{k}_1 = R^{(b)}(k_1), \tilde{k}_2 = bk_2, \cdots, \tilde{k}_d = bk_d$. (iii) Now, consider bond-movings such that k_2 is not moved, but all the others are moved. As we move, for instance, the coupling along the x direction, the only remaining coupling along that direction is $3\tilde{k}_1$. So, we have a coupling along x that is $3\tilde{k}_1$

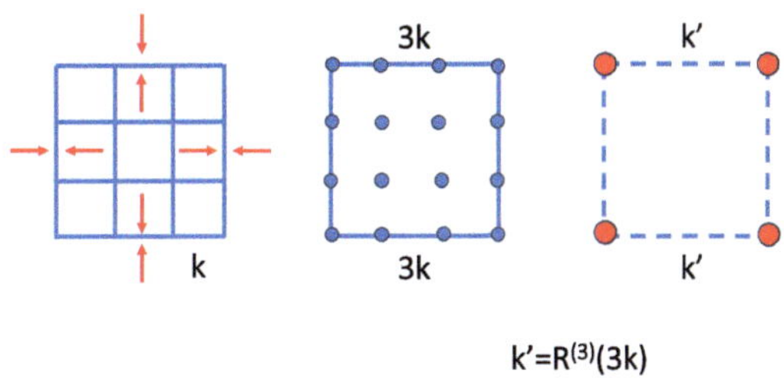

Fig. 18.10 Migdal-Kadanoff transformation

along x between the two edges of the square with side $3a$ and a coupling along y that has a value $3k_2$ between the three lattice sites in the same square. (iv) Decimate now k_2. This eliminates sites along y, leading now to a square that only has spins at its four edges, with a coupling along y given by $k_2' = R^{(3)}(3k_2)$. As a result of the whole process, the coupling along x is now $k_1' = 3R^{(3)}(k_1)$.

Repeating the procedure in all directions, we get for the anisotropic case considered by Kadanoff,

$$k'_\alpha = b^{d-\alpha} R^{(b)} \left(b^{\alpha-1} k_\alpha \right), \tag{18.103}$$

with $\alpha = 1, 2, \cdots, d$. In the case of isotropic couplings (Migdal), we get

$$k' = R^{(b)} \left(b^{d-1} k \right). \tag{18.104}$$

We may make a change of variables of the Migdal case taking $k_\alpha \to b^{\alpha-d} k_\alpha$ and $k'_\alpha \to b^{\alpha-d} k'_\alpha$. Then, we get the same expression as Kadanoff

$$k' = b^{d-\alpha} R^{(b)} \left(b^{\alpha-1} k \right). \tag{18.105}$$

Let us study the transformation in more details. Consider that b is now a continuous parameter, and choose $b = 1 + \delta l$. Inserting in the transformation, considering that the coupling also changes by a small amount $k' = k + \delta k$, and simplifying to the isotropic case, we get

$$\begin{aligned} k + \delta k &= (1+\delta l)^{d-\alpha}\, R^{(1+\delta l)} \left[(1+\delta l)^{\alpha-1} k \right] \\ &= (1+\delta l)^{d-\alpha} \tanh^{-1}\{ \left[\tanh\left((1+\delta l)^{\alpha-1} k \right) \right]^{1+\delta l} \} \\ &= (1+\delta l)^{d-\alpha} \tanh^{-1}\{ \tanh\left[(1 + (\alpha-1)\delta l) \right] [1 + \delta l \ln(\tanh k)] \} + O(\delta l)^2 \\ &= (1+\delta l)^{d-\alpha} \left[(1 + (\alpha-1)\delta l)\, k + \delta l \left(\frac{1}{2} \sinh(2k) \ln(\tanh k) \right) \right]. \end{aligned} \tag{18.106}$$

Consider now the variation $\delta k/\delta l$, and take the continuum limit such that we get a derivative dk/dl. This can be written as

$$\frac{dk}{dl} = (d-1)k + \frac{1}{2} \left[\sinh(2k) \ln(\tanh k) \right]. \tag{18.107}$$

At $d = 1$, we get the equation

$$\frac{dk}{dl} = \frac{1}{2} \sinh(2k) \ln(\tanh k). \tag{18.108}$$

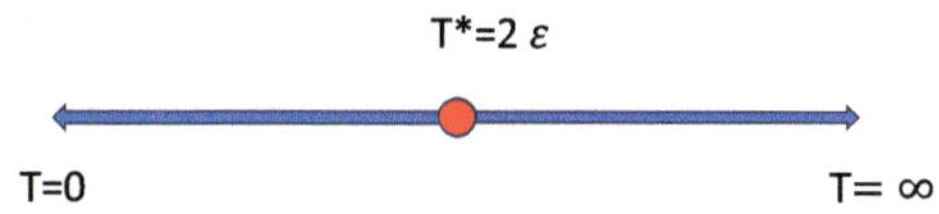

Fig. 18.11 Trajectory flow for the $2d$ Ising model. There is a repulsive fixed point at $T^* = 2\epsilon$, implying that the flow is either to the $T = 0$ solution or to the $T = \infty$ solution. Therefore, there is a phase transition between the two regimes

Away from one dimension, we may define $\epsilon = d - 1$, supposed to be small. Recalling that temperature is $T = k^{-1}$, we may obtain that

$$\frac{dT}{dl} = -\epsilon T + \frac{1}{2}T^2, \tag{18.109}$$

where we look for a solution for small temperature, which means we can expand the functions for large k.

We may now look for fixed points where $dT/dl = 0$. This leads to $\epsilon T = T^2/2$ or $T^* = 2\epsilon$. Since ϵ is small, this is also a small temperature effect. As ϵ vanishes, the fixed point goes to zero, as expected of the $d = 1$ case. Looking at the stability, we see that the fixed point $T^* = 2\epsilon$ is unstable, and so, any point in its vicinity will either flow to $T = 0$ or to $T = \infty$, as shown in Fig. 18.11.

Considering a small deviation from the fixed point $\delta T = T - T^*$, we get that

$$\frac{d(\delta T)}{dl} = (-\epsilon + T^*)\delta T - \epsilon\delta T = \lambda_T \delta T. \tag{18.110}$$

For $T < 2\epsilon$ (which means $\delta T < 0$), we get that $T \to 0$ (since $\lambda_T > 0$). If $T > 2\epsilon$ (which means $\delta T > 0$), we get that $T \to \infty$, as mentioned. The solution may be written as

$$\delta T(l) = e^{\epsilon l}\delta T(0). \tag{18.111}$$

Since $b = 1 + \delta l$, we may write that $b = e^{\delta l}$. Rescaling the lattice constant of the lattice as $a \to e^l a$ leads to a scaling of the correlation length as

$$\xi(l) = e^{-l}\xi(0). \tag{18.112}$$

Defining the critical exponent ν in the usual way

$$\xi(l) \sim (\delta T(l))^{-\nu}, \tag{18.113}$$

we get that

$$\frac{\xi(l)}{\xi(0)} = e^{-l} = \left(\frac{\delta T(l)}{\delta T(0)}\right)^{-\nu} = \left(e^{\lambda_T l}\right)^{-\nu}. \tag{18.114}$$

This implies that

$$\nu = \frac{1}{\lambda_T} = \frac{1}{\epsilon}. \tag{18.115}$$

Extrapolating to two dimensions, we get that for the Ising model in $2d$, we should get $\nu = 1$, since in this case $\epsilon = 1$.

Summarizing the renormalization group procedure in real space, we have to take the following steps:

(i) Apply the Kadanoff transformation to the original Hamiltonian

$$H(0) \to H''. \tag{18.116}$$

(ii) Define the spins of the blocks σ_x, where x is a spatial location.

(iii) Return to the original size of the lattice $x' = x/b$, rescale the spins $\sigma_x \to \lambda_s \sigma_{x'}$, and redefine the Hamiltonian

$$H'(\sigma) = \left[H''(\sigma)\right]_{\sigma_x \to \lambda_s \sigma_{x'}}. \tag{18.117}$$

This leads to a set of new couplings $\mathbf{k}' = R_s\mathbf{k}$. The operations satisfy the property $R_s R_{s'} = R_{ss'}$. The set of operations is a semigroup since it has no inverse.

18.3 Exercises

1. A two-dimensional model has critical exponents $\gamma = 7/6$ and $\delta = 15$. Determine the critical exponents α, β, η, and ν.
2. Show that

$$\delta = \frac{d+2-\eta}{d-2+\eta}.$$

3. Several materials, such as liquid crystals, are very anisotropic. The anisotropy is reflected in the form of the gradient term in the Landau-Ginzburg expansion in terms of the order parameter. In a system in d dimensions, we define the direction parallel to a given axis, $x_{||}$, and $d-1$ perpendicular directions, $\mathbf{x}_\perp$. Considering an order parameter m leads to a Hamiltonian in the presence of an external source h[6]

$$\beta H = \int dx_{||} \int d\mathbf{x}_\perp \left[\frac{K}{2}\left(\nabla_{||} m\right)^2 + \frac{L}{2}\left(\nabla_\perp^2 m\right)^2 + \frac{t}{2}m^2 - hm\right].$$

[6] Exercise adapted from [ChaikinLubensky].

(a) Making a Fourier transform and rewriting the Hamiltonian in terms $m(q_{||}, \mathbf{q}_{\perp})$, construct a renormalization group transformation for βH performing an anisotropic rescaling such that $q'_{||} = bq_{||}$, $\mathbf{q}'_{\perp} = c$, and $\mathbf{q}_{\perp}$ and $m' = m/z$.
(b) The fixed point is obtained imposing that $K' = K$ and $L' = L$. Calculate the eigenvalues y_t and y_h associated with the scalings $t' = tb^{y_t}$ and $h' = hb^{y_h}$.
(c) Write the free energy as

$$f(t, h) = t^{2-\alpha} f_s(h/t^{\Delta}),$$

and determine the exponents α and Δ. Also, determine the exponent γ of the susceptibility $\chi = \partial m/\partial h$.

4. (a) The interaction term with amplitude u in Eq. (18.55), in second-order perturbation theory, in addition to contributing to the renormalization of the interaction in Eq. (18.68), also contributes a term to the renormalization of the parameter, r. Show that such a term, proportional to u^2, involves two integrals over momentum, in the shell with absolute value between Λ/l and Λ. Therefore, this term is proportional to $(\delta l)^2$.
(b) Consider adding to Eq. (18.55) a cubic term of the type

$$\frac{v}{3!} \int_0^{\Lambda} d^d q_1 \int_0^{\Lambda} d^d q_2 \int_0^{\Lambda} d^d q_3 \phi(q_1)\phi(q_2)\phi(q_3)\delta(q_1 + q_2 + q_3).$$

While the quartic term in the coupling u scales as l^{4-d}, the cubic term scales as $l^{(6-d)/2}$. Therefore, at the upper critical dimension of the ϕ^4 theory, $d = 4$, the cubic term scales linearly with l. Therefore, as l increases, the cubic term becomes increasingly larger.
(c) Considering second-order perturbation theory in the coupling v, there is a contribution to the renormalization of r. Find this contribution (note that it involves an integration over momentum).

Chapter 19
Berezinskii-Kosterlitz-Thouless Transition

If the spatial dimension equals or is smaller than the lower critical dimension, there is no phase transition at finite temperature. In the case of a discrete symmetry, such as the Ising model, the lower critical dimension is $d_c = 1$. This was shown explicitly showing that, in the case of one dimension, there is no spontaneous magnetization (see Eq. (15.55)), while in two dimensions there is a phase transition at finite temperatures, as illustrated in Fig. 15.5. These results were also shown using a renormalization group approach, as concluded from Eq. (18.98) (see also Fig. 18.8) and from the analysis of Eq. (18.109) (see also Fig. 18.11). In the case of systems with a broken continuous symmetry, the Hohenberg-Mermin-Wagner theorem states that, for instance, spin systems with continuous rotation symmetry, the lower critical dimension is $d_c = 2$. So, for instance, for the XX model, the Hohenberg-Mermin-Wagner predicts that at $d = 2$ there is no finite temperature phase transition. The result that the theorem proves is that there is no spontaneous symmetry breaking at $d = 2$ and no nonvanishing magnetization. In particular, there is no long-range order at any finite temperature.

However, other correlation functions may signal the transition. This is the content of the Berezinskii-Kosterlitz-Thouless (BKT) transition. It will be shown that there are two regimes, as a function of temperature, where the spatial dependence of the correlation function between two points in space changes. This is associated with the presence of vortices and a change in their density. At low temperatures, there are few vortices. As the temperature increases, one reaches a point when the vortices proliferate and disorder the system and a new regime is found.

19.1 Vortices and Introduction to the BKT Transition

Vortices occur in a superconductor and in a rotating superfluid and may also be defined in spin systems, such as in the two-dimensional XX model. In this system,

P. D. Sacramento, *Introduction to Condensed Matter Theory*,
https://doi.org/10.1007/978-3-031-86251-9_19

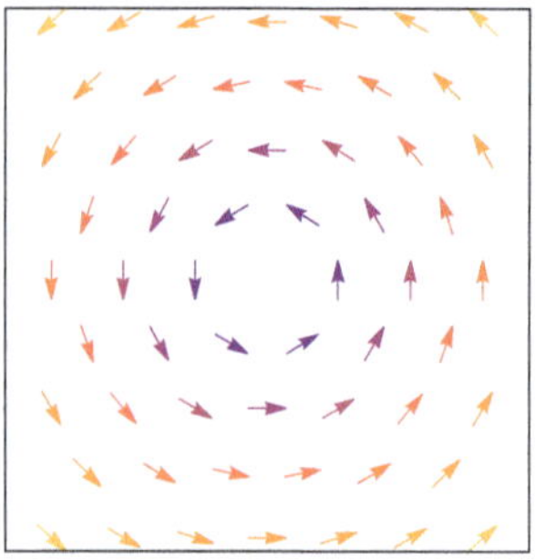

Fig. 19.1 Vortex in a spin system constituted by a rotation of the spins around a given space point, which defines the location of the vortex

the classical spins are vectors defined in a plane, and vortex-like excitations may be considered, taking the orientations of the spins as rotating around a central point, as illustrated in Fig. 19.1.

The classical ferromagnetic two-dimensional XX model previously introduced can be written as[1]

$$H = -\tilde{J} \sum_{\langle n,m \rangle} \mathbf{S}_n \cdot \mathbf{S}_m = -J \sum_{\langle n,m \rangle} \cos(\theta_n - \theta_m), \tag{19.1}$$

in terms of the variable θ_n, which is the angle of the classical spin vector at site n, in relation to a given direction, and where $\tilde{J} > 0$, which implies that in the ground state the spins are aligned. Here, $J = \tilde{J} S^2$, where S is the magnitude of the spin vector.

The partition function is $Z = \int D\theta e^{-S}$. The "action" is given by

$$\begin{aligned} S &= \beta J \sum_{\langle n,m \rangle} \cos\left(\theta_n - \theta_m\right) \\ &= \frac{1}{2} \beta J \sum_{\langle n,m \rangle} \left[e^{i(\theta_n - \theta_m)} + e^{-i(\theta_n - \theta_m)} \right]. \end{aligned} \tag{19.2}$$

We may now consider the limits of low and high temperatures. Even though the Hohenberg-Mermin-Wagner theorem states that no local order parameter may be spontaneously generated and that there is no long-range order, the correlation function

$$\langle e^{i\theta_0} e^{-i\theta_n} \rangle = \frac{1}{Z} \int \left(\prod_m d\theta_m \right) e^{i(\theta_0 - \theta_n)} e^{-\frac{J}{k_B T} \sum_{\langle n',m' \rangle} \cos(\theta_{n'} - \theta_{m'})} \tag{19.3}$$

[1] Here we follow [AraujoSacramento].

will be shown to distinguish the different limiting temperature regimes. In the thermodynamic limit, this correlation function vanishes when the distance between two sites diverges, $|n| \to \infty$, but the asymptotic behavior is different in the limits of vanishing temperature and very high temperature. Consistently, the local correlation functions $\langle e^{i\theta_0} \rangle = \langle e^{-i\theta_n} \rangle = 0$ vanish.

Consider first the regime of high temperatures and expand the Boltzmann factor. Using that

$$\int_0^{2\pi} d\theta_m = 2\pi, \int_0^{2\pi} d\theta_m e^{i\theta_m} = 0,$$

the leading term of the expansion is

$$\langle e^{i\theta_0} e^{-i\theta_n} \rangle \propto \left(\frac{J}{k_B T} \right)^{|n|} = e^{-|n| \log\left(\frac{k_B T}{J}\right)}. \tag{19.4}$$

It vanishes exponentially with the distance, $|n|$.

In the limit of low temperatures, since the system is ferromagnetic, we expect that neighboring spins are almost aligned and that the angles θ_n change slowly. Therefore, the angles between neighboring spins, $\theta_n - \theta_m$, are small. We obtain

$$H - E_0 \propto \frac{1}{2} J \sum_{\langle n,m \rangle} (\theta_n - \theta_m)^2 \propto \frac{1}{2} J \sum_r (\Delta\theta(r))^2 . \tag{19.5}$$

The correlation function is, then, written as a gaussian integral

$$\langle e^{i\theta_0} e^{-i\theta_n} \rangle \propto \frac{1}{Z} \int \prod_m d\theta_m e^{i(\theta_0 - \theta_n)} e^{-\frac{J}{2k_B T} \sum (\Delta\theta)^2} \tag{19.6}$$

and approximated by

$$\langle e^{i\theta_0} e^{-i\theta_n} \rangle \propto e^{\frac{k_B T}{J} \Delta(n)}, \tag{19.7}$$

where for high values of $|n|$,

$$\Delta(n) \propto -\frac{1}{2\pi} \log |n|. \tag{19.8}$$

This result leads to a power law for the decay with distance of the phase-phase correlation function

$$\langle e^{i\theta_0} e^{-i\theta_n} \rangle \propto \left(\frac{1}{|n|} \right)^{\frac{k_B T}{2\pi J}}, \tag{19.9}$$

in contrast to the exponential decrease in the high-temperature regime and with an exponent that depends on the temperature and on the coupling constant between the spins.

The high- and low-temperature regimes are therefore different. The two regimes may be understood in terms of the existence of a few vortices in the low-temperature regime and due to the proliferation of vortices in the high-temperature regime. In addition to understanding the two regimes, we have to determine the evolution from one regime to the other if it takes place at a given temperature or if it is a crossover.

Let us consider the possibility that the excitations that contribute to the low-temperature regime are the vortices illustrated in Fig. 19.1. In a vortex, the spins circle around the center of the vortex, and so, let us use a polar angle, $\varphi = l/\rho$, where l is the distance along the arc and ρ the radius of the circle, to characterize the orientation of the spin. Then,

$$\oint \nabla\varphi \cdot d\mathbf{l} = 2\pi \tag{19.10}$$

if the spins rotate once around the origin. Considering a vortex of size R, we obtain that the excitation energy is given by

$$\begin{aligned} H - E_0 &\propto J \int_a^R \rho d\rho \int_0^{2\pi} d\varphi (\nabla\varphi)^2 \\ &= J \int_a^R \rho d\rho \int_0^{2\pi} d\varphi \frac{1}{\rho^2} \\ &= 2\pi J \log\left(\frac{R}{a}\right), \end{aligned} \tag{19.11}$$

where we identified θ with φ and where a is of the order of the distance between nearest neighbor spins. The number of possible positions to place the vortex in the system is given by $(R/a)^2$. The entropy is, therefore, given by

$$S = k_B \log \Omega = 2k_B \log\left(\frac{R}{a}\right), \tag{19.12}$$

where Ω is the number of configurations. Note that when $R \to \infty$ and/or $a \to 0$ the vortex costs infinite energy and the entropy also diverges in the thermodynamic limit. Evaluating the free energy, we obtain

$$\begin{aligned} F &= U - TS \\ &= 2\pi J \log\left(\frac{R}{a}\right) - T2k_B \log\left(\frac{R}{a}\right) \\ &= 2\left(\pi J - k_B T\right) \log\left(\frac{R}{a}\right). \end{aligned} \tag{19.13}$$

(a)

(b)

Fig. 19.2 (**a**) Pair of vortex and antivortex and (**b**) pair of two vortices

There is a competition between the energy and the entropy terms separated by the condition $F = 0$. In the regime that the energy dominates, the presence of vortices is unfavorable, since they cost energy. In the entropy-dominated regime, it is favorable to have vortices, since it minimizes the free energy. The point $F = 0$ defines a temperature

$$k_B T_c = \pi J, \tag{19.14}$$

suggesting a transition point between the two limiting regimes of low and high temperatures. For $T < T_c$, it is unfavorable to have vortices, while for $T > T_c$ the existence of vortices is favorable. The temperature T_c defines the temperature for which the Berezinskii-Kosterlitz-Thouless transition takes place. The two regimes are distinguished by the presence of few vortices in the regime of low temperatures and by the proliferation of vortices in the high-temperature regime, due to the entropy gain.

When many vortices are thermally excited, the spins get disordered as illustrated in Fig. 19.2. The presence or absence of the vortices separates the two regimes. A complete description of the BKT transition requires an analysis by the renormalization group method, to be detailed in the next section.

19.2 Detailed Analysis of the Transition

Let us look again to the Boltzmann factor $\exp\{-H/(k_B T)\}$ with[2]

$$H = -J \sum_{\langle n,m \rangle} \cos(\theta_n - \theta_m), \tag{19.15}$$

[2] For further reading see J.B. Kogut, Reviews of Modern Physics **51**, 659 (1979), and C. Itzykson and J.M. Drouffe, *Statistical Field Theory*, Cambridge University Press.

where the sites n and m are nearest neighbors. The Boltzmann factor may be rewritten as

$$e^{-\beta[1-\cos(\theta_n-\theta_m)]} = \sum_{l=-\infty}^{\infty} e^{il(\theta_n-\theta_m)} I_l(\beta), \tag{19.16}$$

where $I_l(\beta)$ is the Bessel function with imaginary argument; we have taken $J = 1$ and multiplied by $e^{-\beta}$.

In the regime of low temperatures $T \to 0$, $(\beta \to \infty)$, we obtain the expansion

$$e^{-\beta[1-\cos(\theta_n-\theta_m)]} \sim \frac{1}{\sqrt{2\pi\beta}} \sum_{l=-\infty}^{\infty} e^{il(\theta_n-\theta_m)} e^{-\frac{l^2}{2\beta}}. \tag{19.17}$$

This is the Villain approximation. We get therefore a periodic gaussian model that we will assume is valid for arbitrary temperature. This approximation agrees well with a numerical solution and so validates the approximation.

In a more "elegant" notation (or compact notation), we write

$$e^{-\beta[1-\cos(\Delta_\mu(\mathbf{r})]} \to \frac{1}{\sqrt{2\pi\beta}} \sum_{n_\mu(\mathbf{r})=-\infty}^{\infty} e^{in_\mu \Delta_\mu \theta} e^{-\frac{n_\mu^2}{2\beta}}. \tag{19.18}$$

Here, $\mathbf{r}$ denotes the points on a $2d$ lattice and $n_\mu(\mathbf{r})$ is a vector field of integers. Finally, $\Delta_\mu(\theta(\mathbf{r}) = \theta(\mathbf{r}+\mu) - \theta(\mathbf{r})$, where μ is a direction along the plane (x, y). The partition function may be written as

$$Z = \int \left(\Pi_{\mathbf{r}'} d\theta(\mathbf{r}')\right) \Pi_{\mathbf{r},\mu} \sum_{n_\mu(\mathbf{r})=-\infty}^{\infty} e^{in_\mu \Delta_\mu \theta} e^{-\frac{n_\mu^2(\mathbf{r})}{2\beta}}. \tag{19.19}$$

We may now evaluate the integrals over $\theta(\mathbf{r}')$. These impose a constraint

$$\Delta_\mu n_\mu(\mathbf{r}) = n_\mu(\mathbf{r}+\mu) - n_\mu(\mathbf{r}) = 0. \tag{19.20}$$

We may write

$$Z = \Pi_{\mathbf{r},\mu} \sum_{n_\mu(\mathbf{r})=-\infty}^{\infty} \delta_{\Delta_\mu n_\mu(\mathbf{r}),0} e^{-\frac{n_\mu^2(\mathbf{r})}{2\beta}}. \tag{19.21}$$

The restriction implies that $n_\mu(\mathbf{r}) = \epsilon_{\mu\nu}\Delta_\nu n(\mathbf{r})$, where $n(\mathbf{r})$ is an integer scalar. The vanishing divergence $\nabla \cdot \mathbf{n} = 0$, since $\epsilon_{\mu\nu}$ is antisymmetric, implies that it may be written as

$$\mathbf{n} = \nabla \times \mathbf{A}. \tag{19.22}$$

It is now easy to show that

$$\sum_{\mu} n_{\mu}^{2}(\mathbf{r}) = \sum_{\mu} \left[\Delta_{\mu} n(\mathbf{r})\right]^{2}. \tag{19.23}$$

This leads to the expression for the partition function

$$Z = \sum_{n(\mathbf{r})=-\infty}^{\infty} e^{-\frac{1}{2\beta}\sum_{\mathbf{r},\mu}[\Delta_{\mu} n(\mathbf{r})]^{2}}. \tag{19.24}$$

We have replaced the original XX model at a temperature T, by a model of interactions between integers at a temperature $T^{*} = 1/T$. This is like a duality transformation which maps the high-temperature regime of one problem with the low-temperature regime of another one, and vice-versa.

Let us now replace the integers $n(\mathbf{r})$ by real numbers $\phi(\mathbf{r})$. This is accomplished using the Poisson summation formula

$$\sum_{n=-\infty}^{\infty} g(n) = \sum_{m=-\infty}^{\infty} \int_{-\infty}^{\infty} d\phi g(\phi) e^{i2\pi m\phi}, \tag{19.25}$$

where g is some arbitrary function. Note that the transformation preserves the periodicity. We may now write the partition function as

$$Z = \int_{-\infty}^{\infty} \Pi_{\mathbf{r}} d\phi(\mathbf{r}) \sum_{m(\mathbf{r})=-\infty}^{\infty} e^{-\frac{1}{2\beta}\sum_{\mathbf{r},\mu}(\Delta_{\mu}\phi)^{2}+2\pi i\sum_{\mathbf{r}} m(\mathbf{r})\phi(\mathbf{r})}. \tag{19.26}$$

We may now integrate the ϕ variables and get an expression in terms of the $m(\mathbf{r})$ variables. Anticipating the result, we will conclude that the $\phi(\mathbf{r})$ variables, angular variables, can be understood as spin waves, while the other variables, $m(\mathbf{r})$, are the vortices.

The exponent of the exponential term is given by

$$-\frac{1}{2\beta}\sum_{\mathbf{r},\mu}\left(\Delta_{\mu}\phi \cdot \Delta_{\mu}\phi\right) + 2\pi i \sum_{\mathbf{r}} m(\mathbf{r})\phi(\mathbf{r}). \tag{19.27}$$

Using the result that a gaussian integration (in this case over, the variables $\phi(\mathbf{r})$) is equivalent to considering the saddle point of the argument of the exponential (apart from a normalizing multiplicative factor, which can be absorbed in the partition function), we first consider the variation of the argument of the exponential, varying with respect to the variable ϕ, obtaining the equation

$$\Delta_{\mu}^{2}\phi(\mathbf{r}) = -2\pi i\beta m(\mathbf{r}). \tag{19.28}$$

Let us look for a solution of the type

$$\phi(\mathbf{r}) = \phi_0 + 2\pi i\beta \sum_{\mathbf{r}'} G(\mathbf{r}-\mathbf{r}')m(\mathbf{r}'), \tag{19.29}$$

where

$$\Delta_\mu^2 G(\mathbf{r}-\mathbf{r}') = -\delta(\mathbf{r}-\mathbf{r}'). \tag{19.30}$$

The exponent may be written as (integrating by parts the first term)

$$\frac{1}{2\beta}\sum_{\mathbf{r},\mu} \phi(\mathbf{r})\Delta_\mu^2\phi(\mathbf{r}) + 2\pi i \sum_{\mathbf{r}} m(\mathbf{r})\phi(\mathbf{r}). \tag{19.31}$$

Substituting the form of the solution, we get, taking $\phi_0 = 0$,

$$-2\pi^2\beta \sum_{\mathbf{r},\mathbf{r}'} m(\mathbf{r})G(\mathbf{r}-\mathbf{r}')m(\mathbf{r}'). \tag{19.32}$$

We get, reintroducing normalization factors,

$$Z = Z_{SW} \sum_{m(\mathbf{r})=-\infty}^{\infty} e^{-2\pi^2\beta\sum_{\mathbf{r},\mathbf{r}'} m(\mathbf{r})G(\mathbf{r}-\mathbf{r}')m(\mathbf{r}')}, \tag{19.33}$$

where the function $G(\mathbf{r}-\mathbf{r}')$ takes the role of the correlation function in a lattice for a massless (critical) theory.

We are left with the calculation of the function $G(\mathbf{r})$. Using that $\Delta^2 = \Delta_x^2 + \Delta_y^2$, where, for instance, $\Delta_x^2 f(\mathbf{r}) = f(\mathbf{r}+x) + f(\mathbf{r}-x) - 2f(\mathbf{r})$, we get that in a lattice, and taking a very large system,

$$G(\mathbf{r}) = \int_{-\pi}^{\pi} \frac{dk_x}{2\pi} \int_{-\pi}^{\pi} \frac{dk_y}{2\pi} \frac{e^{i\mathbf{k}\cdot\mathbf{r}}}{4 - 2\cos k_x - 2\cos k_y}. \tag{19.34}$$

For large values of $r = |\mathbf{r}|$, we get that the Fourier transform leads to

$$G(r) \sim -\frac{1}{2\pi}\ln\left(\frac{r}{a}\right) - \frac{1}{4}. \tag{19.35}$$

The exponential term can be written as

$$e^{2\pi^2\beta\sum_{\mathbf{r},\mathbf{r}'} m(\mathbf{r})\left[\frac{1}{2\pi}\ln\left(|\frac{r-r'}{a}|\right)+\frac{1}{4}\right]m(\mathbf{r}')}. \tag{19.36}$$

Therefore, the variables $m(\mathbf{r})$ interact with each other via a logarithmic potential, like vortices. Note that if $sgn(m(\mathbf{r})) = sgn(m(\mathbf{r}'))$ and $|r - r'|$ are large, the Boltzmann factor is large, which implies that the interaction is repulsive. If $sgn(m(\mathbf{r})) \neq sgn(m(\mathbf{r}'))$ and $|r - r'|$ are large, the Boltzmann factor is small, but if $|r - r'|$ is small, the Boltzmann factor gets larger which leads to some attractive interaction between the vortices. So, if the vortices attract, they tend to be very close, which implies that the dominant term is when $r = r'$. Let us write the partition function as

$$Z = Z_{SW} \sum_{m(\mathbf{r})=-\infty}^{\infty} e^{-2\pi^2\beta \sum_{\mathbf{r}\neq\mathbf{r}'} m(\mathbf{r})G(r-r')m(\mathbf{r}') - 2\pi^2\beta \sum_{\mathbf{r}} m(\mathbf{r})G(0)m(\mathbf{r})}. \tag{19.37}$$

The function $G(0)$ diverges. We may regularize it using both a cutoff for the system size and introducing the lattice constant, a. We can write, since the dominant terms are due to small momenta, $k_x, k_y \sim 0$, using polar coordinates in k-space,

$$G(0) = \frac{1}{2\pi} \int_0^{1/a} d\rho_k \rho_k \frac{1}{\rho_k^2 + \left(\frac{1}{R}\right)^2}. \tag{19.38}$$

This leads to the result

$$G(0) \sim \frac{1}{2\pi} \ln\left(\frac{R}{a}\right). \tag{19.39}$$

Recall that the Fourier transform of the type

$$\int d^d k \frac{e^{i\mathbf{k}\cdot\mathbf{r}}}{k^2 + \kappa^2}$$

is in the various spatial dimensions

$$\begin{aligned} 1d &\to \frac{\pi}{\kappa} e^{-\kappa|r|} \\ 2d &\to 2\pi K_0(\kappa|r|) \\ 3d &\to \frac{2\pi^2}{|r|} e^{-\kappa|r|} \end{aligned} \tag{19.40}$$

and that the modified Bessel function behaves logarithmically for small arguments.

The term that we separated in the contribution to the partition function that involves $G(0)$ is of the type $m^2 G(0)$ and is the contribution to the energy of an isolated vortex. We may more explicitly rewrite the various terms as

$$G(r) = G'(r) + G(0), \tag{19.41}$$

where

$$G'(r) = \int_{-\pi}^{\pi} \frac{dk_x}{2\pi} \int_{-\pi}^{\pi} \frac{dk_y}{2\pi} \frac{e^{i\mathbf{k}\cdot\mathbf{r}} - 1}{4 - 2\cos k_x - 2\cos k_y}$$

$$G(0) = \int_{-\pi}^{\pi} \frac{dk_x}{2\pi} \int_{-\pi}^{\pi} \frac{dk_y}{2\pi} \frac{1}{4 - 2\cos k_x - 2\cos k_y}, \tag{19.42}$$

and so we write explicitly

$$Z = Z_{SW} \sum_{m(\mathbf{r})} e^{-2\pi^2 \beta G(0) (\sum_{\mathbf{r}} m(\mathbf{r}))^2} e^{-2\pi^2 \beta \sum_{\mathbf{r},\mathbf{r}'} m(\mathbf{r}) G'(r-r') m(\mathbf{r}')}. \tag{19.43}$$

Since $G(0)$ is very large (it diverges if $R \to \infty$ and $a \to 0$), the Boltzmann factor will vanish unless

$$\sum_{\mathbf{r}} m(\mathbf{r}) = 0. \tag{19.44}$$

This is therefore a restriction on the distribution of vortices. This guarantees that the contributions that are significative to the partition function (and therefore to the sum over states) are those such that the number of vortices equals the number of antivortices or that the system is neutral. This is true if there is no magnetic field that tends to unbalance the number of vortices. So we may rewrite that the partition function is given by

$$Z = Z_{SW} \sum_{m(\mathbf{r})}{}' e^{-2\pi^2 \beta \sum_{\mathbf{r},\mathbf{r}'} m(\mathbf{r}) G'(r-r') m(\mathbf{r}')}, \tag{19.45}$$

where the $'$ in the sum means that only neutral configurations are to be taken into account. Explicitly, we get

$$Z = Z_{SW} \sum_{m(\mathbf{r})}{}' e^{\frac{\pi^2 \beta}{2} \sum'_{\mathbf{r},\mathbf{r}'} m(\mathbf{r}) m(\mathbf{r}')} e^{\pi \beta \sum'_{\mathbf{r},\mathbf{r}'} m(\mathbf{r}) \ln \frac{|r-r'|}{a} m(\mathbf{r}')}, \tag{19.46}$$

where the sums are restricted to $r \neq r'$.

Using now that $\sum_{\mathbf{r}} m(\mathbf{r}) = 0$, we get that

$$\sum_{\mathbf{r},\mathbf{r}'}{}' m(\mathbf{r}) m(\mathbf{r}') = -\sum_{\mathbf{r}} m^2(\mathbf{r}) \tag{19.47}$$

and

$$Z = Z_{SW} \sum_{m(\mathbf{r})}' e^{-\frac{\pi^2\beta}{2}\sum_{\mathbf{r}} m^2(\mathbf{r}) + \pi\beta \sum'_{\mathbf{r},\mathbf{r}'} m(\mathbf{r}) \ln \frac{|r-r'|}{a} m(\mathbf{r}')}. \tag{19.48}$$

This expression for the partition function is formally equivalent to that of a classical model of (unit) charges (positive and negative) of a neutral system in a $2d$ system, like a Coulomb gas, where the $m(\mathbf{r})$ are the charges. There is a chemical potential term and a term that can be seen as the Coulomb interaction in the $2d$ system.

As discussed in the previous section, if $\beta \to \infty$ (or $T \to 0$), the chemical potential term suppresses the number of vortices. Only the spin wave term contributes. If there are vortices, they occur in pairs of vortex-antivortex that are strongly bound, and so are in the vicinity of each other. Far from the location of the pair, their effect cancels out. If now we consider that the temperature is larger or β decreases, the pairs may separate since the entropy allows. Long-range interactions are still present which means the vortices are not free. They behave as a plasma of almost free charges, but in the presence of some screening, that is generated dynamically.

Scaling Analysis of BKT Transition

A proper description of the transition between the two regimes, of few vortices at low temperatures and a proliferation of vortices at high temperatures, can be carried out using a renormalization group approach. Therefore, a low-energy effective theory (long distances) needs to be considered. As before, we may follow a procedure of progressive elimination of high-energy excitations (or small distances of the order of the microscopic scale).

Integrating the degrees of freedom, we expect, from the previous analysis, a term that results from the integration of the spin wave degrees of freedom (or the oscillations in the phases), which leads to the appearance of a chemical potential like term, and also that interactions between the vortex degrees of freedom appear. In the spirit of the renormalization group, terms that are generated by a process of elimination should be included from the start in the action (or Hamiltonian). Therefore, we have to restart and consider the description at a level where the phase degrees of freedom have not been integrated. This leads to a previous expression with the addition of a term that explicitly considers the appearance of the chemical potential term. The starting point for the renormalization procedure is therefore the partition function Eq. (19.26) changed into

$$Z(y) = \int_{-\infty}^{\infty} \prod_{\mathbf{r}} d\phi(\mathbf{r}) \sum_{m(\mathbf{r})} e^{-\frac{1}{2\beta}\sum_{\mathbf{r},\mu}(\Delta_\mu\phi)^2 + \ln y \sum_{\mathbf{r}} m^2(\mathbf{r}) + 2\pi i \sum m(\mathbf{r})\phi(\mathbf{r})}, \tag{19.49}$$

where we have introduced a new variable (activity) that is associated with the chemical potential

$$y = y_0 e^{-\frac{\pi^2\beta}{2}}, \tag{19.50}$$

where at the initial stage of the renormalization procedure we have $y_0 = 1$.

We have to consider both the spin waves and the vortex parts of the partition function. The vortex position correlations contribute to the stiffness of the spin waves. Going back to the formulation of Eq. (19.48), since the chemical potential is large, we get that y is small. A dominant configuration is to consider two charges of opposite signs ($m(\mathbf{r}_1) = -m(\mathbf{r}_2)$, $\sum_{\mathbf{r}} m^2(\mathbf{r}) = 2$). The correlation between vortex positions is of the type

$$\langle m(\mathbf{r}_1)m(\mathbf{r}_2)\rangle \sim -y^2 e^{-2\pi\beta \ln\left(\frac{(\mathbf{r}_1-\mathbf{r}_2|}{a}\right)} \sim -y^2\left(\frac{a}{|\mathbf{r}_1-\mathbf{r}_2|}\right)^{2\pi\beta}. \tag{19.51}$$

This leads to the renormalization[3]

$$\begin{aligned}\frac{1}{\beta_{eff}} &= \frac{1}{\beta} - \frac{\pi^2}{2}\sum_{|r|>a} r^2\langle m(\mathbf{0})m(\mathbf{r})\rangle \\ &= \frac{1}{\beta} + \pi^3 y^2 \int_a^\infty \frac{dr}{a}\left(\frac{a}{r}\right)^{2\pi\beta-3},\end{aligned} \tag{19.52}$$

using polar coordinates. The renormalization procedure is now carried out considering a rescaling of the lengths as $a \to \lambda a$. The variables y and β, associated with the activity (chemical potential) and the inverse temperature (or coupling) have now a dependence on the scale factor indicated as y_λ, β_λ. We consider now that y_λ is small and define $x = r/a$. We have the integral that we separate as

$$\begin{aligned}\int_1^\infty dx x^{3-2\pi\beta} &= \int_1^\lambda dx x^{3-2\pi\beta} + \int_\lambda^\infty dx x^{3-2\pi\beta} \\ &= \frac{\lambda^{4-2\pi\beta}-1}{4-2\pi\beta} + \lambda^{4-2\pi\beta}\int_1^\infty dx x^{3-2\pi\beta},\end{aligned} \tag{19.53}$$

where we have rescaled $x' = x/\lambda$. The integral leads to

$$\sim \ln\lambda + [1 + (4-2\pi\beta)\ln\lambda]\int_1^\infty dx x^{3-2\pi\beta}, \tag{19.54}$$

and, therefore,

$$\frac{1}{\beta_{eff}} = \frac{1}{\beta} + \pi^3 y^2 \ln\lambda + \pi^3 y^2 [1 + (4-2\pi\beta)\ln\lambda]\int_1^\infty dx x^{3-2\pi\beta}. \tag{19.55}$$

As $\lambda \to 1$, we can write that $\ln\lambda \sim \lambda - 1$ is small. This implies that

[3] See C. Itzikson and J. M. Drouffe, *Statistical Physics*, and [ChaikinLubensky].

$$\frac{1}{\beta_{eff}} = \frac{1}{\beta_\lambda} + \pi^3 y_\lambda^2 \int_1^\infty dx x^{3-2\pi\beta_\lambda}, \tag{19.56}$$

where we have defined

$$\frac{1}{\beta_\lambda} = \frac{1}{\beta} + \pi^3 y^2 \ln\lambda \tag{19.57}$$

and

$$y_\lambda^2 = y^2 \left[1 + (4 - 2\pi\beta)\ln\lambda\right]. \tag{19.58}$$

Since $\ln\lambda$ is small, we obtain that $y_\lambda = y + (2 - \pi\beta) y \ln\lambda$.

Summarizing, we have the equations

$$\begin{aligned} \frac{1}{\beta_\lambda} &= \frac{1}{\beta} + \pi^3 y^2 \ln\lambda \\ y_\lambda &= y + (2 - \pi\beta) y \ln\lambda. \end{aligned} \tag{19.59}$$

Taking a derivative with respect to the scaling factor, we get that

$$\begin{aligned} \lambda \frac{d}{d\lambda}\beta_\lambda^{-1} &= \pi^3 y_\lambda^2 \\ \lambda \frac{d}{d\lambda} y_\lambda &= (2 - \pi\beta_\lambda) y_\lambda. \end{aligned} \tag{19.60}$$

Initial conditions for the renormalization flow are

$$\lambda = 1;\ \beta_1 = \beta;\ y_1 = y = e^{-\pi^2\beta/2}. \tag{19.61}$$

Let us now define $x_\lambda = \pi\beta_\lambda - 2$, and redefining $y_\lambda = 2\pi y_\lambda$, we arrive at the equations

$$\begin{aligned} dx &= -y^2 \frac{d\lambda}{\lambda} \\ dy^2 &= -2xy^2 \frac{d\lambda}{\lambda}. \end{aligned} \tag{19.62}$$

The point $x = 0$, $y = 0$ is a fixed point and, expanding in its vicinity, leads to the previous equations. The equations lead to

$$dy^2 = dx^2, \tag{19.63}$$

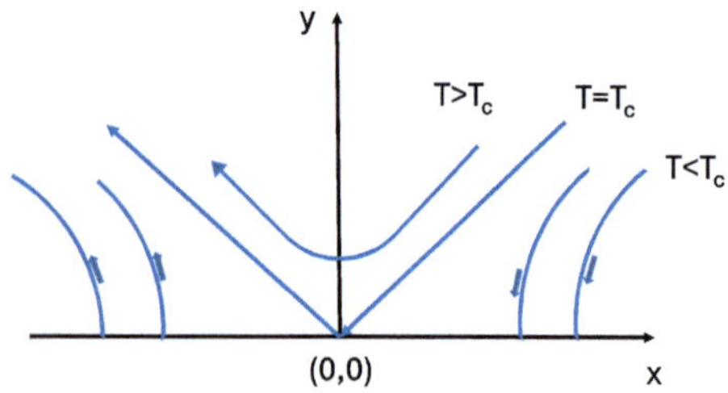

Fig. 19.3 Renormalization group trajectories for the BKT transition

which implies that

$$x^2 - y^2 = C, \tag{19.64}$$

a constant, and are therefore hyperbolae. The trajectories are shown in Fig. 19.3. The point $x = 0$ implies that $\pi\beta = 2$, which fixes the critical temperature, T_c, assuming the coupling constant is fixed. When $\pi\beta > 2$ which means that $k_B T < \pi J/2$, we are in the region of $x > 0$. In this region, the critical line $T = T_c$ is given by $x^2 - y^2 = 0$ or $x = y$, with the fixed point at the origin $x = y = 0$. Along the critical line, we get

$$\frac{\pi J}{k_B T_c} - 2 \sim e^{-\frac{\pi^2 J}{2k_B T_c}}. \tag{19.65}$$

If $T < T_c$, any point converges to $y = 0$ a point on the positive x axis. If $T > T_c$, any point flows to negative values of x. Around the critical flow line that converges to the fixed point, we can write that $x^2 - y^2 = x_0^2$. This leads to $x_0^2 = -ct$, with $c \sim 2.1$ and where we used the standard definition for the deviation from the critical temperature $t = (T - T_c)/T_c$.

Consider the regime where $T > T_c$ equivalent to $t > 0$. In this case, $x_0 = iy_0$, since $x^2 - y^2 = x_0^2$. Near the critical point, $x_0 \sim 0$ and $x \sim y$. Therefore, we get that

$$\frac{dx}{x^2} \sim -\frac{d\lambda}{\lambda}, \tag{19.66}$$

which leads to

$$x \sim \frac{1}{\ln \lambda}. \tag{19.67}$$

We saw that $x \sim t^{1/2}$. Taking the scale λ as the correlation length, we get that the correlation length has the dependency

$$\lambda \sim e^{bt^{-1/2}}, \tag{19.68}$$

which diverges faster than any power and, therefore, the exponent ν is not defined.

19.3 Exercises

1. (a) Determine the critical exponent, η, of the BKT transition, defined by

$$\langle e^{i(\theta_o - \theta_n)} \rangle \sim \left(\frac{1}{|n|} \right)^\eta .$$

   (b) Assuming that hyperscaling holds, determine the critical exponent, δ.

Chapter 20
Quantum Phase Transitions

20.1 Ising Model in a Transverse Field

The Ising model in a transverse field is defined by

$$H = -J \sum_{\langle i,j \rangle} S_i^x S_j^x - h \sum_i S_i^z, \tag{20.1}$$

where the sum i, j is over nearest neighbors, the interactions are among the x component of the spin-$1/2$ operators, and the magnetic field is along the z direction. Since there are two spin components, the model has a quantum nature due to the noncommutativity of the spin components, unlike the zero-field model that is classical. The commutation relations of the spin operators are complicated and is often useful to search for a representation of the spin operators in terms of bosons or fermions or a mixture. Since the Hilbert spaces of spin operators are not the same as those of bosons or fermions, in general, some projection operator is required. One such example was considered before when we used the Holstein-Primakoff transformation.

In $1d$ systems, there is one transformation that does not involve a projection to a physical subspace. This is the Jordan-Wigner transformation considered in the next subsection that maps the spin-$1/2$ problem to a system of spinless fermions. The transformation may be generalized to higher dimensions, but we will limit ourselves to the simpler case of $1d$.

20.1.1 *Jordan-Wigner Transformation*

Spin-$1/2$ operators may be represented using Pauli matrices, σ_i, as $S_i = 1/2\sigma_i$, with $i = x, y, z$ or $1, 2, 3$. The raising and lowering spin operators at site n are expressed

P. D. Sacramento, *Introduction to Condensed Matter Theory*,
https://doi.org/10.1007/978-3-031-86251-9_20

as $S^{\pm}(n) = S^x(n) \pm i S^y(n)$. At different sites, the spin components commute

$$[S^+(n), S^+(m)] = [S^-(n), S^-(m)] = [S^+(n), S^-(m)] = 0 \tag{20.2}$$

if $n \neq m$. At the same site, they obey the usual commutation relations of an angular momentum, and for spin-1/2, there are also anti-commutation relations (as in the Pauli matrices)

$$\begin{aligned} \{S^+(n), S^-(n)\} &= 1 \\ \{S^+(n), S^+(n)\} &= 0 \\ \{S^-(n), S^-(n)\} &= 0. \end{aligned} \tag{20.3}$$

The last two expressions reflect the well-known result that if we have a spin-1/2 operator we cannot raise or lower the spin component twice. Therefore, for instance, $S^+(n)$ can only create one excitation (one spin flip at a given site), and it is forbidden to create two spin excitations. So, if we think of these operators as bosonic, it means that the system behaves as hardcore bosons (since the occupation is restricted to one boson per site), or if we think in terms of fermions, it naturally includes Pauli's principle. So, it seems natural to attempt a fermionic description of the spin-1/2 operators. However, fermions obey anti-commutation relations everywhere and not only at the same site: they also obey anti-commutation relations when different sites are involved. Therefore, a fermionic description must incorporate this difference between the spin operators (that commute at different sites) and the fermions (that also anti-commute at different sites).

Consider the operator

$$K(n) = e^{i\pi \sum_{j=1}^{n-1} S^+(j)S^-(j)} = e^{i\pi \sum_{j=1}^{n-1}\left(S^z(j)+\frac{1}{2}\right)} = i^{n-1} e^{i\pi \sum_{j=1}^{n-1} S^z(j)}. \tag{20.4}$$

The last exponential term is an operator that performs a rotation of the spins from $j = 1, \cdots, n-1$ by π around the z axis. If we take an eigenstate of S^x at every site, we get that

$$\begin{aligned} S^x(n)|\frac{1}{2}, \frac{1}{2}, \cdots, \frac{1}{2}\rangle &= \frac{1}{2}|\frac{1}{2}, \frac{1}{2}, \cdots, \frac{1}{2}\rangle \\ K(n)|\frac{1}{2}, \frac{1}{2}, \cdots, \frac{1}{2}\rangle &= i^{n-1}|-\frac{1}{2}, -\frac{1}{2}, \cdots, -\frac{1}{2}, \frac{1}{2}, \frac{1}{2}, \cdots, \frac{1}{2}\rangle, \end{aligned} \tag{20.5}$$

where the spins are flipped along the x direction up to the $n-1$ location. We conclude that this operator creates a kink in the spin configuration, as shown in Fig. 20.1. This operator has a vanishing expectation value in a state with long-range order, but it may have a nonvanishing expectation value in some disordered state.

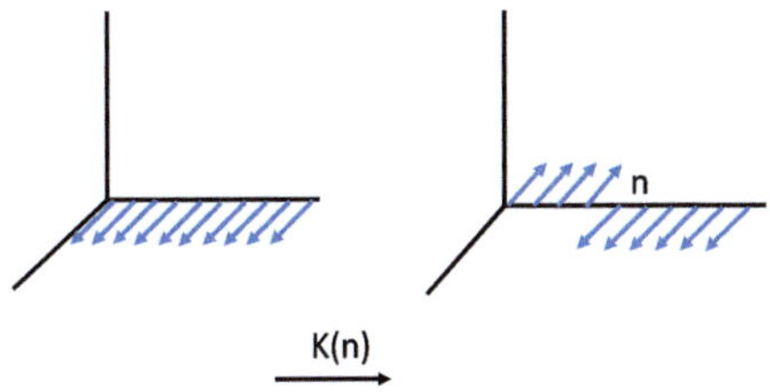

Fig. 20.1 Effect of kink operator on the distribution of spins along the x direction

Let us now define spinless fermionic operators in the following way:

$$c(n) = K(n)S^-(n) = e^{i\pi \sum_{j=1}^{n-1} S^+(j)S^-(j)} S^-(n)$$
$$c^\dagger(n) = S^+(n)K^\dagger(n) = S^+(n)e^{-i\pi \sum_{j=1}^{n-1} S^+(j)S^-(j)}. \tag{20.6}$$

Note that the kink operator is unitary: $K^\dagger(n)K(n) = K(n)K^\dagger(n) = I$. The fermionic destruction operator is related to the lowering operator, and the creation operator is related to the raising spin operator. The role of the kink operator is to give the proper commutation relations of the spin operators at different sites in terms of the anti-commutation relations of the fermionic operators. Or we may consider the inverse result that in terms of the spin operators we obtain fermionic operators.

The fermionic number operator is related with the z component of the spin operator. Indeed,

$$c^\dagger(n)c(n) = S^+(n)K^\dagger(n)K(n)S^-(n) = S^+(n)S^-(n) = \frac{1}{2} + S^z(n)$$
$$c(n)c^\dagger(n) = K(n)S^-(n)S^+(n)K^\dagger(n) = S^-(n)S^+(n) = \frac{1}{2} - S^z(n). \tag{20.7}$$

Since the spin operators commute at different sites, it is easy to see that

$$[S^\pm(n), K(n)] = 0. \tag{20.8}$$

This implies

$$c^\dagger(n)c^\dagger(n) = c(n)c(n) = 0, \tag{20.9}$$

as required of a fermionic operator. Consider now the identity

$$e^{\pm i\pi S^+(n)S^-(n)} = e^{\pm i\pi\left(\frac{1}{2}+S^z(n)\right)} = -2S^z(n). \tag{20.10}$$

Since $\{S^-(n), S^z(n)\} = 0$, it is easy to see that

$$S^-(n)K(m) = -K(m)S^-(n). \tag{20.11}$$

Therefore, for $m \neq n$,

$$
\begin{aligned}
c(n)c(m) &= K(n)S^-(n)K(m)S^-(m) \\
&= -K(n)K(m)S^-(n)S^-(m) \\
&= -K(m)S^-(m)K(n)S^-(n) \\
&= -c(m)c(n), \qquad (20.12)
\end{aligned}
$$

a fermionic anti-commutation relation. Therefore, the operators satisfy the usual fermionic anti-commutation relations

$$\{c(n), c(m)\} = \{c^\dagger(n), c^\dagger(m)\} = 0 \qquad (20.13)$$

and

$$\{c(n), c^\dagger(m)\} = \delta_{n,m}. \qquad (20.14)$$

These relations are the result of the action of the kink operator.

We may also define the inverse Jordan-Wigner transformation written as

$$
\begin{aligned}
S^-(n) &= e^{-i\pi \sum_{j=1}^{n-1} c^\dagger(j)c(j)} c(n) \\
S^+(n) &= c^\dagger(n) e^{i\pi \sum_{j=1}^{n-1} c^\dagger(j)c(j)}. \qquad (20.15)
\end{aligned}
$$

Anisotropic Heisenberg Chain

To illustrate the usefulness of the transformation, let us consider a more general model than the Ising model in a transverse field, such as the anisotropic Heisenberg chain. It can be written in zero magnetic field as

$$
\begin{aligned}
H &= J \sum_{j=1}^{N} \big(S_x(j)S_x(j+1) + S_y(j)S_y(j+1)\big) + \gamma J \sum_{j=1}^{N} S_z(j)S_z(j+1) \\
&= \frac{1}{2} J \sum_{j=1}^{N} \big(S^+(j)S^-(j+1) + S^-(j)S^+(j+1)\big) \\
&\quad + \gamma J \sum_{j=1}^{N} \left(S^+(j)S^-(j) - \frac{1}{2}\right)\left(S^+(j+1)S^-(j+1) - \frac{1}{2}\right). \qquad (20.16)
\end{aligned}
$$

The case of $\gamma = 1$ is the isotropic Heisenberg model and $\gamma = 0$ is the XX model.

Using the Jordan-Wigner transformation, we can obtain some equalities between the spin operators and the spinless fermionic operators. We get

$$
\begin{aligned}
S^+(j)S^-(j+1) &= c^\dagger(j)e^{-i\pi c^\dagger(j)c(j)}c(j+1) \\
&= c^\dagger(j)\left(1-2c^\dagger(j)c(j)\right)c(j+1) \\
&= c^\dagger(j)c(j+1)
\end{aligned} \tag{20.17}
$$

and

$$
\begin{aligned}
S^-(j)S^+(j+1) &= c(j)e^{i\pi c^\dagger(j)c(j)}c^\dagger(j+1) \\
&= c(j)\left(1-2c^\dagger(j)c(j)\right)c^\dagger(j+1) \\
&= c(j)c^\dagger(j+1) - 2c(j)c^\dagger(j)c(j)c^\dagger(j+1) \\
&= c^\dagger(j+1)c(j).
\end{aligned} \tag{20.18}
$$

Then the Hamiltonian can be written in terms of the fermionic operators

$$
\begin{aligned}
H = &\frac{1}{2}J\sum_{j=1}^{N}\left(c^\dagger(j)c(j+1)+c^\dagger(j+1)c(j)\right) \\
&+\gamma J\sum_{j=1}^{N}\left(n(j)-\frac{1}{2}\right)\left(n(j+1)-\frac{1}{2}\right).
\end{aligned} \tag{20.19}
$$

Note that even though the kink operator is a long-range term, the Hamiltonian in terms of the fermionic operators is short-range with only nearest-neighbor terms and the first term is like a hopping Hamiltonian, of tight-binding nature with width controlled by J. If the spin Hamiltonian has further neighbor terms, the fermionic Hamiltonian has a more complicated form since strings of operators will appear connecting the two sites coupled by the further neighbor couplings. In addition, the second term of the Hamiltonian is an interacting term in terms of the spinless fermions, which renders the solution of the problem complicated.

The third component of the spin operator, summed over all sites, is given by

$$
S^z = \sum_{j=1}^{N} S^z(j) = \sum_{j=1}^{N}\left(c^\dagger(j)c(j)-\frac{1}{2}\right) = N_F - \frac{N}{2}, \tag{20.20}
$$

where N_F is the number of fermions. So, a state with total $S^z = 0$ means that the fermions are half-filled $N_F = N/2$ (if N is even). The case of $S^z = 1/2$ means that $N_F = (N+1)/2$, which means N is odd.

Some care with boundary conditions has to be taken in the case of a finite system. For instance, let us suppose that the spin system is periodic (like in a chain closed in a ring). Then, we have that

$$S^i(N+1) = S^i(1) \tag{20.21}$$

for $i = x, y, z$. Consider now the fermionic operator

$$c(N+1) = e^{i\pi \sum_{j=1}^{N} S^+(j)S^-(j)} S^-(N+1) = e^{i\pi \sum_{j=1}^{N}\left(\frac{1}{2}+S^z(j)\right)} S^-(1). \tag{20.22}$$

But we know that $c(1) = S^-(1)$. Therefore, we get

$$c(N+1) = i^N e^{i\pi S^z} c(1) = e^{i\pi N_F} c(1). \tag{20.23}$$

If N is even and $S^z = 0$, we get periodic boundary conditions (or antiperiodic) if $N/2$ is even (or odd).

The case of $\gamma = 0$ (XX model) simplifies in the fermionic language, and we can see it is noninteracting

$$H_{XX} = \frac{1}{2} J \sum_{j=1}^{N} \left(c^\dagger(j)c(j+1) + c^\dagger(j+1)c(j)\right), \tag{20.24}$$

which is just a tight-binding model, expressed in terms of the fermions. It can be diagonalized transforming to momentum space (Fourier transform), and we get a system of free fermions (with no spin). The Hamiltonian is written as

$$H = \int_{-\pi}^{\pi} \frac{dk}{2\pi} \epsilon(k) c^\dagger(k) c(k), \tag{20.25}$$

with $\epsilon(k) = J\cos k$. At zero magnetization, the band is half-filled and the spectrum is gapless. Taking $J < 0$, the band is full for $|k| \leq \pi/2$.

We may also consider a spin-1/2 system that has anisotropy in the xy plane described by the Hamiltonian

$$H_{XY} = -\sum_{j=1}^{N} \left(J_x S_j^x S_{j+1}^x + J_y S_j^y S_{j+1}^y\right). \tag{20.26}$$

Defining $S^x = \left(S^+ + S^-\right)/2$ and $S^y = \left(S^+ - S^-\right)/(2i)$, we can write that

$$\begin{aligned} H_{XY} = &-\frac{1}{4} \sum_j \left[(J_x - J_y)\left(S_j^+ S_{j+1}^+ + S_j^- S_{j+1}^-\right)\right] \\ &-\frac{1}{4} \sum_j \left[(J_x + J_y)\left(S_j^+ S_{j+1}^- + S_j^- S_{j+1}^+\right)\right]. \end{aligned} \tag{20.27}$$

This is an interacting spin quantum problem that is also diagonalizable. A possible way consists in performing the inverse Jordan-Wigner transformation from the spin operators to spinless fermionic operators as above

$$\begin{aligned} S_j^+ &= c_j^\dagger e^{i\pi \sum_{n=1}^{j-1} c_n^\dagger c_n} \\ S_j^- &= e^{-i\pi \sum_{n=1}^{j-1} c_n^\dagger c_n} c_j, \end{aligned} \tag{20.28}$$

leading to

$$\begin{aligned} H_{XY} = &-\frac{1}{4}\sum_j \left[(J_x - J_y)\left(c_j^\dagger c_{j+1}^\dagger + c_{j+1} c_j \right) \right] \\ &-\frac{1}{4}\sum_j \left[(J_x + J_y)\left(c_j^\dagger c_{j+1} + c_{j+1}^\dagger c_j \right) \right]. \end{aligned} \tag{20.29}$$

This model is related to the fermionic Kitaev model, discussed in the next section (at vanishing chemical potential) if we rewrite it as

$$H = -t\sum_j \left(c_j^\dagger c_{j+1} + c_{j+1}^\dagger c_j \right) + \Delta \sum_j \left(c_j c_{j+1} + c_{j+1}^\dagger c_j^\dagger \right), \tag{20.30}$$

choosing $t = (J_x + J_y)/4$; $\Delta = (J_x - J_y)/4$. Therefore, if $J_y = 0$, we get that $\Delta = t$, and if $J_x = J_y$, we get that $\Delta = 0$. If $\Delta = 0$, we will show that the spectrum is gapless, and if $\Delta \neq 0$, there is a gap in the system. Therefore, $\Delta = 0$ is a critical point that separates two gapped phases.

Consider now the addition of a chemical potential term, $\mu \neq 0$. This can be traced back to a magnetic field in the XY model and adds a term of the type $-\mu \sum_j \left(c_j^\dagger c_j - \frac{1}{2} \right)$. Therefore, the chemical potential may be seen in the spin problem as a magnetic field along z. For any $|\mu| > 2t$, the band is either empty or full. For $|\mu| < 2t$, the spectrum is gapless if $\Delta = 0$, and there is a gap if Δ is finite.

If $J_x, J_y > 0$, the spin system is ferromagnetic at zero temperature. If $J_x > J_y$ ($J_x < J_y$), the spins order at zero temperature along the x direction (y direction). If $J_x = J_y$, the system is isotropic and critical with power law correlation functions. In the corresponding Kitaev model, the critical regime is the nonsuperconducting tight-binding model. If $J_x > J_y \to \Delta > 0$ and $J_x < J_y \to \Delta < 0$, in both cases, there is a gap that corresponds in the spin problem to the Landau like quasi-long-range order. Therefore, there is a mapping between an anisotropic XY spin model in a transverse magnetic field and the Kitaev model in a finite chemical potential.

20.1.2 Kitaev Model for a Spinless Triplet Superconductor

Even though superconductivity is not stable in one dimension, due to phase fluctuations, it is interesting to consider the toy model given by the Hamiltonian[1]

$$H = -t \sum_j \left(c_j^\dagger c_{j+1} + c_{j+1}^\dagger c_j \right) + \Delta \sum_j \left(c_j c_{j+1} + c_{j+1}^\dagger c_j^\dagger \right). \tag{20.31}$$

This model was obtained in the previous section from an XY model in zero magnetic field using a Jordan-Wigner transformation. Also, we may consider a one-dimensional system in proximity to a three-dimensional superconductor, such that the Cooper pairs, stable in the three-dimensional superconductor, tunnel into the one-dimensional system, providing the pairing terms. However, in its simpler form, the model actually contains spinless fermions. So, the pairing is not local but involves nearest neighbors. Also, we may understand the spinless nature of the fermions as actually spinful fermions (electrons) that have their spins parallel, and so, effectively, only one spin component is relevant. This means pairing between electrons which is not a spin singlet but part of a spin triplet. This is actually a possibility found in nature even though it is not so common, particularly in low dimension. As we saw before, there are superconductors such that the pairing is a spin singlet but with relative angular momentum other than $l = 0$, as long as l is odd. On the other hand, it can be shown that a one-dimensional triplet superconductor, like the one considered here and proposed by Kitaev, can be found as an effective model for a wire with spin-orbit coupling, in the presence of a conventional s-wave superconductor ($l = 0$, $S = 0$) and under an applied magnetic field [1].

As done for the XX model, the Kitaev model may be written in momentum space defining

$$c_j = \frac{1}{\sqrt{N}} \sum_k e^{ikj} c_k. \tag{20.32}$$

The Hamiltonian may be written as

$$H = \frac{1}{2} \sum_k \begin{pmatrix} c_k^\dagger & c_{-k} \end{pmatrix} \begin{pmatrix} \epsilon_k & -2i\Delta \sin k \\ 2i\Delta \sin k & -\epsilon_k \end{pmatrix} \begin{pmatrix} c_k \\ c_{-k}^\dagger \end{pmatrix}, \tag{20.33}$$

where $\epsilon_k = -2t \cos k$. There is a 2×2 structure as a result of the superconducting nature, that may be traced back to the particle and hole contributions in the BCS theory. Also, there is an off-diagonal term that arises due to the superconducting term. The spectrum is given by

---

[1] See also [AraujoSacramento].

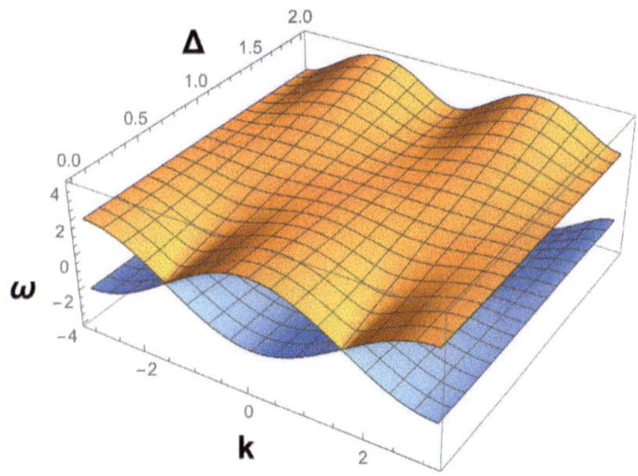

Fig. 20.2 Dispersion relation for the anisotropic XY model or the corresponding Kitaev model as a function of $\Delta = (J_x - J_y)/4$ and for $t = (J_x + J_y)/4 = 1$

$$\omega_k = \pm\sqrt{\epsilon_k^2 + 4\Delta^2 \sin^2 k} \tag{20.34}$$

and is shown in Fig. 20.2. At $\Delta = 0$, the spectrum is gapless if $k = \pm\pi/2$ and describes the transition between phases I and II, at zero chemical potential (in the spin problem at zero magnetic field).

Going back to real space, let us remark that the spinless fermionic operators may be written in terms of fermionic Hermitian, called Majorana, operators, as $c_j = \frac{1}{2}\left(\gamma_{1,j} + i\gamma_{2,j}\right)$. These operators obey anti-commutation rules,

$$\{\gamma_{m,i}, \gamma_{n,j}\} = 2\delta_{ij}\delta_{mn}I, \tag{20.35}$$

where $m, n = 1, 2$ and I is the identity. We use the normalization $\gamma_{m,j}^2 = 1$. In terms of these operators, the Hamiltonian can be written as

$$H = \frac{i}{2}\sum_j \left[(-t + \Delta)\,\gamma_{1,j}\gamma_{2,j+1} + (t + \Delta)\,\gamma_{2,j}\gamma_{1,j+1}\right] - \frac{i}{2}\mu\sum_j \gamma_{1,j}\gamma_{2,j}. \tag{20.36}$$

Note that at some particular point ($\mu = 0, t = \Delta$) the Hamiltonian becomes quite simple

$$H = it\sum_j \gamma_{2,j}\gamma_{1,j+1}. \tag{20.37}$$

Note that if the system is finite with size N and, therefore, $j = 1, \cdots, N - 1$, all operators are linked, two by two, with two exceptions. The operators $\gamma_{1,1}$ and $\gamma_{2,N}$ do not appear in the Hamiltonian and are associated with zero-energy edge states (since they do not appear in the Hamiltonian, they have zero energy). The existence of these zero-energy edge states is linked to a topological nature. Calculating the

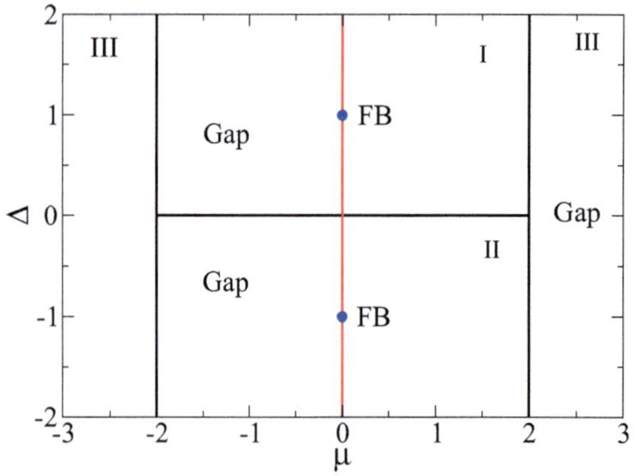

Fig. 20.3 Phase diagram for the Kitaev model (or anisotropic XY model) as a function of Δ and the chemical potential μ and for $t = 1$

winding number for the Kitaev model, at this particular point, $(\mu = 0, t = \Delta)$, one finds a nonzero value for the winding number. Actually, the topology appears in a finite region limited by the chemical potential $|\mu| < 2t$ and any value of Δ. For larger values of $|\mu|$, the winding number vanishes, and there are no edge states. In Fig. 20.3, we show the phase diagram of the Kitaev model. It is left as an exercise at the end of the chapter to show that the three phases are characterized by different winding numbers and zero-energy edge states (or their absence).

If we define nonlocal operators $d_j^\dagger = \frac{1}{2}\left(\gamma_{2,j} - i\gamma_{1,j+1}\right)$, they are related to the other fermionic operators by

$$d_j^\dagger = \frac{1}{2}\left[-i(c_j - c_j^\dagger) - i(c_{j+1} + c_{j+1}^\dagger)\right]. \tag{20.38}$$

We may note that $2d_j^\dagger d_j - 1 = i\gamma_{2,j}\gamma_{1,j+1}$, which leads to

$$H = t\sum_{j=1}^{N-1}\left(2d_j^\dagger d_j - 1\right). \tag{20.39}$$

If we use open boundary conditions (OBC), we see immediately that $d_N^\dagger d_N$ does not appear in the Hamiltonian, which leads to a degenerate ground state $(d_N^\dagger d_N = 0, 1)$ with two Majoranas.

If we use periodic boundary conditions (PBC), we just need to add this term $(d_N^\dagger d_N)$:

$$H = t\sum_{j=1}^{N}\left(2d_j^\dagger d_j - 1\right). \tag{20.40}$$

Note that $d_N = \frac{i}{2}(c_N^\dagger - c_N + c_1 + c_1^\dagger)$. Therefore, $d_N^\dagger + d_N = i(c_N^\dagger - c_N)$ and $d_N - d_N^\dagger = i(c_1 + c_1^\dagger)$. The ground state of the d operators is the ground state of the system, at this special point

$$|GS\rangle = |n_d = 0; n_d = 0; \cdots ; n_d = 0; \cdots\rangle, \tag{20.41}$$

written in terms of the d operators.

20.1.3 Critical Behavior

In the previous section, we considered the regime near the transition between the topological phases I and II, varying Δ. In the spin language, this is a transition at zero magnetic field as the anisotropy changes from a prevalent J_x coupling to a prevalent J_y coupling. We may also consider the transition at finite magnetic field in the spin problem.

Let us return to the 1d Ising model in a transverse field, and let us write it in a different way. In units of the transverse field, let us write the Hamiltonian as

$$H = -\sum_j \sigma^z(j) - \lambda \sum_j \sigma^x(j)\sigma^x(j+1) \tag{20.42}$$

in terms of operators σ_i. This parametrization may be seen as taking the magnetic field $h = 2$ and $J_x = 4\lambda$. Therefore, in the Kitaev model, $\mu = h = 2$, $t = \Delta = \lambda$. Note that with the parametrization chosen, λ is the only relevant parameter, describing the ratio between the spin coupling and the magnetic field. Fixing the spin coupling, we may see the change in λ as due to the change in the transverse magnetic field (or its inverse).

Using the Jordan-Wigner transformation, we obtain

$$H = -2\sum_j c^\dagger(j)c(j) - \lambda \sum_j \left(c^\dagger(j) - c(j)\right)\left(c^\dagger(j+1) + c(j+1)\right). \tag{20.43}$$

Let us again perform a Fourier transform defining

$$a_k = \left(\frac{1}{2N+1}\right)^{1/2} \sum_{j=-N}^{N} e^{ikj} c(j) \tag{20.44}$$

and k takes the values $k = 0, \pm 2\pi/(2N+1), \pm 4\pi/(2N+1), \cdots, \pm 2\pi N/(2N+1)$. The inverse Fourier transform is

$$c(j) = \left(\frac{1}{2N+1}\right)^{1/2} \sum_k e^{-ikj} a_k. \tag{20.45}$$

The Hamiltonian takes the form

$$H = -2\sum_{k} (1 + \lambda \cos k)\, a_k^\dagger a_k - \lambda \sum_{k} \left(e^{-ik} a_k^\dagger a_{-k}^\dagger - e^{ik} a_k a_{-k}\right)$$

$$H = -2\sum_{k>0} (1 + \lambda \cos k) \left(a_k^\dagger a_k + a_{-k}^\dagger a_{-k}\right) + 2i\lambda \sum_{k>0} \sin k \left(a_k^\dagger a_{-k}^\dagger + a_k a_{-k}\right). \tag{20.46}$$

This Hamiltonian can be diagonalized using a Bogoliubov-Valatin transformation and can be written as

$$H = \sum_k \Lambda_k \eta_k^\dagger \eta_k + C, \tag{20.47}$$

where η_k are the destruction operators of the excitations (normal modes) of the system. The band is

$$\Lambda_k = 2\sqrt{1 + 2\lambda \cos k + \lambda^2} \tag{20.48}$$

and is plotted in Fig. 20.4. Naturally, this result is the same as in Eq. (20.34) if we add a chemical potential term $\epsilon_k \to \epsilon_k - \mu$. It ranges from $2|1 - \lambda|$ at $k = \pm\pi$ to $2|1 + \lambda|$ at $k = 0$. (Recall that if $\mu = 0$ the minima are at $k = \pm\pi/2$). Near the minima, it has a quadratic behavior in the momentum shift from $k = \pm\pi$. Therefore, the excitation spectrum is gapped. In terms of the deviation $q = \pi - k$, it behaves like

$$\Lambda_q = 2\sqrt{(1 - \lambda)^2 + \lambda q^2}. \tag{20.49}$$

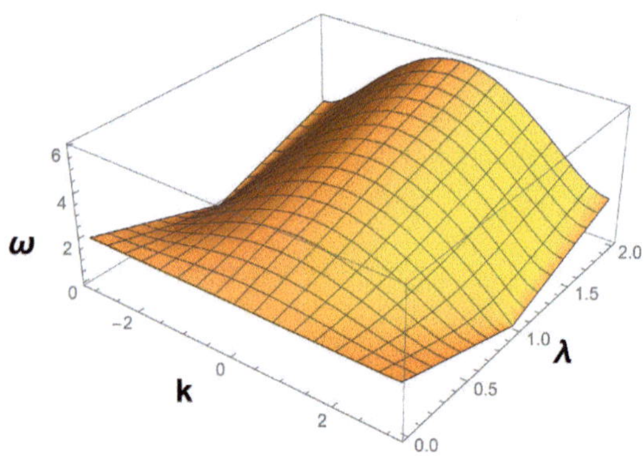

Fig. 20.4 Dispersion relation for the Ising model in transverse field as a function of λ

Near $q \sim 0$, it behaves as

$$\Lambda_q \sim 2|1-\lambda| \left(1 + \frac{1}{2}\frac{\lambda}{(1-\lambda)^2} q^2\right), \tag{20.50}$$

and for $\lambda = 1$, the excitation spectrum is gapless (critical) and

$$\Lambda_q = 2\sqrt{\lambda}|q|, \tag{20.51}$$

and, therefore, we get a linear spectrum.

From Eq. (20.49), we can estimate that

$$\xi^2 \sim \frac{\lambda}{(1-\lambda)^2}. \tag{20.52}$$

Since the critical point is at $\lambda = 1$, we get that $1/\xi \sim |1-\lambda|$. By similarity with a temperature-induced phase transition, we write that $1/\xi \sim |\lambda - \lambda_c|$, which implies that the critical exponent ν defined in the usual way as

$$\xi \sim \left(\frac{1}{|\lambda - \lambda_c|}\right)^{\nu} \tag{20.53}$$

yields $\nu = 1$. It may be shown that

$$m_z \sim (\lambda_c - \lambda)^{\beta}, \tag{20.54}$$

with $\beta = 1/8$.

20.2 Crossover and Quantum Criticality

Let us now generalize the problem to higher dimensions. Taking, for instance, the d-dimensional Ising model in a transverse field,[2] it can be shown that one finds a rich critical behavior. The mean-field phase diagram as a function of the transverse field, Γ, and temperature, T, displays a line of critical points, $T_c(\Gamma)$, that ends at the point $(T = 0, \Gamma = \Gamma_c)$, which is called a multicritical point. For $\Gamma < \Gamma_c$, as we cross the critical line lowering the temperature, the system has a transition that is characteristic of the classical Ising model and such that the transverse field does not change the critical behavior, at the level of the critical exponents. The transition at $T = 0$ as $\Gamma \to \Gamma_c$ displays different critical exponents associated with

[2] It is also a standard notation to write the Hamiltonian as an interaction along the z-axis, with a coupling J_z and a transverse field along the x axis, denoted Γ. We will adopt this notation here.

the quantum critical point. It can be shown that the exponents of the Ising model in a transverse field in d dimensions and $T = 0$ are the same of the classical model at $(d+1)$ dimensions (and $\Gamma = 0$), with Γ playing the role of T (as will be discussed later on). For temperatures close to zero, as we approach the critical line, there is a competition between classical and quantum regimes giving rise to a crossover phenomenon between the two regimes. Let us attempt to clarify the reason for the existence of the two regimes.

20.2.1 Crossover Phenomena

There is a crossover phenomenon when, due to the action of a given parameter p, there is a change in the critical behavior of a system.[3] In a renormalization group analysis, the trajectories in parameter space are determined by the fixed points of the transformations and by their relative stability. At $p = 0$, the most stable fixed point and the respective anomalous dimensions determine the critical behavior of the system. At $p \neq 0$, an extra term in the Hamiltonian appears that we may treat as a perturbation. If the anomalous dimension for this perturbation is negative, then the additional term is irrelevant for the critical behavior. If, on the other hand, it is positive (or null), then the critical behavior will change, corresponding to the passage to a different fixed point, more stable, responsible for different critical exponents.

If the value of the parameter p is small, then the trajectory initially still converges toward the fixed point R^*, finally deviating toward the direction of the new fixed point Q^*. If p has a high value, then the trajectory toward the fixed point R^* is significantly changed and converges fast toward the new fixed point Q^*. The curvature of the trajectory is larger the smaller the value of p. This phenomenon of the transfer of the maximal stability, between fixed points and different trajectories in the parameter space, is called crossover.

Experience shows that the change of behavior occurs at a point that is a function of p. The crossover occurs at a temperature, T^*, given by

$$T^* - T_0(p) \sim p^{1/\phi}, \tag{20.55}$$

where $T_0(p)$ is the point of the phase transition and ϕ is called the crossover exponent. This result is immediately obtained if we assume that the singular part of the free energy, associated with the critical behavior, has the Widom generalized homogeneity form

$$G(\lambda t, \lambda^{\Delta} h, \lambda^{\phi} p) = \lambda^{2-\alpha} G(t, h, p), \tag{20.56}$$

[3] Here, we follow Master's thesis, P. D. Sacramento, FC Universidade de Lisboa (1986).

where h is the magnetic field, $t = (T - T_0(0))/T_0(0)$, and $\Delta = \beta\delta$. In this scaling, relation λ has an arbitrary value. Making the choice $\lambda = t^{-1}$, we obtain

$$G(t, h, p) = t^{2-\alpha} G\left(1, \frac{h}{t^{\Delta}}, \frac{p}{t^{\phi}}\right). \tag{20.57}$$

The important point is that due to the form of the free energy, the thermodynamic functions are going to depend on p/t^{ϕ}, where p is the disturbing field that measures the distance to the fixed point R^*.

It is possible to treat the crossover by adding an extra term to the Hamiltonian of the system. The case of the Ising model in a transverse field is a bit different. First, it is not the temperature that determines the displacement from criticality but the transverse field. The temperature then assumes the role of the field responsible for the crossover since it is the fact that the temperature is nonvanishing or zero that determines the classical or the quantum behavior of the system. The thermodynamic quantities are actually function of T/r^{ϕ}, where $r = (\Gamma - \Gamma_c)/\Gamma_c$, with ϕ the crossover exponent.

20.2.2 Dimensional Crossover

The phenomenon of dimensional crossover is characteristic of quantum systems. The main difference between these and classical systems lies in the non-commutation between two terms in the Hamiltonian (for instance, in the Ising model in transverse field, S^x does not commute with S^z). We may start from the partition function of the system. In the quantum case, the trace of the Boltzmann factor is obtained using some appropriate basis of states, upon which we may act the operators of the Hamiltonian and convert the partition function into a sum, or integral, over the classical variables, obtained from the action of the operators on the states. Typically, matrix elements of the operators are involved. One way to obtain an expression is to factorize the Boltzmann exponential into M equal fractions and to take the limit when $M \to \infty$. Then, we may insert decompositions of unity between each exponential factor, which typically involves calculating matrix elements of the Hamiltonian itself, since in the limit that $M \to \infty$ each exponential may be linearized. This leads to a path integral-like expression where one sums over all possible intermediate states of the used basis.

There is a formal similarity between the Boltzmann exponential of the partition function and the time evolution operator. In the simplest case, when the Hamiltonian is not time-dependent, we can compare the two exponentials $e^{-\beta H}$ and $e^{-(i/\hbar)Ht}$. There is a formal similarity transforming $\beta \to it/\hbar$. We may define an imaginary time, τ, as $\beta\hbar = it = \tau$, or simplifying notation, we may write that $\beta \to \tau$, taking $\hbar = 1$. So, we may establish the correspondence

$$i\int_0^t dt' \to \int_0^{\beta\hbar} d\tau \to \int_0^{\beta} d\tau. \tag{20.58}$$

The trace operation in the partition function implies that $\psi(\tau) = \psi(\tau + \beta)$, initial and final values of the integration over the imaginary time. Decomposing in Fourier series, since τ has a maximal value of β, we can write that

$$\psi(\tau) = \sum_n \psi(\omega_n) e^{i\omega_n \tau}. \tag{20.59}$$

Using the periodicity condition, we get that $e^{i\omega\beta} = 1$, which implies the discrete set $\omega_n \beta = 2\pi n$. These are the so-called bosonic Matsubara frequencies. Note that the spacing between the frequencies is given by $\Delta\omega = 2\pi/\beta = 2\pi k_B T$. Therefore, as $T \rightarrow 0$, the spacing vanishes, and the discrete set of frequencies becomes a continuum. Also, $\beta \rightarrow \infty$.

As we choose a formulation in terms of a path integral, expressed in terms of classical variables, the quantum nature arises in the need to introduce a time ordering of the fields. A time dependence arises, beyond any spatial dependence or on internal degrees of freedom. It is therefore typical the appearance of Matsubara frequencies if we use a finite temperature imaginary time formalism. In this way, some dynamical aspects are automatically included, unlike in a classical system, for which it is possible to separate the statics from the dynamics. The study of the dynamic properties in the classical case is carried out using, for instance, time-dependent Ginzburg-Landau equations. If the system is at a finite temperature, such that the characteristic energies are much smaller than $k_B T$, then the classical statistics is valid, and we obtain a classical critical behavior. At $T = 0$, the previous condition is not satisfied, and therefore, a quantum description is necessary to treat the phase transition that is observed at $\Gamma = \Gamma_c$. It is interesting to note that a classical system described by the Ginzburg-Landau equations does not show dynamics at zero temperature. This is due to the absence of collisions responsible for the dissipation terms that are introduced in the mentioned equations. Such does not happen in a quantum system, since it is exactly in the vicinity of $T = 0$ that the dynamic behavior is richer (from the point of view of the critical behavior).

Let us then consider a functional integral of the form

$$Z = \int \mathcal{D}\phi e^{-\mathcal{H}\{\phi\}}, \tag{20.60}$$

where

$$\mathcal{H}\{\phi\} = \mathcal{H}_0\{\phi\} + \mathcal{H}_I\{\phi\}, \tag{20.61}$$

where $\mathcal{H}_I\{\phi\}$ is the interaction term and the free part is of the type

$$\mathcal{H}_o\{\phi\} = \sum_{k,\omega_l} [r_0 + k^2 + f(k, \omega_l)] |\phi_{k,\omega_l}|^2, \tag{20.62}$$

Table 20.1 Examples of quantum systems and momentum and frequency dependencies

$f(k, \omega_l)$		Quantum systems
		XY model in transverse field
$-i\omega_l$		Bosonic system
		Planar ferromagnet of spin-1/2
$\frac{\lvert\omega_l\rvert^m}{k^{m'}}$	$m = 1, m' = 1$	Paramagnon model for itinerant ferromagnetism
	$m = 1, m' = 0$	Fluctuations in itinerant antiferromagnet
	$m = 1, m' = 0$	Superconductors
	$m = 1, m' = 2$	Doped itinerant ferromagnets
	$m = 2, m' = 0$	Ising model in transverse field
	$m = 2, m' = 0$	Compressible quantum Ising model
	$m = 2, m' = 0$	Structural phase transitions

with ω_l the bosonic Matsubara frequencies. The function $f(k, \omega_l)$ depends on the model under consideration, and some cases of physical interest are listed in Table 20.1.

In order to study the critical behavior, the best way is to use the techniques of the renormalization group. The difference between the treatment of classical and quantum systems lies on the appearance of the Matsubara frequencies, while the other steps of the renormalization group are kept equivalently. Let us first consider the case with $T = 0$. Since $\Delta\omega_l = 2\pi/\beta$, in this limit of zero temperature, the sum in the frequencies is replaced by an integral

$$\frac{1}{\beta}\sum_{\omega} \rightarrow \int \frac{d\omega}{2\pi}. \tag{20.63}$$

According to Wilson, it is necessary to carry out three operations: The first consists in eliminating the fields with higher momenta (and frequencies) keeping the remaining fields fixed. The second consists in performing a rescaling of the momenta and frequencies in such a way that the available set of values is again the initial one. The third step consists in rescaling the fields in such a way as to reestablish the coefficients of k^2 and $f(k, \omega_l)$, in such a way as to recover the Hamiltonian in its original form, possibly with different constants. Then the evolution of these constants in parameter space is studied through the so-called equations of the renormalization group. As we saw earlier, if new terms are generated and are relevant, then they should have been introduced from the start. With this proviso, we expect that the other possible generated terms may be ignored. The difference to the classical case is that in general the rescaling of the momenta and of the frequencies has to be chosen differently in order to obtain a Hamiltonian of the original form. In fact, it is necessary to perform the rescaling

$$\begin{aligned} k' &= kl \\ \omega' &= \omega l^z, \end{aligned} \tag{20.64}$$

Table 20.2 Dynamical critical exponent of various quantum systems

m, m'	z
$m = 2, m' = 0$	1
$m = 1, m' = 0$	2
$m = 1, m' = 1$	3
$m = 1, m' = 2$	4

where z is called the dynamical critical exponent. The value of z depends on the model and is determined by the functional form of the momentum and frequency that appear in the Hamiltonian. At the lowest level of approximation, we obtain that

$$z = \frac{2 + m'}{m}, \tag{20.65}$$

and a set of values is indicated in Table 20.2. Higher-order corrections may change this expression.

The reason for the dimensional crossover may be seen as follows: In the case of $T = 0$, in which the frequencies become a continuum, the application of the renormalization group operations leads to recurrence relations between the Hamiltonian parameters, which are formally similar to those of the classical case, but with the change $d \to d + z$, as illustrated in Fig. 20.5. Using, for instance, Wilson's ϵ-expansion, we can see that the equations are the same but with ϵ defined by

$$\epsilon = d_c - (d + z), \tag{20.66}$$

where d_c is the critical dimension of the relevant interaction term. Therefore, effectively, the dimension of the system is increased by a value that is the dynamical critical exponent. The condition $\epsilon = 0$, which separates the Landau from the Wilson regime, shows that the critical dimension has been lowered, that is, $d_c \to d_c - z$. For values of $d > d_c - z$, the critical exponents are those of Landau, and only for $d \leq d_c - z$, there will be corrections to the mean-field theory. For example, in the case of the Ising model in transverse field, we can see that $z = 1$ (since k^2

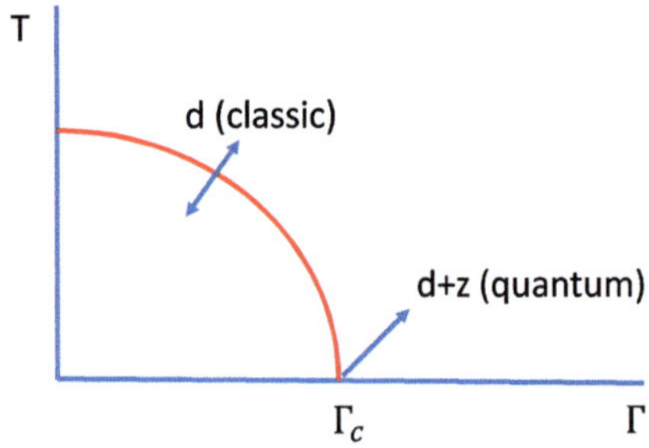

Fig. 20.5 Dimensional crossover from the finite temperature transition to the quantum phase transition

and ω^2 appear symmetrically in the action[4]). In this case, the quantum nature of the problem introduces an extra dimension, which has finite extension in the case of finite temperature and infinite extension in the case of zero temperature. When $T = 0$, it is then natural that the effective dimension of the system is $d + 1$. Since the critical dimension is now $d_c - z = 4 - 1 = 3$, it is then to expect for $d > 3$ the Landau critical exponents. Only for $d \leq 3$ corrections appear due to the fluctuations of the system.

Let us now consider the case of finite temperature. In this case, the Matsubara frequencies no longer are a continuum but are instead a discrete set, with a spacing of $2\pi T$ ($k_B, h = 1$). Since the frequencies and temperature are related, as we rescale ω, we also have to rescale the temperature, T. Then,

$$T(l) = Tl^z. \tag{20.67}$$

Since $\Delta\omega = 2\pi T$, as we take the critical limit ($l \to \infty$), we obtain that, for finite temperatures, the spacing between the levels tends to infinity as we iterate the operations of the renormalization group. As a consequence, only the term $\omega = 0$ survives. For $T = 0$, $\Delta\omega \to 0$, and we obtain the continuum, as said above.

The crossover region is intermediate between the two limits $\Delta\omega = 0$ and $\Delta\omega = \infty$. It may then be characterized by $\Delta\omega \sim 1$, which leads to a particular value $\hat{l}$ given by

$$\hat{l} \sim \left(\frac{1}{T}\right)^{1/z}. \tag{20.68}$$

We have seen (Eq. 18.58) that $r(l) = r_0 l^2$. Since $r \sim \xi^{-2}$, where ξ is the correlation length, when $r(l) \sim 1$ implies that the correlation length has been rescaled to a value of the order of l, which may be treated in perturbation theory. Therefore, the problem of the divergency of ξ has been eliminated, and we may stop the renormalization group operations. If $r(l)$ tends to 1 for a value $l < \hat{l}$, then the transition to the classical behavior cannot take place, and the behavior will be quantum. The transition only takes place if $r(\hat{l}) < 1$, that is, if

$$r_0 < (2\pi T)^{2/z}. \tag{20.69}$$

According to our previous results, the crossover exponent is given by $\phi = z/2$. Since in the case of the Ising model in transverse field $z = 1$, we obtain that $\phi = 1/2$. Also, using the scaling relation[5] $2 - \alpha = \nu(d + z)$, we may get that $\alpha = 0$, and using other scaling relations, we obtain that $\gamma = 7/4$, $\eta = 1/4$, $\delta = 15$.

[4] We will return later to this point, but the relevant Matsubara frequencies are of the order of the dispersion relation energies such as given in Eq. (20.49).

[5] Using Eq. (18.38) and taking $d \to d + z$.

20.3 Zero-Temperature Hertz Dynamical Theory

In this section, let us consider the quantum phase transition at zero temperature between a paramagnet and a ferromagnet, in the context of the Stoner theory of itinerant magnetism.[6] The finite temperature behavior, close to the quantum critical point, may also be considered but will not be presented here.

The Stoner theory based on the Hubbard model was considered in Sect. 16.6. The Hubbard model (Eq. 16.83) may also be written in a momentum space representation (Eq. 16.84). In addition to the kinetic term, the interaction term may be seen as the scattering of two electrons, with momenta $\mathbf{p}, \mathbf{p}'$ and spin components $\uparrow, \downarrow$, respectively, that interchange a momentum $\mathbf{q}$, leading to final states with momenta $\mathbf{p}+\mathbf{q}$ and $\mathbf{p}-\mathbf{q}$. This momentum exchange is provided by the interaction, U. The interaction term may be decoupled and written as in Eq. (16.85) revealing the magnetic nature of the coupling and consequent magnetic properties.

The dynamical susceptibility, such as the transverse susceptibility, was obtained leading to an expression of the type

$$\chi^{-+}(\mathbf{q},\omega) = \frac{\Gamma^{-+}(\mathbf{q},\omega)}{1-U\Gamma^{-+}(\mathbf{q},\omega)}$$
$$\Gamma^{-+}(\mathbf{q},\omega) = \frac{1}{N}\sum_{\mathbf{q}} \frac{f_{\mathbf{q}+\mathbf{q},\downarrow} - f_{\mathbf{p},\uparrow}}{\omega - \left(\bar{\epsilon}_{\mathbf{p},\downarrow} - \bar{\epsilon}_{\mathbf{p}+\mathbf{q},\uparrow}\right) + i\eta}, \tag{20.70}$$

where the f-functions are the Fermi-Dirac distribution. The critical point that separates the paramagnetic from the ferromagnetic phase was obtained determining the singular point of the susceptibility in the paramagnetic phase, considering the static case, $\omega = 0$, and a spatial uniform solution, $q = 0$, leading to the condition $U_c\Gamma(0,0) = 1$ or $U_cN(\epsilon_F) = 1$, where $N(\epsilon_F)$ is the density of states at the Fermi level. Note that we may expand the denominator, since it is a geometric series, leading schematically to

$$\chi = \Gamma\,[1 + U\Gamma + U\Gamma U\Gamma + \cdots]. \tag{20.71}$$

The spin operators of the electronic system may be written as

$$S^{+}(\mathbf{q}) = \sum_{\mathbf{p}} c^{\dagger}_{\mathbf{p}+\mathbf{q},\uparrow} c_{\mathbf{p},\downarrow}$$
$$S^{-}(\mathbf{q}) = \sum_{\mathbf{p}} c^{\dagger}_{\mathbf{p}+\mathbf{q},\downarrow} c_{\mathbf{p},\uparrow}, \tag{20.72}$$

[6] For further reading see J. Hertz, Physical Review B **14**, 1165 (1976), and [Continentino].

describing processes that flip the spin component of an electron. Pictorially, a susceptibility may be seen as an average value of two spin components. Since each spin operator may be seen as a product of two electronic operators, one creation and one destruction operator, to leading order, the susceptibility may be seen as a product of two averages values, of the type of one creation and one destruction electronic operator each. For a given momentum of the spin susceptibility, $\mathbf{q}$, one creates one electron with momentum $\mathbf{p}+\mathbf{q}$ and destroys one with momentum $\mathbf{p}$. The momentum $\mathbf{p}$ is free, and so it must be summed over. In the process, the electron receives the momentum $\mathbf{q}$, and when it is destroyed, it loses the same momentum. The process may be pictorially represented as in Fig. 20.6. Each line in the diagram of the susceptibility represents the average of the product of one creation and one destruction operator of an electron. The final result of the average value of the susceptibility leads to the results in Eq. (20.70). The reader may recall that these results were obtained solving, to some approximation, the equation of motion of the susceptibility. Other methods may be used such as many-body techniques, which are beyond the level of this book. Let us now reformulate some aspects of the problem in order to study the quantum phase transition in some detail.

Let us start by considering that the Hubbard Hamiltonian has a kinetic term and an interaction term written in the usual way $H = H_0 + H'$. The interaction term can be decomposed as

$$H' = U\sum_i n_{i,\uparrow}n_{i,\downarrow} = \frac{U}{4}\sum_i \left(n_{i,\uparrow} + n_{i,\downarrow}\right)^2 - \frac{U}{4}\sum_i \left(n_{i,\uparrow} - n_{i,\downarrow}\right)^2 . \tag{20.73}$$

The first term of the decoupling is a charge term and the second term a spin term. Since we are interested in magnetic properties, we only keep now the spin term. The partition function may be written as

$$Z = Tre^{-\beta H} = Tre^{-\beta(H_0+H')}. \tag{20.74}$$

We want now to consider the interaction as some perturbation, as discussed before when we considered the linear response theory. In that context, we were interested in a time-dependent perturbation, and it is convenient to use the interaction picture.

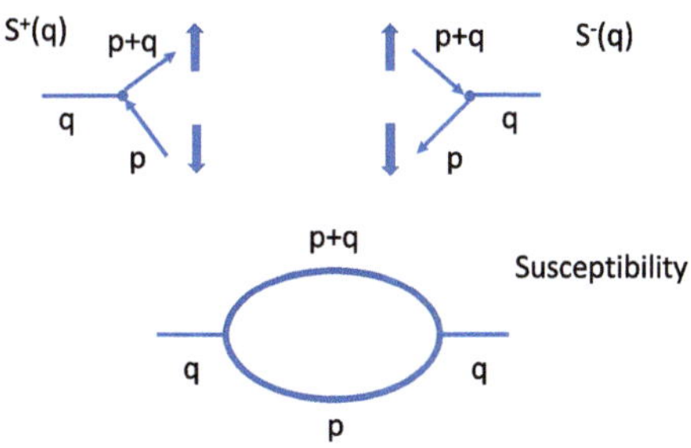

Fig. 20.6 Pictorial representation of the spin operators, $S^+(q)$ and $S^-(q)$, and of the leading term of the transverse susceptibility

Similar to usual procedures in time-dependent perturbation theory, we may then write

$$\begin{aligned} e^{-\beta H} &= e^{-\beta H_0} T_\tau e^{-\int_0^\beta d\tau H'(\tau)} \\ H'(\tau) &= e^{\tau H_0} H' e^{-\tau H_0}, \end{aligned} \tag{20.75}$$

using the interaction picture and where T_τ is the time-ordering operator, in this case in imaginary time.

We may now use the Hubbard-Stratonovich transformation to rewrite the interaction term in terms of an integration over an extra set of variables, $\psi_i(\tau)$. This leads to

$$Z = Z_0 \int D\psi e^{-\frac{1}{2}\sum_i \int_0^\beta d\tau \psi_i^2(\tau)} \langle T_\tau e^{-\sum_{i,\sigma} \int_0^\beta d\tau \sigma V_i(\tau) n_{i,\sigma}(\tau)} \rangle_0. \tag{20.76}$$

Here, the average is defined with the weight the kinetic energy term

$$\langle A \rangle_0 = \frac{1}{Z_0} Tr e^{-\beta H_0} A \tag{20.77}$$

and

$$D\psi = (2\pi)^{N/2} \prod_i d\psi_i(\tau). \tag{20.78}$$

Finally, we introduced the field

$$V_i(\tau) = \sqrt{\frac{U}{2}} \psi_i(\tau). \tag{20.79}$$

This field acts like a time-dependent magnetic field acting on site i and aligned along the z-direction since we may write that

$$\sum_{i,\sigma} \sigma V_i(\tau) n_{i,\sigma}(\tau) \sim \sum_i V_i(\tau) S_i^z(\tau).$$

We may expand

$$\langle T_\tau e^{-\sum_i \int_0^\beta d\tau V_i(\tau) S_{i,z}(\tau)} \rangle_0 = 1 - \sum_i \int_0^\beta d\tau V_i(\tau) \langle S_i^z(\tau) \rangle_0 + \cdots. \tag{20.80}$$

The second-order term is of the type

$$U \langle S_i^z S_j^z \rangle_0 \sim U \chi_{ij}^{zz}. \tag{20.81}$$

Also, recall that, to leading order, $\chi \sim \Gamma$. A rigorous derivation of the terms may be carried out, but here we just need to determine the type of terms that appear. The various steps lead to the following expression for the partition function

$$Z \sim Z_0 \int D\psi e^{-F}, \tag{20.82}$$

where the Landau-Ginzburg-like functional is given by

$$F = \frac{1}{2} \sum_{\mathbf{q},\omega} v_2(\mathbf{q},\omega)|\psi(\mathbf{q},\omega)|^2 + O\left(|\psi|^4\right). \tag{20.83}$$

We have transformed to momentum space, assuming translational invariance. The frequencies ω are actually discrete, and therefore, they are summed over. Finally, one gets that

$$v_2(\mathbf{q},\omega) = 1 - U\Gamma^{zz}(\mathbf{q},\omega). \tag{20.84}$$

The function $\Gamma(\mathbf{q},\omega)$ is related to the Lindhard function that we found in the electromagnetic response of an electron system. Considering a three-dimensional system and considering the limit of small q/q_F and small $\omega/(v_F q)$, one arrives at the expression

$$v_2(\mathbf{q},\omega) = g + q^2 + \frac{|w|}{q}, \tag{20.85}$$

where

$$g = 1 - UN(\epsilon_F) = v_2(0,0) \tag{20.86}$$

is the deviation from the quantum critical point, given by the Stoner criterion.

We have then a critical theory to analyze in the vicinity of the quantum critical point, expressed as

$$\begin{aligned} F = &\frac{1}{2} \int d^d q \int d\omega v_2(q,\omega)|\psi(\mathbf{q},\omega)|^2 \\ &+ \frac{u}{4} \int d^d q_1 \cdots \int d^d q_4 \int d\omega_1 \\ &\cdots \int d\omega_4 \psi(\mathbf{q}_1,\omega_1)\psi(\mathbf{q}_2,\omega_2)\psi(\mathbf{q}_3,\omega_3)\psi(\mathbf{q}_4.\omega_4). \end{aligned} \tag{20.87}$$

There is conservation of momentum and frequency which implies that $\sum_i \mathbf{q}_i = 0$ and $\sum_i \omega_i = 0$. In a lattice, we naturally have a cutoff of the momenta integrations of the sort $\Lambda = q/a$ (in a cubic-like lattice of side a).

In order to study the critical properties, we take steps similar to the renormalization procedure of the critical ϕ^4. We separate

$$\psi(\mathbf{q},\omega) = \psi_0\delta(\omega)\delta_{\mathbf{q},0} + \delta\psi(\mathbf{q},\omega). \tag{20.88}$$

The first term is the uniform and static contribution, and the second term is the dynamical part. The fourth-order term can be written as

$$\psi(\mathbf{q}_1,\omega_1)\psi(\mathbf{q}_2,\omega_2)\psi(\mathbf{q}_3,\omega_3)\psi(\mathbf{q}_4.\omega_4) \sim \psi_0^4 + 6\psi_0^2\delta\psi(\mathbf{q},\omega)\delta\psi(-\mathbf{q},-\omega). \tag{20.89}$$

Assuming that $\omega \sim q^z$, the free energy reduces to leading order to

$$F = L^{d+z}F_0 + \int d^dq \int d\omega \frac{1}{2}\delta\psi(\mathbf{q},\omega)\left(v_2(\mathbf{q},\omega) + 3u\psi_0^2\right)\delta\psi(-\mathbf{q},-\omega), \tag{20.90}$$

with

$$F_0 = \frac{g}{2}\psi_0^2 + \frac{u}{4}\psi_0^4, \tag{20.91}$$

and we may write

$$g = \frac{U}{W} - \left(\frac{U}{W}\right)_c, \tag{20.92}$$

with $W = 1/N(\epsilon_F)$, the bandwidth.

In the spirit of the renormalization group procedure, we want now to focus on the slow modes by integrating out the fast modes. First, we perform a change of variables,

$$q' = \frac{\Lambda}{k}q, \qquad \omega' = \left(\frac{\Lambda}{k}\right)^z \omega. \tag{20.93}$$

In terms of these variables, we may write the free energy as

$$F_< = \left(\frac{k}{\Lambda}\right)^d \left(\frac{k}{\Lambda}\right)^z \int_0^\Lambda d^qq' \int d\omega' \frac{1}{2}|\delta\psi\left(\frac{k}{\Lambda}q', \left(\frac{k}{\Lambda}\right)^z\omega'\right)|^2$$
$$\left(\frac{k^2}{\Lambda^2}\left(q'\right)^2 + g + \frac{\left(\frac{k}{\Lambda}\right)^z\omega'}{\left(\frac{k}{\Lambda}\right)q'} + 3u\psi_0^2\right). \tag{20.94}$$

The fluctuation fields get rescaled as

$$\delta\psi'(q',\omega') = s^{-1}\delta\psi\left(\frac{k}{\Lambda}q', \left(\frac{k}{\Lambda}\right)^z\omega'\right). \tag{20.95}$$

We now fix the scale s such that the gradient term is fixed

$$\left(\frac{k}{\Lambda}\right)^d \left(\frac{k}{\Lambda}\right)^z \left(\frac{k}{\Lambda}\right)^2 s^2 = 1, \tag{20.96}$$

which implies that

$$s = \left(\frac{k}{\Lambda}\right)^{-\frac{d+z+2}{2}}. \tag{20.97}$$

Therefore, in terms of the rescaled fields, the free energy is written as

$$F_< = \int_0^\Lambda d^d q' \int d\omega' \frac{1}{2} |\delta\psi'(q', \omega')|^2 \left((q')^2 + \left(\frac{k}{\Lambda}\right)^{-2} g + \frac{\left(\frac{k}{\Lambda}\right)^{-2} \left(\frac{k}{\Lambda}\right)^z \omega'}{\left(\frac{k}{\Lambda}\right) q'} + 3u \left(\frac{k}{\Lambda}\right)^{-2} \psi_0^2 \right). \tag{20.98}$$

In terms of the scaling factor $b = \Lambda/k$, we obtain that the critical coupling is also rescaled as

$$g' = b^2 g. \tag{20.99}$$

In a thermal phase transition, this would correspond to the rescaling of the temperature deviation from criticality. Now, we choose that ω'/q' remains unrenormalized which leads the dynamical critical exponent to be $z = 3$.

From the uniform part, we get that

$$L^{d+z} g \psi_0^2 = \left(bL'\right)^{d+z} b^{-2} g' \psi_0^2 = \left(L'\right)^{d+z} g' \left(\psi_0'\right)^2 \tag{20.100}$$

and

$$\left(\psi_0'\right)^2 = b^{d+z-2} \psi_0^2, \tag{20.101}$$

which means that defining the exponent η as

$$\left(\psi_0'\right)^2 = b^{d+z-2+\eta} \psi_0^2 \tag{20.102}$$

we get that $\eta = 0$. The interaction is also renormalized as

$$u' = b^{4-(d+z)} u, \tag{20.103}$$

which joins the renormalization

$$g' = b^2 g. \tag{20.104}$$

These transformations have as fixed point the gaussian fixed point with $g^* = 0, u^* = 0$. Also, the effective dimension, given by $d_e = d + z$, since $z = 3$ is such that $d_e > 4$. Therefore, the interaction is an irrelevant perturbation, and the gaussian fixed point dominates the critical behavior. The critical exponents of the gaussian fixed point are well known leading to $\nu = 1/2$ and $\gamma = 1$.

20.4 Fidelity and Phase Transitions

A way to detect a phase transition is to compare the states of a system on the two sides of the transition. Thinking specifically of a quantum phase transition at or near zero temperature, an efficient signature of the transition is to consider the distinguishability between the states of the system, such as the ground state, as some parameter, q, that drives the transition changes. A quantity that quantifies the distinguishability is the fidelity. In the case of pure states, the fidelity is defined by the overlap between the states parametrized by two values of the parameter q and $\tilde{q}$ $|\Phi(q)\rangle$ and $|\Phi(\tilde{q})\rangle$ as

$$F(q, \tilde{q}) = |\langle\Phi(q)|\Phi(\tilde{q})\rangle|. \tag{20.105}$$

If the states are in the same phase, their overlap is close to one, but if the two parameters correspond to states on the two sides of the transition, the fidelity has a minimum.[7] Taking $\tilde{q} = q + \delta(q)$, the minimum of the fidelity signals in an efficient way the transition point.

In the case of mixed states, the system is characterized by a density matrix that may describe a system at a finite temperature or a multiband system. Also, we may be interested in the states of a subsystem that are characterized by a reduced density matrix. In a pure state, the density matrix is simply the projector to that state

$$\rho = |\Phi\rangle\langle\Phi|\,. \tag{20.106}$$

If we divide a system in two parts, A and B, the reduced density matrix of subsystem A is defined by

$$\rho_A = \mathrm{Tr}_B \rho\,. \tag{20.107}$$

[7] See also [AraujoSacramento].

In the general case, the mixed state fidelity between two states a and b is defined by the expression

$$F(\hat{\rho}_a, \hat{\rho}_b) = \mathrm{Tr}\sqrt{\sqrt{\hat{\rho}_a}\hat{\rho}_b\sqrt{\hat{\rho}_a}}\,. \tag{20.108}$$

Its definition is a generalization of the case of classical systems where the classical fidelity F_c is a measure of the distinguishability between two probability distributions $\{p_a(i)\}$ and $\{p_b(i)\}$ and given by $F_c(p_a, p_b) = \sum_i \sqrt{p_a(i)p_b(i)}$. Defining the modulus of an operator $\hat{R}$ as $|\hat{R}| = (\hat{R}\hat{R}^\dagger)^{1/2}$, we obtain that the quantum fidelity is

$$F(\hat{\rho}_a, \hat{\rho}_b) = \mathrm{Tr}\left|\sqrt{\hat{\rho}_a}\sqrt{\hat{\rho}_b}\right| = \mathrm{Tr}\sqrt{\sqrt{\hat{\rho}_a}\sqrt{\hat{\rho}_b}(\sqrt{\hat{\rho}_a}\sqrt{\hat{\rho}_b})^\dagger} = \mathrm{Tr}\sqrt{\sqrt{\hat{\rho}_a}\hat{\rho}_b\sqrt{\hat{\rho}_a}}\,. \tag{20.109}$$

If we consider reduced density matrices, the same definition leads to the fidelity between the partial states $F(\hat{\rho}_A(q), \hat{\rho}_A(\tilde{q}))$, where $\hat{\rho}_A(q) = \mathrm{Tr}_B|\Phi(q)\rangle\langle\Phi(q)|$, and analogously for $\hat{\rho}_A(\tilde{q})$.

Note that the two parts can have different origins, such as a spatial division of the system in two parts or in other spaces such as energy or momentum spaces. If the system has translational invariance and is diagonalizable, the Hamiltonian factorizes in different momentum values, and the fidelity will also factorize. Interesting information may be obtained analyzing the fidelity associated with different momentum values.

Ising Model in Transverse Field

An application of the method to detect a phase transition is on the Ising model in a transverse field written as before

$$H = -\sum_{j=1}^{N}\left(\sigma_j^x\sigma_{j+1}^x + h\sigma_j^z\right), \tag{20.110}$$

where h is the transverse field. The fidelity between two states has been shown to be given by [2]

$$F(h, h') = \prod_{k\geq 0}\cos\left(\theta_k - \theta_k'\right), \tag{20.111}$$

where the angles θ_k are defined (for two values of the transverse field) in the form

$$\cos(2\theta_k) = \frac{\cos k - h}{\sqrt{1 - 2h\cos k + h^2}}$$

$$\sin(2\theta_k) = \frac{\sin k}{\sqrt{1 - 2h\cos k + h^2}}. \tag{20.112}$$

The energy spectrum is given by

$$\epsilon_k = \sqrt{1 - 2h\cos k + h^2}. \tag{20.113}$$

Taking $h = 1$, the spectrum becomes gapless at $k = 0$ and if $h = -1$ at $k = \pi$. Taking $k = 0$, we get that

$$\cos(2\theta_0) = \text{sgn}(1 - h). \tag{20.114}$$

Therefore,

$$\begin{aligned} \cos\theta_0 &= \sqrt{\frac{1}{2}\left(1 + \text{sgn}(1 - h)\right)} \\ \sin\theta_0 &= \sqrt{\frac{1}{2}\left(1 - \text{sgn}(1 - h)\right)}, \end{aligned} \tag{20.115}$$

which leads to

$$\begin{aligned} \cos\left(\theta_0 - \theta_0'\right) = &\sqrt{\frac{1}{2}\left(1 + \text{sgn}(1 - h)\right)}\sqrt{\frac{1}{2}\left(1 + \text{sgn}(1 - h')\right)} \\ &+ \sqrt{\frac{1}{2}\left(1 - \text{sgn}(1 - h)\right)}\sqrt{\frac{1}{2}\left(1 - \text{sgn}(1 - h')\right)}. \end{aligned} \tag{20.116}$$

As a consequence, if we choose two points on the h axis such that the $\text{sgn}(1 - h) = \text{sgn}(1 - h')$, the fidelity at $k = 0$ is one, while if they are different, the fidelity vanishes.

Two-Level System

Let us consider a simple two-level system. The Hamiltonian may be written in terms of Pauli matrices as

$$H = \boldsymbol{h} \cdot \boldsymbol{\sigma}\,. \tag{20.117}$$

A general state may be written as a two-component spinor:

$$\chi = \begin{pmatrix} \chi_1 \\ \chi_2 \end{pmatrix}. \tag{20.118}$$

We may define a density matrix associated with this state as

$$\rho = \chi\chi^\dagger. \tag{20.119}$$

We may also define the density matrix as

$$\rho = \frac{1}{2}\left(I + \mathbf{P}\cdot\boldsymbol{\sigma}\right), \tag{20.120}$$

where I is the identity matrix of dimension 2. The polarization vector is given by

$$\mathbf{P} = Tr\left(\rho\boldsymbol{\sigma}\right). \tag{20.121}$$

Going to the basis that diagonalizes the Hamiltonian and the density matrix, we can show that the density matrix may also be written as

$$\rho = \frac{1}{2}\left(I - \frac{H}{|\boldsymbol{h}|}\right), \tag{20.122}$$

and it acts as a projector onto the lowest energy state. Since $(\boldsymbol{\sigma}\cdot\boldsymbol{n})^2 = 1$, for a unit vector $\boldsymbol{n}$, we get that $\rho^2 = \rho = \sqrt{\rho}$.

We may now calculate the fidelity between two states described by two density matrices $\rho(H_1)$ and $\rho(H_2)$,

$$\begin{aligned} F &= \mathrm{Tr}\sqrt{\sqrt{\rho(H_1)}\rho(H_2)\sqrt{\rho(H_1)}} \\ &= \mathrm{Tr}\sqrt{\frac{1}{2}\left(I - \frac{H_1}{|\boldsymbol{h}_1|}\right)\frac{1}{2}\left(I - \frac{H_2}{|\boldsymbol{h}_2|}\right)\frac{1}{2}\left(I - \frac{H_1}{|\boldsymbol{h}_1|}\right)}, \end{aligned} \tag{20.123}$$

Using $H_1H_2 + H_2H_1 = 2\boldsymbol{h}_1\cdot\boldsymbol{h}_2 I$ and recalling the definition of the versor $\hat{\boldsymbol{h}} = \boldsymbol{h}/|\boldsymbol{h}|$, we may obtain

$$\begin{aligned} F &= \mathrm{Tr}\sqrt{\frac{1}{2}\left(1 + \hat{\boldsymbol{h}}_1\cdot\hat{\boldsymbol{h}}_2\right)\frac{1}{2}\left(I - \frac{H_1}{|\boldsymbol{h}_1|}\right)} \\ &= \sqrt{\frac{1}{2}\left(1 + \hat{\boldsymbol{h}}_1\cdot\hat{\boldsymbol{h}}_2\right)}\mathrm{Tr}\rho(H_1) \\ \Leftrightarrow\ F &= \sqrt{\frac{1}{2}\left(1 + \hat{\boldsymbol{h}}_1\cdot\hat{\boldsymbol{h}}_2\right)}. \end{aligned} \tag{20.124}$$

As this result shows, the fidelity vanishes if $\hat{\boldsymbol{h}}_1\cdot\hat{\boldsymbol{h}}_2 = -1$. In the case of a translational invariant system, where the momentum space description is used, the fidelity is given by the product of the fidelities for each momentum value separately. Therefore, the total fidelity vanishes if for any momentum value it vanishes. If the Hamiltonian depends on some parameter q, $H_q(\boldsymbol{k})$, there may be a set of pairs of parameters $q_1 and q_2$, such that for a given set of momenta, $\boldsymbol{k}$, the fidelity vanishes. If the two parameter points correspond to two situations in the same phase, typically the fidelity does not vanish at any momentum value. However, if the two parameters correspond to two different phases, often, it occurs that the fidelity vanishes at a set of momentum values. Note that this happens even if the two parameters are such that are not near the phase transition interface between the two phases.

It turns out that often there is a relation between the pairs of parameter points (q_1, q_2) for which the fidelity vanishes, $F_{12}(\boldsymbol{k}) = 0$, and the existence of critical gapless points q_c, for which $E_{q_c}(\boldsymbol{k}) = 0$, where $E_q(\boldsymbol{k})$ are the energy eigenvalues of a given momentum value $\boldsymbol{k}$. Under these conditions, the vanishing of the fidelity at a specific momentum and a specific parameter value allows to determine the phase transition point in parameter space.

Kitaev Model

Another example is the Kitaev model. Since there is translational invariance, the fidelity factorizes in momentum space, and the fidelity for each momentum value may be calculated. The full fidelity is then the product of the fidelities for each momentum value, $F = \prod_k F_k$. The momentum space fidelity, F_k, is shown in Fig. 20.7. The Hamiltonian may be written using the Pauli matrices with

$$\boldsymbol{h} = (0, 2\Delta \sin k, \epsilon_k - \mu). \tag{20.125}$$

The eigenvalues are therefore $\pm |\boldsymbol{h}|$, where

$$|\boldsymbol{h}| = \sqrt{4\Delta^2 \sin^2 k + (-2t \cos k - \mu)^2}. \tag{20.126}$$

The transition lines occur for $\mu = 2$ and $k = \pi$, for $\mu = -2$ and $k = 0$, and for $\Delta = 0$ and $\cos k = -\mu/(2t)$. Therefore, for $\Delta = 0$ and $\mu = 0$, the transition occurs at $k = \pi/2$. It is easy to check the vanishing of the fidelity. For this problem, we can write that

$$\frac{\boldsymbol{h}_1 \cdot \boldsymbol{h}_2}{|\boldsymbol{h}_1||\boldsymbol{h}_2|} = \Big(4\Delta_1\Delta_2 \sin^2 k + (2t \cos k + \mu_1)(2t \cos k + \mu_2)\Big)$$

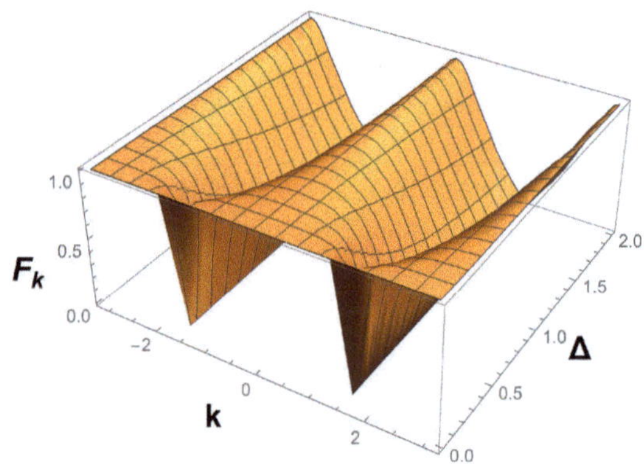

Fig. 20.7 Fidelity as a function of momentum for the Kitaev model as a function of Δ

$$\times\ 1/\sqrt{4\Delta_1^2 \sin^2 k + (2t\cos k + \mu_1)^2}$$
$$\times\ 1/\sqrt{4\Delta_2^2 \sin^2 k + (2t\cos k + \mu_2)^2}. \tag{20.127}$$

Considering, for instance, $\mu_1 = \mu_2 = 0$, it is easily seen that, choosing, for instance, $\Delta_1 > 0$, $\Delta_2 < 0$, the expression reduces to -1 (vanishing fidelity) if $k = \pi/2$, which is the condition for the gapless point.

20.5 Transformations Through a Quantum Critical Point

Let us consider a transformation of the parameters, ξ, of a Hamiltonian such that the system crosses a transition between two phases. To simplify, let us take the transformation to be linear in time such that

$$\xi(t) = \xi(t_0) + r(t - t_0). \tag{20.128}$$

Here, t_0 is the initial time, $r = (\xi(t_f) - \xi(t_0))/(t_f - t_0)$ the rate of time change, and t_f the final time. The evolution of a state is given as before as

$$|\psi_m(\xi(t))\rangle = U(t, t_0)|\psi_m(\xi(t_0))\rangle, \tag{20.129}$$

where $U(t, t_0)$ is the time evolution operator. As time goes by, the Hamiltonian changes, and the eigenstates change continuously in time if the transformation is sufficiently slow. We may decompose the time evolution operator as a path integral

$$U(t_f, t_0) = U(t_f, t_{N-1})U(t_{N-1}, t_{N-2}) \cdots U(t_2, t_1)U(t_1, t_0), \tag{20.130}$$

dividing the time evolution in N intervals. The evolution of any state may be obtained as a set of discrete steps, inserting partitions of unity at each discrete time. If the number of steps is large enough, we may write, to a good approximation,

$$U(t_{i+1}, t_i) = e^{-iH(\xi(\bar{t}))\Delta t}, \tag{20.131}$$

where $\Delta t = (t_f - t_0)/N$ and $\bar{t}$ is an appropriate time in the interval $\{t_i, t_{i+1}\}$. As a matter of convenience, we may take $\bar{t} = t_{i+1}$.

We may now calculate different quantities such as the overlap between the time evolved state and an eigenstate of the Hamiltonian that respects to the parameters of the final time. The overlap amplitude between the initial state $|\psi_m(\xi(t_0))\rangle$ and the state at a given time $t = t_f$, $|\psi_l(\xi(t_f))\rangle$ is defined as

$$A(l, t_f; m, t_0) = \langle\psi_l(\xi(t_f))|\psi_m(\xi(t_0))\rangle. \tag{20.132}$$

It is obtained as

$$A(l, t_f; m, t_0) = \sum_{n_{N-1}} \langle \psi_l(\xi(t_f)) | \psi_{n_{N-1}}(\xi(t_{N-1})) \rangle e^{-iE_{n_{N-1}}(\xi(t_{N-1}))\Delta t} A_{n_{N-1}}, \tag{20.133}$$

where

$$A_{n_i} = \sum_{n_{i-1}} \langle \psi_{n_i}(\xi(t_i)) | \psi_{n_{i-1}}(\xi(t_{i-1})) \rangle e^{-iE_{n_{i-1}}(\xi(t_{i-1}))\Delta t} A_{n_{i-1}}, \tag{20.134}$$

with

$$A_{n_1} = \langle \psi_{n_1}(\xi(t_1)) | \psi_0(\xi(t_0)) \rangle. \tag{20.135}$$

It is interesting to calculate the probability defined by

$$P(t) = |A(l, t_f; m, t_0)|^2. \tag{20.136}$$

If $l = m$, this is called the survival probability of the initial state, m. In the case of an instantaneous transformation (infinite rate) called sudden quench, the parameters of the Hamiltonian only change at the initial time, and the calculation is greatly simplified leading to a final state that in general differs significantly from the original state. In general, for arbitrary rate values, the calculation requires the overlaps between all the intermediate amplitudes, in the limit when $\Delta t \to 0$.

Next, we consider a simplified situation of a two-level system, such that the rate of change is not slow (called adiabatic) but finite, a so-called diabatic regime, which leads to a deviation from the original state and a population of an initial empty state.

20.5.1 Zener-Level Crossings in a Diabatic Regime

Consider a system with two energy states $\psi_1(x)$ and $\psi_2(x)$ such that, for example, the state ψ_2 is the ground state.[8] Consider that the Hamiltonian and, therefore, the states depend on some external parameter, R, that may be changed at a given rate. If the system is originally in the ground state and the transformation is infinitely slow, adiabatic, then, we expect that the system remains in the ground state as R is changed. However, if the transformation is not infinitely slow, that is, diabatic, then, we expect that the final state is a mixture of the two energy states (and in general other higher energy states, which for now we neglect to simplify):

$$\psi(x) = A_1(R)\psi_1(x) + A_2(R)\psi_2(x). \tag{20.137}$$

[8] See C. Zener, Proceedings Royal Society London, Series A **137**, 696 (1932).

Consider now two wave functions that are linear combinations of the energy states. The action of the Hamiltonian may be expressed as

$$\begin{aligned} H\Phi_1 &= \epsilon_1\Phi_1 + \epsilon_{12}\Phi_2 \\ H\Phi_2 &= \epsilon_{12}\Phi_1 + \epsilon_2\Phi_2. \end{aligned} \tag{20.138}$$

Therefore, the Hamiltonian has a structure such as

$$H = \begin{pmatrix} \epsilon_1 & \epsilon_{12} \\ \epsilon_{12} & \epsilon_2 \end{pmatrix}. \tag{20.139}$$

Zener considered that ϵ_{12} is small and that the transition region is small such that

$$\frac{\epsilon_1 - \epsilon_2}{\hbar} = \alpha t; \qquad \dot{\epsilon}_{12} = \dot{\Phi}_1 = \dot{\Phi}_2 = 0. \tag{20.140}$$

A solution of Schrödinger's equation is then searched for in the form

$$\left(H - \frac{\hbar}{i}\frac{\delta}{\delta t}\right)\left(C_1(t)e^{i/\hbar\int\epsilon_1 dt}\Phi_1(x) + C_2(t)e^{i/\hbar\int\epsilon_2 dt}\Phi_2(x)\right) = 0. \tag{20.141}$$

This equation leads to

$$\begin{aligned} \frac{\hbar}{i}\frac{\delta C_1}{\delta t} &= \epsilon_{12}e^{-i/\hbar\int(\epsilon_1-\epsilon_2)dt}C_2 \\ \frac{\hbar}{i}\frac{\delta C_2}{\delta t} &= \epsilon_{12}e^{i/\hbar\int(\epsilon_1-\epsilon_2)dt}C_1. \end{aligned} \tag{20.142}$$

Using the boundary conditions $C_1(t = -\infty) = 0$ and $|C_2(t = -\infty)| = 1$, a solution was found by Zener. Note that the probability for a diabatic transition is obtained calculating

$$P = |C_2(t = \infty)|^2 = 1 - |C_1(t = \infty)|^2 \tag{20.143}$$

to determine the deviation from the adiabatic regime. Eliminating C_2, we obtain an equation for C_1

$$\frac{d^2C_1}{dt^2} + \left(\frac{i}{\hbar}(\epsilon_1 - \epsilon_2) - \frac{\dot{\epsilon}_{12}}{\epsilon_{12}}\right)\frac{dC_1}{dt} + \left(\frac{\epsilon_{12}}{\hbar}\right)^2 C_1 = 0. \tag{20.144}$$

The solution was obtained by Zener

$$|C_1(\infty)|^2 = 1 - e^{-2\pi\gamma}, \tag{20.145}$$

and, therefore, the probability to find the original state at large times is

$$P = e^{-2\pi\gamma}$$
$$\gamma = \frac{\frac{\epsilon_{12}^2}{\hbar}}{\frac{d}{dt}(\epsilon_1 - \epsilon_2)}. \quad (20.146)$$

Using that $d(\epsilon_1 - \epsilon_2)/dt = \hbar\alpha$, we obtain

$$\gamma = \frac{\epsilon_{12}^2}{\hbar^2\alpha}. \quad (20.147)$$

The population of different states is also obtained if we consider a transition across a critical point. In this case, even if the transformation is very slow, adiabatic, due to the critical slowing down at the critical point, different states will be populated leading to a so-called defect production. This will be considered in the following.

20.5.2 Kibble-Zurek Mechanism and Defect Production

When a system suffers a transformation through a phase transition, the neighborhood of the transition point is characterized by a slowing down of its dynamics. In general, the transition point separates two phases characterized by different correlations and a different arrangement of its degrees of freedom. This happens both in transitions that occur at finite temperatures, dominated by thermal fluctuations, and in quantum phase transitions that take place at zero temperature and that are dominated by quantum fluctuations. Topological phase transitions may belong to a different group since in general these are not characterized by changes in symmetry and the consequent appearance of a local order parameter.

Consider again as an example the Ising model in transverse field given by[9]

$$H = -J(t)\sum_{l=1}^{N} \sigma_l^x - W\sum_{l=1}^{N-1} \sigma_l^z \sigma_{l+1}^z, \quad (20.148)$$

and let us consider that the transverse field is a function of time, t. The model may be diagonalized using again the Jordan-Wigner transformation, and the gap to the first excited state is given by

$$\Delta = 2|J(t) - W| = 2W|\epsilon(t)|, \quad (20.149)$$

[9] Note that the interaction is along the z direction and the transverse field is along the x direction. It is equivalent to previous descriptions if we rotate the coordinate axis.

where we are considering an adiabatic, slow, transformation since we are assuming that the eigenstates are those of the static Hamiltonian, at a given instant, t. However, close to the transition point, the adiabaticity condition breaks down.

Associated with the energy scale, there is a timescale, called relaxation time, given by[10]

$$\tau = \frac{\hbar}{\Delta} = \frac{\hbar}{2W|\epsilon(t)|} = \frac{\tau_0}{|\epsilon(t)|}, \tag{20.150}$$

where $\tau_0 = \hbar/(2W)$. The gap closes when $\epsilon(t) = 0$. At this point, $\tau \to \infty$: this result is the slowing down in the neighborhood of the critical point. Associated with this relaxation time, we may define a length, called healing length, defined by $\xi = c\tau$, where c is the sound velocity. Noting that this velocity is the slope of the dispersion relation, we obtain that $c = (2Wa)/\hbar$, where a is the lattice constant. Therefore,

$$\xi = \frac{2Wa}{\hbar}\frac{\hbar}{\Delta} = \frac{2Wa}{\Delta} = \frac{a}{|\epsilon(t)|} = \frac{\xi_0}{|\epsilon(t)|}. \tag{20.151}$$

Note that at the critical point this scale diverges.

Starting at the point $t \to -\infty$ with a high value of J, a state is obtained with all spins pointing in the direction of the x axis. Decreasing the value of the transverse field leads to a situation in which the coupling W dominates and the system tends to a regime such that the spins are aligned along the z axis direction, as a result of the mutual interactions between the spins. Domains of the phase where the symmetry is broken and the spins are aligned appear as we approach the critical point. However, due to the slowing down in the vicinity of the critical point, at the critical point, the system cannot respond since $\tau \to \infty$. In the initial stages of the transformation, the system may follow the values of the external parameters, and the system is found in a quasi-static regime if we are far from the critical point. However, close to the critical point, the regime is different (called impulse regime), and the transformation has two regimes depending if one is close or far from the critical point: this is the so-called Kibble-Zurek mechanism for the transition.

We may define a time, $\hat{t}$, that separates the two regimes. It is expected that it is of the order of the timescale during which the Hamiltonian is changed, τ. One way to define the transformation is to select a linear transformation, as described above

$$\epsilon(t) = \frac{t}{\tau_Q} = rt, \tag{20.152}$$

[10] For further reading see W. H. Zurek, U. Dorner, and P. Zoller, Physical Review Letters **95**, 105701 (2005), and A. Polkovnikov, Physical Review B **72**, 161201(R) (2005).

where r is the rate of change. The change between the two regimes takes place at a time

$$\tau(\hat{t}) = \hat{t}. \tag{20.153}$$

This result leads to

$$\tau(\hat{t}) = \frac{\tau_0}{|\epsilon(\hat{t})|} = \frac{\tau_0 \tau_Q}{\hat{t}} = \hat{t}. \tag{20.154}$$

Therefore,

$$\hat{t} = \sqrt{\tau_0 \tau_Q} = \sqrt{\frac{\hbar}{2Wr}}. \tag{20.155}$$

While for the initial states the regime is quasi-adiabatic and the state of the system follows the value of $J(t)$, after a time $-\hat{t}$, the system cannot follow the instantaneous value of the transverse field. From that instant and until $+\hat{t}$, the system does not follow the value of the transverse field, and only for $t > \hat{t}$ the evolution of the system restarts, having now as initial state the state at the instant $-\hat{t}$.

In the case of phase transitions of thermodynamic systems at finite temperatures, the fluctuations of the order parameter at $\hat{t}$ give origin to domains of size $\hat{\xi}$ given by the healing length at $-\hat{t}$. This length is therefore

$$\hat{\xi} = \frac{\xi_0}{\epsilon(\hat{t})} = a\sqrt{\frac{2W\tau_Q}{\hbar}}. \tag{20.156}$$

In the process and since the process is not quasi-adiabatic, defects may appear in the formation of domains of the stable phase inside the unstable phase. As a consequence one may expect that, if the process is sufficiently slow (r small enough), of the order of one defect per domain should appear, and therefore, the defect density $\hat{\nu}_{KZ}$ may be estimated as

$$\hat{\nu}_{KZ} = \frac{a}{\hat{\xi}} = \sqrt{\frac{\hbar}{2W\tau_Q}} = \sqrt{\frac{\hbar}{2W}r}. \tag{20.157}$$

The reasoning for the case of a quantum phase transition should be similar [3]. Consider again an external perturbation that evolves slowly in time as we cross a second-order quantum phase transition. As discussed above, the adiabaticity property should be violated near the critical point. If the perturbation amplitude is sufficiently small and the process is slow enough, in such a way that the perturbation is small enough when compared to the energy gap between the ground state and the first excited state, the system should remain in the ground state. However, as we approach the critical point, the gap disappears, and some defects, due to the occupation of excited states, may occur. The number of excitations created should tend to zero in the limit of infinitesimal variations.

Consider a Hamiltonian that depends on a parameter, λ, and consider that there is a transition at the point $\lambda = 0$. Consider that the system is in two different phases on the two sides of the transition. Let us suppose that we may diagonalize the Hamiltonian as a function of λ leading to a set of eigenstates labeled by $\Phi_p(\lambda)$. Any state of the system may be expanded in this basis as

$$\psi = \sum_p a_p(t)\Phi_p(\lambda). \tag{20.158}$$

Consider as above $\lambda(t) = rt$. Inserting this expansion in Schrödinger's equation leads to

$$i\frac{da_p}{dt} + ir\sum_q a_q(t)\langle\Phi_p|\frac{d}{d\lambda}|\Phi_q\rangle = \omega_p(\lambda)a_p(t). \tag{20.159}$$

Performing a unitary transformation

$$a_p(t) = \tilde{a}_p(t)e^{-i\int^t \omega_p(\lambda(t'))dt'} = \tilde{a}_p(\lambda)e^{-\frac{i}{r}\int^\lambda \omega_p(\lambda)d\lambda} \tag{20.160}$$

leads to

$$\frac{d\tilde{a}_p}{d\lambda} = -\sum_q \tilde{a}_p(\lambda)\langle p|\frac{d}{d\lambda}|q\rangle e^{\frac{i}{r}\int^\lambda(\omega_p(\lambda')-\omega_q(\lambda'))d\lambda'}. \tag{20.161}$$

Consider as initial state the ground state, $q = 0$. The relative number of excited states is given by

$$n_{exc} = \sum_p{}' |\int_{-\infty}^{\infty} d\lambda\langle p|\frac{d}{d\lambda}|0\rangle e^{\frac{i}{r}\int^\lambda(\omega_p(\lambda')-\omega_0(\lambda'))d\lambda'}|^2, \tag{20.162}$$

where the sum is restricted to the excited states, and we have considered that $\tilde{a}_0(\lambda) \sim 1$.

Let us now try to estimate the density of the resulting excitations. We need to estimate the size of the domains as one approaches the critical point. Therefore, we need to estimate a typical scale of the problem (or a momentum scale). Associated with the time transformation, there is also a timescale (or a frequency scale). Consider a Hamiltonian or Lagrangian of a noninteracting system described by a Φ field that has the form previously discussed

$$H \sim \sum_{k,\omega}\left[\lambda + k^2 + f(k,\omega)\right]|\Phi_{k,\omega}|^2, \tag{20.163}$$

where ω are the frequencies (at finite temperature are the Matsubara frequencies), k is the momentum, and λ gives the deviation with respect the critical point. Recall that the function may be given by $f = -i\omega$ or by

$$f(k, \omega) = \frac{|\omega|^m}{k^{m'}}, \tag{20.164}$$

as in the case of the Ising model in a transverse field with $m = 2, m' = 0$. At the critical point, $\lambda = 0$, let us choose length and timescales (or momentum and frequency) as $k' = kl$ and $\omega' = \omega l^z$, assuming that they may scale differently, as reflected in the dynamical critical exponent z. Near the critical point, we expect that the Hamiltonian should be invariant under a scale transformation: as we approach the critical point, the system becomes self-similar. Introducing these scales imposes that the Hamiltonian invariance is obtained if $zm = 2 + m'$. This restriction implies that

$$z = \frac{2 + m'}{m}, \tag{20.165}$$

as discussed before. In the case of the Ising model in transverse field, it yields $z = 1$. This exponent reflects the different scaling between the momentum and the frequency or the different scaling between lengths and energies. Therefore, $\omega \sim k^z$. Near the transition point, the characteristic length is the correlation length, which scales, by definition, as

$$\xi \sim \lambda^{-\nu}. \tag{20.166}$$

Considering a characteristic momentum as $k \sim \xi^{-1}$, we obtain that the energy gap is of the order of

$$\Delta \sim \left(\frac{1}{\xi}\right)^z \sim |\lambda|^{z\nu}. \tag{20.167}$$

Let us now analyze the expression for the number of excitations and remark that the argument of an exponential has no dimensions. Therefore,

$$r \sim \lambda^{z\nu+1}. \tag{20.168}$$

The number of excitations is therefore of the order of

$$n_{exc} \sim k^d \sim \lambda^{d\nu} \sim r^{\frac{d\nu}{z\nu+1}}. \tag{20.169}$$

Considering a one-dimensional system, $d = 1$, with $z = 1$ and considering that $\nu = 1$, we obtain that the number of excitations is given by

$$n_{exc} \sim \sqrt{r}. \tag{20.170}$$

20.6 Exercises

1. Consider the Jordan-Wigner transformation in Eq. (20.6) and an extension of the chain model Eq. (20.19), such that the hopping terms couple second neighbors. Obtain the corresponding spin Hamiltonian.
2. The Jordan-Wigner transformation may be extended to two dimensions [4]. In terms of spinless fermions, one defines

$$c(n) = e^{-i\varphi(n)} S^-(n). \tag{20.171}$$

The phase is chosen as

$$\varphi(n) = \sum_{m \neq n} c^\dagger(m)c(m)B(n, m). \tag{20.172}$$

(a) Show that $\varphi(n)$ commutes with $c(n)$, $S^-(n)$ and that $[\varphi(n), \varphi(m)] = 0$.
(b) Using the commutation relation between S^+ and S^- at different sites, show that

$$e^{iB(n,m)} = -e^{iB(m,n)}. \tag{20.173}$$

Defining the coordinate of a given spin in the plane as $\mathbf{r}(n) = (x(n), y(n))$ and defining $\tau(n) = x(n) + iy(n)$, we can see that differences of these variables between two points satisfy the previous equation for the B matrix. Therefore, we may take

$$B(n, m) = \Im\left(\ln(\tau(m) - \tau(n))\right) \tag{20.174}$$

and

$$\varphi(n) = \sum_{m \neq n} c^\dagger(m)c(m)\Im\left(\ln(\tau(m) - \tau(n))\right). \tag{20.175}$$

(c) Show that

$$S^z(n) = c^\dagger(n)c(n) - \frac{1}{2}. \tag{20.176}$$

(d) Show that the spin commutation relations are satisfied using this fermionic representation.
(e) Considering the two-dimensional Heisenberg model on a lattice, write the Hamiltonian in terms of the spinless fermionic operators, and show that the hopping term involves the difference of phases between nearest-neighbor sites.

3. (a) Recalling that the Kitaev model can be written in momentum space as in Eq. (20.33), calculate the winding numbers of the phases I, II, and III shown in Fig. 20.3, and show that they may be distinguished by the winding number.

   (b) Associated with a nontrivial winding number, one finds zero-energy edge states at the ends of the Kitaev chain. Diagonalize numerically the Kitaev model in real space (Eq. 20.31) using open boundary conditions (a finite system such that there is an ending point on the right-hand side and an end point at the left-hand side). Relate the existence of the edge states (characterized by wave functions that are localized at the edges and that decay exponentially toward the bulk of the chain) with a nontrivial winding number. Show that if one uses periodic boundary conditions, the edge states disappear.

4. In the case of a bosonic system, the decomposition of Eq. (20.59) led to the bosonic Matsubara frequencies since the trace operation in the partition function implies that $\psi(\tau) = \psi(\tau + \beta)$. It turns out that in the case of the partition function for a fermionic system, written in terms of classical anti-commuting variables, called Grassmann variables, the condition is changed to [5]

$$\psi(\tau) = -\psi(\tau + \beta), \tag{20.177}$$

   where ψ is now a Grassmann variable.

   (a) Show that a similar Fourier decomposition leads to the fermionic Matsubara frequencies of the type

$$\nu_m = \frac{(2m + 1)\pi}{\beta}, \tag{20.178}$$

   with $m = 0, \pm 1, \pm 2, \cdots$.
   As a consequence, bosonic Matsubara frequencies are even multiples of π/β, while fermionic Matsubara frequencies are an odd multiple.

   (b) Noting that the sum (or difference) of two even (or odd) numbers is even and that the sum (or difference) of an even number and an odd number is odd, collisions (interactions) between operators (or fields) of bosonic and/or fermionic nature are conditioned. Considering vertices that involve three operators, show that (i) a collision between two fermions involves an intermediate boson and (ii) a spin operator is described by a bosonic frequency.

5. The elements of the reduced density matrix may be written in terms of correlation functions.

   (a) Consider an s-wave superconductor. Show that the single-site reduced density matrix, using a basis of local states like $|0\rangle, |\uparrow,\downarrow\rangle, |\uparrow\rangle, |\downarrow\rangle$, which denote the four possible states—unoccupied, double occupied, singly occupied with an electron with spin $\uparrow$, and singly occupied with an electron with spin $\downarrow$, respectively—can be defined as

$$\rho_A = \begin{pmatrix} \langle(1-n_\uparrow)(1-n_\downarrow)\rangle & \langle c_\uparrow^\dagger c_\downarrow^\dagger\rangle & 0 & 0 \\ \langle c_\downarrow c_\uparrow\rangle & \langle n_\uparrow n_\downarrow\rangle & 0 & 0 \\ 0 & 0 & \langle n_\uparrow(1-n_\downarrow)\rangle & \langle c_\downarrow^\dagger c_\uparrow\rangle \\ 0 & 0 & \langle c_\uparrow^\dagger c_\downarrow\rangle & \langle(1-n_\uparrow)n_\downarrow\rangle \end{pmatrix}.$$

The spin and charge parts decouple. The spin part couples the two spin orientations, and the charge part couples the empty and doubly occupied cases. The diagonal terms of the matrix describe the number of empty sites, the number of doubly occupied sites, the number of $\uparrow$ spin sites, and the number of $\downarrow$ spin sites, respectively. The sum of the diagonal terms is equal to 1 due to normalization.

(b) Consider now the spinless SSH model. As we saw before, this model describes a dimerized chain of spinless fermions hopping in a tight-binding band. The dimerization is parametrized by η. Due to the dimerization, the unit cell contains two atoms of types A and B. The cells are indexed by j. This model is given by the Hamiltonian

$$\begin{aligned} H = -\mu \textstyle\sum_j & \left(c_{j,A}^\dagger c_{j,A} + c_{j,B}^\dagger c_{j,B}\right) \\ -t \textstyle\sum_j & \Big[(1+\eta)c_{j,B}^\dagger c_{j,A} + (1+\eta)c_{j,A}^\dagger c_{j,B} \\ & + \ (1-\eta)c_{j+1,A}^\dagger c_{j,B} + (1-\eta)c_{j,B}^\dagger c_{j+1,A}\Big]. \end{aligned} \tag{20.179}$$

The operator $c_{j,\alpha}$ destroys a spinless fermion at site j of type $\alpha = A, B$. The amplitude t is the hopping, η the dimerization, and μ the chemical potential.

Show that in the basis of $|n_{j,A}, n_{j,B}\rangle = \{|00\rangle, |01\rangle, |10\rangle, |11\rangle\}$, the reduced density matrix may take the form

$$\rho_j = \begin{pmatrix} u & 0 & 0 & 0 \\ 0 & v & z & 0 \\ 0 & z & v & 0 \\ 0 & 0 & 0 & u \end{pmatrix}, \tag{20.180}$$

and specify the condition(s) used.

6. Consider a noninteracting system with translational invariance. Show that the fidelity factorizes in terms of the fidelities for states of specific momenta.
7. A sudden quench is an extreme instantaneous transformation of the parameters of the various terms of a Hamiltonian. At a given instant, call it $t = 0$, a transformation from a set of parameters, ξ, to a different set of parameters, ξ', takes place. Consider, to simplify, a noninteracting Hamiltonian and that we

know the (single-particle) eigenstates and eigenvalues. Therefore, we know

$$\begin{aligned} H(\xi)|\psi_n(\xi)\rangle &= E_n(\xi)|\psi_n(\xi)\rangle \\ H(\xi')|\psi_{n'}(\xi')\rangle &= E_{n'}(\xi')|\psi_{n'}(\xi')\rangle. \end{aligned} \tag{20.181}$$

After the quench, the eigenstates evolve as

$$|\psi_n(\xi, t)\rangle = e^{-iH(\xi')t}|\psi_n(\xi)\rangle. \tag{20.182}$$

Define the transition amplitudes

$$\begin{aligned} A_{lm}(t) &= \langle\psi_l(\xi)|\psi_m(\xi, t)\rangle \\ B_{l'm}(t) &= \langle\psi_{l'}(\xi')|\psi_m(\xi, t)\rangle. \end{aligned} \tag{20.183}$$

(a) Inserting a decomposition over the eigenstates of the final Hamiltonian, obtain expressions for the two transition amplitudes.
(b) Determine the time dependence of the probability associated with the amplitude $B_{l'm}(t)$.
(c) Obtain an expression for the survival probability of a given initial state defined by

$$P_m = |A_{mm}(t)|^2. \tag{20.184}$$

(d) Consider the Kitaev or SSH models in real space with open boundary conditions, and plot the various probabilities (absolute squares of the amplitudes) as a function of time. Consider different types of quenches (between states in the same or different phases), and compare the evolution of bulk and edge states.

8. The Loschmidt amplitude calculates the overlap amplitude of some initial state, eigenstate of some Hamiltonian H^0, for instance, its ground state $|\psi_0\rangle$, with its time evolution. The unitary time evolution results from the action of a Hamiltonian, H^1, such as due to a sudden quench (evolution with the quenched Hamiltonian, obtained changing abruptly some parameters in the Hamiltonian),

$$\mathcal{G}(t) = \langle\psi_0|\psi_0(t)\rangle = \langle\psi_0|e^{-iH^1t}|\psi_0\rangle, \tag{20.185}$$

or its square $L(t) = |\mathcal{G}(t)|^2$ (here, we take $\hbar = 1$). The exponential is similar to the Boltzmann factor considering a variable $z = it$. In the thermodynamic limit, such an overlap vanishes, due to the orthogonality catastrophe, and it is more convenient to calculate the rate

$$\lambda(t) = -\lim_{N\to\infty} \frac{1}{N} \ln L(t), \tag{20.186}$$

where N is the size of the system.

A way to understand the time dependence of the Loschmidt rate is to recall that traditional phase transitions in statistical mechanics are present when there are singularities in the free energy. These are associated with zeros of the partition function (Lee-Yang zeros). There is a formal similarity between the free energy per particle and the Loschmidt rate if we express the Loschmidt rate in terms of its zeros defined in the complex space [6]. In the case of two-band noninteracting translationally invariant Hamiltonian defined by the matrix $H(k) = \mathbf{d}(k) \cdot \boldsymbol{\sigma}$, where σ are the Pauli matrices, and defining the unit vector $\hat{\mathbf{d}}_k = \mathbf{d}(k)/|\mathbf{d}(k)|$, the Loschmidt amplitude, associated with a quench from some initial parameters, defined by the vector $\hat{\mathbf{d}}_k^0$, to a final state defined by $\hat{\mathbf{d}}_k^1$, may be written as

$$\mathcal{G}(t) = \prod_k \left[\cos(\epsilon_k^1 t) + i\hat{\mathbf{d}}_k^0 \cdot \hat{\mathbf{d}}_k^1 \sin(\epsilon_k^1 t)\right], \tag{20.187}$$

where $\epsilon_k^1 = \pm|\mathbf{d}_k^1|$. This expression holds independently of the space dimension. The zeros in the complex space, also called Fisher zeros, are given by

$$z_n(k) = i\frac{\pi}{\epsilon_k^1}\left(n - \frac{1}{2}\right) - \frac{1}{\epsilon_k^1}\tanh^{-1}\left(\hat{\mathbf{d}}_k^0 \cdot \hat{\mathbf{d}}_k^1\right), \quad n \in N. \tag{20.188}$$

The zeros define a momentum k^* that if $\hat{\mathbf{d}}_{k^*}^0 \cdot \hat{\mathbf{d}}_{k^*}^1 = 0$ this leads to a set of real times

$$t_n = \frac{\pi}{2\epsilon_{k^*}^1}(2n - 1), \tag{20.189}$$

which are associated with the non-analyticities in the Loschmidt rate. These times are obtained taking the real part of the complex zeros to be zero and determining the interception with the imaginary axis.

(a) Consider that the system is in the ground state. The state of the system, for each momentum value (recall that the density matrix factorizes in terms of the momenta), may be represented by the density matrix

$$\rho_k^0 = \frac{1}{2}\left(I - \hat{\mathbf{d}}_k^0 \cdot \sigma\right). \tag{20.190}$$

The Loschmidt amplitude may be calculated as

$$\mathcal{G}(t) = \prod_k \mathcal{G}_k(t), \tag{20.191}$$

where

$$\mathcal{G}_k(t) = Tr\left[\rho_k^0 e^{-i\epsilon_k^1 \hat{\mathbf{d}}_k^1 \cdot \sigma t}\right]. \tag{20.192}$$

Derive Eq. (20.187).

(b) Consider the SSH model Eq. (10.135), and take $t_1 = t(1+\eta)$, $t_2 = t(1-\eta)$. Choose the quenches: (i) $\eta^0 = 0.9, \eta^1 = -0.9$; (ii) $\eta^0 = -0.9, \eta^1 = 0.9$; and (iii) $\eta^0 = 0.9, \eta^1 = 0.5$. For each case, plot the rate as a function of time, and show that it has singularities (sharp peaks). Plot $\Im(z_k) vs. \Re(z_k)$ and verify the cases where the singularities in the rate are associated with the times obtained from Eq. (20.188). Also, note that the various peaks of the rate correspond to the various values of n in Eq. (20.189).

References

1. Roman M. Lutchyn, Jay D. Sau, and S. Das Sarma, Physical Review Letters **105**, 077001 (2010); Yuval Oreg, Gil Refael, and Felix von Oppen, Physical Review Letters **105**, 177002 (2010).
2. P. Zanardi and N. Paunković, Physical Review E **74**, 031123 (2006); S. J. Gu, International Journal of Modern Physics B **24**, 4371 (2010). Note that in this reference the Jordan-Wigner transformation is defined differently: $\sigma_n^+ = e^{i\pi \sum_{j=1}^{n-1} c_j^\dagger c_j} c_n = \prod_{j=1}^{n-1} \sigma_j^z c_n$, $\sigma_n^- = e^{-i\pi \sum_{j=1}^{n-1} c_j^\dagger c_j} c_n^\dagger = \prod_{j=1}^{n-1} \sigma_j^z c_n^\dagger$ and $\sigma_n^z = 1 - 2c_n^\dagger c_n$. As a consequence, the roles of $k = 0$ and $k = \pi$ get reversed.
3. A. Polkovnikov, Physical Review B **72**, 161201(R) (2005).
4. See, for instance, Y. R. Wang, Physical Review B **46**, 151 (1992).
5. See, for example, A. Altland and B.D. Simons, *Condensed Matter Field Theory*, Cambridge University Press.
6. M. Heyl, A. Polkovnikov, S. Kehrein, Physical Review Letters **110**, 135704 (2013).

Chapter 21
Mappings and Dualities

Dualities appear in physics in various contexts, such as dualities in classical electromagnetism (between electric and magnetic fields at the level of Maxwell's equations) and, among others, some dualities that we will consider in this chapter in the realm of condensed matter physics problems. Examples are the Kramers-Wannier identities that relate the partition function of the system in the low- and high-temperature limits, and dual lattices, by establishing a relation between variables and operators defined at the corners or links of some lattice problem. Also, interesting dualities relate quantum and classical problems, and others relate different sets of operators, both with the same statistics or with different statistics, or different commutation relations. Some of these issues will be considered in this chapter.

Having established some relation between operators or physical systems, it may be that the solution of one problem is simplified using the knowledge that may exist on the related problem. A particular useful case is when some exact solution in terms of different variables or operators is possible. In most cases a simplified way to extract valuable information about the behavior of one system and how it translates to some other problem is also useful. Often, some mean-field-like approximation is required.

21.1 Relation Between Classical and Quantum Problems

21.1.1 Classical 1d Ising Model and 0d Quantum Model

Recall the partition function of the one-dimensional Ising model:[1]

[1] For further reading see [Sachdev].

P. D. Sacramento, *Introduction to Condensed Matter Theory*,
https://doi.org/10.1007/978-3-031-86251-9_21

$$Z = \sum_{\{\sigma_l^z = \pm\}} e^{-H} , \tag{21.1}$$

where σ_l^z are classical variables taking the values ± 1 and the Hamiltonian multiplied by $\beta = 1/(k_B T)$ is

$$H = -K \sum_{l=1}^{M_\tau} \sigma_l^z \sigma_{l+1}^z - h \sum_{l=1}^{M_\tau} \sigma_l^z , \tag{21.2}$$

where M_τ is the number of sites. Note that H is dimensionless. As we saw before, the partition function can be written as

$$Z = \sum_{\{\sigma_l^z\}} \prod_{l=1}^{M_\tau} T_1 \left(\sigma_l^z , \sigma_{l+1}^z \right) T_2 (\sigma_l^z) , \tag{21.3}$$

with

$$\begin{aligned} T_1 \left(\sigma_1^z , \sigma_2^z \right) &= e^{K \sigma_1^z \sigma_2^z} \\ T_2 (\sigma^z) &= e^{h \sigma^z} . \end{aligned} \tag{21.4}$$

The partition was then written as

$$Z = Tr \left(T_1 T_2 T_1 T_2 \cdots \right) , \tag{21.5}$$

repeated M_τ times. The transfer matrices are explicitly given by

$$T_1 = \begin{pmatrix} e^K & e^{-K} \\ e^{-K} & e^K \end{pmatrix} \tag{21.6}$$

and

$$T_2 = \begin{pmatrix} e^h & 0 \\ 0 & e^{-h} \end{pmatrix} . \tag{21.7}$$

Recall the expression for the correlation of the spins in two sites, l, l', Eqs. (15.44, 15.47). Therefore,

$$\langle \sigma_l^z \sigma_{l'}^z \rangle = (\tanh K)^{l'-l} . \tag{21.8}$$

Defining the distance, $\tau = la$, with a the lattice constant, and defining the correlation length, ξ, through

$$C(\tau) = \langle \sigma^z(\tau)\sigma^z(0)\rangle = e^{-\frac{|\tau|}{\xi}}, \tag{21.9}$$

leads to

$$\frac{1}{\xi} = \frac{1}{a}\ln(\coth K). \tag{21.10}$$

Therefore, $\xi/a \sim e^{2K}/2$ which is very large if $K \gg 1$, which happens at low temperatures.

The expression for T_1 may be rewritten as

$$\begin{aligned} T_1 &= e^K\left(I + e^{-2K}\hat{\sigma}^x\right) \\ &\sim e^K\left(I + \frac{a}{2\xi}\hat{\sigma}^x\right) \\ &\sim e^{a\left(-E_0 + \frac{1}{2\xi}\hat{\sigma}^x\right)}, \end{aligned} \tag{21.11}$$

where $E_0 = -K/a$; we used that $a/\xi \ll 1$ and $\hat{\sigma}^x$ is a Pauli matrix (here, to distinguish the classical variables, σ^z, from the Pauli matrices, we introduce a hat on the Pauli matrices). Also, defining $\tilde{h} = h/a$ we may write

$$T_2 = e^{a\tilde{h}\hat{\sigma}^z}. \tag{21.12}$$

Note that we may take a as small, since it is the lattice constant. Then we may use that

$$e^{aO_1}e^{aO_2} = e^{a(O_1+O_2)}(1 + O(a^2)) \tag{21.13}$$

and truncate at first order. Call $L_\tau = M_\tau a$ the length of the system and define $T = 1/L_\tau$ and $\Delta = 1/\xi$. Then we may write approximately that

$$T_1T_2 = e^{-aH_Q} \tag{21.14}$$

and

$$Z = Tr\,(T_1T_2)^{M_\tau} \sim Tre^{-\frac{H_Q}{T}}, \tag{21.15}$$

where the "quantum" Hamiltonian is given by

$$H_Q = E_0 - \frac{\Delta}{2}\hat{\sigma}^x - \tilde{h}\hat{\sigma}^z. \tag{21.16}$$

It is a zero-dimensional problem (one site, no spatial dependence), which involves the Pauli matrices (which originated in the 2×2 structure of the transfer matrix in real space of the original classical one-dimensional Ising problem). We have therefore established a mapping between the classical one-dimensional Ising model and a quantum spin (1/2) problem in zero dimensions.

A few comments are in order at this point. The partition function of the "quantum" problem was written as a Boltzmann factor with the parameter, T, seen as the temperature of this related "quantum" problem (written in units with $k_B = 1$). The mapping shows that the temperature is then the inverse of the spatial length of the classical system (in units of the lattice constant). The parameter Δ is the gap in the quantum system and is related to the inverse of the correlation length in the classical system. If the correlation length diverges, the quantum system becomes gapless.

21.1.2 Mapping of a 1*d* Quantum Ising Model on a Classical 2*d* Ising Model

We may now consider an inverse problem that establishes a mapping between a quantum problem in spatial dimension d and a classical problem in a dimension $D = d + 1$. We write the partition function of the quantum problem as

$$Z = Tre^{-\frac{H_Q}{T}} = Tr\left[\left(e^{-aH_Q}\right)^{M_\tau}\right], \tag{21.17}$$

where $L_\tau = M_\tau a = 1/T$ is a size that is associated with the temperature of the system. Since our starting point is the $1d$ quantum system, T is the actual temperature of the reservoir with which the quantum system is in thermal equilibrium and so L_τ is some dimension along the τ direction. If $T = 0$, the extension along this direction is infinite. Otherwise it has a finite extension determined by the temperature of the system.

The system has other dimensions associated with the spatial directions along the distribution of the constituents of the system. If the quantum system is one-dimensional with a lattice constant a_s, then there is only one spatial direction, say with M sites. Therefore, we have an effective two-dimensional system such that there are M_τ sites along one direction (which is associated with temperature) and M along the spatial direction; call it the x direction and the other the τ direction. This effective two-dimensional lattice is shown in Fig. 21.1. Considering the quantum Ising chain, the nearest-neighbor Hamiltonian H_Q couples at most two sites. We may see the problem of calculating the partition function as defining a transfer matrix that couples sites along τ that has a dimension $2^M \times 2^M$. Instead of a lattice model we may as well consider a continuum limit ($a \to 0$) and the small temperature limit ($M_\tau \to \infty$) as long as we keep $M_\tau a = 1/T$ fixed.

To be specific, let us consider once again the transverse Ising model:[2]

$$H_Q = -Jg\sum_i \hat{\sigma}_i^x - J\sum_{\langle i,j\rangle} \hat{\sigma}_i^z\hat{\sigma}_j^z, \tag{21.18}$$

with an exchange interaction between nearest neighbors and a transverse field $h = Jg$. Recall once again that due to the magnetic field the problem has a quantum nature, due to the noncommutativity of $\hat{\sigma}^x$ and $\hat{\sigma}^z$, which are in this context Pauli matrices and not variables, as in the classical Ising model.

We can write that

$$e^{-aH_Q} = T_1T_2 + O(a^2), \tag{21.19}$$

with

$$\begin{aligned} T_1 &= e^{Jga\sum_i \hat{\sigma}_i^x} \\ T_2 &= e^{Ja\sum_i \hat{\sigma}_i^z\hat{\sigma}_{i+1}^z}. \end{aligned} \tag{21.20}$$

The partition function involves a product of exponentials (M_τ times) of the form

$$e^{-aH_Q}e^{-aH_Q}e^{-aH_Q}\dots e^{-aH_Q}.$$

We can now insert a complete set of states between each exponential factor, using as states the eigenstates of $\hat{\sigma}_i^z$, written as $|\{m_i\}\rangle$, with $m_i = \pm 1$ and in total number 2^M, since there are M sites along the spatial direction. The operator T_2 is diagonal

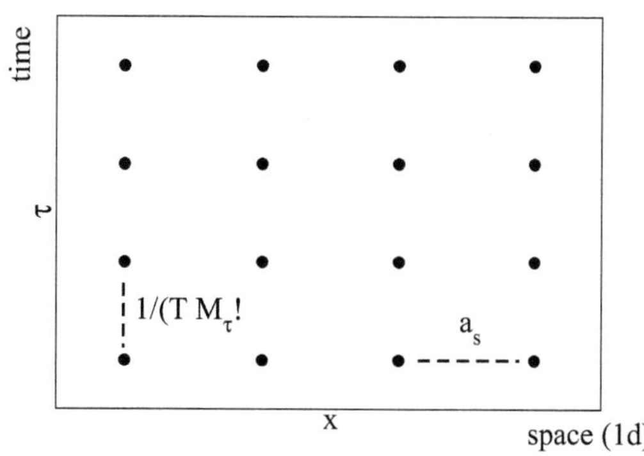

Fig. 21.1 Effective two-dimensional spatial lattice that results from a one-dimensional spatial lattice and an (imaginary) temporal direction

[2] As already mentioned, sometimes one chooses the interaction along the z direction and sometimes along the x direction. A given choice is usually dictated by convenience. If it is convenient to introduce a Jordan-Wigner transformation, a natural choice is to write the interaction along the x direction. In the following it will be useful to introduce a complete set of spin component eigenstates, and so it will be technically easier to consider the interaction along the z-axis.

in this basis and satisfies

$$T_2|\{m_i\}\rangle = e^{Ja\sum_i^M m_i m_{i+1}}|\{m_i\}\rangle. \tag{21.21}$$

Since there are two directions, we consider that $i = 1, \ldots, M$; let us introduce the variable along the direction associated with the temperature, $l = 1, \cdots, M_\tau$. We also have to consider that the eigenvalues have the indices $m_i(l)$. Then we can rewrite the partition function in a more explicit form as

$$Z = \sum_{\{m_i(l)\}} \prod_{l=1}^{M_\tau} \langle\{m_i(l)\}|T_1 T_2|\{m_i(l+1)\}\rangle. \tag{21.22}$$

There are 2^{MM_τ} values of $m_i(l) = \pm 1$, over which we have to sum. Since there are two directions, i associated with the spatial, x, direction and the other l associated with the τ direction; this suggests enlargement of one spatial dimension to effectively two dimensions. This shows that we have mapped the original quantum problem to a two-dimensional spin problem. This model should be classical, because it is determined by the classical variables $m_i(l)$ (eigenvalues of the quantum operator $\hat{\sigma}^z_{i(l)}$ operator) of the Ising type. Therefore, the classical model should be some form of the 2*d* Ising model.

Since the basis used is related with the eigenstates of T_2, all that is left is to determine the action of T_1 on the states:

$$\langle\{m_i(l)\}|T_1|\{m_i(l+1)\}\rangle = \prod_i \langle\{m_i(l)\}|e^{Jga\hat{\sigma}_i^x}|\{m_i(l+1)\}\rangle. \tag{21.23}$$

Each exponential is local in space and the problem reduces to a 2×2 matrix problem. We may write that

$$\langle m|e^{Jga\hat{\sigma}_x}|m'\rangle = Ae^{Bmm'}. \tag{21.24}$$

Considering that $m, m' = \pm 1$ we find that

$$\begin{aligned} e^{-2B} &= \tanh\gamma \\ A^2 &= \frac{1}{2}\sinh(2\gamma), \end{aligned} \tag{21.25}$$

where $\gamma = Jga$, which is of the type of a magnetic field $\gamma \sim h$. We arrive at the result that

$$Z = \sum_{\{m_i(l)\}} e^{\sum_{i,l}[Jam_i(l)m_{i+1}(l) + Bm_i(l)m_i(l+1)]}. \tag{21.26}$$

Since the variables are Ising variables, this is the partition function of a 2*d* anisotropic Ising model.

21.2 Two-Dimensional Ising Model

We may also invert the reasoning and establish a mapping between the classical two-dimensional classical Ising model and a one-dimensional quantum spin model.

As we saw before, the one-dimensional classical Ising model is in a disordered phase (without spontaneous magnetization), except at $T = 0$. At $T = 0$ the model has two ordered phases that correspond to have all spins pointing "up" ($\sigma_i = +1$) or all spins pointing "down" ($\sigma_i = -1$). The system is disordered at any finite temperature. If we consider the classical system in two dimensions, then one expects two phases separated by a finite critical temperature.

Consider the Ising model defined on a two-dimensional spatial lattice. For discussion purposes we will call one direction the spatial direction and the other the "temporal" dimension, already anticipating the mapping to a quantum system in one less spatial dimension plus an (imaginary) temporal direction. The action of the model is defined as

$$S = -\sum_n \left[\beta_\tau \sigma^z(n+\boldsymbol{\tau})\sigma^z(n) + \beta\sigma^z(n+\mathbf{x})\sigma^z(n)\right]. \tag{21.27}$$

The vector $\mathbf{x}$ is the versor of the spatial direction and the vector $\boldsymbol{\tau}$ is the versor of the temporal direction. The Ising variables are in fact the z component of a spin operator $\boldsymbol{\sigma}$ (having in mind the quantum description). The equivalent quantum problem is obtained considering the transfer matrix. The two-dimensional Ising model is equivalent to the Ising model in transverse magnetic field defined by

$$\hat{H} = -\sum_m \hat{\sigma}^x(m) - \lambda\sum_m \hat{\sigma}^z(m+1)\hat{\sigma}^z(m), \tag{21.28}$$

as we just saw in the previous section. The transfer matrix couples the various lines of spins in the spatial direction eliminating the temporal direction and reducing the classical system to a quantum system, with one less dimension. The parameter λ is defined by $\beta = \lambda e^{-2\beta_\tau}$. We leave it to the reader to verify the mapping. Since the Ising model in a transverse field was solved before, this allows an indirect solution to the classical Ising model in two dimensions.

21.2.1 Self-Duality of Ising Model in Transverse Field

An important property of the Ising model is that it is self-dual. The duality transformation was defined by Kramers and Wannier and establishes a relation between the high- and low-temperature properties.

Consider again the Ising model in transverse field in one dimension, Eq. (21.28). Let us associate to each point in the one-dimensional lattice a point in the dual lattice, which corresponds to the bonds that connect the points of the original lattice.

Let us now define the operators of the dual lattice:

$$\mu^x(n) = \sigma^z(n+1)\sigma^z(n)$$
$$\mu^z(n) = \prod_{m<n} \sigma^x(m). \tag{21.29}$$

The operator $\mu^x(n)$ gives information about the state of the bond between the points n and $n+1$ and the operator $\mu^z(n)$ inverts all the spins to the left of point n (this operator is similar to the operator that creates a kink in the Jordan-Wigner transformation). The operators μ also obey the algebra of the Pauli matrices. The Hamiltonian written in terms of the new operators

$$H = \lambda \left[-\sum_n \mu^x(n) - \frac{1}{\lambda} \sum_n \mu^z(n)\mu^z(n+1) \right], \tag{21.30}$$

is related to the Hamiltonian written in terms of the original operators: σ as

$$H(\sigma;\lambda) = \lambda H(\mu; 1/\lambda). \tag{21.31}$$

Since the operators μ and σ have the same algebra, this implies that

$$E(\lambda) = \lambda E(1/\lambda). \tag{21.32}$$

Therefore, the gap must obey the same law $\Delta(\lambda) = \lambda\Delta(1/\lambda)$. If the gap vanishes at a single point, then this should occur when $\lambda = 1$, as concluded from the exact solution. Consider then the Hamiltonian

$$\frac{1}{\lambda} H + h \sum_n \sigma^z(n),$$

where H is the Hamiltonian of the transverse field Ising model and where we have added a magnetic field h along the z-axis. Applying the duality transformation we obtain

$$\frac{1}{\lambda} H(\sigma;\lambda) + h \sum_n \sigma^z(n) = H(\mu; 1/\lambda) + h \sum_n \prod_{m<n} \mu^x(m).$$

We conclude that the average value of the operator $\sum_n \sigma^z(n)$ in the groundstate of the initial Hamiltonian should be equal to the average value of the operator $\sum_n \prod_{m<n} \mu^x(m)$ in the groundstate of the Hamiltonian defined in the dual lattice (evaluated for the value $1/\lambda$). Therefore, since at low temperatures the operator $\sum_n \sigma^z(n)$ has a nonvanishing average value, then at high temperatures the operator $\sum_n \prod_{m<n} \mu^x(m)$ also has an average value that is nonvanishing and, therefore, may be used to characterize the high-temperature phase. Since it is an operator that

disorders the spins, its average value should vanish in the low-temperature phase. Since this operator creates kinks in the spin configuration, the high-temperature phase is then a condensate of kinks, as previously seen. As a consequence of this disorder, the correlation function decays exponentially with distance and, therefore, the order is of short range, corresponding to a correlation length that is finite.

21.2.2 Self-Duality of the 2d Classical Ising Model

Recall that the classical Ising model is described by the partition function:

$$Z(K) = \sum_{\sigma_N=\pm 1} \cdots \sum_{\sigma_1=\pm 1} e^{K \sum_{\langle i,j\rangle} \sigma_i \sigma_j} . \tag{21.33}$$

The interaction is between nearest neighbors and let us consider a square lattice. The parameter $K = J/(k_B T)$, where J is the coupling between the classical variables, and T is the temperature, as in previous sections. N is the number of lattice sites. Using that[3]

$$e^{K\sigma} = \cosh K + \sigma \sinh K, \tag{21.34}$$

we can write that

$$\begin{aligned} Z(K) &= \sum_{\sigma_N=\pm 1} \cdots \sum_{\sigma_1=\pm 1} \prod_{\langle i,j\rangle} \big(\cosh K + \sigma_i \sigma_j \sinh K\big) \\ &= (\cosh K)^{2N} \sum_{\sigma_N} \cdots \sum_{\sigma_1} \prod_{\langle i,j\rangle} \big(1 + \sigma_i \sigma_j \tanh K\big) . \end{aligned} \tag{21.35}$$

The exponent in the prefactor results from the number of pairs as $1/2 \times 4 \times N = 2N$, since we have 4 neighbors and N sites. Let us now determine the behavior in both the high-temperature and the low-temperature regimes.

At high temperature we have that K is small and so we may expand in powers of $\tanh K$. We should note that

$$\begin{aligned} &\sum_{\sigma=\pm 1} \sigma = 0 \\ &\sum_{\sigma=\pm 1} 1 = 2. \end{aligned} \tag{21.36}$$

[3] For further reading see J. Kogut, Reviews of Modern Physics **51**, 659 (1979).

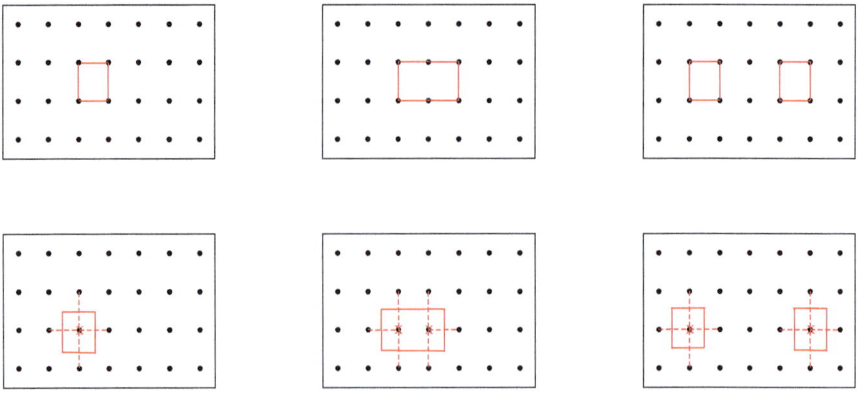

Fig. 21.2 Top row ($T \to \infty$): dominant contributions of closed loops of nearest-neighbor bonds to the partition function. Bottom row ($T \to 0$): dominant contributions to the partition function of broken bonds

Expanding $\prod_{\langle i,j\rangle} \left(1 + \sigma_i \sigma_j \tanh K\right)$ terms with one free variable σ_i give zero. Therefore, only links between neighboring sites that form closed paths contribute. In this case we get that each term that contributes on a set of sites i gives $\sigma_i^2 = 1$, since $\sigma_i = \pm 1$. Terms that contribute are shown in Fig. 21.2 and are of the type of no spin variables, a closed square, rectangle, two separated squares, and so on. Therefore, we can get

$$Z(K)\,(\cosh K)^{-2N}\, 2^{-N} = 1 + N\,(\tanh K)^4 + 2N\,(\tan K)^6 + \frac{1}{2}N(N-5)\,(\tanh K)^8 + \cdots \tag{21.37}$$

where the prefactors of each term are the multiplicities associated to each type of term.

At zero temperature all spin variables have the same value (say all pointing up). For small temperatures some of the spin variables are excited, but in reduced number. The leading term is just one of the spins pointing down. This means four broken links. The next term is two spins down. There are two possibilities: if the flipped spins are nearest neighbors, six links are broken. Otherwise there are eight broken links. We can expand the partition function in increasing number of flipped spins as

$$Z(K)e^{-2NK} = 1 + Ne^{-8K} + 2Ne^{-12K} + \frac{1}{2}N(N-5)e^{-16K} + \cdots \tag{21.38}$$

The various terms are shown in Fig. 21.2 where we represent by crosses the flipped spins. Their graphical representation is similar if we consider the squares or the rectangles. We expect therefore some type of correspondence between the two graphical representations and therefore between the high- and the low-temperature regimes.

Let us define a parameter K^* as

$$\tanh K = e^{-2K^*}. \tag{21.39}$$

In the low-temperature expansion let us replace K by K^*. We get

$$\begin{aligned}\frac{Z(K^*)}{\left(e^{2K^*}\right)^N} &= 1 + Ne^{-8K^*} + 2Ne^{-12K^*} + \frac{1}{2}N(N-5)e^{-16K^*} + \cdots \\ &= 1 + N(\tanh K)^4 + 2N(\tanh K)^6 + \frac{1}{2}N(N-5)(\tanh K)^8 + \cdots \\ &= \frac{Z(K)}{2^N\left(\cosh^2 K\right)^N}. \end{aligned} \tag{21.40}$$

It is easy to show that Eq. (21.39) is equivalent to

$$\sinh(2K)\sinh(2K^*) = 1. \tag{21.41}$$

Therefore, we can write that

$$\frac{Z(K^*)}{[\sinh(2K^*)]^{N/2}} = \frac{Z(K)}{[\sinh(2K)]^{N/2}}. \tag{21.42}$$

The relation between K and K^* shows that if one is large, the other is small. This provides the relation between the high- and the low-temperature regimes. If the model has some unique critical point at a finite temperature, as it was argued before, then the mapping between the two temperature regimes should occur at $K = K^* = K_c$ which is determined by

$$\sinh^2(2K_c) = 1, \tag{21.43}$$

which is equivalent to

$$e^{2K_c} = \sqrt{2} + 1. \tag{21.44}$$

This mapping between the two regimes and the expression for the critical coupling (meaning critical temperature, T_c) was obtained by [7, 8].

21.2.3 Critical Behavior of the 2d Classical Ising Model

As we saw before, there is a mapping between the classical model in D dimensions and a quantum model in $d = D - 1$ dimensions. Also, for a d-dimensional given quantum model, there is some mapping between its critical behavior originated from thermal fluctuations and its quantum critical zero temperature behavior, originated by quantum fluctuations in $d + z$ dimensions, where z is the dynamic critical exponent.

Previously we considered the critical behavior of the quantum Ising model in a transverse field and found its critical exponents, such as $z = 1$, $\nu = 1$ (and therefore $\alpha = 0$, using that $2 - \alpha = \nu(d + z)$, with $d = 1$) and $\beta = 1/8$. The mapping to the classical $d = 2$ dimensions leads to the conclusion that its critical exponents are the same. Using scaling relations we can then list the critical exponents of the 2d classical Ising model. These are $\alpha = 0, \beta = 1/8, \gamma = 7/4, \delta = 15, \nu = 1, \eta = 1/4$.

It is also possible to solve the two-dimensional Ising model in other ways. We will not pursue it here.

21.3 Spin Representations

In addition to the Holstein-Primakoff transformation between spin operators and bosonic operators and the Jordan-Wigner transformation that relates spin operators and spinless fermion operators, other representations of the spin operators in terms of bosons or fermions may be used. In this section we consider two of these transformations.

21.3.1 Schwinger Spin Representation

In the Schwinger representation one introduces two bosons (with destruction operators a_1 and a_2) and we define the spin operators as

$$\begin{aligned} S^+ &= a_1^\dagger a_2 \\ S^- &= a_2^\dagger a_1 \\ S^z &= \frac{1}{2}\left(a_1^\dagger a_1 - a_2^\dagger a_2\right). \end{aligned} \tag{21.45}$$

As seen in the exercises, a fermionic representation is also possible. This representation of the spin operators satisfies the commutation relations of the spin operators such as

$$[S^+, S^-] = 2S^z. \tag{21.46}$$

However, the condition that $S^2 = S(S + 1)$ is only satisfied if a projection to the physical subspace is imposed:

$$n_1 + n_2 = 2S. \tag{21.47}$$

Solving one of the boson operators in terms of the other, through the restriction, the Schwinger representation is equivalent to the Holstein-Primakoff representation, in the physical subspace. We get solving the restriction that

$$a_1^\dagger a_1 = 2S - a_2^\dagger a_2 \tag{21.48}$$

and

$$S^z = S - a_2^\dagger a_2. \tag{21.49}$$

Call $a_2 = a$. We may take

$$a_1^\dagger = a_1 = \sqrt{2S - a^\dagger a} = \sqrt{2S}\sqrt{1 - \frac{a^\dagger a}{2S}}. \tag{21.50}$$

Therefore, it leads to the Holstein-Primakoff relations:

$$\begin{aligned} S^+ &= \sqrt{2S}\sqrt{1 - \frac{a^\dagger a}{2S}}a \\ S^- &= a^\dagger \sqrt{2S}\sqrt{1 - \frac{a^\dagger a}{2S}} \\ S^z &= S - a^\dagger a. \end{aligned} \tag{21.51}$$

21.3.2 Majorana Representation of Spin-$1/2$ Operators

Defining (hermitian) fermionic operators that satisfy the anti-commutation relations

$$\{\eta_a, \eta_b\} = \delta_{ab}, \tag{21.52}$$

with $\eta_a^2 = 1/2$, we can define the spin operator as

$$\mathbf{S} = \frac{-i}{2}\eta \times \eta, \tag{21.53}$$

or explicitly,

$$\begin{aligned} S^x &= -i\eta_y\eta_z \\ S^y &= -i\eta_z\eta_x \end{aligned}$$

$$S^z = -i\eta_x\eta_y. \tag{21.54}$$

It is easy to verify that the commutation relation is satisfied $[S^x, S^y] = iS^z$ and that $\mathbf{S}^2 = S(S+1)$, so apparently no restriction to a physical subspace is needed. However, see the next section where this representation (with a different, inessential, normalization) will be considered in greater detail.

21.4 Unitary Transformations of Fermion and Spin Operators

21.4.1 Fermion, Majorana, and Spin Operators

As mentioned before, a spinful fermion operator at site j may be written in terms of two hermitian operators, $\gamma_{j,\sigma,1}, \gamma_{j,\sigma,2}$, as

$$\begin{aligned} c_{j,\sigma} &= \frac{1}{2}\left(\gamma_{j,\sigma,1} + i\gamma_{j,\sigma,2}\right) \\ c^\dagger_{j,\sigma} &= \frac{1}{2}\left(\gamma_{j,\sigma,1} - i\gamma_{j,\sigma,2}\right). \end{aligned} \tag{21.55}$$

The index σ represents internal degrees of freedom of the fermionic operator, such as spin and/or sublattice index. The γ operators satisfy a Clifford algebra $\{\gamma_m, \gamma_n\} = 2\delta_{mn}I$, where $m, n = j, \sigma, \iota$, with $\iota = 1, 2$, and I is the identity and are normalized as $\gamma_n^2 = 1$. Note that the normalization of the γ_n operators differs from that of the η_i operators.

Also, as seen in the previous section, Majoranas allow a representation of spin-1/2 problems. One needs three Majorana operators to represent local spin operators as

$$S^x = -(i/2)\gamma_2\gamma_3; \qquad S^y = -(i/2)\gamma_3\gamma_1; \qquad S^z = -(i/2)\gamma_1\gamma_2. \tag{21.56}$$

Here, $i = 1, 2, 3 = x, y, z$. It is convenient to define the operators $\mathcal{S}_i = \frac{-i}{2}\epsilon_{ijk}\gamma_j\gamma_k$, related to the standard spin operators as $\mathbf{S} = \frac{1}{2}\vec{\mathcal{S}}$. The reason for the two sets of operators is the different normalizations of the hermitian operators in Eqs. (21.54) and (21.56). Also, it is convenient to introduce the operator $I_3 = -i\gamma_1\gamma_2\gamma_3 = \frac{-i}{6}\epsilon_{ijk}\gamma_i\gamma_j\gamma_k$.

The multiplication rule of the Majorana operators becomes[4]

$$\gamma_i\gamma_j = \delta_{ij}I + i\epsilon_{ijk}\mathcal{S}_k, \tag{21.57}$$

[4] For further reading see P. D. Sacramento and V. R. Vieira, Annals of Physics **391**, 216 (2018) and [AraujoSacramento].

and, as a consequence, $\boldsymbol{\gamma}^2 = 3I$. The multiplication rules of the Majorana and spin operators are given by

$$\gamma_i \mathcal{S}_j = \delta_{ij} I_3 + i\epsilon_{ijk}\gamma_k \tag{21.58}$$

$$\mathcal{S}_i \gamma_j = \delta_{ij} I_3 + i\epsilon_{ijk}\gamma_k \tag{21.59}$$

and

$$\mathcal{S}_i \mathcal{S}_j = \delta_{ij} I + i\epsilon_{ijk}\mathcal{S}_k, \tag{21.60}$$

together with $I_3^2 = I$, $I_3\gamma_i = \gamma_i I_3 = \mathcal{S}_i$, and $I_3\mathcal{S}_i = \mathcal{S}_i I_3 = \gamma_i$. Also, $\vec{\mathcal{S}}^2 = 3I$ and $\boldsymbol{\gamma} \cdot \vec{\mathcal{S}} = \vec{\mathcal{S}} \cdot \boldsymbol{\gamma} = 3I_3$.

Representing fermions or spins using Clifford algebras may lead to an enlargement of the number of states. In the case of fermion systems there is no enlargement of the number of states. The Majorana operators are related to the fermionic operators through $\gamma_1 = c^\dagger + c$ and $\gamma_2 = i\left(c^\dagger - c\right)$. The operator $\mathcal{S}_3 = -i\gamma_1\gamma_2 = \left(2c^\dagger c - 1\right)$ anticommutes with both γ_1 and γ_2, but is not independent of them, and there is no enlargement of the number of states, which continues to be 2. When one considers bilinear representations of spin operators in terms of Clifford operators, an enlargement of the number of states occurs, resulting from the $\mathbb{Z}_2$ symmetry implied by the bilinear representation.

In the case of a single spin operator, there is a duplication of the number of states, with 4 states instead of 2, since we are representing two states by three Majorana operators. Introducing an extra Clifford operator, to pair with the third Majorana, we obtain a second pair of creation and destruction operators, thereby doubling the number of states. Applying this procedure to a set of N spins $\frac{1}{2}$ one has then 2^{2N} states instead of 2^N. In general, if one has N spins $\frac{1}{2}$, the number of states is 2^N. Introducing three Clifford operators for each spin, one has $3N$ Majorana operators. If the number of sites N is even, we can associate them in pairs obtaining $\frac{3N}{2}$ pairs of fermion operators, and the number of states will be $2^{\frac{3N}{2}}$. If the number of states is odd, one introduces then an extra Clifford operator and the number of states is $2^{\frac{3N+1}{2}}$.

In the case of spins interacting in pairs, such as in the Heisenberg interaction, the singlet and triplet states become both doubly represented, and, as a result, the partition function is simply multiplied by this multiplicity factor.

21.4.2 Unitary Transformations on Local Operators

Unitary transformations, U, acting on the spin and Majorana operators in general mix the two types of operators. If they mix the two types of operators (which have different commutation relations), they are called noncanonical and if they do not mix them they are called canonical. Also, it is possible to restrict the unitary transformations by also imposing that they are hermitian. In this case they are of the type $U^2 = I$.

In general, we may look for solutions of the type

$$U = zI + \omega I_3 + \mathbf{x} \cdot \boldsymbol{\gamma} + \mathbf{y} \cdot \vec{\mathcal{S}}. \tag{21.61}$$

Here $z, \omega, \mathbf{x}, \mathbf{y}$ are real numbers. These imply that U is unitary. There are several possible solutions that we will not detail here but it is useful to consider a couple of examples.

One example is the non-hermitian choice $U = \frac{1}{\sqrt{2}}\left(1 + i\mathbf{n} \cdot \vec{\mathcal{S}}\right)$, with $|\mathbf{n}| = 1$. The action of this operator leads to

$$\begin{aligned} U\vec{\mathcal{S}}U^\dagger &= \mathbf{n}\left(\mathbf{n} \cdot \vec{\mathcal{S}}\right) + \mathbf{n} \times \vec{\mathcal{S}} \\ U\boldsymbol{\gamma}U^\dagger &= \mathbf{n}\left(\mathbf{n} \cdot \boldsymbol{\gamma}\right) + \mathbf{n} \times \boldsymbol{\gamma}. \end{aligned} \tag{21.62}$$

The action of this unitary operator does not change the nature of the operators and therefore is an example of a canonical transformation.

Another example is an hermitian noncanonical unitary transformation. A class of transformations like these are given by the choice

$$U = \frac{1}{2}\left[I + I_3 - \mathbf{n} \cdot \left(\vec{\mathcal{S}} - \boldsymbol{\gamma}\right)\right] \tag{21.63}$$

and choosing $\mathbf{n} = \mathbf{e}_z$ we get that

$$\begin{aligned} U &= \frac{1}{2}\left[I + I_3 + \gamma_3 - \mathcal{S}_3\right] \\ &= \frac{1}{2}\left[1 - i\gamma_1\gamma_2\gamma_3 + \gamma_3 + i\gamma_1\gamma_2\right] = U_z. \end{aligned} \tag{21.64}$$

Also, note that

$$U_z\mathcal{S}_zU_z = \mathcal{S}_z \qquad U_z\mathcal{S}_xU_z = \gamma_1 \qquad U_z\mathcal{S}_yU_z = \gamma_2. \tag{21.65}$$

This class of transformations given by $\mathbf{n} = \mathbf{e}_\alpha$ leaves one of the spin operator components invariant, $U_\alpha\mathcal{S}_\alpha U_\alpha = \mathcal{S}_\alpha$ for $\alpha = x, y, z$ (1, 2, 3). The transformation acts only in the perpendicular plane.

One may use the unitary transformations between spin and fermionic operators (here specified in terms of the Majorana operators but, using the relation to the usual fermionic operators, we may establish those transformations) to construct exact mappings between different models. A particular application is a way to understand the Jordan-Wigner transformation as a result of a noncanonical unitary transformation.

Noncanonical Unitary and Jordan-Wigner Transformation

The Jordan-Wigner transformation may also be constructed introducing the local unitary transformation U_z [1]. Consider the XX model ($J_x = J_y = 1$) or the fully anisotropic X model ($J_x = 1$, $J_y = 0$):

$$H_X = \frac{1}{4}\sum_j \left(S_j^+ S_{j+1}^- + S_j^- S_{j+1}^+ + S_j^+ S_{j+1}^+ + S_j^- S_{j+1}^-\right). \tag{21.66}$$

The spin operators may be represented by the Majorana operators as

$$\begin{aligned} S^+ &= \gamma_3 (\gamma_1 + i\gamma_2)/2 \\ S^- &= (\gamma_1 - i\gamma_2)\gamma_3/2. \end{aligned} \tag{21.67}$$

Introduce usual fermionic operators (non-hermitian) as $g = \frac{1}{2}(\gamma_1 - i\gamma_2)$, $g^\dagger = \frac{1}{2}(\gamma_1 + i\gamma_2)$. We obtain that $\gamma_1 = g + g^\dagger$ and $i\gamma_2 = g^\dagger - g$. In terms of these operators we may write the Hamiltonians as

$$\begin{aligned} H_{XX} &= \frac{1}{2}\sum_j \left(g_j^\dagger g_{j+1} + g_j g_{j+1}^\dagger\right)\gamma_{3,j}\gamma_{3,j+1} \\ H_X &= \frac{1}{4}\sum_j \left(g_j^\dagger g_{j+1} + g_j g_{j+1}^\dagger - g_j^\dagger g_{j+1}^\dagger - g_j g_{j+1}\right)\gamma_{3,j}\gamma_{3,j+1}. \end{aligned} \tag{21.68}$$

The presence of quartic terms in the Hamiltonian clearly shows that the Hamiltonian in terms of the Majorana and regular fermions is interacting. A possible way to diagonalize the Hamiltonian is to perform a unitary transformation that eliminates the γ_3 operators. This can be achieved using the local unitary and hermitian operator

$$\begin{aligned} U_{z,j} &= \left(1 - g_j^\dagger g_j\right) + \gamma_{3,j} g_j^\dagger g_j \\ &= \frac{1}{2}\left[1 + i\gamma_{1,j}\gamma_{2,j} + \gamma_{3,j} - i\gamma_{1,j}\gamma_{2,j}\gamma_{3,j}\right]. \end{aligned} \tag{21.69}$$

Defining now an operator that is the product over all sites in the one-dimensional system

$$U_z = \prod_{j=1}^{N} U_{z,j} = U_{z,1}\cdots U_{z,N}, \tag{21.70}$$

we get

$$U_z \gamma_{3,j} g_j^\dagger U_z^\dagger = (-1)^{\sum_{l=1}^{j-1} n_l}\, g_j^\dagger \tag{21.71}$$

and its hermitian conjugate. We see that this unitary transformation gives origin to the so-called strings associated with the occupation of states to the left of a given lattice site, j. This is like in the Jordan-Wigner transformation. Applying to the Hamiltonian

$$U_z H_{XX} U_z^\dagger = -\frac{1}{2}\sum_j \left[g_j^\dagger g_{j+1} + g_{j+1}^\dagger g_j\right]. \tag{21.72}$$

Similarly

$$U_z H_X U_z^\dagger = -\sum_j \left[g_j^\dagger g_{j+1} + g_{j+1}^\dagger g_j + g_j^\dagger g_{j+1}^\dagger + g_{j+1} g_j\right], \tag{21.73}$$

which is Kitaev's model for $t = 1, \Delta = 1$. In contrast to the original model which described interacting spins on a lattice, this model describes spinless fermions that may hop on a lattice and have a nearest-neighbor p-type superconducting pairing, as shown by the appearance of creation and destruction pair operators. A magnetic field term is invariant (see Eq. (21.56)). We can also write

$$H_X = \frac{1}{4}\sum_j \gamma_{2,j}\gamma_{2,j+1}\gamma_{3,j}\gamma_{3,j+1}. \tag{21.74}$$

This quartic Hamiltonian is transformed to a quadratic Hamiltonian by eliminating the γ_3 terms since the action of the unitary operator $U_{z,j}$ is of the form

$$U_{z,j}\gamma_{2,j}\gamma_{3,j}U_{z,j}^\dagger = i\gamma_{1,j}, \tag{21.75}$$

and can be seen as transforming a spin-like operator (bilinear in the Majoranas) into a Majorana operator. We may also see that $U_{z,j}\gamma_{1,j}\gamma_{3,j}U_{z,j}^\dagger = -i\gamma_{2,j}$. Noting now that

$$\left(\prod_{l=1}^{j-1} U_{z,l}\right)\gamma_{1,j}\left(\prod_{l=j-1}^{1} U_{z,l}^\dagger\right) = \gamma_{1,j}\prod_{l=1}^{j-1}\left[i\gamma_{1,l}\gamma_{2,l}\right], \tag{21.76}$$

we get

$$U_z H_X U_z^\dagger = \frac{i}{4}\sum_j \gamma_{2,j}\gamma_{1,j+1}, \tag{21.77}$$

which is a noninteracting fermionic problem.

21.5 Hubbard Model and Operator Representations

As mentioned before, the Hubbard model

$$H = -t \sum_{<i,j>,\sigma} c_{i\sigma}^{\dagger} c_{j\sigma} + U \sum_{i} n_{i\uparrow} n_{i\downarrow}$$

in the limit $U/t \gg 1$ reduces to the $t - J$ model:

$$H = -t \sum_{<i,j>,\sigma} (1 - n_{i\bar{\sigma}}) c_{i\sigma}^{\dagger} c_{j\sigma} (1 - n_{j\bar{\sigma}}) + J \sum_{<i,j>} \mathbf{S}_i \cdot \mathbf{S}_j, \tag{21.78}$$

where $J = 2t^2/U$. The factors $(1 - n_{i\bar{\sigma}})$ exclude double occupancy and, therefore, in this limit the model describes spin operators interacting via an antiferromagnetic Heisenberg interaction and the conditioned motion of the electrons from singly occupied electron sites to empty sites. The kinetic term may be expressed in terms of the Hubbard operators:

$$\begin{aligned} X_i^{\sigma 0} &= c_{i\sigma}^{\dagger} (1 - n_{i\bar{\sigma}}) \\ X_i^{0\sigma} &= c_{i\sigma} (1 - n_{i\bar{\sigma}}), \end{aligned} \tag{21.79}$$

which coincide with the fermionic operators in singly occupied sites. Other operators defined by Hubbard are

$$\begin{aligned} X_i^{2\sigma} &= \sigma c_{i\bar{\sigma}}^{\dagger} n_{i\sigma} \\ X_i^{\sigma\sigma'} &= c_{i\sigma}^{\dagger} c_{i\sigma'} \\ X_i^{20} &= \sigma c_{i\bar{\sigma}}^{\dagger} c_{i\sigma}^{\dagger} \\ X_i^{00} &= (1 - n_{i\uparrow})(1 - n_{i\downarrow}) \\ X_i^{\sigma\sigma} &= n_{i\sigma} (1 - n_{i\bar{\sigma}}) \\ X_i^{22} &= n_{i\uparrow} n_{i\downarrow}. \end{aligned} \tag{21.80}$$

In terms of these operators the fermionic operators may be written as

$$\begin{aligned} c_{i\sigma} &= X_i^{0\sigma} + \sigma X_i^{\bar{\sigma}2} \\ c_{i\sigma}^{\dagger} &= X_i^{\sigma 0} + \sigma X_i^{2\bar{\sigma}}. \end{aligned} \tag{21.81}$$

These operators obey a complicated algebra:

$$[X^{\alpha\beta}, X^{\gamma\delta}] = \delta_{\beta\gamma} X^{\alpha\delta} - \delta_{\alpha\delta} X^{\gamma\beta} \tag{21.82}$$

and standard methods developed to handle bosonic or fermionic systems do not apply directly.

21.5.1 Slave Boson Method

The restriction may be implemented introducing auxiliar bosonic fields defining

$$\begin{aligned} X_i^{00} &= b_i^\dagger b_i \\ X_i^{\sigma\sigma'} &= f_{i\sigma}^\dagger f_{i\sigma'} \\ X_i^{0\sigma} &= b_i^\dagger f_{i\sigma} \\ X_i^{\sigma 0} &= f_{i\sigma}^\dagger b_i, \end{aligned} \tag{21.83}$$

where $f_{i\sigma}$ is a fermionic field that is associated to the spin of the electrons and where b_i is a bosonic field. This method is called the slave boson method [2]. The state occupied by a boson corresponds to a state empty of electrons. This implies the simple transformation

$$c_{i,\sigma} = b_i^\dagger f_{i,\sigma} : \tag{21.84}$$

destroying a fermion at site i with spin projection σ results from creating a bosonic excitation at that site, $b_i^\dagger$, and destroying some auxiliar fermion, $f_{i,\sigma}$, that carries the spin projection of the physical fermion. It is also possible to associate to the empty state a fermionic operator associating then the spin to a boson. This method is called slave fermions.

The slave bosons or slave fermions replace the projection operators such that the Hamiltonian expressed in terms of these operators, as it acts on the states in which double occupancy is excluded, generates states in the same subspace. However, if the Hamiltonian acts on a state occupied by a boson b_i and by a fermion $f_{i\sigma}$ (and the inverse in the slave fermion method), it will create a state with double occupancy. It is therefore necessary to introduce a restriction so that states outside the subspace of the $t - J$ model are not generated. It takes the form

$$\sum_\sigma f_{i\sigma}^\dagger f_{i\sigma} + b_i^\dagger b_i = 1, \tag{21.85}$$

which insures that no double occupancy emerges, since each site either is occupied by a fermion (an electron with spin $\uparrow$ or $\downarrow$) or is occupied by a boson (that is, it is empty of electrons). The restriction may be implemented introducing a Lagrange multiplier λ_i at each lattice site. The Hamiltonian in the slave boson method is then transformed to

$$H = -t \sum_{<i,j>,\sigma} \left(f_{i\sigma}^{\dagger} f_{j\sigma} b_j^{\dagger} b_i + c.c. \right) + J \sum_{<i,j>} \mathbf{S}_i \cdot \mathbf{S}_j$$

$$-\mu \sum_{i\sigma} f_{i\sigma}^{\dagger} f_{i\sigma} + i \sum_i \lambda_i \left(\sum_{\sigma} f_{i\sigma}^{\dagger} f_{i\sigma} + b_i^{\dagger} b_i - 1 \right), \quad (21.86)$$

where μ is the chemical potential. The projection operators, or the Hubbard operators, are difficult to deal with, but the bosonic or fermionic operators allow a standard description, even though one has to worry with the new added field, which has no dynamics. Other transformations have also been used in which other bosons are introduced to describe the various possible states and not only the state of no occupancy. It is also possible to introduce mixed descriptions where both slave bosons and slave fermions are used in a supersymmetric way.

There are other transformations proposed in the literature that lead to a similar decoupling of the electronic degrees of freedom. The main motivation was the study of the large-U limit in the Hubbard model or in the Anderson model for magnetic impurities in a metal, with the intent of controlling in an efficient way the projection to states where double occupancy is restricted (as in the $t-J$ model), but considering a finite value of U, instead of the extreme case of infinite U, usually taken care of by a single slave boson. These representations introduce explicitly operators related to the four possible states associated with each site. Namely, a site may be unoccupied, singly occupied with a given spin projection, or doubly occupied.

In the Kotliar and Ruckenstein procedure four bosonic operators are added, enlarging the operator space, that act as projectors on the original fermionic operators. We will apply this method later on in the context of the Anderson model. In the Zou-Anderson transformation the original electron operators are replaced by two sets of two bosonic and fermionic operators that fulfill the projection. However, both representations lead to an enlargement of the physical Hilbert space and the extra unphysical states have to be projected out. Even though the scheme is not exact, note, however, that the representation introduced by Zou and Anderson has been used to explicitly obtain an exact solution of the Hubbard model in the large U limit.

21.5.2 Zou-Anderson Transformation

The electron operators may be written as [3]

$$c_{i,\sigma} = e_i^{\dagger} S_{i,\sigma} + \sigma S_{i,-\sigma}^{\dagger} d_i$$

$$c_{i,\sigma}^{\dagger} = S_{i,\sigma}^{\dagger} e_i + \sigma d_i^{\dagger} S_{i,-\sigma}, \quad (21.87)$$

where the operators $e_i, d_i, S_{i,\sigma}$ annihilate sites that are unoccupied, doubly occupied, and singly occupied with an electron with spin σ, respectively. Double occupancy is allowed since U is finite. Note that strictly considering the limit $U \to \infty$, we recover the slave boson method. In this limit, double occupancy is strictly forbidden which implies that the operators $d_i, d_i^\dagger$ do not act on the states and the Zou-Anderson transformation just reduces to the first terms, recovering the slave boson method. Since the electron operators are fermionic, we can either choose the operators e, d as bosonic and the operators $S_{i,\sigma}$ as fermionic, or vice versa. Most expressions are the same, formally, in either case. The difference arises when one integrates over degrees of freedom or in the mean-field approach, when the Bose-Einstein or the Fermi-Dirac distributions appear.

The enlargement of the degrees of freedom imposes the constraint,

$$e_i^\dagger e_i + d_i^\dagger d_i + \sum_\sigma S_{i,\sigma}^\dagger S_{i,\sigma} = 1, \tag{21.88}$$

at each site. (This is the completeness relation of the four possibilities: one site is either unoccupied, doubly occupied, or singly occupied by an electron with spin-up or down). The constraint is simply obtained imposing the anticommutation relation $\{c_{i,\sigma}, c_{i,\sigma}\} = 1$ and considering either the slave-boson or slave-fermion commutation or anticommutation relations.

Let us consider the Hubbard model written as

$$H = -t \sum_{i,\delta;\sigma} c_{i,\sigma}^\dagger c_{i+\delta,\sigma} + U \sum_i n_{i,\uparrow} n_{i,\downarrow} - \mu \sum_{i,\sigma} c_{i,\sigma}^\dagger c_{i,\sigma}. \tag{21.89}$$

The first is the hopping term between a site i and its neighbors distant by δ (in general vectors in a d-dimensional space) and μ the chemical potential enforcing the band filling.

The problem to be solved involves the effective Hamiltonian:

$$\begin{aligned} H = &-t \sum_{i,\delta;\sigma} \left(e_i e_{i+\delta}^\dagger - d_i d_{i+\delta}^\dagger \right) S_{i,\sigma}^\dagger S_{i+\delta,\sigma} \\ &-t \sum_{i,\delta} e_i d_{i+\delta} (S_{i\uparrow}^\dagger S_{i+\delta,\downarrow}^\dagger - S_{i\downarrow}^\dagger S_{i+\delta,\uparrow}^\dagger) \\ &-t \sum_{i,\delta} d_i^\dagger e_{i+\delta}^\dagger (S_{i\downarrow} S_{i+\delta,\uparrow} - S_{i\uparrow} S_{i+\delta,\downarrow}) \\ &+U \sum_i d_i^\dagger d_i + \mu \sum_i \left(e_i^\dagger e_i - d_i^\dagger d_i \right) - \mu N \\ &+ \sum_i \lambda_i \left(e_i^\dagger e_i + d_i^\dagger d_i + \sum_\sigma S_{i,\sigma}^\dagger S_{i,\sigma} - 1 \right), \end{aligned} \tag{21.90}$$

where we have introduced at each site a Lagrange multiplier, λ_i, to enforce the constraint. With this transformation, the interacting term of the Hubbard Hamiltonian is now quadratic but the kinetic term becomes now interacting (quartic in the operators). If the dominant term of the Hubbard Hamiltonian is the interacting term (if U is large), we may now consider some expansion due to the kinetic term, since it has a smaller magnitude.

A mean-field approach can be used to decouple the quartic terms in the Hamiltonian. Different choices may be used, depending on the physical properties we expect to be dominant [4]. One possible way uses that the unoccupied and doubly occupied sites are separated, on a first stage, from the singly occupied sites. So a separation of charge degrees of freedom from the spin degrees of freedom is naturally implemented. As we will see later, in one-dimensional systems this type of separation occurs naturally. Also, it has been argued that it may occur in some strongly correlated systems. A natural way to introduce mean-field correlations is to consider that the unoccupied sites and the doubly occupied sites are paired on nearest-neighbor links. On the other hand, the spin states of the singly occupied sites are paired into spin singlets. These decouplings involve correlation functions between two bosonic and between two fermionic fields.

21.5.3 Östlund-Granath Transformation

This exact transformation introduces new operators called quasicharge $\hat{c}_r$ and quasispin operators q_r^i that obey Fermi and Bose statistics [5], respectively:

$$\hat{c}_r = c_{\uparrow,r}^\dagger(1-n_{\downarrow,r}) + (-1)^r c_{\uparrow,r} n_{\downarrow,r} \tag{21.91}$$

$$\begin{aligned} q_r^+ &= (c_{\uparrow,r}^\dagger - (-1)^r c_{\uparrow,r}\,)\, c_{\downarrow,r} \\ q_r^- &= (q_r^+)^\dagger \\ q_r^z &= \frac{1}{2} - n_{\downarrow,r}. \end{aligned} \tag{21.92}$$

Since the transformation is exact, there is no need to enforce a projection to the physical subspace. These operators satisfy the algebra $\{\hat{c}_r, \hat{c}_{r'}^\dagger\} = \delta_{r,r'}$, also $\{\hat{c}_r^\dagger, \hat{c}_{r'}^\dagger\} = 0$ and $\left[\hat{c}_r^\dagger, q_{r'}^i\right] = 0$, $\left[q_r^i, q_{r'}^j\right] = i\delta_{rr'}\sum_k \epsilon_{ijk} q_r^k$. The electron operators are expressed in terms of them as

$$\begin{aligned} c_{\uparrow,r}^\dagger &= \hat{c}_r \left(\frac{1}{2} + q_r^z\right) + (-1)^r\, \hat{c}_r^\dagger \left(\frac{1}{2} - q_r^z\right) \\ c_{\downarrow,r}^\dagger &= q_r^- \,(\hat{c}_r \; - \; (-1)^r\, \hat{c}_r^\dagger\,). \end{aligned} \tag{21.93}$$

The Östlund-Granath representation involves other operators such as the quasicharge operator $n_r^c = \hat{c}_r^\dagger \hat{c}_r$ and the local pseudospin operators $p_r^i = n_r^c q_r^i$, which are the generators of the $SU(2)$ algebra that corresponds to "rotations" between the unoccupied and doubly occupied states.

The following results hold $n_r = 1 - 2n_r^c q_r^z$, $s_r^i = (1 - n_r^c)\, q_r^i$ and $n_r^c = (n_r - 1)^2$ where $s_r^i = \frac{1}{2}\sum_{\alpha,\beta} c_{\alpha,r}^\dagger \sigma_{\alpha\beta}^i c_{\beta,r}$, with σ^i the Pauli matrices. The total z-component of pseudospin can therefore be seen to be half the number of doubly occupied sites minus the number of unoccupied sites, which is precisely the charge relative to half filling.

The Hubbard model can be rewritten in terms of the quasiparticle operators as

$$H = t\,(T_0 + T_1 + T_{-1}) + U h_U \tag{21.94}$$

with $h_U = \frac{1}{2}\sum_r \hat{c}_r^\dagger \hat{c}_r$,

$$T_0 = \frac{1}{2}\sum_{r,r'} (1 + 4\boldsymbol{q}_r \cdot \boldsymbol{q}_{r'})(\hat{c}_r^\dagger \hat{c}_{r'} + CC) \tag{21.95}$$

$$T_1 = \frac{1}{2}\sum_{r,r'} (-1)^r\,(1 - 4\boldsymbol{q}_r \cdot \boldsymbol{q}_{r'})(\hat{c}_r^\dagger \hat{c}_{r'}^\dagger)$$

and $T_{-1} = T_1^\dagger$. Note that the interacting term has also in this representation a simple expression. The price is the complicated nature of the kinetic term, as in the Zou-Anderson transformation.

Even though this transformation also achieves some sort of exact spin-charge separation (actually quasispin and quasicharge), the Hamiltonian has a complicated structure. Although it has quartic interacting terms, since they involve the quasispin operators, these are typically represented in terms of bilinear representations of fermionic or bosonic operators. Usually these representations enlarge the physical Hilbert space and one has to introduce constraints. One can use a Majorana fermion representation but this leads to a Hamiltonian where the leading interacting term involves six operators and, therefore, the analytical treatment is rather complicated.

21.6 Exercises

1. (a) Show that the operators defined in Eq. (21.29) satisfy the same algebra as the Pauli matrices.

   (b) Show that the Hamiltonian may be written as given in Eq. (21.30).

2. Schwinger representation: Consider the following representation of the spin operators:

$$S_r = \frac{1}{2}\mathbf{a}^\dagger \sigma_r \mathbf{a},$$

where σ_r are the Pauli matrices and $r = x, y, z$ and the vector

$$\mathbf{a} = \begin{pmatrix} a_1 \\ a_2 \end{pmatrix},$$

where a_1 and a_2 are bosonic or fermionic operators. Considering both representations in terms of bosons or fermions:

(a) Show that the commutation relations of the spin operators are satisfied.
(b) Calculate $\mathbf{S}^2$. Show that

$$\mathbf{S}^2 = \frac{\hat{N}}{2}\left(\frac{\hat{N}}{2} + 1\right),$$

where $\hat{N} = \hat{n}_1 + \hat{n}_2 = a_1^\dagger a_1 + a_2^\dagger a_2$. Since $\mathbf{S}^2 = S(S+1)$ then to a given value of S corresponds a value of $N = 2S$. As a consequence, the Schwinger representation only describes spin operators $\mathbf{S}$ if we project to the physical subspace where $n_1 + n_2 = 2S$.

3. Another possible representation for spins is the Dyson-Maleev representation:

$$\begin{aligned} S^+ &= \sqrt{2S}a^\dagger\left(1 - \frac{a^\dagger a}{2S}\right) \\ S^- &= \sqrt{2S}a \\ S^z &= a^\dagger a - S. \end{aligned} \tag{21.96}$$

Show that the commutation relation between the spin components is satisfied:

$$[S^+, S^-] = 2S^z. \tag{21.97}$$

The restriction $n = 2S$ has to be imposed to guarantee the proper expression for S^2.

4. Show that the following unitary operators are also solutions of Eq. (21.61):

(a) $U = I_3$.
(b) $U = \frac{1}{\sqrt{2}}(I + i\mathbf{n} \cdot \boldsymbol{\gamma})$; $|\mathbf{n}| = 1$.
(c) $U = \cos\frac{\theta}{2}\mathbf{n}_1 \cdot \boldsymbol{\gamma} + \sin\frac{\theta}{2}\mathbf{n}_2 \cdot \vec{S}$; $|\mathbf{n}_1| = |\mathbf{n}_2| = 1$; $\mathbf{n}_1 \cdot \mathbf{n}_2 = 0$.

5. (a) In the context of the slave boson method, represent schematically the action of the bosonic and fermionic operators in Eq. (21.84) on the states of the fermionic system.
(b) In the context of the Östlund and Grannath method, represent schematically the action of the various operators, quasicharge, quasispin, and pseudospin, on the local eigenstates of the operators n_c, q_z, p_z, s_z.

6. If we have an interacting theory and want to solve it using a mean-field approximation, one convenient way is to introduce Hubbard-Stratonovich transformations to decouple the interaction term. Each Hubbard-Stratonovich term is associated with some average value that one assumes describes the relevant physics. This was done before to consider magnetic or superconducting states in mean field. If there is coexistence of different order parameters, it may happen that the decomposition has to be carried out with some care.

   Consider the example of a spin exchange term of the type $\mathbf{S}_i \cdot \mathbf{S}_j$ between two spin-$1/2$ operators at lattice sites i, j, as, for instance, occurs in the $t - J$ model [6]. We can write the spin operator as $S_i^k = 1/2 f_{i\alpha}^\dagger \sigma_{\alpha\beta}^k f_{i\beta}$, where $k = x, y, z$ and the sum over the spin components α, β is implicit. The f operators are fermionic. We want to decouple the interaction in terms of direct, $f_i^\dagger f_i$, exchange, $f_i^\dagger f_j$, and pairing terms, $f_i f_j$. Let us write

$$\mathbf{S}_i \cdot \mathbf{S}_j = \nu_D + \nu_E + \nu_P, \tag{21.98}$$

   in terms of direct, exchange, and pairing terms, respectively. We take these terms in the form

$$\nu_{D,E,P} = \sum_{k=0}^{3} c_{D,E,P}^k \left(f_{i\alpha}^\dagger \sigma_{\alpha\beta}^k f_{i\beta} \right) \left(f_{j\alpha'}^\dagger \sigma_{\alpha'\beta'}^k f_{j\beta'} \right), \tag{21.99}$$

   where σ^0 is the identity matrix. We want to consider the exchange term in the form $f_{i\uparrow}^\dagger f_{j\uparrow} + f_{i\downarrow}^\dagger f_{j\downarrow}$ and the pairing term in the form $f_{i\uparrow} f_{j\downarrow} - f_{i\downarrow} f_{j\uparrow}$.

   In order to avoid double counting we want to impose that decoupling, for instance, $\nu_D + \nu_P$ into exchange terms, will not generate terms included in ν_E and the same for any sum of two terms. Under these conditions, show that the decomposition is not free and is of the form

$$\begin{aligned}
\nu_D &= -\frac{1}{2} \mathbf{S}_i \cdot \mathbf{S}_j \\
\nu_E &= -\frac{3}{8} \left(f_{i\uparrow}^\dagger f_{j\uparrow} + f_{i\downarrow}^\dagger f_{j\downarrow} \right) \left(f_{j\uparrow}^\dagger f_{i\uparrow} + f_{j\downarrow}^\dagger f_{i\downarrow} \right) + \frac{3}{8} n_i \\
\nu_P &= -\frac{3}{8} \left(f_{i\downarrow}^\dagger f_{j\uparrow}^\dagger - f_{i\uparrow}^\dagger f_{j\downarrow}^\dagger \right) \left(f_{j\uparrow} f_{i\downarrow} - f_{j\downarrow} f_{i\uparrow} \right).
\end{aligned} \tag{21.100}$$

References

1. P. Coleman, E. Miranda, A. Tsvelik, Physical Review B **49**, 8955 (1994).
2. S. E. Barnes, Journal of Physics F: Metal Physics **6**, 1375 (1976); Journal of Physics F: Metal Physics **7**, 2637 (1977); P. Coleman, Physical Review B **29**, 3035 (1984).

3. Z. Zou, P.W. Anderson, Physical Review B **37**, 627 (1988).
4. P. A. Lee, N. Nagaosa and X.-G. Wen, Reviews of Modern Physics **78**, 17 (2006).
5. S. Östlund and M. Granath, Physical Review Letters **96**, 066404 (2006).
6. M. U. Ubbens and P. A. Lee, Physical Review B **46**, 8434 (1992).
7. H. A. Kramers and G. H. Wannier, Statistics of the Two-Dimensional Ferromagnet. Part I. Physical Review **60**, 252 (1941).
8. H. A. Kramers and G. H. Wannier, Statistics of the Two-Dimensional Ferromagnet. Part II. Physical Review **60**, 263 (1941).

Chapter 22
Nonhomogeneous Systems: Effects of Local States

Local perturbations or local states appear in various contexts in extended systems. They may be a result of impurities in a material, that lead typically to a local change in the potential profile that induces scattering on the carriers, with the appearance of possible bound states on the impurity site. When many impurities are present, the interference of multiple scatterings off the impurities may lead to localization of the states. The local states may also be the consequence of states with different energies that are associated with some impurity atom of a different nature (for instance, with a different orbital nature) which has states that, in the free atom limit, have a less extended spatial nature.

In this chapter we will consider the effect of impurities on the properties of otherwise pure systems. We will also consider the limit when the impurities are spread over the material and, in the limiting case, constitute themselves a lattice, leading to a multiband system. The effects of many impurities on transport will be considered in the next chapter.

22.1 Bound States and Scattering Matrix

Let us briefly review a problem in elementary quantum mechanics[1] by considering a potential barrier of height V_0, which may be positive or negative, in a finite one-dimensional region $-a < x < a$. A particle incident from the left goes through three regions and its wave function may be defined as

$$\psi_I(x) = Ae^{ikx} + Be^{-ikx}$$
$$\psi_{II}(x) = Fe^{iqx} + Ge^{-iqx}$$

[1] Here we follow [BicudoSacramento].

P. D. Sacramento, *Introduction to Condensed Matter Theory*,
https://doi.org/10.1007/978-3-031-86251-9_22

$$\psi_{III}(x) = Ce^{ikx}, \tag{22.1}$$

where A is the amplitude of the incident wave (taken as unity $A = 1$), B is the amplitude of the reflected wave, C is the amplitude of the transmitted wave, and F, G are the amplitudes of the waves in the potential barrier region. If we also consider that particles may be incident on the potential region coming from the right, we may add to $\psi_{III}(x)$ a term of the type De^{-ikx}. For a fixed incident energy, E, we have that the kinetic energy plus the potential energy equals the total energy, which implies that $k^2 = 2mE/\hbar^2$ and $q^2 = 2m(E - V_0)/\hbar^2$. Using the boundary conditions that $\psi_I(-a) = \psi_{II}(-a)$, $\psi_{II}(a) = \psi_{III}(a)$, $\psi_I'(-a) = \psi_{II}'(-a)$, and $\psi_{II}'(a) = \psi_{III}'(a)$, where prime is the derivative with respect to the coordinate x, we get that, for $V_0 < 0$,

$$B = \frac{(q^2 - k^2)i\sin(2qa)}{2kq\cos(2qa) - (q^2 + k^2)i\sin(2qa)}e^{-2ika}$$

$$C = \frac{2kq}{2kq\cos(2qa) - (q^2 + k^2)i\sin(2qa)}e^{-2ika}. \tag{22.2}$$

Both coefficients have a pole if

$$2kq\cos(2qa) = i(q^2 + k^2)\sin(2qa). \tag{22.3}$$

This only occurs if $E < 0$ in which case we have that $k = i|k|$. We can then write that

$$\psi_I(x) = Be^{|k|x}$$

$$\psi_{II}(x) = \bar{F}\cos(qx) + \bar{G}\sin(qx)$$

$$\psi_{III}(x) = Ce^{-|k|x}, \tag{22.4}$$

which indicates the existence of a bound state, since the wave function decays exponentially away from the potential region. (The coefficients $\bar{F}$ and $\bar{G}$ are obtained from F and G.) There are two types of solutions, one even, if $\bar{G} = 0$ and with $|k| = q\tan(qa)$, and one odd with $\bar{F} = 0$ and $|k| = -q\,\mathrm{cotg}\,(qa)$. The existence of the even solution implies that any attractive potential gives rise to a bound state. The odd solution only exists if the attractive potential is strong enough. The equation for the poles of the coefficients can also be written as

$$2|k|q\cos(2qa) = (q^2 - |k|^2)\sin(2qa), \tag{22.5}$$

which is identical to the condition for the energies of the bound states.

Considering the more general case where in both external regions, I and III, there are waves propagating in both directions with amplitudes on the left A and B, propagating to the right and left, respectively, and amplitudes on the right C and D, propagating to the right and left, respectively, it is usual to define the S matrix as

$$\begin{pmatrix} B \\ C \end{pmatrix} = S \begin{pmatrix} D \\ A \end{pmatrix} . \tag{22.6}$$

The conditions for a bound state are $D \to 0$ and taking the amplitude of the incident wave from the left $A \to 0$, so that we have real exponentials away from the potential region. This implies that the matrix S has to diverge to give finite B, C. Therefore, the condition that S has poles leads to the condition for bound states, $E < 0$, which have localized wave functions.

In a three-dimensional potential with spherical symmetry, the existence of bound states depends on the depth of the potential. (Since the distance $r \geq 0$, the solution is similar to the odd solution of a one-dimensional problem.) The square potential is a system characterized by a potential of the type

$$V(r) = \begin{cases} -V_0 \ , r < a \\ 0 \qquad , r > a \end{cases} . \tag{22.7}$$

The wave equation in spherical coordinates separates as $\psi(r, \theta, \varphi) = R(r)Y(\theta, \varphi)$. The radial equation for $R(r)$ may be written in the form

$$\frac{\mathrm{d}^2 R}{\mathrm{d}r^2} + \frac{2}{r}\frac{\mathrm{d}R}{\mathrm{d}r} - \frac{l(l+1)}{r^2} R + \frac{2\mu}{\hbar^2}(V_0 + E)R = 0 \ , \ r < a$$

$$\frac{\mathrm{d}^2 R}{\mathrm{d}r^2} + \frac{2}{r}\frac{\mathrm{d}R}{\mathrm{d}r} - \frac{l(l+1)}{r^2} R + \frac{2\mu E}{\hbar^2} R = 0 \ , \ r > a. \tag{22.8}$$

Here μ is the mass of the particle and l the angular momentum. Consider the situation such that the particles form bound states with the potential well, therefore with negative energy, and define the quantities

$$\alpha^2 \equiv -\frac{2\mu}{\hbar^2} E,$$

$$q^2 \equiv \frac{2\mu}{\hbar^2}(E + V_0). \tag{22.9}$$

The solution for $r < a$ is the regular solution corresponding to the spherical Bessel function. In the region $r > a$, the solutions behave as $e^{\mp \alpha r}$. Since we search for a normalizable solution, only the solution that decays exponentially is physically acceptable. Therefore,

$$R_l(r) = \begin{cases} A j_l(qr) \qquad , r < a \\ B h_l^{(+)}(i\alpha r) \ , r > a \end{cases} , \tag{22.10}$$

where h_l is a spherical Hankel function. Demanding the continuity of the wave function and of its derivative at $r = a$, we obtain

$$A j_l(qa) = B h_l^{(+)}(i\alpha a),$$
$$qA\left[\frac{\mathrm{d} j_l}{\mathrm{d}\rho}\right]_{\rho=qa} = i\alpha B\left[\frac{\mathrm{d}h_l^{(+)}}{\mathrm{d}\rho}\right]_{\rho=i\alpha a}. \tag{22.11}$$

Dividing the second equation by the first one

$$q\left[\frac{\frac{\mathrm{d} j_l}{\mathrm{d}\rho}}{j_l}\right]_{\rho=qa} = i\alpha\left[\frac{\frac{\mathrm{d}h_l^{(+)}}{\mathrm{d}\rho}}{h_l^{(+)}}\right]_{\rho=i\alpha a}. \tag{22.12}$$

This is a transcendent equation that in general is very hard to solve and it depends on the variables l, V_0, and E. In the case $l = 0$, the problem simplifies considerably. Consider the function $u(r) = rR(r)$, given in this case by

$$u(r) = r\frac{A\,\mathrm{sen}\,(qr)}{qr} = A\frac{\mathrm{sen}\,(qr)}{q},\ r < a$$
$$u(r) = r\frac{Be^{i(i\alpha r)}}{i(i\alpha r)} = -B\frac{e^{-\alpha r}}{\alpha},\ r > a. \tag{22.13}$$

Demanding the continuity of $u(r)$ and of its derivative at $r = a$, one obtains the following conditions:

$$\frac{A}{q}\,\mathrm{sen}\,(qa) = -\frac{B}{\alpha}e^{-\alpha a},$$
$$A\cos(qa) = Be^{-\alpha a}. \tag{22.14}$$

Dividing the first equation by the second, we obtain the condition for bound states for $l = 0$

$$\tan(qa) = -\frac{q}{\alpha}, \tag{22.15}$$

which are function of the parameters of the well, its width and depth. This equation only has solutions if V_0 is such that

$$V_0\frac{2\mu}{\hbar^2}a > \frac{\pi^2}{4}. \tag{22.16}$$

22.2 Impurity Scattering: Friedel Oscillations

Let us consider now the scattering of a particle by an external potential,[2] such that no bound state is formed. This case includes the scattering of two particles if we use the relative coordinate system and separate the center of mass motion. We consider the case of a free particle that crosses a finite-size region where the potential is nonvanishing and we want to study the outcome of the collision in the region where the particle is once again free.

An incident free particle may be represented by a plane wave. It can be expanded like

$$e^{i\mathbf{k}\cdot\mathbf{r}} \sim \frac{i}{2k}\sum_{l=0}^{\infty}(2l+1)i^l\left(\frac{e^{-i(kr-l\pi/2)}}{r} - \frac{e^{i(kr-l\pi/2)}}{r}\right)P_l(\cos\theta), \tag{22.17}$$

where l is the orbital angular momentum and $r = |\mathbf{r}|$. The functions $P_l(\cos\theta)$ are the Legendre polynomials. After the scattering and far away from the potential center, the shape of the particle wave function is similar. The wave function in the presence of the potential can be expanded asymptotically like

$$\psi(\mathbf{r}) \sim \frac{i}{2k}\sum_{l=0}^{\infty}(2l+1)i^l\left(\frac{e^{-i(kr-l\pi/2)}}{r} - S_l(k)\frac{e^{i(kr-l\pi/2)}}{r}\right)P_l(\cos\theta), \tag{22.18}$$

where due to the conservation of flux $|S_l(k)| = 1$. The wave function can be rewritten as

$$\psi(\mathbf{r}) \sim e^{i\mathbf{k}\cdot\mathbf{r}} + \left(\sum_{l=0}^{\infty}(2l+1)\frac{S_l(k)-1}{2ik}P_l(\cos\theta)\right)\frac{e^{ikr}}{r}. \tag{22.19}$$

Defining the phase shift $\delta_l(k)$ as

$$S_l(k) = e^{2i\delta_l(k)}, \tag{22.20}$$

from Eq. (22.18) we can obtain that the solution of the radial equation is of the type (for $r \gg 1$)

$$R_l(r) \sim \frac{1}{r}\sin\left[kr - \frac{l\pi}{2} + \delta_l(k)\right]. \tag{22.21}$$

Let us now consider the probability flux:

[2] For further reading see [Mahan] and [BicudoSacramento].

$$\mathbf{j} = \frac{\hbar}{2im}\left(\psi^*\nabla\psi - \psi\nabla\psi^*\right). \tag{22.22}$$

In the case of $r \gg 1$

$$\mathbf{j} = \frac{\hbar\mathbf{k}}{m} + \frac{\hbar\mathbf{k}}{m}\mathbf{e}_r\frac{|f(\theta)|^2}{r^2}, \tag{22.23}$$

where

$$\begin{aligned} f(\theta) &= \sum_{l=0}^{\infty}(2l+1)f_l(k)P_l(\cos\theta) \\ f_l(k) &= \frac{S_l(k)-1}{2ik}. \end{aligned} \tag{22.24}$$

The scattering cross section, which is defined as the number of scattered particles that cross the area contained in the solid angle, $d\Omega$, centered in the direction defined by the vector $\mathbf{e}_r(\theta, \varphi)$, divided by the incident flux is

$$d\sigma = \frac{\mathbf{j}\cdot\mathbf{e}_r dA}{\frac{\hbar k}{m}} = |f(\theta)|^2 d\Omega. \tag{22.25}$$

If the spin is introduced, then the potential may depend on φ. The total scattering cross section is given by

$$\begin{aligned} \sigma_{total}(k) &= \int d\Omega\frac{d\sigma}{d\Omega} \\ &= \frac{4\pi}{k^2}\sum_{l=0}^{\infty}(2l+1)\sin^2\delta_l(k) \\ &= \sum_{l=0}^{\infty}\sigma_l(k) \end{aligned} \tag{22.26}$$

where we have used that

$$\int d\Omega P_l(\cos\theta)P_{l'}(\cos\theta) = \frac{4\pi}{2l+1}\delta_{ll'}. \tag{22.27}$$

Notice that the imaginary part of $f(\theta)$ is such that

$$\Im f(0) = \frac{k}{4\pi}\sigma_{total}. \tag{22.28}$$

This result constitutes the so-called optical theorem.

Let us consider a potential of range a. Then we may expect that the maximum value of the angular momentum, l, should be of the order of pa (where p is the momentum). That is, we can restrict the values of l to the range $l \leq \frac{pa}{\hbar} = ka$. Let us consider the simple case of a potential well of depth $-V_0$ and range a, located at the origin. The radial wave function is given by

$$R(r) = Aj_l(qr), r < a$$
$$R(r) = Bj_l(kr) + Cn_l(kr), r > a, \tag{22.29}$$

where j_l and n_l are spherical Bessel and Neumann functions and

$$q^2 = \frac{2m}{\hbar^2}(V_0 + E)$$
$$k^2 = \frac{2m}{\hbar^2}E. \tag{22.30}$$

The phase shift is obtained by solving the continuity for the wave function and its derivative at $r = a$, leading to

$$\tan\delta_l(k) = \frac{kj_l'(ka)j_l(qa) - qj_l(ka)j_l'(ka)}{kn_l'(ka)j_l(qa) - qn_l(ka)j_l'(qa)}. \tag{22.31}$$

For certain energy values the denominator vanishes. This implies that the phase shift goes through $\pi/2$ or that in general it goes through $(n+1/2)\pi$. When the phase shift is $\pi/2$, then

$$\sigma_l(k) = \frac{4\pi(2l+1)}{k^2}\sin^2\delta_l(k) \tag{22.32}$$

has its maximum value. When $\tan\delta_l(k)$ goes through ∞, then goes on through $-\infty$, and then increases again, we have a resonance. One finds that the resonance energies coincide with the eigenenergies of a particle in a $3D$ box. It may be shown that Levinson's theorem holds that states that

$$\delta(0) - \delta(\infty) = N_B\pi, \tag{22.33}$$

where N_B is the number of bound states.

Friedel Oscillations

Let us now consider an impurity at the center of a sphere of radius R (where R is finite but is taken as very large, $R \to \infty$). Since the system is finite, the eigenstates vanish at the boundary.

In the absence of the impurity, the solutions that are regular at the origin have as a radial part a spherical Bessel function $j_l(kr)$ that satisfies $j_l(k_\alpha R) = 0$, where l is the angular momentum. As in other problems, the boundary condition fixes the set of allowed momentum values. For small l and since $R \to \infty$, we may use the asymptotic expansion:

$$j_l(k_\alpha R) \sim \frac{1}{k_\alpha R} \sin\left(k_\alpha R - \frac{l\pi}{2}\right). \tag{22.34}$$

The boundary condition then leads to

$$k_\alpha R = \left(n + \frac{l}{2}\right)\pi. \tag{22.35}$$

Here n is the quantum number for the various modes, for a given angular momentum. In the presence of the impurity, the asymptotic radial function becomes

$$R_l(kr) \sim \frac{C_l}{kr} \sin\left(kr + \delta_l - \frac{l\pi}{2}\right), \tag{22.36}$$

where δ_l are the phase shifts. This form leads to modes that satisfy

$$k_n R + \delta_l(k_n) = \left(n + \frac{l}{2}\right)\pi. \tag{22.37}$$

The impurity introduces additional states. The number of states between k and $k + dk$ is

$$\frac{dn}{dk} = \frac{R}{\pi} + \frac{1}{\pi}\frac{d\delta_l}{dk}. \tag{22.38}$$

The extra states due to the impurity are

$$\frac{d(\delta n)}{dk} = \frac{1}{\pi}\frac{d\delta_l}{dk}. \tag{22.39}$$

We expect that if the potential is attractive the number of states increases and if the potential is repulsive the number of states decreases. In general, it turns out that the sign of $(d\delta)/(dk)$ depends on the momentum.

The total change in the density of states is then

$$\frac{d}{dk}N = \frac{d}{dk}\sum_{m_s,m_l,l}\left(\frac{\delta_{l,m_l,m_s}}{\pi}\right), \tag{22.40}$$

summing over angular momentum and spin. In a metal the electronic states are occupied up to the Fermi level, k_F. Therefore,

$$\int_0^{k_F} dk\frac{dN}{dk} = Z = \sum_{m_s,m_l,l}\left(\frac{\delta_{l,m_l,m_s}(k_F)}{\pi}\right). \tag{22.41}$$

This is the so-called Friedel sum rule. Here Z is the charge of the impurity (in units of e). The change occurs due to the screening exerted by the metal and the requirement of charge neutrality. The electron gas is perturbed in the neighborhood of the impurity charge. The displaced charge exactly cancels the impurity charge (screening charge).

The charge is related to the density difference that takes place in the metal as

$$Z = 4\pi \int_0^\infty dr r^2 [\rho(r) - \rho_0]. \tag{22.42}$$

The density difference is given by

$$\rho(r) - \rho_0 = 2 \int_{k<k_F} \frac{d^3k}{(2\pi)^3} \left(|\psi_k(r)|^2 - |\phi_k(r)|^2 \right). \tag{22.43}$$

In the asymptotic limit this reduces to

$$\lim_{r\to\infty} [\rho(r) - \rho_0] = \frac{8\pi}{(2\pi)^3 r^2} \sum_l (2l+1) \int_0^{k_F} dk \left[\sin^2\left(kr + \delta_l - \frac{l\pi}{2} \right) - \sin^2\left(kr - \frac{l\pi}{2} \right) \right]. \tag{22.44}$$

We can expand the phase shift around the Fermi level like

$$\delta_l(k) = \delta_l(k_F) + (k - k_F)\frac{d\delta}{dk}. \tag{22.45}$$

Performing the integral we get that

$$\lim_{r\to\infty} [\rho(r) - \rho_0] = \frac{1}{4\pi^2 r^3} \sum_l (2l+1)(-1)^l \sin[\delta_l(k_F)] \cos[2k_F r + \delta_l(k_F)] + O\left(\frac{1}{r^4}\right). \tag{22.46}$$

The density oscillates with a period $2k_F$. These are called Friedel oscillations, as shown in Fig. 22.1.

If the phase shift only depends on the angular momentum, we obtain from Eq. (22.41) that

$$Z = \frac{2}{\pi} \sum_l (2l+1)\delta_l(k_F). \tag{22.47}$$

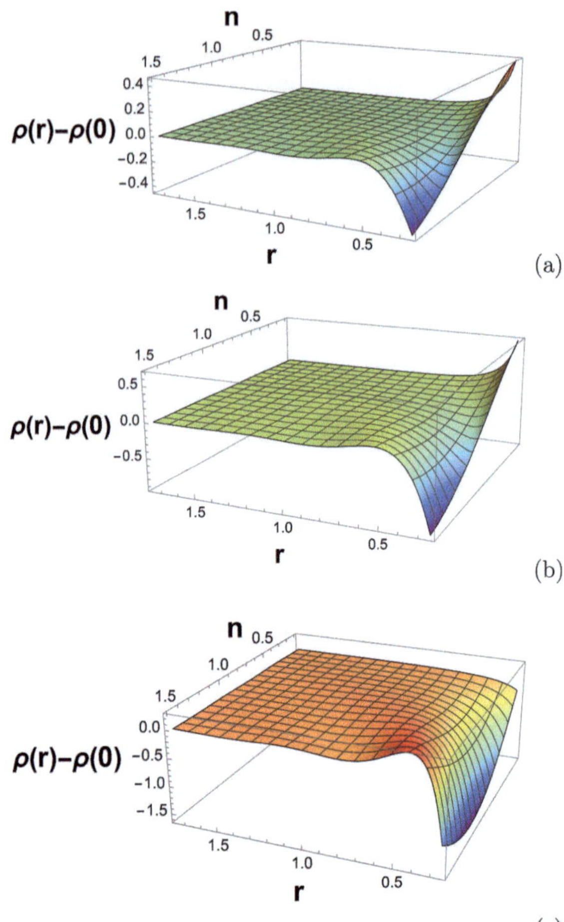

Fig. 22.1 Friedel oscillations as a function of distance, r, and n, where $k_F = \pi n$, for only the $l = 0$ phase shift at the Fermi surface and (**a**) $\delta_0(k_F) = \pi/10$, (**b**) $\delta_0(k_F) = \pi/5$, and (**c**) $\delta_0(k_F) = \pi/2$

The factor 2 is due to the spin degeneracy and the factor $(2l+1)$ is due to the orbital degeneracy. The quantity $\frac{2}{\pi}(2l + 1)\delta_l(k_F)$ is the screening charge in the l angular momentum channel.

22.3 Orthogonality Catastrophe

Let us consider again an impurity at the center of a sphere of radius R (where $R \to \infty$). Also, let us consider again the eigenfunctions of one electron in the sphere with $l = 0$, with and without impurity. With no impurity ($V = 0$) we have seen that

$$\phi_0^n(r) = N_n \frac{\sin k_n r}{k_n r}, \tag{22.48}$$

where $k_n = (\pi n)/R$ and $E_n = (\hbar^2 k_n^2)/(2m) < E_F$. Introducing the impurity at the origin ($V \neq 0$) the wave function is replaced by

$$\psi_0^n(r) \sim N_n' \frac{\sin[k_n r + \delta(E_n)]}{k_n r}. \tag{22.49}$$

Let us consider the overlap between two typical wave functions, one for the free case and the other for the case with impurity.[3] This is given by

$$\begin{aligned} A_{nn'} &= 4\pi \int_0^R drr^2 \phi_0^n(r) \psi_0^{n'}(r) \\ &\sim \frac{2\pi N_n N_{n'}}{k_n k_{n'}} \frac{\sin \delta_{n'}}{k_n - k_{n'} + \frac{\delta}{R}}. \end{aligned} \tag{22.50}$$

By normalization $N_n = k_n/\sqrt{2\pi R}$. Therefore,

$$A_{nn'} = \frac{\sin \delta_{n'}}{\pi(n - n') + \delta}. \tag{22.51}$$

The overlap is however more complicated since the true states are Slater determinants. Therefore,

$$S = \langle \phi | \psi \rangle = det_{E_n < E_F; E_{n'} < E_F} |A_{nn'}|. \tag{22.52}$$

We can write that

$$\psi_{n'} = \sum_{E_n < E_F} A_{nn'} \phi_n + \sum_{E_n > E_F} A_{nn'} \phi_n. \tag{22.53}$$

Normalizing the state implies that

$$1 = \sum_{E_n < E_F} |A_{nn'}|^2 + \sum_{E_n > E_F} |A_{nn'}|^2. \tag{22.54}$$

Since in the determinant only states below the Fermi level are taken into account, we may throw away the second term. The first term in Eq. (22.53) may then be normalized and the overlap may be written as

[3] See P.W. Anderson, Physical Review Letters **18**, 1049 (1967).

$$S = \left[\prod_{n'} \left(1 - \sum_{E_n > E_F} |A_{nn'}|^2 \right)^{1/2} \right] det|A_{nn'} \left(1 - \sum_{E_{n''} > E_F} |A_{n''n'}|^2 \right)^{-1/2} |. \tag{22.55}$$

Now the determinant is smaller or equal than 1. As a consequence we can write that

$$S \leq \prod_{n<n_F} \left(1 - \sum_{E_{n'} > E_F} |A_{nn'}|^2 \right)^{1/2} . \tag{22.56}$$

Taking δ small implies that $A_{nn'}$ is also small. Therefore,

$$\prod_{n<n_F} \left(1 - \sum_{E_{n'} > E_F} |A_{nn'}|^2 \right)^{1/2} \sim \left(1 - \frac{1}{2} \sum_{n<n_F; n'>n_F} |A_{nn'}|^2 \right) . \tag{22.57}$$

The sum can be written as

$$\frac{1}{2\pi^2} \sum_{n<n_F; n'>n_F} \frac{\sin^2 \delta}{(n - n' + \delta/\pi)^2} \sim \frac{\sin^2 \delta}{2\pi^2} \ln M^2, \tag{22.58}$$

where M is some cutoff and here the phase shift δ is calculated at the Fermi surface. We may write that

$$\frac{\hbar^2}{2m} \left(\frac{\pi M}{R} \right)^2 = \Delta, \tag{22.59}$$

where Δ is the energy at which δ is small. Then $M \sim \Omega^{1/3}$, where Ω is the volume. The overlap is then

$$S \leq e^{-\frac{\sin^2 \delta}{\pi^2} \ln \Omega^{1/3}} \sim \Omega^{-\frac{\sin^2 \delta}{3\pi^2}} . \tag{22.60}$$

Considering now $l \neq 0$ we would get

$$\begin{aligned} S &\leq e^{-(\sum_l \frac{2l+1}{\pi^2} \sin^2 \delta_l) \ln \Omega^{1/3}} \\ &\leq e^{-\frac{1}{4\pi^3} k_F^2 \sigma_{total}(k_F) \ln \Omega^{1/3}} \\ &\leq \Omega^{-\frac{1}{12\pi^3} k_F^2 \sigma_{total}(k_F)} . \end{aligned} \tag{22.61}$$

Therefore, as the system size goes to infinity ($\Omega \to \infty$), the overlap goes to zero. The states are therefore orthogonal in the thermodynamic limit. If at least one phase shift is finite, then the states are orthogonal. Note however that due to the exponential behavior, the overlap is quite sensitive to the value of the phase shift. For instance, if δ is very small, then $S \sim 1$, even if $\Omega >> 1$. However, if we are close to the unitarity limit, $\delta \sim \pi/2$ (near a resonance), then $S \sim 0$.

If the potential V is due to the interaction of an electron with the other electrons, then it may happen that, even for small V, the system is qualitatively in a different state as the interaction is turned on. As a consequence, a perturbative description where we use as a starting point free or nearly free electrons in a Fermi surface, with weak interactions, may be altered drastically. Indeed in $1D$ this happens. The phase shifts are not negligible since scattering in a $1D$ system is important, due to the severe restrictions in the available phase space. Two particles moving in a one-dimensional system have no way to avoid each other and their interactions are always important. As we will see, the system is profoundly changed from a Fermi description and the basic excitations are instead bosonic instead of fermionic.

Let us consider again the same problem of the scattering of one free electron from one state $\mathbf{k}$ to another state $\mathbf{k}'$ due to the presence of a short range potential, such that $V_{\mathbf{kk}'} \sim V$. Then the probability that a state below the Fermi surface is scattered to a state $\mathbf{k}'$ above the Fermi surface is, to lowest order,

$$P_{\mathbf{k}\to\mathbf{k}'} = \frac{|V|^2}{|\epsilon_{\mathbf{k}'} - \epsilon_{\mathbf{k}}|^2}. \tag{22.62}$$

Here we have the restrictions that $\epsilon_{\mathbf{k}} < 0$, $\epsilon_{\mathbf{k}'} > 0$ (taking now $\epsilon_{k_F} = 0$). Summing over all values of the initial and final momenta we get that

$$\begin{aligned}\sum_{\mathbf{k},\mathbf{k}'} P_{\mathbf{k}\to\mathbf{k}'} &= |V|^2 \sum_{\epsilon_{\mathbf{k}}<0} \sum_{\epsilon_{\mathbf{k}'}>0} \frac{1}{|\epsilon_{\mathbf{k}'} - \epsilon_{\mathbf{k}}|^2} \\ &\sim |N(\epsilon_F)V|^2 \ln \frac{(\epsilon_{\mathbf{k}'} - \epsilon_{\mathbf{k}})_{\max}}{(\epsilon_{\mathbf{k}'} - \epsilon_{\mathbf{k}})_{\min}}. \end{aligned} \tag{22.63}$$

The minimum possible value for $\epsilon_{\mathbf{k}'} - \epsilon_{\mathbf{k}} \sim 1/\sqrt{\Omega}$ where Ω is the space volume. We see therefore that there is a divergency of the type $\ln \Omega$ in the infinite system limit. The orthogonality catastrophe originates in this divergency.

In $2D$ and in $3D$ the divergency in general disappears because when one particle collides with another one, the momenta are such that $\mathbf{k} \to \mathbf{k}-\mathbf{q}$ and $\mathbf{k}' \to \mathbf{k}'+\mathbf{q}$ and the denominators $\epsilon_{\mathbf{k}'} - \epsilon_{\mathbf{k}}$ no longer vanish in general, and no divergency appears. In $1D$ the phase space is more restricted.

22.4 Anderson Model: Magnetic Moments in Metals

Let us consider an atom with partially filled electronic shells. In general, the angular momentum is nonvanishing. Let us imagine that we insert this atom, as an impurity, in a metal. The electrons of the non-filled shells may delocalize in the metal and, therefore, the total angular momentum at the impurity site may be different, due to the interaction with the metallic environment. We may then expect that, if the eigenfunctions of the electrons in the atom are fairly delocalized (with a large extent with respect to the dimensions of the atom), these electrons probably will mix with the conduction electrons of the metal. In the same way, the conduction electrons may be in states with a spatial distribution close to the impurity site.

On the other hand, electrons in somewhat localized states in the atom, like $4f$ (rare earths: Ce, Yb) or $5f$ (like uranium U), have overlap integrals with the conduction electrons that are much smaller (in particular the $4f$ states are fairly localized). The electrons of the $3d$ type are in an intermediate situation (like Fe, Co, Mn), such that the delocalization is appreciable. A question arises: if an atom is magnetic in the vacuum does it remain magnetic in the metallic environment? That is, what are the conditions that a local magnetic moment exists in a metal and, if it remains finite, what is its interaction with the electrons of the metallic matrix? Both questions are described by the Anderson model (even though its solution, in general, is complex).

The Anderson model is described by a Hamiltonian for an impurity in a metal:

$$H = H_{0s} + H_{0d} + H_{corr} + H_{sd}, \tag{22.64}$$

where[4]

$$H_{0s} = \sum_{\mathbf{k},\sigma} \epsilon_k c^\dagger_{k,\sigma} c_{k,\sigma} \tag{22.65}$$

corresponds to a band of free electrons (originally in the bulk of the metal), which are assumed to have $l = 0$:

$$H_{0d} = E_d \sum_{\sigma} d^\dagger_\sigma d_\sigma, \tag{22.66}$$

where the d electrons are electrons at the impurity with an energy level E_d,

$$H_{corr} = U n_{d,\uparrow} n_{d,\downarrow} \tag{22.67}$$

is the repulsion energy between the localized electrons at the impurity (of the Hubbard type), and the hybridization term

$$H_{sd} = \sum_{\mathbf{k},\sigma} V_k \left(c^\dagger_{k,\sigma} d_\sigma + d^\dagger_\sigma c_{k,\sigma} \right) \tag{22.68}$$

[4] To simplify notation, in this section we will often write $k = \mathbf{k}$.

allows the transfer of electrons from the conduction band to the impurity and vice versa. In general, the electron states at the impurity site have an angular momentum that is $l > 0$. To simplify, here we consider that the degeneracy of these states is only due to the spin component, σ. The matrix element (taken here as real) is given by

$$V_k = \langle \mathbf{k}\sigma|V|d\sigma'\rangle = \int d^3r e^{-i\mathbf{k}\cdot\mathbf{r}} V_{ion}(\mathbf{r})\psi(\mathbf{r})\delta_{\sigma,\sigma'}. \tag{22.69}$$

Here $e^{i\mathbf{k}\cdot\mathbf{r}}$ is the wave function of a free electron and $\psi(\mathbf{r})$ is the wave function localized at the impurity site. In this model the Coulomb interaction between the conduction electrons and any interactions between the electrons and excitations of the system (like phonons, magnons, etc.) are not taken into account.

Let us consider the states of the free impurity:

$$H_{atomic} = E_d \sum_{\sigma} d_\sigma^\dagger d_\sigma + U n_{d,\uparrow} n_{d,\downarrow}. \tag{22.70}$$

Since the degeneracy is only due to the spin, there are 4 possible states with energies:

$$\begin{aligned} |d^2\rangle &\rightarrow E(d^2) = 2E_d + U \\ |d^0\rangle &\rightarrow E(d^0) = 0 \\ |d_\uparrow^1\rangle, |d_\downarrow^1\rangle &\rightarrow E(d^1) = E_d. \end{aligned} \tag{22.71}$$

In order that the ground state is magnetic it is necessary that the state of lowest energy be the doublet $|d^1\rangle$. Therefore, we must have that

$$\begin{aligned} E(d^2) - E(d^1) = E_d + U > 0 \qquad &\rightarrow E_d + \frac{U}{2} > -\frac{U}{2} \\ E(d^0) - E(d^1) = -E_d > 0 \qquad &\rightarrow \frac{U}{2} > E_d + \frac{U}{2}. \end{aligned} \tag{22.72}$$

This is true if

$$\frac{U}{2} > \left| E_d + \frac{U}{2} \right|. \tag{22.73}$$

Then the isolated impurity has a magnetic state as its ground state.

Let us consider now that the impurity is immersed in the metal. Then we may expect that the situation depicted in Fig. 22.2 is observed for the impurity to remain magnetic. If this situation is verified, then we should expect that the impurity level is occupied by one electron and, therefore, that it has a local magnetic moment. If U is high, then the state with double occupancy will have a high energy, and it will be empty. Also, if the hybridization is small, there will be a high probability that

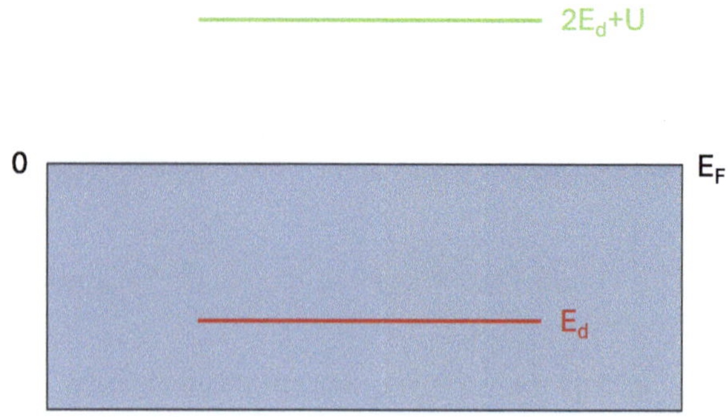

Fig. 22.2 Energy levels of an Anderson impurity such that the local level has a negative value with respect to the Fermi level and the doubly occupied level has a positive energy, above the Fermi level and, therefore, is empty

the level E_d is occupied, since V hybridizes the d electrons with the conduction electrons. Since U is high, the double occupancy in the impurity will be difficult to attain and it will only happen virtually. Therefore, in the small V limit and large U limit, it will make sense to consider that the excitations for the doubly occupied site are eliminated in perturbation theory.

Due to the hybridization between the level d^1 and the conduction electrons, we must expect that the level d should acquire a finite width, given by the Fermi golden rule:

$$\Delta = \Delta(E_d) = \pi \sum_{\mathbf{k}} |V(\mathbf{k})|^2 \delta(\epsilon_{\mathbf{k}} - E_d) = \pi V^2(E_d) N(E_d), \tag{22.74}$$

where $N(\epsilon) = \sum_k \delta(\epsilon_k - \epsilon)$ is the density of states of the conduction electrons.

There will be, as a consequence, a value of $U = U_c$ such that for $U > U_c$ there is a local magnetic moment and for $U < U_c$ there is no local moment (since both spin states may be occupied).

22.4.1 Slave Bosons

In addition to the Hubbard or $t - J$ models previously discussed, other systems with strong correlations, where the motion of the particles is conditioned by some form of projection operators, may be treated in a way similar to those models. This is the case of the Anderson model, here considered. The local Coulomb correlations also play an important role in the behavior of the system and lead to the freezing (partial or total) of the charge fluctuations. The Coulomb repulsion U is the largest energy scale in the problem (typically of the order of 10 eV), while the hybridization is the smaller scale (of the order of $0.1 - 1$ eV). In the case the charge fluctuations are suppressed, the restrictions lead to the Kondo model. The fact that the Coulomb energy is high implies that at the impurity site double occupancy is restricted to virtual fluctuations and, therefore, the limit $U \to \infty$ is usually considered. The finite U case may also be handled, as we will see ahead.

Let us then consider that the state d^2 is inaccessible (that is, we take the limit $U \to \infty$). In this limit we may then write the Hamiltonian like

$$H = \sum_{k,\sigma} \epsilon_k c^{\dagger}_{k,\sigma} c_{k,\sigma} + E_d \sum_{\sigma} d^{\dagger}_{\sigma} d_{\sigma} + V \sum_{k,\sigma} \left(d^{\dagger}_{\sigma} c_{k,\sigma} P_0 + c^{\dagger}_{k,\sigma} d_{\sigma} P_1 \right), \tag{22.75}$$

where P_0 is a projection operator that assures that the state $|d^1_{\sigma}\rangle$ is empty (so that there is no double occupancy that would cost an infinite energy) and P_1 is an operator into the state $|d^1_{\sigma}\rangle$. The use of these operators is complicated when we calculate physical quantities.

It is convenient to use the slave boson method, as discussed before. The operators $d_{i,\sigma}$ are fermionic. That is,

$$\begin{aligned} \{d_{i,\sigma}, d_{j,\sigma'}\} &= \{d^{\dagger}_{i,\sigma}, d^{\dagger}_{j,\sigma'}\} = 0 \\ \{d_{i,\sigma}, d^{\dagger}_{j,\sigma'}\} &= \delta_{ij}\delta_{\sigma,\sigma'}. \end{aligned} \tag{22.76}$$

Let us write the d electron operators like

$$\begin{aligned} d_{i,\sigma} &= b^{\dagger}_i f_{i,\sigma} \\ d^{\dagger}_{i,\sigma} &= f^{\dagger}_{i,\sigma} b_i, \end{aligned} \tag{22.77}$$

the impurity is at site i, b_i is a bosonic operator, and $f_{i,\sigma}$ is a fermionic operator. Let us consider the state $|b^1 f^0\rangle$ which has one boson and no fermion. If the Hamiltonian is defined as

$$H = \sum_{k,\sigma} \epsilon_k c^{\dagger}_{k,\sigma} c_{k,\sigma} + E_d \sum_{\sigma} d^{\dagger}_{\sigma} d_{\sigma} + V \sum_{k,\sigma} \left(f^{\dagger}_{i,\sigma} c_{k,\sigma} b + c^{\dagger}_{k,\sigma} f_{i,\sigma} b^{\dagger}_i \right) \tag{22.78}$$

and it acts on the state $|b^1 f^0\rangle$, then the term $f^{\dagger}_{i,\sigma} c_{k,\sigma} b$ converts this state into the state $|b^0 f_{\sigma}\rangle$. If the Hamiltonian acts on this second state, then the other term $c^{\dagger}_{k,\sigma} f_{i,\sigma} b^{\dagger}_i$ reproduces the original state $|b^1 f^0\rangle$. (The other terms in the Hamiltonian either leave these two states invariant or they annihilate them.) If we consider now the state with double occupancy, the Hamiltonian does not annihilate it, and if we start with the state $|b^1 f_{\sigma}\rangle$, then the forbidden state $|b^0 f^2\rangle$ is created. Therefore, to eliminate this state, a restriction must be introduced such that

$$Q = \sum_{\sigma} f^{\dagger}_{\sigma} f_{\sigma} + b^{\dagger} b = 1. \tag{22.79}$$

We have now a Hamiltonian with a restriction. That is, we should consider instead a description similar to the grand canonical ensemble where only the average number of particles is fixed. We introduce therefore a Lagrange multiplier in the problem, λ, that imposes the restriction $Q = 1$, as mentioned before. The Hamiltonian is then rewritten as

$$H = \sum_{k,\sigma} \epsilon_k c_{k,\sigma}^\dagger c_{k,\sigma} + E_d \sum_\sigma f_\sigma^\dagger f_\sigma$$
$$+V \sum_{k,\sigma} \left(f_\sigma^\dagger c_{k,\sigma} b + c_{k,\sigma}^\dagger f_\sigma b^\dagger \right) + \lambda \left(\sum_\sigma f_\sigma^\dagger f_\sigma + b^\dagger b - 1 \right). \tag{22.80}$$

Note that

$$\sum_\sigma d_\sigma^\dagger d_\sigma = \sum_\sigma f_\sigma^\dagger b b^\dagger f_\sigma = \sum_\sigma f_\sigma^\dagger (b^\dagger b + 1) f_\sigma = \sum_\sigma f_\sigma^\dagger f_\sigma b^\dagger b + \sum_\sigma f_\sigma^\dagger f_\sigma$$
$$= \sum_\sigma f_\sigma^\dagger f_\sigma, \tag{22.81}$$

(in the subspace where the restriction is obeyed, we cannot have at the same time a boson and a fermion at the impurity site). We may also write, defining $\epsilon_d = E_d + \lambda$, that

$$H = \sum_{k,\sigma} \epsilon_k c_{k,\sigma}^\dagger c_{k,\sigma} + \epsilon_d \sum_\sigma f_\sigma^\dagger f_\sigma$$
$$+V \sum_{k,\sigma} \left(f_\sigma^\dagger c_{k,\sigma} b + c_{k,\sigma}^\dagger f_\sigma b^\dagger \right) + (\epsilon_d - E_d)(b^\dagger b - 1). \tag{22.82}$$

Let us now use the mean-field theory. Let us assume that the bosons are condensed. That is, we make the replacement $b \to \langle b \rangle = \sqrt{z}$. Then we obtain the effective mean-field Hamiltonian:

$$H = \sum_{k,\sigma} \epsilon_k c_{k,\sigma}^\dagger c_{k,\sigma} + \epsilon_d \sum_\sigma f_\sigma^\dagger f_\sigma$$
$$+V\sqrt{z} \sum_{k,\sigma} \left(f_\sigma^\dagger c_{k,\sigma} + c_{k,\sigma}^\dagger f_\sigma \right) + (\epsilon_d - E_d)(z - 1). \tag{22.83}$$

Let us use the variational method to determine the parameters of the problem. The energy is therefore stationary with respect to the various parameters, in this case ϵ_d, z, μ, where μ is the chemical potential. The energy is given by $E = \langle H \rangle$. Then, using the Hellman-Feynman theorem,

$$\frac{\partial}{\partial \lambda} \langle H(\lambda) \rangle = \left\langle \frac{\partial H}{\partial \lambda} \right\rangle, \tag{22.84}$$

we obtain that, taking a derivative in order to, ϵ_d

$$\sum_{\sigma} \langle f_{\sigma}^{\dagger} f_{\sigma} \rangle + z - 1 = 0. \tag{22.85}$$

Taking a derivative with respect to z we get

$$\frac{1}{2} \frac{V}{\sqrt{z}} \sum_{k,\sigma} \left(\langle f_{\sigma}^{\dagger} c_{k,\sigma} \rangle + \langle c_{k,\sigma}^{\dagger} f_{\sigma} \rangle \right) + (\epsilon_d - E_d) = 0. \tag{22.86}$$

Remembering that in the grand canonical ensemble $\epsilon_k \to \epsilon_k - \mu$ and the same with the impurity level, taking a derivative with respect to μ we obtain

$$\frac{\partial E}{\partial \mu} = -n = -\sum_{k,\sigma} \langle c_{k,\sigma}^{\dagger} c_{k,\sigma} \rangle - \sum_{\sigma} \langle f_{\sigma}^{\dagger} f_{\sigma} \rangle. \tag{22.87}$$

These mean-field equations have to be solved self-consistently, usually numerically. The averages are understood as thermal averages weighted by the Fermi-Dirac distribution of the eigenenergies of the mean-field hamiltonian (effectively noninteracting).

22.4.2 Excitation Spectrum

To determine the spectrum of excitations it is necessary to diagonalize the Hamiltonian. One impurity should not change appreciably the energies of the continuum and, in addition, we have the energy of the impurity. The level d has now the (renormalized) energy ϵ_d and, using once again the Fermi golden rule, has a width given by

$$\Delta = \pi z V^2(\epsilon_d) N(\epsilon_d). \tag{22.88}$$

Comparing with Eq. (22.74) obtained in perturbation theory, the location of the energy level is shifted and the hybridization is renormalized by the factor z. Then the density of states of the level d is a Lorentzian:

$$\rho_d(\epsilon) = \frac{1}{\pi} \frac{\Delta}{(\epsilon - \epsilon_d)^2 + \Delta^2}, \tag{22.89}$$

as shown in Fig. 22.3.

Then

$$\langle n_d \rangle = \sum_{\sigma} \langle f_{\sigma}^{\dagger} f_{\sigma} \rangle = \int_{-D}^{0} d\epsilon f(\epsilon) \rho_d(\epsilon) = \frac{2}{\pi} \tan^{-1} \left(\frac{\Delta}{\epsilon_d} \right), \tag{22.90}$$

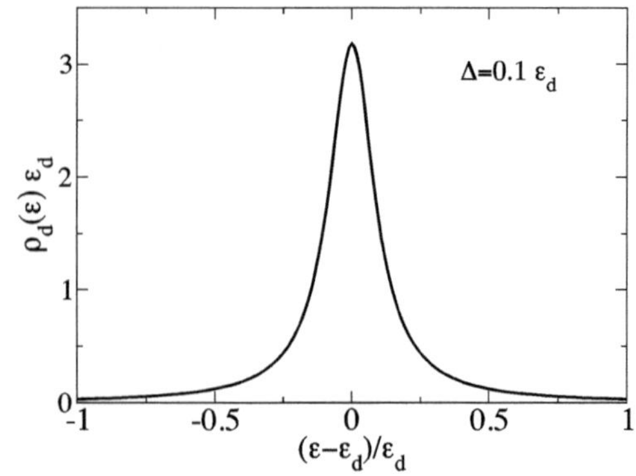

Fig. 22.3 Impurity density of states

at $T = 0$.

Let us write the Hamiltonian Eq. (22.83) in the form[5]

$$H = \sum_{\gamma,\sigma} E_\gamma a^\dagger_{\gamma,\sigma} a_{\gamma,\sigma} + (\epsilon_d - E_d)(z - 1), \tag{22.91}$$

where

$$\begin{aligned} a^\dagger_{\gamma,\sigma} &= \sum_k \alpha_k c^\dagger_{k,\sigma} + \beta f^\dagger_\sigma \\ a_{\gamma,\sigma} &= \sum_k \alpha_k c_{k,\sigma} + \beta f_\sigma . \end{aligned} \tag{22.92}$$

Since these new operators diagonalize the Hamiltonian, we may write that

$$[H, a^\dagger_{\gamma,\sigma}] = E_\gamma a^\dagger_{\gamma,\sigma} . \tag{22.93}$$

Introducing the Hamiltonian in this expression, calculating the commutator, replacing the definition of the operators in terms of the old operators and comparing the coefficients of the same operators, we get in a way similar to other treatments (like the Bogoliubov-de Gennes equations) that

$$\begin{aligned} \beta\epsilon_d + V\sqrt{z}\sum_k \alpha_k &= \beta E_\gamma \\ V\sqrt{z}\beta &= (E_\gamma - \epsilon_k)\alpha_k . \end{aligned} \tag{22.94}$$

Solving this equation gives that

[5] For further reading see [Coleman].

$$E_\gamma = \epsilon_d + \sum_k \frac{V^2 z}{E_\gamma - \epsilon_k}. \tag{22.95}$$

Let us consider the case in which $\epsilon_k = (k + 1/2)\Delta\epsilon$. Here k is an integer that takes $2M$ values between $-M$ and $M - 1$. (With this choice $\epsilon_k \neq 0$.) To simplify let us consider that $\epsilon_d = 0$. Then

$$E_\gamma = \sum_k \frac{V^2 z}{E_\gamma - \epsilon_k}. \tag{22.96}$$

We see that $E_\gamma = 0$ is a solution: the hybridization creates a state in the band (in this case at the origin). The remaining states are translated symmetrically around the new bound state at $E_\gamma = 0$. Each new eigenvalue is translated by a quantity proportional to $\Delta\epsilon$. Then, introducing a phase shift δ_γ,

$$E_\gamma = \epsilon_k - \Delta\epsilon \frac{\delta_\gamma}{\pi}, \tag{22.97}$$

$$E_\gamma = \epsilon_d + \sum_{n=\gamma+1-M}^{\gamma+M} \frac{V^2 z}{\Delta\epsilon (n - \frac{\delta_\gamma}{\pi})}. \tag{22.98}$$

If M is very large, then

$$E_\gamma = \epsilon_d + \frac{V^2 z}{\Delta\epsilon} \sum_{n=-\infty}^{\infty} \frac{1}{n - \frac{\delta_\gamma}{\pi}}. \tag{22.99}$$

The sum gives

$$\sum_{n=-\infty}^{\infty} \frac{1}{n - \frac{\delta_\gamma}{\pi}} = -\pi \cot \delta_\gamma. \tag{22.100}$$

Using the notation that $\delta_\gamma = \delta(E_\gamma)$, we can write that

$$\tan \delta(\epsilon) = \frac{\pi \rho z V^2}{\epsilon_d - \epsilon}, \tag{22.101}$$

where $\rho = \rho(\epsilon) = \frac{1}{\Delta\epsilon}$ is the density of states of the conduction electrons. This implies that the phase shift is

$$\delta(\epsilon) = \tan^{-1}\left(\frac{\Delta}{\epsilon_d - \epsilon}\right), \tag{22.102}$$

where $\Delta = \pi \rho z V^2$. When $\epsilon_d = 0$ the phase shift $\delta(\epsilon = 0) = \pi/2$ (taking $\epsilon = 0$ as the Fermi level). The phase shift goes from zero at negative energies to π at high energies, passing through $\pi/2$ close to the zero energy.

Using that

$$E_\gamma = \epsilon_\gamma - \frac{\Delta\epsilon}{\pi}\delta(E_\gamma), \tag{22.103}$$

we can calculate the change in the density of states. The new density of states is such that

$$\rho^*(E)dE = \rho(\epsilon)d\epsilon. \tag{22.104}$$

Then

$$\rho^*(E) = \rho(0)\frac{d\epsilon}{dE} = \rho + \rho_i(E), \tag{22.105}$$

where the impurity contribution to the density of states is

$$\rho_i(E) = \frac{1}{\pi}\frac{d\delta(E)}{dE} = \frac{1}{\pi}\frac{\Delta^2}{(E-\epsilon_d)^2+\Delta^2}, \tag{22.106}$$

as obtained before from the Fermi golden rule.

22.5 Schrieffer-Wolff Transformation and Kondo Problem

Let us consider once again the Anderson model[6] (considering only $l = 0$ contributions):

$$\begin{aligned}
H &= H_0 + H_1 \\
H_0 &= \sum_{\mathbf{k},\sigma} \epsilon_k c^\dagger_{k,\sigma} c_{k,\sigma} + E_d \sum_\sigma d^\dagger_\sigma d_\sigma + U n_{d,\uparrow} n_{d,\downarrow} \\
H_1 &= \sum_{\mathbf{k},\sigma} V_k \left(c^\dagger_{k,\sigma} d_\sigma + d^\dagger_\sigma c_{k,\sigma} \right).
\end{aligned}$$

Let us eliminate the hybridization term to first order, performing a canonical transformation:[7]

[6] To simplify the notation, in this section we will often write $k = \mathbf{k}$.

[7] See J. R. Schrieffer and P. A. Wolff, Physical Review **149**, 491 (1966).

$$\bar{H} = e^S H e^{-S} = H + [S, H] + \cdots \tag{22.107}$$

Then choosing

$$[H_0, S] = H_1, \tag{22.108}$$

we obtain that

$$\bar{H} = H_0 + \frac{1}{2}[S, H_1] + \frac{1}{3}[S, [S, H_1]] + \cdots . \tag{22.109}$$

The generator S is defined as

$$S = \sum_{\mathbf{k},\sigma,\alpha} \frac{V_k}{\epsilon_k - \epsilon_\alpha} n_{d,-\sigma}^{\alpha} c_{k,\sigma}^{\dagger} d_\sigma - h.c., \tag{22.110}$$

where

$$\begin{aligned} n_{d,-\sigma}^{\alpha} &= n_{d,-\sigma}, \quad \alpha = +, \epsilon_\alpha = E_d + U \\ n_{d,-\sigma}^{\alpha} &= (1 - n_{d,-\sigma}), \quad \alpha = -, \epsilon_\alpha = E_d, \end{aligned} \tag{22.111}$$

corresponding to the situations in which the level d is doubly occupied or not, respectively. Let us call $H_2 = \frac{1}{2}[S, H_1]$ (the other term is of order V^3). Then H_2 can be written as

$$H_2 = H_{ex} + H_{dir} + H_0' + H_{ch}. \tag{22.112}$$

Defining the vectors

$$\psi_k = \begin{pmatrix} c_{k,\uparrow} \\ c_{k,\downarrow} \end{pmatrix} \tag{22.113}$$

and

$$\psi_d = \begin{pmatrix} d_\uparrow \\ d_\downarrow \end{pmatrix}, \tag{22.114}$$

we may write the various terms of H_2 as

$$H_{ex} = -\sum_{k,k'} J_{k',k} \left(\psi_{k'}^{\dagger} \mathbf{S} \psi_k\right) \cdot \left(\psi_d^{\dagger} \mathbf{S} \psi_d\right), \tag{22.115}$$

which is a term of interaction between the spins, where $\mathbf{S} = \boldsymbol{\sigma}/2$, with $\boldsymbol{\sigma}$ Pauli matrices:

$$J_{k',k} = V_{k'} V_k \left(\frac{1}{\epsilon_k - \epsilon_+} + \frac{1}{\epsilon_{k'} - \epsilon_+} - \frac{1}{\epsilon_k - \epsilon_-} - \frac{1}{\epsilon_{k'} - \epsilon_-} \right). \tag{22.116}$$

The direct term is given by

$$H_{dir} = \sum_{k,k'} \left(W_{k',k} + \frac{1}{4} J_{k',k} \psi_d^\dagger \psi_d \right) \left(\psi_{k'}^\dagger \psi_k \right), \tag{22.117}$$

where

$$W_{k',k} = \frac{V_{k'} V_k}{2} \left(\frac{1}{\epsilon_k - \epsilon_-} + \frac{1}{\epsilon_{k'} - \epsilon_-} \right); \tag{22.118}$$

the term H_0' is absorbed in H_0 and is given by

$$H_0' = -\sum_{k,\sigma} \left(W_{kk} + \frac{1}{2} J_{kk} n_{d,-\sigma} \right) n_{d,\sigma} \tag{22.119}$$

and finally a term that involves two creation or two destruction operators of d electrons which is

$$H_{ch} = \frac{1}{4} \sum_{kk'} \sum_{\sigma} J_{k',k} c_{k',-\sigma}^\dagger c_{k,\sigma}^\dagger d_\sigma d_{-\sigma} + (h.c.). \tag{22.120}$$

The condition of a local moment, as previously discussed, is satisfied if $\epsilon_+ > 0$ and $\epsilon_0 < 0$. Considering this case, neglecting H_{ch}, since it lives in the subspace of zero or two d electrons, and using that in this subspace $\psi_d^\dagger \psi_d = 1$, the problem simplifies. The important term is then $H_2 = H_{ex}$. For k and k' near the Fermi momentum the exchange interaction is given by

$$J_{k_f,k_F} = -J_0 = 2V_k^2 \frac{U}{E_d(E_d + U)} < 0. \tag{22.121}$$

This implies that the coupling between the spin of the level d and the spin density of the conduction electrons is antiferromagnetic. This effective model is called Kondo model [1].

We have therefore a gas of (free) conduction electrons whose spin density interacts antiferromagnetically with the impurity spin. Due to the internal degrees of freedom of the impurity spin, the spin of the conduction electrons may change, and vice versa, and a temporal correlation between the conduction electrons is established. The conduction electrons, besides being scattered from $\mathbf{k} \to \mathbf{k}'$ (due to the breaking of the translational symmetry due to the introduction of the impurity), as they collide with the impurity, may also suffer transitions $\sigma \to \sigma'$, due to the exchange interaction of the Heisenberg type.

The Hamiltonian is then given by

$$H_{ex} = \sum_{kk'} J_0 \left(\psi_{k'}^{\dagger}\mathbf{S}\psi_k\right) \cdot \left(\psi_d^{\dagger}\mathbf{S}\psi_d\right)$$

$$H_{ex} = \sum_{kk'} J_0 \left((c_{k',\uparrow}^{\dagger}c_{k,\uparrow} - c_{k',\downarrow}^{\dagger}c_{k,\downarrow})S^z + c_{k',\downarrow}^{\dagger}c_{k,\uparrow}S^+ + c_{k',\uparrow}^{\dagger}c_{k,\downarrow}S^-\right), \tag{22.122}$$

where $\mathbf{S} = \psi_d^{\dagger}\mathbf{S}\psi_d$ is a local spin operator (remember that $J_0 > 0$).

In order to understand the effect of the impurity, let us consider processes in which the state $(\mathbf{k}, \uparrow) \rightarrow (\mathbf{k}', \uparrow)$. The interaction is invariant under rotations and, therefore, it is enough to consider this process. The first-order term in J_0 involves S^z and we will see that it is trivial. Let us then consider the second-order term in J_0. There are 4 possibilities:

(i) The electron transitions to an unoccupied state $\mathbf{k}''$ and then to a state $\mathbf{k}'$. While it is in the state $\mathbf{k}''$, it may have spin up or spin down.
(ii) The electron first goes from an occupied state $\mathbf{k}''$ to $\mathbf{k}'$ and the electron in the state $\mathbf{k}$ then jumps to the hole left in $\mathbf{k}''$. Once again there are two possible spin projections.

In the case of the process $(\mathbf{k}, \uparrow) \rightarrow (\mathbf{k}', \uparrow)$, the second-order contribution, considering (i) and with spin exchange in the intermediate state, is

$$H_{ex,(i)}^{(2)} = \sum_{k''} J_0^2 \frac{c_{k',\uparrow}^{\dagger}c_{k'',\downarrow}c_{k'',\downarrow}^{\dagger}c_{k,\uparrow}}{E_k - E_{k''}} S^- S^+$$

and considering the second type of process (ii), also with spin exchange, is given by

$$H_{ex,(ii)}^{(2)} = \sum_{k''} J_0^2 \frac{c_{k'',\downarrow}^{\dagger}c_{k,\uparrow}c_{k',\uparrow}^{\dagger}c_{k'',\downarrow}}{E_k - (E_k + E_{k'} - E_{k''})} S^+ S^-.$$

The expressions with no spin exchange are

$$H_{ex,(i)}^{(2)} = \sum_{k''} J_0^2 \frac{c_{k',\uparrow}^{\dagger}c_{k'',\uparrow}c_{k'',\uparrow}^{\dagger}c_{k,\uparrow}}{E_k - E_{k''}} S^z S^z$$

$$H_{ex,(ii)}^{(2)} = \sum_{k''} J_0^2 \frac{c_{k'',\uparrow}^{\dagger}c_{k,\uparrow}c_{k',\uparrow}^{\dagger}c_{k'',\uparrow}}{E_k - (E_k + E_{k'} - E_{k''})} S^+ S^-.$$

Considering processes on the Fermi surface $E_k = E_{k'}$. Then the denominators are equal, besides a changed signal. Terms that include products of the type $c_{k'',\sigma}^{\dagger}c_{k'',\sigma}$

appear when in the intermediate state there is an electron with momentum k'' and spin component σ. Therefore, we may include a thermal factor given by the Fermi-Dirac distribution as $f(E_{k''})$. Terms of the type $c_{k'',\sigma}c^{\dagger}_{k'',\sigma}$ mean an intermediate state with a hole at that momentum and spin component, σ, and therefore, we may associate a factor $(1 - f(E_{k''}))$. From the terms with spin flip we get a sum of terms that involve products like

$$(1 - f(E_{k''}))S^{-}S^{+} + f(E_{k''})S^{+}S^{-}. \tag{22.123}$$

Replacing $S^{+}S^{-} = S^{-}S^{+} + 2S^{z}$ we get that the sum of the two processes leads to

$$H^{(2)}_{ex} = \sum_{k''} J_0^2 \frac{c^{\dagger}_{k',\uparrow}c_{k,\uparrow}}{E_k - E_{k''}} \left(S^{-}S^{+} + 2S^{z} f(k'')\right). \tag{22.124}$$

Using that $S^{-}S^{+} = S^2 - (S^z)^2 - S^z$, adding the first-order term in J_0 and the terms with no spin exchange in the intermediate state, we obtain

$$H_{ex} = \left(J_0 S^z + \sum_{k''} J_0^2 \frac{S^2 + S^z[2f(k'') - 1]}{E_k - E_{k''}} \right) c^{\dagger}_{k',\uparrow}c_{k,\uparrow}. \tag{22.125}$$

Then the probability transition for the transition from the state $(\mathbf{k}, \uparrow) \rightarrow (\mathbf{k}', \uparrow)$ involves the term

$$W(\mathbf{k}, \uparrow; \mathbf{k}', \uparrow) \sim J_0^2 S^z \sum_{k''} \frac{2f(k'') - 1}{E_k - E_{k''}} \sim -\ln\left|\frac{E_k - E_F}{E_F}\right|. \tag{22.126}$$

The states considered are within an energy shell which is of the order of the temperature: therefore, $E_k - E_F \sim T$. The transition probability therefore diverges when $T \rightarrow 0$. It implies that the scattering of the electrons diverges as the temperature tends to zero. Since the scattering of the electrons by the phonons increases with temperature (since the number of phonons increases with temperature), this explains the observed low-temperature resistivity minimum in the metals with magnetic impurities, as shown schematically in Fig. 22.4. In Fig. 22.5 we show schematically the full electronic density of states with the appearance of a Kondo peak near the Fermi surface that results from the scattering from the Kondo impurity and consequent binding of the conduction electrons (one bound electron if the impurity has spin $1/2$) with the impurity.

These conclusions were obtained in perturbation theory. However, as $T \rightarrow 0$ the interaction gets to be strong (due to the divergence) and perturbation theory breaks down. Due to the strong scattering of the electrons, due to the impurity spin, the effective coupling gets renormalized to high values. Actually, we will show that the effective coupling diverges and, therefore, a different treatment has to be carried out.

This method is called poor man's scaling [2].

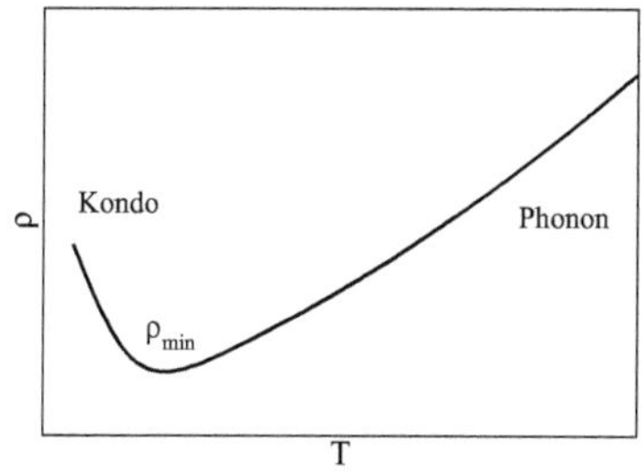

Fig. 22.4 Schematic representation of the resistivity minimum at low temperatures due to the competition between the effect of phonons that decreases as the temperature lowers and the strong scattering at low temperatures due to the Kondo effect

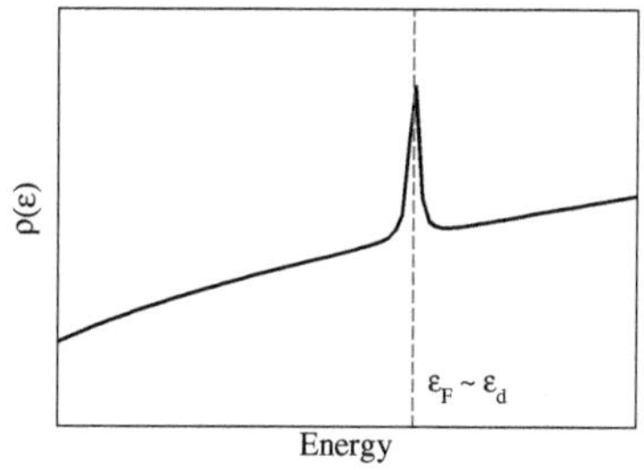

Fig. 22.5 Conduction electron and impurity density of states, with the emergence of the Kondo peak in the vicinity of the Fermi surface

22.5.1 Poor Man's Scaling

It is convenient to analyze the scattering process in terms of the T matrix. If the Hamiltonian may be separated in two parts, $H = H_0 + V$ where V is the interaction that leads to the scattering, then we write Schrödinger equation as

$$(H_0 + V)\,|\phi\rangle = E|\phi\rangle. \tag{22.127}$$

A formal solution is obtained as

$$|\phi\rangle = GV|\phi\rangle + |\xi\rangle. \tag{22.128}$$

Here $G = (E - H_0)^{-1}$ is the Green function (resolvent), and $|\xi\rangle$ is any solution of the homogeneous equation that coincides with the equation for the free particle. The iteration of the solution may be written as the Lippmann-Schwinger series:

$$|\phi\rangle = |\xi\rangle + GV|\xi\rangle + GVGV|\xi\rangle + \cdots . \tag{22.129}$$

The T matrix may be defined as

$$V|\phi\rangle = T|\xi\rangle \tag{22.130}$$

and is obtained by the series

$$T = V + VGV + VGVGV + \cdots = V\frac{1}{1 - GV} = \frac{1}{1 - VG}V. \tag{22.131}$$

The matrix elements of the T-matrix are therefore associated with the scattering from some initial state of a particle described by the Hamiltonian H_0 (typically a free particle) to another state that results from the scattering through the perturbation V. To second order (second Born approximation) we get that

$$T = V + VGV. \tag{22.132}$$

In the Kondo problem H_0 is the kinetic part and the perturbation is the spin exchange term.

Let us rewrite the scattering terms and consider again second-order perturbation theory, as in the discussion of the resistance minimum. We want now, not a full solution, but to consider how the problem scales as we integrate over high-energy degrees of freedom, in the spirit of a renormalization group procedure (as in the Wilson method). We consider that the free electron energy spectrum is bound between $[-D, D]$, where D, the cutoff, is a large scale compared to the energies involved in the scattering process. We consider a narrow region of width δD below the cutoff and want to integrate the energy modes in the interval $[D - \delta D, D]$ (and $[-D, -D + \delta D]$). Since the region is narrow, when converting integrals over momentum to integrals over energies, we take the density of states $\rho(E)$ to be a constant ρ.

The dominant terms to be considered are of the type represented in Fig. 22.6 and can be written as including the various spin components. We are considering a scattering from an electron with momentum $\mathbf{k}$ and spin component α to an electron with momentum $\mathbf{k}'$ and spin component β. In the intermediate step, an electron, or hole, with momentum $\mathbf{k}''$ and spin component λ is involved. This leads to the terms, where all repeated indices are summed over:

$$\left(c^{\dagger}_{k',\beta}\sigma^{n}_{\beta\lambda}c_{k'',\lambda}S^{n}_{\sigma'\sigma''}\right)\left(c^{\dagger}_{k'',\lambda}\sigma^{m}_{\lambda\alpha}c_{k,\alpha}S^{m}_{\sigma''\sigma}\right)\frac{1}{E - E_{k''}};$$

$$\left(c^{\dagger}_{k'',\lambda}\sigma^{n}_{\lambda\alpha}c_{k,\alpha}S^{n}_{\sigma'\sigma''}\right)\left(c^{\dagger}_{k',\beta}\sigma^{m}_{\beta\lambda}c_{k'',\lambda}S^{m}_{\sigma''\sigma}\right)\frac{1}{E - (E_k + E_{k'} - E_{k''})}. \tag{22.133}$$

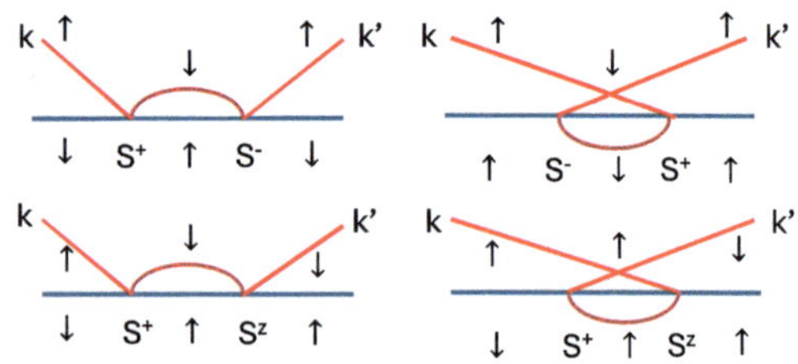

Fig. 22.6 Dominant terms that contribute to the scattering of the conduction electrons with the impurity spin, **S**. The electrons are represented in red and in blue are the states of the impurity spin

The first term involves a particle in the intermediate step, and it is understood that the integration is in the interval $[D - \delta D, D]$, and the second term involves a hole, and it is understood that the integration is carried out in the interval $[-D, -D+\delta D]$. Adding the two contributions, summing over the momenta of the intermediate states, converting the sums into integrations over energy, and recalling that the density of states is taken as a constant in the small integral range, we get for the matrix element of the T matrix

$$T \sim J_0^2 \rho \delta D \frac{1}{E - D} \left(\sigma^n \sigma^m\right)_{\beta\alpha} \left(S^n S^m\right)_{\sigma'\sigma} - J_0^2 \rho \delta D \frac{1}{E - D} \left(\sigma^m \sigma^n\right)_{\beta\alpha} \left(S^n S^m\right)_{\sigma'\sigma}$$

$$\sim -J_0^2 \rho \frac{\delta D}{D} [\sigma^n, \sigma^m]_{\beta\alpha} \left(S^n S^m\right)_{\sigma'\sigma} . \tag{22.134}$$

Summing over the components we get that

$$-\sum_{n,m} [\sigma^n, \sigma_m] S^n S^m = -2i \sum_l \sigma_l \left(\mathbf{S} \times \mathbf{S}\right)_l = 2\boldsymbol{\sigma} \cdot \mathbf{S}. \tag{22.135}$$

Adding the first-order scattering term, we get that the matrix elements of the T matrix lead to an effective interaction, to second order:

$$J(D') = J(D) + 2J^2 \rho \frac{\delta D}{D}, \tag{22.136}$$

where we simplified the notation: $J_0 = J$. D' is the new cutoff, which results from integrating over the width δD. Since the second term is positive, this means that, as a result of the integration over the high-energy degrees of freedom, the coupling constant increases. Repeating the procedure, the coupling constant keeps increasing suggesting that it will grow indefinitely. This means that the weak coupling will not be the stable solution, but instead a solution where the coupling grows to infinity.

An alternative way to reach the same conclusion is to obtain a scaling equation. Defining a coupling $g = \rho J$ we can see that it satisfies the equation

$$\frac{\partial g}{\partial \ln D} = \beta(g) = -2g^2 + O(g^3). \tag{22.137}$$

It carries the same type of information: if the cutoff decreases, the coupling increases and, therefore, the derivative on the left-hand side is negative. Therefore, $\beta(g) < 0$. The solution to this second-order scaling function is

$$g(D') = \frac{g(D)}{1 - 2g(D) \ln \frac{D}{D'}}. \tag{22.138}$$

It is illuminating to introduce a temperature scale as

$$T_K = De^{-\frac{1}{2g(D)}}, \tag{22.139}$$

called the Kondo temperature. The solution can then be written in terms of this scale as

$$2g(D') = \frac{1}{\ln \frac{D'}{T_k}}. \tag{22.140}$$

First, note that the dependence on D is absorbed in the definition of this scale. This is a symptom of the universality of this phenomenon that states that the solution of the scaling equation is independent of the details of the starting point of the renormalization procedure. While $D' \gg T_K$, the coupling is small. As the cutoff D' approaches T_K, which sets the scale of the problem, the coupling grows. This is the result that Kondo obtained, when it was identified that there is a scale at which the perturbation theory diverges. At this order, the scaling equation gives the same result, since the coupling diverges at the scale of the Kondo temperature. One has to include higher-order terms in the perturbation series of the Lippmann-Schwinger equation to improve the result. At the end, one needs to consider a numerical renormalization procedure as carried out by Wilson, or to use the Bethe ansatz to obtain a proper analytical solution. The full solution shows that there are two fixed points: one at $g = 0$, which is a repulsive fixed point, and another at $g \to \infty$, which is an attractive fixed point. It is therefore interesting to consider the infinite coupling limit.

22.5.2 *Infinite Coupling*

Consider a half-chain with an impurity situated at site 0 (the origin) with a spin **S**. Since we are considering only $l = 0$ states, the radial coordinate, r (distance to the impurity site), is enough to describe the process. Since $r \geq 0$, the problem may be seen as one dimensional on a half-line with a coordinate $x > 0$. The electrons may hop between the lattice sites, with an amplitude t. The Hamiltonian is given by

$$H = -t \sum_{j=0,\infty} \left(c_\sigma^\dagger(j+1)c_\sigma(j) + c_\sigma^\dagger(j)c_\sigma(j+1) \right) + Jc_\alpha^\dagger(0)\boldsymbol{\sigma}_{\alpha,\beta}c_\beta(0) \cdot \mathbf{S}. \tag{22.141}$$

The electrons at low energies (or in the regime of low temperatures) will form a singlet state with the impurity spin. If indeed $J \to \infty$, it is favorable energetically to form a state

$$|GS\rangle = \frac{1}{\sqrt{2}} \left(| \uparrow_S \downarrow_c \rangle - | \downarrow_S \uparrow_c \rangle \right), \tag{22.142}$$

where the label S refers to the impurity and the label c to the electrons. Then if $J \gg t$, we may consider the hopping as a perturbation.

The triplet state increases the energy by J with respect to the groundstate and, therefore, the impurity site is effectively out of the way, except for the possibility of virtual processes. The problem reduces to a system of free electrons with one less site available. The effective Hamiltonian is

$$H_{eff} = -t \sum_{j=1,\infty} [c_\sigma^\dagger(j+1)c_\sigma(j) + h.c.] + (\text{weak interaction}), \tag{22.143}$$

where the weak interaction results from the virtual excitations of the electrons, as they jump to the impurity site. Considering one electron far from the impurity, its wave function is of the type $\psi(x) \sim \sin(k_F x)$, where k_F is the Fermi momentum. The exclusion of the electrons from the site of the impurity (site 0) has the effect of introducing a phase shift of one lattice site (the lattice constant is a). We get then that the new wave function is given by $\psi(x) \sim \sin(k_F x - \delta)$, where $\delta = k_F a$. Now we know that at half-filling $k_F = \pi/(2a)$ (recall that U is large and electron densities larger than half-filling are unfavorable energetically). This implies that the phase shift is given by $\delta = \pi/2$. Using the Friedel sum rule implies that the number of bound electrons at the impurity is

$$\sum_\sigma \frac{\delta_{\sigma=\pm 1}}{\pi} = \frac{2\delta}{\pi} = 1. \tag{22.144}$$

22.6 RKKY Interaction

Consider now that there are several impurities diluted in the metallic matrix. Due to the conduction electrons and the Kondo interaction, there will appear an effective interaction between the local spins of the impurities. This is called RKKY interaction (Ruderman, Kittel, Kasuya, Yosida).

Consider two localized impurities at sites $\mathbf{R}_1$ and $\mathbf{R}_2$. The Hamiltonian is then given by

$$\begin{aligned} H &= \sum_{\mathbf{k},\sigma} \epsilon_\mathbf{k} c_{\mathbf{k},\sigma}^\dagger c_{\mathbf{k},\sigma} \\ &+ J \sum_{\mathbf{k},\mathbf{k}'} \sum_{\sigma,\sigma'} c_{\mathbf{k},\sigma}^\dagger \mathbf{s}_{\sigma,\sigma'} c_{\mathbf{k}',\sigma'} \cdot \left(\mathbf{S}_1 e^{i(\mathbf{k}-\mathbf{k}')\cdot \mathbf{R}_1} + \mathbf{S}_2 e^{i(\mathbf{k}-\mathbf{k}')\cdot \mathbf{R}_2} \right). \end{aligned} \tag{22.145}$$

In second-order perturbation theory, there appears a term that couples the spin operators of the two impurities. Let us consider that term that will be a function of $\mathbf{R}_1$ and $\mathbf{R}_2$, and let us integrate over the conduction electrons:

$$H''(\mathbf{R}_1 - \mathbf{R}_2) = \sum_{\mathbf{k},\mathbf{k}';\sigma,\sigma'}^{\prime} \frac{\langle \mathbf{k}, \sigma | H | \mathbf{k}', \sigma' \rangle \langle \mathbf{k}', \sigma' | H | \mathbf{k}, \sigma \rangle}{\epsilon_{\mathbf{k}} - \epsilon_{\mathbf{k}'}}, \tag{22.146}$$

where we exclude $\epsilon_{\mathbf{k}} \neq \epsilon_{\mathbf{k}'}$. The matrix element $\langle \mathbf{k}', \sigma' | H | \mathbf{k}, \sigma \rangle$ is given by

$$\begin{aligned} &= J \sum_{\mathbf{k}_1,\mathbf{k}_2;\sigma_1,\sigma_2} \langle \mathbf{k}', \sigma' | \left(\mathbf{S}_1 e^{i(\mathbf{k}-\mathbf{k}')\cdot\mathbf{R}_1} + \mathbf{S}_2 e^{i(\mathbf{k}-\mathbf{k}')\cdot\mathbf{R}_2} \right) \left(c^{\dagger}_{\mathbf{k}_1,\sigma_1} \mathbf{s}_{\sigma_1,\sigma_2} c_{\mathbf{k}_2,\sigma_2} \right) | \mathbf{k}, \sigma \rangle \\ &= J \sum_{\mathbf{k}_1,\mathbf{k}_2;\sigma_1,\sigma_2} \left(\mathbf{S}_1 e^{i(\mathbf{k}-\mathbf{k}')\cdot\mathbf{R}_1} + \mathbf{S}_2 e^{i(\mathbf{k}-\mathbf{k}')\cdot\mathbf{R}_2} \right) \cdot \langle \mathbf{k}', \sigma' | c^{\dagger}_{\mathbf{k}_1,\sigma_1} \mathbf{s}_{\sigma_1,\sigma_2} c_{\mathbf{k}_2,\sigma_2} | \mathbf{k}, \sigma \rangle. \end{aligned} \tag{22.147}$$

The matrix element of the conduction electron spin density is given by

$$\mathbf{s}_{\sigma_1,\sigma_2} \delta_{\mathbf{k}_2,\mathbf{k}} \delta_{\sigma_2,\sigma} \delta_{\mathbf{k},\mathbf{k}'} \delta_{\sigma_1,\sigma'}. \tag{22.148}$$

Then the matrix element gives

$$\langle \mathbf{k}', \sigma' | H | \mathbf{k}, \sigma \rangle = J \mathbf{s}_{\sigma',\sigma} \cdot \left(\mathbf{S}_1 e^{i(\mathbf{k}-\mathbf{k}')\cdot\mathbf{R}_1} + \mathbf{S}_2 e^{i(\mathbf{k}-\mathbf{k}')\cdot\mathbf{R}_2} \right). \tag{22.149}$$

Using that $\epsilon_{\mathbf{k}} = (k^2)/(2m^*)$, we obtain that

$$\begin{aligned} H''(\mathbf{R}_1 - \mathbf{R}_2) = 2m^* J^2 \sum_{\mathbf{k},\mathbf{k}';\sigma,\sigma'}^{\prime} & \frac{1}{k^2 - (k')^2} \{ (\mathbf{S}_1 \cdot \mathbf{s}_{\sigma',\sigma})(S_1 \cdot \mathbf{s}_{\sigma,\sigma'}) \\ & + (\mathbf{S}_2 \cdot \mathbf{s}_{\sigma',\sigma})(S_2 \cdot \mathbf{s}_{\sigma,\sigma'}) \\ & + (\mathbf{S}_1 \cdot \mathbf{s}_{\sigma',\sigma})(S_2 \cdot \mathbf{s}_{\sigma,\sigma'}) e^{i(\mathbf{k}'-\mathbf{k})\cdot(\mathbf{R}_1-\mathbf{R}_2)} \\ & + (\mathbf{S}_2 \cdot \mathbf{s}_{\sigma',\sigma})(S_1 \cdot \mathbf{s}_{\sigma,\sigma'}) e^{-i(\mathbf{k}'-\mathbf{k})\cdot(\mathbf{R}_1-\mathbf{R}_2)} \}. \end{aligned} \tag{22.150}$$

Define now $\mathbf{x} = \mathbf{R}_1 - \mathbf{R}_2$, and let us convert the summation to an integral over the momentum. Then using that

$$Tr\{(\mathbf{S}_1 \cdot \mathbf{s})(\mathbf{S}_2 \cdot \mathbf{s})\} = \frac{1}{2} \mathbf{S}_1 \cdot \mathbf{S}_2, \tag{22.151}$$

the term that leads to the interaction between the spins $\mathbf{S}_1$ and $\mathbf{S}_2$ is given by

$$[H''(\mathbf{x}) = m^* J^2 \mathbf{S}_1 \cdot \mathbf{S}_2 \frac{V^2}{(2\pi)^6} P \left[\int_0^{k_F} d^3k \int_{k_F}^{\infty} d^3k' \frac{e^{-i\mathbf{x}\cdot(\mathbf{k}-\mathbf{k}')}}{k^2 - (k')^2} \right] + c.c. \tag{22.152}$$

where P is the principal part of the integral. The electron is at the Fermi sphere and is scattered to an accessible state out of the Fermi sphere ($T = 0$). The effective Hamiltonian can be obtained as

$$H''(\mathbf{x}) = m^* J^2 \mathbf{S}_1 \cdot \mathbf{S}_2 \frac{V^2}{(2\pi)^6} P \left[\int_0^{k_F} d^3k e^{-i\mathbf{k}\cdot\mathbf{x}} \int_{k_F}^{\infty} d^3k' \frac{e^{i\mathbf{k}'\cdot\mathbf{x}}}{k^2 - (k')^2} \right] + c.c. \tag{22.153}$$

Using that (approximating $k_F \sim 0$)

$$P \left[\int_0^{\infty} d^3k' \frac{e^{i\mathbf{k}'\cdot\mathbf{x}}}{k^2 - (k')^2} \right] = -\frac{2\pi}{x} \cos kx, \tag{22.154}$$

then we obtain that

$$\int_0^{k_F} d^3k e^{-i\mathbf{k}\cdot\mathbf{x}} \left(\frac{-2\pi^2}{x} \right) \cos kx = \frac{(2\pi)^3}{x^4} \left(\frac{k_F x}{2} \cos 2k_F x - \frac{1}{4} \sin 2k_F x \right). \tag{22.155}$$

Finally, we obtain

$$H_{\text{RKKY}}(\mathbf{x}) = 4m^* J^2 \mathbf{S}_1 \cdot \mathbf{S}_2 \frac{V^2}{(2\pi)^3} k_F^4 F(2k_F x), \tag{22.156}$$

where the function

$$F(x) = \frac{x \cos x - \sin x}{x^4}. \tag{22.157}$$

An interaction between the spins $\mathbf{S}_1$ and $\mathbf{S}_2$ emerges that is proportional to J^2, is antiferromagnetic, but is oscillatory of period $2k_F$, and decays fast, as x^{-3}, as shown in Fig. 22.7. In general it decays as x^{-d}, where d is the space dimensionality.

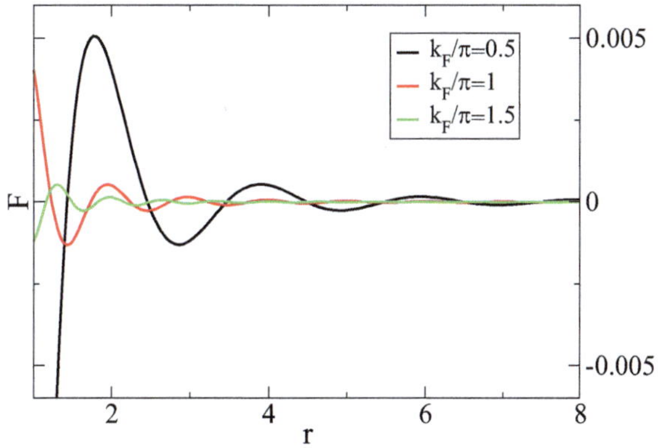

Fig. 22.7 Plot of the function Eq. (22.157) for $F(2k_F r)$, as a function of distance r, for different Fermi wavevectors

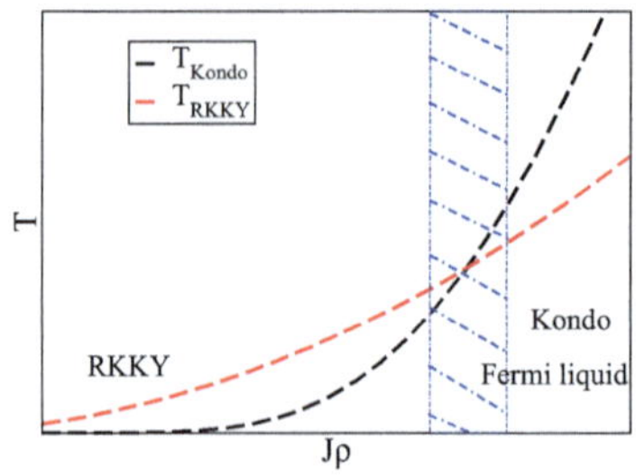

Fig. 22.8 Schematic form of the Doniach phase diagram. At larger values of $J\rho$ the Kondo effect dominates since the Kondo temperature is higher than the ferromagnetic critical temperature that originates from the RKKY coupling between the impurity spins. The Kondo effect gives origin to spin singlets and one may expect some sort of Fermi liquid behavior. At lower values of $J\rho$ the RKKY coupling dominates, the spins may order, and the Kondo effect gets preempted

Doniach has proposed a qualitative diagram of these materials when there is a finite concentration of magnetic impurities. The RKKY coupling constant is of the type $J_{\text{RKKY}} \sim J_{Kondo}^2$ and the Kondo temperature is $T_K \sim D_0 e^{-\frac{1}{J_{Kondo}\rho}}$. For small values of the Kondo coupling, the RKKY interaction dominates and antiferromagnetism is the prevailing order. As the Kondo coupling grows the Kondo effect dominates and there appear spin singlets at each impurity site. At intermediate values of the Kondo coupling, there is a transition between the two regimes. If the Kondo effect dominates, then fluctuations may appear of the type SDW that are the result of instabilities of the Fermi surface. In the other case the RKKY dominates and there is magnetic order if the impurities are regularly distributed. The phase diagram is shown schematically in Fig. 22.8. If they are randomly distributed and the RKKY is dominant, then probably the system will become a spin glass, which is a disordered phase.

22.7 Anderson Lattice

22.7.1 Infinite U Approximation

Consider now a generalization of the Anderson model to a lattice, such that there are now "impurities" at all lattice sites. This occurs in a compound material with two types of atoms such that one provides the s electrons and the other the f electrons, with both sets of electrons near the Fermi energy. We start by considering that U is infinite.

We implement the condition $U = \infty$ within the slave boson formulation in which the empty f-site is represented by a slave boson b_i and the physical operator f_i is replaced with ${b^\dagger}_i f_i$. Condensation of the slave bosons can be described by the replacement $b_i \to \langle b_i \rangle = \langle {b^\dagger}_i \rangle = \sqrt{z}$. We write down the effective Hamiltonian as

$$H_{eff} = \sum_{\mathbf{k},m} \left((\epsilon_{\mathbf{k}} - \mu) c^{\dagger}_{\mathbf{k},m} c_{\mathbf{k},m} + (\epsilon_f - \mu) f^{\dagger}_{\mathbf{k},m} f_{\mathbf{k},m} \right)$$

$$+ \sqrt{z} V \sum_{\mathbf{k},m} (f^{\dagger}_{\mathbf{k},m} c_{\mathbf{k},m} + c^{\dagger}_{\mathbf{k},m} f_{\mathbf{k},m})$$

$$+ (\epsilon_f - \epsilon_0)(z - 1) N_s \,, \tag{22.158}$$

where N_s denotes the number of lattice sites and ϵ_f is the renormalized energy of the f orbitals due to the onsite repulsion. Here the index m encompasses the orbital and the spin degrees of freedom.

In a simplified way, calling $\tilde{V} = \sqrt{z} V$ and ignoring for now the restriction and the internal degrees of freedom (orbital and spin angular momenta), this leads to the Hamiltonian:

$$H = \sum_{\mathbf{k}} \left(\epsilon(\mathbf{k}) c^{\dagger}_{\mathbf{k}} c_{\mathbf{k}} + \epsilon f^{\dagger}_{\mathbf{k}} f_{\mathbf{k}} + \tilde{V} \left(f^{\dagger}_{\mathbf{k}} c_{\mathbf{k}} + c^{\dagger}_{\mathbf{k}} f_{\mathbf{k}} \right) \right) , \tag{22.159}$$

where the narrow band is taken to be flat, with an energy independent of momentum and zero group velocity. The total number of f fermions is

$$N_f = \sum_{\mathbf{k}} f^{\dagger}_{\mathbf{k}} f_{\mathbf{k}} \tag{22.160}$$

and the total number of conduction electrons is

$$N_c = \sum_{\mathbf{k}} c^{\dagger}_{\mathbf{k}} c_{\mathbf{k}} . \tag{22.161}$$

The number of f or c fermions is not conserved, due to the hybridization term that transfers fermions between the f and c bands. Indeed,

$$[H, N_f] = \sum_{\mathbf{k}} \tilde{V} \left(-f^{\dagger}_{\mathbf{k}} c_{\mathbf{k}} + c^{\dagger}_{\mathbf{k}} f_{\mathbf{k}} \right) \tag{22.162}$$

and

$$[H, N_c] = \sum_{\mathbf{k}} \tilde{V} \left(f^{\dagger}_{\mathbf{k}} c_{\mathbf{k}} - c^{\dagger}_{\mathbf{k}} f_{\mathbf{k}} \right) , \tag{22.163}$$

but the total number of fermions is conserved $[H, (N_f + N_c)] = 0$.

The equations of motions for the operators are obtained from

$$i\hbar \frac{\partial c_{\mathbf{k}}}{\partial t} = [c_{\mathbf{k}}, H] = \epsilon(\mathbf{k}) c_{\mathbf{k}} + \tilde{V} f_{\mathbf{k}} \tag{22.164}$$

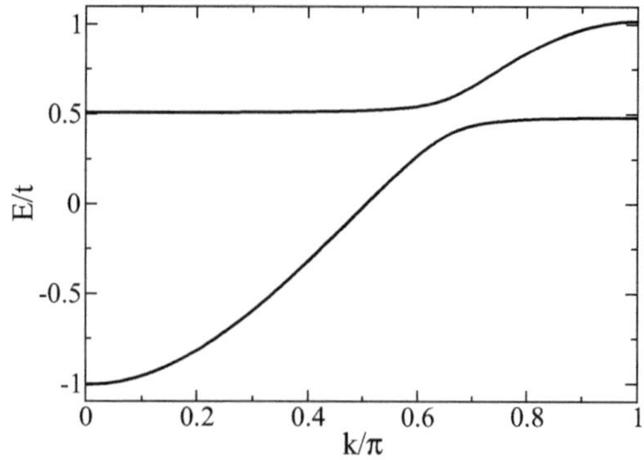

Fig. 22.9 Energy spectrum, Eq. (22.166), for a one-dimensional tight-binding system (to simplify) with hopping $t = 1$, hybridization $V = 0.1$, and $\epsilon = 0.5$. The gap is the consequence of the hybridization between the two bands. Note that the curvature of the bands is small around the gap, leading to high effective masses

and

$$i\hbar \frac{\partial f_{\mathbf{k}}}{\partial t} = \epsilon f_{\mathbf{k}} + \tilde{V} c_{\mathbf{k}}. \tag{22.165}$$

These equations may be solved to give a dispersion relation:

$$E^{\pm}(\mathbf{k}) = \frac{1}{2}\left(\epsilon(\mathbf{k}) + \epsilon \pm \sqrt{(\epsilon(\mathbf{k}) - \epsilon)^2 + 4\tilde{V}^2}\right). \tag{22.166}$$

The spectrum shows a gap, called the hybridization gap, together with bands with small curvature near the gap, that leads to high effective masses and hence the designation of heavy-fermion systems. An example is shown in Fig. 22.9.

In a more careful treatment we need to determine the density of the boson condensate z,which minimizes the free energy of the system, and to determine ϵ_f. This is obtained after imposing local particle (boson+fermion) conservation at the f-sites, in number N_s:

$$z = 1 - \frac{1}{N_s} \sum_{\mathbf{k},m} \langle f^{\dagger}_{\mathbf{k},m} f_{\mathbf{k},m} \rangle , \tag{22.167}$$

$$\epsilon_f - \epsilon_0 = -\frac{V}{2\sqrt{z} N_s} \sum_{\mathbf{k},m} \left(\langle f^{\dagger}_{\mathbf{k},m} c_{\mathbf{k},m} \rangle + \langle c^{\dagger}_{\mathbf{k},m} f_{\mathbf{k},m} \rangle \right). \tag{22.168}$$

Equation (22.167) states that the mean number of electrons at an f-site is $1 - z$. These equations are solved self-consistently and require in general a numerical treatment. It involves the diagonalization of the Hamiltonian, at each iterative step, which allows the calculation of the average values in Eqs. (22.167, 22.168).

22.7.2 Finite U

Consider now the Anderson lattice with a large, but finite value of the local Coulomb repulsion U, at the f sites. The method differs from the infinite U slave boson method since now all local states are allowed. Again, to simplify, we only consider the spin degrees of freedom at the impurities.

There is a Coulomb repulsion U if two f-electrons are located at the same site. We associate four bosons to the various states f-electrons can occupy [3]. The bosons e, d are associated with empty and doubly occupied sites, respectively, and the bosons p_σ with a singly occupied site with spin component σ. There is an enlargement of the Hilbert space and restrictions must be implemented at each site:

$$e_i^\dagger e_i + d_i^\dagger d_i + p_{i,\uparrow}^\dagger p_{i,\uparrow} + p_{i,\downarrow}^\dagger p_{i,\downarrow} = 1$$
$$f_{i,\sigma}^\dagger f_{i,\sigma} = p_{i,\sigma}^\dagger p_{i,\sigma} + d_i^\dagger d_i, \tag{22.169}$$

through Lagrange multipliers λ_i and $\lambda_{i,\sigma}$, respectively. In the physical subspace, the operators $f_{i,\sigma}$ are replaced by $f_{i,\sigma} Z_{i,\sigma}$ where

$$\begin{aligned} Z_{i,\sigma} &= (1 - d_i^\dagger d_{i,\sigma} - p_{i,\sigma}^\dagger p_{i,\sigma})^{-1/2}(e_i^\dagger p_{i,\sigma} + p_{i,-\sigma} d_i) \\ &\times (1 - e_i^\dagger e_{i,\sigma} - p_{i,-\sigma}^\dagger p_{i,-\sigma})^{-1/2}. \end{aligned} \tag{22.170}$$

The usual procedure consists in taking a mean-field approach where we assume the slave bosons to be condensed. It turns out that the mean-field solution shows the possibility of a paramagnetic or a ferromagnetic phase. In the cases of paramagnetic or ferromagnetic solutions, we take $Z_{i,\sigma}^\dagger = Z_{i,\sigma} = Z_\sigma$, $e_i^\dagger = e_i = e$, $d_i^\dagger = d_i = d$, $p_{i,\sigma}^\dagger = p_{i,\sigma} = p_\sigma$. The paramagnetic solution is described by $p_\uparrow = p_\downarrow = p$. As a consequence $\lambda_\uparrow = \lambda_\downarrow = \bar{\lambda}$ and $Z_\uparrow = Z_\downarrow = Z$.

The Hamiltonian is given by

$$\begin{aligned} H &= \sum_{k,\sigma} \epsilon_k c_{k,\sigma}^\dagger c_{k,\sigma} + \sum_{k,\sigma} \epsilon_f f_{k,\sigma}^\dagger f_{k,\sigma} + U \sum_i d_i^\dagger d_i \\ &+V \sum_{i,\sigma} \left(c_{i,\sigma}^\dagger f_{i,\sigma} Z_{i,\sigma} + Z_{i,\sigma}^\dagger f_{i,\sigma}^\dagger c_{i,\sigma} \right) \\ &+ \sum_i \lambda_i \left(e_i^\dagger e_i + d_i^\dagger d_i + p_{i,\uparrow}^\dagger p_{i,\uparrow} + p_{i,\downarrow}^\dagger p_{i,\downarrow} - 1 \right) \\ &+ \sum_{i,\sigma} \lambda_{i,\sigma} \left(f_{i,\sigma}^\dagger f_{i,\sigma} - p_{i,\sigma}^\dagger p_{i,\sigma} - d_i^\dagger d_i \right). \end{aligned} \tag{22.171}$$

In the mean-field approximation the Anderson Hamiltonian can be simplified to

$$\begin{aligned} H_{MF} &= \sum_{k,\sigma}(\epsilon_k - \mu)c^{\dagger}_{k,\sigma}c_{k,\sigma} + \sum_{k,\sigma}(\tilde{\epsilon}_f - \mu)f^{\dagger}_{k,\sigma}f_{k,\sigma} \\ &+ V\sum_{k,\sigma} Z\left(c^{\dagger}_{k,\sigma}f_{k,\sigma} + f^{\dagger}_{k,\sigma}c_{k,\sigma}\right) \\ &+ UN_s d^2 + N_s\lambda\left(e^2 + d^2 + 2p^2 - 1\right) \\ &- N_s\sum_{\sigma}\bar{\lambda}(p^2 + d^2), \end{aligned} \tag{22.172}$$

where $\epsilon_k = -2t\left(\cos k_x + \cos k_y\right)$ in two dimensions, $\tilde{\epsilon}_f = \epsilon_f + \bar{\lambda}$, and we have added the chemical potential μ. The procedure now consists of determining the mean-field values of the four condensed bosons, using the Helmann-Feynmann theorem, by taking derivatives of the Hamiltonian with respect to these parameters. These derivatives lead to mean-field equations that have to be solved self-consistently. Defining

$$F_{V,b} = V\frac{\partial Z}{\partial b}\sum_{k,\sigma}\left(\langle c^{\dagger}_{k,\sigma}f_{k,\sigma}\rangle + \langle f^{\dagger}_{k,\sigma}c_{k,\sigma}\rangle\right), \tag{22.173}$$

where $b = p, d, e$, this leads to

$$F_{V,p} + 2N_s p(\lambda - \bar{\lambda}) = 0 \tag{22.174}$$

$$F_{V,d} + 2N_s d\left(U + \lambda - 2\bar{\lambda}\right) = 0 \tag{22.175}$$

$$F_{V,e} + 2N_s e\lambda = 0, \tag{22.176}$$

together with the restrictions $N_s(e^2 + d^2 + p^2_{\uparrow} + p^2_{\downarrow} - 1) = 0$ and $\sum_k\langle f^{\dagger}_{k,\sigma}f_{k,\sigma}\rangle - N_s(p^2_{\sigma} + d^2) = 0$. In general, the equations have to be solved numerically.

22.8 Fractionalization and Gauge Invariance

At this point some comments may be useful: (i) in strongly correlated systems it is convenient to fractionalize the fermionic operators in terms of auxiliar fields, possibly nonphysical. The description of the systems is carried out more conveniently in terms of these fictitious fields and then a translation to the physical operators is required. In order for the procedure to make sense, the results must be independent of the choice of factorization. There are an infinite number of ways to represent the physical operators, some examples of which were presented in the previous and present chapters. (ii) In general, the problem is not exactly soluble and, therefore, some approximations are usually required. It is a well-known result that the choice of different fractionalizations associated with different approximations leads to different results. (iii) The need and convenience to fractionalize the physical operators occurs independently of spatial dimension. (iv) While we expect that in general these fractionalized degrees of freedom will not be observable, it turns out that in some one-dimensional systems their traces do seem to be detected in experiments. In higher dimensions, most likely these fractionalized degrees of freedom are confined, in such a way that their nature is not revealed in experiments, but there is the possibility of some exotic situations in which the fractionalized excitations remain deconfined. (v) It is easy to see that the introduction of slave bosons (or slave fermions) carries a gauge freedom. Both in the case of the $t - J$ model and in the Anderson model, the $U(1)$ gauge transformation

$$\begin{aligned} b_i &\to b_i e^{i\theta} \\ f_i &\to f_i e^{i\theta} \end{aligned} \tag{22.177}$$

leaves the Hamiltonian invariant. (vi) The most common approximation is a mean-field approximation (Hartree-Fock) in which pairs of operators are replaced by their average values, such that the Hamiltonian is reduced to a quadratic form, which may be diagonalized by a Bogoliubov-Valatin transformation. In the case of the Anderson model for a single impurity, this method gives good results and the agreement with exact results is quite good, particularly if we include fluctuations around the mean field. Such procedure may be implemented using a path integral formalism, but attention to the zero energy modes that result from the gauge invariance must be dealt with. However, these modes may be eliminated by a convenient gauge transformation [4].

Having in mind previous results that antiferromagnetically spins should form bonds with neighboring spins, it is usual to define correlations of the type

$$\chi_{ij} = \langle f_{i\sigma}^\dagger f_{j\sigma} \rangle. \tag{22.178}$$

Taking as example the $t - J$ model, using mean-field theory (in which we may also include superconducting correlations) and integrating exactly the fermionic and bosonic fields (which in this approximation are quadratic), we obtain an effective theory for the correlations χ_{ij}. The solutions of the mean-field equations are in infinite number and, therefore, it is necessary to choose a class of solutions. The simplest choice is reached considering that the correlation is uniform. There are however other interesting solutions and, in particular, the choice $\chi_{ij} = \chi e^{i\theta_{ij}}$ leads to phases that may be energetically competitive. As said above, the local gauge transformation leaves the Hamiltonian invariant:

$$\begin{aligned} f_{i\sigma}^\dagger &\to f_{i\sigma}^\dagger e^{i\varphi_i} \\ \chi_{ij} &\to \chi_{ij} e^{i\varphi_i - i\varphi_j}. \end{aligned} \tag{22.179}$$

Since the system should be invariant to the gauge transformation, the correlations with physical meaning must be gauge invariant. By similarity with electromagnetism and interpreting the phases of the correlation $\chi_{ij} = e^{iA_{ij}}$ as a vector potential defined on the lattice links, the quantity with physical meaning is the flux through a closed path (which will be equivalent to the magnetic field). Defining a cell (plaquette), the physical meaningful quantity is the flux through a plaquette defined by $\theta_{ij}+\theta_{jk}+\theta_{kl}+\theta_{li}$. In this way a connection between the system and a description in terms of a lattice gauge theory is established. This topic is beyond the level of this book and so is left for further study.

22.9 Exercises

1. Solve numerically the mean-field equations Eqs. (22.85), (22.86), and (22.87). The equations have to be solved self-consistently for the various average values[8] as a function of temperature and the parameters E_d, μ. Determine, z, ϵ_d, n.
2. (a) Show that the choice Eq. (22.110) for the generator S satisfies Eq. (22.108).
   (b) Show that the Kondo coupling is antiferromagnetic, that is, show Eqs. (22.116) and (22.121).
3. Consider a Hamiltonian that combines a Kondo interaction with impurity spins located at points $\mathbf{R}_i$, in a $3d$ lattice, and a RKKY interaction between the impurity spins

$$
\begin{aligned}
H = & \sum_{\mathbf{k},\sigma} \epsilon_{\mathbf{k}} c^{\dagger}_{\mathbf{k},\sigma} c_{\mathbf{k},\sigma} \\
& + J \sum_{\mathbf{k},\mathbf{k}'} \sum_{\sigma,\sigma'} \sum_{i} \left(c^{\dagger}_{\mathbf{k},\sigma} \mathbf{s}_{\sigma,\sigma'} c_{\mathbf{k}',\sigma'} \right) \cdot \mathbf{S}_i e^{i(\mathbf{k}-\mathbf{k}')\cdot \mathbf{R}_i} \\
& + \eta J^2 \sum_{i,j} \mathbf{S}_i \cdot \mathbf{S}_j F\left(2k_F |\mathbf{R}_i - \mathbf{R}_j|\right),
\end{aligned} \tag{22.180}
$$

where the function

$$
F(x) = \frac{x \cos x - \sin x}{x^4}. \tag{22.181}
$$

Considering a parabolic dispersion relation or a tight-binding model, use mean-field theory to establish a phase diagram (like Doniach) as a function of the parameters J, η, k_F.

[8] The solution requires finding the zeros of coupled nonlinear equations. One possible way is to use the subroutine hybrd.f, or equivalent subroutines in other languages.

4. Consider a fermionic system with two types of states: (i) f states (heavy) with energy ϵ and (ii) conduction states (light) with dispersion relation $\epsilon_k = k^2/2m$.
   Assume the following Hamiltonian:

$$H = \sum_{\vec{k}} \left(\epsilon f_{\vec{k}}^{\dagger} f_{\vec{k}} + \epsilon_{\vec{k}} c_{\vec{k}}^{\dagger} c_{\vec{k}} + V (f_{\vec{k}}^{\dagger} c_{\vec{k}} + c_{\vec{k}}^{\dagger} f_{\vec{k}}) \right),$$

   where $f_{\vec{k}}$ and $c_{\vec{k}}$ are the respective annihilation operators and V is the hybridization that allows the transfer of fermions between bands. You may consider first spinless fermions and also discuss the realistic case of spinful fermions.

   (a) Show that the Hamiltonian conserves the total fermion number but not the occupation numbers of each band separately (show they are constants of motion).
   (b) Determine the equations of motion of the operators $c_{\vec{k}}$ and $f_{\vec{k}}$.
   (c) Diagonalize the Hamiltonian and show that $E_k = \frac{1}{2}\left(\epsilon_k + \epsilon \pm \sqrt{(\epsilon_k - \epsilon)^2 + 4V^2}\right)$.
   (d) Make a schematic plot of the energy spectrum as a function of momentum k for $\epsilon = 0$. Assume $k < \pi/a$, where a is the lattice constant. Is there a gap in the spectrum?
   (e) If we assume we have exactly one electron per site (recall that there is no spin), where is the Fermi level located? What type of result do you expect for the electric conductivity?

$$i\hbar \frac{dA_H(t)}{dt} = [A_H(t), H(t)] + i\hbar \frac{\partial A_H}{\partial t}$$

$$[A, BC] = [A, B]C - B[C, A] = \{A, B\}C - B\{C, A\}.$$

5. Solve numerically the mean-field self-consistent equations Eq. (22.176) of the Hamiltonian equation (22.171).

   (a) Look for paramagnetic and ferromagnetic solutions as a function of t, V, ϵ_f, μ, T and determine the mean-field parameters.
   (b) Determine the magnetization as a function of U/V, for vanishing temperature and different values of the chemical potential.
   (c) Determine the phase diagram and the various phases.
   (d) Determine the band structure in the various phases.
   (e) Study the behavior of the mean-field parameters and the magnetization as a function of temperature and determine the critical temperature(s) of the phase transition(s).

References

1. A. C. Hewson, *The Kondo Problem to Heavy Fermions*, Cambridge University Press.
2. P.W. Anderson, Journal of Physics C **3**, 2436 (1970); P. Coleman, AIP Conference Proceedings **629**, 79 (2002); [Coleman].
3. G. Kotliar and A. Ruckenstein, Physical Review Letters **57**, 1362 (1986).
4. N. Read and D.M. Newns, Journal of Physics C **16**, 3273 (1983); P. Coleman, Physical Review B **35**, 5072 (1987).

Chapter 23
Nonhomogeneous Systems: Transport and Disorder

The usual starting point to study a macroscopic system is to consider it with infinite size and homogeneous. Most cases we have considered in this book take this simplified and convenient approach. Notable exceptions have also been considered. For instance, when interested in transport properties, often we think of a finite system to which we attach some leads, to allow the flow of some current through a given material, as a result of a gradient of the corresponding intensive quantity through the material, for instance, chemical potential for the flow of particles, temperature gradient for the flow of energy, and so on. The presence of edge states in topological finite systems has also been considered.

Juxtaposing finite systems with different properties allows interesting properties that result from the heterostructures considered, such as the Josephson effect in superconductors, or interfaces of different doped semiconductors. Typically, at the interfaces the properties are new and can be explored in various ways.

The doping of a material with some foreign atoms is also interesting by itself, even in the infinite size limit. In general, perturbations that break the translational invariance lead to a change of the properties of the system, such as its excitations and transport properties. These perturbations may be due to imperfections of the system structure, such as distortions or vacancies, impurities of foreign atoms, etc. These may give origin to localized states near the imperfections, with consequences we have alluded to in previous chapters. For instance, when considering transport, the presence of these imperfections is determinant. Often these impurities are not controlled, implying that their spatial distributions are generally unknown, which means that some averages over their configurations have to be considered.

In this chapter we will revisit some previous results and elaborate some more, along the effects of both nonhomogeneous structures (which may be pure or with defects) and the presence of disorder. Some cases will be considered explicitly and it will be shown that different sets of states allow the understanding of the transport properties, as a consequence of the localized or extended nature of the system eigenstates.

P. D. Sacramento, *Introduction to Condensed Matter Theory*,
https://doi.org/10.1007/978-3-031-86251-9_23

23.1 Transport in Mesoscopic Systems

In an ohmic, resistive, conductor the current density is related to the electric field through a parameter (in general a tensor), called the conductivity, as $\mathbf{j} = \sigma\mathbf{E}$. The current is defined as the integral over a surface of the current density $I = \int_s \mathbf{j}\cdot\mathbf{n}dS$, where $\mathbf{n}$ is the normal to the surface. Standard results for a uniform current density over the surface lead to $I = jS = \sigma SV/L_w$, assuming a constant electric field over the length L_w of a wire, and V is the electrical potential applied at the ends of the wire. The conductance is defined as $I = GV$ which leads to

$$G = \sigma\frac{S}{L_w}. \tag{23.1}$$

If we consider a two-dimensional, rectangular wire, the conductance can be obtained as

$$G = \sigma\frac{W}{L_w}, \tag{23.2}$$

where W is the width of the conductor and L_w its length along the direction of the passage of current. Note that as $L_w \rightarrow 0$ the conductance seems to diverge. However, it is found that it saturates!

Defining the resistance, R, as the inverse of the conductance, and the resistivity, ρ, as the inverse of the conductivity, we recover the expression

$$R = \rho\frac{L_w}{S}. \tag{23.3}$$

Ohm's law holds for system sizes that are larger than typical microscopic lengths, such as the de Broglie wavelength of the transport carriers, but also larger than other scales, such as the average length between collisions (mean free path) of the carriers with imperfections or other carriers. If the system length is smaller than the mean free path, the motion is effectively collisionless and is called ballistic. For these system sizes, the carrier, for instance, an electron under the action of a potential difference between the edges of the conductor, essentially propagates freely. One may wonder then what is the origin of resistance in such a scenario. If there are collisions, we have seen that the steady-state equation of motion of one charge carrier with a given drift velocity, v_d, is of the type

$$\frac{mv_d}{\tau} = eE, \tag{23.4}$$

where τ is the average time between collisions, and E is the electric field. The conductivity has been obtained (Drude model) and is given by

$$\sigma = \frac{ne^2\tau}{m}, \tag{23.5}$$

where n is the electronic density and m is the electron mass. It is convenient to define a quantity which relates the drift velocity with the electric field

$$\mu = \left|\frac{v_d}{E}\right| = \frac{|e|\tau}{m}, \tag{23.6}$$

called the mobility. The conductivity is then written as $\sigma = ne\mu$. As the average time between collisions increases, the mobility grows, and therefore, we expect to find corrections to Ohm's law, if either the systems are very small or the mobility is very high, such as in very pure systems.

23.1.1 Contact Resistance of a Ballistic Conductor

Consider a one-dimensional conductor with length L_w, where the dispersion relation $\varepsilon(\mathbf{k}) = \hbar^2k^2/(2m)$. The electron group velocity is then $v = \hbar k/m$ and, using periodic boundary conditions, the momentum is quantized as $k = 2\pi n/L_w$, with $n \in \mathbb{Z}$. The number of eigenstates in the momentum range dk, or energy range dE, is

$$dn = \frac{L_w}{2\pi}dk = \frac{L_w}{2\pi}\frac{1}{\left(\frac{\partial E}{\partial k}\right)}dE = \frac{L_w}{hv}dE\,. \tag{23.7}$$

The free-electron energy dispersion is chosen since we are interested in the ballistic regime and we ignore any effect of collisions.

A current in the conductor is obtained connecting the wire to two metallic contacts at its (left, L, and right, R,) ends that are at different chemical potentials, as shown in Fig. 23.1. This results in an electric potential difference $V = (\mu_L - \mu_R)/e$ between the ends of the wire. We are going to assume that the left side chemical potential is higher than the one in the right contact. We assume that the contacts are such that have many electronic modes. Also, the conductor is almost one-dimensional if its width, W, is small. Along the direction transverse to the motion of current, only a few energy levels are occupied and one is left with the modes along the longitudinal direction. This leads to the assumption that the contacts are reflectionless, in the sense that any electron propagating along the conductor that

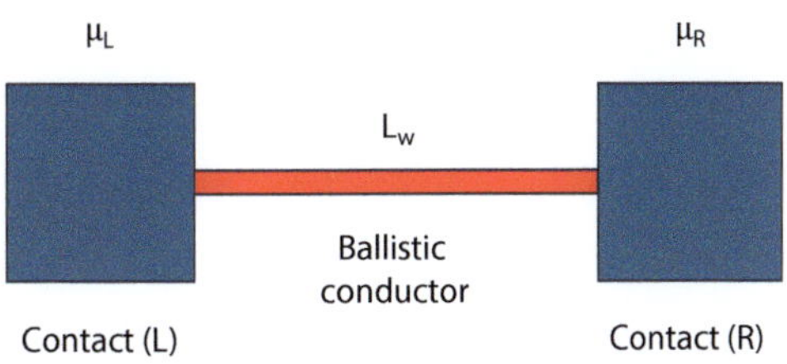

Fig. 23.1 Ballistic conductor with two contacts at its ends with two different chemical potentials

reaches a contact is not reflected and is fully transmitted to the contact. (The reverse is not true, since there are many more modes in the contacts than those available in the conductor and there is a significant reflection of waves incident from a contact to the conductor.) Since the reflection coefficient of waves from the conductor is negligible, it is clear that from the left conductor only modes with $k > 0$ enter the conductor and only modes with $k < 0$ enter the conductor from the right contact. (We are taking positive k from left to right.) Also, since there are no reflections at the interfaces with the contacts, the modes with $k > 0$ and $k < 0$ do not mix.

The current to the right is given by summing over the positive k values, taking into account the occupation of the states, given by the Fermi-Dirac distribution:

$$I_{>} = \frac{e}{L_w} \sum_k v f_R(E), \tag{23.8}$$

since in $1d$ the current density is equal to the current. Considering zero temperature, the Fermi-Dirac distribution is just a step function. The number of injected electrons with $k > 0$ in the energy range $\mu_L - \mu_R$ is then $dn = L_w(\mu_L - \mu_R)/(hv)$ and the energy current density, j, is given by

$$j = ev\frac{dn}{L_w} = \frac{e}{h}(\mu_L - \mu_R)\,. \tag{23.9}$$

In order to make contact with Ohm's law, we take into account that, for one-dimensional systems, the current density is equal to the actual current, $j = I$. On the other hand, Ohm's law can be written as $I = GV = V/R$, and V is the electric potential difference that we identify using $\mu_L - \mu_R = eV$ and obtain

$$G = \frac{e^2}{h}\,, \tag{23.10}$$

which implies that the conductance is quantized. In the above analysis, we did not consider the effect of the spin degeneracy and only consider one transverse mode. Considering M modes and taking into account the spin, we get

$$G = G_c = \frac{2e^2}{h}M. \tag{23.11}$$

This is called contact conductance since its finiteness is only associated with the effect of the interface between the ballistic conductor and the metallic contacts that inject the electrons. The contact resistance is then given by

$$R_c = G_c^{-1} = \frac{h}{2e^2 M} = \frac{12.9k\Omega}{M}. \tag{23.12}$$

23.1.2 Thermal Transport due to Phonons

Previously we discussed the thermal conductivity, due to the energy transport along a given direction, as a result of a temperature gradient. Assuming that the entities that transport the energy move with a given velocity and using that their flux is given by Eq. (6.2), we obtained that the conductivity may be obtained as in Eq. (6.12). A simpler argument for the flux of particles across a surface is usually adopted. The number of particles that cross a given surface, A, in a time interval Δt, is estimated considering the number of particles inside a volume with length perpendicular to the surface $v\Delta t$, where v is the velocity component perpendicular to the surface. Therefore, the volume is $V = A\Delta t v$ and the number of particles inside that volume, and that cross the surface in the time interval Δt, is $N = nV = nA\Delta t v$, where n is the particle density. Therefore, the flux due to the particles that have a positive velocity perpendicular to the surface is $N/(A\Delta t) = nv$. Since there are three spatial dimensions and for each direction there are two senses, the probability that one particle collides is $nv/6$. Therefore, taking similar steps, we obtain the expression usually considered for the thermal conductivity:

$$K = \frac{1}{3}vCl, \tag{23.13}$$

where C is the heat capacity and l the mean free path for collisions. The idea behind Fick's law is that a gradient induces a current that is proportional to the gradient, with a coefficient that is the conductivity. The assumption is that there is a diffusion of the transported quantity, for instance, energy, through collisions, and is through the collisions that the energy is transmitted, from one side to the other side of the material. If there are no collisions ($l \to \infty$ and the conductivity diverges), we face the situation of the ballistic conductor.

Let us consider the thermal conductivity of the phonon gas. The velocity is therefore the phonon velocity. The mean free path is the result of collisions on the surface, with impurities or with other phonons. If the material is harmonic, then there are no interactions with other phonons (independent modes), but anharmonic corrections lead to phonon-phonon scattering.

We may estimate the mean free path due to possible collisions between phonons. The phonon average occupation number with frequency ω, at a given temperature, is given by

$$\langle n \rangle = \frac{1}{e^{\frac{\hbar\omega}{k_B T}} - 1}. \tag{23.14}$$

Considering, to simplify, high temperatures such that $T >> \theta_D$, the Debye temperature, we get that

$$\langle n \rangle \sim \frac{1}{\frac{\hbar\omega}{k_B T}} \sim T. \tag{23.15}$$

Therefore, at high temperatures the number of phonons is proportional to the temperature, which implies that the number of collisions is also proportional to T and the mean free path is then $l \sim 1/T$.

In general, the number of phonons is given by

$$n_{ph} \sim \int_0^{\hbar\omega_D} \frac{w^2 d\omega}{e^{\frac{\hbar\omega}{k_B T}} - 1}. \tag{23.16}$$

Making a change of variable $y = (\hbar\omega)/(k_B T)$ leads to

$$n_{ph} \sim \left(\frac{k_B T}{\hbar}\right)^3 \int_0^{x_D} \frac{y^2 dy}{e^y - 1}, \tag{23.17}$$

with $x_D = (\hbar\omega_D)/(k_B T) = \theta_D/T$. In the opposite limit of small temperatures $T << \Theta_D$, which implies $x_D \to \infty$, we get that the number of phonons is $n_{ph} \sim T^3$, and then the mean free path for small temperatures behaves as $l \sim 1/T^3$ and becomes large at small temperatures.

Consider now, for instance, an insulator such that we may consider that energy transport is only due to lattice vibrations or, in other words, phonons. We may place the left end of the $1d$ system, to simplify, in contact with a thermal reservoir at temperature T_1, and the right end in contact with another thermal reservoir at temperature T_2 and consider $T_1 > T_2$. In the vicinity of the contacts, the phonons are in equilibrium at different temperatures. The phonons will thermalize according to the temperature gradient across the system. Mechanisms that lead to the thermalization of a system, such as in a classical ideal gas, result from the collisions between two sets of molecules at different temperatures that tend to slow down the fastest particles (decreasing their effective temperature) and vice versa for the slower molecules. In the case of the phonons, collisions with impurities or the container walls do not change the energy of the phonons. Simple collisions between the phonons that satisfy the relation between their momenta as

$$\mathbf{k}_1 + \mathbf{k}_2 = \mathbf{k}_3, \tag{23.18}$$

where two phonons collide to give origin to a new phonon with a momentum that satisfies momentum conservation, do not establish equilibrium, since the total phonon momentum is not changed (elastic collisions between electrons behave similarly).

However, there is one process that is allowed in the case of phonons, which is called Umklapp scattering, where a reciprocal lattice vector is involved in the scattering process:

$$\mathbf{k}_1 + \mathbf{k}_2 = \mathbf{k}_3 + \mathbf{g}. \tag{23.19}$$

While energy is conserved, the momentum is not conserved. For instance, if the sum of the momenta is such that the resulting momentum is outside the Brillouin zone, a reciprocal lattice vector brings the momentum to the first Brillouin zone and the momentum decreases. Normal collisional processes ($\mathbf{g} = 0$) imply that the total momentum is conserved and, therefore, the phonon gas moves with constant global momentum. Including the Umklapp processes, the total momentum on average decreases. At high temperatures most phonon modes are excited (including higher energies and momentum) and, therefore, there is an appreciable fraction of collisions of the Umklapp type that lead to significant changes in the total momentum of the phonon gas. This leads to a thermal resistance that scales as T, as we saw above. On the other hand, at low temperatures, when the mean free path is large, a cutoff is introduced when the mean free path is of the order of the system size, and the thermal conductivity has a behavior of the type $K = CvL$, where L is the linear system size. Using dimensional analysis, if the mean free path is effectively very large, the size dimension parameter is the system size. Recalling that the heat capacity behaves as T^3, the thermal conductivity has a similar behavior.

23.1.3 Persistent Charge Currents in 1*d*

Another interesting feature of finite collisionless systems is the appearance of charge currents in a loop pierced by a time-independent magnetic flux. Faraday's law states that if the magnetic flux through a closed loop changes in time, an electromotive force is induced, which leads to a current along the loop with a direction according to Lenz law. If the magnetic flux does not change with time, there is no current induced. Fixing the geometry of the loop and changing the magnetic field, either its direction or amplitude, lead to an induced current. It turns out that in a finite system, of length L, a current exists as long as there is a magnetic flux and the system is small enough. In the thermodynamic limit, when $L \to \infty$, the current vanishes.

Take a circular wire of radius R and a magnetic field perpendicular to the plane of the wire, as shown in Fig. 23.2. Let us consider a single-electron description. The vector potential has the direction $\mathbf{A} = A\mathbf{e}_\theta$, where θ is the angle around the circle. The integral of the vector potential along the wire gives the magnetic flux:

$$\Phi = \int_S \mathbf{B} \cdot \mathbf{n} dS = \oint \mathbf{A} \cdot d\mathbf{l}. \tag{23.20}$$

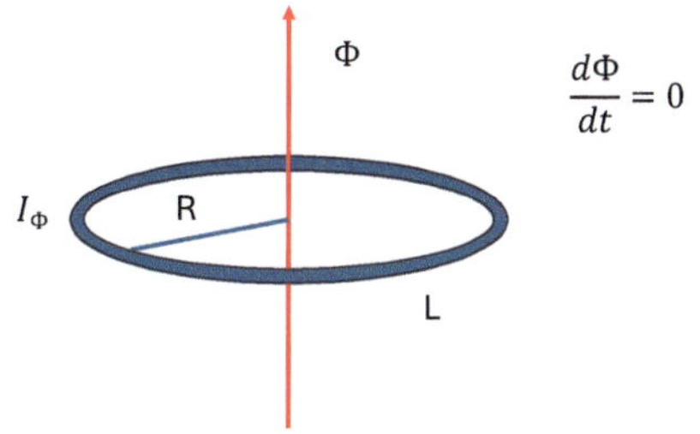

Fig. 23.2 Wire traversed by a time-independent magnetic flux. If the wire is finite, there is an induced electric current

Choosing a gauge with constant A, we get that

$$\Phi = AL = A2\pi R. \tag{23.21}$$

The Hamiltonian for M free electrons, of mass m_e, is

$$H = \sum_{j=1}^{M} \frac{1}{2m_e} \left(-i\hbar\nabla_j + e\mathbf{A}(\mathbf{r}_j)\right)^2. \tag{23.22}$$

Since the electrons are independent, using polar coordinates we can write that the wave function only depends on the polar angle, θ, and satisfies

$$\left(-i\hbar\frac{2\pi}{L}\frac{\partial}{\partial\theta} + \frac{e}{L}\Phi\right)\psi(\theta) = p\psi(\theta), \tag{23.23}$$

where p is the momentum. If the system is small enough and ignoring any effect of impurities, we may assume that there is phase coherence. The approximation applies if the scattering length off impurities is much larger than the system size. If there is phase coherence for the single-electron wave function, we may use periodic boundary conditions:

$$\psi(\theta + 2\pi) = \psi(\theta). \tag{23.24}$$

This implies that there is quantization of the eigenvalues

$$p_n(\Phi) = \frac{\hbar}{L}\left(2\pi n + \phi\right) \tag{23.25}$$

and where

$$\phi = 2\pi\frac{\Phi}{\phi_0}, \qquad \phi_0 = \frac{h}{e}. \tag{23.26}$$

The energy eigenvalues are therefore given by

$$\epsilon_n(\phi) = \frac{\hbar^2}{2m_e L^2}\left(2\pi n + \phi\right)^2. \tag{23.27}$$

The current of each eigenstate is given by

$$i_n = -\frac{e}{\hbar}\frac{\partial\epsilon_n(\phi)}{\partial\phi} = \frac{-e}{L}\frac{p_n}{m_e}. \tag{23.28}$$

The total current associated with a given number of M electrons is obtained calculating

$$I(\phi) = -\frac{e}{\hbar}\frac{\partial E_0(\phi)}{\partial \phi}, \tag{23.29}$$

where E_0 is the groundstate energy of the system given by

$$\begin{aligned} E_0(\phi) &= \sum_{n=-(M-1)/2}^{(M-1)/2} \epsilon_n(\phi), && M, \text{odd} \\ E_0(\phi) &= \sum_{n=-(M-2)/2}^{(M-2)/2} \epsilon_n(\phi) + \epsilon_{\mp M/2}(\phi), && M, \text{even}, \end{aligned} \tag{23.30}$$

which leads to

$$\begin{aligned} I(\phi) &= -\frac{e\hbar}{m_e L^2} M\phi, && M, \text{odd} \\ I(\phi) &= -\frac{e\hbar}{m_e L^2} M(\phi + L), && M, \text{even}. \end{aligned} \tag{23.31}$$

Considering a finite electronic density, $n_e = M/L$, we see that the current vanishes in the thermodynamic limit:

$$\lim_{L\to\infty} I(\phi) = 0. \tag{23.32}$$

The finite current when the system has a finite length is associated with the extended nature of the electron wave functions in the system. This may be tested following arguments that we considered before, in terms of the stiffness of the wave functions. If the states are extended over the system length, we may expect that the wave functions, and the energy, are sensitive to changes at the borders of the system, such as those that result from a change of boundary conditions. If the states are localized, in the sense that their wave functions have a finite spatial extent, then we expect that the states, and energy, are mostly insensitive to changes in boundary conditions (for the bulk states; some states close to the edges will "feel" the changes of the boundary conditions, but these have a small measure). A magnetic flux may also be seen as a change in boundary conditions (through an appropriate gauge transformation the effect of the magnetic flux may be converted to a change of boundary conditions, with respect to a state with no flux present).

We may expand the energy with respect to a given value of the magnetic flux, ϕ_1, as

$$E(\phi) = E(\phi_1) + \frac{1}{2L} D\,(\phi - \phi_1)^2 + \cdots \tag{23.33}$$

where

$$D = L \left(\frac{\partial^2 E_0}{\partial \phi^2} \right)_{\phi=\phi_1} . \tag{23.34}$$

D is called the stiffness to changes of he boundary conditions. It is given by

$$D = \frac{\hbar}{m_e} \frac{M}{L} . \tag{23.35}$$

For a finite system it is finite, indicating that the states are extended and may lead to a finite current. Fixing M and taking $L \to \infty$, the stiffness vanishes. Note that if the concentration n_e is finite, the stiffness is still finite, even though the current vanishes in the thermodynamic limit. The finiteness of the stiffness, however, determines the extended nature of the wave functions. Recall that a finite stiffness means an infinite zero frequency current, as when we discussed the Drude peak of a metallic system.

23.2 Landauer Transport

In a ballistic conductor, an incoming electron has probability one to go across the wire and to enter the other metallic contact. In more realistic systems this probability is less than one. Let us denote by T the average probability that an electron injected at one end of the conductor will transmit to the other end. Therefore, $T = 1$ in the case of a ballistic conductor.[1] In general, there is a transmission that is not $T = 1$, if there are scattering events inside the conductor or if there are interfaces between different conductors arranged in series, even if they are ballistic. The reflectionless condition, assumed for the metallic contacts, does not hold in general between conductors.

Let us then assume a geometry where there are two contacts, one on the left and one on the right, and that are connected by ballistic conductors to a central region composed of a general conductor, as shown in Fig. 23.3. Let us consider that across the conductor there is a transmissibility T, which is in general smaller than one. Each lead supports M modes through which the electrons may flow, the contacts with the leads are reflectionless, and the left contact is at a chemical potential μ_L and the right contact is at a chemical potential μ_R. The difference between the chemical potentials gives origin to a potential difference that leads to a net motion of charges, assumed from left to right. Consider zero temperature.

On the left contact, electrons with positive k that are in equilibrium at the chemical potential, μ_L, are injected, and from the right lead, electrons with $k < 0$

[1] For further reading see S. Datta, *Electronic Transport in Mesoscopic Systems*, Cambridge University Press.

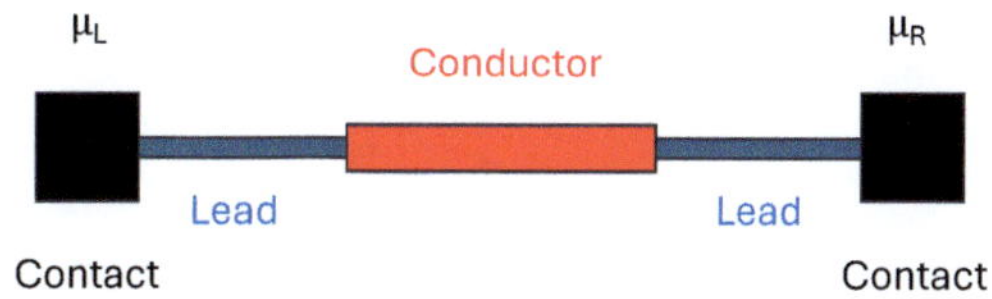

Fig. 23.3 Landauer geometry with a normal conductor connected by two ballistic leads to external contacts that establish a chemical potential difference and an electrical current

are injected, originating from a distribution with chemical potential, μ_R. The current influx from the left lead is

$$I_L^{\rightarrow} = \frac{2|e|}{h} M \left(\mu_L - \mu_R\right). \tag{23.36}$$

Recall Eq. (23.9) and that the spin degeneracy is two. The outflux from the right lead is

$$I_R^{\rightarrow} = \frac{2|e|}{h} MT \left(\mu_L - \mu_R\right). \tag{23.37}$$

On the left lead there is also a flux due to the electrons that are reflected from the interface between the left lead and the conductor. Since there is conservation of current, the reflection coefficient is simply $(1-T)$. Therefore, the current in the left lead that moves towards the left contact is

$$I_L^{\leftarrow} = \frac{2|e|}{h} M(1-T) \left(\mu_L - \mu_R\right). \tag{23.38}$$

The total current on the left lead (or on the right lead) is given by

$$I = I_L^{\rightarrow} - I_L^{\leftarrow} = I_R^{\rightarrow} = \frac{2|e|}{h} MT \left(\mu_L - \mu_R\right). \tag{23.39}$$

By definition, the conductance is then obtained as

$$G = \frac{I}{\frac{\mu_L - \mu_R}{|e|}} = \frac{2e^2}{h} MT. \tag{23.40}$$

Recall that the factor of 2 is due to the spin degeneracy. This is the Landauer formula for the conductance.

We may now generalize this expression to a sequence of conductors in series. Consider two such conductors, as shown in Fig. 23.4. Injecting a current from the left, there is a reflection coefficient, R_i from conductor $i = 1, 2$, and a transmission coefficient, T_i. Let us denote T_{12} the transmission coefficient of the set of two conductors. In the region between the two conductors there may be an infinite

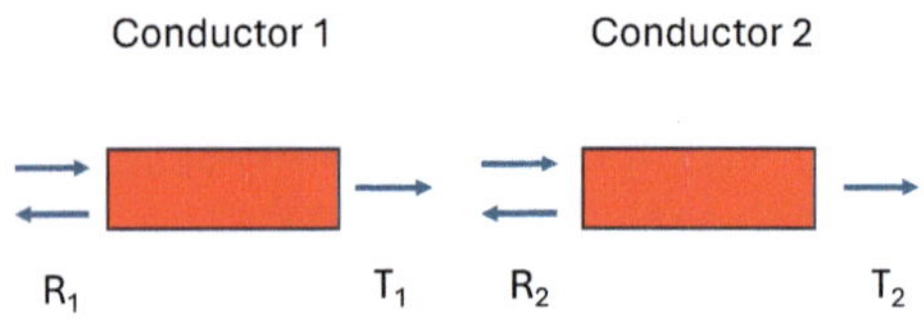

Fig. 23.4 Series of two conductors where T_i are the transmission coefficients across conductor i and R_i are the reflection coefficients from conductor i

sequence of reflections which implies that

$$T_{12} = T_1T_2 + T_1R_2R_1T_2 + T_1R_2R_1R_2R_1T_2 + \cdots = \frac{T_1T_2}{1 - R_2R_1}, \tag{23.41}$$

summing the geometric series. Using that $R_i = 1 - T_i$ we may write that

$$\frac{1 - T_{12}}{T_{12}} = \frac{1 - T_1}{T_1} + \frac{1 - T_2}{T_2} = \frac{R_1}{T_1} + \frac{R_2}{T_2}. \tag{23.42}$$

It shows that the left-hand side is just additive in the contributions of the two conductors and can be generalized to N conductors in series. That is, denoting $T(N)$ as the transmission coefficient of N identical conductors, we can write that

$$\frac{1 - T(N)}{T(N)} = N\frac{1 - T}{T}, \tag{23.43}$$

which can be solved to give

$$T(N) = \frac{T}{N(1 - T) + T}. \tag{23.44}$$

The Landauer formula can be factorized in a convenient way. Let us rewrite the expression in terms of the resistance:

$$R = G^{-1} = \frac{h}{2e^2M}\frac{1}{T} = \frac{h}{2e^2M} + \frac{h}{2e^2M}\frac{1 - T}{T} = \frac{h}{2e^2M} + \frac{h}{2e^2M}\frac{R}{T}. \tag{23.45}$$

If there is no reflection, $R = 0$, as in the ballistic conductor, we are left with only the first term that was identified as the contact resistance. The second term may then be interpreted as the actual resistance of the conductor. Inverting, we get that the conductor has a conductance:

$$G_R = \frac{2e^2M}{h}\frac{T}{R}. \tag{23.46}$$

A more detailed treatment may be obtained generalizing the previous treatment to a case when the various conducting channels may contribute differently.[2] Consider

[2] See D. C. Langreth and E. Abrahams, Physical Review B **24**, 2978 (1981).

a conductor attached to a left and a right lead, as before, and with $a, b = 1, \cdots, M$ distinct channels. Call P_a^L, P_a^R the probability to find one electron on channel a on the left and right leads, respectively. The current density on channel a on the left lead and the electron density on the same channel are given by

$$
\begin{aligned}
J_a^L &= v_a \left(P_a^L - \sum_b R_{ab}^L P_b^L - \sum_b T_{ab}^R P_b^R \right) \\
n_a^L &= \left(P_a^L + \sum_b R_{ab}^L P_b^L + \sum_b T_{ab}^R P_b^R \right),
\end{aligned}
\tag{23.47}
$$

where R_{ab}^L is the reflection coefficient on the left from channel b to channel a, and T_{ab}^R is the transmission probability from channel b on the right to channel a on the left. v_a is the velocity of channel a on the left. In matrix notation these equations may be written as

$$
\begin{aligned}
\frac{J^L}{v} &= P^L - R^L P^L - T^R P^R \\
n^L &= P^L + R^L P^L + T^R P^R.
\end{aligned}
\tag{23.48}
$$

Similarly, we may write that on the right lead we have

$$
\begin{aligned}
\frac{J^R}{v} &= T^L P^L + (R^R - 1) P^R \\
n^R &= T^L P^L + (R^R + 1) P^R.
\end{aligned}
\tag{23.49}
$$

The probabilities may be solved leading to

$$
\begin{aligned}
P^L &= \frac{1}{2}\left(\frac{j^L}{v} + n^L\right) \\
P^R &= -\frac{1}{2}\left(\frac{j^R}{v} - n^R\right).
\end{aligned}
\tag{23.50}
$$

Eliminating the probabilities we get the equations

$$
\begin{aligned}
\frac{1}{2}\left(1 + R^L\right)\frac{j^L}{v} - \frac{1}{2}T^R\frac{j^R}{v} &= \frac{1}{2}\left(1 - R^L\right)n^L - \frac{1}{2}T^R n^R \\
\frac{1}{2}\left(1 + R^R\right)\frac{j^R}{v} - \frac{1}{2}T^L\frac{j^L}{v} &= -\frac{1}{2}\left(1 - R^R\right)n^R + \frac{1}{2}T^L n^L.
\end{aligned}
\tag{23.51}
$$

The coefficients T and R are actually of the form of $T \sim |t|^2$ and $R \sim |r|^2$, where the coefficients t and r are transmission and reflection coefficients, respectively, that are entries of the S-matrix of the scattering off the region where the conductor is located. The S-matrix relates the outgoing waves in terms of the incoming waves and is of the form

$$\begin{pmatrix} L_{out} \\ R_{out} \end{pmatrix} = S \begin{pmatrix} L_{in} \\ R_{in} \end{pmatrix}$$
$$S = \begin{pmatrix} r^L & t^R \\ t^L & r^R \end{pmatrix} . \tag{23.52}$$

Conservation of current (or probability) imposes that the S-matrix is unitary. Also, time-reversal invariance imposes that

$$\left(r^{L,R}\right)^T = \left(r^{L,R}\right), \qquad \left(t^{L,R}\right)^T = \left(t^{R,L}\right)^T , \tag{23.53}$$

where T is the transpose of the reflection and transmission coefficient matrices.

Using unitarity and time-reversal invariance we may conclude that

$$\begin{aligned} 1 &= \sum_b \left(R_{ba}^{L,R} + T_{ba}^{L,R}\right) \\ &= \sum_b \left(R_{ab}^{L,R} + T_{ab}^{L,R}\right) . \end{aligned} \tag{23.54}$$

We may then write that

$$\begin{aligned} \frac{1}{2}\left(1 - R^L\right) n^L - \frac{1}{2} T^R n^R &= \frac{1}{2} T^R \left(n^L - n^R\right) \\ -\frac{1}{2}\left(1 - R^R\right) n^R + \frac{1}{2} T^L n^L &= \frac{1}{2} T^L \left(n^L - n^R\right) . \end{aligned} \tag{23.55}$$

The difference in electronic densities between the left and right leads can be written as

$$n^L - n^R = \delta n = \frac{dn}{dE} e\varepsilon = \frac{1}{\pi \hbar v} e\varepsilon , \tag{23.56}$$

where ε is the electromotive force. We can write then

$$\begin{aligned} J_a^L + \sum_b R_{ab}^L J_b^L - \sum_b T_{ab}^R j_b^R &= \frac{e}{\pi\hbar} \sum_b T_{ab}^R \varepsilon \\ J_a^R + \sum_b R_{ab}^R J_b^L - \sum_b T_{ab}^L j_b^R &= \frac{e}{\pi\hbar} \sum_b T_{ab}^L \varepsilon . \end{aligned} \tag{23.57}$$

In the case of a single channel we get

$$J^R = J^L = G\frac{\varepsilon}{e} = \frac{e}{h}\frac{T}{R}\varepsilon, \tag{23.58}$$

and therefore, we get

$$G = \frac{e^2}{h}\frac{T}{R}. \tag{23.59}$$

In the case of M identical channels we get that $J_a^R = J_a^L = J/M$ and

$$G = \frac{eJ}{\varepsilon} = \frac{e^2}{h}\frac{T}{1-\frac{T}{M}}, \tag{23.60}$$

where $T = 1/2\sum_{a,b}\left(T_{ab}^L + T_{ab}^R\right)$. In the limit of many channels, $M \to \infty$, we get that the conductance is just proportional to T.

23.3 Transfer Matrix in a 1d Lattice

A generalization of the idea of several conductors in series is to consider a single system, defined on a lattice, such that at each point of the lattice there is a given atom, molecule, or unit cell and that electrons propagate along this lattice, injected from a left lead, and emerge from a right lead. As considered before, this may result from the application of some bias potential difference and the consequent appearance of a current. The system may be homogeneous, with all units equivalent, or we may consider any sort of inhomogeneity along the chain. In general, the system will be resistive and, depending on the microscopic details, the system may be conductive or an insulator. In addition to the previously considered difference between conductors and insulators that arise from band theory and the interplay between the position of the chemical potential and the existence of a band structure (with possible gaps in the spectrum), the addition of inhomogeneities, such as impurities distributed randomly, either in their locations or in the magnitudes of the local potential energies, leads to states that may be extended or localized. This may lead, in some cases, to an insulator that is the result of the randomness and not a band insulator.

Let us consider a somewhat general unidimensional system that is described by some Hamiltonian of a tight-binding general class. We consider two leads that are attached to the chain. We may inject a spin up or spin down electron from the left lead into the chain and collect this scattering state on the right lead. The chain is of finite size. In general, at each site we have a spinor with two components, one respecting to spin up and the other to spin down. If we ignore the spin degree of freedom, the spinor has just the component of the wave function of the electron at that site. If we consider a superconductor, in addition to possible spin degrees of freedom, we also have spinor components respecting to the particle and hole components. Other internal degrees of freedom may be considered, leading in general to an M-component spinor.

For instance, considering a normal conductor, such that $M = 2$ corresponds to the two spin components, due to possible spin-orbit couplings and transverse external fields, we must allow for spin flips. The scattering states may be reflected or transmitted along the chain, keeping the same spin component or reversing the spin component. In general, we have a 2×2 matrix structure, with respect to the two spin projections. The scattering states propagate along the chain region. We may choose the left and right leads as simple conductors, using a tight-binding model, with bandwidth large enough to go over the energy range of the central region, where spin-orbit coupling or the effect of possible magnetic fields may be significant. The hopping from the leads to the central region is in general different from the hopping in the leads or in the central region.

At each lattice site, n, consider that we have a spinor, ψ_n, with respect to the two spin projections, leading to an equation of the type

$$H_n \psi_n + T_{n,n+1} \psi_{n+1} + T_{n,n-1} \psi_{n-1} = \epsilon \psi_n. \tag{23.61}$$

Here, ϵ is the energy of the electron coming from the left lead. We may rewrite this equation as

$$\psi_{n+1} = T_{n,n+1}^{-1} (\epsilon - H_n) \psi_n - T_{n,n+1}^{-1} T_{n,n-1} \psi_{n-1}, \tag{23.62}$$

which leads to a transfer matrix, M_n, defined as

$$\begin{pmatrix} \psi_{n+1} \\ \psi_n \end{pmatrix} = M_n \begin{pmatrix} \psi_n \\ \psi_{n-1} \end{pmatrix}$$

$$M_n = \begin{pmatrix} T_{n,n+1}^{-1} (\epsilon - H_n) & -T_{n,n+1}^{-1} T_{n,n-1} \\ 1 & 0 \end{pmatrix}. \tag{23.63}$$

Let us consider the combined system of the two leads and the central region and take the size of the combined system as N. We may consider a point on the left lead, at position N_L, sufficiently far from the central region and a point on the right lead, N_R, also sufficiently far from the central region. We may write that

$$\begin{pmatrix} \psi_{N_R+1} \\ \psi_{N_R} \end{pmatrix} = M_T \begin{pmatrix} \psi_{N_L} \\ \psi_{N_L-1} \end{pmatrix}, \tag{23.64}$$

where

$$M_T = \prod_{n=N_L}^{N_R} M_n. \tag{23.65}$$

23.3.1 Scattering States for Spinless Fermions

To simplify, let us consider first the case of a spinor with just one component. The results are easily generalizable to a spinor with an arbitrary number of components.

In the limit when the number of sites in the central region is very large, the transfer matrix M_T may become numerically unstable, with eigenvalues that either diverge or become very small. It is convenient to define a new matrix:[3]

$$\Gamma = \lim_{L\to\infty} \left(M_T^\dagger M_T\right)^{\frac{1}{2L}} ; \tag{23.66}$$

here $L = N_R - N_L$. This matrix has eigenvalues of the type e^{γ_m} (such that $e^{\gamma_m} < \infty$), with $m = 0, 1$, where γ_m are called the Lyapunov exponents. It turns out that $det[\Gamma] = e^{\gamma_0} e^{\gamma_1} = 1$. Therefore, $\gamma_0 = -\gamma_1$.

We can write that

$$det\left[M_T^\dagger\right] det\left[M_T\right] = \left(\prod_{n=1}^{L} det\left(M_n^\dagger\right)\right)\left(\prod_{n=1}^{L} det\left(M_n\right)\right) = \prod_{n=1}^{L} |T_{n,n+1}^{-1} T_{n,n-1}|^2. \tag{23.67}$$

Considering the simple case that the matrices $T_{n.n+1}$ and $T_{n,n-1}$ are scalars, we get that

$$det[\Gamma] = \lim_{L\to\infty} det\left[\left(M_T^\dagger M_T\right)^{\frac{1}{2L}}\right] = \lim_{L\to\infty} e^{\frac{1}{L}\sum_{n=1}^{L}\left(\log|t_{n,n+1}| - \log|t_{n,n-1}|\right)}. \tag{23.68}$$

If $|t_{n,n+1}| = |t_{n,n-1}|$, it follows that the $det\,[\Gamma] = 1$. If there is disorder, the Hamiltonian is hermitian and we take the infinite size limit, we expect that the average of the hopping terms will be the same for a hopping to the left and a hopping to the right, and we also get that $det\,[\Gamma] = 1$.

It turns out that the Lyapunov exponents are just the inverse of the localization length of the wave functions in the system, since they control the exponential decay of the wave functions, as the distance between two points is considered. A vanishing Lyapunov exponent means that the states are extended, while a finite value implies a localized state. Roughly speaking, in the basis that diagonalizes the matrix Γ, the wave function at a site, distant by L from another site, will be like an exponential of the type $e^{\gamma_m L}$. Considering the exponential decay, we see that the Lyapunov exponent behaves as an inverse localization length.

---

[3] Here we follow personal notes by Miguel Gonçalves, IST, Universidade de Lisboa.

23.3.2 Localized Versus Extended States: Lyapunov Exponent

In the case of a clean system we expect extended states. A simple example is a tight-binding model with no disorder. The Lyapunov exponents must vanish in this case. Introducing disorder, it turns out that one finds localized states, as will be shown later.

Consider then the simple case where $t_{n,n+1} = t_{n,n-1} = t = 1$ and the local potential $H_n = 0$. The transfer matrix is

$$M_n = \begin{pmatrix} \epsilon & -1 \\ 1 & 0 \end{pmatrix}. \tag{23.69}$$

In diagonal form we can write that

$$M_n = U \begin{pmatrix} \tau_1 & 0 \\ 0 & \tau_2 \end{pmatrix} U^{-1}. \tag{23.70}$$

The eigenvalues are given by

$$\tau_{1,2} = \frac{\epsilon}{2} \pm \sqrt{\frac{\epsilon^2}{4} - 1}. \tag{23.71}$$

Therefore,

$$M_T = U \begin{pmatrix} \tau_1^L & 0 \\ 0 & \tau_2^L \end{pmatrix} U^{-1}. \tag{23.72}$$

In the diagonal basis we can write

$$\begin{pmatrix} \phi_{n+1} \\ \phi_n \end{pmatrix} = U^{-1} \begin{pmatrix} \psi_{n+1} \\ \psi_n \end{pmatrix}, \tag{23.73}$$

which implies that, for instance,

$$\phi_{N_R+1} = \tau_1^L \phi_{N_L} = e^{\nu_1 L} \phi_{N_L}, \tag{23.74}$$

with

$$\nu_1 = \log\left(\frac{\epsilon}{2} + \sqrt{\frac{\epsilon^2}{4} - 1}\right). \tag{23.75}$$

Extended states will eventually arise using, for instance, periodic boundary conditions such that $\phi_{N_R+1} = \phi_{N_L}$. This condition cannot be specified for $|\epsilon| > 2$.

(Recall that in a tight-binding model with $t = 1$, the bandwidth is 2.) For $|\epsilon| < 2$ we have that the argument of the square root is negative and

$$\nu_1 = i \arctan\left(\frac{\sqrt{4-\epsilon^2}}{\epsilon}\right). \tag{23.76}$$

This implies that

$$|\tau_{1,2}|^2 = 1. \tag{23.77}$$

Note that we may write

$$M_T^\dagger M_T = U\begin{pmatrix} (\tau_1^*)^L & 0 \\ 0 & (\tau_2^*)^L \end{pmatrix}\begin{pmatrix} \tau_1^L & 0 \\ 0 & \tau_2^L \end{pmatrix}U^{-1} = U\begin{pmatrix} |\tau_1|^{2L} & 0 \\ 0 & |\tau_2|^{2L} \end{pmatrix}U^{-1} \tag{23.78}$$

and, therefore,

$$M_T^\dagger M_T = U\begin{pmatrix} 1 & 0 \\ 0 & 1 \end{pmatrix}U^{-1} = I. \tag{23.79}$$

Therefore,

$$\Gamma = \lim_{L\to\infty}\left(M_T^\dagger M_T\right)^{\frac{1}{2L}} = I = \begin{pmatrix} e^{\gamma_0} & 0 \\ 0 & e^{\gamma_1} \end{pmatrix}, \tag{23.80}$$

with $\gamma_0 = \gamma_1 = 0$ and the localization length diverges, as expected of extended states.

23.3.3 Scattering States for Spinful Fermions

Consider an electron with spin up coming from the left lead, with a given energy ϵ. The spinors have now two components with respecti to the two spin projections. We choose

$$\begin{pmatrix} u_\uparrow(N_L) \\ u_\downarrow(N_L) \end{pmatrix} = \begin{pmatrix} 1 \\ 0 \end{pmatrix} + r\begin{pmatrix} 1 \\ 0 \end{pmatrix} + \bar{r}\begin{pmatrix} 0 \\ 1 \end{pmatrix}, \tag{23.81}$$

where r is the reflection coefficient and $\bar{r}$ the reflection coefficient with spin flip, and

$$\begin{pmatrix} u_\uparrow(N_L-1) \\ u_\downarrow(N_L-1) \end{pmatrix} = \begin{pmatrix} 1 \\ 0 \end{pmatrix}e^{-iq_1} + r\begin{pmatrix} 1 \\ 0 \end{pmatrix}e^{iq_1} + \bar{r}\begin{pmatrix} 0 \\ 1 \end{pmatrix}e^{iq_1}, \tag{23.82}$$

where $q_1 = \cos^{-1}(-\epsilon/(2w_2))$, and w_2 is the hopping on the left (and right) leads.

Considering an electron with spin down coming from the left lead, with a given energy ϵ, leads to

$$\begin{pmatrix} u_\uparrow(N_L) \\ u_\downarrow(N_L) \end{pmatrix} = \begin{pmatrix} 0 \\ 1 \end{pmatrix} + r \begin{pmatrix} 0 \\ 1 \end{pmatrix} + \bar{r} \begin{pmatrix} 1 \\ 0 \end{pmatrix} \tag{23.83}$$

and

$$\begin{pmatrix} u_\uparrow(N_L-1) \\ u_\downarrow(N_L-1) \end{pmatrix} = \begin{pmatrix} 0 \\ 1 \end{pmatrix} e^{-iq_1} + r \begin{pmatrix} 0 \\ 1 \end{pmatrix} e^{iq_1} + \bar{r} \begin{pmatrix} 1 \\ 0 \end{pmatrix} e^{iq_1}. \tag{23.84}$$

On the right lead, we have the transmitted wave. In the case of an electron with spin up coming from the left lead, with a given energy ϵ, we have

$$\begin{pmatrix} u_\uparrow(N_R) \\ u_\downarrow(N_R) \end{pmatrix} = t \begin{pmatrix} 1 \\ 0 \end{pmatrix} + \bar{t} \begin{pmatrix} 0 \\ 1 \end{pmatrix}, \tag{23.85}$$

where t is the transmission coefficient and $\bar{t}$ the transmission coefficient with spin flip, and

$$\begin{pmatrix} u_\uparrow(N_R+1) \\ u_\downarrow(N_R+1) \end{pmatrix} = t \begin{pmatrix} 1 \\ 0 \end{pmatrix} e^{iq_1} + \bar{t} \begin{pmatrix} 0 \\ 1 \end{pmatrix} e^{iq_1}. \tag{23.86}$$

Considering an electron with spin down coming from the left lead, with a given energy ϵ, leads to

$$\begin{pmatrix} u_\uparrow(N_R) \\ u_\downarrow(N_R) \end{pmatrix} = t \begin{pmatrix} 0 \\ 1 \end{pmatrix} + \bar{t} \begin{pmatrix} 1 \\ 0 \end{pmatrix} \tag{23.87}$$

and

$$\begin{pmatrix} u_\uparrow(N_R+1) \\ u_\downarrow(N_R+1) \end{pmatrix} = t \begin{pmatrix} 0 \\ 1 \end{pmatrix} e^{iq_1} + \bar{t} \begin{pmatrix} 1 \\ 0 \end{pmatrix} e^{iq_1}. \tag{23.88}$$

The various coefficients may be obtained in a standard way as briefly explained ahead.

Using the Landauer formula, the charge conductances, for incident electrons with spin components $\uparrow$, $\downarrow$, are obtained as

$$\begin{aligned} \sigma_\uparrow &= 1 - r_\uparrow - \bar{r}_\uparrow = t_\uparrow + \bar{t}_\uparrow \\ \sigma_\downarrow &= 1 - r_\downarrow - \bar{r}_\downarrow = t_\downarrow + \bar{t}_\downarrow, \end{aligned} \tag{23.89}$$

in units of e^2/h. The spin polarization is defined as

$$P = \frac{\left(|t_\uparrow|^2 + |\bar{t}_\uparrow|^2\right) - \left(|t_\downarrow|^2 + |\bar{t}_\downarrow|^2\right)}{\left(|t_\uparrow|^2 + |\bar{t}_\uparrow|^2\right) + \left(|t_\downarrow|^2 + |\bar{t}_\downarrow|^2\right)}. \tag{23.90}$$

As a check we may consider the normalization (conservation of probability/charge) calculating for both spin directions of the incident state as

$$\begin{aligned} 1 &= |r_\uparrow|^2 + |\bar{r}_\uparrow|^2 + |t_\uparrow|^2 + |\bar{t}_\uparrow|^2 \\ 1 &= |r_\downarrow|^2 + |\bar{r}_\downarrow|^2 + |t_\downarrow|^2 + |\bar{t}_\downarrow|^2. \end{aligned} \tag{23.91}$$

The coefficients, for a scattering state in the case of a spin up incident electron, may be obtained solving the equations

$$A_\uparrow \begin{pmatrix} r \\ t \\ \bar{r} \\ \bar{t} \end{pmatrix} = b_\uparrow, \tag{23.92}$$

where

$$b_\uparrow = -\begin{pmatrix} M_{11}^T + M_{13}^T e^{-iq_1} \\ M_{21}^T + M_{23}^T e^{-iq_1} \\ M_{31}^T + M_{33}^T e^{-iq_1} \\ M_{41}^T + M_{43}^T e^{-iq_1} \end{pmatrix} \tag{23.93}$$

and

$$A_\uparrow = \begin{pmatrix} M_{11}^T + M_{13}^T e^{iq_1} & -e^{iq_1} & M_{12}^T + M_{14}^T e^{iq_1} & 0 \\ M_{21}^T + M_{23}^T e^{iq_1} & 0 & M_{22}^T + M_{24}^T e^{iq_1} & -e^{iq_1} \\ M_{31}^T + M_{33}^T e^{iq_1} & -1 & M_{32}^T + M_{34}^T e^{iq_1} & 0 \\ M_{41}^T + M_{43}^T e^{iq_1} & 0 & M_{42}^T + M_{44}^T e^{iq_1} & -1 \end{pmatrix}, \tag{23.94}$$

and similarly for a spin down incident electron. The coefficients, for a scattering state in a spin down incident electron, may be obtained solving the equations

$$A_\downarrow \begin{pmatrix} r \\ t \\ \bar{r} \\ \bar{t} \end{pmatrix} = b_\downarrow, \tag{23.95}$$

where

$$b_{\downarrow} = -\begin{pmatrix} M_{12}^T + M_{14}^T e^{-iq_1} \\ M_{22}^T + M_{24}^T e^{-iq_1} \\ M_{32}^T + M_{34}^T e^{-iq_1} \\ M_{42}^T + M_{44}^T e^{-iq_1} \end{pmatrix} \tag{23.96}$$

and

$$A_{\downarrow} = \begin{pmatrix} M_{12}^T + M_{14}^T e^{iq_1} & 0 & M_{11}^T + M_{13}^T e^{iq_1} & -e^{iq_1} \\ M_{22}^T + M_{24}^T e^{iq_1} & -e^{iq_1} & M_{21}^T + M_{23}^T e^{iq_1} & 0 \\ M_{32}^T + M_{34}^T e^{iq_1} & 0 & M_{31}^T + M_{33}^T e^{iq_1} & -1 \\ M_{42}^T + M_{44}^T e^{iq_1} & -1 & M_{41}^T + M_{43}^T e^{iq_1} & 0 \end{pmatrix} . \tag{23.97}$$

23.3.4 Andreev Reflection

Consider two systems separated by an insulating layer. The whole system may be taken as effectively one-dimensional if we consider transport along the direction perpendicular to the interface and if the dimensions parallel to the interface are much smaller than those along the longitudinal direction. Once again let us establish a potential difference between the two systems so that a current is generated, say from left to right.

As a first example take the two systems as metals and that the insulator has a finite, small, width. We expect that the current versus the potential difference should follow a linear Ohmic relation. If on the left we have a metal and on the right side we place a conventional superconductor, with a gapped spectrum, we expect that for a potential difference smaller than the energy of the spectral gap, the current on the right must be very small. We only expect a finite response for a potential difference that is larger than that scale. In the metallic side the carriers are unpaired electrons, while on the superconductor side the carriers are Cooper pairs. In order to have a current on the right side, one expects that the potential difference is large enough to break the Cooper pairs and the electron may then have a tunneling effect through the insulator and propagate on the right-hand side. This was observed by Giaver in 1960. If we have two superconductors, one expects that Cooper pairs are tunneled from one superconductor to the other, and this is the Josephson effect previously considered.

Other different configurations of materials are possible, such as interfaces that are part of heterostructures that include metals, semiconductors, magnetic materials, superconductors, etc., depending on different functionalities we may want. One such interesting possibility is a conductor/ferromagnet/conductor, or two ferromagnets with magnetizations that form a certain angle between them. These cases are interesting to control transport properties, as a result of magnetization effects on the electron spins. Another possibility is a metal/superconductor/metal heterostructure.

Sending electrons from the metal into the superconductor region, as mentioned above in the Giaver experiment, is worth analyzing with some extra care. Even if the potential difference is smaller than the spectral gap of the superconductor, there is a possibility that some charge is transported into the superconductor. This occurs through a mechanism called Andreev reflection. In this process an electron that is incident on the superconductor pairs with an electron at the interface, leaving a hole behind that propagates into the metal. The pair is now able to propagate into the superconductor.

The propagation into the superconductor is governed by the Bogoliubov-de Gennes (BdG) equations that, as we saw before, take the role the Schrödinger equation has in the case of the transfer matrix method in a normal conductor. Using the BdG equations, the particle and hole components are naturally taken into account, both in the superconductor and in the conductor. The transfer matrix dimension doubles its dimension due to the particle-hole structure, but otherwise the procedure is similar to the one considered in the previous sections. The simplest case of a single-band superconductor involves now a matrix structure that acts on a four-component spinor, associated with the particle and hole degrees of freedom and the two spin projections, $\uparrow$ and $\downarrow$.

The BdG equations determine both the equilibrium states and the scattering states, when a current is passing through. We may use a lattice formulation and define a spinor field at each location in the wire as

$$\psi_i^T = (u(i,\uparrow), v(i,\downarrow), u(i,\downarrow), v(i,\uparrow))^T .$$

We may use a transfer matrix M_i to relate the wave functions in neighboring sites i and $i+1$ as before:

$$\begin{pmatrix} \psi_{i+1} \\ \psi_i \end{pmatrix} = M_i \begin{pmatrix} \psi_i \\ \psi_{i-1} \end{pmatrix} , \tag{23.98}$$

with

$$M_i = \begin{pmatrix} T_i^{-1}(\epsilon - H_i) & -I \\ I & 0 \end{pmatrix} . \tag{23.99}$$

Different heterostructures lead to different Hamiltonians. Since the system is nonhomogeneous, since at the very least we have two or three different regions plus the interfaces, the Hamiltonian is a function of space. Of course, even within a given region the Hamiltonian may differ from point to point, for instance, adding impurities or magnetic impurities, such as parametrized by local magnetic dipoles (or local spin operators). The energy of the incident particle is given by ϵ and T_i represents the kinetic energy on a given link, associated with site i of a given region. The BdG equations are solved self-consistently in a structure of length N, of two normal leads and a superconducting region in the middle, to find the profile of the gap function inside the superconductor, Δ_i. This is then used as input in the transfer matrix (the proximity effect between the two systems is thus taken into account). In general, one takes into consideration the effect of the interface as a potential barrier with a given strength, which may vanish if the interface is transparent to the motion of the electrons. We may let N_L (N_R) label a site in the left (right) lead, sufficiently far from the superconductor to neglect any proximity effect, and take the superconducting region to extend between sites N_{SL} and N_{SR}.

To illustrate the procedure of the calculation of the transport properties, consider the scattering state for a particle with spin up incident from the left, as shown in Fig. 23.5. We write

$$\psi_{N_L}^T = \left(1 + r_z, r_A, \bar{r}, r_A^z\right)^T , \tag{23.100}$$

$$\psi_{N_R}^T = \left(t_z, t_A, \bar{t}, t_A^z\right)^T , \tag{23.101}$$

and

$$\psi_{N_L-1}^T = \left(e^{-iq_1^L a} + r_z e^{iq_1^L a}, r_A e^{-iq_2^L a}, \bar{r} e^{iq_3^L a}, r_A^z e^{-iq_4^L a}\right)^T ,$$

$$\psi_{N_R+1}^T = \left(t_z e^{iq_1^R a}, t_A e^{-iq_2^R a}, \bar{t} e^{iq_3^R a}, t_A^z e^{-iq_4^R a}\right)^T . \tag{23.102}$$

This takes into account reflection of the particle with the same (opposite) spins r_z ($\bar{r}$), and reflection as hole due to Andreev process with the opposite (same) spin r_A (r_A^z). The transmission amplitudes $t_z, t_a, \bar{t}, t_A^z$ have the same meanings. The momenta are obtained far from the superconducting region, where the solution of the BdG equations is easily obtained. Using the product of transfer matrices to relate the points N_L and N_R, one can solve for the various reflection and transmission amplitudes. Identical procedure is carried out for a particle with spin down and for

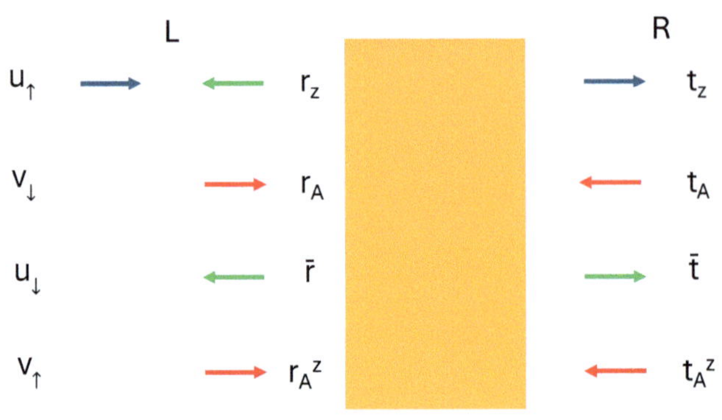

Fig. 23.5 Coefficients for transmission and reflection of an electron incident from the left with spin ↑ from a superconducting region. The states are a spinor with four components (particle-hole and spin ↑, ↓). There are reflections and transmissions without or with spin flip, together with reflections and transmissions that transform a particle into a hole and vice versa. These are Andreev processes

incident holes from the right. For an incident spin up particle the charge current is given by $j_L(\uparrow) = \sin q_1^L a - |r_z|^2 \sin q_1^L a + |r_A|^2 \sin q_2^L a - |\bar{r}|^2 \sin q_3^L a + |r_A^z|^2 \sin q_4^L a$ on the left side (the sine functions are the velocities). The current on the right is given by $j_R(\uparrow) = |t_z|^2 \sin q_1^R a - |t_A|^2 \sin q_2^R a + |\bar{t}|^2 \sin q_3^R a - |t_A^z|^2 \sin q_4^R a$. The differential conductance is given by the currents divided by the velocity of the incident particle.

In experiments, a potential difference, V, is imposed between the two sides of the heterostructure in a standard way. Taking into account the contributions from particles incident from left and holes incident from the right, one gets the expression for the current on the left side:

$$I_\sigma^e + I_\sigma^h = \frac{e}{h} \int_0^\infty d\epsilon \left[f(E - \frac{eV}{2}) - f(E + \frac{eV}{2}) \right]$$
$$\left[1 - R^{ee} + R^{he} - T^{eh'} + T^{hh'} \right], \tag{23.103}$$

where R^{ee} is the reflection coefficient of an electron as an electron, R^{he} the reflection of an electron as a hole, $T^{eh'}$ the transmission of a hole from the right into an electron on the left, and $T^{hh'}$ transmission of a hole from the right into a hole on the left. These coefficients are given by products of reflection or transmission coefficients (absolute value squared) times relative velocities to the incident particle. The expression takes into account the effect of a finite temperature through the weight given by the Fermi-Dirac distributions, $f(E \pm eV/2)$.

23.4 Anderson Localization

23.4.1 Anderson Model

The concept of Anderson localization is well illustrated considering the Anderson model of a disordered lattice system given by the Hamiltonian

$$H = -t \sum_{i,j} c_i^\dagger c_j + \sum_i V_i c_i^\dagger c_i. \tag{23.104}$$

The hopping term is usually taken between nearest-neighbor sites in a d-dimensional lattice and the local potentials V_i are randomly correlated and distributed according to some distribution, $P(V_i)$. This may be taken as a gaussian distribution around some average value (for instance, around $V = 0$) or a box distribution of the type

$$P(V_i) = \frac{1}{W} \Theta \left(\frac{W}{2} - |V_i| \right). \tag{23.105}$$

The model depends on the parameters t and W and, therefore, their ratio, W/t, together with the spatial dimension, determines the properties of the system. Typically, one finds that at small values of $W < W_c$, the states of the system are extended (typically $W_c \to 0$ in one- and some two-dimensional systems) and the system is conducting, while when the disorder is large enough, $W > W_c$, the states become localized and the system may become an insulator. It turns out that the concept of an (energy) mobility edge emerges, such that some states are localized and some states are extended, and the transport properties depend therefore on the location of the chemical potential on the energy spectrum.

Other disorder models may be chosen such as one that represents defects in the lattice that result from a foreign atom at random locations, with a fixed local potential but with a spatial distribution that is random, or we may have a random distribution of vacancies in the system, etc. Also, different symmetries of the Hamiltonian matrix in the presence of disorder lead to different classes of disordered systems with different behaviors.

In general, we may expect a metal-insulator transition due to the presence of disorder. In the insulating regime, characterized by the existence of localized states, by their nature the states are exponentially localized around some defect, which naturally introduces the concept of a localization length, with the physical meaning associated to the degree of localization (or spatial extent) of a state around the defect. The insulating phase may then be characterized by the presence of a finite localization length. In the metallic phase, the states are extended and the localization length diverges. In the metallic phase, a useful parameter is the conductivity of the system that vanishes in the insulating phase. At the transition point, we may expect some scaling that leads to the divergence of the localization length and the vanishing of the conductivity. In the light of other phase transitions previously considered, one may expect some scaling form of the localization transition. This will be addressed later. In addition, at the transition point the states are intermediate between the localized and extended states. Indeed, it is found that the states have a mixed nature, called critical or multifractal states.

In the case of $1d$ all the states are localized as shown in Fig. 23.6, as obtained by direct diagonalization of the Hamiltonian matrix for a finite system. This localization of the eigenstates, which results from the effect of the random potential of the Anderson type, is called strong localization. In the cases of $2d$ and $3d$ the solution is more elusive and will be considered ahead.

The degree of delocalization may be quantified using the participation ratio that is such that it is large if the states are extended and small if the states are localized. The participation ratio for a given state is defined as

$$PR = \frac{L^{-1}}{\sum_j \left(|\psi_j|^2\right)^2}, \tag{23.106}$$

where $|\psi_j|^2$ includes the sum over the two spin components at site j. It is such that is of the order of one if the state is extended and is of the order of $1/L$, where L

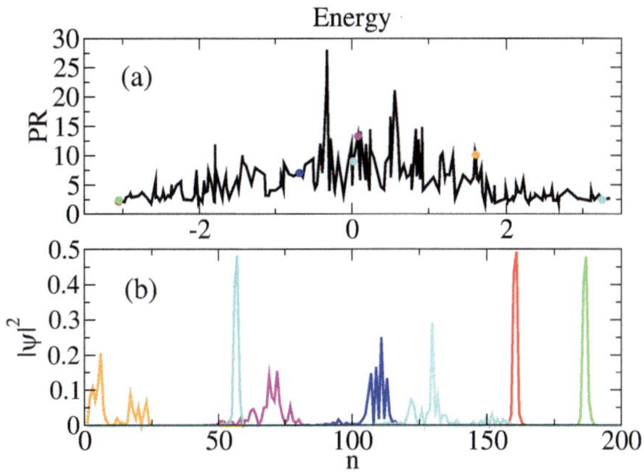

Fig. 23.6 (**a**) Participation ratios of the energy levels of a one-dimensional model with Anderson disorder, with $W = 2$, indicating that all levels are localized. (**b**) Wave functions as a function of site, n, for some states indicated in (**a**). The spatial extent of each wave function is exponentially small

is the system size, if the state is localized. To quantify the effects of disorder, it is also usual to use instead the inverse participation ratio, IPR. For a given eigenstate labeled by m, the IPR is defined as:

$$\mathcal{IPR}_m = \sum_i |\psi_i^m|^4, \tag{23.107}$$

with ψ_i^m the wave function of the eigenstate m at a site i. For perfectly localized states we have that $\mathcal{IPR}_m \sim 1$ and for delocalized states $\mathcal{IPR}_m \sim 1/L$.

Random Matrix Theory and Level Statistics

As we saw before in Chap. 1, the binomial distribution Eq. (1.3) describes the probability distribution to have a given number, N_1, of one of two possible outcomes with a probability, p, out of a given set of N experiments, such that the sequence of events is independent and uncorrelated. It was also shown that the average number of these outcomes is given by Np. Calling $\lambda = Np$ and taking the limit of an infinite number of experiments and the limit $p \to 0$, such that λ is fixed, the binomial distribution converges to the Poisson distribution since

$$\lim_{N\to\infty} \binom{N}{N_1} p^{N_1} (1-p)^{N-N_1} = e^{-\lambda} \frac{\lambda^{N_1}}{N_1!}. \tag{23.108}$$

As shown before, λ is the average value and the variance of the random variable N_1 (note that if $p \to 0$, then $q \to 1$). Note that it is finite for small N_1. This distribution also applies if there are several possible events if they are rare and occur independently of each other.

Consider now a system, for instance, a disordered system, such that its energy eigenvalues are not known and have some random distribution of energy levels.

Suppose the energy levels can be ordered in terms of their magnitude and are labeled by some quantum number, m. If the system is finite, in general there are small gaps between the various energy levels $s_m = E_m - E_{m-1}$. Suppose the energy levels are random and call $\bar{s}$ the average distance between the energy levels. Given a certain energy level, with value E_m, the next energy level will be present with probability $\sim \Delta s/\bar{s}$, within a small energy range of extent Δs. To determine the probability that the energy gap $s_m = E_{m+1} - E_m$ is found in the given range between s and $s + \Delta s$, we may divide the interval s in n small energy ranges, of width Δs, and consider that the next level is not found in the first n intervals but in the next one. The probability is therefore $P(s) = (1 - \Delta s/\bar{s})^n \, \Delta s/\bar{s}$ with $n = s/\Delta s$. Taking the continuum limit, such that $\delta s \to 0$ (or $n \to \infty$), we get the Poisson-like distribution:

$$P(s) = \bar{s}e^{-s/\bar{s}}, \tag{23.109}$$

which decreases monotonically with the size of the gap and is finite as $s \to 0$.

This distribution changes over the spectrum, because the average value of the gaps, $\bar{s}$, may not be constant throughout the spectrum. It is common to use instead the ratio of the gaps between levels as

$$r_m = \frac{min(s_m, s_{m-1})}{max(s_m, s_{m-1})}, \tag{23.110}$$

which eliminates the dependency on the average gap size. If the gaps are uncorrelated, one can show that the distribution of the ratios of the energy gaps satisfies

$$P(r) = \frac{2}{1+r^2}. \tag{23.111}$$

Another approach to the problem of the energy levels of a random system is to take the perspective of total lack of knowledge of the Hamiltonian itself, apart from some possible restrictions that result from some known symmetries of the problem. The classes of global symmetries were already discussed in the context of topological systems: for instance, invariance under time reversal, particle-hole or charge conjugation, or chiral symmetry. Also, there may be local symmetries that result from spatial distributions of the atoms, as in a crystal. Taking as a simple example a two-level system (out of the many possible states), the hermitian Hamiltonian may be written generically as

$$H = \begin{pmatrix} \epsilon_1 & \frac{V}{\sqrt{2}} \\ \frac{V^*}{\sqrt{2}} & \epsilon_2 \end{pmatrix}. \tag{23.112}$$

The eigenvalues are given by

$$E_{1,2} = \frac{\epsilon_1 + \epsilon_2}{2} \pm \frac{1}{2}\sqrt{(\epsilon_1 - \epsilon_2)^2 + 2|V|^2}. \tag{23.113}$$

If $V = 0$, the energy levels are just the diagonal elements, ϵ_1, ϵ_2. If $V \neq 0$, which means that there is a nonvanishing matrix element between the two levels, the new energy levels are further distant from each other. For instance, if $\epsilon_1 = \epsilon_2$ and $V = 0$, the two states are degenerate and, turning on V, the degeneracy is lifted. Therefore, there is what is called level repulsion, if the two states are coupled by V. In some cases this coupling is absent, which may occur in integrable systems. In this case there is no level repulsion, which means that it is possible to find zero gaps between states, which suggests that a Poisson-like distribution may be adequate.

Let us, however, consider that $V \neq 0$ and that the various parameters in the 2×2 Hamiltonian are random. Consider first that there is time-reversal invariance and the Hamiltonian matrix can be written as a real matrix, which implies that $V = V^*$ is a real potential. The randomness of the parameters in the Hamiltonian is chosen as gaussian distributions with zero mean and equal variances, σ^2, for the three random variables $\epsilon_1, \epsilon_2, V$. As before, let us calculate the probability of the energy level separation $s = E_1 - E_2$ given by [1]

$$P(s) = \frac{1}{(2\pi)^{3/2}\sigma^3} \int d\epsilon_1 \int d\epsilon_2 \int dV \delta \left(\sqrt{(\epsilon_1 - \epsilon_2)^2 + 2V^2} - s \right) e^{-\frac{\epsilon_1^2+\epsilon_2^2+V^2}{2\sigma^2}} . \tag{23.114}$$

Write $\epsilon_2 = \epsilon_1 + \sqrt{2}\xi$ and integrate over ϵ_1 to get

$$P(s) = \frac{1}{2\pi\sigma^2} \int d\xi \int dV \delta \left(\sqrt{2\xi^2 + 2V^2} - s \right) e^{-\frac{\xi^2+V^2}{2\sigma^2}} . \tag{23.115}$$

Using polar coordinates $V = e\cos(x)$, $\xi = r\sin(x)$, we can evaluate the remaining integrals yielding, for the case with time-reversal invariance,

$$P(s) = \frac{s}{2\sigma^2} e^{-\frac{s^2}{2\sigma^2}} . \tag{23.116}$$

In the absence of time-reversal symmetry, the matrix element is complex and the real and imaginary parts are independent variables. The three remaining integrals can be evaluated using spherical coordinates, yielding the distribution

$$P(s) = \frac{s^2}{2\sqrt{\pi}\left(\sigma^2\right)^{3/2}} e^{-\frac{s^2}{4\sigma^2}} . \tag{23.117}$$

The distributions Eq. (23.116) and Eq. (23.117) for the level spacing show level repulsion, since as $s \to 0$ both vanish, in contrast with the Poisson distribution. Also, both decay as a gaussian when the level spacing is large. We may compactly write both distributions as

$$P(s) = A_\beta s^\beta e^{-B_\beta s^2} , \tag{23.118}$$

where $\beta = 1$ if there is time-reversal symmetry and $\beta = 2$ in its absence. These types of distributions are called Wigner-Dyson distributions.

These results generalize if we consider larger matrices that involve larger numbers of states that are coupled. One may select Hamiltonian matrices out of a gaussian distribution given by

$$P(H) \sim e^{-\frac{\beta}{2a^2} Tr(H^2)} = e^{-\frac{\beta}{2a^2} \sum_{i,j} H_{ij} H_{ji}}. \tag{23.119}$$

Here, a sets the energy scale. In the case of $\beta = 1$ we have a class of Hamiltonians that are time-reversal invariant, the Hamiltonian terms are real, and $H_{ij} = H_{ji}$. This is called the gaussian orthogonal ensemble of Hamiltonian matrices. The choice $\beta = 2$ applies to systems that do not satisfy the time-reversal invariance, the Hamiltonian terms are complex, and $H_{ij} = H^*_{ji}$. This is called the gaussian unitary ensemble.

23.4.2 Quasidisorder of Aubry-André Type

In contrast with the case of Anderson disorder, that any amount of disorder leads to localized states in a one-dimensional system, quasidisorder (or quasiorder) may lead to extended states as well as localized states. The Aubry-André potential is one such example that leads to two regimes such that for small disorder amplitudes the states remain extended and at high disorder amplitudes (amplitude of the potential) the states are fully localized.

The Aubry-André potential is written as a local chemical potential of the form $\mu_i = V \cos(2\pi\alpha i + \varphi)$, where α is an irrational number, and therefore, the potential is incommensurate with the lattice. A usual choice is $\alpha = (\sqrt{5} - 1)/2$, the inverse golden ratio. We may also use $\alpha = \alpha_n$, where $\alpha_n = F_{n-1}/F_n$ and F_n are the Fibonacci numbers, which are defined recursively as $F_n = F_{n-1} + F_{n-2}$, with $F_1 = F_2 = 1$. Taking the $\lim_{n\to\infty} \alpha_n$ one gets $(\sqrt{5} - 1)/2$. For the case of periodic (open) boundary conditions, one takes for system size $L = F_n$ $(L = F_n + 1)$ and with α_n the local chemical potential becomes periodic, with period L. One can, therefore, simulate the effect of an Aubry-André quasiperiodic (or quasirandom) potential, considering a sequence of Fibonacci sizes and taking $\alpha = \alpha_n$, as above. The transition between the regime of extended states and the regime of localized states occurs at a critical point, $V = 2t$, where critical states appear. These states have a spatial dependence that is intermediate between extended and localized states.

The critical point is the result of a dual transformation between the two regimes, with a fixed point at the critical point. Consider the Hamiltonian

$$H = \sum_n t \left(c_n^\dagger c_{n+1} + h.c \right) + \sum_n V(n) c_n^\dagger c_n, \tag{23.120}$$

where t is the hopping and V the potential, written as $V(n) = V\cos(2\pi\alpha n + \phi)$. The Schrödinger equation for the probability amplitude to find the particle at site n (wave function ψ_n) is written as

$$t\left(e^{ik}\psi_{n+1}^r + e^{-ik}\psi_{n-1}^r\right) + V\cos\left(2\pi\alpha n + \phi\right)\psi_n^r = E\psi_n^r, \tag{23.121}$$

where we introduced a phase e^{ik} between sites n and $n+1$. Define now

$$\psi_n^r = \sum_m e^{i2\pi\alpha m n}\psi_m^d. \tag{23.122}$$

Replacing in the Schrödinger equation we get

$$2t\cos\left(2\pi\alpha m + k\right)\psi_m^d + \frac{V}{2}\left(e^{-i\phi}\psi_{m+1}^d + e^{i\phi}\psi_{m-1}^d\right) = E\psi_m^d. \tag{23.123}$$

There is a duality between the two descriptions if we consider the point $V = 2t$ and $k = -\phi$. Performing the Fourier transform between the wave functions, we find a duality (invariance) at that specific transition point. In the original description, a small or vanishing potential leads to a behavior characteristic of free fermions and therefore extended states if the chemical potential is inside the band. In this case it is inside the band, since the chemical potential is taken as zero, we have a half-filled band with a well-defined Fermi surface. In the other formulation of Schrödinger equation, a vanishing potential leads to a band with no hopping and only local terms, which leads to localized states. The invariant point should be the transition point between the two regimes.

23.4.3 Fractal Analysis

One of the effects of Anderson transitions is the emergence of multifractality, which is characterized by fluctuations of eigenstates. These fluctuations are manifested in the generalized inverse participation ratio. For a given eigenstate labeled by m, the generalized IPR is defined as

$$\mathcal{IPR}(q)_m = \sum_i \left|\psi_i^m\right|^{2q}, \tag{23.124}$$

where, as before, ψ_i^m is the wave function of the eigenstate m at a site i. At criticality, the generalized IPR behaves as

$$\mathcal{IPR}(q)_m \sim L^{\tau(q)}, \tag{23.125}$$

where L is the system size and the exponent $\tau(q)$ is defined in terms of a generalized dimension, $D(q)$, as $\tau(q) = D(q)(q-1)$. In a metallic phase, $D(q) = d$ and for an insulating phase $D(q) = 0$. Wave function multifractality is characterized by a q-dependent value of $D(q)$, whereas the cases of a constant $D(q) \neq d$ are single fractals.

In the case of a one-dimensional system with Anderson disorder, for the clean system the values of $\tau(q)$ closely follow the line $\tau(q) = (q-1)$, indicating that $D(q)$ is $q-$independent and equal to 1. As disorder is introduced, $\tau(q)$ approaches the line $\tau(q) = 0$, where $D(q) = 0$, suggesting the states are localized. In order to reach a proper conclusion it is necessary to evaluate $\tau(q)$ as the system size tends to infinity. To do this, subintervals of L are considered, to which a fit of equation of the form of Eq. (23.125) is done. One finds that as larger and larger values of L are considered, the curves $\tau(q)$ approach $\tau(q) = 0$. This confirms a localization of the bulk states in the thermodynamic limit for small values of disorder.

In the case of Aubry-André disorder one finds that for small values of disorder strength, V, $\tau(q)$ follows the line $q-1$ closely. For larger values of the disorder, the same results as for the case with Anderson disorder are found, with localized states, as concluded by $\tau(q) = 0$. The vicinity of the transition point $V = 2$ is more complex since there is a mixture between extended and localized states. One finds from the exact diagonalization that states have a mixed character, called multicritical states, that have IPR that are intermediate between the values characteristic of extended and localized states, and one finds a multifractal nature since $D(q)$. A similar multifractal behavior is found in the case of Anderson disorder in higher dimensions in the vicinity of the localization transition.

Kitaev Model and Quasidisorder

While in the usual Aubry-André model the multicritical states are found at the transition point, including this type of quasidisorder in the Kitaev model leads to a finite region where multicritical states are found.

We consider a 1D Kitaev chain model of spinless fermions in a lattice of size L, with either open or periodic boundary conditions, whose Hamiltonian reads

$$\hat{H} = \hat{H}_{0J} + \hat{H}_{0\Delta} + \hat{H}_V, \tag{23.126}$$

where $\hat{H}_{0J} = -\sum_i (J\,\hat{c}_i^\dagger \hat{c}_{i+1} + \text{H.c.})$ is the kinetic energy, $\hat{H}_{0\Delta} = -\sum_i (\Delta\,\hat{c}_i^\dagger \hat{c}_{i+1}^\dagger + \text{H.c.})$ is the superconducting p-wave pairing, $\hat{H}_V = \sum_i V_i \hat{c}_i^\dagger \hat{c}_i$ is the Aubry-André potential, and H.c. stands for the hermitian conjugate of the preceding terms. The fermionic creation (annihilation) operator at site i is $\hat{c}_i^\dagger$ ($\hat{c}_i$); J and Δ are the hopping and superconducting p-wave pairing between sites i and $i+1$, respectively. Hereafter, we choose $J = 1$ to set the energy scale of the problem. As above, the potential is of the form $V_i = V\cos(2\pi\alpha i + \varphi)$, where we take α as the inverse golden ratio $(\sqrt{5}-1)/2$, which renders its incommensurability

Fig. 23.7 Phase diagram of one-dimensional Kitaev model in the presence of Aubry-André quasidisorder

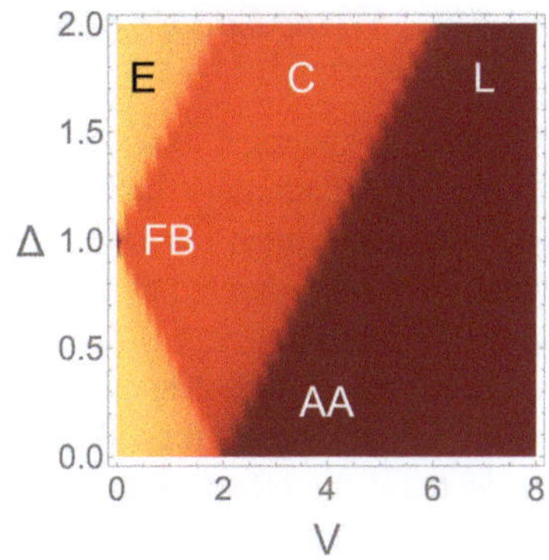

with the lattice. We have also included an additional phase $\varphi \in [0, 2\pi)$ that allows for a "disorder" average, thus reducing the statistical and finite-size effects.

In the absence of superconducting pairing ($\Delta = 0$), the quasi-periodic potential in 1d induces a metal-insulator transition at a finite value of the potential strength ($V = 2$), as discussed above. Now, turning on the pairing term Δ, we notice that the Hamiltonian has a phase diagram as shown in Fig. 23.7. There are essentially three phases, classifying the states in the bulk: (i) localized (L), (ii) critical (C) or multifractal, and (iii) delocalized or extended (E) states. If open boundary conditions are used, regions (ii) and (iii) host Majorana edge modes and are topologically nontrivial. We emphasize two special points in the phase diagram: AA highlights the duality point, where the metal-insulator transition occurs for $\Delta = 0$. FB marks the flat band point, where all the states are degenerate (with energy 2) in a periodic, translationally invariant system, whereas two perfectly localized Majorana zero energy states appear in the open system.

Localization of the bulk states can be studied via the non-ergodic properties of the system's eigenvalues and eigenvectors. To quantify the level of ergodicity, one can use again the inverse participation ratio or the participation ratio (PR) of the eigenvectors of the Hamiltonian. In that case, we define the PR as $\mathcal{R}_m = 1/\sum_i (p_i^m)^2$, where the occupation of the Bogoliubov quasiparticle m on site i is given as $p_i^m = |u_i^m|^2 + |v_i^m|^2$ and u_i^m, v_i^m are the corresponding particle and hole coefficients, respectively. The average PR is then $\overline{\mathcal{R}} = \overline{\langle \mathcal{R}_m \rangle_r}$, where we first average over all the eigenstates m and then take an average over different disorder realizations r. The average PR thus quantifies localization of the eigenvectors in real space. A completely localized state has $\mathcal{R} = 1$, while a perfectly delocalized state (such as a plane wave) has $\mathcal{R} = L$. In contrast, critical states scale with the multifractal dimension of the wave function. The three distinct regions of the phase diagram are thus such that (i) localized with the average normalized PR (NPR) $\overline{\mathcal{R}}/L \sim \mathcal{O}(1/L)$, (ii) critical with $\mathcal{O}(1/L) < \overline{\mathcal{R}}/L < \mathcal{O}(1)$, and (iii) delocalized with $\overline{\mathcal{R}}/L \sim \mathcal{O}(1)$.

23.4.4 Weak Localization and Weak Anti-Localization

The strong Anderson localization considered before was the result of the scattering off the local potentials that lead, in some cases, to either all states being localized (and the consequent insulating behavior) or regions of the energy spectrum being localized (and separated by a mobility edge from the extended states), if the disorder is strong enough.

The presence and effect of the impurities in the system may have a more subtle effect that is directly related to a quantum interference effect, which may enhance or decrease the effect of localization, by favoring or not the backscattering effect due to the impurities. Let us consider that the disorder is weak. Still it may have a significant effect. Unlike the strong disorder we considered above, the disorder is now weak and the states are still delocalized. The tendency of the impurities is to localize states, by binding the electrons to them, but to form true bound states one needs in a $3d$ system a minimum attractive-like potential. However, there is the possibility of interference between delocalized modes that may lead to some localization.

Consider one electron that has multiple scattering effects off impurities. There is some finite probability amplitude that, after some path, it may return to the starting point, as illustrated in Fig. 23.8. In some sense this is similar to a "localization" of the electron, since after multiple steps it returned to its original location. If there is time-reversal invariance, there is the possibility of a corresponding path that is just the same path, but traversed in the opposite sense. Call A the amplitude of one path and $\bar{A}$ the amplitude of the reversed path. The total probability associated with both paths is

$$P = |A + \bar{A}|^2 = |A|^2 + |\bar{A}|^2 + 2Re[\bar{A}^* A]. \tag{23.127}$$

Take $A = \sqrt{p_1} e^{i\phi_1}$ and $\bar{A} = \sqrt{p_2} e^{i\phi_2}$. Then

$$P = p_1 + p_2 + 2\sqrt{p_1 p_2} \cos(\phi_2 - \phi_1). \tag{23.128}$$

If the two paths are the time reversal of one another, then $A = \bar{A}$, which implies that $P = 4p_1$, while without the interference term $P = 2p_1$. The quantum effect of the interference between the two paths is that the return probability to the initial location of the electron has been enhanced. In $1d$ (and in $2d$) this enhancement of the return probability is enough to localize the states. Note that the argument holds

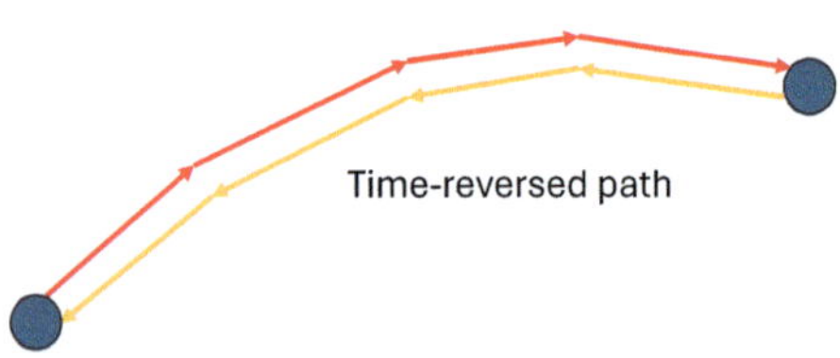

Fig. 23.8 Interference between two time-reversed paths under multiple scattering off impurities

if there is time-reversal invariance. In the presence of a magnetic field, for instance, time-reversal symmetry is broken and this effect is washed out.

The arguments presented above assume a constructive interference that leads to an enhancement of the return probability, which eventually led to the localization of the states. It turns out that under some circumstances the interference may be destructive. This leads to a decreased interference (destructive interference) which leads to a weak anti-localization as a result of the quantum interference effects.

23.4.5 Scaling Theory of Localization

In this section we wish to study in some detail the metal-insulator transition that occurs as the disorder is increased. As mentioned above, it may be studied in various ways, such as looking at the properties of the spectrum and its eigenstates, or the transport properties, like the conductivity or the behavior of the localization length, as one approaches the transition point. Since the localization length is expected to become large at the transition point (if it occurs), we may attempt a real space renormalization group approach looking for a scaling as the size of some typical length increases.

Based on some arguments, a scaling theory of the localization has been proposed some time ago.[4] The physical quantity selected to characterize the transition is the (adimensional) conductance, g_L, that, as mentioned above, clearly distinguishes the conducting from the insulating phases. The proposal consists in the choice

$$g_{bL} = f\left(b, g_L\right) . \tag{23.129}$$

Here L is the linear size of a given characteristic region of the disordered system, b is a rescaling factor, and f is some scaling function. This proposal has been verified by explicit calculations for different systems and different types of disorder, for instance, using numerical methods. Here we will limit ourselves to explore some consequences of the scaling hypothesis, in particular to show that the transition takes place in a $3d$ system and that for lower dimensions, $d = 1, 2$, the disorder localizes the states and the system is insulating.

Let us divide the disordered system into hypercubes with d-dimensions, with linear sizes L and volume L^d. Let us consider that from hypercube to the next, the disorder leads to different sets of states. The argument here is somewhat similar to that used in the qualitative explanation of the Thomas-Fermi approximation, where we considered that near each mesoscopic location there is a given Fermi surface that is displaced from that in neighboring regions due to the application

[4] See E. Abrahams, P.W. Anderson, D.C. Licciardello and T.V. Ramakrishnan, Physical Review Letters **42**, 673 (1979). Also we follow personal notes by Miguel Gonçalves, Instituto Superior Técnico, Universidade de Lisboa (2020).

of a local (nonhomogeneous) potential. Let us consider W' the energy difference between energy levels of neighboring hypercubes (in the vicinity of a given energy E) and let us consider that the coupling between the neighboring cubes is t'. Each cube has a set of L^d eigenstates and we may write that $\rho(E)W' = (L/L_0)^d$, where $\rho(E)$ is the density of states and L_0 is a microscopic scale (like the lattice constant). As the solution of the Anderson model shows, the ratio t/W between the hopping and disorder amplitude (taken here by the hopping t' and the energy level difference between neighboring hypercubes) determines the various regimes such that if $t'/W' << 1$, we expect a localized regime and in the opposite limit we expect a conducting regime. Therefore, the ratio fully determines the properties of the system.

Consider now a rescaling of the hypercube linear dimension by considering a hypercube with rescaled size given by $(bL)^d$. Changing the hypercube size also changes the ratio to $\left(t'/W'\right)_{bL}$. We expect that this new ratio should be a function of the ratio for the size L. Therefore, one assumes that

$$\left(t'/W'\right)_{bL} = f\left(b, \left(t'/W'\right)_L\right). \tag{23.130}$$

The remaining step is the assumption that the adimensional conductance, $g_L = G_L/(e^2/h)$, has the same scaling, since it is expected to be proportional to the ratio t'/W', since they have similar physical behaviors. Note again that while these arguments should be valid, they have indeed been confirmed by explicit calculations carried out in disordered systems.

Similarly to the steps used before, when studying general phase transitions, we take $b = 1 + \delta b$. Consider the logarithm of the adimensional conductance. Then

$$\log g_{bL} = \log f\,(1 + \delta b, g_L) = \log g_L + \left(\frac{\partial \log f\,(b, g_L)}{\partial b}\right)_{b=1} \delta b. \tag{23.131}$$

Using that

$$\delta b = d \log L = \log\left(L(1 + \delta b)\right) - \log L \tag{23.132}$$

and defining the scaling function $\beta(g_L)$, we write

$$\beta(g_L) = \frac{d \log g_L}{d \log L} = \left(\frac{\partial \log f(b, g_L)}{\partial b}\right)_{b=1}. \tag{23.133}$$

Even though we do not have an explicit knowledge of the β scaling function, we have an idea of its form in limiting values. When g is large (metallic regime where the mean free path is large, $l >> L$), we expect Ohm's law and

$$g_L = \sigma L^{d-2}. \tag{23.134}$$

Therefore, we get that in this metallic regime, where g_L is large,

$$\beta(g) = d - 2. \tag{23.135}$$

In the localized regime, when $L >> \lambda$ and the conductance is vanishingly small, we get

$$g_L = g_a e^{-\frac{L}{\lambda}}, \tag{23.136}$$

where λ is the localization length. In this insulating regime we get that

$$\beta(g) = \log\left(\frac{g_L}{g_a}\right). \tag{23.137}$$

In the approach to the critical behavior, we expect that characteristic lengths become very large. One can assume monotonic behavior of the scaling function. When $g <<$ 1 and taking L to be large, we have that the scaling function becomes negative, and we are in the insulating regime. In the limit when g becomes large, we see qualitatively different regimes as a function of the space dimensionality. For $d = 1$ the limiting behavior of the $\beta(g)$ function is such that it tends to $\beta = -1 < 0$, even when $g \to \infty$. Therefore, the β function is always negative, which implies that the system does not scale to a finite conductance and a metallic behavior, as expected. In the case $d = 2$ the β function is also always negative, except at $g \to \infty$ where it vanishes. Therefore, it is also predicted that the system remains an insulator. In the case of $d = 3$ there is a crossover from a regime where $\beta < 0$ to a regime where $\beta > 0$. In this case we may expect a transition between the insulating and the metallic regimes that takes place at the (unstable) fixed point $\beta(g_c) = 0$.

Define a function $F(g)$ as

$$\int_{g_0}^{g} \frac{d\log g}{\beta(g)} = \log\left(\frac{F(g)}{F(g_0)}\right). \tag{23.138}$$

Using the definition of the scaling function $\beta(g_L)$, we get

$$\frac{F(g)}{F(g_0)} = \frac{L}{L_0}. \tag{23.139}$$

In the limiting regimes we get that

$$\begin{aligned} \beta(g) &= \log\frac{g}{g_a}\,; \qquad & F(g) &= A\log\frac{g}{g_a}; \qquad & g &<< g_c \\ \beta(g) &= 1\,; \qquad & F(g) &= Bg; \qquad & g &>> g_c. \end{aligned} \tag{23.140}$$

Near the critical point g_c that satisfies $\beta(g_c) = 0$, we may expand

$$\beta(g) \sim \frac{1}{\nu} \log \frac{g}{g_c}. \tag{23.141}$$

The physical meaning of the parameter ν will become apparent in the following. Assuming that this behavior extends to a finite region around the critical point, we may write that

$$F_{\pm}(g) = C_{\pm} |\log(g/g_c)|^{\nu}. \tag{23.142}$$

Here, the case $+$ applies for the condition $g > g_0 > g_c$ and $-$ applies for the condition $g < g_0 < g_c$. Using that $\log(g/g_c) \sim (g - g_c)/g_c$ we may also write that

$$F_{\pm}(g) = C_{\pm} \left| \frac{g - g_c}{g_c} \right|^{\nu}. \tag{23.143}$$

In the metallic phase we expect the relation

$$\frac{F(g_+)}{F(g_0)} = \frac{B g_+}{C_+ |(g_0 - g_c)/g_c|^{\nu}} = \frac{L}{L_0}. \tag{23.144}$$

This leads to

$$g_+(L) = C'_+ \frac{L}{L_0} \left(\frac{g_0 - g_c}{g_c} \right)^{\nu} = \frac{L}{\lambda_+}, \tag{23.145}$$

which defines a length scale, the correlation length

$$\lambda_+ = \frac{L_0}{C'_+} \left(\frac{g_0 - g_c}{g_c} \right)^{-\nu}, \tag{23.146}$$

which identifies ν as the correlation length critical exponent. The conductivity is then

$$\sigma = \frac{e^2}{h} \frac{g}{L} = \frac{e^2}{h} \frac{1}{\lambda_+}. \tag{23.147}$$

In the localized phase we have instead

$$\frac{F(g_-)}{F(g_0)} = \frac{A \log(g_-/g_a)}{C_- |(g_c - g_o)/g_c|^{\nu}} = \frac{L}{L_0}, \tag{23.148}$$

which implies a localization length λ_- defined by

$$g_-(L) = g_a e^{-C'_- \left(\frac{g_c - g_0}{g_c} \right)^{\nu} \frac{L}{L_0}} \sim e^{-\frac{L}{\lambda_-}}, \tag{23.149}$$

with

$$\lambda_{-} = \frac{L_0}{C'_{-}} \left(\frac{g_c - g_0}{g_c} \right)^{-\nu} . \tag{23.150}$$

The localization length also diverges at the critical point with the same critical exponent, ν.

23.5 Quenched and Annealed Disorder

In this section we want to consider the effect of disorder on a magnetic system. The disorder may take place in different ways, such as in the interactions between the spins, leading to so-called spin glasses, or due to the presence of an external magnetic field that changes randomly from site to site. This case is simpler and it is the aim of this discussion to show how the randomness affects the behavior of the system. In particular, we will see that, in a simplified description, the effect of the random magnetic field may be seen as an effective interaction between the spins, adding to the existing exchange.

A possible description of a set of magnetic moments is obtained from the isotropic Heisenberg model in a magnetic field:

$$H = -\frac{1}{2} \sum_{i,j} J_{ij} \mathbf{S}_i \cdot \mathbf{S}_j - \sum_j \mathbf{h}_j \cdot \mathbf{S}_j . \tag{23.151}$$

Here J_{ij} is the coupling between the spins on sites i, j (usually taken as nearest neighbors) and $\mathbf{h}_j$ is the local magnetic field that may take random values. The spin operators satisfy the usual commutation relations and are described in various ways, as seen before. Convenient representations that do not require a projection to a physical subspace are the Majorana representation and the representation in terms of spin coherent states. Using a suitable representation, we may obtain an expression for the partition function in terms of effective c-numbers, as explained in the following. The representation of the quantum problem in terms of these c-numbers (which may be matrix elements of the operators) allows a study of the thermodynamics of the system, using the partition function as a generator of thermal correlation functions, as discussed before.

In the presence of disorder and associated, in this example, with the uncertainty of the values of the local magnetic fields, one needs to consider some sort of average over different configurations of the local magnetic fields. This average procedure is carried out in general over the set of configurations, of whichever parameters take random values. The way the average is taken leads to different physical situations. Two typical ways to consider the average over disorder configurations are the so-called quenched and annealed disorders. The disorder is called quenched when the probability distribution of the random parameter is temperature independent and is called annealed when the random parameter is such that its spatial distribution is self-adjusted to minimize the system free energy, with the constraint that its global distribution is temperature independent. Quenched disorder is more realistic and the one usually found in experiments. However, from the calculational point of view, it is hard to implement. The average over disorder configurations has to be carried over the physical quantities, such as the free energy. On the other hand, in the case of annealed disorder, the average is taken over the partition function itself, which is easier to evaluate. Since the free energy is a logarithm of the partition function, one has to take averages on the logarithm of the partition function, in the realistic case of quenched disorder.

To be specific, consider the case of a random local magnetic field and assume that the distribution of its values is a gaussian:[5]

$$\begin{aligned} \overline{h_i^a} &= 0 \\ \overline{h_i^a h_j^b} &= \delta_{ab}\Delta_{ij}. \end{aligned} \tag{23.152}$$

In the case of quenched disorder, the free energy is obtained considering a disorder average of the type

$$\sim \overline{\log Z}. \tag{23.153}$$

In the case of annealed disorder the free energy is obtained as

$$\sim \log \overline{Z}. \tag{23.154}$$

Also, in the quenched case, the average value of an observable is obtained as

$$\overline{\langle O\rangle} = \overline{\left[\frac{Tr e^{-\beta H} O}{Tr e^{-\beta H}}\right]}. \tag{23.155}$$

In the annealed case it is obtained as

$$\overline{\langle O\rangle} = \frac{\overline{Tr e^{-\beta H} O}}{\overline{Tr e^{-\beta H}}}, \tag{23.156}$$

which is simpler since it involves the separate averages on the numerator and denominator.

The first step is to obtain an expression for the partition function of the spin quantum problem.

[5] Here we follow the Master degree thesis of Iveta Pimentel, Lisboa 1986.

23.5.1 Partition Function and Spin Coherent States

The path integral expression for the partition function can be obtained using a description of the spin, S, degrees of freedom in terms of spin coherent states. These are discussed in Appendix E. Using the spin coherent states, an expression is obtained for the evolution operator and from that expression we can obtain a similar expression for the partition function, considering imaginary time instead of real time and taking periodic boundary conditions between the initial and final points of the trajectory. It leads to

$$Z = \int \mathcal{D}^2\mu(t) e^{-\frac{1}{\hbar}\int_0^{\beta\hbar} d\tau [-\hbar \frac{S}{1+\alpha^*\alpha}(\frac{d\alpha^*}{d\tau}\alpha - \alpha^*\frac{d\alpha}{d\tau}) + \mathcal{H}_c(\alpha^*,\alpha)]}, \tag{23.157}$$

where periodic boundary conditions in the imaginary time have been used and by similarity with the discussion on the transition amplitude in Appendix E.

The free term in the Lagrangian is geometrically the area in the sphere defined by the trajectory under consideration and by two great arcs from the north pole (of the Bloch sphere where the classical spin vector evolves) to the initial and final points of the trajectory. It can be rewritten as

$$\mathcal{L} = \frac{d\mathbf{S}}{dt} \times \mathbf{S} \cdot \boldsymbol{\varphi}, \tag{23.158}$$

where $\boldsymbol{\varphi} = \frac{\mathbf{u}}{S+\mathbf{S}\cdot\mathbf{u}}$, with $\mathbf{u} = \mathbf{e_z}$, reflecting the choice of the z axis as the quantization axis in the construction of the coherent states. In general, $\mathbf{u}$ can be any arbitrary unit vector. Here $\mathbf{S}$ is the spin vector.

The Lagrangian may be totally rewritten in terms of the vector $\mathbf{S}$. The integration on the sphere can also be written as $1/S^2 d^3 S\delta(\mathbf{S}^2 - S^2)$. We arrive then at the path integral for spins

$$\int \mathcal{D}\mathbf{S}\delta(\mathbf{S}^2 - S^2) e^{i\int \mathcal{L}dt}. \tag{23.159}$$

The matrix elements of the spin operators $\hat{S}_\pm$ and $\hat{S}_z$ on the spin coherent states, needed for the construction of the expression for the partition function, are given by $\langle\alpha|\hat{\mathbf{S}}|\alpha\rangle = S\mathbf{m}$, where $\mathbf{m}$ is the direction of the spin average on the state $|\alpha\rangle$.

The partition function is therefore written in terms of a (path) integral over configurations of the spins (converted to average values or c-numbers), with an exponential that results from a Boltzmann factor. This factor involves the Hamiltonian written in terms of the averages of the spin operators, plus a term called kinetic term (which is the equivalent in the Lagrangian formulation of the Légendre transform between the Lagrangian and the Hamiltonian). Adding the term of the external magnetic field to this kinetic term and considering that the spins do not interact, the partition function reduces to that of a spin in the presence of a magnetic field. This partition function can be calculated by elementary methods, as we have seen before. Therefore, the complicated term is the one that results from the interactions. This term can be eliminated using a Hubbard-Stratonovich transformation.

The interaction term can be written as

$$e^{\frac{1}{2}\int_0^\beta d\tau \sum_{i,j} \mathbf{S}_i J_{ij} \cdot \mathbf{S}_j} = \int \frac{D\phi}{(det 2\pi J)^{1/2}} e^{\int_0^\beta d\tau \left(-\frac{1}{2}\sum_{i,j} \boldsymbol{\phi}_i \cdot J_{ij}^{-1} \boldsymbol{\phi}_j + \sum_i \boldsymbol{\phi}_i \cdot \mathbf{S}_i\right)}, \tag{23.160}$$

in terms of a set of bosonic fields $\boldsymbol{\phi}_i$. In this way the problem of interacting spins in the presence of a magnetic field is replaced by a set of free spins in the presence of a bosonic field, with correlation J_{ij}. The partition function is then written as a double integral over the initial variables and over the bosonic fields introduced by the exact Hubbard-Stratonovich transformation. The original fields may be integrated over, since the integral is gaussian leading to an expression for the partition function as

$$Z = \int \frac{D\phi}{(det 2\pi J)^{1/2}} e^{-\int_0^\beta d\tau \frac{1}{2}\sum_{i,j} (\boldsymbol{\phi}_i - \mathbf{h}_i) \cdot J_{ij}^{-1} (\boldsymbol{\phi}_h - \mathbf{h}_j) - \beta \sum_i \overline{F}(\boldsymbol{\phi}_i)}, \tag{23.161}$$

where we have made a translation $\boldsymbol{\phi} \to \boldsymbol{\phi} - \mathbf{h}$ and where $\overline{F}(\boldsymbol{\phi})$ is the free energy of one spin in a space and time-dependent effective field $\boldsymbol{\phi}$. This result was previously obtained in Eq. (15.69).

The average values over the bosonic fields are related to the average values of spin operators through exact relations, obtained taking derivatives with respect to the magnetic field, $\mathbf{h}$. Recall that we got that

$$\begin{aligned} \langle \boldsymbol{\phi} \rangle &= \mathbf{h} + J \langle \mathbf{S} \rangle \\ \langle \boldsymbol{\phi}\boldsymbol{\phi} \rangle_c &= J + J \langle \mathbf{SS} \rangle_c J, \end{aligned} \tag{23.162}$$

where $\langle \cdots \rangle_c$ denotes a connected correlation function (in this case the correlation function of two fields subtracted by the product of the average values of each field, separately). These averages are the thermal averages. In the presence of disorder we also have to consider the averages over the disorder configurations, as discussed above.

23.5.2 Random Magnetic Field

The problem of a random magnetic field is in general complicated and, even in its absence, it is a good approximation to consider a mean-field approach to simplify the problem. One way to do that is to consider a saddle-point approximation of the partition function, looking for a variational solution of the argument of the Boltzmann factor. This leads to a solution such that the average of the bosonic field is stationary and homogeneous and is oriented along a given direction (call it the

z-direction). Due to the exact relation to the average of the spin operator, that is, the magnetization, we consider a magnetization that is aligned along a given direction, z. The problem is then simplified to an equivalent paramagnet (in the presence of the bosonic field that takes the role of an effective magnetic field) or, if there is indeed a homogeneous external magnetic field, **h**, the fields just add. As we have seen before and considering a spin-1/2, we get that the average value of the bosonic field, call it $\langle \phi^z \rangle$, is of the form of a self-consistent Brillouin function for spin-1/2:

$$\langle \phi^z \rangle = J_0 \frac{1}{2} \tanh \frac{\beta \langle \phi^z \rangle}{2}. \tag{23.163}$$

Here,

$$J_0 = J_{\mathbf{k}=0} \tag{23.164}$$

is the zero momentum Fourier component of the exchange coupling between the spins, where we assumed translation invariance such that J_{ij} only depends on $\mathbf{r}_i - \mathbf{r}_j$. Taking into account disorder configurations and using the mean-field approximation implies taking now averages on the free energy itself (quenched disorder) or on the partition function (annealed disorder).

One way to carry out the average disorder in the quenched case amounts to use the so-called replica trick where one uses the mathematical formal expression:

$$\overline{\log Z} = \lim_{n \to 0} \overline{\frac{Z^n - 1}{n}}. \tag{23.165}$$

In the case of annealed disorder we take the disorder average of the partition function directly. This last case is simpler from the technical point of view and will be considered next.

Let us consider again the original partition function written in terms of the spin variables and consider that the distribution of magnetic fields is a gaussian with correlation Δ_{ij}. The magnetic field is coupled to the spin variables linearly. The average over disorder is then taken into account performing the gaussian average on the partition function of a time independent random magnetic field distribution:

$$\int D\mathbf{h} e^{-\frac{1}{2}\sum_{i,j} \mathbf{h}_i \cdot \Delta_{ij} \mathbf{h}_j + \sum_j \mathbf{h}_j \cdot \int d\tau \mathbf{S}_j} = e^{-\frac{1}{2}\sum_{i,j} \Delta_{ij} \int d\tau \mathbf{S}_i \cdot \int d\tau \mathbf{S}_j}. \tag{23.166}$$

The integration over the distribution of the random magnetic field leads to a *time-dependent* correction term to the coupling between the spins. The spin coupling term becomes

$$-\frac{1}{2} \int_0^\beta d\tau \int_0^\beta d\tau' \sum_{i,j} \tilde{J}_{ij}(\tau, \tau') \mathbf{S}_i(\tau) \cdot \mathbf{S}_j(\tau'), \tag{23.167}$$

where the effective interaction is given by

$$\tilde{J}_{ij}(\tau, \tau') = J_{ij}\delta(\tau - \tau') + \Delta_{ij}. \tag{23.168}$$

We conclude that the effect of the random (static) magnetic field is to induce an interaction that is nonlocal in time, in addition to the original nonlocal in space interaction. The interaction term may now be eliminated using a Hubbard-Stratonovich transformation (which is now nonlocal in time).

A leading approximation is obtained considering, in a similar way, a mean-field approximation (saddle point of the functional written in terms of auxiliar bosonic fields, as before). In mean field one assumes that the leading solution of the time- and space-dependent problem is time-independent and space-independent:

$$\langle \phi^z \rangle = \frac{1}{2} (J_0 + \beta \Delta_0) \tanh \frac{\beta \langle \phi^z \rangle}{2}, \tag{23.169}$$

where

$$J_0 = \sum_j J_{ij}; \qquad \Delta_0 = \sum_j \Delta_{ij}. \tag{23.170}$$

Therefore, the effective field in the Brillouin function is changed due to the presence of the random magnetic field. Recalling the connection between the bosonic field and the magnetization, we get that the magnetization is given by the self-consistent set of equations:

$$\begin{aligned} \langle S^z \rangle &= \frac{1}{2} \tanh \frac{\beta \overline{\phi}^z}{2} \\ \overline{\phi}^z &= (J_0 + \beta \Delta_0) \langle S^z \rangle. \end{aligned} \tag{23.171}$$

The critical temperature for the appearance of a nonvanishing magnetization satisfies

$$\frac{\beta_c}{4} (J_0 + \beta_c \Delta_0) = 1, \tag{23.172}$$

leading to

$$T_c = \frac{J_0}{8k_B} \left(1 + \sqrt{1 + 16 \frac{\Delta_0}{J_0^2}} \right). \tag{23.173}$$

Therefore, the annealed field leads to an increase of the critical temperature.

23.6 Exercises

1. A simple model for a disordered system, such as a glass, consists in taking that the atoms are distributed in some random way by two energy levels of the type

$$E_i = -\delta, \delta,$$

where we consider that for each atom only the two lowest energy levels, to simplify. However, the energies δ take random values between 0 and ∞. Assume that the distribution of these values is such that all values are equally probable. Therefore, the probability distribution $P(\delta)$ is a constant. Assume that the two-level systems are independent. The specific heat of one two-level system is given by

$$C_\delta = k_B \left(\frac{\delta}{k_B T}\right)^2 \frac{1}{\cosh^2\left(\frac{\delta}{k_B T}\right)}.$$

Calculate the temperature dependence of the specific heat of glass in the limit of low temperatures.

2. The Kubo formula for the dynamic conductivity has a term that involves the commutator of the charge current operator. Consider a one-dimensional tight-binding model. The various currents may be obtained in the usual way using their conservation laws in the general schematic form $\frac{\partial \rho}{\partial t} + \nabla \cdot j = 0$ where ρ is the charge density operator. In a one-dimensional lattice the divergence takes the form

$$(\nabla \cdot j)_i = J_{i\to i-1} + J_{i\to i+1},$$

where the current $J_{i\to i-1}$, for instance, has a hopping term that takes an electron from $i \to i-1$ and another term with opposite sign that takes an electron from $i-1 \to i$.

Using the equation of motion for the charge density operator, show that the charge current can be obtained from

$$J_c = -ei\frac{t}{\hbar}\sum_\sigma \sum_i \left(c^\dagger_{i,\sigma} c_{i+1,\sigma} - c^\dagger_{i+1,\sigma} c_{i,\sigma}\right).$$

Taking the Fourier transform to momentum space obtain an expression for the zero momentum charge operator.

3. Use the relaxation time approximation to the Boltzmann equation to obtain expressions for the electrical conductivity, the Hall coefficient, and the thermal conductivity of a metal, considering the electrons as a highly degenerate gas of particles that move freely, except for collisions with lattice impurities and

phonons. Assume that the chemical potential in the background distribution function f_0 is constant.

4. Consider a one-dimensional lattice, described by a tight-binding model with hopping t and a local potential V_i that fluctuates from site to site with a box distribution of amplitude W, as in Eq. (23.105). Consider a finite extent chain with periodic boundary conditions or open boundary conditions.

   (a) Diagonalize the Hamiltonian and obtain the energy eigenvalues and the eigenfunctions. Plot some characteristic wave functions and determine if they are localized or extended.
   (b) Calculate the participation ratio or the inverse participation ratio to confirm the nature of the eigenstates.
   (c) Consider now the Aubry-André model, Eq. (23.120). Determine the phase diagram as a function of V/t and take $\phi = 0$, to simplify. Plot some characteristic eigenfunctions for different values of V/t. Confirm that there is a transition from a regime of extended states to another regime of localized states, as V/t increases. Plot some characteristic eigenfunctions at the critical point.

5. Show that

$$\log Z = \lim_{n \to 0} \frac{Z^n - 1}{n}$$

Reference

1. L. D'Alessio, Y. Kafri, A. Polkovnikov, M. Rigol, Advances in Physics **3**, 239 (2016).

Chapter 24
One-Dimensional Systems

In many three-dimensional systems, the effect of interactions between fermionic particles is taken into account using the so-called Fermi liquid theory, previously discussed. In this theory, the excitation spectrum is fermionic, but the quasiparticle parameters are renormalized with respect to the free system. As shown in Eq. (14.45) the quasiparticles have residual interactions between them and finite lifetimes. The spectrum is a continuum of low energy excitations due to the excitations of particles and holes of arbitrarily small energies and momenta around the Fermi surface, in addition to the plasma mode at finite frequency.

The one-dimensional case is special due to the fact that the Fermi surface has only two points. An important consequence is that any instability of momentum $2k_F$ couples the two states at the Fermi surface and may lead to a gap in the spectrum (Peierls instability). In general, in systems of higher dimension, any instability that couples two points of the Fermi surface has a null measure and therefore is not relevant to the behavior of the system (with the exception of nesting where a dense set of points is coupled). The Peierls instability in one dimension suggests that the excitations of the system may have a different nature and indeed can be described in terms of collective bosonic excitations.

24.1 Luttinger Liquid and Bosonization

Due to the Pauli exclusion principle, at low temperatures, only the electrons close to the Fermi surface are important. In the case of a one-dimensional system, it is usual to linearize the dispersion relation of a free electron gas in the form $\epsilon_r(k) = v_F(rk - k_F)$ introducing two branches ($r = \pm 1$) around the two Fermi points, with velocities $\pm v_F$. It is then necessary to introduce a cutoff k_0. Such a procedure leads to the Tomonaga model. It is also possible (and convenient) to consider a dispersion relation without cutoff, extending the bands to $\pm\infty$ (note that considering

P. D. Sacramento, *Introduction to Condensed Matter Theory*,
https://doi.org/10.1007/978-3-031-86251-9_24

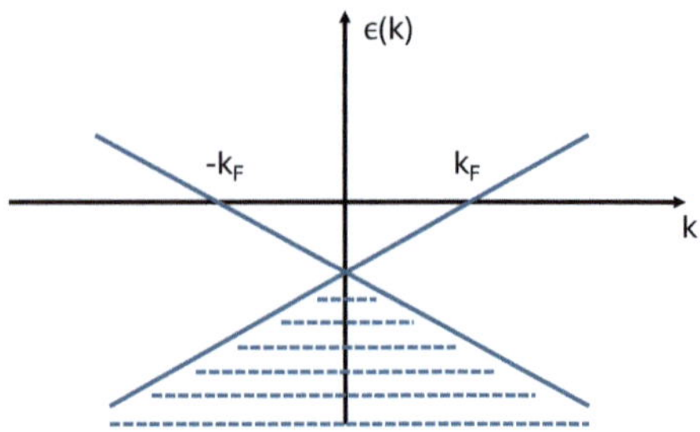

Fig. 24.1 Linearization of the spectrum around the Fermi momentum as in the Luttinger model

only the regime of small energies, these additional states are not expected to affect the behavior). This choice corresponds to the Luttinger model.[1] This is shown in Fig. 24.1.

Let us now add interactions. Focusing on the Luttinger case, the model describes two types of fermions corresponding to the two branches $r = \pm 1$, whose densities interact via two interactions: one term corresponds to interactions between electrons in the same branch (g_2) and the other between electrons in different branches (g_4). The Luttinger model is defined by $H = H_0 + H_2 + H_4$ where,

$$
\begin{aligned}
H_0 &= \sum_{r,k,s} v_F(rk - k_F) : c^\dagger_{rks} c_{rks} : , \\
H_2 &= \frac{1}{L} \sum_{p,s,s'} \left(g_{2,||}(p)\delta_{s,s'} + g_{2,\perp}(p)\delta_{s,-s'} \right) \rho_{+,s}(p)\rho_{-,s'}(-p) , \\
H_4 &= \frac{1}{2L} \sum_{r,p,s,s'} \left(g_{4,||}(p)\delta_{s,s'} + g_{4,\perp}(p)\delta_{s,-s'} \right) : \rho_{r,s}(p)\rho_{r,s'}(-p) : . \qquad (24.1)
\end{aligned}
$$

The first term is the usual kinetic operator, and in the interacting terms, the spin, s, dependence on the interactions is considered in the forms $||$ or $\perp$. The operators have to be normal ordered to eliminate the infinite number of states introduced in the model, by subtracting the average value in the groundstate. We define

$$
\begin{aligned}
\rho_{r,s}(p) &= \sum_k : c^\dagger_{r,k+p,s} c_{r,k,s} := \\
&= \sum_k \left(c^\dagger_{r,k+p,s} c_{r,k,s} - \delta_{p,0} \langle c^\dagger_{r,k+p,s} c_{r,k,s} \rangle_0 \right) , \qquad (24.2)
\end{aligned}
$$

subtracting the average value of the groundstate.

[1] For further reading see, for instance, J.M.P. Carmelo, P.D. Sacramento, D. Bozi and L.M. Martelo in "Strongly Correlated Systems, Coherence and Entanglement", Chapter 1, Eds. J.M.P. Carmelo, P.D. Sacramento, J.M.B. Lopes dos Santos and V. Rocha Vieira (World Scientific, Singapore, 2007).

Besides these terms that only include low-momentum scattering, the Luttinger model may be extended, considering additional terms that take into account the possibility of finite momentum excitations, where the two branches are coupled, allowing backscattering (g_1) or considering Umklapp processes (g_3).

It is usual to consider two new terms

$$H_{1,\perp} = g_{1\perp} \sum_s \int_0^L dx : \psi^\dagger_{+,s}(x)\psi_{-,s}(x)\psi^\dagger_{-,-s}(x)\psi_{+,-s}(x) : \tag{24.3}$$

and

$$H_{3,\perp} = \frac{g_{3\perp}}{2} \sum_s \int_0^L dx : \psi^\dagger_{+,s}(x)\psi^\dagger_{+,-s}(x)\psi_{-,-s}(x)\psi_{-,s}(x) + h.c. : . \tag{24.4}$$

The first term corresponds to an exchange of momentum of the order of $2k_F$ and the second term an exchange of momentum of the order of $4k_F$ (the corresponding term $g_{1||}$ is equivalent to $g_{2||}$, and the terms given by $g_{3||}$ or $g_{4||}$ do not give contributions).

The model without these additional terms is exactly solvable (without the backscattering term and without the Umklapp term). The Hamiltonian conserves the total charge, the spin, and also these quantities separately in each branch. Therefore the charge and spin currents are also conserved.

The interactions present in the Luttinger model also occur in problems of higher dimensionality. However, while in $1d$, these interactions are marginal; for instance, in $3d$, they are irrelevant. A consequence of the interactions in $1d$ is that the eigenmodes that correspond to charge and spin degrees of freedom move with propagation velocities that are different, leading to a separation of the degrees of freedom, as we will see. Since the treatment is somewhat technical, we will present the results without explicit derivations.

One way to solve the Luttinger model results from the property that the commutator of the density operators is given by,

$$[\rho_{r,s}(p), \rho_{r',s'}(-p')] = -\delta_{r,r'}\delta_{s,s'}\delta_{p,p'}\frac{rpL}{2\pi}, \tag{24.5}$$

where L is the system length. Therefore the density operators satisfy bosonic commutation relations. Defining charge and spin density operators in each branch

$$\begin{aligned} \rho_r(p) &= \left(\rho_{r,\uparrow}(p) + \rho_{r,\downarrow}(p)\right)/\sqrt{2} \\ \sigma_r(p) &= \left(\rho_{r,\uparrow}(p) - \rho_{r,\downarrow}(p)\right)/\sqrt{2}, \end{aligned} \tag{24.6}$$

and using Kronig's identity, where the kinetic term of the Hamiltonian can be written in a bilinear form in the density operators

$$H_0 = \frac{2\pi}{L} v_F \times \sum_{p>0} [\rho_+(p)\rho_+(-p) + \rho_-(-p)\rho_-(p) + \sigma_+(p)\sigma_+(-p) + \sigma_-(-p)\sigma_-(p)], \tag{24.7}$$

one obtains that the Luttinger model can be diagonalized via a Bogoliubov-Valatin transformation leading to a separation of the charge and spin degrees of freedom $\tilde{H} = \tilde{H}_\rho + \tilde{H}_\sigma + \tilde{H}_c$,

$$\begin{aligned} \tilde{H}_\nu &= \frac{2\pi v_\nu}{L} \sum_{p\neq 0} : (\tilde{\nu}_+(p)\tilde{\nu}_+(-p) + \tilde{\nu}_-(p)\tilde{\nu}_-(p)) :, \\ \tilde{H}_c &= \frac{\pi}{2L} \sum_{\nu=\rho,\sigma} \left(\frac{v_\nu}{K_\nu}(N_{+,\nu} + N_{-,\nu})^2 + v_\nu K_\nu (N_{+,\nu} - N_{-,\nu})^2 \right), \end{aligned} \tag{24.8}$$

($\nu = \rho, \sigma$) where the charge and spin velocities are given by,

$$v_\nu = \sqrt{\left(v_F + \frac{1}{2\pi}(g_{4\parallel} \pm g_{4\perp})\right)^2 - \left(\frac{1}{2\pi}(g_{2\parallel} \pm g_{2\perp})\right)^2}, \tag{24.9}$$

($+, -$ corresponds to ρ, σ respectively), $N_{r,\nu} = \rho_{r,\nu}(p = 0)$, and the parameters K_ν are given by,

$$K_\nu = \sqrt{\frac{\pi v_F + \frac{1}{2}(g_{4\parallel} \pm g_{4\perp}) - \frac{1}{2}(g_{2\parallel} \pm g_{2\perp})}{\pi v_F + \frac{1}{2}(g_{4\parallel} \pm g_{4\perp}) + \frac{1}{2}(g_{2\parallel} \pm g_{2\perp})}}. \tag{24.10}$$

The operators of the charge and spin degrees of freedom commute among themselves and are separately conserved, as said above. Moreover, they propagate with different velocities, leading to an effective separation in real space.

Collective charge and spin modes that propagate with different velocities also occur in systems of higher dimensionality. The difference from the one-dimensional systems is that in one dimension, the excitation spectrum does not include a continuum (the Stoner continuum) at low energies. While in a Fermi liquid the collective modes may decay into quasiparticles in the continuum, in the one-dimensional systems, there is no such continuum, and therefore, the collective modes remain as the only excitations of the system. This is shown in Fig. 24.2. The non-existence of fermionic excitations implies that there are no quasiparticles at the Fermi surface. The density of states vanishes at the Fermi surface, and therefore, we find a spectral weight reduction near that surface.

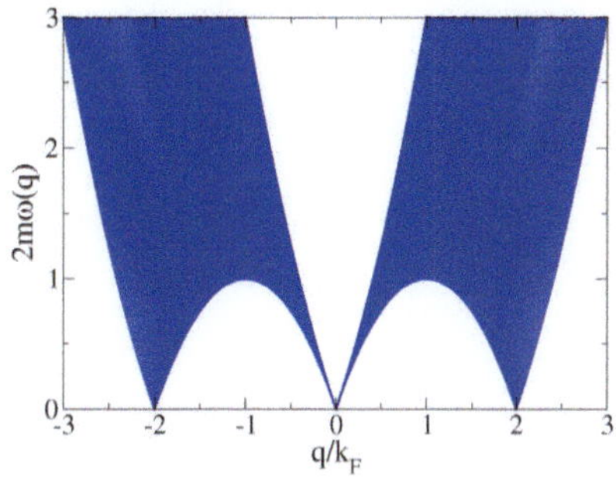

Fig. 24.2 Low-energy excitation spectrum in a one-dimensional fermionic system. Note the absence of low-energy modes, except sharp features at $k = \pm 2k_F$

The diagonalization of the Luttinger model shows that the Hamiltonian and its excitations are described by bosonic modes. However, the calculation of arbitrary correlation functions requires products of fermionic operators. The full solution of the model including all its correlation functions involves the representation of the fermionic operators in terms of the bosonic operators. This procedure is called bosonization.

The charge density and the current, $j_s(x)$, satisfy the following commutator:

$$[\rho_s(x), j_s(x')] = -\frac{i}{\pi}\frac{\partial}{\partial x}\delta(x - x'), \tag{24.11}$$

where $j_s(x) =: \rho_{+,s}(x) - \rho_{-,s}(x) :$. Let us then consider two conjugate fields, $\Phi_s(x)$ and $\Pi_s(x)$, satisfying the canonical commutation relation $[\Phi_s(x), \Pi_s(x')] = i\delta(x - x')$. Identifying the operators $\rho_s(x) = -\frac{1}{\pi}\frac{\partial}{\partial x}\Phi_s(x)$ and $j_s(x) = \Pi_s(x)$ satisfies the commutation relation. This result suggests that we may represent the density by a bosonic field of the form $\frac{\partial \Phi_s(x)}{\partial x} = -\pi\rho_s(x)$.

The introduction of a fermion at site x creates a kink of amplitude π in the bosonic field. The introduction of one particle at point x implies that the rest of the particles have to adjust to accept the new particle at x. We may therefore express the fermionic operator in the form of a translation operator (plus a phase needed to yield the anti-commutation relations),

$$\psi_{r,s}(x) \sim e^{irk_F x - ir\Phi_{r,s}(x) + i\pi\int_{-\infty}^{x} dz \Pi_{r,s}(z)}. \tag{24.12}$$

This relation establishes, therefore, a representation of a fermionic field operator in terms of bosonic fields and is called bosonization of the original fermionic theory.

The Luttinger model may also be expressed as $H = H_\rho + H_\sigma$, where,

$$H_\nu = \int dx \left(\frac{\pi v_\nu K_\nu}{2}\Pi_\nu^2 + \frac{v_\nu}{2\pi K_\nu}(\partial_x \phi_\nu)^2 \right), \tag{24.13}$$

with $\nu = \rho, \sigma$. Such a transformation allows the calculation of all the correlation functions of the Luttinger model (since it is a gaussian model). The correlation functions are characterized by critical exponents which are non-universal. The calculation of the correlation functions is reduced in this context to averages over a gaussian distribution. For instance, within bosonization it can be shown that

$$\langle \rho(x)\rho(0)\rangle = \frac{K_\rho}{(\pi x)^2} + A_1 \frac{\cos(2k_F x)}{x^{1+K_\rho}} \frac{1}{\ln^{-3/2}(x)} + A_2 \frac{\cos(4k_F x)}{x^{4K_\rho}}, \tag{24.14}$$

and

$$\langle \mathbf{S}(x)\cdot\mathbf{S}(0)\rangle = \frac{1}{(\pi x)^2} + B_1 \frac{\cos(2k_F x)}{x^{1+K_\rho}} \frac{1}{\ln^{1/2}(x)}. \tag{24.15}$$

Also, the quantity K_ρ determines the singularity of the momentum distribution,

$$\rho_k \sim \frac{1}{2} - \frac{k-k_F}{|k-k_F|}|k-k_F|^\alpha, \tag{24.16}$$

where $\alpha = (K_\rho + 1/K_\rho - 2)/4$ is the exponent of the single-particle density of states $D(\omega) \sim |\omega|^\alpha$ as well. The behavior of the momentum distribution in the vicinity of the Fermi momentum is shown in Fig. 24.3. One needs to determine the parameter K_ρ for each specific model.

Even though the Luttinger model is very simplified, it describes in a qualitatively correct way the low energy properties of many interacting one-dimensional systems. A broadly used nomenclature, classifying interacting one-dimensional systems whose low-energy behavior falls in the universality class of the Luttinger model, is that of a Luttinger liquid. Another class of one-dimensional systems is the Luther-Emery class, which groups systems with gaps in the spectrum. In particular, for instance, the addition of the backscattering term to the Luttinger model leads in some regimes (for an attractive interaction) to a gap in the spin excitations. The Umklapp term, on the other hand, leads to a gap in the charge excitations.

Luther-Emery Model and 2*d* Coulomb Gas

Let us consider the extended Luttinger model that includes the backscattering and Umklapp terms. These may be written in terms of bosonic operators associated with

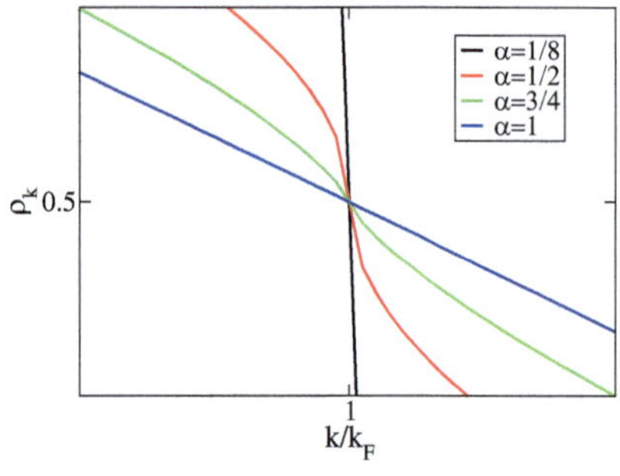

Fig. 24.3 Behavior of ρ_k around the Fermi momentum for different values of the exponent α. Unlike in a Fermi liquid, there is no discontinuity at the Fermi momentum

the spin and charge operators, similarly to Eq. (24.6), as

$$H_{1\perp} = \frac{2g_{1\perp}}{(2\pi\alpha)^2}\int dx\cos[\sqrt{8}\Phi_\sigma(x)]$$
$$H_{3\perp} = \frac{2g_{3\perp}}{(2\pi\alpha)^2}\int dx\cos[\sqrt{8}\Phi_\rho(x)], \tag{24.17}$$

where

$$\Phi_\rho(x) = -\frac{i\pi}{L}\sum_{p\neq 0}\frac{e^{-\alpha|p|/2-ipx}}{p}(\rho_+(p)+\rho_-(p)) - \left(N_{+,\rho}+N_{-,\rho}\right)\frac{\pi x}{L}$$
$$\Phi_\sigma(x) = -\frac{i\pi}{L}\sum_{p\neq 0}\frac{e^{-\alpha|p|/2-ipx}}{p}(\sigma_+(p)+\sigma_-(p)) - \left(N_{+,\sigma}+N_{-,\sigma}\right)\frac{\pi x}{L}. \tag{24.18}$$

The Umklapp term is only important at half-filling. However, the backscattering term is in general important and for a particular value of $g_{1||} = -\frac{3}{5}\frac{1}{2\pi v_F}$ Luther and Emery solved the problem exactly and concluded that the spin part has a gap. At this special point, the model is not in the class of the Luttinger model (which has a gapless spectrum for both the charge and spin degrees of freedom and therefore is critical). In the general case, the model is not solved exactly, but by similarity with the sine-Gordon model, solved, for instance, by the renormalization group method, conclusions about the extended Luttinger model may be inferred.

Chui and Lee showed that the spin part of this model is equivalent to a problem of a two-dimensional Coulomb gas, also by itself equivalent to the sine-Gordon model and the XY model in two-dimensions. Additional information on the phase diagram was obtained using bosonization, supplemented by direct application of the renormalization group method on the extended Luttinger model. The phase diagram was elaborated, identifying the dominant instabilities in each region of the space of parameters. One remark that is in order is to note that there are no true phase transitions in one dimension; only dominant instabilities are found that may determine the phase of the system, if small couplings between one-dimensional systems are allowed in a material of higher spatial dimension.

Even though we will not enter into any details of the derivation of the mapping, let us note that the equivalent Coulomb model establishes a relation between the one-dimensional conductor at $T = 0$ and a set of charges q and $-q$ that interact via the two-dimensional Coulomb potential, with an effective temperature $T = \bar{\beta}^{-1}$ where

$$\bar{\beta}q^2 = 2\sqrt{\frac{1+g_{1||}/2\pi v_F}{1-g_{1||}/2\pi v_F}}, \tag{24.19}$$

and in a chemical potential

$$e^{-\bar{\beta}\mu} = \frac{g_{1\perp}}{(2\pi)^2}. \tag{24.20}$$

This mapping is somewhat related to others discussed in this book. The analysis of the renormalization group of the Coulomb model allows the construction of the phase diagram. The transition temperature, T_c, defined in the attractive regime by $g_{1||} = |g_{1\perp}|$, separates a regime in which for $T \leq T_c$ ($g_{1||} \geq |g_{1\perp}|$), the positive and negative charges are paired, $g_{1\perp}$ scales to zero, and the backscattering term is irrelevant. For $T > T_c$, the gas is ionized, the Coulomb gas screens a test charge, a Yukawa-like potential appears, and a gap in the excitation spectrum appears as well. In this regime, the system is in the universality class of the Luther-Emery model, and the correlation functions of charge and spin are different from those of the Luttinger model. The correlation functions that involve parallel spins are no longer divergent. Also, the spin susceptibility is now exponentially activated at low temperatures, due to the finite gap.

24.2 Spin-1/2 Heisenberg Chain

The anisotropic spin-1/2 Heisenberg chain is an interesting quantum problem. As we have seen, when changing the anisotropy, there are some limits that we have solved: the XX model is gapless, while the Ising model has a gap. It may be expected that interpolating between the XX model, by adding an interaction term like ZZ, that grows from zero to the Ising limit, we may find some transition point where the gap opens. It turns out that this happens at the isotropic point. This result may be obtained explicitly because the problem of a system of spin-1/2 operators on a chain may be solved exactly using the Bethe ansatz method (1931) and is therefore integrable. We will show in the following that the spectrum is indeed gapless.

Since the problem is integrable, one expects the Hamiltonian to exhibit an extensive number of nontrivial conserved quantities or, equivalently, to possess a complete set of eigenstates that have a particular simple structure, characterized by a fixed set of momenta and scattering S-matrices. In non-integrable models on the other hand, the scattering of particles usually leads to particle production and decay; thus, only the total momentum is expected to be conserved. But in a model solvable by Bethe ansatz, the multi-particle scattering can be factorized into two-particle scatterings, and each of the bare momenta is conserved. Even though we will diagonalize the model and obtain its spectrum, the calculation of correlation functions is, in general, a hard problem, particularly dynamic correlation functions. The thermodynamics is more accessible, but here we will only consider the zero temperature case and will focus on the excitation spectrum.

24.2.1 Solution by the Bethe Ansatz Method

The Hamiltonian of the isotropic Heisenberg model may be written as[2]

$$H = J\sum_{i=1}^{N} \mathbf{S}_i \cdot \mathbf{S}_{i+1}, \tag{24.21}$$

where J is the coupling constant and N is the number of lattice sites. If $N = 2$ the problem is easily solved. It is enough to define the total spin $\mathbf{S}_T = \mathbf{S}_1 + \mathbf{S}_2$ and to note that $2\mathbf{S}_1 \cdot \mathbf{S}_2 = \mathbf{S}_T^2 - \mathbf{S}_1^2 - \mathbf{S}_2^2$. If $N = 3$ and we use periodic boundary conditions, the problem is also soluble, but when N increases, the solution is hard. As long as N is finite, the diagonalization problem may be solved in a computer, but quickly the dimension of the matrices to diagonalize becomes too large to solve. In particular, in the thermodynamic limit, the problem is complex.

In any case, let us consider the case where $S = 1/2$. When $J < 0$, the groundstate (state of minimal energy) is obtained if all spins are parallel. The state is infinitely degenerate, since the spins may point together along any space direction. Let us consider, with no loss of generality, that the direction along which the spins point is the z axis. The ferromagnetic state is then given by

$$|F\rangle = |\uparrow\uparrow\cdots\uparrow\rangle, \tag{24.22}$$

where all the spins in number N point "upward." Let us consider the action of the Hamiltonian on this state. It is easy to see that this state is an eigenstate of the Hamiltonian:

$$\begin{aligned} H|F\rangle &= J\sum_{i=1}^{N} \mathbf{S}_i \cdot \mathbf{S}_{i+1}|\uparrow\uparrow\cdots\uparrow\rangle \\ &= E_F|F\rangle = N\frac{J}{4}|F\rangle. \end{aligned} \tag{24.23}$$

Let us now look for the excitations of the system. Consider first that one of the spins points "down." We define a state in the form

$$|\Omega\rangle = \sum_{n} \psi(n) S_n^- |F\rangle \tag{24.24}$$

and consider the action of the Hamiltonian on this state. We obtain that

[2] For further reading see, for instance, the PhD thesis of Kong-Ju Bock Lee, Temple University (1990).

$$H|\Omega\rangle = \sum_n H S_n^- |F\rangle = E|\Omega\rangle, \tag{24.25}$$

if the state is an eigenstate. Using that

$$S^\pm = S^x \pm i S^y,$$

$$[S^x, S^y] = i S^z,$$

$$[S^+, S^-] = 2S^z,$$

$$[S^z, S^\pm] = \pm S^\pm,$$

$$\mathbf{S}_i \cdot \mathbf{S}_{i+1} = \frac{1}{2}\left(S_i^+ S_{i+1}^- + S_i^- S_{i+1}^+\right) + S_i^z S_{i+1}^z,$$

$$[AB, C] = A[B, C] + [A, C]B,$$

then

$$[H, S_n^-] = J\left(\frac{1}{2}S_{n+1}^- + \frac{1}{2}S_{n-1}^- - \frac{1}{2}S_n^- - \frac{1}{2}S_n^-\right). \tag{24.26}$$

Using this result in the equation for the eigenvalues we obtain

$$\begin{aligned} H|\Omega\rangle &= \sum_n \psi(n)\left(S_n^- H + [H, S_n^-]\right)|F\rangle \\ &= E_F|\Omega\rangle + \frac{J}{2}\sum_n \psi(n)\left(S_{n+1}^- + S_{n-1}^- - 2S_n^-\right)|F\rangle \\ &= E|\Omega\rangle. \end{aligned} \tag{24.27}$$

We get that

$$2(E - E_F)\psi(n) = J\left(\psi(n-1) + \psi(n+1) - 2\psi(n)\right). \tag{24.28}$$

Let us try a solution of the form

$$\psi(n) = Ae^{ikn}. \tag{24.29}$$

Replacing, we obtain

$$E = E_F + J\left(\cos k - 1\right). \tag{24.30}$$

This is the energy of one excitation, in which a single spin has changed its orientation. Note that the excitation was searched in the form of a collective excitation in which the change of the component along the quantization axis is split among all the system spins, in the form of a wave of momentum k. This type of excitation was called before a magnon (spin wave). Also note that if the momentum vanishes, the energy of the excitation also vanishes.

Consider now higher energy excitations that correspond to two spins changing their orientation. We then look for solutions to the eigenvalue equation in the form

$$|\Omega\rangle = \sum_{n_1,n_2} \psi(n_1, n_2) S_{n_1}^- S_{n_2}^- |F\rangle. \tag{24.31}$$

Suppose that $n_2 > n_1 + 1$. In this case we obtain that the action of the Hamiltonian leads to

$$\begin{aligned} H|\Omega\rangle &= E|\Omega\rangle \\ &= \sum_{n_1,n_2} \psi(n_1, n_2) H S_{n_1}^- S_{n_2}^- |F\rangle \\ &= \sum_{n_1,n_2} \psi(n_1, n_2) \left(S_{n_1}^- H + [H, S_{n_1}^-]\right) S_{n_2}^- |F\rangle. \end{aligned} \tag{24.32}$$

Then,

$$\begin{aligned} &2(E - E_F) \sum_{n_1,n_2} \psi(n_1, n_2) S_{n_1}^- S_{n_2}^- |F\rangle \\ &= J \sum_{n_1,n_2} \psi(n_1, n_2) \\ &\left(S_{n_1}^- S_{n_2+1}^- + S_{n_1}^- S_{n_2-1}^- + S_{n_1+1}^- S_{n_2}^- + S_{n_1-1}^- S_{n_2}^- - 4 S_{n_1}^- S_{n_2}^-\right) |F\rangle. \end{aligned} \tag{24.33}$$

Therefore,

$$\begin{aligned} 2(E - E_F)\psi(n_1, n_2) = J(&\psi(n_1, n_2 - 1) + \psi(n_1, n_2 + 1) + \psi(n_1 - 1, n_2) \\ &+ \psi(n_1 + 1, n_2) - 4\psi(n_1, n_2)). \end{aligned} \tag{24.34}$$

Let us try a solution of the form

$$\psi(n_1 < n_2) = A e^{ik_1 n_1 + ik_2 n_2} + B e^{ik_1 n_2 + ik_2 n_1}. \tag{24.35}$$

Assume that ψ is symmetric under the permutation of n_1 and n_2. If $n_1 > n_2$ we obtain that

$$\psi(n_1 > n_2) = B e^{ik_1 n_1 + ik_2 n_2} + A e^{ik_1 n_2 + ik_2 n_1}. \tag{24.36}$$

When we change conditions from $n_1 < n_2$ to $n_1 > n_2$, the coefficients A and B exchange. Let us then replace Eq. (24.35) in Eq. (24.34). We obtain that

$$E = E_F + J\left(\cos k_1 + \cos k_2 - 2\right). \tag{24.37}$$

Actually, the coefficients A and B are not independent. Consider now the case $n_2 = n_1 + 1$. Retracing the same steps, we obtain that

$$\begin{aligned} 2(E - E_F)\psi(n_1, n_1+1) = J\big(&\psi(n_1-1, n_1+1) + \psi(n_1, n_1+2) \\ &-2\psi(n_1, n_1+1)\big). \end{aligned} \tag{24.38}$$

Since we cannot change twice the spin orientation in the same point (because $S = 1/2$), we have that

$$\psi(n_1, n_1) = 0.$$

We get that

$$\frac{B}{A} = \frac{\cos\frac{1}{2}(k_1+k_2) - e^{\frac{i}{2}(k_1-k_2)}}{e^{-\frac{i}{2}(k_1-k_2)} - \cos\frac{1}{2}(k_1+k_2)}, \tag{24.39}$$

which can be written as

$$\frac{B}{A} = \frac{\frac{1}{2}\left(\cot\frac{k_1}{2} - \cot\frac{k_2}{2}\right) - i}{\frac{1}{2}\left(\cot\frac{k_1}{2} - \cot\frac{k_2}{2}\right) + i} = e^{2i\phi(k_1,k_2)}. \tag{24.40}$$

Therefore,

$$\cot\phi(k_1, k_2) = \frac{1}{2}\left(\cot\frac{k_1}{2} - \cot\frac{k_2}{2}\right). \tag{24.41}$$

Let us call $\lambda_j = 1/2\cot\frac{k_j}{2}$. This quantity is called rapidity. We see then that $\phi(k_1, k_2)$ is function of $(\lambda_1 - \lambda_2)$.

Consider now that we have M spins that change their orientation. We have then a M-particle problem, in the sense that we have M excitations with momenta k_i propagating in the spin system. Let us look for a solution as

$$|\Omega\rangle = \sum_{n_1<\cdots<n_M} \psi(n_1, \cdots, n_M) S^-_{n_1}\cdots S^-_{n_M}|F\rangle, \tag{24.42}$$

where we choose the amplitudes in the form

$$\psi(n_1, \cdots, n_M) = \sum_P e^{i \sum_{j=1}^{M} k_{P_j} n_j + i \sum_{i>j}^{M} sgn(n_i - n_j)\phi(k_{P_i}, k_{P_j})} \tag{24.43}$$

and where we sum over all the momenta permutations P. In the same way, we get

$$E = E_F + J \sum_{j=1}^{M} \left(\cos k_j - 1 \right). \tag{24.44}$$

We have therefore M particles on a line. Define

$$T_j = S_{j,j+1} \cdots S_{j,M} S_{j,1} \cdots S_{j,j-1}, \tag{24.45}$$

where the operator $S_{j,j+1}$ exchanges the particles j and $j+1$. In the simple case of the Heisenberg model we are considering, $S_{j,j+1}$ is just a phase shift. In general it is a matrix. In the case at hand we have that

$$S_{j,j+1} = e^{2i\phi(\lambda_j - \lambda_{j+1})}. \tag{24.46}$$

The action of the operator T_j is to exchange particle j with all the remaining ones and to return it to the same lattice site. Using periodic boundary conditions, the system should return to the same state (minus a phase of the type $e^{ik_j L}$, due to Bloch's theorem, where $L = Na$ is the lattice size). We want now to know the eigenvalues of the type

$$T_j \xi = e^{ik_j L} \xi. \tag{24.47}$$

It can be obtained directly that

$$e^{ik_j L} = \prod_{h=1, h \neq j}^{M} e^{2i\phi(\lambda_j - \lambda_h)}. \tag{24.48}$$

Taking logarithms, we obtain

$$k_j(\lambda_j) L = 2\pi I_j + 2 \sum_{h=1, h \neq j}^{M} \phi(\lambda_j - \lambda_h), \tag{24.49}$$

where I_J are integers (if N is even). Knowing that

$$e^{2i\phi(\lambda_j - \lambda_h)} = \frac{(\lambda_1 - \lambda_2) - i}{(\lambda_1 - \lambda_2) + i} \tag{24.50}$$

and using that

$$e^{-ik} = \frac{\cot\frac{k}{2} - i}{\cot\frac{k}{2} + i},$$

we finally reach the Bethe ansatz equations

$$\left(\frac{\lambda_j + \frac{i}{2}}{\lambda_j - \frac{i}{2}}\right)^M = \prod_{l \neq j} \left(\frac{\lambda_j - \lambda_l + i}{\lambda_j - \lambda_l - i}\right), \tag{24.51}$$

where $j = 1, \cdots, M$, $\lambda_j = \frac{1}{2}\cot(k_j/2)$ or

$$e^{ik_j} = \frac{\lambda_j + \frac{i}{2}}{\lambda_j - \frac{i}{2}}. \tag{24.52}$$

These equations have solutions for the rapidities that may be real or complex (designated strings). The states of lowest energies are obtained considering the solutions of real rapidities. This is the case we consider next.

The energy is given by

$$\begin{aligned} E - E_0 &= J \sum_{j=1}^{M} \left(\cos k_j - 1\right) \\ &= -2J \sum_{j=1}^{M} \frac{1}{(2\lambda_j)^2 + 1}. \end{aligned} \tag{24.53}$$

Taking logarithms we obtain

$$2N \tan^{-1} 2\lambda_j = 2\pi I_j + 2 \sum_{l=1}^{M} \tan^{-1}(\lambda_j - \lambda_l). \tag{24.54}$$

The quantum numbers are contained in the interval $[-(N - M - 1)/2, (N - M - 1)/2]$. Defining the pseudo-momentum

$$q_j = \frac{2\pi}{N} I_j, \tag{24.55}$$

we may write the logarithmic equations of the Bethe ansatz in the continuum in the form

$$q = 2\tan^{-1}(\lambda(q)) - \frac{1}{\pi} \int_{-k_{F,\uparrow}}^{k_{F,\uparrow}} dq' N(q') \tan^{-1}\left(\frac{\lambda(q) - \lambda(q')}{2}\right), \tag{24.56}$$

where we have introduced the distribution $N(q)$ of the values of q that are occupied. Consider that the magnetization of the system is positive or zero. If all the spins point "up", we have that $M = 0$ (in the ferromagnetic, $J < 0$ case). In the antiferromagnetic case ($J > 0$), the magnetization vanishes in the groundstate, $M = N/2$, and the pseudo-momenta are limited to the interval $[-\pi/2, \pi/2]$. If a magnetic field is applied, the magnetization takes intermediate values, and the maximal value of the pseudo-momenta is designated as $k_{F,\uparrow}$ and by $k_{F,\downarrow}$ the maximum value that is occupied in the groundstate. The energy is given by

$$E = -|J| \frac{N}{2\pi} \int_{-k_{F,\uparrow}}^{k_{F,\uparrow}} dq N(q) \frac{1}{2} \frac{1}{\left(\frac{1}{2}\right)^2 + \lambda^2(q)}. \tag{24.57}$$

The entities that correspond to the occupation of the available pseudo-momentum values are called pseudo-particles, and correspondingly, the empty momenta values are called pseudo-holes.

The number of spins that point "down" is obtained

$$\frac{N}{2\pi} \int_{-k_{F,\uparrow}}^{k_{F,\uparrow}} dq N(q) = M. \tag{24.58}$$

In the groundstate we have that

$$N(q) = N^0(q) = \theta\left(|q| - k_{F,\downarrow}\right), \tag{24.59}$$

as mentioned before. This notation is applicable in the case of a finite magnetic field (or finite magnetization) and identifies the maximum value of the quantum numbers I_j as the limit of the Brillouin zone ($q = k_{F,\uparrow}$), and the maximal quantum number occupied is identified as the Fermi surface ($q = k_{F,\downarrow}$). In zero magnetic field and considering the antiferromagnetic case, $k_{F,\uparrow} = k_{F,\downarrow}$, and all the quantum numbers are occupied in the groundstate (the band is full).

Recall that the Bethe ansatz quantum numbers are defined in the interval

$$-\frac{N-M-1}{2} \leq I_\gamma \leq \frac{N-M-1}{2}. \tag{24.60}$$

A triplet spin excitation may be specified by $M = M_0 - 1$, $S = N/2 - M = 1$. As a consequence $I_\pm = \pm N/4$, and there are $N/2 + 1$ available spaces. Since $M = N/2 - 1$, there are two remaining spaces that are not occupied (holes). The spin excitation therefore gives rise to two holes in the state occupation.

Also, recall that

$$k_j(\lambda_j)L = 2\pi I_j + 2 \sum_{h=1, h\neq j}^{N} \phi(\lambda_j - \lambda_h). \tag{24.61}$$

The total momentum is given by

$$P = \sum_j k_j = \sum_j (2\pi I_j). \tag{24.62}$$

Using the anti-symmetry of $\phi(\lambda_j - \lambda_h)$ we obtain

$$P = \sum_j \left(2\tan^{-1} 2\lambda_j - \pi\right)(mod 2\pi). \tag{24.63}$$

24.2.2 Theory of the Expansion in the Number of Pseudo-Particles

We are interested in considering excited states with a small deviation with respect to the groundstate. Specifically, we want to consider states that differ from the groundstate by small deviations in the quantum numbers of the pseudo-momenta distribution [1]. Therefore, let us consider situations where

$$N(q) = N^0(q) + \delta(q), \tag{24.64}$$

where $\delta(q)$ is small. Note that this condition implies small deviations in the number of changes of the quantum numbers, but does not imply necessarily small energies, since we may make changes in the occupation numbers of states that are far from the Fermi surface (where the zero energy of each pseudo-particle is defined).

The change in the distribution $N(q)$ implies a change in the rapidities—that we are going to expand around the rapidities $\lambda^0(q)$ in powers of the deviation $\delta(q)$ as

$$\lambda(q) = \lambda^0(q) + \lambda^1(q) + \lambda^2(q) + \cdots \tag{24.65}$$

The procedure is now to insert the two expansions in the expression for the energy and in the Bethe ansatz equations. The expansion of the energy leads to an expansion in terms of the deviation $\delta(q)$ that we may write as

$$\begin{aligned} E &= E^0 + E^1 + E^2 + \cdots \\ E^1 &= \frac{N}{2\pi}\int_{-k_{F,\uparrow}}^{k_{F,\uparrow}} dq\delta(q)\epsilon(q) \\ E^2 &= \frac{N}{(2\pi)^2}\int_{-k_{F,\uparrow}}^{k_{F,\uparrow}} dq \int_{-k_{F,\uparrow}}^{k_{F,\uparrow}} dq'\delta(q)\frac{1}{2}f(q,q')\delta(q'). \end{aligned} \tag{24.66}$$

Here E^0 is the energy of the groundstate, $\epsilon(q)$ is the band of the states that the pseudo-particles may occupy, and $f(q,q')$ are the f functions of the interactions between the pseudo-particles (by similarity with the Landau Fermi-liquid theory Eq. 14.45). As already mentioned, let us define the zero energy of the bands of the pseudo-particles as

$$\epsilon(\pm k_{F,\downarrow}) = 0. \tag{24.67}$$

Using that

$$\tan^{-1}(x_0 + y) = \tan^{-1} x_0 + \frac{1}{1+x_0^2} y - \frac{x_0}{\left(1+x_0^2\right)^2} y^2 + \cdots \tag{24.68}$$

we may obtain that

$$E^0 = -|J| \frac{N}{2\pi} \int_{-k_{F,\uparrow}}^{k_{F,\uparrow}} dq \frac{1}{2} N^0(q) \frac{1}{\left(\frac{1}{2}\right)^2 + (\lambda(q))^2}, \tag{24.69}$$

$$\begin{aligned} E^1 = &-|J| \frac{N}{2\pi} \int_{-k_{F,\uparrow}}^{k_{F,\uparrow}} dq \frac{1}{2} N^0(q) \left(-2 \frac{\lambda^0(q)}{\left(\left(\frac{1}{2}\right)^2 + \left(\lambda^0(q)\right)^2\right)^2} \lambda^1(q) \right) \\ &- |J| \frac{N}{2\pi} \int_{-k_{F,\uparrow}}^{k_{F,\uparrow}} dq \frac{1}{2} \delta(q) \frac{1}{\left(\frac{1}{2}\right)^2 + \left(\lambda^0(q)\right)^2} \end{aligned} \tag{24.70}$$

and

$$\begin{aligned} E^2 = &-|J| \frac{N}{2\pi} \int_{-k_{F,\uparrow}}^{k_{F,\uparrow}} dq \frac{1}{2} N^0(q) \left(-2 \frac{\lambda^0(q)}{\left(\left(\frac{1}{2}\right)^2 + \left(\lambda^0(q)\right)^2\right)^2} \lambda^2(q) \right. \\ &\left. + \left(\lambda^1(q)\right)^2 \frac{3\left(\lambda^0(q)\right)^2 - \left(\frac{1}{2}\right)^2}{\left(\left(\frac{1}{2}\right)^2 + \left(\lambda^0(q)\right)^2\right)^3} \right) \\ &- |J| \frac{N}{2\pi} \int_{-k_{F,\uparrow}}^{k_{F,\uparrow}} dq \frac{1}{2} \delta(q) \left(-2 \frac{\lambda^0(q)}{\left(\left(\frac{1}{2}\right)^2 + \left(\lambda^0(q)\right)^2\right)^2} \right) \lambda^1(q). \end{aligned} \tag{24.71}$$

We need now to determine the functions $\lambda^0(q)$, $\lambda^1(q)$, $\lambda^2(q)$. These functions may be obtained using the Bethe ansatz equations. Inserting the expansions in the Bethe ansatz equation for the pseudo-momentum as

$$
\begin{aligned}
q &= 2\tan^{-1}\left(2\left(\lambda^0(q)+\lambda^1(q)+\lambda^2(q)\right)\right) \\
&\quad -\frac{1}{2\pi}\int_{-k_{F,\uparrow}}^{k_{F,\uparrow}} dq'\left(N^0(q')+\delta(q')\right) \\
&\quad 2\tan^{-1}\left(\lambda^0(q)-\lambda^0(q')\lambda^1(q)-\lambda^1(q')\lambda^2(q)-\lambda^2(q')\right),
\end{aligned} \tag{24.72}
$$

we obtain

$$
q = 2\tan^{-1}\left(2\lambda^0(q)\right) - \frac{1}{2\pi}\int_{-k_{F,\uparrow}}^{k_{F,\uparrow}} dq' N^0(q') 2\tan^{-1}\left(\lambda^0(q)-\lambda^0(q')\right), \tag{24.73}
$$

which determines $\lambda^0(q)$, and

$$
\begin{aligned}
0 &= \frac{4}{1+\left(2\lambda^0(q)\right)^2}\lambda^1(q) \\
&\quad -\frac{1}{2\pi}\int_{-k_{F,\uparrow}}^{k_{F,\uparrow}} dq' N^0(q')\frac{2}{1+\left(\lambda^0(q)-\lambda^0(q')\right)^2}\left(\lambda^1(q)-\lambda^1(q')\right) \\
&\quad -\frac{1}{2\pi}\int_{-k_{F,\uparrow}}^{k_{F,\uparrow}} dq' \delta(q') 2\tan^{-1}\left(\lambda^0(q)-\lambda^0(q')\right).
\end{aligned} \tag{24.74}
$$

We may obtain the expression that determines $\lambda^2(q)$, but is not shown here, for simplicity.

To determine $\lambda^1(q)$ we write

$$
\lambda^1(q) = \frac{d\lambda^0(q)}{dq} Q^1(q). \tag{24.75}
$$

First, it is necessary to obtain the derivative of the rapidity of the groundstate. This can be obtained taking the derivative of q in order to $\lambda^0(q)$ and then inverting. We obtain

$$
1 = \frac{d\lambda^0(q)}{dq}\left(\frac{4}{1+\left(2\lambda^0(q)\right)^2} - \frac{1}{2\pi}\int_{-k_{F,\uparrow}}^{k_{F,\uparrow}} dq' N^0(q')\frac{2}{1+\left(\lambda^0(q)-\lambda^0(q')\right)^2}\right). \tag{24.76}
$$

Using this result in the equation for $\lambda^1(q)$, one obtains an equation for $Q^1(q)$ as

$$Q^1(q) = \frac{1}{2\pi}\int_{-k_{F,\uparrow}}^{k_{F,\uparrow}} dq'\delta(q')2\tan^{-1}\left(\lambda^0(q) - \lambda^0(q')\right)$$
$$-\frac{1}{2\pi}\int_{-k_{F,\uparrow}}^{k_{F,\uparrow}} dq' N^0(q')\frac{2}{1+\left(\lambda^0(q) - \lambda^0(q')\right)^2}\frac{d\lambda^0(q')}{dq')}Q^1(q'). \quad (24.77)$$

We may now introduce another function called phase shift $\phi(q, q')$

$$Q^1(q) = \int_{-k_{F,\uparrow}}^{k_{F,\uparrow}} dq'\delta(q')\phi(q, q'). \quad (24.78)$$

Inserting this expression in the previous equation for $Q^1(q)$ we obtain that

$$\phi(q, q') = \frac{1}{\pi}\tan^{-1}\left(\lambda^0(q) - \lambda^0(q')\right)$$
$$-\frac{1}{\pi}\int_{-k_{F,\uparrow}}^{k_{F,\uparrow}} dq'' N^0(q'')\frac{1}{1+\left(\lambda^0(q)-\lambda^0(q'')\right)^2}\frac{d\lambda^0(q'')}{dq''}\phi(q'', q'). \quad (24.79)$$

We may now carry out a transformation of variables

$$\frac{d\lambda^0(q)}{dq}dq = dr \quad (24.80)$$

and change the notation defining $\lambda = r$. We may also take $\lambda^0(q^{\pm}_{FS}) = \pm B$ that, in the limit of zero magnetic field, takes the value $B = \infty$. Also, we define

$$\phi(q, q') = \bar{\phi}\left(\lambda^0(q), \lambda^0(q')\right) = \bar{\phi}(\lambda, \lambda'). \quad (24.81)$$

We obtain then

$$\bar{\phi}(\lambda, \lambda') = \frac{1}{\pi}\tan^{-1}(\lambda - \lambda') + \int_{-\infty}^{\infty} d\lambda'' S(\lambda - \lambda'')\bar{\phi}(\lambda'', \lambda'), \quad (24.82)$$

where a kernel has been defined

$$S(\lambda) = -\frac{1}{\pi}\frac{1}{1+\lambda^2}. \quad (24.83)$$

Let us now turn our attention to the expression for E^1. Substituting the expression for $\lambda^1(q)$ we may obtain

$$\epsilon(q) = -|J|\frac{\frac{1}{2}}{\left(\frac{1}{2}\right)^2 + \left(\lambda^0(q)\right)^2} + |J|\int_{-\infty}^{\infty} d\lambda \frac{\lambda}{\left(\left(\frac{1}{2}\right)^2 + \lambda^2\right)^2}\bar{\phi}(\lambda, \lambda^0(q)). \tag{24.84}$$

The first term is the undressed term, and the second term, which involves the phase shift, is the feedback term. This is the pseudo-particles band.

We are left with the calculation of the phase shifts. Solving the expression for the rapidity of the groundstate $\lambda^0(q)$ and the phase shifts, we may obtain the energy bands of the pseudo-particles. However, it turns out that it is more convenient to use an alternative method.

It is convenient to establish a connection between the description of the Bethe ansatz equations we adopt here and a description in terms of the rapidity densities that is usually considered in the literature. The usual density is defined by

$$\sigma(\lambda) = \frac{dq}{d\lambda^0}. \tag{24.85}$$

This function satisfies the equation

$$\sigma(\lambda) = 4\frac{1}{1+4\lambda^2} - \frac{1}{\pi}\int_{-\infty}^{\infty} d\lambda' \frac{1}{1+(\lambda-\lambda')^2}\sigma(\lambda'). \tag{24.86}$$

We may see that

$$\int_0^B d\lambda \sigma(\lambda) = k_{F,\downarrow} \tag{24.87}$$

and

$$q = \int_0^{\lambda^0(q)} d\lambda \sigma(\lambda). \tag{24.88}$$

Let us now define another function $\eta(\lambda)$ using

$$\epsilon(q) = \int_B^{\lambda^0(q)} d\lambda \eta(\lambda). \tag{24.89}$$

Therefore,

$$\frac{d\epsilon(q)}{d\lambda^0(q)} = \eta(\lambda). \tag{24.90}$$

We may then obtain as equation for this new function taking a derivative of the pseudo-particles band. We obtain

$$\eta(\lambda) = |J| \frac{\lambda}{\left(\left(\frac{1}{2}\right)^2 + \lambda^2\right)^2} + |J| \int_{-\infty}^{\infty} d\lambda' \frac{\lambda'}{\left(\left(\frac{1}{2}\right)^2 + (\lambda')^2\right)^2} \frac{d\bar{\phi}(\lambda', \lambda)}{d\lambda}. \tag{24.91}$$

To calculate the derivative of the phase shift, we use its integral equation. As the term out of the integral is only a function of the difference and the integral is a convolution, the phase shift is only a function of the difference of its arguments. After some manipulations, we obtain an integral equation for $\eta(\lambda)$ in the form

$$\eta(\lambda) = |J| \frac{\lambda}{\left(\left(\frac{1}{2}\right)^2 + \lambda^2\right)^2} + \int_{-\infty}^{\infty} d\lambda' \left(-\frac{1}{\pi}\right) \frac{1}{1 + (\lambda - \lambda')^2} \eta(\lambda'). \tag{24.92}$$

We can solve this equation using Fourier transforms. We obtain

$$\eta(\omega) = |J| \pi i e^{-\frac{1}{2}|\omega|} \omega - e^{-|\omega|} \eta(\omega) \tag{24.93}$$

and therefore

$$\eta(\omega) = \frac{|J| \pi i \omega}{2 \cosh \frac{\omega}{2}}. \tag{24.94}$$

The inverse may be calculated leading to

$$\eta(\lambda) = |J| \frac{\pi^2}{2} \frac{\sinh \pi \lambda}{\cosh^2 \pi \lambda}. \tag{24.95}$$

Therefore, we obtain that the pseudo-particles band is given by

$$\epsilon(q) = \int_{\infty}^{\lambda^0(q)} d\lambda \eta(\lambda)$$

$$\epsilon(q) = -|J| \frac{\pi}{2} \frac{1}{\cosh \pi \lambda^0(q)}. \tag{24.96}$$

This band represents the dispersion relation of the pseudo-particles associated with the degrees of freedom of the system. The state of the system is characterized by a given occupation of this band. The excitations are now obtained, changing the occupation of these states. A simple excitation is obtained if the band is not fully occupied (finite magnetic field), creating pairs of particle-hole and, in general, introducing holes in the distribution of the occupied states. The energy of these holes typically involves energies of the type $-\epsilon(q)$, which are therefore positive, as in the case of the excitations of a Fermi gas.

We may now determine the phase shifts. We obtain that

$$\frac{d\bar{\phi}(\lambda)}{d\lambda} = \frac{1}{\pi}\frac{1}{1+\lambda^2} - \frac{1}{\pi}\int_{-\infty}^{\infty} d\lambda' \frac{1}{1+(\lambda-\lambda')^2}\frac{d\bar{\phi}(\lambda')}{d\lambda')}. \tag{24.97}$$

Defining

$$f(\lambda) = \frac{d\bar{\phi}(\lambda)}{d\lambda}, \tag{24.98}$$

we may solve this equation using Fourier transforms

$$f(\omega) = \frac{1}{1+e^{|\omega|}} \tag{24.99}$$

and then inverting to obtain that

$$f(\lambda) = \frac{1}{4\pi}\left(-Hn\left[-\frac{1}{2}-\frac{i\lambda}{2}\right] + Hn\left[-\frac{i\lambda}{2}\right] + Hn\left[\frac{i\lambda}{2}\right] - Hn\left[-\frac{1}{2}+\frac{i\lambda}{2}\right]\right). \tag{24.100}$$

Using that $H[n] = \gamma + \psi[n+1]$ where γ is Euler's constant and ψ is the digamma function, we obtain that

$$f(\lambda) = \frac{1}{4\pi}\left(-\psi[\frac{1}{2}-\frac{i\lambda}{2}] + \psi[1-\frac{i\lambda}{2}] + \psi[1+\frac{i\lambda}{2}] - \psi[\frac{1}{2}+\frac{i\lambda}{2}]\right). \tag{24.101}$$

Integrating, we obtain that the phase shift is given by

$$\bar{\phi}(\lambda) = \frac{i}{2\pi}\log\frac{\Gamma\left(\frac{1}{2}+\frac{i\lambda}{2}\right)\Gamma\left(1-\frac{i\lambda}{2}\right)}{\Gamma\left(\frac{1}{2}-\frac{i\lambda}{2}\right)\Gamma\left(1+\frac{i\lambda}{2}\right)}. \tag{24.102}$$

This expression is in agreement with the result obtained by other methods and justifies its designation as phase shift. A possible way to define the phase shift consists in identifying the deviation of the pseudo-momenta of two pseudo-particles that collide, in comparison to the momenta of the free particles, defined in Eq. (24.55). Recalling Eq. (24.63) and Eq. (24.54), we may use a definition for the momentum of the type

$$p = \frac{2\pi}{N}I_j + \frac{2\pi\bar{\phi}}{N}, \tag{24.103}$$

where $\bar{\phi}$ is the phase shift that results from the deviation of the momentum from that of a free particle in a ring of size N, due to the collisions with other particles, in this case the M spin excitations.

All that is left now is to calculate the rapidity density. This satisfies

$$\sigma(\lambda) = 4\frac{1}{1+4\lambda^2} - \frac{1}{\pi}\int_{-\infty}^{\infty} d\lambda' \frac{1}{1+(\lambda-\lambda')^2}\sigma(\lambda'). \tag{24.104}$$

Again, it can be solved using Fourier transforms and we obtain

$$\sigma(\omega) = \pi \frac{1}{\cosh\frac{|\omega|}{2}}. \tag{24.105}$$

Inverting we get

$$\sigma(\lambda) = \pi \frac{1}{\cosh \pi\lambda}. \tag{24.106}$$

Therefore,

$$\begin{aligned} q &= \int_0^{\lambda^0(q)} d\lambda \sigma(\lambda) \\ &= 2\left[\tan^{-1}\tanh\frac{x}{2}\right]_0^{\pi\lambda^0(q)} \end{aligned} \tag{24.107}$$

We obtain then

$$\tanh\frac{\pi}{2}\lambda^0(q) = \tan\frac{q}{2}. \tag{24.108}$$

The expression for the energy band may be rewritten using

$$\cosh 2x = \frac{1+\tanh^2 x}{1-\tanh^2 x}, \tag{24.109}$$

leading to

$$\epsilon(q) = -|J|\frac{\pi}{2}\frac{1-\tan^2\frac{q}{2}}{1+\tan^2\frac{q}{2}} = -|J|\frac{\pi}{2}\cos q, \tag{24.110}$$

and the velocity may be obtained as

$$v(q) = |J|\frac{\pi}{2}\sin q. \tag{24.111}$$

As mentioned above, the energy second order term, E^2, may be obtained, but it involves a long and fastidious calculation that is not shown here. The result may be expressed in terms of the phase shifts and the velocities of the pseudo-particles in the form

$$E^2 = \frac{N}{\pi}\int_{-k_{F,\uparrow}}^{k_{F,\uparrow}} dq \int_{-k_{F,\uparrow}}^{k_{F,\uparrow}} dq'\delta(q)\delta(q')$$
$$[v(q)\phi(q,q') + v(q')\phi(q',q) + v\phi(k_{F,\downarrow},q)\phi(k_{F,\downarrow},q')$$
$$-v\phi(-k_{F,\downarrow},q)\phi(-k_{F,\downarrow},q')], \tag{24.112}$$

where $v = v(q = k_{F,\downarrow})$.

24.2.3 Low-Energy Excitation Spectrum

The excitation momentum satisfies

$$q = 2\left[\tan^{-1}\tanh\frac{x}{2}\right]_0^{\pi\lambda^0(q)}$$
$$= 2\tan^{-1}\tanh\frac{\pi}{2}\lambda^0(q), \tag{24.113}$$

which implies that

$$\tan\frac{q}{2} = \tanh\frac{\pi}{2}\lambda^0(q). \tag{24.114}$$

The excitations are obtained creating two holes in the quantum number distribution. The momenta are limited to the interval $[-\pi/2, \pi/2]$. The momenta of the two holes are designated q_1 and q_2. A way to calculate the excitation energy is obtained maintaining one of the holes at the Fermi surface (energy zero) and allowing the other hole to go around the Brillouin zone. In general, the excitation spectrum is a continuum if we allow both holes to move around the Brillouin zone, as shown in Fig. 24.4. The hole at the Fermi surface has momentum $\pm\pi/2$. The energy of the hole is the symmetric of the energy of the pseudo-particle that originated it and the same for its momentum. Therefore the energy of the excitation with one hole on the Fermi surface is

$$\epsilon(q) = |J|\frac{\pi}{2}\cos q \tag{24.115}$$

and the momentum is

$$k = \frac{\pi}{2} - q. \tag{24.116}$$

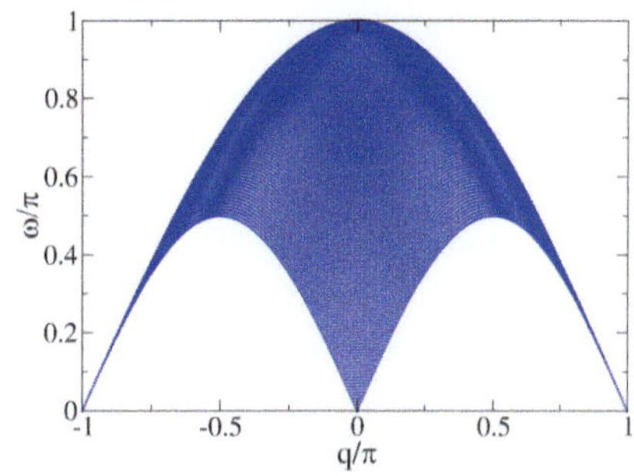

Fig. 24.4 Continuum of excitations that result from exciting two holes in the band $\epsilon(q)$

We have then as examples

$$
\begin{aligned}
q &= -\frac{\pi}{2}; k = \pi; \epsilon = 0 \\
q &= 0; k = \frac{\pi}{2}; \epsilon = |J|\frac{\pi}{2} \\
q &= \frac{\pi}{2}; k = 0; \epsilon = 0.
\end{aligned}
\tag{24.117}
$$

The results show that the excitation spectrum is gapless.

24.2.4 Integrability

Integrable systems have sufficient conservation laws such that, all degrees of freedom can be eliminated in terms of constants of integration in classical system or in terms of a set of quantum numbers in a quantum system. In a one-dimensional system, the set of momenta is conserved in every scattering process, in addition to the usual total momentum conservation. The only changes that may occur are permutations of the momenta between the particles in the scattering process. If two particles scatter and have initial momenta k_1 and k_2, then the final momenta will be (k_1, k_2) or (k_2, k_1) (forward scattering only).

The necessary and sufficient condition for the integrability of a many-body system is the factorization of every N particle scattering matrix into the product of $N(N-1)/2$ two-particle scattering matrices, in such a way that the scattering process can be seen as a sequence of pair collisions, with the momenta kept constant or interchanged. The mathematical description of this condition are the so-called Yang-Baxter equations

$$
S^{b_1c_1}_{b_2c_2}(k_1, k_2)S^{a_1b_1}_{b_3c_3}(k_1, k_3)S^{a_2b_2}_{a_3b_3}(k_2, k_3) = S^{b_2c_2}_{b_3c_3}(k_2, k_3)S^{b_1c_1}_{a_3b_3}(k_1, k_3)S^{a_1b_1}_{a_2b_2}(k_1, k_2),
\tag{24.118}
$$

where $S^{a_1b_1}_{a_2b_2}(k_1, k_2)$ is the two-particle scattering matrix for particles of momenta k_1, k_2 and internal degrees of freedom (a_1, a_2) and (b_1, b_2)i, with respect to spin, orbital, chiral, color, etc. of the particles of the incoming and outgoing states in the scattering process. The sum over all possible intermediate states has to be taken into account. The Yang-Baxter relations reflect the invariance of the two possible sequences of two-particle scatterings, such that first particles 2 and 3 scatter, then 1 and 3 scatter, and finally 1 and 2, or the other possibility such that first particles 1 and 2 scatter, then 1 and 3, and finally 2 and 3. The processes are equivalent since they have the same initial and final states, guaranteeing the single-valuedness of the wavefunction. It turns out that the condition that these equations impose is very strict and only a few models satisfy it, meaning only a subset of models is called integrable.

The case of the Heisenberg chain satisfies this condition trivially, since the scattering matrix is just a phase shift

$$S(k_1, k_2) = e^{2i\phi(k_1,k_2)} \tag{24.119}$$

and the commutativity implies the Yang-Baxter equations. In more complicated models, with internal degrees of freedom, the scattering matrix is not just a phase shift, and the scattering matrices in general do not commute. In the case of the Heisenberg chain or other models with no internal degrees of freedom, like spin-zero bosons, the factorization condition of the N-particle scattering matrix implies that the total phase shift is just the sum of two-particle scattering phases.

For integrable systems with no internal degrees of freedom, the Yang-Baxter equations are simply satisfied leading, as we saw earlier in the explicit example of the spin-1/2 isotropic Heisenberg chain, to the N-particle wavefunction (the Bethe ansatz)

$$\psi(x_1, \cdots, x_N) = \sum_P e^{i\sum_{j=1}^N k_{pj}x_j + i\sum_{i>j}^N sgn(x_i - x_j)\phi(k_{pi},k_{pj})}, \tag{24.120}$$

where $P = (P1, \cdots, PN)$ is a permutation of the integers $(1, \cdots, N)$, x_j are the coordinates on the lattice, k_j the momenta of the particles, and $\phi(k_i, k_j)$ is the two-particle phase shift.

In the case where the system has internal degrees of freedom, the wavefunction takes the form

$$\psi_{a_1,\cdots,a_N}(x_Q) = \sum_P A_{a_1,\cdots,a_N}[Q, P]\, e^{i\sum_{j=1}^N k_{pi}x_{Qj}}, \tag{24.121}$$

where $Q = (Q1, \cdots, QN)$ and $P = (P1, \cdots, PN)$ are permutations of the integers $(1, \cdots, N)$. Q represents a given sequence of positions of the N particles. If we consider two regions X_Q and $X_{Q'}$ that differ by the permutation of particles i

and j, then

$$A_{a_1,\cdots,a_i,\cdots,a_j,\cdots,a_N}\left[Q,P\right] = S^{a_i b_i}_{a_j b_j} A_{a_1,\cdots,b_i,\cdots,b_j,\cdots,a_N}\left[Q',P'\right]. \qquad (24.122)$$

In the Bethe ansatz wavefunction, the momenta are somewhat free with the only restriction that a given set of momenta is invariant under all scattering processes. However, the momenta are not strictly free since they are determined by the boundary conditions imposed on the system. A common case are periodic boundary conditions, but twisted boundary conditions may also be imposed, if some flux goes through the system, such as due to a magnetic field going through a ring of the one-dimensional system. Considering the case of periodic boundary conditions we impose, on a chain of size L, for every j location,

$$\psi(x_1,\cdots,x_j,\cdots,x_N) = \psi(x_1,\cdots,x_j+L,\cdots,x_N). \qquad (24.123)$$

Translating a particle from site x_j around the ring implies carrying out a permutation of the particle with all the others and in the order of their arrangement, in a given spatial distribution. This is illustrated in Fig. 24.5. This involves a product of $N-1$ two-particle scattering matrices. Imposing periodic boundary conditions, this fixes the value of possible momenta, since the problem reduces to an eigenvalue equation

$$e^{ik_j L}\xi = T_j \xi, \qquad (24.124)$$

where

$$T_j = S_{j,j+1}\cdots S_{j,N} S_{j,1}\cdots S_{j,j-1}. \qquad (24.125)$$

In the case that there are internal degrees of freedom, the diagonalization problem becomes complicated, due to the matrix structure of the scattering matrices. The simpler case of the Heisenberg model was considered before, leading to the Bethe-ansatz equations. In more complicated models, the procedure is the same.

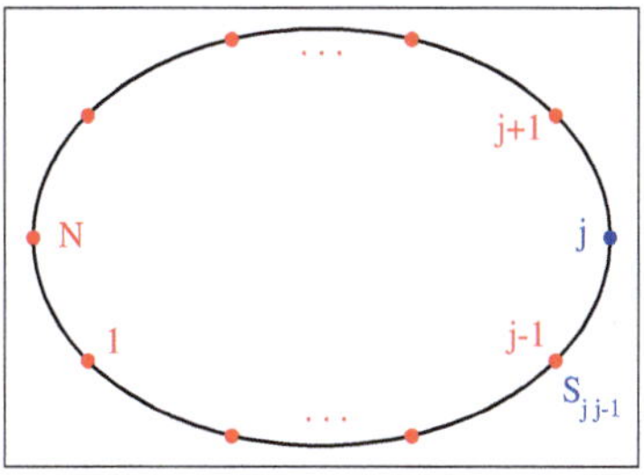

Fig. 24.5 Schematic representation of the sequence of scatterings of particle j with all other particles in a closed chain. Each scattering process involves the scattering matrix $S_{j,i}$ of particle j with particle i

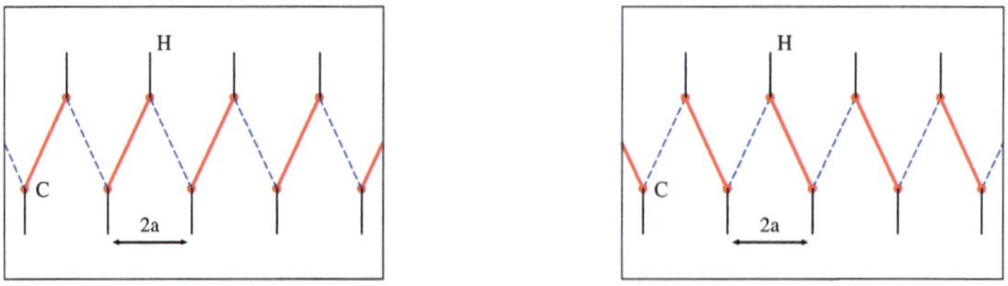

(a) (b)

Fig. 24.6 Bond structure of polyacetylene (SSH model). There is an alternancy of single and double bonds along the chain, with two possible sequences (**a**) and (**b**). The thicker lines represent double bonds, and the dashed bonds represent single bonds. To each carbon, C, atom, one hydrogen atom, H, is bound

24.3 Solitons in Polyacetylene

Polyacetylene, $(CH)_x$, has a very simple structure. The electronic structure of the σ electrons leads to a dimerized structure, where single and double bonds between the C atoms alternate along the chain (this is the SSH model, previously discussed, if the atoms are frozen in a dimerized structure[3]). There are two possible phases A (associated with the choice (a)) and B (associated with the choice (b)), as shown in Fig. 24.6. One phase is obtained from the other by a translation of one lattice unit. The positions of the C atoms are not fixed, and the atoms' oscillations around their equilibrium positions (phonons) give origin to a term in the Hamiltonian

$$H_{ph} = \frac{1}{2}\sum_n K\left(u_{n+1} - u_n\right)^2 + \frac{1}{2}\sum_n M\dot{u}_n^2, \tag{24.126}$$

where M is the mass of the nuclei, K is the force constant, and the variables u_n represent longitudinal displacements, as shown in the figure.

The electrons of the π orbitals move in this dimerized structure according to a tight-binding model, leading to a total Hamiltonian

$$H = H_{ph} - \sum_{n,s} t_{n+1,n}\left(c^\dagger_{n+1,s}c_{n,s} + c^\dagger_{n,s}c_{n+1,s}\right) \tag{24.127}$$

and $t_{n+1,n} = t_0 - \alpha(u_{n+1} - u_n)$ is the transition amplitude between two lattice sites n and $n+1$. Minus a constant t_0, it has an electron-phonon coupling. In a perfectly dimerized structure (either A or B), the displacements are of the type

$$u_{n0} = \pm(-1)^n u_0,$$

[3] For further reading see W. P. Su, J. R. Schrieffer and A. J. Heeger, Physical Review Letters 42, 1698 (1979); A.J. Heeger, S. Kivelson, J. R. Schrieffer and W. P. Su, Reviews of Modern Physics 60, 781 (1988).

where (+) corresponds to the phase A and (−) to the phase B. In these cases we have that

$$\begin{aligned} t_{n+1,n} &= t_0 - t_1, \\ &= t_0 + t_1, \end{aligned} \tag{24.128}$$

where in the first case we have a single bond and in the second case we have a double bond. Due to the dimerization, the spectrum shows a gap for the momentum values $k = \pm\pi/2a$. If the chemical potential vanishes, it is located in the gap, and the system is an insulator.

Soliton and Zero Energy Mode

Consider now that the system is divided in two domains, such that on the left side, we have phase B, and on the right side, we have phase A (recall that the two phases are degenerate). The separation between the two domains defines a soliton (kink), which may be considered as localized at the origin, $n = 0$. Defining an order parameter $\psi_n = (-1)^n u_n$ that reduces to $\pm\, u_0$ far from the wall that separates the two phases and that is zero at the origin, we may estimate the domain wall width between the two phases using a simple variational method. Looking for a solution of the type

$$\psi_n = u_0 \tanh\left(\frac{n}{l}\right)$$

and determining the energy as a function of the variational parameter l, it can be shown that $l \sim$ 5–9 and that the mass of the soliton is quite small, of the order of $M_s \sim 6m_e$. This result means the soliton has a high mobility. The solution of the problem shows that a zero energy mode emerges, due to the presence of the soliton, for each spin projection value.

The density of states changes due to the presence of the soliton. From the valence and conduction bands half a state disappears (integrating the density of states either in the valence or in the conduction band one finds one less state), which appears in the zero energy mode. Since the total number of electronic states is conserved,

$$\int_{-\infty}^{\infty} \Delta\rho(E)dE = 0,$$

where $\rho(E)$ is the density of states. Since the valence band is fully occupied, having the solitonic excitation present implies that the valence band has one electron missing (half a state for each spin value). That is, in the case of a neutral soliton, the missing electron occupies the state at the origin. Note, however, that the valence band remains with the spins paired, while the electron in the zero energy mode has a given spin ($\uparrow$ or $\downarrow$). Therefore, the result is that a neutral soliton (where the number of electrons of the system has not changed) has spin-$1/2$. Other possibilities are obtained, either removing an electron from the zero mode or adding another electron

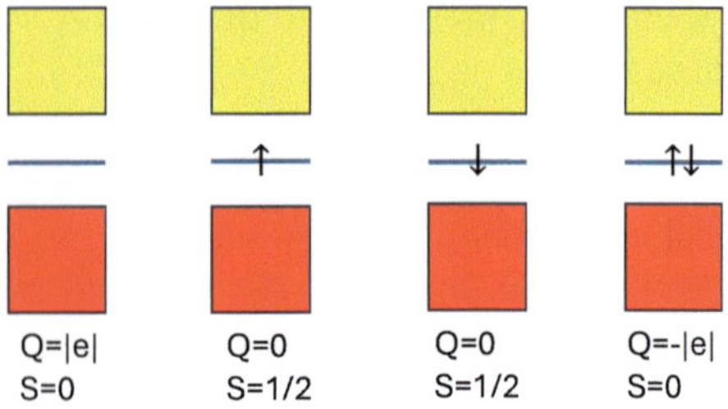

Fig. 24.7 Unusual charge-spin relations associated with different fillings of the zero energy mode at the soliton location

on the same mode (with the opposite spin polarization). Therefore, a charged soliton, with charge $+e$ or with charge $-e$, respectively, has zero spin. As a consequence, one finds exotic relations between charge and spin that do not respect Kramers' theorem, since a spin-1/2 system with an even number of electrons should have integer spin. In short, the excitations are of the type

$$Q_0 = 0 ; \qquad s_0 = \frac{1}{2}$$
$$Q_{\pm} = \pm e ; \qquad s_{\pm} = 0.$$

They are schematically represented in Fig. 24.7. The effect of the addition of the soliton may be understood in a way similar to the Kondo problem. The presence of the soliton gives rise to a phase shift in the electrons, such that the density of states is affected, leading to a decrease of the number of states in the vicinity of the soliton, integrating to the disappearance of half a state for each spin value.

The apparent violation of Kramers' theorem implies that there must exist another spin that compensates the spin of the soliton. Indeed, topological restrictions imply that, as a soliton S forms, an anti-soliton, $\bar{S}$, should form as well. In this case there is a separation between the phase A and the phase B, such that the system returns to the phase A on the right side. If the pair soliton-antisoliton is found at a short distance from each other, the pair will respect Kramers' theorem, but in the case that the two members of the pair are far apart, we may consider the soliton (and the antisoliton) as an excitation of the system with an unusual charge-spin relation. (Note that the term soliton is being used in a loose way; in fact these excitations are not solitons but solitary waves. If they are allowed to move along the chain, as they cross, the excitations interfere, while true solitons do not interact and remain with the same shape as they collide, such as the solitons of the sine-Gordon model).

Fractional Charges

It turns out that the problem has other interesting properties. Let us consider spinless fermions instead of electrons. In this case, and considering a charged soliton, such that the zero mode is empty, then, since half a state is missing from the valence band, the charge of the soliton is $+\frac{1}{2}|e|$, which means a fractional charge. It is the fact that electrons have spin-1/2 that gave origin to unusual charge-spin relations but that preserve integer charges. This type of problem was considered by Jackiw and Rebbi in the context of a relativistic theory in $(1+1)$ dimensions, where they identified the charge of a soliton in a bosonic field coupled to a spinless fermionic field, as being a fermion of number $1/2$.

The case we have considered up to now is that of a system with electronic density that corresponds to half-filling. Let us now take a situation in which the band is one-third full. In this case it will be favorable that the system trimerizes, instead of dimerizing. There will be now two gaps in the system. There are now three degenerate states. These states can be called A, B, and C. The displacements for the perfectly trimerized lattice are given by

$$u_n = u \cos\left(\frac{2}{3}\pi n - \theta\right).$$

The three states are then obtained taking $\theta - \frac{1}{6}\pi = 0, \frac{2}{3}\pi, \frac{4}{3}\pi$, respectively. Once again, the three states may be obtained from each other by translations. The hopping of one electron takes two values as $-(t_0+\delta)$ and $-(t_0+2\delta)$, respectively. There are two types of excitations. One type is such that the degenerate states are in the sequence $A \to B \to C \to A$ from left to right, centered at the points $n_1, n_2.n_3$. Another sequence is $A \to C \to B \to A$. The first type corresponds to a value of $\theta - \frac{1}{6}\pi = 0$; then we go through values of $\frac{2}{3}\pi$, $\frac{4}{3}\pi$, and $\frac{6}{3}\pi$. Since the total phase shift is 2π, the system returns to phase A for high values of n. In this process, a charge of $-2|e|$ has been transferred. Since the three kinks that separate the various degenerate phases must be equivalent, the charge transported by each kink must be equal. That is, on each kink, we should find an accumulated charge of $+\frac{2}{3}|e|$. As a consequence, each kink has a fractional charge. The second type of sequence transports a charge $-\frac{2}{3}|e|$, since the phase shift is now -2π and a charge of $-2|e|$ is now accumulated in the region of the three kinks. In general, in a problem with spin degeneracy n_s and a commensurability n_c, each soliton will have a fractional charge of $|e|n_s/n_c$.

Returning to the case of commensurability $n_c = 3$, consider first the kinks of type I. There are two states in the gaps called ϕ_l and ϕ_u. The state localized in the lower gap, ϕ_l, is the result of one third of a state being removed from the lower band and two thirds of a state being removed from the middle band. A symmetric situation occurs for the state ϕ_u. If the state ϕ_l is unoccupied, we have a kink with a charge $\frac{2}{3}|e|$. All the spins are paired, and the total spin is therefore 0. If this state is occupied with one electron, then the kink has charge $-\frac{1}{3}|e|$ and spin $1/2$. Finally if the state is doubly occupied, the spin is 0, and the charge is $-\frac{4}{3}|e|$.

In the case of sequences of type II, something similar occurs. The state ϕ_l is the result of two thirds of a state being removed from the lower band and one third removed from the middle band. If the state is unoccupied, the kink has charge $+\frac{4}{3}|e|$, and the spin is 0. If it is occupied with one electron, the charge is $+\frac{1}{3}|e|$, and the spin is $1/2$, and if it is doubly occupied, the charge is $-\frac{2}{3}|e|$, and the spin vanishes.

Kivelson and Schrieffer showed that the fractional charge of the solitons has physical significance, since they showed that it is the eigenvalue of a charge operator, and not just the average value of different situations, in which the charges are integers, and only the average value is fractionary. The fractional charge is therefore an observable, even though the charge of a system is an integer if we include all the solitons (and antisolitons). This proof was also carried out by Jackiw and Rebbi.

Schrieffer's Argument

The mechanism that generates fractional charges was clarified by a simple argument due to Schrieffer. Let us take the trivial example of an infinite chain. Let us associate a charge q to each alternate lattice site. For example, let us place the charges in the odd numbered sites, starting from $-\infty$ until a site labeled 1. Next, we change the sequence introducing a charge on site 2 and then charges on all even sites up to ∞. In this way we have created two domains, such that the frontier is found between sites 1 and 2. Let us now move the charge at site 2 to site 3. The location of the border between the two domains moved to a position between sites 3 and 4. That is, the charge was moved by one position, but the position of the frontier moved two positions. Let us calculate now the charge of the frontier between the two domains. To do that, we calculate the change of the electric dipole moment, caused by the displacement of the charge q. The change of the electric dipole moment is 1. The change due to the motion of the frontier with charge Q (to be determined) is 2. Since the change of the dipole moment must be independent of the description used, we must have that $Q = q/2$, which means that if the charges are integers, the charge of the frontier (or domain wall) must be fractionary. (Obviously, if we had placed a charge q every three sites, the charge of the domain wall would be $Q = q/3$).

Solitons and Other Excitations

The discussion of the conditions under which the solitons carry fractional charges, or that excitations have an unusual charge-spin relation, is based on the assumption that the charge carried by the soliton (or domain wall) is mostly located at the interface between the domains. In fact, the zero energy mode has a wavefunction that is relatively extended in space. However, in the limit of a very large system, it is possible to have a situation in which the various solitons are far apart, such that they preserve their individual nature. However, the whole discussion carried out up to this point assumes that the possible states are degenerate. If the degeneracy between the states is lifted, the behavior of these excitations is different. Considering, for instance, the dimerized case, suppose that phase A is the more stable. Take an interface such that on the left side we have phase A and on the right side we have phase B. Since phase A is favorable energetically, the energy cost of a region with phase B is proportional to its extension. Therefore, if $\Delta E/l$ is the energy difference per site between phases A and B, then the energy cost is given by $\sim d(\Delta E/l)$, where d is the extension of phase B. Therefore, if we have a pair soliton-antisoliton, the energy increases linearly with the distance between them (the soliton corresponds to a change from phase A to phase B and the antisoliton is the opposite). As a consequence, there is a confinement of the pair of solitons that tends to keep them close to each other, giving rise to an excitation called bipolaron, which is a confined pair of solitons $(S + \bar{S})$.

One of the initial arguments in favor of the existence of solitons was the comparison between the energy required to excite electrons across the gap (for instance, for the dimerized case) and the energy necessary to create one soliton. Indeed, it was found that the energy necessary to create a soliton was favorable, and therefore solitons are possible excitations. The solitonic excitation may be expressed in the continuous version of the Su, Schrieffer, and Heeger model as an external potential $\Delta(x)$, which appears in the equation

$$\epsilon_n \psi_n(x) = \left(-i v_F \sigma_3 \frac{\partial}{\partial x} + \Delta(x)\sigma_1 \right) \psi_n(x), \tag{24.129}$$

where σ_i are Pauli matrices that act on the spinor $\psi_n^\dagger = (u_n^*, v_n^*)$, where the amplitudes u_n and v_n pertain to the electrons that move to the right and to the left, respectively. The soliton excitation corresponds to a profile such that $\Delta(x) = \Delta_o \tanh(x/l)$, where Δ_0 is half the gap of the dimerized system.

There are other excitations called polarons that do not have a topological origin and may be described by the choice

$$\Delta(x) = \Delta_0 - K_0 v_F\{\tanh[K_0(x + x_0)] - \tanh[K_0(x - x_0)]\}.$$

This excitation is represented in Fig. 24.8 and compared to a soliton. Instead of interpolating between two phases (and therefore distinct boundaries at infinity, which give a topological nature to the solitonic excitation), it is just a local deformation of the bosonic configuration (recall that the origin of the lattice deformation lies with the phonons). Since the solitons have to be created in pairs (unless non-trivial boundary conditions are imposed externally), the energy to create a pair soliton-antisoliton is given by $2(2/\pi)\Delta_0$, while the energy to create a polaron is smaller and given by $\sim 0.90\Delta_0$ and is therefore energetically preferred. In general, we have to consider both possibilities. On the other hand, solitons are excitations that require that the dimerization extends over large regions. In general, in the real materials, this does not happen, and the experimental observation is difficult.

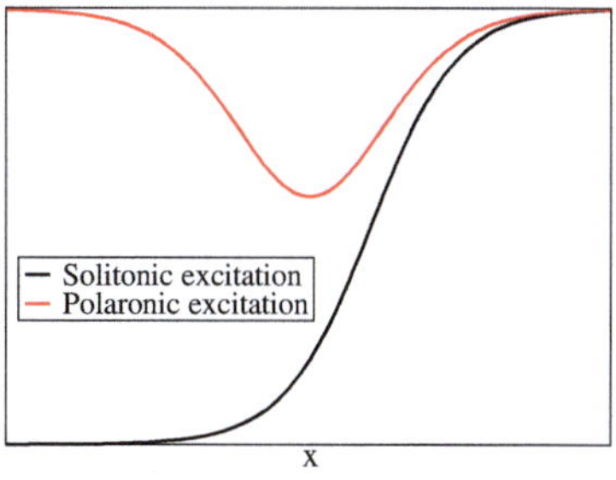

Fig. 24.8 Comparison of solitonic and polaronic excitations

Soliton with Fermionic Number $1/2$

The Jackiw and Rebbi solution for the case of spinless fermions allows a simple way to obtain a solution for the zero energy mode. Consider a fermionic field, $\psi(x,t)$, coupled to a scalar field, $\Phi(x,t)$, in a space $(1+1)$ (the authors also considered a space $(3+1)$ with similar conclusions), with a Lagrangian density

$$\begin{aligned} L &= L_B + L_F \\ L_B &= \frac{1}{2g^2}\left[\left(\frac{\partial\Phi}{\partial t}\right)^2 - \left(\frac{\partial\Phi}{\partial x}\right)^2 - \frac{1}{2}\left(\Phi^2-1\right)^2\right] \\ L_F &= \bar{\psi}\left(i\partial_\mu\gamma^\mu - m\Phi(x,t)\right)\psi. \end{aligned} \tag{24.130}$$

Note that the fermionic field has just two components in a space $(1+1)$. Consider a kink solution of the type

$$\Phi_s(x) = \pm\tanh(x/\sqrt{2}).$$

The solution of the Dirac equation for the fermionic field in the presence of this bosonic profile is written as

$$\left(-i\alpha\partial_x + \beta m \tanh\frac{x}{\sqrt{2}}\right)\psi(x) = E\psi(x).$$

Beyond the usual solutions of a free particle, one also finds a zero energy solution

$$\eta_0 = \begin{pmatrix} A\exp\{-m\int^x dy\tanh(y/\sqrt{2})\} \\ 0 \end{pmatrix}.$$

This is a fermionic mode that has the shape of an exponential of a bosonic field, which is somewhat similar to the bosonization we found previously. The fermionic operator may then be expanded as

$$\psi(x,t) = \sum_{k\neq 0}[b_k\eta_k(x)e^{-iE_kt} + d_k^\dagger\tilde{\eta}_k(x)e^{iE_kt}] + a\eta_0(x), \tag{24.131}$$

where a is the destruction operator for the zero energy mode and the other terms correspond to the usual decomposition, in positive and negative energy modes. Note that the zero energy mode is its own conjugate (in the thermodynamic limit). As in the case of polyacetylene, there are two degenerate states. They may be represented by $|s_+\rangle$ and $|s_-\rangle$ and satisfy

$$a|s_-\rangle = b_k|s_-\rangle = d_k|s_-\rangle = 0$$

and

$$|s_+\rangle = a^\dagger |s_-\rangle; a|s_+\rangle = |s_-\rangle.$$

The states are energetically degenerate but have different charges. The charge is given by

$$\begin{aligned} Q &= \frac{1}{2} \int dx [\psi^\dagger(x,t), \psi(x,t)] \\ &= \sum_k (b_k^\dagger b_k - d_k^\dagger d_k) + a^\dagger a - \frac{1}{2}, \end{aligned} \tag{24.132}$$

which shows once again that the charge is half-integer, fractionary. Moreover, this treatment confirms that the eigenvalues of the charge correspond to a fractionary charge. The states have charges

$$\begin{aligned} Q|s_-\rangle &= -(1/2)|s_-\rangle \\ Q|s_+\rangle &= +(1/2)|s_+\rangle. \end{aligned} \tag{24.133}$$

24.4 Topology of Spin-1 Chain

One expects that an antiferromagnetic chain of half-integer spins should be either a spin singlet state with a gapless spectrum or to be doubly degenerate with a gap to the first excited state, corresponding to a breaking of translation symmetry. This is the Lieb, Shultz, and Mattis theorem. While the spin-$1/2$ case is gapless, as shown by the Bethe ansatz solution presented in this chapter, the model with higher values of the spin, including the integer values, is not integrable, and in general no exact solution is known, except in some specially chosen models such as the Takhtajan-Babujan model.

Haldane conjectured that antiferromagnetic spin chains of integer and half-integer spins belong to two separate classes [2]: undimerized systems with half-integer spins have gapless spectra, but the integer spin systems have a gap in the spectrum and are said to be in the Haldane phase. This conjecture has been confirmed extensively, both with numerical calculations and experimentally.

A model proposed by Affleck, Kennedy, Lieb, and Tasaki for a spin-1 chain (AKLT model) [3] provides an example of the Haldane phase.[4] In this model, the spin-1 operators are coupled to their nearest neighbors. Each link may be in a state with total spin $\mathbf{S} = \mathbf{S}_i + \mathbf{S}_{i+1}$ and may take the values $S = 0, 1, 2$. Defining the

[4] For further reading see, for instance, [AraujoSacramento].

Hamiltonian in terms of projectors to some total spin value and identifying some state that has no projection on that spin value ensures that the state has zero energy. If this is the lowest energy, then it is the groundstate or belongs to the manifold of groundstates.

The Hamiltonian is written as

$$H = \sum_i P_2 \left(\mathbf{S}_i + \mathbf{S}_{i+1}\right), \tag{24.134}$$

where the projector P_2 ensures that each link is in the subspace of $S = 2$. The choice

$$P_2 \left(\mathbf{S}_i + \mathbf{S}_{i+1}\right) = \frac{1}{2}\mathbf{S}_i \cdot \mathbf{S}_{i+1} + \frac{1}{6}\left(\mathbf{S}_i \cdot \mathbf{S}_{i+1}\right)^2 + \frac{1}{3} \tag{24.135}$$

implements the desired projection and leads to

$$\begin{aligned} P_2((\mathbf{S}_i + \mathbf{S}_{i+1})\,|S = 2\rangle &= |S = 2\rangle \\ P_2((\mathbf{S}_i + \mathbf{S}_{i+1})\,|S = 1\rangle &= 0 \\ P_2((\mathbf{S}_i + \mathbf{S}_{i+1})\,|S = 0\rangle &= 0, \end{aligned} \tag{24.136}$$

which implies that the state has zero or positive energy. AKLT constructed a state that is contained in the subspaces of $S = 0$, 1, and, therefore, it is a groundstate. It was shown that the spectrum has a gap, as wanted, and that the correlation functions decay exponentially with the distance between sites on the lattice.

The state proposed relies on the representation of each spin-1 operator as the symmetric coupling of two spin-$1/2$ operators, assumed as occupying each lattice site. In an antiferromagnet, it is natural that spins tend to form singlets. It is imposed that one of the spin-$1/2$ operators of a given site forms a spin singlet $S = 0$ state with one spin-$1/2$ operator of a neighboring site. The two remaining spin-$1/2$ operators of the two neighbors may then couple to give rise to a spin $S = 0$ state or a spin 1 state. Therefore, states with $S = 2$ are naturally excluded. One can construct a state under these conditions. If the system is infinite, this state is called a valence bond solid (VBS).

If the system has finite extent, there are two spin-$1/2$ operators, one at each edge, which are unpaired. Since each spin-$1/2$ has two states, there are four degenerate states, with zero energy. The groundstate has therefore a degeneracy of 4. The existence of these degenerate edge states illustrates the topological nature of the state. One may also show the topological nature of this state through a calculation of the Berry phase.

References

1. J. Carmelo and A. A. Ovchinikov, Journal of Physics Condensed Matter **3**, 757 (1991); J. M. P. Carmelo and H. Q. Lin, Modern Physics Letters B **7**, 1387 (1993).
2. F. D. M. Haldane, Physics Letters **93A**, 464 (1983); F. D. M. Haldane, Physical Review Letters **50**, 1153 (1983).
3. I. Affleck, T. Kennedy, E. H. Lieb and H. Tasaki, Physical Review Letters **59**, 799 (1987).

Chapter 25
Short Tour of Materials and Applications

In this chapter, some new or more specific material properties will be considered, together with some applications. The discussion will be mainly descriptive, and no derivations or detailed discussions of the physical mechanisms will be presented. On the other hand, several references are added. In some cases, the original works will be referred, but for the most part, the references are recent reviews on the subjects presented. The reader is encouraged to consult these reviews and, if interested, to go through the extensive list of references in each review paper. There is a very large list of different phenomena in condensed matter physics. Here a small subset is introduced, with no particular motivation on completeness.

Electronic Structure

Real materials are in general complex. In previous chapters we were mainly concerned with minimal or simplifying descriptions of different phenomena. The understanding of the properties of a given system involves an experimental study, using various techniques and methods, and/or some theoretical study. This usually requires some detailed study of the electronic structure, together with the detailed understanding of the material structure and the lattice dynamics.

Detailed, first principle methods, such as density functional theory (DFT) or other methods, that allow a determination of the band structure of some material, are usually a good and necessary starting point [1], in particular if the system is not strongly correlated. The full solution leads to a myriad of bands. Examples for some semiconductors are shown in Fig. 25.1. Typically, from these bands, those that are assumed to be most relevant, such as those in the neighborhood of the Fermi surface, are then modelled in some simplified way, expected to reproduce as best as possible the most relevant physical effects.

Adding interactions, the DFT method has been extended [2], and other methods have also been used [3] It is the usual procedure to include the effect of interactions on the simplified model that describes the behavior of the system.

P. D. Sacramento, *Introduction to Condensed Matter Theory*,
https://doi.org/10.1007/978-3-031-86251-9_25

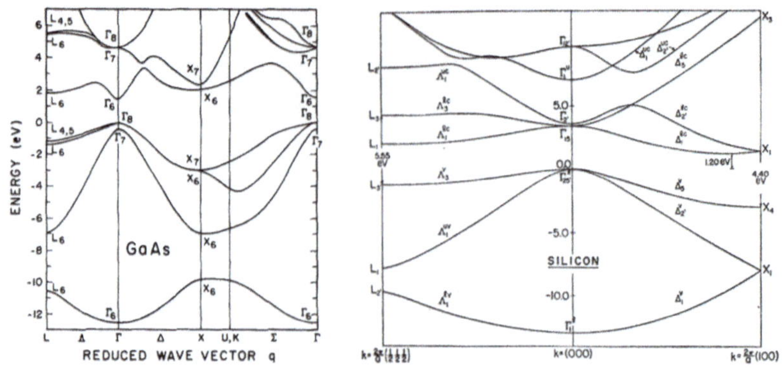

Fig. 25.1 Band structure of GaAs (direct gap) and Si (indirect gap) semiconductors. Reproduced with permission from J. S. Blakemore, J. Appl. Phys. **53**, R123 (1982) and from Manuel Cardona and Fred H. Pollak, Physical Review **142**, 530 (1966)

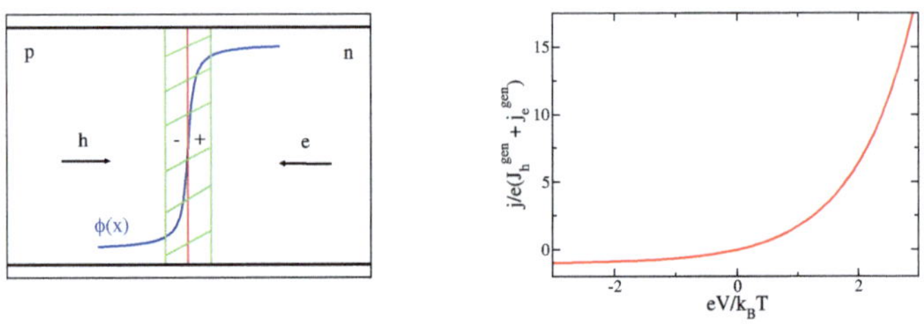

Fig. 25.2 $p - n$ junction and effect of an external potential, V, on the current across the junction

Even though these detailed descriptions of a given material are essential as starting points, simplified models also have their merits and in the remainder of this chapter we will maintain a simplified description, as followed in other chapters.

Junctions of Semiconductors

In n-type semiconductors, there is an excess of electrons, and in p-type semiconductors, there is an excess of holes. If we place several semiconductors in contact, there is in general some motion of the charges across the junction. This leads to an accumulation of charge of negative and positive values that then serves as a potential energy barrier with respect to further motion of the charges.

One example is a $p - n$ junction,[1] as shown in Fig. 25.2. Since there is a deficiency of holes on the right there is a current of holes to the right and the opposite for the negative charges. As the holes (electrons) accumulate on the right (left), a non-homogeneous potential, $\phi(x)$, develops around the junction ($x = 0$). This gives rise to a charge profile and a potential barrier that prevents further net migration of charges across the barrier. If we now add a positive potential, V, on the p side,

[1] For further reading see, for instance, [Kittel].

this has the effect of decreasing ϕ and, if the potential is negative, it increases ϕ. If $V = 0$ there is no net current after equilibrium is reached. If $V \neq 0$ currents are established and there is a significant difference as the sign of the external potential changes, as shown in Fig. 25.2. The currents are mainly due to the thermal excitation of holes in the n side in the region around the barrier and that are repelled to the p side by the potential barrier and the same for electrons generated on the p side. The total current is approximately given by

$$j = e\left(j_h^{gen} + j_e^{gen}\right)\left(e^{\frac{eV}{k_B T}} - 1\right), \tag{25.1}$$

neglecting the small currents that are the result of the motion of holes (electrons) from the p (n) region into the other regions, implying some recombination of charges. This junction serves as a rectifier in the sense that it changes the current, depending on the potential difference established across the junction. If alternating voltage is applied, the current will flow mainly in one direction, and, therefore, the junction “rectifies” the current.

The existence of the potential barrier and the associated region where it is located (called depletion layer) has applications such as in solar cells and photovoltaic detectors. An incident photon on the junction creates an electron-hole pair that diffuses into the depletion layer and adds to the potential. The electron moves to the right and the hole to the left leading to a current that may be collected in a circuit. In this way the solar energy in terms of the photons may be converted into electrical energy.

We may also consider heterostructures with different semiconductors of the same charge type. In general, they have different crystalline structures, different band gaps, and different Fermi levels. This implies that, as they come together, potential barriers form at the interface. One example is a $n - N$ junction of two n-type semiconductors (such as a junction $Ge/GaAs$), where N has the larger gap and higher Fermi level. When they form a junction, the top of the valence and the bottom of the conduction band have to adjust so that a common Fermi level is established in equilibrium. This is shown in Fig. 25.3. In this case, one finds a potential well at the bottom of the conduction band that may be occupied by electrons that are then localized at the interface forming a quasi-two dimensional electron system, if the Fermi level is located in the potential well. Considering an heterostructure such as

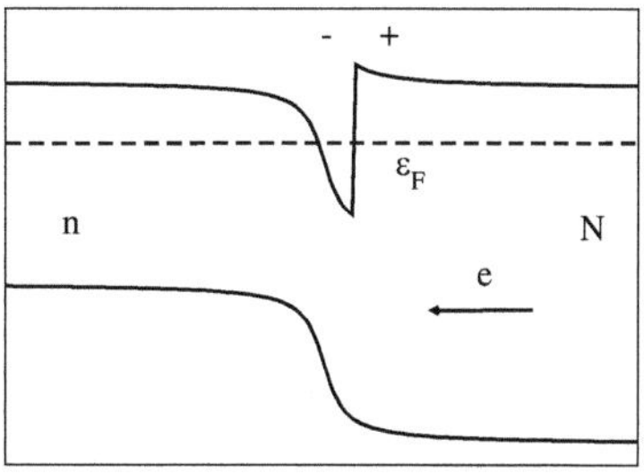

Fig. 25.3 $n - N$ junction with the formation of a potential well at the interface

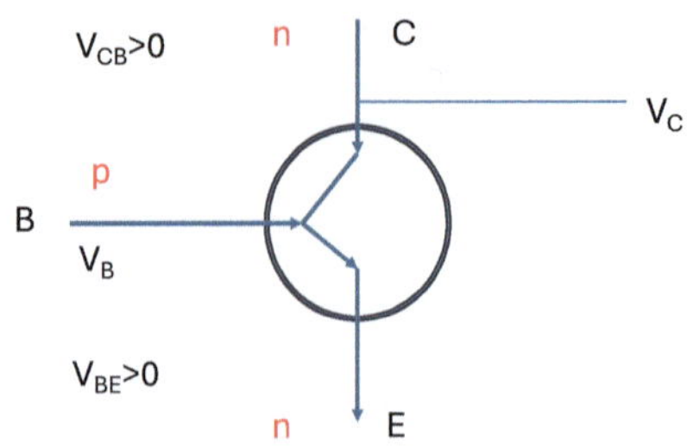

Fig. 25.4 $n-p-n$ junction: the n-type semiconductors are called the collector (C) and the emitter (E). The p-type semiconductor is the base (B). With the polarities shown in the figure, a current enters the junction from the collector and the base and leaves from the emitter

a metal, an oxide SiO_2, and Si (MOSFET, metal-oxide-semiconductor field-effect transistor), imposing a gate voltage with positive polarity on the Si side, allows an accumulation of charge at the oxide and a two-dimensional conducting system.

There are many heterostructures with different functionalities. One example is a $n-p-n$ junction, as schematized in Fig. 25.4. One functionality is that it works as an amplifier. Inserting a resistance in the wire of the collector, call it R_C, and a resistance in the emitter wire, R_E, and establishing a potential at the base, V_B, it can be shown that any change, ΔV_B, leads to a change, ΔV_C, at the collector that is $|\Delta V_C| = |\Delta V_B| R_C/R_E$, which may be tuned choosing the two resistances appropriately.

Two-Dimensional Materials

Two-dimensional systems have received recently considerable attention, since a strictly two-dimensional system is not expected to be stable (see Eq. (15.196)), except under special conditions. Graphene is a significant example together with transition metal dichalcogenides (TMD). The experimental control on obtaining single-layer graphene has fueled the field. Its honeycomb structure (Fig. 8.9) leads to its band structure (Fig. 10.15), with the presence of Dirac cones and their interesting consequences, such as relativistic-like dispersion relation, its significative quantum Hall effect (Eq. 13.149), Hall conductance (Eq. 13.152), and sensitivity to strain (Eq. 13.193) that leads to high synthetic magnetic fields. As a consequence of its band structure [4], many other properties have proven interesting and useful in different contexts. The dispersion relations around the two Dirac valleys can be written as in Eq. (10.186).

We may focus on its relativistic-like dispersion relation. In the context of relativistic problems, both in the context of the Klein-Gordon equation and Dirac equation, when a particle collides with a potential barrier, if the height of the barrier is very high, there is a possibility of a propagating mode (with real momentum) inside the barrier (Klein paradox). This occurs if the height of the barrier, V_0, is such that $V_0 > E + mc^2$, where E is the energy of the incident particle. This occurs because the potential may create pairs of particle-antiparticle, and the antiparticle may propagate inside the potential barrier. In the context of the Dirac equation, this leads to a reflection and transmission coefficients that read

$$R = \frac{(1-r)^2}{(1+r)^2}$$
$$T = \frac{4r}{(1+r)^2}, \tag{25.2}$$

with $R + T = 1$ and

$$r = \sqrt{\frac{\left(V_0 - E + mc^2\right)\left(E + mc^2\right)}{\left(V_0 - E - mc^2\right)\left(E - mc^2\right)}}. \tag{25.3}$$

If the mass vanishes, as in graphene, $r = 1$, and $T = 1$. Therefore, there is perfect transmittance. One way to understand such effect in graphene is to notice that, defining the helicity operator as $\hat{h} = 1/2\boldsymbol{\sigma} \cdot \mathbf{p}/|\mathbf{p}|$, it has eigenvalues $\pm 1/2$ on the energy eigenstates of the wave functions near the two Dirac cones, $\mathbf{K}, \mathbf{K}'$. It is therefore conserved if we take the linear approximation around the Dirac cones. Also, the two Dirac cones have opposite helicities. As a consequence, if there is some scattering on the vicinity of a given Dirac point, if the two Dirac points are not mixed, backscattering is not allowed since it would imply a reversal of helicity, which is forbidden, since it is conserved. Considering a barrier of finite extent with width D and height V_0, it can be shown that for an electron with incidence angle ϕ, if $Dq_x = n\pi$, with q_x the momentum component perpendicular to the potential barrier interface, $T(\phi) = 1$. Also, if the incidence is normal ($\phi = 0$), there is perfect transmittance, $T = 1$.

Another related property has to do with a minimal conductivity [5, 6] through a graphene region of length L and width W. The length is along the direction of a passing current. Consider a neutral graphene system, with the Fermi energy at the Dirac point (zero energy). Since the density of states is given by $\rho(E) = 2|E|/(\pi\sqrt{3}t^2)$, with t the hopping between nearest neighbors (between the two sublattices), it vanishes at the Dirac point and, therefore, at the Fermi level. Therefore, one might expect a vanishing conductivity. However, one may recall that to calculate the conductivity, the effect of disorder is to be taken. It turns out that an effect of disorder is to give origin to a finite density of states and, therefore, it allows a finite conductivity. Considering the effect of a random potential it may be shown that the conductivity has a minimal value, independent of disorder, if disorder is small, and given by

$$\sigma_{\min} = \frac{4e^2}{\pi h}. \tag{25.4}$$

The same result has been obtained calculating the conductivity due to the evanescent modes through a graphene strip inserted between two leads, in the limit that $W >> L$.

Another interesting result is the transmissivity of light through a graphene monolayer. Either using a perturbative approach or using Kubo's formula, it has been shown that the transmissivity is given by Nair et al. [7]

$$T = 1 - \pi\alpha. \tag{25.5}$$

Here, α is the fine structure constant and, therefore, the result can be expressed solely in terms of fundamental constants. Absorbing light leads to transitions between bands, keeping the momentum fixed (since the speed of light is much larger than the electrons velocities). The transmissivity in the optical regime may be compared between an area with air, one layer of graphene, and two layers of graphene. It is interesting that the differences are visible to the naked eye [5]. The transmissivity is calculated by subtracting the absorbed part from the incident light.

Stacking graphene layers leads to interesting changes and varied properties. One example is a bilayer graphene. In this case, the band structure is such that the number of bands doubles, and the spectrum remains gapless, but the bands have a quadratic dispersion relation at the gapless point. Adding an external electric field perpendicular to the layers leads to the lifting of the degeneracy between the two sublattices and a gap opens, which can be tuned varying the electric field [8]. Other heterostructures of graphene layers and TMDs lead to interesting physics that will be mentioned later in this chapter.

Graphene is also notable for many other properties [9] despite (and/or because) it being a one-atom layer of carbon atoms. It has a very strong material, very large charge mobilities and current densities, and high thermal conductivity, and it is impermeable to gases. Even though spin does not enter in the usual discussion of its basic properties, the pseudospin associated with the two Dirac valleys is an important characteristic to be explored. Since graphene has many coexisting properties that in other materials are usually exclusive, it seems as a material that is potentially useful in a number of different fields. One aspect that is however to be noted is the small spin-orbit coupling, which seems to hinder applications that combine orbital and spin degrees of freedom, for instance, the use of electric fields to control spin degrees of freedom. In that respect, TMDs seem more interesting due to their larger intrinsic spin-orbit coupling. We will return to this point.

Topological Electronic Systems

From a realization that topology has a role in understanding and characterizing some physical systems, there has been an increasing awareness that it pervades a multitude of problems, such as electronic systems ([10]; suggestively entitled "All topological bands of all nonmagnetic stoichiometric materials") and many others. Here we illustrate some topological electronic systems.

An early example where topology is helpful is the integer quantum Hall effect, where the transverse Hall conductivity takes values $\sigma_{xy} = Ne^2/h$ (see Eq. (13.122), where N is the number of filled Landau levels and e^2/h is the conductance quantum, Eq. (23.10). Associated with bulk topological properties, as discussed in Chap. 10, one usually finds edge states due to the bulk-edge correspondence. These are often

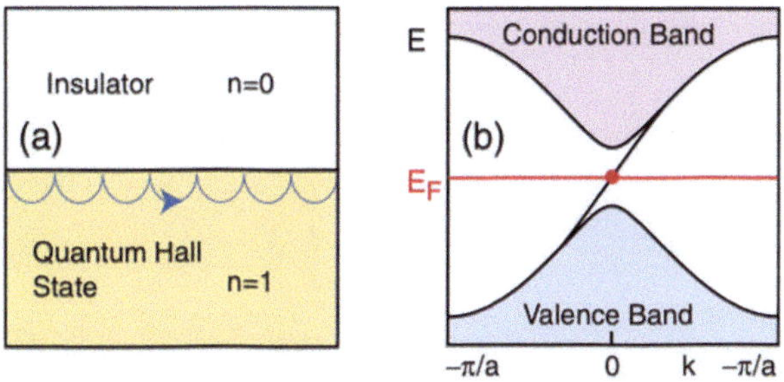

Fig. 25.5 Chiral edge mode due to the presence of an external high magnetic field. The mode has a given velocity. On the opposite edge, the velocity is reversed. If the edges are far apart, backscattering due to imperfections at the edges is inhibited. Reproduced with permission from M. Z. Hasan and C. L. Kane, Reviews of Modern Physics **82**, 3045 (2010)

used to identify the presence of topologically non-trivial phases. A perpendicular magnetic field on a two-dimensional electronic system leads to circulating Landau orbits that, near the edges of a finite width sample, give origin to modes that propagate along the edge in a given direction, as illustrated in Fig. 25.5. These are called chiral modes, present in the case time reversal invariance is broken, in this case due to the external magnetic field. Considering the effect of the electronic spin and the presence of a spin-orbit coupling, together with time reversal symmetry, one finds helical edge modes, where there is momentum-spin locking, as shown in Fig. 25.6. This constitutes a quantum spin Hall insulator. At each edge, instead of a single mode, as in the chiral case, there are two counterpropagating modes with opposite spin components. Imperfections at the edges that conserve spin, such as potential scattering impurities, cannot give rise to backscattering, and the edge modes are topologically protected.

An early experimental verification of topological modes in a topological insulator considered a heterostructure [11] $CdTe/HgTe/CdTe$, with a variable thickness, d, of $HgTe$. In $HgTe$, there is a band inversion responsible for the non-trivial behavior. As the width d increases, two edge states appear that are responsible for the quantized $2e^2/h$ conductance. Some of the samples shown in the figure have a width that is small enough that the edge states are absent, while samples called III

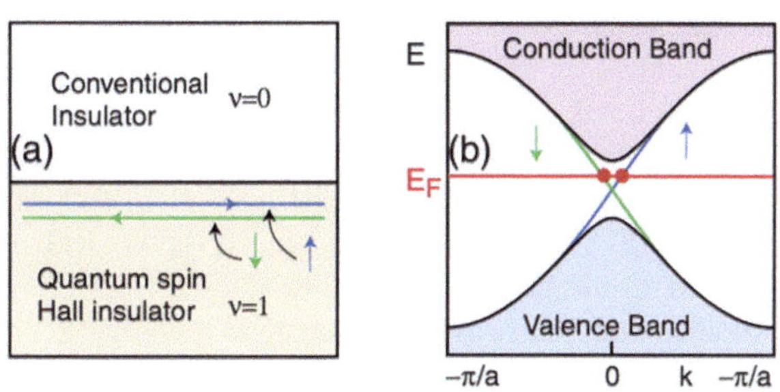

Fig. 25.6 Helical edge modes in a quantum spin Hall insulator. Reproduced with permission from M. Z. Hasan and C. L. Kane, Reviews of Modern Physics **82**, 3045 (2010)

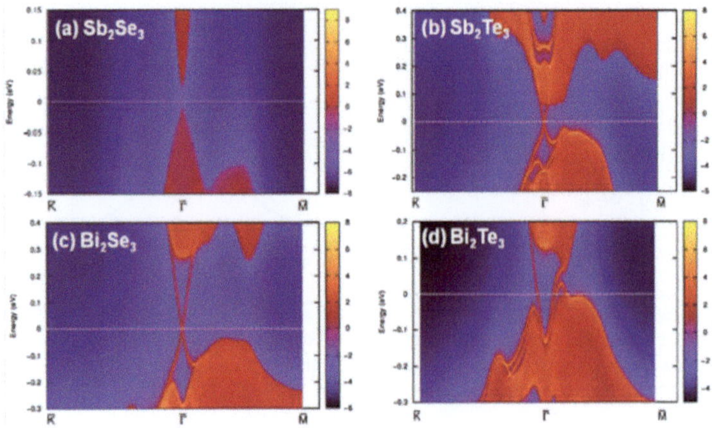

Fig. 25.7 Edge states of several materials. Reproduced with permission from Xiao-Liang Qi and Shou-Cheng Zhang, Reviews of Modern Physics **83**, 1057 (2011)

and IV have an appropriate width that leads to the appearance of two edge modes that contribute each e^2/h to the conductance.

Later several topological insulators have been identified in crystals [12], such as Bi_2Se_3, with the experimental detection of helical momentum-spin locked edge [13] states.

The method used was angle-resolved photoemission electron scattering (ARPES) in which an incident photon extracts an electron (via the photoelectric effect) that is then detected determining its energy and momentum [14]. In Fig. 25.7 theoretical results are shown for other topological insulators, such as those in the Bi_2Se_3 class, like Bi_2Te_3, Sb_2Te_3, $BI_{1-x}Sb_x$. In addition to systems that are protected by point symmetries, there are also systems that are protected by crystalline symmetries, called topological crystalline insulators. One example [15] is $SnTe_2$ in which the states are protected by mirror symmetry.

Other topological effects have been found such as in cold atom systems [16], whose parameter tunability by laser lattices enables the study and construction of various states. Also, returning to electronic systems, topological states in $3d$ Weyl semimetals have been shown experimentally [17] such as in the $TaAs$ class, $NbAs, TaP, NbP$, as well as in TMD materials such as $WTe2, MoTe_2, W_xMo_{1-x}Te_2$, in Heusler topological insulators $LnPtBi$ with $Ln = Y, La, Lu$, and in Heusler magnetic Weyl semimetals with $Ln = Gd, Nd$.

Materials in the class of the quantum anomalous Hall effect (Hall effect in the absence of a magnetic field) have also been considered. There are several classes [18] that result from doping a topological insulator with magnetic ions, such as Cr, V dopings [19]. Examples are $(Bi, Si)_2Te_3$ doped with Cr or V. Another example is intrinsic magnetic topological insulators such as $MnBi_2Te_4$. Other examples, to be addressed later, are the result of heterostructures. In general, topological magnets [20] constitute an interesting class of systems. These systems, such as antiferromagnetic Weyl semimetals, display interesting phenomena such as anomalous Hall effect, anomalous Nernst effect, and magneto-optical and magnetic spin Hall effects.

Altermagnets

We have considered different types of magnetically ordered systems, with the common characteristic of either the presence of finite magnetic moments in real space or some form of band magnetism with spin splitting. The magnetic orderings we have considered can be understood as either commensurate with the crystal lattice or incommensurate, such as spin-density waves. Two examples were ferromagnets (FM) and ferrimagnets, which are called commensurate not crystal-symmetry compensated, and antiferromagnets (AF), which are crystal-symmetry compensated. Recently [21] a novel type of magnetic commensurate crystal-symmetry compensated phases has been proposed, called altermagnets. One may say that the significant distinction between antiferromagnets and altermagnets is that antiferromagnets display Kramers' degeneracy (time reversal symmetry), while altermagnets do not (time reversal symmetry broken). The altermagnets have properties that are between ferromagnets and antiferromagnets, sharing some properties with one or the other type of magnetic systems. Specifically, ferromagnets (FM) may have net magnetization, while antiferromagnets (AF) and altermagnets (AM) have zero magnetization. FM and AM do not display Kramers' degeneracy, while AF do. Consistently, FM and AM show an anomalous Hall response, while AF do not.

In Fig. 25.8 a comparison of some characteristics of ferromagnets, antiferromagnets, and altermagnets is presented, focusing on the band structure and on the symmetry properties, both with respect to real space and spin space. The spatial symmetry of an altermagnet is of the d-wave type and is the corresponding magnetic system of the d-wave superconductors. Candidate materials are metallic RuO_2 and insulators FeF_2, MnF_2.

Different models describe an altermagnet. Considering Pauli matrices acting on spin space leads to the consequence of a general spin splitting of the bands, in the absence of a magnetic field. The bands are such that even though the two spin components have different energies, the magnetization of the bands vanishes.

Two simple two-dimensional examples are

$$H = 2t \cos k_x \cos k_y + 2t_j \sin k_x \sin k_y \sigma_z, \tag{25.6}$$

where σ_z is a Pauli matrix, and

$$H = \pm 2t \left(\cos k_x - \cos k_y\right) \sigma_z. \tag{25.7}$$

Another model in $3d$ space is given by Amundsen et al. [22]

$$H(k) = \left(\frac{\hbar^2 k^2}{2m} - \mu\right) \sigma_0 + a_k \sigma_z - (\alpha\,(\mathbf{z} \times \mathbf{k})) \cdot \boldsymbol{\sigma}, \tag{25.8}$$

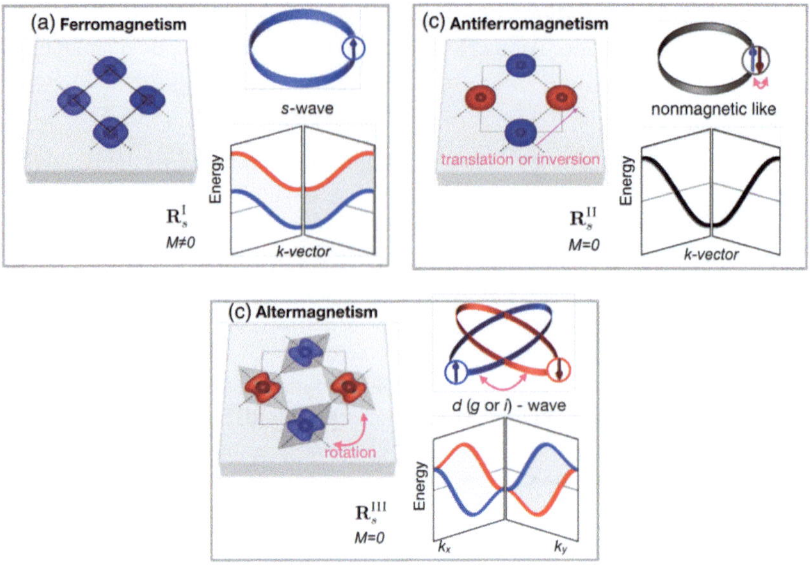

Fig. 25.8 Comparison between ferromagnets, antiferromagnets, and altermagnets. Reproduced with permission from Libor Smejkal, Jairo Sinova, and Tomas Jungwirth, Physical Review X **12**, 040501 (2022)

where a_k is the d-wave-like magnetic term, given by

$$a_k = \frac{\hbar^2 \beta \left(k_x^2 - k_y^2\right)}{2m}, \tag{25.9}$$

α is the amplitude of the Rashba spin-orbit coupling.

We may also consider a two-dimensional tight-binding form

$$\begin{aligned} H(k) = & \left[-2t\left(\cos k_x + \cos k_y\right) - \mu\right]\sigma_0 + \beta\left(\cos k_x - \cos k_y\right)\sigma_z \\ & -\alpha\left(-\sin k_y \sigma_x + \sin k_x \sigma_y\right). \end{aligned} \tag{25.10}$$

In Fig. 25.9 the two energy bands are shown as a function of momentum. There is spin splitting, except at the center and corners of the Brillouin zone. However, integrating over momentum, the total magnetization of the bands vanishes. This is also shown in the same figure where the momentum dependence of the magnetization is presented for the two bands.

Spintronics

Spin and orbital degrees of freedom (or in general spin space and spatial degrees of freedom) are often coupled. We have seen the interplay of the symmetry of the spin states with the symmetry of the spatial states, how in the Kondo and RKKY effects

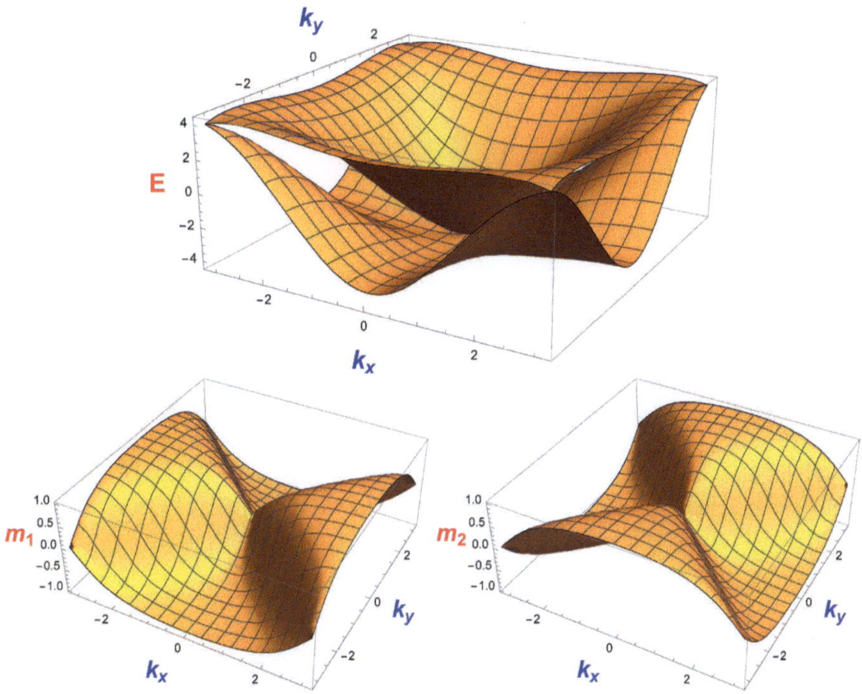

Fig. 25.9 Top panel: Energy bands of an altermagnet given by Eq. (25.10) for $t = 1, \beta = 1.8, \alpha = 1, \mu = 0$. Lower panels: momentum dependent magnetization for the two bands

the spin interactions are mediated by orbital degrees of freedom, or how magnetic states arise in the context of itinerant magnetism.

Another standard example is the relativistic spin orbit interaction in atoms. For instance, in the hydrogen atom, the charged electron, in its reference frame, perceives the effect of the moving proton as a magnetic field, in addition to the electric field due to the charge of the proton. If S' is the reference frame of the moving electron, moving with velocity $\mathbf{v}$ with respect to the reference frame of the proton, S, we know that the electric and magnetic fields in the two reference frames are related by

$$\mathbf{E}' = \gamma\left(\mathbf{E} + c\boldsymbol{\beta} \times \mathbf{B}\right) - \frac{\gamma^2}{\gamma + 1}\boldsymbol{\beta}\left(\boldsymbol{\beta} \cdot \mathbf{E}\right)$$

$$\mathbf{B}' = \gamma\left(\mathbf{B} - \frac{\boldsymbol{\beta}}{c} \times \mathbf{E}\right) - \frac{\gamma^2}{\gamma + 1}\boldsymbol{\beta}\left(\boldsymbol{\beta} \cdot \mathbf{B}\right), \qquad (25.11)$$

where

$$\gamma = \frac{1}{\sqrt{1 - \left(\frac{v}{c}\right)^2}}$$
$$\boldsymbol{\beta} = \frac{\mathbf{v}}{c}. \tag{25.12}$$

Considering that $v << c$, we may write that

$$\mathbf{B}' = -\frac{1}{c^2}\mathbf{v} \times \mathbf{E}. \tag{25.13}$$

The magnetic field then interacts with the spin of the electron, via a Zeeman coupling.

In the case of the electron in the hydrogen atom $\mathbf{E} = -\nabla\phi$, where ϕ is the Coulomb potential of the proton. The coupling to the spin of the electron then leads to the usual[2] spin-orbit coupling

$$V_{so} = \frac{e^2}{8\pi\epsilon_0 m^2 c^2}\frac{1}{r^3}\left(\mathbf{S} \cdot \mathbf{L}\right), \tag{25.14}$$

where $\mathbf{S}$ is the electron spin and $\mathbf{L}$ the orbital angular momentum.

In a solid we may have a similar effect if an electron of momentum $\mathbf{k}$ is propagating along a region where an electric field is present. We may write a contribution to the Hamiltonian as

$$H_{so} = \lambda\boldsymbol{\sigma} \cdot \left(\mathbf{k} \times \mathbf{E}\right), \tag{25.15}$$

where $\boldsymbol{\sigma}$ are the Pauli matrices.

We have discussed before the effect of the presence of a magnetic moment due to both the intrinsic spin of the electrons and the angular momentum due to the orbital part; how the spin-orbit coupling, V_{so}, takes a role in selecting the total angular momentum; and how it is manifested in the Hund's rules. An effect of H_{so} has been shown in the context of spin selective transport due to the passage of a current in chiral molecules [23]. This leads to a current that is at least partially spin polarized. Another current spin polarizing method consists in passing a current through a ferromagnetic region, where there is a predominance of one spin component with respect to the other. A spin-orbit term may also arise as a consequence of the breaking of inversion symmetry, such as in noncentrosymmetric systems, or by the presence of an interface. Considering a two-dimensional system in the $x - y$ plane, this gives rise to a Rashba spin-orbit term with coupling α_R that may be written as

$$H_R = \alpha_R\left(\sin k_y\sigma_x - \sin k_x\sigma_y\right), \tag{25.16}$$

[2] Including a $1/2$ factor due to the Thomas precession.

or to a Dresselhaus spin-orbit coupling of the type

$$H_D = \alpha_D \left(\sin k_x \sigma_x - \sin k_y \sigma_y \right) . \tag{25.17}$$

These terms lift the degeneracy of the spin bands (except at some points in the Brillouin zone). The presence of a spin-orbit coupling implies that the spin components are coupled and that spin is not a conserved quantity (in contrast to charge, with the exception of a superconductor).

The existence of spin degrees of freedom together with their transport properties, and possible manipulation, suggested the field of spintronics [24], as a complementary field to electronics, where the manipulation of charge is the aim. The manipulation of magnetic degrees of freedom allows, for instance, the manipulation of the spatial distribution of magnetic configurations, such as magnetic memories, with the possibility of accessing and changing the distributions (reading/writing) and transport of information in the form of spin degrees of freedom. An important aim is the motion of domain walls as a way to change a magnetic configuration. This may be achieved with local magnetic fields but is in general a slow process. A faster way that is common to electronics is to use the passage of currents (in the case of spintronics spin polarized) and through the action of a spin torque that can affect magnetic configuration. Due to the coupling of the orbital and spin degrees of freedom, this may be achieved in a fast way using electric fields to induce currents through the coupling to the charge degree of freedom of the electrons, which associated with its spin degrees of freedom affects the spin distribution in space. Often heterostructures lead to enhanced spin-orbit torques [25].

Assuming a spin polarized current passing through a region with a magnetic configuration, such as different magnetic domains in succession, the spin density of the current electrons, **s**, will interact with the magnetic moments of the magnetic configuration, **S**, typically via an exchange interaction of the type $\mathbf{s} \cdot \mathbf{S}$. This interaction leads to a torque exerted on the magnetic moments. There are also effects resulting from the scattering off the interfaces between the domains. The torque leads to the dynamics of the magnetic moments that is usually described by the Landau-Lifshitz-Gilbert (LLG) equation

$$\frac{d\mathbf{S}}{dt} = -\gamma \mathbf{S} \times \mathbf{H} + \boldsymbol{\tau} + \frac{\alpha}{S} \mathbf{S} \times \frac{d\mathbf{s}}{dt}, \tag{25.18}$$

where γ is the gyromagnetic ratio, a damping term with amplitude α has been added, **h** is a magnetic field, and the torque has the general form

$$\boldsymbol{\tau} = \frac{j_0}{e} \left[\eta \mathbf{S} \times (\mathbf{S} \times \mathbf{s}) + \eta \mathbf{S} \times \mathbf{s} \right], \tag{25.19}$$

where j_0 is the charge current associated with the injected spin polarized current. The solution of the LLG equation (usually numerically) describes the time evolution of the spin configuration.

Even in the case of a non-polarized spin current, as the electrons go through different materials, in particular with some type of magnetic ordering, the two spin component electrons scatter differently depending on the spin dominance of each region. Recalling the Landauer expression for the conductance, Eq. (23.40), in terms of the transmission coefficient, this coefficient actually depends on the spin projection. Going through regions of different ferromagnetic magnetization orientations leads to different charge transport and different resistivities. In this case, it is the magnetic configuration that affects the charge transport, instead of the charge transport affecting the magnetization.

One interesting phenomenon is the giant magnetoresistance effect (GMR). Magnetoresistance is the property of changing the resistance of a material or system (like a heterostructure) by applying a magnetic field. In particular, one is interested in the relative change of the resistance as a field is applied.

The change of the bands and density of states as a magnetic field is applied or as a result of the spontaneous magnetization in a ferromagnet, which has been discussed before. In the absence of magnetization, the spin up and spin down bands are degenerate, but the magnetization splits the bands, and one spin orientation is dominant over the other. Consider now a heterostructure where we have two ferromagnets with opposite magnetizations and an interlayer barrier that is non-magnetic. Two cases are considered, where (in zero external magnetic field, $H = 0$) the magnetization orientations are opposite and (in finite external magnetic field) they are parallel. Sending a current through the heterostructure, one finds that the resistances in the cases of $H = 0$, and finite H are different. Or in other words, the currents that go through the heterostructure differ, being larger when the field is applied. In the first case, there is a spin $\uparrow$ majority on the left side and a spin $\downarrow$ majority on the right side, while at finite H, there is a spin $\uparrow$ majority on both sides. This means that the dominant spin $\uparrow$ current on the left side carries over to the right side with less scattering, since there is no spin flip needed.

The incident current that enters the left ferromagnet is not spin polarized, containing both spin components, $\uparrow$ and $\downarrow$. There are therefore two channels (associated with the two spin components) across the heterostructure. If $H = 0$, the channel for both the up and the down spin components has a total resistance of $R_\uparrow + R_\downarrow$, where R_σ is the resistance offered by each ferromagnet to the motion of spin σ electrons. Therefore, the heterostructure is like two resistances in parallel with a total resistance $R_0 = (R_\uparrow + R_\downarrow)/2$. On the other hand, applying a magnetic field and aligning the magnetizations implies that the spin-up electrons find a resistance $2R_\uparrow$ and the down-spin electrons find a resistance $2R_\downarrow$. The parallel resistance is therefore $R_H = 2R_\uparrow R_\downarrow/(R_\uparrow + R_\downarrow)$. (In addition to the resistances inside the ferromagnetic systems, there is also a contribution to the resistance from the various interfaces). Therefore, the two situations are different. The difference is

$$\Delta R = R_H - R_0 = -\frac{1}{2}\frac{\left(R_\uparrow - R_\downarrow\right)^2}{R_\uparrow + R_\downarrow} \tag{25.20}$$

and is negative and nonvanishing, if $R_\uparrow \neq R_\downarrow$. Choosing a heterostructure of the type $Fe/Cr/Fe$, a change of about 10% has been achieved, and considering many layers of Fe/Cr, a drop in the resistance of close to 50% has been achieved [26]. Similar behavior is found if one considers a half-metal such as CrO_2. In this case, due to the magnetizations on the left, only spin $\downarrow$ electrons are located at the Fermi level, and if $H = 0$, the spin $\downarrow$ band is full and below the Fermi level. Therefore, there is no current flowing to the right. Applying a magnetic field $H \neq 0$, a current is obtained.

It is also interesting to have spin currents in non-magnetic systems due to intrinsic effects, like strong spin orbit coupling. One example is semiconductors. Two magnetic semiconductors $Ga_{1-x}Mn_xAs$ function as source and drain of spin currents that go through non-magnetic semiconductors, $GaAs$ with spin-orbit coupling, and separated by tunnel layers of $AlAs$. There is a significant difference in the resistances when the magnetizations are parallel or antiparallel. Due to its high mobility, one would hope that graphene would also be useful in spintronics. As mentioned before, the spin orbit coupling is however quite small. Ways to increase the spin-orbit coupling include doping with heavy atoms (with large spin-orbit coupling), by some proximity effect with a system with intrinsic spin-orbit coupling or considering hydrogenated graphene [27]. Since TMDs have larger spin-orbit coupling, they may be expected to provide more suitable candidates for high-mobility spin carriers.

Along the same line of reasoning, antiferromagnetic spintronics has attracted considerable attention [28]. As discussed in this reference, they are to some extent shielded from external magnetic fields, do not give origin to outside field lines and are able to give origin to significant magnetotransport effects. Extensive lists of materials are presented in the same reference, such as metallic antiferromagnets, usually Mn based, insulating antiferromagnets (usually oxides), and semiconductor and semimetal antiferromagnets.

The use of superconductors (SC) to transport the current, in FM/SC/FM heterostructures, has also been considered [29]. In these systems, the contribution of Andreev scattering is significant, leading in some cases to larger currents as compared to a non-superconductor. Andreev scattering occurs when an electron enters a superconducting region: the electron captures an extra electron, giving rise to a hole that propagates back from the superconducting region, and the Cooper pair propagates inside the superconductor. However, a hole propagating away from the superconducting region is equivalent to an electron propagating inside the superconductor and, therefore, adds to the current on the right side [30].

There are many applications of spintronics [24] such as reading and writing in magnetic memories, sensors, spin diodes, and spin transistors.

Strongly Correlated Systems and Unconventional States

In various materials, correlations play an important role. Often they are the result of narrow bands such that the presence of interactions between the particles cannot be taken in perturbation theory. One model that describes this type of materials to some approximation is the Hubbard model. Another typical case involves several bands where at least one of them is a narrow band. This case is described by the lattice Anderson model or the lattice Kondo model. Also, one finds that these materials often have complex phase diagrams, since due to the presence of strong interactions/correlations, various phases compete and even coexist. The theoretical treatment of these materials is often complicated, and the results are usually not completely settled. For instance, when superconductivity is present in some part of the phase diagram, it is usually of some unconventional type, with Cooper pairing symmetries that are not the simpler s-wave, spin-singlet type. Having strong interactions, such as local couplings of the Hubbard-U type ($U > 0$), one might expect that pairing would not be of s-wave type ($L = 0$), to avoid the local repulsion. Therefore, we may anticipate higher angular momentum values, and if the pairing is of the spin-singlet type, the lowest available angular momentum is $L = 2$ or d-wave. A pairing with $L = 1$ is also possible and is associated with spin-triplet pairing, but it has proven hard to confirm experimentally. Also, the mechanism is not always completely understood, and it often seems to differ from the phonon exchange considered in the standard (conventional) BCS theory.

The use of detailed experimental methods has been important to characterize these materials, such as high-resolution ARPES or scanning tunneling microscopy (STM) [31], among many other standard thermodynamic and transport measurements, together with other experimental methods. For instance, in STM measurements, a metallic tip is placed near the surface of a material, and as one moves the tip along the surface, important information is obtained, such as the local density of states of the material. Integrating over space, one obtains the density of states. (In ARPES, we study the energy momentum dispersion relations of the quasiparticle excitations, and with STM, we obtain information about the energy and space quantities. Integrating either over momentum or over space, respectively, we obtain information on the energy dependence of the density of states). In STM, establishing a potential difference, V, between the tip (T) and the sample (S), one finds a current that is given by

$$I(eV) = \frac{4\pi e}{h} \int dE \left[f(E_f - eV + E) - f(E_f + E) \right] |M|^2 \\ N_T (E_F + E) \, N_S (E_F - eV + E) \, , \tag{25.21}$$

where M is the matrix element for the tunneling between the tip and the sample and f are Fermi-Dirac distributions and a differential conductance that gives information on the sample density of states

$$\frac{dI}{dV}(eV) \sim N_s (E_F - eV) \, . \tag{25.22}$$

In Fig. 25.10, the phase diagrams of several strongly correlated materials are presented. These are in the classes of heavy-fermion systems, high-temperature superconductors of the cuprate type, and a metallic ferromagnet. In common, they display several phases that compete or coexist. The phases are tuned using pressure

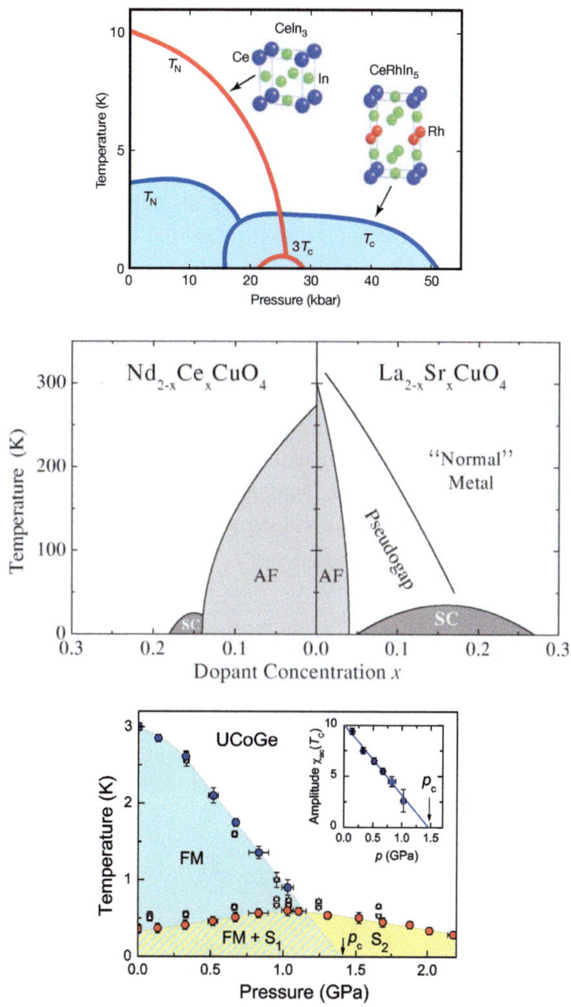

Fig. 25.10 Phase diagrams of several superconductors. Heavy-fermion materials: Reproduced with permission from P. Monthoux, D. Pines, and G. G. Lonzarich, Nature **450**, 1177 (2007). Cuprates: reproduced with permission from Andrea Damascelli, Zahid Hussain, and Zhi-Xun Shen, Reviews of Modern Physics **75**, 473 (2003). Phase diagram of $UCoGe$: reproduced with permission from E. Slooten, T. Naka, A. Gasparini, Y. K. Huang, and A. de Visser, Physical Review Letters **103**, 097003 (2009)

or chemical pressure, by doping. In addition to superconductivity, there are magnetic phases that may be antiferromagnetic or ferromagnetic, and in some cases, the appearance of superconductivity expels the magnetic order, but in other cases, they may coexist, even in the case of superconductivity and ferromagnetism.

In the first row, heavy fermions are considered. In the second row, we consider the phase diagrams of cuprates, with hole doping on the right side and electron doping on the left side, with respect to the "parent," undoped ($x = 0$) antiferromagnet. La_2CuO_4 is an antiferromagnet insulator whose magnetism arises from the spins on the copper ions. As can be seen from the hole doped case, there is a region that in the normal phase is similar to a normal metal, while if the system is underdoped (optimal doping is the value of x such that the superconducting critical temperature is maximal), the system behaves as a non-Fermi liquid, with linear in-temperature resistivity and several unusual features, including some pseudogap in the spectrum. In the third row, we consider the case of $UCoGe$, to highlight the similarities of the phase diagrams of the several classes of materials.

These are not the only materials displaying complex phase diagrams. Among others, we may mention one-dimensional systems. Even though there is no long-range order in these systems, several phases and quantum critical points may be identified at zero temperature through the dominance of different instabilities in the systems.

The first heavy-fermion superconductor $CeCu_2Si_2$, discovered in 1979, started a series of discoveries of strongly correlated systems, displaying different phases and non-trivial normal phases. The heavy-fermion behavior is due to the presence of $4f$ or $5f$ states, which lead to narrow bands and enhanced correlations, with high effective masses (hence the name heavy fermions) (recall Fig. 22.9). These materials typically have low critical temperatures [32]. Several phase diagrams are shown in Fig. 25.11. Using the Anderson lattice model Eq. (22.158) or Eq. (22.171) and adding pairing and antiferromagnetic terms in a self-consistent mean-field approach,

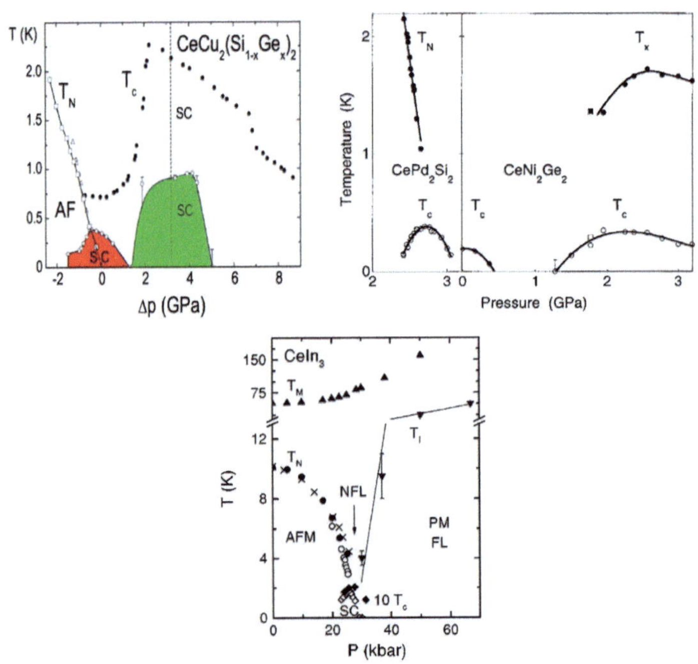

Fig. 25.11 Phase diagrams of several heavy fermions. Reproduced with permission from Christian Pfleiderer, Reviews of Modern Physics **81**, 1551 (2009), from F M Grosche, P Agarwal, S R Julian, N J Wilson, R K W Haselwimmer, S J S Lister, N D Mathur, F V Carter, S S Saxena and G G Lonzarich, Journal of Physics Condensed Matter **12**, L533 (2000) and from G. Knebel, D. Braithwaite, P. C. Canfield, G. Lapertot, and J. Flouquet, Physical Review B **65**, 024425 (2001)

the competition and coexistence of the two types of order found in the phase diagrams can be obtained, and a qualitative agreement with experimental results may be achieved, particularly if the pairing is d-wave.

Even though the critical temperatures are small, their overall properties are interesting, often displaying non-Fermi liquid behavior [33]. The heavy fermion materials (and the cuprates) display several features that do not fall under the classification of Fermi liquids. Several thermodynamic and transport properties have power law or logarithmic dependences on temperature or external fields with exponents that show that these materials may be classified as singular or non-Fermi liquids [34]. Focusing on the heavy fermion materials, possible explanations were considered in terms of the overcompensated multichannel Kondo model [35] but were, to some extent, discarded since some observed power laws exponents were not in the appropriate range, considering dilute Kondo impurities (a multichannel Kondo lattice is a hard problem to solve). In particular, a determination in the case of $U_x Y_{1-x} Pd_3$ yields $\beta = 1.3$, which is larger than the maximal value of one, suggesting that at least the single-impurity multichannel Kondo model is not appropriate. A more general approach may be taken considering the vicinity to a quantum critical point, which may be responsible for the non-Fermi liquid behavior [36].

As mentioned before, non-Fermi liquid behavior is expected in one-dimensional strongly correlated systems. For instance, in the context of the Luttinger model, we showed that one expects the absence of fermionic states at the Fermi surface and spin-charge separation. The same is expected from an exact Bethe ansatz solution and considering other approaches beyond the Luttinger liquid regime [37]. In Fig. 25.12 experimental results are shown for a one-dimensional chain, TTF-TCNQ, and for the chain $SrCuO_2$. As the Bethe ansatz solution for the Heisenberg chain shows, while in two dimensions one finds a coherent mode of excitations [38], such as in the case of $Sr_2CuCl_2O_2$, in a $1d$ problem, the excitations involve two elementary spin excitations and therefore a continuum of states, in accordance with the solution we discussed in the previous Chapter. In addition to the fractionalization in one-dimensional systems, including the case of polyacetylene, the existence of fractional degrees of freedom has also been discussed in other contexts in previous chapters, such as in Chap. 21, and also notably in the context of the fractional quantum Hall effect. In this case charges are expected to be fractional with respect to the free electron charge. This has been confirmed experimentally [39].

The transition to two-dimensional or higher-dimensional systems may be taken by coupling chains. The coupling between chains is in general a relevant parameter, and the nature of the low energy states changes. In the groundstate, one may find some ordered phase, but considering the effect of temperature, the ordered phase is washed away, and we find a disordered phase, but with characteristics that are different from the starting one-dimensional problem, at least at small energies. Considering higher energies, one may expect that the chains effectively decouple and some signatures of the one-dimensional behavior may be found. This is particularly obtained as a result of frustration, such as due to geometrical frustration. The quasi-$2D$ triangular spin system Cs_2CuCl_4 has an excitation spectrum similar

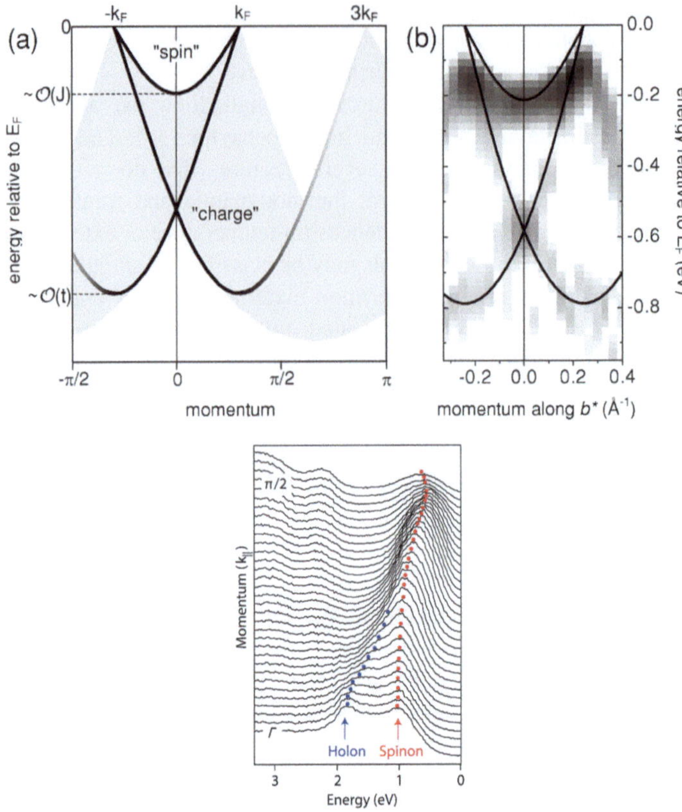

Fig. 25.12 (Top) Spin-charge separation in TTF-TCNQ and (bottom) spin fractionalization in $SrCUO_2$. Reproduced with permission from M. Sing, U. Schwingenschlögl, R. Claessen, P. Blaha, J. M. P. Carmelo, L. M. Martelo, P. D. Sacramento, M. Dressel, and C. S. Jacobsen, Physical Review B **68**, 125111 (2003); and from Jonathan A. Sobota, Yu He, and Zhi-Xun Shen, Reviews of Modern Physics **93**, 025006 (2021)

to the chain, with a continuum originating from fractionalized pairs of spin-1/2 spinons. Other materials show similar properties, such as CPC, $KCuF_3$, and copper benzoate. These features are mainly observable by photoemission experiments, where the finite energy excitation spectrum may be probed. This fractionalization of the degrees of freedom, and its origin in the one-dimensional behavior, has been explained using exact solutions for the chains and the random phase approximation to handle the coupling between chains, both for spin chains and the Hubbard model [40]. We will return later to the appearance of fractional excitations in frustrated magnetic systems.

An important characteristic of the structure of the cuprates is the existence of the two dimensional planes of CuO_2, where the spins are located. Also, the expected higher angular momentum pairing symmetry is natural even in the normal state,

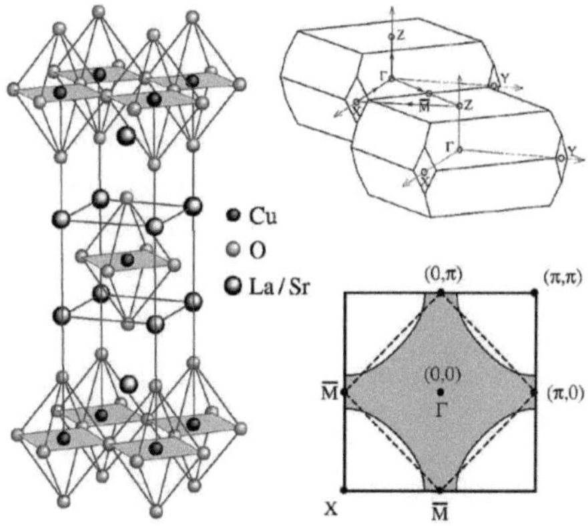

Fig. 25.13 Crystal structure and Fermi surface of LSCO. Reproduced with permission from Andrea Damascelli, Zahid Hussain, and Zhi-Xun Shen, Reviews of Modern Physics **75**, 473 (2003)

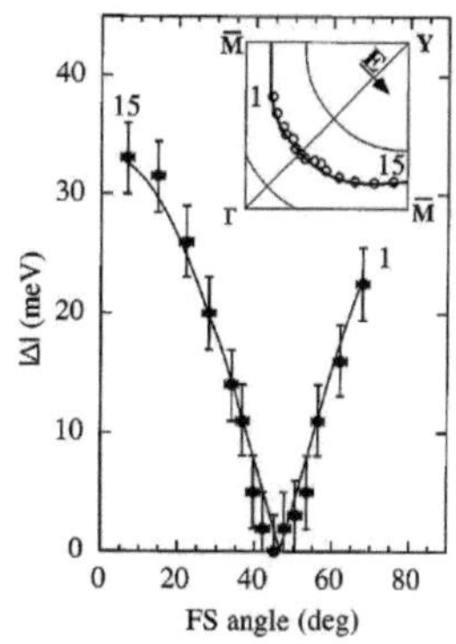

Fig. 25.14 Cuprates gap as a function of the Fermi surface angle. Reproduced with permission from H. Ding, M. R. Norman, J. C. Campuzano, M. Randeria, A. F. Bellman, T. Yokoya, T. Takahashi, T. Mochiku, and K. Kadowaki, Physical Review B **54**, R9678 (1996)

and the $d_{x^2-y^2}$ component has a large content in the vicinity of the Fermi level. The crystal structure and Fermi surface topology are shown in Fig. 25.13. In the superconducting regime the gap function becomes finite and has a symmetry that is consistent with a $d_{x^2-y^2}$ symmetry, as shown in Fig. 25.14. This symmetry, on the square lattice of the CuO_2 planes, translates into a gap function that has a momentum dependence of the type $\cos k_x - \cos k_y$. Therefore, it is gapless along the lines $k_x = \pm k_y$. Taking into account the Fermi surface of Fig. 25.13, this implies the presence of four nodal points at the Fermi surface. In Fig. 25.14 the cuprates gap is shown. As one changes the angle across one quarter of the Brillouin zone, we cross the nodal point into the anti-nodal regions near $(\pi, 0)$ and equivalent points.

Rather than just looking at the points where the gap function vanishes, the symmetry of the gap function may be understood in greater detail using phase-sensitive methods [41]. Different symmetries may give rise to nodal points, and so a direct check of the d-wave symmetry is helpful.

If we have a Josephson junction, the phase difference at the junction is given by

$$\gamma = \phi_L - \phi_R + \frac{2\pi}{\Phi_0} \int_L^R \mathbf{A} \cdot d\mathbf{l}, \tag{25.23}$$

where ϕ_l, ϕ_R are the phases of the pairing function on the L, R sides of the junction, respectively, Φ_0 is the flux quantum, and $\mathbf{A}$ is some vector potential that may be present. The Josephson current is therefore of the type $I_J = I_c \sin\gamma$. It turns out that the symmetry of the pairing affects the current across the junction if the crystallographic axis on the two sides of the junction, with respect to the interface, defining angles θ_L, θ_R do not coincide. If we now consider a superconducting ring, eventually traversed by some external flux, Φ_A, and composed of several junctions, we may write that

$$\Phi_A + I_J L + \frac{\Phi_0}{2\pi}\sum_{i,j}\gamma_{ij} = n\Phi_0. \tag{25.24}$$

Here, L is the inductance of the ring, and γ_{ij} are phase differences between sides i and j of the various junctions and n is an integer. In this case, in general we have that, for each junction, $I_J = I_c^{ij}(\theta_i, \theta_j)\sin\gamma_{ij}$. Considering just one junction, the energy as a function of the flux is given by

$$U = \frac{1}{2}\frac{\Phi^2}{L} - \frac{1}{2\pi}|I_c|\Phi_0 \cos\left(\frac{2\pi\Phi}{\Phi_0} + \varphi\right). \tag{25.25}$$

The phase difference φ is the result of the non-alignment of the crystallographic axis across the junctions and can take the values $\varphi = 0, \pi$. Minimizing with respect to the flux, one finds that if $\varphi = 0$ the minimum of the energy is such that $\Phi/\Phi_0 = 0$. However, if $\varphi = \pi$ (so-called π-ring) the minimum is such that $\Phi/\Phi_0 = \pm 1/2$. Therefore, one finds a half-quantum flux. In general, one finds

$$\begin{aligned} \Phi &= n\Phi_0 \qquad Neven, \\ \Phi &= \left(n + \frac{1}{2}\right)\Phi_0 \qquad Nodd, \end{aligned} \tag{25.26}$$

where N is the number of π junctions in the ring.

A set of three regions of YBCO (assumed to have a $d_{x^2-y^2}$ symmetry) with different crystallographic axes was considered experimentally. These regions define interfaces such that different types of junctions are possible. The central ring is a π ring, while the others are zero rings. Therefore, the zero rings should show integer flux quantum vortices and the π ring a half-flux quantum vortex. In zero magnetic field, the zero rings should show no flux quantization. It has been shown that the π ring shows a spontaneously generated half-flux quantum [41].

In the case of heavy-fermions, while d-wave symmetry seems to be a strong possibility, such as in the original material $CeCu_2Si_2$ (believed to be a d-wave superconductor with a nodal structure), this result has been questioned, and it appears that instead it is a gapped superconductor, either a two-band s-wave or a two-band d wave superconductor [42]. Also, the spin-triplet pairing, proposed for the candidate UTe_2 discovered in 2018, has recently been questioned, and the

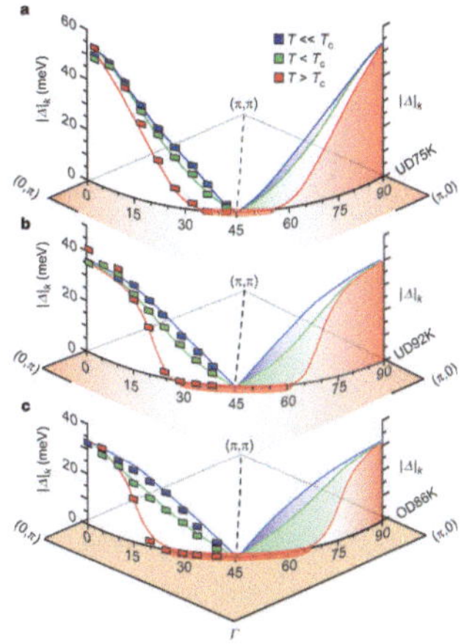

Fig. 25.15 Spectral gap in Bi2212 including the pseudogap in the anti-nodal region. Reproduced with permission from W. S. Lee, I. M. Vishik, K. Tanaka, D. H. Lu, T. Sasagawa, N. Nagaosa, T. P. Devereaux, Z. Hussain, and Z.-X. Shen, Nature **450**, 81 (2007)

issue is not settled [43]. Also, Sr_2RuO_4, proposed to be a spin-triplet chiral p-wave material, is also under discussion [44], even though the material was discovered in 1994.

In Fig. 25.15, the temperature and momentum dependence of the spectral gap of Bi2212 are shown for two underdoped samples and on overdoped sample. In a metallic system, the spectral weight at the Fermi surface is usually finite. In a conventional superconducting phase, usually, there is a finite gap, with exceptions such as in the presence of disorder or in a very large magnetic field, in the Landau regime. If the pairing function has nodes, we may find a gapless regime, if there are nodes at the Fermi surface. In Fig. 25.15, one finds that at low temperatures, and at a temperature smaller than the critical temperature, $T < T_c$, the gap closes at the nodal point. For $T > T_c$, we see an extended gapless region around the nodal point, since the superconductivity is absent, however, close to the anti-nodal points, one gap remains. This is the so-called pseudogap and is found for temperatures above the superconducting critical temperature. Its nature has been controversial, and there are several proposals. It appears that for temperatures slightly above T_c, it may be associated with the pre-formation of the Cooper pairs, but phase fluctuations prevent the existence of superconducting order. At higher temperatures, the amplitude fluctuations seem to be absent as well, and the pseudogap may be due to spatial density fluctuations, such as the formation of density stripes [45].

One of the significant properties of the cuprates was the discovery of high critical temperatures, in some systems close to 100 K. The search for room temperature superconductors and its potential applications, due to its dissipation free nature, have captured the attention and imagination since the discovery of superconductivity. In early 2008 another group of superconductors with relatively high critical temperatures, up to 55 K, was identified. Moreover, these materials have Fe, which would seem less likely to give rise to superconductivity and high T_c. These materials can be organized in different families [46] (Fig. 25.16).

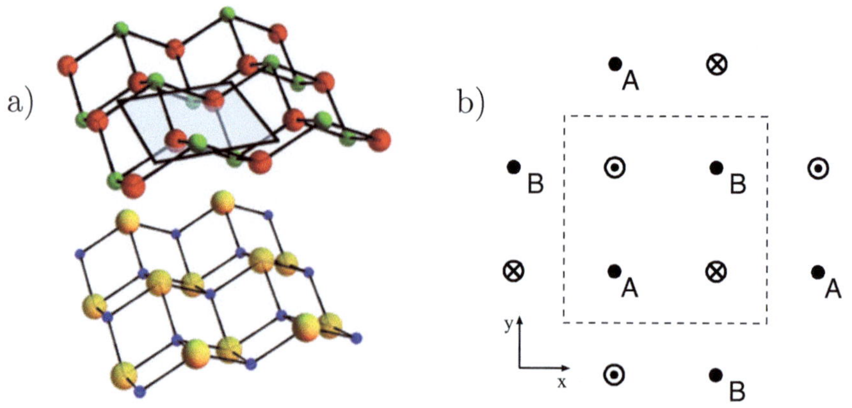

Fig. 25.16 Structure of an iron compound with the $FeAs$ layer on top and the rare earth-oxygen layer below (the Fe atoms are in red, the As atoms are in green, the rare earth is in yellow and O in blue). The Fe atoms can be separated into two groups, A and B, as indicated on the right panel. The As atoms are displaced up or down with respect to the Fe atoms. Reproduced with permission from V. Cvetkovic and Z. Tesanovic, Europhysics Letters **85**, 37002 (2009)

A relevant feature of these pnictide materials is that they have to be considered as multiband, in the sense that several bands have significant contributions at the Fermi surface. One finds that both electron and hole bands cross the Fermi surface. A very simplified model should at the very least consider two bands, one electron, and one hole band. Near the M point in the Brillouin zone, the bands have an electron nature, and near the Γ point, they have a hole nature.

This band structure opens the interesting possibility that, in addition to possible pairing between two electrons or between two holes, we may consider the possibility of pairing between one electron and one hole (meaning one electron in the electron dominated band and one electron in the hole dominated band; the pairs are Cooper pairs of two real electrons). This pairing would imply a boson of momentum $\mathbf{Q} = (\pi, \pi)$. The vicinity to an antiferomagnetic phase implies that magnons with this momentum are readily available, and so it would be reasonable to assume that the mechanism to provide the binding of a Cooper pair would be the exchange of magnons, instead of the conventional phonons. Indeed this possibility seems to explain experimental results. Moreover, the interaction does not have to be attractive and may be taken as repulsive. The mechanism is illustrated in Fig. 25.17.

Recall the gap equation written as we saw in Eq. (17.64)

$$\Delta(\mathbf{k}) = -\sum_{\mathbf{k}'} V\left(\mathbf{k}, \mathbf{k}'\right) \frac{\Delta(\mathbf{k}')}{2E(\mathbf{k}')} \tanh\left(\frac{1}{2}\beta E(\mathbf{k}')\right). \tag{25.27}$$

It may be shown that taking the interaction as repulsive but selecting different signs for the gap functions in the hole and electron bands, one finds a convergent finite solution. This is called alternating s-wave pairing symmetry, as illustrated in

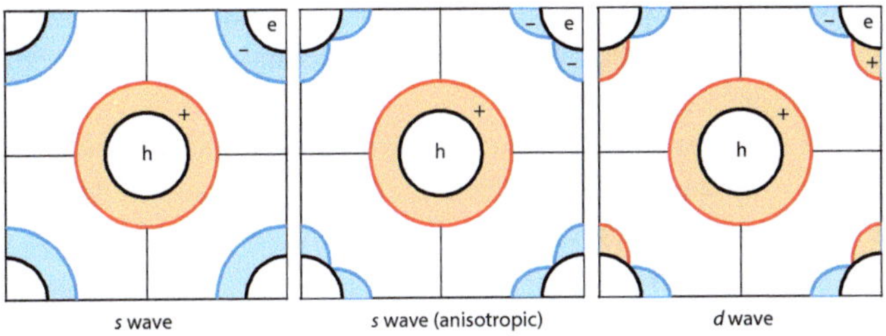

Fig. 25.17 Possible pairing symmetries in pnictides. Reproduced with permission from Johnpierre Paglione and Richard L. Greene, Reviews of Nature Physics **6**, 645 (2010)

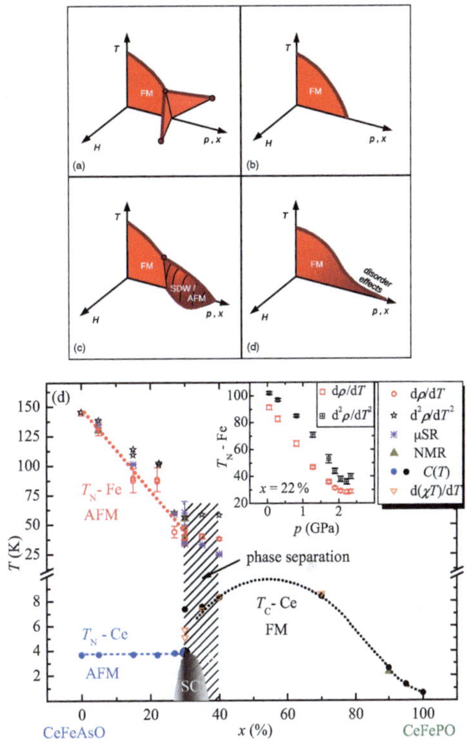

Fig. 25.18 Scheme of different cases of coexistence or competition of phases in a metallic ferromagnet. Reproduced with permission from M. Brando, D. Belitz, F. M. Grosche, and T. R. Kirkpatrick, Reviews of Modern Physics **88**, 025006 (2016) (reproduced from Hisashi Kotegawa, Valentin Taufour, Dai Aoki, Georg Knebel, and Jacques Flouquet, Journal of Physical Society of Japan 80, 083703 (2011)). Phase diagram of $CeFeAs_{1-x}P_xO$. Reproduced with permission from A. Jesche, T. Forster, J. Spehling, M. Nicklas, M. de Souza, R. Gumeniuk, H. Luetkens, T. Goltz, C. Krellner, M. Lang, J. Sichelschmidt, H.-H. Klauss, and C. Geibel, Physical Review B **86**, 020501 (2012)

Fig. 25.17. Other possibilities arise, but the alternating s-wave seems to be a good (and interesting) choice that describes the observed properties. At the very least, this possibility is interesting because it reveals that multiband systems may offer a large variety of mechanisms.

Finally, we show in Fig. 25.18 further examples of metallic ferromagnets, in which a series of coexistence and competition of phases is found, including the vicinity of ferromagnetic and antiferromagnetic phases, separated by second- and first-order transitions. The point here is to stress, once more, the complexity and richness of behaviors of these types of systems. There is a large variety of metallic ferromagnets [47] such as elemental Fe, and that includes transition-metals, uranium- and lanthanide-based compounds, and strontium ruthenate, either clean or in the presence of disorder.

The search for higher superconducting critical temperatures has been a continuous goal [48]. To some extent, the materials are organized in families, from the conventional superconductors with relatively low critical temperatures, heavy fermions with lower critical temperatures, to the pnictides and cuprates. Higher critical temperatures have been achieved considering the materials under large pressures. A class of materials that have quite high critical temperatures are the hydrides but at very large pressures. These critical temperatures reach room temperature values but require very special conditions, due to the high pressures involved [48]. It would be interesting to find situations of room temperature superconductivity, without requiring high pressures. As we saw before, chemical pressure or doping plays a somewhat equivalent role in some materials, by compressing the electron orbitals and increasing wavefunction overlaps. Confinement may be achieved in different ways, and it seems that considering interfaces between different materials may be another path to generate and manipulate superconductivity, albeit at the moment with low critical temperatures.

Interface Phases

At interfaces we may find phases with enhanced or changed behavior, including phases that are not present in either of the materials at the sides of the interface. Consider, for instance, the superconductor $FeSe$ with its critical temperature around 8 K. Constructing an interface $FeSe/SrTiO_3$, one finds that the critical temperature is significantly enhanced [49], reaching a critical temperature of about 80 K. The role of the interface has been interpreted as providing a coupling between the spin-fluctuation pairing mechanism and high-energy phonons localized in the vicinity of the interface, enhancing the critical temperature.

The utility of heterostructures, combining different materials with different properties, is, for instance, highlighted in the case of complex oxide heterostructures. These materials display various properties that are very sensitive to small changes of structure or chemistry, ranging between different types of materials, such as paraelectric insulators, ferroelectrics, metals, and even superconducting [50], such as the perovskite compound $SrTiO_3$. Mixing different transition metal oxides, one may have an endless variety of systems with shared interfaces. Manganites are also interesting with a complex phase diagram, as one evolves from two antiferromagnetic insulators, $LaMnO_3$ and $SrMnO_3$, as one changes the Sr concentration in $La_{1-x}Sr_xMnO_3$. This evolution is shown in Fig. 25.19.

An interesting example of the appearance of new phases at interfaces is the system $LaAlO_3/SrTiO_3$. Both systems are gapped, but the interface is conducting, and with appropriate change of parameters, it may be superconducting. So two oxide insulators lead to an oxide interface superconductor [51]. Interfacing two superconductors may lead to an interface that is also superconducting, but with an enhanced critical temperature. Specifically, bringing together an underdoped cuprate and an overdoped cuprate, it is predicted that it may give origin to enhanced interface

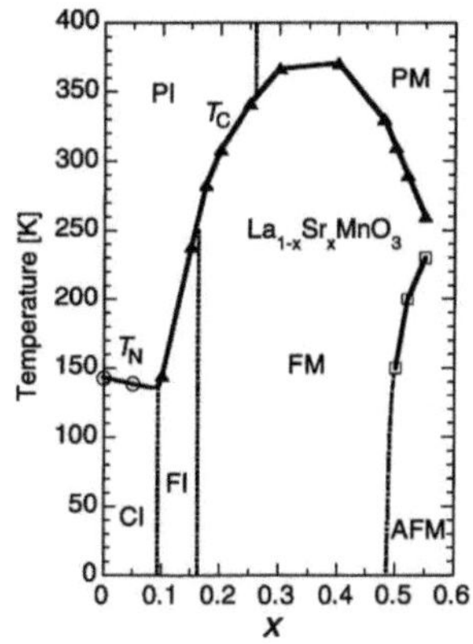

Fig. 25.19 Phase diagram of $La_{1-x}Sr_xMnO_3$, showing a variety of phases as x changes. Reproduced with permission from Y Tokura, Y Tomioka, Journal of Magnetism and Magnetic Materials **200**, 1 (1999)

superconductivity [52]. The reasoning is that one combines the high pairing energy of the underdoped regime with the large superfluid density typical of the overdoped regime, which, together with a proximity effect across the interface, mixes the two characteristics of the two differently doped superconductors. In addition to a metallic or a superconducting interface phase, magnetic order is also possible and even the coexistence of the two types of order, with the possibility that, in this case, there is phase separation [53].

Interface physics finds many other applications, such as magnetic phases that are the result of interface phenomena. At the interface, there is spatially symmetry breaking that leads to various phenomena, as we have seen before, such as anomalous magnetic effects, which may result from interface induced spin-orbit coupling, Dzyaloshinskii-Moriya couplings, etc., [54].

In addition to the interesting properties we get when we construct heterostructures with different materials, interesting effects also arise when we have layered materials of the same type. The way various monolayers are stacked also leads to interesting consequences. For instance, in the case of graphene, a typical single-layer material, we saw before that a bilayer has electronic properties that are qualitatively different from the monolayer graphene. This occurs in different contexts, as, for instance, coupling two spin chains (and constructing a spin ladder) leads to different properties. For instance, if the spins are $S = 1/2$, coupling two chains leads to a system that is in some cases more related to a spin-1 chain. Recalling previous discussions, half-integer and integer chains behave differently; in the first case, the spectrum is gapless, and in the second case, the spectrum is gapped. Also, considering multilayers of graphene, one finds behaviors different from the monolayer, and the same happens with multilayers of TMDs [55]. Moreover, the sequence of the various monolayers that is chosen to construct the multilayer system also leads to different spectra and properties. One may also consider that the sequence is such that there is a rotation between the various monolayers.

In the case of twisted bilayer graphene, a Moiré like-pattern that results from the twist of one monolayer with respect to the other, as shown in Fig. 25.20, leads

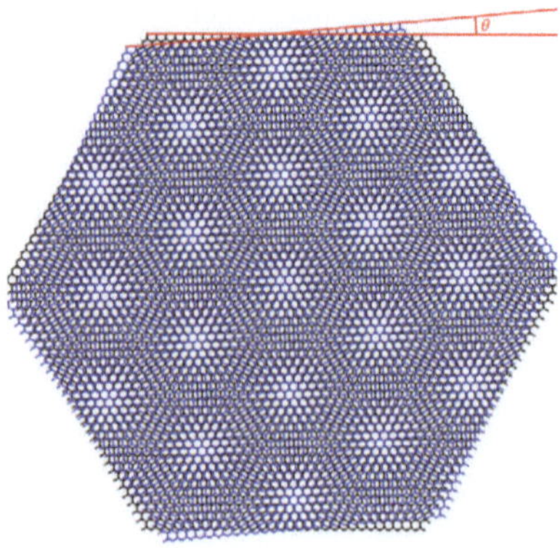

Fig. 25.20 Twisted bilayer graphene with a Moiré like-structure. Reproduced with permission from Cui-Zu Chang, Chao-Xing Liu, and Allan H. MacDonald, Reviews of Modern Physics **95**, 011002 (2022)

to an electronic spectrum that depends on the twist angle [56]. As a function of the twist, the electron velocity changes considerably, and in the vicinity of some "magic" angles, the bands become very flat, with nearly vanishing electron velocity [57]. Since the bands are very flat, the role of interactions becomes very relevant, and a large variety of phases can be found [58]. These include superconducting phases. Another example is the existence of the quantum anomalous Hall effect, including the fractional effect [18]. The effective magnetic field is provided by orbital effects, due to the spontaneous valley polarization and time reversal breaking, which, combined with the enhanced effect of the interactions, leads to a fractional quantum Hall effect, as found in (pentalayer) graphene [59]. This effect has also been found in TMDs bilayers, such as in twisted $MoTe_2$. In these systems, there is a range of angles where flat bands are found, instead of the magic angles found in graphene.

Quantum Circuits

With the aim of achieving quantum computation and taking advantage of the quantum superposition principle, in order to access and manipulate a much larger amount of information, quantum circuits have been considered. The basic unit for a quantum circuit is a qubit. This is a two-level system, like a spin-$1/2$ system with its two projections or a photon with its two polarization states. Physical realizations are a real spin-$1/2$, like an atom, nucleus, or electron, a quantum dot with one spin-$1/2$ particle, whose spin degree of freedom or charge (occupied or empty dot) may be taken as a two-level system. Manipulating the state of a qubit may be achieved using operators like Pauli matrices or their combinations. The interest for a quantum circuit is to have a series of qubits and to manipulate according to some desired operation. Therefore, it is also useful and important to operate on more than one qubit simultaneously and to take advantage of their entanglement. Typical simple operations (gates) that act on one qubit are a X gate (σ_x Pauli matrix), a Z gate (σ_z Pauli matrix), and the Hadamard gate $H = (\sigma_x + \sigma_z)/\sqrt{2}$ defined by their actions on a spinor

$$X\begin{pmatrix}\alpha\\ \beta\end{pmatrix}=\begin{pmatrix}0&1\\ 1&0\end{pmatrix}\begin{pmatrix}\alpha\\ \beta\end{pmatrix}=\begin{pmatrix}\beta\\ \alpha\end{pmatrix}$$
$$Z\begin{pmatrix}\alpha\\ \beta\end{pmatrix}=\begin{pmatrix}1&0\\ 0&-1\end{pmatrix}\begin{pmatrix}\alpha\\ \beta\end{pmatrix}=\begin{pmatrix}\alpha\\ -\beta\end{pmatrix}$$
$$H\begin{pmatrix}\alpha\\ \beta\end{pmatrix}=\frac{1}{\sqrt{2}}\begin{pmatrix}1&1\\ 1&-1\end{pmatrix}\begin{pmatrix}\alpha\\ \beta\end{pmatrix}=\frac{1}{\sqrt{2}}\begin{pmatrix}\alpha+\beta\\ \alpha-\beta\end{pmatrix}. \tag{25.28}$$

Representing the eigenstates of the two-level system as $|0\rangle$ and $|1\rangle$, another gate that is usually considered, and that acts on two qubits, is the $C - not$ gate defined by

$$|00\rangle \to |00\rangle,$$
$$|01\rangle \to |01\rangle,$$
$$|10\rangle \to |11\rangle,$$
$$|11\rangle \to |10\rangle. \tag{25.29}$$

Condensed matter systems provide a way to scale up the number of qubits. Examples are solid-state systems where one can grow quantum dots, taking advantage, for instance, of the technological developments in the semiconducting industry. One such case is semiconducting spin qubits [60]. Also, local states that originate from impurities, like interstitial impurities or the absence of some atom in a given lattice location, may constitute, to a good approximation, two-level systems. Another physical realization are qubits in superconductors [61] such as Josephson junctions, with the presence or absence of flux through them as the two-level system, which may be manipulated using optical methods or small electric currents. Cooper-pair boxes can also be used as a charge degree of freedom qubit.

Due to quantum decoherence problems, associated with fluctuations and the write/read process, the issue of errors (and error correction) is an important one. One way to try to minimize errors is to keep the circuit at low temperatures, and so the use of superconducting based circuits is a natural and convenient choice. Also, it is convenient to consider systems such that the qubit states have some sort of protection, and topological systems are an interesting possibility. In this context, it has captured the attention, in the context of superconductors, on the use of the Majorana modes that were introduced in Chap. 20 and in Chap. 21. One theoretical possibility is the Majorana modes that are found in vortices of some two-dimensional superconductors. Another possibility is the Majorana modes at the ends of a chain described by the Kitaev model, Eq. (20.31) (and at the flat band point as Eq. (20.37)).

It has been shown [62] that vortices in a bidimensional superconductor with triplet pairing have an excitation spectrum with a Majorana fermion. The wave function is given by [3] [see Eq. (17.195)]

$$\psi(\mathbf{r}) = e^{i\varphi}\left[d_x\left(|\uparrow\uparrow\rangle + |\downarrow\downarrow\rangle\right) + id_y\left(|\uparrow\uparrow\rangle - |\downarrow\downarrow\rangle\right) + d_z\left(|\uparrow\downarrow\rangle + |\downarrow\uparrow\rangle\right)\right] \times \left(k_x + ik_y\right), \tag{25.30}$$

where the vector $\mathbf{d}$ characterizes the triplet pairing and the phase of the condensate is φ. $\psi(\mathbf{r})$ remains invariant under the transformation $(\varphi, \mathbf{d}) \rightarrow (\varphi + \pi, -\mathbf{d})$. Therefore, a half phase rotation accompanied by the vector $\mathbf{d}$ rotating to $-\mathbf{d}$ in a turn around the vortex. Choosing $d_z = 0$ and $d_x + id_y = de^{i\varphi(\mathbf{r})}$ satisfies this result. Also, only one spin component is affected by the change

$$\psi(\mathbf{r}) = d(r)\left(e^{2i\varphi(\mathbf{r})}|\uparrow\uparrow\rangle + |\downarrow\downarrow\rangle\right)\left(k_x + ik_y\right). \tag{25.31}$$

The solutions are obtained transforming to the operators $\hat{\gamma}^\dagger = u\hat{\psi}^\dagger + v\hat{\psi}$. A change of phase by a value ϕ implies that $(u, v) \rightarrow (ue^{i\phi/2}, ve^{-i\phi/2})$. If $\phi = 2\pi$ we obtain that $\gamma \rightarrow -\gamma$.

If we have two vortices, each one having an associated Majorana mode, and one of them, γ_1, is transported around the other, the rotation exchanges the signal of the fermion associated with the vortex, leaving invariant the Majorana fermion, γ_2, that was at rest. We obtain the non-Abelian transformation

$$\begin{aligned} \gamma_1 &\rightarrow -\gamma_2, \\ \gamma_2 &\rightarrow \gamma_1. \end{aligned} \tag{25.32}$$

The unitary transformation

$$U_{12} = e^{\frac{\pi}{4}\gamma_1\gamma_2} = \frac{1}{\sqrt{2}}\left(1 + \gamma_1\gamma_2\right), \tag{25.33}$$

leads to

$$\begin{aligned} U_{12}\gamma_1 U_{12}^\dagger &= -\gamma_2 \\ U_{12}\gamma_2 U_{12}^\dagger &= \gamma_1, \end{aligned} \tag{25.34}$$

as desired. Defining a non-Hermitian fermion $f = 1/2(\gamma_1 + i\gamma_2)$, the action of the unitary operator on the states $|n_f = 0, 1\rangle$, where n_f is the occupation number of the operator f, is given by

$$U_{12}|n_f\rangle = e^{i\frac{\pi}{4}(1-2n_f)}|n\rangle, \tag{25.35}$$

[3] Here we follow [AraujoSacramento].

since $\gamma_1\gamma_2 = i(1 - 2f^\dagger f)$, and, therefore, is simply a phase factor and two Majoranas cannot be taken as a qubit.

If we have two pairs of Majorana fermions, the action of the unitary operators U_{12} and U_{34} is given by

$$U_{12}|n_1 n_3\rangle = e^{i\frac{\pi}{4}(1-2n_1)}|n_1 n_3\rangle ,$$
$$U_{34}|n_1 n_3\rangle = e^{i\frac{\pi}{4}(1-2n_3)}|n_1 n_3\rangle . \tag{25.36}$$

The permutation of Majoranas of different pairs takes the form

$$U_{23}|n_1 n_3\rangle = \frac{1}{\sqrt{2}}\left(|n_1 n_3\rangle + i|1 - n_1, 1 - n_3\rangle\right) , \tag{25.37}$$

and, therefore, mixes different states, even though it preserves the fermionic parity. Importantly, we obtain that

$$[U_{12}, U_{23}] = \gamma_1\gamma_3, \tag{25.38}$$

the commutator is non-vanishing and the order of the fermionic permutation operations affects the final result, allowing non-trivial operations. Note that

$$U_{12} = e^{i\frac{\pi}{4}S_z} ,$$
$$U_{23} = e^{i\frac{\pi}{4}S_x} , \tag{25.39}$$

may be interpreted as a rotation by an angle $\pi/2$ in a space of spin-$1/2$ (the unitary operator U_{34} is of the same form as U_{12}) and, therefore, corresponds to the algebra of one qubit (a spin-$1/2$ operator).

However, while theoretically appealing, it is not clear if the model of a bidimensional superconductor with triplet pairing has an experimental realization in nature. Another possibility, while still to be fully established, is the presence of Majorana modes at the ends of a semiconducting wire with strong spin-orbit coupling, in proximity with a conventional superconductor and in the presence of a magnetic field [63], or the presence of Majorana modes at the ends of a chain of magnetic impurities on top of a superconductor [64].

One may consider networks of wires that host Majorana modes at their ends [65] and take advantage of their non-Abelian statistics to construct non-trivial gate operations. In Fig. 25.21 it is illustrated how one may construct a sequence of wires with Majorana modes at their ends (γ modes) and that, by applying voltages appropriately (or currents), we may move the locations of the Majorana modes [66]. Also, it shows how to extend a topological region or shrinking it, as the Majorana modes converge to the same location and give rise to a normal fermion. Also, it is shown in Fig. 25.21 one way how to exchange Majorana modes using a T junction, by either exchanging the Majoranas at the ends of a single wire (from a-d) or exchanging the Majoranas at the left and right edges of two topological regions.

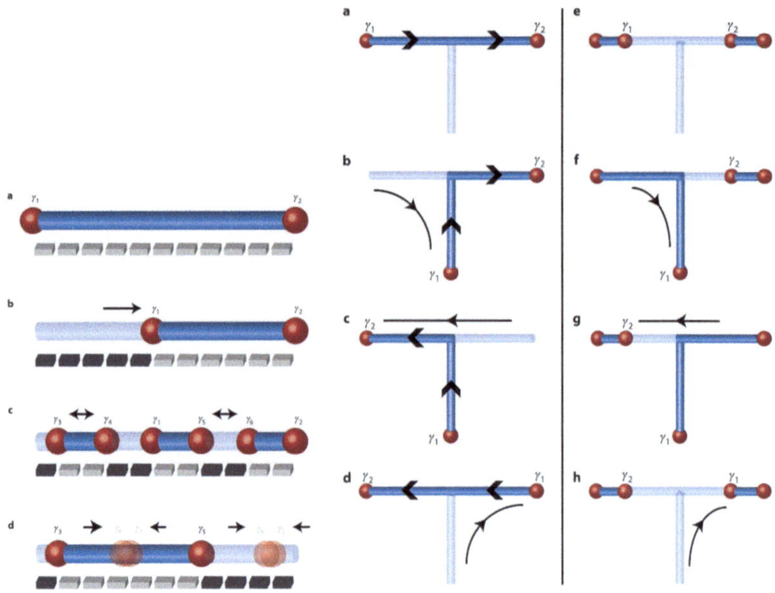

Fig. 25.21 Braiding of Majorana modes at the edges of topological wires. Reproduced with permission from Jason Alicea, Yuval Oreg, Gil Refael, Felix von Oppen, and Matthew P. A. Fisher, Nature Physics **7**, 412 (2011)

As explained before, the zero-energy-edge Majorana modes in the Kitaev chain may combine non-locally to form a usual fermion whose occupations, occupied and empty, are energy degenerate. Therefore, the two states, while energy degenerate with zero energy, can be distinguished by their parity. Charging effects due to Coulomb repulsion may distinguish parity, and therefore, information may be obtained on the occupation of the Majorana modes. Considering a system of Cooper-pair boxes and including the effect of Coulomb interactions allow the identification of zero energy modes due to the change in parity [67] and therefore the even-odd effect in the total fermionic number.

Frustrated Magnets

If we have competing interactions in magnetic system, this may lead to a degeneracy of the lowest energy state. Considering antiferromagnetic interactions, in either a triangle or in a tetrahedron, one finds that, due to the geometry, the system becomes frustrated, in the sense that not all bonds can be satisfied. A similar situation occurs in a chain, if we consider first- and second-neighbor antiferromagnetic interactions.

Frustration may also arise if we have random interactions, which may be either antiferromagnetic or antiferromagnetic and ferromagnetic. One example is a system where RKKY interactions take place, due to the oscillating nature of the interaction and its dependence on the distance between the spins. This occurs in diluted magnetic alloys where the substrate is metallic. If there is randomness associated with the locations of the impurities, this gives rise to a situation where there are many possible groundstates of the magnetic system, with many metastable states of very close energies and no single groundstate emerges. Moreover, as one lowers the temperature, the system may reach some stable configuration such that locally the spin is oriented along some direction, but as one moves from one spatial location to another, the preferred orientation is randomly changed. This phase is called a spin glass [68]. This type of system provides an alternative class of disorder to that discussed in Sect. 23.5.2, where we had a random magnetic field. An example of metallic alloys is provided by $CuMn$, but we can also have insulators with frustration, such as $Eu_xSr_{1-x}S$, where the origin is longer range exchange couplings.

Due to the geometric frustration or the disorder effects or the combination of the two and the existence of many low energy states, these systems do not order magnetically as one lowers the temperature, at least they do not show periodic magnetic order. A designation of spin liquid, as a magnetic system that does not order no matter how low the temperature is, was adopted. The definition has recently been more restricted to include systems that, indeed do not order, but have fractionalized excitations, have some sort of (long-range) topological order and whose main characteristic is the presence of geometric frustration. Several systems are candidates, as discussed recently [69]. Candidate materials in two-dimensional systems are the jarosites, Kitaev spin liquids such as the iridates, and several others, as discussed in this reference.

Here we will just focus on the disorder-free three-dimensional pyrochlore lattice [70], of the type $A_2B_2O_7$, such as the oxides $Dy_2Ti_2O_7$ and $Ho_2Ti_2O_7$, where there is an Ising anisotropy [71] and the systems have some type classical behavior. There are other materials that have XY anisotropy and are, therefore, expected to display quantum effects, such as $Yb_2B_2O_7$ and $Er_2B_2O_7$ and other families [72]. The pyrochlore lattice shown in Fig. 25.22, intercalating triangular and Kagome $2d$ lattices, may be seen as corner-sharing tetrahedra. This type of materials have properties that make them good candidates to being a spin liquid. They do not have magnetic order, have excitations that are fractionalized, and have topological order.

Consider the oxides $Dy_2Ti_2O_7$ and $Ho_2Ti_2O_7$: the interactions between the spins is of a long-range dipolar type, but the system does not order. There is a finite entropy per spin, associated with the geometric frustration. The interactions

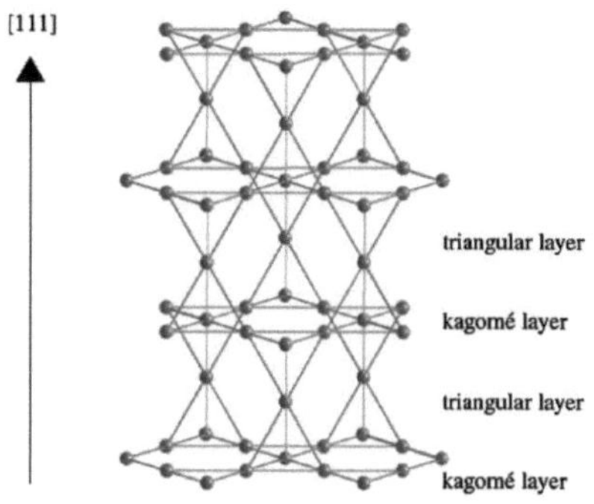

Fig. 25.22 Pyrochlore structure. Reproduced with permission from Jason S. Gardner, Michel J. P. Gingras, and John E. Greedan, Reviews of Modern Physics **82**, 53 (2010)

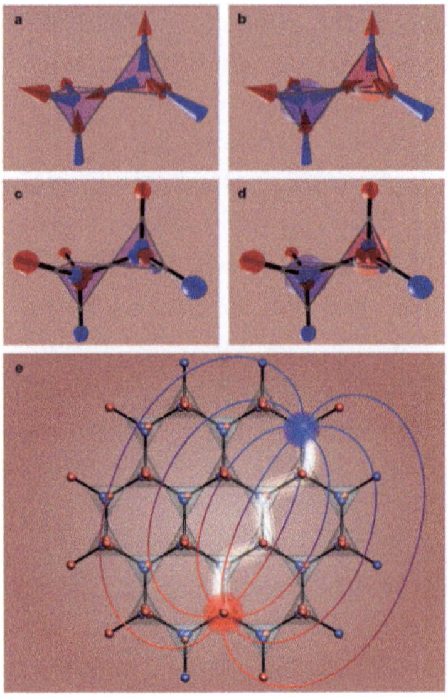

Fig. 25.23 Spin ice and water ice structures. In the spin case, the spins point inward or outward, and in water ice, the hydrogen atoms are further away or closer to the central oxygen atom. (**a**) Spin ice rule and (**b**) flipping one spin to create an excitation. (**c**) and (**d**) represent the same but using dumbbells instead of vectors with arrows. The red end gives the arrow direction. (**e**) Magnetic monopoles in spin ice and the equivalent to a Dirac string connecting them. In the dumbbell representation, it is easier to represent a monopole and an anti-monopole together with the line of flipped spins ("Dirac string") that connects the monopoles. Reproduced with permission from C. Castelnovo, R. Moessner, and S. L. Sondhi, Nature **451**, 42 (2008)

are such that the spins tend to point, either toward or away, along the bonds of the tetrahedra, in such a way that two spins point inward and two spins point outward. This is similar to what happens in ice, and this system has been denoted a spin ice. This is illustrated in Fig. 25.23. Out of the 16 states of a tetrahedron, only 6 are allowed by this restriction, and since a lattice with N spins has $N/2$ tetrahedra, the entropy per spin is $S/k_B = 1/2 \log(3/2)$. The ice restriction is related to the restriction of a divergence-free magnetic field, since from each tetrahedron, the "magnetic lines" (in this case the magnetic momenta or spins) have a vanishing flux across the surface of the tetrahedron. This suggests that some gauge field is associated with these degenerate states. It turns out that the excitations of this gauge field are alike magnetic monopoles, in the sense that flipping one spin and breaking the ice rule at a given tetrahedron is compensated overall by flipping another spin, inducing a line of flipped spins that may be seen as a Dirac-like string. So flipping one spin requires two excitations (like a monopole and an antimonopole), that are fractionalized excitations. These materials provide therefore a condensed matter realization of the looked after magnetic monopoles in vacuum [71]. Recalling the prediction of Majorana fermions (or actually Majorana modes) we discussed before, we see that in condensed matter, one finds realizations of "particles" that have been theorized in an empty space but that have proved hard to find experimentally. The excitations in terms of monopoles and the strings that connect them are shown in Fig. 25.23.

Other Topics

There are many other topics that are interesting because of fundamental science and/or applications. The list is endless.

- As mentioned in the preface, liquids were not considered for the most part in the book. Here we summarize some examples of liquids or liquid-like systems.
  In the context of superfluids, a system of interacting bosons was shown to have the property of superfluidity, but this state was not described in any detail, since it is beyond the level of the bulk of the book. Considering a dilute bosonic system, it was shown that there is the effect of Bose-Einstein condensation. The addition of interactions changes the spectrum of the system, leading to superfluidity. It was necessary for the experimental facilities to improve, due to technological advances and the development of several experimental methods, to detect a Bose condensate. A very dilute gas of bosons had to be studied, with a consequent very low critical temperature and optical confinement methods, to guarantee the virtual absence of interactions [73]. Superfluidity is also present in a fermionic system due to the effect of interactions. This occurs in ^3He. The pairing symmetry is of triplet p-wave type [74] and has a complex phenomenology. While the system does not seem to be useful in applications (as mentioned by A. J. Legget in his Nobel address, for instance, due to the very low critical temperature of the order of mK), it has proven as a source of inspiration for fundamental physics [75].
  A liquid-like set of systems that has had a significant impact on applications are liquid crystals, being at the interface between liquids (with no translational or rotational invariance due to the motion of the molecules or atoms in the system) and a crystal (where usually one finds both translational and rotational invariant structures). In liquid crystals there are many phases that have in common the preservation of some sort of symmetry, either rotational, at least on average, or translational, even if only in certain directions [76]. Among other applications, we can mention its use in liquid crystal displays [77].
  In liquid crystals, the direction of the average orientation of the molecules at a given location in space $\mathbf{r}$ is called the director field, $\hat{\mathbf{n}}(\mathbf{r})$. It is not a polar vector field, since both senses in a given direction are equivalent. Different possible phases of liquid crystals are shown in Fig. 25.24. Associated with fluctuations around the average directions shown in Fig. 25.24, we may find distortions of the director field such as bend, splay, or twist These distortions, if stable, may give rise to different phases with interesting properties and the appearance of defects [78]. Also, some molecule shapes give rise to new phases, such as bent-core and pear-shaped molecules [78].
  Interestingly, a mapping between liquid crystal phase and some superconductors has been established, both with conventional superconductors, including the presence of an external magnetic field and the presence of vortices in a type-

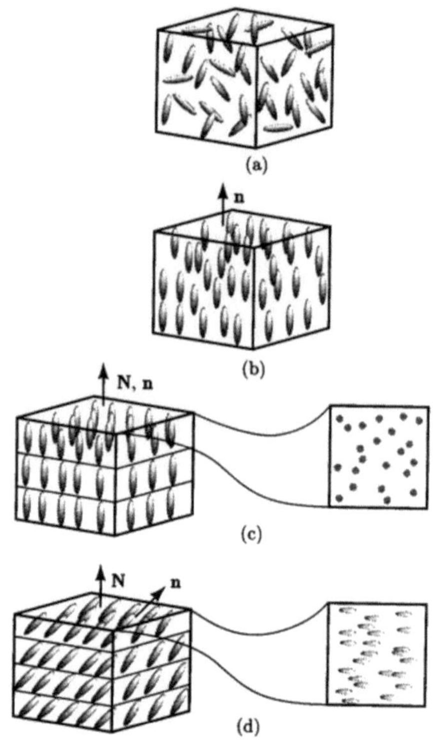

Fig. 25.24 Different phases in liquid crystals. (**a**) Disordered liquid phase, (**b**) nematic, (**c**) smectic A, and (**d**) smectic C. In the nematic phase, rotation symmetry is broken on average, but there is still translational symmetry. In the smectic phases, there is some translational symmetry along the N direction, but the molecules are free to move in the planes. Reproduced with permission from [ChaikinLubensky], Cambridge University Press

II superconductor, and with unconventional d-wave superconductors, in this case considering bent-core molecule phases [79]. For instance, there is a mapping (at the level of the free energy symmetries) between a smectic liquid crystal and a conventional superconductor. In the smectic, one may define an order parameter that distinguishes the phase from a disordered phase considering a density wave of the type $\psi(\mathbf{r}) = \rho(\mathbf{r})e^{i\Phi(\mathbf{r})}$, where the phase is associated with the spatial modulation due to the presence of the planes. This is formally equivalent to a wave function that describes the wave function of a superconductor in the context of the Landau-Ginzburg functional. Comparing in detail the free energy of the two systems, which in the liquid crystal is associated with the various possible deformations of the average structure, one finds that, in addition to the similarity in the order parameter, the deviation of the perpendicular component to the planes of the director field is like a vector potential in the superconductor, the elastic constant of the deformations is like the inverse mass of the Cooper pairs, the periodic modulation associated with the planes takes the role of the charge, the gradient energy is the kinetic energy, and the Meissner effect in the superconductor finds its match in the expulsion of bent and twist deformations.

Another situation where one finds a mixture between a solid and a liquid is superionic conductors. In these solid systems, there are two sublattices such that, at some temperature, one of the sublattices melts while keeping the other solid, such that the ions of the melted sublattice may flow through the solid structure of the unmelted sublattice. This occurs if the initial lattice (usually with a complex structure) has channels through which ions may flow or if the atoms of the melted sublattice have small diameters. These systems have found interesting applications since they have ionic conductivities that are high and somewhat comparable to ionic solutions, with applications in batteries, since the superionic conductors are essentially solid stable systems. Examples are Li_2SO_4 (Li has a small diameter), AgI, and the compound sodium β-alumina. The cations that are usually interesting are Li^+, Ag^+, Na^+, H^+.

- Another interesting topic is the interplay of condensed matter systems and quantum information. Using quantum information methods such as entanglement and fidelity [80], interesting and complementary understanding of condensed matter systems and their phases and phase transitions has been achieved in recent years. In particular, in systems where interactions and in general correlations are significant, the use of quantum information techniques has proven useful, for instance, in the light of entanglement of the degrees of freedom.
- The dynamic generation of new phases has also attracted considerable attention. In addition to imposing externally a transition between a phase and another one, for instance, by promoting a change from a given groundstate to an excited state with different properties, the method also leads to the creation of non-equilibrium phases, which disappear when the perturbation is turned off and that have no counterparts in equilibrium situations. One example is photoinduced transitions [81].
- Another vast topic not discussed here is the wide range of energy applications of materials (batteries, solar cells, medical applications, etc.), even though a few cases were commented on.
- As discussed before, many advances in experimental results have been a consequence of the significant technological advances that enabled better resolutions in established methods and the appearance of new methods. The improved experimental results have also allowed to better clarify the physical mechanisms, polishing the theoretical ideas and methods and enabling a better cooperation between experimental and theoretical efforts. One example is the development of metamaterials to create properties and functionalities not present in materials found in nature. Another such example is the development of attosecond spectroscopy [82]. The possibility to follow in real-time microscopic mechanisms enables a better understanding of the physical phenomena and will most likely uncover unexpected properties.

References

1. See, for instance, P. Hohenberg and W. Kohn, Physical Review **136**, B864 (1964); W. Kohn and L. J. Sham, Physical Review **140**, A1133 (1965); "Electronic structure: Basic theory and practical methods", R. M. Martin (Cambridge Univ. Press, 2004); the computer code WIEN2K,

that uses the augmented plane wave method (APW) to determine the band structure; KITE: high-performance accurate modelling of electronic structure and response functions of large molecules, disordered crystals and heterostructures.

2. Vladimir I. Anisimov, Jan Zaanen, and Ole K. Andersen, Physical Review B **44**, 943 (1991); John P. Perdew, Kieron Burke, and Matthias Ernzerhof, Physical Review Letters **77**, 3865 (1996).
3. See, for instance, the dynamical mean field theory in Antoine Georges, Gabriel Kotliar, Werner Krauth, and Marcelo J. Rozenberg, Reviews of Modern Physics **68**, 13 (1996).
4. A. H. Castro Neto, F. Guinea, N. M. R. Peres, K. S. Novoselov, and A. K. Geim, Reviews of Modern Physics **81**, 109 (2009).
5. N. M. R. Peres, Reviews of Modern Physics **82**, 2673 (2010).
6. S. Das Sarma, Shaffique Adam, E. H. Hwang, and Enrico Rossi, Reviews of Modern Physics **83**, 407 (2011).
7. R. R. Nair, P. Blake, A. N. Grigorenko, K. S. Novoselov, T. J. Booth, T. Stauber, N. M. R. Peres, and A. K. Geim, Science **320**, 1308 (2008).
8. Eduardo V. Castro, K. S. Novoselov, S. V. Morozov, N. M. R. Peres, J. M. B. Lopes dos Santos, Johan Nilsson, F. Guinea, A. K. Geim, and A. H. Castro Neto, Physical Review Letters **99**, 216802 (2007).
9. See, for instance, A. K. Geim, Science **324**, 1530 (2009); K. S. Novoselov, Reviews of Modern Physics **83**, 837 (2011).
10. See, for instance, M. G. Vergniory, B. J. Wieder, L. Elcoro, S. S. P. Parkin, C. Felser, B. Andrei Bernevig, and N. Regnault, Science **376**, 816 (2022).
11. M. König, S. Wiedmann, C. Brüne, A. Roth, H. Buhmann, L. W. Molenkamp, X.-L. Qi, and S.-C. Zhang, Science **318**, 766 (2007); Xiao-Liang Qi and Shou-Cheng Zhang, Reviews of Modern Physics **83**, 1057 (2011).
12. See, for instance, Y. Xia, D. Qian, D. Hsieh, L. Wray, A. Pal, H. Lin, A. Bansil, D. Grauer, Y. S. Hor, R. J. Cava and M. Z. Hasan, Nature Physics **5**, 398 (2009).
13. See, for instance, M. Z. Hasan and C. L. Kane, Reviews of Modern Physics **82**, 3045 (2010).
14. See, for instance, Jonathan A. Sobota, Yu He, and Zhi-Xun Shen, Reviews of Modern Physics **93**, 025006 (2021).
15. Y. Tanaka, Zhi Ren, T. Sato, K. Nakayama, S. Souma, T. Takahashi, Kouji Segawa and Yoichi Ando, Nature Physics **8**, 800 (2012).
16. N. R. Cooper, J. Dalibard, and I. B. Spielman, Reviews of Modern Physics **91**, 015005 (2019).
17. M. Zahid Hasan, Su-Yang Xu, Ilya Belopolski, and Shin-Ming Huang, Annual Review of Condensed Matter Physics **8**, 289 (2017); Binghai Yan, and Claudia Felser, Annual Review of Condensed Matter Physics **8**, 337 (2017).
18. Cui-Zu Chang, Chao-Xing Liu, and Allan H. MacDonald, Reviews of Modern Physics **95**, 011002 (2022).
19. C.-Z- Chang, J. Zhang, X. Feng, J. Shen, Z. Zhang, M. Guo, K. Li, Y. Ou, P. Wei, L.-L. Wang, Z.-Q. Ji, Y. Feng, S. Ji, X. Chen, J. Jia, X. Dai, Z. Fang, S.-C. Zhang, K. He, Y. Wang, L. Lu, X.-C. Ma, and Q.-K. Xue, Science **340**, 167 (2013); J. Zhang, C.-Z. Chang, P. Tang, Z. Zhang, X. Feng, K. Li, L.-L, Wang, X. Chen, C. Liu, W. Duan, K. He, Q.-K. Xue, X. Ma, and Y. Wang, Science **339**, 1582 (2013).
20. Satoru Nakatsuji and Ryotaro Arita, Annual Review of Condensed Matter Physics **13**, 119 (2022).
21. Igor Mazin and The PRX Editors, Physical Review X **12**, 040002 (2022); Libor Smejkal, Jairo Sinova, and Tomas Jungwirth, Physical Review X **12**, 031042 (2022); Libor Smejkal, Jairo Sinova, and Tomas Jungwirth, Physical Review X **12**, 040501 (2022).
22. Morten Amundsen, Arne Brataas, and Jacob Linder, Physical Review B **110**, 054427 (2024).
23. Zouti Xie, Tal Z. Markus, Sidney R. Cohen, Zeev Vager, Rafael Gutierrez, and Ron Naaman, Nano Letters **11**, 4652 (2011); Ron Naaman, and David H. Waldeck, Annual Review of Physical Chemistry **66**, 263 (2015).
24. See, for instance, Igor Zutic, Jaroslav Fabian, and S. Das Sarma, Reviews of Modern Physics **76**, 323 (2004).

25. See, for instance, A. Manchon, J. Zelezny, I. M. Miron, T. Jungwirth, J. Sinova, A. Thiaville, K. Garello, and P. Gambardella, Reviews of Modern Physics **91**, 035004 (2019).
26. Albert Fert, Reviews of Modern Physics **80**, 1517 (2008); Peter A. Grünberg, Reviews of Modern Physics **80**, 1531 (2008).
27. Jayakumar Balakrishnan, Gavin Kok Wai Koon, Manu Jaiswal, A. H. Castro Neto, and Barbaros Özyilmaz, Nature Physics **9**, 284 (2013).
28. V. Baltz, A. Manchon, M. Tsoi, T. Moriyama, T. Ono, and Y. Tserkovnyak, Reviews of Modern Physics **90**, 015005 (2018).
29. Jacob Linder and Jason W. A. Robinson, Nature Physics **11**, 307 (2015).
30. G. E. Blonder, M. Tinkham, and T. M. Klapwijk, Physical Review B **25**, 4515 (1982); Guy Deutscher, Reviews of Modern Physics **77**, 109 (2005).
31. Ady Stern, Annual Review of Condensed Matter Physics **7**, 11 (2016).
32. See, for instance, Christian Pfleiderer, Reviews of Modern Physics **81**, 1551 (2009) and for plutonium based materials see E.D. Bauer, and J.D. Thompson, Annual Review of Condensed Matter Physics **6**, 137 (2015).
33. For extensive lists of materials and properties see, for instance, G. R. Stewart, Reviews of Modern Physics **73**, 797 (2001); G. R. Stewart, Reviews of Modern Physics **78**, 743 (2006).
34. C.M. Varma, Z. Nussinov, Wim van Saarloos, Physics Reports **361**, 267 (2002).
35. P. Schlottmann and P. D. Sacramento, Advances in Physics **42**, 641 (1993); D. L. Cox and A. Zawadowski, Advances in Physics **47**, 599 (1998).
36. See, for instance, [Continentino], [Sachdev], [Coleman].
37. Adilet Imambekov, Thomas L. Schmidt, and Leonid I. Glazman, Reviews of Modern Physics **84**, 1253 (2012).
38. C. Kim, A. Y. Matsuura, Z.-X. Shen, N. Motoyama, H. Eisaki, S. Uchida, T. Tohyama, and S. Maekawa, Physical Review Letters **77**, 4054 (1996).
39. V. J. Goldman and B. Su, Science **267**, 1010 (1993).
40. R. Coldea, D. A. Tennant, A. M. Tsvelik, and Z. Tylczynski, Physical Review Letters **86**, 1335 (2001); Masanori Kohno, Oleg A. Starykh, and Leon Balents, Nature Physics **3**, 790 (2007); P. Ribeiro, P. D. Sacramento, and K. Penc, Physical Review B **84**, 045112 (2011).
41. See, for instance, C. C. Tsuei and J. R. Kirtley, Reviews of Modern Physics **72**, 969 (2000).
42. Michael Smidman, Oliver Stockert, Emilian M. Nica, Yang Liu, Huiqiu Yuan, Qimiao Si, and Frank Steglich, Reviews of Modern Physics **95**, 031002 (2023).
43. D Aoki, J-P Brison, J Flouquet, K Ishida, G Knebel, Y Tokunaga and Y Yanase, Journal of Physics Condensed Matter **34**, 243002 (2022).
44. Yoshiteru Maeno, Shingo Yonezawa, Aline Ramires, arXiv:2402.12117.
45. Takeshi Kondo, Yoichiro Hamaya, Ari D. Palczewski, Tsunehiro Takeuchi, J. S. Wen, Z. J. Xu, Genda Gu, Jörg Schmalian, and Adam Kaminski, Nature Physics **7**, 21 (2011).
46. See, for instance, Hai-Hu Wen, and Shiliang Li, Annual Review of Condensed Matter Physics **2**, 121 (2011).
47. M. Brando, D. Belitz, F. M. Grosche, and T. R. Kirkpatrick, Reviews of Modern Physics **88**, 025006 (2016).
48. See, for instance, Chris J. Pickard, Ion Errea and Mikhail I. Eremets, Annual Review of Condensed Matter Physics **11**, 57 (2020).
49. Dung-Hai Lee, Annual Review of Condensed Matter Physics **9**, 261 (2018).
50. Pavlo Zubko, Stefano Gariglio, Marc Gabay, Philippe Ghosez, and Jean-Marc Triscone, Annual Review of Condensed Matter Physics **2**, 141 (2011).
51. Stefano Gariglio, and Jean-Marc Triscone, Comptes Rendus Physique **12**, 591 (2011); E. Cappelluti, C. Grimaldi, and F. Marsiglio, Physical Review Letters **98**, 167002 (2007).
52. S. A. Kivelson, Physica B Condensed Matter **318**, 61 (2002); Erez Berg, Dror Orgad, and Steven A. Kivelson, Physical Review B **78**, 094509 (2008); Satoshi Okamoto and Thomas A. Maier, Physical Review Letters **101**, 156401 (2008).
53. Lu Li, C. Richter, J. Mannhart, and R. C. Ashoori, Nature Physics **7**, 762 (2011).
54. F. Hellman, A. Hoffmann, Y. Tserkovnyak, G. S. D. Beach, E. E. Fullerton, C. Leighton, A. H. MacDonald, D. C. Ralph, A. A Arena, H. A. Dürr, P. Fischer, J. Grollier, J. P. Heremans,

T. Jungwirth, A. V. Kimel, B. Koopmans, I. N. Krivorotov, S. J. May, A. K. Petford-Long, J. M. Rondinelli, N. Samarth, I. K. Schuller, A. N. Slavin, M. D. Stiles, O. Tchernyshyov, A. Thiaville and B. L. Zink, Reviews of Modern Physics **89**, 025006 (2017).

55. Rafael Roldán, Jose A. Silva-Guillén, M. Pilar López-Sancho, Francisco Guinea, Emmanuele Cappelluti, Pablo Ordejón, Annalen der Physik **526**, 347 (2014).
56. J. M. B. Lopes dos Santos, N. M. R. Peres, and A. H. Castro Neto, Physical Review Letters **99**, 256802 (2007).
57. Rafi Bistritzer and Allan H. MacDonald, Proceedings National Academy USA **108**, 12233 (2011).
58. Eva Y. Andrei and Allan H. MacDonald, Nature Materials **19**, 1265 (2020); Eva Y. Andrei, Dmitri K. Efetov, Pablo Jarillo-Herrero, Allan H. MacDonald, Kin Fai Mak, T. Senthil, Emanuel Tutuc, Ali Yazdani, and Andrea F. Young, Nature Reviews Materials **6**, 201 (2021).
59. Zhengguang Lu, Tonghang Han, Yuxuan Yao, Aidan P. Reddy, Jixiang Yang, Junseok Seo, Kenji Watanabe, Takashi Taniguchi, Liang Fu and Long Ju, Nature **626**, 759 (2024).
60. Guido Burkard, Thaddeus D. Ladd, Andrew Pan, John M. Nichol, and Jason R. Petta, Reviews of Modern Physics **95**, 025003 (2023).
61. Morten Kjaergaard, Mollie E. Schwartz, Jochen Braumüller, Philip Krantz, Joel I.-J. Wang, Simon Gustavsson, and William D. Oliver, Annual Review of Condensed Matter Physics **11**, 369 (2020).
62. D. A. Ivanov, Physical Review Letters **86**, 268 (2001).
63. Anindya Das, Yuval Ronen, Yonatan Most, Yuval Oreg, Moty Heiblum, and Hadas Shtrikman, Nature Physics **8**, 887 (2012); V. Mourik, K. Zuo, S. M. Frolov, S. R. Plissard, E. P. A. M. Bakkers, and L. P. Kouwenhoven, Science **336**, 1003 (2012).
64. Stevan Nadj-Perge, Ilya K. Drozdov, Jian Li, Hua Chen, Sangjun Jeon, Jungpil Seo, Allan H. MacDonald, B. Andrei Bernevig, and Ali Yazdani, Science **346**, 602 (2014).
65. Jason Alicea, Yuval Oreg, Gil Refael, Felix von Oppen, and Matthew P. A. Fisher, Nature Physics **7**, 412 (2011).
66. Ching-Kai Chiu, M. M. Vazifeh and M. Franz, Europhysics Letters **110**, 10001 (2015).
67. Yuval Oreg, and Felix von Oppen, Annual Review of Condensed Matter Physics **11**, 397 (2020).
68. K. Binder and A. P. Young, Reviews of Modern Physics **58**, 801 (1986); Giorgio Parisi, Reviews of Modern Physics **95**, 030501 (2023).
69. J. Knolle, and R. Moessner, Annual Review of Condensed Matter Physics **10**, 451 (2019).
70. Jason S. Gardner, Michel J. P. Gingras, and John E. Greedan, Reviews of Modern Physics **82**, 53 (2010).
71. C. Castelnovo, R. Moessner, and S.L. Sondhi, Annual Review of Condensed Matter Physics **3**, 35 (2012).
72. Alannah M. Hallas, Jonathan Gaudet, and Bruce D. Gaulin, Annual Review of Condensed Matter Physics **9**, 105 (2018).
73. E. A. Cornell and C. E. Wieman, Reviews of Modern Physics **74**, 875 (2002).
74. Anthony J. Leggett, Reviews of Modern Physics **47**, 331 (1975); **76**, 999 (2004).
75. "The Universe in a Helium Droplet", G. E. Volovik, Oxford Univ. Press.
76. "The Physics of Liquid Crystals", P. G. de Gennes and J. Prost, Oxford University Press.
77. See for instance, J. C. Jones, Liquid Crystals Today **27**, 44 (2018).
78. J. V. Selinger, Annual Review of Condensed Matter Physics **13**, 49 (2022).
79. See for instance, P. Tolédano and A. M. Figueiredo Neto, Liquid Crystals Today **10**, 5 (2010).
80. Luigi Amico, Rosario Fazio, Andreas Osterloh, Vlatko Vedral, Reviews of Modern Physics **80**, 517 (2008); S.-J. Gu, International Journal of Modern Physics **24**, 4371 (2010).
81. See for instance, A. Cavalleri, Comtemporary Physics **59**, 31 (2017); Takashi Oka, and Sota Kitamura, Annual Review of Condensed Matter Physics **10**, 387 (2019).
82. Renate Pazourek, Stefan Nagele, and Joachim Burgdörfer, Reviews of Modern Physics **87**, 765 (2015); Rocío Borrego-Varillas, Matteo Lucchini, and Mauro Nisoli, Reports Progress Physics **85**, 066401 (2022).

Appendix A
Many-Body Systems and Second Quantization

In first quantization physical quantities are replaced by (Hermitian) operators that, in general, do not commute. Physical quantities are associated with average values of these operators, in some states. Therefore, to quantize a theory, a prescription is to replace the classical variables by corresponding operators and to impose commutation relations between them. Associated with the non-trivial commutation relations between operators are associated uncertainty rules between so-called canonically conjugate variables. These are variables that are related through the Lagrangian of the system and classically have non-trivial Poisson brackets. Associated with a given Lagrangian (or action) we may find the equations of motion through the Euler-Lagrangian equations. Therefore, if we have the equations of motion of a given classical variable or field, we may find a Lagrangian from which the Euler-Lagrange equations give the equations of motion. Then from the Lagrangian we may identify canonically conjugate variables and impose commutation relations between their operator correspondents, or just calculate their Poisson brackets and transform them into commutators.

One way to introduce the concept of second quantization is to regard Schrödinger equation as an equation of motion for some field, in this case the wave function.[1] Consider the time-dependent Schrödinger equation, for instance for a one-dimensional system

$$-\frac{\hbar^2}{2m}\frac{\partial^2\psi(x,t)}{\partial x^2}+V\psi(x,t)-i\hbar\frac{\partial\psi(x,t)}{\partial t}=0.$$

The Euler-Lagrange equation for a classical field is written as

[1] Here we follow [BicudoSacramento].

P. D. Sacramento, *Introduction to Condensed Matter Theory*,
https://doi.org/10.1007/978-3-031-86251-9

$$\frac{\partial}{\partial x}\frac{\partial L}{\partial\left(\frac{\partial\psi}{\partial x}\right)}+\frac{\partial}{\partial t}\frac{\partial L}{\partial\left(\frac{\partial\psi}{\partial t}\right)}-\frac{\partial L}{\partial\psi}=0.$$

Let us try to identify the Lagrangian density from which Schrödinger equation is derived, considering the Euler-Lagrange equation. The Lagrangian density may be written as

$$L=i\hbar\psi^*\frac{\partial\psi}{\partial t}-\psi^*V\psi-\frac{\hbar^2}{2m}\frac{\partial\psi^*}{\partial x}\frac{\partial\psi}{\partial x},$$

or in the symmetrized form $-\frac{i\hbar}{2}\left(\frac{\partial\psi^*}{\partial t}\psi+\psi^*\frac{\partial\psi}{\partial t}\right)$, where we are taking the potential, V, as real. The canonically conjugate moment of the field ψ, defined in the usual way, is given by

$$\Pi=\frac{\partial L}{\partial\left(\frac{\partial\psi}{\partial t}\right)}=i\hbar\psi^*,$$

and therefore the canonical conjugate is the field itself (with the complex conjugate). So the conjugate field of the wave function $\psi(\mathbf{r})$ is $i\hbar\psi^*(\mathbf{r})$. Let us then quantize the wave function imposing the commutation relation

$$\left[\hat{\psi}(\mathbf{r},t),i\hbar\hat{\psi}^\dagger(\mathbf{r}',t)\right]=i\hbar\delta(\mathbf{r}-\mathbf{r}')$$

and, therefore,

$$\left[\hat{\psi}(\mathbf{r},t),\hat{\psi}^\dagger(\mathbf{r}',t)\right]=\delta(\mathbf{r}-\mathbf{r}').$$

The wave function has been transformed into an operator field.

In general, the wave function may be expanded in a complete basis of orthogonal functions $\Phi_\lambda(\mathbf{r})$, with coefficients $a_\lambda(t)$

$$\psi(\mathbf{r},t)=\sum_\lambda a_\lambda(t)\Phi_\lambda(\mathbf{r}).$$

In the same way

$$\psi^*(\mathbf{r},t)=\sum_\lambda a_\lambda^*(t)\Phi_\lambda^*(\mathbf{r}).$$

As we impose commutation relations on the wave function, this is transformed into an operator, hence the name second quantization. In the first quantization we quantize observables into operators and in the second quantization we quantize the wave functions into operators. Since the wave function is now an operator, the coefficients $a_\lambda(t)$ are themselves operators. The commutation relation imposes commutation relations between these operators in the form

$$\left[a_\lambda(t), a^\dagger_{\lambda'}(t)\right] = \delta_{\lambda,\lambda'}$$
$$[a_\lambda(t), a_{\lambda'}(t)] = 0$$
$$\left[a^\dagger_\lambda(t), a^\dagger_{\lambda'}(t)\right] = 0,$$

which are similar to those of a set of harmonic oscillators.

Let us now consider the energy and the operator in first quantization that is the correspondent one, the Hamiltonian $\mathcal{H}$, of a set of independent particles. The energy of a state described by the wave function $\psi(\mathbf{r})$ in first quantization is given in the usual way by

$$E = \int \mathrm{d}^3 r \psi^*(\mathbf{r})\mathcal{H}\psi(\mathbf{r}).$$

Let us now define a Hamiltonian operator in second quantization, $\hat{H}$, with the transformation

$$\hat{H} = \int \mathrm{d}^3 r \hat{\psi}^\dagger(\mathbf{r})\mathcal{H}\hat{\psi}(\mathbf{r}).$$

We obtain

$$\begin{aligned}\hat{H} &= \int \mathrm{d}^3 r \sum_\lambda a^\dagger_\lambda(t)\Phi^*_\lambda(\mathbf{r})\mathcal{H}\sum_{\lambda'} a_{\lambda'}(t)\Phi_{\lambda'}(\mathbf{r})\\ &= \sum_{\lambda,\lambda'}\langle\lambda|\mathcal{H}|\lambda'\rangle a^\dagger_\lambda(t)a_{\lambda'}(t),\end{aligned}$$

where

$$\langle\lambda|\mathcal{H}|\lambda'\rangle = \int \mathrm{d}^3 r \Phi^*_\lambda(\mathbf{r})\mathcal{H}\Phi_{\lambda'}(\mathbf{r}).$$

If we choose as complete basis the eigenfunctions of the Hamiltonian

$$\mathcal{H}\Phi_\lambda = \epsilon_\lambda\Phi_\lambda,$$

the Hamiltonian will take the simplified form

$$\hat{H} = \sum_\lambda \epsilon_\lambda a^\dagger_\lambda a_\lambda.$$

In the case of free particles, $\lambda = \mathbf{k}$,

$$\hat{H} = \sum_{\mathbf{k}} \epsilon_{\mathbf{k}} a_{\mathbf{k}}^{\dagger} a_{\mathbf{k}}.$$

We may then interpret $a_{\mathbf{k}}^{\dagger} a_{\mathbf{k}}$ as a number of particle operator with momentum $\mathbf{k}$. The energy of a free system is therefore the sum over all momenta of the energy of each momentum, times the number of particles with that momentum. In general, the operator $a_{\lambda}^{\dagger} a_{\lambda}$ counts the number of particles in the state λ.

This resembles the relation between the Planck interpretation of the black-body radiation, as resulting from states in the material described as harmonic oscillators, and Einstein interpretation of the states, as states of the radiation field. In the harmonic oscillator problem raising and lowering operators were introduced, from which a number operator with eigenvalues that are related to the energy and occupation of a given energy level appears. In Einstein's proposal the same operators, with the same algebra, are interpreted as counting numbers of particles (the photons) on a given state (a given frequency mode).

With these considerations in mind let us take a system of N identical particles, contained in a box of volume Ω, with periodic boundary conditions.[2] To describe the state of the system we can either use a wave function $\psi(\mathbf{r}_1, \mathbf{r}_2, \ldots, \mathbf{r}_N)$, or else we can specify the state of the system indicating the number of particles, $n_{\mathbf{p}}$, in a given one-particle state, described by a plane wave of momentum $\mathbf{p}$. We can then characterize the state of the system by the vector[3]

$$|n_1, \ldots, n_p, \ldots\rangle.$$

This state is given by a symmetrical or antisymmetrical product of plane waves, one for each particle. Using the symmetrization of the wave functions is a complex problem if the number of particles is large and it is not a very convenient description to carry out calculations.

Let us define an operator, called creation operator, by

$$a_p^{\dagger}|\ldots, n_p, \ldots\rangle = \sqrt{n_p + 1}|\ldots, n_p + 1, \ldots\rangle.$$

Similarly we can define a destruction operator by

$$a_p|\ldots, n_p, \ldots\rangle = \sqrt{n_p}|\ldots, n_p - 1, \ldots\rangle.$$

Let us consider first the case of bosons. Consider the product of two creation operators acting on the vacuum state (state with zero occupation numbers)

[2] See also [PinesNozieres].

[3] To simplify notation, here we represent $n_{\mathbf{p}}$ as n_p, and the same for destruction and creation operators.

$$a_p^\dagger a_{p'}^\dagger|0,\ldots,0\rangle = |0,\ldots,1,\ldots,1,\ldots 0\rangle,$$

indicating that a particle was created with momentum $\mathbf{p}$ and another with momentum $\mathbf{p}'$. Since the wave function is symmetrical with respect to the permutation of particles, the final state is the same if instead we act with the operators in reverse order like $a_{p'}^\dagger a_p^\dagger|0\rangle$, using a simplified notation to represent the vacuum state. Therefore we conclude that

$$a_p^\dagger a_{p'}^\dagger - a_{p'}^\dagger a_p^\dagger = 0,$$

meaning that the operators commute. We may also conclude that

$$[a_p, a_{p'}] = [a_p^\dagger, a_{p'}^\dagger] = 0,$$

$$[a_p, a_{p'}^\dagger] = \delta_{p,p'}.$$

The last result is easily shown:

$$a_p a_{p'}^\dagger|\ldots,n_p,\ldots,n_{p'},\ldots\rangle = \sqrt{n_{p'}+1}\sqrt{n_p}|\ldots,n_p-1,\ldots,n_{p'}+1,\ldots\rangle$$
$$a_{p'}^\dagger a_p|\ldots,n_p,\ldots,n_{p'},\ldots\rangle = \sqrt{n_{p'}+1}\sqrt{n_p}|\ldots,n_p-1,\ldots,n_{p'}+1,\ldots\rangle.$$

Therefore if $p \neq p'$ we get that

$$(a_p a_{p'}^\dagger - a_{p'}^\dagger a_p)|\ldots,n_p,\ldots,n_{p'},\ldots\rangle = 0.$$

If $p = p'$, then

$$a_p a_p^\dagger|\ldots,n_p,\ldots\rangle = \sqrt{n_p+1}\sqrt{n_p+1}|\ldots,n_p,\ldots\rangle$$
$$a_p^\dagger a_p|\ldots,n_p,\ldots\rangle = \sqrt{n_p}\sqrt{n_p}|\ldots,n_p,\ldots\rangle$$

and

$$(a_p a_p^\dagger - a_p^\dagger a_p)|\ldots,n_p,\ldots\rangle = (n_p+1-n_p)|\ldots,n_p,\ldots\rangle,$$

which proves the above result. The operator $n_p = a_p^\dagger a_p$ is a number operator since the state $|\ldots,n_p,\ldots\rangle$ is an eigenvector with eigenvalue n_p.

Let us now consider the case of fermions. In the same way, we define creation and destruction operators. We have to keep in mind now that the wave function is antisymmetric and that the exchange of two particles involves a minus sign. Then

$$c_p^\dagger c_{p'}^\dagger|\ldots\rangle = -c_{p'}^\dagger c_p^\dagger|\ldots\rangle.$$

This implies that these operators anticommute instead of commuting:

$$\left(c_p^\dagger c_{p'}^\dagger + c_{p'}^\dagger c_p^\dagger\right) = \left\{c_p^\dagger, c_{p'}^\dagger\right\} = 0.$$

Also, we can show that

$$\{c_p, c_{p'}\} = 0$$
$$\{c_p^\dagger, c_{p'}\} = \delta_{p,p'}.$$

In the same way, $n_p = c_p^\dagger c_p$ is a number operator. As a consequence, $c_p^2 = 0$ and $n_p^2 = n_p$. We have then to define that

$$c_p|\ldots, n_p, \ldots\rangle = (-1)^{-\sum_{j<p} n_j} n_p|\ldots, n_p - 1, \ldots\rangle$$
$$c_p^\dagger|\ldots, n_p, \ldots\rangle = (-1)^{-\sum_{j<p} n_j}(1 - n_p)|\ldots, n_p + 1, \ldots\rangle.$$

We have to assign a specified order to the one-particle states. Any order is allowed since only relative signs have physical meaning. For instance we may choose to define

$$|\ldots, n_p = 1, \ldots, n_{p'} = 1, \ldots\rangle = c_p^\dagger c_{p'}^\dagger|.., 0., , , .0, \ldots\rangle.$$

We have overlooked so far that the fermions have half-integer spin. Let us then consider spin-1/2 fermions and add the spin index to the operators. The anticommutation relations are then properly rewritten as

$$\{c_{p\sigma}, c_{p'\sigma'}\} = \left\{c_{p\sigma}^\dagger, c_{p'\sigma'}^\dagger\right\} = 0$$
$$\left\{c_{p\sigma}, c_{p'\sigma'}^\dagger\right\} = \delta_{\sigma\sigma'}\delta_{pp'}.$$

Field Operators

A normalized plane wave (eigenfunction of the momentum operator) is written as

$$\phi_p(\mathbf{r}) = \frac{1}{\sqrt{\Omega}} e^{i\mathbf{p}\cdot\mathbf{r}}.$$

As a consequence, $\phi_p(\mathbf{r})a_\mathbf{p}$ is the probability amplitude to destroy a boson (or a fermion, if we replace the boson destruction operator by the fermion destruction operator) of momentum $\mathbf{p}$, at the point $\mathbf{r}$. Let us then define a field operator $\hat{\psi}(\mathbf{r})$ by the Fourier transform of the destruction operator

$$\hat{\psi}(\mathbf{r}) = \frac{1}{\sqrt{\Omega}} \sum_{\mathbf{p}} a_{\mathbf{p}} e^{i\mathbf{p}\cdot\mathbf{r}}$$

$$a_{\mathbf{p}} = \frac{1}{\sqrt{\Omega}} \int d^3r \psi(\mathbf{r}) e^{-i\mathbf{p}\cdot\mathbf{r}}.$$

This field operator destroys a particle at site **r**. Clearly the conjugate is an operator that creates a particle at point **r** defined by

$$\hat{\psi}^\dagger(\mathbf{r}) = \frac{1}{\sqrt{\Omega}} \sum_{\mathbf{p}} a_{\mathbf{p}}^\dagger e^{-i\mathbf{p}\cdot\mathbf{r}}.$$

It is easy to show, using the commutation and anticommutation relations of the momentum creation and destruction operators, that for bosons the field operators obey

$$[\hat{\psi}(\mathbf{r}), \hat{\psi}(\mathbf{r}')] = [\hat{\psi}^\dagger(\mathbf{r}), \hat{\psi}^\dagger(\mathbf{r}')] = 0,$$

$$[\hat{\psi}(\mathbf{r}), \hat{\psi}^\dagger(\mathbf{r}')] = \delta(\mathbf{r} - \mathbf{r}'),$$

and for fermions obey

$$\{\hat{\psi}_\sigma(\mathbf{r}), \hat{\psi}_{\sigma'}(\mathbf{r}')\} = \{\hat{\psi}_\sigma^\dagger(\mathbf{r}), \hat{\psi}_{\sigma'}^\dagger(\mathbf{r}')\} = 0,$$

$$\{\hat{\psi}_\sigma(\mathbf{r}), \hat{\psi}_{\sigma'}^\dagger(\mathbf{r}')\} = \delta_{\sigma\sigma'}\delta(\mathbf{r} - \mathbf{r}').$$

Let us now consider the fermion gas as an example. The number operators in momentum and in real spaces are defined by

$$n_{p\sigma} = c_{p\sigma}^\dagger c_{p\sigma}$$

$$\rho(\mathbf{r}, \sigma) = \hat{\psi}_\sigma^\dagger(\mathbf{r})\hat{\psi}_\sigma(\mathbf{r}),$$

respectively. Clearly the total number of particle operator is given by

$$N = \sum_{\mathbf{p},\sigma} c_{p\sigma}^\dagger c_{p\sigma} = \sum_{\sigma} \int d^3r \hat{\psi}_\sigma^\dagger(\mathbf{r})\hat{\psi}_\sigma(\mathbf{r}).$$

The kinetic energy is given by

$$T = \sum_i \frac{\mathbf{p}^2}{2m}$$

$$= \sum_{\mathbf{p},\sigma} \frac{\mathbf{p}^2}{2m} n_{p\sigma},$$

which is straightforwardly written in second quantization replacing the number of particles, in a given momentum and with a given spin projection, by the corresponding number operator. Therefore, in second quantization the kinetic energy is written as

$$T = \sum_{\mathbf{p},\sigma} \frac{\mathbf{p}^2}{2m} c^{\dagger}_{p\sigma} c_{p\sigma}.$$

We can as well recall that the kinetic energy in first quantization is written as

$$T = -\frac{\hbar^2}{2m} \sum_{\sigma} \int d^3 r \psi^*_\sigma(\mathbf{r}) \nabla^2 \psi_\sigma(\mathbf{r}),$$

where now $\psi_\sigma(\mathbf{r})$ is the wave function and interpret this wave function as a field operator. Then if we Fourier transform it is easy to see that we recover the above expression.

The potential energy is more complicated to handle. Let us consider a two-body factorizable potential energy like

$$V = \frac{1}{2} \sum_{i \neq j} V(\mathbf{r}_i - \mathbf{r}_j))).$$

The particle density is given by

$$\rho(\mathbf{r}) = \sum_i \delta(\mathbf{r} - \mathbf{r}_i).$$

Then the potential energy is written as

$$V = \frac{1}{2} \int d^3 r \int d^3 r' V(r - r')[\rho(r)\rho(r') - \rho(r)\delta(r - r')].$$

The last term eliminates the interaction of a particle with itself. Recalling that $\rho(r) = \sum_\sigma \rho(r, \sigma)$ and interpreting the density functions as operators, we can write the potential energy as

$$V = \frac{1}{2} \int d^3 r \int d^3 r' V(r - r') \left[\sum_{\sigma,\sigma'} \hat{\psi}^\dagger_\sigma(r) \hat{\psi}_\sigma(r) \hat{\psi}^\dagger_{\sigma'}(r') \hat{\psi}_{\sigma'}(r') \right.$$
$$\left. - \sum_\sigma \hat{\psi}^\dagger_\sigma(r) \hat{\psi}_\sigma(r) \delta(r - r') \right].$$

Using that

$$[A, BC] = [A, B]C - B[C, A]$$
$$= \{A, B\}C - B\{C, A\},$$

where A, B, C are operators, we can rewrite the potential energy in so-called normal order, where all the creation operators are to the left and all the destruction operators are to the right,

$$V = \frac{1}{2}\int d^3r \int d^3r' V(r - r') \sum_{\sigma,\sigma'} \hat{\psi}^{\dagger}_{\sigma}(r)\hat{\psi}^{\dagger}_{\sigma'}(r')\hat{\psi}_{\sigma'}(r')\hat{\psi}_{\sigma}(r).$$

In momentum space it can be written as

$$V = \frac{1}{2\Omega}\sum_{\sigma,\sigma'}\sum_{\mathbf{k},\mathbf{p},\mathbf{q}} V_q c^{\dagger}_{p+q,\sigma} c^{\dagger}_{p-q,\sigma'} c_{k,\sigma'} c_{p,\sigma}.$$

Here V_q is the Fourier transform of the two-body potential energy.

Due to the potential, there will be scattering of the particles in two-body collisions. One particle with momentum $\mathbf{k}$ and spin projection σ' will scatter off a particle with momentum $\mathbf{p}$ and spin projection σ. After the collision, a particle with momentum $\mathbf{k} - \mathbf{q}$ and spin projection σ' will emerge and another one with momentum $\mathbf{q} + \mathbf{p}$ and spin projection σ. Since the particles are indistinguishable, any of the two particles will be in either final state leading to two possible outcomes of the collision. The scattering matrix element of a pair of fermions $(\mathbf{p}, \sigma; k, \sigma')$ into a new state $(\mathbf{q} + \mathbf{p}, \sigma; \mathbf{k} - \mathbf{q}, \sigma')$ is given by

$$\langle \ldots; \mathbf{q} + \mathbf{p}, \sigma; \ldots; \mathbf{k} - \mathbf{q}, \sigma'; \ldots |V| \ldots; \mathbf{p}, \sigma; \ldots; \mathbf{k}, \sigma'; \ldots \rangle.$$

Replacing the expression for the potential energy, we have to evaluate the matrix element

$$\left\langle \mathbf{p} + \mathbf{q}, \sigma; \mathbf{k} - \mathbf{q}, \sigma' | c^{\dagger}_{p'+q',\sigma_1} c^{\dagger}_{k'-q',\sigma_2} c_{k',\sigma_2} c_{p',\sigma_1} | \mathbf{p}, \sigma; \mathbf{k}, \sigma' \right\rangle$$

and then sum over the spin projections σ_1, σ_2 and over the momenta $\mathbf{k}', \mathbf{p}', \mathbf{q}'$ and multiply by $V_{q'}$. There are two possibilities: we may act with the destruction operator c_{k',σ_2} on the single-particle state $|\mathbf{p}, \sigma\rangle$ and the destruction operator c_{p',σ_1} in the single particle state $|\mathbf{p}, \sigma\rangle$ or vice versa (recall that the order of the operators is important when we are dealing with fermions). The matrix element yields for fermions $V_q - V_{p+q-k}\delta_{\sigma,\sigma'}$ and for bosons it yields $V_q + V_{p+q-k}$. The different relative sign between the boson and fermion cases is very important. The first term V_q is called the direct term and the other term is the exchange term. Note that the exchange term in the fermion case only appears if the spins are equal.

To finish, it is easy to see that

$$\rho_q = \int d^3 r \rho(r) e^{-iq\cdot r}$$

$$\rho(q) = \sum_{\mathbf{p}\sigma} c^\dagger_{p\sigma} c_{p+q,\sigma}$$

and that the current operator is given in second quantization by

$$\mathbf{J}_q = \sum_{\mathbf{p}\sigma} \frac{\mathbf{p}}{m} c^\dagger_{p\sigma} c_{p+q,\sigma}.$$

Exercises

1. Consider the one-dimensional tight-binding model. In this model the electrons hop between nearest-neighbor sites with a probability amplitude t. Write this model in second quantization in real space. Next, make a Fourier transform of the creation and destruction operators in the form

$$c_{i,\sigma} = \frac{1}{\sqrt{N}} \sum_k c_{k,\sigma} e^{ikx_i},$$

   where we have N lattice sites, $\sigma = \uparrow, \downarrow$, the operator $c_{i\sigma}$ destroys an electron of spin σ at site i, and the operator $c_{k,\sigma}$ destroys on electron of spin σ and momentum k. Show that the problem is diagonal

$$H = \sum_{k,\sigma} \epsilon_k c^\dagger_{k\sigma} c_{k,\sigma},$$

   where

$$\epsilon_k = -2t \cos ka,$$

   with a the lattice constant.
2. Show that

$$A = \left\langle 0 | c_2 c_1 c_3^\dagger c_4^\dagger | 0 \right\rangle$$

   is given by the determinant of the anticommutators between the fermionic operators. That is, by the determinant of

$$\begin{pmatrix} \left\{c_3^\dagger, c_1\right\} & \left\{c_3^\dagger, c_2\right\} \\ \left\{c_4^\dagger, c_1\right\} & \left\{c_4^\dagger, c_2\right\} \end{pmatrix}.$$

Therefore, verify that

$$A = \delta_{3,1}\delta_{4,2} - \delta_{3,2}\delta_{4,1}.$$

3. Imagine a system constituted by two wires, 1 and 2, and a one-dimensional system, of small dimensions (unidimensional box). The wires are unidimensional conductors. The energy levels of the system with reduced dimensions may be ordered in the way $\epsilon_1, \epsilon_2, \cdots$.

   Using second quantization write a Hamiltonian for the system considering that:

   (a) The charge carriers in the wires are free particles and that as they approach the box tunnel effect may occur between each wire and the box.
   (b) Redo the same problem considering that the charge carriers interact with each other, on the same wire, in the form of a two-particle potential V.

Appendix B
Discrete Symmetries

Symmetries play an important role in many fields. The quantization of the Berry phase as a result of the spatial inversion symmetry in the SSH chain and the quantization of the Hall conductance are examples of the importance of symmetries on the topological properties. Various types of symmetries, both spatial and local symmetries, lead to different universality classes, with distinct properties. In addition to the spatial symmetry inversion, we may consider mirror symmetries and crystal groups. Also, in the absence of spatial symmetries, local symmetries also play an important role, such as time-reversal symmetry, chiral symmetry, and charge-conjugation or particle-hole symmetry. To simplify, let us briefly consider some symmetries like spatial inversion and local symmetries.[1]

The symmetries of a given problem are reflected on the symmetries of the Hamiltonian that describes it. Sometimes, the states of the system may lose symmetries that are in the Hamiltonian, and we say the symmetries are broken. This occurs in the context of phase transitions. For now, let us consider the symmetries of the Hamiltonian. In some situations it may be useful to consider a real space description, but often we use a momentum space description, if the system is invariant under translations. Operations on the real space Hamiltonian (with momentum space Hamiltonian matrix $H(\mathbf{k})$), such as transpose, complex conjugate, and Hermitian conjugate, translate in momentum space to

$$H \to H(\mathbf{k})\,,\ H^T \to H^T(-\mathbf{k})\,,\ H^* \to H^*(-\mathbf{k})\,,\ H^\dagger \to H^\dagger(\mathbf{k})\,,$$

respectively. Given a Hamiltonian, either in real or momentum space, we must search for an operator that satisfies the condition defining the symmetry. In its absence the symmetry does not exist.

A symmetry that is often present is parity or inversion symmetry. The inversion operation in real space is written as

[1] For further reading see, for instance, [AraujoSacramento].

P. D. Sacramento, *Introduction to Condensed Matter Theory*,
https://doi.org/10.1007/978-3-031-86251-9

$$P\mathbf{r} = -\mathbf{r}.$$

That is, the symmetry is defined by the condition: if ψ satisfies Schrödinger equation, then $P\psi(-\mathbf{r})$ also satisfies, where P is a suitable matrix. Inverting the coordinates in Schrödinger equation, including inside the matrix H, one can always write

$$H\psi(-\mathbf{r}) = i\hbar\partial_t\psi(-\mathbf{r}) \Leftrightarrow PHP^\dagger P\psi(-\mathbf{r}) = i\hbar\partial_t P\psi(-\mathbf{r}),$$

and, therefore, the symmetry imposes that, in real space, $PH(-\mathbf{r}, -\mathbf{r}')P^\dagger = H(\mathbf{r}, \mathbf{r}')$. In momentum space we write

$$PH(-\mathbf{k})P^\dagger = H(\mathbf{k}).$$

Consider now the usual discussion of time-reversal symmetry. There is a time-reversal operator, T, that satisfies

$$T\mathbf{r} = \mathbf{r};\ T\mathbf{k} = -\mathbf{k};\ t \to -t$$
$$T\psi(\mathbf{r}, t) = \psi^*(\mathbf{r}, -t).$$

Time-reversal symmetry may be stated as if ψ satisfies the Schrödinger equation, then $T\psi^*$ satisfies the Schrödinger equation with the substitution $t \to -t$. T is a unitary matrix to be determined. From the time-dependent Schrödinger equation we get

$$\begin{aligned} i\hbar\partial_t\psi = H\psi &\Leftrightarrow -i\hbar\partial_t\psi^* = H^*\psi^* \\ &\Leftrightarrow -i\hbar\partial_t(T\psi^*) = TH^*T^\dagger(T\psi^*). \end{aligned} \tag{B.1}$$

The time-reversal operation, $\mathcal{T}$, satisfies

$$\begin{aligned} \psi &\to T\psi^*, \\ H &\to TH^*T^\dagger, \end{aligned} \tag{B.2}$$

where it is assumed that ψ and H are expressed in real space. But if $T\psi^*$ satisfies the Schrödinger equation with $t \to -t$, then $-i\hbar\partial_t(T\psi^*) = H(T\psi^*)$. From the time-dependent Schrödinger equation we get

$$TH^*T^\dagger = H, \qquad TH^*(-\mathbf{k})T^\dagger = H(\mathbf{k}).$$

In the absence of spin, T is a symmetric matrix. Considering the physical spin, T must be an antisymmetric matrix. This leads us to the result: $T = \sigma_y$. The notation $\mathcal{T} = i\sigma_y K$ is often used, where K denotes complex conjugation.

The operator $\mathcal{T}$, either with or without physical spin, is antiunitary ($\mathcal{T}\psi \cdot \mathcal{T}\phi = (\psi \cdot \phi)^*$) and for spin $\frac{1}{2}$ we have $\mathcal{T}^2 = -1$. This implies that the two states, ψ and $\mathcal{T}\psi$, are orthogonal and have the same energy, making up a Kramers pair. For momentum $\boldsymbol{k} = 0$ and other time invariant momenta, the two eigenstates ψ and $\mathcal{T}\psi$ are two orthogonal spinors.

Particle-hole or charge-conjugation symmetry is defined by the following condition: there is a unitary matrix, C, such that $C\psi^*$ satisfies Schrödinger equation. This implies $CH^*C^\dagger = -H$. The operation of charge conjugation or particle hole is defined as

$$\mathcal{C} : H \to CH^*C^\dagger .$$

Chiral or sublattice symmetry is defined as

$$\exists_{\mathcal{O}} : \mathcal{O}H\mathcal{O}^\dagger = -H \, , \; \mathcal{O}^2 = 1 \, .$$

This symmetry implies that the Hamiltonian can be brought to off-diagonal form. Indeed, expressing $\mathcal{O}$ in the basis of its eigenvectors,

$$\mathcal{O} = \begin{pmatrix} 1 & 0 \\ 0 & -1 \end{pmatrix} = \tau_z \implies H = \begin{pmatrix} 0 & D \\ D^\dagger & 0 \end{pmatrix} .$$

Rotating $\hat{H}$ to the off-diagonal form, introducing a unitary matrix, U,

$$UHU^\dagger = \begin{pmatrix} 0 & D \\ D^\dagger & 0 \end{pmatrix},$$

we can write

$$\tau_z U H U^\dagger \tau_z = - \begin{pmatrix} 0 & D \\ D^\dagger & 0 \end{pmatrix} = -UHU^\dagger \Leftrightarrow U^\dagger \tau_z U H U^\dagger \tau_z U = -H \, ,$$

and therefore it leads to $\mathcal{O} = U^\dagger \tau_z U$. Chiral symmetry must be present whenever the Hamiltonian enjoys the symmetries $\mathcal{T}$ and $\mathcal{C}$. Starting with the Hamiltonian H and applying the symmetry operators in succession, we get

$$H \xrightarrow{\mathcal{T}} TH^*T^\dagger = H \xrightarrow{\mathcal{C}} CH^*C^\dagger = -H \, ,$$

which means that we have rotated H into $-H$ through the operator $\mathcal{O} = \mathcal{C}\mathcal{T}$. Chiral symmetry may also exist alone, without any of the symmetries $\mathcal{T}$ or $\mathcal{C}$.

Appendix C
Translation Operator in a Lattice

In the continuum, the momentum operator is defined as the generator of infinitesimal translations. Let us recall the derivation of this result.

Consider a translation such that $\mathbf{r}' = \mathbf{r} + \mathbf{a}$. Let U_T be the translation operator by a vector a. Then, by definition, we have

$$U_T|\alpha\rangle = |\alpha'\rangle$$
$$\langle\alpha'|\mathbf{r}|\alpha'\rangle = \langle\alpha|\mathbf{r}|\alpha\rangle + \mathbf{a},$$

where $|\alpha\rangle$ represents the initial state and $|\alpha'\rangle$ represents the state that results from the translation. If there is conservation of momentum,

$$\langle\alpha'|\mathbf{p}|\alpha'\rangle = \langle\alpha|\mathbf{p}|\alpha\rangle.$$

Assuming that U_T is of the type

$$U_T = e^{-\frac{i}{\hbar}\mathbf{a}.\mathbf{G}},$$

where $\mathbf{G}$ is called the translation generator, we have that

$$\langle\alpha'|\mathbf{r}|\alpha'\rangle = \langle\alpha|U_T{}^{+}\,\mathbf{r}\,U_T|\alpha\rangle = \langle\alpha|\mathbf{r}|\alpha\rangle + \mathbf{a} = \langle\alpha|\mathbf{r}|\alpha\rangle + \mathbf{a}\langle\alpha|\alpha\rangle,$$

where we used that $\langle\alpha|\alpha\rangle = 1$. We conclude that

$$\begin{cases} U_T{}^{+}\,\mathbf{r}\,U_T = \mathbf{r} + \mathbf{a} \\ U_T{}^{+}\,\mathbf{p}\,U_T = \mathbf{p}. \end{cases}$$

Consider now that $\mathbf{a} = \delta\mathbf{a}$ ($\delta \ll 1$). Then

P. D. Sacramento, *Introduction to Condensed Matter Theory*,
https://doi.org/10.1007/978-3-031-86251-9

$$U_T = e^{-\frac{i}{\hbar}\delta\mathbf{a}.\mathbf{G}} \approx \hat{1} - \frac{i}{\hbar}\,\delta\mathbf{a}.\mathbf{G} + O(\delta a)^2,$$

where $\delta\mathbf{a}.\mathbf{G} = \sum_{j=1}^{3} \delta a_j\, G_j \equiv G_j\, \delta a_j$. We get that

$$[r_i, G_j] = i\hbar\delta_{ij}.$$

Also, from the momentum invariance we get

$$[p_i, G_j] = 0.$$

Finally, from the two previous expressions, we conclude that $G_j = p_j$, which means that $\mathbf{p}$ is the generator of translations.

In a lattice, the definition of the operator that takes the role of the generator of translations has to be changed.[1] The lattice momentum operator, Π, may be defined by $U = e^{i\Pi}$, $\Pi = \Pi^\dagger$, and $[H, \Pi] = 0$. Here, U is a one-site shift operator of the entire lattice. Defining the set of wave vectors in, say a $1d$ fermionic system, as ϕk, with $k = 0, \cdots, L-1$ and $\phi = 2\pi/L$, the expected definition of the momentum operator

$$\Pi = \phi \sum_{k=1}^{L-1} k c_k^\dagger c_k,$$

where c_k is the destruction operator of one fermion of wave vector ϕk, is a suitable definition if the fermionic system is non-interacting. However, if the system is interacting, in general this definition of the momentum operator does not commute with the Hamiltonian. An example is the Hubbard model.

We may define for complex numbers α the function

$$g(\alpha) = \sum_{k=0}^{L-1} i e^{-i\phi k a} = i\frac{1 - e^{-i\phi L\alpha}}{1 - e^{-i\phi\alpha}}.$$

This implies that

$$g'(m) = \sum_{k=1}^{L-1} \phi k e^{-i\phi k m},$$

for $m = 1, \cdots, L$. Applying an inverse Fourier transform, we get that

[1] For further reading see F. H. L. Essler, H. Fram, F. Göhmann, A. Klümper, and V. E. Korepin, *The One-Dimensional Hubbard Model*, Cambridge University Press.

$$\phi k = \frac{1}{L} \sum_{m=1}^{L} g'(m) e^{i\phi k m},$$

for $k = 0, \cdots, L-1$. For the set $m = 1, \cdots, L-1$ we may differentiate the function $g(\alpha)$ to obtain

$$g'(m) = \frac{\phi L}{e^{-i\phi m} - 1}.$$

The particular value

$$g'(L) = \frac{1}{2} \phi L(L-1)$$

may be used to rewrite that

$$\phi k = \phi \sum_{m=1}^{L-1} \left(\frac{1}{2} + \frac{e^{i\phi k m}}{e^{-i\phi m} - 1} \right),$$

for $k = 0, \cdots, L-1$.

We impose that the spectrum of the momentum operator is of the form of ϕZ. Substituting Π for ϕk implies that the momentum operator is defined $mod(2\pi)$. We get, by hypothesis,

$$\Pi = \phi \sum_{m=1}^{L-1} \left(\frac{1}{2} + \frac{U^m}{e^{-i\phi m} - 1} \right),$$

which implies that this definition is of the form of a polynomial on the translation (shift) operator U. This means that, for a N-particle state,

$$e^{i\Pi} |N\rangle = e^{i\phi(k_1 + \cdots + k_N)} |N\rangle.$$

Appendix D
Non-Hermitian Hamiltonian

In elementary quantum mechanics, the probability that a particle in a state, given by the wave function $\psi(x, t)$, is found between the positions x and $x + \mathrm{d}x$ at time t, $P(x, t)\mathrm{d}x$, is given by the expression

$$P(x, t)\mathrm{d}x = |\psi(x, t)|^2\mathrm{d}x = \psi^*(x, t)\psi(x, t)\mathrm{d}x.$$

The time variation of the probability density $P(x, t)$ is given by

$$\frac{\partial P(x, t)}{\partial t} = \frac{\partial \psi^*(x, t)}{\partial t}\psi(x, t) + \psi^*(x, t)\frac{\partial \psi(x, t)}{\partial t}.$$

In a standard way, using Schrödinger equation, considering a real potential and defining the probability current

$$j(x, t) = \frac{\hbar}{2im}\left(\psi^*\frac{\partial \psi}{\partial x} - \frac{\partial \psi^*}{\partial x}\psi\right),$$

we get the continuity equation (or conservation of probability)

$$\frac{\partial P(x, t)}{\partial t} + \frac{\partial j(x, t)}{\partial x} = 0.$$

This equation written for the case of a one-dimensional system is easily generalized to three dimensions

$$\frac{\partial P(\mathbf{x}, t)}{\partial t} + \nabla \cdot \mathbf{j}(\mathbf{x}, t) = 0,$$

P. D. Sacramento, *Introduction to Condensed Matter Theory*,
https://doi.org/10.1007/978-3-031-86251-9

where the vector current is given by

$$\mathbf{j}(\mathbf{x}, t) = \frac{\hbar}{2im}\left(\psi^* \nabla \psi - \nabla \psi^* \psi\right).$$

The meaning of the continuity equation is clear: if the probability to find the system in a given volume changes, then there is a corresponding flux of the vector current through the surface of the volume.

We may consider now a potential that in general may be complex. In this case, the continuity equation generalizes to the expression

$$\frac{\partial}{\partial t} P(x,t) + \frac{\partial}{\partial x} j(x,t) = \frac{2}{\hbar} \mathrm{Im} V(x) P(x,t),$$

where $\mathrm{Im} V(x)$ is the imaginary part of the potential. The extra term may be seen as a source or sink of probability. In particular, it may describe the creation or destruction of particles, with the associated increase or removal of probability. An interesting case that will be considered ahead is a situation where there is an incoming and an outgoing probability flux and that they balance each other. This may happen as the result of a spatial integration of the flux, for instance, a situation such that at some end of a system there is incoming flux and at another end there is an outgoing flux such that they compensate. In the following, we will briefly review some Hamiltonians that are non-Hermitian [1].

Consider a simple one-dimensional tight-binding model given by

$$H = \sum_n \left(J_L c_n^\dagger c_{n+1} + J_R c_{n+1}^\dagger c_n \right),$$

with J_L, J_R real numbers. If $|J_l| \neq |J_R|$, the Hamiltonian is not Hermitian. Transforming to momentum space, the Hamiltonian may be diagonalized, and the energy spectrum is

$$E(k) = (J_L + J_R)\cos k + i\,(J_L - J_R)\sin k.$$

Since the Hamiltonian is not Hermitian, the spectrum is not real, in general. If $J_L = J_R$, the imaginary part vanishes, and the spectrum is real. In the general case, there is a preferred flux along one direction or the other depending on the hopping amplitudes to the left and right. If the fluxes are compensated, the spectrum is real. Defining a winding number as

$$w = \frac{1}{2\pi i} \int_{-\pi}^{\pi} dk \partial_k \ln E(k),$$

we can find that when $|J_L| - |J_R| < 0$, the winding number $w = -1$, and in the opposite case, when $|J_L| - |J_R| > 0$, the winding number $w = 1$; therefore, the two regimes are characterized by the value of w. Also, there must be (at least) a point where the spectrum vanishes when the transition takes place for $|J_L| = |J_R|$ and the spectrum is real. In this case, it occurs when $\cos k = 0$.

While in general if the Hamiltonian is not Hermitian the spectrum is not real, there are cases where it is. Consider the symmetry operation of parity, defined by its action on the wave function as

$$P\psi(\mathbf{r}, t) = \psi(-\mathbf{r}, t),$$

with $P^2 = I$. Also, recall the time reversal operation

$$T\psi(\mathbf{r}, t) = \psi^*(\mathbf{r}, -t),$$

also with $T^2 = 1$. Consider now a Hamiltonian that is invariant under the combined action of the two symmetry operations. We say that the system has PT-symmetry leading to

$$[PT, H] = 0, \qquad H = PTHPT.$$

Consider the eigenvalue equation $H\psi = E\psi$. Then if there is PT-symmetry, we expect that there are common eigenstates of the PT operator with the Hamiltonian, which may be written as $PT\psi = \lambda\psi$, with $\lambda = e^{i\varphi}$. We obtain that

$$\begin{aligned} PT\,(H\psi) &= PT\,(E\psi) \\ H\,(PT\psi) &= E^*\,(PT\psi) \\ H\lambda\psi &= E^*\lambda\psi \\ H\psi &= E^*\psi, \end{aligned}$$

which implies that $E = E^*$ and the spectrum is real, even though the Hamiltonian is non-Hermitian. This is called a problem where the Hamiltonian is PT-symmetric and that the symmetry is unbroken, so that the states have the same symmetry. If the symmetry is broken, then complex eigenvalues probably appear, together with their complex conjugates.

Consider a simple one-particle Hamiltonian in a one-dimensional potential given by

$$U(x) = V(x) + iW(x).$$

Let us impose that the Hamiltonian is PT-symmetric. This implies

$$\begin{aligned} V(x) &= V(-x) \\ W(x) &= -W(-x). \end{aligned}$$

Consider an example of a PT-symmetric potential

$$U(x) = (x - i\alpha)^2,$$

with α a real number. The problem can be solved, and the spectrum is that of a harmonic oscillator, $E_n = 2n + 1$ (a real number). The wavefunctions are simple extensions of the wave functions of the usual harmonic oscillator like

$$\psi_n(x) = H_n(x - i\alpha)e^{-\frac{(x-i\alpha)^2}{2}},$$

that result from the direct substitution $x \to x - i\alpha$. It may then be argued that the existence of the PT-symmetry leads to a real spectrum.

Consider now another potential of the form

$$U(x) = -(ix)^N.$$

This potential is also PT-symmetric. The diagonalization of the problem leads to the conclusion that when $N \leq 2$, the spectrum is real, and all energies are positive. When $N = 2$, the potential is real and is just the usual harmonic potential with a real spectrum. However, it has been found that when $1 < N < 2$, there is a finite number of real positive energy eigenvalues and an infinite number of pairs of complex conjugate eigenvalues. Further, when $N < 1$, there are no real solutions. So in this problem, the PT-symmetry is broken.

Consider now a two-band system described by the Hamiltonian

$$H = k\sigma_1 + i\gamma\sigma_3,$$

with $\gamma \leq 0, k \leq 0$ two real numbers, acting on a two-component spinor. The operation of PT-symmetry is acted by choosing $P = \sigma_1$, and T is just the operation of complex conjugation. We can see easily that $PTHPT = H$. The spectrum of the Hamiltonian is given by

$$E_\pm = \pm\sqrt{k^2 - \gamma^2}$$

and the eigenstates by

$$\psi_\pm = \begin{pmatrix} i\frac{\gamma}{k} \pm \sqrt{1 - \frac{\gamma^2}{k^2}} \\ 1 \end{pmatrix}.$$

If $\gamma \leq k$, the PT-symmetry is unbroken, and the spectrum is real. If $\gamma > k$, the symmetry is broken: the spectrum has no real values, and all eigenvalues appear as complex conjugate pairs. The case $\gamma = k$ is special: the spectrum is doubly degenerate with a zero eigenvalue, and the state is unique and given by

$$\psi_0 = \begin{pmatrix} i \\ 1 \end{pmatrix}.$$

This is called an exceptional point with only one linearly independent eigenvector. At this point, the geometric multiplicity (number of linearly independent eigenvectors) is smaller than the algebraic multiplicity (degree of degeneracy of the characteristic polynomial).

Another illustrative example of a non-Hermitian two-band Hamiltonian is written as

$$H = \begin{pmatrix} 0 & \alpha \\ 1 & 0 \end{pmatrix},$$

where α is a complex number. The spectrum is

$$E_{\pm} = \pm\sqrt{\alpha}.$$

The eigenspectrum is not analytic since $|\partial_\alpha E(\alpha)| \to \infty$ as $\alpha \to 0$. There are right and left eigenvectors of the Hamiltonian given by

$$\begin{aligned} H\psi_{R,\pm} &= E_{\pm}\psi_{R,\pm} \\ \psi_{L,\pm}H &= \psi_{L,\pm}E_{\pm}, \end{aligned}$$

that are obtained as

$$\begin{aligned} \psi_{R,\pm} &= \begin{pmatrix} \pm\sqrt{\alpha} \\ 1 \end{pmatrix} \\ \psi_{L,\pm} &= \begin{pmatrix} 1 & \pm\sqrt{\alpha} \end{pmatrix}, \end{aligned}$$

which are not orthogonal if $\alpha \neq 1$. Furthermore, if $\alpha = 0$, the Hamiltonian is not diagonalizable, which is the point where the exceptional point is found, with a doubly degenerate zero-energy solution and one left eigenvector and one right eigenvector.

The band touching exceptional points are topologically protected. Consider a generic two-band Hamiltonian

$$H(\mathbf{k}) = \big(b_\mu(\mathbf{k}) + id_\mu(\mathbf{k})\big)\,\boldsymbol{\sigma}_\mu$$

and consider a two-dimensional space. Here $\mu = 0, 1, 2, 3$. The eigenvalues of the Hamiltonian are given by

$$E_{\pm} = b_0 + id_0 \pm \sqrt{b^2 - d^2 + 2i\mathbf{b}\cdot\mathbf{d}}.$$

In order to have band touching, we impose the conditions

$$b^2 - d^2 = 0, \qquad \mathbf{b} \cdot \mathbf{d} = 0,$$

and the energy takes the value $E_0 = b_0 + id_0$.

We can define an invariant, designated vorticity, by considering a contour enclosing the exceptional point, such that

$$\nu = \frac{1}{4\pi i} \oint d^2k \nabla_{\mathbf{k}} \ln det\left(H(\mathbf{k}) - E_0 I\right),$$

which is equivalent to

$$\nu = \frac{1}{2\pi} \oint d^2k \nabla_{\mathbf{k}} arg\left(E_+(\mathbf{k}) - E_-(\mathbf{k})\right).$$

The reasoning for the existence of these invariants may be understood considering the mapping of the non-Hermitian Hamiltonian to an Hermitian matrix that is chiral symmetric

$$\tilde{H} = \begin{pmatrix} 0 & H(\mathbf{k}) - E_0 I \\ H^\dagger(\mathbf{k}) - E_0^* I & 0 \end{pmatrix}.$$

The exceptional point, $\mathbf{k}_0$, is obtained as the solution of

$$det\left(H(\mathbf{k}_0) - E_0 I\right) = 0,$$

which is also the solution for the extended Hermitian matrix, $\tilde{H}$. The fact that the extended matrix was explicitly constructed as chiral symmetric implies that there is a matrix, $\Sigma = 1 \otimes \rho_3$, that allows the definition of a winding number as a topological invariant, given by

$$\nu_w = \frac{1}{4\pi i} \oint d^2k Tr\left(\Sigma \tilde{H}^{-1} \nabla_{\mathbf{k}} \tilde{H}\right),$$

that coincides with the vorticity defined above, confirming its topological nature. If the winding number or vorticity are non-vanishing, we expect a topological protection of the existence of the exceptional point.

Reference

1. C.M. Bender and S. Boettcher, Physical Review Letters **80**, 5243 (1998); T. Yoshida, R. Peters, N. Kawakami, Y. Hatsugai, Progress of Theoretical and Experimental Physics **2020**, 12A109 (2020).

Appendix E
Spin Coherent States and Stereographic Projection

Before we turn to the case of spin coherent states, let us briefly review the standard case of bosonic coherent states. These are defined as the eigenstates of the destruction operator (or lowering operator of the simple harmonic oscillator)

$$a|\alpha\rangle = \alpha|\alpha\rangle,$$

where α is complex. These states may be expanded in a complete basis, such as that of the eigenstates of the number operator, as

$$|\alpha\rangle = \sum_{n=0}^{\infty} c_n|n\rangle.$$

The coefficients satisfy the recurrence relation

$$c_{m+1}\sqrt{m+1} = \alpha c_m.$$

Normalizing the states implies that we get

$$|\alpha\rangle = e^{-\frac{1}{2}|\alpha|^2} \sum_{n=0}^{\infty} \frac{\alpha^n}{\sqrt{n!}}|n\rangle = e^{-\frac{1}{2}|\alpha|^2} e^{\alpha a^\dagger}|0\rangle.$$

These states may also be constructed using the so-called displacement operator

$$|\alpha\rangle = D(\alpha)|0\rangle,$$

where the operator $D(\alpha)$ is given by

$$D(\alpha) = e^{a^\dagger \alpha - \alpha^* a}.$$

P. D. Sacramento, *Introduction to Condensed Matter Theory*,
https://doi.org/10.1007/978-3-031-86251-9

Using

$$e^{A+B} = e^A e^B e^{-\frac{1}{2}[A,B]},$$

valid if $[A, [A, B]] = [B, [A, B]] = 0$, we may write

$$D(\alpha)|0\rangle = e^{-\frac{1}{2}|\alpha|^2} e^{\alpha a^\dagger} e^{-\alpha^* a}|0\rangle = e^{-\frac{1}{2}|\alpha|^2} e^{\alpha a^\dagger}|0\rangle = |\alpha\rangle.$$

These states, even though normalized, are not orthogonal

$$\langle\alpha|\beta\rangle = e^{-\frac{1}{2}|\alpha|^2} e^{-\frac{1}{2}|\beta|^2} e^{\alpha^*\beta}.$$

The decomposition of the identity may be written as

$$\int \frac{d^2\alpha}{\pi}|\alpha\rangle\langle\alpha| = 1,$$

where $d^2\alpha = d\alpha d\alpha^*$.

One significant point is that these states minimize the Heisenberg uncertainty relation and, therefore, are the quantum states as classical as possible. Recalling the relation with the position and momentum variables of a one-dimensional operator

$$x = i\sqrt{\frac{\hbar}{2m\omega}}\left(a - a^\dagger\right)$$

$$p = \sqrt{\frac{m\hbar\omega}{2}}\left(a + a^\dagger\right),$$

we can show that

$$\Delta x\Delta p = \frac{\hbar}{2}.$$

(see the Exercises).

In the same way that we define the coherent states for the bosonic case, we may introduce spin coherent states for the spin operators. The spin coherent states for a system of spin J are defined by[1]

$$|\Omega\rangle = \hat{R}(\theta, \varphi)|JJ\rangle,$$

[1] For further reading see V. R. Vieira and P. D. Sacramento, Annals of Physics **242**, 188 (1995), V. R. Vieira and P. D. Sacramento, Nuclear Physics B **448**, 331 (1995) and [BicudoSacramento].

where

$$\hat{R}(\theta,\varphi) = e^{i\theta \mathbf{n}.\hat{\mathbf{J}}}$$
$$= e^{i\theta(\sin\varphi \hat{J}_x - \cos\varphi \hat{J}_y)}$$
$$= e^{\zeta \hat{J}_- - \zeta^* \hat{J}_+}$$

and

$$\zeta = \frac{\theta}{2} e^{i\varphi}.$$

The operator $\hat{R}(\theta,\varphi)$ is a rotation through an angle θ about the axis

$$\mathbf{n} = (\sin\varphi, -\cos\varphi, 0),$$

normal to the z-axis and to the vector

$$\mathbf{m} = (\sin\theta\cos\varphi, \sin\theta\sin\varphi, \cos\theta),$$

as represented in Fig. E.1.

This definition is the analog of the definition of the boson coherent states obtained taking the limit $J \to \infty$

$$|\alpha\rangle = \hat{D}(\alpha)|0\rangle,$$

where $\hat{D}(\alpha) = e^{a^\dagger \alpha - \alpha^* a}$ is the displacement operator and $|0\rangle$ is the vacuum state. The boson coherent state $|\alpha\rangle$ is a (right) eigenvector of the destruction operator a, with eigenvalue α. In the case of spins, it is possible to write eigenvalue equations for the coherent states, but they involve three or, at least two, of the

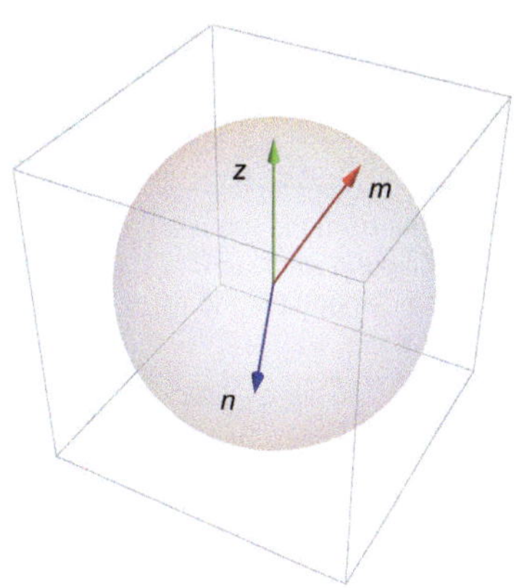

Fig. E.1 Representation of the unit vectors $\mathbf{e}_z$, $\mathbf{n}$, $\mathbf{m}$. The operator $\hat{R}(\theta,\varphi)$ rotates the vector $\mathbf{e}_z$, through a rotation around the vector $\mathbf{n}$ into the vector $\mathbf{m}$

operators $\hat{J}^-$, $\hat{J}^+$, $\hat{J}^z$ since the spin commutation relations imply that, in general, three operators are involved, instead of two, except when $J \to \infty$. To clarify the parallelism to the boson expressions, we compare some of our results to those obtained in the boson case. This can be done using the Holstein-Primakoff transformation $\hat{J}^- = a^\dagger\sqrt{2J - \hat{N}}$, $\hat{J}^+ = \sqrt{2J - \hat{N}}a$, and $\hat{J}^z = J - \hat{N}$, where $\hat{N} = a^\dagger a$, with $a, a^\dagger$ Boson operators. Taking the limit $J \to \infty$, the spin operators $\hat{J}^-$ and $\hat{J}^+$, suitably rescaled, reduce to $\hat{J}^-/\sqrt{2J} \to a^\dagger$ and $\hat{J}^+/\sqrt{2J} \to a$. It is convenient to rescale, together with the operators, the complex variables α like $\sqrt{2J}\alpha \to \alpha$.

We can write $\hat{R}(\theta, \varphi)$ with a different parametrization, which is very convenient for the calculations

$$\hat{R}(\theta, \varphi) = e^{\alpha \hat{J}^-} e^{\ln \frac{1}{1+|\alpha|^2} \hat{J}^z} e^{-\alpha^* \hat{J}^+},$$

where

$$\alpha = \tan \frac{\theta}{2} e^{i\varphi}.$$

This last equation defines the stereographic projection from the south pole of the sphere to the plane passing through the equator with complex coordinates α and α^*. A phase point lying on a sphere with radius 1 can be projected to the x, y plane from either the north or the south pole. The projection from the south pole is defined by the complex variable

$$\alpha = \tan \frac{\theta}{2} e^{i\varphi}$$

and from the north pole by

$$\alpha' = \frac{1}{\alpha^*}.$$

This projection is shown in Fig. E.2.

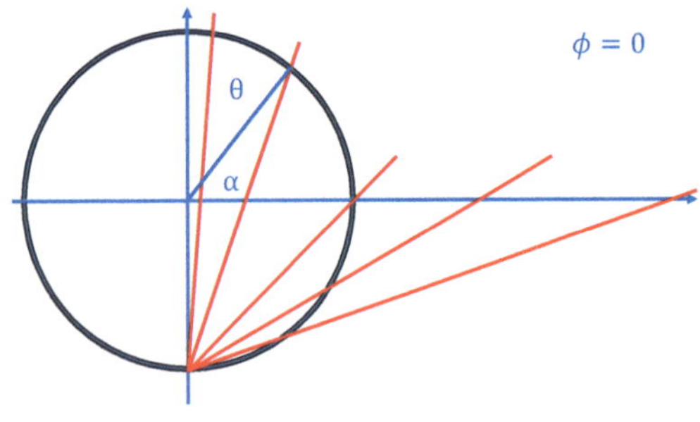

Fig. E.2 Stereographic projection from the south pole of a point in the unit sphere into a point in the equator point, with coordinates interpreted as a complex number

The coherent states can then be labeled as

$$|\alpha\rangle = \frac{1}{(1+|\alpha|^2)^J} e^{\alpha \hat{J}^-} |JJ\rangle,$$

These states are not orthogonal, and their overlap is

$$\begin{aligned}\langle\alpha'|\alpha\rangle &= \frac{(1+\alpha'^*\alpha)^{2J}}{(1+|\alpha'|^2)^J(1+|\alpha|^2)^J} \\ &= e^{i\Phi}\left[\frac{1}{2}(1+\cos\Theta)\right]^J,\end{aligned}$$

where Φ is the area of the spherical triangle with vertices $\mathbf{m}$, $\mathbf{m}'$, and $\mathbf{e}_z$, and Θ is the angle between the two directions defined by $\mathbf{m}$ and $\mathbf{m}'$. These states are overcomplete, and the identity can be decomposed as

$$\frac{2J+1}{4\pi}\int d\Omega|\Omega\rangle\langle\Omega| = \hat{1},$$

where

$$\begin{aligned}d\Omega &= \sin\theta d\theta d\varphi \\ &= \frac{4d^2\alpha}{(1+|\alpha|^2)^2}.\end{aligned}$$

The last equation gives the element of area appropriate for the plane of the stereographic projection of the sphere. The partition of unity can then also be written as

$$\frac{2J+1}{\pi}\int \frac{d^2\alpha}{(1+|\alpha|^2)^2}|\alpha\rangle\langle\alpha| = \hat{1}.$$

The trace of an operator is given by

$$Tr\hat{F} = \frac{2J+1}{\pi}\int \frac{d^2\alpha}{(1+|\alpha|^2)^2}\langle\alpha|\hat{F}|\alpha\rangle.$$

The coherent state $|\alpha\rangle$ depends on the variables α and α^*, which should be considered independent. As we will see in the sequence, it is convenient to define matrix elements depending only on two complex variables (and not four). The same happens in the case of bosons where the so-called holomorphic representation is used. The easiest way to use this type of representation is to define the new states

$$\|\alpha\rangle = e^{\alpha\hat{J}^-}|JJ\rangle,$$

which depend only on α. Their overlap is

$$\langle\beta\|\hat{1}\|\alpha\rangle = (1+\beta^*\alpha)^{2J},$$

and the decomposition of the identity becomes

$$\frac{2J+1}{\pi}\int d^2\alpha \frac{\|\alpha\rangle\langle\alpha\|}{(1+|\alpha|^2)^{2(J+1)}} = \hat{1}.$$

The transition amplitude

$$\bar{U}(\alpha_f^*, t_f; \alpha_i, t_i) =< \alpha_f||T_t e^{-\frac{i}{\hbar}\int_{t_i}^{t_f} dt\hat{\mathcal{H}}(t)}||\alpha_i\rangle,$$

can be obtained by factorizing the finite evolution operator in elementary ones and inserting the resolution of unity between them. We obtain that

$$\bar{U}\ (\alpha_f^*, t_f; \alpha_i, t_i) = \int \left(\prod_{k=1}^{N-1} d^2\mu_k\right)\left(\prod_{k=0}^{N-1} e^{-\frac{i}{\hbar}\Delta t\mathcal{F}_c(\alpha_{k+1}^*,\alpha_k,t_k)}\right)(1+\alpha_N^*\alpha_{N-1})^J$$

$$\prod_{k=1}^{N-1}\left(\frac{1+\alpha_{k+1}^*\alpha_k}{1+\alpha_k^*\alpha_k}\right)^J\left(\frac{1+\alpha_k^*\alpha_{k-1}}{1+\alpha_k^*\alpha_k}\right)^J(1+\alpha_1^*\alpha_0)^J,$$

where $\Delta t = \frac{t_f-t_i}{N}$,

$$d^2\mu_k = \frac{2J+1}{\pi}\frac{d^2\alpha_k}{(1+|\alpha_k|^2)^2},$$

is the integration measure and

$$e^{-\frac{i}{\hbar}\Delta t\mathcal{F}_c(\alpha_{k+1}^*,\alpha_k,t_k)} = \frac{<\alpha_{k+1}||e^{-\frac{i}{\hbar}\Delta t\hat{\mathcal{H}}}||\alpha_k\rangle}{\langle\alpha_{k+1}||1||\alpha_k\rangle}$$

is the exact normalized matrix element. The appropriate boundary conditions are given by $\alpha_0 = \alpha_i$, $\alpha_N^* = \alpha_f^*$.

Taking the continuum limit $\Delta t \to 0$ and using the Euler-McLaurin formula, we obtain that

$$\bar{U}(\alpha_f^*, t_f; \alpha_i, t_i) = \int \mathcal{D}^2\mu(1+\alpha_f^*\alpha(t_f))^J e^{\frac{i}{\hbar}\int_{t_i}^{t_f} dt\mathcal{L}}(1+\alpha^*(t_i)\alpha_i)^J,$$

where

$$\mathcal{L} = \frac{\hbar}{i}\frac{J}{1+\alpha^*\alpha}\left(\frac{d\alpha^*}{dt}\alpha - \alpha^*\frac{d\alpha}{dt}\right) - \mathcal{H}_c(\alpha^*, \alpha, t),$$

is the Lagrangian and $\mathcal{H}_c(\alpha^*, \alpha, t)$ is the normalized matrix element of the Hamiltonian. We note the two powers reflecting the boundary conditions $\alpha(t_i) = \alpha_i$ and $\alpha^*(t_f) = \alpha_f^*$. Note that $\alpha(t_f)$ is not the complex conjugate of α_f^* and that $\alpha^*(t_i)$ is not the complex conjugate of α_i. They are the values taken by the variables $\alpha(t)$ and $\alpha^*(t)$, satisfying these boundary conditions, at the times $t = t_f$ and $t = t_i$, respectively. In particular, in the saddle-point approximation, they are obtained solving the classical equations of motion for $\alpha(t)$ and $\alpha^*(t)$ and evaluating their values at t_f and t_i, respectively.

A similar procedure can be carried out for the partition function. The final result is shown in the main text.

Exercises

1. Consider the harmonic oscillator or a problem of bosons. We may define the eigenstates of the destruction operator (called coherent states) in the form

$$a|\alpha\rangle = \alpha|\alpha\rangle,$$

where α is a complex variable.

(a) These states may be written in the basis of the eigenstates of the harmonic oscillator in the form

$$|\alpha\rangle = \sum_{n=0}^{\infty} c_n|n\rangle.$$

Show that

$$c_{n+1}\sqrt{n+1} = \alpha c_n.$$

(b) Defining the states as normalized $\langle\alpha|\alpha\rangle = 1$; show that

$$|\alpha\rangle = e^{-\frac{1}{2}|\alpha|^2} \sum_{n=0}^{\infty} \frac{\alpha^n}{\sqrt{n!}}.|n\rangle.$$

(c) Show that

$$\langle\alpha|\beta\rangle = e^{-\frac{1}{2}|\alpha|^2} e^{-\frac{1}{2}|\beta|^2} e^{\alpha^*\beta}.$$

(d) Show that the partition of unity is written as

$$\int \frac{d^2\alpha}{\pi} |\alpha\rangle\langle\alpha| = 1.$$

(e) Show that in these states we have the minimization of the Heisenberg uncertainty relation

$$\Delta x \Delta p = \frac{\hbar}{2},$$

where we define as usual

$$\Delta A = \sqrt{\langle\alpha|\,(A - \langle\alpha|A|\alpha\rangle)^2\,|\alpha\rangle}.$$

Index

P. D. Sacramento, *Introduction to Condensed Matter Theory*,
https://doi.org/10.1007/978-3-031-86251-9

N

O

P

Q

R

MIX
Papier aus verantwortungsvollen Quellen
Paper from responsible sources
FSC® C105338

If you have any concerns about our products,
you can contact us on
ProductSafety@springernature.com

In case Publisher is established outside the EU,
the EU authorized representative is:
Springer Nature Customer Service Center GmbH
Europaplatz 3, 69115 Heidelberg, Germany

Printed by Libri Plureos GmbH
in Hamburg, Germany